IFMBE Proceedings

CW00833259

Volume 68/3

Series editor

Ratko Magjarevic

Deputy Editors

Fatimah Ibrahim
Igor Lacković
Piotr Ładyżyński
Emilio Sacristan Rock

The International Federation for Medical and Biological Engineering, IFMBE, is a federation of national and transnational organizations representing internationally the interests of medical and biological engineering and sciences. The IFMBE is a non-profit organization fostering the creation, dissemination and application of medical and biological engineering knowledge and the management of technology for improved health and quality of life. Its activities include participation in the formulation of public policy and the dissemination of information through publications and forums. Within the field of medical, clinical, and biological engineering, IFMBE's aims are to encourage research and the application of knowledge, and to disseminate information and promote collaboration. The objectives of the IFMBE are scientific, technological, literary, and educational.

The IFMBE is a WHO accredited NGO covering the full range of biomedical and clinical engineering, healthcare, healthcare technology and management. It is representing through its 60 member societies some 120.000 professionals involved in the various issues of improved health and health care delivery.

IFMBE Officers
President: James Goh, Vice-President: Shankhar M. Krishnan
Past President: Ratko Magjarevic
Treasurer: Marc Nyssen, Secretary-General: Kang Ping LIN
http://www.ifmbe.org

More information about this series at http://www.springer.com/series/7403

Lenka Lhotska · Lucie Sukupova
Igor Lacković · Geoffrey S. Ibbott
Editors

World Congress on Medical Physics and Biomedical Engineering 2018

June 3–8, 2018, Prague, Czech Republic
(Vol. 3)

IUPESM
PRAGUE 2018

Editors
Lenka Lhotska
CIIRC
Czech Technical University in Prague
Prague
Czech Republic

Lucie Sukupova
Institute of Clinical and Experimental
   Medicine
Prague
Czech Republic

Igor Lacković
Faculty of Electrical Engineering
   and Computing
University of Zagreb
Zagreb
Croatia

Geoffrey S. Ibbott
Department of Radiation Physics
The University of Texas MD Anderson
   Cancer Center
Houston, TX
USA

ISSN 1680-0737   ISSN 1433-9277   (electronic)
IFMBE Proceedings
ISBN 978-981-10-9022-6   ISBN 978-981-10-9023-3   (eBook)
https://doi.org/10.1007/978-981-10-9023-3

Library of Congress Control Number: 2018940876

Printed on acid-free paper

This Springer imprint is published by the registered company Springer Nature Singapore Pte Ltd. part of Springer Nature
The registered company address is: 152 Beach Road, #21-01/04 Gateway East, Singapore 189721, Singapore

Preface

This book presents the proceedings of the IUPESM World Congress on Biomedical Engineering and Medical Physics, a triennially organized joint meeting of medical physicists, biomedical engineers, and adjoining healthcare professionals. Besides the purely scientific and technological topics, the 2018 Congress will also focus on other aspects of professional involvement in health care, such as education and training, accreditation and certification, health technology assessment, and patient safety. The IUPESM meeting is an important forum for medical physicists and biomedical engineers in medicine and healthcare to learn and share knowledge, and to discuss the latest research outcomes and technological advancements as well as new ideas in both medical physics and biomedical engineering field.

Biomedical engineering and medical physics represent challenging and rapidly growing areas. Building on the success of the previous World Congresses, the aim of the World Congress 2018 is to continue in bringing together scientists, researchers, and practitioners from different disciplines, namely from mathematics, computer science, bioinformatics, biomedical engineering, medical physics, medicine, biology, and different fields of life sciences, so that they can present and discuss their research results. We hope that the World Congress 2018 will serve as a platform for fruitful discussions between all attendees, where participants can exchange their recent results, identify future directions and challenges, initiate possible collaborative research, and develop common languages for solving problems in the realm of biomedical engineering and medical physics.

Following a thorough peer-reviewed process, we have finally selected 498 papers. The Scientific Committee would like to thank the reviewers for their excellent job. The articles can be found in the proceedings and are divided into the main tracks and special sessions. The papers show how broad the spectrum of topics in biomedical engineering and medical physics is.

The editors would like to thank all the participants for their high-quality contributions and Springer for publishing the proceedings of the World Congress.

Prague, Czech Republic   Lenka Lhotska
Prague, Czech Republic   Lucie Sukupova
Zagreb, Croatia   Igor Lacković
Houston, USA   Geoffrey S. Ibbott
March 2018

Congress Coordinating Committee

Chair

Kin Yin Cheung, Chair, IOMP, Hong Kong

Members

Howell Round, IOMP, New Zealand
Slavik Tabakov, IOMP, UK
Virginia Tsapaki, IOMP, Greece
James Goh, IFMBE, Singapore
Kang Ping Lin, IFMBE, Chinese Taipei
Herb Voigt, co-opted, IFMBE, USA †
Monique Frize, co-opted, IFMBE, Canada

Congress Organizing Committee

Co-chairs

Jaromir Cmiral, Czech Republic, BME
Libor Judas, Czech Republic, MedPhys

Secretaries

Frantisek Lopot, Czech Republic, BME
Karel Nechvil, Czech Republic, MedPhys

Members

Financial Committee Co-chairs

Martin Mayer, Czech Republic, BME
Vít Richter, Czech Republic, MedPhys

Scientific Committee Co-chairs

Lenka Lhotska, Czech Republic, BME
Lucie Sukupova, Czech Republic, MedPhys

Publicity Committee Chair

Martina Novakova, Czech Republic, BME, MedPhys

Education Committee Co-chairs

Jiri Hozman, Czech Republic, BME
Irena Koniarova, Czech Republic, MedPhys

Financial Committee

Co-chairs

Martin Mayer, Czech Republic, BME
Vit Richter, Czech Republic, MedPhys

Members

Jan Hanousek, Czech Republic, BME
Anchali Krisanachinda, IOMP
Mark Nyssen, IUPESM, IFMBE
Vaclav Poljak, Czech Republic, MedPhys

Scientific Committee

Co-chairs

Geofrey Ibbott, IOMP
Lenka Lhotska, Czech Republic, BME
Igor Lackovic, IFMBE
Lucie Sukupova, Czech Republic, MedPhys

Members

Magdalena Bazalova-Carter, Canada
Carmel Caruana, Malta
Tomas Cechak, Czech Republic
Paul Chang, Singapore
Kin Yin Cheung, Hong Kong, China
Jiri Chvojka, Czech Republic
Matej Daniel, Czech Republic
Marie Davidkova, Czech Republic
Harry Delis, IAEA
Pavel Dvorak, Czech Republic
Ludovic Ferrer, France
Martin Falk, Czech Republic
Christian Gasser, Sweden
Csilla Gergely, France
Susanna Guatelli, Australia
Jens Haueisen, Germany
Jan Havlik, Czech Republic
Jiří Hozman, Czech Republic
Marjan Hummel, The Netherlands
Leonidas D. Iassemidis, USA
Tom Judd, USA
Peter Knoll, Austria

Radim Kolar, Czech Republic
Jana Kolarova, Czech Republic
Christian Kollmann, Austria
Irena Koniarova, Czech Republic
David Korpas, Czech Republic
Vladimir Krajca, Czech Republic
Jan Kremlacek, Czech Republic
Petr Marsalek, Czech Republic
Deborah Van Der Merwe, IAEA
Karel Nechvil, Czech Republic
Chris Nuggent, UK
Pirkko Nykänen, Finland
Hakan Nystrom, Sweden
Pawel Olko, Poland
Maria Perez, WHO
Vaclav Porod, Czech Republic
Kevin Prise, UK
Ivo Provaznik, Czech Republic
Jaroslav Ptáček, Czech Republic
Vladimir Rogalewicz, Czech Republic
Simo Saarakkala, Finland
Ioannis Sechopoulos, The Netherlands
Milan Sonka, USA
Olga Stepankova, Czech Republic
Lucie Sukupova, Czech Republic
Krystina Tack, USA
Annalisa Trianni, Italy
Jiri Trnka, Czech Republic
Virginia Tsapaki, Greece
Adriana Velazquez, WHO
Frantisek Vlcek, Czech Republic
Jan Vrba, Czech Republic
Kevin Warwick, UK
Martha Zequera Diaz, Columbia

Program Committee

Co-chairs

Jaromir Cmiral, Czech Republic, BME
Libor Judas, Czech Republic, MedPhys

Members

Lenka Lhotska, Czech Republic, BME
Frantisek Lopot, Czech Republic, BME
Karel Nechvil, Czech Republic, MedPhys
Martina Novakova, Czech Republic, BME
Lucie Sukupova, Czech Republic, MedPhys

Publicity Committee

Chair

Martina Novakova, Czech Republic

Members

Pavla Buricova, Czech Republic
Michele Hilts, Canada
Jeannie Hsiu Ding Wong, Malaysia
Akos Jobbagy, Hungary
Luiz Kun, USA
Pavla Novakova, Czech Republic
Magdalena Stoeva, Bulgaria
Tae-Suk Suh, Republic of Korea
Jaw-lin Wang, Chinese Taipei

Education Committee

Co-chairs

Jiri Hozman, Czech Republic, BME
Irena Koniarova, Czech Republic, MedPhys

Members

Jiri Kofranek, Czech Republic, BME
Vladimír Krajca, Czech Republic, BME
Simo Saarakkala, Finland, BME
Ioannis Sechopoulos, The Netherlands, MedPhys

Professional Standards Committee

Co-chairs

David Korpas, Czech Republic, BME
Libor Judas, Czech Republic, MedPhys

Members

John Damilakis, Greece, MedPhys
Michele Hilts, Canada
Jeannie Hsiu Ding Wong, Malaysia
Tomas Kron, Australia, MedPhys
Siew Lok Toh, Singapore, BME
Yakov Pipman, USA, MedPhys
Kang Ping Lin, Chinese Taipei, BME
Jaroslav Ptacek, Czech Republic, MedPhys
Christoph Trauernicht, South Africa, MedPhys

Publication Committee

Chair

Vladimir Marik, Czech Republic

Co-chair

Igor Lacković, Croatia

International Advisory Board (BME)

Co-chairs

James Goh, Singapore
Ratko Magjarevic, Croatia

Members

Guillermo Avendano C., Chile
Paulo De Carvalho, Portugal
Jaromir Cmiral, Czech Republic
Fong Chin Su, Chinese Taipei
David Elad, Israel
Yubo Fan, China
Mário Forjaz Secca, Portugal
Monique Frize, Canada
Birgit Glasmacher, Germany
Peter Hunter, New Zealand
Ernesto Iadanza, Italy
Fatimah Ibrahim, Malaysia
Timo Jämsä, Finland
Akos Jobbagy, Hungary
Eleni Kaldoudi, TBA
Peter Kneppo, Czech Republic
Shankar Krishnan, USA
Eric Laciar Leber, Argentina
Igor Lackovic, Croatia
Piotr Ladyzynski, Poland
Lenka Lhotska, Czech Republic
Nigel Lovell, Australia
Alan Murray, UK
Marc Nyssen, Belgium
Leandro Pecchia, UK
Kang Ping Lin, Chinese Taipei
Ichiro Sakuma, Japan
Maria Siebes, The Netherlands
Nitish Thakor, Singapore
Herbert F. Voigt, USA
Min Wang, Hong Kong

International Advisory Board (MP)

Co-chairs

Kin Yin Cheung, Hong Kong, China
Slavik Tabakov, UK

Members

Laila Al Balooshi, UAE
Abdullah Al Hajj, KSA
Huda Al Naemi, Qatar
Rodolfo Alfonso Laguardia, Cuba
Supriyanto Ardjo Pawiro, Indonesia
Eva Bezak, Australia
Marco Brambila, Italy
David Brettle, UK
Arun Chougule, India
John Damilakis, Greece
Catherine Dejean, France
Ludovic Ferrer, France
Michelle Hilts, Canada
Jeannie Hsiu Ding Wong, Malaysia
Amaury Hornbeck, France
Geoffrey Ibbott, USA
Ahmed Ibn Seddik, Morocco
Taofeeq Ige, Nigeria
Petro Julkunen, Finland
Simone Kodlulovich, Brazil
Dimitri Kostylev, Russia
Anchali Krisanachinda, Thailand
James C. L. Lee, Singapore
Melissa Martin, USA
Rebecca Nakatudde, Uganda
Herke Jan Noordmans, The Netherlands
Fridtjof Nüsslin, Germany
Yakov Pipman, USA
Jaroslav Ptacek, Czech Republic
Magdalena Rafecas, Germany
Madan Rehani, USA
Jose L. Rodriguez, Chile
Howell Round, New Zealand
Magdalena Stoeva, Bulgaria
Tae-Suk Suh, Republic of Korea
Lucie Sukupova, Czech Republic
Virginia Tsapaki, Greece
Graciela Velez, Argentina
Ulrich Wolf, Germany

Acknowledgements

Organizing Societies

International Societies

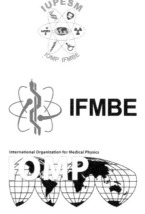

Collaborating Institution

Sponsors

About IFMBE

The International Federation for Medical and Biological Engineering (IFMBE) is primarily a federation of national and transnational societies. These professional organizations represent interests in medical and biological engineering. IFMBE is also a non-governmental organization (NGO) for the United Nations and the World Health Organization (WHO), where we are uniquely positioned to influence the delivery of health care to the world through biomedical and clinical engineering.

The IFMBE's objectives are scientific and technological as well as educational and literary. Within the field of medical, biological, and clinical engineering, IFMBE's aims are to encourage research and application of knowledge and to disseminate information and promote collaboration. The ways in which we disseminate information include the following: organizing World Congresses and Regional Conferences, publishing our flagship journal Medical and Biological Engineering and Computing (MBEC), our Web-based newsletter—IFMBE News, our Congress and Conference Proceedings, and books. The ways in which we promote collaborations are through networking programs, workshops, and partnerships with other professional groups, e.g., Engineering World Health.

Mission
The mission of IFMBE is to encourage, support, represent, and unify the worldwide medical and biological engineering community in order to promote health and quality of life through the advancement of research, development, application, and management of technology.

Objectives
The objectives of the International Federation for Medical and Biological Engineering shall be scientific, technological, literary, and educational. Within the field of medical, clinical, and biological engineering, its aims shall be to encourage research and the application of knowledge and to disseminate information and promote collaboration.

In pursuit of these aims, the Federation may, in relation to its specific field of interest, engage in any of the following activities: sponsorship of national and international meetings, publication of official journals, cooperation with other societies and organizations, appointment of commissions on special problems, awarding of prizes and distinctions, establishment of professional standards and ethics within the field, or in any other activities which in the opinion of the General Assembly or the Administrative Council would further the cause of medical, clinical, or biological engineering. It may promote the formation of regional, national, international, or specialized societies, groups or boards, the coordination of bibliographic or informational services, and the improvement of standards in terminology, equipment, methods and safety practices, and the delivery of health care.

In general, the Federation shall work to promote improved communication and understanding in the world community of engineering, medicine, and biology.

Contents

Part VIII   Dosimetry and Radiation Protection

Part I
Micro- and Nanosystems, Active Implants, Biosensors

Semiconductor Ethanol Sensor Inducted with Visible Light

Yu. Dekhtyar, M. Komars, and M. Sneiders

Abstract

Modern semiconductor sensors are based on an MOS (metal–oxide–semiconductor) structure and their further development is tending to combine them together with UV irradiation. In this paper, a system is presented that can pave the way towards gas analyzers that rely on visible spectrum light. The article presents a semiconductor gas sensor induced with safe optical irradiation delivered from the LED matrix to the semiconductor surface. Different irradiation wavelengths (440, 530, 600, 710 nm) under constant flux were used separately to find the condition for the best sensor response on gas sorption. An output signal was recorded in zero-level emission and in the presence of ethanol. Changes in sensor response and signal rise/relaxation time constant in the presence of saturated ethanol vapor were observed. Sensor response to the ethanol vapor was detected for each used irradiation wavelength and it is approximated by the non-linear falling regression curve with the highest sensor response at 440 nm (R = 104%). In addition, the different rise and relaxation time constant of the signal trace was detected depending on the irradiation wavelength. The time constant ratio is approximated by the non-linear rising regression curve with a maximum value at 710 nm for both signal rise ($\tau = 0.61$) and relaxation ($\tau = 2.08$) parts.

Keywords

LED stimulation • Time constant • Ethanol sensing
Room temperature

1 Introduction

Nowadays, there is increasing concern about the environmental pollution which increases each year, thereby causing irreparable damage and influencing every inhabitant on Earth. Air pollution from industrial gases, from fuel combustion products and other sources, inevitably increases the incidence of diseases, including: lung cancer, asthma, heart attacks and allergies [1, 2].

Modern methods are directed to prevent and even foresee the problem. Studies show that the determination of biomarkers in exhaled air can become a powerful tool for medical diagnostics. Currently, there are certain biomarkers in exhaled air that are accompanied by diseases such as obesity, lung cancer, kidney failure, heart attacks and diabetes mellitus [3, 4].

Along with this, the ethanol concentration in exhaled air may correspond as an indicator of the glucose level in blood [5]. Moreover, the potential endogenous source of ethanol is intestinal bacterial flora [6]. This is a challenging area, as the ethanol concentration in exhaled air is normally much lower than the concentration level in human breath after alcohol ingestion [7].

In the past decade, semiconductor gas sensors have been rapidly developed. The sensors are cheap, easy to use, able to detect a variety of different gases and are stable. Because of this, the sensor gained widespread use. Gas sensor technologies are providing significant progress in the detection of hazardous substances and exhaled biomarkers [8].

Among the actual technologies widely distributed, MOS sensors have several significant drawbacks. They are not appropriate for the detection of inflammable gases with a low autoignition temperature as the sensing area must be heated up to 500 °C. Hence, heating to a higher temperature demands higher power consumption [9, 10]. More often the sensors are combined with UV irradiation. As a result, gases adhere to the MOS surface and the current via the MOS

Yu. Dekhtyar (✉) · M. Komars · M. Sneiders
Institute of Biomedical Engineering and Nanotechnologies,
Riga Technical University, Riga, Latvia
e-mail: Jurijs.Dehtjars@rtu.lv

© Springer Nature Singapore Pte Ltd. 2019
L. Lhotska et al. (eds.), *World Congress on Medical Physics and Biomedical Engineering 2018*,
IFMBE Proceedings 68/3, https://doi.org/10.1007/978-981-10-9023-3_1

transistor is influenced [11–13]. However, UV light is potentially harmful for humans and the environment.

As an alternative, optically induced semiconductor gas sensors were designed to provide both high sensitivity without heating the sensing area and a possibility to differentiate the presence of different gases by measuring the output signal and processing time constant at different supply voltages [14]. Furthermore, the electrophysical properties of an optically induced semiconductor gas sensor depending on the irradiation type and measurement of low acetone concentration under UV stimulation at room temperature were shown previously [15]. Despite the progress already made, several knowledge gaps in the working principle still exist.

2 Materials and Methods

The aim for the current research was to conduct sensor response measurements and time constant ratio calculations using optical stimulation under different irradiation wavelengths of visible spectrum light.

For those purposes, a test setup (Fig. 1) was built. The LED matrixes (wavelengths equal to 440, 530, 600, 710 nm) were used to provide homogeneous optical irradiation on the semiconductor sensor surface. By using an LED power supply, the equal photon flux was adjusted for each irradiation wavelength. Arduino controller was assembled to carry out measurements, process signal trace and export data for further processing to a PC.

2.1 Sensor Response Measurement

During each measurement in zero-level emission and in the presence of saturated ethanol vapor, an output signal was recorded. Finally, the sensor response on ethanol was calculated:

$$R = S(ethanol)/S(ambient) - 1 \qquad (1)$$

where S(ethanol)—is the signal peak value in an ethanol environment; S(ambient)—is the zero-level emission signal peak value.

2.2 Signal Rise and Relaxation Time

Trace of the sensor response was recorded and then processed to the time constant through the Arduino microcontroller (Fig. 2).

Signal rise and relaxation time can be characterized using the following definition:

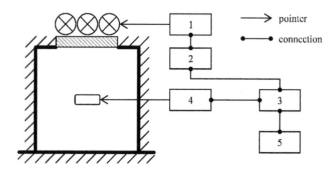

Fig. 1 Experimental test setup for measurements: (1) LED matrix; (2) LED power supply; (3) Arduino controller, used to power the sensor, measure the output signal and adjust LED output power; (4) semiconductor sensor inside gas chamber; (5) PC, used to process the signal and to adjust the Arduino controller. The LED matrix was used as an optical irradiation source to provide homogeneous irradiation (photon flux is 10^{20} cm^{-2}s^{-1}) on the sensor surface

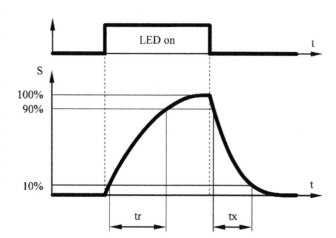

Fig. 2 Graphical interpretation of the signal rise time (tr) and relaxation time (tx) constants

- Rise time (tr)—the interval of time it takes an output signal to increase from 10 to 90% of its peak value;
- Relaxation time (tx)—the time interval required for an output signal to decrease from 90 to 10% of its peak value.

Time constant ratio defined as the ratio between the time constant of the signal in the presence of ethanol and the time constant of the signal in zero-level emission was calculated:

$$\tau = \tau(ethanol)/\tau(ambient) \qquad (2)$$

where τ(ethanol)—is the time consonant measured in an ethanol environment; τ(ambient)—is the time constant measured in a zero-level emission.

3 Results and Discussion

3.1 Sensor Response Measurement

The obtained result indicates the non-linear dependence of the semiconductor gas sensor response to the optical irradiation wavelength and can be approximated by a falling regression curve. The maximal signal response was detected at 440 nm and equals R = 104% and the minimum sensor response at 710 nm, R = 39% (Fig. 3).

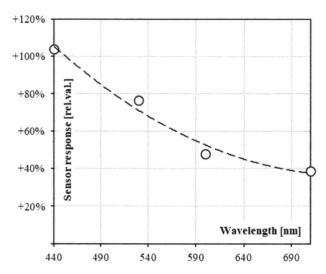

Fig. 3 Semiconductor gas sensor response to the ethanol vapor induced with different irradiation wavelengths

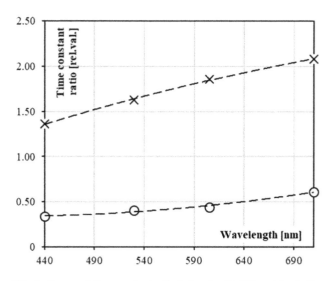

Fig. 4 Processed time constant ratio induced with different irradiation wavelengths: rise time (o) and relaxation time (x)

3.2 Signal Rise and Relaxation Time

Figure 4 shows the processed signal rise and relaxation time from the signal shapes recorded at a different irradiation wavelength.

Obtained results indicate non-linear dependence of processed time constant ratio to optical irradiation wavelength for both rise and relaxation time. Both results can be approximated by a rising regression curve. For both time constant ratios, the minimal value was detected at 440 nm and maximal value at 710 nm.

4 Conclusions

Visible spectrum LED light may be used for an optically induced semiconductor gas sensor to carry out the sensor response measurement of ethanol vapor. Highest sensor response was detected at the lowest tested irradiation wavelength of 440 nm.

Processed values of signal rise and relaxation time ratio may be used to sense the presence of ethanol vapor in air.

A correlation between the sensor response and time constant ratio was not observed and, therefore, both methods can be used jointly to increase the detection quality.

Although results demonstrate potential application of visible light usage to detect presence of ethanol, current results are achieved for saturated ethanol vapor and, therefore, concentration comparison to modern progress cannot be done.

Conflict of Interest The authors declare that they have no conflict of interest.

References

1. Kampa,M., Castanas,E.: Human health effects of air pollution. Environmental Pollution 151(2), 362–367 (2008).
2. Katsouyanni,K.: Ambient air pollution and health. British Medical Bulletin 68(1), 143–156 (2003).
3. Risby,T.H., Sehnert,S.S.: Clinical application of breath biomarkers of oxidative stress status. Free Radical Biology and Medicine 27 (11–12), 1182–1192 (1999).
4. de Zwart,L.L., Meerman,J.H., et al.: Biomarkers of free radical damage: Applications in experimental animals and in humans. Free Radical Biology and Medicine 26(1–2), 202–226 (1999).
5. Galassetti,P.R., Novak,B., et al.: Breath ethanol and acetone as indicators of serum glucose levels: an initial report. Diabetes Technology & Therapeutics 7(1), 115–123 (2005).
6. Cope,K., Risby,T., Diehl,A.M.: Increased gastrointestinal ethanol production in obese mice: Implications for fatty liver disease pathogenesis. Gastroenterology 119(5), 1340–1347 (2000).

7. Miekisch,W., Schubert,J.K., Noeldge-Schomburg,G.F.: Diagnostic potential of breath analysis—focus on volatile organic compounds. Clinica Chimica Acta 347(1–2), 25–39 (2004).

8. Arshak,K., Moore,E., et al.: A review of gas sensors employed in electronic nose applications. Sensor Review 24(2), 181–198 (2004).

9. Kim,H.J., Lee,J.H.: Highly sensitive and selective gas sensors using p-type oxide semiconductors: Overview. Sensors and Actuators, B: Chemical 192, 607–627 (2014).

10. Karmakar,M., Mondal,B., et al.: Acetone and ethanol sensing of barium hexaferrite particles: A case study considering the possibilities of non-conventional hexaferrite sensor. Sensors and Actuators B: Chemical 190, 627–633 (2014).

11. Herrán,J., Fernández-González,O., et at.: Photoactivated solid-state gas sensor for carbon dioxide detection at room temperature. Sensors and Actuators B: Chemical 149(2), 368–372 (2010).

12. Comini.E., Faglia,G., Sberveglieri,G.: UV light activation of tin oxide thin films for NO2 sensing at low temperatures. Sensors and Actuators B: Chemical 78(1–3), 73–77 (2001).

13. Chen,H., Liu,Y., et al.: A comparative study on UV light activated porous TiO2 and ZnO film sensors for gas sensing at room temperature. Ceramics International 38(1), 503–509 (2012).

14. Dekhtyar,Y., Sneiders,M., et al.: Towards Optically Induced Semiconductor Human Exhalation Gas Sensor. In: XIV Mediterranean Conference on Medical and Biological Engineering and Computing 2016, IFMBE Proceedings, vol. 57, pp. 482–485. Springer, Cham (2016).

15. Dekhtyar,Y., Sneiders,M., et al.: Optically Induced Semiconductor Gas Sensor: Acetone Detection Range using Continuous and Cyclic Optical Irradiation Types. In: EMBEC & NBC 2017. EMBEC 2017, NBC 2017. IFMBE Proceedings, vol. 65, pp. 330–333. Springer, Singapore (2018).

Study of the Influence of the Molecular Weight of the Polymer Used as a Coating on Magnetite Nanoparticles

Christian Chapa, Diana Lara, and Perla García

Abstract

Designing coated magnetic nanoparticles for nanomedicine applications, such as magnetic resonance imaging contrast enhancement, hyperthermia, and drug-delivery, has been in the focus of scientific interest for the last decade. Biocompatible polymers are used as nanoparticles coating for its physical and chemical properties that are very useful for biomedical applications. The aim of this contribution was to prepare the magnetite nanoparticles stabilized with poly (ethylene glycol) (PEG) and poly (ethylene glycol) methyl ether (mePEG) to elucidate the influence of the molecular weight on the corresponding amount of coating. The X-ray diffraction studies determined inverse spinel structure of magnetite nanoparticles, and field–emission scanning electron microscopy indicated the formation of quasi-spherical nanostructures with the final average particle size of 88–136 nm depending on the type of polymer coating. The bonding status of different polymers on the magnetite nanoparticles was confirmed by the Fourier transform infrared spectroscopy. According to the thermogravimetric analysis polymer amount in nanocomposites is related to molecular weight in the PEG-modified MNPs. The results of this study indicate the possibility of controlling the properties of theranostics nanomaterials, starting from the molecular weight of the polymer used as a coating.

Keywords

Nanomedicine • Magnetite • Nanocomposite

C. Chapa (✉) · D. Lara · P. García
Autonomous University of Ciudad Juarez,
32310 Juarez City, CHIH, Mexico
e-mail: christian.chapa@uacj.mx

1 Introduction

Magnetite nanoparticles (MNPs) can be used for both diagnosis and therapy in the nanomedicine field [1, 2]. However, the naked magnetite nanoparticles are barely suitable for biomedical applications due to the aggregation near the physiological pH in the absence of electrostatic or steric stabilization [3]. Huge efforts are made to find a coating for MNPs, which fulfills the serious criteria of biocompatibility, and the chemical and the colloidal stability as well [4]. An alternative for overcoming such difficulties is to modify the MNPs with biocompatible polymers.

Poly (ethylene glycol) (PEG), is one of the main coatings that is used in nanomedicine because of its effectiveness to inhibit the absorption of proteins in the blood and the capture of nanoparticles by phagocytic cells as well as its long history of safety in humans and classification as Generally Regarded as Safe (GRAS) by the FDA [5]. To date, the synthesis of PEG-modified magnetite, including those derivatives of PEG, have been attempted to investigate the effects of individual parameters on the properties of modified NPs. However, it is difficult to manipulate the amount of polymer adsorbed to the particles. In addition, it is necessary to study the influence of molecular weight on the amount of polymer associated with MNP.

The molecular weight of the polymer used as a coating for nanoparticles is a very important variable because it is directly related to the physical properties of the nanomedicine system. Currently, the molecular weight of the polymer used as a nanoparticle coating is rarely considered within the criteria to control the properties of the final material. In this work, we quantify a critical parameter that influences the efficiency of the nanomedicine systems, the amount of PEG and PEGME associated to MNPs in relation to the molecular weight of the polymer used.

© Springer Nature Singapore Pte Ltd. 2019
L. Lhotska et al. (eds.), *World Congress on Medical Physics and Biomedical Engineering 2018*,
IFMBE Proceedings 68/3, https://doi.org/10.1007/978-981-10-9023-3_2

2 Materials and Methods

2.1 Preparation of Magnetite Nanoparticles

The magnetite nanoparticles were synthesized by coprecipitation method as previously reported elsewhere [6]. Briefly, the precursor solutions ferric chloride (10 mM) and ferrous sulfate (5 mM) were mixed with magnetic stirring for 1 h at 60 °C, then a strong base solution (NH_4OH) was added dropwise until pH 13. After that, several washes were made until reaching pH 7, the obtained precipitated was repeatedly collected by centrifugation (10 min, 9000 G). The precipitated product was dried in oven at 100 °C and stored to further modification and characterization.

2.2 Preparation of PEG and PEGME-Modified Magnetite

For modification of MNPs, 25 mg of the collected magnetite resuspended in 2% triton. Meanwhile, four different polymer solutions, poly (ethylene glycol) (PEG400 and PEG3350) and poly (ethylene glycol) methyl ether (PEGME400 and PEGME2000), were prepared using 50 mg of the polymer in 30 mL of distilled water. For each coating reaction, the polymer solution was added directly to stirred magnetite. The pH was then adjusted to 9, by addition of NH_4OH or HCl, as required, and the reaction mixture was stirred by ultrasound for 5 min. After that, several washes were made with distilled water until reaching neutral pH. Finally, the product was placed in an oven at 80 °C for 24 h and store for further characterization.

2.3 Characterization of Nanoparticles

The magnetite sample was analyzed in PANalytical X'Pert MRD PRO equipment with a Cu-kα source ($\lambda = 1.5406$ Å) operated at 40 kV and 30 mA and at a scanning rate of 0.1° $2\theta s^{-1}$ from 10 to 80° 2θ. The morphology of the bared MNPs and surface-modified MNPs were evaluated by Scanning Electronic Microscopy (Hitachi SU5000). To determine the particle size of the nanoparticles diameter of 100 individual nanoparticles was measured directly from the images using the line tool, the results are presented as mean ± SD and statistical comparisons between groups were carried out using one-way ANOVA followed by the Student's t-test. Infrared spectra were obtained by using Fourier Transform Infrared (FTIR) with attenuated total reflection (ATR) spectrometer (Nicolet 6700/Thermo Electron). The infrared spectra were recorded with a resolution of 4 cm^{-1}, and the scan range was set from 4000 to 600 cm^{-1}.

The results are presented as the average of 32 scans. Thermogravimetric analysis (TGA) were performed on a Q600 thermobalance from TA Instruments Inc. The samples were heated from ∼25 to 600 °C, at a heating rate of 10 °C/min. The content of polymer on the modified-MNPs was determined by the weight loss of the samples.

3 Results and Discussion

The X-ray diffraction pattern of MNPs is shown in Fig. 1. Positions and relative peaks intensities point to pure magnetite phase. Positions and relative intensities of all diffraction peaks can be indexed to a Fe_3O_4 phase. The XRD pattern was indexed in a spinel-like structure and there was no evidence of any extra phases, indicating that pure magnetite had been produced under the experimental conditions.

Figure 2 shows selected SEM images with their corresponding particle size distribution histograms for modified-MNPs. SEM images of nanoparticles illustrate that the nanoparticles consist of roughly spherical and almost uniform in size nanoparticles with an average diameter of about 254 ± 169.5 nm for M-PEG400, 58 ± 18.9 nm for M-PEG3350, 85 ± 31.2 nm for M-PEGME550, and 102 ± 35.4 nm for M-PEGME2000. For the bare nanoparticles the main size was 17 nm ± 3.9 nm. The coating with polymeric materials did significantly ($P > 0.05$) alter mean size. The differences in size indicates that the nanoparticles were not individually coated, but that a nanocomposite was formed in which several magnetite NPs were immobilized in the polymeric matrix. For all four polymers there is a distribution of chain lengths. This could be important for the short-chain ones. PEG is OH end-capped while PEGME is

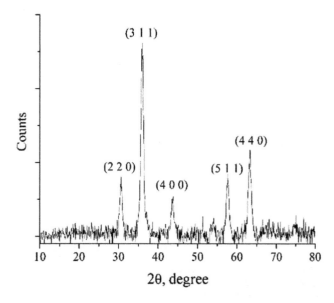

Fig. 1 XRD pattern of magnetite

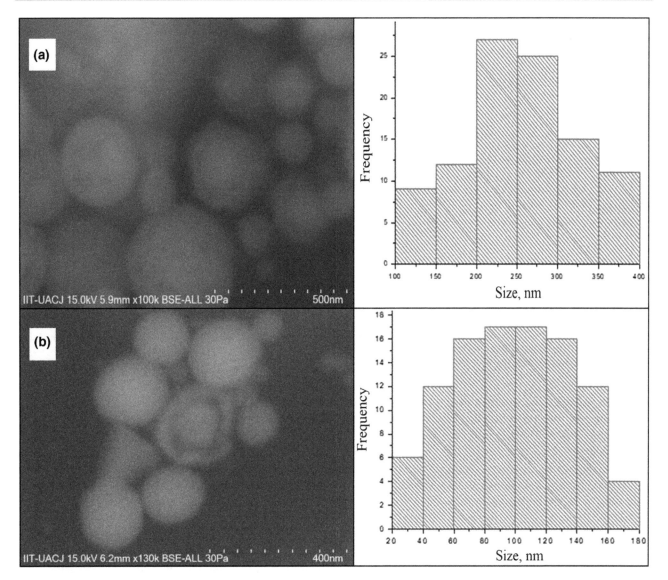

Fig. 2 Selected SEM micrographs for M-PEG400 (**a**) and M-PEGME2000 (**b**) and their corresponding histogram of size distribution based on SEM images

capped with CH_3 groups. The OH-terminals in the short PEG chains facilitate the formation of hydrogen bonds between the molecules around the magnetite nanoparticles which causes the formation of larger particles. While the terminal CH_3 in PEGME does not form hydrogen bonds with the polymer backbone, this prevent the association of a greater number of molecules, which explains the formation of nanoparticles with smaller size.

The chemical signature of polymer molecules associated to magnetite nanoparticles was demonstrated by IR spectroscopy (Fig. 3). The IR spectra of Poly(ethylene glycol), (PEG, linear formula: $H(OCH_2CH_2)_nOH$) and Poly(ethylene glycol) methyl ether (PEGME, linear formula: $CH_3(OCH_2CH_2)_nOH$) have very few differences between them. PEG and PEGME consist of the monomer (-O-CH_2-CH_2-), differing only in the chain length and end groups. It can be noticed that both, PEG and PEGME, exhibit bands around 2877 and 1460 cm^{-1} attributed to C-H stretching and bending modes of methylene groups, respectively. The absorption peak at 1355 cm^{-1} is attributed to O-H bending mode and the absorption peak at 1100 cm^{-1} is due to the C-O stretching vibration mode. It can be noticed that the two bands near to 1100 cm^{-1} (1147 and 1061 cm^{-1}) in both, PEG and PEGME, are revealed when the molecular weight increases because, when increasing the molecular weight, there are more C-O bonds that contribute to the stretching vibration mode. New absorption peak can be found at about 1730 cm^{-1} by comparing the IR spectra of the polymers

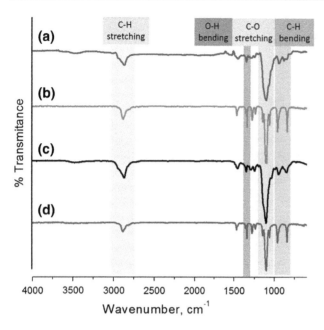

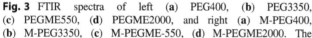

Fig. 3 FTIR spectra of left (**a**) PEG400, (**b**) PEG3350, (**c**) PEGME550, (**d**) PEGME2000, and right (**a**) M-PEG400, (**b**) M-PEG3350, (**c**) M-PEGME-550, (**d**) M-PEGME2000. The

absorption bands of representative functional groups are indicated in colored bars (Color figure online)

(left) with those of the modified magnetite nanoparticles (right). The new absorption band can be assigned to C = O, this suggests that -CH$_2$–OH group has been oxidized to COO$^-$ group during the reaction. The strong absorption band at about 700 cm^{-1} is due to Fe–O stretching vibration for the polymer-modified magnetite nanoparticles.

In order to confirm the content of polymeric material in the nanocomposites the thermal behavior of the modified-MNPs, M-PEG and M-PEGME nanocomposites

were measured via thermogravimetric analysis (Fig. 4). The mass reduction below 100 °C corresponding to the loss of residual water. The thermal behavior of the polymeric materials was previously measured, their decomposition starts at around 200 °C, therefore, the weight loss that occurs after 200 °C is attributed to the decomposition of polymeric material in the nanocomposite. The weight loss of the nanocomposites after 200 °C was 5.82% for M-PEG400, 8.15% for M-PEG3350, 16.40% for M-PEGME550, and 3.74% for M-PEGME2000. The thermal decomposition of PEGME matrix is presumably due to the breakdown of organic skeleton in a complex degradation process.

4 Conclusion

Magnetite nanoparticles which were synthesized by coprecipitation method were modified with PEG and PEGME to form the nanocomposites. The XRD patterns show pure phase magnetite. According to SEM measurements bare magnetite nanoparticles has a mean particle size of 17 ± 3.9 nm. After coating with PEG and PEGME, the M-polymer nanocomposites have significantly bigger mean particle size with polydispersity attributed to the distribution of chain lengths. The FTIR shows the vibrational modes of the magnetite and the polymeric material onto the nanocomposites and a IR band at 1730 cm^{-1} indicates the union between the polymer and the magnetite. There is a relationship between the molecular weight of the PEG and

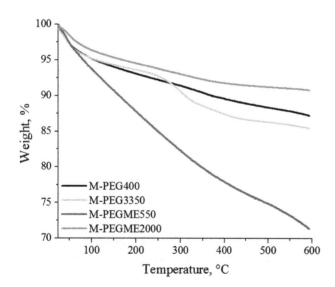

Fig. 4 TGA of M-PEG400, M-PEG3350, M-PEGME550, and M-PEGME2000 nanocomposites

the amount of weight loss from 200 °C; however, this relationship is not observed when the PEGME is used.

Acknowledgements The authors would like to thank the CONACYT for supporting this research.

Conflict of Interest All authors declare no conflict of interests in this work.

References

1. Golovin, Y. I., Klyachko, N. L., Majouga, A. G., Sokolsky, M., Kabanov, A. V. Theranostic multimodal potential of magnetic nanoparticles actuated by non-heating low frequency magnetic field in the new-generation nanomedicine. J Nanoparticle Res 19:63 (2017)

2. Bañobre-López, M., Piñeiro, Y., López-Quintela, M. A., Rivas, J. Magnetic Nanoparticles for Biomedical Applications. In: Handbook of Nanomaterials Properties. Springer Berlin Heidelberg, Berlin, Heidelberg, pp 457–493 (2014)

3. Favela-Camacho, S. E., Pérez-Robles, J. F., García-Casillas, P. E., Godinez-Garcia, A. Stability of magnetite nanoparticles with different coatings in a simulated blood plasma. J Nanoparticle Res 18:176 (2016)

4. Mandal, S., and Chaudhuri, K. Magnetic Core-Shell Nanoparticles for Biomedical Applications. In: Complex Magnetic Nanostructures. Springer International Publishing, Cham, pp 425–453 (2017)

5. Suk, J. S., Xu, Q., Kim, N., Hanes, J., Ensign, L. M. PEGylation as a strategy for improving nanoparticle-based drug and gene delivery. Adv Drug Deliv Rev 99:28–51 (2016)

6. Flores-Urquizo, I. A., García-Casillas, P., Chapa-González, C. Development of magnetic nanoparticles $Fe_2^{+3}X^{+2}O_4$ (X = Fe, Co y Ni) coated by amino silane. Rev Mex Ing Biomed 38 (2017)

Classification Algorithm Improvement for Physical Activity Recognition in Maritime Environments

Ardo Allik, Kristjan Pilt, Deniss Karai, Ivo Fridolin, Mairo Leier, and Gert Jervan

Abstract

Human activity recognition using wearable sensors and classification methods provides valuable information for the assessment of user's physical activity levels and for the development of more precise energy expenditure models, which can be used to proactively prevent cardiovascular diseases and obesity. The aim of this study was to evaluate how maritime environment and sea waves affect the performance of modern physical activity recognition methods, which has not yet been investigated. Two similar test suits were conducted on land and on a small yacht where subjects performed various activities, which were grouped into five different activity types of static, transitions, walking, running and jumping. Average activity type classification sensitivity with a decision tree classifier trained using land-based signals from one tri-axial accelerometer placed on lower back and leave-one-subject-out cross-validation scheme was 0.95 ± 0.01 while classifying the activities performed on land, but decreased to 0.81 ± 0.17 while classifying the activities on sea. An additional component produced by sea waves with a frequency of 0.3–0.8 Hz and a peak-to-peak amplitude of 2 m/s^2 was noted in sea-based signals. Additional filtration methods were developed with the aim to remove the effect of sea waves using the least amount of computational power in order to create a suitable solution for real-time activity classification. The results of this study can be used to develop more precise physical activity classification methods in maritime areas or other locations where background affects the accelerometer signals.

Keywords

Physical activity classification
Human activity recognition • Sea wave filtration
Maritime environment • Wearable systems

1 Introduction

Physical activity classification is used for human activity recognition, which provides information about user's movement and activity levels and can be used to develop more precise energy expenditure models [1]. Physical activity classification is usually based on acceleration signals, for which acceleration could be measured in various locations on the human body [2]. For this purpose it is possible to use the accelerometers inside smart phones [3] and smart watches [4] or integrate them into necklaces [5] and garments [6].

In activity recognition studies the accelerometer signal is often separated into static and dynamic components. The static component (Acc$_S$) is affected by gravity and gives information about body posture, the dynamic component (Acc$_D$) is based on motion and captures the body movement information. Different filtration methods have been used to separate these components, such as applying low-pass filter to separate Acc$_S$ and subtracting Acc$_S$ from acceleration signals to find Acc$_D$ [7, 8] or using two different filters [5, 9]. While acceleration signals measured on land are mostly only affected by gravity and body movement, accelerometers on sea also capture the fluctuation caused by sea waves, which could be decreased by applying appropriate filters.

The aim of this study was to evaluate how physical activity classifiers perform in maritime environment where sea waves affect the accelerometer signals and to develop a filter to remove acceleration induced by waves.

A. Allik (✉) · K. Pilt · D. Karai · I. Fridolin · M. Leier · G. Jervan
Tallinn University of Technology, Ehitajate tee 5, 19086 Tallinn, Estonia
e-mail: ardo.allik@ttu.ee

© Springer Nature Singapore Pte Ltd. 2019
L. Lhotska et al. (eds.), *World Congress on Medical Physics and Biomedical Engineering 2018*,
IFMBE Proceedings 68/3, https://doi.org/10.1007/978-981-10-9023-3_3

2 Methods

2.1 Instrumentation

Tri-axial acceleration signals were measured using Shimmer3 sensor platform accelerometer. Sensor was placed on the back of the lower trunk. This location was chosen because it is suitable for integrating the accelerometers inside garments and previous studies have had good results with accelerometers attached near the center of the body mass [10]. Signals were recorded using Wide Range accelerometer setting with dynamic range of ± 16 g and sampling rate of 512 Hz.

2.2 Test Overview

Two different test suits were carried out to evaluate the differences between signals measured on land and signals measured in maritime environments. The land-based signal measurements were done in a sports facility with ample space for movement. Accelerometer signals were measured during standing, sitting, lying, picking up objects, transitions between standing, sitting and lying, hopping on one leg and on both legs, walking and running. During walking and running subjects were asked to alternate their pace. For activity recognition and classification these signals were divided into different activity types of static, transitional, walking, running and hopping. The collected signals and their length are shown in Table 1.

The sea-based test suit was carried out in the cabin of a sailing yacht on an open gulf. During the experiments there was constant moderate wind of 10–11 m/s and the height of the waves was between 0.5 and 1 m. This test suit was very similar with the land-based test suit, with minor differences in time schedule. The size of the test area was about 6 m^2

and the height of the cabin was about 2 m. The measured signals are shown in Table 1.

2.3 Study Group

The study group for the land test suit consisted of 12 subjects, of which 10 were male and 2 female. The study group for the sea test suit consisted of 4 males. The subjects' anthropometric parameters are shown in Table 2.

2.4 Signal Resampling and Filtration

Tri-axial accelerometer signals measured with sampling frequency of 512 Hz were resampled to 100 Hz using MATLAB function *resample* without changing the signal spectrum.

Two different filtration methods (filtration method 2 and 3) were used in this study to test their ability to filter out acceleration caused by sea waves from the accelerometer signals and their effect on classification performance compared to a filter unable to filter out sea waves (filtration method 1). Butterworth type IIR filter was chosen, since IIR filters require less computational power than comparable FIR filters and are thus more suitable for real-time wearable systems. The filtration methods are shown in Table 3.

Following filtration signals were fragmented into fragments of 3 s with no inter-window gaps and without overlap, which has also been found suitable for classification in a previous work [11]. Afterwards each fragment was labelled with the corresponding activity type.

2.5 Feature Extraction

Machine learning based physical activity classification uses features which are extracted from signal fragments. These features are used as an input for training and evaluating the classifier and need to capture the specifics of human body movement and posture to increase the classification performance.

The feature set of 19 features used in this study was achieved after using feature selection methods on various sets adopted from previous studies [11]. Only time-domain features were used in order to keep the required computational power minimal.

2.6 Classifier Training and Evaluation

A machine learning based decision tree classifier was chosen, since it has been found to have a good performance

Table 1 Classified activity types and length of conducted activities

Activity type	Activities carried out in land test suit (in minutes)	Activities carried out in sea test suit (in minutes)
Static	Standing (1), sitting (1), lying (1)	Standing (2), sitting (2), lying (2)
Transitional	Picking up objects (1), sit-stand transitions (1), lie-sit transitions (1), lie-stand transitions (1)	Picking up objects (1), sit-stand transitions (1), lie-sit transitions (1), lie-stand transitions (1)
Walking	Walking (3)	Walking (3)
Running	Running (3)	Running (3)
Hopping	Hopping on one leg (1), hopping on both legs (0.5)	Hopping on one leg (1), hopping on both legs (0.5)

Table 2 Subjects' anthropometric parameters

Test suit	Number of subjects	Age (years) mean ± SD; range	Height (cm) mean ± SD; range	Weight (kg) mean ± SD; range
Land	12	30.3 ± 10.1; 15–46	176.8 ± 7.5; 167–189	71.1 ± 19.0; 45–115
Sea	4	36.3 ± 8.4; 26–57	181.8 ± 8.3; 171-190	87.0 ± 22.0; 65–115

Table 3 Filtration methods and filter parameters

Filtration method no.	Acceleration component	Filter type	Filter parameters (filter order, passband, stopband, passband ripple, stopband ripple)
1	Acc_S	Low-pass	3, 0.15 Hz, 2 Hz, 1 dB, 20 dB
	Acc_D		Found by subtracting Acc_S component from accelerometer signals
2	Acc_S	Low-pass	4, 0.1 Hz, 0.3 Hz, 1 dB, 20 dB
	Acc_D		Found by subtracting Acc_S component from accelerometer signals
3	Acc_S	Low-pass	6, 0.1 Hz, 0.2 Hz, 1 dB, 20 dB
	Acc_D	High-pass	6, 1.2 Hz, 0.8 Hz, 1 dB, 20 dB

with small classification time [10] and is thus also suitable for wearable systems. Classifier was trained using MATLAB's function *fitctree*, which returns a fitted binary decision tree.

To reduce overfitting errors, land signals were classified using a leave-one-subject-out cross-validation scheme, where each test subject's signals were classified using a classifier trained based on the signals of the other subjects. When classifying sea signals the classifier was trained based on all the land signals to determine how classifiers developed based on land signals perform on sea.

The results were evaluated using statistical measure sensitivity, which shows the ratio of true positives in relation to real positives [12].

3 Results

Figure 1 shows the Z-axis accelerometer values (axis perpendicular to the Earth) during lying of one test subject after applying the filtration methods shown in Table 3. Lying should be the most static position and so is chosen to illustrate the effect of the sea waves in the acceleration signals. The sea waves induced an additional component with a frequency of 0.3–0.8 Hz and a peak-to-peak amplitude of 4 m/s² in the sea-based Acc_D component signals and 2 m/s² in Acc_S component signals.

The average classification sensitivities using different filtration classifying land and sea signals are shown in Table 4. With all filtration methods the average classification sensitivity for land signals was higher than for sea signals.

4 Discussion

The average activity classification sensitivity of land signals was about 0.95 in this study, which is comparable to the results achieved by other researchers [2, 10]. No large difference could be noted in classification performance with different filtration methods. Figure 1 shows that while the Acc_D and Acc_S acceleration signals were more affected by sea waves with the first filtration method than with other filtration methods, the results of classifying sea signals were not considerably lower compared to other filtration methods based on the classification sensitivities in Table 4. Filtration method 2 was able to filter out sea wave induced acceleration from Acc_S, while filtration method 3 decreased it in both Acc_D and Acc_S. Also, with every filtration method sea signals were classified with lower performance than land signals.

There are several possibilities why filtering out the acceleration produced by the sea would not increase the classification sensitivities of sea signals to the same level as land signals. Some of the chosen features, such as mean of the signal, are only slightly affected by sea waves, which would not have an effect on the classification performance. The difference between sensitivities classifying land and sea signals could have been caused by using different testing areas—the size of the cabin of the sailing yacht might have restricted the subjects' movement, which could have caused the larger difference seen in classifying running and hopping activities compared to other activity types. Additionally, the study groups used in this study could have been too small, in which case additional measurements might be needed for definitive results.

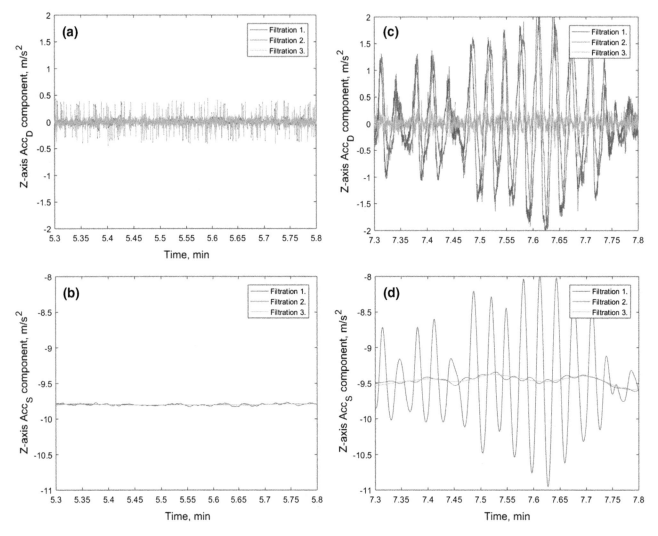

Fig. 1 Z-axis accelerometer signal values during lying from one test subject after applying different filtration methods. **a** Land, Acc_D component, **b** sea, Acc_D component, **c** land, Acc_S component, **d** sea, Acc_S component

Table 4 Classification sensitivities of different activity types with different filtration methods

Filtration method no.	Classified signals	Static	Transitional	Walking	Running	Hopping	Average
1	Land	0.97	0.94	0.94	0.94	0.96	0.95 ± 0.01
	Sea	0.82	0.94	1.00	0.56	0.72	0.81 ± 0.17
2	Land	0.97	0.94	1.00	0.82	0.93	0.93 ± 0.03
	Sea	0.92	0.89	0.99	0.66	0.81	0.85 ± 0.12
3	Land	0.97	0.95	0.99	0.92	0.84	0.94 ± 0.06
	Sea	0.89	0.89	1.00	0.47	0.72	0.79 ± 0.21

It is also important to note that the 0.3–0.8 Hz frequency of sea waves could contain important information for human activity recognition. In this case filtering out the sea wave induced noise would also lower the physical activity classification results, but in this study no large difference was found between classifying land signals when using different filtration methods. Only simple filtration methods using IIR filters were tested in the study, which are suitable for using in real-time wearable systems. Better results could be achieved with more complex filtration methods, but they would also require more computational power.

5 Conclusion

In this study it was evaluated how sea waves in maritime environment affect the performance of modern physical activity recognition methods. Even though the three different filtration methods used in this study removed the accelerations caused by sea waves in various degrees, no large differences could be noted between classification results. With every filtration method the classification sensitivities classifying activities performed on sea were lower compared to activities performed on land. This study helps to understand how sea waves affect the human activity recognition and is a good basis for further research.

Compliance with Ethical Standards

The research was funded partly by the Estonian Ministry of Education and Research under institutional research financing IUTs 19-1 and 19-2, by Estonian Centre of Excellence in IT (EXCITE) funded by European Regional Development Fund and supported by Study IT programme of HITSA.

The authors declare that they have no conflict of interest. All procedures performed in studies involving human participants were in accordance with the ethical standards of the institutional and/or national research committee and with the 1964 Helsinki declaration and its later amendments or comparable ethical standards. This article does not contain any studies with animals performed by any of the authors.

References

1. Altini, M., Penders, J., Vullers, R., Amft, O.: Estimating Energy Expenditure Using Body-Worn Accelerometers: A Comparison of Methods, Sensors Number and Positioning. IEEE Journal of Biomedical and Health Informatics 19 (1), 219–226 (2015).
2. Awais, M., Mellone, S., Chiari, L.: Physical activity classification meets daily life: Review on existing methodologies and open challenges. In: Proceedings of the 37th Annual International Conference of the IEEE Engineering in Medicine and Biology Society (EMBC), IEEE, Milan, Italy, pp. 5050–5053 (2015).
3. Lu, Y., Wei, Y., Liu, L., Zhong, J., Sun, L., Liu, Y. Towards unsupervised physical activity recognition using smartphone accelerometers. Multimedia Tools and Applications 76 (8), 10701–10719 (2017).
4. Weiss, G. M., Timko, J. L., Gallagher, C. M., Yoneda, K., Schreiber, A. J.: Smartwatch-based Activity Recognition: A Machine Learning Approach. In: Proceedings of the IEEE-EMBS International Conference on Biomedical and Health Informatics, IEEE, Las Vegas, USA, pp. 426–429 (2016).
5. Altini, M., Penders, J., Vullers, R.: Combining wearable accelerometer and physiological data for activity and energy expenditure estimation. In: Proceedings of the 4th Conference on Wireless Health, ACM New York, Baltimore, USA (2013).
6. Curone, D., Tognetti, A., Secco, E. L., Anania, G., Carbonaro, N., De Rossi, D., Magenes, G.: Heart Rate and Accelerometer Data Fusion for Activity Assessment of Rescuers During Emergency Interventions. IEEE Transactions on Information Technology in Biomedicine 14 (3), 702–710 (2010).
7. Moncada-Torres, A., Leuenberger, K., Gonzenbach, R., Luft, A., Gassert, R.: Activity classification based on inertial and barometric pressure sensors at different anatomical locations. Physiological Measurement 35 (7), 1245–1263 (2014).
8. Wang, J., Redmond, S. J., Voleno, M., Narayanan, M. R., Wang, N., Cerutti, S., Lovell, N. H.: Energy expenditure estimation during normal ambulation using triaxial accelerometry and barometric pressure. Physiological Measurement 33 (11), 1811–1830 (2012).
9. Tapia, E. M.: Using Machine Learning for Real-time Activity Recognition and Estimation of Energy Expenditure. PhD Thesis, Massachusetts Institute of Technology (2008).
10. Altun, K., Barshan, B., Tuncel, O.: Comparative study on classifying human activities with miniature inertial and magnetic sensors. Pattern Recognition 43 (10), 3605–3620 (2010).
11. Allik, A., Pilt, K., Karai, D., Fridolin, I., Leier, M., Jervan, G.: Activity Classification for Real-time Wearable Systems: Effect of Window Length, Sampling Frequency and Number of Features on Classifier Performance. In: Proceedings of the IEEE-EMBS Conference on Biomedical Engineering and Sciences, IEEE, Kuala Lumpur, Malaysia, pp. 460–464 (2016).
12. Powers, D. M. W.: Evaluation: from precision, recall and F-factor to ROC, informedness, markedness and correlation. Journal of Machine Learning Technologies 2 (1), 37–63 (2011).

First Steps Towards an Implantable Electromyography (EMG) Sensor Powered and Controlled by Galvanic Coupling

Laura Becerra-Fajardo and Antoni Ivorra

Abstract

In the past it has been proposed to use implanted electromyography (EMG) sensors for myoelectric control. In contrast to surface systems, these implanted sensors provide signals with low cross-talk. To achieve this, miniature implantable devices that acquire and transmit real-time EMG signals are necessary. We have recently in vivo demonstrated electronic implants for electrical stimulation which can be safely powered and independently addressed by means of galvanic coupling. Since these implants lack bulky components as coils and batteries, we anticipate it will be possible to accomplish very thin implants to be massively deployed in tissues. We have also shown that these devices can have bidirectional communication. The aim of this work is to demonstrate a circuit architecture for embedding EMG sensing capabilities in our galvanically powered implants. The circuit was simulated using intramuscular EMG signals obtained from an analytical infinite volume conductor model that used a similar implant configuration. The simulations showed that the proposed analog front-end is compatible with the galvanic powering scheme and does not affect the implant's ability to perform electrical stimulation. The system has a bandwidth of 958 Hz, an amplification gain of 45 dB, and an output-referred noise of 160 μV_{rms}. The proposed embedded EMG sensing capabilities will boost the use of these galvanically powered implants for diagnosis, and closed-loop control.

Keywords

Galvanic coupling • Microsensors • Microstimulator

1 Introduction

Apart from diagnosis, electromyography (EMG) has been extensively used as a source of control for exoskeletons and prostheses [1, 2], and has been proposed for closed-loop control in electrical stimulation. Implanted EMG sensors are an alternative to superficial EMG as they provide signals with lower cross-talk for myoelectric control, potentially helping to control devices with multiple degrees of freedom, which is not possible nowadays with superficial EMG [3]. These sensors have been tried as implantable central units wired to electrodes [4], which present surgical complexity; and as wireless implant capsules (e.g. IMES) [2]. Even though these capsules have accomplished high miniaturization levels, they are still too stiff and bulky to be massively deployed in tissues, hampering the development of a network of microdevices with high selectivity for stimulation and sensing. This can be explained as they require voluminous components as coils and batteries for power transfer/generation to produce the current magnitudes required for neuromuscular stimulation. In the case of inductive coupling, the implant's diameter is limited by the coil used; while in existing battery technologies, the implant's size is limited by their energy density.

In [5] we proposed a heterodox method to create ultrathin implants that lack coils and batteries. The implants act as rectifiers of innocuous high frequency (HF) current bursts (≥ 1 MHz) supplied to the tissues by galvanic coupling using superficial electrodes (Fig. 1). We in vivo demonstrated microcontrolled injectable stimulators that could be galvanically powered and that could deliver low frequency (LF) currents capable of stimulating excitable tissues [6]. These implants (diameter = 2 mm), made of commercially available components, are the first step towards future

L. Becerra-Fajardo (✉) · A. Ivorra
Department of Information and Communication Technologies,
Universitat Pompeu Fabra, 08018 Barcelona, Spain
e-mail: laura.becerra@upf.edu

A. Ivorra
Serra Húnter Fellow, Universitat Pompeu Fabra, 08018 Barcelona,
Spain

© Springer Nature Singapore Pte Ltd. 2019
L. Lhotska et al. (eds.), *World Congress on Medical Physics and Biomedical Engineering 2018*,
IFMBE Proceedings 68/3, https://doi.org/10.1007/978-981-10-9023-3_4

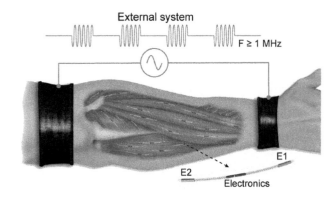

Fig. 1 Basic scheme of the method to galvanically power implants. An external system delivers high frequency current (≥ 1 MHz) bursts, which flow through the tissues by galvanic coupling. The implants are envisioned as elongated, flexible and ultrathin devices with two electrodes at opposite ends (E1 and E2) to pick up the high frequency current and rectify it for power and for electrical stimulation

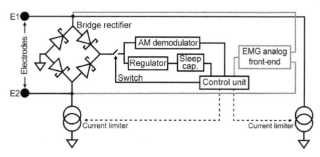

Fig. 2 Core architecture of the implantable device based on commercially available components. The analog front-end for EMG acquisition proposed here is shown in gray. The sleep capacitor powers the electronics, including the analog front-end, when no HF current is delivered by the external system

ultrathin and flexible implants (diameter < 1 mm) based on an application-specific integrated circuit (ASIC). In [7] we in vitro demonstrated that bidirectional communication is feasible in this method, allowing closed-loop control. The aim of this work is to demonstrate a signal conditioning electronics architecture (i.e. an analog front-end) for embedding EMG sensing capabilities in our galvanically powered implants. This will boost their use in diagnosis, closed-loop control in neuroprostheses, and man-machine interfaces as those used for prostheses control.

2 Methods

2.1 Proposed Electronic Architecture

The core architecture of the microcontrolled injectable stimulators was described in [6]. Briefly, implant electrodes E1 and E2 pick up a portion the HF current delivered to the tissues by the external system (Fig. 2). This HF current is full-wave rectified by a bridge rectifier. A regulator subcircuit stabilizes the rectified voltage to power the control unit (CU) and the rest of the electronics, and it is followed by a capacitor that powers the electronics during sleep mode (i.e. when no HF current is delivered by the external system). Current consumption is limited in this mode to keep the sleep capacitor's size small. The circuit includes a demodulator that is capable of extracting information amplitude modulated in the same HF current used for galvanic powering. This information is used to independently address each implant. The CU drives two current limiters that deliver LF current pulses for electrical stimulation. These current limiters are also used to modulate the HF current to send information from the implantable circuit to the external system [7].

In here we propose to add an analog front-end for EMG acquisition to this core architecture (Fig. 3a). The implant senses the EMG signals as the difference between the voltage obtained from the two implant electrodes E1 and E2 when no HF current is delivered by the external system. The first stage consists on analog filters: (1) a first-order low-pass filter ($F_c = 342$ kHz) used to protect the front-end's amplifier from the HF current used for powering and (2) a first-order high-pass filter to prevent possible dc components seen across the implant electrodes ($F_c = 73$ Hz). An operational amplifier (OPAMP) configured as a differential amplifier is used to amplify the voltage difference and suppress voltages that are common to both E1 and E2. A single-supply, low power OPAMP is used for this purpose as the implant's regulator delivers a single-supply voltage (VCC), and the front-end powers from the sleep capacitor. Additionally, the OPAMP must ensure low noise, very low input bias current, and very small package. The ADA4691-2 (by Analog Devices Inc.) is a $1.21 \times 1.22 \times 0.4$ mm dual OPAMP that provides all these features and has a shutdown mechanism for further power reduction. A voltage divider is used to set a steady state of VCC/2 V at the output of the differential amplifier when no EMG signal is picked-up by the implant electrodes. The second OPAMP of the package is used to further amplify the obtained voltage, and it is followed by a low-pass filter ($F_c = 1.25$ kHz) to avoid aliasing during digitalization.

2.2 Computer Simulations

Computer simulations were performed in a SPICE simulator (LTspice XVII by Linear Technologies) using the setup shown in Fig. 3b. A Thévenin equivalent is used to model the coupling between the tissue and implant electrodes E1 and E2, which have a diameter D and a separation distance L. If the implant is aligned with the electric field ($E(t)$), then $V_{Th} = LE(t)$. The equivalent resistance (R_{Th}) is the

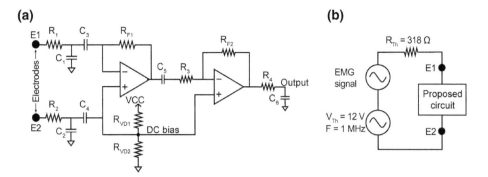

Fig. 3 **a** Analog front-end scheme added to the core architecture of the implants (Fig. 2). The circuit includes a filtering stage, differential amplification, and antialiasing filter. **b** Simulation setup used to evaluate the front-end proposed here. A Thévenin equivalent is used to model the coupling between muscle tissue and the implant electrodes E1 and E2. EMG signals are applied through the EMG voltage source, and a sinusoidal signal of 1 MHz is used as the HF current delivered for galvanic coupling (V_{Th})

resistance of the dipole formed by the two electrodes. If $L \gg D$, $R_{Th} = 1/\pi\sigma D$, where σ is the electrical conductivity of the tissue (S/m). The impedances of the implant electrodes are neglected for simplicity. For this study, we supposed the same length and diameter of the implant we demonstrated in [6] (L = 5 cm; D = 2 mm), an electric field of 240 V/m, and an electrical conductivity of 0.5 S/m which approximately corresponds to the admittivity magnitude of skeletal muscle at 1 MHz [8]. A voltage source connected in series with the Thévenin equivalent is used to simulate the EMG signals that appear across the implant electrodes. They are based on a 500 ms sample of a digitized intramuscular EMG signal obtained from an analytical infinite volume conductor model that used an implant configuration with two electrodes at opposite ends [3].

Transient, ac and noise analyses were used to evaluate the behavior of the proposed circuit, including if the front-end affected the ability of the implant to perform electrical stimulation, and to assess its gain and bandwidth.

3   Results

The simulations show that the proposed front-end does not unbalance the load of the circuit on any electrode, and the circuit is able to deliver symmetrical pulses of 2 mA, as those obtained in previous demonstrations [6].

Figure 4a shows the intramuscular EMG signal applied (top) and the output of the proposed front-end (bottom). The picked-up EMG signal was effectively amplified and biased over the 1.65 V established in the differential amplifier (VCC = 3.3 V). To analyze the delay between the picked-up EMG signal and the output of the analog front-end, the signals were normalized and resampled and a cross correlation was applied between them. The results show that the lag between the output and the original EMG signal was 58 μs.

Figure 4b shows the results obtained from the ac analysis. The acquisition system has a bandwidth of 958 Hz. The

Fig. 4 **a** Sample of simulated intramuscular EMG signal detected by a myoelectric sensor [3], and corresponding output signal of the analog front-end proposed. The lag between signals is 58 μs. **b** Frequency response of the analog front-end proposed

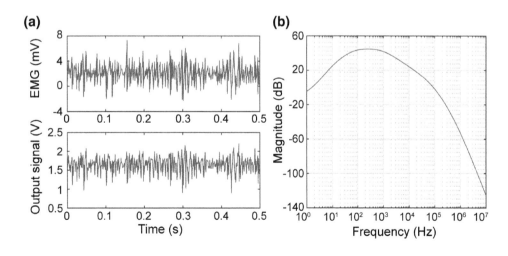

maximum gain of the EMG acquisition system is 45 dB. The output-referred noise of the analog front-end is 160 μV_{rms} (from 1 Hz to 10 MHz). Having in mind that the output signal without dc bias is 0.21 V_{rms}, the calculated signal-to-noise ratio (SNR) for the signal conditioning circuit is 62.4 dB.

4 Conclusion

In here we have proposed an analog front-end architecture for an existing microcontrolled injectable stimulator based on commercially available components and that is galvanically powered using HF currents. According to simulations, the proposed signal conditioning circuit is compatible with the galvanic powering scheme proposed and demonstrated in the past. The simulations also show that the circuit filters and amplifies the expected intramuscular EMG signals that are going to be picked up by the implant electrodes, as those obtained from an analytical model that used a similar implant configuration [3]. Additionally, the gain, bandwidth and noise parameters obtained with the simulations demonstrate that the proposed circuit has a similar behavior to that previously reported by implantable wireless EMG capsules [2]. In contrast to these capsules based on inductive coupling, our future galvanically powered implants based on ASICs will be potentially thinner and more flexible. In the near future, we plan to develop the first implantable prototypes for their evaluation in vivo.

The proposed embedded EMG sensing capabilities will boost the use of these galvanically powered implants for diagnosis, closed-loop control in neuroprostheses, and man-machine interfaces as those used for prostheses control.

Acknowledgements This project has received funding from the European Research Council (ERC) under the European Union's Horizon 2020 research and innovation programme (grant agreement No 724244).

Conflict of Interest The authors declare no conflict of interest.

References

1. Pons, J.L. Rehabilitation Exoskeletal Robotics. IEEE Eng. Med. Biol. Mag 29(3), 57–63 (2010).
2. Weir, R.F., et al. Implantable Myoelectric Sensors (IMESs) for Intramuscular Electromyogram Recording. IEEE Trans. Biomed. Eng. 56(1), 159–171 (2009).
3. Lowery, M.M., Weir, R.F., Kuiken, T.A. Simulation of Intramuscular EMG Signals Detected Using Implantable Myoelectric Sensors (IMES). IEEE Trans. Biomed. Eng. 53(10), 1926–1933 (2006).
4. Memberg, W.D., et al. Implanted Neuroprosthesis for Restoring Arm and Hand Function in People With High Level Tetraplegia. Arch Phys Med Rehabil. 95(6), 1201–1211 (2014).
5. Ivorra, A. Remote Electrical Stimulation by Means of Implanted Rectifiers. PLoS One 6(8), e23456 (2011).
6. Becerra-Fajardo, L., Schmidbauer, M., Ivorra, A. Demonstration of 2-mm-Thick Microcontrolled Injectable Stimulators Based on Rectification of High Frequency Current Bursts. IEEE Trans Neural Syst Rehabil Eng. 25(8), 1343–1352 (2016).
7. Becerra-Fajardo, L., Ivorra, A.: Bidirectional communications in wireless microstimulators based on electronic rectification of epidermically applied currents. In: 2015 7th International IEEE/EMBS Conference on Neural Engineering (NER), pp. 545–548. IEEE (2015).
8. Gabriel, S., Lau, R.W., Gabriel C. The dielectric properties of biological tissues: III. Parametric models for the dielectric spectrum of tissues, Phys. Med. Biol., 41(11), 2271–2293 (1996).

Powering Implants by Galvanic Coupling: A Validated Analytical Model Predicts Powers Above 1 mW in Injectable Implants

Marc Tudela-Pi, Laura Becerra-Fajardo, and Antoni Ivorra

Abstract

While galvanic coupling for intrabody communications has been proposed lately by different research groups, its use for powering active implantable medical devices remains almost non-existent. Here it is presented a simple analytical model able to estimate the attainable power by galvanic coupling based on the delivery of high frequency (>1 MHz) electric fields applied as short bursts. The results obtained with the analytical model, which is in vitro validated in the present study, indicate that time-averaged powers above 1 mW can be readily obtained in very thin (diameter < 1 mm) and short (length < 20 mm) elongated implants when fields which comply with safety standards (SAR < 10 W/kg) are present in the tissues where the implants are located. Remarkably, the model indicates that, for a given SAR, the attainable power is independent of the tissue conductivity and of the duration and repetition frequency of the bursts. This study reveals that galvanic coupling is a safe option to power very thin active implants, avoiding bulky components such as coils and batteries.

Keywords

Galvanic coupling • Active implant • Wireless power transfer

1 Introduction

Miniaturization of electronic medical implants has been hampered because of the use of batteries and inductive coupling for power. Both mechanisms require bulky and rigid parts which typically are much larger than the electronics they feed.

As we have recently shown in vivo [1], galvanic coupling can be an effective power transfer method which can lead to unprecedented implant miniaturization. Remarkably, although galvanic coupling for intrabody communications has been proposed lately by different research groups [2], its use for powering implants has remained almost non-existent. Reluctance to use galvanic coupling for power transfer may arise from not recognizing two facts. First, large magnitude high frequency (>1 MHz) currents can safely flow through the human body if applied as short bursts. Second, to obtain a sufficient voltage drop across its two intake (pick-up) electrodes, the implant can be shaped as a thin and flexible elongated body suitable for minimally invasive percutaneous deployment (Fig. 1a).

2 Methods

2.1 The Analytical Model and Its Rationale

Safety standards for human exposure to electromagnetic fields identify two general sources of risk regarding passage of radiofrequency (RF) currents through the body. On the one hand, the standards indicate risk of thermal damage due to the Joule effect, which roughly can be considered as frequency independent. On the other hand, the standards recognize risks caused by unsought electrical stimulation of excitatory tissues. In this case, safety thresholds increase with frequency. In particular, for frequencies above 1 MHz and short bursts, the IEEE standard [3] specifies limitations related to heating which are more restrictive than those

M. Tudela-Pi · L. Becerra-Fajardo · A. Ivorra
Department of Information and Communication Technologies,
Universitat Pompeu Fabra, 08018 Barcelona, Spain
e-mail: marc.tudela@upf.edu

A. Ivorra (✉)
Serra Húnter Fellow,
Universitat Pompeu Fabra, 08018 Barcelona,
Spain
e-mail: antoni.ivorra@upf.edu

© Springer Nature Singapore Pte Ltd. 2019
L. Lhotska et al. (eds.), *World Congress on Medical Physics and Biomedical Engineering 2018*,
IFMBE Proceedings 68/3, https://doi.org/10.1007/978-981-10-9023-3_5

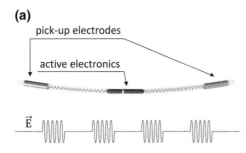

(a)

(b)

(c)

Fig. 1 **a** We envision thin and flexible implants powered by innocuous high frequency current bursts through tissues. **b** Simplified model of the implant; D, electrode diameter; L, implant length. **c** Equivalent

Thévenin circuit for the conductive medium and the field, connected to the implant circuitry modeled as a resistive load

related to stimulation. Therefore, here only the thermal limitation is considered as we deem that frequencies between 1 and 10 MHz will be adequate for galvanic coupling. (Because of the skin effect, frequencies above 10 MHz may not be convenient as at that frequency the effect becomes significant [4] and the operation of implants at deep locations would be hindered.)

The limitations specified by the standards regarding heating are indicated as a limitation to the so-called Specific Absorption Rate (SAR), which can be calculated as:

$$SAR = \frac{\sigma(E_{rms})^2}{\rho} \qquad (1)$$

where σ (S/m) is the electrical conductivity of the tissue, ρ (kg/m^3) is the mass density of the tissue and E_{rms} is the root mean square value of the applied electric field (V/m). For occupational exposure or persons in controlled environments —as would be the case considered here—this limit is 10 W/kg.

If the field is applied as sinusoidal bursts (duration = B, repetition rate = F):

$$SAR = \frac{\sigma(E_{peak})^2}{2\rho} FB \qquad (2)$$

where E_{peak} is the amplitude of the sinusoidal burst. Then, assuming a homogeneous medium and a uniform electric field, the maximum peak voltage across two points a and b at a separation distance L, is:

$$V_{ab\_peak} = \overrightarrow{E_{peak}} \cdot \overrightarrow{r_{ab}} = E_{peak}L \cos(\theta)$$
$$= \sqrt{\frac{2\rho SAR_{max}}{\sigma FB}} L \cos(\theta) \qquad (3)$$

where $\overrightarrow{r_{ab}}$ is the vector defined by the points a and b and θ is the angle between this vector and the field. If these two points correspond to the location of the two intake (pick-up) electrodes of the implant—here simply modeled as spheres

—then the rms open circuit voltage (i.e. the voltage across the electrodes in Fig. 1b if $R_{Load} = \infty$) is:

$$V_{OC\_rms} = \frac{V_{ab\_peak}}{\sqrt{2}} \sqrt{FB} = \sqrt{\frac{\rho SAR_{max}}{\sigma}} L \cos(\theta) \qquad (4)$$

And assuming that the implant is aligned with the electric field ($\theta = 0$):

$$V_{OC\_rms} = \sqrt{\frac{\rho SAR_{max}}{\sigma}} L \qquad (5)$$

If the implant circuitry is simply modeled as a load (R_{Load}), then it is straightforward to compute the maximum power that it will dissipate if the Thévenin resistance (R_{Th}) of the Thévenin equivalent circuit (Fig. 1c) is known:

$$P_{Load\_max} = P_{Load}(if \ R_{Load} = R_{Th}) = \frac{V_{OC\_rms}^2}{4R_{Th}} \qquad (6)$$

For an infinite medium (i.e. the electrodes that deliver the field are far away), R_{Th} is the resistance across the two implant electrodes. If the implant electrodes are modeled as spheres with a separation distance much larger than the diameter ($L \gg D$), then [5]:

$$R_{Th} = \frac{1}{\sigma \pi D} \qquad (7)$$

Therefore, the maximum average power for a given SAR limit (SAR_{max}) that can be drawn by the implant is:

$$P_{Load\_max} = \frac{V_{OC\_rms}^2}{4R_{Th}} = \frac{\pi}{4} SAR_{max} \rho DL^2 \qquad (8)$$

2.2 Experimental Setup

An in vitro experimental setup that replicates the assumptions made to generate the previous model was developed to validate the analytical expression in (8), (Fig. 2).

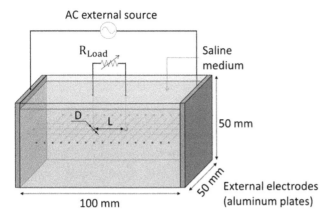

AC external source

R_{Load}

Saline medium

D L

50 mm

50 mm

External electrodes (aluminum plates)

100 mm

Fig. 2 Schematic representation of the in vitro setup developed to validate the analytical model (see text for details). For geometrical reference, a 0.5 cm × 0.5 cm grid made of cotton thread was sewed across the plates, 2.5 cm from the bottom of the structure

The field was delivered by two 5 cm × 5 cm parallel aluminum plates held at a distance of 10 cm using two polycarbonate plates. 1 MHz voltage bursts across these two electrodes were generated by the combination of a function generator (4060 Series by BK Precision) and a high voltage amplifier (WMA 300 by Falco systems).

The above electrode structure was placed inside a 19 cm × 14 cm × 6.3 cm glass container, which was filled with a saline solution. Three different concentrations were tried: 0.3, 0.6 and 0.9% NaCl. The conductivity of these solutions at 20 °C, as measured with a conductivity tester (HI 98312 by Hanna), was 0.58 S/m, 1.1 S/m and 1.56 S/m respectively.

The magnitude of the applied voltage was adjusted for a SAR of 10 W/kg according to expression 2.

The pick-up electrodes of the implants were modeled by stainless steel spherical electrodes (SAE 316) with four different diameters: 0.5, 1, 1.5, and 2 mm. Each electrode was laser welded to a 10 cm piece of 32 AWG enameled copper wire.

In each trial, a pair of electrodes with the same diameter was connected to a 1 kΩ high-precision potentiometer (R_{Load} in Fig. 2).The potentiometer was adjusted in advance to the computed R_{Th} value in order to drawn the maximum possible power.

Instant power dissipated at the potentiometer was computed as the square of the recorded voltage across it, using an oscilloscope (TPS2014 by Tektronix Inc.), divided by the

value of R_{Load}. Power was then time averaged for the burst repetition period (i.e. average power) and for the duration of the burst (i.e. peak power).

3 Results and Discussion

Figure 3 shows a set of numerical results from the analytical model (expression 8) together with the corresponding experimental results. As it can be observed, the experimental results fit the analytical model and powers above 1 mW are obtained for all the diameters when the inter electrode distance (L) is larger than 2 cm.

Expression 8 indicates that the maximum time-averaged power than can be drawn by the implant is independent of the duration (B) and the repetition frequency (F) of the bursts. However, the same does not apply for the maximum power that can be drawn during the burst (peak power). This is illustrated in Fig. 4d where it is modeled the delivery of three different burst patterns (Fig. 4a–c) with a rms value of 141 V/m (D = 1 mm, SAR = 10 W/kg, and ρ = 1000 kg/m³). Since the power increases with the square of the applied voltage, peak powers of hundreds of mW can be obtained when short burst (B < 1 ms) are applied.

Remarkably, the analytical model (expression 8) indicates that, for a given SAR, the attainable power is independent of the tissue conductivity. This is validated in Fig. 4e where experimental results are displayed for three different conductivities ($\sigma_{0.9\%}$ = 1.56 S/m, $\sigma_{0.6\%}$ = 1.1 S/m, and $\sigma_{0.3\%}$ = 0.58 S/m).

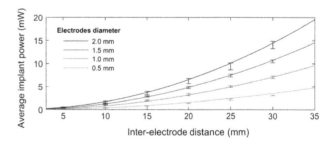

Fig. 3 Time-averaged power dissipated at R_{Load} for different electrodes diameters (D) and inter-electrode distances (L). Solid lines correspond to the analytical model (expression 8). Bars correspond to the 95% confidence interval of 10 experimental measurements. Experimental parameters: σ = 0.58 S/m (0.3% NaCl), ρ = 1000 kg/m³ (water), V_{rms} = 13.1 V, R_{Load} = 549 Ω

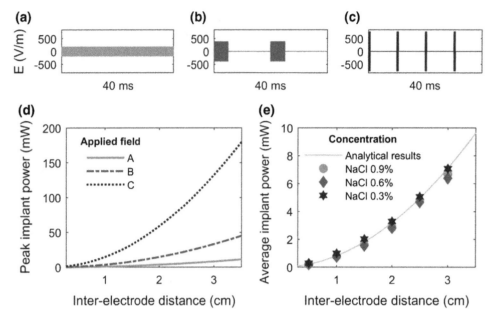

Fig. 4 Dependences on the burst pattern and the media conductivity. **a** Continuous 1 MHz 141 V_{rms}/m sinusoidal field, E_{Peak} = 200 V/m. **b** Bursts of 1 MHz 141 V_{rms}/m sinusoidal field B = 5 ms, F = 50 Hz, E_{Peak} = 400 V/m. **c** Bursts of 1 MHz 141 V_{rms}/m sinusoidal field B = 625 μs, F = 100 Hz, E_{Peak} = 800 V/m. **d** Predicted peak power attainable in an implant (D = 1 mm) for the burst patterns of subfigures A, B and C. **e** Modeled and experimentally obtained time-averaged powers for three conductivities (D = 1 mm, SAR = 10 W/kg)

4 Conclusions

The results of this study indicate that it should be possible to safely supply powers well above 1 mW to thin elongated (D ≤ 1 mm, L ≥ 20 mm) implants using galvanic coupling. For that, the energizing electric field—generated by an externally delivered current or voltage—can consist in a high frequency (>1 MHz) sinusoidal wave applied in short bursts resulting in a SAR value below 10 W/kg.

The predicted time-averaged powers (>1mW) are one to three orders of magnitude above the requirements of some implantable technologies for sensing and stimulation [6], including conventional pacemakers (10 to 50 μW).

Remarkably, the developed analytical model predicts that, for a given SAR, the attainable power is independent on the tissue conductivity and on the duration and repetition frequency of the bursts. These predictions are experimentally confirmed by the in vitro model that replicates the assumptions made to generate the analytical model.

Peak powers in the order of tens or hundreds of mW are attainable when bursts are very short in comparison to their repetition period. This suggests that galvanic coupling may be particularly useful in applications requiring large amounts of power in short intervals, such as is the case of neuromuscular stimulation [1].

Acknowledgements This project has received funding from the European Research Council (ERC) under the European Union's Horizon 2020 research and innovation programme (grant agreement No 724244).

Conflict of Interest The authors declare no conflict of interest.

References

1. Becerra-Fajardo, L., Schmidbauer, M., Ivorra, A.: Demonstration of 2 mm thick microcontrolled injectable stimulators based on rectification of high frequency current burst. IEEE Trans Neural Sysy. Rehabil. Eng. 25(8), 1342–52 (2017).
2. Seyedi, M., Kibret, B., Lai, D.T.H., Faulkner, M.: A survey on intrabody communications for body area network applications. IEEE Trans. Biomed. Eng. 60(8), 2067–79 (2013).
3. IEEE Standard for Safety Levels With Respect to Human Exposure to Radio Frequency Electromagnetic Fields, 3 kHz to 300 GHz. 1–238 (2006).
4. Vander Vorst, A., Rosen, A., Kotsuka, Y.: RF/Microwave Interaction with Biological Tissues. New Jersey: John Wiley & Sons, Inc. (2006).
5. Grimnes, S., Martinsen, Ø.G., Geometrical Analysis, in: Bioimpedance & Bioelectricity basics. 2nd edn. Elseiver, Oxford (2008).
6. Ho, J.S. et al., Supplementary material in: Wireless power transfer to deep-tissue microimplants. Proc. Natl. Acad. Sci. 111, 7974–7979 (2014).

Microfluidic Diamond Biosensor Using NV Centre Charge State Detection

Marie Krečmarová, Thijs Vandenryt, Michal Gulka, Emilie Bourgeois, Ladislav Fekete, Pavel Hubík, Ronald Thoelen, Vincent Mortet, and Miloš Nesládek

Abstract

In this work we develop DNA sensors that are based on charge switching in colour centres in diamond. The presented method allows the combination of luminescence sensor and electrochemical sensor working on the principle of electrochemical impedance spectroscopy (EIS). The sensor employs specifically designed diamond structures grown by the means of chemical Vapour deposition (CVD). This diamond structure consists of highly boron doped diamond electrode on which an intrinsic diamond layer is deposited. This intrinsic layer is about 15 nm thick and it contains NV colour centres. The device is then embedded in polydimethylsiloxane (PDMS) microfluidic flow cell and covered by a transparent indium tin oxide (ITO) coated electrode. The switching of the NV centre charge state as a response, on diamond surface termination, is crucial tool for the sensitive charged molecules sensing. First we demonstrated high sensitivity of the near surface NV centres on a diamond biosensor surface charge termination. The measured data are supported by band bending modelling. Negative O- terminated surface results in a preferable NV centre charge state of NV^0 or NV^-, whereas positive H- terminated surface leads to mostly non-PL NV^+ charge state. By this principle any charged molecule, such as polymer on DNA, can be detected by a customized surface functionalization. Functionality of the microfluidic diamond device is also verified by the EIS.

Keywords

Nitrogen vacancy • Diamond • Microfluidics

1 Introduction

Diamond as a wide gap (5.5 eV) material semiconducting or insulating has unique combination of physical and chemical properties for biosensing applications. We present a novel concept of microfluidic biosensor based on combination of exceptional electrochemical properties of boron doped diamond and luminescence properties of quantum colour NV centre defects in diamond. The novel design allows electrochemical detection together with NV centre charge state detection. The NV centre can exhibit in negative NV^- and neutral NV^0 charge state or in dark NV^+ charge state without photoluminescence (PL) [1], depending on number of electrons co-occupying the centre. The conversion of $NV^{0/+/-}$ charge state can be achieved passively by an interaction of NV centre at the surface with electric field at (surface termination [2]) or actively (electrically [1], optically [3], and electrochemically [4]) manipulation of the Fermi level position. The NV centre sensitivity, placed close to a specifically terminated diamond surface can be used as a tool for detection of charged molecules, such as DNA [5]. In this work we attempt to develop a microfluidic system for combining electrochemical and NV optical detection of DNA attachment.

M. Krečmarová (✉) · M. Gulka · V. Mortet · M. Nesládek
Faculty of Biomedical Engineering, Czech Technical University in Prague, Nám. Sítná 3105, 272 01 Kladno, Czech Republic
e-mail: krecmarova.marie@gmail.com

T. Vandenryt · M. Gulka · E. Bourgeois · R. Thoelen
Material Physics Division, Institute for Materials Research, University of Hasselt, Wetenschapspark 1, B 3590 Diepenbeek, Belgium

T. Vandenryt · E. Bourgeois · R. Thoelen
IMOMEC Division of IMEC, Wetenschapspark 1, B 3590 Diepenbeek, Belgium

L. Fekete · P. Hubík · V. Mortet
Institute of Physics, Academy of Sciences Czech Republic V.V.I, Na Slovance 1999/2, 182 21 Praha 8, Czech Republic

© Springer Nature Singapore Pte Ltd. 2019
L. Lhotska et al. (eds.), *World Congress on Medical Physics and Biomedical Engineering 2018*,
IFMBE Proceedings 68/3, https://doi.org/10.1007/978-981-10-9023-3_6

2 Methods

2.1 Preparation of the Microfluidic Diamond Biosensor

Metallically boron doped homoepitaxial diamond electrode was grown on (100) oriented single crystal diamond substrate using the Seki Technotron AX5010 CVD system in hydrogen rich gas including 1% CH_4 and 6000 ppm of trimethylboron (TMB), with 550 W microwave power, 100 mbar working pressure and 1100 °C substrate temperature. Diamond surface was re-polished after the growth to obtain a very flat surface. The B-doped electrode was then overgrown by thin un-doped diamond thin film using the Astex AX6500 CVD system with 0.4% of CH_4 in hydrogen gas, 900 W microwave power, 93 mbar working pressure and 940 °C substrate temperature. NV centres were incorporated in the un-doped film during the growth from a residual nitrogen chamber background. The diamond device structure includes ~ 20 μm metallically B-doped layer and ~ 15 nm thin N- film with NV centres on top. Oxygen and hydrogen surface termination was made by O_2 and H_2 plasma exposure. Ohmic titanium/gold (Ti/Au = 20/80 nm) contacts were fabricated by classical photolithography followed by thermal annealing. The polydimethylsiloxane (PDMS) flow cell was fabricated by creation of the microchannel in a PDMS film followed by laser cutting of the desired design. The PDMS flow cell is placed on the diamond substrate embedded in the electronic board and covered by indium tin oxide (ITO) coated glass slide.

2.2 Characterizations of the Biosensor

Raman spectroscopy measurement was performed using the Horiba Jobin Yvon T64000 Raman with excitation wavelengths of 488 nm. AFM was measured in semi contact mode using the NTEGRA-Prima. Resistivity, carrier density and mobility of the B- doped layer were determined by van der Pauw method at room temperature. NV PL was measured by a home built confocal microscope excited with 500 mW Gem laser from Laser Quantum producing 532 nm CW excitation with 8mW (spectroscopy) and 50 μW (mapping) laser power, directed by Gaussian beam optics to $100\times$ Long Working Distance M Plan Semi-Apochromat—LMPLFLN Olympus objective with N.A. of 0.8 and WD of 3.4 mm. The PL was detected using Excelitas single photon counter and 650 nm long pass filter to remove the laser light and NV^0 PL. Band bending under a different surface termination of the diamond device (B- layer: [acceptor] = 1e21 cm^{-3} and N- layer: [donor] = 1e18 cm^{-3}) was simulated using the AMPS-1D software (Analysis of Microelectronic and Photonic Structures) by numerical calculations of the Poisson's and the electron and hole continuity equations. The electrochemical

impedance spectroscopy (EIS) was measured using the Hewlett Packard 4248 Precision LCR meter. The impedance was measured with 10 mV AC voltage applied to the flow cell at room temperature. As an electrolyte we used BupHTM phosphate PBS with pH 7.2. The impedance was measured for 31 frequencies in a range from 100 to 100 kHz. The electrical circuit parameters were determined by fitting the impedance data to the equivalent circuit (Fig. 5a—inset) using Zview software from Scribner Associates.

3 Results and Discussions

3.1 Concept of the Microfluidic Diamond Device

Highly boron doped film with thickness of 20 μm, conductivity of 333 S/cm, carrier concentration of $1.1 \cdot 10^{21}$ cm^{-3} and mobility of 1.9 cm^2/Vs was deposited on (100) oriented diamond substrate, followed by polishing to obtain a very smooth surface. AFM image shows RMS surface roughness of 1.2 nm (Fig. 1a). Raman spectra of the boron doped layer is shown in Fig. 1b, showing distinct Fano resonances of boron electronic state with 1329 cm^{-1} Raman line [6], corresponding to metallically doped diamond. The narrow bandwidth of the line confirms extreme quality of B-doped epilayer. The metallically boron doped layer was further overgrown by 15 nm thin un-doped film with NV centres incorporated during the growth from residual nitrogen background.

The biosensor consists of diamond device embedded in the electronic board gap and covered with PDMS flow cell and ITO coated glass slide on top (Fig. 2). Total thickness of the PDMS/ITO structure is 2.5 mm, which is smaller than working distance of the used objective (WD = 3.4 mm). Volume of the flow cell cavity is 20 μl. The NV PL is detected by confocal microscopy through transparent ITO glass in the microfluidic cavity (Fig. 2a). The diamond device is equipped with 4 electrodes. Source electrode on B-layer and gate electrode on N-layer can be used for active control of NV centre charge state by an applied electric field. Another electrodes are used for electrochemical control or readout. Working electrode is located directly on the B- layer and reference electrode on the N-layer. Counter electrode represents optically transparent conductive ITO coated glass which covers the microfluidic PDMS flow cell (Fig. 2b).

3.2 NV Centre Charge State Control and Electrochemical Readout

3.2.1 NV Centre Manipulation by Diamond Surface Termination

The construction of sensor was investigated first by passive charge control of NV centres by oxygen and hydrogen

Fig. 1 AFM surface morphology of the diamond device (**a**), Raman spectra of the highly boron doped diamond (**b**)

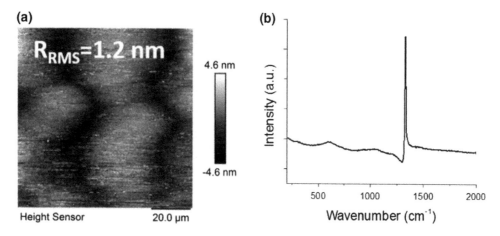

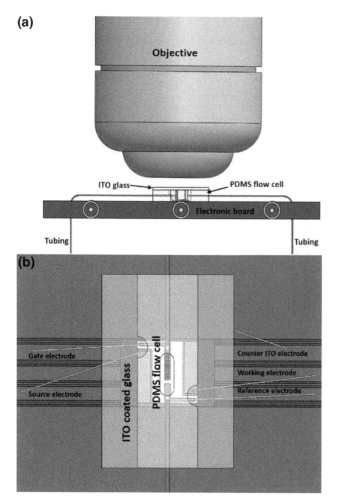

Fig. 2 Scheme of the sensing setup (**a**) and detail of the biosensor including the diamond device embedded in the electronic board and covered by PDMS flow cell and/TO glass slide (**b**)

NV PL. It means that NV centers in thin N-layer are affected from both side, by B-layer from bottom and by surface termination from top. With H-termination (Fig. 3a), the Fermi level is positioned below the NV$^-$ and NV^0 transition level and thus NV centres are in non-PL NV$^+$ charge state mostly. Even after O-termination (Fig. 3b) the NV centres are mainly in NV^0 charge state.

Figure 4a–d shows PL cross section and surface maps for both surface terminations measured with 650 nm long pass filter removing NV^0 PL. Very low counts were observed on hydrogen terminated surface, suggesting that the partial depletion of the N-layer due to the band bending from B-doped side is further increased by upward bending at the H-terminated surface. PL spectroscopy (Fig. 4e) shows predominantly NV^0 charge state with O-termination, which is quenched by H-termination to more preferably non-PL NV$^+$ state. In case of oxygen terminated surface PL counts increased dramatically by more than an order of magnitude (2 times higher) compared to H-terminated surface, suggesting stabilization of the NV$^-$ due to restoration of downward band bending at the surface. This is a significant improvement compared to previous research where the changes were $\sim 30 - 50\%$ [7].

The idea of the proposed biosensor is using of the close surface (few nm) NV centre charge state sensitivity for detection of biomolecules caring electrical charge, for instance DNA. The detection principle can be based on switching of NV PL as a response on the diamond functionalization. Positive surface functionalization quenches the NV PL, after immobilization of negative DNA molecules, the NV PL can be reversed to negative NV$^-$ charge state by captured electron and higher signal of PL detected.

3.2.2 Electrochemical Properties of the Biosensor

Further on we verified electrochemical functionality of the biosensor by electrochemical impedance spectroscopy (EIS) for the H- terminated diamond surface. The EIS system includes counter electrode (conductive ITO coated

surface termination. Figure 3 shows modeling of the band bending of the diamond device with H- and O- surface termination. The highly boron doped under-layer tunes the Fermi level position close to the valence band and quenches

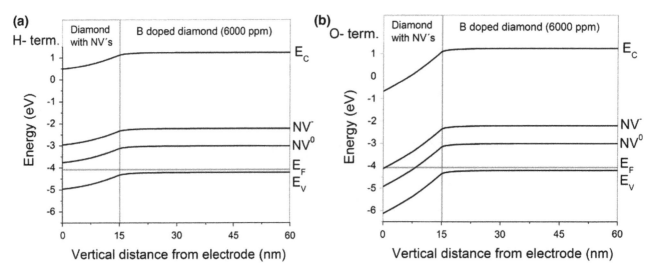

Fig. 3 Band bending of the diamond device including highly B doped layer and thin diamond film with NV centres on top with hydrogen (**a**) and oxygen (**b**) surface termination

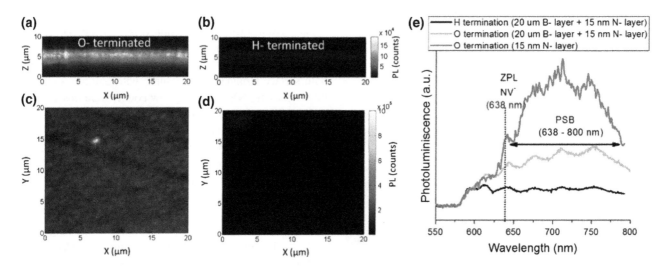

Fig. 4 PL intensity cross section (**a**, **b**) and surface (**c**, **d**) maps measured with 650 nm long pass filter removing NV^0 PL of diamond device showing app. 15 times higher intensity for oxygen terminated diamond surface (**a**, **c**) than for hydrogen terminated diamond surface (**b**, **d**). PL spectra shows app. 2 times higher intensity for oxygen termination with predominantly NV^0 charge state (red line) then hydrogen termination with predominantly NV^+ charge state (black line) (**e**). For illustration is shown PL spectra of oxygen terminated N- layer without presence of the B-doped under-layer showing higher occupation of the NV^- charge state (blue line) with zero phonon line (ZPL) at 638 nm and broad phonon sidebands (PSB) from 638–800 nm

glass slide), working electrode (highly boron doped diamond electrode with 15 nm thin diamond NV centre containing film on top) and electrolyte between the electrodes. The electrochemical system represents equivalent electrical circuit (Fig. 5a—inset), which was used for fitting of the impedance data (Fig. 5a). The circuit consists of resistance of the electrolyte R_S, double layer capacitance C_{dl}, charge transfer resistance R_{cr} and Warburg impedance W. Nyquist and Bode plot of the complex

impedance and phase is shown in Fig. 5a, b. The measured EIS data fitted to the equivalent electrical circuit shows R_{cr} of 2.89 kΩ cm^2 and low C_{dl} of 0.67 μF/cm^2 comparable with good quality boron doped electodes [8, 9]. Due to the high sensitivity of the EIS, biochemical reactions on a diamond surface can be monitored, such as DNA hybridization in real time. By the concept of the biosensor the EIS can effectively support the very sensitive NV centre charge state readout.

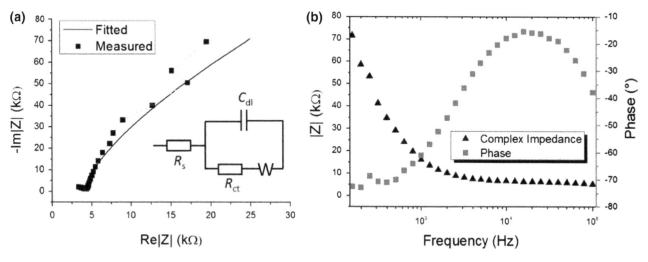

Fig. 5 Nyquist plot with inset an equivalent electrical circuit for fitting of parameters (**a**) and Bode plot of the complex impedance and phase (**b**) of H-terminated diamond device

4 Conclusions

In conclusion, we fabricated diamond microfluidic NV centre charge state biosensor for charge state molecule detection. The detection principle is based on close surface NV centres charge state manipulation by a surface functionalization. Moreover the developed biosensor allows an active control of the NV centre charge state electrochemically or electrically for biosensing applications. The diamond chip consists of metallic boron doped diamond and thin (~ 15 nm) un-doped diamond film with NV centres on top. The inter conversion of $NV^{-/0/+}$ charge state was investigated by positive H- and negative O- termination as a pre-step for further detection of charged molecules. The H-termination results in a preferable non-PL NV^+ charge state, whereas O- termination leads to NV^0 or NV^- charge state. Presence of the charged molecules as DNA on the surface of the diamond chip can alter this switching effect for the shallow NV centres. The electrochemical functionality of the biosensor was verified by electrochemical impedance spectroscopy. In futher work we plan to demonstrate simultaneous measurement of DNA hybridization by both optical and electrochemical means.

Acknowledgements The authors acknowledge the institutional resources of the Department of Biomedical Technology FBMI CTU; CTU grant SGS14/214/OHK4/3T/17; the Czech Science Foundation (GACR) Grant ID: GAČR 16-16336S; EU–FP7 research grant DIA-DEMS, No. 611143, FWO (Flanders) G.0.943.11.N.10.; the Erasmus Student Mobility Grant and the J.E. Purkyne fellowship awarded by Academy of Sciences of the Czech Republic.

Conflict of Interest The authors declare that they have no conflict of interest.

References

1. C. Schreyvogel, V. Polyakov, R. Wunderlich, J. Meijer and C. E. Nebel. Active charge state control of single NV centres in diamond by in-plane Al-Schottky junctions. *Scientific Reports* 5 (2015) 12160.
2. V. Petrakova, I. Rehor, J. Stursa, M. Ledvina, M. Nesladek, P. Cigler. Charge-sensitive fluorescent nanosensors created from nanodiamonds *Nanoscale* 7 (2015) 12307.
3. X. Chen, Ch. Zou, Z. Gong, Ch. Dong, G. Guo and F. Sun. Subdiffraction optical manipulation of the charge state of nitrogen vacancy center in diamond. Light: *Science and Applications* 4(1) (2015) e230.
4. S. Karavelia, O. Gaathona, A. Wolcotta, R. Sakakibaraa, O. A. Shemeshe, D. S. Peterkag, Edward S. Boydene, J. S. Owend, R. Yusteg and Dirk Englund. Modulation of nitrogen vacancy charge state and fluorescence in nanodiamonds using electrochemical potential. Proceedings of the National Academy of Sciences of the United States of America 113(15) (2016) 3938–3943.
5. V. Petrakova, V. Benson, M. Buncek, A. Fiserova, M. Ledvina, J. Stursa, P. Cigler and M. Nesladek. Imaging of transfection and intracellular release of intact, non-labeled DNA using fluorescent nanodiamonds. *Nanoscale* 8 (2016) 12002–12012.
6. V.D. Blank, V.N. Denisov, A.N. Kirichenko, M.S. Kuznetsov, B.N. Mavrin, S.A. Nosukhin, S.A. Terentiev. Raman scattering by defect-induced excitations in boron-doped diamond single crystals. *Diamond & Related Materials* 17(11) (2008) 1840–1843.
7. V. Petráková, M. Nesládek, A. Taylor, F. Fendrych, P. Cígler, M. Ledvina, J. Vacík, J. Štursa, and J. Kučka. Luminescence properties of engineered nitrogen vacancy centers in a close surface proximity. *Phys. Status Solidi* 208(9) (2011) 2051–2056.
8. R. F. Teófilo, H. J. Ceragioli, A. C. Peterlevitz, L. M. Da Silva, F. S. Damos, M. M. C. Ferreira, V. Baranauskas, L.T. Kubota. Improvement of the electrochemical properties of "as-grown" boron-doped polycrystalline diamond electrodes deposited on tungsten wires using ethanol. *J Solid State Electrochem* 11 (2007) 1449–1457.
9. B. P. Chaplina, D. K. Hublerb, J. Farrell. Understanding anodic wear at boron doped diamond film electrodes. *Electrochimica Acta* 89 (2013) 122–131.

UWB Platform for Vital Signs Detection and Monitoring

Ivana Čuljak, Hrvoje Mihaldinec, Zrinka Kovačić, Mario Cifrek, and Hrvoje Džapo

Abstract

In this paper a non-invasive method for vital signs detection and monitoring employing ultrawide bandwidth (UWB) technology is proposed. The idea behind the proposed approach is to use UWB technology to measure the variations in RF communication channel characteristics to detect vital signs. The feasibility of the proposed approach was experimentally tested with custom developed software and hardware platform, based on Decawave DW1000 M module. The platform was specifically designed and optimized to enable data acquisition of physical parameters with high sampling rate. The experiment consisted of placing UWB transmitter and receiver units in predetermined positions on the anterior and posterior thoracic wall where the transmitter generates an ultra-short UWB pulse with a minimum bandwidth of 500 MHz. From the channel impulse response (CIR) of the UWB channel measured at the UWB receiver the information about the heart muscle contraction is extracted. The heart muscle contraction detection algorithm exploits on the fact that the heart movements are periodic and therefore suitable for detection in frequency domain. The algorithm for feature extraction processes the sampled signal frequency spectrum, in order to estimate the heart rate. The obtained results showed the validity of the proposed approach and the performance of the proposed method was evaluated in comparison with commercial ECG device.

Keywords

Ultrawide bandwidth (UWB) • Vital signs monitoring
Heart rate detection • Biomedical instrumentation

I. Čuljak (✉) · H. Mihaldinec · Z. Kovačić · M. Cifrek · H. Džapo
Faculty of Electrical Engineering and Computing, University of
Zagreb, Unska 3, 10 000 Zagreb, Croatia
e-mail: ivana.culjak@fer.hr

1 Introduction

Early detection of the cardiac abnormalities can provide timely patient's diagnosis and as such could prevent further progression of the potential pathogenesis condition. UWB communication is based on sending very short pulses (duration from 100 ps to 1 ns) which occupy the minimum bandwidth of 500 MHz [1]. The Federal Communications Commission (FCC) specifies spectral and power limitation of the UWB technology in medical applications [2]. Features of ultrawideband (UWB) communication systems provide its application in numerous research fields, including monitoring and detecting human physiological parameters [3]. One of the first studies in the detection of human breathing and heartbeat through the wall by radar was proposed in [4] by Bugaev et al. After that there were also some further studies [5, 6] on the same topic. Related studies [7–9] proposed similar measuring technique but with different approaches in the implementation of feature extracting algorithm. In [7] four methods were proposed for detection of simulated heart rate using ultrawideband signals (based on variance, Fast Fourier Transformation (FFT), Wavelet and on Power Spectrum Density (PSD)). An algorithm for heart and respiration rate with UWB radar based on detection of movement energy in a specified band of frequency using wavelet and filter banks that contain other motion is proposed in [8]. Researchers proposed the harmonic path (HAPA) algorithm for vital signs monitoring based on UWB [9].

The existing approaches to the HR detection by means of UWB signals are based on a radar working principle where the electromagnetic energy propagated towards and through the body and it is reflected back from the tissue interfaces due to different relative dielectric constants of the organs [10]. In this paper we propose a different approach where a transversal method of measuring the changes in characteristics of the electromagnetic wave propagation path due to heart muscle activity are detected. We use the solution based on the commercially available UWB chip and module [11]. We employ similar algorithms for features extraction based

© Springer Nature Singapore Pte Ltd. 2019
L. Lhotska et al. (eds.), *World Congress on Medical Physics and Biomedical Engineering 2018*,
IFMBE Proceedings 68/3, https://doi.org/10.1007/978-981-10-9023-3_7

on variance and FFT, as described in [7]. The measurement method and algorithm are elaborated in Sect. 2. The preliminary experimental results of the proposed measurement method are presented in Sect. 3.

2   Measurement Method Description

2.1   Theoretical Background

The proposed method is based on the assumption that the heart motion is continuously changing the communication channel parameters and thus modulates the observed signal power on the receiver side; under the assumption that the transmitted signal power and mutual position of transmitter and receiver units do not change over the time. It has been shown that the signal propagation in a human body is significantly attenuated due to the muscle tissue layer variations in all frequency bands [12]. The UWB module [11] used as a basis for our custom-designed measurement system provides a real-time information about the channel impulse response (CIR), which is described in details in [13]. By measuring CIR we can detect variations in reflection and absorption rates at the receiving end for the uniformly transmitted signal. With continuous measurements in the observed time interval we should be able to detect the motion of the heart muscle and with its rate of change. To ensure the maximum signal absorption rate (SAR), the antenna was placed no less than 1 cm from the human body, as it was shown in [14]. Due to the significant difference of the dielectric properties of an inflated lung and deflated lung in the proposed method, to minimize this effect measurements were executed during the retained exhale phase.

2.2   System Description

The prototype hardware module shown in Fig. 2 is based on the Decawave DWM1000 M module [11]. The DW1000 chip [11] is IEEE 802.15.4-2011 compliant and provides high communication data rate (up to 6.8 Mbps), with optional simultaneous measurement of transmit and receive timestamps with a resolution of 15.6 ps. Furthermore, the chip provides large bank of memory that holds the accumulated channel impulse response (CIR) data. Data contains 992 or 1016 samples with complex values, a 16-bit real integer and a 16-bit imaginary integer for nominal pulse repetition frequency (PRF) of 16 or 64 MHz respectively [13]. The size of the printed circuit board (PCB) is 60 x 25 mm s shown in Fig. 1, which makes it suitable for on-body placement and measurements. Software stack is based on FreeRTOS operating system. The sampling rate of the system was set at 50 ms, where each sample represents the current CIR reading.

Fig. 1 Developed UWB platform based on Decawave DWM1000 M module

The ECG waveform was recorded simultaneously with the commercially available ECG device Shimmer3 [15], which provides a configurable digital front-end for 5-lead ECG measurements with sampling rate set at 512 Hz. ECG vector from the right arm (RA) position to the LA (left arm) position was measured on Lead I. To synchronize the measurements between our test system and ECG reference device, the trigger signal from module was used and directly connected to the ECG lead. The pin was triggered when the UWB module started the measurement and stopped at the end of the predetermined measurement window (Fig. 2).

2.3   Experimental Setup

The UWB transmitter and receiver units were placed on the anterior and posterior thoracic wall of the test subject. Measurements were conducted on a 25-year-old female subject. To minimize non-related effects, the subject was asked to exhale and sit still for 10–30 s while readings were captured. The antenna placement is shown on Fig. 3. Chip antenna of DW1000 M module was placed no less than 1 cm from the skin. The experiments were executed with variable channel settings given in Table 1 to ensure the validity of the platform and the proposed algorithm. The CIR accumulator was sampled every 50 ms and logged on the SD card for offline processing.

2.4   Algorithm for Heart Rate Detection

In this study we applied a basic algorithm for heart motion estimation, like the method described in [7]. The first step was to remove background clutter by subtracting the average value of the CIR accumulator with the originally received CIR data. Then we applied the moving average filter to increase a dynamic effect of the heart motion. Finally, the

Fig. 2 Block diagram of the developed UWB module

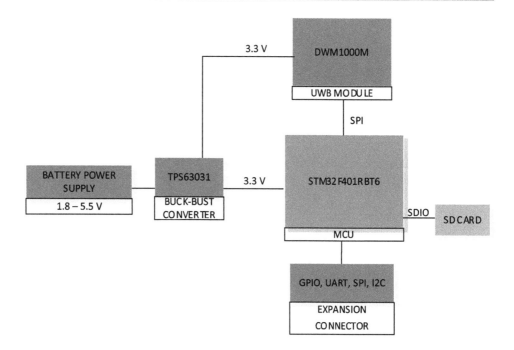

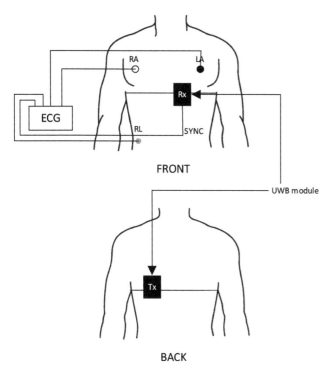

Fig. 3 The experimental setup for monitoring heart rate activity with UWB transmitter and receiver

Table 1 Parameters of the DWM1000 used in experiments

Channel	f_c (MHz)	BW (MHz)
2	3993.6	499.2
4	3993.6	1331.2
5	6489.6	499.2
7	6489.6	1081.6

total energy in time was calculated to show absorption due to the cardiac cycle. We expect that the maximum energy expressed at the end of the systole when heart is contracted, and the minimum value of the energy at the end of the diastole when the heart is full of the blood. Then we apply FFT on all calculated total energies in one measurement. In the frequency domain we estimate the HR and compare it with the commercial ECG device to validate our measurement setup. The R peaks were detected by means of Pan Tompkins algorithm [16]. The error rate is expressed in

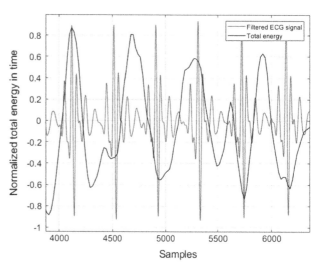

Fig. 4 Comparison of the normalized total energy in time $E(t)$ with the filtered ECG signal: for UWB parameters channel 4, PRF 64 MHz; BT 6.8 Mbps

Table 2 Heart rate error rate obtained by experiment

Meas. No.	Meas.duration (s)	UWB Parameters [11]	Ref. HR frequency [Hz]	Estimated HR frequency [Hz]	ER (%)
1	15.16	CH5, 16 MHz, 6.8 Mbps	1.31	1.25	4.99
2	13.49	CH7, 16 MHz, 6.8 Mbps	1.18	1.18	0.00
3	20.50	CH2, 16 MHz, 6.8 Mbps	1.21	1.26	4.00
4	25.08	CH4, 16 MHz, 6.8 Mbps	1.31	1.11	15.15

percentage as a difference in the HR frequency estimated by our device and the reference ECG system.

3 Results

We conducted several series of measurements following the measurement procedure described in the Sect. 2. In the Fig. 4 the $E(t)$, calculated from sampled CIR variations in time, is plotted along with simultaneous measurements acquired by means of reference ECG device.

The results shown in Fig. 4 show comparison of the normalized signal of the total energy in time E(t) with the filtered ECG signal for one case of UWB communication parameters. The results are very promising because one can visually notice very good agreement of the waveforms in terms of following the activity related to heart rate. Our results also exhibit good heart motion estimation for all tested UWB channels. The y-axis provides the normalized signal in reference to maximum values of the total energy in time. Additionally, error rates in HR detection were calculated presented in Table 2.

4 Conclusion

In this paper we proved that it is feasible to detect heart muscle motion and heart rate relatively accurately by using off-the-shelf UWB chips, employing the transversal method of measuring the changes in characteristics of EM wave propagation media due to the heart muscle activity. Previous studies have shown that it is possible to use a radar principle in conjunction with UWB technology to detect heart rate but such approach requires costly laboratory equipment. The proposed method is implemented in custom-designed prototype device which is suitable for low–power, low–profile, safe, contactless and inexpensive vital signs monitoring system solution, that can be tailored to achieve the high performance for various specific applications. The future research will focus on further investigation on the most useful parts of the information and efficient algorithms for extracting the information contained in the CIR accumulator.

Acknowledgements This research has been supported by the European Regional Development Fund under the grant KK.01.1.1.01.0009 (DATACROSS).

Conflict of Interest The authors declare that they have no conflict of interest.

References

1. Di Benedetto, Maria-Gabriella. UWB Communication Systems: A Comprehensive Overview. New York, NY [u.a.]: Hindawi Publ (2006)
2. Code of Federal Regulations: Part 15 Subpart F Ultra-Wideband Operation. Federal Communications Commission (May 2002)
3. Christine N. Paulson, John T. Chang, Carlos E. Romero, Joseph Watson, Fred J. Pearce, Nathan Levin, Ultra-wideband radar methods and techniques of medical sensing and imaging, Proc. SPIE 6007, Smart Medical and Biomedical Sensor Technology III, 60070L (11 November 2005)
4. Bugaev, A. A. et al., Through wall sensing of human breathing and heart beating by monochromatic radar, Proceedings of the Tenth International Conference on Grounds Penetrating Radar, 2004. GPR 2004. 1 (2004)
5. Lai, Chieh-Ping et al., Hilbert-Huang Transform (HHT) Analysis of Human Activities Using Through-Wall Noise Radar, 2007 International Symposium on Signals, Systems and Electronics (2007)
6. Yan, Jiaming et al., Through-Wall Multiple Targets Vital Signs Tracking Based on VMD Algorithm, Sensors (2016)
7. Amjad Hashemi, Alireza Ahmadian, Mehran Baboli, An Efficient Algorithm for Remote Detection of Simulated Heart Rate Using Ultra-Wide Band Signals, American Journal of Biomedical Engineering, Vol. 3 No. 6 (2013)
8. Baboli, Mehran, Olga Boric-Lubecke, and Victor Lubecke, A New Algorithm for Detection of Heart and Respiration Rate with UWB Signals, Conference proceedings: Annual International Conference of the IEEE Engineering in Medicine and Biology Society. IEEE Engineering in Medicine and Biology Society. Annual Conference 2012 (2012)
9. Nguyen, Van et al., Harmonic Path (HAPA) algorithm for non-contact vital signs monitoring with IR-UWB radar, 2013 IEEE Biomedical Circuits and Systems Conference (BioCAS) (2013)
10. E. Priidel, R. Land, V. Sinivee, P. Annus and M. Min, Comparative measurement of cardiac cycle by means of different sensors, 2017 Electronics, Palanga (2017)
11. "DWM User Manual". [Online]. Available: http://www.decawave.com/ (2018)

12. Dove Ilka, Analysis of Radio Propagation Inside the Human Body for in-Body Localization Purposes, University of Twnte, Faculty of Electrical Engineering, Mathematics & Comupter Science, M. Sc. Thesis (August 2014)

13. McLaughlin, Michael, Ciaran McElroy, Sinbad Wilmot, and Tony Proudfoot. RECEIVER FOR USE IN AN ULTRA-WIDEBAND COMMUNICATION SYSTEM. DECAWAVE LIMITED, Dublin (IE), assignee. Patent US 2014/0133522 A1. 15 May 2014. Print

14. M. Klemm and G. Troester, " EM energy absorption in the human body tissues due to UWB antennas," Progress in Electromagnetics Research, Vol. 62, 261–280 (2006)

15. "Shimmer User Manual". [Online]. Available: http://www. shimmersensing.com/ (2018)

16. Pan, Jiapu and Willis J. Tompkins, A Real-Time QRS Detection Algorithm, IEEE Transactions on Biomedical Engineering BME-32 (1985)

A Novel Hybrid Swarm Algorithm for P300-Based BCI Channel Selection

Víctor Martínez-Cagigal, Eduardo Santamaría-Vázquez, and Roberto Hornero

Abstract

Channel selection procedures are essential to reduce the curse of dimensionality in Brain-Computer Interface systems. However, these selection is not trivial, due to the fact that there are 2^{N_c} possible subsets for an N_c channel cap. The aim of this study is to propose a novel multi-objective hybrid algorithm to simultaneously: (i) reduce the required number of channels and (ii) increase the accuracy of the system. The method, which integrates novel concepts based on dedicated searching and deterministic initialization, returns a set of pareto-optimal channel sets. Tested with 4 healthy subjects, the results show that the proposed algorithm is able to reach higher accuracies (97.00%) than the classic MOPSO (96.60%), the common 8-channel set (95.25%) and the full set of 16 channels (96.00%). Moreover, these accuracies have been obtained using less number of channels, making the proposed method suitable for its application in BCI systems.

Keywords

Brain-computer interface • Multi-objective optimization
Swarm intelligence • Electroencephalography
Channel selection

1 Introduction

Brain-Computer Interfaces (BCI) have proven to be able to establish an effective communication system that allows users to control applications using their own brain signals [1]. Due to the non-invasiveness, portability and low cost characteristics of the electroencephalography (EEG), brain signals are usually registered by placing several electrodes (i.e., channels) on the users' scalp [1]. In order to identify the users' intentions in real time, it is essential to employ a recognizable control signal, such as the P300 evoked potentials. These potentials are positive peaks of the EEG mainly produced in the parietal cortex in response to infrequent and particularly significant stimuli at about 300 ms after their onset [1]. The most common setup, known as P300 Speller, allows users to spell words or select certain commands [2]. The user just need to focus attention on one of the character cells of a displayed matrix, while its rows and columns are randomly flashing. Whenever the target's row or column are intensified, a P300 potential is produced in the user's scalp. Thus, the desired character can be determined by computing the intersection where those P300 responses were found [2].

However, the inter-session variability and the low signal-to-noise-ratio that are present in these event-related responses make it difficult to obtain a reliable P300 potential. Thus, it is necessary to compute an average of several sequences, which may produce an over-fitting of the classifier, resulting in a spoiled system performance [3]. The curse of dimensionality may be reduced by using a channel selection procedure, which also reduces the power consumption in wireless EEG caps, increases the users' comfort and assures suitable performances [3]. Nevertheless, this selection is not trivial, owing to the fact that there are 2^N possible combinations for an N-channel cap, making the exhaustive search intractable [3]. For this reason, most P300-based studies omit this stage and use a combination of 8 typical channels in parietal and occipital positions as a general rule of thumb [4]. Nonetheless, due to the intrinsic inter-subject variability of the EEG, an optimization is required for leveraging the system performance.

In this regard, metaheuristics based on evolutionary computation have demonstrated excellent performances solving complex optimization problems. Even though

V. Martínez-Cagigal (✉) · E. Santamaría-Vázquez · R. Hornero
Biomedical Engineering Group, University of Valladolid,
Valladolid, Spain
e-mail: victor.martinez@gib.tel.uva.es

© Springer Nature Singapore Pte Ltd. 2019
L. Lhotska et al. (eds.), *World Congress on Medical Physics and Biomedical Engineering 2018*,
IFMBE Proceedings 68/3, https://doi.org/10.1007/978-981-10-9023-3_8

several methods have been successfully applied to P300-based BCI systems, most of them have used single-objective strategies, ignoring the trade-off between the final number of selected channels and the system performance [5, 6], or merging both trade-off objectives using aggregation approaches [7–9]. However, a practical BCI channel selection algorithm should simultaneously optimize a two-fold purpose: (i) to maximize the performance of the system, and (ii) to minimize the required number of channels. Multi-objective optimization algorithms, such as Non Sorting Genetic Algorithm II (NSGA-II) [7, 10] or Multi-Objective Particle Swarm Optimization (MOPSO) [10] have been applied in this regard. Although all of them have been proved to be suitable to its application in this field, their inter-trial variability, as well as their lack of deterministic approaches, hinder their full adaptation to binary BCI systems.

In this study, a novel multi-objective hybrid algorithm that merges the key aspects of MOPSO and forward selection (FS) is proposed for selecting the optimal channel sets in BCI applications. The method, which performs a dedicated local search over each channel, provides a set of Pareto optimal solutions that minimizes the error of the system and the required number of channels.

2 Subjects

The subject pool was composed of 4 male healthy subjects (mean of 26.25 ± 5.19 years) that were asked to spell a total of 200 characters with the P300 Speller in 2 sessions [2] (i.e., half for training and half for testing). EEG signals were recorded using a 16-channel cap with a g.USBamp amplifier (g.Tec, *Guger Technologies*, Austria). Sampling rate was fixed at 256 Hz and bandpass (0.1–60 Hz), notch (50 Hz) and common average reference filters were applied.

3 Methods

In order to evaluate the usefulness of our proposed algorithm for BCI systems, the method has been compared with the traditional MOPSO, described in [11].

3.1 Processing Pipeline

The signal processing pipeline is detailed in the Fig. 1a. In the feature extraction stage (i) a 0–700 ms window from the stimuli onset was selected; and a (ii) sub-sampling to 20 Hz was computed. Then, the multi-objective algorithm is applied to the training subset, returning sets of optimal

combinations of channels. Finally, these sets are evaluated in the testing dataset using a Linear Discriminant Analysis (LDA) classifier and final accuracies are calculated. This pipeline is repeated 20 times in order to discard stochastic effects.

3.2 Two-Fold Objective

Multi-objective metaheuristics involve the optimization of multiple conflicting objective functions at the same time. In this case, our two-fold objective is:

$$\min F(\boldsymbol{x}) = \begin{cases} f_1(\boldsymbol{x}) = 1 - AUC(\boldsymbol{x}) \\ f_2(\boldsymbol{x}) = \sum \boldsymbol{x} \end{cases}, \quad (1)$$

where $\boldsymbol{x}$ denotes a solution (i.e., particle position, a specific set of channels where $\boldsymbol{x}_c = \{0, 1\}$ with $c = 1, \ldots, N_c$) and AUC denotes the area under ROC curve, derived from a 5-fold cross-validated LDA that is trained and tested with the same solution $\boldsymbol{x}$. Therefore, $f_1(\boldsymbol{x})$ involves the minimization of the error rate, whereas $f_2(\boldsymbol{x})$ the minimization of the required number of channels.

3.3 δMOPSO/FS Algorithm

The proposed algorithm, dedicated MOPSO with FS (δMOPSO/FS), whose pseudo-code is detailed in Fig. 1c, was developed to overcome the inter-trial variability and the lack of search depth that binary MOPSO experiments when is applied to BCI systems [10]. In order to achieve this objective, the method provides a set of novel concepts, such as (i) deterministic initialization, (ii) dedicated particle subgroups for each channel, (iii) leader selection based on binary tournament, and (iv) three-fold mutation, which are detailed below.

Deterministic initialization Forward selection is applied in order to reduce the inter-trial variability due to stochastic effects. Starting from an empty set, the algorithm tests each channel for its inclusion based on the metric $f_1(\boldsymbol{x})$. Then, the repository is filled with the non-dominated solutions.

Dedicated particles In order to perform a depth local search and favor the convergence, MOPSO/FS dedicates subgroups of N_{sp} particles focused on each possible number of channels $c \in 1, \ldots, N_c$. Thus, each particle's position is randomly initialized as long as $\sum \boldsymbol{x} = c$, where c is the number of channels that belongs to its subgroup.

Leader selection Each particle should point to a repository leader that has the same number of channels than the particle's subgroup. Thus, the leader selection is based on binary tournament odds: each particle selects its

Fig. 1 **a** Detailed signal processing pipeline. **b** Leader selection based on distances. Each particle should select its corresponding leader with probability p, whereas the rest have decreasing probabilities as distance increases. **c** Pseudo-code of the proposed MOPSO/FS algorithm

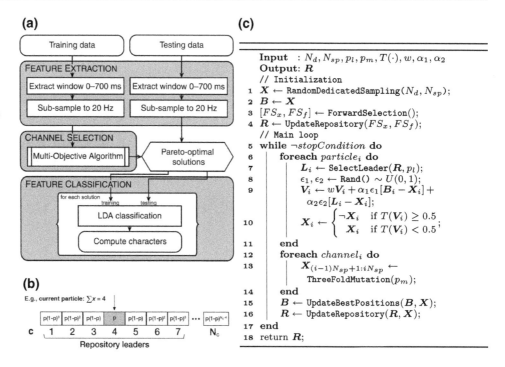

corresponding leader with probability p; and the rest of them with probabilities $p(1-p)^d$, where d is the distance from its leader. The procedure is illustrated in Fig. 1b.

Position updating Due to the dichotomous nature of the problem, once the velocity of each particle is calculated, the positions are updated on a transfer function basis. If the transformed velocity in each dimension $T(v_i)$ exceeds a threshold t, the position is inverted, otherwise is maintained. The transfer function is v-shaped: $T(v) = \left| v/\sqrt{v^2 + 1} \right|$, based on [12].

Three-fold mutation The mutation operator is similar to [11], but applied to each subgroup of particles. Therefore, each subgroup is divided in three parts, in which: (1) no mutation is applied, (2) uniform mutation with probability P_m is applied, and (3) non-uniform mutation is applied (i.e., probability $P_n = 1 - \left[(\text{gen} - 1)/\max_{\text{gen}} \right]^5$).

Repository update The repository (i.e., the set of Pareto optimal solutions) is updated in each generation of the algorithm as follows: (i) the current population X and the repository R are merged into a new population R_n, (ii) fitness is calculated in R_n according to Eq. (1), and (iii) non-dominated solutions of R_n are stored into the new repository. Due to the discrete nature of the channel selection problem, the repository keeps a maximum of N_c solutions, which provides a range of combinations to choose from, depending on the number of channels that the user would want to use.

4 Results

The proposed method has been compared with a traditional binary MOPSO, described in [11]. Both of them have run a fixed number of 500 generations in order to facilitate the comparison between them. Optimal training phase Pareto Fronts of both methods, as well as testing final accuracies of the optimal channel sets, are depicted in Fig. 2. Moreover, the highest reached accuracies using MOPSO/FS, MOPSO, the classical 8-set [4] and the 16-channel full set are shown in Fig. 3a. Finally, the Fig. 3b displays the averaged number of times that each channel has been selected as a Pareto Optimal solution across subjects.

5 Discussion and Conclusion

Although multi-objective metaheuristics have been proved to be suitable for the channel selection procedure in BCI systems, there is still room for improvement. Their lack of deterministic approaches cause the algorithms to suffer from stochastic effects, making necessary the computation of several runs in order to reach the optimal solutions [7, 10]. As can be observed in Fig. 2, not only δMOPSO/FS reaches more optimal Pareto Fronts than MOPSO, but also its convergence is faster (mean of 96.36 generations for δMOPSO/FS, and 417.70 for δMOPSO). In addition, the

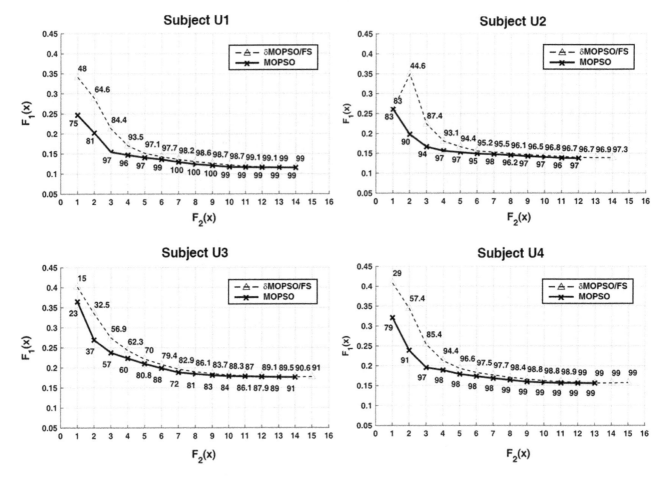

Fig. 2 Optimal Pareto solutions for δMOPSO/FS (solid line) and MOPSO (dashed line) for each user. The curves, composed by the optimal solutions returned by both methods, depict the trade-off between both objectives in training data. Final testing accuracies are also show next to each solution

(a)

	δMOPSO/FS		MOPSO		8-set	16-set
	Acc.	Ch.	Acc.	Ch.	Acc.	Acc.
U1	100.0 ± 0.0	7	99.1 ± 0.3	11	99.0	99.0
U2	98.0 ± 0.0	7	97.3 ± 0.5	14	94.0	97.0
U3	91.0 ± 0.0	14	91.0 ± 0.0	15	90.0	89.0
U4	99.0 ± 0.0	8	99.0 ± 0.0	12	98.0	99.0

(b)

Fig. 3 a Highest reached accuracies and their required number of channels using different approaches. **b** Averaged normalized channel ranks for the obtained Pareto optimal solutions (e.g., a value of 1 indicates that the channel is selected in all the solutions that belongs to the Pareto Front of every single user)

results indicate that the final δMOPSO/FS accuracies are higher than that obtained with MOPSO and, furthermore, the solutions use less number of channels. These accuracies are also higher than that obtained using the common 8-set [4] or the entire full set of channels, which reinforces the idea that channel selection is beneficial for the system's performance. It is also noteworthy to mention that the standard deviation of the final accuracies across trials for δMOPSO/FS is null, which means that every single trial has converged to the same set of solutions. This fact, in conjunction with the rapid convergence of the algorithm, demonstrates that δMOPSO/FS has successfully avoided the stochastic effects and thus, it can assure the identification of global optima in a single run, saving a high amount of computation time. Moreover, the Fig. 3b shows that there are certain channels that have been repeatedly selected along the Pareto Fronts of the subject pool. These channels are mainly distributed over the parietal and occipital regions, which reinforces the study of Krusienski et al. [4], who stated that the P300 potentials are mainly generated in those positions. It can also be noticed that the dispersion of the selected channels by MOPSO is smoother than by δMOPSO/FS, which indicates again that δMOPSO/FS successfully finds the same global optima in each run.

Even though this new technique has been proved to be suitable for use in P300-based BCI systems channel selection, we can point out several limitations. Firstly, the algorithm requires several hyperparameters to be fixed, whose optimization lies on the user experience. Moreover, both MOPSO and δMOPSO/FS use a transfer function for adapting them into binary-based approaches. In order to overcome this limitations, we contemplate the following future research lines: (i) to implement a dynamic fixation of the hyperparameters, and (ii) to apply these novel concepts to binary objected algorithms, such as the ones that are based on genetic algorithms.

In conclusion, a novel swarm-based algorithm has been proposed for selecting the optimal channel set in BCI applications. The proposed algorithm, δMOPSO/FS, has been tested with 4 healthy subjects and compared with the traditional binary MOPSO. Results show that δMOPSO/FS is not only able to converge more faster than MOPSO, but also to reach higher accuracies (mean of 97.00%) than that obtained by using MOPSO (mean of 96.60%), the common 8-channel set (mean of 95.25%) and the full set of 16 channels (mean of 96.00%). Moreover, these accuracies are obtained using less number of channels than MOPSO, approximately the half of the full set. For these reasons, we conclude that δMOPSO/FS is suitable for use in P300-based BCI systems channel selection procedures.

Acknowledgements This study was partially funded by projects TEC2014-53196-R of 'Ministerio of Economía y Competitividad' and FEDER, the project "Análisis y correlación entre el genoma completo y la actividad cerebral para la ayuda en el diagnóstico de la enfermedad de Alzheimer" (Inter-regional cooperation program VA Spain-Portugal POCTEP 2014–202) of the European Commission and FEDER, and project VA037U16 of the 'Junta de Castilla y León' and FEDER. V. Martínez-Cagigal was in receipt of a PIF-UVa grant of the University of Valladolid. The authors declare no conflict of interest.

References

1. Wolpaw, J.R., Birbaimer, N., *et al.*: Brain-computer interfaces for communication and control. Clin. Neurophysiol., 113(6), pp. 767–791 (2002).
2. Farwell, L.A., Donchin, E.: Talking off the top of your head: toward a mental prosthesis utilizing event-related brain potentials. Electroen. Clin. Neuro., 70(6), pp. 510–523 (1988).
3. Cecotti, H., Rivet, B., *et al.*: A robust sensor-selection method for P300 brain-computer interfaces. J. Neural Eng., 8(1), p. 016001 (2011).
4. Krusienski, D.J., Sellers, E.W., *et al.*: Toward enhanced P300 speller performance. J. Neurosci. Meth., 167(1), pp. 15–21 (2008).
5. Jin, J., Allison, B.Z., *et al.*: P300 Chinese input system based on Bayesian LDA. Biomed. Tech., 55(1), pp. 5–18 (2010).
6. Perseh, B., Sharafat, A.R.: An Efficient P300-based BCI Using Wavelet Features and IBPSO-based Channel Selection. J. Med. Signals Sens., 2(3), pp. 128–143 (2012).
7. Kee, C., Ponnambalam, S.G., *et al.*: Multi-objective genetic algorithm as channel selection method for P300 and motor imagery data set. Neurocomputing, 161, pp. 120–131 (2015).
8. Martínez-Cagigal, V., Hornero, R.: P300-Based Brain-Computer Interface Channel Selection using Swarm Intelligence. Rev. Iberoam. Autom. In., 14(4), pp. 372–383 (2017).
9. Martínez-Cagigal, V., Hornero, R.: A Binary Bees Algorithm for P300-Based Brain-Computer Interfaces Channel Selection. In: Advances in Computational Intelligence, LNCS, IWANN2017, pp. 453–463. Cádiz, Spain (2017).
10. Martínez-Cagigal, V., Hornero, R.: Multi-Objective Optimization for P300-Based Channel Selection. In: Proceedings of the 9th CEA symposium, pp. 73–78. Barcelona, Spain (2017).
11. Sierra, M.R., Coello, C.A.: Improving PSO-Based Multi-objective Optimization Using Crowding, Mutation and E-Dominance. Lect. Notes Comput. Sc., 3410, pp. 505–519 (2005).
12. Mirjalili, S., Lewis, A.: S-shaped versus V-shaped transfer functions for binary Particle Swarm Optimization. Swarm Evol. Comput, 9, pp. 1–14 (2913).

The Probablistic Random Forest Clinico-Statistical Regression Analysis of MER Signals with STN-DBS and Enhancement of UPDRS

Venkateshwarla Rama Raju

Abstract

In this study, we present classification and regression analysis to predict the UPDRS score and its enhancement after the microelectrode STN signal recording (MER) with DBS surgery (implantation of the micro-electrode). We hypothesized that a data informed grouping of features extrapolated from MER signals of STN can envisage restore (by decreasing the tremor) and functioning the motor improvement in Parkinson's disease (PD) patients. A random—forest is used to account for unbalanced datasets and multiple observations per PD subject, and showed that only five features of STN-MER signals are sufficient and account for prognosting UPDRS advancement. This finding suggests that STN signal characteristics are maximum correlated to the extent of improvement motor restoration and motor behavior observed in STN DBS.

Keywords

Microelectrode-recording (MER)
Parkinson's disease (PD) • STN-DBS • Classification and prediction • Random forest

V. Rama Raju (✉)
Department of Computer Science and Engineering, CMR College of Engineering & Technology (UGC Autonomous), Kandlakoya, Medchal Rd, 501401 Hyderabad, India
e-mail: drvrr@cmrcet.org; system@ou.ernet.in; idcoucea@hd1.vsnl.net.in

V. Rama Raju
Jawaharlal Nehru Technological University Hyderabad JNTUH, Hyderabad, India

V. Rama Raju
Nizam's Institute of Medical Sciences, Biomedical, Neurology and Neurosurgery, Hyderabad, India

V. Rama Raju
Biomedical Engineering Department, Osmania University College of Engineering (Autonomous), Hyderabad, OU, India

1 Introduction

1.1 Parkinson's Disease

Parkinson's disease (PD) is a chronic complex progressive neurodegenerative brain disorder that belongs to a larger class of disorders called movement disorders. PD is one of the most common neurologic disorders that elders' experience with "severe health hazard" is a devastating diagnosis affecting circa ~ 2 of every 1000 (2/1000) older adults [1–8]. The causes are unknown and so far no cure [1, 2] and the search for optimal cure is on for the past 2 centuries since the time it was first described by James Parkinson [18]. In PD, one particular population of brain cells those that produce a chemical messenger termed dopamine become impaired and lost over time. The loss of these brain cells causes circuits (basal ganglia circuits) in the brain to function bizarrely and those uncharacteristic circuits effect in movement problems [1]. Basal ganglion is an important organ of the brain mainly mean for our movement and control. Present healings for PD are meant for alleviating the symptoms rather than the disease's progression (For instance, Levedopa—a chemical building block that converts human body into dopamine. It replaces the dopamine that is lost in Parkinson's. However, there are more side effects with this drug), hence fresh hope lies in new research and findings, such a latest classification and prediction of clinical enhancements with microelectrode STN recording with DBS (MER with STN-DBS) [1]. The early signs of the disease may help us understand the progress of the disease because it is more than just these dopamine cells in the brain; it affect other cells as well that we are learning more and more about every day [2]. Therefore, prediction is one of the most significant factors in the detection of PD features at very early stage (say two decades in advance). In this paper, we present the classification and prediction of clinical and/or diagnostic setup in deep brain stimulation (DBS) by using with the help of electro-neuro-physiological MER recordings of STN signals.

The PD is characterized by its four classes of cardinal motor features (or symptoms), namely, tremor, postural instability, bradykinesia and rigidity [1–4]. PD is caused by damage to the central nervous system (CNS) [5]. Symptoms similar to PD have been mentioned as "Kampavata" in ancient Indian Hindi documents [6]. The search for optimal cure is on for the last two hundred years ever since it was first discovered by James Parkinson a way back in 1817 [7]. Since then, the disease has become the pathfinder for other neurodegenerative disorders, starting with the discovery of dopamine (in PD, one particular population of the brain cells that produce a chemical messenger to communicate with other cells) deficiency within the basal ganglia, which led to the development of first effective treatment for a progressive neurodegenerative condition [8]. However, it is possible that PD was present long before this landmark description. A disease known as Kampa Vata consisting of *shaking* (*kampa*) and lack of muscular movement (*vata*), existed in ancient India as long as 4500 years ago [9]. Deep brain stimulation (DBS) of subthalamic-nuclei (STN) is a surgical technique proving better results not only for the detection of PD features—symptoms but also significantly reducing tremors and restoring the motor function highly which was invented by the two neuroscientists, namely Benabid and Delong [10, 11]. Its mechanisms are not fully elucidated quantitatively—objectively, though the technique was clinically established. However, the clinical outcome is determined by many factors. Microelectrode-recordings (MER) of subthalamic-nucleus (STN) intraoperatively for targeting during DBS procedures are most useful for deducing inferences. This is because anatomical structural organization provide some clues as to what might be the function of basal ganglia circuits, the inference of function from anatomical structure is exploratory [12, 13]. So far quantitative work was done MER with STN-DBS but subject specific enhancement was not performed. In this study we attempted to quantify also predict the UPDRS subject specific enhancement. Objective PD scale can provide a more complete picture of the neurophysiological basis for PD.

2 Methods

The process for DBS was a one-stage bilateral stereotactic approach using a combined electrode for both MER and macrostimulation. Up to five micro/macro-electrodes were used in an array with a central, lateral, medial, anterior, and posterior position. Final target location was based on test stimulation (intraoperatively). Bilateral STN-DBS performed in our tertiary-care center NIMS hospital Hyderabad (South India).

2.1 MR Image-Targeting

One of the major problem with the targeting subthalamic nucleus is that it is a small biconvex lens diamond structure almond shaped and not clearly detected on MRI due to lack of contrast between the STN and the surrounding structures [1–18]. The STN can be visualized on MRI but other methods such as Lozano's technique where a position 3 mm lateral to the superolateral border of the red nucleus is targeted have been studied and found to be effective areas for stimulation. As the MRI techniques are not absolutely perfect, use of electrophysiological techniques such as microelectrode recording from the subthalamic nucleus as well as intraoperative stimulation have assisted in clearly demarcating the STN. Microelectrode recording can identify subthalamic neurons by their characteristic bursting pattern and their signals clearly identify the nucleus form the surrounding structures. On table stimulation is studied to ensure that the there is optimal benefit with the least side effects and this is the final test to ensure the correct targeting of the STN. All these techniques are normally used in combination during targeting, albeit, the individual role of each modality is still not known.

2.2 MER Signal Acquisition—Recording

Five electrodes (Medtronic maker) were placed in an array with a central, lateral, medial, posterior, and an anterior position placed 2 mm apart, to delineate the borders of the nucleus. The targeting was performed according to Lozano's technique—2 mm sections are taken parallel to the plane of anterior comissure-posterior commissure line and at the level with maximum volume of red nucleus, STN is targeted at 3 mm lateral to the anterolateral border of red nucleus. The co-ordinates are entered into stereocalc software which gives the co-ordinates of the STN. Another neuro navigation frame-link-software is also used to plot the course of the electrodes and to avoid vessels. The surgery is performed with two burr holes on the two sides based on the co-ordinates. Five channels that are introduced with the central channel representing the MRI target while medial and lateral are placed in the X-axis while anterior and posterior are placed in the Y-axis to cover an area of 5 mm diameter. Intra-operative recording was performed in all 5 channels. For performing microelectrode-signal-recording of STN with DBS, five microelectrodes are slowly passed through the STN and recording is performed from ±10 mm (10 mm above to 10 mm below) the STN calculated on the MRI. STN is identified. Extracellular MER was performed with Medtronic-micro-electrodes having an input-impedance of 1.1 ± 0.4 mega-Ohms (MΩ's) which was calculated at 220 Hz. Signals were recorded with biological-amplifiers signal average's (10,000 times-amplification) of the Medtronic Lead-point system, by employing bootstrapping method. Signals were filtered using analog band pass filters with lower and upper cut-off frequencies between 0.5 and 5 kHz (amplifier bandwidth). The signal was sampled at 10 kHz, by using 12-bit-resolution analogue-to-digital converter (A/DC) card (2^N, $N = 12$ dynamic-range giving 4096 sample values) and then later up-sampled to 20-kHz at off-line. The channel with maximum recording and the earliest recording were recorded on both sides. Intraoperative test stimulation was performed in all channels from the level at the onset of MER recording. Stimulation was done at 1mv, 3mv to assess the improvement in bradykinesia, rigidity and tremor. Appearance of dyskinesias was considered to be associated with accurate targeting. Side effects were assessed at 5mv and 7mv to ensure that the final channel chosen had maximum improvement with least side effects. Correlation was assessed between the aspects of MER and the final channel chosen in 20 PD-subjects (40 sides).

2.3 Microelectrode Signal Processing and Feature—Selection

In MER signal processing, local field potentials(LFP) and multi-unit-activity(MUA) signals were gathered by low pass and high pass filtering techniques at cut-off frequency 200 and 500 Hz. Spike-detection was performed by MUA voltage-thresholding. Spike-related-features were assessed by common spike-train-metrics [14, 15]. To examine the behavior of local neuronal populations, the BUA was extracted from the MUA following the procedure [18]. In the same studies, it was suggested that the rationality between the MUA-BUA signal-envelopes and LFP may reveal coherent-activity of small or large neuronal-populations. From every signal, we extracted 89 features. A list of investigative-features and their corresponding-metaphors is given in Table 1.

2.4 Random—Forests

A professional way to alleviate above-fitting is by imparting —training several uncorrelated trees in an ensemble-learner referred to as random-forest (RF), which can be applied for classification and regression. RF's can handle highly nonlinear interactions and they can cope with small observations and large-number of predictors. During training phase, each tree in RF is trained using a different subset of data "bootstrap-aggregation" and features "random-subspace

method" randomly-sampled with replacement. The data that are left out during construction of each tree are used for validation purposes (Fig. 1).

As building the forest advances, the system generates an internal unbiased-estimate of the generalization error (OOB error) which is then used to identify most important variables. The final OOB prediction for a given observation is the average score attained over-all-trees(regression) or choosing majority within forest(classification), without trees that included this observation during their training-phase. In this study, we used RFs both for classification and regression. In the former case, we extracted features that heuristically classified "good" and "poor" STN-DBS responders, defined as patients that exhibited an "off"-state UPDRS-enhancement above or below 38% [17].

2.5 Model Training and Corroboration

RFs were trained using subject-wise bootstrapping, intriguing separately into account the left-hemisphere(LH) and right-hemisphere(RH) STN-feature vectors of each subject Fig. 1a. Each RF consisted of 300-trees. For classification, each tree was created by choosing randomly with replacement 7/9 "good" responders and 7/11 "poor" responders. Therefore, the training pool envisaged 14 feature-vectors (7 PD-subjects $\times$ 2 hemi-spheres) labeled as one (1 meant for "good"), and 14 feature-vectors labeled as zero (0, meant for "poor"). The left-over feature-vectors (2 are "good" responders and 4 are "poor" responders) were used as the OOB set (Fig. 1). A PD-subject was classified as a "good" responder if and only if the average predicted as with "good" response and the probability attained for the LH-STN and RH-STN feature-vectors was ≥ 0.5. The predicted UPDRS improvement (%) was computed as the average prediction obtained from the L and R-STN.

2.6 The Performance of Model

In the case of classification, we used the Matthews Cor-relation Coefficient (MCC) for the OOB data as a performance metric, which is a class skew insensitive measure given by

$$MCC = \frac{TP.TN - FP.FN}{(TP + FP)(TP + FN)(TN + FP(TN + FN)}$$

where, TN (TP) and FN (FP) are the numbers of correctly and incorrectly predicted "poor" ("good") response observations, respectively. An MCC value of 1 corresponds to a perfect prediction, while a value of -1 indicates a total

Table 1 Name of features and their corresponding metaphors

Name	Metaphor	Name	Metaphor
PowerX$_W$	Power-band ratio of signal-X in frequency Band-W	RR	Bursting-rate
PKX$_W$	Peak-to-average power-ratio of signal-X in frequency band-W	PB	Percentage-of-bursts
FmaxPKX$_W$	Frequency corresponding to maximum peak to average power ratio of signal X in frequency band W	FR	Firing-rate
CVX$_W$	Coefficient of variation of signal X in frequency band W	stim$_E$	coordinates of the stimulation contact on axis E, where E corresponds to x (lateral–medial),y (posteri or–anterior), or z (ventral–dorsal)
PAFC$_{WZ}$	LFP phase–amplitude cross frequency coupling index betwese phase in band **W and voltage in band Z**	stimd	Euclidean-distance of stimulation contact from the STN center
PPFC$_{WZ}$	LFP phase–phase cross frequency coupling index between the phase in band W and amplitude in band Z	dist	Euclidean-distance of STNMER from the stimulation contact
ZeroCrossX	Percentage of electrical-baseline i.e., zero-line crossings in signal X	distpeak$_B$	Distance between the maximum aggregate beta LFP peak and the stimulation contact
SNRX	$20\log_{10}\frac{\sigma X}{\sigma n}, \sigma X = SD(X), \sigma n = \frac{median(X)}{0.7645}$	**hemi**	Hemisphere (Left or Right)
maxCohXY	Maximum coherence between signals X and Y	prep	Preponderance (L/R: most affected body side is the right/left, controlled by the left/right hemisphere)
maxCohXY$_W$	Maximum coherence between signals X and Y in frequency band W	HY	Hoehn and Yahr PD scale
max_PL$_W$	Maximum phase locking index in band W for the LFP signal	levpre	Preoperative LED
MISI	Mean interspike interval	age	Age
SISI	Interspike interval standard deviation	years	Disease duration
CVISI	Interspike-interval-coefficient-of-variation	sex	sez (female/male coded as 1/2)
PS	Percentage of spikes in the spike signal		

MER-Signals X and Y correspond to LFP, EMUA, or EBUA. The frequency bands W and Z are defined as follows: delta (D; 1–4 Hz), theta (T; 4–10 Hz), beta (B; 10–45 Hz), gamma (G; 45–100 Hz), and high gamma (HG; 100–200 Hz). For example, maxCohXY$_W$ refers to the maximum coherence between LFP and EMUA, LFP and EBUA, or EMUA and EBUA in one of the aforementioned frequency bands

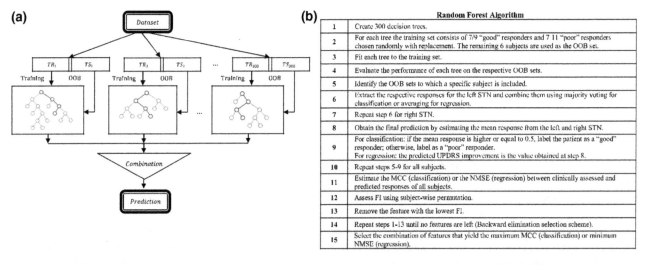

Fig. 1 **a** 300-trees of random forest, in the group, every tree uses a different-training (TR) and testing-set (TS). **b** Sequence of algorithmic steps pursued to foresee for each-subject

disagreement between prediction and observation. Random —classification gives values closure to zero (0). In case of a tie in terms of the MCC value, we chose the classifier that given the minimum cross entropy loss function (*J*) defined mathematically can be expressed in different ways as follows:

$$J = -\frac{1}{N}\left[\sum_{k=1}^{N} y_k \ln(p_k) + \sum_{k=1}^{N}(1-y_k)\ln(1-p_k)\right] \quad (1)$$

$$= -\frac{1}{N}\sum_{k=1}^{N}[y_k \ln(p_k) + \ln(1-p_k) - y_k\ln(1-p_k)] \quad (2)$$

$$= -\frac{1}{N}\sum_{k=1}^{N}\{\ln(1-p_k) + y_k[\ln(p_k) - \ln(1-p_k)]\} \quad (3)$$

$$= -\frac{1}{N}\sum_{k=1}^{N}\left[\ln(1-p_k) + yk.\ln\left(\frac{p}{(1-p_k)}\right)\right] \quad (4)$$

Here, N is total number of patients, y_k is prognostically and/or diagnostically assessed response of subject k, P_k—predicted-response (i.e. average predicted-probability of good response from right and left of subthalamic-nuclei. In connection with the regression—model, the performance is assessed by using the correlation co-efficient (the Pearson's correlation-coefficient, ρ) and the NMSE between the predicted and the clinically assessed UPDRS enhancement (%) output vector for the OOB data. For the classification, Matthews Correlation Coefficient (MCC) is used for the out of bag data as performance-metric as a system of standard measurement [16].

Feature—selections

Feature-significance (FS) is expressed as reduce in the predicted classification or augments in the predicted-regression if the values of this feature is randomly shuffled during the regression phase. This measure was computed for every-tree, averaged and then divided by the following standard deviation (SD) over the whole forest.

$$s^2 = \sqrt{s^2} = \sqrt{\frac{\sum_{i=1}^{n}(x - \bar{x})^2}{n-1}} \quad (5)$$

where, $\sigma = \sqrt{\sigma^2}$ is the population (subjects N = 20) standard deviation (SD).

The SD of a sample is square-root-of-variance computed as given in Eq. (5). The combined features given the maximum classification or minimum normalized MSE-regression was chosen.

3 Results

In this study, the STN DBS prediction was considered as a classification—problem. A backward removal feature—selection method and found that four-features attained a maximal MCC-value is 0.9045. Important features found in this study are PKLFP$_{HG}$, power-BUA$_T$, max-PL$_B$, and max-PL$_{HG}$ with FIs 0.1495, 0.9142, 0.3899, and 0.5982 correspondingly.

4 Conclusions

UPDR scale is not a rationale but to some extent hypothetical means rationally or scientifically not accepted scale. It is based on clinician's choice scale. Hence for objective—scientific evidence computer simulation and statistical modeling for disease symptoms—or features prediction at early stage be conducted. Then it can be compared with the UPDR scale in terms of the performance improvement after the DBS. Significancy of the work and its importance to the medical physics and biomedical eng: The approach can employ a small number of the signal features inside the STN to predict, separately for each subject, the behavioral outcome of STN DBS, justifying further investigation and, clinical applications possibly. This work has broad implication and innovation of newer statistical and electrophysiological techniques and improving currently available MRI DBS machines for evaluating all types of neurological disorders in particular Parkinson's disease and other movement disorders. It will be of great interest to the Scientists/engineers involved in medical physics and biomedical research in the fields of biomedical instrumentation and signal processing applications to neuroelectrophysiology.

References

1. Juliann, Schaeffer., New Hope in Neuroprotection? A Parkinson's disease Update. Neuroprotection has become the focus of efforts aimed at slowing the progression of Parkinson's disease, Nature.-com 1–4 (2017).
2. Benabid, AL., Koudsie, A., Benazzouz, A., Vercueil, L., Fraix V., Chabardes, S., LeBas, JF., Pollak, P.: Deep brain stimulation of the corpus luysi (subthalamic nucleus) and other targets in Parkinson's disease. Extension to new indications such as dystonia and epilepsy. J Neurol 248(3), III37–III47 (2001).
3. Benabid, AL., Krack, PP., Benazzouz, A., Limousin, P., Koudsie, A., Pollak, P.: Deep brain stimulation of the subthalamic nucleus for Parkinson's disease: methodologic aspects and clinical criteria. Neurology 55:S40–S44 (2000).

4. J, Jankovic.: Parkinson's disease: Clinical features and diagnosis. J. Neu-rol. Neurosurg. Psychiatry 79(4), 368–376 (2008).

5. Tugwell, C.: Parkinson's disease in focus. Pharmaceutical Press, London (2008).

6. Yashar, Sarbaz., Hakimeh, Pourakbari.: A review of presented mathematical models in Parkinson's disease: black and gray box models. Med Biol Eng Comput 54, 855–868 (2016).

7. Parkinson, J.: An Essay on Shaking Palsy, Sherwood, Neely and Jones: London (1817).

8. K, Ray Chaudhuri., Victor, SC Fung., Fast Facts: Parkinson`s Disease. Fourth Edition, Health Press Ltd. London (2016).

9. Ronald, F. Pfeiffer., Zbgkiew, K. Wazolek., Manuchair, Ebadi.: Parkinson`s Disease, CRC Press, Taylor & Francis Group (2013).

10. Alim, Luis Benabid.: Laskar Award Winner, Nature Medicine. 10 (20), 1–3 (2014).

11. Mahlon, DeLong.: Laskar Award Winner, Nature Medicine. 10 (20), 4–6 (2014).

12. Larry, Squire., Darwin, Berg., Floyd, E. Bloom., Sascha, du. Lac., Anirvan, Ghosh., Nicholas C. Spitzer.: Fundamental Neuroscience, 4th Edition. *AP* Academic Press (2012).

13. Duane E. Haines., Gregory A. Mihailoff.: Fundamental Neuroscience for Basic and Clinical Applications. E-Book 5th Edition. ELSEVIER (2018).

14. Ashkan, K., Blomstedt, P., Zrinzo, L., Tisch, S., Yousry, T., Limousin-Dowsey, P.: Variability of the subthalamic nucleus: the case for direct MRI guided targeting. Br J Neurosurg. 21 (2), 197–200 (2007).

15. Patel, NK., Khan, S., Gill, SS.: Comparison of atlas- and magnetic-resonance-imaging-based stereotactic targeting of the subthalamic nucleus in the surgical treatment of Parkinson's disease. Stereotact Funct Neurosurg. 86(3), 153–61 (2008).

16. Andrade-Souza, YM., Schwalb, JM., Hamani, C., Eltahawy, H., Hoque, T., Saint-Cyr, J., Lozano, AM.: Comparison of three methods of targeting the subthalamic nucleus for chronic stimulation in Parkinson's disease. Neurosurgery. 62 (2), 875–83 (2008).

17. B. W. Mathews.: Comparison of the predicted and observed secondary structure of T4 phage lysozyme. Biochim. Biophys. Acta. Protein Struct. 405(2), 442–451 (1975).

18. Parkinson J.: An Essay on Shaking Palsy, Sherwood, Neely and Jones: London (1817).

A Computer Simulation Test of Feedback Error Learning-Based FES Controller for Controlling Random and Cyclic Movements

Takashi Watanabe and Naoya Akaike

Abstract

Feedback control of movements by functional electrical stimulation (FES) can be useful for restoring motor function of paralyzed subjects. However, it has not been used practically. Some of possible reasons were considered to be in designing a feedback FES controller and its parameter determination, and nonlinear characteristics with large time delay in muscle response to electrical stimulation, which are different between subjects. This study focused on the hybrid controller that consists of artificial neural network (ANN) and fuzzy feedback controller. ANN was trained by feedback error learning (FEL) to realize a feedforward controller. Although FEL can realize feedforward FES controller, target movement patterns are limited to those similar to patterns used in the training. In this paper, FEL-FES controller was tested in learning both random and cyclic movements through computer simulation of knee joint angle control with 4 different training data sets: (1) sinusoidal patterns, (2) patterns generated by low pass filtered random values, (3) using both the sinusoidal and the LPF random patterns alternatively and (4) patterns that consisted of 3 random sinusoidal components. Trained ANNs were evaluated in feedforward control of sinusoidal and random angle patterns. Training with data set (1) caused delay in controlling random patterns, and training with data set (2) caused delay in controlling sinusoidal patterns. Training with data set (3) showed intermediate performance between those with data set (1) and (2). Training with data set (4) could control adequately both random and sinusoidal patterns. These results suggested that generating movement patterns using sinusoidal components would be effective for various movement control by FEL-FES controller.

Keywords

FES • Feedback error learning • Ankle Tracking control

1 Introduction

Functional electrical stimulation (FES) can be useful for restoring or assisting paralyzed motor function due to a spinal cord injury or a cerebrovascular disease [1, 2]. However, feedback FES control has not been used practically, although it can be effective for restoring paralyzed movements, while a method of using pre-determined stimulation data were practical [3]. Some of possible reasons are considered to be in difficulties of designing a feedback FES controller and its parameter determination because the musculoskeletal system has nonlinear, time-variant characteristics with large time delay in muscle response to electrical stimulation, which are different between subjects, and redundancy in stimulation intensity determination.

In our previous work, a multichannel proportional- integral- derivative (PID) controller was developed to control the redundant musculoskeletal system that involves an ill-posed problem in stimulus intensity determination [4, 5] and the PID controller was applied to Feedback Error Learning (FEL) controller [6–10]. In the FEL controller for FES (FEL-FES controller), an artificial neural network (ANN) was trained by the FEL to develop the inverse dynamics model (IDM) of electrically stimulated musculoskeletal system, which can be used as a feedforward controller. Although FEL can realize feedforward FES controller for each subject, target movement patterns are limited to those similar to patterns used in the training such as sinusoidal angle patterns or randomly generated angle patterns of two-point reaching movement. In addition, there are few studies on FES control of movements of nonlinear musculoskeletal system [11–13]. Therefore, an FES control

T. Watanabe (✉) · N. Akaike
Graduate School of Biomedical Engineering, Tohoku University,
Sendai, Miyagi, Japan
e-mail: t.watanabe@tohoku.ac.jp

of various movements for paralyzed subjects has been desired.

In this study, FEL-FES controller was developed using fuzzy feedback controller. Fuzzy controller is considered to be useful to realize a practical feedback FES controller. The FEL-FES controller was tested in learning both random and cyclic movements through computer simulation of knee joint angle control with 4 different training data sets.

2 Methods

2.1 Outline of FEL-FES Controller

The block diagram of the FEL-FES controller used in this study is shown in Fig. 1, which is composed of a fuzzy feedback controller and ANN. The ANN was trained to realize the inverse dynamics model (IDM) of electrically stimulated musculoskeletal system, which can be used as a feedforward controller. The fuzzy controller consisted of 2 sub-fuzzy controllers. One calculates change of stimulation intensity from error and the other calculates stimulation intensity proportional to the error. Stimulation intensity u_{fb} is the sum of the previous stimulation intensity and the calculated intensities. The sum of stimulation outputs from the ANN and the fuzzy controller is applied to a muscle.

2.2 Computer Simulation Method

The FEL-FES controller was tested in computer simulation of knee joint angle control by stimulating the rectus femoris. A three-layered ANN was used as a feedforward controller. The inputs of the ANN were time series of angles, angular velocities and angular accelerations of target movements at continuous 6 times, from n to $n + 5$ (sampling frequency of 30 Hz). The numbers of neurons were 18, 18 and 1 for the input, hidden and output layers, respectively.

The ANN was trained by the FEL under 4 different training data sets. Learning of the ANN was performed after a single control trial. In each control trial, movement was controlled for 24 s, and the first 4 s was not used for the learning. The 4 training data sets were as follows:

(1) sinusoidal patterns
(2) patterns generated by low pass filtered random values
(3) using both the sinusoidal and the LPF random patterns alternatively
(4) patterns consisted of 3 random sinusoidal components.

For the data set (1), sinusoidal pattern was determined for each control trial, in which cycle period and amplitude was selected randomly from 2, 3, 4, 5, and 6 s of cycle period and 2, 4, 6, 8, and 10° of amplitude with an offset of 5°. For the data set (2), random value between 0 and 1 was generated for 24 s data and the data was low pass filtered with cut off frequency of 0.2 Hz. Angle of the LPF random pattern was adjusted to be between 5 and 25°. Pattern of data set (4) was generated by the following for each control trial:

$$f(n\Delta t) = \sin\frac{2\pi}{T_1}n\Delta t + \sin\frac{2\pi}{T_2}n\Delta t + \sin\frac{2\pi}{T_3}n\Delta t \qquad (1)$$

Here, Δt shows sampling interval and n is the sample number. Cycle period T_1, T_2 and T_3 were determined randomly in order to satisfy the following relation:

$$1.8 < T_1 < 3.2 < T_2 < 4.6 < T_3 < 6.0 \qquad (2)$$

Amplitude was adjusted to be between 5 and 25°.

ANN learning was performed more than 10000 control trials, and stopped as mean error ME converged.

$$ME = \frac{1}{N}\sum_{n=1}^{N}|e(n\Delta t)| \qquad (3)$$

where N is the total number of sampled data used for learning in a single control trial. The 4 ANNs trained with all data sets were tested in feedforward control of 3 sinusoidal angle patterns (cycle period of 2 s with amplitude of 10°, cycle period of 4 s with amplitude of 8° and cycle period of 6 s with amplitude of 6°). The 3 ANNs trained with data set (1), (2) and (3) were tested in feedforward control of LPF-random patterns and the ANN trained with data set (4) was tested in controlling random sinusoidal component patterns.

The model of electrically stimulated muscle was represented by a second order system with time delay. Gain of the model was determined by a cubic polynomial approximation of measured input-output characteristics of the rectus femoris of a healthy subject.

3 Results and Discussions

Examples of feedforward control by trained ANNs are shown in Figs. 2 and 3. ANN trained with sinusoidal patterns (data set (1)) could control properly all 3 sinusoidal patterns.

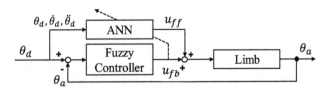

Fig. 1 Block diagram of the FEL controller for FES. θ_d and θ_a represent the desired and the measured joint angles. The ANN learns with the outputs of the fuzzy controller while controlling limbs

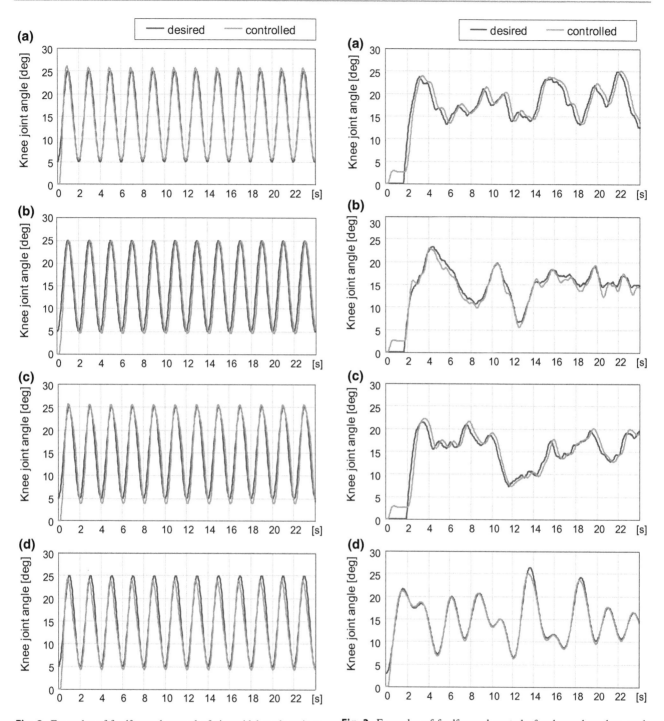

Fig. 2 Examples of feedforward control of sinusoidal angle trajectories by trained ANNs (cycle period of 2 s with amplitude of 10°). From the top, results of ANN trained with data set (**a**), (**b**), (**c**) and (**d**) are shown

Fig. 3 Examples of feedforward control of unlearned random angle trajectories by trained ANNs. From the top, results of ANN trained with data set (**a**), (**b**), (**c**) and (**d**) are shown

However, time delay was caused in controlling LPF-random patterns. On the other hand, ANN trained with LPF-random patterns (data set (2)) could control LPF-random patterns well, while it caused larger time delay in controlling sinusoidal patterns than ANNs trained with other data sets. ANN trained

with data set (3) showed intermediate results between those by the ANN trained with data set (1) and those with data set (2). The ANN trained with random sinusoidal component patterns (data set (4)) could control adequately both sinusoidal and random sinusoidal component patterns.

Table 1 Average *ME* of feedforward control of 3 sinusoidal angle patterns and 3 unlearned random angle patterns by trained ANNs using ANN connection weights after 10000 learnings

Data set	Sinusoidal	Random
(1)	0.52 ± 0.34	1.16 ± 0.08
(2)	1.23 ± 0.94	0.72 ± 0.12
(3)	0.91 ± 0.82	0.90 ± 0.07
(4)	0.68 ± 0.79	0.37 ± 0.04

Table 1 shows average mean error (*ME*) for 3 sinusoidal patterns and 3 random patterns. Values of *ME* were calculated for 20 s of each control trial from 4 s after the beginning of control. The error for tracking of sinusoidal angle patterns was smallest by the ANN trained with data set (1), although the ANN trained with data set (4) also showed small error. Training with data set (3) sometimes caused increase of error after decrease as the learning progresses. The ANN trained with the data set (4) showed small error for both patterns. Especially, 2 sinusoidal patterns with larger cycling periods and smaller amplitudes were controlled with better performance than the ANN trained with sinusoidal patterns (data set (1)).

The computer simulation tests suggested that generating movement patterns using sinusoidal components would be effective for various movement control by FEL-FES controller. As shown in our previous study [10], it is considered that using various target positions and movement velocities rather than repeated training with same target would be effective, even if training data is used only one time.

The model of electrically stimulated musculoskeletal system used in this study was a simple second order system although it included time delay and nonlinear gain. Therefore, data sets (1) and (2), which were basically effective for angle patterns similar to those used in ANN training, are considered not to be suitable for FES application. Using both patterns alternatively could not improve significantly the ANN learning. Although the computer simulation tests suggested that random sinusoidal component patterns (data set (4)) could be effective for learning various movement patterns for FES, target angle trajectories used in this paper did not include constant angle as used in moving between 2 points [10]. Random patterns for evaluation of data set (4) were different from others. It is necessary to test the trained ANN with various angle patterns including keeping constant angles.

The ANN used in this study was a 3-layered network with the fixed numbers of neurons. Learning coefficients were determined by a trial and error manner. Therefore, further examinations to determine parameters of ANN would be required.

4 Conclusion

In this paper, FEL-FES controller was tested in learning both random and cyclic movements through computer simulation of knee joint angle control with 4 different training data sets. Trained ANNs were evaluated in feedforward control of sinusoidal and random angle patterns. The ANN trained with random sinusoidal component patterns could control both sinusoidal and random patterns appropriately, while training with sinusoidal patterns and using both patters alternatively caused delay in controlling random angle patterns, and training with the LPF-random patterns caused delay in controlling sinusoidal patterns. These suggested that generating movement patterns using sinusoidal components would be effective for various movement control by FEL-FES controller.

Acknowledgements This work was supported in part by the Ministry of Education, Culture, Sports, Science and Technology of Japan under a Grant-in-Aid for Scientific Research (B).

References

1. Handa, Y.: Current Topics in Clinical Functional Electri-cal Stimulation in Japan. J. Electromyogr. Kinesiol. 7(4), 269–274 (1997).
2. Lyons, G. M., Sinkjær, T., Burridge, J. H. and Wilcox, D. J.: A Review of Portable FES-Based Neural Orthoses for the Correction of Drop Foot. IEEE Trans. Neural Sys. Rehabil. Eng., 10(4), 260–279 (2002).
3. Hoshimiya, N., Naito, A., Yajima, M., Handa, Y.: A multi-channel FES system for the restoration of motor functions in high spinal cord injury patients: a respiration-controlled system for multijoint upper extremity. IEEE Trans. Biomed. Eng., 36(7), 754–760 (1989).
4. Watanabe, T., Iibuchi, K., Kurosawa, K. and Hoshimiya, N.: Method of Multichannel PID Control of 2-Degree of Freedom of Wrist Joint Movements by Functional Electrical Stimulation. Systems and Computers in Japan, 34(5), 25–36 (2003).
5. Kurosawa, K., Watanabe, T., Futami, R., Hoshimiya, N. and Handa, Y.: Development of a closed-loop FES system using 3-D magnetic position and orientation measurement system. J. Automatic Control, 12(1), 23–30 (2002).
6. Kurosawa K, Futami R, Watanabe T, Hoshimiya N.: Joint angle control by FES using a feedback error learning controller. IEEE Trans Neural Syst Rehabil Eng. 13(3), 359–371 (2005).
7. Watanabe, T., Kurosawa, K. and Yoshizawa, M.: An Effective Method of Applying Feedback Error Learning Scheme to Functional Electrical Stimulation Controller. IEICE Transactions on Information and Systems, E92-D(2), 342–345 (2009).
8. Watanabe, T. and Sugi, Y.: Computer Simulation Tests of Feedback Error Learning Controller with IDM and ISM for Functional Electrical Stimulation in Wrist Joint Control., J. Robotics, 2010, 908132, https://doi.org/10.1155/2010/908132 (2010).
9. Watanabe, T and Fukushima K.: An approach to applying feedback error learning for functional electrical stimulation (FES) controller: Computer simulation tests of wrist joint control. Advances Artificial Neural Systems, 2010, 814702, https://doi.org/10.1155/2010/814702 (2010).

10. Watanabe, T., Fukushima, K.: A Study on Feedback Error Learning Controller for Functional Electrical Stimulation: Generation of Target Trajectories by Minimum Jerk Model. Artificial Organs, 35(3), 270–274 (2011).
11. Lynch, C.L., Popovic, M.R.: A comparison of closed-loop control algorithms forregulating electrically stimulated knee movements in individuals with spinal cord injury. IEEE Trans. Neural Syst. Rehabil. Eng., 20(4), 539–548 (2012).
12. Alibeji, N., Kirsch, N., Farrokhi, S., Sharma, N.: Further Results on Predictor-Based Control of Neuromuscular Elec-trical Stimulation. IEEE Trans. Neural Syst. Rehabil. Eng., 23(6), 1095–1105 (2015).
13. Oliveira, T.R., Costa, L.R., Catunda, J.M.Y., Pino, A.V., Barbosa, W., Souza, M.N.: Time-scaling based sliding mode control for Neuromuscular Electrical Stimulation under uncertain relative degrees. Med. Eng. Phys., 44, 53–62 (2017).

A Hybrid BCI-Based Environmental Control System Using SSVEP and EMG Signals

Xiaoke Chai, Zhimin Zhang, Yangting Lu, Guitong Liu, Tengyu Zhang, and Haijun Niu

Abstract

The paper developed a hybrid Brain–computer interface (hBCI) home environmental control system for paralytics' active and assisted living, by integrating single channel Electromyography (EMG) of occlusal movement and steady state visual evoked potentials (SSVEP). The system was designed as three-level interface, besides the idle state interface, for work state there are one main interface and five sub-interfaces. The main interface included five visual stimulus corresponding to different devices such as nursing bed, wheelchair, telephone, television and lamps, the sub-interfaces present control function of those devices. Gazing at stimuli at different frequencies corresponding to a certain function can select a device or device action. Several particular occlusal patterns respectively are used to confirm the selected function, return from sub-interface to main interface and switch on/off the system. Ten healthy subjects without any training completed the virtual system verification experiment, the averaged target selection accuracy based on SSVEP achieved 96.3%. Moreover with a simple clench action for target confirmation, the false positive rate was minimized to zero, which improved the control accuracy. This indicated that Combining SSVEP and EMG can effectively enhance the security and interactivity of the environmental control system.

Keywords

Hybrid brain computer interface • Environmental control
Electromyography (EMG)
Steady state visual evoked potential (SSVEP)
Paralytics

1 Introduction

Brain–computer interfaces is an alternative channel of neuro-muscular pathway, which allows users to directly control a device by brain activity [1]. The interest in BCI research is increasing as the rapid development of brain cognition and neuroscience, computer science and biomedical engineering. Particularly in rehabilitation engineering, BCI technology is potentially useful for people who are severe paralyzed after amputation, stroke or spinal cord injuries (SCIs), by reconstructing their communication with daily living environment [2].

Multiple Electroencephalography (EEG) paradigms have been used in BCIs, such as Motor Imagery (MI), P300 and Steady State Visual Evoked Potentials (SSVEP) [3]. Among those, BCI based on SSVEP received much attention as it can achieve higher information transmission rate (ITR) without training, and has been used for the control of nursing bed and wheelchair using SSVEP-BCI [4–6]. Those BCIs based on single EEG paradigm were used to control a single device, however communication with a variety of devices in home environment system is complicated. Moreover, single modal BCIs are usually synchronous and cannot distinguish idle state with work state automatically which is likely to cause wrong operation.

Considering those problems, researchers have proposed hBCI, combining two or more EEG paradigms to control complex devices. Pfortscheller et al. added an event related desynchronization (ERD) potential to SSVEP-BCI to realize the switch of the system [7]; Allison et al. used mu rhythm and SSVEP to achieve asynchronous control [8]; Li et al. combined SSVEP and P300 to reduce the false positive rate

X. Chai (✉) · Z. Zhang · Y. Lu · G. Liu · T. Zhang · H. Niu (✉)
School of Biological Science and Medical Engineering, Beihang University, Beijing, China
e-mail: chaiXiaoke@buaa.edu.cn

H. Niu
Beijing Advanced Innovation Center for Biomedical Engineering, Beihang University, Beijing, China
e-mail: hjniu@buaa.edu.cn

T. Zhang
National Research Center for Rehabilitation Technical Aids, Beijing, China

© Springer Nature Singapore Pte Ltd. 2019
L. Lhotska et al. (eds.), *World Congress on Medical Physics and Biomedical Engineering 2018*,
IFMBE Proceedings 68/3, https://doi.org/10.1007/978-981-10-9023-3_11

during idle state [9]. Other researchers developed hBCIs using different physiological signals from electrocardiogram (ECG), electrooculogram (EOG) or electromyogram (EMG), Lin K et al. detected the EMG of hand movement to choose the location region of target in a SSVEP-BCI speller [10–12]. Nevertheless, for severe paralysis or amputees, only muscles above neck remain function, especially EMG of their occlusal movement is obviously suitable as a control signal [13].

In this paper, EMG signals from occlusal movement were integrated into SSVEP-BCI to build the hBCI home environmental control system. To validate the reliability of the system function, a virtual environmental control experiment was designed, the performance of target selection and confirmation, return and switch were evaluated.

2   Structure and Method

2.1   System Modules

As shown in Fig. 1, the hBCI home environmental control system includes visual stimulation, signals processing and results feedback modules. Visual stimulation was presented at the LCD monitor. The stimulation and signal processing program are written using MATLAB. The recognition results were encoded sending to the environmental control module.

2.2   Functional Interface

As shown in the Fig. 2, the system interface consists of three levels, which can be switched between each level. After switching from the idle state to work state, there were one main interface and five sub-interfaces. For wheelchair interface, the target corresponds to moving forward/backward, turning left/right and switching the posture to stand up or lay down. For

nursing bed, the action includes rise/fall of the bed, flipping left/right and adjusting the angle of back and leg. For television interface, users can select the menu or turn back to the home-page. For telephone interface, there are three emergency numbers. For environment interface, there are two lamps and a curtain.

2.3   Control Method

The system combined the recognition of three EMG patterns and SSVEP to design the control mode. The control flow is illustrated as Fig. 3, during idle state, EMG pattern 3, a long-time clench was used as a switch to turn on the system. Two stages were needed from main interface to sub-interfaces, the of the target selection stage lasts three seconds, target confirmation stage duration was two second. In device-selection stage, devices on the main interface was chosen by gazing at the flicker. And the flicker corresponding to the certain function was labeled green when its SSVEP feature was recognized. In device-confirmation stage, EMG pattern 1, a one-time clenching was used to get into the sub-interface. Each sub-interface include different number of targets, functions of devices on the sub-interface also can be chosen through SSVEP recognition, then be confirmed by EMG pattern recognition. In all the sub-interfaces, EMG pattern 2, a two-time clenching was used to return to the main interface.

2.4   Signal Processing

The signal processing includes preprocessing and feature recognition. The classification method of SSVEP is Classical Correlation Analysis (CCA). EEG signals from occipital region channel of the brain (PO4, PO3, O1, O2) were selected to calculate the characteristics. The reference signals in CCA were composed of sinusoids and cosinusoids pairs at

Fig. 1 System modules

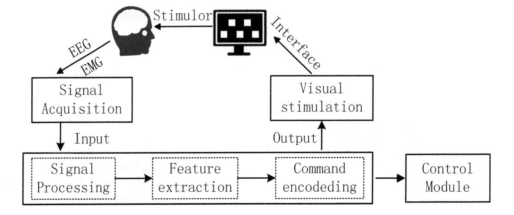

Fig. 2 Main interface and
sub-interfaces

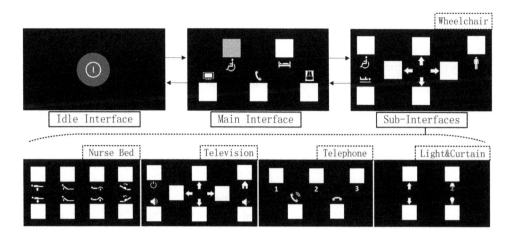

Fig. 3 Control method

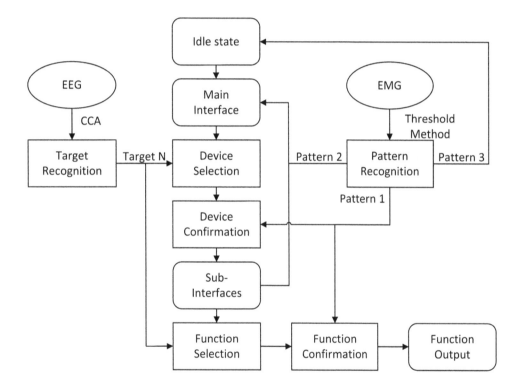

the same frequency of the stimulus and its second harmonics. Different frequencies of SSVEPs correspond to flicker in different position.

The pattern detection of EMG signals are based on the feature threshold method. Integrated EMG (iEMG) calculated in a window of a certain length was selected as the feature, and the appropriate threshold were used to detect the number of exceed points upon the threshold. Combing the threshold of amplitude and time to detect the clench action pattern.

3 Experiment and Results

3.1 Subject and Experiment

Ten healthy volunteers (five males and five females, mean age 22.8 ± 2.3 years) were recruited to participate in system verification experiment, all subjects had normal or corrected to normal vision. EEG and EMG signals were simultaneously collected using the Neuroscan signal acquisition system with a sampling rate of 1000 Hz. Each participant was instructed to complete the control task following the indicated icon in accordance with the prescribed procedures. There were 37 times of target selection and confirmation commands, five return and two switch commands in the experiment.

3.2 Evaluation

The correct control output was recorded only once the target selection and the target confirmation were both correct. The Selection Accuracy is the ratio of the command number to the total times of target selection operation. The Confirmation Accuracy is the ratio of the confirmation times to the total number of correct selection times. The Control Accuracy is the ratio of the correct command output number to the number of target commands. The Return Accuracy is the ratio of the correct return number to the total number of return operation times. The Switch Accuracy is the ratio of the correct switch number to the total number of switch operation times.

3.3 Results

As shown in Table 1, the averaged selection accuracy for ten subjects was 96.3%, and there were four subjects who obtained correct target selection and confirmation completely. The averaged confirmation accuracy of all subjects was 99.2%, and there were seven subjects whose target-confirmation accuracy rate was 100%. There were five subjects who had incorrect target-selection, but there was no confirmation after all the wrong selection, so the control accuracy of the system was 100%. Four subjects failed to

correctly control all the return commands, the accuracy of two-time clenching twice for return command averaged as 92%. Three subjects failed to correctly control all the switch commands, the accuracy of long-time clench for switch command averaged as 90.0%.

4 Discussion

The interface of the system is simple, the graphic logo and visual feedback improves interactivity of the system without interfering the stimulus. From the experiment of healthy young subjects who were first time use BCI system, the average recognition of SSVEP reached a high accuracy. After selecting the target, users can judge whether it is the current operation intention through the presented recognition result. Therefore no one made wrong confirmation, which effectively avoids the wrong output and ensure safety of the device control.

In the system, one channel EMG signal and four channels of EEG signals are used to build an hBCI which can implement complex control in home environment with less input signal. The EMG patterns recognition of temporal muscle also achieved high accuracy, effectively using a simple clench motion achieved the confirmation of selected target. The introduction of EMG pattern enriched the function of the system, realized the switch between different interfaces. And the system is Plug&Play which can be closed at any time and woke up at idle state, thereby also reducing the possibility of wrong operation.

The function of this home environment control system is easy to expand, which can be used in different environment, by adjusting function according to the actual need. The hBCI system can achieve hand-free home environment control, its high flexibility and security in control also provides the reliability for the application to paralyzed patients.

Table 1 Accuracy of commands

	Selection accuracy (%)	Confirmation accuracy (%)	Control accuracy (%)	Return accuracy (%)	Switch accuracy (%)
N1	100.0	100.0	100.0	100.0	100.0
N2	92.5	100.0	100.0	83.3	100.0
N3	97.4	100.0	100.0	100.0	100.0
N4	100.0	100.0	100.0	100.0	66.7
N5	97.4	97.4	100.0	83.3	100.0
N6	100.0	100.0	100.0	100.0	100.0
N7	100.0	100.0	100.0	100.0	66.7
N8	88.1	97.4	100.0	100.0	100.0
N9	94.9	97.4	100.0	71.4	66.7
N10	92.5	100.0	100.0	83.3	100.0
Average	96.3	99.2	100.0	92.0	90.0

Acknowledgements This work was supported by the National Science and Technology Ministry of Science and Technology Support Program (Grant No.2015BAI06B02) and the National High Technology Research and Development Program of China (Grant No.2015AA042304).

Conflict of Interest The authors declare that they have no conflict of interest.

References

1. Wolpaw, J. R., Birbaumer, N., Heetderks, W. J., Mcfarland, D. J., Peckham, P. H., Schalk, G.: Brain-computer interface technology: a review of the first international meeting. IEEE Transactions on Rehabilitation Engineering 8(2), 164–73 (2000).
2. Daly, J. J., Huggins, J. E.: Brain-computer interface: current and emerging rehabilitation applications. Archives of Physical Medicine & Rehabilitation 961(3), S1-S7 (2015).
3. Abdulkader, S. N., Atia, A., Mostafa, M. S. M.: Brain computer interfacing: applications and challenges. Egyptian Informatics Journal 16(2), 213–230 (2015).
4. Bin G, Gao X, Yan Z, Hong B, Gao S.: An online multi-channel SSVEP-based brain-computer interface using a canonical correlation analysis method. Journal of neural engineering 6(4), 046002 (2009).
5. Rebsamen, B., Guan, C., Zhang, H., Wang, C., Teo, C., Jr, A. M., et al.: A brain controlled wheelchair to navigate in familiar environments. IEEE Transactions on Neural Systems & Rehabilitation Engineering 18(6), 590–598 (2010).
6. Shyu, K. K., Chiu, Y. J., Lee, P. L., Lee, M. H., Sie, J. J., Wu, C. H., et al.: Total design of an fpga-based brain–computer interface control hospital bed nursing system. IEEE Transactions on Industrial Electronics 60(7), 2731–2739 (2013).
7. Pfurtscheller, G., Solis-Escalante, T., Ortner, R., Linortner, P., Muller-Putz, G. R.: Self-paced operation of an SSVEP-based orthosis with and without an imagery-based "brain switch:" a feasibility study towards a hybrid BCI. IEEE Transactions on Neural Systems & Rehabilitation Engineering 18(4), 409–414 (2010).
8. Allison, B. Z., Brunner, C., Altstätter, C., Wagner, I. C., Grissmann, S., Neuper, C.: A hybrid ERD/SSVEP BCI for continuous simultaneous two dimensional cursor control. Journal of Neuroscience Methods 209(2), 299 (2012).
9. Rui, Z., Wang, Q., Kai, L., He, S., Si, Q., Feng, Z.: A BCI-based environmental control system for patients with severe spinal cord injuries. IEEE Transactions on Biomedical Engineering PP(99), 1–1 (2017).
10. Scherer, R., Mã¼Ller-Putz, G. R., Pfurtscheller, G.: Self-initiation of EEG-based brain-computer communication using the heart rate response. Journal of Neural Engineering 4(4), L23 (2007).
11. Shinde, N., George, K.: Brain-controlled driving aid for electric wheelchairs. In: IEEE, International Conference on Wearable and Implantable Body Sensor Networks, pp. 115–118. IEEE (2016).
12. Lin, K., Cinetto, A., Wang, Y., Chen, X., Gao, S., Gao, X.: An online hybrid BCI system based on SSVEP and EMG. Journal of Neural Engineering 13(2), 026020 (2016).
13. Chang, B. C., Bo, H. S.: Development of new brain computer interface based on EEG and EMG. In: IEEE International Conference on Robotics and Biomimetics, pp. 1665–1670. IEEE Computer Society (2009).

Principal Component Latent Variate Factorial Analysis of MER Signals of STN-DBS in Parkinson's Disease (Electrode Implantation)

Venkateshwarla Rama Raju

Abstract

Although clinical benefits of deep brain stimulation (DBS) in subthalamic-nuclei (STN) neurons have been established, albeit, how its mechanisms improve the motor features of PD have not been fully established. DBS is effective in decreasing tremor and increasing motor-function of Parkinson's disease (PD). However, objective methods for quantifying its efficacy are lacking. Therefore, we present a principal component analysis (PCA) method to extract-features from microelectrode-recording(MER) signals of STN-DBS and to predict improvement of unified Parkinson's disease rating scale(UPDRS) following DBS (applied on 12 PD patients). Hypothesis of this study is that the developed-method is capable of quantifying the effects-of-DBS "on state" in PD-patients. We hypothesize that a data informed combination of features extracted from MER can predict the motor improvement of PD-patients undergoing-DBS-surgery. This shows the high-frequency-stimulation in diseased-brain did not damage subthalamic-nuclei (STN) neurons but protect. Further, it is safe to stimulate STN much earlier than it was accepted so far. At the experimental level, high-frequency-stimulation of the STN could protect neurons in the subsstantia-nigra (SN, an important element of the brain). Therefore, to test this hypothesis in humans, we need to perform STN stimulation at the very beginning of the disease so that we can predict the disease at an early-stage. The latent-variate-factorial is a statistical-mathematical technique PCA based tracking method for computing the effects of DBS in PD. Ten parameters capturing PD characteristic signal-features were extracted from MER-signals of STN. Using PCA, the original parameters were transformed into a smaller number of PCs. Finally, the effects-of-DBS were quantified by examining the PCs in a lower-dimensional-feature-space. This study showed that motor-symptoms of PD were effectively reduced with DBS.

Keywords

Microelectrode-recording (MER) • Parkinson's disease (PD) • STN-DBS • Principal component analysis (PCA) Latent variate factorial analysis

V. Rama Raju (✉)
Department of Computer Science and Engineering, CMR College of Engineering & Technology (UGC Autonomous), Kandlakoya, Medchal Rd, 501401 Hyderabad, India
e-mail: drvrr@cmrcet.org; system@ou.ernet.in; idcoucea@hd1.vsnl.net.in

V. Rama Raju
Jawaharlal Nehru Technological University Hyderabad JNTUH, Hyderabad, India

V. Rama Raju
Nizam's Institute of Medical Sciences, Biomedical, Neurology and Neurosurgery, Hyderabad, India

V. Rama Raju
Biomedical Engineering Department, Osmania University College of Engineering (Autonomous), Hyderabad, OU, India

1 Introduction

Parkinson's disease (PD) is a chronic disorder characterized by four primary cardinal symptoms, namely, resting tremor, bradykinesia, postural instability and rigidity. Currently, the diagnosis is based on the presence of clinical-features (i.e., symptoms and signs) and the response to antiparkinsonian medications [1–3].

The most established scale for assessing disability and impairment in PD-disease is the "Unified Parkinson's disease Rating Scale" (designated as UPDRS) [4] is based on subjective clinical evaluation of features. Hence, it is required to quantify PD characteristics objectively in order to enhance the prognosis, and prognostic diagnosis to identify disease sub forms, observe disease progression and explain healing treatment-curing efficacy [5, 6]. Deep brain stimulation (DBS) is a well established surgical technique for PD treatment that uses high frequency electrical pulses to stimulate the subthalamic nuclei(STN) and associated brain-regions. However, large outcome disparity exists

L. Lhotska et al. (eds.), *World Congress on Medical Physics and Biomedical Engineering 2018*,
IFMBE Proceedings 68/3, https://doi.org/10.1007/978-981-10-9023-3_12

Fig. 1 The signal patterns of single and multi-unit subthalamic-nuclei neurons at range of depths

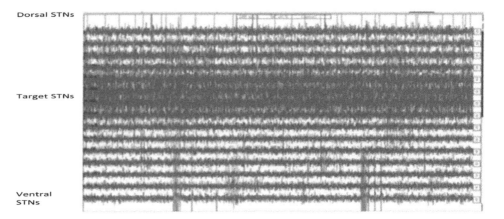

among recipients due to varied standards for postoperative management, particularly concerning DBS programming optimization [6, 7]. Though, mechanisms of DBS act are not clear properly and correct electrode placement, the effectiveness of lead-point position and stimulation might advance motor features and restore increase motor function and allows for a reduction in antiparkinsonian medication doses [1–8]. Further, DBS stimulation parameters are set by subjective evaluation of PD symptoms, and no physiological-based quantitative measures are used to optimize the efficacy of DBS in reducing motor disorders [8, 9]. Hence, a tracking method (PCA based) is worth or objective reasoning. The objective of this study is to quantify the efficacy of DBS.

2 Methods

12subjects with PD diagnosis having more than 6 years as per united kingdom (UK) PD society brain bank criteria with good response to a precursor to dopamine cells "Levodopa" and Hoehn and Yahr score of less than 4 with normal cognition were included in this study. The signal-acquisition—microelectrode recording was performed in all patients. To perform MER in STN-DBS, five MER/macrostimulation electrodes were placed in an array with a central, lateral, medial, posterior, and an anterior position placed 2 mm apart, to define the borders of the nucleus. Depending on the pre-operative magnetic resonance imaging (MRI), it was decided in some cases to acquire with three or four microelectrodes rather than five. Usually, 3–4 channel-recordings were performed in the central, medial, posterior, and lateral channel. The anterior channel was included rarely. Extracellular single and multi-unit MER was performed with small (10 μm-micron meters width) polyamide-coated tungsten microelectrodes (Medtronic maker; microelectrode 291; input impedance was 1.1 ± 0.4 Mega Ohm; measured

at 220 Hz, at the beginning of each trajectory) mounted on a sliding burrow/or cannula. Signals were recorded with the amplifiers (10,000 times amplification) of the Lead-Point system (Medtronic), using a bootstrapping principle and were filtered with analog Band-Pass filters between 0.5 and 5 kHz (−3 dB;12 dB/Oct). The signal was sampled at 12 kHz, by employing a 16 bit analog to digital (A/D) converter (ADC). Later using Nyquist criteria sampled up to 24 kHz. Following a two seconds signal stabilization period after electrode movement cessation, multi-unit segments were recorded for five to twenty seconds. For STN, 8 and 12 mm above the MRI-based target, the microelectrodes were advanced in steps of 500 μm towards the target by a manual microdrive. When the needles were inside the STN, at each depth, the spiking activity of the neurons lying close to the needle (pick-up area up to 200 μm) could be recorded. Depending on the neuronal density not more than 3–5 units were acquired—recorded concurrently. More distant units could not be distinguished from the background level. The following Fig. 1 obtained with MER. The subthalamic-nuclei was detected by interfered noise with a bigger electrical zero line and asymmetrical discharge patterns of multiple-frequencies.

The STN was clearly distinguished from the dorsally located zona incerta and lenticular fasciculus (H$_2$-field) by an abrupt amplify (signal amplitudes) in background-noise level and augment in discharge-rate typically characterized by rhythmic bursts of activity with a burst frequency. All five microelectrodes were passed through STN and signal-recording was performed from dimensions ±10 mm, and STN calculated on MRI. The STN was detected by a high noise with a large electrical baseline and an irregular discharge patterns with multiple frequencies. Figure 2 shows the microelectrode recording which was obtained from the STN.

The STN MER signals features (Table 1). PCA is used to transform originally correlated variables into uncorrelated and to reduce number of variables.

Single unit (single
channel) STN

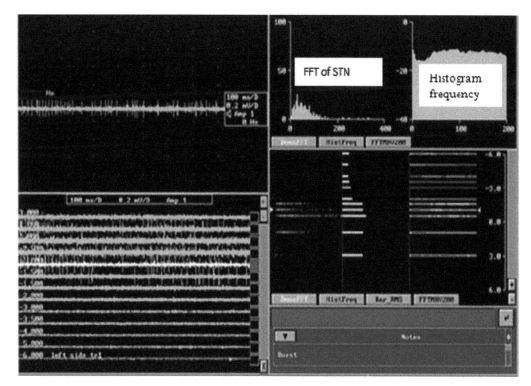

Central Channel
(11 mm) typical
firing pattern
with irregular
firing and broad
baseline (from –
1.00 level)

Fig. 2 MER-signal recording. At the posterior level, STN was encountered lower than expected at that location

Table 1 Feature-values: Mean ± SD) for Parkinson's disease subjects with DBS "ON" and "OFF"

Signal feature	PD patients with DBS on	PD patients with DBS off
$D_{2,r}$	6.1 ± 1.0	5.4 ± 1.6
$D_{2,l}$	5.7 ± 1.4	5.4 ± 2.1
% REC_r	9.2 ± 5.1	14.9 ± 13.6
% REC_l	12.7 ± 7.6	15.3 ± 14.9
R M S_r 10^3	0.8 ± 0.4	2.5 ± 4.6
R M S_l 10^3	1.0 ± 0.9	6.0 ± 12.4
Samp En_r	0.9 ± 0.3	0.7 ± 0.3
Samp En_l	1.0 ± 0.4	0.8 ± 0.4
$Coherence_r$	1.3 ± 0.7	1.5 ± 1.4
$Coherence_l$	1.4 ± 0.8	2.2 ± 1.5

The ten calculated signal parameters were sequentially placed and normalized (to zero mean and unit SD of subjects) to form feature vectors for all subjects. The dimension of the feature vectors was then reduced by applying the PC approach. In that approach, the feature vectors were decomposed into weighted sums of orthogonal basis vectors where the scalar weights were called the principal components. These PCs were the new uncorrelated features.

3 Results

In our results, the UPDRS motor score was lower with DBS "ON" comparing with DBS "OFF" in 12 subjects but the decrease rate was patient character. The clinical scores describing disabilities in hand movements decreased for all subjects in either side of the body. The principal components were computed (in Mat Lab tool) for 12 subjects using computed Eigen-vectors. We observed that the first three PC scores are sufficient to account the variance approximately 80% (in our scatter plot). These PCs are good enough to discriminate between the DBS "ON" and "OFF" states, and between the PD patients. The first Eigen vector is the best mean-square fit for the feature-vectors. Consequently, first highest magnitude value is the PC1, second highest magnitude is PC2, and third highest magnitude is PC3 in the order of decreasing. These magnitudes are amplitudes of 12 subjects MER signals of STN. By visually inspecting the morphology of the third Eigen-vector, we could recognize, that PC3 emphasizes differences between right and left side variables. Indeed, the unilateral onset continual and unrelenting asymmetries of symptoms support the diagnosis of PD in relation to other similar diseases. The distances between the DBS "ON" and "OFF" states in the feature

Table 2 Characteristics' of Parkinson's diseased subjects

P#.	Age	Gender	UPDRS off	UPDRS on	Medications	Co-morbidities
1	70	F	56	43	Clozapine, Diazepam, Fludrocortisone Acetate, Namenda, Senna Plus, Seroquel	None
2	70	F	64	48	Levodopa	Chronic obstructive pulmonary disease
3	82	F	59	40	Mirapex	none
4	71	M	34	14	Lipitor, Glipizide, Lisinopril, Aspirin	Diabetes, angioplasty in 2000
5	61	M	71	42	Sinemet, Omeprazole, Aspirin	none
6	61	M	38	31	Sinemet	Hypertension
7	66	F	47	28	Azilect, Amitriptyline, Evista	None
8	63	M	57	33	Sinemet Tasmar, Pantprozol	None
9	50	M	43	34	Mirapex, Trazadone, Metformin, Glimepiride Advair, Metolazone	Diabetes
10	59	M	43	24	Levodopa, Detrol LA	Diabetes
11	63	M	44	30	Sinemet, Simvastatin	None
12	64	M	62	38	Carbidopa, Naproxen, Ropinirole	Hypertension

space were highly individual. Similarly, the improvements in clinical scores were individual. However, strong changes in the total UPDRS motor score did not result in strong changes in the analyzed principal components. This could be the fact that the total UPDRS motor score [4] is a complicated score that consists of a large number of sub-scores. These sub-scores are defined for different areas of the body in different movement conditions (Table 2).

We presented a principal component analysis (PCA) based tracking method for quantifying the effects of DBS in. It was observed that the PC-based tracking method was more sensitive to PD with associated tremor.

4 Conclusions

In this study, we applied PCA-based latent variate-factorial tracking method. The hypothesis of the study was that the developed method is capable of quantifying the effects of DBS "ON" patients with PD objectively. Our findings suggest that electrophysiological STN signal characteristics are strongly correlated to the ex tent of motor behavior improvement observed in STN-DBS. In the future studies, multivariate PCA approach can be tested in serving the adjustment of DBS settings. In addition, the sensitivity of the presented method to different types of PD should be estimated more carefully in further experimental studies. The exciting systems approach and computational framework can be prepared performed in our future studies. The proposed approach can employ few signal features within the STN to compute prognostics, separately for each PD subject,

the performance and behavioral outcome of STN-DBS, justifying further investigation and, possibly, clinical-experimental applications. Lastly, only few neurophysiologically interpretable MER signal features are sufficient to account for predicting UPDRS improvement. Future study also involve applying nonlinear time domain analysis of average amount of mutual information (AAMI) technique with controls for better prediction and understanding of Parkinson's disease through MER with STN-DBS.

References

1. Kyriaki Kostoglou, Konstantinos P. Michmizos. Classification and prediction of clinical improvement in deep brain stimulation from intraoperative microelectrode recordings. IEEE Transactions on Biomedical Engineering. 2017; 64(5): 1123–1130.
2. Jan L. Bruse, Maria A. Zuluaga, Abbas. Khushnood. Detecting Clinically Meaningful Shape Clusters in Medical Image Data: Metrics Analysis for Hierarchical Clustering Applied to Healthy and Pathological Aortic Arches. IEEE Transactions on Biomedical Engineering. 2017; 64(10): 2373–2382.
3. Lorenzo Livi, Alireza Sadeghian, Hamid Sadeghian. Discrimination and Characterization of Parkinsonian Rest Tremors by Analyzing Long-Term Correlations and Multifractal Signatures. IEEE Transactions on Biomedical Engineering. 2016; 63(11), 2243–2249.
4. Fahn, S.; Elton, RL. The Unified Parkinson's Disease Rating Scale. In: Fahn, S.; Marsden, CD.; Calne, DB.; Goldstein, M., editors. Recent developments in Parkinson's disease. Florham Park, N.J: Macmillan Healthcare Information; 1987. p. 153–63.
5. Dustin A. Heldman, Christopher L. Pulliam, Enrique Urrea Mendoza. Computer guided deep brain stimulation programming for Parkinson's disease, Neuromodulation. 2016; 19 (2): 127–132.

6. Viswas Dayal, Patricia Limousin, Thomas Foltynie. Subthalamic nucleus deep brain stimulation in Parkinson's disease: the effect of varying stimulation parameters. Journal of Parkinson's disease. 2017; 7: 235–245.

7. Jankovic J. Parkinson's disease: clinical features and diagnosis. J Neurol Neurosurg Psychiatry. 2008; 79:368–376. [PubMed: 18344392].

8. Antoniades CA, Barker RA. The search for biomarkers in Parkinsons disease: a critical review. Expert Rev. 2008; 8 (12):1841–1852.

9. Morgan JC, Mehta SH, Sethi KD. Biomarkers in Parkinson's disease. Curr Neurol Neurosci Rep. 2010; 10:423–430. [PubMed: 20809400].

Functional State Assessment of an Athlete by Means of the Brain-Computer Interface Multimodal Metrics

Vasilii Borisov®, Alexey Syskov®, and Vladimir Kublanov®

Abstract

The estimation in real time of the functional and mental state level for the athlete during the loads is essential for management of the training process. New multimodal metric, obtained by means of the brain-computer interface (BCI), is proposed. The paper discusses the results of the joint usage of data from Emotiv EPOC+ mobile wireless headset. It includes motion sensors (accelerometer) and EEG channels. The features of the Emotiv EPOC+ interface allow to record the deviation of the head from the body axis, which provides an additional channel of information about the physical and mental (psycho-emotional) state of the athlete. Based on this data a new multimodal metric is calculated. Approbation of the metric was performed for functional stress studies on group of 10 volunteer subjects, including evaluations of the TOVA-test and the hyperventilation load. The joint application of different signals modalities allows to obtain estimates level of attention for these functional studies.

Keywords

Brain-computer interface • Functional study
Multimodal interaction

V. Borisov (✉) · A. Syskov · V. Kublanov
Ural Federal University, Yekaterinburg, 620002, Russian
Federation
e-mail: v.i.borisov@urfu.ru

A. Syskov
e-mail: a.m.syskov@urfu.ru

V. Kublanov
e-mail: v.s.kublanov@urfu.ru

1 Introduction

Traditionally, the assessment of the functional state of an athlete using EEG devices is carried out in the laboratory conditions. The progress of mobile data transmission technologies, the emergence of mobile and energy saving computing technologies allows to create a mobile functional status monitoring systems for personal use (including applications in sport tasks) [1]. The development of user EEG systems brain-computer in the last decade has shown a fundamental difference in approaches for laboratory studies and functional state assessment in everyday life [2]. In this work we use the 14-channel wireless EEG headset Emotiv EPOC+ [3]. Studies have shown that this headset is a worthy replacement for EEG laboratory instruments in everyday life [4, 5].

Multimodal signals paradigm allows to increase the accuracy of classification the functional state of a person in real world conditions. So, the accelerometer signal indicates the daily activity type and allows to increase accuracy of classification in HRV feature space [6]. On the other hand, human motion data depends of vestibular system in almost all aspects of life [7]. In this way, the information on the movements of the head can be used to assess the cognitive status: self-perception of movement, spatial perception, including moving objects.

In [8], EEG and accelerometer signals from Emotiv Epoc+ are used for assessing the functional state in integrated multimodal feature space. Integrated feature space was constructed and verified for telemedicine and workout applications with multimodal intelligence user interface [9, 10]. The averaged data for a stage (3–5 min duration) was used for feature extraction and classification. It is more useful to have a metric for continuous athlete functional state assessment during each stage.

The purpose of this work is the development of metrics, the detection of physiological patterns with changes in the functional state of the athlete in the training process. An example of a metric is the calculated values of the level of focus, stress, etc. at certain points in time.

2   Materials and Methods

2.1   Stages of the Experiment

To receive signals during the experiments, Emotiv Epoc+ headsets with Community SDK for data processing were used. The Emotiv EPOC+ headset contains 14 EEG channels and a three-axis accelerometer. The location of the electrodes of the EEG corresponds to the standard scheme 10–20: AF3, F7, F3, FC5, T7, P7, Pz, O2, P8, T8, FC6, F4, F8, AF4. The sample rate is 128 Hz. The number of digits of the ADC are 14. As additional software, Pebl was used to monitor the training program, Matlab was used for analysis and data processing.

The program of the experiment was sharpened as follows in order to simulate changes in the functional state during the training:

- Rest state (RS) during 300 s;
- TOVA test (T1) during 180 s;
- Hyperventilation load (HL) during 180 s;
- TOVA test (T2) during 180 s;
- Aftereffect (AE) during 300 s.

At the stage of functional rest, the subject sits opposite the monitor of the personal computer and looks at the black screen. Stage of TOVA test is an intellectual test for the variability of attention. It is a mental test to evaluate the function of active attention and control reactions. During the test, squares and circles appear alternately in the upper and lower parts of the computer screen. The task of the subject is to press a space on the keyboard when the square appears at the top of the screen [11]. At the stage of hyperventilation, the subject often breathes during the whole time, imitating breathing during heavy loads.

For functional state control purpose, ECG signal during all stages was registered. The spectral characteristics of HRV are investigated in the frequency bands indexes. The changes in VLF index were significant during the stages [3]. It is known [12], that fluctuations of VLF index of HRV signal with periods in range (0.04–0.003) Hz is complex and is due to the influence on the heart rhythm of the suprasegmental regulation level, since the amplitude of these waves is closely related to the mental stress and the functional state of the cerebral cortex.

2.2   Formation of Feature Space

2.2.1   Primary Signal Processing

Signals of motion modalities. During the experiments, the data was saved from the built-in three-axis accelerometer of the Emotiv EPOC+ headset; it provides data: time; acceleration values along the 3-axis.

The signal measured by the accelerometer is a linear sum of three components:

- Body Acceleration Component is acceleration resulting from body movement;
- Gravitation Acceleration Component is acceleration resulting from gravity;
- Noise inherent to the measuring system.

Changes in acceleration along the axes caused by human movements correspond to frequencies from 0 to 20 Hz in the signal spectrum. The gravitational component can be isolated in the range from 0 to 0.3 Hz. A component containing instrument noise is usually in the range above 20 Hz. To isolate the BA component, a second-order Butterworth window filter with frequencies from 0.3 to 20 Hz is used [13]. The study used the characteristics of the accelerometer signal from [6].

Signals of bioelectrical activity modalities. In the first step, all EEG data were transformed to the frequency domain. To separate EEG—rhythms from the signal, a second-order Butterworth bandpass filter were applied. Rhythms borders were: Theta (4–7) Hz, Alpha (7–15) Hz, Beta-Low (15–25) Hz, Beta-High (25–31) Hz. Discrete Fourier transform method was used for frequencies magnitudes extraction. As result, four coefficients are calculated for each of 14-th channel. Each coefficient is sum of magnitudes for one of the rhythms. Thus, EEG data in frequency domain are described as 56-dimension feature space [8].

2.2.2   The Model of the Integrated Feature Vector

Further, in order to build feature space and exclude artifacts components, the methods of the principal components (PCA) [14, 15], and linear discriminant analysis (LDA) [16] are used. Pairs of main components with the maximum explained variance and better classification accuracy are used to identify the most informative characteristics, for this purpose, the information on loads for EEG channels is used [8]. Data sets for pairs of states (RS and HL; RS and T1; T1 and HL) are used for building cross all stages feature space. The Fig. 1 contains data sets units, data processing unit descriptions and example of classification as described above.

Fig. 1 The diagram of data processing for feature selection purpose

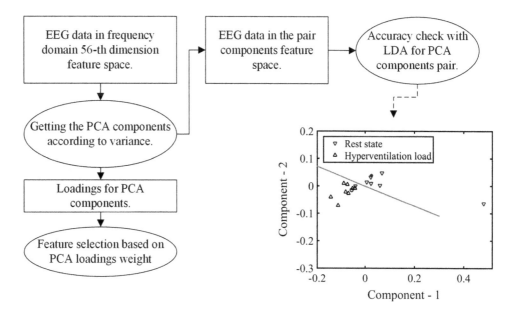

Fig. 2 The process of integrated feature vector creation

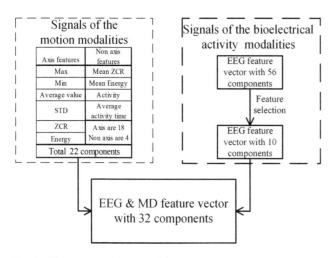

The result of feature selection is EEG feature vector, which contains AF3, T7, O1, T8, AF4 channels with Theta and Alpha rhythms components, correspondingly. Therefore, the new EEG vector size is 10 versus 56 in initial EEG data.

After that, new EEG feature space was expanded by adding the 22 components of accelerometer features as it depicted in Fig. 2.

2.3 Calculating a New Metric

The LDA method was used for creation the metric. Data sets for analysis contained EEG and MD recordings in integrated feature space for all subjects and the following pairs of stages: RS and HL, RS and T1. Test data were obtained from the experiment data using a cross-validation test (leave-one-out-cross-validation).

In our case, the accuracy obtained was 100% for all test samples. To determine the level of the athlete functional state, the formula derived in [17] is used to estimate the distance of each subject from the hyperplane PD separating the two load tests:

$$PD = \frac{\sum_{i=1}^{32} k_i x_i + C}{\sqrt{\sum_{i=1}^{32} k_i^2}},$$ (1)

where

- k_i are the coefficients of the hyperplane separating the two functional states;
- C is constant;
- x_i are the coordinates of the state of the subject in the characteristic space.

3 Results

The metric values were calculated on the basis of the averaged data for each stage and each subject. Boxplots were obtained for the states of the RS-T1 metric (as shown in Fig. 2)

This metric describes statistically significant state changes for the RS and T1 stages. For the stages HL, T2 and AE, the changes are not significant. In Fig. 3 shows scatter graphs for the RS-HL metric. From these graphs it is clear that:

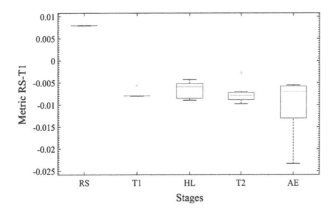

Fig. 3 The metric values calculated for RS-T1 stages

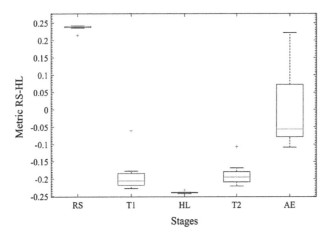

Fig. 4 The metric values calculated for RS-HL stages

- There are statistically significant differences in the metric values for the following stages: RS, T1, HL, and AE levels;
- There are no statistically significant changes for stages T1 and T2;
- There is a dynamic reflecting changes in the metric in the state AE towards recovering RS values after cognitive and physical exertion (Fig. 4).

4   Conclusions

One of the problems in developing an intelligent, multimodal interface is to obtain a model for combining modalities to calculate various metrics for assessing a person functional state. In this work, metrics to determine the physiological patterns of changes in the functional state of athletes in the process of training were obtained.

To calculate each metric, the coefficients of the LDA model were used. The coefficients of LDA models were

obtained on training samples in the integrated feature space of EEG modalities and an accelerometer for pairs of functional states:

- Rest state (RS) and TOVA test (T1) is the metric of RS-T1;
- Rest state (RS) and Hyperventilation load (HL) is the metric of RS-HL.

The value of the metrics was calculated for the following functional states: Rest state (RS); TOVA test (T1); Hyperventilation load (HL); TOVA test (T2); Aftereffect (AE). The calculation was made for the averaged characteristics of the modalities of each athlete. As a result, boxplots of metric values were obtained for each stage of the experiment.

Based on the results of the boxplot diagram analysis, the following conclusions can be drawn:

1. for the RS-HL metric, it is possible to obtain statistically significant changes in the assessment of the athlete functional state for the stages: Rest state (RS); TOVA test, Hyperventilation load (HL); Aftereffect (AE);
2. for the RS-HL metric, there is a dynamic that reflects the changes in the metric in the state AE towards the recovery of RS values after cognitive and physical exertion.

The revealed dynamics of the RS-HL metric in our opinion is associated with a change in the controlling effect of the cerebral cortex. What causes changes in the activity of alpha rhythm and changes in patterns of head movements. Changes in the patterns of movement of the head are associated with the search for the most stable position of the head relative to the gravitational vector.

Acknowledgements The work was supported by Act 211 Government of the Russian Federation, contract № 02.A03.21.0006.

Conflict of Interest The authors declare that they have no conflict of interest.

References

1. Silva, H.P. da et al.: Biosignals for Everyone. IEEE Pervasive Comput. 13, 4, 64–71 (2014).
2. Mcdowell, K. et al.: Real-world neuroimaging technologies. IEEE Access. 1, 131–149 (2013).
3. Borisov, V. et al.: Mobile brain-computer interface application for mental status evaluation. In: 2017 International Multi-Conference on Engineering, Computer and Information Sciences (SIBIRCON). pp. 550–555 (2017).
4. David, H. et al.: Usability of four commercially-oriented EEG systems. J. Neural Eng. 11, 4, 046018 (2014).

5. Ries, A. et al.: A Comparison of Electroencephalography Signals Acquired from Conventional and Mobile Systems. (2014).

6. Wu, M. et al.: Modeling perceived stress via HRV and accelerometer sensor streams. Presented at the Proceedings of the Annual International Conference of the IEEE Engineering in Medicine and Biology Society, EMBS (2015).

7. Danilov, Y.P. et al.: Vestibular sensory substitution using tongue electrotactile display. In: Human Haptic Perception: Basics and Applications. pp. 467–480 Birkhäuser Basel (2008).

8. Alexey Syskov et al.: Feature Extraction and Selection for EEG and Motion Data in Tasks of the Mental Status Assessing. Presented at the BIOSTEC 2018: 11th International Joint Conference on Biomedical Engineering Systems and Technologies (2018).

9. Dumas, B. et al.: Multimodal interfaces: A survey of principles, models and frameworks. Lect. Notes Comput. Sci. Subser. Lect. Notes Artif. Intell. Lect. Notes Bioinforma. 5440 LNCS, 3–26 (2009).

10. Syskov, A.M. et al.: Intelligent multimodal user interface for telemedicine application. In: 2017 25th Telecommunication Forum (FOR). pp. 1–4 (2017).

11. Mueller, S.T., Piper, B.J.: The Psychology Experiment Building Language (PEBL) and PEBL Test Battery. J. Neurosci. Methods. 222, 250–259 (2014).

12. N.I. Shlyk, R.M.B.: Rhythm of the heart and type of vegetative regulation in assessing the level of health of the population and the functional preparedness of athletes. Udmurt University, Izhevsk (2016).

13. Mathie, M.: Monitoring and Interpreting Human Movement Patterns Using a Triaxial Accelerometer. (2003).

14. Jolliffe, I.: Principal Component Analysis. In: Wiley StatsRef: Statistics Reference Online. John Wiley & Sons, Ltd (2014).

15. Wolpaw, J., Wolpaw, E.W.: Brain-Computer Interfaces: Principles and Practice. Oxford University Press, USA (2012).

16. McLachlan, G.J.: Discriminant Analysis and Statistical Pattern Recognition: McLachlan/Discriminant Analysis & Pattern Recog. John Wiley & Sons, Inc., Hoboken, NJ, USA (1992).

17. Kublanov, V.S. et al.: Classification of the physical training level by heart rate variability and stabilography data. In: 2017 Siberian Symposium on Data Science and Engineering (SSDSE). pp. 49–54 (2017).

18. Machado, I.P. et al.: Human activity data discovery from triaxial accelerometer sensor: Non-supervised learning sensitivity to feature extraction parametrization. Inf. Process. Manag. 51, 2, 201–214 (2015).

Experimental Setup for the Systematic Investigation of Infrared Neural Stimulation (INS)

Paul Schlett, Celine Wegner, Thilo Krueger, Thomas Buckert, Thomas Klotzbuecher, and Ulrich G. Hofmann

Abstract

Infrared neural stimulation (INS) gains growing interest both in electrophysiological research and for potential clinical applications, promising advantages like contactless operation, superior focality and hence spatial selectivity, and lack of electrical stimulation artifacts for nerve stimulation. We established an experimental setup for systematic investigation of relevant INS parameters, since little quantitative research for deeper understanding has been performed yet. Our customized setup facilitates the use of multiple fiber-based infrared laser systems of different wavelengths for remote stimulation, multi-site low-noise EMG recording and automated laser beam characterization. Hence, this setup simplifies upcoming systematic studies on both technical and physiological conditions for laser-induced neural activation. Determination of safe margins for reliable stimulation will help to understand the underlying physiological mechanisms and establish INS as alternative method for neural activation. Here, we present our experimental setup and preliminary results of our ongoing work.

Keywords

Infrared neural stimulation · Peripheral nerve Laser stimulation · Rat sciatic nerve · CMAP EMG

P. Schlett (✉) · U. G. Hofmann
Section for Neuroelectronic Systems, Uniklinik Freiburg, Freiburg, Germany
e-mail: paul.schlett@uniklinik-freiburg.de

C. Wegner · T. Krueger
inomed Medizintechnik GmbH, Emmendingen, Germany

T. Buckert
Arges GmbH, Wackersdorf, Germany

T. Klotzbuecher
Fraunhofer ICT-IMM, Mainz, Germany

© Springer Nature Singapore Pte Ltd. 2019
L. Lhotska et al. (eds.), *World Congress on Medical Physics and Biomedical Engineering 2018*,
IFMBE Proceedings 68/3, https://doi.org/10.1007/978-981-10-9023-3_14

1 Introduction

Infrared neural stimulation (INS) was introduced in 2005 by Wells in order to investigate alternative methods for neural activation [1, 2]. It found with work by Izzo [3] and Teudt [4] its path into auditory respectively facial nerve stimulation, yet little was published for stimulation of the central nervous system. Three different theories compete to elucidate the modus operandi of INS as there are: transient photothermal gradients [1], change of cell membrane capacitance due to heating [5] or nanoporation meaning short-time opening of the cell membrane [6] leading to depolarization of neurons. However, to the best of our knowledge no final conclusion was presented and the existing publications are hard to independently reproduce due to a lack of technical details reported. The underlying bio-physiological mechanisms remain unclear. We therefore reached out to re-evaluate infrared nerve stimulation with our experimental approach, which is not only to qualitatively reproduce INS, but also to enable novel explanation based on quantitative observations in the animal model. In contrast to literature, we try to maximize the working distance between laser fiber and illuminated tissue to maintain the touch-free properties of laser stimulation with regard to potential clinical application. Here, we depict our experimental approach to explore the promising field of INS.

2 Methods

2.1 Infrared Laser Stimulation

Laser devices for INS stimulation are commonly being operated in pulsed rather than continuous mode [1, 2, 7–9]. Illuminating a target with a pulsed laser, the *radiant exposure* H_p caused by one pulse is defined as

$$H_p = \frac{E_p}{A} = \frac{4E_p}{\pi d^2} \quad and \quad [H] = \frac{J}{m^2} \quad , \quad (1)$$

where E_p is the radiant energy of the pulse and A is the effective illuminated surface area assuming a circular laser spot of diameter d, commonly referred to as *spot size*. The *pulse energy* E_p depends on both average laser power and the *pulsewidth* τ_p. Although reported laser pulses for INS stimulation in the current literature are mainly described by means of the radiant exposure H, however, this does not account for the exposure time at all [1, 2, 7–9]. Considering the radiant energy E_p of a single pulse being delivered to a target within a spot of area A and a discrete exposure time $\tau = \tau_p$, we describe the radiant load using the *irradiance I*, which is given by

$$I = \frac{E}{\tau A} = \frac{4}{\pi} \frac{E_p}{\tau_p d^2} \quad and \quad [I] = \frac{J}{s \cdot m^2} = \frac{W}{m^2} \quad . \quad (2)$$

For a pulse train of duration τ_{tr} containing a fast burst of n identical pulses, the radiant exposure and irradiance have to be modified to be H_{tr} and I_{tr} respectively:

$$H_{tr} = \frac{n \cdot E_p}{A} \quad resp. \quad I_{tr} = \frac{n \cdot E_p}{\tau_{tr} A} \quad . \quad (3)$$

For the application of laser radiation into the rat sciatic nerve, we use an infrared diode laser system (1470 nm, 12 W, DILAS Diodenlaser GmbH, Mainz, Germany), coupled into a multi-mode fiber of 200 μm core diameter and a numerical aperture NA = 0.22. The laser fiber is connected to a customized focusing optics, composed of a matched set of five spherical lenses mounted on a vertical C-mount rail (Fig. 1a). Shifting of the last lens allows the variation of the laser spot size diameter down to $d = 0.9$ mm while maintaining minimum 15 mm depth of field and a fixed focus position of about 50 mm below the rail mount. The optical assembly supports the use of any other fiber-coupled laser system for future studies with different (infrared) wavelengths.

2.2 Laser Characterization

To determine the radiant energy E_p of single laser pulses and fast pulse trains of short (<2 ms) duration, we use a pyro-electric energy meter (ES145C, Thorlabs, Newton, New Jersey, US). For beam profiling, we employ the commonly known knife-edge method [10]: a razor blade is moved orthogonally into the continuous beam by a motorized micromanipulator (Luigs und Neumann, Ratingen, Germany), while measuring the downstream radiant power (S310C, Thorlabs) with regard to the blade position (Fig. 1a). Beam diameter $d(z)$ as a function of the position z along the

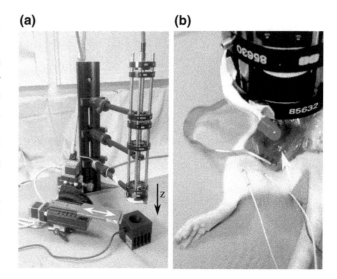

Fig. 1 Experimental setup. **a** Vertically mounted focusing optics for INS studies. Automated beam characterization (knife-edge) with micromanipulator and power meter. **b** Rat sciatic nerve under stimulation with laser pulses (white arrow). EMG is recorded by means of intramuscular needle electrodes (Color figure online)

optical axis can be extracted, the beam profile characterized. We use a laser beam profiler (LBP-1-usb, Newport Corporation, Irvine, California, US) to rapidly determine the beam diameter or *spot size* during experiments.

2.3 Parametric Laser Control

The ability of our setup to cover a wide range of defined stimulation conditions like *pulse energy* E_p, *exposure time* τ and the irradiated area A (*spot size*), holds great potential for detailed research of INS. To this end, we use a set of parameters for pulse train definition, which are depicted in Fig. 2. Assuming a fixed average laser power $\bar{P}$, the pulse energy E_p scales linearly with the *pulsewidth* τ_p. Considering n identical pulses as shown in Fig. 2, the period between two pulses is set by the Interval time τ_i. Hence, the pulse train is defined and can be iterated specified by the *Repeat* time and number of *Runs* to execute.

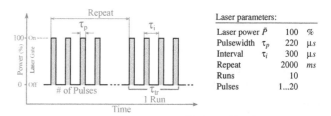

Fig. 2 Laser parameters for INS stimulation: Definition and experimental settings

The free choice of those parameters allows arbitrary laser stimulation sequences and ensures maximum versatility for upcoming studies. With regard to Eq. (3), differently composed pulse trains that result in identical irradiance I_{tr}, however, may still feature different internal patterns, which might be physiologically relevant when optically activating nerves.

The inherent constraints of the laser's proprietary control software require a more sophisticated approach for external laser control to facilitate advanced flexibility. Hence, a customized LabVIEW user interface operates the entire experiment. All parameters and commands are passed to an Arduino Nano microcontroller and electronic circuit (Fig. 3). On the Arduino, a local program interprets the parameters to render pulse trains and—after a firing command—accordingly generates pin output voltages, which are then fed into the circuitry and drive the laser gate. On the left side in Fig. 3, a PWM output voltage is stabilized using a combination of two integrating capacitors and an operational amplifier to provide a steady analog voltage (0…10 V) for setting the power value $\bar{P}$ as a fraction of full laser power (12 W). Analog inputs constantly monitor the laser status for safety and prevent unintended firing. On the right side, the laser's internal gate voltage is controlled by means of a digital output controlling a set of two serial fast switching transistors (Darlington) to supply a gate voltage above 18 V. Further, a trigger signal envelopes each pulse train according to Fig. 2 for data synchronization.

2.4 Data Acquisition

Laser-induced nerve activation by means of INS can result in visible muscle activity like twitching [11]. The elicited neural signal travels along the nerve leading to muscle contraction, termed compound muscle action potential (CMAP). Depending on the strength of the neural signal, limb or muscle movement can be weak and therefore remain unobserved. For the reliable detection of any emerging muscle activity, we record the electromyogram (EMG) by means of intramuscular needle electrodes attached to a commercial EMG recording system (ISIS, inomed Medizintechnik GmbH, Emmendingen, Germany). For each experiment, all laser parameters are saved into a text file.

2.5 Animal Preparation and Electrode Placement

For preliminary studies of INS on the rat sciatic nerve in vivo, we used female Sprague-Dawley rats (n = 11, 240–350 g). All animal experiments conducted in this study were performed with approval from the locally responsible Animal Welfare Committee with the Regierungspraesidium Freiburg in accordance with the guidelines of the European Union Directive 2010/63/UE (permit G17/80). Rats were initially anesthetized with Isofluran prior to intraperitoneal injection of a mixture of ketamine (100 mg/kg) with xylazine (8 mg/kg) and accordingly readministered depending on the depth of anesthesia. The skin was shaved and removed over both thighs. The sciatic nerves were then exposed by incision of the muscular fascia and mostly blunt dissection of *m. gluteus superficialis* and *m. biceps femoris*. For INS, the rat was placed under the laser setup and the wound was held open using a surgical retractor and moistened frequently with saline solution. A reference electrode was placed close to the tail and a pair of EMG electrodes was inserted into *m. biceps femoris* for preliminary studies. Proper nerve function was verified

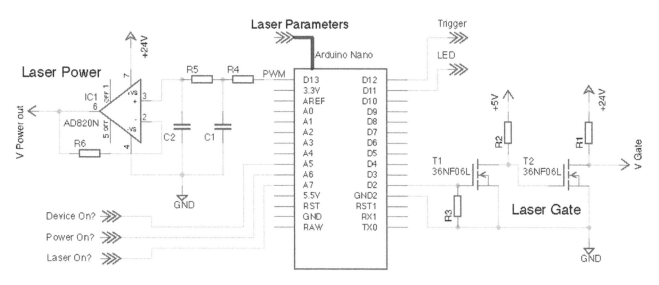

Fig. 3 Schematics of the customized circuit for laser operation with Arduino Nano controlling laser power and gate voltages according to laser parameters

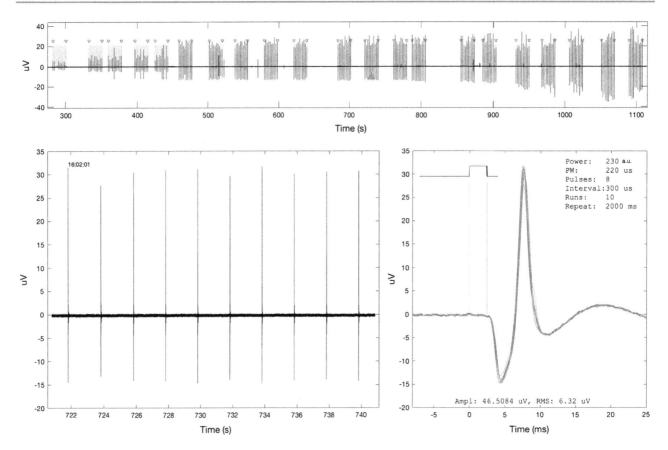

Fig. 4 Preliminary results. Top: long-term EMG recording under INS with increasing number of stimulation pulses. Left: Detailed extract (top green arrow, 8 pulses) of distinct response to laser stimulation with parameters according to Fig. 2. Right: Overlay of the 10 laser-evoked CMAPs with regard to the stimulation onset (trigger). CMAPs were highly consistent and do not show stimulation artifacts (Color figure online)

using an electrical nerve stimulation device (Neurostimulator, inomed Medizintechnik GmbH, Emmendingen, Germany). After experiment termination, both sciatic nerves were extracted for histological analysis.

2.6 Experimental Procedure

For proof of concept, the laser was focused at various locations on the main trunk of the sciatic nerve, up to about 15 mm proximal to the first branching point (Fig. 1b). Data recording was started and the laser was fired according to the pulse parameters (see Fig. 2). Overall radiant exposure per pulse train H_{tr} was then successively increased with the number of pulses n.

3 Results

In all of $n = 11$ preliminary experiments, we were able to evoke muscle activity from the rat sciatic nerve. However, successful stimulation occurred primarily at specific locations susceptible for INS and is therefore considered to be highly location-dependent. At locations that did not reveal a clear response to INS at low to medium radiation levels, the increase of radiant exposure for stimulation partially led to visible thermal damage. Tissue damage by INS will therefore be addressed in upcoming dosimetry histo-pathological studies.

In Fig. 4, an exemplary data set is shown. The top graph depicts a recorded EMG under INS. The laser trigger signals are shown in yellow. The bottom left graph shows one selected pulse train with laser-evoked CMAP signals, in the right graph an overlay plot is depicted. In general, the majority of the elicited CMAP responses rated "successful" occurred in a synchronous, rhythmical manner according to the laser stimulus, within 3–20 ms after stimulus onset ($t = 0$). Also, muscle twitching was observed at higher CMAP amplitudes. Apart from few outliers, the shape of the evoked responses was highly consistent. No stimulation artifact was observed through all experiments.

When considering the approximation for the train time $\tau_{tr} \approx n \cdot \tau_i$, we note that the irradiance I_{tr} remains constant with fixed single pulse energy which was determined to be

$E_p = 1.9 - 2$ mJ. In upcoming studies, different stimulation patterns, and therefore alteration of the irradiance, will be performed by variation of laser power $\bar{P}_i$, exposure times τ (τ_p, τ_i) and different spot sizes.

4 Conclusion

In this paper, we presented our novel experimental setup, that allowed us to consistently reproduce infrared nerve stimulation in all preliminary experiments on the rat sciatic nerve. In the future we will continue our research on different optical stimulation parameters and wavelengths, exposure thresholds and limits for reliable and safe stimulation, and bio-physiological aspects of INS.

Conflict of Interests The authors declare that they have no conflict of interests. This work was funded by the German Federal Ministry of Education and Research (BMBF), project *NeuroPhos*, FKZ 13GW0155C.

References

1. Wells, Jonathon; Kao, Chris; Mariappan, Karthik; Albea, Jeffrey; Jansen, E D.; Konrad, Peter; Mahadevan-Jansen, Anita: Optical stimulation of neural tissue in vivo. In: *Optics letters* 30 (2005), Nr. 5, S. 504–506

2. Wells, Jonathon; Kao, Chris; Jansen, E D.; Konrad, Peter; Mahadevan-Jansen, Anita: Application of infrared light for in vivo neural stimulation. In: *Journal of biomedical optics* 10 (2005), Nr. 6, S. 064003–064003

3. Izzo, Agnella D.; Richter, Claus-Peter; Jansen, E D.; Walsh, Joseph T.: Laser stimulation of the auditory nerve. In: *Lasers in surgery and medicine* 38 (2006), Nr. 8, S. 745–753

4. Teudt, Ingo U.; Nevel, Adam E.; Izzo, Agnella D.; Walsh, Joseph T.; Richter, Claus-Peter: Optical stimulation of the facial nerve: a new monitoring technique? In: *The Laryngoscope* 117 (2007), Nr. 9, S. 1641–1647

5. Shapiro, Mikhail G.; Homma, Kazuaki; Villarreal, Sebastian; Richter, Claus-Peter; Bezanilla, Francisco: Infrared light excites cells by changing their electrical capacitance. In: *Nature communications* 3 (2012), S. 736

6. Beier, Hope T.; Tolstykh, Gleb P.; Musick, Joshua D.; Thomas, Robert J.; Ibey, Bennett L.: Plasma membrane nanoporation as a possible mechanism behind infrared excitation of cells. In: *Journal of neural engineering* 11 (2014), Nr. 6, S. 066006

7. Wells, Jonathon D.; Thomsen, Sharon; Whitaker, Peter; Jansen, E D.; Kao, Chris C.; Konrad, Peter E.; Mahadevan-Jansen, Anita: Optically mediated nerve stimulation: Identification of injury thresholds. In: *Lasers in surgery and medicine* 39 (2007), Nr. 6, S. 513–526

8. Izzo, Agnella D.; Walsh, Joseph T.; Ralph, Heather; Webb, Jim; Bendett, Mark; Wells, Jonathon; Richter, Claus-Peter: Laser stimulation of auditory neurons: effect of shorter pulse duration and penetration depth. In: *Biophysical journal* 94 (2008), Nr. 8, S. 3159–3166

9. Richter, C.-P.; Matic, A.I.; Wells, J.D.; Jansen, E.D.; Walsh, J.T.: Neural stimulation with optical radiation. In: *Laser & Photonics Reviews* 5 (2011), Nr. 1, 68–80. http://dx.doi.org/10.1002/lpor.200900044. – https://doi.org/10.1002/lpor.200900044. – ISSN 1863–8899

10. Khosrofian, John M.; Garetz, Bruce A.: Measurement of a Gaussian laser beam diameter through the direct inversion of knife-edge data. In: *Applied Optics* 22 (1983), Nr. 21, S. 3406–3410

11. McCaughey, Ryan G.; Chlebicki, Cara; Wong, Brian J.: Novel wavelengths for laser nerve stimulation. In: *Lasers in surgery and medicine* 42 (2010), Nr. 1, S. 69

Chronically Monitoring of Optogenetic Stimulation-Induced Neural and Hemodynamic Response

Chun-Wei Wu and Jia-Jin Chen

Abstract

Understanding the link between neuronal activity and cerebral hemodynamics, known as neurovascular coupling (NVC), is critical for studying cerebral vascular dysfunctions. Optogenetics, which involves using light to control electrical activity of opsin-expressing neurons, is an approach to monitor and modulate functions of specific neurons in neuronal networks. Recent studies suggested that optogenetic stimulation induced hemodynamic response could be a good model for studying the mechanism of NVC. The aim of the study is to develop an optogenetic platform for long-term monitoring of the cerebral hemodynamics. An implantable optrode for optical stimulation and neural recording in addition to near-infrared spectroscopy (NIRS) recording have been developed. Through fiber optics, the optogenetic stimulation was introduced into ChR2-positive glutamatergic neuron of primary motor cortex (M1). Neural and hemodynamic responses were evaluated using local field potentials (LFPs) and NIRS recordings. An in-line optical filter was applied to filter off the blue light collected from M1 and to avoid interfering the NIRS measurement. The results showed the classic waveform of evoked hemodynamic response was obtained alongside of LFPs during optogenetic stimulation. Overall, our innovative optogenetic-NIRS interface could provide long-term observation on cerebral hemodynamics during optogenetic modulation.

Keywords

Neurovascular coupling • Hemodynamic response
Optogenetic stimulation • Channelrhodopsin
Near infrared spectroscopy

1 Introduction

Neurovascular Coupling (NVC) is the process that regulates local cerebral blood flow (CBF) to meet the oxygen and nutrient demands from activated brain area [1]. Many functional brain imaging techniques work depending on principle of NVC, such as fMRI measuring blood-oxygenation level dependent (BOLD) signals, positron emission tomography (PET) measuring CBF changes, and functional near infrared spectroscopy (fNIRS) measures changes in hemoglobin concentrations [1]. Impaired NVC has been reported in neurodegenerative diseases such as stroke, Alzheimer's disease and Parkinson's disease [2, 3]. However, the mechanisms that modulate NVC are not fully understood.

Optogenetics, which involves using light to control cellular function in genetic modified cell, has become a powerful tool in neuroscience animal study [4]. When a light sensitive ion channel was introduced into neuron and activated by certain wavelength of light, it regulates the ion conductance of the membrane and alters the electrophysiological activity of neuron. Optogenetics can achieve fast, spatially localized and tissue-type specifically neural modulation [5]. Unlike conventional electrical and magnetic stimulation, optogenetic stimulation does not generate significant electromagnetic artifact, which allows investigators to collect electrophysiological signals tightly coupled with stimuli [6].

NVC dynamics evoked by optogenetic stimulation has been studied using laser Doppler flowmetry, intrinsic optical imaging and fMRI [7–11]. In the current study we developed a method that adopting fNIRS to chronically measure optogenetic stimulation-induced hemodynamic response in ChR2-expressed rat. By integrating fNIRS and neural recording, we aimed to monitor long-term NVC dynamics at high temporal resolutions.

C.-W. Wu (✉) · J.-J. Chen
National Cheng Kung University, Tainan, 701, Taiwan
e-mail: chenjj@mail.ncku.edu.tw; chunwei0521@gmail.com

© Springer Nature Singapore Pte Ltd. 2019
L. Lhotska et al. (eds.), *World Congress on Medical Physics and Biomedical Engineering 2018*,
IFMBE Proceedings 68/3, https://doi.org/10.1007/978-981-10-9023-3_15

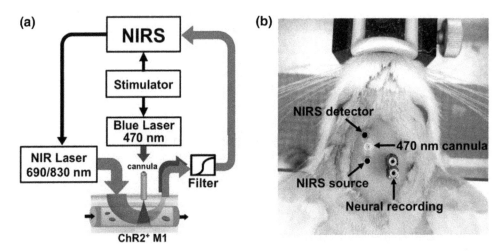

Fig. 1 System set-up of optogenetic-induced hemodynamic recording. **a** Block diagram of the NIRS system integrated with blue laser stimulation. **b** Chronic implantation of the optrode and NIRS cannula on rat's head

2 Materials and Methods

2.1 Animal Preparation

All experiments were approved by the National Cheng Kung University Medical College Animal Use Committee. Adult male SD rats with body weight 300–350 gw were obtained from the Animal Center of National Cheng Kung University Medical College. They were housed in standard cages at a temperature of 25 ± 1 °C with a 12/12-h light/dark cycle and free food and water access.

2.2 Lentivirus Preparation

A second-generation lentivirus system was used for transduction of ChR2. The packaging plasmid psPAX2 and envelope plasmid pMD2.G were transfected into HEK293FT cell with ChR2 expressing plasmid: pLenti-CaMKIIa-ChR2(H134R)-EYFP-WPRE, which was a gift from Karl Deisseroth in Optogenetics Resource Centre, Stanford University, Stanford, CA, USA (Addgene plasmid # 20944). The medium was collected and concentrated to harvest virus particle in 1,000 folds concentrated solution.

2.3 Stereotaxic Surgery

Rats were first anesthetized with 5% isoflurane in oxygen and then maintained on 2% isoflurane in oxygen. Rats were anchored in a stereotaxic frame. An incision was made on scalp, and the skull was exposed to locate bregma. A hole was drilled over M1 corresponding to right forelimb (AP: +1.5 mm, ML: −3.0 mm). Lentivirus containing pLenti-CaMKIIa-ChR2(H134R)-EYFP-WPRE was injected in M1 at two different depths: 2.0 and 2.5 mm. Optrode composed of a 200 μm-optical fiber and Pt wire electrode

was implanted into M1. Two of the 18 gauge stainless steel cannulas (10 mm in length) were placed near optrode to guide optical fibers when applying NIRS measurement. After implantation, the optrode and cannulas were covered with dental cement (Fig. 1b). Wounds were closed with tissue glue; rats were given lidocaine (2%) and 5 ml of normal saline (i.p.). Optical stimulation and hemodynamic monitoring were performed after 4 weeks of recovering.

2.4 Optogenetic Stimulation, Neural Recording and NIRS Measurement

Figure 1a shows a simplified block diagram of the optogenetic-NIRS system. Optical stimuli were generated by stimulator (Model 2100, A-M Systems) and sent out as TTL controlling signal to a blue laser device (10 mW/mm²; MBL-III-473-50, Ultralasers Inc.). The optical stimuli were guided into ChR2-expressed tissue through implantable optrode. Neural signals are picked up by Pt wire electrode, and then amplified (20,000X), filtered with 60 Hz notch and 10–600 Hz bandpass filter, finally sampled at 10 kHz (MP36, BIOPAC Systems Inc.). NIRS measurements were performed with a frequency domain photon migration (Imagent, ISS). NIR lasers of two wavelengths (690 and 830 nm) were guided into ChR2 adjacent tissue through optical fibers. Optical signals were collected by optical fibers connected to an optical filter to remove blue light. After filtration, NIR signals were detected by PMT sensors, and analyzed to obtain the concentration changes of oxygenated ([HbO]) and deoxygenated ([Hb]) hemoglobins based on modified Beer-Lambert's Law.

3 Results

The function of ChR2 was first evaluated under blue laser stimulation after four weeks of expression. The local field potentials (LFPs) were evoked by optical pulses, which were

Fig. 2 Electrophysiological recording of neural responses induced by optogenetic stimulation. **a** Local field potential (LFP) trace before, during and after 20 Hz optical stimulation with 1 ms pulse-width. **b** Averaging waveforms of optogenetic-evoked potentials under blue laser stimulation with various pulse-widths

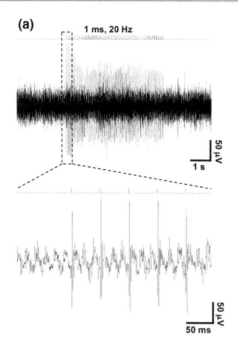

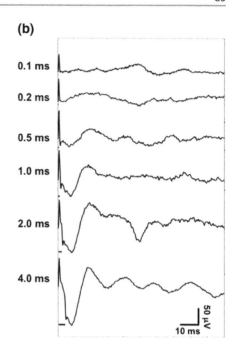

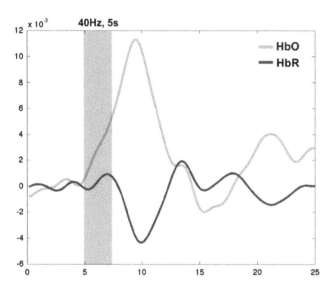

Fig. 3 Hemodynamic response under high frequency optogenetic stimulation (1 ms pulse width at 40 Hz for 5 s, power density: 10 mW/mm^2)

A significant elevation was observed in [HbO] while a drop was seen in [HbR] after 5 s of optical pulses at 40 Hz.

4   Discussion and Conclusions

In the current study we developed a novel NIRS platform to monitor hemodynamics and neural activity during optogenetic modulation. Using optical filter to remove 470 nm wavelength components before NIRS recording allowed the quantification of hemoglobin under the blue laser stimulation without stimulating artifacts. This set-up is feasible for long-term hemodynamic measuring when using optogenetic approach to investigate NVC mechanism in small rodent model.

Acknowledgements The authors would like to thank the Ministry of Science and Technology (MOST) of R.O.C. for supporting (Grand number: MOST 104-2314-B-006-007-MY3).

Conflict of Interest The authors declare that they have no conflict of interest.

delivered through optical fiber tethered from laser device to the implanted optrode. The corresponding evoked potentials are visible at each epoch (Fig. 2a). The classic waveforms were calculated by averaging 60 epochs. Traces in Fig. 2b indicated the classic waveforms of optogenetic-evoked potentials induced by various durations of pulse (0.1, 0.2, 0.5, 1.0, 2.0 and 4.0 ms). By applying high-frequency optogenetic stimulation, the changes in [HbO] and [HbR] in ChR2-expressed brain are quantified using NIRS. Figure 3 shows the hemodynamic waveforms averaged by 11 times.

References

1. Raichle, M.E.: Behind the scenes of functional brain imaging: a historical and physiological perspective. Proc Natl Acad Sci U S A. 95(3), 765–72 (1998).
2. Hamilton, N.B., et al.: Pericyte-mediated regulation of capillary diameter: a component of neurovascular coupling in health and disease. Front Neuroenergetics. 2 (2010).
3. Lourenco, C.F., et al.: Neurovascular-neuroenergetic coupling axis in the brain: master regulation by nitric oxide and consequences in

aging and neurodegeneration. Free Radic Biol Med. 108668–82 (2017).

4. Zhang, F., et al.: Optogenetic interrogation of neural circuits: technology for probing mammalian brain structures. Nat Protoc. 5 (3), 439–56 (2010).

5. Boyden, E.S., et al.: Millisecond-timescale, genetically targeted optical control of neural activity. Nat Neurosci. 8(9), 1263–8 (2005).

6. Richner, T.J., et al.: Optogenetic micro-electrocorticography for modulating and localizing cerebral cortex activity. J Neural Eng. 11(1), 016010 (2014).

7. Scott, N.A., et al.: Hemodynamic responses evoked by neuronal stimulation via channelrhodopsin-2 can be independent of intra-cortical glutamatergic synaptic transmission. PLoS One. 7(1), e29859 (2012).

8. Kahn, I., et al.: Optogenetic drive of neocortical pyramidal neurons generates fMRI signals that are correlated with spiking activity. Brain Res. 151133–45 (2013).

9. Vazquez, A.L., et al.: Neural and hemodynamic responses elicited by forelimb- and photo-stimulation in channelrhodopsin-2 mice: insights into the hemodynamic point spread function. Cereb Cortex. 24(11), 2908–19 (2014).

10. Li, N., et al.: Study of the spatial correlation between neuronal activity and BOLD fMRI responses evoked by sensory and channelrhodopsin-2 stimulation in the rat somatosensory cortex. J Mol Neurosci. 53(4), 553–61 (2014).

11. Iordanova, B., et al.: Neural and hemodynamic responses to optogenetic and sensory stimulation in the rat somatosensory cortex. J Cereb Blood Flow Metab. 35(6), 922–32 (2015).

System for Motor Evoked Potentials Acquisition and Analysis

V. Čejka, A. Fečíková, O. Klempíř, R. Krupička, and R. Jech

Abstract

Biological signal acquisition is a fundamental part of the following signal processing methods. This study is focused on hardware and software solution for an electrophysiological measurement in neurological patients and healthy controls. This paper deals with a design and an implementation of the system for transcranial magnetic stimulation (TMS) applied over the human motor cortex, which has the diagnostic and potential therapeutic effect, respectively. The system was successfully used for examinations of 22 neurological patients (mean age 51 ± (SD) 17 years) suffering from dystonia of various distribution and etiology treated by chronic deep brain stimulation of globus pallidus interna (GPi DBS). Established values of the motor-evoked potential's (MEP) parameters are in line with the current literature. Designed system for TMS examination is an effective tool for studying the pathophysiology of neurological diseases.

Keywords

Transcranial magnetic stimulation
Motor evoked potential • Data acquisition

V. Čejka (✉) · O. Klempíř · R. Krupička
Faculty of Biomedical Engineering, Department of Biomedical Informatics, Czech Technical University in Prague, Prague, Czech Republic
e-mail: vaclav.cejka@fbmi.cvut.cz; vaclav.cejka@lf1.cuni.cz

V. Čejka · A. Fečíková · R. Jech
Department of Neurology and Centre of Clinical Neuroscience, First Faculty of Medicine and General University Hospital, Charles University, Prague, Czech Republic

V. Čejka · A. Fečíková · R. Jech
Faculty of Medicine, General University Hospital in Prague, Prague, Czech Republic

1 Introduction

Over the past decades, TMS has become a tool of significant importance in both basic and clinical neurosciences [1]. Gradually coupled with stimulation by single TMS pulses, techniques of paired-pulse stimulation, repetitive stimulation, and various hybrid protocols appeared [2, 3]. The cost of measurement is rising quickly, so it is essential to use efficient HW and SW resources to shorten the time of the electrophysiological examination. The measuring chain is a pivotal element that, in a convenient configuration, facilitates the work and allows part of operations to be partially or fully automated.

Instrumentation technique routinely used to record motor-evoked potentials (MEPs) in EMG laboratories usually allows only to display the potential waveform after the arrival of the stimulation TMS pulse. This approach is insufficient for our purposes as it does not allow to randomize the order of pulses varying in stimulation parameters. Moreover, it is unable to subsequently sort and process data according to these criteria. Most of these amplifiers are not equipped for communication with the connected stimulator as they only synchronize the moment of recording with the stimulation discharge. Therefore, we selected and implemented a system with its own programming language to enable us to divide our experimental measurements on a dystonic patient group into blocks, to automatically randomize the pulse rate differing by intensity and time span, and to display the resulting data on-line.

As Albanese defines: 'Dystonia is a movement disorder characterized by sustained or intermittent muscle contractions causing abnormal, often repetitive, movements, postures, or both' [4]. Much of the current literature on pathophysiology of dystonia pays particular attention to three general abnormalities: loss of inhibition, sensory dysfunction and a derangement of plasticity [5]. A well-established approaches to test intracortical inhibition and plasticity in humans in a noninvasive way are short interval intracortical inhibition (SICI) and paired associative stimulation (PAS) [5].

© Springer Nature Singapore Pte Ltd. 2019
L. Lhotska et al. (eds.), *World Congress on Medical Physics and Biomedical Engineering 2018*,
IFMBE Proceedings 68/3, https://doi.org/10.1007/978-981-10-9023-3_16

2 MEP Acquisition

MEPs are obtained using TMS. TMS is an indirect and non-invasive method used to induce excitability changes in a motor cortex via wire coil generating a magnetic field that passes through the scalp [6]. In general, single-pulse and paired-pulse TMS are used to explore brain functioning, whereas repetitive TMS (rTMS) is used to induce changes in brain activity that can last beyond the stimulation period.

TMS applied over the motor cortex leads to an activation of pyramidal cells evoking descending volleys in the pyramidal axons projecting on spinal motoneurons. Motoneuron activation in response to corticospinal volleys induced by TMS leads to contraction in the target muscle evoking MEP on electromyography (EMG). This potential is recorded by the constructed TMS system [6]. Its latency [7], peak-to-peak amplitude [8, 9], duration [10], number of phases [11] or area under curve [12] can be used to describe and estimate parameters of motor cortex and whole corticospinal tract [6].

2.1 System Requirements

The proposed system is designed to address three major challenges. The first part is for the motor threshold (resting MT and action MT). The other two parts are for SICI and PAS. During the determination of AMT and RMT, the acquisition of potentials is controlled by the moment of the stimulator discharge. The length of the recorded epoch is 100 ms. A request is made to display the last 10 MEPs on the screen and to calculate their amplitude on-line. There is also a requirement to determine the number of pulses greater than the stated limit (50 uV for RMT, 200 uV for AMT).

During SICI measurement we require that MEPs are sorted separately after paired pulse and especially after simple TMS stimulation. By that, it is possible to visually assess the effect of paired stimulation immediately. It is important to automatically adjust stimulus intensities and interstimulation intervals and randomize single/paired TMS pulses.

For PAS, we want to measure sets of curves showing the size of MEP depending on the pacing pulse power. Using these curves the PAS effect is measured before the intervention and then at 0, 15 and 30 min after stimulation.

2.2 Data Flow

SICI (Fig. 1) is obtained with paired-pulse stimulation and reflects interneuron influence in the cortex. The measurement is divided by the intensity of the conditioning pulse into blocks. In each block, 15 simple and 15 paired pulses

are randomly mixed. These pulses are filtered by bandpass and averaged. From each average curve, the latency (ms) and its amplitude (mV) are determined. From the amplitude ratio, the magnitude of the intracortical inhibition is finally calculated.

PAS is a combination of repetitive transcranial magnetic stimulation (rTMS) and repetitive electrical peripheral nerve stimulation (rENS) [13]. This protocol produces a long-lasting and somatotopically specific increase in corticospinal excitability. PAS effect (Fig. 2) can be accessed in a variety of ways. One of possible options is to measure the so-called 'Stimulus response curve' (SRc) before and after PAS and evaluate its change.

2.3 System Testing

The system has been tested on 22 patients (13F, 9 M, mean age 51 ± (SD)17 years) with dystonia of various distribution (15 generalized, 7 cervical) and etiology treated by chronic pallidal deep brain stimulation (DBS). For comparison, we included 22 age- and gender-matched healthy controls (13F, 9 M, aged 51 ± (SD)17 years) with no history of neurological or psychiatric disorder. All subjects gave their informed consent to participate and the study was approved by the local ethics committee of the General Faculty Hospital in Prague in compliance with the Declaration of Helsinki.

After the examination, latencies and amplitudes measured by MEP were evaluated off-line. A magnitude of inhibition was calculated from the MEP amplitude ratio, whereas the PAS effect was evaluated by SRc.

3 TMS System Description

3.1 Hardware Solution

The created system (see Fig. 3) measures the MEPs elicited by TMS applied over motor cortex. The signal is recorded using surface electrode (Alpine Biomed Denmark, REF: 90,13L0453, connector 1.5 mm TPC, cable 50 cm) in belly-tendon montage from the targeted hand muscle (abductor pollicis brevis—APB, abductor digiti minimi—ADM), amplified (Quad System 1902, Cambridge Electronics Design), converted to the digital representation (ADC, Power 1401 mk II, CED) and then saved on the computer (Notebook Lenovo, Think Pad T530, Intel Core i7, OS Windows 7 Professional, 8 GB RAM) for later offline analysis.

Individual parts of the system communicate together using USB and RS-232 interface. Synchronized pulses are

Fig. 1 Short interval intracortical inhibition (SICI) diagram

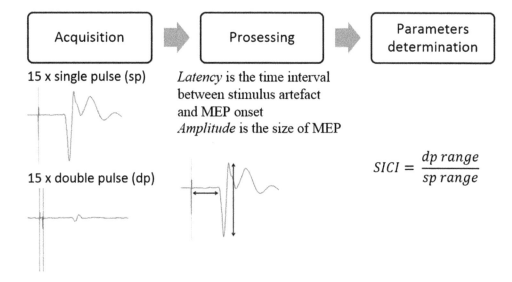

Fig. 2 Paired associative stimulation (PAS) diagram

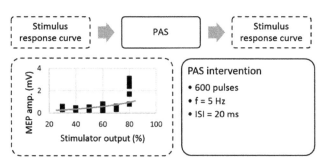

propagated through coaxial cable (RG 58 C/U, 50 O, Ø 0, 9/19 × 0,18 mm, PE) in TTL (transistor-transistor-logic) values.

Measuring chain is for better functionality and data reproducibility completed by an optical tracking system (OTs, BrainSight Frameless, Rogue Research Inc) and stereotactic frame. OTs is a means of determining in real-time the position of an object (stimulating coil) by tracking the positions of either active or passive infrared markers attached to the object [14]. The position of the point of reflection is determined using special camera system (Polaris). This system is necessary for better monitoring of

Fig. 3 Block diagram of the TMS system

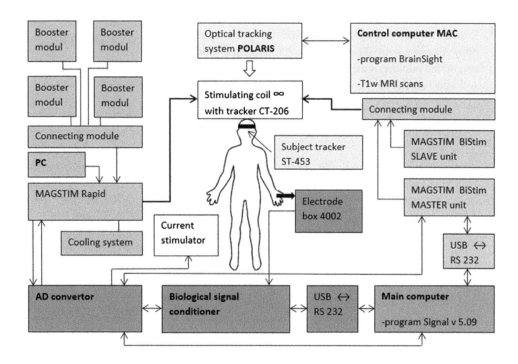

Fig. 4 Script algorithm

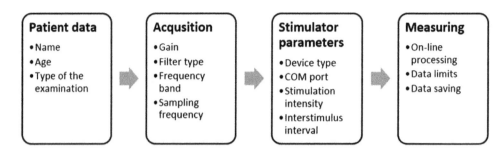

the magnetic field position and non-invasive stereotactic frame minimizes patients' head movements.

3.2 Software Solution

Whole process is driven by scripts written in Signal programming language (version 5.09, CED). Signal is a sweep-based data acquisition and analysis package. Its use ranges from a simple storage oscilloscope to complex applications requiring stimulus generation, data capture, control of external equipment and analysis. These scripts ensure an adjustment of parameters of the ADC (sampling frequency: 5 kHz, resolution: 16 bit), signal conditioner (frequency band: 5–2000 Hz, filter type: B'Worth 3rd order) and TMS stimulator (number of pulses, stimulation intensity, interstimulus interval).

Script algorithm description:

1. Termination of all previous measurements and deletion of the current configuration file.
2. Display of information about the upcoming measured protocol.
3. Loading of new configuration file that consists of: sampling frequency of the AD converter, recorded frame length, the number of used channels and their parameters:
   a. Gain value (1000x), voltage offset (0 mV), filter type (high pass filter: 3-pole B'Worth, low pass filter: B'Worth), cut-off frequency (5 Hz for the high pass filter, 2 kHz for the low pass filter) and notch filter switch on/off.
   b. Activation of an output port, through which a control TTL pulses for the magnetic and electric stimulator are generated.
   c. Selection of the number of extra states and the number of repetitions of each of them. Specification of the exact type of connected device (Magstim BiStim) and the COM port used for mutual communication.
   d. Creating of a file name template for a data storage. Setting a path for data saving, and sampling limits on how much data should be stored in one file.

4. Creating the dialog for entering patient identification (surname, year of the birth), threshold values (AMT, RMT) and other data according to the current protocol.
5. Establishing a folder for storing measured data. The folder name contains the patient's name and year of birth. Folders are further sorted into a structure according to specific examinations.
6. At the end of acquisition storing data and close windows (Fig. 4).

4 System Validation

The system was used for examination of dystonic patients group and healthy control subjects. Found values (see Table 1) are comparable to studies handling with MEP analysis. The average RMT value was 53% while Kojovic [15] in his study reported 51.3% of the magnetic stimulator output. The action threshold is then lower in both cases by about 10% of the stimulator output. Latencies of MEP evoked by single TMS pulse measured from the APB muscle are 21.5 ± 1.1 ms and ADM muscle 21.7 ± 1.1 ms. In his work, Livingston [16] reported 20.7 ± 1.4 ms for n.medianus (APB) and 20.1 ± 2.1 ms for n.ulnaris (ADM).

5 Discussion

The purpose of this study was to describe the technical solution and software for the motor-evoked potential acquisition system. According to defined requirements, we have developed a system that enables us to examine a group of dystonic patients and of control subjects in the electrophysiological examination protocol. It allows the determination of AMT and RMT values, to investigate the influence of intracortical inhibition and to evaluate the effect of PAS. The system was implemented both technically and programmatically. In the Signal programming environment, we wrote scripts to control the function of individual cells in the measuring chain (stimulators, amplifier, AD converter) and data acquisition. The collected data were analyzed in the Signal environment as well.

Table 1 MEP parameters

Parameter	Found value (hc)	Similar systems (hc)
Resting threshold	53.3% (IQR 42–65%)	51.3% [15]
Action threshold	37.0% (APB)	40.0% [15]
MEP amplitude (mV)	1.2 ± 1.2 mV (APB)[a] 1.5 ± 1.3 mV (ADM)[a]	
MEP latency (ms)	21.5 ± 1.1 ms (APB) 21.7 ± 1.1 ms (ADM)	20.7 ± 1.4 ms (APB) [16] 20.1 ± 2.1 ms (ADM) [16]

hc = healthy control, APB = abductor pollicis brevis, ADM = abductor digiti minimi, [a]= stimulation intensity 1,3 × RMT

The values of MEPs parameters are in line with the literature (Table 1). Since each TMS study has a group of control subjects that serves as a source of reference values, the deviations in our findings from other authors' ones are not significant.

The main advantages of the system are the ability to automatically control pacing parameters via the control computer. It allows the examination to be divided into blocks within which pulses are randomized with no need to manually interfere with the process. Therefore, the attending staff can pay full attention to the position of the stimulation coil which is most important during TMS examination.

We plan to expand the system for measuring the transcallosal inhibition in the group of patients after stroke.

6 Conclusion

We designed, developed and described TMS system, which was successfully used for examination of 22 patients (mean age 51 ± (SD) 17 years) suffering from dystonia and 22 age- and gender-matched healthy controls. Using the system, we found significant differences between the patients and healthy controls in MEP parameters (MEP onset latency, amplitude). Designed system for TMS examination is an effective tool for studying the pathophysiology of neurological diseases.

Acknowledgements Supported by research grant projects offered by the Czech ministry of education: IGA NT 12282, by the Czech Science Foundation GAČR 16-13323S and by the Charles University, Prague, Czech Republic: Progres Q27/LF1.

Statement The authors declare that they have no conflict of interest.

References

1. Herbsman, T., et al., *Motor threshold in transcranial magnetic stimulation: the impact of white matter fiber orientation and skull-to-cortex distance.* Hum Brain Mapp, 2009. **30**(7): p. 2044–55.
2. Stefan, K., et al., *Induction of plasticity in the human motor cortex by paired associative stimulation.* Brain, 2000. **123 Pt 3**: p. 572–84.
3. Wittenberg, G.F. and M.A. Dimyan, *How do the physiology and transcallosal effects of the unaffected hemisphere change during inpatient rehabilitation after stroke?* Clin Neurophysiol, 2014. **125**(10): p. 1932–3.
4. Albanese, A., et al., *Phenomenology and classification of dystonia: a consensus update.* Mov Disord, 2013. **28**(7): p. 863–73.
5. Quartarone, A. and M. Hallett, *Emerging concepts in the physiological basis of dystonia.* Mov Disord, 2013. **28**(7): p. 958–67.
6. Klomjai, W., R. Katz, and A. Lackmy-Vallee, *Basic principles of transcranial magnetic stimulation (TMS) and repetitive TMS (rTMS).* Ann Phys Rehabil Med, 2015. **58**(4): p. 208–13.
7. Kallioniemi, E., et al., *Onset Latency of Motor Evoked Potentials in Motor Cortical Mapping with Neuronavigated Transcranial Magnetic Stimulation.* Open Neurol J, 2015. **9**: p. 62–9.
8. Rosler, K.M., D.M. Roth, and M.R. Magistris, *Trial-to-trial size variability of motor-evoked potentials. A study using the triple stimulation technique.* Exp Brain Res, 2008. **187**(1): p. 51–9.
9. Vaseghi, B., M. Zoghi, and S. Jaberzadeh, *Inter-pulse Interval Affects the Size of Single-pulse TMS-induced Motor Evoked Potentials: A Reliability Study.* Basic Clin Neurosci, 2015. **6**(1): p. 44–51.
10. Kalupahana, N.S., et al., *Abnormal parameters of magnetically evoked motor-evoked potentials in patients with cervical spondylotic myelopathy.* Spine J, 2008. **8**(4): p. 645–9.
11. Chowdhury, F.A., et al., *Motor evoked potential polyphasia: a novel endophenotype of idiopathic generalized epilepsy.* Neurology, 2015. **84**(13): p. 1301–7.
12. Leon-Sarmiento, F.E., et al., *A new neurometric dissection of the area-under-curve-associated jiggle of the motor evoked potential induced by transcranial magnetic stimulation.* Physiol Behav, 2015. **141**: p. 111–9.
13. Quartarone, A., *Rapid-rate paired associative stimulation of the median nerve and motor cortex can produce long-lasting changes in motor cortical excitability in humans*, in *Jurnal of Physiology*. 2006. p. 657–670.
14. Hernandez-Pavon, J.C., et al., *Effects of navigated TMS on object and action naming.* Front Hum Neurosci, 2014. **8**: p. 660.
15. Kojovic, M., et al., *Secondary and primary dystonia: pathophysiological differences.* Brain, 2013. **136**(Pt 7): p. 2038–49.
16. Livingston, S.C., H.P. Goodkin, and C.D. Ingersoll, *The influence of gender, hand dominance, and upper extremity length on motor evoked potentials.* J Clin Monit Comput, 2010. **24**(6): p. 427–36.

Comparison Between Support Vector Machine with Polynomial and RBF Kernels Performance in Recognizing EEG Signals of Dyslexic Children

A. Z. Ahmad Zainuddin, W. Mansor, Khuan Y. Lee, and Z. Mahmoodin

Abstract

Dyslexia is seen as learning disorder that causes learners having difficulties to recognize the word, be fluent in reading and to write accurately. This is characterized by a deficit in the region associated with learning pathways in the brain. Activities in this region can be investigated using electroencephalogram (EEG). In this work, Discrete Wavelet Transform (DWT) with Daubechies order of 2 (db2) based features extraction was applied to the EEG signal and the power is calculated. The differences between beta and theta band with responding to learning activities were explored. Multiclass Support Vector Machine (SVM) was used to classify the EEG signal. Performance comparison of Polynomial and Radial Basis Function (RBF) kernel recognizing EEG signal during writing word and non-word is presented in this paper. It was found that SVM with RBF kernel performance was generally higher than that of the polynomial kernel in recognizing normal, poor and capable dyslexic children. The SVM with RBF kernel produced 91% accuracy compared to the polynomial kernel.

Keywords

Dyslexia • Electroencephalogram • Support vector machine • RBF kernel • Polynomial kernel

A. Z. A. Zainuddin · W. Mansor (✉) · K. Y. Lee · Z. Mahmoodin
Faculty of Electrical Engineering, Universiti Teknologi MARA, 40450 Shah Alam, Selangor, Malaysia
e-mail: wahidah231@salam.uitm.edu.my

A. Z. A. Zainuddin · W. Mansor · K. Y. Lee · Z. Mahmoodin
Computational Intelligent Detection RIG, Pharmaceutical and Life Sciences CORE UiTM, 40450 Shah Alam, Selangor, Malaysia

A. Z. A. Zainuddin · Z. Mahmoodin
Medical Engineering Technology Section, Universiti Kuala Lumpur, 53100 Gombak, Selangor, Malaysia

1 Introduction

Dyslexia is a neurological disorder that causes learners having difficulties to decode a word, read and write despite receiving the adequate level of academic education [1]. Generally, the dyslexic children intelligent quotient (IQ) is normal or above average even though they have the problem to acquire smooth skill in reading and writing [2]. Schools in Malaysia screen children with dyslexia through an assessment that consists of measuring capability in spelling, reading, writing and as well as children strength and weakness in learning [3]. According to the report from Malaysia Ministry of Education, approximately 53,613 children enrolled the special program for learning disability in 2016 in which 8.35% expected to have dyslexia [4]. Another report shows that dyslexic children that enrol the intervention program have increased from 1,679 in 2014 to 10,329 in 2017 in which 5,806 is at primary level (age 7–12 years old) [5]. This number is increasing every year.

Looking into brain function, the cerebral cortex is the part of the brain that consists of four lobes which associated with a different function known as a frontal, temporal, parietal and occipital lobe. When an activity is carried out, the bioelectrical signal is generated in the area that related to its function which can be recorded using EEG. Compared to other imaging technique to identify dyslexia such as fMRI, PET and MEG [6], EEG has advantages as it can record higher temporal resolution of the signal where time and frequency domains of the signal are kept, is portable, easy to use, low cost, noninvasive and practical to be applied during learning activities [7].

A few studies have been conducted using EEG to determine area associates with brain functions such as sleep studies [8], epileptic [9], mental task, mental imaginary, motor imaginary, brain-computer interface [10] and learning disabilities [11]. This EEG signal is extracted to find good features for classification. Some of the features that can be

© Springer Nature Singapore Pte Ltd. 2019
L. Lhotska et al. (eds.), *World Congress on Medical Physics and Biomedical Engineering 2018*,
IFMBE Proceedings 68/3, https://doi.org/10.1007/978-981-10-9023-3_17

extracted from EEG signal are power, skewness, variance, energy, entropy and standard deviation [12].

Dyslexia information in EEG signal can be obtained by extracting the features of the signal and then classified the signal using a suitable classifier. SVM is one of the well-known classifiers that can produce accurate results [13]. It is based on statistical learning theory and can work in small sample size, nonlinear and multiple classifications [14]. Choosing different kernel function of SVM may produce different performance [15]. Polynomial and RBF were widely used nonlinear kernel that projected data into infinite dimensional feature space [16]. The SVM performance using both kernels in classifying EEG of dyslexic children has not been reported.

This paper describes the classification of EEG signals of normal, poor and capable dyslexic children using SVM with Polynomial and RBF kernels. In this work, the performance of Polynomial and RBF kernel through writing known word and non-word is examined for suitability in identifying dyslexia.

2 Research Methodology

In this work, the classification of EEG signal of dyslexic children was carried out in several stages which include signal acquisition, subject identification and processing, features extraction and SVM classification using Polynomial and RBF Kernels.

2.1 Signal Acquisition, Subject Identification, and Processing

EEG signals were acquired using wireless bio-signal acquisition system called g.Nautilus with 8 active electrodes placed on the subject scalp in accordance with the International 10/20 System. These electrodes act as a sensor to pick up brain waves. Eight (8) electrode locations were chosen with reference to the areas associated with reading and writing pathways. At the left hemisphere of the brain, the electrodes are positioned at C3, P3, T7, and FC5 while at the right hemisphere of the brain, the electrodes are located at C4, P4, T8, and FC6.

There were four tasks carried out by each subject while EEG signal was recorded. The subject was asked to sit comfortably on a chair with a piece of paper and a pencil. A screen monitor was placed on a table in front of the subject. In the first task, the subject has to write 3 simple words and in the second task the subjects are required to write 3 complex words, these words are the words that have a specific meaning and can be understood. While in the third task, the subject has to write 3 simple non-words and in the

fourth task, the subject must write 3 complex non-words. These non-words are the words that have no specific meaning. Each word and non-word was shown on the monitor screen one by one.

These sets of words were prepared according to age-appropriate academic level. Set A is for the subjects aged 7–8 which comprises 3 alphabets, set B is for the subjects aged 9–10 which contains 4–5 alphabets and set C is for the subjects aged 11–12 and have 5–8 alphabets. The choice of words and non-words were based on the assessment used by Dyslexia Association of Malaysia.

In this study, EEG data were recorded from 8 normal control subjects, 17 poor dyslexic children, and 8 capable dyslexic children. Normal control subjects are children from public school that can read and write smoothly. Poor dyslexic is referred to children that could not read and write correctly compared with normal control subject with the same age group level while capable dyslexic children refer to children that are able to read and write after they went through a dyslexia intervention program. The subject age was in the range of 7–12 years old since at this stage they start to receive formal learning activity at school where the symptom of dyslexia can be clearly seen from reading and writing. These subjects were first screened to identify the level of learning disorder which is poor or capable dyslexic with the assistance from Dyslexia Association of Malaysia and Rakan Dyslexia Malaysia group. During the assessment, physiological background, medical history, right and left hand dominant and IQ were recorded to ensure conformity of data.

EEG signals were recorded using g.Nautilus wireless biosignal acquisition system that has a built-in amplification and provides 24bit resolution with 500 Hz sampling rate. Noise embedded in the signal was removed using 2 types of filter. A notch filter was used to eliminate artifacts from power lines frequency at 50 Hz and a high pass filter with cut off frequency at 0.5 Hz was employed to remove noise from dc source. Once the artifacts were removed, features extraction was carried out.

2.2 Features Extraction

EEG signals are divided into five frequency bands known as delta δ (up to 4 Hz), theta θ (4–8 Hz), α alpha (8–13 Hz), beta β (13–30 Hz) and gamma γ (above 31 Hz). The delta is associated with deep sleep, theta is related to drowsiness, alpha indicates relaxed awareness, beta refers to the concentration or active attention and finally, gamma is simultaneous processing of information from different brain areas. Learning activities such as reading and writing, are mental activities which associated with the beta band frequency. While in theta band, the brain focusing is withdrawn.

Since EEG signal has non-stationary properties, time-frequency domain approaches using DWT was used for extracting the signal features. Daubechies of order 2 (db2) was employed to provide time-frequency scale representation due to its ability to localize features and provide smooth EEG signals [12]. Hence, db2 decomposes EEG signal into 5, however, in this work, only beta (13–30 Hz) and theta (4–8 Hz) bands were considered.

The power features were computed from reconstructed signal detail coefficient and the power was calculated from the sum of squared reconstructed signal values (x) divided by the signal length (L) as shown in using Eq. (1).

$$Power = \sum x^2 / L(x) \tag{1}$$

The beta band power and the ratio of theta/beta band power are the two statistical feature vectors used as input to the classifier.

2.3 Classification

As mentioned previously, SVM with polynomial and RBF kernels were used to classify the three categories of EEG signals; normal, poor and capable dyslexic. SVM performs classification by finding maximum separation boundary by optimizing the spaces between two classes. In the linear case, a straightforward separation can be done using linear kernel but in nonlinear condition, the data need to be placed in features space where the separation is carried out in hyperspace. Nonlinear separation is accomplished by employing Radial Basis Function (RBF) and Polynomial kernel. Multiclass SVM with one versus one was employed in this work to classify normal, poor and capable dyslexic children. One versus one mechanism was carried out by separating each pair of classes against each other and using majority voting scheme to determine the output.

The SVM classifier equation used in the work is shown in (2).

$$f(x) = \sum_{i}^{N} \alpha_i y_i k(x_i, x) + b \tag{2}$$

where b is the bias, $k(x_i, x)$ is the kernel used in SVM, α_i is the weight vector, y_i is the target vector and N is the size of training data. While maximizing the margin of the data separation, the SVM minimizes the misclassification to zero. The trade-off between the misclassification and the margin is controlled by a parameter called box constraint. For the polynomial kernel, the order of polynomial kernel is determined by d as shown in Eq. (3). Here, the parameter d was set to 3. The RBF kernel projects vectors into an infinite dimensional space to compute the inner product between two projected vectors. The RBF equation used in the work is

shown in Eq. (4) where the tuned parameter, σ that specifies the kernel width was set to 1. Both parameters were selected since it gives the lowest error from ten-fold cross-validation.

$$k(x_i, x) = (x_i.x + 1)^d \tag{3}$$

$$k(x_i, x) = exp\left(-\frac{||x_i - x||^2}{2\sigma^2}\right) \tag{4}$$

To select the optimum kernel, the box constraint was varied from 0.001 to 1000. The performance of each kernel was then evaluated and the accuracy, sensitivity, and specificity were determined using Eqs. (5), (6) and (7) respectively. Confusion matrix for multiclass was then employed to verify the performance of the classification models.

$$Accuracy, A_c = \frac{T_N + T_P}{T_P + T_N + F_P + F_N} \tag{5}$$

where T_N is the true negative, T_P is the true positive, F_P is the false positive and F_N is the false negative.

$$Sensitivity, S_e = T_{PR} = \frac{T_P}{T_P + F_N} \tag{6}$$

$$Specificity, S_p = T_{NR} = \frac{T_N}{T_N + F_P} \tag{7}$$

3 Results and Discussion

In this study, one dataset refers to total features obtained from a recording of EEG signals from 8 channels (C3, C4, P3, P4, FC5, FC6, T7 and T8) during performing a task. Since two features which are beta band power and theta/beta band ratio were computed for a task, one dataset gives 16 features. As each subject completes a total of 4 tasks, the accumulative dataset is 132 for 33 subjects. Therefore, the total data used is 2112. The datasets later were divided into 64% for training and 36% for testing. As mentioned

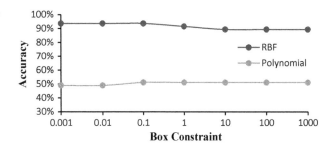

Fig. 1 Accuracy for SVM with Polynomial and RBF kernels when box constraint is varied

Table 1 Classification Performance of SVM with both kernels using box constraint = 1

Kernel	Type	Sensitivity	Specificity	Accuracy
Polynomial	Normal	1.00	0.67	
	Poor Dyslexic	0.44	1.00	0.51
	Capable Dyslexic	0.75	0.79	
RBF	Normal	1.00	0.95	
	Poor Dyslexic	0.92	0.88	0.91
	Capable Dyslexic	0.75	0.98	

previously, the optimum parameter for RBF and polynomial kernel of SVM were selected using K-Fold cross-validation.

Figure 1 shows the accuracy of SVM in identifying normal, poor and capable dyslexic when box constraint is varied. The results show that RBF kernel provides high accuracy (94%) when the box constraint is between 0.001 and 0.1 whereas the polynomial kernel maintains high accuracy (51%) when the box constraint is in the range of 0.1–1000.

Table 1 shows the classification performance of SVM with polynomial and RBF kernels. The SVM with polynomial kernel provides the highest sensitivity when classifying the normal subjects and have the highest specificity when recognizing poor dyslexic children. It is also found that using the polynomial kernel, the SVM provides an accuracy of 51% in classifying the normal, poor and capable dyslexic children.

It can be seen that the SVM with RBF kernel gives good performance when classifying EEG signals of normal, poor and capable dyslexic children. It provides 91% accuracy in classifying all subjects. The highest sensitivity which is 100% is obtained when classifying the normal subjects and the highest specificity (98%) is achieved when distinguishing the capable dyslexic. Comparing the performance of these two types of kernel at box constraint is 1, it is obvious that the RBF kernel is the most accurate kernel since it produces the highest classification accuracy which is 91% whereas the polynomial kernel only gives 51%. The RBF kernel performs better than the polynomial kernel since it uses Gaussian curve with infinite dimensionality in separating data points which offers more predictive efficiency.

4 Conclusion

The performance of SVM with polynomial and RBF kernels in recognizing EEG signals of dyslexic children has been described in this paper. The sensitivity, specificity, and accuracy of each kernel were determined to select the optimum kernel. It was found that the SVM with RBF kernel performance is much better than that of polynomial kernel since it produces an accuracy of 91% in classifying all

subjects. The SVM with polynomial kernel was unable to identify poor dyslexic correctly compared to normal and capable dyslexic. Therefore, the SVM with RBF Kernel is proposed to be used in recognizing EEG signals of normal, poor and capable dyslexic.

Acknowledgements This work was supported by Fundamental Research Grant Scheme (FRGS), Malaysia (600-RMI/FRGS 5/3 (137/2015)). The authors would like to thank Ministry of Higher Education, Malaysia, Research Management Institute and Faculty of Electrical Engineering, Universiti Teknologi MARA, Shah Alam, for financial support, facilities and various contributions, and to Dyslexia Association Malaysia for their assistance.

Conflict of Interest The authors declare that they have no conflict of interest.

References

1. E. S. Norton, S. D. Beach, and J. DE Gabrieli, "Neurobiology of dyslexia," *Curr. Opin. Neurobiol.*, vol. 30, pp. 73–78, Feb. 2015.
2. U. Goswami, "Dyslexia, Developmental," in *International Encyclopedia of the Social & Behavioral Sciences*, vol. 6, Elsevier, 2015, pp. 727–730.
3. Ministry of Education, "Instrumen Senarai Semak Disleksia," 2011.
4. B. P. K. K. P. Malaysia, "Data Pendidikan Khas 2016," 2016.
5. Z. Mahfuzah, "Statistik Murid Disleksia di Malaysia," 2017. [Online]. Available: https://www.mahfuzahzainol.com/single-post/2017/12/06/Statistik-Murid-Disleksia-di-Malaysia. [Accessed: 19-Jan-2018].
6. Y. Sun, J. Lee, and R. Kirby, "Brain Imaging Findings in Dyslexia," *Pediatr. Neonatol.*, vol. 51, no. 2, pp. 89–96, Apr. 2010.
7. S. Mohamad, W. Mansor, L. Y. Khuan, C. W. N. F. C. W. Fadzal, N. Mohammad, and S. Amirin, "Development of computer-based assessment for brain electrophysiology technique of dyslexic children," in *2016 IEEE Symposium on Computer Applications & Industrial Electronics (ISCAIE)*, 2016, pp. 79–83.
8. S. Qureshi and S. Vanichayobon, "Evaluate different machine learning techniques for classifying sleep stages on single-channel EEG," in *2017 14th International Joint Conference on Computer Science and Software Engineering (JCSSE)*, 2017, pp. 1–6.
9. S. Siuly and Y. Li, "Designing a robust feature extraction method based on optimum allocation and principal component analysis for epileptic EEG signal classification," *Comput. Methods Programs Biomed.*, vol. 119, no. 1, pp. 29–42, 2015.

10. X. Li, X. Chen, Y. Yan, W. Wei, and Z. J. Wang, "Classification of EEG signals using a multiple kernel learning support vector machine," *Sensors (Basel).*, vol. 14, no. 7, pp. 12784–12802, 2014.

11. D. C. Hammond, "What is Neurofeedback: An Update," *J. Neurother.*, vol. 15, no. 4, pp. 305–336, 2011.

12. T. Gandhi, B. K. Panigrahi, and S. Anand, "A comparative study of wavelet families for EEG signal classification," *Neurocomputing*, vol. 74, no. 17, pp. 3051–3057, 2011.

13. X. Liu, C. Gao, and P. Li, "A comparative analysis of support vector machines and extreme learning machines," *Neural Networks*, vol. 33, pp. 58–66, 2012.

14. Y. Ma, X. Ding, Q. She, Z. Luo, T. Potter, and Y. Zhang, "Classification of Motor Imagery EEG Signals with Support Vector Machines and Particle Swarm Optimization," *Comput. Math. Methods Med.*, vol. 2016, no. 5, pp. 667–677, 2016.

15. E. A. Zanaty, "Support Vector Machines (SVMs) versus Multilayer Perception (MLP) in data classification," *Egypt. Informatics J.*, vol. 13, no. 3, pp. 177–183, 2012.

16. C. K. I. Williams, "Learning With Kernels: Support Vector Machines, Regularization, Optimization, and Beyond," *J. Am. Stat. Assoc.*, vol. 98, no. 462, pp. 489–489, Jun. 2003.

Photothermal Inhibition of Cortex Neurons Activity by Infrared Laser

Qingling Xia and Tobias Nyberg

Abstract

Some brain diseases are caused by neurons being abnormally excited, such as Parkinson's disease (PD) and epilepsy. The aim of this study was to investigate the feasibility and the efficacy of infrared laser irradiation for inhibiting neuronal network activity. We cultured rat cortex neurons, forming neural networks with spontaneous neural activity, on multi-electrode arrays (MEAs). To inhibit the activity of the networks we irradiated the neurons using different intensity of 1550 nm infrared laser light. A temperature model was created using COMSOL Multiphysics software to predict the temperature change at different laser intensity irradiation. Our initial result shows that the wavelength of 1550 nm infrared laser can be used to inhibit the network activity of cultivated rat cortex neurons directly and reversibly. The degrees of network inhibition can be manipulated by changing the laser intensity. The optical thermal effect is considered the primary mechanism during infrared neural inhibition (INI). These results demonstrate that INI could potentially be useful in the treatment of neurological disorders and that temperature may play an important role in INI.

Keywords

Brain diseases • Cortex neurons • Temperature model
Infrared neural inhibition (INI)

Q. Xia (✉) · T. Nyberg
School of Engineering Sciences in Chemistry,
Biotechnology and Health, KTH Royal Institute of Technology,
Huddinge, Sweden
e-mail: qingling@kth.se

Q. Xia
Bioengineering College, Chongqing University, Chongqing,
China

L. Lhotska et al. (eds.), *World Congress on Medical Physics and Biomedical Engineering 2018*,
IFMBE Proceedings 68/3, https://doi.org/10.1007/978-981-10-9023-3_18

99

1 Introduction

Neural modulation techniques that can inhibit neural activity safely and reversibly are valuable for treating nervous system diseases, such as Parkinson's disease and epilepsy. Electrical stimulation (ES) is a standard stimulation method that has been widely used to modulate neural activities in the nervous system [1]. However, it is hard using ES to achieve inhibition of neural activity without contact, electrochemical junction and excitation, which restrict further control of the nervous systems [2, 3]. Recent advances in optogenetics [4] and nanoparticle-enhanced near-infrared laser (NIR) techniques [5–7] have revealed the potential of precise inhibition of neural activity. However, both techniques require the introduction of exogenous additives to neuronal targets, which hinder their clinical application.

In the last decade, Infrared neural stimulation (INS) has been widely applied for evoking neurons response in the peripheral nervous system (PNS) [8–10] and central nervous system (CNS) [11, 12]. Recently, it is shown that IR light can also inhibit neural the activity of neurons in vitro [13, 14]. The advantages of using IR compared with ES are that the technique is contact-free and has high spatial selectivity [8]. In contrast to optogenetics and nanoparticle-enhanced NIR optical methods, INI is a direct stimulation without genetically modifying the neuronal tissue and cells or adding exogenous transducer materials. However before INI could potentially be used as a clinical method for treating nervous system diseases we need to investigate the efficacy and safety of the method.

In this study, we investigate how infrared laser light, $\lambda = 1550$ nm, can inhibit the activity of in vitro cortex neural network. The spikes rate and spike rate change (SRC) were calculated to test the degree of INI. A temperature model was used to simulate the temperature elevation, during INI, which was subsequently used for safety evaluation of the laser irradiation.

2 Method

2.1 Experimental Setup of Primary Cortex Neurons Stimulation and Recording

Figure 1c illustrates the schematic of IR irradiation and recording system based on micro-electrode arrays (MEAs). Neurons were cultivated on MEAs having 60 electrodes with an electrode diameter of 30 μm and an inter-electrode distance of 200 μm using silicon nitride to insulate the leads (Multichannel Systems, Reutlingen, Germany).

To irradiate neurons a 1550 nm continuous wave (CW) infrared laser (Modulight, Tampere, Finland) with a 200 μm multimode fiber with an attached collimator (F220SMA-1550, Thorlabs, Newton, NJ) was used Fig. 1c. Laser power was varied from 90, 180, 270 to 340 mW. Before the stimulation experiments, MEAs were taken out from incubator to check for activity in the form of spontaneous bursts and single action potentials. Laser irradiation experiments were only conducted on MEAs having spontaneous activity.

Neural activity was recorded by the MEA at the age of 19–27 DIV. The MEA was placed in an MEA-1060 Headstage (Multichannel Systems, Reutlingen, Germany) and heated to 32 °C. The electrodes signal was amplified by a 16 channel amplifier (Micro Electrode Amplifier, model 3600, AM Systems) and digitized using DataWave software (SciWorks, Loveland, CO v. 6.0).

2.2 Cell Culturing

Primary cortex neurons were prepared from an embryonic day 18 (E18) Sprague Dawley (SD) rat cortex (Brainbits UK). The dissociation procedure followed was the BrainBits Primary Neuron Culture Protocol. MEAs were coated with 100 μl of 1% polyethyleneimine (Sigma-Aldrich, USA) overnight. Before use, the MEAs were rinsed with sterilized water and allowed to dry. Dishes were covered by 20 μl mouse laminin (L2020, Sigma-Aldrich) solution, at a concentration of 20 μg per ml medium, for 1 to 2 h in the incubator (37 °C, 5% CO2) before adding the cell suspension. MEA membrane lids (ALA Scientific Instruments, USA) was used to limit medium evaporation during cultivation.

Cells were cultivated following the Brainbits culturing protocol with an addition of pencillin resulting in a concentration of 100 U/ml penicillin and 100 μg/ml streptomycin (Sigma-Aldrich, USA.), to the medium. Half of the medium was changed every 3 days.

2.3 The Heating Model

To model the temperature during the laser irradiation, a finite element model (FEM, Fig. 1b) was adapted from the model of Liljemalm et al. [14] using COMSOL Multi-physics (COMSOL Multi-physics, 2017, version 5.3, Stockholm,

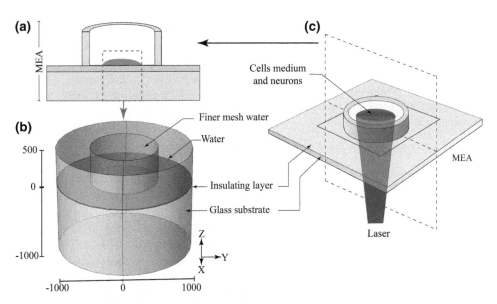

Fig. 1 The schematic of IR irradiation and recording and the 3D geometry of temperature model. **a** The side view of the MEA chamber during IR irradiation of neurons (From the dark dash box of **c**), not to scale), the small red area represents the laser spot. **b** The 3D geometry used in the model represents the red dash rectangular box of **a** and consists of glass, an insulating layer (SiN, 2 μm) and water. Unit: μm. A finer mesh was used for the central water region to achieve a more detailed spatial solution in the volume of interest. **c** The illustration of the experimental setup (Color figure online)

Sweden) to present the region of interest (ROI, the red dash box in Fig. 1a). We discretized the model as 20 μm × 20 m × 20 μm cubic elements and considered the interface non-isothermal flow, a multiphysics coupling of fluid flow and heat transfer. The two main components of heat transport in the model are conduction and convection. The normalized laser intensity distribution was assumed as an ideal Gaussian function with a beam diameter of 6 mm.

2.4 Data Analysis

To analyze and extract the spikes, digitized signals were filtered with Butterworth band-pass filter (20–2000 Hz) and the detection threshold was set to two times the noise level. For the analysis of neural activity, average spike rates were calculated before, after, and during laser stimulation (bin = 1 s). To evaluate the degree of the neural activity inhibition, we chose the channels which had average spike rate more than 1 spikes/sec before the laser stimulation. The spike rate change (SRC) with or without the IR irradiation was estimated by the following Eq. 1.

$$\text{SRC}(\%) = \left(\frac{\text{spike count during stim} - \text{spike count before stim}}{\text{spike count before stim}} \right) \times 100\%$$

$$(1)$$

2.5 Safety Analysis

We evaluated the safety of our laser experiments by calculating the damage signal ratio (DSR) using the method and presented in [15]. The highest spatial temperature, in the center of the MEA, was used for the calculations.

3 Results

3.1 Temperature Changes at the Cell Layer

The temperature model predicted the temperature changes over the time in the center of the laser beam at output powers of 90–340 mW (Fig. 2a). The temperatures rose immediately following the IR exposure and reached the equilibrium temperature after about 2 s (Fig. 2a). To save simulation time we simulated the first 3 s of the 60 s long laser pulse and used the temperature plateau value at 2 s as our maximum temperature. Figure 2b shows how the temperature space distribution at 2 s under the different laser power.

3.2 Infrared Neural Stimulation Inhibited the Neural Network Activities

To better illustrate the strong correlation between laser stimulus and cellular response and allow the firing of the neurons to recover, we repeatedly stimulated the neurons 3 times by turning the laser on 60 s and off 180 s (Fig. 3a). The results show this stimulation is repeatable and reversible (Fig. 3b). Spike rate plots show the mean spikes per second prior to, during 60 s and post 120 s of the infrared light exposure (Fig. 4). The spike rate decreased with increasing laser power (90–340 mW) (Fig. 4).

Neural inhibition occurred instantly after the IR irradiation, inset Fig. 4c, d. The first 5 s the spikes were suppressed most severely and then the degree of inhibition was slightly decreased. Both Figs. 3 and 4 demonstrate spiking firing was fully recovered about 90 s post-IR exposure.

To quantify the degree of inhibition, spike rate change (SRC) was calculated using Eq. 1. When the laser power increased from 0 to 340 mW, for MEA 1 and MEA2, the degree of suppression rose monotonically from 0 to 88.18% and from 0 to 62.96% respectively (Fig. 5a). The temperature increase, for the equilibrium plateau in Fig. 2, was slightly non-linear with the power increase (Fig. 5b).

3.3 Damage Signal Ratio (DSR) at the Different Laser Power

The DSR for the maximum power was 0.00028. For an extended laser irradiation of 10 min at the power 340 mW, the DSR was 0.0028.

4 Discussion

In our present work, we show that the rapid, repeatable and reversible block of cortex neurons was obtained with 1550 nm infrared laser. In addition, the different degree of inhibition could be achieved by tuning the laser power.

The inhibition is presumed to be a heat block phenomena [16] or a heat sensitive membrane channel [5]. As can be seen from, Fig. 2b, the temperature is not uniform over the MEA surface. The inhibition may thus be lesser at the edges of the illuminated area. Thus it is difficult to make a quantitative evaluation of the inhibition for one specific temperature. However as we have used the peak temperature in the center of the MEA we are constantly underestimating the degree of inhibition, if the lowest temperatures on the MEA had been used we would be overestimating the efficacy of the method.

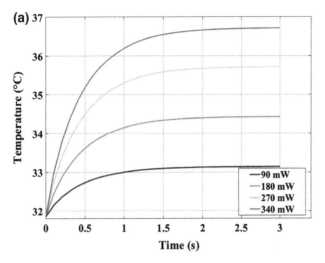

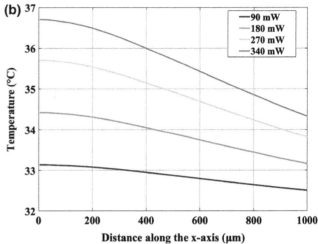

Fig. 2 Temperature distribution on the cell layer (0.1 μm above the MEAs substrate) at the different laser power (Laser power: 90, 180, 270 and 340 mW; Laser wavelength: 1550 nm; Initial temperature 32 °C). **a** Temperature distribution in the center of the laser beam at the different laser power over the time. **b** Temperature distribution at the distance from the center along the x-axis at 2 s

Fig. 3 Laser irradiation scheme and single channel suppression illustration. **a** The schematic of 3 times repeated laser irradiation. **b** The raw data of repeating suppression of spontaneous neural firing from a single electrode (laser power 270 mW, MEA 1)

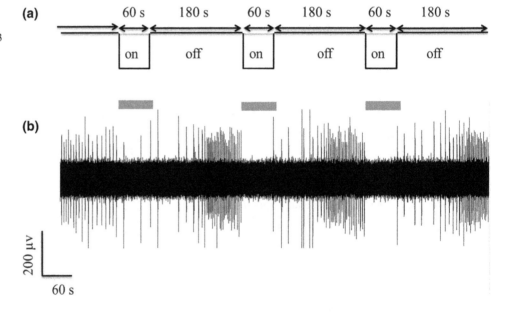

Comparing the maximum temperature change with laser power (Fig. 5b) and the degree of inhibition with laser power (Fig. 5a) indicated that higher temperature increased the suppression of neuron activity. The inhibition is not linear with the temperature increase but is increasing more per degree for a low laser power and then levels out (Fig. 5). The degree of inhibition is different for the two MEAs (Fig. 5a). We attribute this to differences in the network layout where parameters such as synaptic connections, local cell density, and the neuron type, may all contribute to differences in activity of the network both before and during inhibition.

The maximum inhibition 88%, spike data from all electrodes, was obtained for MEA1 using 340 mW resulting in an irradiance of 14 mW/mm². These inhibition results are in accordance with reports of the photothermal suppression effect induced by different nanoparticle-enhanced NIR resulting in a degree of inhibition of 80–96% for 15 mW/mm² at 785 nm [5, 7]. Thus we manage to reach a similar degree of inhibition without exogenous additives. This is an advantage of INI as there is no need to consider any cytotoxicity of nanomaterials if considering the technique for a clinical purpose.

The rapid and stable neuron response is important to consider when translating the modulation technique of INI to

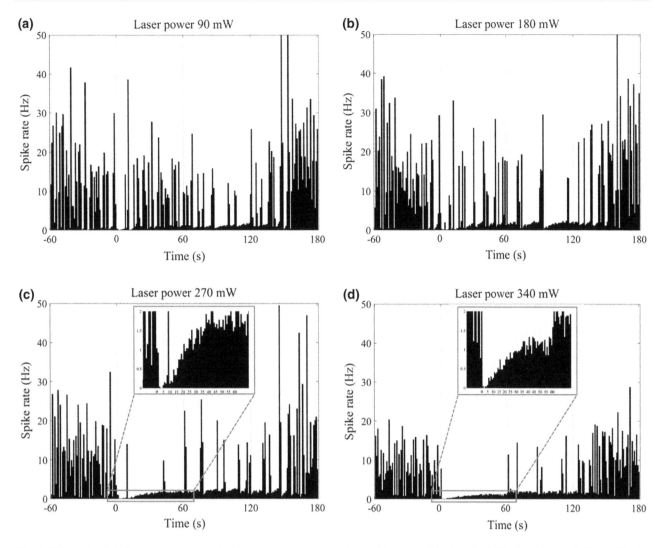

Fig. 4 Spike rate plots for neurons exposed to 60 s continuous wave infrared laser at **a** 90, **b** 180, **c** 270 and **d** 340 mW (recording electrodes: n = 9, repeat stimulation: 3 times, red words 0 and 60: laser start and laser end, bin size: 1 s.). The inset in **c** and **d** magnifies the time period 0–60 s during the IR (Color figure online)

clinical application. In our experiments, inhibition was accrued within one second after the laser was turned on (Fig. 4). This is roughly the time it takes for the temperature to reach equilibrium, Fig. 2a. Inhibition was obvious from the start of the laser pulse, especially the first 5 s, and continued after the end of the pulse for roughly 60 s (Fig. 4). Spike rate slowly increased even during the IR irradiation (the insert of Fig. 4c, d) and spikes slowly recovered after almost 90 s laser exposure. To allow the activities of the neurons have enough time to recovery, we thus waited for 180 s before the next irradiation. Inhibition is present for the 90 mW and increases most fast up to 270 mW. These spikes were fully reversible and highly reproducible across the repeated experiments (Fig. 3). These repeated experiments

show the reliability of INI and the potential to be used as the clinical tool.

In the current experiment, we used a baseline temperature of 32 °C. As the maximum temperature increase is only about 5 °C we reach a modest maximum temperature of 37 °C. We do not expect this temperature to inflict any damage on the neurons when they are exposed for a limited time. Calculating the DSR for our maximum irradiance at 340 mW gave a value of 0.00028 which is far below the damage threshold for cerebral cortex cells of 0.051 [15]. If the temperature baseline would have been increased we cannot rule out potential damage. This problem could possibly be used by pulsing the laser during the stimulation as pulses may also inhibit the activity without inflicting damage [14, 15].

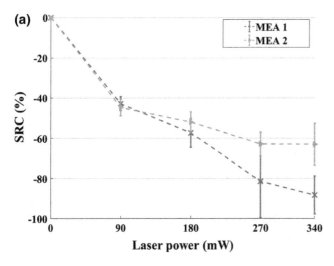 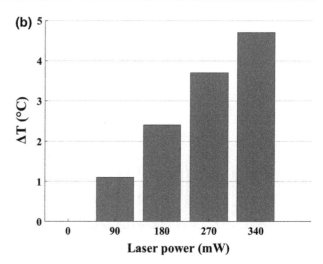

Fig. 5 Photothermal inhibition of the neural network activities and peak temperature changes. **a** Quantification of spike rate changes (SRC, mean ± SD) of two MEA samples at the different laser power (the value of blue line was calculated from Fig. 4). **b** Maximum temperature changes at the different laser power (Color figure online)

5 Conclusion

In this study, we showed the high efficiency of INI and a different degree of INI could be achieved by tuning the laser power, which shows the promising to be used to treat some neurological diseases. However, for evaluating potential clinical use of INI further research is needed to find more specific parameters for laser pulses to be used at physiological temperatures.

Acknowledgements The authors are grateful for the financial support provided by the KTH-CSC Programme.

Conflict of Interest The authors declare no conflict of interest.

References

1. Edwards, C.A., Kouzani, A., Lee, K.H., Ross, E.K.: Neurostimulation Devices for the Treatment of Neurologic Disorders. Mayo Clin Proc 92, 1427–1444 (2017).
2. Jensen, A.L., Durand, D.M.: High frequency stimulation can block axonal conduction. Exp Neurol 220, 57–70 (2009).
3. Kilgore, K.L., Bhadra, N.: Reversible nerve conduction block using kilohertz frequency alternating current. Neuromodulation 17, 242–254; discussion 254–245 (2014).
4. Zhang, F., Wang, L.P., Brauner, M., Liewald, J.F., Kay, K., Watzke, N., Wood, P.G., Bamberg, E., Nagel, G., Gottschalk, A., Deisseroth, K.: Multimodal fast optical interrogation of neural circuitry. Nature 446, 633–639 (2007).
5. Yoo, S., Hong, S., Choi, Y., Park, J.-H., Nam, Y.: Photothermal inhibition of neural activity with near-infrared-sensitive nanotransducers. ACS nano 8, 8040–8049 (2014).
6. Yoo, S., Kim, R., Park, J.H., Nam, Y.: Electro-optical Neural Platform Integrated with Nanoplasmonic Inhibition Interface. ACS Nano 10, 4274–4281 (2016).
7. Lee, J.W., Jung, H., Cho, H.H., Lee, J.H., Nam, Y.: Gold nanostar-mediated neural activity control using plasmonic photothermal effects. Biomaterials 153, 59–69 (2018).
8. Wells, J., Konrad, P., Kao, C., Jansen, E.D., Mahadevan-Jansen, A.: Pulsed laser versus electrical energy for peripheral nerve stimulation. J Neurosci Methods 163, 326–337 (2007).
9. Peterson, E.J., Tyler, D.J.: Motor neuron activation in peripheral nerves using infrared neural stimulation. J Neural Eng 11, 016001 (2014).
10. Teudt, I.U., Nevel, A.E., Izzo, A.D., Walsh, J.T., Jr., Richter, C.P.: Optical stimulation of the facial nerve: a new monitoring technique? Laryngoscope 117, 1641–1647 (2007).
11. Cayce, J.M., Friedman, R.M., Jansen, E.D., Mahavaden-Jansen, A., Roe, A.W.: Pulsed infrared light alters neural activity in rat somatosensory cortex in vivo. Neuroimage 57, 155–166 (2011).
12. Cayce, J.M., Friedman, R.M., Chen, G., Jansen, E.D., Mahadevan-Jansen, A., Roe, A.W.: Infrared neural stimulation of primary visual cortex in non-human primates. Neuroimage 84, 181–190 (2014).
13. Mohanty, S.K., Thakor, N.V., Jansen, E.D., Walsh, A.J., Tolstykh, G.P., Martens, S.L., Ibey, B.L., Beier, H.T.: Short infrared laser pulses block action potentials in neurons. 10052, 100520 J (2017).
14. Liljemalm, R., Nyberg, T., von Holst, H.: Heating during infrared neural stimulation. Lasers Surg Med 45, 469–481 (2013).
15. Liljemalm, R., Nyberg, T.: Damage criteria for cerebral cortex cells subjected to hyperthermia. Int J Hyperthermia 32, 704–712 (2016).
16. Duke, A.R., Lu, H., Jenkins, M.W., Chiel, H.J., Jansen, E.D.: Spatial and temporal variability in response to hybrid electro-optical stimulation. J Neural Eng 9, 036003 (2012).

Automated Atlas Fitting for Deep Brain Stimulation Surgery Based on Microelectrode Neuronal Recordings

Eduard Bakštein, Tomáš Sieger, Daniel Novák, Filip Růžička, and Robert Jech

Abstract

Introduction: The deep brain stimulation (DBS) is a treatment technique for late-stage Parkinson's disease (PD), based on chronic electrical stimulation of neural tissue through implanted electrodes. To achieve high level of symptom suppression with low side effects, precise electrode placement is necessary, although difficult due to small size of the target nucleus and various sources of inaccuracy, especially brain shift and electrode bending. To increase accuracy of electrode placement, electrophysiological recording using several parallel microelectrodes (MER) is used intraoperatively in most centers. Location of the target nucleus is identified from manual expert evaluation of characteristic neuronal activity. Existing studies have presented several models to classify individual recordings or trajectories automatically. In this study, we extend this approach by fitting a 3D anatomical atlas to the recorded electrophysiological activity, thus adding topological information. Methods: We developed a probabilistic model of neuronal activity in the vicinity the subthalamic nucleus (STN), based on normalized signal energy. The model is used to find a maximum-likelihood transformation of an anatomical surface-based atlas to the recorded activity. The resulting atlas fit is compared to atlas position estimated from pre-operative MRI scans. Accuracy of STN classification is then evaluated in a leave-one-subject-out scenario using expert MER annotation. Results: In an evaluation on a set of 27 multi-electrode trajectories from 15 PD patients, the proposed method showed higher accuracy in STN-nonSTN classification (88.1%) compared to the reference methods (78.7%) with an even more pronounced advantage in sensitivity (69.0% vs 44.6%). Conclusion: The proposed method allows electrophysiology-based refinement of atlas position of the STN and represents a promising direction in refining accuracy of MER localization in clinical DBS setting, as well as in research of DBS mechanisms.

Keywords

Deep brain stimulation • Anatomical atlas fitting Microelectrode recordings

1 Introduction

The deep brain stimulation (DBS), targeting the basal ganglia is a symptomatic treatment technique, applied routinely to late-stage Parkinson's disease (PD) and other movement disorders, such as dystonia or essential tremor. In case of the PD, chronic electrical stimulation is most commonly applied to the subthalamic nucleus (STN), which is small (ca 10 mm along its longest axis) and located in subcortical structures, which makes it a challenging target to implant an electrode into. Moreover, brain shift, electrode bending and other influences during the surgery introduce additional inaccuracies into the process.

As highly accurate electrode placement within the nucleus is crucial for achieving a good clinical outcome, most centers use manually evaluated microelectrode recordings (MER) for additional electrophysiological verification of optimal target position. Over more than a decade, successful efforts have been made to provide automatic MER classification to ease the process using various signal-derived features and machine-learning models(e.g. [1, 2]).

In this paper, we extend on the work of Lujan et al. [3], who suggested fitting of a 3D atlas to manually-labeled

E. Bakštein (✉) · T. Sieger · D. Novák
Faculty of Electrical Engineering, Department of Cybernetics, Czech Technical University in Prague, Prague, Czech Republic
e-mail: eduard.bakstein@fel.cvut.cz
URL: http://neuro.felk.cvut.cz/

E. Bakštein
National Institute of Mental Health, Klecany, Czech Republic

T. Sieger · F. Růžička · R. Jech
First Faculty of Medicine, Department of Neurology, Center of Clinical Neuroscience, Charles University, and General University Hospital, Prague, Czech Republic

© Springer Nature Singapore Pte Ltd. 2019
L. Lhotska et al. (eds.), *World Congress on Medical Physics and Biomedical Engineering 2018*,
IFMBE Proceedings 68/3, https://doi.org/10.1007/978-981-10-9023-3_19

MER locations. Using a probabilistic framework, which we described previously in [4], we develop a model that allows fully automatic fitting of a surface STN atlas directly to raw MER data, without the need for manual annotation.

2 Methods

The proposed model is based on finding a maximum likelihood fit of a surface STN model to neuronal background activity, assuming different probability distribution of neuronal activity level inside and outside the STN. The aim is then to find transformation of the STN atlas, which maximizes the likelihood of STN position with respect to the measured MER data. We use the surface atlas by Krauth et al. [5] but any STN atlas can be used in general. The model is described in more detail below, further technical details can be found in the thesis [6].

To extract an estimate of the neuronal background activity level from raw MER signal, we used the normalized root-mean-square (NRMS) measure proposed in [1], which sets the mean RMS value of the first five recording positions of each trajectory equal to one. This approach compensates for variability in electrode impedance.

2.1 The 3D Atlas Transformation Procedure

We define the 3D transformation used in this study as a matrix operation with 9 degrees of freedom (DOF), allowing translations t_x, t_y and t_z, scaling s_x, s_y and s_z along the x, y and z axis respectively and also rotation around the three axes, given by the angles γ_x, γ_y and γ_z.

The transformation is given by the vector r and can be completely characterized as:

$$r = [t_x, t_y, t_z, s_x, s_y, s_z, \gamma_x, \gamma_y, \gamma_z]. \quad (1)$$

2.2 Model Structure

The model assumes two states with different NRMS levels: (i) Inside the STN (*IN*) and (ii) outside the STN (*OUT*). The probability distribution of the NRMS values in each state is modeled by the log-normal distribution in what we call the *emission probabilities*. Additionally, we incorporate smooth transition between states around the boundary, modeled by logistic (sigmoid) function, which we call the *sigmoid membership function*. This provides smooth gradient for more convenient optimization, as well as a more realistic representation of the

electrophysiological boundary of the STN, which is fuzzy especially at the lateral end (see Fig. 2). The emission probabilities, as well as parameters of the sigmoid membership functions are estimated during model *training phase* on data from the training set and form the parameter vector Θ.

The atlas fitting is then done during the *evaluation phase*, typically on unseen test data. The aim is to find a transformation vector r^* which maximizes the likelihood of producing a set of observations (i.e. NRMS values) $x = \{x_1, \ldots, x_N\}$ recorded at locations $L = \{l_1, \ldots, l_N\}$, where l_i are the 3D recording site coordinates corresponding to observation x_i. The transformation using the parameters r is then applied to the STN atlas vertices v at the initial position. All parameters from the vector Θ are held fixed during the whole evaluation phase.

Emission probabilities The emission probabilities represent how likely a background activity (NRMS) level x_i is to be observed in the respective state. The emission probabilities are modeled using the log-normal distribution, whose parameters $\{\hat{\mu}_{OUT}, \hat{\sigma}_{OUT}, \hat{\mu}_{IN}, \hat{\sigma}_{IN}\}$ are estimated during the training phase using standard maximum-likelihood estimation. Example of trained emission probabilities can be found in Fig. 1.

In the evaluation phase, the emission probability $p(x_i|s, \Theta)$ of observing NRMS value x_i in a state s given model parameters Θ is calculated using formula for probability density function of the log-normal distribution.

Membership probabilities The transition between states is modeled by the membership sigmoid function S, which also represents the fuzzy electrophysiological boundary of the STN, as observed on real data (see Fig. 2). As the slope of the transition is steeper at the proximal boundary (where the electrode enters the STN) the training NRMS data aligned with respect to the STN entry, combined with mirrored data aligned with respect to the STN exit are used to fit a single sigmoid function S, defined by two parameters: shift β_0 and slope β_1.

In the *evaluation phase*, the sigmoid transition function depends only on the distance from the model surface, rotated using vector r and is computed according to:

$$S(d_i|\Theta) = \left(1 + \exp -(\beta^0 + \beta^1(d_i))\right)^{-1}, \quad (2)$$

where d_i is the euclidean distance between the MER measurement location l_i and the nearest point on the surface of the STN model. Additionally, the distance d_i is multiplied by -1 if the location l_i lies outside of the model and by $+1$ when inside.

The membership probabilities for trained model parameters Θ and anatomical model transformed by the vector r are then computed according to:

Fig. 1 Fitted emission probabilities: histograms of observed NRMS values inside (red area) and outside (blue area) the STN, with fitted log-normal probability density functions (dashed curves) and their parameters (vertical lines) (Color figure online)

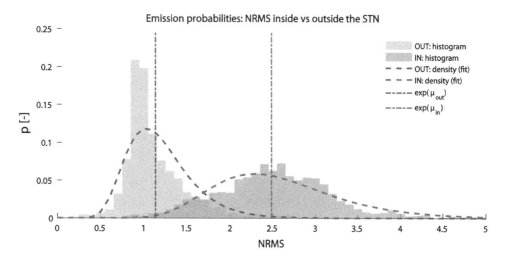

Fig. 2 The membership logistic sigmoid function S (red) fitted to measured NRMS data around the STN entry (blue circles) and exit (green circles, depth-flipped/negative) data. The fitted sigmoid S can be compared to separate entry and exit sigmoid S_{en} and S_{ex}, fitted on STN entry or exit data separately (Color figure online)

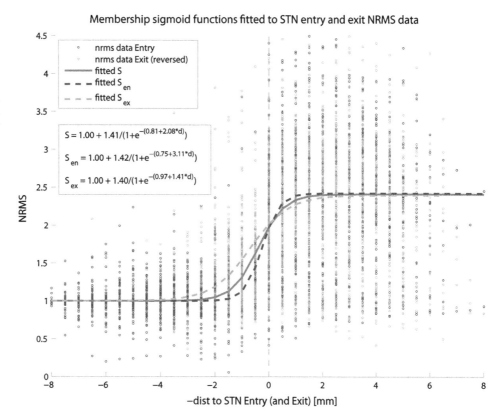

$$p(l_i \in IN | r, \Theta) = S(l_i | r, \Theta) \tag{3}$$

for the state IN and:

$$p(l_i \in OUT | r, \Theta) = 1 - p(l_i \in IN | r, \Theta) \tag{4}$$

for the state OUT.

The trained model is fully characterized by the parameter vector.

$\Theta = \{\hat{\mu}_{OUT}, \hat{\sigma}_{OUT}, \hat{\mu}_{IN}, \hat{\sigma}_{IN}, \beta^0, \beta^1\}$, comprising parameters of the emission probability densities and those of the sigmoid function.

Likelihood function and MLE estimation The aim of optimization in the evaluation phase is to find transformation vector r^*, which maximizes the likelihood given the observed data:

$$r^* = \underset{r}{argmax} \, \mathcal{L}(r | \{x, L\}, \Theta) = \underset{r}{argmax} \, p(\{x, L\} | r, \Theta) \tag{5}$$

Where p is the joint probability of observation sequence x at locations L, given trained model parameters Θ and transformation vector r. When decomposed, the probability of being in state s (i.e. IN or OUT) and observing a NRMS value x_i at position l_i is computed as a product of the emission and membership probability functions according to the Bayes' theorem:

$$p(\{x_i, l_i \in s\}|r, \Theta) = p(x_i|l_i \in s, r, \Theta) \cdot p(l_i \in s|r, \Theta) \quad (6)$$

The joint probability for a single observation is then computed as a marginalization over both states:

$$\begin{aligned} p(\{x_i, l_i\}|r, \Theta) &= p(\{x_i, l_i\}|r, \Theta) = \\ &= p(\{x_i, l_i \in IN\}|r, \Theta) \\ &+ p(\{x_i, l_i \in OUT\}|r, \Theta) \end{aligned} \quad (7)$$

To compute the joint probability of the whole observation sequence $x = \{x_1, \ldots, x_N\}, L = \{l_1, \ldots, l_N\}$, we naïvely assume conditional independence given model parameters and compute the joint probability as:

$$p(\{x, L\}|r, \Theta) = \prod_{i=1}^{N} p(\{x_i, l_i\}|r, \Theta) \quad (8)$$

For numerical stability, we use the equivalent task and minimize the negative log-likelihood instead:

$$r^* = \underset{r}{argmin} \sum_{i=1}^{N} -\ln(p(\{x_i, l_i\}|r, \Theta)), \quad (9)$$

where r^* is the MLE estimate of optimal transformation parameters, given the model parameters and the observation sequence. The minimization is performed using general purpose constrained optimization (the *active set* algorithm as implemented in MathWorks Matlab `fmincon` function). To prevent the model from diverging from clinically reasonable scaling and rotation, we set the maximum shift to ± 5 mm in any direction, maximum scaling $\pm 25\%$ in each direction and rotation maximum $\pm 15°$ around each axis, hence the model abbreviation *nrmsCon*, used below.

2.3 Reference Methods

In order to evaluate performance of the proposed method, we implemented three reference methods, based solely on anatomical landmarks, identified manually by neurologists in the pre-operative MRI images:

1. **target**—the method consists in finding a translation $[t_x, t_y, t_z]$, which shifts central point of the atlas model to the planned target point without any scaling or rotation. This method is also used as the initalization for NRMS-based fitting, as it requires no additional information apart from planned target coordinates, which is the result of standard pre-surgical planning procedure.

2. **acpc**—this method represents a simple atlas fitting approach, based on two significant brain landmarks: the anterior commisure (AC) and the posterior commisure (PC). The method analytically finds a full 9-DOF transformation which maps the vector given by AC and PC points in the atlas to the vector given by AC-PC points, identified in patient's MRI scans.

3. **allpoints**—additionally to the AC-PC points, this method uses 12 landmarks on the STN boundaries, defined previously in the supplement of [7]. The method than finds a full 9-DOF transformation to minimize the least-square distance between the characteristic points on the atlas and in manually annotated patient MRI data.

2.4 Data Collection and Preprocessing

The MER signals were recorded intra-operatively from five parallel electrode trajectories, spaced 2 mm apart in a "ben-gun" configuration around the central electrode. The sampling frequency was 24 kHz, signals were filtered by a bandpass filter in the range 500–5000 Hz upon recording and stored for offline processing. At each of the recording positions, spaced apart by 0.5 mm, a typically ten seconds of MER signal were recorded using each electrode. In order to eliminate artifact-bearing segments of the signals, we used our automatic artifact classifier, presented previously in [8]. Manual intra-operative expert annotation of the MER signals has been stored, labeling each signal as coming either from inside or outside the STN.

2.5 Performance Evaluation

In order to estimate the out of sample performance of the proposed method and due to the relatively small sample size (in terms of whole patient sets), we employed the leave one subject out (LOSO) procedure. In each iteration we kept one subject's data (maximum two 5-electrode trajectories for bi-laterally implanted patients) for model fitting and evaluation, while all other data were used to obtain the parameters Θ.

To evaluate quality of the model fit, we used the machine-learning based approach used also in [3]: the MER recording sites, expert-labeled as STN, were expected to be encapsulated inside the fitted atlas (true positives), while other recordings were expected to lie outside. The accuracy, sensitivity, specificity and Youden J-index ($J = sensitivity + specificity - 1$) were computed.

Fig. 3 Comparison of classification performance across methods (correctly included/excluded recorded NRMS points): the proposed electrophysiology-based method *nrmsCon* (yellow) showed higher STN identification accuracy than the reference MRI-based methods (Color figure online)

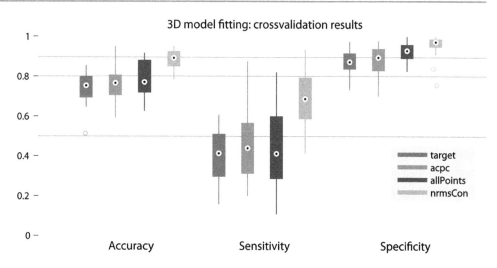

Table 1 Overall 3D STN model fitting crossvalidation results on the 27 validation trajectories for all methods

Method	Accuracy		Sensitivity		Specificity		Youden J	
	Mean (%)	Std (%)	Mean (%)	Std	Mean (%)	Std	Mean (%)	Std (%)
Target	**74,3**	7,6	**40,7**	12,3	**87,3**	5,7	**28,0**	17,0
acpc	**75,7**	8,9	**44,7**	17,2	**87,6**	8,2	**32,3**	21,1
All Points	**78,7**	8,7	**44,6**	19,8	**92,3**	4,9	**36,8**	21,3
nrmsCon	**88,1**	5,2	**69,0**	14,2	**95,5**	5,4	**64,5**	13,6

3 Results and Discussion

3.1 Collected Data

The dataset contained data from 27 explorations in 15 PD patients with complete 3D information and additional 8 explorations from 4 patients with measured and annotated MER signals but without information on spatial recording locations. The latter small set was included for estimation of model parameters (Θ) but was excluded from validation. Each exploration consisted of 5 electrode trajectories with 25.9 recording positions on average. In total, the data included 35 explorations from 19 patients, leading to 175 electrode trajectories and 4538 recorded MER signals.

3.2 Performance Evaluation

Classification performance (i.e. the proportion of correctly included/excluded recording sites) was evaluated for each of the fitting methods on the 27 exploration trajectories, the

results are shown in the Fig. 3 and Table 1. As seen from the results, it is apparent that the presented *nrmsCon* method provided substantially better fit to the measured MER sites than any of the other methods. The results further show, that the main difference is driven especially by the higher sensitivity, i.e. the proportion of correctly included STN points inside the model. This is even more clearly seen from the tabulated values of the Youden J statistic, where the proposed method surpasses the reference methods by a factor of two. It has to be considered that the dataset is highly imbalanced dataset with only 27% of signals coming from the STN.

To provide additional insight into the results, we evaluated the fitted values of the transformation parameters individually. Results of the proposed *nrmsCon* method showed similar distribution to the landmark-based *allPoints* method, except for a relatively large ca 2 mm shift in the *y* direction. According to previous studies [9], this is the main direction of the brain shift occuring during surgery and this preliminary evaluation thus provides promising results for intra-operative brain-shift compensation. An example visualization of atlas fit can be found in Fig. 4.

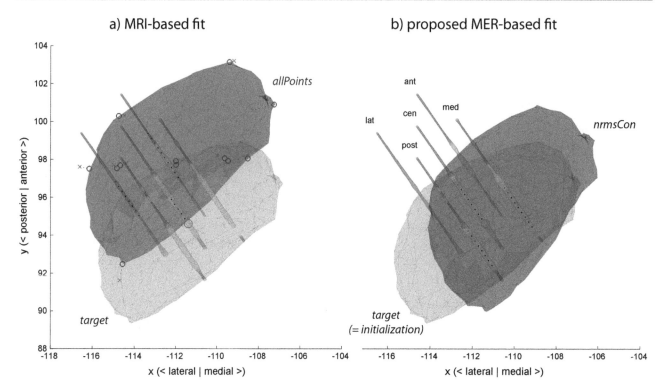

Fig. 4 Examples of model fit using the **a** *allPoints* method based on characteristic MRI points in patient pre-operative scans (red x) and on the atlas (blue o) and **b** the proposed *nrmsCon* method based solely on electrophysiology on a single five-electrode trajectory. The final model position after fitting is shown in purple, the initial position (*target* method) is shown in grey. The width of the five microelectrode trajectory cyllinders denotes the NRMS value, while colors denote manual labels: STN in yellow, non-STN in grey. MER positions inside the resulting model are denoted by black points, planned target by red o (Color figure online)

4 Conclusion

We proposed a probabilistic model for automatic direct fitting of a STN atlas to multi-electrode explorative DBS MER data. The presented results indicate that the proposed MER-based system may potentially bring increased accuracy in intra-operative MER localization and thus contribute to higher efficacy in DBS research and potentially also in therapy.

Acknowledgements The work presented in this paper was supported by the Czech Science Foundation (GACR), under grant no. 16-13323S and by the Ministry of Education Youth and Sports, under NPU I program Nr. LO1611.

References

1. Anan Moran, Izhar Bar-Gad, Hagai Bergman, and Zvi Israel. Real-time refinement of subthalamic nucleus targeting using Bayesian decision-making on the root mean square measure. *Mov Disord*, 21(9):1425–1431, September 2006.

2. Adam Zaidel, Alexander Spivak, Lavi Shpigelman, Hagai Bergman, and Zvi Israel. Delimiting subterritories of the human subthalamic nucleus by means of microelectrode recordings and a Hidden Markov Model. *Mov Disord*, 24(12):1785–1793, sep 2009.

3. J. Luis Lujan, Angela M. Noecker, Christopher R. Butson, Scott E. Cooper, Benjamin L. Walter, Jerrold L. Vitek, and Cameron C. McIntyre. Automated 3-Dimensional Brain Atlas Fitting to Microelectrode Recordings from Deep Brain Stimulation Surgeries. *Stereotactic and Functional Neurosurgery*, 87(4):229–240, jan 2009.

4. Eduard Bakstein, Tomas Sieger, Daniel Novak, and Robert Jech. Probabilistic Model of Neuronal Background Activity in Deep Brain Stimulation Trajectories. In *Information Technology in Bio- and Medical Informatics - 7th International Conference, (ITBAM)*, 2016.

5. Axel Krauth, Remi Blanc, Alejandra Poveda, Daniel Jeanmonod, Anne Morel, and Gabor Szekely. A mean three-dimensional atlas of the human thalamus: Generation from multiple histological data. *NeuroImage*, 49(3):2053–2062, 2010.

6. Eduard Bakstein. *Deep Brain Recordings in Parkinson's Disease: Processing, Analysis and Fusion with Anatomical Models*. doctoral thesis, Czech Technical University in Prague, 2016.

7. Tomas Sieger, Tereza Serranova, Filip Ruzicka, Pavel Vostatek, Jiri Wild, Daniela Stastna, Cecilia Bonnet, Daniel Novak, Evzen Ruzicka, Dusan Urgosik, and Robert Jech. Distinct populations of neurons respond to emotional valence and arousal in the human subthalamic nucleus. *Proceedings of the National Academy of Sciences of the United States of America*, 112(10):3116–21, 2015.

8. Eduard Bakstein, Tomas Sieger, Jiri Wild, Daniel Novak, Jakub Schneider, Pavel Vostatek, Dusan Urgosik, and Robert Jech. Methods for automatic detection of artifacts in microelectrode recordings. *Journal of Neuroscience Methods*, 290:39–51, 2017.

9. Srivatsan Pallavaram, Benoit M. Dawant, Michael S. Remple, Joseph S. Neimat, Chris Kao, Peter E. Konrad, and P. F. D'Haese. Effect of brain shift on the creation of functional atlases for deep brain stimulation surgery. *International Journal of Computer Assisted Radiology and Surgery*, 5(3):221–228, 2010.

Applying Weightless Neural Networks to a P300-Based Brain-Computer Interface

Marco Simões, Carlos Amaral, Felipe França, Paulo Carvalho, and Miguel Castelo-Branco

Abstract

P300-based Brain Computer Interfaces (BCI) are one of the most used types of BCIs in the literature that make use of the electroencephalogram (EEG) signal to convey commands to the computer. The efficiency of such systems depends drastically on the ability of correctly identifying the P300 wave in the EEG signal. Due to high inter-subject and inter-session variability, single-subject classifiers must be trained every session. In order to achieve fast setup times of the system, only a few trials are available each session for training the classifier. In this scenario, the capacity to learn from few examples is crucial for the performance of the BCI and, therefore, the use of weightless neural networks (WNN) is promising. Despite its possible added value, there are no studies, to our knowledge, applying WNNs to P300 classification. Here we compare the performance of a WNN against the state-of-the-art algorithms when applied to a P300-based BCI for joint-attention training in autism. Our results show that the WNN performs as good as its competitors, outperforming them several times. We also perform an analysis of the WNN hyperparameters, showing that smaller memories achieve better results most of the times.

This study demonstrates that the adoption of this type of classifiers might help increase the prediction accuracy of P300-based BCI systems, and should be a valid option for future studies to consider.

Keywords

P300 • Brain-Computer interfaces (BCI) Weightless neural networks (WNN)

1 Introduction

P300-based Brain Computer Interfaces (BCI) represent a widely used type of electroencephalographic (EEG) BCI in the literature. Those BCIs work based on the identification of the P300 wave in the EEG signal, elicited by paying attention an infrequent stimulus. Although the most common application in the P300-speller [1], an interface for selecting letters from a matrix based on flashes, more complex interfaces have been proposed [2]. Although they base their functioning in the same neurological process related to attention, different waveforms are elicited by different interfaces [3]. In this sense, it is important to have strong classification algorithms to correctly identify the target stimulus.

Due to high inter-subject and inter-session variability, single-subject classifiers must be trained every session. In order to achieve fast setup times to the system, only a few trials are available each session for training the classifier.

In this scenario, being able to learn from few cases is crucial and, therefore, the use of weightless neural networks (WNN) is promising. The WNN are a family of classifiers that do not require weight optimization, being trained only with a forward step [4]. This way, the WNN are usually faster to train than its usual competitors [5]. Despite their foreseeable merits, the applications of WNN remain underexplored [6], being this the first work applying WNN to BCI, to our knowledge.

M. Simões (✉) · C. Amaral · M. Castelo-Branco
CIBIT, Coimbra Institute for Biomedical Imaging and Translational Research, ICNAS, University of Coimbra, Coimbra, Portugal
e-mail: msimoes@dei.uc.pt

M. Simões · C. Amaral · M. Castelo-Branco
Faculty of Medicine, University of Coimbra, Coimbra, Portugal

M. Simões · P. Carvalho
CISUC, Center for Informatics and Systems, University of Coimbra, Coimbra, Portugal

F. França
PESC-COPPE, Universidade Federal do Rio de Janeiro, Rio de Janeiro, Brazil

© Springer Nature Singapore Pte Ltd. 2019
L. Lhotska et al. (eds.), *World Congress on Medical Physics and Biomedical Engineering 2018*,
IFMBE Proceedings 68/3, https://doi.org/10.1007/978-981-10-9023-3_20

In this paper, we compare the performance of a WNN against the state-of-the-art algorithms when applied to a P300-based BCI clinical trial for joint-attention training in autism. We further study how the WNN hyperparameters influence the classifier in this task.

2 Methods

In this section we describe the methods used in the paper, starting with the dataset used, the feature extraction procedure and then the classifiers used for performance comparison.

2.1 Dataset

We used the EEG data from 13 subjects using the gTec g. Nautilus system while performing an innovative P300-based BCI task for joint-attention training in autism, published in [2]. This innovative system uses the BCI in a virtual reality setting, where the participants train to follow the gaze of a virtual avatar. 13 subjects performed one session of BCI with 3 systems. The paper concluded that the best performing EEG system was the g.Nautilus, in terms of use for a clinical trial in autism. Here, we also use the g.Nautilus data for the comparison of classifiers. For a detailed explanation of the task, please refer to the paper in [2].

2.2 Signal Processing and Classification Pipeline

The signal processing procedure follows the traditional feature extraction and selection using two independent datasets (one for training, one for testing). This process is repeated for each session of each participant.

2.2.1 Feature Extraction and Selection

The feature extraction procedure follows the algorithm of [7], where a two-step spatial filter is applied to the signal, one using the Fisher Criteria to maximize the differences between target/non-target signals and another one to maximize the Signal-to-Noise Ratio (SNR) (see Fig. 1). The filters are applied in cascade. To a deeper explanation of the procedure, please refer to the original paper [7].

From the final signal generated, we select the features that maximize the difference between the target and non-target signals (measured by Pearson R correlation). Only the features whose p-value are below 0.01 are selected. A minimum of 50 features is enforced, meaning that if not enough features have a p-value below threshold, the 50 features with lowest p-value are selected.

2.2.2 Classification Procedure

In order to select the best parameters for each classifier, the train dataset is split in train (70% of the samples) and validation (30% of the samples) sets. Then, we iteratively select hyperparameter combinations to test. For each hyperparameter combination we train a classifier with the train set and evaluate its performance with the test set. The classifier with the best accuracy in the validation set is selected for evaluation with the test set.

The metric selected for evaluation of the classifier performance was the object detection accuracy. In the virtual scenario, the avatar directs the attention of the participant for one out of eight objects. For each blink of the eight objects, the classifier must perform a binary classification (target vs non-target), generating a target score for each object. For the eight objects, the one with the highest target score is selected as target, while the other are labeled as non-target. Chance-level is, therefore, fixed as 12.5% (1 out of 8).

2.3 Classifiers

In this paper, we compared the most used classifiers for this purpose [8] (Fisher Linear Discriminant Analysis (**fisher**), Support-Vector Machine (**svm**) and Naïve Bayes Classifier (**nbc**)) with a Weightless Neural Network (Wilkes, Stonham and Aleksander Recognition Device—**WiSARD** variant). The WiSARD is a network of discriminants, each composed by n-tuples of RAM memories. The input feature vector is transformed in a binary representation. That binary representation is randomly mapped to the memories. For training, each sample goes to the discriminant of its class, increasing the counter of each memory location. For evaluation, the feature vector is evaluated in both discriminants (see Fig. 2), selecting the discriminant with higher sum of counts, after a bleaching procedure (for a more in-depth description, see [9]). We adapted the WiSARD algorithm to consider prior class frequencies in its prediction, to deal with the unbalanced characteristics of this problem. So, after the training, the counts of each memory of each discriminant is divided by the total samples of its class, used in training.

The hyperparameters optimized for the WiSARD classifier were the number of bits per memory (memory size) and the number of bits to represent each feature (input size). For the SVM, we optimized the BoxContraint C value. The NBC and Fisher classifier do not need hyperparameter optimization.

We also registered the training time needed for each classifier and compare them in addition to the accuracy metric. It should be noted that while the SVM, Fisher and Naïve Bayes classifiers make use of precompiled C routines, the WNN uses an in-house developed software fully scripted in Matlab, which makes the training times comparison biased against the WNN.

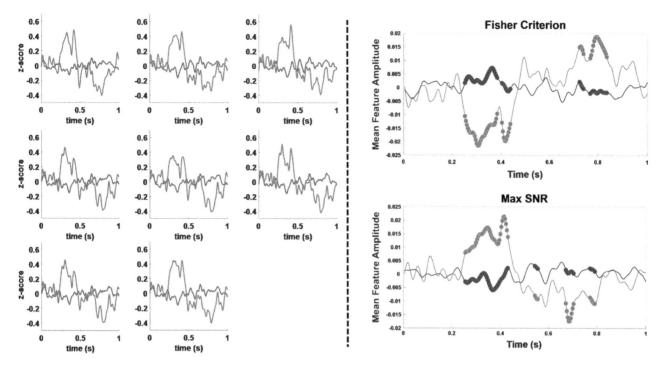

Fig. 1 Left—Event-related potentials (ERPs) from the 8 original electrodes, showing the grand averages of all target stimuli (orange) and all non-target stimuli (blue), for a randomly selected subject. Right—Signal processed for feature extraction, after application of Fisher Criterion filter (top) and max-SNR filter (bottom). Most discriminant features are selected and highlighted in the figures (Color figure online)

Fig. 2 WiSARD schematic, explaining the process from the signal to the discriminants. During training, each sample is mapped to the correspondent discriminant. For the prediction part, the sample is submitted to both discriminants and the resulting scores compared

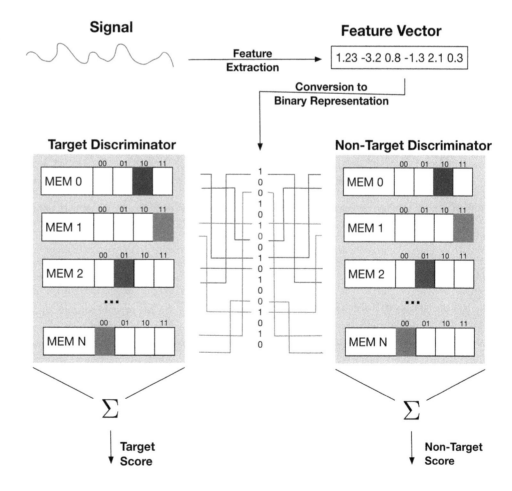

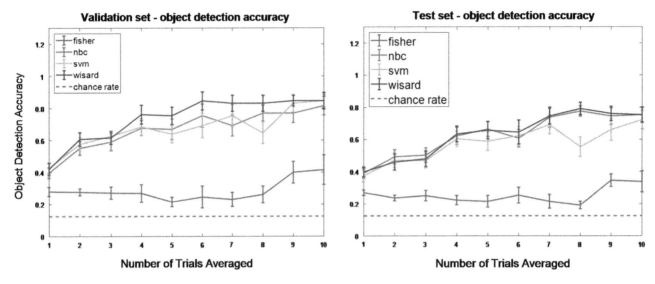

Fig. 3 Object detection accuracy for the four classifiers tested across number of averaged trials. At the left, the results for the validation accuracy and at the right, the results for the test set

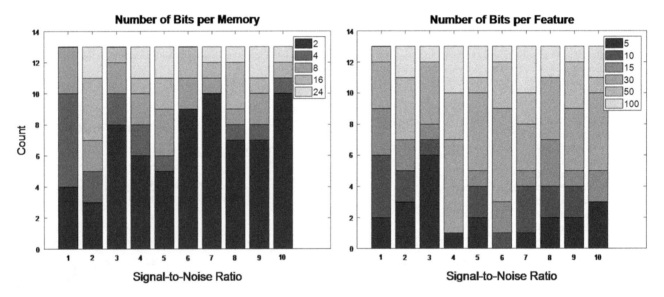

Fig. 4 Histogram of hyperparameter selection in the validation set by SNR. At the left, the memory size and, at the right, the number of bits used to represent a feature

3 Results

Due to the intrinsic noise characteristic of event-related potential signals, the SNR increases when we average several responses together. In this sense, we present in Fig. 3 the results of the performance of the classifiers across the number of averaged trials.

As expected, we see a monotonic increase in accuracy across trials for almost every classifier. We see that, for the validation set, the WiSARD present the best results for every SNR level. At the test set, we see that its performance is, at least, as good as the best classifier of the common used solutions, outperforming it several times. We see that in several cases it outperforms the SVM, possibly the most used classifier in the literature for this type of problem.

Regarding the hyperparameters (Fig. 4), we see that for most of the times, smaller memories are selected (2 bits). When looking to the histogram of the number of bits used to represent each feature, 30 bits is the most common, but without a clear superiority, showing that the memory size has a greater influence in the performance of the classifier than the number of bits per feature.

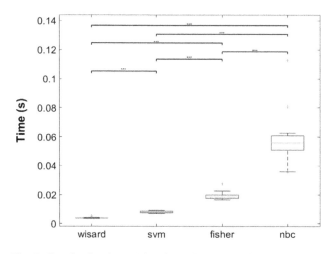

Fig. 5 Boxplot showing training times of each classifier. *** represent statistically significant differences with p < 0.001 after correction for multiple comparisons with the Bonferroni method

Regarding the training duration for each classifier, Fig. 5 shows the time needed to train each classifier. Due to some training times not to be normally distributed, we compared the training times with a non-parametric Friedman test, showing statistically significant differences between the classifiers $\chi^2(3) = 39$, p < 0.001. Post hoc tests (pair-wise Wilcoxon signed-rank tests, corrected for multiple comparisons with the Bonferroni method) showed strong statistical differences between all methods.

4 Discussion

In this paper we compared the most used classifiers in the P300 detection field with a weightless neural network, which, to our knowledge, was never tested with this purpose. We used an innovative BCI application that aims to train joint-attention skills to people with autism spectrum disorder.

The results achieved show that the presented WNN is able to, at least, perform at the level of the best classifier, outperforming it several times. Additionally, the WNN shows the fastest training time of the classifiers sampled, even without the use of pre-compiled routines. The WNN faster training times and ability to generalize responses using small memory sizes worked specially well for the characteristics of the problem, where inter-trial variability presents a big challenge for any learning algorithm. Further exploration is needed to assess if other configurations of WNN

can improve even more the accuracy results achieved by the WiSARD algorithm here tested.

This paper demonstrates that the adoption of WNN classifiers might help to increase the prediction accuracy of P300-based BCI systems. Therefore, future studies should consider its adoption when choosing the classifier to include in the P300 detection systems.

Acknowledgements This work was supported by FTC—Portuguese national funding agency for science, research and technology [Grant PAC—MEDPERSYST, POCI-01-0145-FEDER-016428], fellowship SFRH/BD/77044/2011, and the BRAINTRAIN Project FP7-HEALTH-2013-INNOVATION-1-602186 20, 2013.

Conflict of Interest The authors declare no conflict of interests.

References

1. Pires G, Nunes U, Castelo-Branco M (2012) Comparison of a row-column speller vs. a novel lateral single-character speller: Assessment of BCI for severe motor disabled patients. Clin Neurophysiol 123:1168–1181. https://doi.org/10.1016/j.clinph.2011.10.040.
2. Amaral CP, Simões MA, Mouga S, et al (2017) A novel Brain Computer Interface for classification of social joint attention in autism and comparison of 3 experimental setups: A feasibility study. J Neurosci Methods 290:105–115. https://doi.org/10.1016/j.jneumeth.2017.07.029.
3. Amaral CP, Simões M a., Castelo-Branco MS (2015) Neural Signals Evoked by Stimuli of Increasing Social Scene Complexity Are Detectable at the Single-Trial Level and Right Lateralized. PLoS One 10:e0121970. https://doi.org/10.1371/journal.pone.0121970.
4. Aleksander M de G, França FMG, Lima PM V, Morton H (2009) A brief introduction to Weightless Neural Systems. ESANN'2009 proceedings, Eur Symp Artif Neural Networks - Adv Comput Intell Learn 22–24.
5. Cardoso DO, Carvalho DS, Alves DSF, et al (2016) Financial credit analysis via a clustering weightless neural classifier. Neurocomputing 183:70–78. https://doi.org/10.1016/j.neucom.2015.06.105.
6. França FMG, de Gregorio M, Lima PM V., de Oliveira WR (2014) Advances in Weightless Neural Systems. 22th Eur Symp Artif Neural Networks 497–504. https://doi.org/10.13140/2.1.2688.6403.
7. Pires G, Nunes U, Castelo-Branco M (2011) Statistical spatial filtering for a P300-based BCI: tests in able-bodied, and patients with cerebral palsy and amyotrophic lateral sclerosis. J Neurosci Methods 195:270–81. https://doi.org/10.1016/j.jneumeth.2010.11.016.
8. Manyakov N V, Chumerin N, Combaz A, Van Hulle MM (2011) Comparison of Classification Methods for P300 Brain-Computer Interface on Disabled Subjects. Comput Intell Neurosci 2011:1–12. https://doi.org/10.1155/2011/519868.
9. Grieco BPA, Lima PMV, De Gregorio M, França FMG (2010) Producing pattern examples from "mental" images. Neurocomputing 73:1057–1064. https://doi.org/10.1016/j.neucom.2009.11.015.

How Does Cell Deform Through Micro Slit Made by Photolithography Technique?

Shigehiro Hashimoto, Yusuke Takahashi, and Haruka Hino

Abstract

Several slits sort cell according to the deformability in vivo. A micro slit (0.87 mm of width, 0.010 mm height) was newly designed between a micro ridge on a transparent polydimethylsiloxane plate and micro ridges on a borosilicate glass plate. These ridges made by photolithography technique make contact each other in the perpendicular position to make the slit between the ridges. A one-way flow system was designed to observe each cell passing through the slit in vitro. Four kinds of cells were used in the test: C2C12 (mouse myoblast cell), HUVEC (human umbilical vein endothelial cell), Hepa1-6 (mouse hepatoma cell), and Neuro-2a (mouse neural crest-derived cell). The suspension of each kind of cells was injected to the slits. The deformation of each cell passing through the slit was observed with an inverted phase-contrast microscope. At the microscopic images, the outline of each cell was traced, and the area (S) was calculated. The deformation ratio was calculated as the ratio (S_2/S_1) of the projected area of each cell before the slit (S_1) and that in the slit (S_2). The velocity of the cell passing through the slit was calculated by the trace at the microscopic movie. The experimental results show that each cell deforms to the flat circular disk and passes through the micro slit. Hepa1-6 is flattened with the increase of the passing velocity, and HUVEC is elongated along the flow. The designed slit between micro ridges is effective to evaluate the deformability of cells.

Keywords

Cell deformation · Micro machining · Micro slit

1 Introduction

The deformability is important for the biological cell in vivo [1, 2]. An erythrocyte, for example, deforms and passes through capillary. Some cells deforms and passes through a slit in vivo. Several systems sorts cells according to the deformability in vivo.

In the previous studies, several kinds of micro-channels were made to simulate deformation of erythrocytes through the micro capillary. The photolithography technique is useful to make a micro-channel [3–5]. The previous studies show that the slit of width between 0.010 and 0.025 mm has capability to sort several kinds of floating cells as the filter. In the present study, the micro slit of 0.010 mm has been manufactured by photolithography technique to simulate deformation of cells through the micro-slit.

2 Methods

2.1 Micro Slit

Micro ridges are made by photolithography technique. A micro slit (0.87 mm of width, 0.010 mm of height, 0.10 mm of length) was newly designed between a micro ridge on a transparent polydimethylsiloxane plate and micro ridges on a borosilicate glass plate. These ridges make contact each other in the perpendicular position to make the slit between the ridges (Fig. 1a). The dimension of the surface micro morphology on each plate was confirmed with a laser microscopic measurement (Fig. 3b). The slit was placed in the center part of a flow channel (height of

S. Hashimoto (✉) · Y. Takahashi · H. Hino
Kogakuin University Shinjuku, Tokyo, 1638677, Japan
e-mail: shashimoto@cc.kogakuin.ac.jp

© Springer Nature Singapore Pte Ltd. 2019
L. Lhotska et al. (eds.), *World Congress on Medical Physics and Biomedical Engineering 2018*,
IFMBE Proceedings 68/3, https://doi.org/10.1007/978-981-10-9023-3_21

(a)

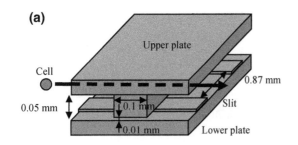

(b)

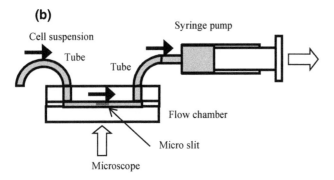

Fig. 1 **a** A micro slit between ridges. **b** A one-way flow (from left to right) system

0.05 mm, width of 5 mm, and length of 30 mm). A one-way flow system (Fig. 1b) was designed to observe each cell passing through the slit in vitro.

2.2 Flow Test

Four kinds of cells (3 < passage < 10) were used in the test: C2C12 (mouse myoblast cell line originated with cross-striated muscle of C3H mouse), HUVEC (human umbilical vein endothelial cell), Hepa1-6 (mouse hepatoma cell line of C57L mouse), and Neuro-2a (a mouse neural crest-derived cell line). Each kind of cells suspended in each culture medium (3×10^{-4} cell/cm^3) was introduced into the slits by suction at the outlet by the syringe pump. The constant flow rate ($<6 \times 10^{-10}$ m^3/s) was adjusted by the pump. The deformation of each cell passing through the slit was observed with an inverted phase-contrast microscope.

At the microscopic images, the contour of each cell was traced, and the area (S) was calculated. The deformation ratio (Rs) was calculated as the ratio of the projected area of each cell before the slit (S_1) and that in the slit (S_2).

$$Rs = S_2/S_1 \qquad (1)$$

The velocity (v) of the cell was measured by the trace of the centroid of the cell at the microscopic movie. The velocity (Rv) ratio was calculated from the velocity passing through the slit (v_2), and the velocity before the slit (v_1).

$$Rv = v_2/v_1 \qquad (2)$$

The contour of the two dimensional projection image of each cell in the slit was approximated to the ellipsoid, and the angle (φ) between the major axis and the flow direction was measured. The angle is zero, when the cell elongates along the stream line.

3 Results

Figure 2 shows the scanning electron microscopic images of the upper (left) and lower (right) ridges. Each top surface (almost flat) contacts together to make the slit between them. Figure 3 exemplifies the tracing across the ridge (dotted line in Fig. 2 left) on the upper plate measured by the laser microscope.

Figure 4 shows the microscopic image during the flow test. Table 1 shows the range of the deformation ratio and the velocity of cell passing through the slit. Each cell deforms to the flat circular disk and is passing through the micro slit.

Figure 5 shows the relationship between the deformation ratio and the velocity ratio of Hepa1-6. The deformation ratio of several cells exceeds 1.5 at the velocity ratio higher than 2.

Figure 6 shows distribution of HUVEC according to the direction of the major axis of each cell. Before the slit, every ratio of cells is around 0.17, which shows that the direction of cell is random. In the slit, most of cells are included in the range between 0° and 15°, which shows that HUVEC is elongated along the flow.

Fig. 2 Scanning electron microscopic images of the upper (left) and lower (right) ridges. Each top surface contacts together to make the slit between them

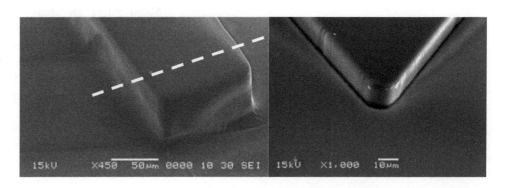

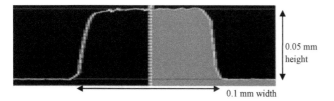

Fig. 3 Tracing across the ridge on the upper plate measured by laser microscope

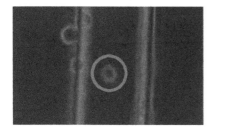

0.1 mm

Fig. 4 A deformed cell passing through the slit (in the red circle); cell flows from left to right

Table 1 Deformation ratio and velocity of cell passing through slit. Data range of twenty cells at each kind of cell

Cell	Deformation ratio (Rs)	Velocity (v_2) [mm/s]
C2C12	1.0–1.5	0.1–1.1
HUVEC	1.0–1.4	0.3–0.6
Hepa1-6	1.0–1.7	0.05–0.9
Neuro-2a	1.0–1.5	0.05–0.4

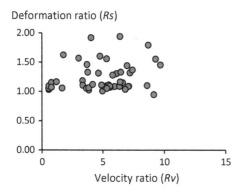

Fig. 5 Relationship between deformation ratio and velocity ratio of Hepa1-6

4 Discussion

Four kinds of cells has been selected in the present study. HUVEC is the typical cell, which is exposed to the flow at the inner surface of the vessel wall in vivo. The previous

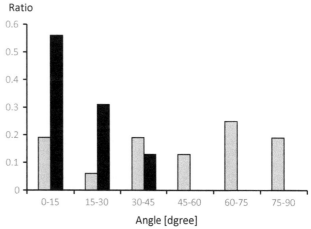

Fig. 6 Distribution of HUVEC according to direction (angle (φ) between major axis of each cell and flow): gray bar, before slit: black bar, in slit: n = 16

study shows that C2C12 tilts perpendicular to the flow in vitro [6]. HUVEC, on the other hand, makes orientation along the streamline in vitro [7]. Neuro-2a elongates the neurite along the direction of the excess gravitational force field [8]. Cancer cells might tend to penetrate through slits. The viability of cells has been confirmed by the culture of the residual cells after the present flow test. C2C12, for example, was able to differentiate to myotube, which showed repetitive contraction by stimulation of electric pulses. In the present experimental system, the mean velocity of flow through the slit increases 5.7 times as that in the flow channel, which is inversely proportional to the cross sectional area of the flow path. The higher velocity ratio corresponds to the cell, which flows near wall before the slit. The slit technique can be applied to cell sorting, which may contribute to make tissue in vitro [9] in regenerative medicine.

5 Conclusion

The micro slit was able to be manufactured by the photolithography technique. Each cell flattened to be a circular disk during passing through the micro slit. Hepa1-6 becomes thinner at the higher passing velocity, and HUVEC is elongated to the direction of the flow. The designed slit is useful for evaluation of the deformability of cells.

Acknowledgements This work was supported by a Grant-in-Aid for Strategic Research Foundation at Private Universities from the Japanese Ministry of Education, Culture, Sports and Technology.

Conflict of Interest This paper has no conflict of interest.

References

1. Hashimoto, S., Otani, H., Imamura, H., et al.: Effect of aging on deformability of erythrocytes in shear flow. Journal of Systemics Cybernetics and Informatics 3(1), 90–93 (2005).
2. Hashimoto, S.: Detect of sublethal damage with cyclic deformation of erythrocyte in shear flow. Journal of Systemics Cybernetics and Informatics 12(3), 41–46 (2014).
3. Takahashi, Y., Hashimoto, S., Hino, H., Azuma, T.: Design of slit between micro cylindrical pillars for cell sorting. Journal of Systemics, Cybernetics and Informatics, 14(6), 8–14 (2016).
4. Takahashi, Y., Hashimoto, S., Hino, H., Mizoi, A., Noguchi, N.: Micro groove for trapping of flowing cell. Journal of Systemics, Cybernetics and Informatics 13(3), 1–8 (2015).
5. Mizoi, A., Takahashi, Y., Hino, H., Hashimoto, S., Yasuda, T.: Deformation of cell passing through micro slit. In: Proceedings of the 19th World Multi-Conference on Systemics Cybernetics and Informatics, Vol. 2, pp. 270–275. International Institute of Informatics and Systemics, Orlando (2015).
6. Hashimoto, S., Okada, M.: Orientation of Cells Cultured in Vortex Flow with Swinging Plate in Vitro. Journal of Systemics Cybernetics and Informatics 9(3), 1–7 (2011).
7. Hashimoto, S., Hino, H., Sugimoto, H., Takahashi, Y., Sato, W.: Endothelial cell behavior after stimulation of shear flow. In: Proceedings of the 21th World Multi-Conference on Systemics, Cybernetics and Informatics, Vol. 2, pp. 197–202. International Institute of Informatics and Systemics, Orlando (2017).
8. Tamura, T., Hino, H., Hashimoto, S., Sugimoto, H., Takahashi, Y.: Cell Behavior After Stimulation of Excess Gravity. In: Proceedings of the 21th World Multi-Conference on Systemics, Cybernetics and Informatics, Vol. 2, pp. 263–268. International Institute of Informatics and Systemics, Orlando (2017).
9. Takahashi, Y., Sugimoto, K., Hino, H., Takeda, T., Hashimoto, S.: Electric stimulation for acceleration of cultivation of myoblast on micro titanium coil spring. In: Proceedings of the 20th World Multi-Conference on Systemics, Cybernetics and Informatics, Vol. 2, pp. 153–158. International Institute of Informatics and Systemics, Orlando (2016).

Creation of Bio-Roots with Usage of Bioengineered Periodontal Tissue—a General Overview

Ahmed Osmanović, Sabina Halilović, and Naida Hadžiabdić

Abstract

Teeth are crucial for health and appearance. Loss of teeth leads to functional, psychological and esthetic issues. For many years, osseointegrated dental implants have been successfully used as a popular prosthetic restoration method for missing teeth. These implants have a direct connection with the alveolar bone, which can cause damage and affect the implant itself and the temporomandibular joint. Thus, those implants should be inspired by natural teeth, which possess periodontal ligament fibers—a connective tissue structure that has supportive, remodeling, sensitive and nutritive function. With advancement in the fields of tissue engineering and dental implantology, a great number of experiments is performed to reconstruct the periodontium around the titanium implants. The aim of this study was to examine studies published between 2000 and 2017, and the clinical benefits of such bioengineered implants. Research is based on full-length papers retrieved from PubMed/Medline electronic database using the key words 'dental implants', 'regenerative dentistry, 'tissue engineering', 'bioengineered periodontal tissue', 'tooth replantation'. After application of inclusion and exclusion criteria, 14 papers were selected and critically reviewed. In the following articles it was found that bioengineered dental tissue could be used as a successful therapy method with a focus on significant improvement in the quality of a patient's life. Further studies are needed for the development of these novel approaches as they cannot be easily applied clinically for various reasons due to the complexity of wrapping periodontal ligament fibers around the dental titanium implants. In addition to this, aggravating factors for usage of such tissue engineering implants on the daily basis are most suitable in terms of costs and time required for practical applications.

Keywords

Dental implants • Regenerative dentistry
Tissue engineering • Bioengineered periodontal tissue

1 Introduction

In the population, approximately 2–10% of the people are affected by tooth loss and the ratio increases with age [1]. Dental implants are used more than 50 years with an aim to restore normal function, comfort, aesthetics, speech, and health to individuals who are missing teeth. Those implants include osseointegrated dental implants that are used as a substitute for the root of the tooth. Osseointegration is a term first used by Branemark in the 1960s [2] representing a direct connection between the implant and bone tissue without periodontium anchorage. Because of this connection, teeth do not have the same mobility as natural teeth, and cause damage to the alveolar bone, the implant itself, and even the temporomandibular joint [2]. For many years, osseointegrated implants represent a clinical challenge because of localized bone loss. The success of the implant depends on the osseointegration between the implant and the alveolar bone, unlike the periodontal ligament (PDL) around the roots of the teeth [3].

With development of the field of tissue engineering and dental implantology, a great number of experiments are performed to develop periodontal ligament around an implant [4]. The aims of these experiments are to create a bio-root, which would provide ideal conditions for periodontio-integrated implant treatment in the future [4].

According to Arunachalam et al. [5], the periodontal ligament has the following roles: (a) It permits forces elicited during masticatory function and other contact movements to

A. Osmanović (✉) · S. Halilović
Genetics and Bioengineering, International Burch University,
Sarajevo, Bosnia and Herzegovina
e-mail: ahmed_osmanvic@hotmail.com

N. Hadžiabdić
Department of Oral Surgery, School of Dentistry, University of
Sarajevo, Sarajevo, Bosnia and Herzegovina

Fig. 1 Diagrams of approaches
for peri-implant periodontium
reconstruction [2]

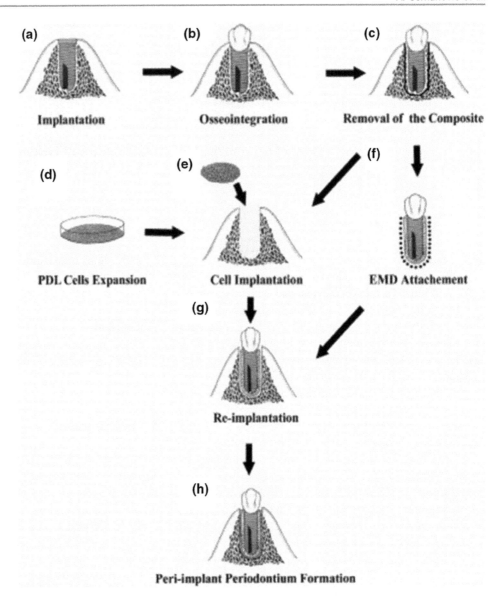

be distributed towards the alveolar via the alveolar bone, (b) It acts as a shock absorber giving the tooth a degree of movement in the socket. It also provides proprioception, (c) The periodontal ligament also has an important interaction with the adjacent bone, playing the role of the periosteum, at the bone side facing the root, (d) It homes vital cells such as osteoclasts, osteoblasts, fibroblasts, cementoblasts, cementoclasts, and most importantly, the undifferentiated mesenchymal stem cells. These cells are important in the dynamic relationship between the tooth and the bone. Due to this, there is a large interest in implants associated with periodontal tissues.

The process of periodontal tissue formation around the dental implant is shown in Fig. 1. In this figure, the following steps are visible: (a) An implant is placed into the empty space of the missing tooth, (b) Three months later, osseointegration is formed and an upper artificial tooth is fabricated, and then

(c) The dental implant with a 2-mm thickness of surrounding bone tissue is harvested, (d) PDL cells derived from the periodontium of autogenous extracted teeth and cultured in vitro are (e) transferred to this alveolar socket which is properly prepared for reimplantation. Then, (f) The bone-implant complex attached by Enamel matrix derivative (Emdogain) is (g) Re-implanted into the socket. (h) The peri-implant periodontal tissues will be observed several months later.

2   Methods

To locate sufficient resources for writing of this review paper, a literature search was conducted using online resource Medline/PubMed. The aim of this search was to examine studies published between 2000 and 2017, and to study clinical benefits of bioengineered implants.

Fig. 2 Process of application of inclusion and exclusion criteria

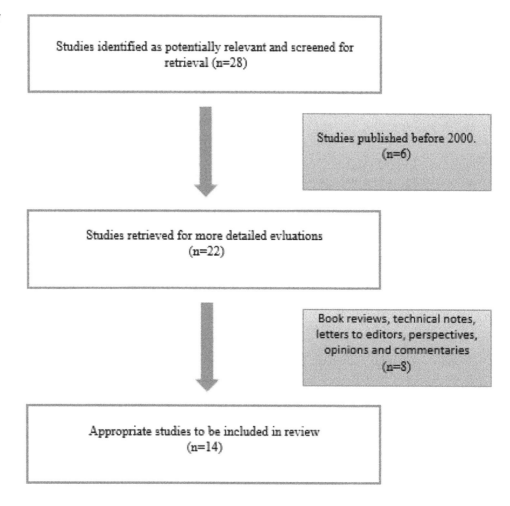

Research is based on full-length papers retrieved from electronic databases using the key words 'dental implants', 'regenerative dentistry, 'tissue engineering', 'bioengineered periodontal tissue'. A literature search was performed according to three main inclusive criteria: articles published only in English, human and animal experiments and articles published in last 17 years. Book reviews, technical notes, letters to editors, perspectives, opinions and commentaries were excluded.

After application of inclusion and exclusion criteria, 14 papers were selected, critically reviewed and considered appropriate to be included in this review. At the end of our study selection process, 14 relevant publications were included, as depicted in Fig. 2.

3 Results and Discussion

In the last 17 years, many experiments and studies were conducted (Fig. 3a) to investigate the usage of bioengineered periodontal tissue in dentistry. These studies included both human and animal cases (Fig. 3b).

In 2000, Urabe et al. [6] performed an animal study to assess whether the nature of implant material affects the migration, proliferation and differentiation of the progenitor cells for periodontium formation. From their study, it was revealed that bioactivity of the implant material strongly affects cell differentiation but does not influence the migration of periodontium-derived cells.

In the same year, a group of scientists headed by Guarnieri et al. [7], published a case report in which they evaluated histological characteristics of the tissue present between a titanium implant and a retained root in a 40-year old man. It was found that there is neoformation of cementum and collagen fibers on an implant in the presence of root residues. As this was the first case for this event, they concluded that additional studies are required to come to a strong conclusion.

Another animal study by Choi [8], also conducted in 2000, aimed to investigate new PDL attachment on titanium implants. This study included teeth from 3 dogs. Examination after three months concluded that cultured PDL cells can form tissues that appear similar to PDL fibers wrapped around implants.

(a)

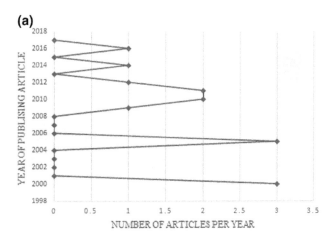

(b)

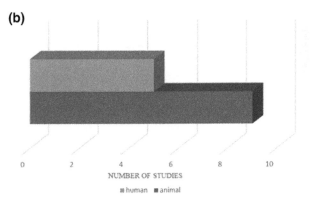

Fig. 3 **a** Distribution of published articles related to bioengineered dental implants from 2000 to 2017 and **b** Comparison of a number of published articles related to bioengineered dental implants of human and animal studies

In 2005, researchers continued to work on developing periodontal tissues around titanium implants. In this year, three animal studies are conducted by Miyashita et al. [9], Parlar et al. [10] and Jahangiri et al. [11]. Their studies showed potential of the formation of a new cementum from PDL which helps in the development of adequate mechanical strength. Also, it was revealed that periodontal tissue has adequate capacity for new formation at sites where this tissue was previously lacking [9–11].

Usage of bone marrow-derived mesenchymal stem cells for the formation of periodontal structure around titanium implants was completed by Marei et al. [12] in 2009. This animal study showed that undifferentiated mesenchymal stem cells are capable of forming three critical tissues required for periodontal tissue regeneration: Cementum, bone and periodontal ligament around the titanium implants.

One of the key experiments in this area was explored by Gault et al. [13]. The aim of their study was to describe the technical development and the clinical application of the so-called "ligaplants", the combination of PDL cells with implant biomaterial. This study was also unique as cells isolated from PDL and cultured in a bioreactor on titanium

pins are implanted in the dental alveoli of humans. As a result, the scientists included revealed that ligaplants have potential advantages over standard osseo-integrated implants due to their capability of true, functional loading. Out of the eight implants inserted, one implant was still in place and functioned even after 5 years, and even exhibited substantial bone regeneration in the adjacent bone defect 2 years after implantation. This implies that future clinical use of ligaplants might also be able to avoid bone grafting, its expenses, inconvenience and discomfort to the patient [13].

Rinaldi and Arana Chavez [14] proved that a thin cementum-like layer is formed after implantation in the areas in which the implant was in contact with PDL. They confirmed that the titanium surface with its well-known biocompatibility exerts an effect on PDL to lay down a cementum-like layer on the implant surface.

In 2011, Lin et al. [15], performed an animal study to validate the possibility of formation of bioengineered periodontal tissue on titanium dental implants. This study performed on rats reveal the potential to replace missing teeth in humans with dental implants consisting of bioengineered periodontal tissues. The remaining PDL tissue around the extracted sockets can regenerate bone and PDL-like tissues on hydroxyapatite (HA) coated tooth-shaped implants. Occlusal loads to the HA-coated implants may induce regeneration of PDL-like tissue in the peri-implant tissue [16].

In contrast to the above-mentioned authors, Oshima et al. [17] presents a new concept in bioengineering of dental tissues. This study was followed by studies conducted by the same author [18] and Nakajima et al. [19]. It revealed a novel bioengineering method including a functional biohybrid implant that is combined with an adult-derived periodontal tissue and attached with bone tissue as a substitute for the cementum. This principle restored physiological function, such as orthodontic movement through bone remodeling and appropriate responsiveness to noxious stimuli. This approach represents the potential for a next-generation bio-hybrid implant for tooth loss as a future bio-hybrid artificial organ replacement therapy [17–19].

4 Conclusion

Today tissue engineering is widely used in many branches of medicine with interesting potential applications in dentistry for the treatment of missing natural teeth. Data obtained from the present review of 14 clinical studies demonstrate that bioengineered dental tissue could be used as a successful therapy method with a focus on the significant improvement in the quality of a patient's life. Further studies are needed for the development of these novel approaches as they cannot be easily applied clinically for various reasons such as the complexity of procedure for wrapping of periodontal ligaments around dental titanium implants. In addition to this, aggravating factors for usage of such tissue engineering implants on the daily basis are the costs and time required for practical applications. This should not be observed as demotivation as based on previous experiences, such as the human genome project, which took 13 years and cost more than 2.7 billion of dollars, had prices which decreased significantly after 14 years. This fact can be a driving force as new developed methods will bring reduced costs and improved speed of application.

Declaration of Interest The authors have no conflict of interest to declare.

References

1. Kennedy DB. Orthodontic management of missing teeth. Canadian Dental Association 1999; 65(10), 548–550.
2. Lin C, Dong Q-S, Wang L, Zhang J-R, Wu L-A, Liu B-L. Dental implants with the periodontium: A new approach for the restoration of missing teeth. Medical Hypotheses. 2009; 72(1):58–61.
3. Lemmerman KJ, Lemmerman NE. Osseointegrated dental implants in private practice: a long-term case series study. Journal of periodontology. 2005 Feb 1; 76(2):310–9.
4. Minkle Gulati VA, Govila V, Jain N, Rastogi P, Bahuguna R, Anand B. Periodontio-integrated implants: A revolutionary concept. Dental research journal 2014; 11(2), 154.
5. Arunachalam L, Janarthanan A, Merugu S, Sudhakar U. Tissue-engineered periodontal ligament on implants: Hype or a hope? Journal of Dental Implants. 2012; 2(2):115.
6. Urabe M, Hosokawa R, Chiba D, Sato Y, Akagawa Y. Morphogenetic behavior of periodontium on inorganic implant materials: An experimental study of canines. Journal of Biomedical Materials Research. 2000; 49(1):17–24.
7. Guarnieri R, Giardino L, Crespi R, Romagnoli R. Cementum formation around a titanium implant: a case report. International Journal of Oral & Maxillofacial Implants 2002; 17(5).
8. Choi BH. Periodontal ligament formation around titanium implants using cultured periodontal ligament cells: a pilot study. International Journal of Oral & Maxillofacial Implants 2000; 15(2).
9. Miyashita A, Komatsu K, Shimada A, Kokubo Y, Shimoda S, Fukushima S, et al. Effect of Remaining Periodontal Ligament on the Healing-up of the Implant Placement. Journal of Hard Tissue Biology. 2005; 14(2):198–200.
10. Parlar A, Bosshardt DD, Unsal B, Cetiner D, Haytac C, Lang NP. New formation of periodontal tissues around titanium implants in a novel dentin chamber model. Clinical Oral Implants Research. 2005; 16(3):259–267.
11. Jahangiri L, Hessamfar R, Ricci JL. Partial generation of periodontal ligament on endosseous dental implants in dogs. Clinical Oral Implants Research. 2005; 16(4):396–401.
12. Marei MK, Saad MM, El-Ashwah AM, El-Backly RM, Al-Khodary MA. Experimental Formation of Periodontal Structure Around Titanium Implants Utilizing Bone Marrow Mesenchymal Stem Cells: A Pilot Study. Journal of Oral Implantology. 2009; 35(3):106–129.
13. Gault P, Black A, Romette J-L, Fuente F, Schroeder K, Thillou F, et al. Tissue-engineered ligament: implant constructs for tooth replacement. Journal of Clinical Periodontology. 2010; 37(8), 750–758.
14. Rinaldi JC, Arana-Chavez VE. Ultrastructure of the Interface between Periodontal Tissues and Titanium Mini-Implants. The Angle Orthodontist. 2010; 80(3):459–65.
15. Lin Y, Gallucci G, Buser D, Bosshardt D, Belser U, Yelick P. Bioengineered Periodontal Tissue Formed on Titanium Dental Implants. Journal of Dental Research. 2010; 90(2):251–256.
16. Kano T, Yamamoto R, Miyashita A, Komatsu K, Hayakawa T, Sato M, et al. Regeneration of Periodontal Ligament for Apatite-coated Tooth-shaped Titanium Implants with and without Occlusion Using Rat Molar Model. Journal of Hard Tissue Biology. 2012; 21(2):189–202.
17. Oshima M, Mizuno M, Imamura A, Ogawa M, Yasukawa M, Yamazaki H, et al. Functional Tooth Regeneration Using a Bioengineered Tooth Unit as a Mature Organ Replacement Regenerative Therapy. PLoS ONE. 2011 Dec; 6(7).
18. Oshima M, Inoue K, Nakajima K, Tachikawa T, Yamazaki H, Isobe T, et al. Functional tooth restoration by next-generation bio-hybrid implant as a bio-hybrid artificial organ replacement therapy. Scientific Reports. 2014; 4(1).
19. Nakajima K, Oshima M, Yamamoto N, Tanaka C, Koitabashi R, Inoue T, et al. Development of a Functional Biohybrid Implant Formed from Periodontal Tissue Utilizing Bioengineering Technology. Tissue Engineering Part A. 2016; 22(17–18):1108–15.

In Vitro and in Vivo Hemolysis Tests of a Maglev Implantable Ventricular Assist Device

Keqiang Cai, Lianqiang Pan, Yuqian Liu, Guanghui Wu, and Changyan Lin

Abstract

Objective: Implantation of a ventricular assist device (VAD) is a seminal therapeutic option for patients with terminal cardiac failure. A growing number of VAD patients are successfully bridged to transplantation, or can even live permanently with the device. However, the success is restricted by frequent severe complications. Haemolysis is a relevant adverse effect of several VAD types, which is the result of destruction of red blood cells, reduced by wall shear stress, flow acceleration and interaction with artificial surfaces. The CH-VAD, a small implantable continuous-flow blood pump, featuring a magnetically levitated impeller and enough hydrodynamic performance, was under development and completed for a 60-days animal implantation experiment in 6 sheep. The goal of this study is to validate the hemolysis of the pump through in vitro and in vivo studies. Methods: A series of in vitro tests was quantified experimentally by using in vitro circulation loop system according to ASTM F1841, the standard practice for the assessment of hemolysis in continuous-flow blood pumps. The hemolysis test in vivo was performed during a 60-days ovine model implantation, which was being conducted under the Institutional Animal Care and Use Committee (IACUC) protocol 05-0600 1. Results in vitro tests showed that the average normalized index of hemolysis (NIH) value of the VAD was 0.007 mg/l. The hemolysis in vivo was evaluated based on the amount of free hemoglobin in the plasma, and which showed that the free hemoglobin level in plasma peaked at 0.95 mg/l on the fifth postoperative day and then returned to an acceptable range of 6.0 mg/dL. Conclusion The magnetic levitation left ventricular assist device has good hemolytic performance. These acceptable performance results supported proceeding initial clinical trail conditions.

Keywords

CH-VAD • Maglev blood pump Hemolysis performance • In vivo test

1 Introduction

Heart failure (HF), a serious threat to human life and health, has become a global social and public health problem [1]. Every year, about 20% patients with heart disease in the world develop into end-stage heart failure. Clinically, the effect of internal medicine on HF is not ideal to provide the quality life of patients, the rate of disability and mortality is still high. Heart transplantation is also faced with donor shortage, adaptation difficulties, high cost, many complications and other deficiencies [2]. With the development of science and technology, implantation of a ventricular assist device (VAD) is a seminal therapeutic option for patients with terminal cardiac failure. A growing number of VAD patients are successfully bridged to transplantation, which can be weaned or can even live permanently with the device. However the success is restricted by frequent severe complications, hemolysis is a relevant adverse effect of several VAD types. Destruction of red blood cells (RBCs) is the result of wall shear stress, flow acceleration and interaction with artificial surfaces [3]. It contributes, together with the specific bleeding problems [4] to debilitating anaemia in VAD patients. Hemolysis is usually measured by plasma free hemoglobin (FHB), and when FHB content is more than 40 mg/l, it is considered hemolysis occurred [5].

In China, some medical scientific research institutions have begun to devote themselves to the development and initial testing of VADs, however there is no mature commercial product of VAD made so far, because the hemolytic

K. Cai (✉) · L. Pan · Y. Liu · G. Wu · C. Lin
Beijing Anzhen Hospital, Capital Medical University, Beijing, 100029, China
e-mail: llbl@sina.com

G. Wu · C. Lin
Beijing Institute of Heard Lung and Blood Vessel Diseases, Beijing, 100029, China

© Springer Nature Singapore Pte Ltd. 2019
L. Lhotska et al. (eds.), *World Congress on Medical Physics and Biomedical Engineering 2018*,
IFMBE Proceedings 68/3, https://doi.org/10.1007/978-981-10-9023-3_23

ability of VAD is one key to limit its development. Therefore the hemolyis test in vitro and in vivo is very important to develop a high quality VAD and to reduce the incidence of clinical complications.

CH-VAD, the third generation ventricular assistive device was developed by Suzhou Tongxin Medical Devices Co., Ltd. The core of the blood pump is designed by magnetic suspension centrifugal scheme without wear of mechanical bearings. The structure of the pump body is profiled by computational fluid dynamics (CFD) and made of titanium alloy material with good blood compatibility, which has excellent hemolysis performance in theory. In this study, we will test the hemolysis performance of CH-VAD in vitro and in vivo experiments, which provides a foundation for further long-term animal survival experiment and clinical trial.

2   Materials and Methods

2.1   Animals

Eight male sheep, weight 54–60 kg, were inoculated. All of them were in good health condition without cardiovascular system and blood related diseases. The sheep were purchased from Beijing Pinggu Simulation Hospital (license No. SYXK(Beijing)2010-0019). This study was commissioned by the Ethics Committee of Beijing Anzhen Hospital, Capital Medical University.

2.2   CH-VAD

The CH-VAD system consists of the implantable components, external power sources and an external controller. The implantable components include a blood pump, an outflow graft with reinforced tubing and the cable of the percutaneous driveline. The cable of the percutaneous driveline is 6 mm in diameter and extends from the pump through the skin to an external controller with a cable connector cap, which controls the operation of the device, sending power and operating signals to the pump. The external controller is connected to the power source, an AC or DC adapter and a rechargeable battery.

The implantable blood pump (Fig. 1) was designed as a centrifugal pump with a fully electromagnetic suspended impeller, which avoids the problems of mechanical bearing wear and thrombi seen with second-generation blood pumps. The pump's cannula, with 28 mm long and 16-mm outer diameter, is designed to be implanted into the apex without cardiopulmonary bypass (CPB).

As an LVAD from the left ventricle to the aorta, the pump's hemodynamic output is designed to have a flow rate

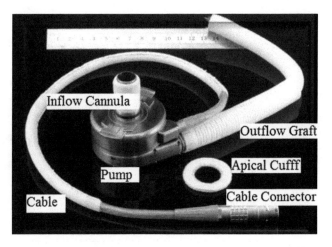

Fig. 1 Pump of CH-VAD is shown

of 5.0 L/min against 100 mmHg pressure when the impeller rotates at 3000 rpm. The 350-g blood pump is 56 mm in diameter and 31 mm long (without the inflow cannula), has a titanium alloy surface and has a displaced volume of 45 ml.

2.3   Experimental Instruments

Specially made sheep cages (self-made), anesthesia apparatus (Aeon 7200), ACT measuring instrument (ACT II, Medtronic Company, USA), Flowmeter (SM6000, Yi Fomen Electronics Co., Ltd), Pressure gauge (MMBPTSA20, Beijing Tiandi Hehe Technology Co., Ltd). Heparinized Cardiopulmonary Bypass conduit (Beijing Weijinfan Medical instrument Technology Limited), Pressure measuring Catheter (Shenzhen Yixinda Medical New Technology Co., Ltd.), High speed centrifuge (Eppendorf, Germany), ECG monitor (RSM-4101 K, NIHON Kohdene, Japan), and Ultraviolet Spectrophotometer (752 z. Beijing Optical instrument Factory).

2.4   Methods

Hemolysis Test in Vitro. Two adult male sheep, weighing 56.0 and 58.0 kg, were used in this test. Fresh blood was collected from sheep jugular vein aseptically and treated with heparin for 1.5 mg/kg. The whole blood Activated Clotting Time (ACT) was kept above 450 s. Put the fresh blood into a special blood storage bag of 500 ml and connect blood pumps, pressure gauges, flowmeters and damping valves in cyclic closed circuit according to ASTM F1841, the protocol for the assessment of the hemolytic properties of continuous flow blood pumps used in extracorporeal or implantable circulatory assist. The assessment was made

Table 1 FHB values in vitro hemolysis test (n = 6)

	0 min	60 min	180 min	240 min
FHB(mg/l)	0.058 ± 0.008	0.070 ± 0.009	0.082 ± 0.006	0.094 ± 0.009

Table 2 FHB values of hemolysis test in vivo (n = 6)

Time	Preoperative	1st day	Fifth day	Tenth day	Twentieth day	Thirtieth day	Sixtieth day
FHB (mg/l)	0.59 ± 0.03	0.84 ± 0.19	0.95 ± 0.16	0.65 ± 0.09	0.60 ± 0.08	0.58 ± 0.06	0.57 ± 0.02

based on the pump's effects on the erythrocytes over a certain period of time. For this assessment, a recirculation test is performed with a pump for 4 h.

Adjusting the pump rotation speed and damping valve made the output of pump was 5.0 L/min flow rate and 100 mm Hg pressure. Samples of 2 ml blood were collected at 0, 60, 120, 180 and 240 min after pump operated. The plasma free hemoglobin (FHB) and hematocrit values in every sample were measured.

Ten days later, the same test was repeated and for a total of 6 times.

Finally, the normalized index of hemolysis (NIH) [6] was calculated during the operation of the blood pump, which represents the amount of FHB produced by the blood pump delivers 100 L hematocrit-normalized blood per unit time.

Hemolysis Test in Vivo. Six adult male sheep, weighing from 54.0 to 60.0 kg, were used in this study. The surgical technique of the VAD implantation and postoperative care of animals were performed in accordance with the Institutional Animal Care and Use Committee (IACUC) protocol 05-0600 1 and the literature [7].

The experiment endpoint was reached when the sheep survived to 60 days after device implantation. The plasma free hemoglobin (FHB) was measured preoperatively to measure baseline levels. These tests were repeated at the first day and then every 5 days postoperatively until study termination.

3 Results

3.1 Hemolysis in Vitro

In the six hemolysis tests in vitro, the blood pump rotated stably with basically normal temperature. The levels of FHB measured at 0, 60, 120, 180 and 240 min are shown in Table 1.

According to the hematocrit values of the samples, the NIH of the blood pump in vitro was calculated as 0.007 mg/l.

3.2 Hemolysis in Vivo

The levels of FHB measured before CH-VAD implantation and at postoperative time points are shown in Table 2.

The level of FHB began to rise rapidly after the CH-VAD implanted, reached a peak of 0.95 mg/l on the fifth day postoperatively, and then decreased gradually. At the 20th days postoperatively, the FHB value was close to the normal level before operation, and tended to be stable at the 30th days after operation.

4 Discussion and Conclusion

Hemolysis is an important index to evaluate the effect of VAD on blood composition and physical and chemical properties. It is of great significance to evaluate the blood pump by hemolysis tests, because it is a necessary step before the later clinical trials. In this study, the hemolytic performance of CH-VAD tested both in vitro and in vivo showed satisfactory results.

In hemolysis test in vitro, the NIH value of CH-VAD was 0.007 mg/l, which was superior to other blood pumps made in China, and it is also superior to some foreign third generation blood pumps such as European DuraHeart centrifugal pump [8]. However, there is still a certain gap compared with the successful blood pump used in clinic [9].

The hemolysis tests in vitro and in vivo showed that CH-VAD had good hemolytic ability and could be used for long term chronic survival test and further clinic trails.

Disclosure Statement The authors declare that they have no conflict of interest.

Funding This study was supported by the National Science Foundation of China (81670371), and the Capital Public Health Project (Z161100000116086).

References

1. Khatibzadeh S, Farzadfar F, Oliver J, Ezzati M, Moran A. Worldwide risk factors for heart failure: a systematic review and pooled analysis. Int J Cardiol 2:1186–1194(2013).
2. Gopalan RS, Arabia FA, Noel P, et al. Hemolysis from aortic regurgitation mimicking pump thrombosis in a patient with a heartMate II left ventricular assist device: a case report[J]. ASAIO J 58(3):278–280(2012).
3. McLarty A. Mechanical Circulatory Support and the Role of LVADs in Heart Failure Therapy[J]. Clin Med Insights Cardiol 9 (Suppl 2):1–5(2015).

4. Hernandez AF, Grab JD, Gammie JS, et al. A decade of shortterm outcomes in post cardiac surgery ventricular assist device implantation: data from the Society of Thoracic Surgeons' National Cardiac Database. Circulation 116:606–612(2009).

5. Luckraz H, Woods M, Large SR, et al. And hemolysis goes on: ventricular assist device in combination with veno-venous hemofiltration. Ann Thorac Surg 73(2):546–548(2002).

6. Nishinaka T, Schima H, Roethy W, et al. The DuraHeart VAD, a magnetically levitated centrifugal pump: the University of Vienna bridge-to-transplant experience. Circulation 70(11):1421–1425(2006).

7. Tuzun E, Roberts K, Cohn WE, et al. In vivo evaluation of the HeartWare centrifugal ventricular assist device[J]. Tex Heart Inst J 34:406–411(2007).

8. Carl A, Johnson Jr, et al. Biocompatibility assessment of the first generation PediaFlow pediatric ventricular assist device. Artif Organs 35(1):9–21(2011).

9. Vermeulen Windsant IC, de Wit NC, Sertorio JT, et al. Blood transfusions increase circulating plasma free hemoglobin levels and plasma nitric oxide consumption: a prospective observational pilot study. Critical Care 16(3):R95(2012).

Therapeutic Embolization by Cyanoacrylate Liquid Glues Mixed with Oil Contrast Agent: Time Evolution of the Liquid Emboli

Yongjiang Li, Dominique Barthes-Biesel, and Anne-Virginie Salsac

Abstract

Glue embolization is a minimally invasive treatment used to block the blood flow to specific targeted sites. Cyanoacrylate liquid glues, mixed with radiopaque iodized oil, have been widely used for vascular embolization owing to their low viscosity, rapid polymerization rate, good penetration ability and low tissue toxicity. In this study, we have conducted an in vitro study to quantitatively investigate the polymerization kinetics of two n-butyl cyanoacrylate (nBCA) glues (Glubran 2 and Histoacryl) mixed with an iodized oil (Lipiodol) at various concentrations. The polymerization process of the glue-oil mixture is systematically characterized upon contact with a protein ionic solution mimicking plasma and compared to the case without protein. The results provide essential information for interventional radiologists to help them understand the glue behavior upon injection, and thus control embolization.

Keywords

Glue embolization • Cyanoacrylate glue • Polymerization kinetics

1 Introduction

Glue embolization is a therapeutic treatment technique used to block the blood flow to specific targeted sites. It is carried out under X-ray by introducing an embolic glue, mixed with a radiopaque iodized oil, into the circulation through a microcatheter. The technique can be performed as a definitive treatment or an adjunct to the management of arterio-venous malformations, tumors, trauma or hemorrhage [1, 2]. Cyanoacrylate liquid glues are widely used as embolic agents owing to their low viscosity, rapid polymerization rate and low tissue toxicity. Histoacryl (B. Braun, Melsungen, Germany), a pure n-butyl cyanoacrylate (nBCA) glue, has been the only glue available for external use in Europe for many years, but it is still not approved for intravascular use by the European Community (EC). It has, nevertheless, been tested for more than 10 years on patients [3]. One commonly used liquid adhesive that has the EC approval for endovascular use is Glubran 2 (GEM, Viareggio, Italy), which consists of nBCA mixed with metacryloxysulpholane (MS) as comonomer. The addition of MS allows to lower the polymerization temperature to about 45 °C and to thus reduce cytotoxicity [4]. Upon injection in the blood flow, a glue-oil mixture simultaneously polymerizes and flows with blood, leading to vessel occlusion. However, the procedure is difficult to control, because very little information exists on the polymerization kinetics of the glue-oil mixture. One empirical technique consists of dropping a small quantity of glue-oil mixture onto a plasma substrate and visualising its change in opacity [5]. This procedure provides empirical information on the initial stage of polymerization inside a thin sheet of glue mixture, only. We have designed a novel experimental setup to characterize precisely the polymerization kinetics inside a glue-oil mixture upon contact with an ionic solution containing (or not) protein concentrations similar to blood. The objective of the study is to use this technique to analyze and compare the polymerization process of Glubran 2 and Histoacryl, mixed with a radiopaque oil (Lipiodol, Guerbet, Aulnay-sous-Bois, France) at various concentrations, and identify the influence of proteins on it.

2 Materials and Methods

Glubran 2-Lipiodol (G-L) and Histoacryl-Lipiodol (H-L) mixtures are prepared at glue concentrations $C_G = 100\%$, 50% and 25% by means of a female luer connector attached

Y. Li · D. Barthes-Biesel (✉) · A.-V. Salsac
Biomechanics and Bioengineering Laboratory (UMR CNRS 7338), Université de Technologie de Compiègne-CNRS, Sorbonne Universités, CS 60319, 60203 Compiègne, France
e-mail: dbb@utc.fr

to two 1-ml syringes, one loaded with glue and the other with Lipiodol. The mixing process consists of passing the content back and forth, from one syringe to the other at high speed. Two model solutions are used as substitutes of human blood plasma: an ionic solution (IS) consisting of PBS with 0.08% glucose and a protein solution consisting of PBS mixed with bovine serum albumin (BSA) at concentrations 80 g/L (8%) or 40 g/L (4%). The two latter solutions are referred to as IS-BSA8 and IS-BSA4, respectively.

The polymerization reaction is studied under static conditions in a vertical glass capillary tube (internal diameter $D_t = 1.06 \pm 0.01$ mm) by following the procedure described by Li et al. [6, 7]. The lower end of the tube, filled with the glue-oil mixture to be tested, is put in contact with the reacting solution (IS or IS-BSA): this creates a sharp, well-defined interface between the two liquids. As the polymerization proceeds, the glue mixture density increases, which leads to an increase of opacity. This change in opacity of the fluids is monitored with an imaging system consisting of a high-speed camera (SA3, Photron, USA) coupled to a back illumination source (Schott-Fostec, LLC, USA) (Fig. 1a). An upwards vertical z-axis is defined along the tube with origin $z = 0$ at the bottom of the capillary tube: the progression of the polymerization reaction is then evaluated from the change in image grey level $G_p(t, z)$ of the glue mixture at measuring points equally distributed along the z-axis with an interval $0.5D_e$, where D_e is the diameter of the liquid region measured on the image. Grey levels are averaged within boxes of width $0.7D_e$ and height $0.4D_e$ centered on each test point. The progression of the polymerization reaction is monitored with two recording phases: a continuous recording to capture the beginning of the polymerization process at a frame rate of 50 fps over 217.8 s or 435.7 s, followed by a time-lapse mode to monitor the long-term polymerization process at a frame rate of 0.5 fps. The

duration of the time-lapse mode ranges from 60 to 240 min depending on the glue concentration.

3 Results and Discussions

3.1 Polymerization Phases

A typical polymerization process of a G-L mixture ($C_G = 50\%$) on contact with a protein solution (8%) is shown in Fig. 2a. As soon as the two liquids are in contact, a darkening appears at the tube bottom indicating the polymerization of the mixture. The darkening front propagates upwards and stops at a distance z_f at time t_f, thus creating a polymerized glue plug. In the case shown in Fig. 2a, $z_f = 2.1 \pm 0.3$ mm for $t_f = 90 \pm 60$ s: the polymerization process is thus fast. Scanning electron microscope (SEM) observations of the bottom surface of the polymerized glue plug show a complex network of connected polymerized structures with interstices filled with oil. Qualitatively, the resulting structure is hard and resists compression. Some thirty five minutes after the first polymerization has stopped, a second polymerization phase takes place. The polymerization front propagates upwards from the upper boundary of the glue plug and reaches the top of the mixture column after ~ 1 h from the moment of contact (Fig. 2a). The final grey level is much lighter than the one in the plug, indicating that the new polymerized bulk is less opaque and thus less dense than the plug. SEM images of a section of the polymerized column show micro oil droplets encapsulated by polymerized glue. Qualitatively, the column is still a hard solid that resists compression, like the plug. Similar phenomena are observed for a 50% H-L mixture, as shown in Fig. 2b. Note that in this case the top of the Histoacryl plug is not as sharp as that of Glubran 2: the corresponding plug height is $z_f = 2.9 \pm 0.3$ mm for a time $t_f = 245 \pm 60$ s. In conclusion, the polymerization reaction proceeds in two different phases which are referred to as slow and fast volumetric polymerization, respectively, and which are discussed in the following.

3.2 Polymerization Reactions and Kinetics

A careful analysis of the evolution of the grey level $G_p(z, t)$, as described in detail in [7], allows us to evaluate z_f and t_f. Fast polymerization results are shown for a variety of conditions in Fig. 3a. Since the polymerization process is random, there is some scatter of the results. As expected, t_f increases with z_f. For the 50% G-L mixture, the fast polymerization altogether propagates over an average distance $z_f = 2.1 \pm 0.4$ mm over an average time $t_f = 132 \pm 73$ s,

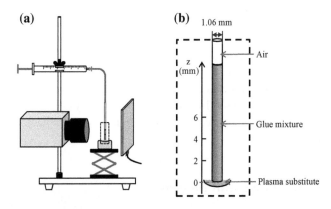

Fig. 1 **a** Experimental setup to study the polymerization of glue-oil mixtures on contact with a reacting solution: a camera monitors the changes in grey level of a glue sample contained in the tube and lighted from behind. **b** Detail of the capillary tube and of the coordinate system

Fig. 2 At time $t = 0$, a glue-oil mixture $(C_G = 50\%)$ is put in contact with IS-BSA8. The darkening of the glue solution indicates that polymerization is occurring. A fast polymerization reaction over the two first minutes is followed by a slow polymerization some 30 min later. **a** G-L mixture; **b** H-L mixture

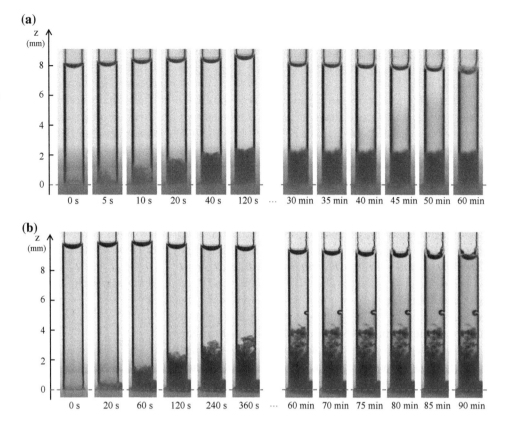

Fig. 3 a Time t_f necessary to polymerize a distance z_f during the fast polymerization. **b** Average front propagation velocity V_p during the slow polymerization for various mixtures upon contact with different protein solutions

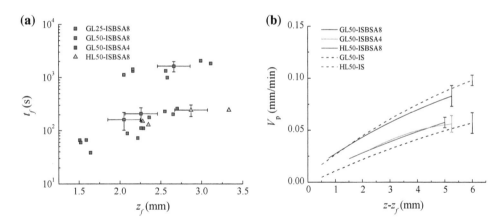

leading to an average propagation velocity $z_f/t_f \sim 0.95$ mm/min. We note that there are no significant differences between 4 and 8% BSA concentrations, which means that there is a saturation of BSA molecules. However, when C_G is reduced to 25%, t_f increases significantly to about 20 min, and the average propagation velocity reduces to $z_f/t_f \sim 0.11$ mm/min, which is an order of magnitude slower than for 50% mixtures. Histoacryl tends to create slightly larger plugs $z_f = 2.7 \pm 0.5$ mm, but the difference with Glubran 2 is not significant (Fig. 3a).

The second phase of polymerization is slow enough to allow us to measure the propagation velocity V_p of the reaction front [7]. As shown in Fig. 3b, V_p increases with z, because the reaction is exothermic. For all cases of glue-oil mixtures on contact with protein solutions, V_p varies between 0.02 and 0.08 mm/min, which is much smaller than z_f/t_f. The values and tendencies of V_p are comparable with those obtained for glue-oil mixtures on contact with pure IS (Fig. 3b). It should be pointed out that no slow polymerization phase is observed for a 25% G-L mixture on contact with either IS-BSA8 or IS. For pure Glubran 2 or Histoacryl polymerizing with IS-BSA8, both fast and slow polymerization reactions are observed. However, the fast phase is very quick and extends over a short distance $(z_f \sim 1$ mm,

$t_f \sim 5$ s), which makes it difficult to measure it with a good precision. In addition, the slow phase velocity is very difficult to assess from the image grey levels. Nevertheless, a glue column of 5 mm in height, completely polymerizes within 10 min.

The fast polymerization is triggered by the BSA molecules which have about 583 side chains of amino acids, and thus many possible sites for a zwitterionic polymerization [8, 9]. It is fast because the concentration of BSA molecules is high and thus provides a large number of potential initiation sites. This results in the formation of star polymeric structures, branching out from one BSA molecule [10], which would explain the compact structure of the fast reaction plug at the bottom of the tube. When the fast polymerization stops, the glue mixture above the plug still contains non-polymerized monomers. The slow polymerization is probably triggered by charges on the surface of the polymer in the plug. These charges lead to an anionic polymerization, similar to what is observed when the glue is put in contact with a pure ionic solution. The polymer structure is then composed of linear chains of monomers and is then less dense than the structure resulting from the zwitterionic polymerization. Increasing the oil concentration leads to an increase of the mean distance between the monomer molecules and thus of the chain formation time, as observed empirically [5]. The two tested nBCA glues differ only by the addition of the co-monomer metacryloxysulpholane (MS) to Glubran 2. The addition of MS does not modify significantly the polymerization phases and kinetics. It leads however, to a dispersion of results which is quite larger for Glubran 2 than for Histoacryl, as can be surmised from the relative size of the error bars in Fig. 3b.

4 Conclusions

The experimental setup we designed allows us to make a detailed analysis of the polymerization process when an nBCA glue, mixed with a radiopaque oil, is put in contact with a protein solution analogous to blood plasma. It thus allows to monitor what happens inside the bulk of a glue volume, when it is injected into blood. The main findings are that (i) the polymerization proceeds on two steps: a fast zwitterionic reaction leading to the formation of compact structures over a couple of minutes, is followed by a slow anionic reaction leading to less compact structures over tens of minutes; (ii) the addition of a radiopaque oil, which is necessary for intravascular applications, has a major effect on the reaction kinetics: the higher the oil concentration, the slower the polymerization; (iii) there is no significant difference between Histoacryl and Glubran 2, as regards to the polymerization kinetics. The typical time that it takes for Histoacryl and Glubran 2 to polymerize over a 1-mm thickness varies from 5 s for pure glue to about 1 min for a 50% glue concentration, and 10 min for a 25% glue mixture. Such information can help interventional radiologists understand the glue behavior upon injection, and thus control embolization.

Acknowledgements The authors thank the Chinese Scholarship Council, GEM S.r.l. and Guerbet S.A. for their support, and Dr. A. Fohlen for bringing the clinical perspective.

Conflict of Interest The authors declare that they have no conflict of interest.

References

1. Rosen R.J., Contractor S., Semin. Interven. Rad. 21, 59–66 (2004).
2. M. A. Lazzaro, A. Badruddin, O. O. Zaidat, Z. Darkhabani, D. J. Pandya, J. R. Lynch, Front. Neurol. 2, 64 (2011).
3. Bellemann, N., Stampfl, U., Sommer, C.M., Kauczor, H.U., Schemmer, P., Radeleff, B.A., Dig. Surg. 29, 236–242 (2012).
4. M. Leonardi, C. Barbara, L. Simonetti, R. Giardino, N. N. Aldini, M. Fini, L. Martini, L. Masetti, M. Joechler, F. Roncaroli, Interv. Neuroradiol. 8, 245–250 (2002).
5. C. Takasawa, K. Seiji, K. Matsunaga, T. Matsuhashi, M. Ohta, S. Shida, K. Takase, S. Takahashi, J. Vasc. Interv. Neuroradiol. 23, 1215–1221 (2012).
6. Li Y.J., Barthès-Biesel D., Salsac A.-V., J. Mech. Behav. Biom. 69, 307–317 (2017).
7. Li Y.J., Barthès-Biesel D., Salsac A.-V., J. Mech. Behav. Biom. 74, 84–92 (2017).
8. D. C. Pepper, Polymer J. 12, 629–637 (1980).
9. I. C. Eromosele, Micromol. Chem. Physics 190, 3085–3094 (1989).
10. Kim S., Evans K., Biswas A., Coll. Surf. B: Biointerfaces. 107, 68–75 (2013).

The Biomaterial Surface Nanoscaled Electrical Potential Promotes Osteogenesis of the Stromal Cell

Yuri Dekhtyar, Igor Khlusov, Yurii Sharkeev, Nataliya Polyaka, Vladimir Pichugin, Marina Khlusova, Fjodor Tjulkin, Viktorija Vendinya, Elena Legostaeva, and Larisa Litvinova

Abstract

The calcium phosphate coating was provided onto the titanium substrate because of the nanoarc coatings technology. Both surface morphology and electrical charge of the coating were measured at the nano/micro-scaled lateral resolution. The negative electrical potential was typical for sockets, however the positive one to the peaks of the roughness. The cells were mainly attached at the negatively charged sockets. The cells expressed both osteocalcin and alkaline phosphatase that are the osteoblastic molecular markers.

Keywords

Surface morphology • Surface electrical potential Human mesenchymal cells • Oeteogsteogenesis

1 Introduction

Stem cells are the fundamental units to produce/regenerate tissue. Therefore, a surfaces of the biomaterials, that are in use for implants, are mimicked to control the stem cells. To reach this a morphology of the surface and its physical/chemical properties are engineered. However, the surface is able to control the cell, if the latter is attached to the biomaterial, the surface properties influencing attachment of the cells are being studied intensively. A significant growth of the related publications started from 90th of the previous century and succeeded totality 1722 in January 2018 (SCOPUS, keywords: surface-influence-cell-attachment [1]).

The cell adheres to the implant surface via a specific proteins layer that coats an implant shortly after the implant is inserted into a living organism [2]. Fundamentally, adhesion of the protein molecule obeys the theory [3] that considers dispersion and electrostatic interactions between the molecule and the substrate. The dispersive interaction potentials decrease very fast in dependence on a distance ($\sim distance^{-6}$) [4] against the electrostatic potential ($\sim distance^{-1}$). Therefore, the latter is expected to deliver a stronger impact to trap an electrically charged or polarized "tail" of the adhering molecule. The research focusing the electrostatic factor significantly started from 2005 and currently is at the very beginning contributing just around 4% of the above number of the publications (SCOPUS, keywords: surface-electrical charge/potential-cell attachment [1]).

The present article identifies an influence of the widely exploited calcium phosphate surface having electrically charged structures on the mesenchymal stem cells (MSCs).

2 Materials and Methods

2.1 Preparation of the Specimens

Commercially pure titanium (99.58 Ti, 0.12 O, 0.18 Fe, 0.07 C, 0.04 N, 0.01 H wt%) plates ($10 \times 10 \times 1$ mm^3) were used as substrates for deposition of calcium phosphate coatings. The samples were cleaned ultrasonically by Elmasonic S10 (Elma Schmidbauer GmbH, Sigen, Germany) for 10 min in distilled water immediately before deposition. The coating was prepared in the anodal regime as described in [5]. An aqueous solution consisted of 20 mass% phosphoric acid, 6 mass% dissolved synthetic hydroxyapatite, and 9 mass% dissolved calcium carbonate was used to

Y. Dekhtyar (✉) · N. Polyaka · F. Tjulkin · V. Vendinya
Riga Technical University, Riga, Latvia
e-mail: jurijs.dehtjars@rtu.lv

I. Khlusov · M. Khlusova
Siberian State Medical University, Tomsk, Russia

Y. Sharkeev · E. Legostaeva
Institute of Strength Physics and Materials Science of SB RAS, Tomsk, Russia

V. Pichugin
National Research Tomsk Polytechnic University, Tomsk, Russia

L. Litvinova
Immanuel Kant Baltic Federal University, Kaliningrad, Russia

© Springer Nature Singapore Pte Ltd. 2019
L. Lhotska et al. (eds.), *World Congress on Medical Physics and Biomedical Engineering 2018*,
IFMBE Proceedings 68/3, https://doi.org/10.1007/978-981-10-9023-3_25

fabricate calcium phosphate. Micro-arc oxidation process was performed with initial current densities in the range of 0.2–0.25 A/cm2. Micro-arc parameters were: pulse frequency of 100 Hz, pulse duration of 100 μs, process duration in the range of 5–10 min; voltage 150–400 V. The fabricated specimens were dried in the dry-heat manner with Binder FD53 (Binder GmbH, Tuttlingen, Germany) at 453 K for 1 h.

2.2 Surface Characterization

Surface topography was analyzed with a scanning electron microscope (SEM) (AG-EVO® 50 Series (Carl-Zeiss, Oberkochen, Germany). The phase composition was determined by the X-ray diffraction technique (XRD, Bruker D8 Advance) as described elsewhere [6].

Atomic force microscopy (AFM) was applied in the contact mode to characterize the roughness of the samples on the nanoscale. The Solver–PRO47 microscope (NT-MDT Co., Zelenograd, Russia) was applied. The measured roughness was characterized with an average nanoroughness index R_{an} calculated by the by AFM software. Kelvin probe atomic force spectroscopy measurements were employed for the same area as the AFM ones to identify the surface electrical potential (V_k). The latter was calculated by the AFM software as the average one per each scannes area.

To characterize the electrical potential at the macro scale an electron work function (φ) of the specimen surface was measured because of the photoelectron emission detection in the vacuum conditions 10^{-4} Pa. The value of φ is a minimal energy provided to an electron to escape it from the solid. The value of φ increases, when the surface charge becomes more negative. To excite a photoemission current (I) the specimens were irradiated by a soft ultraviolet light beam (diameter 5 mm) increasing an energy of the photons from 3 to 6 eV. The value of φ was identified as the energy of the photons, when I = 0. The spectrometer was described in details in [7].

Before the biological testing, the samples were dry-heat sterilized as described above. The prenatal stromal cells from the human lung (HLPSCs) with CD34-CD44 + OCN (osteocalcin)- phenotype (Stem Cell Bank Ltd., Tomsk, Russia) were used to study the MSCs osteogenic differentiation and maturation induced by the calcium phosphate coatings. The details were described in [8]. The HLPSCs suspension was freshly prepared with a concentration of 3×104 viable karyocytes/mL of the following culture medium: 80% DMEM/F12 (1:1) (Gibco Life Technologies; Grand Island, NY, USA), 20% fetal bovine serum (Sigma-Aldrich, St. Louis, MO, USA), 50 mg/l gentamicin (Invitrogen, UK) and freshly added L-glutamine sterile solution in a final concentration of 280 mg/L

(Sigma-Aldrich). After the cells were thawed the viability of 92% of cells was identified with the ISO 10993-5 test; 0.4% trypan blue was used. Each specimen was placed in a separated plastic well of a 24-well plate (Orange Scientific, Belgium).

To identify the osteogenic potency of the surface, the culture medium was not saturated by osteogenic supplements such as β-glycerophosphate, dexamethasone, and ascorbic acid. The cell suspension was added in a volume of 1 mL per well. The cell culture was incubated for four days in a humidified atmosphere of 95% air and 5% CO2 at 37 °C.

The specimens with adherent stromal cells were continuously air-dried, fixed for 30 s in formalin vapor and stained with alkaline phosphatase (ALP). Naphtol AS-BI phosphate ($C18H15NO6P$, molecular weight (m.w.) 452.21) and fast garnet GBC salt ($C14H14N4O4S$, m.w. 334.35) (both from Lachema, Brno Czech Republic) were used. The brown sites of enzymatic activity served as cellular ALP staining criteria [8].

Some calcium phosphate coated specimens were fixed in formalin vapor as described above. Primary antibodies (rabbit polyclonal anti-human IgG (1:100), Epitomics Inc., Burlingame, CA, USA) to OCN, and universal immunoperoxidase antirabbit and antimouse polymer (Histofine Simple Stain MAX PO MULTY, Nichirei Biosciences Inc., Tokyo, Japan) were used as described previously [8]. The brown sites of colored cells served as the OCN staining criteria.

The morphometry method was used to recognize quantitative parameters of the cells [9]. The ImageJ 1.43 software (http://www.rsb.info.nih.gov/ij) was employed to process digital images of OCN or ALP stained cells. Ten randomly selected images were computed for each sample. The squares of stained cells as well squares of both cell-free and seeded by stained cells valleys and sockets were estimated in μm^2.

3 Results and Discussion

The prepared coatings were amorphous; the X-ray diffraction maxima corresponded to calcium phosphate and calcium oxide compounds, i.e. $CaTi_4(PO_4)_6$, β-$Ca_2P_2O_7$, TiP_2O_7, TiO_2 (anatase) were observed after annealing of the specimens at 1073 K for 1 h (Fig. 1).

The calcium phosphate coating had spherulite peaks sized at the baseline to 20–30 μm in a diameter and single and interconnected pores sized to 1–10 μm in a diameter revealed in both spherulites and valleys (Fig. 2a). The latter had sockets (Fig. 2b).

The AFM measurements (Fig. 2c) identified that the coating was superimposed with submicron particles of 500–

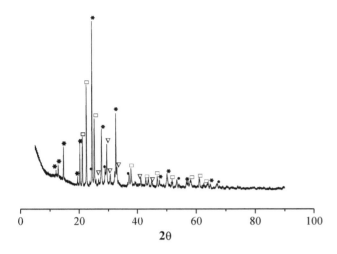

Fig. 1 X-ray diffraction spectrum of the calcium phosphate coating. Maximus corresponding to the compounds: *—$CaTi_4(PO_4)_6$, □—TiP_2O_7, ▽—β-$Ca_2P_2O_7$, ●—TiO_2(anatase)

1000 nm. The particles were assembled in globules, 1–2 μm in diameter with 30 nm height. The pores (of 1–2 μm diameter) were located in between the globules. The R_{an} was in a range 400–1300 nm for different specimens and correlated (Fig. 3) with φ (r = 0.74; significance 95%). This means that the nanoscaled roughness R_{an} had an influence on the electrical potential φ measured at the macro scale. On the other hand the potential V_k measured at the nanoscale was also connected with φ (r = 0.97; significance 95%; Fig. 3). This means that the roughness contributed also to the nanoscaled surface electrical potential. The potential V_k at the valleys was equal to −0.024 V that was the negative value (significance 95%) against the potential +0.021 V at the peaks.

The cells were not absorbed in the pores having the dimensions ∼1–10 μm that were less that MSCs sizes. MSCs adhered preferentially to the spherulites and valleys of the coating. However the HLPSCs located in the sockets belonged to the negatively charges valleys expressed ALP and OCN proteins as the markers of synthesizing osteoblasts and as the humoral component of the bone matrix. Approximately 84% of ALP- or OCN-positive cells were revealed in the surface sockets, but only 16% ones were placed on the spherulites.

The squares covered with the ALP- and OCN-stained cells correlated with the surrounding sockets area seeded by HLPSCs (r = 0.99 and 0.91; significance >99% for ALP and OCN, correspondingly). At the same time the total square of the sockets correlated with the electrical potential of the surface (r = −0.77; significance 98%). This meant that the electrical potential at the sockets had an influence on connected with them cells.

Fig. 2 Images of the coating surface. SEM—**a** (× 1100 magnification), **b** (× 8400 magnification); AFM—(**c**)

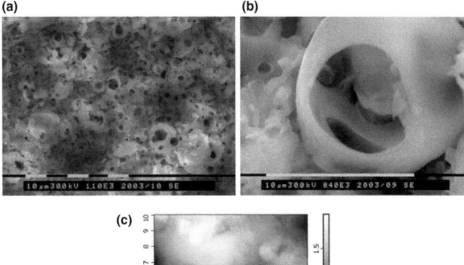

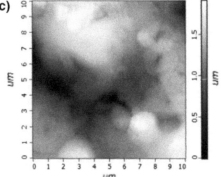

Fig. 3 The correlations of: φ with R_{an} (**a**) and V_k (**b**) of the coatings

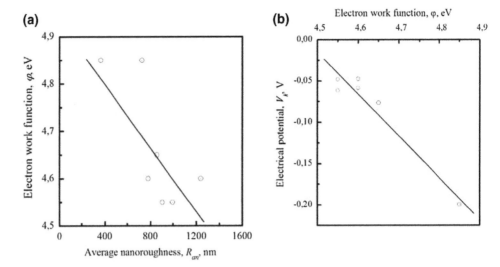

4 Conclusions

1. The calcium phosphate surface roughness has an influence on the surface electrical potential. The valleys having the sockets of the coating deliver the negative electrical potential.
2. The sockets of the calcium phosphate coating capture HLPSCs and promotes expression of ALP and OCN proteins as the markers of synthesizing osteoblasts and as the humoral component of the bone matrix.

Acknowledgements The authors wish to thank K.V. Zaitsev (Stem Cell Bank Ltd., Tomsk, Russia) for the cell line provision and Rafael Manory for assistance in preparing this article for submission. The work has been partially financially supported by the Russian Science Foundation, Project No. 16-15-10031 (in part of cell culturing in vitro).

Declaration of the Conflict Interest The authors do not have the conflict of interests.

References

1. SCOPUS, 2018.29.01.
2. B. D. Ratner et. al Biomaterials science, Academic press 1996, 484.
3. Derjaguin B.V., Landau L.D. Acta Physic Chimica, USSR, 1941, 14, 633–642.
4. London F., Properties' and application of molecular forces. – Ztschr. Phys. Chem., 1930, Bd 11, 222–251.
5. Sharkeev Yu.P., Legostaeva E.V., Eroshenko A.Yu., Khlusov I.A., Kashin O.A. The structure and physical and mechanical properties of a novel biocomposite material, nanostructured titanium-calcium -phosphate coating. Compos Interfac. 2009; 16: 535–46.
6. Legostaeva E.V., Kulyashova K.S., Komarova E.G., Epple M., Sharkeev Y.P., Khlusov I.A. Physical, chemical and biological properties of micro-arc deposited calcium phosphate coatings on titanium and zirconium-niobium alloy. Mat -wiss u Werkstofftech 2013; 44: 188–97. https://doi.org/10.1002/mawe.201300107.
7. Akmene R.J., Balodis A.J., Dekhtyar Yu.D., Markelova G.N., Matvejevs J.V., Rozenfelds L.B., Sagaloviąs G.L., Smirnovs J.S., Tolkaąovs A.A., Upmiņš A.I. Exoelectron emission specrometre complete set of surface local investigation. Poverhnost, Fizika, Himija, Mehanika (in Russian), 1993; 8:125–8.
8. Khlusov I.A., Shevtsova N.M., Khlusova M.Y. Detection in vitro and quantitative estimation of artificial microterritories which promote osteogenic differentiation and maturation of stromal stem cells. Methods Mol Biol. 2013; 1035:103–19. https://doi.org/10.1007/978-1-62703-508-8_9.
9. Culture of animal cells: A manual of basic techniques, 5th edition, Ed. R. Ian Freshney. -N.Y.: John Wiley & Sons, 2005, p. 580.

Removal of Vascular Calcification Inducer Phosphate in Different Dialysis Treatment Modalities

Jana Holmar, Ivo Fridolin, Merike Luman, Joachim Jankowski, Heidi Noels, Vera Jankowski, and Setareh Alampour-Rajabi

Abstract

Approximately 8–10% of the adult population in Europe suffers from kidney diseases. Cardiovascular complications are the leading cause of death in chronic kidney disease (CKD) patients, and vascular calcification is prevalent. High serum phosphate (P) level is a trigger of higher prevalence of vascular calcification in CKD patients. Phosphate is removed from the blood of end-stage renal disease (ESRD) patients regularly by extracorporeal renal replacement therapy, called dialysis. This paper aims to evaluate the calcification capability of CKD phosphate levels and compare the removal of phosphate during the different dialysis modalities. Human vascular smooth muscle cells and rat aortic rings were incubated in a medium containing CKD levels of phosphate. Both, calcium content measurements and histochemical staining proofed significantly increased calcification. Ten uremic patients, five males, and five females mean age 59 ± 16 years, were followed during 40 chronic midweek hemodialysis sessions. Four dialysis modalities with different settings were used once for each patient: hemodialysis (HD), high-flux hemodialysis (HF1, HF2) and postdilutional online hemodiafiltration (HDF). Total removed phosphate (TRP) was calculated by using phosphate concentration and the weight of the total spent dialysate collection. Phosphate reduction ratio (RR) was calculated by using patients' pre- and post-dialysis phosphate concentrations in serum. Patients' mean pre-dialysis serum phosphate levels were 1.72 ± 0.57 mmol/L, which is higher than in healthy subjects (0.81–1.45 mmol/L). Phosphate serum reduction ratios achieved during HD procedures were significantly lower from the ratios achieved during HDF and HF2 procedures. The mean total removed phosphate (TRP) values for HD were significantly lower than TRP values of other modalities (HF1, HF2, and HDF). Differences in removal values between HF1, HF2, and HDF were not significant. The results are indicating that phosphate levels presented in CKD increase vascular calcification and it is possible to remove phosphate more effectively by adjusting the dialysis treatment parameters.

Keywords

Vascular calcification • Phosphate • Dialysis adequacy

1 Introduction

Around 8–10% of the adult population suffers from kidney damage, and cardiovascular complications are the leading cause of death in chronic kidney disease (CKD) patients [1]. One of the serious and prevalent co-morbidities in CKD patients is vascular calcification (VC). The key trigger of higher prevalence of vascular calcification in CKD is high serum phosphate (P) level [2, 3]. It makes proper and sufficient removal of phosphate crucial for end-stage renal disease (ESRD) patients to prevent cardiovascular disease in this population. During the renal replacement therapy, called dialysis, phosphate is removed from the blood regularly. However, phosphate removal is more complicated than the removal of other small molecular weight uremic toxins, e.g. urea. One of the reasons is that phosphate is removed mainly from the plasma space during the dialysis while urea is also

J. Holmar (✉) · J. Jankowski · H. Noels · V. Jankowski
S. Alampour-Rajabi
RWTH Aachen University/University Hospital, Institute for
Molecular Cardiovascular Research, Aachen, Germany
e-mail: jholmar@ukaachen.de

J. Holmar · I. Fridolin
Department of Health Technologies, Tallinn University
of Technology, Tallinn, Estonia

M. Luman
North Estonian Medical Centre, Centre of Nephrology, Tallinn,
Estonia

J. Jankowski
CARIM School for Cardiovascular Diseases, University
of Maastricht, Maastricht, Netherlands

© Springer Nature Singapore Pte Ltd. 2019
L. Lhotska et al. (eds.), *World Congress on Medical Physics and Biomedical Engineering 2018*,
IFMBE Proceedings 68/3, https://doi.org/10.1007/978-981-10-9023-3_26

removed from red blood cell water [4]. Secondly, phosphate is negatively charged, and the fact that some dialysis membranes have a negative charge or they become negatively charged during dialysis (due to an accumulation of negatively charged proteins) makes phosphate diffusion across the membrane difficult [5]. It has demonstrated that increasing the dialysis time has a strong effect on phosphate removal, but also dialysis settings, i.e., filter type, blood- and dialysate flow rates are essential for the removal of phosphate [5]. Estimating the removed phosphate could be done by using serum samples before and after the dialysis, for elementary reasons it is not done in every session. Current guidelines are suggesting to perform the monitoring every 1–3 months [6]. Still, an easy to perform, an indirect optical method for estimating the phosphate concentration and removal during each dialysis has been proposed earlier [7–9]. These approaches are using UV- absorbance, and fluorescence of the spent dialysate and could simplify the estimation of treatments quality regarding phosphate removal.

The aim of the study was to evaluate the calcification capability of elevated phosphate levels presented in CKD and compare the removal of phosphate during the different dialysis modalities.

2 Materials and Methods

2.1 In Vitro Studies

Study protocols were approved by the local ethics committees.

Human aortic smooth muscle cells HAoSMCs (Promo-Cell; Germany) were cultivated in "Smooth Muscle Cell Growth Medium 2" (PromoCell; Germany) at 37 °C in a humidified atmosphere of 5% CO_2.

Cells in passage 5–8 were used in 3 independent experiments. Per experiment four wells (25,000 cells in each) of cells were used (N = 12).

Normal medium (NM) consisted of high glucose Dulbecco's Modified Eagle's Medium (DMEM) (Sigma–Aldrich, Germany) containing 0.9 mM phosphate, 1.8 mM calcium, 2.5% fetal bovine serum (Biochrom, Germany) and 1% penicillin-streptomycin 10,000 U/mL (ThermoFisher, Germany). In the calcifying medium (CM) phosphate was included to reach 1.6 mM concentration. The cells were cultured in NM and CM for 7 days; medium was changed every 2 days.

Calcium content was quantified using the o-cresolphthalein complexone method with the Randox Calcium Kit (Randox, UK) and was normalized to protein concentration measured using the Micro BCA Protein Assay Kit (ThermoFisher, Germany).

In addition, the thoracic aortas of 8–13 weeks old male Wistar-rats were isolated and cut in 1–2 mm rings, and the endothelial layer of the rings was damaged. Aortic rings were incubated in normal- and calcifying medium. Two independent experiments, in both four rings per condition, were used (N = 8). Incubation conditions and used mediums were the same as described in the previous section except for the phosphate concentration of CM which was fixed to 2.0 mmol/L. After 7 days, cultured rings were fixed with 4% paraformaldehyde, embedded in paraffin and 3 μm slices were stained using von Kossa method. The results were compared using paired two-tailed Student T-test. GraphPad Prism 5 (GraphPad Software, USA) was used for analysis. *p < 0.05 was considered statistically significant.

2.2 Dialysis Studies

Ten uremic patients (five males, five females) (mean age 59 ± 16 years) were followed during 40 chronic midweek hemodialysis sessions in North Estonian Medical Centre, Estonia. The study was performed after the approval of the protocol by the national ethics committee. The goal of the analysis was to compare the removal of phosphate during the different dialysis modalities. Four dialysis modalities with different settings were used once for each patient in a cross-over design (Table 1).

Patients' pre- and post-dialysis phosphate (P_{start} and P_{end}) and calcium level in the serum were measured to quantify the initial pre-dialysis serum concentrations and for calculating phosphate reduction ratios.

The outcome of the analysis was measured by comparing the phosphate removal values achieved by different dialysis treatments. During the dialysis sessions, spent dialysate was collected in the dialysate collection tank. At the end of the procedure the tank was weighed (W_{tank}) and one sample (P_{tank}) was taken from it after careful stirring. Concentrations of serum and dialysate samples were measured in a clinical chemistry laboratory (Synlab Eesti OÜ, Estonia). The total removed phosphate (TRP) was calculated by multiplying the weight of the tank (W_{tank}) and its phosphate concentration (P_{tank}).

$$TRP = P_{tank} * W_{tank} \tag{1}$$

The reduction ratio (RR) of phosphate was calculated based on the phosphate concentration in the patients' serum at the start (P_{start}) and the end (P_{end}) of the procedure.

$$RR = \frac{P_{start} - P_{end}}{P_{start}} * 100\% \tag{2}$$

For two procedures out of 40, RR was not possible to calculate due to missing P_{end} sample value.

Table 1 Dialysis treatment parameters

Treatment	Blood flow (ml/min)	Dialysate flow (ml/min)	Dialyser (Fresenius Medical Care, Germany)
Hemodialysis (HD)	300	500	FX8
High-flux hemodialysis (HF1)	300	800	FX1000
High-flux hemodialysis (HF2)	350	500	FX1000
Postdilutional online hemodiafiltration (HDF)	350	800	FX1000
All treatments lasted 240 min			

The linear correlation values (R) between pre-dialysis serum phosphate concentrations and TRP and RR values were calculated.

The TRP and RR values for different dialysis modalities were compared using 1-way ANOVA with Newman-Keuls multiple comparison test and paired two-tailed Student T-test. GraphPad Prism 5 (GraphPad Software, USA) was used for analysis. $*p < 0.05$ was considered statistically significant.

3 Results

Patients' mean pre-dialysis serum phosphate levels were 1.70 ± 0.58 mmol/L, the values were in the range of 0.61–2.72 mmol/L. In case of 7 patients out of ten, the pre-dialysis serum phosphate levels were higher than in healthy subjects (0.81–1.45 mmol/L) [10]. Patients' serum calcium levels were 2.34 ± 0.14 mmol/L, which is in the range of the values of the healthy subjects (2.1–2.6 mmol/L) [10].

Calcium content measurements in cultured HAoSMCs demonstrated significantly increased calcification in cells incubated in calcification medium (Fig. 1).

Further, von Kossa staining demonstrated increased calcification in aortic rings incubated in calcification medium (Fig. 2).

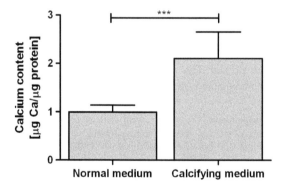

Fig. 1 Calcium content in human aortic smooth muscle cells incubated for 7 days in the normal and calcifying medium. ***($p < 0.001$); N = 12

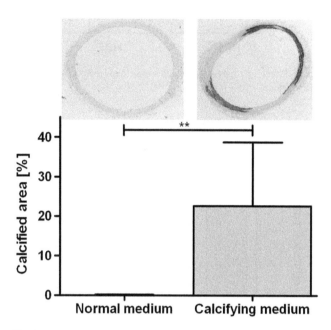

Fig. 2 Von Kossa staining and calcified area of rat aortic rings incubated for 7 days in the normal and calcifying medium. ** ($p < 0.01$); N = 8

The mean $\pm$ standard deviation (SD) of total removed phosphate (TRP) values for different dialysis modalities are presented in Fig. 3. Four dialysis modalities were compared: hemodialysis (HD), high-flux hemodialysis with different blood- and dialysate flow rates (HF1, HF2), and postdilutional online hemodiafiltration (HDF). The mean TRP for HD was 33.6 ± 9.9 mmol, which was significantly lower than TRP values achieved during HF1 (45.7 ± 15.3 mmol), HF2 (45.8 ± 11.4 mmol) and HDF (43.9 ± 8.1 mmol). TRP values of HF1, HF2, and HDF were not significantly different.

The linear correlation value between pre-dialysis serum phosphate concentration and TRP was 0.77.

The mean $\pm$ SD phosphate reduction ratios (RR) achieved by different dialysis modalities are presented in Fig. 4. The mean phosphate reduction ratio during HD procedures was $46.7 \pm 10\%$, which was significantly lower than the values achieved during HF2 ($62.0 \pm 16.9\%$) and HDF ($59.4 \pm 14.5\%$) but not from HF1 ($60.4 \pm 20.5\%$), reduction ratios of HF1, HF2, and HDF were not

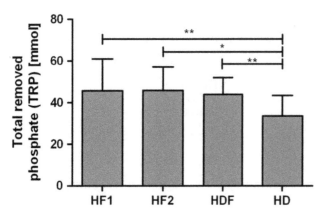

Fig. 3 Effect of dialysis modalities on total removed phosphate (TRP) amount. *($p < 0.05$); **($p < 0.01$); N = 40

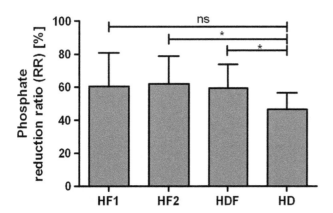

Fig. 4 Effect of dialysis modalities on phosphate reduction ratio (RR). *($p < 0.05$); N = 38

significantly different. The linear correlation value between pre-dialysis serum phosphate concentration and RR was 0.47.

4 Discussion and Conclusion

It has been demonstrated that CKD patients are more endangered for vascular calcification compared to the general population [11, 12]. Also, the vascular calcification inducing effect of phosphate has been shown previously [2, 3]. The latter was confirmed in the current paper: CKD levels of phosphate increased vascular calcification in human aortic smooth muscle cells and rat aorta (Figs. 1 and 2). Interestingly, phosphate is currently not listed as an uremic toxin in the European Work Group on Uremic Toxins (EUTox) database [13]. For 70% of patients included in this study, serum phosphate levels were elevated compared to healthy controls; these patients were also prescribed for phosphate-binders. Patients' serum calcium levels were in

the normal range for all patients. It shows that calcium levels are well maintained, but the effective removal of phosphate during dialysis is crucial for these patients.

It has been demonstrated that due to kinetic limitations, phosphate removal during dialysis differs from the removal of other small marker molecules, e.g. urea. There are two main reasons that do not allow very effective phosphate removal during dialysis. Firstly, the intradialytic phosphate concentration is low. Secondly, phosphate is negatively charged, and many dialyzer membranes have a negative charge originally or accumulate the coating of negatively charged proteins during dialysis. Prolonged dialysis or offering more frequent dialysis could help to remove larger amounts of phosphate [5].

Due to the above mentioned complex kinetics of phosphate removal, the reduction ratio is probably not the best way to estimate the efficiency of procedures regarding phosphate removal. Still, the results for both, reduction ratios and total removed amounts follow the same trend (Figs. 3 and 4). The both, RR and TRP values are presented in the paper since the RR is giving direct information of the phosphate removal from the blood side, and the TRP is giving the values for total removed phosphate from the spent dialysate side. Estimating total removed phosphate values should be preferred and optical methods could make the estimations easily feasible [7–9].

The results of the current paper show that total removed phosphate and phosphate reduction ratios are dependent on dialysis modalities. Total removed phosphate values for hemodialysis (HD) procedures were significantly lower than in case of other modalities (Fig. 3). Phosphate reduction ratios were also lowest for HD procedures (Fig. 4). Removal values between other modalities were not significantly different. It demonstrates the possibility to remove more phosphate during dialysis by using high flux filters and adjusting the dialysis treatment parameters.

The limitation of the study is a low number of monitored procedures, and follow-up studies are planned.

In summary, phosphate is an inducer of vascular calcification, and most of the studied CKD patients had elevated phosphate levels. This makes the effective removal of phosphate crucial. This study demonstrated the possibility to increase phosphate removal by dialysis settings.

Acknowledgements The authors want to thank all dialysis patients, technical assistant, and nurses. The research was funded partly by the Transregional Collaborative Research Center TRR219, Estonian Ministry of Education and Research under institutional research financing IUT 19-2; by Estonian-Norwegian cooperation programme Green Industry Innovation Estonia project EU 47112; the European Union through the European Regional Development Fund; by Estonian Centre of Excellence in IT (EXCITE) funded by European Regional Development Fund, and by Estonian Research Council grant PUTJD66.

Conflict of Interest The authors declare that they have no conflict of interest.

References

1. Webster, A.C., et al.: Chronic Kidney Disease. The Lancet 389 (10075), 1238–1252 (2017).
2. Gross, P., et al.: Vascular toxicity of phosphate in chronic kidney disease: beyond vascular calcification. Circ J 78(10), 2339–46 (2014).
3. Levin, N.W. and N.A. Hoenich: Consequences of hyperphosphatemia and elevated levels of the calcium-phosphorus product in dialysis patients. Curr Opin Nephrol Hypertens 10(5), 563–568 (2001).
4. Descombes, E., et al.: Diffusion kinetics in blood during haemodialysis and in vivo clearance of inorganic phosphate. Blood Purif 19(1), 4–9 (2001).
5. Daugirdas, J.T.: Removal of Phosphorus by Hemodialysis. Seminars in Dialysis 28(6), 620–623 (2015).
6. Kidney Disease: Improving Global Outcomes (KDIGO) CKD-MBD Update Work Group. KDIGO 2017 Clinical Practice Guideline Update for the Diagnosis, Evaluation, Prevention, and Treatment of Chronic Kidney Disease–Mineral and Bone Disorder (CKD-MBD). Kidney International Supplements 7(1), 1–59 (2017).
7. Enberg, P., et al.: Utilization of UV absorbance for estimation of phosphate elimination during hemodiafiltration. Nephron Clin Pract 121(1–2), c1-9 (2012).
8. Holmar, J., et al. Removal Estimation of Uremic CVD Marker Phosphate in Dialysis Using Spectrophoto-and Fluorimetrical Signals. In: H. Eskola, et al. (eds.) EMBEC & NBC 2017. Tampere, Finland 2017, vol. 65, pp. 358–361: Springer, Singapore (2018).
9. Holmar, J., et al.: An Optical Method for Serum Calcium and Phosphorus Level Assessment during Hemodialysis. Toxins 7(3), 719–727 (2015).
10. Kidney Disease: Improving Global Outcomes (KDIGO) CKD-MBD Work Group. Clinical practice guideline. Kidney Int Suppl. 76(113), S1-S130 (2009).
11. Covic, A., et al.: Vascular calcification in chronic kidney disease. Clin Sci (Lond) 119(3), 111–21 (2010).
12. Rattazzi, M., et al.: Aortic valve calcification in chronic kidney disease. Nephrol Dial Transplant 28(12), 2968–76 (2013).
13. The European Uremic Solutes Database (EUTox-db), http://www.uremic-toxins.org/DataBase.html, last accessed 2018/01/10.

DNA Intracellular Delivery into 3T3 Cell Line Using Fluorescence Magnetic Ferumoxide Nanoparticles

Ondrej Svoboda, Josef Skopalik, Larisa Baiazitova, Vratislav Cmiel, Tomas Potocnak, Ivo Provaznik, Zdenka Fohlerova, and Jaromir Hubalek

Abstract

Gene delivery is a widespread strategy in current experimental medicine. In this work, we report a method for low-toxic intracellular DNA vector delivery and post transfection localisation of this vector in mouse embryonic fibroblast cell lines. The surface of modified ferumoxide nanoparticles conjugated with Rhoda-mine B isothiocyanate (FeNV-Rh) was modified with linear polyethyleneimine and medium molecular weight chitosan to increase Accelerated Sensor of Action Potentials DNA vector adhesion. The size of the FeNV-Rh/DNA transfection complex was studied using dynamic light scattering (DLS) and scanning electron microscopy (SEM) techniques. The transfection complex internalisation of plasmid expression and FeNV-Rh, and stability of rhodamine fluorescence in intracellular space were observed at time periods 6, 12, 24 and 48 h post transfection. Results showed high transfection complex intracellular biocompatibility—cell viability after Rh-MNP labelling was higher than 97% 24 h after transfection, and higher than 95% after the next 24 h. Selective FeNV-Rh localisation in the lysosomes was quantified. More than 82% of nanoparticles were localised in the lysosomes 12 h post transfection and 94% of lysosomes had a significant and long-term deposit of nanoparticles. DNA vector expression was visible in >65% of the cells and precise protein localisation on the cell membrane was confirmed using confocal microscopy.

Keywords

DNA • Delivery • 3T3 cells • Fluorescence Ferumoxide magnetic nanoparticles • FeNV-Rh

1 Introduction

Genetically modified cells can express functional ion channels and produce growth factors, for example, as a new direction of cardiac therapy or express bioactive molecules such as gene therapy or tumour and inflammatory disease. In genetic modifications, transfection, which introduces foreign nucleic acid, is widely used. The very promising transfection approach for intracellular delivery of biomolecules is magnetofection. This method, based on the particle internalization principles formed in the 1970s was first described in 2000 [1] and is still not fully understood [2]. Magnetofection increases transfection efficiency and biocompatibility in comparison to the traditional transfection methods, such as lipofection or electroporation. It applies a magnetic field, which enhances the contact of the transfection complex with the cell membrane in combination with the low toxicity of the magnetic nanoparticle, given that iron oxide is biodegradable. Nanoparticles (NPs) can play the role of a "Trojan horse". They take on the role of components and the size tuneable delivery system of the molecule, and the parallel role of cell labelling.

Transfection efficiency enhancement has been the objective for the past few years [3]. However, the transfer of the transfection complex across the cell mem-brane, the NPs and plasmid distribution in intracellular organelles and the

O. Svoboda (✉) · J. Skopalik · L. Baiazitova · V. Cmiel
T. Potocnak · I. Provaznik
Faculty of Electrical Engineering and Communication,
Department of Biomedical Engineering, Brno University
of Technology, Brno, Czech Republic
e-mail: svobodao@feec.vutbr.cz

O. Svoboda · Z. Fohlerova · J. Hubalek
Central European Institute of Technology, Brno University
of Technology, Brno, Czech Republic

Z. Fohlerova · J. Hubalek
Faculty of Electrical Engineering and Communication,
Department of Microelectronics, Brno University of Technology,
Brno, Czech Republic

J. Hubalek
Faculty of Electrical Engineering and Communication,
Department of Microelectronics, SIX Centre, Brno University
of Technology, Brno, Czech Republic

release of plasmids from the NPs during the hours and days post-application needs to be studied carefully, because these have an impact on fine cell homeostasis [4] or proliferation [5].

In this work, we present the transfection method, which uses agent-coated ferumoxide magnetic nanoparticles conjugated with rhodamine B isothiocyanate for intracellular delivery of DNA plasmids. The results show spontaneous cell uptake of the transfection nanocomplex, with high transfection efficiency (82%), long-term biocompatibility (97% viable cells at 24 h post-transfection) and nanoparticle colocalization with lysosomes (91% of lysosomes had deposits of nanoparticles).

2 Materials and Methods

2.1 Cell Culture

Mouse embryonic fibroblast cell line (3T3) cells were cultured in high glucose Dulbec-co's Modified Eagle's Medium (DMEM; Sigma-Aldrich) containing 10% FBS (Sigma-Aldrich), 1% Penicillin/Streptomycin (Sigma-Aldrich), and 1% L-glutamine (Sigma-Aldrich) and at 37 °C, 5% CO2. Only cells with a low passage number (<20) were used for the experiments. Forty-eight hours before the transfection, the cells were seeded in an 8-well confocal plate (C8-1.5H-N, Cellvis) with a density of 5·103 cells/well and cultured to achieve optimal confluency of 50–60%.

2.2 Synthesis of Magnetic Nanoparticles

Ferumoxide nanoparticles (FeNV) were synthesised by borohydride reduction of ferric chloride $FeCl_3 \cdot 6H_2O$ (37 mmol; Sigma-Aldrich, USA) at room temperature. After the reduction reaction, the temperature of the mixture was increased to 100 °C and held constant for 2 h. Thereafter the surfaces of the bare FeNV nanoparticles were functionalised by Rhodamine B isothiocyanate (details in [8]) to obtain the final FeNV-Rh.

2.3 Transfection

The 3T3 cells were transfected using FeNV-Rh coated with linear polyethyleneimine (PEI; MAX 40 K, Polysciences) or medium weight chitosan (Sigma-Aldrich) and with conjugated Accelerated Sensor of Action Potentials 1 (ASAP1) DNA plasmids (pcDNA3.1/Puro-CAG-ASAP1, Addgene, 7.5 kpb). The three transfection complex combinations, with varying components, were established (Table 1). The transfection reagent concentrations were selected from routine protocol at the point of enhanced transfection efficiency.

The FeNV-Rh coating was made in 30 μl of serum-free DMEM by PEI or chitosan, with 20 min incubation at RT followed by the addition of the DNA plasmids and a further 20 min incubation at RT. The positive control consisted of a Matra-A reagent (IBA GmbH)/DNA complex incubated for 20 min at RT. The negative controls contained FeNV-Rh/PEI or FeNV-Rh/chitosan with 20 min incubation, or bare FeNV-Rh, PEI, chitosan and DNA. The final concentrations of FeNV-Rh, coating agents and DNA are listed in Table 1.

After the incubation, each sample was purified of free transfection components by magnetic field (10 s, 103mT) and an exchange of the solving agent. Thereafter, the complex was filled to 200 μl, with a complete culture media added to the cells in the 8-well confocal plate (8WP). The 8WP was then transferred to the magnet and kept for 20 min at 37 °C and 5% CO_2. Then, the 8WP was removed from the magnet (103mT) and incubated for 48 h.

2.4 Transfection Complex Characterization

Dynamic light scattering (DLS; Zetasizer Nano ZS, Malvern Instruments) was used for the size analysis of the FeNV-Rh transfections (I–III) and the negative control transfections (I–VI).

Scanning electron microscopy (SEM; Lyra 3, Tescan Orsay Holding) was used to analyse the FeNV-Rh size distribution.

2.5 Quantification of Nanoparticle Intracellular Infiltration, Plasmid Infiltration and Final DNA Transcription

A confocal laser scanning microscope, Leica TCS SP8 X, equipped with gateable hybrid and PMT detectors and a white light laser was used to take fluorescent images. The FeNV-Rh complex was detected using confocal setting 530/650-670 (exc./em.). GFP was detected at wavelengths 488/517 nm and DAPI at 405/461 nm (exc./em.).

Transfection efficiency. Transfection efficiency images were taken at 6, 12, 24 and 48 h post transfection. Transfection efficiency was analysed using the Matlab 2013b (Mathworks) script. The analysis was based on the comparison of the quantity of ASAP1 expressing cells (GFP), to the total quantity of cells (DAPI stained).

Additional analysis. After 12 h of FeNV-labelling, lysosomes were stained with high fluorescence specific marker LysoTracker® (exc./em = 490/530 nm) and the

Table 1 The working concentrations of transfection reagents

Type	Nanoparticles [μg·ml^{-1}]	Coating agent [μg·ml^{-1}]	DNA [ng·ml^{-1}]
Protocol (I)	6.00	–	2.20
Protocol (II)	6.00	3.00 PEI	2.20
Protocol (III)	6.00	6.00 Chitosan	2.20
Positive control	2.20 (Matra-A)	–	2.20
Negative control (I)	6.00	–	–
Negative control (II)	6.00	3.00 PEI	–
Negative control (III)	6.00	6.00 Chitosan	–
Negative control (IV)	–	3.00 PEI	–
Negative control (V)	–	6.00 Chitosan	–
Negative control (V)	–	–	2.20

same cells were scanned in the red spectrum of FeNV-Rh (exc./em = 530/650-670 nm). The colocalization of green and red pixel channels was quantified.

2.6 Viability

Cell viability was determined using the LIVE/DEAD Viability/Cytotoxicity Kit for detection of viable and dead cells (Molecular Probes) and a flow cytometry assay based on propidium iodide (PI) staining (BD FACSCanto).

3 Results and Discussion

Herein we present the results of using PEI or FeNV-Rh for intracellular delivery of the ASAP1 DNA vector. ASAP1 is a fluorescent membrane voltage sensor, with circularly permuted GFP. It was originally developed for encoding neural communication [6], though the ASAP1's application is much more extensive. It had already been expressed in Xenopus laevis oocytes [7] or HEK293 [6].

Results from the DLS technique showed a significant difference in complex size (p < 0.05) between all transfection protocol types (I–III) and the negative controls (I–III) in the cells. No fractions were detected in the negative transfections (IV–VI), as these fractions were under the DLS detection limit. The zeta potential measurements show significant changes from -32 ± 1.0 mV for bare FeNV-Rh

(Negative control I) to -40 ± 3.2 mV (Protocol I), -22 ± 2.5 mV (Protocol II), -25 ± 2.9 mV (Protocol III), -17 ± 1.7 mV (Negative control II) and -19 ± 2.3 mV (Negative control III). These results con-firm that the transfection nanocomplex contains the DNA vector. SEM analysis (Fig. 1b) of bare FeNV-Rh nanoparticles illustrated the size distribution of the nanoparticle clusters 30 min from sonication (30 mW, 3 s inter-pulse interval, 15 min) and confirmed the magnetic properties of the FeNV-Rh nanoparticles. The apparent polydispersity of FeNV-Rh on Fig. 1b is the effect of self-aggregation of nanoparticles; the polydispersity of bare FeNV-Rh was not observed (Fig. 1a) immediately after sonication.

The confocal analysis showed that $82 \pm 9\%$ of rhodamine positive pixels were localised in the lysosomes 12 h after the start of cell incubation with FeNV-Rh-DNA, and $94 \pm 4\%$ of lysosomes inside the cells have significant and long-term deposits of nanoparticles. The statistical results were computed from 500 randomly selected cells (Fig. 2).

Additional colocalization analysis showed that $91 \pm 7\%$ of "orange-red" pixels (FeNV-Rh-DNA) had incidence with the "green" pixels area (lysosomes) 12 h after the end of labelling. This confirms selective distribution of almost all the FeNV-Rh-DNA nanocomplex in lysosomes or early endosomes.

These results show that DNA plasmids and NPs are connected in first 12 h after transfer into the cells, and that the nanocomplex is mostly redistributed into the lysosomes. The exact mechanism of dissociation of FeNV-Rh and plasmids will be the focus of future studies. However, the GFP protein propagation in the cell membrane supports the argument that a minimal part of the complex can be dissociated in a specific lysosome microenvironment and expressed plasmid is functional and well transcribed (Fig. 3a). The best transfection efficiency occurred in protocol (II), in which the transfection complex containing PEI (free PEI was removed) was higher than 65%, with very precise ASAP1 transcription on the membrane (Fig. 3a). Study of cell viability using the LIVE/DEAD kit (up to 96 h) demonstrated a difference in cell viability (T-test, p < 0.05) between the control and the FeNV-Rh exposed cells (Fig. 3b). The cell viability determined by flow cytometry showed comparable results in the same period of time.

The presented method was focused on the preparation of FeNV-Rh-agent-DNA nanocomplex. The evaluation gives positive results about the colloidal stability of the nanocomplex in a water medium, confirms spontaneous cell uptake, and confirms excellent long-term biocompatibility of nanoparticles. The effectiveness of transfection (% of GFP positive cells from all rhodamine positive cells or all cells in culture) is now comparable with common techniques [3].

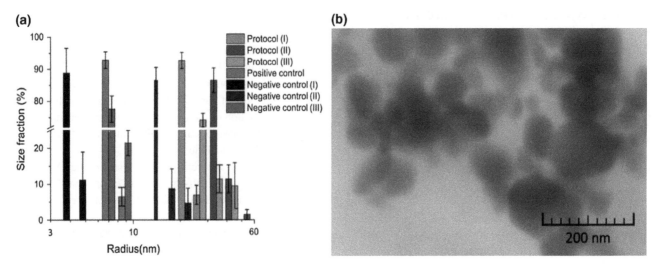

Fig. 1 FeNV-Rh and transfection complex size characteristic from DLS **a** (n = 3) and SEM technique (**b**). Scale bars as percentage fractional representation ± SD

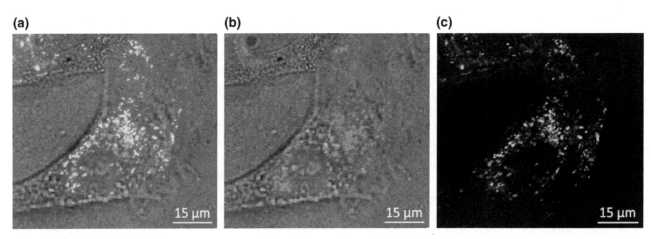

Fig. 2 **a** Intracellular distribution of FeNV-Rh; **b** lysosomes stained with LysoTracker® Green; **c** colocalization of lysosomes and intracellular distribution of FeNV-Rh in 3T3 cells (Color figure online)

Fig. 3 **a** Confirmation ASAP1 plasmid expression, localization in the 3T3 cells (GFP) and correlation with FeNV-Rh distribution; **b** cell viability determined using LIVE/DEAD. Black bars represent control samples, grey bars represent cells incubated with FeNV-Rh nanoparticles at the desired time (**b**)

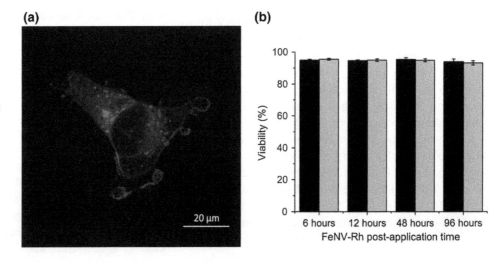

Future analysis will focus on improving nanocomplex incorporation and plasmid transcription in other types of cells (especially cardiomyocytes of animals, or patients with ion channel abnormalities, which should be the main target of future genetic medicine). The next analysis will be focused on enhancing transfection effectiveness (the nanocomplex metal core gives the opportunity of magneto-fection by static or oscillating magnetic fields). This "met-alofluorescence nanovector" gives potential to the wide-scale application for future clinical studies, including precise optical selection of cells before implantation, precise analysis of cell distribution in tissue, specific force manipulation and cell de-livery during cell therapy or gene therapy.

Acknowledgements Research described in this paper was financed by Czech Ministry of Education in frame of National Sustainability Program under grant LO1401. For research, infrastructure of the SIX Center was used.

Conflicts of Interest The authors declare that they have no conflict of interest.

References

1. Dobson J (2006) Gene therapy progress and prospects: magnetic nanoparticle-based gene delivery. Gene Ther 13:283–287. https://doi.org/10.1038/sj.gt.3302720.
2. Bregar VB, Lojk J, Šuštar V, et al (2013) Visualization of internalization of functionalized cobalt ferrite nanoparticles and their intracellular fate. Int J Nanomedicine 8:919–931.
3. Durán MC, Willenbrock S, Barchanski A, et al (2011) Comparison of nanoparticle-mediated transfection methods for DNA expression plasmids. J Nanobiotechnology 9:47.
4. Duan J, Yu Y, Yu Y, et al (2014) Silica nanoparticles enhance autophagic activity, disturb endothelial cell homeostasis and impair angiogenesis. Part Fibre Toxicol 11:50.
5. Popov A, R. Popova N, Selezneva I, et al (2016) Cerium oxide nanoparticles stimulate proliferation of primary mouse embryonic fibroblasts in vitroSt-Pierre F, Marshall JD, Yang Y, et al (2014) High-fidelity optical reporting of neuronal electrical activity with an ultrafast fluorescent voltage sensor. Nat Neurosci 17:884–9.
6. Lee EEL, Bezanilla F (2017) Biophysical Characterization of Genetically Encoded Voltage Sensor ASAP1: Dynamic Range Improvement. Biophys J 113:2178–2181.
7. Cmiel V, Skopalik J, et al (2017) Rhodamine bound maghemite as a long-term dual imaging nanoprobe of adipose tissue-derived mesenchymal stromal cells. Eur Biophys J 46:433–444.

Modern Semi-automatic Set-up for Testing Cell Migration with Impact for Therapy of Myocardial Infarction

Larisa Baiazitova⬤, Josef Skopalik, Vratislav Cmiel, Jiri Chmelik, Ondrej Svoboda, and Ivo Provaznik

Abstract

Ischemic heart disease and resulting acute myocardial infarction (AMI) is one of the main causes of morbidity and mortality in industrial countries. The idea for the modern therapeutic strategy, which should activate the migration of stem/progenitor cells or reduce the migration of inflammatory cells in AMI regions, has emerged in the last 15 years, mainly as a result of physiological observation and post-mortem histology. Published data from direct measurements of cell migration are very limited. We prepared a universal set-up that can be used for the testing of cell migration in AMI micro-environment. Mesenchymal stromal cells (MSCs), the most commonly used stem/progenitor cells in experimental cellular therapy for AMI, were used in the recent set-up tests. The cells, which should be tested for their migration potential, were injected into the starting point in a special micro-chamber on the substrate, and optics of the microscope allowed a time-lapse recording of cells in micrometre resolution every 2 min. Our software tools provided precise 2D and 3D tracking of moving cells and data export for statistical analysis. Set-up should be upgraded to a fully-automatic preclinical screening tool in the future.

Keywords

Patient screening • Stem cell analysis • Cell motility
Time-lapse image analysis

1 Introduction

Cell migration plays a major role in several physiological processes. Cell migration is affected by factors of the micro-environment, and by the actual state of the cell (affected by the age of the donor, genetic disorder, etc.). Historically, the first analysis and simulation of cell migration focused on tumour and epithelial cells [1], 'Transwell migration assay' and 'wound healing assay' are standard methods for quantification of these cell migrations. These assays need only a low-cost microscope and manual camera. However, there also exists another family of migratory cells: 'stem cells and progenitor cells'. The stem cells migrate commonly towards apoptotic or inflammatory regions [2]. Stem cell migration displays lower velocity and more complicated trajectories. Monitoring of these cells requires more precise time-lapse hardware and more sophisticated software for individual motility visualisation and quantification.

Currently, there are many algorithms for 2D and 3D cell segmentation, which can be implemented to analyse stem cell migration [3, 4]. Usually, the cells are considered either as a cluster or as separate objects. In the first case, direction of displacement of the cluster's centre of mass is detected [5]. In the second case, the migration of each cell is studied individually. An overview of some currently available software tools for the visualisation and validation of microscopic image data of migrating cells is presented in [6]. In our case, we programmed a custom algorithm in the MATLAB program environment for automatic cell segmentation and migration tracking using fluorescent images (more advantageous than phase contrast images).

L. Baiazitova (✉) · J. Skopalik · V. Cmiel · J. Chmelik
O. Svoboda · I. Provaznik
Faculty of Electrical Engineering and Communication,
Department of Biomedical Engineering,
Brno University of Technology, Brno, Czech Republic
e-mail: baiazitova@feec.vutbr.cz

O. Svoboda
Central European Institute of Technology,
Brno University of Technology, Brno, Czech Republic

© Springer Nature Singapore Pte Ltd. 2019
L. Lhotska et al. (eds.), *World Congress on Medical Physics and Biomedical Engineering 2018*,
IFMBE Proceedings 68/3, https://doi.org/10.1007/978-981-10-9023-3_28

2   Materials and Methods

2.1   Preparation of the Cells

MSCs were isolated, harvested, and cultured, as previously described [7]. Three donors aged 50–52 were selected. All collections of cells were performed after approval by the ethics committee and after gaining informed consent. Cells were cultivated at 37 °C and 5% CO_2 in medium IMDM with 1% Penicillin-Streptomycin (all Sigma-Aldrich) and with FBS (Invitrogen).

The migration experiment itself was started by removal of adherent MSCs by trypsin and by transfer of the cells into one inlet of a cultivating micro-chamber (Ibidi μ-Slide I Luer), which was intended to simulate a micro-area of tissue (details in Sect. 2.2). Sixty microliters of a 20,000 cell/ml solution were pipetted into the left inlet of the micro-chamber, and after 30 min, the micro-chamber was completely filled with the cultivation medium. The next day, the medium was exchanged (pure IMDM without FBS) and MSCs were labelled with CellTracker™ Green CMFDA fluorescent dye (Invitrogen). In the case of the advance chemotaxis experiment, the chemoattractant [5] was pipetted into the opposite inlet (Fig. 1).

2.2   Preparation of the Gas Control Chamber

It is necessary to observe the incubation conditions for long-term experiments. Our setup consisted of a cell micro-cultivation chamber (Ibidi μ-Slide I Luer) and a custom-made frame made of PLA thermoplastic using a 3D printer (FELIX 3.1) equipped with a transparent glass lid (Fig. 1). This chamber includes CO_2 and N_2 inlets connected to a control system (Arduino, Italy) and gas sensors (Tele-dyne, U.S.A.) that allow the possibility for precise settings of

different normoxic or hypoxic conditions, including the setting of the gas concentration oscillation in time. The levels of CO_2 and O_2 were set at 5 and 10% in our experiment.

2.3   Confocal Microscopy

The data were acquired using a Leica TCS SP8 X confocal microscope equipped with White Light Laser (WLL). The samples in the micro-chamber were observed using a lens with 10X magnification. MSCs were labelled with Cell-Tracker™ Green CMFDA fluorescent dye, which can be retained in living cells through several generations (72 h). Due to this property, we were able to observe the migration of cells continuously for several hours. Excitation wavelength was set to 490 nm and emission range to 500–540 nm, corresponding to the dye spectral properties.

To follow the MSC migration, we acquired 2D and 3D time-lapse data, which were obtained every 2 min over a period of 6 to 9 h. The size of the image stack object is 1.16×1.16 mm with a spatial resolution of 1024×1024 pixels and 3 μm depth. When scanning the micro-chamber temperature, CO_2 and N_2 levels were controlled during this period.

2.4   Data Analysis

Initially, our algorithm determined x- and y-coordinates of the centre of mass for each cell. If there were several slices on the z-axis, the image stacks at every hour interval were merged together as a sum of pixels. To reduce the noise in the background and preserve cell edges, a median filter was applied. Then, simple thresholding of a pre-processed image was applied, resulting in a binary image, where the

(a)

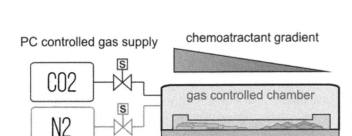

(b)

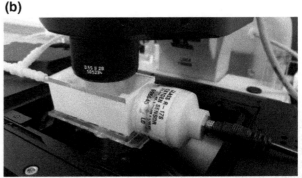

Fig. 1 a Schematic overview of the micro-chamber. MSCs are injected into the area at the right end of the chamber. In the basic variant of the experiment, cell motility can be monitored under absence of any chemoattractant (zero gradient). In the advanced variant, the

source of chemoattractant is added to the left end of chamber (blue spindle-shaped cells); after that, the cells move towards higher concentrations of chemoattractant. **b** Micro-chamber on microscope holder

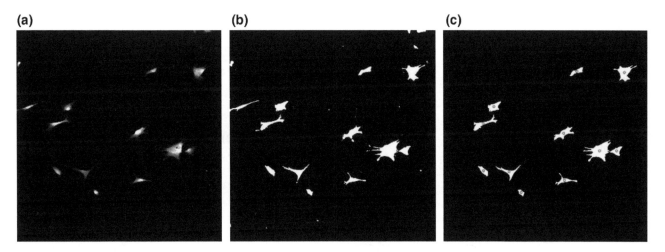

Fig. 2 Segmentation of cells. **a** Median filter applied to the merge image stack on the z-axis. **b** Simple thresholding. **c** Centre of mass (red) on the binary image with fill holes and removal of small objects

segmented cells were marked with a white colour on a black background. The image post-processing was done using the 'fill holes' morphological operation, followed by a morphological 'open' that separates potentially touching cells. Objects that were connected to the image border were excluded. Finally, small false positive objects were removed based on their area size, and the mass centres of the rest of cells were determined (Fig. 2).

This algorithm was used for image stacks in each time interval. Migration tracks were calculated from coordinates and then visualised as an additional layer in the original microscope image (Fig. 3). The tracks of cell division and the left side of the scan field were considered redundant and manually removed.

Our algorithm is suitable for low cell density. For high cell density, an alternative method of cell marker is desirable

because the cells can merge, causing complications in cell segmentation.

3 Results and Discussion

Our basic results after developing the final version of the micro-chamber were: (i) excellent stability of a focused micro-chamber region in the field of view of the microscope; (ii) stability of humidity and gas concentration in the micro-chamber is sufficient (short-time oscillation was less than 2% of set values, data not shown) for safe survival of the cells; (iii) viability of the cells reached 90–97% after 6 h of the migration experiment and their motility under constant external factors seemed constant during the whole time interval; (iv) fluorescence markers (CellTracker™ Green CMFDA) displayed minimal photo bleaching.

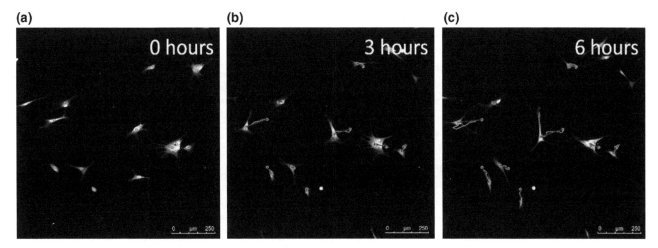

Fig. 3 The first and last image of the confocal image sequence were acquired by a Leica TCS SP8 X over 6 h. MSCs were labelled with CellTracker™ Green CMFDA fluorescent dye. **a** The first image with the starting point marked red. **b** and **c** The image after three and 6 h with migration track marked red (accumulated distance) and Euclidean distance marked blue

Table 1 Results from nine used tracks from one patient, from the duration of six hours

Measured values	Min	Max	Mean
Accumulated distance [μm]	48.2305	226.0950	126.2190
Euclidean distance [μm]	10.6604	131.5080	70.9598
Velocity [μm/min]	0.1340	0.6280	0.35061

Fig. 4 Angular diagram of vectors without (**a**) or with chemoattractant (**b**)

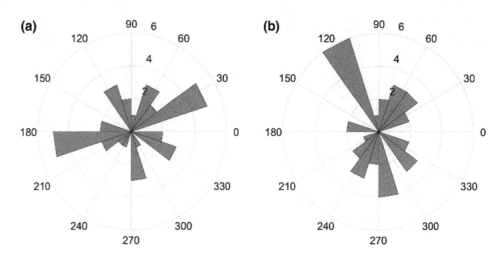

Further results gave quantification of velocity for MSCs in micro-environments without chemoattractant. At first, for quantification of cell motility, it was important to determine the parameters for our software, such as forward migration indices, displacement of centre of mass, and directness. For this, it was necessary to calculate Euclidean and accumulated distance [6].

The cells from all three donors (50–52 years old) showed very similar speeds of migration. The illustrative value of cell motility for one donor is shown in Table 1. Measured values for results from nine used tracks from one patient, for the duration of 6 h.

The quantification of different migration properties of MSCs from patients of different ages was not the aim of the present study. However, this research question will be important in our subsequent studies. The speed of migration, represented by Euclidean distance in Table 1, was approximately 12 μm/h. This speed demonstrated a good correlation with the speed 11 μm/h, which was quantified from a histological study of MSC migration in an animal heart [8].

In the advanced variant of migration experiment, the chemoattractant gradient was set within the micro-chamber (Fig. 1; details in Methods and [5]). The movement of the cells displayed no change in the mean velocity of cell migration, but there was change in the preferential direction of cell movement. Quantitative angular diagram of cell movements is displayed by software output in Fig. 4.

Recent all presented data was obtained from experiments on a polymer coverslip culture surface (bottom of micro-chamber). For a more precise simulation of cell migration in a real micro-environment, we would need to

upgrade the bottom of the surface, for example, to a collagen coating or a real heart collagen insert.

Modern therapeutic strategies using the stem/progenitor cell application and/or different stimulants for the migration of stem/progenitor cells is very limited because of the absence of objective and user-friendly methods for quantification of the migration potential of stem cells. This work has provided a presentation and evaluation of a newly developed complex setup that is suitable for measurement of the migration potential of cells, quantification of their trajectories with our software tools, and export of quantitative results in a user-friendly form. Our future aim is also to upgrade the method for testing the migration of several types of cells at the same time (MSCs, neutrophil, etc.).

Conflict of interest The authors declare that they have no conflicts of interest.

Ethical approval All procedures performed in studies involving human participants were in accordance with the ethical standards of the institutional and/or national research committee and with the 1964 Helsinki declaration and its later amendments or comparable ethical standards.

References

1. Bredin, C. G., Liu, Z., Hauzenberger, D. and Klominek, J.: Growth-factor-dependent mi-gration of human lung-cancer cells. Int. J. Cancer, 82(3), 338–345 (1999).
2. Skopalik, J., Pasek, M., Rychtarik, M.,et al.: Formation of Cell-To-Cell Connection between Bone Marrow Cells and Isolated Rat Cardiomyocytes in a Cocultivation Model. Journal of Cell Science & Therapy, 5(5), (2014).

3. Hodneland, E., Kögel, T., Frei, D. M., et al.: CellSegm-a MATLAB toolbox for high-throughput 3D cell segmentation. Source code for biology and medicine, 8(1), 16, (2013).
4. Wang, Y., Zhang, Z., Wang, H., and Bi, S.: Segmentation of the clustered cells with op-timized boundary detection in negative phase contrast images. PloS one, 10(6), e0130178. (2015).
5. Baiazitova, L., Skopalik, J., Cmiel, V., et al.: Characterization of cells migration through cardiac tissue using advanced micros-copy techniques and matlab simulation. In: Computing in Cardiology Conference (CinC), pp. 1125–1128, IEEE, Nice (2015).
6. Masuzzo, P., Huyck, L., Simiczyjew, A., et al.: An end-to-end software solution for the analysis of high-throughput single-cell migration data. Scientific Reports, 7, 42383 (2017).
7. Havrdova, M., Polakova, K., Skopalik, J., et al.: Field emission scanning electron microscopy (FE-SEM) as an approach for nanoparticle detection inside cells. *Micron*, *67*, 149–154 (2014).
8. Kim, Y. J., Huh, Y. M., Choe, K. O., et al.: In vivo magnetic resonance imaging of injected mesenchymal stem cells in rat myocardial infarction; simultaneous cell tracking and left ventricular function measurement. *The international journal of cardiovascular imaging*, *25*(1), 99–109 (2009).

Recent Research Progress on Scaffolds for Bone Repair and Regeneration

Stefano Nobile and Lucio Nobile

Abstract

Currently, the major areas of research in nanotechnology with potential implications in ostearticular regeneration are: nano-based scaffold construction and modification to enhance biocompatibility, mechanical stability, and cellular attachment/survival. Nanotechnologies can be used to form scaffolds and to deliver drugs and growth factors in the lesion site in order to enhance bone formation. The aim of this paper is to give an overview of some recent advances of osteoarticular tissue engineering allowed by the application of nanotechnologies.

Keywords

Nanothecnology • Bone tissue engineering
Cartilage • Scaffolds

1 Introduction

Recent advances in the research community on nanotechnology, nanomaterials and nanomechanics have stimulated research activities in medicine and bone tissue engineering devoted to their development and their applications. Applications of nanomaterials in medicine involve diagnostic and therapeutic processes for several diseases affecting different organs [1, 2].

Research on nanomaterials for medical applications has been carried out in the last 30 years; most studies focused on their safety and toxicity on human cell and tissue function. Several critical issues involve the ability to translate inorganic nanoparticle from academic studies to industrial scaling processes that comply with commercial quality systems, governmental standards, and regulatory contexts for human use. In particular, inorganic nanoparticle size and shape, their physicochemical properties and, most importantly, surface and interfacial properties in biological systems that result in formation of protein corona on particle surfaces are critical parameters to consider. In vitro tests may therefore provide only a partial indication of possible toxicity potential, compared to in vivo exposures. Animal models have been used for preclinical studies, and so far few nanodrugs have been approved for human use.

Tissue engineering is another field in which nanotechnologies appear to have a promising role. In particular, osteoarticular reconstruction following bone fracture is of interest due to the increasing number of elderly people and bone fractures requiring reconstruction with tissue transplants. Other potential applications are trauma, congenital bone malformations, osteoarticular diseases and tumor resections. Osteoarticular reconstruction requires three components: a biocompatible scaffold, cells replacing tissue and biochemical mediators (such as growth factors) to guide cell function. Nanotechnologies can be used to form scaffolds and to deliver drugs and growth factors in the lesion site in order to enhance bone/cartilage formation thanks to their high surface-to-volume ratio.

In this paper the recent research progress on scaffolds for bone repair and regeneration is reviewed, with particular reference to their mechanical properties.

2 Progress of Bone Tissue Engineering Scaffolds

The extracellular matrix of musculoskeletal tissue is mainly composed of collagens, elastin, proteoglycans, and glycoproteins; however, the structural organization of the extracellular matrix is highly specific in order to meet the tissue-specific function. Each component of the

S. Nobile
Neonatal Intensive Care Unit, "G. Salesi" Children's Hospital,
Via Corridoni 11, 60123 Ancona, Italy
e-mail: stefano.nobile@ospedaliriuniti.marche.it

L. Nobile (✉)
Department DICAM-University of Bologna-Campus of Cesena,
Via Cavalcavia 61, 47521 Cesena, Italy
e-mail: lucio.nobile@unibo.it

© Springer Nature Singapore Pte Ltd. 2019
L. Lhotska et al. (eds.), *World Congress on Medical Physics and Biomedical Engineering 2018*,
IFMBE Proceedings 68/3, https://doi.org/10.1007/978-981-10-9023-3_29

musculoskeletal system (such as bone, cartilage, meniscus, ligament, tendon, and muscles) has its specific architecture, composition, and functions.

Bone is a composite of collagen fibers reinforced with nanocrystals of calcium phosphate arranged in a semiregular pattern. Most of the organic matrix of bone is formed by collagen fibers of 50 nm size. Highly ordered carbonated apatite crystals are present at the nucleation sites on the collagen fibers. These mineralized collagen fibers are aligned and organized in different patterns to form different kinds of bone.

Cartilage is composed of specialized cells (chondrocytes) that produce collagenous extracellular matrix, rich in proteoglycan and elastin fibers, and does not contain blood vessels nor nerves. Cartilage is classified in three types, elastic cartilage, hyaline cartilage and fibrocartilage, which differ in relative amounts of collagen and proteoglycan.

Cartilage bears frictional, compressive, shear and tensile loading, and has limited repair capabilities mainly due to the lack of blood supply [3]. Since nanoparticles have significant limitations in bearing compressive loading, they have been mainly studied in cartilage pathology as signal delivery substrate to enhance chondrogenic differentiation of pluripotent cells [4].

A three-dimensional scaffold is an extracellular biodegradable matrix that guides tissue regeneration. An ideal scaffold should be both biocompatible and biodegradable, allowing for complete replacement with functional tissue. The structure, chemical composition, and biologic signals delivered by the scaffolds are key factors to guide the cellular behavior and to promote regeneration of tissues. Simple biodegradable polymeric materials or ceramics have been investigated as bone tissue engineering scaffolds; however, these materials have some limitations, including insufficient mechanical resistance. In contrast, nanoscale organic and inorganic materials incorporated into polymeric scaffolds may provide more favorable properties necessary for mechanical support, as well as release of bioactive elements and cellular adhesion, differentiation, and integration into the surrounding environment [5].

Some studies have demonstrated that tissue regeneration is faster in nanofibers than other scaffold types, such as solid wall (i.e. collagen) scaffolds [6]. Nanofiber scaffolds should have sufficient porosity to allow cell infiltration, nutrient transfer, blood supply, and the mechanical stability to sustain the neo-tissue formation.

Many approaches such as sacrificial fibers (selective leaching of co-electrospun fibers), low-density nanofibers, and use of salt or other particles that can be selectively dissolved, laser irradiation, or ultrasonication have been studied to enhance porosity, but the best approach needs to be tailored to the damaged tissue [7, 8]. Nanofibers mineralization by immersion in solutions containing calcium and phosphate or by loading with hydroxyapatite or polypeptide-polyester blends can be performed to mimic the chemical composition of biological tissues [9–11].

Among the scaffolds for cartilage and bone tissue engineering applications, injectable hydrogels have demonstrated great potential for use, owing to their high water content, similarity to the natural extracellular matrix, porous structure for cell transplantation and proliferation, and ability to repair irregular defects [12]. Natural biomaterial-based injectable hydrogels (i.e. chitosan, collagen/gelatin, alginate, fibrin, elastin, heparin, chondroitin sulfate, and hyaluronic acid) as well as synthetic biomaterials (mainly composed of polyethylene glycol) have been developed and recently reviewed in [12].

The process of scaffold construction is a complex issue. The main techniques available for the synthesis of nanofibers are electrospinning, self-assembly, and phase separation; other processes include meltblowing, flash spinning, bicomponent spinning, forcespinning, and drawing. In most of these processes, the fibers are collected as nonwoven random fiber mats known as nanowebs, consisting of fibers having diameters from several nanometers to hundreds of nanometers [13].

Of these, electrospinning is the most widely studied technique and also seems to exhibit the most promising results for tissue engineering applications thanks to the possibility to fabricate nanofibrous assemblies of various materials (i.e. polymers, ceramics and metals) with possible control of the fiber fineness, surface morphology, orientation and cross-sectional configuration. Moreover, limited reports about self-assembly and phase separation synthesis of nanofibers have been published [14]. Electrospinning involves converting a polymer into a viscous solution with the addition of a solvent, charging the polymer with a high voltage source to create a Taylor cone, extending the polymer into a thin jet stream across an electrostatic field, and then collecting the produced nanofibers on a grounded collector [15]. Electrically conductive nanofibers may provide topographical signals that facilitate scaffold degradation while simultaneously guiding skeletal tissue repair. These scaffolds can stimulate osteogenesis and trigger little immunogenic response. The process of fabricating scaffolds with electrospinning has seen many innovations. There have been new techniques to overcome the hydrophobic nature of polymeric materials using surfactants, new structural formations like that of cotton wool, new composite combinations like chitosan and silk fibroin, and new confirmatory studies of the bone stimulating effects of known pro-osteogenic proteins [16, 17]. Amiri et al. [18] showed that the combination of willemite (Zn_2SiO_4) nanoparticles and electrospun fibers is able to provide a suitable and efficient matrix to support stem cells differentiation for bone tissue engineering applications.

Traditional tissue engineering methods use a "top-down" approach, in which cells are seeded onto a scaffold with biocompatible and biodegradable properties, and are expected to populate in the scaffold and create their own extracellular matrix. However, the fabrication of complex larger functional tissues (i.e. liver, kidney) with high cell densities and complex metabolic activity is still difficult to achieve, mainly because of the limited diffusion properties of biomimetic scaffolds. An emerging, alternative "bottom-up" method to face this problem focuses on the fabrication of microscale tissue building blocks with a specific microarchitecture and assembling these units to engineer larger tissue constructs from the bottom up. Fabrication of tissue building blocks can be achieved via multiple approaches, including cell-encapsulating microscale hydrogels (microgels), self-assembled cell aggregation, generation of cell sheets, and direct printing of cells [19].

Some authors reported the possibility of integrating three different techniques (sponge replica method, freeze-drying and electrospinning) for a successful scaffold fabrication with potential application in osteochondral tissue engineering [20].

Silva et al. [21] developed honeycomb-like scaffolds by combining poly (D, L-lactic acid) with a high amount of graphene/multi-walled carbon nanotube oxides, and performed in vitro and in vivo tests in rats showing high mechanical performance and promotion of osteogenesis without toxicity. Radha et al. [22] reported on the preparation of nano-hydroxyapatite incorporated poly(methylmethacrylate) scaffolds by conjugated thermal induced phase separation and wet-chemical approach, with resulting enhancement of mechanical and biological properties of the scaffolds. Dong and colleagues [23] developed microporous graphene oxide (GO) modified titanate nanowire scaffolds with tunable mechanical properties prepared through a simple hydrothermal process followed by electrochemical deposition of GO nanosheets. In vitro tests showed that these scaffolds, in particular the one terminated with -OH groups, had improved cell viability, and proliferation, differentiation and osteogenic activities.

A critical factor influencing cell differentiation and proliferation in the context of artificial scaffolds is oxygen and metabolites supply for the differentiating cells. Vessel sprouting and growth can be promoted with the aid of recombinant vascular inductive growth factors such as vascular endothelial growth factor or angiopoietins. Angiogenic factors in combinations with biocompatible materials or to make biofunctionalized scaffolds are widely tested in preclinical model systems. Schimke et al. [24] tested nano-diamond particles bound to angiogenic factors and found significantly increased rates of angiogenesis one month after implantation in an in vitro model.

In another study, Sagar et al. [25] developed a hybrid lyophilized polymer composite blend of anionic charged sodium salt of carboxymethyl chitin and gelatin reinforced with nano-rod agglomerated hydroxyapatite with enhanced biocompatibility and tunable elasticity. In a rabbit model, this nanosystem displayed improved activity of bone regeneration in comparison to self-healing of control groups.

Another useful technique to enhance scaffold biocompatibility and osteoinductivity is osteoblast conditioning of nano-hydroxyapatite/gelatin scaffolds, as shown by Samadikuchaksaraei et al. [26] in a rat model.

Other preclinical studies showed further techniques to enhance scaffold osteogenicity: Çakmak et al. [27] described a co-culture model based on a trilayered silk fibroin-peptide amphiphile scaffold cultured with human articular chondrocytes and human bone marrow mesenchymal stem cells in an osteochondral cocktail medium and found in vitro evidence of osteogenic differentiation. Dodel and colleagues [28] used the electrospinning technique and vapor phase polymerization combination method with freeze-drying to produce a ligament construct of silk fibroin/PEDOT/Chitosan nanocomposite scaffold, which was coated with chitosan. Somatic human stem cells were cultured on the scaffold and underwent electrical stimulation with resulting facilitation of cell seeding and promotion of cell proliferation and differentiation.

Tracheal cartilage injuries, either congenital or acquired, are other potential targets for tissue engineering. Wang et al. [29] fabricated a core-shell nanofibrous scaffold to encapsulate bovine serum albumin plus a growth factor (recombinant human transforming growth factor-β3) into the core of the nanofibers for tracheal cartilage regeneration. This scaffold promoted the chondrogenic differentiation ability of mesenchymal stems cells derived from Wharton's jelly of human umbilical cord and could be useful for tracheal repair. In a pilot preclinical study, Maughan et al. [30] compared tracheal graft scaffolds to synthetic nanoscaffolds and discussed pros and cons of each material for tracheal replacement in children and adults.

In conclusion, the application of nanotechnologies for the repair of osteoarticular system diseases is showing promising results. The simultaneous demonstration of low toxicity and good efficacy will enhance the clinical use of nanoparticles in the next future.

Conflict of Interest The authors declare that they have no conflict of interest.

References

1. Nobile L and Nobile S: Recent advances of nanotechnology in medicine and engineering. AIP Conference Proceedings 1736 (1), 020058/1-020058/4(2016).
2. Nobile S and Nobile L.: Nanotechnology for biomedical applicaitons: recent advances in neurosciences and bone tissue engineering, Polym Eng Sci 57, 644–650(2017).

3. Sophia Fox AJ, Bedi A, Rodeo SA: The basic science of articular cartilage: structure, composition, and function. Sports Health 1(6), 461–8(2009).

4. Toyokawa N et al.: Electrospun synthetic polymer scaffold for cartilage repair without cultured cells in an animal model. Arthroscopy 26(3), 375–83(2010).

5. Walmsley GG, McArdle A, Tevlin R, Momeni A, Atashroo D, Hu MS, Feroze AH, Wong VW, Lorenz PH, Longaker MT, Wan DC: Nanotechnology in bone tissue engineering. Nanomedicine 11 (5), 1253–63(2015).

6. Woo KM, Chen VJ, Jung HM, Kim TI, Shin HI, Baek JH, Ryoo HM, Ma PX: Comparative evaluation of nanofibrous scaffolding for bone regeneration in critical-size calvarial defects. Tissue Eng Part A 15(8), 2155–62(2009).

7. Wang, Y.Z., Wang, B.C., Wang, G.X., Yin, T.Y., Yu, Q.S: A novel method for preparing electrospun fibers with nano-/ micro-scale porous structures. Polym Bull 63:259–265(2009).

8. Baker BM, Shah RP, Silverstein AM, Esterhai JL, Burdick JA, Mauck RL: Sacrificial nanofibrous composites provide instruction without impediment and enable functional tissue formation. Proc Natl Acad Sci U S A. 109(35), 14176–14181(2012).

9. Wang P, Zhao L, Liu J, Weir MD, Zhou X, Xu HH: Bone tissue engineering via nanostructured calcium phosphate biomaterials and stem cells. Bone Res. 30;2:14017, 1–13(2014).

10. Lu LX, Zhang XF, Wang YY, Ortiz L, Mao X, Jiang ZL, Xiao ZD, Huang NP: Effects of hydroxyapatite-containing composite nanofibers on osteogenesis of mesenchymal stem cells in vitro and bone regeneration in vivo. ACS Appl Mater Interfaces 5(2), 319–330(2013).

11. Wang J, Valmikinathan CM, Liu W, Laurencin CT, Yu X: Spiral-structured, nanofibrous, 3D scaffolds for bone tissue engineering. J Biomed Mater Res A 93(2), 753–62(2010).

12. Liu M, Zeng X, Ma C, Yi H, Ali Z, Mou XB, Li S, DengY, He NY: Injectable hydrogels for cartilage and bone tissue engineering. Bone Res 5,17014(2017).

13. Leach MK, Feng ZQ, Tuck SJ, Corey JM: Electrospinning fundamentals: optimizing solution and apparatus parameters. J Vis Exp 21(47), 2494(2011).

14. Kuo YC, Hung SC, Hsu SH: The effect of elastic biodegradable polyurethane electrospun nanofibers on the differentiation of mesenchymal stem cells. Colloids Surf B Biointerfaces.122, 414–422(2014).

15. Poologasundarampillai G et al.: Cotton-wool-like bioactive glasses for bone regeneration. Acta Biomater 10(8), 3733–46(2014).

16. Lai GJ, Shalumon KT, Chen SH, Chen JP: Composite chitosan/silk fibroin nanofibers for modulation of osteogenic differentiation and proliferation of human mesenchymal stem cells. Carbohydr Polym 111, 288–97(2014).

17. Kim BR, Nguyen TB, Min YK, Lee BT: In vitro and in vivo studies of BMP-2-loaded PCL-gelatin-BCP electrospun scaffolds. Tissue Eng Part A 20(23–24), 3279–89(2014).

18. Amiri B, Ghollasi M, Shahrousvand M, Kamali M, Salimi A: Osteoblast differentiation of mesenchymal stem cells on modified PES-PEG electrospun fibrous composites loaded with Zn2SiO4 bioceramic nanoparticles. Differentiation 92(4), 148–158(2016).

19. van de Weert M, Hennink WE, Jiskoot W: Protein instability in poly(lactic-co-glycolic acid) microparticles. Pharm Res 17(10), 1159–67(2000).

20. Nie H, Ho ML, Wang CK, Wang CH, Fu YC: BMP-2 plasmid loaded PLGA/HAp composite scaffolds for treatment of bone defects in nude mice. Biomaterials 30(5):892–901(2009).

21. Silva E et al.: PDLLA honeycomb-like scaffolds with a high loading of superhydrophilic graphene/multi-walled carbon nanotubes promote osteoblast in vitro functions and guided in vivo bone regeneration. Mater Sci Eng C Mater Biol Appl 73, 31–39 (2017).

22. Radha G, Balakumar S, Balaji Venkatesan, Elangovan Vellaichamy: A novel nano-hydroxyapatite — PMMA hybrid scaffolds adopted by conjugated thermal induced phase separation (TIPS) and wet-chemical approach: Analysis of its mechanical and biological properties. Materials Science and Engineering: C 75, 221–228(2017).

23. Dong W, Hou L, Li T, Gong Z, Huang H, Wang G, Chen X, Li X: A dual role of graphene oxide sheet deposition on titanate nanowire scaffolds for osteo-implantation: Mechanical hardener and surface activity regulator. Scientific Reports 5, Article number: 18266 (2015).

24. Schimke NM et al.: Biofunctionalization of scaffold material with nano-scaled diamond particles physisorbed with angiogenic factors enhances vessel growth after implantation.Nanomed.: Nanotechnol Biol Med 12 (3), 823–833(2016).

25. Sagar N et al.: Bioconductive 3D nano-composite constructs with tunable elasticity to initiate stem cell growth and induce bone mineralization. Materials Science and Engineering: C 69, 700–714 (2016).

26. Samadikuchaksaraei A et al.: Fabrication and in vivo evaluation of an osteoblast-conditioned nano-hydroxyapatite/gelatin composite scaffold for bone tissue regeneration. J Biomed Mater Res A. 104 (8), 2001–10(2016).

27. Çakmak S, Çakmak AS, Kaplan DL, Gümüşderelioğlu M: A Silk Fibroin and Peptide Amphiphile-Based Co-Culture Model for Osteochondral Tissue Engineering. Macromol Biosci 16(8), 1212– 26(2016).

28. Dodel M et al.: Electrical stimulation of somatic human stem cells mediated by composite containing conductive nanofibers for ligament regeneration. Biologicals 46, 99–107(2017).

29. Wang J et al.: Evaluation of the potential of rhTGF- β3 encapsulated P(LLA-CL)/collagen nanofibers for tracheal cartilage regeneration using mesenchymal stems cells derived from Wharton's jelly of human umbilical cord. Mater Sci Eng C Mater Biol Appl 70, 637–645(2017).

30. Maughan EF et al.:A comparison of tracheal scaffold strategies for pediatric transplantation in a rabbit model. Laryngoscope 127(12), E449–E457(2017).

μCT Based Characterization of Biomaterial Scaffold Microstructure Under Compression

Markus Hannula, Nathaniel Narra, Kaarlo Paakinaho, Anne-Marie Haaparanta, Minna Kellomäki, and Jari Hyttinen

Abstract

Scaffolds are often designed with progressive degradation to make way for cell proliferation of seeded cells for native tissue. The viability of the scaffold has been shown to depend on, among other things, the microstructure. Common parameters, that are used to describe microstructure, are porosity, material thickness, pore size and surface area. These properties quantify the suitability of the scaffold as a substrate for cell adhesion, fluid exchange and nutrient transfer. Bone and cartilage scaffolds are often placed or operated under loads (predominantly compression). This can alter the structural parameters depending on the stiffness of the scaffold and applied deformation. It is important to know, how scaffolds' parameters change under deformation. In this study, two scaffolds (PLCL-TCP and collagen-PLA) intended for use in bone and cartilage applications, were studied through micro computed tomography based imaging and in situ mechanical testing. The scaffolds were subjected to uniaxial compressive deformation up to 50% of the original size. The corresponding changes in the individual scaffold bulk characteristics were analyzed. Our results show an expected decrease in porosity with increasing deformation (with PLCL-TCP scaffold 52% deformation resulted in 56% decrease in porosity). Especially in the sandwich constructs of collagen-PLA, but also in PLCL-TCP composites, it was evident that different materials are affected differently which may be of significance in applications with mechanical loading. Our results are a step towards understanding the changes in the structure of these scaffolds under loading.

Keywords

X-ray microtomography • Biomaterials • Compression In situ imaging • Porosity

1 Introduction

Scaffolds are usually created for inducing cell proliferation with progressive degradation to make way for native tissue. In bone applications, this hold particularly true.

Porosity, pore size and material distribution are convenient features to gauge the suitability of a particular scaffold structure for its intended purpose. Other features include material and pore size thickness. In case of elastic materials, changes in porosity or surface areas, localized strains which can have a bearing on cell proliferation. In case of plastic deformation or fracture, changes in the mechanical rigidity of the overall structure can occur.

Micro computed tomography (μCT) based study are prevalent, in particular for in vitro experimental and small animal studies. In situ loading devices provide the response of static mechanical testing—compression, deformation, tensile test etc. With μCT it is possible to monitor the internal microstructure of these constructs and validate their behavior with respect to the required ideal under loading.

This study presents a preliminary application of the concept in observing the changes within two foamed PLCL based and one collagen-PLA scaffolds under deformation. Foamed PLCL-TCP scaffolds are interesting under compression due to the TCP component. Pure PLCL scaffold will recover from the compression and permanent changes are minor. On the other hand, TCP particles in the composite can break the PLCL structure that can modify the pore size distribution.

M. Hannula (✉) · N. Narra · K. Paakinaho · A.-M. Haaparanta
M. Kellomäki · J. Hyttinen
BioMediTech Institute and Faculty of Biomedical Sciences and Engineering, Tampere University of Technology, Tampere, Finland
e-mail: markus.hannula@tut.fi

N. Narra
Division of Biomedical Engineering, Faculty of Health Sciences, University of Cape Town, Cape Town, South Africa

© Springer Nature Singapore Pte Ltd. 2019
L. Lhotska et al. (eds.), *World Congress on Medical Physics and Biomedical Engineering 2018*,
IFMBE Proceedings 68/3, https://doi.org/10.1007/978-981-10-9023-3_30

2 Materials and Methods

2.1 Sample Details

2.1.1 PLCL-TCP Scaffold

The composites were manufactured by melt-mixing PLCL-polymer (70L/30CL; Purasorb PLC7015, Corbion Purac Biomaterials, Gorinchem, The Netherlands) with 60 wt% of β-tricalcium phosphate (β-TCP) having the particle size range between 100 and 300 μm (Plasma Biotal Ltd., Buxton, United Kingdom). The formed composites were foamed using supercritical carbon dioxide ($ScCO_2$) in a mold assembly. Thereafter, the foamed composite blocks were cut into cylindrical shape (Ø = 10 mm, h = 4.7 mm), and used as such in the deformation.

2.1.2 Collagen-PLA Composite

The collagen-PLA composite samples were manufactured as described in [1]. Briefly, type I bovine dermal collagen (PureCol®, Nutacon B.V., Leimuiden, the Netherlands) was gelled into collagen gel with concentration of 0.5 wt%. Medical grade polymer poly(L/D)lactide 96/4 (Purac Biochem, Gorinchem, The Netherlands) was used for fiber manufacture. The polymer was melt-spun into fibers (Ø ~ 20 μm), using a Gimac microextruder (Gimac, Gastronno, Italy). The fibers were cut to staple fibers (length of ~ 10 cm), and carded and needle punched into felt. The felt was then cut with a puncher (Ø = 8 mm). The collagen solution was loaded into Teflon molds (Ø = 8 mm, h = 4 mm) together with the felts (felts at the bottom and at the top of the molds). The samples were frozen for 24 h at −30 °C prior to freeze-drying for 24 h. The samples were cross-linked with 95% ethanol solution with 14 mM EDC (N-(3-Dimethylaminopropyl)-N'-ethylcarbodiimide hydrochloride, Sigma-Aldrich, Helsinki, Finland) and 6 mM NHS (N-Hydroxysuccinimide, Sigma-Aldrich, Helsinki, Finland) for 4 h RT. The samples were washed with deioniced water and re-freeze-dried as described earlier.

2.2 Compression and Imaging

Due to differences in the rigidity of the samples, different approaches were used for the sample compression. The PLCL-TCP scaffold was significantly more rigid than the collagen-PLA scaffold. Thus relatively greater force had to be applied to achieve and hold the deformation. The collagen-PLA scaffold, on the other hand, was pliable and very low force was required to induce and hold deformation. The deformation magnitudes for both the samples were chosen arbitrarily. The samples were first deformed and then

the magnitude calculated from observing the difference in sample in the corresponding projection images.

The PLCL-TCP sample was compressed with a mechanical compression device designed to be operated in situ in the μCT device [2]. The sample was placed within a polycarbonate tube, between opposing pistons. The compression of a sample was performed by mechanically moving the top piston with the stationary bottom piston acting as a sample stage. The top piston assembly (80 gms) was screwed-in until it was resting on the sample and the position was fixed, thus providing an initial static loading force in the range of 0.8 ± 0.1 N. The applied forces were not monitored during the deformation process. The sample was compressed at three levels—13% (deformation: 1.3 mm), 35% (deformation: 3.5 mm) and 52% (deformation: 5.2 mm).

The collagen-PLA sample was compressed in a syringe (diameter: 4.6 mm) between two pistons. In order to get a better image quality the sample was cut to 3.6 mm diameter to fit in a smaller syringe. The sample height was 3 mm and it was compressed by 33% (deformation: 1 mm).

Compression of the both samples is shown in Fig. 1.

The imaging procedure for both the samples was such that pre and post compression tomographic image volumes were obtained with a Zeiss high-resolution μCT device (Xradia MicroXCT-400, Zeiss, Pleasanton, CA, USA). Thus the samples were first placed in their respective sample holders between the opposing pistons and image data collected with zero deformation. Subsequently, they were deformed in situ and image data collected again. The imaging parameters for both samples are presented in Table 1. The reconstruction was performed by the Xradia XMReconstructor software native to the device.

2.3 Deformation Analysis

The reconstructed image volumes obtained from the μCT device were imported into Avizo 9.4 image processing software (Thermo Fisher Scientific, Waltham, MA, USA). Analysed volumes, 3.8 × 3.8 × 4.2 mm for PLCL-TCP scaffold and 2.5 × 2.5 × 2 mm for collagen-PLA composite, were selected from the center of samples and non-local means filtering was used to reduce the noise levels. The scaffolds were segmented in the image volumes using simple thresholding procedures. As the samples had high contrast with respect to the background, this simple image processing task was sufficient for extracting the region of interest. Segmented binary image stacks were imported into Fiji software [3] where porosity analysis was done by using the BoneJ plugin [4]. Calculated parameters for both samples

Fig. 1 Above PLCL-TCP scaffold, below Collagen-PLA composite. Scale bars above 5 mm, below 0.5 mm

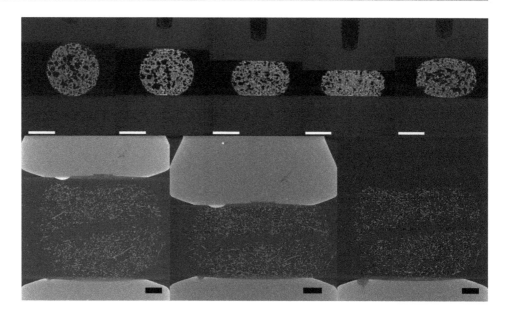

Table 1 Imaging parameters

Sample	Source voltage (kV)	Source current (µA)	Exposure time (s)	Projections	Pixel size (µm)
PLCL-TCP	80	125	2	1600	5.63
Collagen-PLA	40	250	3	1600	2.29

were mean material and pore thickness of analysed volumes. All image processing and analysis tasks were performed on a Windows© 7 based desktop with an Intel® Xeon® 3.4 Ghz CPU, NVIDIA® Quadro® K6000 graphics card and 128 GB of installed RAM.

3 Results

A summary of the changes observed in the bulk parameters of the scaffolds under uniaxial compression are listed in Table 2 (PLCL-TCP) and Table 3 (collagen-PLA). Preliminary study was done with PLCL scaffold without TCP. Those results are also presented in Table 2.

PLCL-TCP scaffold analysis results are shown in Fig. 2.

4 Discussion and Conclusions

Here we used in situ compression to assess the porosity of materials using µCT- imaging. The results show that in both the scaffold constructs the material parameters (thickness) remain unaffected. This indicates that the compression load applied to induce deformation in the structure does not compress the material itself to any significant degree. On the other hand, the changes in the void parameters indicate that porosities and pore sizes decrease under compression but recover nearly to the initial level after compression with the collagen-PLA sample. In ambient environment (21 °C) the PLCL-TCP sample is not recovering fully to its original shape, but remaining 25% smaller. This is most likely due to the polymers glass transition temperature that is 22–23 °C and the high concentration of elastic recovery hindering ceramic particles embedded in the polymer phase. PLCL sample had only 5% difference after compression compared to the original size. It seems that TCP particles are breaking PLCL structure and detaching from the polymer matrix during the compression and the sample cannot recover fully anymore at ambient temperature the same rate as the pure polymer scaffold, i.e. the PLCL scaffold. However, as the temperature rises to the body temperature, e.g. during implantation, the elastic recovery rate of the polymer chains is likely to increase significantly. TCP particles will break out from PLCL walls and particles will be available for cells to interact. On the other hand, TCP particles might affect pore sizes. Although in this case the deformation was more than 50%, there were no major fractures in the PLCL-TCP sample.

The results of this study are a step towards understanding the changes in the structure of these scaffolds under expected operation. Thus when using these scaffolds in a clinical setting, they may be subjected to compression to achieve a snug fit. Aside from bulk properties, it may also be beneficial to understand the 3D distribution of the scaffold properties

Table 2 PLCL-TCP and PLCL samples, calculated parameters with different compressions

		Porosity (%)	Material thickness mean (μm)	Material thickness std. (μm)	Pore thickness mean (μm)	Pore thickness std. (μm)	Deformation (%)
PLCL-TCP	Before compression	52.8	132.2	45.5	528.4	257.0	0
	With 13% compression	49.7	131.2	45.8	468.7	258.8	13
	With 35% compression	38.8	138.9	48.8	267.0	171.0	35
	With 52% compression	23.0	158.6	56.7	163.9	128.8	52
	After compression	49.0	143.8	53.0	344.8	200.0	25
PLCL	Before compression	71.7	236.1	84.8	975.8	443.8	0
	With 23% compression	69.1	232.2	84.1	823.3	408.9	23
	With 54% compression	56.2	236.0	85.6	480.9	297.8	54
	After compression	71.0	234.6	85.6	912.3	425.8	5

Table 3 Collagen-PLA composite sample, calculated parameters of different parts before, during and after compression

		Porosity (%)	Material thickness mean (μm)	Material thickness std. (μm)	Pore thickness mean (μm)	Pore thickness std. (μm)
Whole sample	Before compression	86.4	18.9	5.6	48.7	17.2
	With 33% compression	82.4	18.9	5.4	39.6	13.9
	After compression	86.3	19.0	5.3	57.3	23.9
PLA upper layer	Before compression	84.3	20.3	4.0	74.9	34.4
	With 33% compression	81.2	20.0	3.7	65.3	30.7
	After compression	82.8	19.1	3.5	66.0	30.6
Collagen	Before compression	94.7	6.0	1.2	46.0	16.7
	With 33% compression	91.8	6.1	1.3	33.1	13.4
	After compression	94.6	6.0	1.2	45.1	16.5
PLA lower layer	Before compression	83.3	20.3	3.9	69.8	34.2
	With 33% compression	79.2	20.2	3.9	57.3	25.2
	After compression	82.8	19.3	3.7	66.1	34.7

under compression that mimic the use environment. This would enhance our understanding of the realistic distribution of porosity during in situ functional environment.

Acknowledgements The work of M. Hannula has been supported by the Human Spare Parts Project funded by Finnish Funding Agency for Technology and Innovation (TEKES). Nathaniel Narra's work was in part supported by the South African Research Chairs Initiative of the

Fig. 2 PLCL-TCP scaffold
analysis results

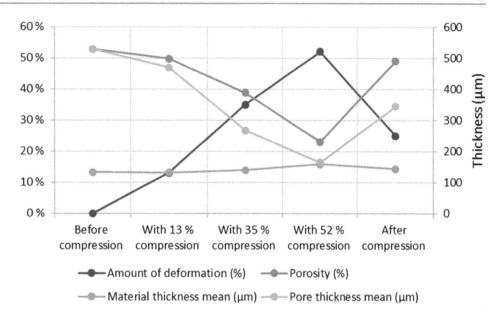

— Amount of deformation (%)   — Porosity (%)

— Material thickness mean (µm)   — Pore thickness mean (µm)

Department of Science and Technology and the National Research
Foundation (grant no 98788).

Ethical Statements The authors declare that they have no conflict of
interests.

References

1. Haaparanta A., Järvinen E., Cengiz I.F., Ellä V., Kokkonen H.T.,
Kiviranta I., Kellomäki M.: Preparation and characterization of
collagen/PLA, chitosan/PLA, and collagen/chitosan/PLA hybrid
scaffolds for cartilage tissue engineering. J Mater Sci Mater Med 25
(4):1129–1136 (2014)
2. Narra N., Blanquer S.B., Haimi S.P., Grijpma D.W., Hyttinen J.:
µCT based assessment of mechanical deformation of designed
PTMC scaffolds. Clin Hemorheol Microcirc 60(1):99–108 (2015)
3. Schindelin J., Arganda-Carreras I., Frise E., Kaynig V., Longair M.,
Pietzsch T., Preibisch S., Rueden C., Saalfeld S., Schmid B.: Fiji: an
open-source platform for biological-image analysis. Nature methods
9(7):676–682 (2012)
4. Doube M., Kłosowski M.M., Arganda-Carreras I., Cordelières F.P.,
Dougherty R.P., Jackson J.S., Schmid B., Hutchinson J.R.,
Shefelbine S.J.: BoneJ: free and extensible bone image analysis in
ImageJ. Bone 47(6):1076–1079 (2010)

Multi-gaussian Decomposition of the Microvascular Pulse Detects Alterations in Type 1 Diabetes

Michele Sorelli, Antonia Perrella, Piergiorgio Francia,
Alessandra De Bellis, Roberto Anichini, and Leonardo Bocchi

Abstract

Among diabetic patients, microangiopathy represents a relevant cause of morbidity and mortality. Diabetes induces detrimental changes in the biomechanical characteristics of blood microvessels, and fuels the development of a dysfunctional vascularization. Since the structural properties of the circulatory system affect the microvascular pulse, the aim of this study was to detect these vascular alterations through a model-based quantitative analysis of its waveform. Baseline microvascular perfusion was recorded on the hallux with a laser Doppler flowmeter. 54 healthy subjects (age: 34 ± 26 years) and 22 type 1 diabetic (T1D) patients without known cardiovascular complications and smoking history (age: 34 ± 17 years) were compared. A novel multi-Gaussian decomposition algorithm was applied to reconstruct the heartbeat-related oscillations, which were evaluated according to normalized and physiologically-motivated shape descriptors. Eight out of the nine properties assessed significantly differed between the groups ($p < 0.001$), indicating that the proposed pulse modeling method is sensitive to the effects of T1D on the peripheral perfusion.

Keywords

Pulse decomposition • Laser Doppler flowmetry Diabetes • Microangiopathy

1 Introduction

Microangiopathy is deemed an important cause of morbidity and mortality among patients affected by diabetes mellitus. In these subjects, the long-term exposure to high glucose concentrations, and the consequent tissue oxidative stress, determine a range of biochemical, structural and functional alterations involving the microvascular endothelium, smooth muscle cells, and capillary pericytes. In particular, this complication is fuelled by the excessive formation of advanced glycation end-products, and the build-up of inelastic matrix materials, which lead to a disordered and inefficient vascularization, and to an abnormal permeability and stiffness of blood vessels [1]. Despite being established that vascular stiffness directly affects the pulse wave velocity (PWV) within the arterial tree, according to the Moens-Korteweg model [2], several studies have highlighted also an indirect effect on the profile of the peripheral pulse [3, 4], and identified specific features which hold a close correlation with the arterial PWV and the tone of peripheral vessels [5]. Therefore, the aim of this study was to assess the possibility to detect these diabetes-related biomechanical changes of the circulation through a model-based, detailed, quantitative analysis of the microvascular pulse, evaluated non-invasively by laser Doppler flowmetry (LDF).

2 Materials and Methods

Research activities were carried out in accordance with the guidelines of the Declaration of Helsinki of the World Medical Association. The enrolled volunteers received a detailed explanation of the adopted study protocol and its

M. Sorelli (✉) · A. Perrella · L. Bocchi
Department of Information Engineering, University of Florence, Via di S. Marta 3, 50139 Florence, Italy
e-mail: michele.sorelli@unifi.it

L. Bocchi
e-mail: leonardo.bocchi@unifi.it

P. Francia
Department of Clinical and Experimental Medicine, University of Florence, Viale Pieraccini 6, 50139 Florence, Italy

A. De Bellis · R. Anichini
Department of Internal Medicine, Diabetes Unit, General Hospital of Pistoia, Via Ciliegiole 98, 51110 Pistoia, Italy

© Springer Nature Singapore Pte Ltd. 2019
L. Lhotska et al. (eds.), *World Congress on Medical Physics and Biomedical Engineering 2018*,
IFMBE Proceedings 68/3, https://doi.org/10.1007/978-981-10-9023-3_31

purpose, and signed an informed consent form prior to the start of the measurement sessions. Acquisitions were made in resting conditions on the pulp of the right hallux with a Periflux 5000 LDF system (Perimed, Sweden), with the subjects lying supine in a comfortable position, in a temperature-controlled environment (T ≈ 23 °C). Microvascular perfusion was recorded from 54 healthy controls (age: 34 ± 26 years) and 22 type 1 diabetic (T1D) patients without known cardiovascular complications (age: 34 ± 17 years). All the included participants were non-smokers. Perfusion signals were sampled at 32 Hz, and the minimum time constant available for the instrument output low-pass filter, i.e. 0.03 s, was set in order to properly retain the heartbeat-related frequency content.

The methodology adopted in the present work for analysing the LDF signals was translated from recent research on the assessment of the digital volume pulse (DVP) of photoplethysmographic signals [6–8]. More specifically, a pulse decomposition algorithm (PDA) evolved from a previous method recently presented in [9], was implemented so as to reconstruct each heartbeat oscillation extracted from the LDF recordings. Gaussian functions represent the basic modeling components of the developed PDA, as reported in literature by others [6, 8, 10, 11]. In detail, following the segmentation of the separate cardiac cycles and the identification of the *incisura* reference, as schematized in Fig. 1a, the PDA decomposes each pulse wave into four separate Gaussians: one for the forward travelling systolic beat and three devoted to the reconstruction of the diastolic profile on the right side of the *incisura*, associated with the reflection

and re-reflections of the main pulse. The identification of the multi-Gaussian model was carried out with a non-linear optimization algorithm, tuning the initialization and the boundaries of the model parameters according to temporal and amplitude properties specific of each detected pulse wave. After fitting each signal, the rate of false detections and mismodeled waves is decreased by imposing 3-σ limits on the estimated cardiac cycle duration, and the α_1 and σ_1 parameters of the systolic component. Two sets of 20939 and 7233 cardiac pulse models were thus obtained for the control and T1D groups, respectively.

In order to characterize the pulse contour, the percentage amplitude ratios α_2/α_1, α_3/α_1, α_4/α_1 and the delays Δt_{1-2}, Δt_{1-3}, Δt_{1-4} of the diastolic components to the systolic one were derived for each identified waveform model (Fig. 1b). Moreover, these normalized morphological features were combined with the Crest Time (CT), i.e. the duration of the ascent of the primary wave, and two other physiologically-motivated parameters: the Stiffness Index (SI), and the Reflection Index (RI). The former was originally defined in [12] as the ratio of subject height to the time interval between the systolic and the diastolic DVP maximum (or inflection point), ΔT_{DVP}, and has been positively correlated with arterial stiffness and PWV; the latter, instead, has been linked to the tone of small peripheral arteries [13], and is generally evaluated as the ratio between the diastolic and systolic DVP amplitudes [6]. In the present work, the SI was estimated as the time delay between the forward wave and the centroid of the three diastolic Gaussian components, while the RI was obtained from the percentage ratio of the

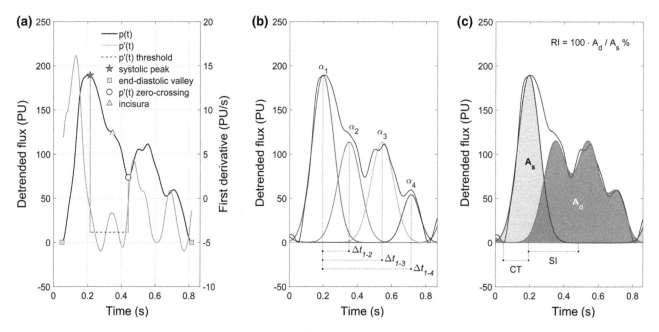

Fig. 1 Multi-Gaussian PDA. **a** detection of the *incisura*: $p'(t)$ maximum point exceeding the mean signal slope in the interval between the systolic peak and the first $p'(t)$ zero-crossing (if no such maximum exists, the latter is adopted as reference); **b** normalized morphological properties; **c** physiologically-motivated features

areas beneath the diastolic (A_d) and systolic (A_s) pulse profiles, as shown in Fig. 1c.

3 Results and Conclusion

The modeling accuracy of the multi-Gaussian PDA was verified on the basis of goodness-of-fit parameters: on the whole set of 28172 waveform models, the algorithm exhibited a coefficient of determination (R^2) of 0.97 ± 0.03 (mean $\pm\sigma$); furthermore, the average normalized root mean square error achieved over the analysed LDF signals was 0.89 ± 0.03, where a value of 1 would mean a perfect data fit. These outcomes thus strongly suggest that the proposed approach can adequately reconstruct the profile of the LDF pulse waves.

The sample distributions of the extracted waveform features were assessed for normality by means of the Shapiro-Wilk test. Since all of them significantly deviated from a normal distribution ($p < 0.001$), the non-parametric Mann-Whitney test for independent samples was used to compare the control and T1D groups. Table 1 summarizes the results of the statistical analysis.

Except for the CT, all the analyzed pulse features significantly differ among the compared groups. Specifically, higher median Δt_{1-3} and Δt_{1-4} delay times are observed in T1D subjects, whereas their median Δt_{1-2} is decreased to a moderate extent. Overall, the median SI derived from the T1D group results to be mildly higher. In [12], a significant positive correlation ($r = 0.70$, $p < 0.001$) has been identified between the systolic-diastolic peak delay, ΔT_{DVP}, and the aorto-femoral transit time, which is known to decrease with increasing vascular stiffness and PWV. Accordingly, the detected augmentation of the SI might be interpreted as a reduction in the rigidity of large elastic arteries, which would contrast with the assumption of an increased mechanical

stiffness, due to the T1D-related exacerbation of inter- and intra-molecular cross-linking of collagen fibers. However, it is relevant to consider that the model-based approach applied in this study differs from the original method, which relies entirely on the detection of local maxima. Also, the latter is based on the analysis of photoplethysmography signals, that relate to local blood volumetric changes, whereas LDF provides a non-absolute measure of perfusion. These dissimilarities might explain this inconsistent outcome. Undoubtedly, novel studies devoted to address more thoroughly the definition and interpretation of this pulse-derived feature of stiffness would be of great value. Furthermore, the relative amplitude of all the three diastolic Gaussians is increased in patients with T1D, which are also associated with a consistent rise of the RI. Millasseau et al. have shown that the relative height of the diastolic DVP peak is increased in a dose-dependent manner by angiotensin II [14], which is known to have a vasoconstrictive action on peripheral arteries. Therefore, these data appear to be in accordance with diabetes-associated vascular stiffening.

On the whole, the presented results indicate that the proposed multi-Gaussian PDA might be sensitive to the effects of T1D on the peripheral circulation. Further studies are required in order to better characterize the relation between the morphological features of the peripheral LDF pulse and the biomechanical properties of the circulatory system, and demonstrate the suitability of this technique for the screening of diabetic-related vascular complications.

Conflict of Interest This study was supported by Ente Cassa di Risparmio di Firenze, Florence, Italy (grant number 2015.0914). The authors declare that they have no potential conflict of interest in relation to the study in this paper.

Table 1 LDF pulse wave features: control versus T1D patients

Waveform	Features	Median (IQR)		p
		Control	T1D	(Mann-Whitney)
RI	%	59.3 (37.5–94.1)	84.0 (55.1–127.8)	<0.001
SI	ms	281 (245–317)	293 (256–333)	<0.001
CT	ms	155 (131–180)	153 (128–183)	0.335
α_2/α_1	%	29.9 (19.0–44.6)	40.2 (26.4–59.5)	<0.001
α_3/α_1	%	28.5 (19.3–38.7)	35.0 (25.0–47.1)	<0.001
α_4/α_1	%	18.7 (12.4–27.3)	24.3 (16.7–34.1)	<0.001
Δt_{1-2}	ms	168 (134–203)	157 (127–191)	<0.001
Δt_{1-3}	ms	292 (253–334)	305 (264–346)	<0.001
Δt_{1-4}	ms	431 (374–511)	477 (408–559)	<0.001

References

1. Madonna, R., Balistreri, C. R., Geng, Y., De Caterina, R.: Diabetic microangiopathy: pathogenic insights and novel therapeutic approaches. Vascular Pharmacology 90, pp. 1–7 (2017). https://doi.org/10.1016/j.vph.2017.01.004
2. Nichols, W. W., O'Rourke, M. F., Vlachopoulos, C.: McDonald's Blood Flow in Arteries, Sixth Edition: Theoretical, Experimental and Clinical Principles, CRC Press, 2011.
3. Alastruey, J., Passerini, T., Formaggia, L., Peiró, J.: Physical determining factors of the arterial pulse waveform: theoretical analysis and calculation using the 1-D formulation. Journal of Engineering Mathematics 77, pp. 19–37 (2012). https://doi.org/10.1007/s10665-012-9555-z
4. Alty, S. R., Angarita-Jaimes, N., Millasseau, S. C., Chowienczyk, P. J.: Predicting arterial stiffness from the digital volume pulse waveform. IEEE Transactions on Biomedical Engineering 54, pp. 2268–2275 (2007). https://doi.org/10.1109/tbme.2007.897805
5. Millasseau, S. C., Ritter, J. M., Takazawa, K., Chowienczyk, P. J.: Contour analysis of the photoplethysmographic pulse measured at the finger. Journal of Hypertension 24, pp. 1449–1456 (2006). https://doi.org/10.1097/01.hjh.0000239277.05068.87

6. Rubins, U.: Finger and ear photoplethysmogram waveform analysis by fitting with Gaussians. Medical & Biological Engineering & Computing 46, pp. 1271–1276 (2008). https://doi.org/10.1007/s11517-008-0406-z

7. Baruch, M. C., Warburton, D. E. R., Bredin, S. S. D., Cote, A., Gerdt, D. W., Adkins, C. M.: Pulse Decomposition Analysis of the digital arterial pulse during hemorrhage simulation. Nonlinear Biomedical Physics 5 (2011). https://doi.org/10.1186/1753-4631-5-1

8. Couceiro, R., Carvalho, P., Paiva, R. P., Henriques, J., Quintal, I., Antunes, M., Muehlsteff, J., Eickholt, C., Brinkmeyer, C., Kelm, M., Meyer, C.: Assessment of cardiovascular function from multi-Gaussian fitting of a finger photoplethysmogram. Physiological Measurement 36, pp. 1801–1825 (2015). https://doi.org/10.1088/0967-3334/36/9/1801

9. Sorelli, M., Perrella, A., Bocchi, L., Cardiac pulse waves modeling and analysis in laser Doppler perfusion signals of the skin microcirculation. In: CMBEBIH 2017: Proceedings of the International Conference on Medical and Biological Engineering, pp. 20–25, Springer, Singapore (2017). https://doi.org/10.1007/978-981-10-4166-2_4

10. Liu, C., Zheng, D., Murray, A., Liu, C.: Modeling carotid and radial pulse pressure waveforms by curve fitting with Gaussian functions. Biomedical Signal Processing and Control 8, pp. 449–454 (2013). https://doi.org/10.1016/j.bspc.2013.01.003

11. Wang, L., Lisheng, X., Feng, S., Meng, M., Wang, K.: Multi-gaussian fitting for pulse waveform using weighted least squares and multi-criteria decision making method. Computers in Biology and Medicine 43, pp. 1661–1672 (2013). https://doi.org/10.1016/j.compbiomed.2013.08.004

12. Millasseau, S. C., Kelly, R. P., Ritter, J. M., Chowienczyk, P. J.: Determination of age-related increases in large artery stiffness by digital pulse contour analysis. Clinical Science 103, pp. 371–377 (2002). https://doi.org/10.1042/cs1030371

13. De Loach, S. S., Townsend, R. R.: Vascular stiffness: its measurement and significance for epidemiologic and outcome studies. Clinical Journal of the American Society of Nephrology 3, pp. 184–192 (2008). https://doi.org/10.2215/cjn.03340807

14. Millasseau, S. C., Kelly, R. P., Ritter, J. M., Chowienczyk, P. J.: The vascular impact of aging and vasoactive drugs: comparison of two digital volume pulse measurements. American Journal of Hypertension 16, pp. 467–472 (2003). https://doi.org/10.1016/s0895-7061(03)00569-7

GOAL (Games for Olders Active Life): A Web-Application for Cognitive Impairment Tele-Rehabilitation

Leonardo Martini, Federica Vannetti, Laura Fabbri, Filippo Gerli,
Irene Mosca, Stefania Pazzi, Francesca Baglio, Leonardo Bocchi,
and GOAL Project Group

Abstract

Vascular Dementia (VaD) and Alzheimer's Disease (AD) are the major causes of advanced cognitive decline. Serious Games (SGs) are computer games, recently proposed in the healthcare sector, specifically for the evaluation and rehabilitation of psychiatric and neurological disorders. The main objective of the GOAL project is to test a suite of SGs on a group of subjects with Mild Cognitive Impairment (MCI) and Vascular Cognitive Impairment (VCI), conditions at risk of frank dementia, with the aim to characterize and quantify their functional, cognitive and motor abilities. The games were implemented using "ad hoc" ICT tools, for longitudinal monitoring and rehabilitation management directly from home. In this context, the Web-Application GOAL-App was developed for allow patients to access to scheduled physical and cognitive trainings. HTML5, JavaScript and CSS have been used to create a clear, intuitive and extremely easy to use UI. The back-end is JAVA-based. The preliminary results showed good feedback from the subjects, who regularly practiced the proposed scheduled trainings.

Keywords

Dementia • Mild cognitive impairment
Web-application • Rehabilitation

1 Introduction

During the aging process, older people often encounter difficulties in cognitive skills (memory, language, executive functions), ranging from the isolated memory disorder to the established dementia, which involves serious functional repercussions in the skills of everyday life. Dementia is one of the major causes of disability in the world and epidemiological studies indicate that in 2020, only in the countries of the European Union (EU), people with dementia will be over 15 million, with a prevalence rate that increases with age, where values ranging from 1.4% for the age-class 65–69 years and 28.5% for the 85–89 years group [1]. Vascular dementia (VaD), in association with Alzheimer's disease (AD), are two of the main causes of cognitive impairment in old age. In this scenario it's easy to understand how the prevention and treatment of these conditions in the initial phases are fundamental. For this reason, many studies have been initiated with the aim of identifying preventative approaches to be applied in the Mild Cognitive Impairment (MCI) and Vascular Mild Cognitive Impairment (VCI), two conditions at high risk of conversion into frank dementia. Indeed, several studies indicate that about 15% of patients with MCI/VCI evolve into dementia within 2 years [2, 3]. Other studies reveal that the pharmacological treatments currently available for this type of condition are not effective and indicate that clinicians should recommend regular exercise, cognitive training and monitor of cognitive status of these subjects over time [3, 4].

Serious Games (SGs) are computer games used in many contexts such as education, training, simulation, fitness. More recently SGs have also been proposed in the health

GOAL Project Group: Sandro Sorbi, Claudio Macchi, Tommaso Migliazza, Andrea Stoppini, Giulia Lucidi, Federica Savazzi.

L. Martini · L. Bocchi (✉)
University of Florence, Florence, Italy
e-mail: leonardo.bocchi@unifi.it

F. Vannetti · L. Fabbri · F. Gerli · I. Mosca
Don Gnocchi Foundation IRCCS, Florence, Italy

S. Pazzi
CBIM, Pavia, Italy

F. Baglio
Don Gnocchi Foundation IRCCS, Milan, Italy

sector, in particular in the evaluation and rehabilitation of psychiatric and neurological diseases [5]. A recent study, published in Nature [6] shows that the deficit of cognitive skills can be slowed down by using SGs as tools for cognitive enhancement. Furthermore, physical activity has also been demonstrated contributing to the maintenance of cognitive functions (in particular of executive functions) and reduces the risk of developing neurodegenerative diseases such as AD [7]. Therefore, the proposal that seems to be more promising to counteract the progression of these degenerative conditions, is the implementation of combined rehabilitative treatments, working on both the cognitive and motor domains [8–10].

The GOAL (Games for Olders Active Life) project has the aim to develop a platform useful for the clinicians to assess and monitoring the clinical status of patients and for the patient to take advantage from a combined rehabilitation program, both cognitive and physical. The project is the result of a collaboration between the Don Carlo Gnocchi Foundation, the Bioengineering and Medical Informatics Consortium (CBIM), the University of Florence and has been financed by Regione Toscana (FAS Salute 2014). All these activities are implemented and integrated with specific ICT tools, for a longitudinal monitoring and rehabilitation management of cognitive and motor skills in these subjects, directly at home. Patients, provided with a computer with touchscreen, can access the cognitive and physical activities training using the web application.

2   Materials and Method

GOAL-App is a web-application specifically designed to implement a weekly training program, consisting of combined cognitive training sessions, physical training sessions and leisure activities (to be performed with the caregiver, spouse, sons, etc.), according to clinicians indications. Scheduling and monitoring of activities are performed on an external portal accessible by the web-application administrator. Each activity is implemented in independent modules. The cognitive module integrates a collection of brain training exercises from BrainHQ, a third-party platform developed by Posit Science, that makes available serious games for multidimensional stimulation (memory, speed, attention, intelligence etc.). The proposed SGs are adaptive type, i.e. the difficulty varies in relation to the user performance. Difficulty is indeed maintained immediately over the user comfort threshold, which, according to several studies, efficaciously stimulate the neural plasticity [11]. The caregiver module includes suggestions of leisure activities to be carried out with the caregiver during the weekend, such as visit museums, gardening, watching film, etc. The newly developed physical activities module includes a training program

of APA (Adapted Physical Activity) exercises, delivered through guided video, and specific questionnaires about the performed exercises and the daily activities.

The process of GOAL web-application requirements analysis, design and implementation is described below.

2.1   Requirement Analysis

The first step was the analysis of the architectural, functional and safety requirements at global level of the web-app and the various integrated modules, in particular regarding the new physical activity module. The main application requirements are:

- user-friendly interface with large and intuitive buttons, considering the user-target and the poor or absent familiarity with ICT;
- communication with administration portal for the scheduling of all activities;
- integration of the physical activity, cognitive and caregiver modules. A single home page where the patient can access to the daily scheduled activity;
- security: authentication system and use of non-sensitive data: GOAL-App doesn't use any sensitive data present on the portal;
- easy updating of contents and high maintainability.

The resulting architecture, as shown in Fig. 1, thanks to its modularity, allows to integrate or replace one or more activity modules.

Main requirements for the physical activity module are:

- pre and post physical training questionnaires.
- automatic playback of full-screen video with touch controls, maximizing the view;
- all the answers to the questionnaires and the time spent on each web page are automatically saved, providing a real-time tracking of the patient's activities.

2.2   Design and Implementation

GOAL-App has a three-tier architecture characterized by 3 communicating layers: a presentation layer realized through HTML and CSS pages, a business logic layer, consisting of the back-end present on the application server, which receives, processes and satisfies the client's requests and a data layer consisting of the database management system, where the data reside. The home page, shown in Fig. 1, is designed to maintain the same layout, while only the large central button changes according to the type of activity to be

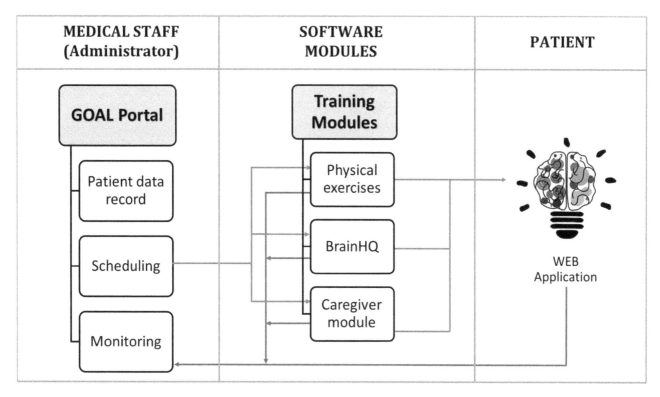

Fig. 1 Functional architecture

Fig. 2 Home page: different central button for different type of activity

performed. In this way, access to the various modules is extremely simple and intuitive.

Furthermore, HTML5 and JavaScript allows the playback of videos directly on the browser without any plugins and the creation of a full-screen video player with intuitive controls, as shown in Fig. 3. User can easily play/pause videos touching the screen in any point.

The back-end is completely Java-based, whose servlets allow to manage the authentication of the users, the initialization of the daily training session and the data loading/saving on the database. The used DBMS is PostgreSQL.

The strength of the application and in particular of the physical activity module, is the ability to dynamically

Fig. 3 Fullscreen video-player with touch play/pause controls

Fig. 4 Dynamic creation of web-page content from JSON Schema

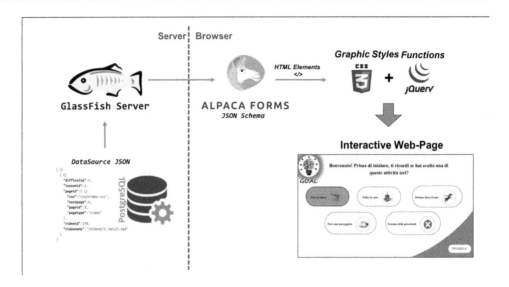

generate the elements present in the web page using fetched data from database. Thus, the content of the web pages can be completely managed, modified and updated from the database, without editing the HTML code or the Java backend, leading to high maintainability. The generation process is shown in Fig. 4. For this purpose, we used Alpaca Forms, an open source suite of javaScript tools able to generate forms and HTML elements using a JSON schema; JSON data is loaded directly from the database, using a REST interface. Once the JSON Schema is specified and filled with the retrieved JSON data, Alpaca renders them in the corresponding HTML structures. CSS style sheets and JavaScript functions were used for the page graphics and interactivity.

The following is a sample UML Sequence Diagram of the operations of web pages generation and the data saving on the database. The browser, through a HTTP Request with GET method, requests and obtains from the server the HTML page to be displayed, within which the Alpaca code is executed. Alpaca then retrieves the necessary datasets from the database in JSON format and translates them into the HTML elements of the page, which are rendered by the browser and presented to the user (Fig. 5).

The user answers to the questionnaire or visualizes the videos of the proposed physical exercises; collected data are sent to the backend using the next HTTP POST Request; a java servlet manages the data saving and redirects to the next page.

The system security is guaranteed by a java filter, specially implemented to check and reject unauthenticated connections. All the attempts to access any page without verified credentials are immediately blocked by redirecting the user to the login page.

3 Results and Discussion

The GOAL-App was tested on a first group of 12 patients with MCI or VCI (Mean: 74.6 yo, SD: 3.6 yo) from a sample of 33, appropriately evaluated and blocks randomized. In this pilot test 8 weeks of rehabilitation have been programmed and globally the utilization results by the patients showed a high compliance of the proposed activities and feedback in terms of usability, measured by means of collected data on database and ad hoc questionnaires. The possibility of performing cognitive and physical activity trainings comfortably at home was particularly appreciated. As shown in Fig. 6, 75% expressed perceived improvements in maintaining attention and retaining memory, combined with perceived physical well-being and renewed energies. Activities, on the whole, were highly appreciated, so much that many patients expressed displeasure in interrupting the rehabilitation program. Also, 83% did not experience particular obstacles in interacting with a computer, despite a referred lack of familiarity with the technology. This could be the result of a deep customization of the operating system and of the attention to the usability issues in the design of the application.

These characteristics allow older people to quickly familiarize with widely diffused and low-cost ICT technologies. In addition, the direct connection of all clients to the administration portal, allows the clinician not only to completely customize the training program, but to follow in real time the progresses of several patients. All the collected data generate a data base that also can be of considerable importance for research purposes. In future, the sample enlargement with further patient groups, will allow to collect additional data useful to validate the Web-App system architecture.

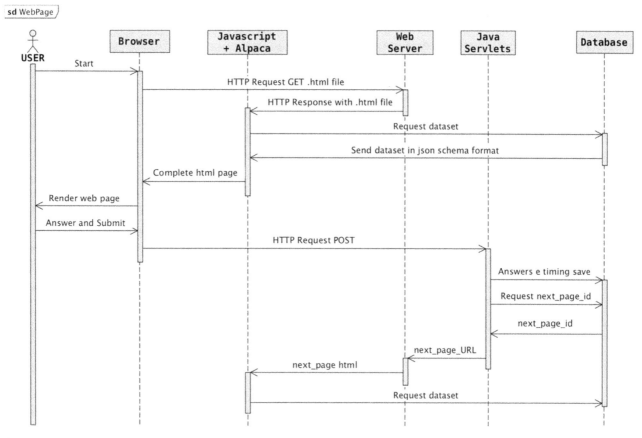

Fig. 5 Sequence diagram of web page loading and data saving on DB

Fig. 6 Compliance and user feedback

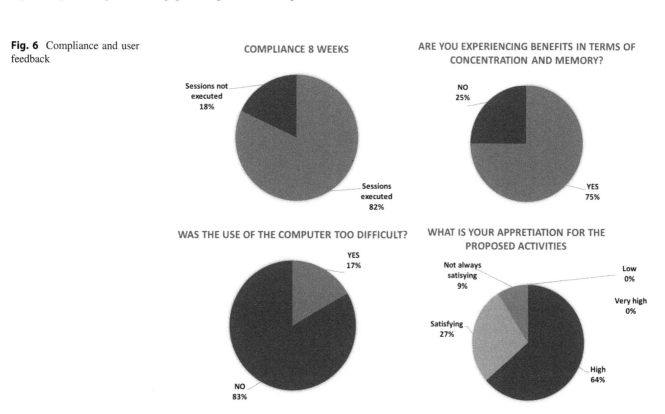

References

1. http://www.alzheimer-europe.org/Policy-in-Practice2/Country-comparisons/2013-The-prevalence-of-dementia-in-Europe.

2. Perri, R., Serra, L., Carlesimo, GA., Caltagirone, C., Early Diagnosis Group of Italian Interdisciplinary Network on Alzheimer's Disease: Preclinical dementia: an Italian multicentre study on amnestic mild cognitive impairment. Dement Geriatr Cogn Disord, 23(5), pp 289–300 (2007).

3. Petersen R.C., Lopez, O., Armstrong, M., Getchius, T., Ganguli, M., Gloss, D., Gronseth, G., Marson, D., Pringsheim, T., Day, G., Sager, M., Stevens, J., Rae-Grant, A.: Practice guideline update summary: Mild cognitive impairment. Report of the Guideline Development, Dissemination, and Implementation Subcommittee of the American Academy of Neurology. Neurology, Vol. 90, N°3, pp 126–135 (2018).

4. Karakaya, T., Fabian Fußer, F., Schröder, J., Pantel, J.: Pharmacological Treatment of Mild Cognitive Impairment as a Prodromal Syndrome of Alzheimer's Disease. Current Neuropharmacology, 11, pp 102–108 (2013).

5. Mahncke, H., Connor, B., Appelman, J., Ahsanuddin, O., Hard, J., Wood, R., Joyce, N., Boniske, T., Atkins, S., Merzenich, M.: Memory enhancement in healthy older adults using a brain plasticity-based training program: A randomized, controlled study. PNAS, vol. 103 no. 33, pp 12523–12528 (2006).

6. Anguera, JA., Boccanfuso, J., Rintoul, JL., Al-Hashimi, O., Faraji, F., Janowich, J., Kong, E., Larraburo, Y., Rolle, C., Johnston, E., Gazzaley, A: Video game training enhances cognitive control in older adults. Nature; 501, pp 97–101 (2013).

7. Hillmann, CH., Erickson, KI., Kramer, AF.: Be smart, exercise your heart: exercise effects on brain and cognition. Nar Rev Neurosci. 9, pp 58–65 (2008).

8. Karssemeije, E.G.A., Aaronson, J.A., Bossers, W.J., Smits, T., Olde Rikkert, M.G.M, Kessels, R.P.C.: Positive effects of combined cognitive and physical exercise training on cognitive function in older adults with mild cognitive impairment or dementia: A meta-analysis. Ageing Research Reviews, 40, pp 75–83 (2017).

9. Gagliardi, C., Papa, R., Postacchini, D., Giuli, C.: Association between Cognitive Status and Physical Activity: Study Profile on Baseline Survey of the My Mind Project. Int. J. Environ. Res. Public Health, 13, 585 (2016).

10. Cancela, J., Ayán, C., Varela, S., Seijo, M.: Effects of a long-term aerobic exercise intervention on institutionalized patients with dementia. Journal of Science and Medicine in Sport, 19, pp 293–298 (2016).

11. Lumsden, J., Edwards, E.3, Lawrence, N., Coyle, D., Munafò, M.: Gamification of Cognitive Assessment and Cognitive Training: A Systematic Review of Applications and Efficacy. JMIR Serious Games, vol. 4, iss. 2, e11, pp 1–14 (2016).

Tango™ Wellness Motivator for Supporting Permanent Lifestyle Change

Antti Vehkaoja, Jarmo Verho, Mikko Peltokangas, Teppo Rantaniva, Vala Jeyhani, and Jari Råglund

Abstract

We present a system designed for assisting people in obtaining healthier lifestyle. The system includes a monitoring device worn on the chest and a web portal that visualizes the measured parameters and provides the user motivating tips for healthier lifestyles. The monitored parameters include heart rate, step count, calorie consumption, activity level, heart rate variability and sleep quality. A unique feature of the system is that the communication from the wearable unit to the backend server is arranged via direct mobile network connection, thus avoiding the need for a separate gateway device. The measured data can be viewed with a web browser user interface. We evaluated the beat-to-beat heart rate estimation performance with ten subjects in a controlled exercise protocol and with three subjects in 24-h free-living conditions. The average mean absolute error of the R-R interval estimation was 8.0 ms and 6.4 ms in the two test scenarios, respectively and the corresponding coverages of the obtained R-R intervals 76% and 94%.

Keywords

Wearable monitoring • Lifestyle coaching

1 Introduction

Lifestyle has the single most important role in maintaining or affecting general health condition of a person. Sedentary lifestyle and poor dietary habits are the main causes of increase in prevalence of cardiovascular diseases, obesity, type-2 diabetes, and metabolic syndrome in Western countries and recently also in Middle East and Asia [1].

While general health awareness is increasing and there is a growing trend of self-monitoring of wellness, these trends only touch a small portion of the people, i.e. the ones who are already active and interested in their wellbeing. Those people who suffer from the aforementioned problems and who would really benefit from a lifestyle change are easily left out and do not get support for it. Wellness trackers and activity monitors coming on the market are mainly designed for people who are already active. The devices are not designed from the perspective of a person who is just trying to start a lifestyle change and who would need external motivating and encouraging. Some motivating systems exist for facilitating physical rehabilitation, such as the one described in [2] but what has been lacking are tools that the patients could use on their own and that would provide coaching in several aspects of life habits.

Healthy lifestyle consists of three main pillars: activity, sleep, and nutrition. All these together affect the overall wellbeing of a person. There are technological solutions developed for assisting in quantifying the behavior of the user with respect to any of these three areas. Activity is the one that has been gained the biggest attention from users and technology developers and the range of functionality of activity tracking devices has increased from the first ones that just used to show the average heart rate or count steps. For example, many of the newer wrist-worn devices also include sleep monitoring and sleep quality assessment in their service. In addition, there are devices dedicated exclusively for sleep monitoring and a lot of research has been done with different monitoring methods. One example of a commercial solution that uses a flexible mattress sensor has been developed by Beddit company [3].

Current solutions for nutrition monitoring are mostly based on food diary applications, where the user manually enters the meal content and the application breaks it up into nutrients. Automatic image based solutions for the detection of food content and computation of the energy and

A. Vehkaoja (✉) · J. Verho · M. Peltokangas · V. Jeyhani
BioMediTech Institute and Faculty of Biomedical Sciences and Engineering, Tampere University of Technology, Tampere, Finland
e-mail: antti.vehkaoja@tut.fi

T. Rantaniva · J. Råglund
Health Care Success Ltd, Helsinki, Finland

© Springer Nature Singapore Pte Ltd. 2019
L. Lhotska et al. (eds.), *World Congress on Medical Physics and Biomedical Engineering 2018*,
IFMBE Proceedings 68/3, https://doi.org/10.1007/978-981-10-9023-3_33

nutritional content have been studied and developed, too [4]. Another approach for assisting in changing one's dietary habits is not to try to quantify of what is eaten but rather try to directly assist in making better dietary choices by providing educative tips towards healthier diet.

In this paper we present our solution named Tango™ wellness motivator, which is intended for people who have not paid attention to their living habits earlier but are looking for a lifestyle change. The Tango™ system is based on a holistic approach considering all the three pillars that affect on physical welfare and gives support on improving them.

2   Tango™ Wellness Motivator System

The proposed system consists of a chest-worn monitoring device that is attached to a regular heart rate monitoring belt, a back-end server system for storing the measurement data, and a front-end user interface that is accessed through a web browser. A special feature of the system is that the monitoring device communicates with the back-end directly using the GSM cellular network, thus avoiding the need for a separate gateway device.

As said earlier, the purpose of the system is not to work as a basic activity monitor, but rather as a lifestyle coach that motivates the user to pay attention on his/her daily habits; i.e. generally encourages to move more, adjust the amount of sleep, and optimize dietary choices. Therefore, a decision was made not to provide the user real time information to e.g. a wrist device but to encourage the user to use the dedicated web portal to see the daily activity results and at the same time be provided with tips on how to improve the general wellbeing. The components of the system and their most important features are presented next.

2.1   The Monitoring Device

The heart of the chest-worn monitoring device is nRF52832 SoC (System on a Chip) that includes an ARM Cortex M4F processor core and several versatile peripheral devices, e.g. Bluetooth Low Energy (BLE) radio. The monitoring device uses an ADS1292R analog front-end from Texas Instruments for measuring the ECG signal and an MPU-9255 9D motion sensor from TDK Invensense for detecting steps and monitoring sleep. Communication with the cellular network is handled with SARA-G350 2G GSM modem from u-Blox. The device is charged with a custom charging dock powered by a normal micro USB charger. The BLE connection provided by the SoC could be used for transmitting the measured information to a local, e.g. a wrist-worn device but this has not yet been implemented and is reserved for future use.

Operating modes
The monitoring device has two operating modes. In the daytime mode it measures the ECG signal and calculates beat-to-beat heart rate from it, detects step counts and measures activity and posture of the user. The other operating mode is low-power sleep monitoring mode or nighttime mode. In nighttime mode, the device is worn on the wrist with in a dedicated wrist cradle. The ECG measurement and step counting are disabled and an algorithm for detecting sleep/wake states based on acceleration measurement is used. The presence of the nighttime measurement cradle is detected by the signal obtained from the ECG measurement channel.

Communication with the back-end server
The communication with the back-end is initialized and executed in a following way: the device initializes the GSM modem on start-up, sets up a GPRS data connection, obtains the IP address of the back-end server via DNS, opens a TCP socket connection to it and then requests the initial configuration using a simple custom protocol. The configuration contains the initial time; the daytime mode data transmit interval (one minute by default) and information about the latest available firmware.

Assuming the device is in the daytime mode (measuring ECG and counting steps), it keeps the socket connection to the back-end server alive and sends new data periodically in one minute intervals by default. The server, in turn, acknowledges each received data block. The acknowledges are not strictly necessary, but they make the detection of data loss simpler and more reliable. The main drawback of this is the increased current consumption, which is, however, minimal in comparison to the power required simply to keep the GPRS data connection open. The monitoring device has on-board memory for storing up to two hours of data in case of a loss of network connection.

When the device is in the nighttime mode (measuring sleep quality), the GPRS data connection is normally shut down. The device registers to the GSM network every 15 min, sends the collected sleep quality data and then deregisters from the network. While the network registration and deregistration are slow operations consuming lots of power, the net effect is still a significant reduction in average power consumption. The average current consumption of the device was measured to be 16 mA in daytime operating mode and 4.4 mA in nighttime mode.

2.2   Signal Processing

Some of the monitored parameters are computed on the device in order to decrease the amount of transmitted data and the rest, which are derived from the previous ones, are calculated in the back-end system. The heart rate and step counting algorithms were developed specifically for this system and for the other parameters, algorithms proposed in the literature were used. Minute by minute energy expenditure is estimated through the average heart rate by using method proposed by Keytel et al. in [5]. The model-based estimation algorithm considers the age, weight, and the gender of the person. The energy expenditure values are calculated in the back-end side, which is a natural choice, as the user information is stored there. Sleep/wake detection is based on the algorithm proposed by Cole et al. in [6].

The developed step counting algorithm exploits a two-phase approach by first classifying the type of activity, and then, if walking or running activity is detected, the steps are being counted. This approach enables using simple, adaptive threshold approach in the step counting phase. Proprietary heart rate estimation algorithm uses the slopes and the amplitude of the R-peak as features and compares the features of new R-peak candidates to the already detected R-peaks. The algorithm does not report an R-peak if the signal has too much noise. A relatively low 100 Hz sampling frequency is used for the ECG to minimize the computational load of the microprocessor. This sampling rate has still been found adequate for beat-to-beat interval estimation [7].

2.3 Web Portal User Interface

User can observe the recorded data via a web browser user interface partially shown in Fig. 1. When signing-in, the UI first provides the user with an intuitive summary view of user's activity status. This summary view takes into account the user behavior from the past seven days and presents the status with changing background color. All the monitored parameters are presented below the summary view. These are divided in four categories: activity, heart, rest and nutrition.

Heart rate tachogram is shown from a desired period together with a heart rate variability (HRV) index (not shown in Fig. 1). The HRV parameter displayed is the RMSSD (root-mean-square of successive differences) and features an automated function for obtaining comparable values of this index i.e. recognizes when the user is still and in relaxed orientation for long enough.

Daytime activity levels are categorized in four classes based on the heart rate and the activity information obtained with the movement sensor. The commonly used definitions of 55 and 70% of the maximum heart rate are used as thresholds between low and moderate, and moderate and high intensities. Sedentary time is indicated when the heart rate and activity level stay low for too long time. Activity category also includes steps taken and calories burned during the day.

Other features of the user interface, not shown in Fig. 1, include sleep/wake hypnogram and other derived sleep parameters: sleep latency, sleep efficiency and total sleep time. In addition, the dietary and nutrition section is not shown in Fig. 1. Due to the lack of convenient and easy-to-use solutions for food intake, we have currently taken the approach of providing the user with motivational dietary tips and recommendations.

3 Validation of the Heart Rate Algorithm

We evaluated the performance of the beat-to-beat heartrate estimation algorithm in two tests. The first HR test setup had a controlled test protocol of approximately 30-minute duration and the other one was a 24-hour recording in free-living conditions. The following test protocol was implemented in each controlled measurement: (1) normal walking and going stairs up and down, (2) laying on a bed, (3) riding an exercise bike (12–16 km/h), (4) running on a treadmill (speeds of 3, 5, 8 and 11 km/h), and (5) standing up. The duration of each phase was 3 min, and there was a break of 30 s between each phase. The data were analyzed for 10 subjects in controlled measurements and for 3 subjects in 24-hour measurements. The heart rate belt was not moisturized before the beginning of the tests to mimic the intended usage condition, which is long-term use throughout the day not just when going for a walk or to do some other exercise.

The subjects were young healthy adults. The subjects were informed about the study and they signed informed consent forms. The principles outlined in the Helsinki Declaration of 1975, as revised in 2008, were followed in the study. Reference data was recorded with Ambu Blue Sensor R-00-S disposable stress test electrodes placed under the right clavicle and on the left hip to measure approximately the Lead II ECG signal. The signal was recorded with Faros 360 ECG monitoring device from Bittium Biosignals Ltd. Faros is certified as Class IIa medical device. The reference ECG data was analyzed with professional Holter ECG analysis software, Cardiac Navigator from Bittium Biosignals. The detected R-peaks were visually verified and corrected. Some periods of the measurement had so high noise level in the reference ECG that the true R-peaks were not visually observable. These segments were marked as noise. In this algorithm evaluation test the reference results were compared to our own heart rate algorithm by loading the data to Matlab and running the algorithm.

R-R-intervals were found in both signals and the corresponding intervals were assigned to each other. The assignment was visually verified and corrected. If an extra beat was detected in the belt ECG, this R-R-interval was compared with the previous detected R-R-interval of the

Fig. 1 Tango™ web portal user interface

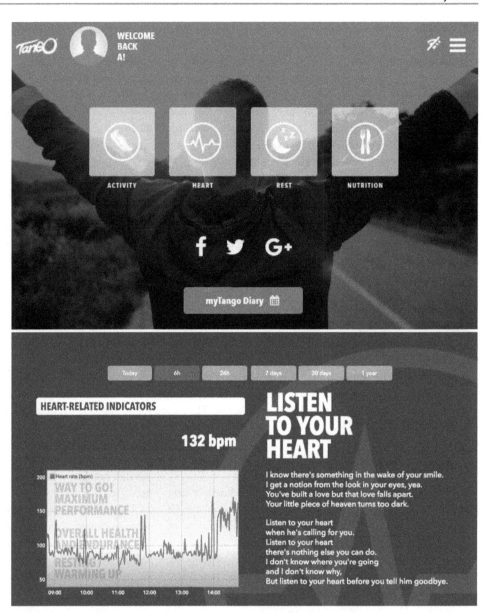

reference ECG. In case the belt ECG was missing an R-peak, the erroneous (too long) R-R-interval was assigned to both of the corresponding intervals in the reference ECG.

4 Results and Discussion

Table 1 shows the results of the algorithm evaluation. The performance of the system was approximately equal during both, the controlled test and the free-living conditions in terms of RRI estimation error. The coverage was significantly better during the 24-hour test due to generally smaller amount of movement during the recording. In the controlled measurement, two test subjects had the coverage of the belt ECG less than 30% due to the high-amplitude artefacts in the measurement signal. They both had the largest R-R

estimation error as well. With the other subjects, the belt ECG coverage was higher than 72% with mean absolute error less than 10.3 ms and rot mean square error less than 26 ms. In one test subject, the belt ECG provided higher coverage (82.88%) than the reference ECG (68.64%). The majority of the large beat detection errors occurred during short periods of RRI-series interrupted by bad-quality ECG or just before or after a break caused by bad-quality ECG. The sweating during the treadmill exercise improved the skin-electrode contact: all subjects had good-quality ECG at the end of the measurement, even though some of them had major artifacts in the ECG during the other phases of the measurement. The choice of not moisturizing the heart rate belt before starting of the recording was made on purpose for trying to mimic the intended usage conditions of the system. We assumed that the users would not be moisturizing the

Table 1 R-R-interval estimation performance during the two test protocols

	Controlled measurements (average (range)), (N = 10)	24-hour measurements (average (range)), (N = 3)
Mean absolute error (ms)	7.99 (4.69...26.34)	6.34 (5.79...7.30)
Root mean square error (ms)	19.10 (6.00...61.56)	20.67 (11.88...23.84)
Mean error (bias) (ms)	−0.53 (−10.34...1.83)	1.33 (−1.49...3.53)
Coverage (%)	76.24 (24.19...99.67)	94.17 (90.61...99.82)
Reference coverage (%)	95.64 (68.64...100)	99.84 (99.53...99.99)

heart rate belt electrodes when putting the device on for a day. However, it can be concluded based on the test results that moisturizing the electrodes is needed to obtain signal with adequate quality, at least if there will be excessive amount of movement before the skin-electrode contact improves naturally by sweating.

5 Conclusion

We have presented a system intended to be used as a motivator tool for people trying to improve their living habits. In wellbeing point of view, the system addresses the three main components of health: activity, sleep, and nutrition by supporting healthier lifestyle and providing tips for improving it. The future work will include evaluation of the system with subjects belonging to the actual target group. Also, the subjective perceptions will be evaluated with interviews. Technical development of the future will include implementing the respiration monitoring to gain better insight to users' fitness.

Conflict of Interest JR and TR are employees of Health Care Success Ltd, producer of Tango™ system.

References

1. Ramahi T M: Cardiovascular disease in the Asia Middle East region: global trends and local implications, Asia Pacific Journal of Public Health 22, 83S–89S (2010).
2. Vehkaoja A et al: System for ECG and heart rate monitoring during group training, Proc. 30th Annu Int Conf IEEE Eng Med and Biol Soc, pp. 4832–35, Vancouver, Canada (2008).
3. Paalasmaa J et al: Unobtrusive online monitoring of sleep at home, Proc. 34th Annu Int Conf IEEE Eng Med and Biol Soc, pp. 3784–88, San Diego, USA (2012).
4. Hassannejad H et al: Automatic diet monitoring: a review of computer vision and wearable sensor-based methods, Int. J Food Sci and Nutr, 68(6), 656–670 (2017).
5. Keytel L R et al: Prediction of energy expenditure from heart rate monitoring during submaximal exercise, J Sports Sci, 23(3), 289–297 (2005).
6. Cole R J et al: Automatic sleep/wake identification from wrist activity, Sleep, 15(5), 461–469 (1992).
7. Mahdiani S et al: Is 50 Hz high enough ECG sampling frequency for accurate HRV analysis?, Proc. 37th Annu Int Conf IEEE Eng Med Biol Soc, pp. 5948–51, Milan, Italy (2015).

Measuring the Response of Patients with Type I Diabetes to Stress Burden

Vit Janovsky, Patrik Kutilek, Anna Holubova, Tomas Vacha, Jan Muzik, Karel Hana, and Pavel Smrcka

Abstract

The article presents measurement methodology and systems for measuring and evaluating the response of patients suffering from type I diabetes to stress burden. The proposed methodology and systems have been developed to recognize the effects of strain on patients' reaction time and their work performance. Stress burden was measured by a multi-sensory monitoring system designed for this purpose. The methods used were based on monitoring physiological stress symptoms by measuring pulse rate, respiratory rate, temperature, galvanic skin resistance, and electrical activity of muscles. It was suggested to monitor the measured parameters in waveform, from initiation of stress stimuli, throughout the period of growth, and up to the point of decline. The research was conducted under the supervision of psychologists, and the proposed methodology for measuring stress burden and its impact on physiological functions of type I diabetics involved a group of patients and a control group of healthy subjects. The proposed measurement methodology would be beneficial not only in the design of systems for detecting mental stress, but also in the treatment of patients suffering from diabetes and the assessment of their physical/mental state while performing demanding work tasks.

Keywords

Type I diabetes • Stress burden • Physiological variables
Reaction time

1 Introduction

Relation between the stress burden of type I diabetes is studied primarily from the perspective of long-term development and the deterioration of health problems in patients with type I diabetes. Although several studies refer to the potential impact of long-term stress on the emergence of type I diabetes, [1, 2], others report almost no conclusive results and reject this relation. The same results were achieved in the past when stress-related research was conducted in patients with type II diabetes [3].

The research of glycemic changes control proves a correlation between glycemic control and mental strain if patients with type 1 diabetes suffer from long-term stress, [4, 5]. The relationship between short-term stress burden and glycemic control was investigated in [6, 7]. Although in most of the observed cases the correlation was confirmed, its extent varies greatly and can be considered a rather individual matter.

The quoted studies have not yet presented the research of short term stress measurements and comparison between diabetics and healthy subjects by means of physiological data monitoring. For this reason, this study focuses on methods and systems developed for measuring and evaluating short-term stress developed in diabetics and healthy subjects. A multi-sensory device for scanning physiological data and related research of mental stress has been developed specifically for

V. Janovsky · P. Kutilek (✉) · A. Holubova · J. Muzik
K. Hana · P. Smrcka
Faculty of Biomedical Engineering, Czech Technical
University in Prague, Sitna sq. 3105, Kladno, Czech Republic
e-mail: kutilek@fbmi.cvut.cz

V. Janovsky
e-mail: vit.janovsky@fbmi.cvut.cz

A. Holubova
e-mail: anna.holubova@fbmi.cvut.cz

J. Muzik
e-mail: muzik@fbmi.cvut.cz

K. Hana
e-mail: hana@fbmi.cvut.cz

P. Smrcka
e-mail: smrcka@fbmi.cvut.cz

T. Vacha
University Centre for Energy Efficient Buildings, Czech Technical
University in Prague, Třinecká 1024, Bustehrad, Czech Republic
e-mail: tomas.vacha@uceeb.cz

© Springer Nature Singapore Pte Ltd. 2019
L. Lhotska et al. (eds.), *World Congress on Medical Physics and Biomedical Engineering 2018*,
IFMBE Proceedings 68/3, https://doi.org/10.1007/978-981-10-9023-3_34

this purpose. The designed device seems to be applicable not only as a technology in monitoring systems for mental stress detection, but also in the treatment process of subjects with diabetes for the assessment of their physical/mental condition while performing labor-demanding tasks.

2 Methods

Based on the mentioned insufficient research in area of mental strain influence on people suffering from type I type diabetes, a pilot study focused on measuring the strain in I type 1 diabetics and healthy subjects has been proposed. The measurement methodology and an appropriate multi-sensory biomedical data monitoring system have been designed for the same purpose.

2.1 Participants

In order to analyze the subjects of the pilot study, five healthy subjects (control group—C) (aged 26 (SD 3.5)) and five type I diabetes patients (Pt) (aged 28 (SD 5.6)) were subjected to measuring. The diagnostic evaluation of the subjects included a detailed anamnesis, aneurological examination and routine laboratory testing. The subjects were selected randomly and on different days. The study was performed in accordance with the Helsinki Declaration. The study protocol was approved by the local Ethical Committee

and the Ethical Committee of the First Faculty of Medicine of Charles University, and informed consent was obtained from all subjects.

2.2 Measurement Equipment

To carry out measurements, the VLVlab Wireless Pocket Poligraphy (FlexiCare s.r.o., Czech Republic) was used, see Fig. 1. It is a device for the online monitoring of physiological data.

The system, which allows for measuring synchronized physiological data, was developed at a joint workplace of the Faculty of Biomedical Engineering of the Czech Technical University (CTU) and the First Faculty of Charles University in Prague. The selected version of the system enables the following measurements:

- temperature (TEMP)—measured in °C by an electronic thermometer on the surface of the body,
- respiratory rate (EIMP) measured in Ω units by electrodes, and determined on the basis of changes in bioimpedance of the chest,
- heart rate (HR) measured in Hz by standard adhesive, disposable ECG electrodes and chest strap heart rate monitor, and determined on the basis of channel ECG signal,
- galvanic skin resistance (GSR) measured in Ω units by an external finger probe.

Fig. 1 VLVlab pocket polygraph application

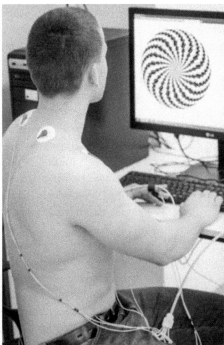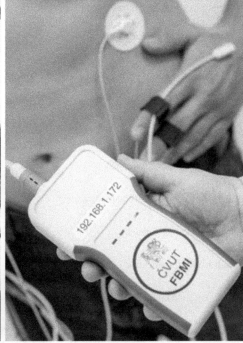

2.3 Test Procedure

The proposed VLVlab system is used for simultaneous TEMP, EIMP, HR and GSR measurements. The VLVlab system sensors are placed on each subject prior to measuring. The VLVlab system monitors changes in biomedical data over time and particularly in response to experimentally applied stimuli. Subsequently, monitored responses are compared between the groups of participants. The measurement methodology consists of a preparatory and testing phase. The objective of individual steps is to induce mental strain and to monitor physiological parameters affecting mental state, and further development of these parameters. The measurements are taken in the following phases:

(A) The preparatory phase (5 min.), consists of:

- initial explanation of the experiment and mounting the sensors of measuring equipment onto the subjects bodies
- measurement of biomedical data
- placing the participants at a table on the edge of the room

(B) The testing phase consists of:
- 3 min of individual preparation for the presentation within the simulated job interview: the participant is allowed to take notes but cannot use them
- 5 min. presentation in front of the jury. The jury shows no emotion. If a participant stops talking, the jury asks him supportive questions
- After the presentation has been completed, the jury will ask the proband to subtract the constant 17 from the number 2023. If he fails, the jury corrects the mistake, and let him start from the beginning.

The measured biomedical data is subsequently stored and prepared for pre-processing before the next analysis.

2.4 Data Processing Method

Three blocks of data were subjected to measuring. One block from the preparatory resting phase, and the other two blocks from the testing phases: one taken during preparation for the presentation and the other during the presentation itself. Data measured during the preparatory resting phase are used to determine the baseline of biomedical data, i.e. the data from this phase provide the values for determining median TEMP, EIMP, HR, and GSR. The advantage of using the median is that it is not easily distorted by extreme values caused by

artifacts, as opposed to the mean value, which is extremely sensitive to them. Subsequently, the median of the measured data is determined for the testing phase of presentation for preparation and the presentation itself. Since the researchers were interested in the changes of biomedical data over time, and particularly in those responding to experimental stimuli, and subsequent comparisons between groups of participants, the different median values are set for the resting phase and for individual testing phases. The mentioned difference in median values is set for all the measured subjects.

2.5 Statistical Analysis

First, the difference between the medians for the resting and testing phases was calculated for each subject. This was followed by the Jarque–Bera test testing the normal distribution of calculated median values relating to biomedical data in a group of patients and a group of healthy subjects, [8]. The median (Mdn), minimum (Min), maximum (Max), the first quartile (Q1) and the third quartile (Q3) of the biomedical data were calculated in the next step. The first quartile is defined as the middle number between the smallest number and the median of the data set. The third quartile is the middle value between the median and the highest value of the data set. These indicators serve for the statistical presentation of the results and possible future use. Another, the Wilcoxon test, was used to assess the significance of the differences between individual phases of measurements (preparatory phase for the presentation and the phase of presentation) and differences between the control group (C) and patients (Pt). The level of significance was set at $\alpha = 0.05$. The statistical methods were used according to [8] and the statistical analysis was performed by MatLab software.

3 Results

The Jarque-Bera test displayed the not normal distribution of data in some data sets. The statistical data served to illustrate the differences between the measurement phases in groups of patients and healthy subjects. The plot (Fig. 2) displays the Min, Max, Mdn, Q1, and Q3 for the calculated values in different phases and a deviation from the TEMP, EIMP, HR, GSR baseline in the resting phase. Neither the comparison of biomedical data values measured during the different phases of the test—nor the comparison of biomedical data values measured during the same testing phase showed significant statistical differences between the group of patients (Pt) and the control group (C). The value p calculated by the

Fig. 2 Comparison of the biomedical data measured in the resting and testing phase of preparation for the presentation (T) and those for the presentation itself (P). The measurements were taken in a group of patients (Pt) and a control group (C)

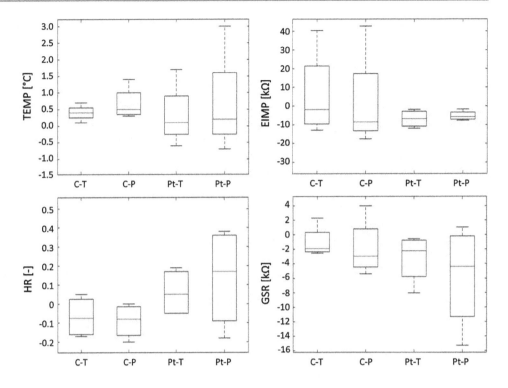

Wilcoxon test was always higher than the pre-set significance level $\alpha = 0.05$, i.e. the calculated values were higher than 0.25 in all the cases where statistically significant differences between groups of data were identified.

Future research should supplement the measured data with: ECG analysis, breathing curve parameters and motion activity data. It is also recommended that the groups consist of more subjects and that the observed subjects suffer from a different type of diabetes.

4 Discussion

In the preliminary research into the area of type I diabetics' response to stress, the authors presented the modules of the monitoring system, as well as the methodology for measurement and data processing. Preliminary results did not reveal statistically significant differences in physiological data between Pts and C, which contradicted their expectations, and will need to be subjected to a more in-depth medical analysis. The issues considered for follow up research are primarily focused on the study of further physiological data changes. Since the main objective of the current study was to design a system/methodology for monitoring physiological data of diabetics in stress situations, the goals of the research task have been met. Methodology functionality was tested in a laboratory environment. While performing measurements on 10 participants, the system worked without any loss of signal. The system can provide information about the changes in physiological data which may be essential for the optimisation of future measurements. As diabetic patients suffer from autonomic nervous disorders, it is difficult to compare autonomic nerve responses in stress responses with healthy subjects.

5 Conclusion

The article presented the application of the system for monitoring biomedical data and evaluation of their changes after the patients with type I diabetes are put under stress. The results did not show any significant differences between the biomedical data of healthy subjects and diabetics during the tests. The findings described, however, prove the ability of the proposed methods to identify differences in the physical and mental conditions of the subjects under observation.

Preliminary results would be beneficial for further research into the behaviour of type I diabetics and for the adjustment and extension of the monitoring system by involving more subjects in the experiment.

Acknowledgements This work has been supported by the Ministry of Education, Youth and Sports within National Sustainability Programme I, project No. LO1605. Also, this work has been supported by project SGS17/108/OHK4/1T/17 sponsored by Czech Technical Uni-versity in Prague.

Conflict of Interest The authors declare that they have no conflict of interest.

References

1. Surwit, R.S., Schneider, M.S., Feinglos, M.N.: Stress and diabetes mellitus. Diabetes Care 15(10), 1413–1422 (1992).
2. Hägglöf, N., Blom, L., Dahlquist, H., Lönnberg, G., Sahlin, B.: The Swedish childhood diabetes register: indication of severe psychological stress as a risk factor for Type 1 (insulin dependent) diabetes mellitus in childhood. Diabetologia 34(8), 579–583 (1991).
3. Trovato, G.M., Catalano, D., Martines, G.F., Spadaro, D., Di Corrado, D., Crispi, V., Garufi, G., Di Nuovo, S.: Psychological stress measure in type 2 diabetes. Eur Rev Med Pharmacol Sci. 10 (2), 69–74 (2006).
4. Delamater, A.M., Patiño-Fernández, A.M., Smith, K.E., Bubb, J.: Measurement of diabetes stress in older children and adolescents with type 1 diabetes mellitus. Pediatr Diabetes14(1), 50–6 (2013).
5. Kamody, R.C., Berlin, K.S., Hains, A.A., Kichler, J.C., Davies, W. H., Diaz-Thomas, A.M., Ferry, R.J.: Assessing Measurement Invariance of the Diabetes Stress Questionnaire in Youth With Type 1 Diabetes. Journal of Pediatric Psychology 39(10), 1138–1148 (2014).
6. Kramer, J.R., Ledolter, J., Manos, G.N., Bayless, M.L.: Stress and metabolic control in diabetes mellitus: Methodological issues and an illustrative analysis. Annals of Behavioral Medicine 22(1), 17–28 (2000).
7. Eom, Y.S., Park, H.S., Kim, S.H., Yang, S.M., Nam, M.S., Lee, H. W., Lee, K.Y., Lee, S., Kim, Y.S., Park, I.B.: Published online 2011 Apr 30. Evaluation of Stress in Korean Patients with Diabetes Mellitus Using the Problem Areas in Diabetes-Korea Questionnaire. Diabetes Metab J. 35(2), 182–187 (2011).
8. Kutilek, P., Volf, P., Cerny, R., Hejda, J.: The application of accelerometers to measure movements of upper limbs: Pilot study. Acta Gymnica 47(1), 24–32 (2017).

Wearable Sensors and Domotic Environment for Elderly People

Sergio Ponce, David Piccinini, Sofía Avetta, Alexis Sparapani,
Martín Roberti, Nicolás Andino, Camilo Garcia, and Natalia Lopez

Abstract

Technology is an important instrument for a new concept of home care and continuous monitoring. New and smaller devices and communication advances have constructed a more humanitarian model of assisted alternative to hospital. For this purpose an integral scheme must to consider physiological, emotional, social and environmental conditions that promote the multidisciplinary support for the patients. In this work we present an Integral Assistive Home Care System, specially designed for elderly or chronic illness people. The approach proposed comprises a user wearable device, a domotic system's core installed in a personal computer (PC) and an ichnographic software (SICAA) that allows the interaction of the patient with the environment and peripheral devices. Wearable sensors system have a master module that deals with data acquisition, synchronization and wireless transmission, connected to sensors or slaves which acquire biological signals and process them to minimize the amount of data to be transmitted by Bluetooth. The biologic variables (each with its own specific acquisition and preprocessing module) acquired are temperature, heart rate and pulse oximetry, and kinetics measurements through an inertial sensor IMU. The domotic SICAA soft and control hardware was designed to achieve some automatic tasks through an ichnographic software. The programmed functions comprises: house control (that comprises blinds, lights, orthopedic bed, air conditioner, television, and intercom); medication alarm; career communication (nurse call, voice synthesizer), and computer access (internet, chat, games, and text processors). The entire system is low cost, modular and adaptable for different user's capabilities and pathologies.

Keywords

Wearable • Sensor • IoT • Domotic • Elderly

1 Introduction

Medical technology is a key element in global problems concerning health, as it can improve the quality of life or extend it through the appropriate use and prevention. Its value is calculated not only for its economic cost but also for its impact in practice and the success of medical treatments. Due to medicine improvements, life expectancy has increased. Along with this, the caring for the elderly and people with chronic diseases have also improved, emphasizing the need of preventive medicine and monitoring of these patients. Also, the concept of Home Care appears as a promising alternative for these patients, preserving the social and familiar support and relations. Domotic is a growing area of commercial devices, due to the comfort search and home automation, but not specific for patients home care. Several attempts have been proposed, for monitoring daily life activities [1] and vital signs [2], between others, focusing the attention in alarms and emergency help. On the other side, new trends in Internet of Things have introduced the remote control of home appliances and electronic devices in order to optimize energy consumption and comfort. The control input is through cell phones, voice commands or remote access, through Wi-Fi or Bluetooth communication [3–5]. However, all these approaches are adapted to elderly or chronic people and not fully designed for them. The main objective of this work is the design and implementation of a non-obtrusive system, that accomplish with technical requirements of wearable systems, like low weight, small

S. Ponce (✉) · D. Piccinini · S. Avetta · A. Sparapani · M. Roberti
N. Andino · C. Garcia · N. Lopez
Universidad Tecnológica Nacional, F. R. San Nicolás, GADIB,
San Nicolás de los Arroyos, Argentina
e-mail: sponce@rec.utn.edu.ar

N. Lopez
Universidad Nacional de San Juan, GATEME, CONICET, San
Juan, Argentina

© Springer Nature Singapore Pte Ltd. 2019
L. Lhotska et al. (eds.), *World Congress on Medical Physics and Biomedical Engineering 2018*,
IFMBE Proceedings 68/3, https://doi.org/10.1007/978-981-10-9023-3_35

size, battery autonomy and versatility to different patients, pathologies and geographic situation.

2 Materials and Methods

The system has the following main blocks: wearable sensors network, home automation network, scenarios analyzer and information management (Fig. 1).

2.1 Wearable Sensors Network

Health monitoring is accomplished through the wearable sensors network, that is unobtrusive, lightweight and low consumption. The physiological variables chosen for this application were corporal temperature (T°), blood oxygen saturation (SpO2), heart rate (HR), kinetic measurements (KM) and voice. The sensor nodes are based in the microcontroller CC2650, with Bluetooth 4.0 Low Energy (LE) of Texas Instruments® [6]. The processor is a M3Cortex 32bits, at 48 MHz. The benefit of this configuration is the low size and low energy consumption that allows the autonomy for long time periods with a CR2450 battery (3 V and 600 mAh).

The SpO2 node uses the AFE4403 embedded circuit, (Texas Instruments®), which consist in an analogical Front-End. For more information, see [7]. For temperature acquisition the sensor used was the LMT70, Texas Instruments ®, small size such as 1×1 mm and 0.05 °C of accuracy in the range of 20–42 °C [8]. To voice register the

choice was a microcontroller with WiFi radio embedded. The microphone signal is filtered and digital codified by a TLV320 codec through Pulse Code Modulation (PCM) sampled at 22 kHz and with a resolution of 16bits [9]. This codec send the data under the I2S communication protocol to the microcontroller, which is in turn connected to the WiFi net for the Audio Over IP (AOIP) distribution using the User Datagram Protocol (UDP). This slaves or sensor nodes are schematized in Fig. 2.

The sensor node communication is achieved by a wireless link Bluetooth LE and concentrated in a collector that allows the relation between this net and another WiFi net, that pickup the information from all the sensors, including the other blocks sources. The battery level is also included in the information and all the data packages are sent to the scenario analyzer block in order to decide the interaction with the control actions, such as familiar alarm, emergency alarm, domotic control, and so on.

2.2 Domotic Network

Home-User interaction is done through a network of sensors and automated devices communicated through the MQTT message protocol [10]. This protocol proposes a Publicant-Suscriptor model between computers via WiFi, and can be established with topics that is represented by a hierarchical chain. The topics are created by the publicant and the nodes must be subscripted and communicated with him (Fig. 3). The basis of the MQTT is the TCP/IP protocol, but is a much better option due to the small size and the

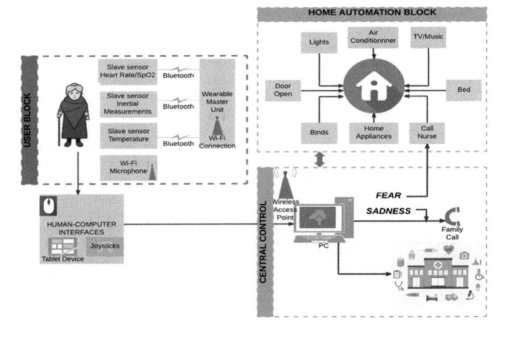

Fig. 1 General diagram: The user block comprises the wearable sensors network and the relation with the input HCI; the home automation block and the Central Control, for scenario analysis and information management

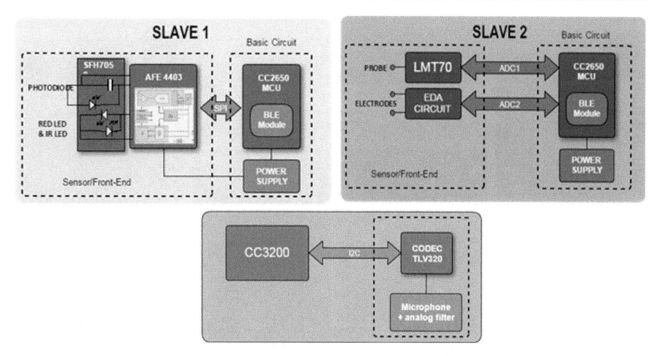

Fig. 2 **a** Block diagram for SpO2 and Heart Rate Sensor. The operation is based in the classical principles of IR and Visible Light absortion. Heart rate is obtained from signal processing. **b** Block diagram for Body Temperature Sensor. The data are sampled one per minute, for energy saving. EDA block is Electrodermic Activity, not tested yet. **c** Block diagram for Voice Register Node

secure message delivery. The Broker chosen was the EMQ, because is open source, is scalable, allow coworking and secure connections by SSL, and can be used with Big Data applications (>1 M package per second).

Several nodes are distributed in home, in order to acquiere variables such as room temperature, luminosity, carbon monoxide, doors open and movement detection on circulation zones. Also, devices nodes are necessary to lights, air conditioner, blinds, and TV control. An important node is the call to nurse. The user commands the home

appliances with different Human–Computer Interfaces (HCI), i.e. a tablet [11] (Fig. 4).

2.3 Scenario Analyzer

The determination of patient's situation is the ultimate objective of the work. For this purpoe, the sensor information is analyzed in order to interact with home appliances and activate the notification and alarms mechanisms if it is necessary. For this purpose a graphic web interface (Node-RED [12]) was designed to interconnect easily the blocks that represents the nodes distributed in the home. The information is analyzed and the system decides the action to pursuive.

2.4 Information Management

Sensors information is stored in a database of time series, that allows the analysis of the evolution of one variable or the multivariable study, which is our concern. The impact of habits changes in physiological and behavioral variables can be reflected in these time series. A time series database greatly facilitates the process of searching, collecting and validating information. For all the above, it is the choice for solutions where it is necessary to manage large volumes of

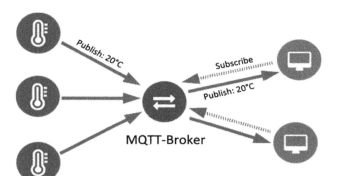

Fig. 3 Message exchange between nodes using MQTT protocol

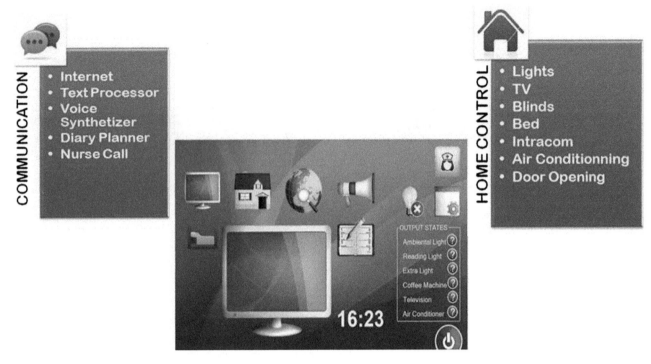

Fig. 4 Domotic UI -User Interface SICAA for TV, PC or mobile tablet

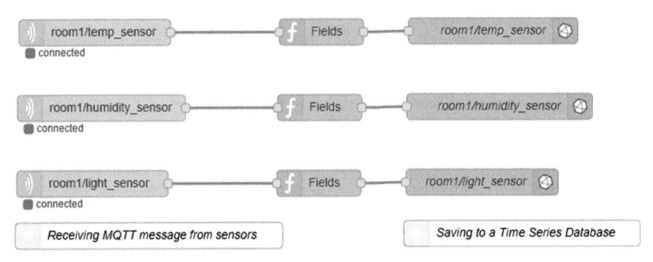

Fig. 5 Node-RED as easy way for interact with sensors

Fig. 6 Node Wearable modules. Master (left) and Slaves (right) modules for Hearth Rate/SpO2, IMU, Temperature, and microphone. Is important to note the size related with a wristwatch

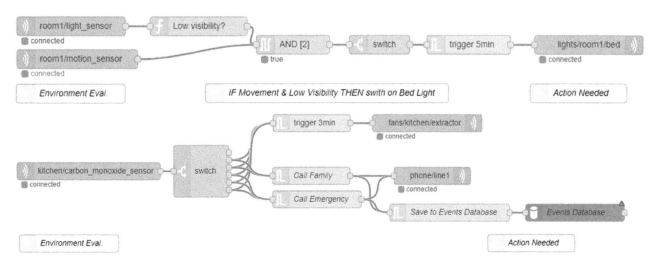

Fig. 7 **Up**: Example for movement detection in bedroom at night. **Down**: Example for Carbon Monoxide detection and alarm trigger

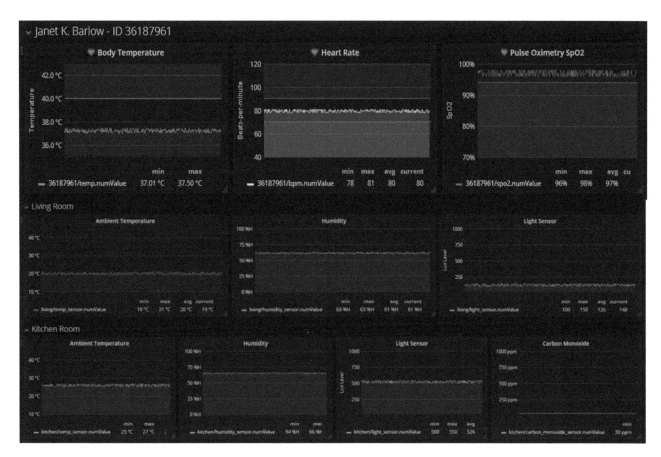

Fig. 8 **Up**: Real-time graph showing wearable sensors' values for a specific patient. **Down**: Real-time graph showing patient's environment variables

information and analysis in real time. In our platform, we chose to use InfluxDB, from InfluxData [13]. On the other hand, the variables are monitored and analyzed using Grafana [14]. This solution allows us to obtain real-time graphs of all the variables collected by the system and visualize them in an interactive web interface (Fig. 5).

2.5 Blocks Connection

The whole system is constructed above a virtual structure based in XenServer [15]. This is a flexible and modular scheme that allows the addition of new services, computing power and hardware changes without problems or reconfigurations (even in fail cases). Each virtualized system behaves as if it was a standalone PC, and each one is called Virtual Machine (VM) (Fig. 6).

3 Results

All sensors were designed and implemented in small size, for a first validation and test. The oxymeter and heart rate data were compared with a commercial device (Medix OXi-3), which was taken as a standard to calculate the absolute measurement error in the 10 volunteers. The values obtained (error average of 2%) show an acceptable performance [16]. Related to the graphical programming of the action plan for different situations with Node-RED, Fig. 7 shows two posible situations and the solutions proposed. Graphical interface and time series visualization is presented in Fig. 8. The Grafana interface allows us to obtain real-time graphs of all the variables collected by the system and visualize them in an interactive web interface.

4 Conclusions

The system presented is a complete solution for home and patient monitoring, constituting a suitable alternative for elderly and chronic people. All electronic designs were tested and validated with acceptable performance and the communication protocols and graphical interfaces are according with the main objective of the work. In the next stage of the project, home automation of volunteers must be done in order to conclude in the social impact of the system.

References

1. Xiang Y, Tang Y-p, Ma B-q, Yan H-c, Jiang J, Tian X-y (2015) Remote Safety Monitoring for Elderly Persons Based on Omni-Vision Analysis. PLoS ONE 10(5): e0124068. https://doi.org/10.1371/journal. pone.0124068.
2. Morris ME, Adair B, Miller K, Ozanne E, Hansen R, et al. (2013) Smart-Home Technologies to Assist Older People to Live Well at Home. Aging Sci 1: 101. https://doi.org/10.4172/jasc.1000101.
3. Ziyu Lv; Feng Xia; Guowei Wu; Lin Yao; Zhikui Chen. iCare: A Mobile Health Monitoring System for the Elderly. Green Computing and Communications (GreenCom), 2010 IEEE/ACM Int'l Conference on & Int'l Conference on Cyber, Physical and Social Computing (CPSCom).
4. Bilal Ghazal; Khaled Al-Khatib. Smart Home Automation System for Elderly, and Handicapped People using XBee. International Journal of Smart Home 9(4):203–210 • April 2015. https://doi.org/10.14257/ijsh.2015.9.4.21.
5. Akanbi C.O.; Oladeji D.O. Design of a Voice Based Intelligent Prototype Model for Automatic Control of Multiple Home Appliances. Transactions on Machine Learning and Artificial Intelligence. Vol.4 Issue 2. 2016.
6. http://www.ti.com/lit/gpn/cc2650.
7. http://www.ti.com/lit/gpn/afe4403.
8. http://www.ti.com/lit/gpn/lmt70.
9. http://www.ti.com/lit/gpn/tlv320aic23b-q1.
10. http://mqtt.org/.
11. Natalia López, David Piccinini, Sergio Ponce, Elisa Perez. Martín Roberti. From Hospital to Home Care. Creating a domotic environment for elderly and disabled people. IEEE Pulse. 2016 May–Jun; 7(3):38–41. https://doi.org/10.1109/mpul.2016.2539105. ISSN 2154-2287.
12. https://nodered.org/.
13. https://www.influxdata.com/.
14. https://grafana.com/.
15. https://xenserver.org/.
16. S. Ponce, N. Lopez, D. Piccinini, M. Roberti, M. Caggioli, S. Avetta, N. Andino, A. Sparapani C. Garcia and O.L. Quintero. Sistema vestible para detección de estados fisiológicos y emocionales en entornos industriales. XVII Latin American Conference of Automatic Control (CLCA). Medellín, Colombia. 13t al 15t de Octubre, 2016.

Videogame Implementation for Rehabilitation in Patients with Parkinson Disease

F. Chacha, J. P. Bermeo, M. Huerta, and G. Sagbay

Abstract

The present paper is focused on the rehabilitation of patients with Parkinson Disease (PD), using video games that respond to the movements of the player. It is important, to begin reviewing the state of the art, to delimit the characteristics of the system. Subsequently, the system was designed and implemented for use with software Unity and Arduino. Finally, the validation tests of the system were performed with patients of the PD and healthy patients. First, a review of the theoretical framework of Parkinson's disease, rehabilitation, and videogames present today is carried out, analyzing advantages and disadvantages of the different systems existing in the market. Then, the characteristics that the game must fulfill are defined, such as the transmission technologies, methodologies to be implemented, equipment and software that will be used for the design. Next, the videogame design is done using the Unity software, the same one that allows to export it to the Android platform. The system consists in acquiring patient data using a Bluetooth module and an accelerometer, whose data results from the game are sent wirelessly to the mobile device, for the project were used a smartphone. The "Rehabilitation" application contained PlaneGame with three levels and BallGame with two levels of participation. Finally, the validation of the system and analysis of results were executed, where the tests are carried out at the "Adulto Mayor" (elderly) University. In conclusion, the results at the end of the rehabilitation sessions showed that the developed system improves the motor movements of the upper extremities, which favors raising the life quality of patients.

Keywords

Android • Rehabilitation • Parkinson disease
Videogames

1 Introduction

Parkinson's disease continues to be one of the most common disorders in society that affects a greater number of women than men of adult age, causing to progressively lose their functional abilities and motor skills, since there is an absence of dopamine in the mescenfalo of brain.

This disease is characterized by movement disorders, such as muscle stiffness, tremors, slowness of movement (bradykinesia) and, in extreme cases, a loss of physical movement [1–3], since commonly patients with PD tend to be inactive [4] and through inactivity, secondary disorders may arise, which include decreased aerobic capacity, decreased muscle function, decreased joint mobility and decreased bone quality. Actions such as reaching, grasping, manipulating, and replacing objects often cause problems when performing activities such as dressing and eating causing the reduction of joint speed. [5]

For the control of the symptoms medical treatments, surgical and conventional techniques of rehabilitation are established, this often this include expensive equipment in specialized centers, where the patient must to go daily to improve their life conditions, which causes to lose interest or feel tired and do not go anymore. For this reason, there is another type of rehabilitation called Exergaming, which consist of strength, balance, and flexibility activities for patients to exercise while playing video games integrating physical effort or sedentary movements in response to the

F. Chacha · J. P. Bermeo (✉) · M. Huerta · G. Sagbay
Universidad Politécnica Salesiana, Cuenca, 010105, Ecuador
e-mail: jbermeo@ups.edu.ec

F. Chacha
e-mail: fchachac@est.ups.edu.ec

M. Huerta
e-mail: mhuerta@ups.edu.ec

© Springer Nature Singapore Pte Ltd. 2019
L. Lhotska et al. (eds.), *World Congress on Medical Physics and Biomedical Engineering 2018*,
IFMBE Proceedings 68/3, https://doi.org/10.1007/978-981-10-9023-3_36

demands of the game in a fun, entertaining, and it can be used with patients of any age. [6, 7]

Various studies have been developed to determine the importance of Exergames for the prevention and rehabilitation of patients with neurodegenerative diseases [8]. In the treatment of PD, consoles of Nintendo Wii and Microsoft Kinect have been used to improve certain patient conditions, such as the WuppDi application, which consists of a mini-game to improve three main aspects: movement, movement coordination and improvement of the concentration in patients with PD. [9]

This paper presents the design of a low-cost system that will improve the rehabilitation of the upper extremities in mature adult people who have EP and people healthy. The design is detailed in [10] and the tests are done with a mobile phone, a tablet and finally the system is compared with conventional therapy.

2 Methodology

2.1 System Design

Figure 1 shows the system design. Figure 1a shows the block diagram of the system, which consists of four stages: the stage of detection and acquisition of data using the developed device, the transmission of data through the Bluetooth module, the stage of video games that were developed in Unity and finally the stage of analysis of results.

2.2 Design Device

The developed device has low cost elements, among them: Arduino, Bluetooth module, Accelerometer sensor and batteries. It is easy to use since it has a wristband so that the participant feels comfortable. In Fig. 1b can see the elements of the device and its connection.

2.3 Detection and Acquisition

In the detection phase, the accelerometer allows to detect the values of each movement made by the participant.

Then, in the acquisition of data, the device is placed in the participant's hand, and the rehabilitation exercises are carried out: supination, pronation, flexion, and extension, as shown in Fig. 2a.

2.4 Videogame

The video game was developed on the Unity3D platform, and it contents two minigames FunnyPlaneGame y RedBall Game which are described in [10], after to the first tests carried out with the system in [10], improvements were added with three menus: Start Session, (Ss), Therapist Parameters (TP) and Results (R). The Fig. 2b shows the menu. The Ss button allows each participant to register with their name, so that the game data is saved as they play. The TP button allows a health person to control the

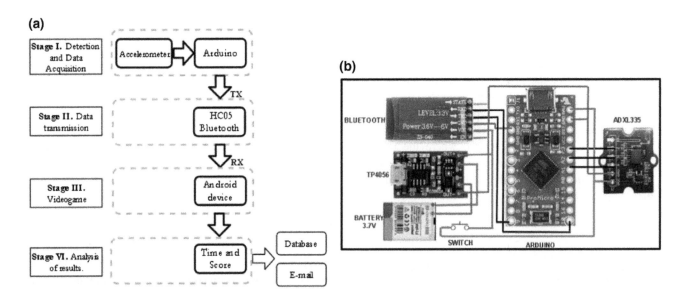

Fig. 1 **a** Diagram of system, **b** Connection diagram of the device elements

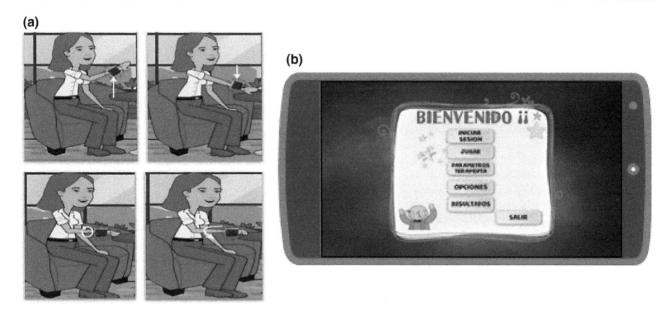

Fig. 2 **a** Rehabilitation exercises for patients with PD, **b** Main menu of video game

participant's playing time, where they must enter the time data they wish to reach. Finally, the R button allows you to generate the participant's data in a .txt file for its better visualization and later it can be sent by email to the person in charge.

3 Results

A study was conducted in 26 older adults between sixty-two and eighty-five, where six have Parkinson's disease and 20 were healthy adults. 16 women (Four with PD) and ten men (two with PD).

3.1 Data Obtained with Mobile Phone

The consideration that was had was that each participant is placed in a chair, the sensor device is placed, first in the right hand and then in the left, at 30 cm from the Mobile device or the tablet. The participants had 5 test sessions twice per day, using an Android phone, they practiced during two minutes with the device in each hand,

The results indicate the time and score obtained from the first session until the last session, during the two mini-games, both for patients with Parkinson's and healthy people. The Table 1 shows the results for time and score during the FunnyPlaneGame with the PD patients, when the test sessions end, an increase in score and a decrease in playing time is observed.

Table 1 Data obtained in FunnyPlaneGame by PD Patients

Hand	1st Session	5th Session	Percentage change (%)
Time [sec.]			
Right	61.88	49.33	−20.3
Left	65.66	50.33	−23.3
Average	63.77	49.83	−21.9
Score [points]			
Right	37.5	75	100
Left	27.5	51.26	86.4
Average	32.5	63.13	94.2

The Table 2 shows the results during the RedBallGame, the game time decreases by about 20% with respect to the first session, while the score increases by 93.5%.

In the same way, the results are presented with healthy patients, the Table 3 shows the data of time and score for the FunnyPlaneGame where there is a time decrease by 12.3% and the score increase about 136.4%.

For the RedBallGame, the data are different as seen in the Table 4 where the average of time when finish the sessions is about 23.7%, while the score increases considerably.

3.2 Comparison with Conventional Therapy

Tests were conducted to compare and validate the proposed system in relation to conventional therapy, so that there are five patients with PD and eight healthy patients. The Table 5 shows the characteristics of participants.

Table 2 Data obtained in RedBallGame by PD Patients

Hand	1st Session	5th Session	Percentage change (%)
Time [sec.]			
Right	60.99	47.57	−22.0
Left	60.08	49.41	−17.8
Average	60.53	48.49	−19.9
Score [points]			
Right	37.5	77.5	106.7
Left	40	72.5	81.3
Average	38.75	75	93.5

Table 3 Data obtained in FunnyPlaneGame by Healthy People

Hand	1st Session	5th Session	Percentage change (%)
Time [sec.]			
Right	50.12	46.13	−8.0
Left	55.04	46.05	−16.3
Average	52.58	46.09	−12.3
Score [points]			
Right	42.5	100	135.3
Left	40	95	137.5
Average	41.25	97.5	136.4

Table 4 Data obtained in RedBallGame by Healthy People

Hand	1st Session	5th Session	Percentage change (%)
Time [sec.]			
Right	54,1	37,83	−30.1
Left	49,04	40,88	−16.6
Average	51,57	39,35	−23.7
Score [points]			
Right	42,5	110	158.8
Left	40	95	137.5
Average	41,25	102,5	148.5

The fifteen participants had two daily sessions for ten days with the proposed rehabilitation system and conventional therapy, at the end of which a motor test was performed using tweezers for two minutes on each hand.

The motor test last for two minutes in each arm, where each one had to catch the largest number of pompoms with the indicated color and place them inside the container, and the score is the amount of pompoms collected correctly. The average results obtained from the test after conventional therapy with the 15 participants and after videogame therapy are shown in Fig. 3.

Table 5 Information of Participants for comparison with conventional therapy

	State	Sex	Age		State	Sex	Age
Patient 1	PD	F	82	Patient 9	Healthy	F	71
Patient 2	PD	M	71	Patient 10	Healthy	M	72
Patient 3	PD	F	73	Patient 11	Healthy	F	66
Patient 4	PD	F	83	Patient 12	Healthy	F	58
Patient 5	PD	M	74	Patient 13	Healthy	F	70
Patient 6	Healthy	M	67	Patient 14	Healthy	M	65
Patient 7	Healthy	M	71	Patient 15	Healthy	F	80
Patient 8	Healthy	F	73				

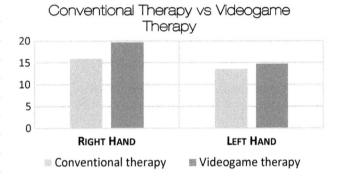

Fig. 3 Motor test

4 Conclusions

The proposed system is a good option for a new type of therapy using video games in patients with PD, the tests performed with the mobile phone show good results improving the time and score of the game, the results show improvements in the decrease of time and increase of score since it has bigger screen size which allows each participant to observe the video game more clearly.

The participants who used conventional therapy obtained an average of 15.93, while the participants who used the proposed system of videogame therapy obtained an average of 19.67 using the right hand. For the left hand, an average of 13.53 is obtained with conventional therapy and 14.73 with the proposed system, denoting better performance with videogame than conventional therapy.

Acknowledgements The authors gratefully acknowledge the support of the Project named "SISMO-NEURO: Análisis del Movimiento Corporal en Enfermedades Neurológicas" (Analysis of the body movement in neurological diseases) of Universidad Politécnica Salesiana from Ecuador.

Conflict of Interest statement The authors have no conflict of interest.

References

1. S. Date, T. Tashiro, K. Nozaki, H. Nakamura, S. Sakoda, and S. Shimojo, "A grid-ready clinical database for Parkinson's disease research and diagnosis," *Proc.-IEEE Symp. Comput. Med. Syst.*, pp. 483–488, 2007.
2. H. Mohammadi-Abdar, A. L. Ridgel, F. M. Discenzo, and K. A. Loparo, "Design and Development of a Smart Exercise Bike for Motor Rehabilitation in Individuals with Parkinson's Disease," *IEEE/ASME Trans. Mechatronics*, vol. 21, no. 3, pp. 1650–1658, 2016.
3. "Homepage | National Parkinson Foundation." [Online]. Available: http://www.parkinson.org/. [Accessed: 28-Aug-2017].
4. C. Tranchant, M. Koob, and M. Anheim, "Parkinsonism and Related Disorders Parkinsonian-Pyramidal syndromes : A systematic review," *Park. Relat. Disord.*, 2017.
5. E. Fertl, A. Doppelbauer, and E. Auff, "Physical activity and sports in patients suffering from Parkinson's disease in comparison with healthy seniors.," *J. Neural Transm. Park. Dis. Dement. Sect.*, vol. 5, no. 2, pp. 157–161, 1993.
6. F. Li *et al.*, "Tai Chi and Postural Stability in Patients with Parkinson's Disease," *N. Engl. J. Med.*, vol. 366, no. 6, pp. 511–519, Feb. 2012.
7. C. J. Winstein, P. S. Pohl, and R. Lewthwaite, "Effects of physical guidance and knowledge of results on motor learning: support for the guidance hypothesis.," *Res. Q. Exerc. Sport*, vol. 65, no. 4, pp. 316–323, Dec. 1994.
8. M. Pirovano, E. Surer, R. Mainetti, P. Luca, and N. A. Borghese, "Exergaming and rehabilitation : A methodology for the design of effective and safe therapeutic exergames q," *Entertain. Comput.*, vol. 14, pp. 55–65, 2016.
9. O. Assad *et al.*, "WuppDi !–Supporting Physiotherapy of Parkinson's Disease Patients via Motion-based Gaming," no. January, 2011.
10. F. Chacha, J.P. Bermeo and M. Huertas, "A rehabilitation system for upper limbs in Adult Patients using video games" of 2017 IEEE 37th Central America and Panama Convention (CONCAPAN XXXVII), pp. 1–6, 2017.

Design of a Multisensory Room for Elderly People with Neurodegenerative Diseases

F. Duchi, E. Benalcázar, M. Huerta, J. P. Bermeo, F. Lozada, and S. Condo

Abstract

Multisensory stimulation in older persons is an effective practice that helps to train the mind and motor skills through elements that stimulate the senses of the people. The spaces where specialists work the sensory stimulation are the so-called multisensory rooms "Snoezelen". Initially, the use of these rooms was mainly intended for children with learning difficulties, those that had difficulties to explore their environment. Recently, last few years, there have been realized investigations of the implementation of these rooms in persons who present cognitive deteriorations from moderate to severely and neurodegenerative pathology as Parkinson's, Dementia, Alzheimer's, Huntington's, Bipolar disorder, among others. This project developed and implemented a Multisensory Black Room for older patients with neurodegenerative diseases and cognitive impairment. Different sensory stimulus were used to help the cognitive and functional sphere of older people. The implemented room has some elements: stairs of colors, star curtain, fiber optic shower, textures path, virtual reality glasses and sound therapy. We test our room with a group of older people, some of them form a control group without room stimulation and the others used the room during 3 months. The results indicate that it has been possible to reduce the

F. Duchi · E. Benalcázar · M. Huerta (✉) · J. P. Bermeo
Universidad Politécnica Salesiana, Cuenca, Ecuador
e-mail: mhuerta@ups.edu.ec; mhuerta@ieee.org

F. Duchi
e-mail: hduchi@est.ups.edu.ec

E. Benalcázar
e-mail: ebenalcazarp@est.ups.edu.ec

J. P. Bermeo
e-mail: jbermeo@ups.edu.ec

F. Lozada · S. Condo
Centro Gerontológico Hogar Miguel León, Cuenca, Ecuador
e-mail: pscfher@gmail.com

S. Condo
e-mail: disetocuenca@gmail.com

aggressiveness pattern and to evolve in the functional part (fine-gross motor) and focus the attention that was dispersed in patients with neurodegenerative diseases and cognitive impairment in the early and late stages. In addition, patients improve the relationship with their social and personal environment, since the initiative of the room is to provide an atmosphere of wellness and relaxation, for the patient and the specialist.

Keywords

Snoezelen room • Multi-sensory room • Dementia Elderly • Therapy

1 Introduction

Multisensory stimulation in older persons is an effective practice that helps to train the mind and motor skills through elements that stimulate the senses of the people. The spaces where specialists work the sensory stimulation are the so-called multisensory rooms "Snoezelen" [1, 2]. The functions that can be promoted in a Snoezelen room are: (1) Relaxation; (2) Development of self-confidence; (3) Achieve sense of self control; (4) Encourage exploration and creative activities; (5) Establish rapport with care takers; (6) Provide leisure and enjoyment; (7) Promote choice; (8) Improve attention span, and (9) Reduce challenging behaviors [3–5].

Initially, the use of these rooms was mainly intended for children with learning difficulties, those that had difficulties to explore their environment [6–9]. In 1987, in Whittington (United Kingdom), the first Snoezelen facility was created in a center for mentally disabled adults, with six different multisensory environments. Recently, last few years, there have been growing the number of realized investigations of the implementation of these rooms in persons who present cognitive deteriorations from moderate to severely and neurodegenerative pathology as Parkinson's, Dementia,

© Springer Nature Singapore Pte Ltd. 2019
L. Lhotska et al. (eds.), *World Congress on Medical Physics and Biomedical Engineering 2018*,
IFMBE Proceedings 68/3, https://doi.org/10.1007/978-981-10-9023-3_37

Alzheimer's, Huntington's, Bipolar disorder, among others [10–13]. In therapy, participants have no obligation to learn new things and are given the opportunity to feel, relax, explore and experience [14].

The benefits of a multisensory room are not only reflected in patients. Because the care of patients with dementia or brain damage are accompanied by a high dependency on attention and behavioral disorders, caregivers suffer a heavy workload. Studies have shown that there are significant effects in relation to workload, stress, job satisfaction and exhaustion of caregivers [15]. Its usefulness has meant that today the therapy is extended to more groups of people such as the disabled centers that have widely adopted the use of them, as well as the geriatric centers [2, 14]. Currently in Ecuador 7% of its inhabitants are over 65 years old, this part of society is associated with an increase in the probability of presenting cognitive deterioration due to aging.

This project developed and implemented a Multisensory Black Room for older patients with neurodegenerative diseases and cognitive impairment. Different sensory stimulus were used to help the cognitive and functional sphere of older people. The stimulation elements included were: visuals (fiber optic shower, stair od colors, star curtain, virtual reality glasses), tactile (textures path), auditory (sound therapy) and an interactive lighting system for the environment.

We test our room with a group of older people, some of them form a control group without room stimulation and the others used the room during 12 weeks, 5 days a week, 30 min each day. The results comparing control group and stimulated group indicate that it has been possible to reduce the aggressiveness pattern and to evolve in the functional part (fine-gross motor) and focus the attention that was dispersed in patients with neurodegenerative diseases and cognitive impairment in the early and late stages. In addition, patients improve the relationship with their social and personal environment, since the initiative of the room is to provide an atmosphere of wellness and relaxation, for the patient and the specialist.

2 Dementia

Dementia is not a disease in itself but the symptomatic result of various lesions in the central nervous system. Also, is accompanied by the deterioration of the cognitive and functional faculties of the patient. It produces language and memory disorders, which leads to the inability to do activities of daily life. Such symptoms are combated by pharmacological therapies or with other types of therapies such as the realized ones in multisensory rooms [16].

We used a short test of mental status that can be performed in about 5 min in an outpatient setting. It is quick

tool that allows the psychologist to quantitatively measure mental functions: orientation, learning, attention, mathematical calculations, abstraction, information, construction, remembering. The scores are the following: normal (38 – 28) points), mild cognitive impairment (27 – 25 points) [17].

Other test used in this paper was Barcelona test. That is an instrument for neuropsychological exploration to evaluate the cognitive status of the persons. Different aspects were evaluated such as: temporal orientation, spatial orientation and personal orientation [18].

3 Methods

3.1 Multi-sensory Room

The dimensions of the room are dimensions of 5 × 5 m in which the different sensory stimulation elements are distributed. In our case it is considered a black room, with white walls to allow the reflection of LED light that was used for the environment. It is composed of an array of multi-sensory equipment that provide stimulation in different modes to each participant (see Fig. 1).

- **Star Curtain.** The idea with the star curtain is to stimulate the attention and focus of older adults who perform activities such as touching, observing and manipulating it, they achieve a visual and tactile stimulus.
- **Stair of colors.** Therapists encourage patients to perform vocalizations and control the volume of their voice, reflecting it with visual effects. It seeks to stimulate the vision, the control of breathing, as well as to work the relation cause and effect.

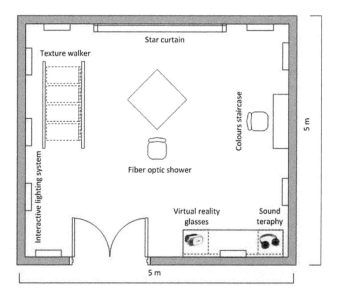

Fig. 1 Location of the elements in the multi-sensory room

- **Fiber optic shower**. Elderly people perform follow-up exercises, tactile stimulation, fine and gross motor skills when placed and passed through the body. Their color changes help with attention and memory exercises.
- **Sound Therapy**. Increases attention, concentration and improves verbal and communication skills in the person.
- **Textures Path**. People experience different tactile sensations when walking over different types of textures.
- **Virtual reality glasses**. Help in the stimulation of higher cognitive areas (memory, orientation, perception).

3.2 Development of Therapies

A research study was conducted in which participants were categorized according to their cognitive and functional status, in activities and instrumentals of daily living.

Participants. The population was selected among the residents of "Centro Gerontológico Hogar Miguel León" in the city of Cuenca (Ecuador). The inclusion criterion was a diagnosis of dementia (Alzheimer's, Parkinson's, Huntington's), diagnoses backed by a doctor specializing in Gerontology and corroborated by the Department of Psychology using the Assessment Cognitive Clinical Mayo which classifies dementia as mild, moderate and severe levels (25–38).

The treated universe was 60 patients, the clinical psychologist verified the eligibility of the participants according to the inclusion criterion, of which a number of 12 people were selected for the multisensory room. Informed consent was obtained from all legal representatives of patients included in the study.

Table 1 shows the sociodemographic characteristics of the participants of the Multisensory room.

Procedure. The group selected for the Multisensorial room participated in different focal sessions of senso-perceptive stimulation through: fiber optic shower, stair of colors, star curtain, textures path, sounds, virtual reality glasses and an interactive lighting system.

To develop focused therapies according to their reality, a field study was carried out to know the reality of the older adults, their activities, customs, traditions of which they formed part. It was concluded that the great majority come from the rural areas of the country.

Table 1 Sociodemographic characteristics of the patients

	Age	Gender	Education level
Patient 1	101	Male	No formal education
Patient 2	84	Female	Primary
Patient 3	70	Female	No formal education
Patient 4	91	Female	Primary
Patient 5	66	Female	Secondary
Patient 6	83	Male	Primary
Patient 7	89	Female	Primary
Patient 8	95	Female	No formal education
Patient 9	82	Female	No formal education
Patient 10	85	Female	No formal education
Patient 11	91	Female	No formal education
Patient 12	41	Female	Primary

The design of the multisensorial group sessions was based on the protocol of neuropsychological examination Test de Barcelona. All patients in this group participated in 5 weekly sessions of 30 min, for 12 weeks. All sessions were performed by professionals (clinical psychologist, occupational therapist, graduate in therapy, therapist and social worker) with training in the thematic.

4 Results

The following figures show the average of the evaluations made before and after the sessions in the multisensory room over the twelve weeks. Significant improvements in the cognitive and functional field of patients are observed. The data shown in the following charts are arranged as follows: X-axis: treated patients; Y-axis: disease degree. Being: 5 (very serious), 4 (severe), 3 (moderate), 2 (discrete), 1 (Normal).

After the sessions certain patients show more spontaneous speech and a better relationship with the people in their environment (see Fig. 2). Other encouraging results presented by the patients were in their motor skills, improving their muscular strength and visual-motor coordination (see Fig. 3). One of the factors that has impacted on the expected results has been the type of personality, which goes from the apassible to the euphoric. In order for the

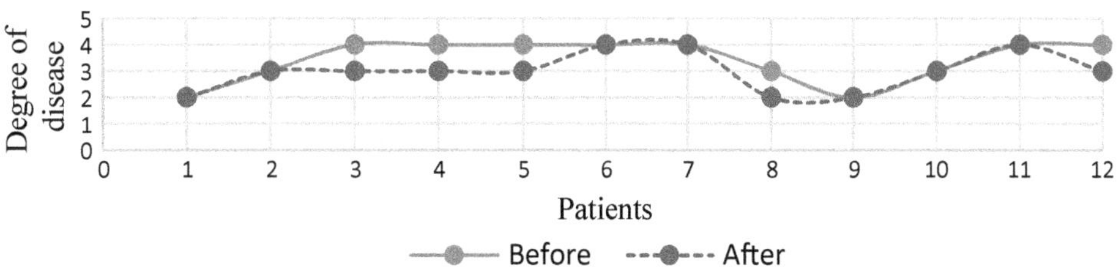

Fig. 2 Results of behavior and social interaction

Fig. 3 Motor coordination results

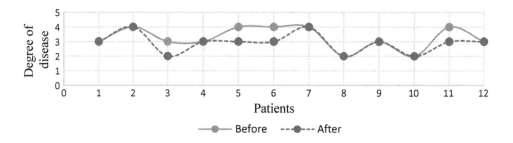

results to become permanent, there has been a constancy in the therapeutic process, talking about the mood and behavioral state of the patients.

5 Conclusions

After the study conducted on the efficacy of a multisensory room aimed at older adults, it is possible to see an improvement in the cognitive functions of the people which results in the reduction of the pattern of aggression of the people. The therapies have managed to improve their functional part, as well as a significant improvement in the relationship of these people with their social and personal environment. As for fields such as vision and hearing there has not been breakthrough, this may be due to the advanced degree of deterioration that they already present. It is important to take special care about the use of some implements in people with severe cognitive impairment, as it may present as something threatening and confusing.

Acknowledgements The authors gratefully acknowledge the support of the Project named "SISMO-NEURO: of Universidad Politécnica Salesiana from Cuenca Ecuador, UAM Universidad del Adulto Mayor and Centro Gerontológico Hogar Miguel León and all patients who participated in this research.

Conflict of Interest The authors declare that they have no conflict of interest.

References

1. A. M. Tinga, J. M. A. Visser-Meily, M. J. van der Smagt, S. V. der Stigchel, R. van Ee, and T. C. W. Nijboer, "Multisensory Stimulation to Improve Low- and Higher-Level Sensory Deficits after Stroke: A Systematic Review," *Neuropsychol. Rev.*, vol. 26, no. 1, pp. 73–91, Mar. 2016.
2. A. L. Lázaro, S. Blasco, and A. Lagranja, "La integración sensorial en el Aula Multisensorial y de Relajación: estudio de dos casos," *Rev. Electrónica Interuniv. Form. Profr.*, vol. 13, no. 4, pp. 321–334, 2010.
3. J. Hulsegge, *Snoezelen: Another World : a Practical Book of Sensory Experience Environments for the Mentally Handicapped.* Rompa, 1987.
4. S. B. N. Thompson and S. Martin, "Making Sense of Multisensory Rooms for People with Learning Disabilities," *Br. J. Occup. Ther.*, vol. 57, no. 9, pp. 341–344, Sep. 1994.
5. H. W. M. Kwok, Y. F. To, and H. F. Sung, "The application of a multisensory Snoezelen room for people with learning disabilities-Hong Kong experience," *Hong Kong Med. J. Xianggang Yi Xue Za Zhi*, vol. 9, no. 2, pp. 122–126, Apr. 2003.
6. G. Perhakaran *et al.*, "SnoezelenCAVE: Virtual Reality CAVE Snoezelen Framework for Autism Spectrum Disorders," in *Advances in Visual Informatics*, 2015, pp. 443–453.
7. N. Castelhano and L. Roque, "The 'Malha' project: A game design proposal for multisensory stimulation environments," in *2015 10th Iberian Conference on Information Systems and Technologies (CISTI)*, 2015, pp. 1–5.
8. H. A. Caltenco and H. S. Larsen, "Designing for Engagement: Tangible Interaction in Multisensory Environments," in *Proceedings of the 8th Nordic Conference on Human-Computer Interaction: Fun, Fast, Foundational*, New York, NY, USA, 2014, pp. 1055–1058.
9. G. A. Hotz, A. Castelblanco, I. M. Lara, A. D. Weiss, R. Duncan, and J. W. Kuluz, "Snoezelen: a controlled multi-sensory stimulation therapy for children recovering from severe brain injury," *Brain Inj.*, vol. 20, no. 8, pp. 879–888, Jul. 2006.
10. A.-J. Bianchi, H. Guépet-Sordet, and P. Manckoundia, "[Changes in olfaction during ageing and in certain neurodegenerative diseases: up-to-date]," *Rev. Med. Interne*, vol. 36, no. 1, pp. 31–37, Jan. 2015.
11. S. D. Berkheimer, C. Qian, and T. K. Malmstrom, "Snoezelen Therapy as an Intervention to Reduce Agitation in Nursing Home Patients With Dementia: A Pilot Study," *J. Am. Med. Dir. Assoc.*, vol. 18, no. 12, pp. 1089–1091, Dec. 2017.
12. B. S. Strøm, S. Ytrehus, and E.-K. Grov, "Sensory stimulation for persons with dementia: a review of the literature," *J. Clin. Nurs.*, vol. 25, no. 13–14, pp. 1805–1834, Jul. 2016.
13. C. Muller and M. Gillet, "Maladie d'Alzheimer: stratégies de communications et d'interventions. Une approche sans mots," *Kinesither. Rev.*, vol. 159, no. 15, pp. 65–69, 2015.
14. S. Chan, M. Y. Fung, C. W. Tong, and D. Thompson, "The clinical effectiveness of a multisensory therapy on clients with developmental disability," *Res. Dev. Disabil.*, vol. 26, no. 2, pp. 131–142, Mar. 2005.
15. J. C. M. van Weert, A. M. van Dulmen, P. M. M. Spreeuwenberg, J. M. Bensing, and M. W. Ribbe, "The effects of the implementation of snoezelen on the quality of working life in psychogeriatric care," *Int. Psychogeriatr.*, vol. 17, no. 3, pp. 407–427, Sep. 2005.
16. I. M. Pérez and F. J. M. Martínez, *Demencia : qué es y cómo puede tratarse*, 1ª ed., 1ª imp. edition. Madrid: Editorial Síntesis, S.a., 2014.
17. E. Kokmen, J. M. Naessens, and K. P. Offord, "A Short Test of Mental Status: Description and Preliminary Results," *Mayo Clin. Proc.*, vol. 62, no. 4, pp. 281–288, Apr. 1987.
18. M. Quintana Aparicio and J. Peña Casanova, *Test Barcelona abreviado.* Universitat Autònoma de Barcelona, 2010.

Feasibility Study of a Methodology Using Additive Manufacture to Produce Silicone Ear Prostheses

Barbara Olivetti Artioli, Maria Elizete Kunkel, and Segundo Nilo Mestanza

Abstract

In the current years, many kinds of prostheses have been developed to replace parts with deformity or absent from the human body. Facial prosthesis production methods have undergone little change in the last 40 years. Today, manual techniques are used to produce prostheses, which are usually cast in wax over a replica of the patient's anatomy. Silicone is the most appropriate structural material used in the production of atrial prostheses due to its mechanical and chemical characteristics. Recently, new technologies, such as 3D scanning and additive manufacturing techniques are being applied in the production of auricular prostheses with better cost/quality relation when compared to the manual manufacturing process. In this research, four methods to acquire the structure of a human ear were investigated (photogrammetry, 3D scanning, 3D reconstruction of computed tomography images and parameterized 3D modeling) to produce auricular prosthesis molds by additive manufacture using the fused deposition modeling (FDM) technique. The best silicone ear prosthesis produced shows an excellent esthetic result with 0.1–3% of dimensional error. Mechanical analysis of the silicone (tensile strength and hardness tests) shows that the prostheses produced have excellent mechanical resistance that is not altered by the pigmentation process.

The research demonstrates the feasibility of an accessible methodology to produce ear prostheses using free software, technologies and supplies available in the market.

Keywords

Auricular prosthesis • Additive manufacturing
Auricular pavilion • Ear • 3D scanning
3D reconstruction

1 Introduction

The ear is an organ of great importance in the aesthetics of the facial contour, not being evident when its size, shape, location and positioning are the usual. However, any changes in the auricle cause significant aesthetic differences [1]. Aesthetic modifications of the face provide patients with noxious psychosocial alterations, leading to difficulties in interpersonal and social relationships, and consequently compromising their quality of life [2, 3]. Ear deformities can be caused by various pathologies and situations, from congenital diseases such as microtia to automotive accidents and burns [4, 5].

The absence or partial absence of the human auricle can be restored by atrial reconstruction surgery or prosthetic rehabilitation [6–8]. In recent years, handcrafted techniques have been used as a method of producing maxillofacial prostheses. This process, in addition to being long and costly, restricts the quality of the prosthesis to the manual skill of the prosthetist [9]. Aiming at the brevity and the reduction of the manual processes, new methodologies of auricular prosthesis production using additive manufacturing technology (AM) emerged. The AM allows the production of physical parts based on three-dimensional(3D) computer models, without the need for accessories, cutting tools, or other ancillary resources. The AM allows the optimization of design and the production of custom parts on demand [10].

B. O. Artioli · S. N. Mestanza
Center for Engineering, Modeling and Applied Social Sciences, Federal University of ABC, Sao Bernardo, Brazil
e-mail: barbara@figmentface.com

M. E. Kunkel (✉)
Science and Technology Institute, Federal University of Sao Paulo, São José dos Campos, Brazil
e-mail: elizete.kunkels@gmail.com

© Springer Nature Singapore Pte Ltd. 2019
L. Lhotska et al. (eds.), *World Congress on Medical Physics and Biomedical Engineering 2018*,
IFMBE Proceedings 68/3, https://doi.org/10.1007/978-981-10-9023-3_38

1.1 Objective

To develop and validate an effective methodology to produce silicone auricular prosthesis by additive manufacture considering the cost-benefit for the current emerging market.

2 Literature Review

The conventional method used to produce an atrial prosthesis is usually performed by a dental surgeon or a professional called anaplastologist. This is the specialty that unites the art and science of restoring parts with deformity or absent from the human body through artificial means. The technique involves five steps [11]: 1. Physical obtaining of the patient's contralateral ear structure (by molding the ear with a fluid mixture of alginate or silicone); 2. Manufacture of an ear model in wax to sculpt the anatomy of interest; 3. Production of a final alginate ear mold; 4. Pigmentation and silicone application and 5. Removal of the silicone auricular prosthesis from the mold.

In the last decade, some researchers have developed new methodologies to produce auricular prosthesis from ear molds produced by AM. These studies represent attempts to automate the prosthesis production process reducing the final cost and improving the quality of the product.

In some studies, AM was introduced in the manufacturing process of auricular prosthesis to obtain the contralateral ear anatomy to generate the structure of the wax model, abbreviating the first step of the conventional method. The first report was published in 1999, magnetic resonance imaging (MRI) was used to acquire images of the ear and stereolithography (SLA) technology was used to produce a photopolymer resin model of this structure [12]. In 2000, other research compared four AM technologies (SLA, selective laser sintering (SLS), fused deposition modeling (FDM) and laminate object manufacturing (LOM)) being the first to design and produce an ear prosthesis mold by AM [13]. The image acquisition of the ear was performed through 3D laser scanning, and the final prosthesis was produced in polyurethane through a vacuum casting system using as master model of the mold obtained with AM.

Some studies developed a direct method, where it is not necessary to perform 1 to 3 steps of the original procedure, since the final template of the auricular prosthesis is digitally modeled and prototyped by AM. [14] reported the usage of the AM ear prosthesis production methodology with SLS technology in 10 cases of patients with unilateral absence of the auricular pavilion. [15] reported a clinical case demonstrating that the combination of AM technologies positively influences the production of an ear prosthesis. The study points out that through this new method it is possible to reduce the preliminary procedures of the conventional manufacturing methodology.

In 2011, [16] developed two case studies applying multiJet modeling and SLA technology for direct mold printing (obtained through computed tomography (CT) and Computer-aided design). The study confirmed that this methodology can be used in the production of prostheses with greater realism and facial harmony, besides requiring less processing time. It also highlighted the need for studies to evaluate more accessible methodologies, expanding the fields of use.

Underdeveloped and emerging countries need a more accessible methodology, based on free software, technology, widespread materials and a methodology with less waste, that hasn't been highlighted and evaluated in previous studies. The development and corroboration of the ear prostheses production phases aimed at these markets are essential for the correct implementation and applicability of the prostheses in this population.

3 Materials and Methods

The methodology adopted in this research to produce an atrial prosthesis was based on a previous study [17]. The use of different methodologies to produce a silicone auricular prosthesis by AM was investigated considering the current market. The volunteer was a 24-year-old woman without any maxillofacial deformities, with both normal ears morphology. Four structures were detached: helix-antihelix complex, conchal complex, the lobe and tragus (Fig. 1).

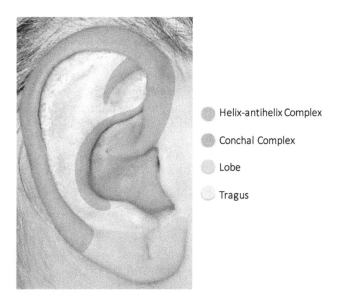

Fig. 1 Anatomical details of the volunteer's ear used as base to produce an ear prosthesis

Five procedures were investigated to acquire the volunteer's ear structure: Photogrammetry, two 3D scanning techniques, 3D reconstruction of the ear from computed tomography (CT) images and parameterized 3D modeling with a software of human models. The auricle was segmented and reconstructed with the free software Meshmixer and Tinkercad (Autodesk, USA), excluding any artifacts and expendable structures, to generate the 3D model of the structure. Through this ear model, a mold was generated to be produced through AM utilizing three different FDM equipments. The preparation and pigmentation of the silicone (Silpuran 2400 (Polisil, Brazil)) used to produce the prothesis was carried out and it was spilled to the mold. After the time defined by the silicone the complete prosthesis is easily removed from the mold and can be fixed to the user with the use of a suitable adhesive for this purpose. Mechanical analysis was performed on pure and pigmented silicone, with the purpose of comparing their behavior and analyzing the mechanical interference generated using the pigment in the manufacture of the auricular prostheses. The Fig. 2 shows an assembly of the materials and processes applied in this research.

4 Results

The investigation about the four procedures to acquire the volunteer's ear structure showed that the photogrammetry, Intel 3D scanning and parameterized 3D modeling are inefficient due to meager definition of ear structures and reproductions with open and irregular meshes. Nevertheless, the final structure obtained from Zeiss 3D scanning and the segmentation of CT images presents the morphologies of the helix-antral complex, conchal complex, the lobe and tragus compatible with the anatomy of the volunteer ear, enable the use of these reconstruction for 3D modeling of the auricular prosthesis. Four mold types and five silicone ear prosthesis were produced using the technique of FDM by AM (Fig. 3).

The anatomical structures of the helix-antral complex, conchal complex, lobule and tragus were correctly represented in the auricular prosthesis produced from the Zeiss 3D Scanner. However, there is a minimal presence of striations on the entire surface. The prosthesis differs 2.86% of the size of the volunteer's ear. The auricular prosthesis produced through CT using a 60 μm 3D print resolution and ABS filament without the application of the planning

Fig. 2 Descriptive framework of the methodology adopted to produce ear prosthesis in this research. # software, CT—Computed tomography, PLA—Polylactic Acid, ABS—Acrylonitrile Butadiene, FDM—Styrene fused deposition modeling

PHASE	METHOD	EQUIPMENT AND SOFTWARE#	DESCRIPTION
3D structure Acquisition of the ear	Photogrammetry	IPhone 6 Plus Trnio #	3D reconstruction from 34 photos
	3D scanning	Intel® RealSense™ R200 ReconstructMe #	4 types of procedures
		Zeiss® Comet L3D	5 types of procedures
	3D reconstruction of CT images	Siemens Emotion 6 / Invesalius #	3D reconstruction from 634 CT images
	Parameterized 3D modeling	Make Human #	11 types of modifications on ear pavilion
3D modeling of the ear	Ear segmentation	Meshmixer #	-
	Mold 3D modeling	Tinkercad Beta #	
Prothesis mold prototyping	Additive Manufacturing FDM technique	Stella 3D printer	PLA filament, 1.75 mm 100 μm resolution
		3D Machine ONE printer	ABS filament, 1.75 mm 60 μm resolution
		Manufactured 3D printer	ABS filament, 3.00 mm 150 μm resolution
Production of the silicon ear prosthesis	Mold filling	-	Silicone Silpuran 2400 8 ml Silpuran A plus 8 ml Silpuran B and 0.75ml of liquid pigment
Ear prosthesis - mechanical analysis	Tensile strength	Instron 3369	8 test specimens 100 N Speed 25 mm/min
	Hardness test	HT-6510A dipole measuring apparatus	20 test specimens 25x25 mm Thickness: 6 - 9 mm

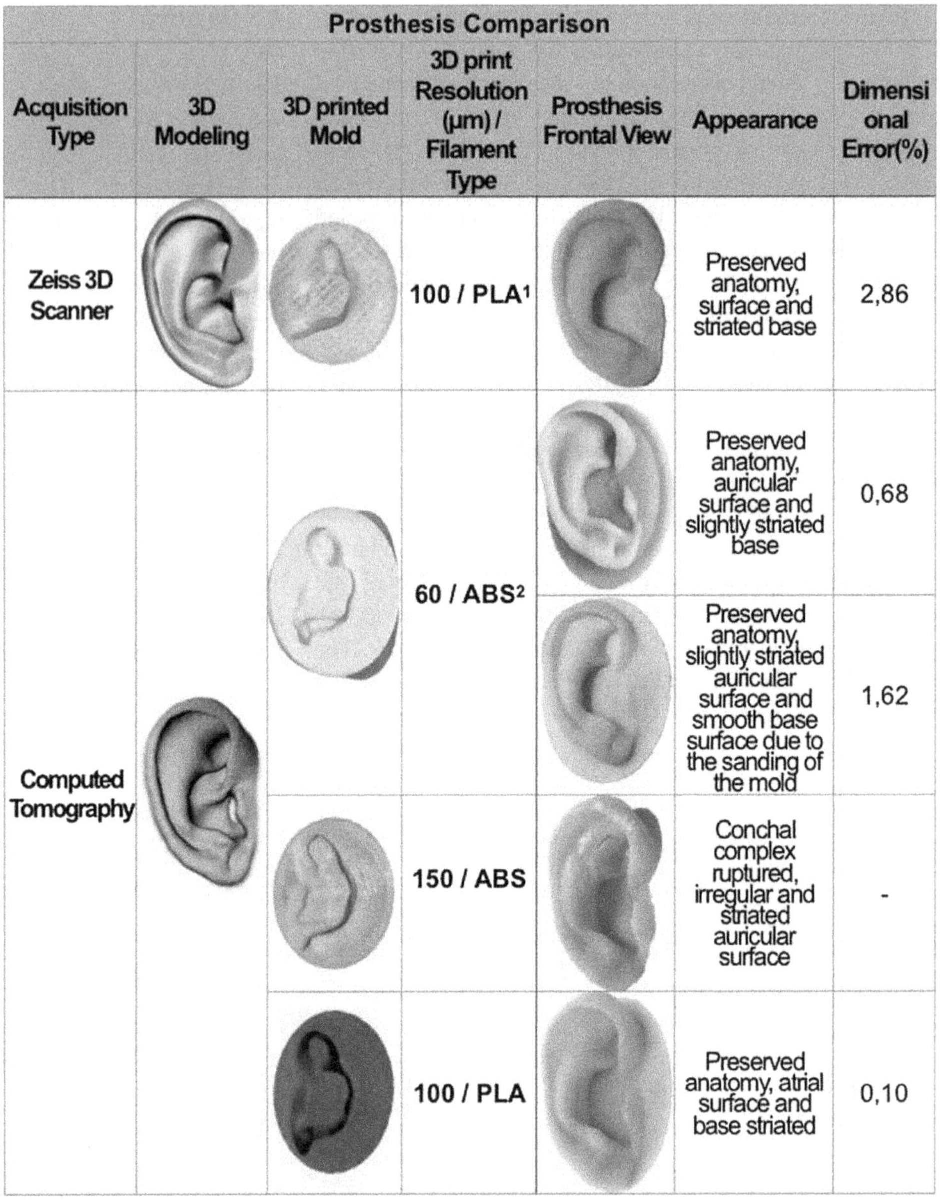

Fig. 3 Descriptive framework of the best ear prothesis produced by additive manufacturing technology with fused deposition modeling technique. PLA—Polylactic Acid, ABS—Acrylonitrile Butadiene

Acquisition Type	3D Modeling	3D printed Mold	3D print Resolution (µm) / Filament Type	Prosthesis Frontal View	Appearance	Dimensional Error(%)
Zeiss 3D Scanner			100 / PLA[1]		Preserved anatomy, surface and striated base	2,86
Computed Tomography			60 / ABS[2]		Preserved anatomy, auricular surface and slightly striated base	0,68
					Preserved anatomy, slightly striated auricular surface and smooth base surface due to the sanding of the mold	1,62
			150 / ABS		Conchal complex ruptured, irregular and striated auricular surface	-
			100 / PLA		Preserved anatomy, atrial surface and base striated	0,10

process of the external structures was effective, presenting the morphologies of the helix-antihelix complex, conchal complex, the lobe and tragus with precision, and similar measures diverging in 0, 68% of the objective ear. However, the striated appearance of the base structure of the part impaired its final appearance, reducing the likelihood of using it. After the superficial optimization, the prototype of the auricular prosthesis was produced obtaining better superficial appearance in the corresponding areas, which were scorched with sandpaper, obtaining the difference of 0.68% of the ear of the volunteer. The ear prosthesis produced by CT images with 150 µm 3D print resolution and ABS filament did not result in an applicable atrial prosthesis because of the error occurring during the impression of the part (high velocity), resulting in failures that makes it impossible to remove the prosthesis from the mold without rupturing the structure. The auricle prosthesis produced from CT images and manufactured with 100 µm 3D print resolution and PLA filament, reproduced the structures of the helix-antral complex, conchal complex, the lobe, and tragus efficiently, obtaining the smallest difference of measurements with the objective ear, only 0.10%. However, its surface had a more striated appearance when compared to prosthesis produced through CT using a 60 µm 3D print resolution and ABS as expected, because of the higher printer resolution.

The mechanical analysis (tensile strength and hardness tests) of the silicone used to produce the protheses showed that they have a mechanical resistance consisted with the need that is not altered by the pigmentation process. In addition, the average hardness of 7.70 Shore A scale also consonant to the skin demanded.

5 Discussion and Conclusion

The study investigated the use of different methodologies to produce an auricular prosthesis, proving the use of the 3D reconstruction by the acquisitions of the Zeiss 3D scanner and CT images to obtain the 3D structure of the ear. The 3D modeling of the structures and the mold were elaborated only with the aid of free software. The feasibility of performing an atrial prosthesis produced by MA with the FDM technique using the ABS and PLA inputs was confirmed. Through the mechanical tests in conjunction with the adhesion analysis, the applicability of Silpuran 2400 silicone was applied to produce atrial prostheses, as well as the use of capillary adhesives as a means of fixing the product to the user. The research finally proved the use of technologies and materials available and accessible in emerging locations, sustaining the goal of achieving a methodology accessible to the current market.

References

1. Avelar, J.M.: Importance of ear reconstruction for the aesthetic balance of the facial contour. Aesthetic plastic surgery, v. 10, n. 1, p. 147–156 (1986).
2. Horlock, N. et al.: Psychosocial outcome of patients after ear reconstruction: a retrospective study of 62 patients. Annals of plastic surgery, v. 54, n. 5, p. 517–524 (2005).
3. Tam, C.K. et al.: Psychosocial and quality of life outcomes of prosthetic auricular rehabilitation with CAD/CAM technology. International journal of dentistry, v. 2014 (2014).
4. Sakae, E., Aki, F.: Estudos das Complicações na Reconstrução de Orelha. São Paulo, Diss. Universidade de São Paulo (2007).
5. Giot, J.P. et al.: Prosthetic reconstruction of the auricle: indications, techniques, and results. In: Seminars in plastic surgery, pp. 265–272, Thieme Medical Publishers (2011).
6. Guttal, S.S. et al.: Rehabilitation of a missing ear with an implant retained auricular prosthesis. The Journal of the Indian Prosthodontic Society, v. 15, p. 70 (2015).
7. Leach, J., Biavati, M.: Ear Reconstruction. University of Texas Southwestern Medical School (2015).
8. Mevio, E. et al.: Osseointegrated implants: an alternative approach in patients with bilateral auricular defects due to chemical assault. Case reports in otolaryngology, v. 2016 (2016).
9. Eggbeer, D., Bibb, R., Evans, P.: The appropriate application of computer aided design and manufacture techniques in silicone facial prosthetic. Proceedings of the 5th National Conference on Rapid Design, Prototyping and Manufactore. p. 45–52, (2004).
10. Huang, S.H. et al.: Additive manufacturing and its societal impact: a literature review. The International Journal of Advanced Manufacturing Technology, p. 1–13 (2013).
11. Swaminathan, A.A. et al. Fabrication of a silicone auricular prosthesis-a case report. Nitte University Journal of Health Science, v. 6, n. 1 (2016).
12. Coward, T.J.; Watson, R.M.; Wilkinson, I.C. Fabrication of a wax ear by rapid-process modeling using stereolithography. International Journal of prosthodontics, v. 12, n. 1 (1999).
13. Chee Kai, et al. Facial prosthetic model fabrication using rapid prototyping tools. Integrated Manufacturing Systems, v. 11, n. 1, p. 42–53 (2000).
14. Turgut, et al. Use of rapid prototyping in prosthetic auricular restoration. Journal of Craniofacial Surgery, v. 20, n. 2, p. 321–325, (2009).
15. Liacouras, et al. Designing and manufacturing an auricular prosthesis using computed tomography, 3-dimensional photographic imaging, and additive manufacturing: a clinical report. The Journal of prosthetic dentistry, v. 105, n. 2, p. 78–82 (2011).
16. Karatas, M.O. et al. Manufacturing implant supported auricular prostheses by rapid prototyping techniques. European Journal of Dentistry, v. 5, n. 4, p. 472–477 (2011).
17. Artioli, B.O.; Maluf, N.N.; Roxo, V.G.L. Produção de um protótipo de prótese de pavilhão auricular. Monografia Pontifícia Universidade Católica de São Paulo (2014).

Plantar Pressure Measurement Transformation Framework

Dusanka Boskovic, Iris Kico, and Abdulah Aksamovic

Abstract

Pedobarography measurements of pressure distribution across the plantar surface can be a source of valuable information for gait analysis in context of injury prevention, improvement in balance control, diagnosing disease, and gait analysis. Different applications demand different measurement types: platforms in the gait labs or in-shoe smart soles, and subsequent analysis methods vary with the application domain. Although the pedobarography is considered experimental, technology advancements in the field of IoT, and popularity of collecting different data linked to human activities and behavior, are contributing to increase in pedobarography research. Comprehensive analysis of research results is impeded since the data collecting is not standardized, and differ in volume and structure, thus not facilitating comparative analyses, as it is a case with other biomedical signals as ECG or EEG. In our research we have implemented software framework in Python language with objective to extract relevant pedobarography information using foot segmentation and data aggregation algorithms. In order to validate our solution we processed data from public plantar pressure data set and transformed it in data sets comprising pressure signals covering selected number of foot segments, down to one sensor signal. The proposed data transformation application can help in data sharing and comparison of different approaches in pedobarography.

Keywords

Pedobarography • Gait analysis

D. Boskovic (✉) · I. Kico · A. Aksamovic
University of Sarajevo, Sarajevo, Bosnia and Herzegovina
e-mail: dboskovic@etf.unsa.ba

I. Kico
e-mail: ikico1@etf.unsa.ba

A. Aksamovic
e-mail: aaksamovic@etf.unsa.ba

1 Introduction

Since ancient times, people have been interested in movement analysis and gait analysis. Bipedal gait is typical form of movement in humans and one of the recognizable features of our species. Walk refers to the style of movement of an individual. Research has shown that people can recognize each other by walking, even from poor representations, indicating presence of identity data in "gait signature". Human gait is an attractive way for recognizing people who are far way, but it is also possible to identify certain difficulties. Numerous neurological diseases are associated to changes in gait. Measuring foot loading and distribution of plantar pressure can help identify these diseases. These measurements are facilitated and improved thanks to the advancement and development of technology.

The importance of plantar pressure measurement is that provides an insight into force and/or weight distribution. This is important for several different fields of applications, first of all as indicator if there is a proper weight distribution. Plantar pressure distribution data can be used for different objectives as: identification of gait [1] and posture features [2], monitoring deviations of the center of pressure [3], monitoring weight overload [4], and for different purposes as research, in health care [5], for athletes' medical and performance examinations [6]. Data can be obtained using platforms in labs and also in real environment using shoe insoles.

The paper is organized as follows: the next Section provides background information on plantar pressure measurement and what knowledge relevant for gait analysis can be obtained from monitoring plantar pressure distribution over time. Section 3 describes the gait analysis framework and results obtained when transformation algorithms were applied to public plantar pressure data set to extract gait parameters. Last section presents results discussion, conclusions and recommendations for the future work.

© Springer Nature Singapore Pte Ltd. 2019
L. Lhotska et al. (eds.), *World Congress on Medical Physics and Biomedical Engineering 2018*,
IFMBE Proceedings 68/3, https://doi.org/10.1007/978-981-10-9023-3_39

2 Plantar Pressure

Analysis of pedobarography measurements provide data on foot pressure distribution. Analysis of plantar pressure distribution requires proper foot segmentation. It is important to notice there is no formal procedure for foot segmentation, but frequently used sub-segment schematic is described in [7].

Plantar pressure measurements provide several parameters of interest: peak pressure, contact area, contact time, maximum force, force-time integral, and can be recorded when the subject is barefoot or using in-shoe system.

Systems for plantar pressure measurements available in the market or in research laboratories have different sensor configuration to meet different application requirements. There are various systems for plantar pressure measurements, but all of them can be divided into two groups: platform systems and insoles. Key requirements for designing devices for plantar pressure measurements are: spatial resolution, sampling frequency, accuracy, sensitivity and calibration.

It is easy to see there is a need for a system that will meet all of these requirements, but there is also a need for algorithms for processing and analysis of the collected data and for extracting gait features from these data in a manner that results can be compared across different research studies.

Gait analysis is an important diagnostic process with applications in rehabilitation, therapy and exercising. Changes in gait reveal information about person's life quality. This is of particular interest when reliable information about progress of disease is required, such as neurological diseases, multiple sclerosis or Parkinson's disease, system illnesses such as cardiopathies affecting the gait, changes in dynamic motion due to the effects of stroke, age related illnesses. Knowledge about gait characteristics at particular moment in time, and more important, the follow-up and evaluation over time, will provide early diagnosis of disease and selection of the best possible treatment.

Gait represents series of gate cycles. One gait cycle is also known as a stride. It represents time period since one foot touches the surface until the same foot touches the surface again.

Every stride has two phases: stance phase—foot is in contact with the surface and swing phase—foot has no contact with the surface. Stance phase is approximately 60% of gait cycle while swing phase is about 40%.

Some of the main events in gait cycle based on time of contact of different left and right foot positions are: (a) initial contact/heel strike, (b) opposite toe off, (c) opposite initial contact, (d) toe off. By measuring time spans between two events it is possible accurately calculate time parameters: Gait cycle duration, Stance duration, Swing phase and Single support [2].

3 Plantar Pressure Measurement Transformation

In this paper we have used data from research database for plantar pressure [8] containing measurements obtained following strict protocol. The experiment included 16 healthy subjects and measurements at different walking speeds and corresponding duration.

Recording was performed at different walking speeds, while duration of walking at one speed was 5 min. Plantar measurement system uses a treadmill with sensors, Zebris FDM-THM. Values of the pressures was recorded every 10 ms and stored into matrix with dimensions 56×128 elements. One element of the matrix responds to pressure value at one point and to the cell surface 8.46 mm $\times$ 8.46 mm. The measurements were successively stored in the file, and one matrix will be referred as a "frame". 5-minute walking measurements were recorded as a series of 30.000 frames. Matrices contain a lot of zero elements,

Analysis of plantar pressure distribution is linked to identification of main events in gait cycle which correspond to the maximum pressure within specific foot segments. In order to identify these events it is necessary to aggregate this long sequence of pressure data: (1) in time—to obtain specific precision of gait events, and (2) for contact surface —to focus on foot segment or segments of interest.

Time interval (t1, t2) has to be chosen so it is possible to identify foot segments with the highest values of pressure, so the characteristics gait events can be identified as well (see Fig. 1). There are three cases:

1. If time interval has only one value, t1 = t2, then only measurements at the moment t1 would be analyzed, and changes between measurements are too small to identify significant events easily.
2. If time interval is too big then measurements include several contact areas and it is not possible to identify corresponding event.
3. If time interval is adequate contact area will be appropriate to identify significant gait events.

Choice of the time interval depends on number of frames that are aggregated, as illustrated (see Fig. 2) with aggregation results for 1, 10 and 100 frames.

The number of 10 frames is chosen for aggregating frames into one matrix. Aggregation function used is maximum, so it is possible to get maximum pressure distribution.

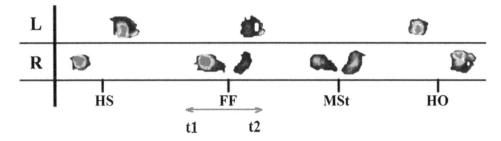

Fig. 1 Gait events and time interval t1 to t2 selection (based on a diagram from [9])

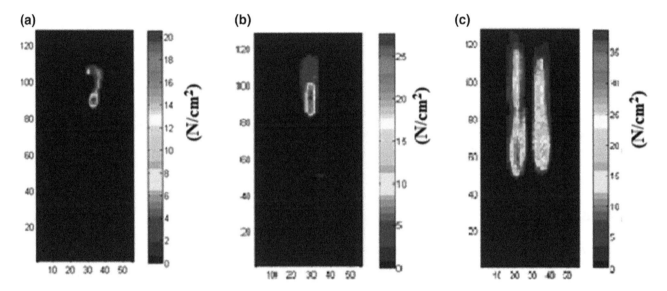

Fig. 2 Plantar pressure distribution depend on time interval: **a** the real measurement—the original frame, **b** aggregated 10 frames and **c** aggregated 100 frames

This process could be generalized so algorithm finds the corresponding values of these parameters. It is important to notice that the result matrix is divided into two parts so the displayed results correspond to tracking of one foot and they are more understandable for interpretation.

The next step is spatial segmentation, when maximum pressure is determined for foot segment of interest. Again aggregation function used is maximum, so we obtain maximum pressure for the selected segment. Algorithm for extracting the segment in focus returns the maximum value of the pressure. Selection of the segment is generalized by defining set of indexes corresponding to specific foot segment.

In this way a set of 10 frames dimension 56×128 elements is transformed into one measured value corresponding to specific foot segment. When described steps are repeated and all data frames processed, resulting output is plantar pressure signal corresponding to foot segment of interest and with selected time precision. In this paper the results for the forefoot segment will be presented.

4 Gait Analysis

In the example used to illustrate the plantar measurement transformation results, selected foot segment corresponds to forefoot. By successive aggregation of measured pressure values into one value, the maximum value vector for the foot segment in focus is obtained. This vector represents plantar pressure as time series and is pressure signal projected at one point. This signal is further analyzed so the important gait events can be identified.

The process of the transformation of plantar pressure data can be applied to more segments, where the result would be several pressure signals. This would require a different number of frames for aggregation and shorter time intervals (t1 and t2).

From the signal shown in Fig. 3 it is possible to determine number of steps in certain time interval. Change of pressure from zero to non-zero value is recognized and marked as step identifications. Based on these markers it is

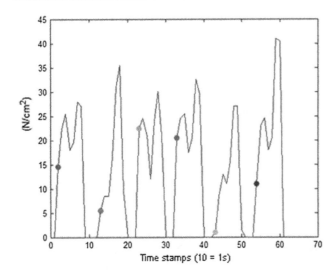

Fig. 3 Changes of maximum pressure and event identification

possible to determine number of steps and also duration of a step. In the above example in 6 s subject made 6 steps, which corresponds to the pace of the subject in the original experiment.

5 Conclusions

In this paper we described a framework for transformation of plantar pressure data measurements with objective to facilitate gait analysis. We have used public data set from research database for plantar pressure, and transformed huge series of plantar pressure matrices into forefoot pressure signal which we used for stride identification and measurement. We transformed the set of 10 frames with dimension 56×128 elements into one measured value corresponding to specific foot segment—in this case forefoot. The process can be applied to more segments, resulting in a set of pressure signals. Implemented algorithms are generic and associated with specific parameters aimed to produce data sets with desired time and space pressure resolutions. In the

process of transformation we have retained information relevant for gait analysis and identification of gait cycle events.

In our future work we will implement user interface enabling users to change transformation parameters—as time intervals and area used for aggregation and compare different results. In this way we will provide our research teams with flexible system facilitating future research in this promising area.

References

1. Sejdic E, Lowry K, Roche J, Brach J.: Extraction of Stride Events From Gait Accelerometry During Treadmill Walking, IEEE Journal of Translational Engineering in Health and Medicine, (2015).
2. Das R, Kumar N: Investigations on postural stability and spatiotemporal parameters of human gait using developed wearable smart insole, J Med Eng Technol 39(1), 75–78, (2015).
3. Ruhe A., Fejer R., Walker B.: Center of pressure excursion as a measure of balance performance in patients with non-specific low back pain compared to healthy controls: a systematic review of the literature, Eur Spine J (20), 358–368, (2011).
4. Hellstrom, P., Folke, M., Ekström, M.: Wearable Weight Estimation System, Procedia Computer Science 64, 146–152, (2015).
5. Muro-de-la-Herran, A., Garcia-Zapirain, B., Mendez-Zorrilla, A.: Gait Analysis Methods: An Overview of Wearable and Non-Wearable Systems, Highlighting Clinical Applications. Sensors 14(2), 3362–3394, (2014).
6. Petry, V.K.N., Paletta, J.R.J., El-Zayat, B.F., Efe, T., Michel, N.S. D., Skwara, A.: Influence of a training session on postural stability and foot loading patterns in soccer players. Orthopedic Reviews, vol. 8(1), 6360, (2016).
7. Martinez-Nova, A.J.C., Cuevas-Garcia, J.C., Pascual-Huerta, J., Sanchez-Rodriguez, R.: BioFoot® in-shoe system: Normal values and assessment of the reliability and repeatability, The Foot 17, 190–196, (2007).
8. McClymont J, Pataky TC, Crompton RH, Savage R, Bates KT, Data from: The nature of functional variability in plantar pressure during a range of controlled walking speeds. Dryad Digital Repository. http://dx.doi.org/10.5061/dryad.09q2b, (2016) Accessed on January 15, 2017.
9. Zayegh A, Woulfe J, Begg R, Identification of Foot Pathologies Based on Plantar Pressure Asymmetry, Sensors (Basel), 15(8), 20392–408, (2015).

Body Tracking Method of Symptoms of Parkinson's Disease Using Projection of Patterns with Kinect Technology

Raquel Torres, Mónica Huerta, Roger Clotet, and Giovanni Sagbay

Abstract

The analysis of the body movement is relevant in different areas, such as therapy, rehabilitation, bioinformatics and medicine. The Parkinson's disease (PD) is a progressive degenerative process of the central nervous system that primarily affects the movement. To measure motor disorders, body sensor networks and portable technologies are the trend for tracking and monitoring symptoms in PD. Through the use of technological tools, such as sensors, whether sensors for movement acquisition (accelerometers, gyroscopes, inclinometers) or environment sensors (sensors that record physiological properties), it is possible to track the symptoms of Parkinsonism in a person. A system has been designed using a Kinect sensor, that uses the projection of patterns technology for monitoring change in body posture, obtaining information for a set of points or joints, and variation that could have during the observed period. The designed Kinect sensor system consists of four modules: the first acquisition of the body movement of the patient with the Kinect sensor V1.0, the second feature extraction module to process captured scene by Kinect V1.0, the third recognition of the skeleton module and finally the acquired data processing module, developed with MatLab. The acquisition of the center of mass with the presented methodology, through projection of patterns used by the Kinect sensor technology, is non-invasive a method and convenient to use in people.

Keywords

Body tracking • Kinect sensor • Motor disorders Parkinson disease

1 Introduction

The Parkinson's disease (PD) is a degenerative and progressive process of the central nervous system that primarily affects the movement. In presences of symptoms of Parkinsonism there is a need to use some sort of device to capture signals or indicators of the body movement of the patient in non-invasive approach, in order to assist diagnosis, knowledge, and evolution of the disease with impartiality of trial, which brings convenience and low cost, both to the patient as to the physician. To measure motor disorders, body sensor networks and portable technologies set the trend for tracking and monitoring symptoms in PD.

In this research has been studied the movement of the body of a group of healthy people, a through a system that uses a sensor Kinect as a method of data acquisition, for the purpose of establishing a descriptive platform position and displacement of key joints of the human skeleton, so as a basis to compare the position and movement of joints of patients who have neurodegenerative diseases and recognize some symptoms with motor impairment.

R. Torres (✉)
University of the Armed Forces, Caracas, Venezuela
e-mail: rtorres@unefa.edu.ve

R. Torres · R. Clotet
Universidad Simón Bolívar, Caracas, Venezuela
e-mail: clotet@usb.ve

M. Huerta · G. Sagbay
Universidad Politécnica Salesiana, Cuenca, Ecuador
e-mail: mhuerta@ups.edu.ec

G. Sagbay
e-mail: gsagbay@ups.edu.ec

© Springer Nature Singapore Pte Ltd. 2019
L. Lhotska et al. (eds.), *World Congress on Medical Physics and Biomedical Engineering 2018*,
IFMBE Proceedings 68/3, https://doi.org/10.1007/978-981-10-9023-3_40

2 The Extraction of Information of an Image on a 3D Space

In systems that analyze and follow the body movement are considered the parameters that determine the speed according to the movement of the subject is generated by observing body joints [1].

2.1 Body Tracking

There are different algorithms for body tracking; H.Weiming et al. in [2] studied some of them. Figure 1 show a resume of these algorithms. In a Kinect system the detection of the

position is based on dynamic models of the human body, these models are taken as a basis for the movement of the body to the skeleton mapping. In the tracking algorithm based on the model of the skeletal figure, it is considered that the essence of the movement is typically limited to the movements of the head, torso and extremities.

The extraction of information of an image on a 3D space, starting from one or more projections that represent it [3], consists of two large processes, first to acquire an image of the real event and is followed by another stage for the analysis of acquired images to extract information.

The capture stage, considers everything related with the formation process of the image and the mechanism whereby the information of the physical world (scene) is transmitted.

Fig. 1 Methods for monitoring according to [2]

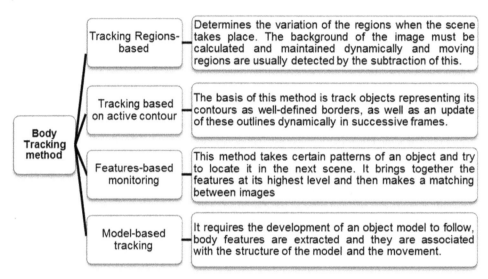

Fig. 2 The process of analysis or imaging components

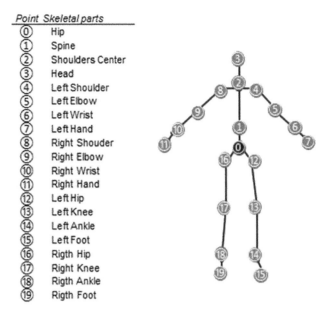

Point	Skeletal parts
(0)	Hip
(1)	Spine
(2)	Shoulders Center
(3)	Head
(4)	Left Shoulder
(5)	Left Elbow
(6)	Left Wrist
(7)	Left Hand
(8)	Right Shouder
(9)	Right Elbow
(10)	Right Wrist
(11)	Right Hand
(12)	Left Hip
(13)	Left Knee
(14)	Left Ankle
(15)	Left Foot
(16)	Rigth Hip
(17)	Right Knee
(18)	Rigth Ankle
(19)	Rigth Foot

Fig. 3 Dot structure of Skeletal Tracking for the Kinect (Color figure online)

To extract information, once it is captured or acquires the image of the scene in study, proceeds in four essential steps: preprocessed, segmentation, description and recognition [4], presented in Fig. 2.

2.2 Technology of Projection of Patterns with Kinect Sensor

The Kinect system uses the technology of projection of patterns, where a known pattern of points of infrared light hits on the surface whose distances you want to calculate, determining its geometry through the analysis of differences in the image obtained with the original pattern is projected [5–7]. Dot structure of Skeletal Tracking for the Kinect, it is shown in Fig. 3. The red point (0) represents center hip.

3 Case Study

In this project, a system composed of a Kinect sensor, was used to follow the change of body stance, finally obtaining information about the location a set of point or systems of joints on the Kinect sensor and variation that they might have had during the observed period [12, 13].

For the body tracking method for different age groups, it followed the process shown in Fig. 4. The designed system using Kinect sensors consists of four modules: first one is an acquisition of the body movement of the patient with the Kinect sensor V1.0, the second one features the extraction module to process the scene captured by Kinect V1.0, third

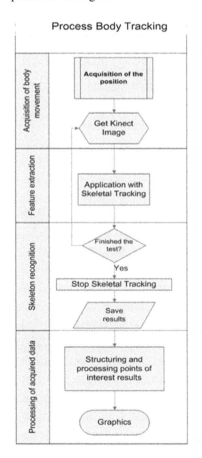

MODULES

1. Adquisition

In order to get the movement captures, it has to be used a motion alteration test, like Romberg test, Forefinger test or Barany indicator evaluation (it measures the postural balance), UPDRS scale (measures the degree of movement alterations in Parkinson Disease), or any other medical tests.

2. Feature extraction

When the signal is totally prepared, it is started NUI API (Natural User Interface Application Programming Interface) which contains the interface of Skeletal Tracking . Each frame contains a set of data that provides information of 20 points that make up a structure of skeletal figure for the measured instant [8] as shown in Figure 3.

3. Reconigtion of skeleton

In our proposed system, some groups of joints are considered as one. In Figure 3 are shown each joint on the skeleton. It has been associated a coordinate to each joint in the skeleton. These coordinates are obtained in this phase, which can be stored in txt files. The txt files allow to draw skeletal figure later.

4. Acquired data processing

The obtained joint coordinates are structured and processed using a Matlab routine. The obtained points are grouped in ten (10) different branchs, from three different matrices associated with the coordinates x, y and z from each frame.

Fig. 4 Process for the body tracking method for different age groups and modules of designed system

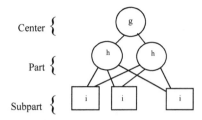

Center {
Part {
Subpart {

Fig. 5 Hierarchy model [11]

one works on the recognition of the skeleton module, and finally, the acquired data processing module, which is based on a routine in MatLab.

4 Results

It was realized a series of measurements to a group of people with the purpose of establishing a model with a group of patients from different groups of ages [8], in order to establish a comparative basis of motor signs for the evaluation of the position with the use of sensors, as it has been reported in [9]. In this case study, has been considered the track to body mass center, located in the middle of the hip, it is taken as premise computer vision algorithms presented in [10] and the hierarchical model for visual objects categorization presented in [11], see Fig. 5.

Table 1 Data acquired of center of mass on the x axis (Volunteers, age)

Age (years)	Gender	X_1	X_2	X_3	X_4	X_5	X_6	X_7	X_8	X_9	X_{10}
22	M	0.0355	0.03555	0.0356	0.03565	0.03571	0.03577	0.03583	0.03588	0.03594	0.036
22	F	0.10656	0.10664	0.10671	0.10678	0.10684	0.10689	0.10694	0.10699	0.10704	0.10709
26	M	0.07521	0.07522	0.07522	0.07523	0.07523	0.07524	0.07524	0.07524	0.07524	0.07524
29	M	0.11948	0.11933	0.11919	0.11906	0.11892	0.11879	0.11868	0.11857	0.11846	0.11835
29	M	−0.12748	−0.12744	−0.12741	−0.12739	−0.12737	−0.12735	−0.12734	−0.12733	−0.12733	−0.12732
31	F	0.0901	0.0901	0.09011	0.09012	0.09015	0.09017	0.0902	0.09021	0.09023	0.09024
46	M	0.06309	0.06337	0.06364	0.06388	0.0641	0.06431	0.0645	0.06468	0.06484	0.06499
49	M	0.09574	0.09617	0.09666	0.09715	0.0977	0.09823	0.09874	0.09923	0.09967	0.10009
50	F	0.18671	0.18342	0.17875	0.17316	0.16616	0.15662	0.14781	0.1406	0.13301	0.12542
57	F	0.01341	0.01351	0.0136	0.01368	0.01376	0.01382	0.01388	0.01393	0.01397	0.01401
64	F	0.15188	0.15588	0.15912	0.16182	0.16382	0.16458	0.1642	0.16303	0.16205	0.16039
70	F	0.22714	0.23424	0.23832	0.23855	0.23621	0.22942	0.22403	0.22056	0.21651	0.21357
79	M	−0.04057	−0.04078	−0.04098	−0.04117	−0.04135	−0.04153	−0.0417	−0.04187	−0.04205	−0.04226
75	F	0.25321	0.08545	0.0602	−0.07075	−0.07991	−0.13775	−0.13499	−0.14351	−0.16498	−0.16256

Table 2 Data acquired of center of mass on the y axis (Volunteers, age)

Age (years)	Gender	y_1	y_2	y_3	y_4	y_5	y_6	y_7	y_8	y_9	y_{10}
22	M	−0.2833	−0.283	−0.283	−0.283	−0.283	−0.283	−0.284	−0.284	−0.28356	−0.28358
22	F	−0.22107	−0.22099	−0.22092	−0.22085	−0.22078	−0.22071	−0.22064	−0.22057	−0.22049	−0.22041
26	M	−0.16732	−0.1672	−0.16709	−0.16699	−0.1669	−0.16682	−0.16674	−0.16667	−0.16661	−0.16655
29	M	0.23773	0.23767	0.23762	0.23758	0.23745	0.23734	0.23719	0.23693	0.23669	0.23639
29	M	−0.02209	−0.02207	−0.02204	−0.02202	−0.02201	−0.02199	−0.02198	−0.02197	−0.02197	−0.02196
31	F	−0.42804	−0.42776	−0.42751	−0.42716	−0.42673	−0.42633	−0.42598	−0.42559	−0.42514	−0.42473
46	M	−0.13417	−0.1342	−0.13423	−0.13427	−0.13431	−0.13435	−0.13439	−0.13444	−0.13448	−0.13452
49	M	−0.03881	−0.03858	−0.03832	−0.03803	−0.0377	−0.03738	−0.03705	−0.03674	−0.03644	−0.03615
50	F	0.15829	0.15575	0.15332	0.14966	0.14595	0.13978	0.13308	0.12988	0.12689	0.12474
57	F	0.07208	0.0721	0.07212	0.07213	0.07215	0.07217	0.07219	0.0722	0.07222	0.07224
64	F	−0.16	−0.16172	−0.1625	−0.16205	−0.16061	−0.15847	−0.15473	−0.15022	−0.14399	−0.13665
70	F	0.22714	0.23424	0.23832	0.23855	0.23621	0.22942	0.22403	0.22056	0.21651	0.21357
79	M	0.13999	0.11304	0.10631	0.06886	0.05767	−0.04716	0.01002	−0.0112	−0.02394	−0.02526
75	F	0.13999	0.11304	0.10631	0.06886	0.05767	−0.04716	0.01002	−0.0112	−0.02394	−0.02526

Fig. 6 Shaft displacement in axes x of center of mass

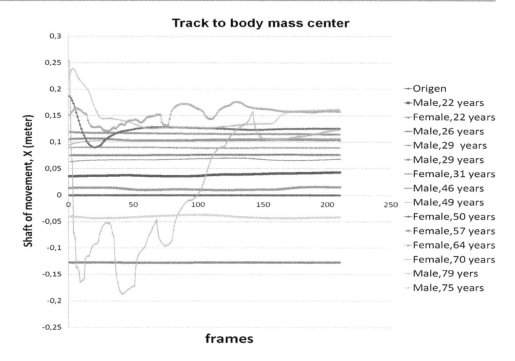

Fig. 7 Shaft displacement in axes y of center of mass

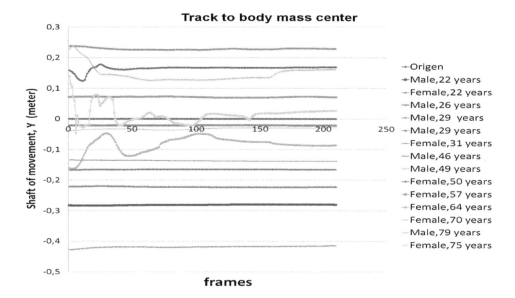

With a series of measurements to a group of people of different ages, we can establishing a model of reference, to classify them in different age groups as [8] suggest. Data acquired center of mass on the X axis is show in Table 1 and data acquired center of mass on the Y axis is show in Table 2.

The displacement in axes X of center of mass is show in Fig. 6 and displacement in axes Y of center of mass is show in Fig. 7.

5 Discussion

The results obtained with the methodology used in this proposal, for the monitoring of body center of mass of a group of volunteers from different age groups, shown in Fig. 6, exposed that the group of volunteers with greater variation from body center of mass with regard to the origin, is represented by people with different ages both for the axis

X as in the shaft Y. To obtain information about the center of mass of a body, will allow analyzing the positions of equilibrium, as well as also could work in the appropriate increase of stability with the support and caring by a specialist. The acquisition of the center of mass with the presented methodology, through projection of patterns used by the Kinect sensor technology, a method is non-invasive and convenient to use in people. Although, is important to consider the constraints of the sensor (maximum distance from the camera to the human body in observation).

Acknowledgements The authors gratefully acknowledge the support of the NEURO-SISMO project, Universidad Politécnica Salesiana from Ecuador.

Conflict of Interest statement The authors have no conflict of interest.

References

1. Cappozzo A., Cappello A.; Croce U.D., Pensalfini F., "Surface-marker cluster design criteria for 3-D bone movement reconstruction"; Biomedical Engineering, IEEE Transactions on, vol.44, pp. 1165– 1174, 1996.
2. Weiming H., Tieniu T.; Liang W.; Maybank S., "A survey on visual surveillance of object motion and behaviors," Systems, Man, and Cybernetics, Part C: Applications and Reviews, IEEE Transactions on, vol. 34, no. 3, pp. 334, 352, Aug. 2004.
3. Watt A., Policarpo F., "The image computer", England, 1998.
4. González A., "Técnicas y algoritmos básicos de visión artificial", Universidad de la Rioja, 2006.
5. Shotton J., Fitzgibbon A., Cook M., Sharp T., Finocchio M., Moore R., Kipman A., Blake A., "Real-time human pose recognition in parts from single depth images," Computer Vision and Pattern Recognition (CVPR) 2011, IEEE Conference on, pp. 1297, 1304, June 2011.
6. Zhang Z., "Microsoft Kinect Sensor and its effect", IEEE Computer Society, 2012.
7. Skeletal Tracking, available in http://msdn.microsoft.com/en-us/library/hh973074.aspx. Fecha de consulta: March 2018.
8. Los factores definitorios de los grandes grupos de edad de la población: tipos, subgrupos y umbrales, available in http://www.ub.edu/geocrit/sn/sn-190.htm. March 2018.
9. Torres R., Huerta M., González R., Clotet R., Bermeo J., Vayas G., "Sensors for Parkinson's disease evaluation", Devices, Circuits and Systems (ICCDCS), 2017 International Caribbean Conference on, 2017.
10. Szeliski R., "Computer Vision,Algorithms and Applications", Springer-Verlag London Limited 2011.
11. Bouchard G. and Triggs B., "Hierarchical part-based visual objects categorization", CVPR, Springer,Volumen 1, 2005.
12. Torres R., Huerta M., Clotet R., González R., Sánchez L., Erazzo M,, y otros. (2014). A Kinect based approach to assist in the Diagnosis and Quantificaction of Parkinson's Disease. VI Congreso Latinoamericano de Ingeniería Biomédica CLAIB2014. Paraná-Argentina.
13. Torres R. et al., "Diagnosis of the corporal movement in Parkinson's disease using kinect sensors," in IFMBE Proceedings, 2015, vol. 51, pp. 1445–1448.

A Parametrization Approach for 3D Modeling of an Innovative Abduction Brace for Treatment of Developmental Hip Dysplasia

Natalia Aurora Santos, Barbara Olivetti Artioli, Ellen Goiano, Maraisa Gonçalves, and Maria Elizete Kunkel

Abstract

Developmental hip dysplasia (DHD) is frequently encountered in the pediatric orthopedic practice. DHD is characterized by dislocation of the femoral head in the acetabulum. In Brazil are diagnosed three times more cases than the world average (5–8 cases for 1,000 births). The lack of treatment leads to long-term morbidity, abnormal gait, chronic pain and arthritis. Early detection and treatment with a Pavlik harness results in improved outcomes. After 6 months of age, closed or open reduction with spica casting is required for 4 months to treat a persistent hip dislocation. Plaster is used for orthopedic immobilization due the low cost, moldability and good mechanical resistance. However, there are several risks and complications due to the use of spica cast in DHD treatment: Skin problems due to lack of adequate hygiene (itching, ulceration, dermatitis and infection), formation of pressure areas, plaster fracture (11% of cases) and fever. Digitization techniques have been explored for production of customized hip abduction brace by additive manufacturing. However, it is not possible to keep a child standing still to perform 3D scanning of the hip and legs region. The goal of this research was to develop an alternative approach for acquisition of the external geometry of the infant to create 3D model of an abduction brace. The parameterization technique created includes: The creation of a virtual 3D model of a child's body using the MakeHuman software; Articulation of the hip region of the model to the position required in the treatment of DHD with the Blender software; Definition of the parameters required for the modeling of a hip abduction brace. A DHD pediatric orthopaedist approved the methodology created. Innovations in the area of assistive technology can bring many benefits to the user in the process of rehabilitation.

Keywords

3D modeling • Parametrization • Developmental hip dysplasia • Hip abduction brace

1 Introduction

Several clinical situations in orthopedics require immobilization as a form of treatment, such as developmental hip dysplasia (DHD), Legg Calvé Perthes disease, fractures and others [1]. DHD occurs in infants in the first months of life and is characterized by dislocation of the femoral head of the acetabule. In Brazil, about 5–8 cases are diagnosed for every 1,000 births. Lack of DHD treatment leads to long-term morbidity, abnormal gait, chronic pain, and arthritis in adult life. If DHD is diagnosed in the first month of life, treatment is performed with the Pavlik harness. If the treatment does not work or if the diagnosis of DHD is performed after the sixth month of life, it is necessary to replace the femoral head in the acetabular region surgically and maintain the infant with legs and hip in abduction and flexion with the use of spica casting during a period of 4 months.

N. A. Santos · M. E. Kunkel (✉)
Science and Technology Institute, Federal University of Sao Paulo, São Paulo, Brazil
e-mail: biomecanica.unifesp@gmail.com; elizete.kunkels@gmail.com

B. O. Artioli · M. Gonçalves
Center for Engineering, Modeling and Applied Social Sciences, Federal University of ABC, Sao Bernardo, Brazil

E. Goiano
Pediatric Orthopedics, Hospital Municipal Dr. José de Carvalho Florence, São José dos Campos, Brazil

M. Gonçalves
Science and Technology Institute, Federal University of Sao Paulo, São José dos Campos, Brazil

© Springer Nature Singapore Pte Ltd. 2019
L. Lhotska et al. (eds.), *World Congress on Medical Physics and Biomedical Engineering 2018*,
IFMBE Proceedings 68/3, https://doi.org/10.1007/978-981-10-9023-3_41

Plaster is the most commonly used material for orthopedic immobilization due to its low cost, easy moldability and good mechanical strength. However, there are several risks and complications for to the use of plaster in DHD treatment: skin problems due to lack of proper hygiene, itching, ulceration, dermatitis, infection, formation of pressure areas, plaster fracture in 11% of cases and increase in body temperature.

Some types of orthosis have been investigated to replace the spica casting in the immobilization of the hip and legs in the treatment of DHD [2, 3]. The technology used in this orthotics is outdated, the material is thermoformed in the region to be immobilized, or a pre-cast with plaster is done directly on the patient's body, production is laborious and time consuming procedure. These orthoses are expensive, the risks and complications due to the use are the same as those found in the spica casting, and there are few studies on their effectiveness [2–7].

In recent years, the application of new technologies such as 3D modeling and additive manufacturing (MA) have been explored in the health area in the production of customized orthoses [8, 9], mainly of upper limbs [10–14]. The MA allows the production of solid components from digital models with a 3D printing machine. It allows the manufacture of parts with good mechanical strength, low cost and in a relatively short period of time [15].

An alternative to immobilization in the treatment of DHD might be a custom orthosis constructed from high technology with low cost that can replace the spica casting efficiently. Some digitizing techniques have been explored for 3D modeling and production of a custom infant bracing for hip abduction produced by additive manufacturing (MA). In [16], the external surface of the hip region and legs of a puppet was digitized by a photogrammetry technique and by 3D scanning [8, 9]. Photogrammetry allows the 3D reconstruction of an object from photographs, being a lower cost alternative to laser scanning. The scanning technique was tested only on a static doll, however, 3D scanning of the hip region and legs of a baby are not feasible; because this procedure requires the child to remain in a static position for a few minutes [17].

A virtual model of the infant can be created by means of a parametrization technique, without the need of plaster cast or 3D scanning. This approach may allow the creation of an orthosis for the treatment of DHD without the need to digitize its structure. In this study, a methodology will be described for the acquisition of the geometry of the infantile hip and that can be substituted for the techniques of photogrammetry and scanning. The objective of this research is to develop an alternative approach for the acquisition of the external geometry of a baby, through the creation of a virtual model of an infant, for the 3D modeling of an abduction orthosis for DHD, through of a parameterization technique.

2 Materials and Methods

The MakeHuman open source software is a database of 3D humanoid models based on anthropometric measurements. The software allows the creation of parameterized customized models and some functions allows to export a humanoid model to the free software Blender so that it can be articulated. Thus it would be possible to position a virtual model in the required position in the DHD treatment. In cases of DHD, each child needs to be immobilized in a different position with the angle of hip flexion from 40° to 45°, abduction angle of the legs between 100° and 110° and angle of knee flexion varying from 80° to 90°.

In this study, a parametrization technique was developed to transfer patient measurements to a virtual humanoid model. The technique ensures that the patient's measurements are more accurate than those obtained by scanning from the technique of photogrammetry or scanning that require the baby not to move. A parametrization approach was developed in three stages:

(1) Creation of an articulated 3D model of the infantile body (avatar) with the software MakeHuman and a skeletal mesh deformation: A humanoid infant model was selected (Age: 1, Muscle: 50%, Weight: 100%, Height: 50.22 cm (Fig. 4) and also the skeleton model most appropriate to the project ("Cmu mb" model with only 31 bones) was selected.

(2) Articulation of the hip and legs of the avatar to the position required in the treatment of DHD using the free software Blender: The positioning of the thigh and legs required in the treatment of DHD was represented with the help of the tool plugin BlenderTools. The template file exported in MakeHuman was then imported with the extension ".mhx". In Blender the position of the object was modified using the rotation cursor in the "pose mode" option. The bone corresponding to the member that was to be modified was selected. The "bone" corresponding to the thigh was selected and the green line of the rotation cursor moved to the desired abduction angle, in this case approximately 110°. The "bones" corresponding to the legs were selected, so the green line of the rotation cursor was moved, thus leading to knee flexion, positioned at 90°. Other adjustments such as rotation on the Y-axis of the thigh were required. Finally, an articulated doll model was created and could be positioned as needed.

(3) 3D modeling of the child's avatar and definition of a minimum number of parameters for the 3D modeling of a bracing: Once the doll is in the desired position, it was possible to determine the measures of the hip region and legs, which will be the measures taken of a

patient. With the help of the Meshmixer software, some functions analyze and correct where there is a vector mesh error, closing the cloud of open points. With only the part of interest of the hip, in Blender, the hip object was solidified in the option "edit" with the function "make solid". After this step the object could be modeled and parameterized. The dummy object is placed in the desired position with the help of the Blender tool, in case it was placed in the treatment position for DHD to exemplify. In this way, a huma-noid model was created articulated and positioned to be produced in the future, a customized orthosis model for a patient.

3 Results

The humanoid model created in the MakeHuman software was dimensioned with age, weight and height close to an infant, to represent its body geometry (Figs. 1, 2, 3). A set of

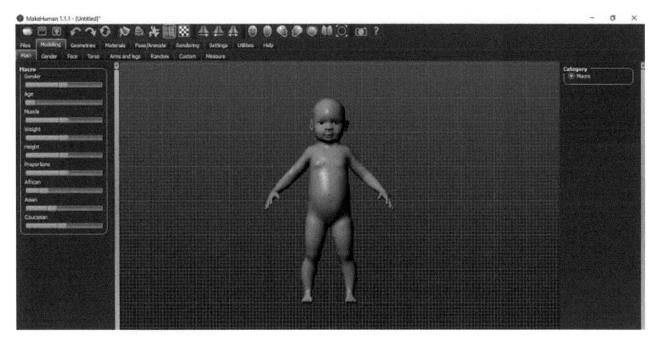

Fig. 1 Selected template in MakeHuman representing the geometry of a child

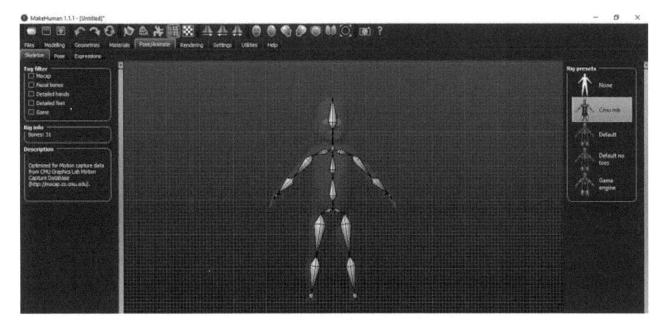

Fig. 2 Skeleton selected for posterior articulation of the humanoid model

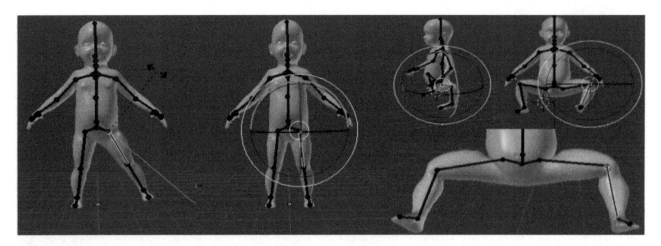

Fig. 3 Articulated doll with hip in the DHD treatment position

bones was added to the model (Fig. 1), to create a skeleton in which it was possible to articulate the doll later. In the Blender software, the articulating doll (Fig. 2) was imported. With the aid of the skeleton previously inserted with a doll, the modifications can be made (Fig. 3), starting with the bone representing the femur, following the bone that moves the leg.

The virtual doll in anatomical position and the virtual doll already positioned for DHD treatment were parameterized (Fig. 4). The 3D model of a created infant allows the high fidelity representation of the kinematics of the whole body using the definition of anatomical and reproducible bone segment.

4 Discussion

The plaster used in orthopedics is efficient in its immobilization function, however, the burdens entailed by its use are far greater than the benefits. The child who uses the plaster can not sit down, take a bath, it goes through anesthetic precedence for the placement of the plaster and other problems. There are studies and technologies capable of creating customized orthoses to eliminate the problems generated by the use of plaster. These technologies can be

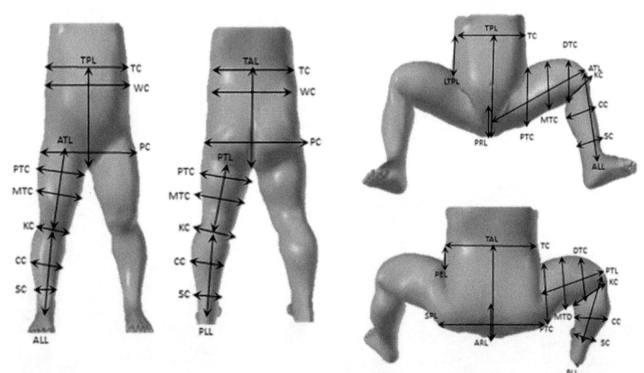

Fig. 4 Circumference of: TC (thorax), WC (waist), PC (pelvis), PTC (proximal thigh), MTC (mean thigh), DTC (distal thigh), KC (Knee), CC (calf), SC (supramaleolar). Length of: TPL (thoraco-perineal), TAL (thoraco-anal), ATL (anterior thigh), PTL (posterior thigh), ALL (anterior leg), PLL (posterior leg), SPL (special pelvic), LTPL (lateral thoraco-pelvic), PRL (perineal region), ARL (anal region)

used in an even more efficient and practical way with the use of the parametrization technique developed in this research.

MakeHuman has proven to be a good tool for acquiring ready-made models, without the need for digitization, either from dolls or from the patient itself. MakeHuman is commonly used in computer animation for games and movies. However, no study was found using the software in the creation of a parameterized virtual model for the development of personalized orthotics. Based on literature searches, it was possible to observe that there is not much information on anthropometric parameterization. The difficulty of making a personalized orthosis in children who move unconsciously motivated the creation of an innovative methodology. This technique was created for the development of hip orthoses but can be used for any structure of the body acting in the creation of orthoses by additive manufacture.

The acquisition of human geometry, for the production of orthoses, through preexisting models is possible. But the positioning of the 3d puppet is a difficult task because each clinical case requires a different and crucial pose for a favorable prognosis of the treatment. Subsequently, it is practicable to manipulate this geometry, to create a bracing. This methodology can aid in new protocol research, so that specific, precise and specialized methodologies for the production of orthoses can arise, as well as the creation of orthoses for various regions of the human body, as well as for animals. This research is being developed and evaluated with the participation of a medical team from the orthopedics of a large public hospital. The approach requires little contact time with the infant, and the time to produce a hip abduction orthosis can be optimized and can bring many benefits for the babies in the DHD rehabilitation process.

5 Conclusion

Personalized technological innovations in the area of orthosis can bring many benefits for users and caregivers, generally aiding in the process of rehabilitation. It is possible to use pre-created humanoid models to efficiently represent the geometry of the external anatomy of the human body. The acquisition of anatomical hip geometry and modeling may allow the production of orthoses for the treatment of DHD without the need for 3D scanning. Using the parametrization technique in humanoid models to create scripts of the geometry of an infant can allow the creation of personalized products as an abductor orthosis. These technologies allow you to produce low-cost orthoses, so that they fit perfectly to the hip with a safety margin for comfort.

References

1. Campion JC, Benson MK. Developmental dysplasia of the hip. Surgery (Oxford), 2007;25(4):176–180.
2. Wilkinson AG, et al. The efficacy of Pavlik harness, the craig splint and the von Rosen splint in the management of neonatal dysplasia of the hip. JBone & Joint Surgery. 2002;84(5):716–719.
3. Hedequist D, et al. Use of an abduction brace for developmental dysplasia of the hip after failure of Pavlik harness use. Journal of Pediatric Orthopedics. 2003;23(2):175–177.
4. Ibrahim DA, et al. Abduction bracing after Pavlik harness failure. JPediatric Orthop. 2013;33(5):536.
5. Uras I, et al. The efficacy of semirigid hip orthosis in the delayed treatment of developmental dysplasia of the hip. Journal of Pediatric Orthopaedics B. 2014, 23(4):339–342.
6. Dyskin E, Ferrick M. Semirigid abduction bracing is effective treatment of reducible developmental dysplastic hips after failure of Pavlik harness. Annals of Orthopedics & Rheumatology. 2015;3 (2):1045.
7. Wahlen R, Zambelli P. Treatment of the developmental dysplasia of the hip with an abduction brace in children up to 6 months old. Advances in orthopedics. 2015:1–6.
8. Toledo I, et al. Metodologia para produção de órteses por meio de fotogrametria, modelagem 3D e manufatura aditiva. XVII Congresso Brasileiro de Biomecânica CBB e I Encontro Latino Americano de Biomecânica, 2017, Porto Alegre. p. 564–465.
9. Mavroidis C et al. Patient specific ankle-foot orthoses using rapid prototyping. Journal of Neuroengineering and Rehabilitation. 2011;8(1):1–11.
10. Jumani MS, et al. Fused deposition modelling technique (FDM) for fabrication of custom-made foot orthoses: a cost and benefit analysis. Scientific International (Lahore). 2014;26 (5):2571–2576.
11. Paterson AMJ, Bibb RJ, Campbell RI. A review of existing anatomical data capture methods to support the mass customisation of wrist splints. Virtual and Physical Prototyping. 2010;5(4):201–207.
12. Paterson A. Digitisation of the splinting process: exploration and evaluation of computer aided design approach to support additive manufacture [thesis]. Leicestershire: Loughborough University; 2013.
13. Paterson AM, Donnisson E. Computer-aided design to support fabrication of wrist splints using 3D printing: A feasibility study. Hand Therapy. 2014;19(4):102–103.
14. Paterson AM et al. Comparing additive manufacturing technologies for customised wrist splints. Rapid Prototyping Journal. 2015; 21(3):230–3.
15. Cano APD. Parametrização e produção de órtese termomoldável para imobilização de punho produzida por manufatura aditiva. 102p. Trabalho de Conclusão de Curso de Engenharia Biomédica (TCC) - Universidade Federal de São Paulo. 2017.
16. Meidanshahy T. Feasibility of using 3D printing in the manufacture of orthotics componentry. Diss. Flinders University, Adelaide, Australia 2014.
17. Munhoz R, Moraes CADC, Tanaka H, Kunkel ME. A digital approach for design and fabrication by rapid prototyping of orthosis developmental dysplasia of the hip. Research on Biom Eng. 2016;32(1),63–73.

ApOtEl: Development of a Software for Electroporation Based Therapy Planning

Luisa Endres Ribeiro da Silva, Marcos Tello, Dario F. G. de Azevedo, and Ana Maria Marques da Silva

Abstract

The objective of this paper is to describe the implementation of a software application, called *ApOtEl*, developed for needle electrodes positioning optimization for electrochemotherapy procedures in the treatment of cutaneous and subcutaneous tumors. The software was developed using MATLAB®, and it optimizes the needle-type electrodes positioning configurations, through the study of the analytical electric field, using Laplace equation in a homogeneous bi-dimensional environment. The optimization was based on requirements chosen to guarantee the tumor total coverage and to minimize healthy neighboring tissues damage. An optimization function was created to orientate the electrochemotherapy application, and the best option available in all configurations generated by the distance variations between electrodes and positioning orientations was determined. The software provides, by the entry of tumor dimensions, the optimized distances for positioning needle-type electrodes, as well as the representation of the electric field distribution and intensity. Representation of the electrodes positioning and instructions for the procedure to facilitate the procedure planning are provided.

Keywords

Electrochemotherapy • Software development
Electric field

L. E. R. da Silva (✉) · A. M. M. da Silva
Electrical Engineering Graduate Program, PUCRS,
Porto Alegre, Brazil
e-mail: luisa.endres@acad.pucrs.br

A. M. M. da Silva
e-mail: ana.marques@pucrs.br

M. Tello
State Electrical Energy Company, CEEE, Porto Alegre, Brazil

D. F. G. de Azevedo · A. M. M. da Silva
Laboratory of Medical Imaging, Science School, PUCRS,
Porto Alegre, Brazil

1 Introduction

Electrochemotherapy (ECT) is an effective and safe form of local antitumor treatment in which short and intense electrical pulses are applied to the target volume in conjunction with chemotherapeutic agents [1, 2]. Due to exposure to electrical pulses of sufficiently high intensity, the cell membrane becomes temporarily permeable, a process known as reversible electroporation, allowing large molecules, such as chemotherapeutics, to cross the cytoplasmic membrane barrier, potentializing its effects [3].

Studies on electroporation date back to the 1960s [1]. Over the years, several experiments, both in vitro and in vivo, have been performed. In 2002, the European Standard Operating Procedures for Electrochemotherapy and Electrogenetherapy (ESOPE) was created to define standard operating procedures for the technique. The study reported complete tumor regression in 73.7% of nodules treated with only one ECT application [1, 4]. Many other clinical studies have been performed, and the regression rate followed by a single ECT application is 60% (85% of objective response), although this percentage varies between tumor types [5].

Among the main ECT advantages, the technique has the ability to preserve tissues, functions and sensitive structures, as well as, to produce minimal and tolerable side effects. Moreover, ECT has the ability to treat tumors of any histology, it is a quick (taking only a few minutes) and a cost-effective treatment, that greatly improves the quality of life of the treated patients [4, 5].

The number of patients benefiting from ECT as a form of treatment has been increasing rapidly, with more than 1,500 patients treated in 2011, in more than 100 hospitals around the world [1]. By the end of 2015, approximately 13,000 cancer patients were treated with ECT and currently it is used routinely in about 140 European centers [5]. In Brazil, research with electroporation began in 2008 in a university veterinarian hospital. Currently, the treatment is part of the routine of this animal hospital [3].

© Springer Nature Singapore Pte Ltd. 2019
L. Lhotska et al. (eds.), *World Congress on Medical Physics and Biomedical Engineering 2018*,
IFMBE Proceedings 68/3, https://doi.org/10.1007/978-981-10-9023-3_42

However, ECT success is highly dependent on the electric field distribution and parameters related with its distribution and intensity, such as type, quantity and electrodes positioning, whose incorrect choice may compromise the treatment.

The objective of this paper is to describe the implementation of a software application, developed for needle electrodes positioning optimization for ECT procedures in the treatment of cutaneous and subcutaneous tumors.

2 Methods

2.1 Electric Field

The distribution of the electric field is one of the fundamental aspects for the effectiveness of electroporation treatments. For a successful treatment, all tumor cells must be destroyed and it is extremely important that the electric field is as uniform as possible, covering the entire tumor with intensity that exceeds the electroporation threshold (E_{rev}), around 600 V/cm for most neoplasms [3]. In order to obtain the electric field intensity, some parameters such as type and configuration of electrodes, their positioning should be carefully chosen, since they have a great influence on the electric field intensity and distribution. If these parameters are not chosen appropriately, the electric field may be insufficient or inadequate, leading to an ineffective treatment and tumor resection due to the insufficient magnitude of the local electric field or to non-coverage of parts of the tumor.

The analytical solution for the electric field distribution has limitations, such as being only possible for regular geometries, usually confined to 2D and it assumes that the electrical properties of the tissues are uniform, which rarely occurs in practice. However, it is a quick and convenient way for analyzing the electric field distribution in the tumor, providing an overview of the treatment area, as well as its effectiveness.

For the development of this work, mathematical models and analytical calculations, based on Čorovič (2010), were incorporated in the software application to calculate the electric field between needle electrodes, positioned in the region of interest. According to this method, electric field intensity and distribution are estimated by Laplace equation, considering the electric potential as a sum of the multipoles of all electrodes, in a homogenous environment.

2.2 Electrode Positioning Optimization

The distribution of the electric field can be controlled by the applied voltage, number, electrodes distances and how they are placed in the target tissue. We sought an optimization of the main parameters that dictate the field distribution for different tumor sizes.

The simplifications assumed include the analytical calculations for 2D representations of superficial cutaneous tumors measuring 1 cm × 1 cm to 15 cm × 15 cm, including tumors with elliptical shapes. The area of interest is represented as a 2D plane of the tumor in 3D space. This assumption is that the electric field is homogeneous along the electrode's axis of insertion, perpendicular to 2 D plane of the tumor.

The electric field is calculated by assuming 1 V between electrodes, allowing the use of several E_{rev} thresholds in the optimization, with electroporation occurring at 1 V/cm. However, in order to guarantee a greater safety, it is assumed that the area where the electroporation occurs is equal or greater than 1.20 V/cm, to provide a 20% safety margin. This safety margin is justified by model simplifications and tumors irregularity.

Electric field distribution is analyzed for 1, 2 and 3 pairs of electrodes positioned parallel in line with the configuration center, coincident with tumor center. Such linear arrangements are commonly used in ECT for minimizing the inhomogeneity of the electric field distributions generated by needle electrodes. The possibilities of distance between the electrodes (starting in a distance of 1 cm and using steps of 0.1 cm) and two orientations lead to a total of 1360 configurations available. It is important to mention that a previous evaluation was carried out to eliminate redundant configurations.

For the electrodes positioning optimization some requirements are chosen, such as the electric field with an intensity superior or equal to E_{rev} must cover the entire tumor; neighboring healthy tissue should not be exposed to an excessively high field, being as small as possible; and as few applications as possible should be carried out.

Considering these requirements, an "optimization" function was developed to evaluate the available electrode configurations, appropriate for each tumor size, which takes into account the tumor coverage by the electric field for electroporation and the conformation of the electric field to the tumor shape.

First the tumor coverage fraction by the electric field configurations is evaluated through the calculating of the coverage fraction (CF), given by Eq. (1):

$$CF = \frac{AC}{AT} \tag{1}$$

where AC refers to the total area covered by the setting within the criterion of intensity greater than or equal to E_{rev}, and AT is the total area of the tumor. CF values below 1 indicate that the area of the tumor covered by the electric field threshold that generates electroporation is less than necessary; if CF is greater than 1, electroporation occurs in a

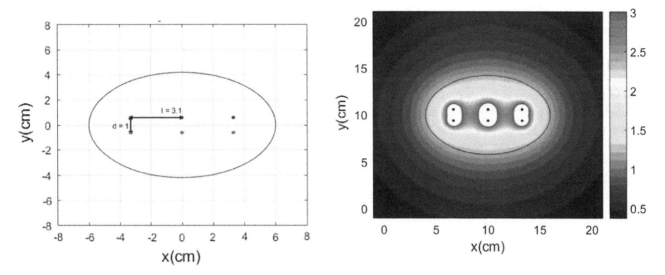

Fig. 1 (Left) Representation of electrodes positioning for a 12 cm × 8.4 cm tumor. (Right) Electric field distribution for the optimized configuration for a 12 cm × 8.4 cm tumor

region larger than the tumor area, reaching the healthy neighborhood.

However, the comparison between areas is not enough to optimize the positioning, since some electrode configurations can generate asymmetric areas and tumors can be irregular. To solve this problem, the tumor maximum dimensions in vertical and horizontal directions and the electric fields thresholds generated by the configurations are determined. For optimization, a penalty is established if tumor dimensions are greater than the electric field distribution area with intensity superior or equal to E_{rev} of a given configuration, in any direction. The penalty is calculated as follows:

$$\begin{cases} P_x = W_T - W_E, \text{if } W_T \leq W_E \\ \quad P_x = 100, \text{if } W_T > W_E \end{cases} \quad (2)$$

$$\begin{cases} P_y = L_T - L_E, \text{if } L_T \leq L_E \\ \quad P_y = 100, \text{if } L_T > L_E \end{cases} \quad (3)$$

$$P = Px + Py \quad (4)$$

P_x is the penalty applied to height, P_y is the penalty applied to width, P is the total penalty applied to the configuration, W_T is the tumor height, W_E is the height of the configuration, L_T is the tumor width and L_E is the configuration width.

Thus, the total optimization of the configuration is given by Eq. (5):

$$Op = CF + P \quad (5)$$

The closer the *Op* function value is to 1, better the criteria are met by the configuration. If the configuration for one application is not optimized for tumor size or shape, a larger

number of applications optimized for different tumor areas should be performed.

The software application was developed using MATLAB® tools, language and compiler, which generated an executable for installation. The software application is restricted to Windows operational systems.

3 Results

The software application, called *ApOtEl*, which stands for Application for Optimization of Electrodes, can be installed through the executable (.EXE) provided by the authors.

Through the entry of tumor dimensions, optimized parameters for electrodes positioning are provided, with the electric field covering the entire tumor with intensity higher than E_{rev} threshold. In addition, *ApOtEl* provides an electric field distribution representation, and a brief explanation of the technique application procedure.

When starting the simulation panel, user should provide the 2D tumor dimensions (in cm). *ApOtEl* fills in the results —Tumor Dimensions, Procedure and Optimization.

ApOtEl will initiate the optimization process for the inserted tumor size. After optimizing or choosing the ideal parameters. Dimensions section presents tumor dimensions and electroporated volume; volume calculation is estimated, assuming the smaller dimension as the tumor depth. Procedure section provides instructions on how electrodes should be placed, the depth they should be inserted and whether more than one application will be needed. In Optimization section, ECT optimum parameters for the tumor are presented, such as number and distances between the electrodes. *ApOtEl* also generates an electric field distribution representation for the optimized parameters, and a representation of electrodes positioning in the tumor.

An example of *ApOtEl* in a hypothetical case study with a 12 cm × 8.4 cm tumor is presented (see Fig. 1). By informing the tumor dimensions, the system will optimize and provide the most appropriate parameters for a safe and efficient procedure.

For a total coverage of the tumor with an electric field with intensity above E_{rev} and with the minimum of healthy tissue affected, the system determined the use of six electrodes in two parallel lines with a distance of 1 cm between the electrodes of different polarity (d) and distance of 3.1 cm for the electrodes with the same polarity (l). Finally, the system provided a chart schematically showing where these electrodes should be positioned and the distribution of the electric field generated by them.

4 Conclusion

The main result of this work is the software application *ApOtEl*, desktop-installable, that provides the visualization of the optimized electric field distribution of needle electrodes for ECT.

The oncology professionals are not familiar with the concept of electric field distribution, which is the ECT base. Thus, the developed software will help these professionals to visualize what happens during ECT application, as well as providing an optimization of the treatment, seeking greater effectiveness. The visualization of the electric field distribution will give oncology professionals a better understanding of the physical phenomena involved in ECT, in addition to providing a greater "confidence" in the application of the technique.

This software application will contribute to the evaluation and planning of treatments involving ECT in a safe and efficient way, through studies using mathematical models (in silico studies). These analyzes may benefit in decision-making and planning of pre-application procedures.

The preliminary application of the developed software has been shown to be suitable for the planning of cancer treatment in animals. Further studies will integrate this software application with optical medical images for the evaluation of tumor dimensions and subsequent choice of the appropriate positioning for each situation.

Compliance with ethical standards

Conflicts of interest: The authors declare that they have no conflict of interest.

References

1. Haberl, S., et al.: Cell membrane electroporation–part: the applications. IEEE Electr Insul Mag 29 (1), 29–37 (2013).
2. Pavšelj, N., Miklavčič, D.: Numerical modeling in electroporation-based biomedical applications. Radiol Oncol 42 (3), 159–68 (2008).
3. Telló, M.: Uso da corrente elétrica no tratamento do câncer. 1st ed. EDIPUCRS, Porto Alegre (2004).
4. Čorovič, S.: Modélisation et visualisation de l'électroperméabilisation dans des tissus biologiques exposés à des impulsions électriques de haut voltage. PhD (dissertation). Paris Sud: Paris (2010).
5. Calvet, C. Y., Mir, L. M.: The promising alliance of anti-cancer electrochemotherapy with immunotherapy. Cancer Metastasis Rev 35 (2), 165–177 (2016).

Physical Analysis of Pulse Low-Dynamic Magnetic Field Applied in Physiotherapy

Aleš Richter, Miroslav Bartoš, and Želmíra Ferková

Abstract

This article makes an effort to explain some physical and energy aspects of practice using magneto therapy in the treatment of the musculoskeletal system (orthopaedic surgery, physiotherapy and the rehabilitation). We are presenting the principles of electromagnetic induction in muscle tissue as typical example of the parts of human body. The main accent of presenting theory is put on macroscopic physical behaviour of low-frequency electromagnetic field in living body parts. The problems are indicated with modelling. This approach use simplified model of tissue conductivity. One of the goals is to warn about different distribution of magnetic field in parts of body caused by different position, spatial orientation and metal implants too. The each metal part implanted into human body has strongly influence on distribution physical effects into. Another analyse introduces physical and energy differences among individual types of power sources of magnetic field and their dynamic behaviour.

Keywords

Magneto therapy
Electromagnetic induction in living tissue
Metal implants

A. Richter (✉)
Institute of Mechatronics and Computer Engineering, Technical University of Liberec, Liberec, Czech Republic
e-mail: ales.richter@tul.cz

M. Bartoš
Department of orthopedics, First Faculty of Medicine Charles University, Central Military Hospital, Prague, Czech Republic

Ž. Ferková
Department of Electrical Engineering and Mechatronics, Technical University of Košice, Košice, Slovakia

1 Introduction

Every living creatures by own activities produce electromagnetic field. Does external low frequency magnetic field have influence on living structure?

1.1 Medical Aspects

This "historical" magnet therapy, magnetic therapy, or magneto therapy, based on the use of the static magnetic fields is considered by most of the doctors' pseudoscientific alternative medicine practice.

Pulsed electromagnetic field therapy (PEMFT), also known as *pulsed magnetic field therapy (PMFT)* or *low field magnetic stimulation (LFMS)*, was introduced as a therapeutic method in many areas of medicine at the end last century and especially during last decade and it is considered as an effective procedure especially for the treatment of the musculoskeletal disorders (orthopedic surgery, physiotherapy and the rehabilitation) [1, 2].

Clinical experience has shown that pulsed electromagnetic field therapy may be used for a number of conditions and issues, and that the benefits include:

- Reduction of pain and inflammation
- Improve energy/circulation, blood/tissue oxygenation
- Regulate blood pressure and cholesterol levels as well as the uptake of nutrients [3, 4]
- Increase cellular detoxification and the ability to regenerate cells
- Accelerates repair of bone and soft tissue/relaxes muscles, [4] etc.

Many theories about the healing principles of the pulsed electromagnetic field on the living human tissue were published, but no of them has brought the clear explanation, so

© Springer Nature Singapore Pte Ltd. 2019
L. Lhotska et al. (eds.), *World Congress on Medical Physics and Biomedical Engineering 2018*,
IFMBE Proceedings 68/3, https://doi.org/10.1007/978-981-10-9023-3_43

many doctors don't accept this method as an "evidence based medicine" procedure [3, 5].

PEMFT is considered very safe method. There are no contraindications to PEMFT except in cases of haemorrhage or where electrical implants are in use. In contrast to chemical medicaments, there is no over dosage, at least within the field range that are presently used for treatments. The PMF therapy is a heatless therapy, therefore, all implants (except heart pacemakers) can be treated. Fractures can be treated even through a plaster cast, since magnetic fields permeate all materials. Its advantages are relatively low cost and simple use. It does not require any skilled operator. Furthermore, in order to get the maximum efficiency of the method, its basic principle and important parameters should be known.

1.2 Theoretical Premises

The fundamental description of magnetic effects on living organism is described in many articles and books [1]. These texts, particularly medical literature, very often present the studies in different physical units. The electromagnetic field is completely described by four Maxwell equations [6]. This theory is appropriate to description of inner electromagnetic field in human bodies and in the tissues. The tissues and human and animal bodies mainly consist of diamagnetic substances. The magnetic fields penetrate through all parts of the body without difficulty and external magnetic field strength H_m is slightly attenuated. The fractional reduction of the magnetic field in living tissue is caused by a variation of electron orbits of atoms or molecules and it is less than 10 millionths from range. The frequency of external magnetic field is low and therefore dielectric losses are not taken into account too therefore the following table (Table 1) presents conductivity of selected typical tissues only [7]. The examples of tissue conductivity show that human or animal body is strongly heterogeneous surrounding from electromagnetic view. We can speculate about low intensity of external electromagnetic field so the conductivity of individual tissue is linear.

The ones of tissues (particularly muscle and brain) are anisotropic. The magnetic field direction has substantial influence on physical effect in each point tissue.

For unified description it is very important to present the definition of physical notions and units in International System of Units (SI), (Table 2) e.g. [6, 7].

1.3 Energy or Power Balance in Tissue

The variable external magnetic field is transformed into the strength of electric field E_t in tissue. According to Ohm's law the strength of electric field produces current density

Table 1 The characteristic conductivity of representative tissues. Electromagnetic properties of tissues in frequency range up to 1 MHz

Conductivity	γ_t [S/m]
Blood	1.23
Cartilage	1
Muscle along	0.43
Muscle across	0.17
Porous bone	0.17
Subcutaneous fat	0.057
Brain Grey Matter (average)	0.23
Brain Grey Matter—according to direction	0.19–0.25
Metal Implants: Stainless steel	1.1×10^6
Titanium	1.8×10^6

J because the tissue is adequately conductive γ_t, e.g. [1, 6, 8]. The product of electric field strength E and current density J indicates power density w_t which is absorbed in tissue. We suppose that external magnetic field is low frequency and therefore dielectric losses are not taken into account [7].

One of the important parameter how to define the effect of variable magnetic field in tissue is electric field strength E_t (see Eq. 1).

Medical literature defines this biophysical effect with the current density in tissue J_t [A/m^2] or [A/cm^2], 1[A/m^2] correspond to 100 [μA/cm^2] or 0.1 [mA/cm^2] (3). This parameter is significant for statement of non-hazardous levels of magnetic field e.g. [9–13]. The local current density in the tissue we can describe as follows (2, 3):

$$E_t = \left(\frac{\mu}{\gamma_t} H_m \frac{\partial H_m}{\partial t}\right)^{\frac{1}{2}} \quad (1)$$

$$J_t = \left(\mu \gamma_t H_m \frac{\partial H_m}{\partial t}\right)^{\frac{1}{2}} \quad (2)$$

The important parameter how to evaluate the impact of external magnetic field is energy density absorbed in tissue. This situation is presented by the following Eq. (4). The density of energy in tissue w_t (see Eq. 5) is a local parameter valid for actual time intervals and space in the part of body.

The variable external magnetic field is transformed into the strength of electric field E_t in tissue. The product of electric field strength E_t and current density J_t indicates power density w_t which is absorbed in tissue. This follows from 2nd Maxwell Equation or Faraday's Induction law. According to Ohm's law the strength of electric field produces current density J_t because the tissue is adequately conductive γ_t, e.g. [6].

$$J_t = \gamma_t E_t \quad (3)$$

Table 2 Determination of Physical Units

γ_t—conductivity of tissue	[S/m], [m^{-3} kg^{-1}s^{3}A^2]
J_t—current density in the tissue	[A/m^{-2}]
E_t—electric field strength in the tissue	[V/m], [m.kg.s^{-3}A]
w_m—density of magnetic energy coming into tissue	[J/m^3], [m^{-1} kg. s^{-2}]
H_m—external magnetic field strength	[A/m]
w_t—density of energy absorbed by tissue	[J/m^3], [m^{-1} kg. s^{-2}]
B_m—external magnetic induction in tissue	[T], [kg.s^{-2}A^{-1}]
W_m—total magnetic energy coming into tissue	[J]
$P_{m\,max}$—top of magnetic power coming into tissue	[W]
p_t—density of power absorbed by tissue	[W/m^3]

Table 3 The circuit parameters of magneto therapy coil

Inner Inductance	L = 0.05	[H]
DC resistance	R_{DC} = 15	[Ω]
Number of turns	n = 335	
Winding length	l = 340	[mm]
Ellipse axis	600 × 560	[mm]

$$w_t = \int_{t_1}^{t_2} \boldsymbol{J}_t \cdot \boldsymbol{E}_t dt = \gamma_t \int_{t_1}^{t_2} E_t^2\, dt = \frac{1}{\gamma_t} \int_{t_1}^{t_2} J_t^2\, dt \quad (4)$$

$$w_t = \mu \int_{t_1}^{t_2} H_m \frac{\partial H_m}{\partial t} dt = \frac{1}{2}\mu \left[H_m^2 \right]_{t_1}^{t_2} \quad (5)$$

$$W_t = \int w_t dV \quad (6)$$

We expect that this power-producing parameter would be more important for evaluation of magnetic field efficiency. It is evident that impact of magnetic field depends on the content of tissue which is inserted into the space of operating external field. This situation is described by Eqs. (7, 8).

$$P_t = \int p_t dV \quad (7)$$

$$P_t = \int_V \boldsymbol{J}_t \cdot \boldsymbol{E}_t\, dV = \frac{d}{dt} \int_V w_m dV \quad (8)$$

$$W_t = \iiint_V w_t dx dy dz \quad (9)$$

This relation represents the total energy loss and it is given by cubic integral (7), e.g. [7, 8]. Other way how to define impact of magnetic field into human body is expression of instantaneous power. This situation is described by Eq. (8).

The Eqs. (7, 8) is derived from the integral representation of energy conversion principle which is presented in Eq. (1). The Eq. (2) is transformed by this way.

2 Limitations of Magneto Therapy Using

Physiotherapy use several types and sizes of magnetic applicators for application of magneto therapy. These magnetic adapters can be classified as the applicators with open or closed magnetic circuit. Comparatively great cylindrical coil are very popular in medical offices that is why we chose these types for our experiments. These coils do not include ferromagnetic core and the magnetic field is continuous.

2.1 Sources of Magnetic Field

This presented type of magnetic applicator is the biggest which is used in magneto therapy and rehabilitation care. The large profile along with relatively short length of open air coil does not allow creating inside sufficient large space of homogenous magnetic field. The magnetic applicator (according to Fig. 1) is open air coil without ferromagnetic pole extension. The design of device is frameless and therefore the profile of coil is slightly ellipsoidal.

In this case the maximum current in coil (Fig. 1) is chosen 2 A. The distribution of magnetic induction is shown in the Fig. 2.

Violet curve shows distribution along x axis on periphery of the coil and red curve is according x axis in centre. The simulation verifies our assumption that the lowest intensity of field is right in the point of intersection between axis y and axis z in the center. The level of magnetic induction **B** can range in units [mT]. In comparison with induction of geomagnetic field ($\mathbf{B}_{geomg} \sim 30 \div 50$ [μT]) this application of magneto therapy used more than hundredfold more powerful magnetic induction, [6].

2.2 Influence of Location and Orientation in Magnetic Field

In our simulation is used simple model of leg which is placed off-center (Fig. 3). This approximation is imperative to indicating energy effects in the model of leg. The position

Fig. 1 Magnetic field of elliptic magnetic coil (Table 3)

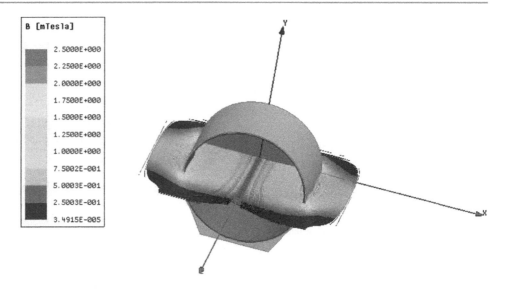

of treated body part in the center of coil is not optimal for therapy but we can see this positioning in medical praxis very often. Low-frequency magneto therapy uses particular sequences of rectangular voltage pulses generated on input of a coil. For this reason the simulation is implemented on one pulse of therapeutic sequence only.

Low-frequency magneto therapy uses particular sequences of rectangular voltage pulses generated on input of a coil. For this reason the simulation is implemented on only one pulse of therapeutic sequence.

Magnetic field is driven by current of the coil only. The Fig. 4 presents waveform, one of representative current pulse (red line). The energy effect in tissue is realised only in dynamic changes of magnetic field. The model of leg is reflecting the conductivity of muscle $\gamma_t = 0.67$ [S/m] only, e.g. [7, 8]. The duration of behaviour of induction power loss is too short (blue waveform on Fig. 4) with maximum pulse power about $P_{m\ max} \sim 1.3 \times 10^{-6}$ [W]. We obtain the total energy losses of one magnetic pulse by integration

but these energy losses will be insignificant, approximately $W_m \sim 1.3 \times 10^{-9}$ [J].

In medical literature this biophysical effect is classified with the current density in tissue $\mathbf{J}_t$, e.g. [9, 10]. Consequently the attention will be put on the distribution of current density in the model of leg. The Fig. 5 show our results.

Maximal top power in longitudinal axis is 1.25 µW and by rotation 30° is 1.75 µW. That is represented increasing of power possibly energy approximately 1.4 times.

2.3 Influence of Metal Implants and Their Orientation in Magnetic Field

Magneto therapy is a treatment method that is used quite often as an additional treatment for bone healing disorders after various traumatic conditions and orthopedic operations where metal implants can be implanted in the limbs. Metal implants are mostly made of materials that are non-magnetic, but these metals are usually highly conductive with conductivity substantially higher than any other types of tissue found in the human body. The following figure (Fig. 6) shows a part of the x-ray image of the lower limbs with metal implants. The patient was in this case advised to undergo magneto therapy.

The x-ray image shows the condition after a complicated fracture of the upper part of the lower leg that was treated by surgery—combined osteosynthesis by bolts and secured intramedullary nail. Basically, the intramedullary nail (hollow metal rod) longitudinally passes through the entire luminal cavity of the tibia bone. Based on this real case, a simplified numerical model has been compiled, which includes only a longitudinal rod without bolts. The orientation of the implanted rod is in the direction of the exciting

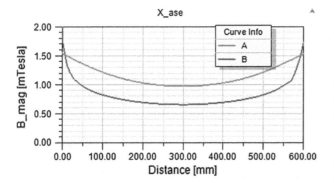

Fig. 2 Distribution of magnetic induction B inside cylindrical magnetic coil (input current 2A)

Fig. 3 Location of the leg in modelling and magnetic vectors specification, **a** The leg is oriented in direction z axis and is placed in center bottom quarter, **b** The leg is rotated 30° in direction z axis and is placed in center bottom quarter Model of leg in accordance with 166 cm stature and 66 kg weight of woman figure

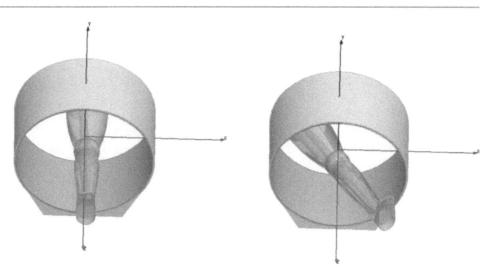

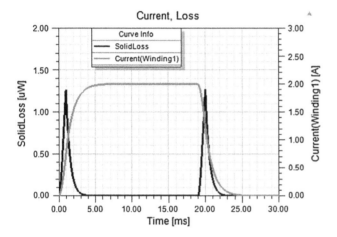

Fig. 4 Location of the leg in modelling and magnetic vectors specification

the direction of the magnetic field force (in the x-axis direction). It can be expected that the metal screws connecting the fragments of the upper part of the tibia bone with its position at the center of the applicator and approximately perpendicular orientation to the field lines will also have a considerable effect on the amount of stored energy.

2.4 Influence of Volume Exposed Tissue and Their Composition

The theoretical preface, chapter 1.3 indicates that therapeutic impact will depend on the content of tissue which is inserted into the magnetic field. This situation stems from Eqs. (7, 8, 9).

The total energy loss is given by the degree of filling into the coil and is presented by cubic integral (9), e.g. [6].

The introduced example uses the same driving current pulse as is presented in (Fig. 4) but volume of tissue is plays major importance. The aim is to show how the induced current density will increase in the same shape model of leg but with leg 50% wider which presents 2.25× enlarged volume.

The duration of pulse induction power loss is too short again but with six times higher maximum power which is about $P_{m\ max} \sim 8 \times 10^{-6}$ [W] and total energy losses of one magnetic pulse $W_m \sim 8 \times 10^{-9}$ [J]. The surface current density is about $\mathbf{J}_t \sim 0.022$ [A.m^{-2}] which is twice.

Main effect of magneto-therapy is the production of currents in the part of human body subjected to magnetic field. According to Reference of ICNIRP committee, e.g. [10], the value of induced current density range from 0.01 to 0.1 [A.m^{-2}] in human body and can influence nervous system. The range from 0.1 to 1 [A.m^{-2}] is able to cause health hazard and the changes in stimulation of nervous system were detect.

PEMFT. This means that the minimum induced current in the metal implant can be expected. In the simulation, both limb sizes are used, as shown in previous cases (Fig. 3a). At the same time, two diameters (8 and 15 mm) of implanted reinforcement were chosen for better understanding of the physical nature of the reinforcement. The leg model corresponds to a female figure of about 166 cm in height and 66 kg in weight.

Conducted numerical simulations confirm the expected theoretical assumptions that the metal implant will significantly influence the distribution of magnetic field in the tissue (Table 4). It is not possible to think about its influence separately. It will always depend on the total volume of tissue and the volume of metallic implants. In this case, the implant is in the direction of the magnetic field line, that is, the induced energy into its volume will be the lowest, but it will still significantly outweigh the energy induced into other tissues! The most induced energy is applied if the metal implant is oriented in its largest dimension perpendicular to

Fig. 5 Distribution of current density in the model of leg along longitudinal axis z

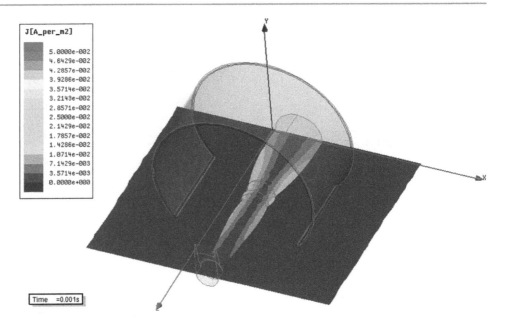

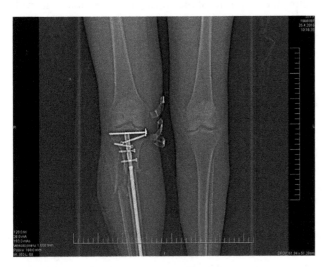

Fig. 6 X-ray image of the lower limbs with metal implants

Table 4 Influence of metal implants on maximum pulse power

$P_{m\ max}$ [μW]	Implant less	ϕ 8 mm implant	ϕ15 mm implant
	1.3	8.2	98.5
Model of lower limb with 2.25× enlarged volume			
	8	15.97	110.2

3 Conclusion

Approximate solution of magneto therapy describes the induced electric field in the body. It produces currents that are under the safety limits. The location and orientation body, volume exposed tissue and their composition has strong effect on distribution induced field and its energy impact. It means that the standard use of magneto-therapy is safe. The metal implants in body pose health hazard.

Acknowledgements This work was supported from institutional support for long term strategic development of the Ministry of Education, Youth and Sports of the Czech Republic.This work was supported by the Slovak Research and Development Agency under the contract No. APVV-16-0270 and project VEGA No.1/0283/18 of Scientific Grant Agency of the Ministry of Education, Science, Research and Sport of the Slovak Republic.

References

1. MALMIVUO, Jaakko. a Robert. PLONSEY. *Bioelectromagnetism: principles and applications of bioelectric and biomagnetic fields*. New York: Oxford University Press, 1995. ISBN 0195058232.
2. Hannemann PF[1], Mommers EH, Schots JP, Brink PR, Poeze M.: The effects of low-intensity pulsed ultrasound and pulsed electromagnetic fields bone growth stimulation in acute fractures: a systematic review and meta-analysis of randomized controlled trials. Arch Orthop Trauma Surg. 2014 Aug; 134(8):1093–106.
3. Bassett, C. A. (1987). "Low energy pulsing electromagnetic fields modify biomedical processes." Bioessays 6(1): 36–42.
4. Shi et al. (2013). "Early application of pulsed electromagnetic field in the treatment of postoperative delayed union of long-bone fractures: a prospective randomized controlled study." BMC Musculoskelet Disord 14: 35.
5. Watson, T. (2010). "Narrative Review: Key concepts with electrophysical agents." Physical Therapy Reviews 15(4): 351–359.
6. S. U. Inan, S. A. Inan: Engineering Electromagnetics, Addison-Wesley, Menlo Park, CA, USA, 1999 (ISBN 0-201-47473-5).
7. https://www.itis.ethz.ch/virtual-population/tissue-properties/database/low-frequency-conductivity/.

8. A.Richter, Ž. Ferková: Physical and energy analysis of therapy applying low-dynamic magnetic fields, Conference Paper · May 2017,https://doi.org/10.1109/ecmsm.2017.7945889   Conference: 2017 IEEE.

9. Brodeur, P.: Annals of Radiation: The hazards of electromagnetic fields. Parts 1–3. New Yorker, 12 June, 51–88; 19 June, 47–73; 26 June, 39–68, 1989. (Later published as Currents of death: Power lines, computer terminals and the attempt to cover up their threat to your health. New York: Simon and Schuster, 1989).

10. Reference of ICNIRP committee: Guidelines for Limiting Exposure to Time-Varying Electric, Magnetic, and Electromagnetic Fields (up to 300 GHz). Health Physics 74/ 4: 494–522, 1998.

11. IEC/TS 60479-1, Effect of current on human beings and livestock, Part 1: General aspects, 1987.

12. IEC/TS 60479-2, Effect of current on human beings and livestock, Part 2: Special aspects, 1998.

13. IEC/TS 60479-3, Effect of current on human beings and livestock, Part 3: 1998.

Analyzing Energy Requirements of Meta-Differential Evolution for Future Wearable Medical Devices

Tomas Koutny and David Siroky

Abstract

Recent advances in clinical engineering include development of physiological models to deliver optimized healthcare. Physiological model comprises a number of equations to relate biomedical signals. Each equation contains a set of coefficients. Determining the coefficients is a complex task as the models are non-linear. Therefore, development of the models must be accompanied by a development of methods to determine model coefficients. With the advent of wearable medical devices, we have to consider energy requirements of the models and the methods. Considering an illustrative case of type-1 diabetes mellitus patients, we already demonstrated that Meta-Differential Evolution outperforms analytical methods, when determining coefficients of glucose dynamics. In this paper, we analyze convergence of the Meta-Differential Evolution, running time and associated power consumption on a single board computer with a system-on-a-chip—Cortex-A8 AM335x. Based on the analysis, we recommend splitting the process of determining the coefficients into two phases. First phase determines the initial, per-patient optimized coefficients. Second phase is an energetically efficient update of these coefficients with new, continuously measured signal of the patient. Meta-Differential Evolution searches for optimal coefficients by evolving a number of generations of candidate coefficients, using a number of evolutionary strategies. We demonstrate that the proposed approach significantly reduces the number of candidate coefficients to evaluate, while achieving the desired accuracy. This positively reflects in the lifetime of wearable device's battery.

T. Koutny (✉)
NTIS – New Technologies for Information Society,
University of West Bohemia, Univerzitni 8, 306 14,
Pilsen, Czech Republic
e-mail: txkoutny@kiv.zcu.cz

D. Siroky
Department of Computer Science and Engineering, University of
West Bohemia, Univerzitni 8, 306 14, Pilsen, Czech Republic

Specifically, calculating coefficient's update took 0.05 Ws only. It shows the feasibility of using Meta-Differential Evolution with its improved accuracy for blood glucose calculations in a wearable device.

Keywords

Diabetes • Meta-differential evolution
Energy requirements

1 Introduction

To optimize delivered healthcare, a physiological model processes continuously measured biomedical signals. Considering an illustrative example of type 1 diabetes mellitus (T1D), we have demonstrated that Meta-Differential Evolution (Meta-DE) is a promising method to determine coefficients of glucose-dynamics model for T1D patients [1]. Meta-DE improves accuracy of blood glucose level (BG) calculation. In addition, more approaches that are evolutionary [2] follow, with respect to study [1]. Therefore, we analyze Meta-DE to develop guidelines how to translate the evolutionary knowledge, which we learned with high-performance computing, to low-power devices—a single board computer with a system-on-a-chip.

While a processor with increased computational power provides ample computing capabilities, it requires a great deal of power consumption, which causes creating a thermal hot spot and putting pressure on the energy resource in a mobile device [3]. A specialized device can deliver desired power consumption, but we need an advanced signal processing [4]. Offloading the computation, e.g. to a mobile phone, is a possibility [5]. Nevertheless, gaming or multimedia application can unexpectedly deplete mobile phone battery. Therefore, we do not consider mobile phone as attractive as having a single system-on-a-chip with balanced power-consumption and computational capabilities.

© Springer Nature Singapore Pte Ltd. 2019
L. Lhotska et al. (eds.), *World Congress on Medical Physics and Biomedical Engineering 2018*,
IFMBE Proceedings 68/3, https://doi.org/10.1007/978-981-10-9023-3_44

2 Diabetes Technology

Diabetes mellitus is a silent, civilization disease. It does not hurt until it has developed. Specifically, T1D pathogenesis involves autoimmune destruction of pancreas cells, which produce insulin [6]. Insulin promotes glucose utilization by cells to produce energy. Without insulin, glucose accumulates in the body. Excessive amounts of glucose binds to multiple organs, hence damaging them, eventually leading to their failure. Therefore, T1D patient needs insulin therapy to survive. Insulin therapy comprises artificial insulin delivery. Insulin bolus must be calculated precisely as more complications may arise from deficient or excessive dosing of insulin.

Technology plays a key role in managing T1D. There is a minimally invasive continuous glucose monitoring system (CGMS), which monitors interstitial-fluid glucose level (IG) and can be paired with insulin pump. Due to physiological reasons, BG can differ considerably from IG. Therefore, CGMS recalculates IG to BG with a varying degree of success [7]. We demonstrated that our algorithm (i.e., the model and the method to determine its coefficients) outperforms CGMS in this task [1, 8–10] and therefore it would be desirable to bring our algorithm into the practice. Equation (1) describes the model.

$$p \times i(t) + cg \times b(t) \times [b(t) - i(t)] + c = i(t + \Delta t) \quad (1)$$

Let $b(t)$ and $i(t)$ denote BG and IG at time t, respectively. Then, let us represent BG gain with p, effect of concentration gradient and capillary membrane permeability with cg and residual IG with c [8]. The model (1) relates present BG and IG to future (Δt ahead) IG, to calculate BG by exploiting the IG time-gradient [9].

3 Meta-Differential Evolution

Differential Evolution (DE) is an evolutionary optimization technique. It utilizes a vector of model coefficients and a fitness function that reduces this vector into a single scalar value. Less value indicates better coefficients. In the beginning, DE generates a number of random vectors. It is called a population and DE evolves it in subsequent iterations. In each iteration, DE combines each vector with another vector, while applying crossbreeding and mutation. A specific strategy governs how DE combines the vectors. DE keeps the resulting vector, if it produces less fitness scalar than the original vector. This way, DE updates entire population and proceeds with next iteration, until it meets a stopping condition. Stopping condition is e.g., a fixed number of iterations or reduction of the fitness scalar below a specific threshold.

As DE is not analytic method, it can fail to find sufficiently good coefficients. Specifically, DE population can cease to evolve or it may converge pre-maturely [1, 11]. To mitigate these risks, (1) we initialize the population with analytically determined coefficients; (2) apply partial derivatives to escape local extreme; (3) use Mersenne Twister and chaos random number generators; and (4) complement each candidate coefficients with private cross-breeding and mutation factors and DE strategy [1]. This turns DE into Meta-DE, as these vector-private settings are generated randomly on a failure to produce an improved offspring.

4 Reducing the Computational Complexity

4.1 First Phase

As Eq. (1) is a linear combination, Linear Least Squares (LLS) represents computationally inexpensive way to determine p, cg and c parameters for a given Δt coefficient [9]. As a result gramian matrix does not need to change with respect to Δt, only the right-side vector must be calculated for a specific Δt. Based on our previous work, let us constrain the Δt-parameter to range from 0 to 40 min. Discretizing this range by 10 s, we have to calculate the fitness function 240 times only. This gives us 240 candidate coefficients with calculated fitness scalar. Eventually, we select the best coefficients by the fitness scalar [9].

4.2 Second Phase

Clinical practice has adopted relative error to estimate accuracy of BG estimation. Relative error is absolute difference between measured and calculated BG, divided by measured BG. It implies a tolerance of increasing absolute difference between measured and calculated BG with increasing measured BG. Hence, we can apply simple linear regression to optimize BG, which we obtain with the LLS-determined coefficients. Specifically, we use Eq. (2) [12] to calculate (i.e., to correct) the calculated BG—$BG_{corrected}$. Let $b_{calculated}(t)$ denote calculated BG at time t, according to the proposed model of glucose dynamics—Eq. (1) [9].

$$BG_{Corrected}(t) = BG_{Slope} \times b_{calculated}(t) + IG_{Slope} \times i(t) + Intercept \quad (2)$$

Equation (2) is simple and therefore it requires less energy than Eq. (1). In addition, calculating relative error with Eq. (2) can be vectorized easily.

Table 1 Efficiency of Meta-DE strategies for population sizes 100; 40; 20

Strategy [1]	Average fitness	Average iterations to best coefficients	Standard deviation of iterations to best coefficients
All	0.245; 0.247; 0.251	5297; 5109; 4560	3151; 3081; 3015
Current to umBest and Rand1	0.245; 0.247; 0.247	5230; 5058; 5662	3199; 3069; 3116
Current to P-Best	0.244; 0.245; 0.246	5808; 5507; 5293	3209; 3126; 3106
Current to umP-Best	0.242; 0.244; 0.251	6432; 6225; 4599	2974; 3041; 3034
Best to Bin	0.247; 0.246; 0.249	7352; 7910; 6271	2160; 2274; 2621
Current to umBest1	0.245; 0.249; 0.249	5041; 6564; 6017	3173; 2478; 2596
Current to Rand 1	0.247; 0.246; 0.254	5067; 5776; 4549	3185; 3155; 2969

Table 2 ECDF of relative error of calculated BG

Cumulative probability of less than or equal relative error (%)	Relative error			
	CGMS (%)	LLS BG (%)	Study [10] (%)	Optimized BG (%)
10	2.6	1.5	1.3	0.7
20	5.3	3.3	2.8	2.5
30	8.0	5.1	4.5	4.2
40	11.0	7.0	6.6	6.3
50	14.6	9.2	8.7	8.6
60	18.7	12.0	11.4	11.3
70	23.5	15.5	14.8	14.8
80	30.6	20.8	19.7	19.3
90	41.1	30.2	29.0	27.5
95	52.1	39.4	37.8	36.4
100	641.8	213.0	201.2	166.7

4.3 Meta-DE

As the fitness function of the second phase is linear combination, we removed (2) partial derivatives, (3) chaos random number generator and (4) reduced the number of DE strategies. In addition, we investigated effect of population size.

In the clinical practice, mean relative error (MARD) with 0.1 precision is a well-established marker to determine accuracy of glucometer—be it BG meter or CGMS. This implies that Meta-DE can stop once the MARD, which we calculate with Eq. (2), stops improving beyond the desired precision.

5 Results

From NCT01591681 study (see the US Clinical Trials database), we extracted 78 CGMS profiles with 3516 measured BG to verify the proposed approach [1].

Table 1 gives Meta-DE strategies evaluation for the original computation [1], with population sizes 100, 40 and 20 respectively. These results indicate that Current to umBest and Rand1 represent a best balance between energy requirements and achieved accuracy. Energy requirement is a function of iterations needed to obtain the coefficients. As Meta-DE is non-deterministic method, consumed energy, i.e., number of iterations, may vary. Table 1 gives standard deviation to capture this variability.

To describe accuracy of BG calculation, let us use empirical cumulative distribution function (ECDF) of relative error of calculated BG [13]. It represents the probability that relative error of calculated BG is less than or equal a given relative error.

Table 2 gives ECDF for IG considered as BG (CGMS accuracy), BG calculated with LLS determined coefficients, recent study [10], and Eq. (2) optimized BG.

To optimize BG, we used the Current-to-umBest and Rand1 strategies, with population size 40. On average, Meta-DE required 170 iterations (with standard deviation of 253) to find Eq. (2) coefficients.

To measure power consumption with Cortex-A8 AM335x, we connected it with 0.1 Ω resistor (*Rm*) in series to 5 V power supply (U). Then, we measured voltage (U_{Rm}) across R_m. Specifically, we measured the voltage at no load ($U_{Rm\text{-}no\ load}$) and with Meta-DE running ($U_{Rm\text{-}MetaDE}$). The voltages were 15.5 mV and 26.1 mV, respectively. According to Ohm's Law, Meta-DE increased the electric current by 106 mA.

To make this measurement robust, we magnified the computational problem. Specifically, we forced Meta-DE to evolve 10000 generations. Then, the execution took 107.3 s (a time period p). According to Eq. (3), we obtain power that Meta-DE consumed for the period p. It was 53.2 Ws. Thus, 170 iterations required 0.9 Ws only.

$$E = U \times \frac{(U_{Rm-MetaDE} - U_{Rm-no\,load})}{R_m} \times p \qquad (3)$$

In addition, we artificially expanded the number of calculated BG to 500. In nor-mal case, T1D patient takes 2–3 BG samples a day. Lifetime of present CGMS sensor span approximately across 10 days. Considering 30 BG samples, we have to scale the power consumption down by 30/500–0.05 Ws. This result is feasible for a low-power device.

6 Conclusion

The results confirm feasibility of the proposed reduction of computational complexity. Optimized BG even slightly outperformed results, which we reported in study [10]. Therefore, the results encourage us to push the proposed model of glucose dynamics, and the method to determine its coefficients, into the practice—to a future wearable medical device.

Acknowledgements This publication was supported by the project LO1506 of the Czech Ministry of Education, Youth and Sports and university specific research project SGS-2016-013.

Conflict of Interest The authors declare no conflict of interest.

References

1. Koutny, T.: Using meta-differential evolution to enhance a calculation of a continuous blood glucose level. Comput Methods Programs Biomed 133, 45–54 (2016).
2. Choi, J., Jung, B., Choi, Y., Son, S.: An adaptive and integrated low-power framework for multicore mobile computing. Mobile Information Systems 2017 (2017).
3. Tuominen, J., Lehtonen, E., Tadi, M.J., Koskinen, J., Pankaala, M., Koivisto, T.: A miniaturized low power biomedical sensor node for clinical research and long term monitoring of cardiovascular signals. In: 2017 IEEE International Symposium on Circuits and Systems (ISCAS), pp. 1–4. IEEE (2017).
4. Lane, N.D., Bhattacharya, S., Georgiev, P., Forlivesi, C., Jiao, L., Qendro, L., Kawsar, F.: Deepx: A software accelerator for low-power deep learning inference on mobile devices. In: 15th ACM/IEEE International Conference on Information Processing in Sensor Networks (IPSN), pp. 1–12. IEEE (2016).
5. De Falco, I., Della Cioppa, A., Koutny, T., Scafuri, U., Tarantino, E., Krcma, M.: An evolutionary approach for estimating the blood glucose by exploiting interstitial glucose measurements. In: BIOSTEC 2018. Springer (2018).
6. Longo, D., Fauci, A., Kasper, D., Hauser, S., Jameson, J., Loscalzo, J.: Harrison's Principles of Internal Medicine: Volumes 1 and 2, 18th Edition, 18 edn. McGrawHill Professional (2011).
7. Thabit, H., Leelarathna, L., Wilinska, M.E., Elleri, D., Allen, J.M., Lubina-Solomon, A., Walkinshaw, E., Stadler, M., Choudhary, P., Mader, J.K., Dellweg, S., Benesch, C., Pieber, T.R., Arnolds, S., Heller, S.R., Amiel, S.A., Dunger, D., Evans, M.L., Hovorka, R.: Accuracy of Continuous Glucose Monitoring During Three Closed-Loop Home Studies Under Free-Living Conditions. Diabetes Technol. Ther. 17(11), 801–807 (2015).
8. Koutny, T.: Prediction of interstitial glucose level. IEEE Trans Inf Technol Biomed16(1), 136–142 (2012).
9. Koutny, T.: Blood glucose level reconstruction as a function of transcapillary glucose transport. Comput. Biol. Med. 53, 171–178 (2014).
10. Koutny, T.: Crosswalk–a time-ordered metric. In: EMBEC & NBC 2017, pp. 884–887. Springer (2017).
11. Hu, Z., Xiong, S., Su, Q., Zhang, X.: Sufficient conditions for global convergence of differential evolution algorithm. Journal of Applied Mathematics 2013 (2013).
12. Koutny, T.: Validating temporal concentration gradient to predict blood glucose level. Submitted to JAMIA (2017).
13. Koutny, T.: Modelling of glucose dynamics for diabetes. In: International Conference on Bioinformatics and Biomedical Engineering, pp. 314–324. Springer (2017).

Energy Consumption Profiles of Common Types of Medical Imaging Equipment in Clinical Settings

Anthony Easty, Linda Varangu, J. J. Knott, Shawn Shi, and Kent Waddington

Abstract

Imaging equipment such as MRIs, CT scanners and general radiography equipment consume significant amounts of energy while operating. This study describes a series of detailed energy consumption studies on these devices during clinical use at three major health care centres in Canada, one in British Columbia and two in Ontario. The study was conducted by the Canadian Coalition for Green Health Care [1], with funding provided by Natural Resources Canada [2] and BC Hydro [3]. The primary goal of the study was to accelerate the development of ENERGY STAR specifications for medical imaging equipment. Natural Resources Canada is assisting the United States Environmental Protection Agency (US EPA) [4], by collecting these data from the field. Eight testing events were undertaken, providing energy consumption data for low power energy modes, standby/idle power energy modes and active/patient scanning energy modes. Energy consumption was measured over periods ranging from three to eleven days, to provide rich information about when and how frequently the equipment was used and what the associated energy consumption profiles were. Data acquisition rates were varied to gain a detailed understanding of the temporal variations in energy consumption profiles during each use mode. Results from this study showed that there were variations in the low power mode energy consumption of greater than 25% in some cases, and that non-scanning energy consumption, either low power or stand-by modes, in some cases accounted for up to 80% of the total energy consumption of the system at some hospitals. These findings indicate that there is considerable scope for manufacturers to reduce the energy consumption levels of their devices, and for users to reduce energy consumption during clinical use through practices such as placing the system into a lower energy mode or shutting it down while not in use, where possible.

Keywords

Imaging devices • Energy consumption • Use profiles

1 Introduction

The purpose of this study was to obtain energy consumption data from hospital medical imaging equipment (MIE) which would assist government sponsors of the ENERGY STAR® program to determine if a new category of ENERGY STAR products should be developed for MIE. Partnering hospitals, as well as BC Hydro, also wanted to gain a better understanding of MIE energy consumption at hospitals so that strategies could be developed to help reduce energy consumption and costs.

Three hospital partners took part in this project:

- Nanaimo Regional General Hospital (NRGH), Island Health (VIHA), in British Columbia;
- The Hospital for Sick Children (SickKids) in Ontario; and
- University Health Network (UHN) in Ontario.

Energy consumption data were obtained from three types of MIE:

1. Computed Tomography (CT)
2. General Radiography (X-ray)
3. Magnetic Resonance Imaging (MRI)

Eight testing events were undertaken using calibrated energy data-loggers with varying sample intervals, resulting

A. Easty (✉)
IBBME, University of Toronto, Toronto, ON, Canada
e-mail: tony.easty@utoronto.ca

L. Varangu · J. J. Knott · S. Shi · K. Waddington
Canadian Coalition for Green Health Care, Branchton, ON, Canada

© Springer Nature Singapore Pte Ltd. 2019
L. Lhotska et al. (eds.), *World Congress on Medical Physics and Biomedical Engineering 2018*,
IFMBE Proceedings 68/3, https://doi.org/10.1007/978-981-10-9023-3_45

in large numbers of data points for energy consumption. Data were collected for low power energy modes, standby/idle power mode and active/scanning energy modes. The MIEs were measured mostly over periods ranging from three to 11 days. Longer measurement periods allowed richer information about when and how frequently the equipment was used and enabled the calculation of estimated annual energy consumption and energy costs. Extended measurement periods also provided insights into energy reduction strategies that would not be available through the original testing protocol.

The European Coordination Committee of the Radiological, Electromedical and Healthcare IT Industry (COCIR) developed the Self Regulatory Initiative (SRI) for Medical Imaging Equipment. This Initiative, launched in 2009 and officially acknowledged by the European Commission in 2012, aims at reducing the environmental impacts of medical imaging devices, and represents the proactive approach of COCIR towards sustainable health care and the circular economy. COCIR have produced six status reports to date, the latest one released in September 2016 [5]. The major international equipment manufacturers participate in this initiative. For example, from the MRI manufacturing sector participants include GE, Phillips, Siemens, Toshiba and Hitachi.

COCIR developed standardized testing protocols for MIE which were used as the basis for the test methods developed for the ENERGY STAR MIE testing process—'Final Draft Test Method For Determining Medical Imaging Equipment Energy Use—Rev. Aug 2014' [6] that was used as the basis for product testing for the Canadian study.

The COCIR SRI reports significant reduction in annual energy consumption for MIE such as MRIs from 2011–2015. Manufacturers have collectively decreased the energy consumption of the MRIs. In 2015, the daily average energy consumption per unit for MRI equipment decreased to 176.91 kWh/day, showing a 21% reduction compared to 2011 (225.92 kWh/day) and 5% compared to 2014 (see Fig. 1). These aggregate figures for the products from multiple manufacturers make it clear that significant efforts are being made by manufacturers to reduce the daily power consumption of their MRI systems.

Comparing energy consumption values for MRIs manufactured by five companies between 2011 and 2015 shows there is a greater than 25% spread in the average daily energy consumption over that time period, indicating that some manufacturers are having more success than others in reducing daily energy consumption (see Fig. 2).

Thus there is a general trend toward reduced energy consumption, and it appears that some manufacturers are focusing more on this issue than others, or have technology that is more amenable to reduced energy consumption.

2 The Study

Imaging equipment energy consumption testing occurred with three MIE types and eight separate testing events, all conducted on operational units in clinical settings. The facilities engineering teams connected the data loggers to the power feeds for each system and verified their functionality prior to testing. The following three types of equipment were tested:

a. Magnetic Resonance Imaging (MRI)
b. Computed Tomography (CT)
c. General Radiography (X-ray)

Partner hospitals wanted to use longer testing periods to gain a better understanding of total energy consumption from MIE at their facilities. This resulted in many data points (instead of the 36 originally agreed to) and much more reliable data for off mode, low power mode and standby modes.

Most of the data measurements took place over periods of between three and eleven days. These protocols are listed below, and the data acquisition intervals for the data loggers were varied as shown, to help to gain an understanding of the timescale of various fluctuations in energy consumption during use. During each sampling interval, the energy consumption recorded represents the accumulated total during that time. Since scans with these devices typically last a number of minutes, most of the data acquisition intervals were of one-minute duration, or longer. However, we also chose to sample a CT system at a one-second interval, to determine whether very short-term variations in energy consumption were occurring as well (see Fig. 3).

Fig. 1 MRI energy reduction achievements: calculated values for 2011–2015 with projections to 2017

	Sold units[8]	Total daily energy consumption (kWh)[9]	Average daily energy consumption per unit (kWh/d)	Beyond BAU (kWh/d)	BAU (kWh/d)
2011	✓394	✓89.011	✓225,92		
2012	✓446	✓100.038	✓224,30	225,61	230,39
2013	✓454	✓95.148	✓209,58	225,30	234,87
2014	✓513	✓95.572	✓186,30	225,00	239,34
2015	✓604	✓106.855	✓176,91	224,69	243,81
2016				224,38	248,29
2017				224,07	252,76

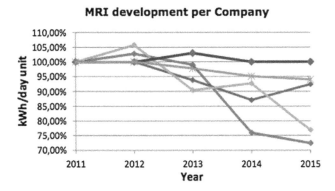

Fig. 2 COCIR members reports of average daily energy consumption for MRIs over a five-year period from five different manufacturers

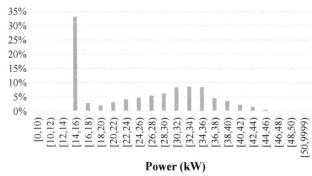

Fig. 4 MRI energy consumption (kW) percentage at each power range at Site 1

The collected data are too numerous to include here in their totality, but there are some key features that stood out in subsequent analysis. Looking at the distribution of MRI energy consumption at Sites 1 and 2, the first clear finding is that the largest percentage of power consumption occurs in the idle mode between 12 to 14 and 14 to 16 kW respectively. Further, idle mode accounts for over 80% of total energy consumption at Site 1, compared with over 30% at Site 2. This wide difference is likely accounted for by a higher utilization for patient scans at Site 2 (see Figs. 4 and 5).

Looking at typical CT energy consumption, a similar pattern emerges, where over 70% of total energy consumption occurs in the standby mode between 2 and 3 kW (see Fig. 6). A deeper examination of the energy consumption profile for twenty-four hours reveals that for short periods of time, the system is put into a low energy idle mode that reduces the power consumption to below 1 kW (see Fig. 7).

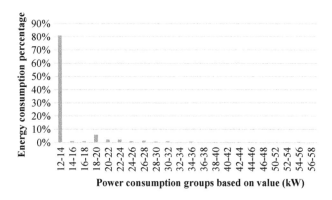

Fig. 5 MRI energy consumption (kW) percentage at each power range at Site 2

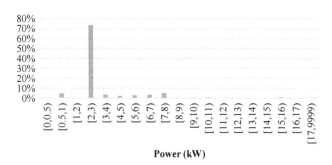

Fig. 6 CT energy consumption (kW) percentage at each power range at Site 1

3 Discussion

This study demonstrates that it is feasible to monitor medical imaging equipment energy consumption in the long-term, and that interesting patterns of performance and system usage emerge. Given the high energy consumption of the

Fig. 3 Power consumption testing events, test length and data acquisition intervals

MEI Type	Testing Event and Site	Test Length	Data Acquisition Intervals (mins or secs)
MRI	1. Site 1	11 days	6 mins
MRI	2. Site 2	7 days	1 min
CT	3. Site 1	7 days	6 mins
CT	4. Site 1	7 days	6 mins
CT	5. Site 2	8 days	1 min
CT	6. Site 3	2 min	1 sec
X-ray	7. Site 1	6 days	6 mins
X-ray	8. Site 1	3 days	6 mins

Fig. 7 CT power consumption (kW) over twenty-four hours showing drop in kW at Site 1

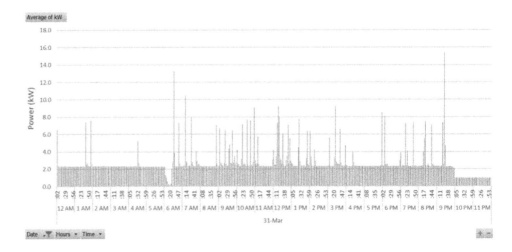

equipment itself and ancillary systems such as computer monitors and processors and HVAC systems, there are significant savings to be had in reducing overall system energy consumption. This can be achieved in part by focusing on manufacturers' specifications and including energy consumption as a consideration during purchase selection. It is clear that manufacturers are paying attention to this issue since the overall energy consumption profiles for these systems are dropping in each new model year. Further, there are opportunities during clinical operation to reduce the standby power consumption of at least some of the major items of medical imaging equipment, and users are encouraged to review these opportunities as part of their protocol for the safe and efficient operation of these systems. Specifically, it was observed that users often failed to put systems into lower standby modes once a scan was complete, despite the fact that these modes were accessible on the system. If there is no clinical disadvantage to this, it offers the opportunity to significantly reduce overall system power consumption, where available. It is also worth noting that a reduction in primary power consumption can in turn lead to reduced energy consumption for room cooling systems, compounding the effect of the reduction in energy

consumption of the system itself, since less system power dissipation in turn requires less cooling to maintain a given ambient temperature.

Conflicts of Interest The authors have no conflicts of interest to report regarding the contents of this paper.

References

1. Canadian Coalition for Green Health Care Homepage, greenhealth-care.ca, last accessed 2017/12/13.
2. Natural Resources Canada Homepage, https://www.canada.ca/en/natural-resources-canada.html, last accessed 2017/12/13.
3. BC Hydro Homepage, https://www.bchydro.com/, last accessed 2017/12/13.
4. United States Environmental Protection Agency Homepage, https://www.epa.gov/, last accessed 2017/12/13.
5. COCIR Self-Regulatory Initiative for the Ecodesign of Medical Imaging Equipment Status Report2015,www.cocir.org/fileadmin/6_Initiatives_SRI/SRI_Status_Report_2015_-21_September_2016.pdf , last accessed 2017/12/13.
6. https://www.energystar.gov/sites/dfault/files/ENERGY%20STAR%20Medical %20Imaging%20Equipment%20Final%20Draft%20Test %20Method.pdf, last accessed 2017/12/13.

Automated Sunglasses Lens Exposure Station and the Preliminary Effects of Solar Exposure

Leonardo Mariano Gomes, Artur Duarte Loureiro, Guilherme Andriotti Momesso, Mauro Masili, and Liliane Ventura

Abstract

The Laboratory of Ophthalmic Instrumentation (LIO) from the University of Sao Paulo—Brazil, is involved in research about Sunglasses and its standards, and has already contributed for changing parameters in the previous Brazilian sunglasses standard, the ABNT NBR 15111-2013. The focus of the work presented in this paper is to investigate on the long-term solar exposure effects in sunglasses lenses and to conduct this study, we have developed an automated exposure system dedicated to expose sunglasses lenses towards the sun. The system also measures the dose of ultraviolet radiation which the lenses were subjected to and other weather variables, like temperature and relative humidity of the air. In this paper, we discuss about the materials used to manufacture sunglasses lenses, about the machine we developed to exposes the lenses and the methods used to measure the lenses transmittance characteristics over the time and to determine if and how long-term solar exposure may affect the samples.

Keywords

ISO 12312-1 • ABNT NBR ISO 12312-1
Resistance to solar radiation • Sunglasses
Solar simulator

1 Introduction

Sunglasses may assure ultraviolet (UV) safety when adequate UV protection filters are used. There are evidences that sunglasses UV protection can degrade with exposure to the Sun, but such an experiment has never been done. In our current research, we propose to evaluate the changes of luminous and UV transmittances in sunglasses lenses after long periods of controlled solar exposition, by using an automated sunglasses lens exposure station that we developed.

1.1 Sunglasses Standards

The Brazilian standard for sunglasses for general use (ABNT NBR ISO 12312-1:2015) is a translation of the international one (ISO 12312-1:2013) [1, 2]. According to the standard used in Brazil, depending on luminous spectral transmittance, lenses are classified into different categories and for each category there is a required UV protection (Table 1). In this standard is established a lens aging test that involves exposing sunglasses lenses in solar simulator for 50 h and analyze transmittance difference before and after the test.

Some sunglasses lenses show luminous transmittance changes after long solar exposure, becoming lighter. There are evidences that solar exposure can increase sunglasses UV transmittance and there is no experimental study in the literature showing how much solar exposure time corresponds to 50 h in solar simulator.

1.2 Previous Works

In a previous study conducted in LIO [3], we discussed the parameters for the solar simulator test (distance between lamp and samples, and exposure time) to make the test equivalent to the real solar exposure conditions that population are subjected to when wearing sunglasses. The present test parameters, as specified by the Brazilian standard, should be revisited to establish safe limits for UV filters of sunglasses.

L. M. Gomes · A. D. Loureiro · G. A. Momesso · M. Masili
L. Ventura (✉)
Ophthalmic Instrumentation Laboratory – LIO, Department of
Electrical Engineering, Sao Carlos School of Engineering
University of Sao Paulo, Sao Paulo, Brazil
e-mail: lilianeventura@usp.br

© Springer Nature Singapore Pte Ltd. 2019
L. Lhotska et al. (eds.), *World Congress on Medical Physics and Biomedical Engineering 2018*,
IFMBE Proceedings 68/3, https://doi.org/10.1007/978-981-10-9023-3_46

Table 1 Transmittance requirements for sunglasses lenses for general use [1]

Lens category	Visible spectral range		UV spectral range	
	Range of luminous transmittance (τ_V)		Maximum value of solar UVB transmittance ($\tau_{SU\ VB}$)	Maximum value of solar UVA transmittance ($\tau_{SU\ VA}$)
	From over (%)	To (%)	280–315 nm	315–380 nm
0	80	100	$0.05\ \tau_V$	τ_V
1	43	80		
2	18	43	1.0% absolute or $0.05\ \tau_V$ (the greater)	$0.5\ \tau_V$
3	8	18	1.0% absolute	1.0% absolute or $0.25\ \tau_V$ (the greater)

2 Materials and Methods

The automatic solar exposition station for sunglasses is a machine which the main function is to expose a set of sunglasses lenses to the sun while favorable weather conditions and, in the absence of such conditions, to protect the lenses. The favorable weather conditions are during the daylight, i.e. after sunrise and before sunset, and when it is not raining. It is desired to avoid rain on the samples because it would facilitate the deposition and accumulation dirt particles on the lens surface and compromise the action of UV radiation on them.

The system mechanics was designed to protect the set of five acrylic panels that houses 60 lenses (12 each). It was designed a metallic frame with four of each sides covered by UV protected polycarbonate sheet. The box could slide through a metallic rail, with double of the box length, to cover or uncover the panels when needed, and the motion is driven by a garage door motor. After the manufacture of the system, we installed it on the rooftop terrace because it was a restricted access place; in addition to permit the samples would be free of shadows during all day. Along with the automatic solar exposition station, we installed an IP camera for monitoring the system's functioning, weather sensors for providing information about the conditions that lenses would be exposed and to alert when rain begins and the UV sensors for registering the total radiation dose that lenses would be exposed (Fig. 1).

The analysis of lens transmittance changes after exposition to the Sun depends on measures of the transmittance spectrum of each lens. The visible (380–780 nm) and UV

Fig. 1 The automatic solar exposition station for sunglasses installed in the terrace of a building

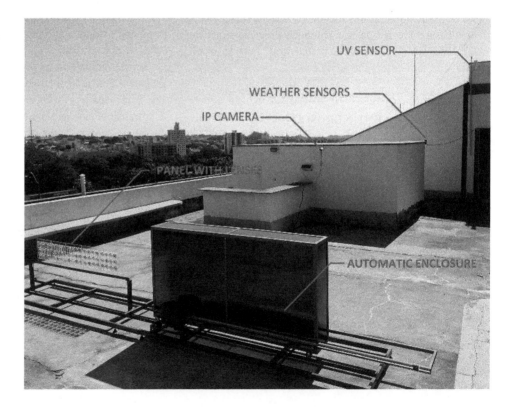

(280–400 nm) transmittances are measured in 5 points of lenses: one central point and 4 points located 5 mm above, below, on the left and right of the central point. Transmittance measures are taken from 780 to 280 nm, with steps of 5 nm. For doing these measures, it was used the VARIAN Cary 5000 UV-Vis-NIR spectrophotometer. A mechanical device is used for holding the lens, which should be placed with its central point aligned with a mark on the device. The device is placed inside the spectrophotometer and measures of transmittance in the central point are taken. After this step, the mechanical device is adjusted to move the lens 5 mm up, 5 mm down and so on. Measures of transmittance in all 5 points of each lens are recorded in a text file, with 5 nm step for each spectral response, in compliance with ISO 12312-1 [2].

3   Results and Discussion

After setting up the machine and the controlling hardware and software, we tested the functioning of the machine in the real conditions it would work when exposing the sunglasses lenses. The system was programmed to detect the current time and open or close the enclosure to expose or protecting the lenses respectively. By the online control panel (Fig. 2) it was possible to manually open or close the enclosure if any problems were detected, in addition to live monitoring the machine through the IP camera video stream. The control panel also shows the opening and closing time for the current day, the current UV index level and other information regarding the total time of lenses exposure to the sun and weather conditions report.

The closing time length for protecting the lenses is about 18 s, which is enough for protecting the most rains in São Carlos city. We tested the machine for 5 months before installing the first set of 60 lenses to be exposed, and by monitoring the machine functioning and the system reports, we verified that the system worked as expected, with minor bugs that were easily repaired.

With the evaluation of the automated panel system concluded, the next step of the research is to expose the lenses to the natural sunlight and measure the effects of the exposition to the changes in UV and visible transmittances. The lenses of our sample set are manufactured of the following materials: 80% polycarbonate, 2% polymethyl methacrylate (PMMA), 5% CR-39 (with polarizing filter inside), 12.8% polyamide and 0.2% glass. Our research team is engaged in determining if the parameters of the tests of the certification standards for sunglasses are enough for the safety of users and ultraviolet radiation levels in Brazil [4], and we've already exposed the lenses set in a solar simulator. The test was conducted for a total of 3000 h and we've already detected transmittance changes, which will be compared with the future results after exposition in the automated sunglasses lens exposure station.

Fig. 2 The automatic solar exposition station for sunglasses installed in the terrace of a building

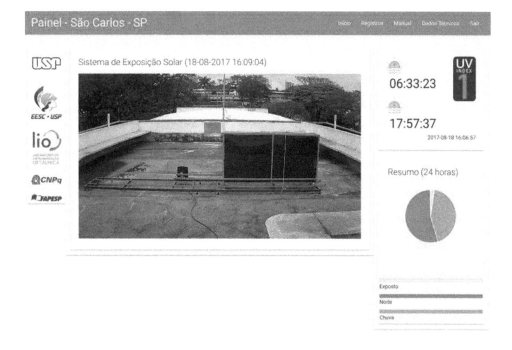

4 Conclusion

In this paper we presented the automatic solar exposure station for sunglasses lenses, which was developed in our laboratory (LIO—EESC/USP). From the construction of the machine, a study about the behavior of sunglasses lenses after the long periods of exposure to solar radiation will be carried out to contribute to the redefinition, if necessary, of the parameters of the Brazilian standard regarding lenses stress after solar exposition. Lenses are being exposed towards geographic north, and total exposure time is recorded. It is expected to verify whether the requirements of the ISO 12312-1:2013 and ABNT NBR ISO 12312-1:2015 standards are still met or whether new certification items need to be provided for ensuring sunglasses protection for population.

References

1. ASSOCIAÇÃO BRASILEIRA DE NORMAS TÉCNICAS. NBR ISO 12312-1: Proteção dos olhos e do rosto – óculos para proteção solar e óculos relacionados – parte 1: óculos para proteção solar para uso geral. Rio de Janeiro, 2015. 26 p.
2. INTERNATIONAL STANDARD ORGANIZATION. ISO 12312-1: Eye and face protection – sunglasses and related eyewear – part 1: Sunglasses for general use. August. 2013. 23 p.
3. MASILI, M., VENTURA, L. Equivalence between solar irradiance and solar simulators in aging tests of sunglasses. Biomedical Engineering Online (Online), p. 86–98, n. 2016.
4. MAGRI, R., MASILI, M., DUARTE, F. O., VENTURA, L. Building a resistance to ignition testing device for sunglasses and analysing data: a continuing study for sunglasses standards. Biomedical Engineering Online (Online), 16:114, 2017.

Experience in the Design of Temporary External Pacemaker in the Case of the Mexican Institute of Social Security (IMSS)

Gustavo Adolfo Martinez Chavez

Abstract

The present paper shows the design and development of the external pacemaker circuit, for use in diagnose or research of diseases than affect the heart. In this case they used lineal circuit, existent in the Mexican National Market. We design and used a simple prototype for generation pulse for heart stimulation by employing the oscillators with the circuit LM555. Previous results obtained in HGR1 "Carlos Macgregor" hospital displayed good operation in the electrical impulses from the heart muscle cause your heart to beat (Contract). Build up this type of pacemaker is the objective, in this paper we show how was the development of a circuit that cold sense the heart rhythm and could supply the impulses necessaries to live. The most relevant is your use in medical persons of the hospital as diagnose and research tool.

Index Terms

Pacemaker • Heart • Oscillators circuit • LM555

1 Introduction

Arrhythmias of the heart can be very detrimental to heart function. Such abnormal ties are often treated with pacemakers. Pacemakers can function in many ways, depending on the mode of operation and the design choices for your components of the pacemakers. The technological advances that have developed over the past three decades, especially integrated circuit technology, allow for pacemakers to be titrated for specific patient needs, to serve as a diagnostic tool o some the agree, and to optimize cardiac output in rate-responsive pacing.

Vital to all cardiac function is the spontaneous and repetitive generation of electrical impulses by the heart. These impulses control the sequence of muscle contraction of each heartbeat. The pattern and timing of these impulses determines the heart rhythm. Abnormalities of this rhythm impair the hearts ability to pump blood as the body demands.

The Fig. 1 shows the heart in cross section, normally, the heartbeat begins in the right atrium when the sinoatrial (SA) node, a special group of cells, transmits an electrical signal across the heart. This signal spreads throughout the atria and to the atrioventricular (AV) node. The AV node connects a group of fibers in the sinoatrial node is the heart's pacemaker, the ventricles that conducts the electrical signal and sends the impulse to all parts of the ventricles. This exact route must be followed to ensure that the heart pumps properly.

As the electrical impulse cross through the heart contracts; this normally occurs about 60–100 times per minute, with each contraction equaling a single heartbeat. The atria contract about one-fifth of a second before the ventricles, allowing them to vacate their blood into the ventricles before the ventricles contract.

The cardiac pacemaker is an electric stimulator that produces periodic electric pulses conducted to electrodes located on the surface of heart (epicardium), within the heart muscle (myocardium), or within the cavity of the heart or the lining o the heart (endocardium). The stimulus thus conducted to the heart causes it to contract, this effect can be used prosthetic ally in disease states in witch the heart is not stimulated at a proper rate on its own [2].

Pacemaker can help to pace the heart in cases of slow heart rate, fat and slow heart rate, or a blockage in the heats electrical system.

There are different pacemakers; one of them sends pulses to the heart so that it beats to a rhythm hat have been determined of fixed rhythm. Their name is asynchronous.

Another class can feel the heart's rhythm and turn them selves off when the heartbeat is above a certain level. They will turn on again when the heartbeat is too slow.

G. A. Martinez Chavez (✉)
Mexican Institute of Social Security, South Delegation D.F,
Mexico City, Mexico
e-mail: gamartinezch@yahoo.com; gustavo.martinezc@imss.gob.mx

© Springer Nature Singapore Pte Ltd. 2019
L. Lhotska et al. (eds.), *World Congress on Medical Physics and Biomedical Engineering 2018*,
IFMBE Proceedings 68/3, https://doi.org/10.1007/978-981-10-9023-3_47

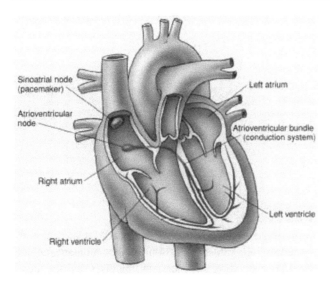

Fig. 1 Show the heart in cross section and node sensitive the hearts pacemaker

These types of pacemakers are called demand pacemakers. The prototype that was developed works as a demand pacemaker and it is in the capacity to feel if exists or not heart rhythm.

2 Materials and Methods

2.1 Description Circuit

The circuit consists of two integrate circuit timer (LM55) which control the opening and closing one start switch, The first circuit is used in astable operation and determines the heart rate witch may be altered between 30 a 320 beats/minute by adjusting the variable resistor. The second circuit regulates the pulse width, which determines the ratio between systole and diastole (duty cycle) this situation is controlled by one second resistor, alters this ratio to produce a change in systolic time which results is the change in stroke volume of the ventricle.

As equal forms, the circuit they have a two light emitting diodes (LED), which are used as visual indicators of the duration the pulse width (ventricular ejection) and operation the voltage power, through the regulator circuit whose

magnitude is controller with three resistor provides +5 voltage to the timing circuit.

The block diagram of pacemaker circuit is illustrated in the Fig. 2.

The design of the experimental pacemakers described in the above diagram depicts a simple asynchronous pacemaker where the uniform pulsing rates produce by his oscillator. The pulse output circuit generates pulses in either a constant current or constant voltage, which are transmitted to the heart via the electrodes.

2.2 Pulse Generation (Stimulation)

The amplitude and pulse duration of the pacemaker depended on whether to the pacemaker is operating in constant-current or constant-voltage [3]. The constant current mode, the pulse current is constant, typically from 8 to 10 mA with a duration of 1.0–1.2 ms. In this developed in the principal characteristic is the constant-Voltage mode, the voltage is maintained at 5.0–5.5 Volts with duration of 0.5 a 0.5 ms. these threshold values are half of the chronic values. The diagram below shows pulses of varying widths (Fig. 3).

2.3 Clinical Measurements Protocol

To verify the performance of the pacemaker proposed in real conditions, measurements were performed over medical procedures in the HGR1 "Carlos Macgregor" Hospital City, previous the authorization of the medical protocol.

Whose results are registered through an electrocardiography monitor of high-speed in such a way that can reproduce the wave form stimulating the pacemaker what constitutes an advance step to prove the results the circuit pacemaker in real measures.

2.4 Patient

Feminine patient with 35 year-old, was studied and with previous laboratory studies, she has been practiced surgery for the inserted electro catheter via the femoral arterial, what enables the connection of the external pacemaker.

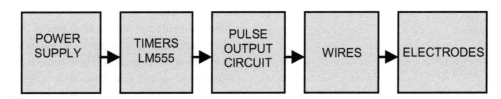

Fig. 2 Diagram of a block the electronic pacemaker

Fig. 3 The shown standards valued the pulse generator, correspond to the chronic values. The pulses sent out by the pacemaker to the heart cause the heart to contract. This is referred to as stimulation. A pulse is characterized by pulse width and pulse amplitude

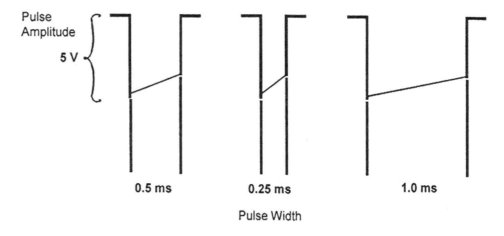

Fig. 4 Show the characteristics out signal the circuit pacemaker indicated to pulse amplitude at 5.0–5.5 Volts with duration of 0.5 a 0.5 ms, to effect of producing the stimulation signal

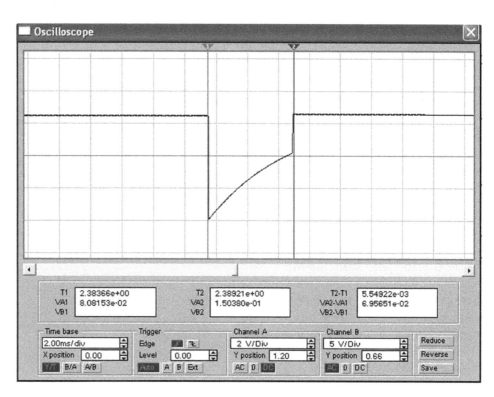

3   Results

3.1   Experimental System (Simulated)

Previous to the assembling of electronic components and testing of the circuit pacemaker, the design was simulated through the program Electronics Workbench in that first stay was looked to obtain the pulses generation, as well as the variability of pulse according to the characteristics the pacemaker mode (constant-Voltage), indicated to maintained at 5.0–5.5 Volts with duration of 0.5 a 0.5 ms, to effect of producing the stimulation sign of the pacemaker. Weaker

pulses either not lasting long enough (less pulse width) or lesser strength (less amplitude) may not cause stimulation. The results shown in the Fig. 4.

3.2   The Test of Circuit Pacemaker

The results obtained with pacemaker system were coherent with to simulate. The analysis in the mensurations show pulse amplitude is 4.89 V and pulse width is 0.486 ms. whose graphics in the oscilloscope are presented in the Fig. 5. We can observe that the results in the simulated and testing pacemaker circuit indicate good correlation.

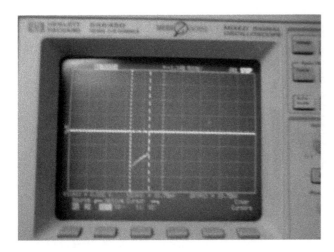

Fig. 5 Graphics in the oscilloscope are presented by the output pulses of a circuit pacemaker; with correlation respect to simulate circuit. The analysis in the mensurations show pulse amplitude is 4.89 V and pulse width is 0.486 ms

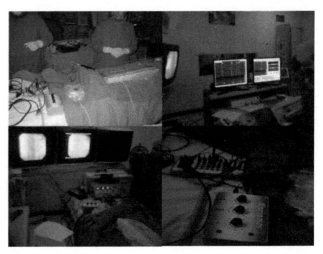

Fig. 7 Photographs the use of circuit pacemaker during the surgery in patient

3.3 The Test of Circuit Pacemaker in Patients

The circuit pacemaker was proven under real conditions, where all the stimuli generated by the same one were detected, what constitute an important step in the validation the circuit. Figure 6. Show the photograph the pacemaker system. Figures 7 and 8. Show the results in patient during the surgery in HGR1 "Carlos Macgregor"

4 Discussion and Conclusions

The main goal of this paper is to show that electronic hardware design is a valid alternative to be taken into account when deciding on the implementation of a certain development or the technical—economical possibility of a certain product in our context.

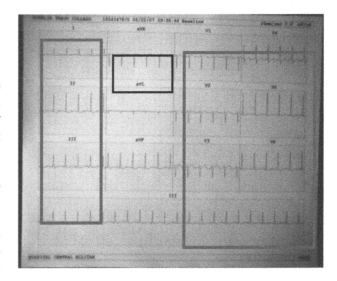

Fig. 8 Results the EKG sign in patient as the use of external pacemaker. Blue color show the beats are marked for caused the stimulate circuit pacemaker and red color the normal beats among the vectors V1–V6

Many factors have contributed to the feasibility of the electronic hardware design in our country. The use of state of the art technologies is being made possible by the reduction in prototyping and production, fixed costs the affordable cost of powerful CAD stations and the simplicity of communications and access to information from providers and manufacturers, some examples of these techniques that they have been used in design and implementation the external pacemaker system.

A series of experiences done in recent in HGR1 HGR1 "Carlos Macgregor" Hospital are presented in this paper what was live and expressed in design circuit pacemakers.

The results obtained in this circuit confirm the proposed the development of external pacemaker they are used

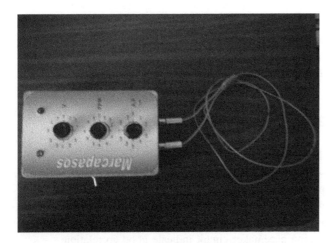

Fig. 6 Photograph the pacemaker and electro cater

electronic circuit, the low price and high benefits existent in the Mexican National Market.

The statistical results between laboratory analysis and prototype show a direct dependence between both systems, considering measures in patients, the dispersion is elevated. This dispersion lowers if we consider homogeneous depending the particularly situation of the patient. This parameter can help to adjust the settings improvements the future design of pacemaker.

References

1. Belott, P., Sands, S., et al.: Resetting of DDD pacemakers due to EMI. PACE, 7:169, 1984.
2. K.A.Ellenbogen "Clinical Cardiac Pacing", 1985.
3. AAMI ECG Committee, Performance Standard for Ambulatory Electrocardiographs, Publication EC38-1994, August 1994.
4. Dodinot, B., Godenir, J., et al.: Electronic article surveillance: A possible danger for pacemaker patients, PACE, 16:46–53, 1993.
5. Carrillo, R., Preliminary observations on cellular telephones and pacemakers, PACE, 1995, 18: 863 (EMI-168).
6. Hayes, D. EMI update. 9th Annual National Symposium on Pacing and Arrhythmia control, Feb. 1992.
7. Irnich W., Lazica M., et al.: Pacemaker patients and extracorporal shock wave lithotripsy. Cardiac Pacing and Electrophysiology - Belhassen B., Feldman S., Copperman Y. Eds. R & L Creative Communications Ltd., Pub. Jerusalem 1987: P 221–226.
8. Irnich, W.,: Interference in pacemakers. PACE, 7:1021, 1984.
9. Kuan, P., Kozlowski, J., et al.: Interference with pacemaker function by cardiokymographic testing. Am. J. Cardiol., 58:362, 1986.
10. Medtronic: Pacemaker patients who use transcutaneous electrical nerve stimulators. Cardiovascular Tech Note, Issue No: 82-4, 1982.
11. Moore, S., Firstenberg, M.: Long term effects of radiocatheter ablation on previously implanted pacemakers, PACE, 16:947, 1993.
12. Parker, B., Furman, S., et al.: Input signals to pacemakers in a hospital environment. Annals New York Academy of Sciences, pp. 823–34.
13. Rahn, R., Zegelman, M., et al.: The influence of dental treatment on rate response pacemakers. PACE, 12:1300, 1989.
14. Telectronics Technical Note: Effects of Radiation on pacemakers. 1990.
15. Telectronics Technical Note: Electromagnetic Interference and the Pacemaker Patient: an Update. 1996.
16. Ivan Gelder, L.M., Bracke Fale, El Gamal, IH: ECG monitoring and minute ventilation rate adaptive pacing. Cardiologie, 1:301, 1994.
17. Vanerio, G., Rashidi, R., et al.: Effect of catheter ablation procedures on permanent pacemakers. Circulation, Abstract 0717, 1990.
18. Villforth, J.: Cardiac pacemaker warning signs near microwave ovens. Dept. of HEW/FDA, Bureau of Radiological Health, Reference Report No. 5, 1976.
19. Warnowicz-Papp, M.: The pacemaker patient and the electromagnetic environment. Clin. Prog. in Pacing and Electrophysiol., 1:166.1983.

Prototype Device for Driving Suitability Tests in Sunglasses

Artur D. Loureiro⊙ and Liliane Ventura⊙

Abstract

Wearing inadequate sunglasses while driving may lead to dangerous misunderstandings in objects and traffic lights recognition. Sunglasses standards propose transmittance requirements that sunglasses should fit to be classified as suitable for driving. Transmittance tests are time-consuming and laborious. Also, it requires a spectrophotometer and a skilled technician to be performed. The aim of this study was to develop and to build an easy-to-use, quick and accurate device for luminous and traffic lights transmittance tests which runs the tests by itself in a way anyone can operate it without any training. A microcontrolled prototype was developed and built using a white LED and a four-channel sensor combination. This combination generated luminous and traffic lights weighting functions similar to standard ones. Using our prototype and a gold standard (VARIAN Cary 5000 spectrophotometer), luminous transmittance and relative attenuation quotients for traffic lights were measured in 128 sunglasses lenses. Bland-Altman method was used to assess concordance between both measurement methods. The bias was insignificant for all measurement and the limits of agreement were broad for luminous transmittance and for relative attenuation quotient for blue light detection, and narrow for the others. Thus, within the predefined tolerance, prototype measurements are equivalent to gold standard ones for relative attenuation quotients for red, yellow and green light detection. Despite not all prototype measurements being equivalent to gold standard ones, results were accurate; only 5 from 128 lenses were defectively classified as to suitability for driving (2 for luminous transmittance, 1 for red light

quotient and 2 for blue light quotient). Our prototype creates means to general public to assess characteristics of their own sunglasses including whether they are suitable for driving according brazilian and ISO standards.

Keywords

ABNT NBR ISO 12312-1 • ISO 12312 • Sunglasses
Luminous transmittance • Traffic lights transmittance
Transmittance measurement

1 Introduction

International standard ISO 12312-1:2013 defines requirements that sunglasses lenses must comply with in order to be suitable for driving [1].

Luminous transmittance, τ_V, and traffic lights signal transmittances, τ_{signal}, are calculated by Eq. (1) where $\tau_F(\lambda)$ is the lens spectral transmittance and $W(\lambda)$ is a known weighting function. Traffic lights signal transmittances are calculated for red, yellow, green and blue signals [1].

$$\tau = \frac{\int_{380}^{780} \tau_F(\lambda) W(\lambda) d\lambda}{\int_{380}^{780} W(\lambda)\, d\lambda} \quad (1)$$

The relative visual attenuation quotients for signal lights detection, Q_{signal}, are calculated using Eq. (2) and they are also calculated for red, yellow, green and blue signals [1].

$$Q_{signal} = \frac{\tau_{signal}}{\tau_V} \quad (2)$$

Sunglasses with dark lenses or that attenuate excessively traffic light signals are inappropriate for driving as they could lead to perilous misunderstandings [2, 3]. To be suitable for driving, sunglasses lenses must have luminous transmittance greater than 8% (0.08), red signal detection quotient greater or equal to 0.8 and yellow, green and blue ones greater or equal to 0.6.

A. D. Loureiro · L. Ventura (✉)
Ophthalmic Instrumentation Laboratory - LIO/Sao Carlos School of Engineering - EESC, University of Sao Paulo - USP, São Carlos, SP, Brazil
e-mail: lilianeventura@usp.br

© Springer Nature Singapore Pte Ltd. 2019
L. Lhotska et al. (eds.), *World Congress on Medical Physics and Biomedical Engineering 2018*,
IFMBE Proceedings 68/3, https://doi.org/10.1007/978-981-10-9023-3_48

To certify compliance with the standard it is required a laborious, time-consuming test performed with a spectrophotometer by a skilled technician.

After long solar exposure, sunglasses lens transmittance may vary spectrally unpredictably [4]. Thus, even lenses approved in traffic lights tests need to be retested after years of use to assure that they are still suitable for driving.

The purpose of this study is to develop a portable device capable to perform transmittance measurements related to suitability for driving in a fast and automatic way.

2 Materials and Methods

Using a white LED and a four-channel sensor, four weighting functions were obtained.

These functions were linearly combined to produce weighting functions similar to standard ones for luminous transmittance and traffic lights transmittance (red, yellow, green and blue).

We measured 128 sunglasses lenses with our device and a gold standard (VARIAN Cary 5000 spectrophotometer). By Bland-Altman method, we assess concordance between both measurement methods. For luminous transmittance values, it was adopted 0.5 and 6% as values above which bias absolute value is significant and 95% limits of agreement amplitude, wide, respectively. For traffic lights visual attenuation quotient values, 0.1 and 0.4, respectively.

All calculations were performed using the free software GNU Octave v.4.0.0 [5].

3 Results and Discussion

The five weighting functions (for luminous, red, yellow, green and blue light transmittances) we obtained by sensor-LED combination were plotted overlapped with respective standard ones and they are shown in Figs. 1, 2, 3, 4 and 5.

From the 128 measured sunglasses lenses, 19 have luminous transmittance inferior to 8% and for theses, signal detection quotients were not measured.

3.1 Luminous Transmittance Measurement

Mean-difference Tukey plot for 128 sunglasses luminous transmittance values obtained using our prototype and a gold standard is presented in Fig. 6. Two lenses were classified in wrong categories. The difference between methods tends to rise with the average. The greatest absolute error, the bias and the 95% limits of agreement are 12.939%, −0.4994% and [−7.0984%; 6.0996%], respectively.

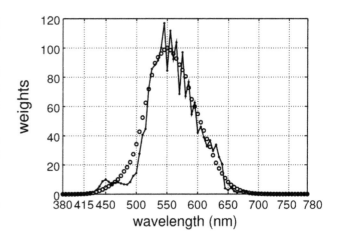

Fig. 1 Our weighting function (continuous line) and the standard one (empty circles) for luminous transmittance

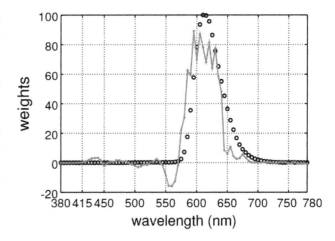

Fig. 2 Our weighting function (continuous line) and the standard one (empty circles) for red signal transmittance (Color figure online)

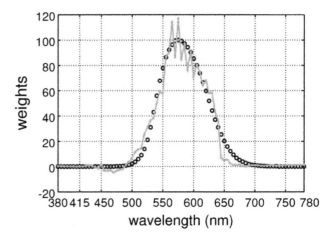

Fig. 3 Our weighting function (continuous line) and the standard one (empty circles) for yellow signal transmittance (Color figure online)

Bias is not significant and 95% limits of agreement interval is wide. Consequently, within the predefined tolerance, our method is not equivalent to gold standard one for luminous transmittance measurement.

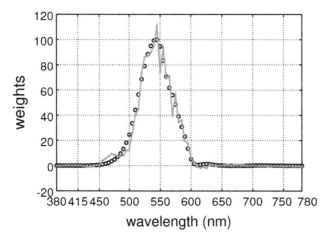

Fig. 4 Our weighting function (continuous line) and the standard one (empty circles) for green signal transmittance (Color figure online)

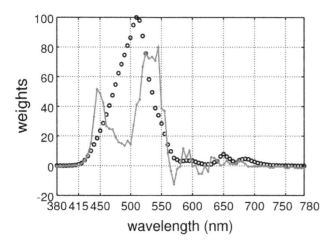

Fig. 5 Our weighting function (continuous line) and the standard one (empty circles) for blue signal transmittance (Color figure online)

3.2 Red Signal Detection Quotient Measurement

Mean-difference Tukey plot for 109 sunglasses red signal detection quotient values obtained using our prototype and a gold standard is presented in Fig. 7. One lens was wrongly classified as to suitability for driving (Qred $\geq$ 0.8). The greatest absolute error, the bias and the 95% limits of agreement are 0.268, 0.0536 and [−0.1415; 0.2487], respectively.

Bias is not significant and 95% limits of agreement interval is narrow. Consequently, within the predefined tolerance, our method is equivalent to gold standard one for red signal detection quotient measurement.

3.3 Yellow Signal Detection Quotient Measurement

Mean-difference Tukey plot for 109 sunglasses yellow signal detection quotient values obtained using our prototype and a gold standard is presented in Fig. 8. All lenses were correctly classified as to suitability for driving (Qyellow $\geq$ 0.6). Prototype measures tend to be inferior to gold standard ones. The greatest absolute error, the bias and the 95% limits of agreement are 0.256, 0.0909 and [−0.0406; 0.2223], respectively.

Bias is not significant and 95% limits of agreement interval is narrow. Consequently, within the predefined tolerance, our method is equivalent to gold standard one for yellow signal detection quotient measurement.

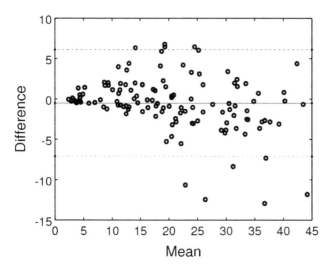

Fig. 6 Mean-difference Tukey plot for luminous transmittance measures in 128 sunglasses lenses

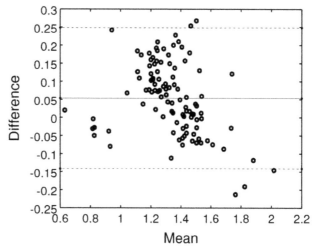

Fig. 7 Mean-difference Tukey plot for red signal detection quotient measures in 109 sunglasses lenses

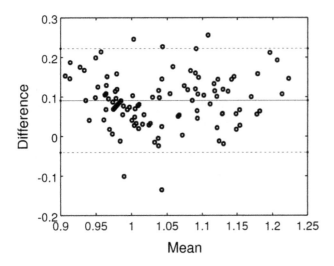

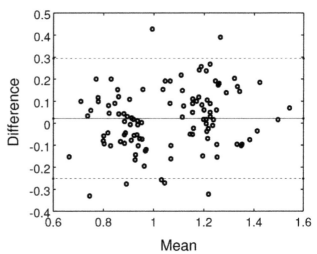

Fig. 8 Mean-difference Tukey plot for yellow signal detection quotient measures in 109 sunglasses lenses

Fig. 10 Mean-difference Tukey plot for blue signal detection quotient measures in 109 sunglasses lenses

3.4 Green Signal Detection Quotient Measurement

Mean-difference Tukey plot for 109 sunglasses green signal detection quotient values obtained using our prototype and a gold standard is presented in Fig. 9. All lenses were correctly classified as to suitability for driving (Qgreen $\geq$ 0.6). Prototype measures tend to be greater to gold standard ones. The greatest absolute error, the bias and the 95% limits of agreement are 0.159, −0.0523 and [−0.1377; 0.0330], respectively.

Bias is not significant and 95% limits of agreement interval is narrow. Consequently, within the predefined tolerance, our method is equivalent to gold standard one for green signal detection quotient measurement.

3.5 Blue Signal Detection Quotient Measurement

Mean-difference Tukey plot for 109 sunglasses blue signal detection quotient values obtained using our prototype and a gold standard is presented in Fig. 10. Two lenses were wrongly classified as to suitability for driving (Qblue 0.6). The greatest absolute error, the bias and the 95% limits of agreement are 0.427, 0.0216 and [−0.2512; 0.2944], respectively.

Bias is not significant and 95% limits of agreement interval is wide. Consequently, within the predefined tolerance, our method is not equivalent to gold standard one for blue signal detection quotient measurement.

4 Conclusion

The proposed method for luminous and traffic lights measurements in sunglasses lenses uses simple components and provides accurate results.

Within the predefined tolerance, prototype measurements are equivalent to gold standard ones for relative attenuation quotients for red, yellow and green light detection.

From 128 measured lenses, only 5 were incorrectly classified as to suitability for driving; 2 for luminous transmittance measure errors; 1 for Qred measure error and 2 for Qblue measure errors.

Our device aims to provide to people a mean to obtain informations about their own sunglasses and the importance to use suitable sunglasses while driving.

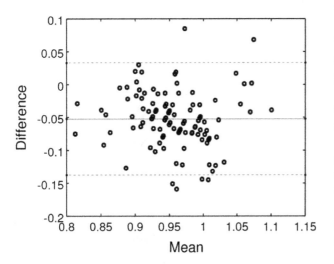

Fig. 9 Mean-difference Tukey plot for green signal detection quotient measures in 109 sunglasses lenses

Acknowledgment The authors declare no conflict of interest. The authors acknowledge ABIOPTICA, CNPq (130755/2015-0) and FAPESP (2014/16838-0).

References

1. International Organization for Standardization. Eye and face protection - sunglasses and related eyewear Part 1: sunglasses for general use. Geneva: ISO 12312-1:2013; (2013).

2. Dain, S. J., "Sunglasses and sunglass standards," Clin Exp Optom, 86(2), 77–90 (2003).

3. Hovis, J. K., "When yellow lights look red: tinted sunglasses on the railroads," Optom Vis Sci, 88(2), 327–333, (2011).

4. Loureiro, A D; Gomes, L M; Ventura, L. Transmittance Variations Analysis in Sunglasses Lenses Post Sun Exposure. Journal of Physics. Conference Series (Online), v. 733, p. 012028, (2016).

5. John W. Eaton, David Bateman, Søren Hauberg, Rik Wehbring (2015). GNU Octave version 4.0.0 manual: a high-level interactive language for numerical computations, available online at: http://www.gnu.org/software/octave/doc/interpreter/.

Using the Monte Carlo Stochastic Method to Determine the Optimal Maintenance Frequency of Medical Devices in Real Contexts

Antonio Miguel-Cruz, Pedro Antonio Aya Parra,
Andres Felipe Camelo Ocampo, Viena Sofia Plata Guao,
Hector H. Correal O, Nidia Patricia Córdoba Hernández,
Angelmiro Núñez Cruz, Jefferson Steven Sarmiento Rojas,
Daniel Alejandro Quiroga Torres, and William Ricardo Rodríguez-Dueñas

Abstract

The purpose of this study was to implement and validate a Monte Carlo Algorithm (MCA) to determine the best T value (the time between two preventative maintenances) that optimizes the achieved availability of equipment types. In doing so, we (1) collected 796 maintenance works orders from 16 medical devices installed in a 900-bed hospital; (2) we fitted the probability distributions for each of the inputs of the achieved availability mathematical model (the mean preventative and corrective service time (in hours)); (3) we generated a set of random inputs following a Weibull distribution of the achieved availability mathematical model; (4) we calculated the achieved availability for every random input generated; this process was repeated for "m" iterations (an accuracy of 1%, 95% CI, alpha = 0.05); (5) the trends of the mean achieved availability for the different maintenance T intervals versus mean time to failure (MTTF) for all the equipment types were plotted; finally, (6) the best T value with the maximum value of the achieved availability of a medical device type for a specific MTTF was the optimal target. The mean simulation time for all the cases was 12 min. The MCA was able to determine the best T value, optimizing the achieved availability in 81.25% of cases. In conclusion, the results showed that, on average, the T maintenance intervals determined by the MCA were statistically significantly different from the original T values suggested either by the clinical engineering department or third-party maintenance providers (MCA_{Tmean} = 1.68 times/yr, $Actual_{Tmean}$ = 2.56 times/yr, p = 0.008).

Keywords

Monte carlo simulation • Maintenance optimization Preventative maintenance frequency • Biomedical engineering • Clinical engineering

1 Introduction

The aim of this study was to implement and validate a Monte Carlo algorithm (MCA) as a method of determining the optimal maintenance intervals, i.e. T (or maintenance frequency, $F_{pm} = 1/T$), of medical devices. The motivation for conducting this research to provide additional evidence for the clinical engineering community when determining whether the maintenance frequency should be established *"periodically in accordance with the manufacturers' recommendations"* [1], or whether they can use their own maintenance records. In doing so, the authors believe they are contributing to filling the gap in the existing knowledge for implementing maintenance optimization models, specifically the optimal maintenance intervals of medical devices [2], thus adding another point of view to the current debate (named following the *manufacturers' recommendations* or not) about how to establish the maintenance frequencies of medical devices.

1.1 Maintenance Policies and Analysis

A maintenance policy is the implementation of a certain maintenance model to improve the reliability of a system [1]. According to the authors' experience, the most commonly

A. Miguel-Cruz (✉) · P. A. A. Parra · A. F. C. Ocampo · V. S. P. Guao · J. S. S. Rojas · D. A. Q. Torres · W. R. Rodríguez-Dueñas
Biomedical Engineering Program, School of Medicine and Health Sciences, Universidad del Rosario, Calle 63D#24-31 Barrio 7 de Agosto, Bogotá, Colombia
e-mail: antonio.miguel@urosario.edu.co

H. H. Correal O · N. P. C. Hernández · A. N. Cruz
Biomedical Engineering Maintenance Department, Universidad del Rosario-Hospital Universitario Mayor Méderi, Cl. 24#29-45, Bogotá, Colombia
e-mail: hector.correal@mederi.com.co

© Springer Nature Singapore Pte Ltd. 2019
L. Lhotska et al. (eds.), *World Congress on Medical Physics and Biomedical Engineering 2018*,
IFMBE Proceedings 68/3, https://doi.org/10.1007/978-981-10-9023-3_49

used maintenance policy for the maintenance of medical devices is the periodic preventative maintenance (PM) policy [3]. In this policy, the parts of a medical device are preventively maintained at fixed time intervals k*T (k = 1, 2,...n, and T is constant) independently of the failure history of the medical device (frequency of failures, λ), and repaired at intervening failures. The cornerstone of this policy is determining the T value (or the maintenance frequency, f_{pm} = 1/T). Ideally, if the T value is adequate (i.e. $f_{pm} \approx \lambda$) the device will never fail because the PM is always performed first, after which the availability of the device is maximized. Therefore, the next problem to be solved is: what is the best T value that "optimizes" the availability of a certain device, knowing that maintenance tasks have a cost? The mathematical model of the achieved availability to be optimized is shown in Eq. 1 [2]

$$A_a = \frac{1}{1 + \lambda * (\text{MCMT} + \text{MCVT}) + f_{pm} * (MPMT + MPVT)} \tag{1}$$

where:

A_a: Achieved availability
λ: Failure rate (MTBF = 1/λ)
MCMT: Mean time to perform a corrective maintenance (in hours)
MCVT: Mean time to perform a verification or quality inspection of corrective maintenance (in hours)
f_{pm}:   PM frequency (times/year, 1/T)
MPMT: Mean time to perform a PM (in hours)
MPVT: Mean time to perform a verification or quality inspection of PM (in hours)

To solve this problem, the authors implemented and validated an algorithm using a Monte Carlo simulation [4]. To implement and run the Monte Carlo algorithm it is necessary to generate random variables that follow a statistical distribution. In this research, the Weibull distribution (see Fig. 1) was used to generate the random numbers of MCMT, MPMT, and λ

2 Materials and Methods

2.1 Materials

The MCA was implemented using the Scilab v.5.5.2 software package licensed by GPL [1]. The SPSS® V 22.0 statistics package was used to generate descriptive, univariate, and bivariate statistics. The programming and validation of the simulation were conducted on an Inter Core i7 personal computer, 3,40 Ghz, 6 Gb RAM, with the Windows 7, 64 bits operative system.

2.2 Methods

In this study, we followed the methodology proposed by [2]. The data were extracted from a primary source of maintenance records from the clinical engineering department. The maintenance records corresponded to the medical devices located in the Intensive Care Units (ICUs) and imaging services. Two research assistants extracted from a secondary information source the preventative and corrective maintenance data records from 2009–2017 including the mean time to failure (MTTF) (or frequency of failures λ, times/yr), the actual maintenance frequencies (f_{pm}, times/yr) established by either an in-house or external service provider, and the corrective and preventative service time (in hours). In addition, whether or not the maintenance task was conducted by an in-house or an external service provider was determined. With the extracted maintenance data, the Weibull probability distributions for the MCMT, MPMT, and λ variables were fitted. We selected the Weibull probability distribution because it has the great advantage in reliability work that by adjusting the distribution parameters it can be made to fit many life distributions [5, p. 78]. Next, the MCA was implemented using Scilab (see the MCA implementation for more detail). Then, the MCA was validated. To validate the implemented MCA, the authors first ran one experiment per device type (i.e. electrocardiographs, infusion pumps, ECG monitors, etc.); next they verified that for each availability curve obtained from the different scenarios of f_{pm}, a typical parabolic curve was obtained (side-opening parabola, see Eq. 1); finally they verified that for a specific λ value, if $f_{pm} \approx \lambda$ (i.e. ±5%) the achieved availability curve at this point was maximum (i.e. the f_{pm} was optimum or "tuned" with failures). Descriptive statistics to summarize the demographic data of the medical device population and a test for differences between the groups were conducted. The Mann–Whitney U test for differences between the groups was conducted with the aim of determining whether the preventative maintenance frequency of the medical devices as determined by the implementation of the MCA was statistically significantly different from the actual preventative maintenance frequency established by either an in-house or external service provider. The alpha level of significance was set at $p \leq 0.05$.

Monte Carlo algorithm implementation. Monte Carlo simulation is a method for iteratively evaluating a deterministic model using sets of random numbers as inputs. It is a simple mathematical procedure, with random inputs and random outputs [5, p. 108]. To run a Monte Carlo simulation the generation of random variables that follow an arbitrary statistical distribution is needed. The inputs are randomly generated from probability distributions to simulate the process of sampling from an actual population. Therefore a

Fig. 1 Diagram flow of the implemented MCA

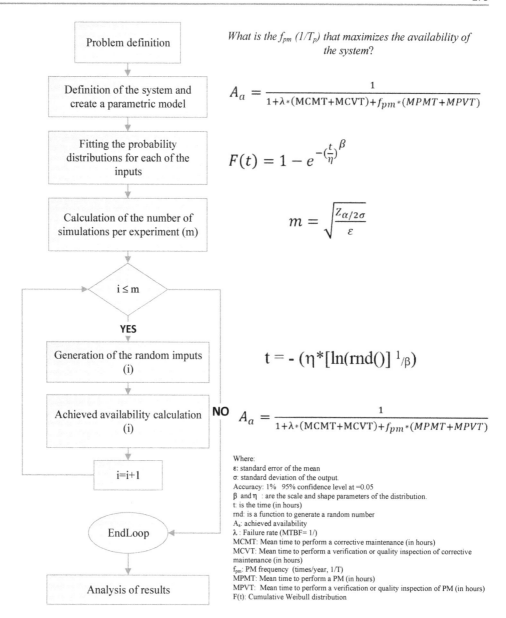

Problem definition

Definition of the system and create a parametric model

Fitting the probability distributions for each of the inputs

Calculation of the number of simulations per experiment (m)

$i \leq m$

YES

Generation of the random imputs (i)

Achieved availability calculation (i)

NO

$i=i+1$

EndLoop

Analysis of results

What is the f_{pm} ($1/T_p$) that maximizes the availability of the system?

$$A_a = \frac{1}{1+\lambda*(MCMT+MCVT)+f_{pm}*(MPMT+MPVT)}$$

$$F(t) = 1 - e^{-(\frac{t}{\eta})^{\beta}}$$

$$m = \sqrt{\frac{Z_{\alpha/2\sigma}}{\varepsilon}}$$

$$t = - (\eta*[\ln(rnd()] \, ^{1}/_{\beta})$$

$$A_a = \frac{1}{1+\lambda*(MCMT+MCVT)+f_{pm}*(MPMT+MPVT)}$$

Where:
ε: standard error of the mean
σ: standard deviation of the output.
Accuracy: 1% 95% confidence level at =0.05
β and η : are the scale and shape parameters of the distribution.
t: is the time (in hours)
rnd: is a function to generate a random number
A_a: achieved availability
λ : Failure rate (MTBF= 1/)
MCMT: Mean time to perform a corrective maintenance (in hours)
MCVT: Mean time to perform a verification or quality inspection of corrective maintenance (in hours)
f_{pm}: PM frequency (times/year, 1/T)
MPMT: Mean time to perform a PM (in hours)
MPVT: Mean time to perform a verification or quality inspection of PM (in hours)
F(t): Cumulative Weibull distribution

distribution for each input that best represents the current state of knowledge is chosen. Figure 1 shows the diagram flow of the MCA implemented in this study.

3   Results

Table 1 shows the results obtained from the simulations and the maintenance data per equipment type, including the mean MPMT (SD), MCMT (SD) (in hours), the equipment type brand name, the actual failure frequency (λ, in times/yr), the maintenance frequency (f_{pm}, times/yr) at which the optimum achieved availability (A_v) was found using the implemented MCA, and the actual maintenance frequency (f_{pma}, times/yr) established by either an in-house or external service provider. Table 1 also shows, highlighted in italic, the types of medical device where the optimum achieved availability was determined by the implemented MCA. The mean simulation time for all the cases was 12 min.

The results obtained from the data analyses showed that the actual average failure frequency (λ, times/yr) for every type of device (5.56 times/yr SD 3.99) was higher than either

Table 1 Achieved availability versus mean time to failure (in months) for each medical device type

Equipment type	Brand name	MPMT		MCMT		Actual λ (t/yr)		Monte Carlo f_{pm} (t/yr)		Actual f_{pma} (t/yr)	
		Mean	SD	Mean	SD	Mean	MTTF	f_{pm}	A_v	f_{pm}	A_v
Angiography U	Siemens	7.33	2.03	5.32	3.51	15.63	0.77	4	0.940	4	0.935
Angiography U	Toshiba	7.34	2.51	4.68	2.88	7.83	1.53	2	0.974	3	0.973
Mobile C-arm	Siemens	2.75	0.96	2.69	1.53	4.38	2.74	2	0.989	2	0.980
Fluoroscopic U	Siemens	4.86	2.60	4.01	2.83	2.88	4.17	1	0.985	4	0.983
Gamma Camera	Siemens	7.93	2.82	4.23	3.11	5.5	2.18	2	0.969	4	0.967
Mammography	Siemens	4.00	1.81	2.16	1.53	1.83	6.56	1	0.965	2	0.955
ECG monitor	Nihon K	0.52	0.06	0.21	0.06	6	2.00	2	0.998	2	0.998
TAC	Toshiba	6.72	1.79	3.95	3.02	10.13	1.18	2	0.976	3	0.977
Mobile X Ray	Siemens	2.23	0.62	2.84	2.09	6.38	1.88	3	0.987	2	0.986
Mobile X Ray	Toshiba	2.5	0.68	3.03	1.88	2	6.00	1	0.995	2	0.996
MRI	Siemens	7.15	1.65	4.32	3.3	10.75	1.12	2	0.963	4	0.961
X Ray U	Siemens	4.00	1.58	2.99	2.27	6.75	1.78	1	0.985	2	0.984
X Ray U	Toshiba	2.86	0.83	2.08	0.82	1	12.0	1	0.999	2	0.998
X Ray U	Universal	4.31	2.05	2.04	1.16	2.5	4.80	1	0.995	2	0.994
Tourniquet	Pangas	2.67	1.30	2.00	1.06	1.43	9.60	1	0.994	1	0.994
Surgical lamp	Drager	2.27	1.42	1.16	0.85	5.5	2.18	1	0.987	2	0.986

the average of the optimal maintenance frequencies determined by the MCA (MCA_{Tmean} = 1.68 times/yr SD 0.87) or the actual maintenance frequencies suggested either by the clinical engineering department or third-party maintenance providers ($Actual_{Tmean}$ = 2.56 times/yr SD 0.96). The latter result indicates that the optimal maintenance frequencies determined by the MCA are not always well-tuned with the failure frequencies (times/yr) for every type of device. On average, the actual maintenance frequencies suggested either by the clinical engineering department or third-party maintenance providers was 1.5 times/yr compared with the optimal maintenance frequencies determined by the MCA. The results also showed that, on average, the optimal maintenance frequencies determined by the MCA were statistically significantly different from the actual maintenance frequencies suggested either by the clinical engineering department or third-party maintenance providers (Mann–Whitney U test two-tailed p = 0.008, power of the test = 62%). The results also showed that the average achieved availability obtained from the MCA (MCA_{av} = 98.13% SD 1.60) was higher compared with the actual average achieved availability ($Actual_{av}$ = 96.98% SD 1.76) for all the medical devices, although these differences were not statistically significantly different (Mann–Whitney U test two-tailed p = 0.694, power of the test = = 39%). The MCA

was able to determine the best T value (or the maintenance frequency), optimizing the achieved availability in 81.25% of cases (Pearson-Chi-Square = 16.0, df = 1, p = 0.000).

4 Discussion

This study aimed to implement and validate an MCA to determine the optimal maintenance intervals (or the maintenance frequency) of types of medical device. In doing so, we extracted a set of maintenance records data from a 900-bed hospital.

Although the MCA was able to determine in most cases the optimal maintenance frequency, some opportunities for improving this study existed. One is to improve the power of the statistical tests. The powers of all the tests were below 80%, an unacceptable value that increases the risk of Type II error [3]. This can be solved by increasing the data sample of the types of medical device involved in future studies. The standard deviation of the failure frequencies mean was too high (i.e. 71.76% on average), with such a high variation in the failure frequencies that generating the random number of failure frequencies in the implemented MCA was problematic. Future research should focus on collecting data with less standard deviation.

5 Conclusions

On average, the T maintenance intervals (or the maintenance frequency) determined by the MCA were statistically significantly different from the original T values suggested either by the clinical engineering department or external maintenance providers.

References

1. Ridgway, M. Manufacturer-recommended PM intervals: Is it time for a change? Biomedical Instrumentation and Technology 43(6), 498–500 (2009).

2. Jamshidi, A., Ait-kadi, A., Bartolome, A. Medical Devices Inspection and Maintenance; A Literature Review. In the Proceedings of the 2014 Industrial and Systems Engineering Research Conference, Montréal, Canada (2014).

3. Pham, H., Wang, H. Reliability and optimal maintenance, NY: Springer (2007).

4. Cruz, A., Dueñas, W. Estimation of the optimal maintenance frequency of medical devices: A Monte Carlo simulation approach. In: 7th Latin American Congress on Biomedical Engineering, CLAIB 2016, Bucaramanga (2017).

5. S. Enterprises, Scilab: Free and Open Source software for numerical computation (OS, Version 5.XX) [Software], http://www.scilab.org, last accessed 2017/11/21.

The Supportability of Medical Devices

M. Capuano

Abstract

Clinical Engineering (CE) and Health Technology Management (HTM) appears to be spending more time trying to get medical device manufacturers to provide support for inhouse servicing than ever before. Surveys conducted by AAMI in 2015 and CMBES in 2016 revealed that model-specific technical training and documentation were the top two priorities for this group of respondents. Many manufacturers and even healthcare institutions seem to minimize the value or even the existence of CE/HTM programs. It is important to note that these programs operate to save hospitals and healthcare money and also serve to provide quick and necessary support for healthcare technology in the clinical setting. They are now being challenged by many of their commercial partners. Almost every other acquisition of medical equipment now requires the need to negotiate support for inhouse services and is met with a balance of success and failure. It appears manufacturers are designing equipment without considering the customer's option to service it. These customers include medium to large hospitals that have the capacity, economies of scale, and know-how to create and sustain CE/HTM departments. At the same time, there are many companies that provide good support for inhouse servicing. Their examples of appropriate support strategies may serve as a baseline for most other companies to make their products serviceable and to ensure CE/HTM is qualified and properly equipped to perform the required service. This paper highlights most of the issues surrounding the notion of Supportability in the CE/HTM world. These issues affect independent service organizations (ISOs) in a similar way. There are efforts to manage the supportability issue and ideas on how certain barriers might be dealt with. The paper attempts to recognize these

and the rationale behind certain behaviors. There are standards and regulations, or an absence of them, which either help or hinder the issue.

Keywords

Supportability • Serviceability • Service manual
Training • Clinical engineering • OEM • HTM
ISO

1 Early Efforts

In 2012, Mike Capuano CBET/CCE, Manager of Hamilton Health Sciences Biomedical Technology department, began to look at the situation as an issue worth investigating. He knew the challenges clinical engineers faced on a day to day basis. One of those challenges—the support of medical devices inhouse—seemed to take on a definitive persona. As a long-time member of AAMI (the Association for Advancement of Medical Instrumentation) and active contributor on its various boards and committees, he began to think that these activities may provide an opportunity to bring the issue to the forefront. There were no existing efforts on supportability at the time. Capuano decided that he would submit a work proposal to the standards board of AAMI just to see what would happen. It was not approved for development but it did result in the publication of an AAMI Leading Practice document on the Supportability of Medical Devices [1]. This was the first action taken on the topic ever. It coined the first definition of supportability as follows: The degree to which a medical device or system can be effectively and economically supported, in terms of its design features and product support (information, training, technical support, tools, and spare parts), throughout its lifecycle. Capuano extends this definition as being executed by 'entities other than representatives or direct agents of the original equipment manufacturer (OEM)' (e.g. CE/HTM, ISOs, etc.).

M. Capuano (✉)
Hamilton Health Sciences, Biomedical Technology, Hamilton,
ON L8S4J9, Canada
e-mail: capuamik@hhsc.ca

© Springer Nature Singapore Pte Ltd. 2019
L. Lhotska et al. (eds.), *World Congress on Medical Physics and Biomedical Engineering 2018*,
IFMBE Proceedings 68/3, https://doi.org/10.1007/978-981-10-9023-3_50

If there is something the company can do in the field but prevents the customer's technical personnel or ISOs from being able to do it; this is a supportability barrier. Factors that may contribute to supportability include: equipment design, vendor policies; availability of education, phone assistance, and service aids; bundling practices/contract options, etc. (see Table 1). Since this early effort, AAMI continued to address the issue by various means and from a wider scope of contributors. Articles on supportability started to pop up in the form of contributed articles, blogs, conference sessions, and even a cover story in AAMI's journal, 'Biomedical Instrumentation and Technology (BIT)' [2].

Other organizations and their publishers also began broaching the topic. On top of this, another phase of action began to take shape. This came in the form of organized groups working to address the issue in teams and sub-committees. The AAMI Technology Management Council (TMC) assigned a sub-committee in 2015 known to be the Supportability Task Force. The task force began acting on the need to address some of the polarized viewpoints coming from both CE/HTM and industry sectors. It is important to note that AAMI's breadth and influence comes from its diverse membership including biomedical technicians, clinical engineers, and single/corporate memberships coming from the OEMs. The AAMI Forum on the Supportability of Medical Devices was held in November of 2015 at AAMI Headquarters in Arlington, VA. It was one of the first events of its type where, by invitation, CE/HTM and OEM representatives came together to identify what was driving the perceptions on supportability and try to break them down into actionable pieces. This strategy had some success and sparked the creation of sub-groups under the TMC's Supportability Task Force. These included the creation of a comprehensive service level agreement template,

Table 1 Factors affecting supportability

Access to:
Service manuals
Technical training
Diagnostic codes
Error codes
Event logs
Test equipment
Service updates
Phone support
Other:
Bundling practices
Service options
Equipment design
Availability of replacement parts

an AAMI reference on competencies required for the support of medical devices, and recommendations on the sourcing of replacement parts. These actions stem directly from the prioritized breakouts facilitated at the forum in Arlington.

2 Remaining Issues

Despite the progress made to address the issue, the concern remains and appears to be of particular concern in Canada. Based on transcribed comments from the 2015 World Congress held in Toronto, several countries came forward at the CMBES-hosted World Summit on the Supportability of Medical Devices and stated that they too face the challenge of obtaining supports especially from OEMs originating abroad. In Canada, most of the technologies supported inhouse originate in the United States and are distributed through either a separate vendor, subsidiary, or a Canadian arm of the OEM situated in our country. The concerns, as indicated above, come in the form of various limitations or barriers that may be as a result of marketing strategy, technological force, procurement strategy, competency, weak standards, and the interpretation of regulatory requirements. These barriers can get in the way of providing safe, cost-effective, and expedient service to medical equipment in the field. They drive increases in the cost of service and also limit the ability of competent inhouse or independent servicers to conduct prompt, and sometimes crucial, on-site services.

Viewpoints from both CE/HTM and OEMs cover common themes. Some of the CE/HTM community feels that OEMs limit support for inhouse programs in order to affect revenue. Several OEMs state that increasing supports for inhouse may affect reliability of their product and that designing supportability may limit its level of sophistication. Hospitals may take serious heed in OEM claims of heightened complexity and risk. Lofty warranty options and service agreements that capture software upgrades tend to sell the client and take CE/HTM out of the equation. Group purchasing strategies tend to sideswipe CE/HTM because they satisfy the many smaller centers that lack CE/HTM support and pass over the opportunities that are available to the larger centers that use CE/HTM. This is a common occurrence in Canada.

In light of the increased awareness of companies that do not provide adequate supportability, there is still a base of companies that still provide excellent support for inhouse programs. They appear to support the notion that good supportability means increased uptime for their product, adds value, and is representative of good business practice.

Today's devices and systems are becoming increasingly similar from hardware and software perspectives. The level of distinction among competing products and vendors is shrinking. As a result, values and options provided to the customer is becoming more so a critical component of any transaction.

The customer pays for and owns the equipment. They have a right to obtain support mechanisms that are compatible with their abilities and infrastructure (e.g. CE/HTM). However, for reasons not completely understood or thought out, the issue remains contentious.

3 Survey Data

Surveys from both AAMI (2015) and CMBES (2016) have helped to shed light on the topic. The surveys included a request for respondents to prioritize the survey statements (questions). The AAMI 2015 survey showed that for CE/HTM, the top three priorities out of 12 statements were:

1. **Training is inaccessible (not affordable, not available, etc.)**
2. **Service documentation is not readily available**
3. **Products are not designed with supportability in mind**

And for OEMs, they were:

1. **Aptitude of non-OEM individuals working on equipment is unclear/unverifiable/insufficient**
2. **Non-OEM individuals working on equipment make changes (improper replacement parts, etc.) to equipment, thus making it unsafe to use**
3. **Non-OEM technicians in the field don't keep up with technology changes**

This was the first marker to show the polarizing perspectives from each of these stakeholders on the question of supportability. Other issues from the CE/HTM response were: unrealistic manufacturer recommendations for maintenance, the usefulness of service documentation, and the ability to reach phone/human support. Other issues from the OEM response were: concern for repairs being done right and clarity around what customers want in terms of supportability. This has led to some of the AAMI efforts underway which focuses on competency and the type of service level agreement between the OEM the customer.

From the CMBES survey conducted in 2016, the top three statements affecting respondents were:

1. **Access to comprehensive device-specific technical training**
2. **Access to comprehensive service documentation**
3. **Access to diagnostics**

Out of 232 respondents (mostly CE/HTM), responses to statements gauging adequacy of these resources are as follows:

1. **Access to comprehensive device-specific technical training is adequate.**
   Agree 47%
   Neutral 25%
   Disagree 29%
2. **Access to comprehensive service documentation is adequate.**
   Agree 42%
   Neutral 21%
   Disagree 37%
3. **Access to equipment diagnostics (without having to purchase passwords, dongles, service agreements or training) is adequate.**
   Agree 24%
   Neutral 23%
   Disagree 53%

There were other very important responses from the CMBES survey worth mentioning (see Table 2). There is some agreement that Remote Phone Support is adequate (54%) but disagreement that Test Equipment (47%) and disagreement that Regulatory Oversight of Supportability (50%) appears to be significant.

53% agree that obtaining inhouse support for equipment purchased through Group Purchasing Organizations (GPOs) requires more effort than with regular capital purchases. This places a light on how hospital-vendor acquisition processes can affect supportability. CE/HTM's involvement can be impeded based on a central entity that makes decisions in isolation of the larger members of the GPOs most likely to have CE/HTM services.

74% agree that, in the last 5–10 years, changes in technology and design has contributed to a change in the supportability of medical devices. This implies that there may be a certain element in the factor of supportability that will remain constant regardless of any effort undertaken to improve it.

78% of agree that, in the last 5–10 years, business practices of the OEMs (vendors) has contributed to a change in the supportability of medical devices. What may drive this could possibly be linked with the technology question. However, many dispute this as indicated in several of the comments submitted with the survey.

70% agree that, in the last 5–10 years, hospital and procurement practices has contributed to a change in the supportability of medical devices. Decision-makers and trends taking place in purchasing departments may be leaving out CE/HTM in their evolving processes. Some onus is placed on CE/HTM to provide advice and influence on these

Table 2 Responses from the CMBES survey (2016)

Survey statement	Agree (%)	Disagree (%)
Vendors currently provide adequate access to comprehensive technical training	47	**29**
Vendors currently provide adequate access to comprehensive service documentation	42	**37**
Vendors currently provide adequate access to remote phone support	54	20
Vendors currently provide adequate access to equipment diagnostics	24	**53**
Vendors currently provide adequate access to specialized test equipment	27	**47**
Regulatory bodies or standards play a sufficient role to ensure vendors support inhouse service of medical devices	26	**50**
Obtaining inhouse support for equipment purchased through Group Purchasing Organizations (GPOs) requires more effort than with regular capital purchases	**53**	10
Changes in technology/design has contributed to a change in the supportability of medical devices over the last 5–10 years	74	12
Changes in business practices by vendors has contributed to a change in the supportability of medical devices over the last 5–10 years	**78**	6
Changes in business/purchasing practices by hospitals has contributed to a change in the supportability of medical devices over the last 5–10 years	**70**	13
BMETs working inhouse have the abilities and knowledge to support most equipment (assuming support from vendors is adequate)	86	6
BMETs working inhouse continually keep up with technology changes (assuming support from vendors is adequate)	80	9
A list of vendors good at supporting inhouse service should be posted online to improve the supportability of medical devices	**87**	5
A list of vendors bad at supporting inhouse service should be posted online to improve the supportability of medical devices	**78**	8
It is currently difficult to comprehensively support medical devices inhouse	**41**	28
The supportability of medical devices inhouse is becoming more difficult	**69**	16
Prior to the survey, I was not aware there were efforts under way to address the issue (CMBES, AAMI, ACCE, etc.)	54	29

decision-makers. However, when the effort is made on a continuous basis and with no policy change, getting heard becomes a frustrating and futile process.

83% agree that clinical engineers, biomedical technicians, and technologists continue to keep up with changes in technology and are able to maintain ability and knowledge to support medical devices inhouse. There is a bias element to this question. However, the response is likely not an inaccurate representation of the confidence and competency of the respondents.

An especially interesting response was to the question of having an on-line rating system to help improve supportability (from OEMs). A notable 83% of respondents agreed. As to the form in which such a system would take is yet to be discussed. This bold mechanism would directly impact the perception of OEMs and in real time.

A dynamic indicator was built into the CMBES survey in the form of two survey questions. These relate to CE/HTM's ability to comprehensively support medical devices inhouse. 41% agree that it *is* currently difficult. 69% agree that it is becoming more difficult. This indicates where we are and where we may be in the near future. The ability to comprehensively support medical devices inhouse depends heavily on the OEM. It is important to realize that unless mechanisms are instituted now, the benefits of CE/HTM services are at risk.

50% of respondents were not aware of the efforts currently underway in dealing with the supportability question. Publicity on the topic has been relatively minimal considering its importance. Several responses were Neutral because many vendors do a good job on supportability. This quandary left respondents not knowing how to respond. Nevertheless, an average of 38% agreed that these three elements of supportability are adequate.

Despite efforts from the many OEMs that make supportability a priority, results of the AAMI Survey and the significant top three responses that Disagree from the CMBES Survey (Avg 40%), points to a weighty situation. Current efforts, although noble, are having to work fast to catch up with what appears to be a developing tide.

On the international front, the survey did end up in the hands of respondents in thirteen countries outside of Canada. None came from the United States. Although it is a small segment (in the order of single digits) of the overall response (9%), about half of these felt supportability is an issue (see Table 3). This, in combination with a large turnout at the 2015 World Congress—Summit on the Supportability of Medical Devices and comments from that event, indicates that this is likely a global issue.

The World Summit was followed by two additional follow up events in 2017. One at CMBEC40 in Winnipeg, Canada and an open CMBES Webinar held in October. These efforts were organized by Mike Capuano of Hamilton Health Sciences and co-presented by Jean Ngoie, now at the Ninewells Hospital in New Dundee, Scotland. All of these events involved a panel of clinical engineers representing the

Table 3 International response

Survey statement	Agree (%)	Comment
Brazil	1	Issue
Bhutan	1	Issue
Canada	177	Issue
Ghana	1	Issue
Hungary	1	Issue
India	1	No issue
Iran	2	Somewhat
Pakistan	1	Somewhat
Peru	1	No issue
Saudi Arabia	1	Somewhat
South Africa	1	Somewhat
Spain	1	Somewhat
UK	4	Issue
Yemen	1	Issue
Total	194 (out of 232)	

CMBES and in Winnipeg, a panel of OEM representatives (GE, Spacelabs, Drager, and BD) were present to provide their perspectives. The CMBES panel consists of CMBES President, Martin Poulin (Victoria), Kelly Kobe (Calgary), Mario Ramirez (Toronto), Andrew Ibey (Vancouver), Murray Rice (Toronto), and Marco Carlone (Toronto).

Although metrics do well in providing a figurative picture of the issue, the survey comments are much more descriptive and direct. Probably the most profound set of comments in that survey are those directed at hospitals and purchasing departments. Several institutions have seen a change from when everything automatically went to 'Biomed' to now seeing every device or system requiring a project assignment and intense negotiation to obtain the required supports.

The message appears to be that hospital administration needs to better-recognize that CE/HTM is good for the institution and that vendors must be holistically accountable to the healthcare organization.

4 The OEM Lobby

In 2016, the FDA opened a docket for comments (Docket N-0436) on the 'Refurbishing, Reconditioning, Rebuilding, Remarketing, or Remanufacturing of Medical Devices.' The purpose of the docket was to obtain comments and feedback pertaining to the risk of third party service on medical devices. It is believed that it was initiated by representatives of certain OEMs regarding the quality of work performed by these third parties. Nevertheless, the docket, which closed in June of 2016, contained important information on the supportability of medical devices. In a presentation at the 2017

AAMI conference in Austin, Texas; Capuano and Binseng Wang shared the results of an analysis of the docket responses [3]. Out of 171 responses to the docket, 83 recommended improved supportability from OEMs (49%) (see Fig. 1a). This comment category had the highest number of responses compared with all other comment categories (see Fig. 1b).

Conversely, responses in the docket addresses the question of regulation but from a different perspective. Where comments from the CMBES survey centered on OEMs being regulated to provide better supportability, the docket responses focused more on regulating third party servicers. These came mostly from the OEM community looking to achieve an improved method of capturing failure data on their products. It was also seen as a means to give the FDA the ability to register all third party entities that provide services on medical devices. This desire on the part of the OEMs and the FDA has advanced to the political level. With Bill 2118, Congress has asked the FDA to produce a report on how the FDA might implement a process requiring all independent service organizations to register with the FDA, file adverse event reports, and maintain a complaint-handling system. The FDA must submit this report by May 15, 2018. The intent of the bill is to give the FDA 'more oversight and patient protections to the third-party servicing process.' [4]. They want to know how many businesses are 'engaged in servicing medical equipment' and a 'better handle on adverse events to ensure that they never happen again.'

Certain groups are opposed to this indicating that it is costly to the business side and may jeopardize many independent service organizations. Robert J. Kerwin, general counsel for IAMERS (International Association of Medical Equipment Resellers and Servicers), told a Congressional subcommittee that 'this is a solution for which there has been no evidence of a problem.' [4]. Members of the American College of Clinical Engineering (ACCE) and the ECRI Institute have cited their opposition to it as well. Although AAMI remains neutral because of their diverse representation, they did provide an informative response to Docket N-0436 citing the benefits of a well-supported CE/HTM programs (and ISOs). They continue to promote the idea that CE/HTM/ISO and OEMs need to work on resolving issues together. This is evident based on their collaborative efforts thus far.

5 Service Information, Training, and Access to Diagnostics

It is common now to see equipment, traditionally made serviceable in the field, now severely limited in supportability. This is found to be true based on product design, sales

Fig. 1 **a** FDA Docket N-0436
responses recommending
improved supportability from
OEMs. **b** FDA Docket N-0436
responses recommending per
comment category

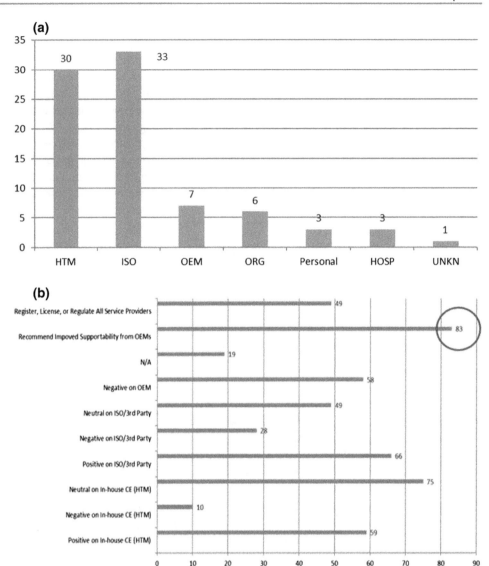

policy, and limits on exactly what information can be obtained. Notations such as, '…not field serviceable,' '… limited number of serviceable parts,' and '…training [is] required to be eligible to receive the service tool' are becoming much more common within the limited service information we are beginning to see (see Figs. 2, 3 and 4). Regarding the training required to get the service tool, the cost of the training mentioned above is very high. In this example, the training was 'on-line' at a cost of $900 CDN per person. For this device, a medium sized inhouse program would likely train 10–12 people, thus accumulating a cost of about $10,000 CDN. To bring someone on-site to do the training it would have costed $12,000 CDN plus $2,000 per person. For 10 people it would have costed $32,000 CDN.

An issue more relevant in Canada is that of obtaining good quality service training at a reasonable cost. Since most of the devices acquired in Canada are manufactured or distributed in the United States or abroad, access to appropriate technical training is more of a challenge to obtain or it comes at relatively higher cost. This is due primarily to the need for Engineers and Biomedical Engineering/Equipment Technicians and Technologists (BMETs) to travel to the United States for factory training or to attend a custom service school in one of Canada's bigger cities. Some companies add a premium (per attendee) taking advantage of customers wishing to train several of their staff. For example, one manufacturer of a common medical device charged about $737 CDN per attendee (no cap). For 19 attendees, it came to $14,000 CDN.

Access to service diagnostics for troubleshooting and quality assurance testing is also a control factor some OEMs employ in order to keep the reigns on their product. Schemes such as requiring a separate license for each device, requiring a purchase order for a one-time use of a passcode, and keeping access away from competent BMETs unless they take an expensive training course.

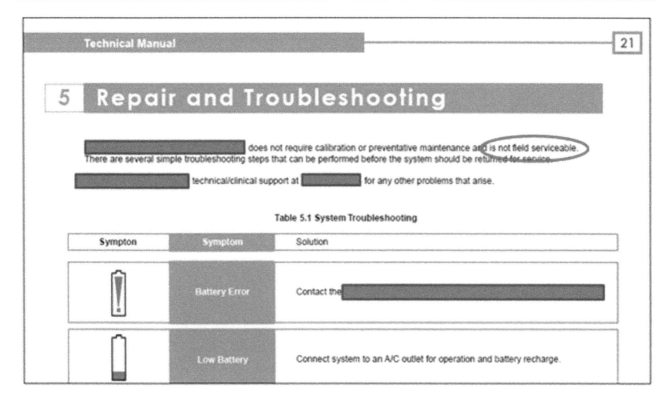

Fig. 2 Limits on field serviceability

There are several reasons that many company representatives will typically cite when attempting to resolve the supportability issue. A popular one is the risk of revealing trade secrets or intellectual property; for example if too much information were to be released. This rationale is seldom an issue especially if the technology is appropriately patent-protected.

Liability concerns are also cited. OEMs continue to cite the potential for litigation should its product be maligned in the hands of an unqualified individual. At the end of the day, every wrongful injury or death suit is based on the unique and individual merits of the case and rarely on any biased presumptions of guilt. In almost all circumstances of equipment related litigations, the forensic evidence usually points to the root cause whether it be a faulty component of design or to the work of an unqualified person.

Another attempt at justification is the reference to FDA requirements (21CFR 820). One should not accept this as a reason an OEM cannot provide service information. Simply put, there is no such requirement.

6 Standards/Regulations

The **NFPA 99 Healthcare Facilities Code (2012)**, used primarily in the United States, provides recommendations concerning service and maintenance of equipment in healthcare institutions. The code indicates what manufacturers 'shall furnish' with the sale of their products [appliances]. In Sect. 10.5.3 it states that documents shall contain at least a technical description, instructions for use, and a means to contact the manufacturer. It also states that illustrations showing locations of controls and step by step procedures for testing and proper use be provided; as well as schematics, wiring diagrams, and repair procedures. This standard, although held in high regard and used religiously throughout the United States, appears to be loosely-followed by a growing number of manufactures. The elements mentioned above are congruent with what we know as a 'service manual;' a term not used in this standard or other similar standards. For example, **CAN/CSA-C22.2 No. 60601-1:08 Medical electrical equipment—Part 1: General requirements for basic safety and essential performance, Section 7.9.2.16** refers to a technical description indicating that 'instructions for use shall contain the information specified in 7.9.3 or a reference to where the material specified in 7.9.3 is to be found (e.g. 7.9.3.3 Circuit diagrams, component part lists, etc.). It also states that 'The technical description shall contain a statement that the MANUFACTURER will make available on request circuit diagrams, component part lists, descriptions, calibration instructions, or other information that will assist SERVICE PERSONNEL to repair those parts of

Fig. 3 A limited number of serviceable parts

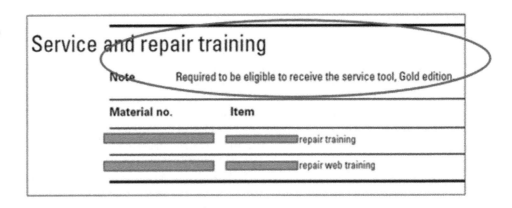

Fig. 4 Training [is] required to be eligible to receive the service tool

ME EQUIPMENT that are designated by the MANU-FACTURER as repairable by SERVICE PERSONNEL.'

Not mentioning the term 'Service Manual' is a missed opportunity. Clearly indicating the need for this type of information would certainly help non-OEM service communities. The standards do little to make manufacturers comply with the need to provide information comprehensive enough to adequately service their product.

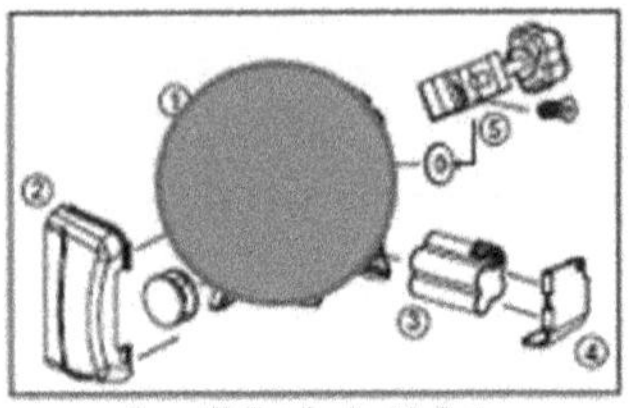

As mentioned previously, the United States Food and Drug Administration (FDA) does not prevent manufacturers from providing service information to their customers. However, it does not require them to provide it either. This does little to help non-OEM entities obtain what they need. They do require OEMs to provide service information for medical lasers as indicated in **Section 21 CFR 1040.10.** It states as follows, 'To servicing dealers and distributors and to others upon request at a cost not to exceed the cost of preparation and distribution, adequate instructions for service adjustments and service procedures for each laser product model.'

Table 4 Effectiveness of standards and regulations

Reference	Effective?
NFPA 99	No
CSA 22.2	No
FDA CFR 21 040.10	No
ISO 13485	No
JC—EC.01.01.01, EP 3	Somewhat

Health Canada has no requirement to make service information available. Its requirement to satisfy **ISO 13485 Quality Standard for Medical Devices** does not address servicing in the field. As indicated in comments from the CMBES survey, some stakeholders wish they did.

At both the 2015 World Congress Summit and the AAMI Forum on Supportability, one of the gaps identified was the absence of supportability covered in the evaluations that ECRI Institute were conducting. ECRI Institute is a well-known and well-utilized nonprofit organization dedicated to the non-biased evaluation of medical devices. In around December of 2015, ECRI Institute began publishing 'Service and Maintenance' as a criteria in their evaluations. This, along with the Joint Commission's latest requirement for HTM programs to house a 'library of information,' [5] may indicate that some of the recent efforts on supportability are paying off (see Table 4). Organizations such as AAMI, CMBES, ACCE, the Joint Commission, and ECRI Institute continue to address the issue of supportability in CE/HTM.

7 Summary

Information from a combination of survey results; industry, organizational and government forums; device documentation, standards, regulations, and home-grown examples of barriers to supportability; all point to an authentic issue. CE/HTM and ISOs are facing challenges more prominent now than once perceived. At this point in time, the issue appears to be at a cusp where Supportability will continue to be increasingly evasive or, due to efforts currently under way, it will subside or improve from its current state. The CMBES is an advocate for the efficient utilization of Clinical Engineers and BMETs to effectively manage and support medical devices in Canada's healthcare institutions. One of its aims is to advance and promote the theory and practice of engineering sciences and technology to medicine. Their members do this by providing safe, cost-effective, and expedient technical and technological services to medical devices in the field. They believe that a healthy presence of relevant CE/HTM and ISO entities greatly benefits healthcare. A stable backing from manufacturers and healthcare institutions is crucial to maintaining the unique and necessary services of the Clinical Engineering/Healthcare Technology department.

Conflict of Interest The author declares that he has no conflict of interest.

References

1. Vockley, M., AAMI Supportability Task Force: Flexible solutions for device supportability, a guide for manufacturers and healthcare technology management professionals. *AAMI Leading Practice* publication, pp. 3 (2014), http://s3.amazonaws.com/rdcms-aami/files/production/public/FileDownloads/HTM/LP_Supportability_2014.pdf, last accessed 2018/01/25.
2. Sherwin, J., Have you seen me?, AAMI *Biomedical Instrumentation and Technology*, Nov-Dec 2012.
3. AAMI News & Views Website, Comments to FDA highlight fault lines in third-party service debate, Posted June 12, 2017, http://www.aami.org/newsviews/newsdetail.aspx?ItemNumber=4808.
4. AAMI News Website, Lawmakers dive into device service debate, June 2017, http://www.aami.org/productspublications/articledetail.aspx?ItemNumber=4765, last accessed 2018/01/25.
5. AAMI News & Views Website, Joint commission's Mills addresses fight over service manuals, Posted June 10, 2017, http://www.aami.org/newsviews/newsdetail.aspx?ItemNumber=4802.

Smart Tourniquet System for Military Use

Erdem Budak, Faruk Beytar, Aytekin Ünlü, and Osman Eroğul

Abstract

Extremities are the most frequently injured regions of body encountered with the combat casualties. The extremity hemorrhages constitute the leading cause of preventable deaths in the first aid period. Thus, tourniquets are indispensable devices for combat casualty care. There are some military tourniquets, which are produced worldwide and can be manually applied by the wound to prevent blood loss. However, in military applications, there is no tourniquet system comprising these features that can be used with one hand, can be applied quickly and transmits information. We have developed a tourniquet system which applies the required pressure to the extremity of the person by moving a belt connected to the pulley with a motor. When the arm or leg buttons on the device are pressed, the system is activated. Once the belt is fitted to the extremity, the system automatically starts the tourniquet process and is continued until the bleeding is stopped. The information of the blood flow and force applied are acquired via the feedback from the motor encoder and the force sensor. The system starts the tourniquet process and the bluetooth transmits the location and application time of the tourniquet. The receiver informs the headquarters via the military communication standard. In this respect, it is possible to be informed about exact location of the injured soldiers in the hot zone. In order to test the developed tourniquet, we have produced the leg phantoms which consist of femur bones and plastics similar to in actual dimensions of the human leg and artificial veins. Tourniquet operation was applied to the point where the tourniquet operated blood flow stopped. It is thought that the developed system will be used in military applications and internal security.

Keywords

Extremity injury • Hemorrhage • Preventable death
Tourniquet

1 Introduction

1.1 Tourniquet Structure and Demand of the Tourniquet Systems

The tourniquet is a medical device that has been used for a century and a half ago and that is being used to stop blood flow in amputee operations and to stop bleeding in the limb injuries. The tourniquet is applied to the proximal part of the limb by a tightened belt or a bandage wider than 3 cm [1]. Although the application of tourniquet seems simple, the correct application of tourniquet is critical to prevent mortality. In the terrorist incidents that took place in our country between 2006 and 2014, 677 security force employees were killed, and 1925 personnel were injured [2]. Data from a combat support hospital shows that 70% of casualties inflicted extremity injuries of which 16.4% required tourniquet application due to concomitant vascular injuries [3]. Combat application tourniquet (CAT) is among the most widely used tourniquets in United States Army's with a reported success rate of 79% [4]. Similar tourniquets are being used in Turkish Armed Forces (TAF). However, in two separate prospective randomized trials with 102 and 145 participants, the efficacy in the lower limb was 50–88% and 70%, respectively [5, 6]. It was observed that if the efficacy was not controlled by an objective pressure threshold, the subjective feeling of pressure by the applier leads to 30% application failure [5]. It is therefore possible that military personnel may still to bleed to death in Turkish Armed Forces (TAF) due to extremity injuries.

E. Budak (✉) · F. Beytar · O. Eroğul
TOBB University of Economics and Technology,
06560 Ankara, Turkey
e-mail: erdembme@gmail.com; edudak@etu.edu.tr

A. Ünlü
Healthcare University, 06018 Ankara, Turkey

© Springer Nature Singapore Pte Ltd. 2019
L. Lhotska et al. (eds.), *World Congress on Medical Physics and Biomedical Engineering 2018*,
IFMBE Proceedings 68/3, https://doi.org/10.1007/978-981-10-9023-3_51

1.2 Tourniquet Systems and Tourniquet Patents in Literature

There are different types of tourniquets developed for military purposes to stop the bleeding in extremity injuries. The Combat Application Tourniquet (CAT) and the Special Operations Forces Tactical Tourniquet (SOFTT) are applied by manually twisting the belt with the windlass. At the Emergency and Military Tourniquet (EMT), the hand pump is manually inflated to create pressure and stop bleeding. Success rates of these tourniquets are 92% in EMT, 76% in CAT and 66% in SOFTT [7]. There are also tourniquets produced for commercial purposes which are Ratcheting Medical Tourniquet (RMT), Mechanical Advantage Tourniquet (MAT) and Combat Ready Clamp (CRoC). There are number of patents such as Tourniquet and Method of Using Same, Tourniquet Timer, Electromechanical Tourniquet for Battlefield Application, Electric Automatic Tourniquet System. Considering tourniquet systems commercially available on the market, there is no system that can be applied with one hand to completely stop the blood flow especially on the lower extremity [8]. It is necessary that tourniquet can reach a constant pressure value and remain at that value for successful application. When the soldier is injured and rapidly loses blood, he tries to wear and squeeze the tourniquet, but it is difficult to adjust the tourniquet with required pressure. Once the tourniquet is applied, it should be loosened for 5 min at intervals of 1 h and let the blood flow of the injured extremity should be ensured. Thus, the risk of possible gangrene is reduced [9]. With this motivation, an intelligent tourniquet system which is named Military Smart Tourniquet (MST) has been produced that can be used for the military field to stop the blood flow on both the lower and upper extremities, can be applied to the arm or leg alone, can show and send the tourniquet's application time, applied position and pressure. Moreover, the MST has been passed the system functionality tests which are tourniquet application between proper time both tightening and loosening, system using in dark environment, stopping blood sample using with a leg phantom and working the MST at least 6 h with a full chargerd battery.

2 Materials and Methods

2.1 Mechanical Design

MST is a mechanism which has a bandage to surround extremities via a pulley. MST is used to block the bleeding on the limbs and besides it applies certain force on extremities. In order to transfer power and movement a worm gear mechanism is used in MTS. The mechanism design was started first evaluating system needs then it was modeled in computer with 3D solid design software (Fig. 1a). MST is an intelligent system that can apply the required force via tourniquet operation on both upper and lower extremities. In humans, the average arm circumference is 288.4 mm and the average circumference of the leg is 349.7 mm [10]. MTS's tourniquet belt has 47.6 mm wide. Hence, it is possible to calculate the application area of the tourniquet with the extremity known for its width and length. For successful application of the tourniquet, a certain pressure has to apply on extremites which is at least 140 mmHg on the upper limb and at least 229 mmHg on the lower limb [11]. There are no other criteria in the literature about the required pressure (mmHg) to stop the flow of blood in extremity injuries. Using with average leg circumference and tourniquet belt width can be reached the area of tourniquet applying region. Therefore, some cascade equations which are given below between 1 and 4, estimated tourniquet applying force can be derived.

$$A = 47.6 \times 10^{-3}.349.7 \times 10^{-3} m^2 \tag{1}$$

$$1\,\text{mmHg} = 133.33\,\text{Pa}, \, 1\,\text{Pa} = 7.5006 \times 10^{-3}\,\text{mmHg} \tag{2}$$

$$P = 229\,\text{mmHg} \times \frac{1\,\text{Pa}}{7.5006 \times 10^{-3}\text{mmHg}} \tag{3}$$

$$F = \frac{229}{7.5006 \times 10^{-3}} \frac{N}{m^2} \times 47.6 \times 10^{-3}.349.7 \times 10^{-3} m^2 \tag{4}$$

According to calculations, the desired force must be applied on the average leg circumference has determined as 508 N in the tourniquet application. Moreover, to tourniquet application a motor is specified which can be applied 2288.28 N in MTS. In this case, the selected motor can be able to apply a force of 1.96 times the targetted force to the muscle with the MST [8]. The mechanical design of the MST (Fig. 1b) is completed by the addition of the MST's electronic control card, user information display, system user buttons, tourniquet belt, tourniquet belt lock mechanism.

2.2 Electronics Design

While the MST electronic design was being made, the requirements of the system were determined first, then the electronic materials needed for the system design were procured and the PCB of the system control board was produced. Simultaneously with the hardware design, the software development process was completed by embedding

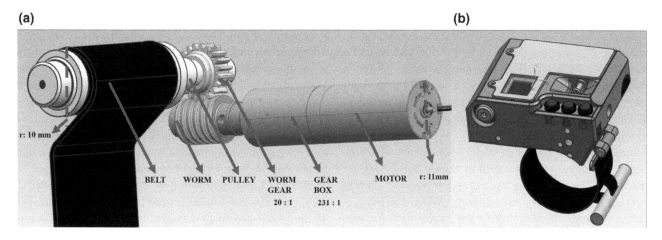

Fig. 1 **a** MST's worm gear mechanism 3D design, **b** MST's 3D design

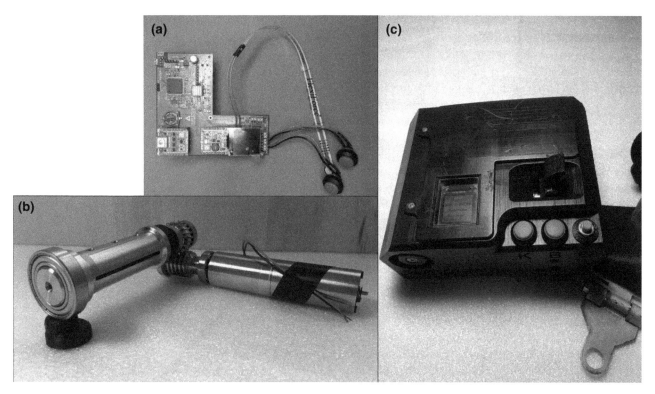

Fig. 2 **a** MST's control card, **b** MST's worm gear mechanism assembly, **c** MST

the software in the designed PCB and ensuring the optimum operation with the peripheral elements. MST comprises of; electronic control card (Fig. 2a), tourniquet mechanism, tourniquet belt's lock mechanism, system battery, user information display and user system control buttons. The ARM Cortex-M4 based STM32F407VGT microcontroller is used in the controller of the MST. The controller takes input from the output of the encoder which calculates the number of rotations of the motor, digital signals are produced by user buttons, the output of the force sensor,

which provides feedback for the user to make the tourniquet process happen. The controller uses different communication protocols during tourniquet operation. It controls the user information display via SPI, the Bluetooth module via USART, the motor driver using with TIMER and GPIO peripherals. The controller, which uses its own FLASH memory to keep the data from the encoder in memory, uses the TIMER interrupt so that the system algorithm can take place at a certain time. The tourniquet time information applied by the injury, the force information applied by the tourniquet taken from the force sensor are sent as a text message to the CENKER system with the Bluetooth module. CENKER is a system designed by ASELSAN, which is integrated with MTS. The CENKER system, which is equipped with hightech products, designed for military personal in the hot zone. With the GPS module mounted on the CENKER system, latitude and longitude information of the injured person, who applies the tourniquet system, is sent to the headquarters with other vital data of the injured via the special military communication standart by means of CENKER. The MST also has safety protocole to prevent applying over force on the extremity. When the MST is started, running algorithm checks the increasing force sensor's data. If algorithm doesn't catch any slope, motor will disable, and system will give an error which indicates the tourniquet doesn't apply properly.

2.3 System Manufacturing

According to the production speed, precision and durability, additive manufacturing the best solution of to produce MST main body. 3D design of MST main body generates vectors based on the spatial proximity of all points which are modeled as rigid bodies. The merging of vectors exists a data cloud file. Slicer software divides the model, which is.stl format, into layers according to the printing quality of the 3D printing device. Coordinates of the points to which the laser beam is transmitted for each layer separated by the stereolithography (SLA) layered manufacturing tech-nique are also created by the resolving functions that works in the background of the Slicer software. All 3D parts of the MST were produced by selecting a layer thickness of 50 μm with downward SLA technique. MST's some parts, which should have high mechanical resistance, has been produced with machining (Fig. 2b). The pulley and worm parts manufactured from stainless stell 304 and worm gear manufactured from CuAl11Ni materials. As a final step, all subcomponents of the MST were combined according to fit the 3D dimensional assembly file. The system integration is completed in Fig. 2c. The whole tourniquet system weight is 1606 gr and it has 75 × 125 × 155 mm dimensions. The MST works with 8.4 V 2S Li-ion battery to drive 6 V DC motor and other electronic components. Aid of internal charging circuit, the MST can be charged only a 8.4 V switching power supply from outside of the cage.

3 Conclusion

In this study, military purpose MST system has been developed. The developed system was electronically tested in computer environment and the system passed all the hardware and software tests. The most important test of tourniquet systems is stopping bleeding. For this, we have developed a limb phantom which is nearly the same with human tissue in terms of mechanical characteristicsm that can be used in flow and pressure tests [12]. With the developed leg phantom, the tourniquet system was tested and the blood flow was stopped with MST. In order to test the MST on humans, a medical device ethics committee permission must be obtained. If necessary permissions are obtained, the system will be tested on human and the accuracy of operation of the system will be determined statistically. Moreover, the MST has been tested in both cold and hot condition according to MIL-STD-810G Method 502.5 and the system has been accopmlished the test in −33 °C and 43 °C stabilize temperature. Thanks to the improved tourniquet system, severe extremity injuries and blood loss can be prevented on the battlefield, resulting in gun injuries and explosions. Accordingly, even if the wounded does not know how to apply the tourniquet, he will be able to easily and correctly apply the tourniquet by putting on the MTS and pressing a button on its wounded extremity. In addition, since the location information of the injury will be reported to the headquarter, the medics will be able to intervene as soon as possible. The smart tourniquet system is not only use in the military field but also it can use in civilian area to prevent emergency injuries. Some researches, which are about on emergency smart tourniquet system, still continue in the Medical Device Design Laboratory where it is in the TOBB University of Economics and Technology Biomedical Engineering Department.

Acknowledgements The developed tourniquet system has been funded by The Scientific and Technological Research Council of Turkey as a science—industry project which has "Military and Emergency Field Smart Tourniquet (Harp ve Acil Yardım Akıllı Turnikesi—HAYAT)" name and "0932.STZ.2015" project no. Besides, we thanks to project industry partner of ASELSAN A.Ş. which is a Turkish Armed Forces Foundation company.

References

1. Klenerman, L. (1962). The tourniquet in surgery, Journal of Bone & Joint Surgery, British Volume 44.4 937–943.
2. The Gendarmerie General Command activity reports in 2007/2014, http://www.jandarma.gov.tr/duyurular/faaliyet_raporu/FLASH/index.html, last accessed 2018/01/28.
3. Role 2 military hospitals: results of a new trauma care concept on 170 casualties. Ünlü A, Cetinkaya RA, Ege T, Ozmen P, Hurmeric V, Ozer MT, Petrone P. Eur J Trauma Emerg Surg. 2015 Apr;41(2):149–55.
4. Kragh J., J. F., vd. (2008). Practical use of emergency tourniquets to stop bleeding in major limb trauma, Journal of Trauma and Acute Care Surgery, 64.2 S38–S50.
5. Unlu, A., vd. (2014). An evaluation of combat application tourniquets on training military personnel: changes in application times and success rates in three successive phases, Journal of the Royal Army Medical Corps jramc-2014.

6. Unlu A, Petrone P, Guvenc I, Kaymak S, Arslan G, K.aya E, Yılmaz S, Cetinkaya RA, Ege T, Ozer MT, Kilic S. Combat application tourniquet (CAT) eradicates popliteal pulses effectively by correcting the windlass turn degrees: a trial on 145 participants. Eur J Trauma Emerg Surg. 2017 Oct;43(5):605–609.

7. Kragh J., J. F., Walters, T. J., Baer, D. G., Fox, C. J., Wade, C. E., Salinas, J. & Holcomb, J. B. (2008). Practical use of emergency tourniquets to stop bleeding in major limb trauma. Journal of Trauma and Acute Care Surgery, 64(2), S38–S50.

8. Budak, E. İ. (2017). Development of Tourniquet System For Military Use (master thesis). TOBB University of Economics and Technology, Ankara, TURKEY.

9. Sapega, A. A., vd. (1985). Optimizing tourniquet application and release times in extremity surgery. A biochemical and ultrastructural study, The Journal of Bone & Joint Surgery 67.2 303–314.

10. Gavan, J. A. (1950). The consistency of anthropometric measurements. American journal of physical anthropology, 8(4), 417–426.

11. McEwen, J. A. & Inkpen, K. (2004). Surgical tourniquet technology adapted for military and prehospital use, BRITISH COLUMBIA UNIV VANCOUVER, CANADA.

12. Budak, E., Beytar, F., Özdemir, M., Susam, B. N., Göker, M., Ünlü, A., & Eroğul, O. (2017). Lower limb phantom design and production for blood flow and pressure tests, The EuroBiotech Journal, volume 1, issue 4, 278–284, October 2017.

Using Data Mining Techniques to Determine Whether to Outsource Medical Equipment Maintenance Tasks in Real Contexts

Antonio Miguel-Cruz⊙, Pedro Antonio Aya-Parra⊙,
William Ricardo Rodríguez-Dueñas⊙, Andres Felipe Camelo-Ocampo,
Viena Sofia Plata-Guao, Hector H. Correal O.,
Nidia Patricia Córdoba-Hernández, Angelmiro Nuñez-Cruz,
Jefferson S. Sarmiento-Rojas⊙, and Daniel Alejandro Quiroga-Torres⊙

Abstract

The purpose of this study was to determine whether the maintenance of medical equipment should be outsourced (or not). For this, we used data mining techniques called decision trees. We (1) collected 2364 maintenance works orders from 62 medical devices installed in a 900-bed hospital; (2) then we randomly selected 90% of the maintenance works orders to train 8 different decision tree schemas (J48 (pruned and unpruned), Naive Bayes tree, random tree, alternating decision tree, logistic model tree, decision stump, REP tree); (3) next, the remaining 10% of the works orders were used to test the decision tree schemas. The relative absolute error was used to evaluate what the tested decision tree schemas had learned; finally (4), we chose the decision tree schema with the lowest relative absolute error. Overall, the decision tree schemas performed well. 62.5% (5/8) of the decision tree schemas had less than 20% relative absolute error. 87.5% (7/8) of the decision tree schemas had more than 90% in the correct classification (whether to outsource maintenance tasks or not). The different tested decision tree schemas showed that the most important variables when making the decision whether to outsource maintenance tasks or not were: medical device, risk class (I, IIA, IIB, III), complexity, obsolescence, maintenance frequency, service time and outsourcing. The best decision tree schema was the logistic model tree (LMT) with 14.6628% relative absolute error and 94.7034% in the correct classification.

Keywords

Clinical engineering • Outsourcing
Maintenance management • Data mining • Decision tree

1 Introduction

Outsourcing is defined as "contracting with a third-party supplier for the management and completion of a certain amount of work, for a specified length of time, cost and level of service" [1, p. 8]. According to KPMG's Shared Services and Outsourcing Advisory, in the first quarter of 2017, the defense, telecommunications and healthcare sectors increased the number of their outsourcing contracts by 25% compared to the last quarter of 2016. According to a report on medical device outsourcing by Grand View Research, outsourcing was valued at US$ 33.2 billion in 2016 and is expected to grow by 11.5% by 2025 [2].

Although outsourcing has grown in recent years, making the decision to outsource a process is not a simple one. Several factors such as cost, quality, safety, the availability of spare parts, skilled personnel and resource optimization must be taken into consideration [3, 4].

For this reason, it is essential to make use of tools that allow better decisions to be made when outsourcing maintenance tasks. Some of these tools are mathematical models such as the analytical hierarchy process technique (AHP),

A. Miguel-Cruz (✉) · P. A. Aya-Parra · W. R. Rodríguez-Dueñas ·
A. F. Camelo-Ocampo · V. S. Plata-Guao · J. S. Sarmiento-Rojas ·
D. A. Quiroga-Torres
Biomedical Engineering Program. School of Medicine and Health Sciences, Universidad Del Rosario, Calle 63D # 24-31, 7 de Agosto, Bogotá D.C, Colombia
e-mail: Antonio.miguel@urosairo.edu.co; miguelcr@ualberta.ca

A. Miguel-Cruz
Department of Occupational Therapy. Faculty of Rehabilitation Medicine, University of Alberta, 2-64 Corbett Hall, Edmonton, AB T6G 2G4, Canada

H. H. Correal O. · N. P. Córdoba-Hernández · A. Nuñez-Cruz
Biomedical Engineer Maintenance Department, Corporación Hospitalaria Juan Ciudad, Calle 24 # 29 45, Bogotá D.C, Colombia

L. Lhotska et al. (eds.), *World Congress on Medical Physics and Biomedical Engineering 2018*,
IFMBE Proceedings 68/3, https://doi.org/10.1007/978-981-10-9023-3_52

linear programming (LP) or data envelopment analysis (DEA) [5, 6]. On the other hand, there are also decision-making methods by machine learning tools or data mining such as decision trees, clustering or neural networks [5, 7]. However, there is almost no research on the application of some of these tools in the medical device maintenance field [8]. For this reason, the aim of this research was to implement and determine the best tree-based prediction model that allows the managers of clinical engineering departments in healthcare institutions to make decisions about outsourcing maintenance tasks.

2   Methodology

The methodology proposed by Miguel-Cruz et al. in [9] was used to develop this research with regard to the data collection, and to the selection and operationalization of the variables.

2.1   Data Set

The unit of analysis for this investigation was the maintenance tasks obtained from a data set in a hospital (now referred to as Hospital "A"). Hospital "A" is a third-level high-complexity private hospital with more than 900 beds and with a total of 2295 medical devices in its inventory. A total of 2394 maintenance tasks were carried out on 62 medical devices in the hospital. The devices included in our sample and its maintenance tasks originated from two different areas in the hospital: diagnostic imaging and the intensive care unit. We selected these hospital areas because the technical complexity, equipment types, and equipment acquisition costs involved in these areas makes them the most likely to have maintenance services contracted to an external provider [10].

2.2   Data Collection Procedure

The data were collected by two research assistants. The following variables were collected for each maintenance task: equipment type, obsolescence, technological complexity, equipment acquisition cost, date of acquisition, equipment risk, frequency of maintenance, whether the maintenance was outsourced (by an external provider) or not (in-house), the maintenance type (corrective or preventive) and the service time (in hours). To avoid bias, all the technology users, clinical engineers, external maintenance service providers and research assistants were trained at the beginning of the study on how to collect the maintenance data.

2.3   Selection and Operationalization of the Variables

Dependent variable (outcome variable). The output or dependent variable of this study was whether the maintenance was outsourced (*OutSource*). This is a nominal variable. It was operationalized as '1' if the maintenance was outsourced, otherwise the code was set to '0'.

Independent variables. The independent variables used in this research and its operationalizations are shown in Table 1.

2.4   Learning Algorithms: The Decision Trees Selected

We used WEKA's machine learning library [6] as the processing tool for the algorithms applied in this research. Eight algorithms were used: AdTree, Decision Stump, REPTree, NBTree, J48 (pruned and unpruned), LMT and Random Tree. We selected decision trees as learning schemas because they can handle data of different types, including continuous, categorical, ordinal, and binary, and uses non-parametric statiscial methods, as a result do not require normality assumptions of the data [6].

2.5   Experimental Setup and Assessment of the Decision Trees' Performance

We evaluated the learning algorithms using the method proposed by Witten and Frank. We used two data sets, one for training and one for validation. We randomly divided the data set using 90% of the maintenance tasks for training, and the remaining 10% to validate the 8 decision tree algorithms used [10]. We used the cross-validation test to avoid bias in the algorithms. The characteristics used to evaluate the decision trees were the correct classification percentage and relative absolute error.

3   Results

3.1   Descriptive Statistics

In total, 62 medical devices from two areas in hospital "A" were reported to have had maintenance tasks performed on

Table 1 Variables and values

Variables (codes)	Values
Technological complexity (TComplex)	High, medium and low
Medical device risk (MedDRisk)	Class I, IIa, IIb and III
Whether the maintenance type task was unscheduled/corrective (MaintType)	Corrective, Preventive
Technological obsolescence (TObsolec)	Exploitation time/Useful life ([0,n] n∈ℜ+)
Maintenance frequency transactions (MainTFreq)	Times/year [0,n], n∈ℜ+
Service time in hours (ST)	[0,n], n∈ℜ+

Table 2 Results of the decision tree algorithms

Tree name	Parameters	Correct classification (%)	Relative absolute error (%)
LMT	-I 1 M 15 -W 0.0	94.7034	14.6628
RandomTree	-K 0 -M 1.0 -V 0.001 -S 1	93.8983	15.7941
J48 Unpruned	-U -M 2	94.322	17.6848
REPTree	-M 2 -V 0.001 -N 3 -S 1 -L 1 -I 0.0	93.9831	18.9753
NBTree	N/A	93.7977	19.8825
J48	-C 0.25 -M 2	93.9407	20.27
AdTree	-B 10 -E 3 -S 1	91.8644	42.5171
DecisionStump	N/A	75.4237	66.0995

them. 78.3% of the devices were classified as medium-high and high-risk in their class. 96.6% of the devices were classified as medium or high technological complexity. The results also showed that 39.2% of the maintenance tasks were corrective maintenance, whereas 60.8% were preventive maintenance. The mean obsolescence of the medical equipment was 0.7354 ± 0.4442. Overall, the maintenance frequency had a mean of 2.34 SD 0.806 times/year. The mean service time was 2.8599 SD 9.65 h. Finally, 47% of the maintenance tasks were not outsourced, while 53% were performed by an external service provider.

3.2 Assessment of Decision Tree Algorithms

Table 2 shows the results of the performance of the decision tree algorithms used in the research. The best decision tree schema for this study is highlighted in gray (i.e. the LMT algorithm)

Figure 1 shows highlighted in gray the decision tree with the smallest relative absolute error and the best correct classification percentage (i.e. the LMT algorithm). It is not possible to show the mathematical models of the LMT equations resulting from the decision tree algorithm in Fig. 1 due to a lack of space.

4 Discussion

This document explores the implementation of different decision tree algorithms to determine whether medical device maintenance should be outsourced. The best algorithm found in this case was the logistic model decision tree

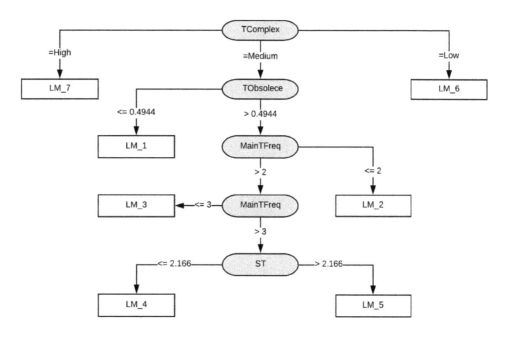

Fig. 1 Decision tree from the LMT algorithm. Note: the LMT equations are available on request

(LMT) algorithm. This is consistent with investigations [11] which use logistical regression models to define subcontracting in industry. Although the logistic model decision tree algorithm was able to determine whether maintenance tasks should be outsourced in 94.7034% of cases, there are some ways this study could be improved. To achieve higher levels of validity and generalization, future research should focus on testing the decision tree algorithms on hospital data sets other than those used in training and validating a model.

5 Conclusion

It was found that the best algorithm to determine whether maintenance tasks should be outsourced or not is the logistical model tree algorithm (i.e. the LMT algorithm), thus demonstrating that it is possible to use data mining tools for decision making in the clinical engineering field.

Conflicts of Interest The authors report no conflicts of interest.

References

1. Oshri, I., Kotlarsky, J., Willcocks, L.P. The handbook of global outsourcing and offshoring: the definitive guide to strategy and operations, Third edn., Hampshire: Palgrave Macmillan, 2015, p. 8.
2. Grand View Research. Medical Device Outsourcing Market Analysis By Application, By Service, And Segment Forecasts, 2014–2025. Grand View Research Inc., 2017.
3. Bertolini, M., Bevilacqua, M., Frosolini, M. An analytical method for maintenance outsourcing service selection. International Journal of Quality & Reliability Management 21(7), 772–788 (2004).
4. Falsini, D., Fondi, F., Schiraldi, M.M. A logistics provider evaluation and selection methodology based on AHP, DEA and linear programming integration. International Journal of Production Research 50(17), 4822–4829 (2012).
5. Gerdes, M. Decision trees and genetic algorithms for condition monitoring forecasting of aircraft air conditioning. Expert Systems with Applications 40, 5021–5026 (2013).
6. The University of Waikato: Weka 3: Data Mining Software in Java, Machine Learning Group at the University of Waikato, 2016. Available at: https://www.cs.waikato.ac.nz/ml/weka/, last accessed 2018/01/22.
7. Chen, R.Y., Sheu, D.D., Liu, M.Y. Vague knowledge search in the design for outsourcing using fuzzy decision tree. Computers & Operations Research 34, 3628–3637 (2007).
8. Miguel-Cruz, A., Rios-Rincón, A.M. Medical device maintenance outsourcing: Have operation management research and management theories forgotten the medical engineering community? A mapping review. European Journal of Operational Research 221, 186–197 (2012).
9. Miguel-Cruz, A., Rios-Rincón, A.M., Haugan, G.L. Outsourcing versus in-house maintenance of medical devices: a longitudinal, empirical study: Rev Panam Salud Publica, 35(3) 193–199 (2014).
10. Cruz, A.M., Haugan, G.L., Rincon, A.M.R. The effects of asset specificity on maintenance financial performance: An empirical application of Transaction Cost Theory to the medical device maintenance field. European Journal of Operational Research 237 (3) 1037–1053(2014).
11. Witten, I.H., Frank, E., Hall, M.A. Data mining practical machine learning tools and techniques, Burlington: Morgan Kaufmann Publications (2011).

Set Up for Irradiation and Performing Spectroscopy for Human Lenses

Fernanda Oliveira Duarte, Márcio Makiyama Mello, Mauro Masili, and Liliane Ventura

Abstract

Introduction: In the environment with natural solar irradiation, the part of the electromagnetic spectrum that should be most attentive when it comes to eye health is the ultraviolet. Studies with both animals and cells have shown that chronic exposure of the eyes to ultraviolet radiation (UVR) causes significant damage to the eyes structures, particularly the cornea, lens, and retina. Unfortunately, still today, there is controversy regarding the harm caused to the human eye due to exposure of the ocular environment to UVR. There is great methodological difficulty with the use of human lenses, which leads to the increase of controversial results. Aim: Development of a device for that the human lens can be irradiated in a solar simulator and analyzed in spectrophotometer for determination of the UVR effects in the eye. Methodology: A lens holder has been developed using a mathematical model and a 3D printer. Results: The result obtained after ZMorph 3D printer was a very light and effective holder for fitting the lenses as well as the cuvette. Conclusion: The development of this holder will allow different experimental protocols to be performed with the human lens, once this support decreases the need for excessive manipulation of the lens in relation to other supports found in the literature, avoiding the degradation of the tissue. We believe that the development of this holder will contribute in a promising way for future research with human lens, assuring reliable results. Financial support: FAPESP (2013/08038-7 and 2014/16938-0).

Keywords

Human lenses • Solar simulator • Lens holder
Ultraviolet radiation

1 Introduction

The lens of the eye performs specific and complex functions that depend on involving different biological pathways. It is an avascular, transparent, and delicate tissue that have all cells derived from proliferation and differentiation of lens epithelial cells [1]. The most part of the lens is comprised of uniquely elongated fiber cells, which can reach lengths of up to 1000 microns. Fiber cells produce abundant levels of crystalline (about 90% of the water-soluble proteins of the mammalian) essential for transparency and refractive properties by a gradient of uniform concentration on the lens [2].

The lens grows throughout entire life even with lens cell division on a very limited scale, just in epithelial cells, and with a pattern of greater growth in the embryo than in adults. During the late stage of fibrous cells differentiation, there is degradation of the endoplasmic reticulum, Golgi apparatus, mitochondria and nuclei, what is also imperative for lens transparency. Any alteration or destruction of this process will imply in the dispersion of light and, in the last case, in the development of the cataract. Considering that, the differentiated cells do not have the capacity to produce new proteins. That way, maintenance of lens proteins is crucial for ocular health and for the prevention of lens opacities [1–3]. However, due to the lens location along the optic axis of the eye, this structure is chronically exposed to environmental stress from solar radiation. In this context, in the environment with natural solar irradiation, the part of the electromagnetic spectrum that should be most attentive when it comes to eye health is the ultraviolet. Studies with both animals and cells have shown that chronic exposure of the eyes to ultraviolet radiation (UVR) causes significant damage to the eyes structures, particularly the cornea, lens, and

F. O. Duarte · M. M. Mello · M. Masili · L. Ventura (✉)
Department of Electrical Engineering, University of Sao Paulo (EESC), Av. Trabalhador Saocarlense 400, Sao Carlos, SP 13566-590, Brazil
e-mail: lilianeventura@usp.br

retina [4–8]. The spectral energy increases with decreasing wavelength and the potential damage associated with energy increases exponentially. In this way, the solar UVR exposure, many times inevitable and routine, makes it one of the most relevant environmental agents capable of damaging the DNA to which humans are exposed. In addition, the chromophores present in the cells also absorb UVA and UVB photons. However, still today there is controversy regarding the harm caused to the human eye due to exposure to UVR. In this context, it is noteworthy that human lenses are quite different in many important ways from their laboratory animal counterparts [9]. Nevertheless, most of the studies involving the lens are carried out with animals creating a gap in relation to the results for humans. Besides that, extraction methods and manipulation of the lens can affect the lens proteins, resulting in tissue opacification, which also limits the studies. Hence, we believe that studies that mimic the natural environment should be conducted in order to verify the real effect of UVR on the human lens. For such studies to be developed, it is necessary to eliminate the possible degradation and over-manipulation of human lenses. Therefore, the objective of this study was to develop a specific holder for the human lens that facilitates the lens handling during the conduct of different experimental protocols.

2 Materials and Methods

2.1 Human Lens Prototype

A lens prototype was created to be used as a geometrical reference to the lens holder. To create the prototype geometry of the lens, a mathematical model was used, proposed by Kasprzak [10]. The method can be used for modelling optical properties of the lens and is based in hyperbolic cosine type function. It consists in an approximation of the anterior and posterior lens profile with a hyperbolic cosine function and is given in polar coordinates. Consider the hyperbolic cosine function in polar coordinates as in Eq. (1), with a, b and d arbitrary constants:

$$(\varphi) = [\cos h(b\varphi) - 1] + d, \qquad (1)$$

where d denotes the axial distance between the origin and the curve, Kasprzak [10] developed the equations and found that:

$$a = d\,R_0(R_0 - d)/[3(R_0 - d)(2R_0 - d) - R_0^2], \qquad (2)$$

$$b = (1/R_0)\sqrt{[3(R_0 - d)(2R_0 - d) - R_0^2]}, \qquad (3)$$

where R_0 is the central radius of curvature. The posterior profile of the lens is given by:

$$\rho_P(\varphi) = (a_P/2)[\cos h(b_P\varphi) - 1][\tan h(m(s_P - \varphi)) + 1] + d \qquad (4)$$

and the anterior profile by:

$$\rho_A(\varphi) = (a_A/2)[b_A \cos h(\pi - \varphi) - 1][1 - \tan h(m(s_A - \varphi))] + d, \qquad (5)$$

where m describes the slope of the hyperbolic tangent, s_P and s_A describe the respective shifts of the hyperbolic tangents in relation to the origin. The final lens profile is:

$$(\varphi) = \rho_P(\varphi) + \rho_A(\varphi) - d. \qquad (6)$$

Thus, to create our lens prototype we used the constants [10]: $R_{0P} = 6$ mm, $R_{0A} = 10.2$ mm, $= 2$ m, $m = 8$, $s_P = 1.68$, and $s_A = 1.715$. The model was implemented in MATLAB®, in which the lateral profile, plotted in x-z plane, is shown in Fig. 1a. The model was reconstructed in Solidworks®, shown in Fig. 1b.

2.2 Cuvette and Spectrophotometer Optical Path

The analysis of the lens will be performed by a spectrophotometer CARY 5000 from Varian. The lens will be in the vertical position inside the holder, which in turn was designed to fit inside a cuvette. The lens holder must place the optical axis of the lens coincident with the optical path in the spectrophotometer. The standard quartz cuvette, 10 mm optical path, and 3.5 mL, from Varian, was used as a reference. The spectrophotometer irradiates the cuvette with a monochrome slit.

2.3 Solar Simulator Temperature Test

The solar simulator internal temperature test was performed to ensure that the material resists the heat and is appropriate to hold the lenses without damaging the tissue. The test was done in the 450 W, ozone-free xenon arc lamp solar simulator (LEMA, Italy). According to previous studies by our research group [11], the irradiance provided by the solar simulator is 460 W/m^2, at a distance of 30 cm from the lamp. A digital thermometer was placed inside the solar simulator at 5, 12, 20, and 30 cm from the lamp. This thermometer has an accuracy of ±0.5 °C for the actual temperature range. At each position, the simulator was switched on for 25 h and the temperature was recorded at 5-s intervals. The recorded maximum temperatures were 61.9 ± 0.5 °C at 5 cm from the lamp and 40 ± 0.5 °C at 30 cm from the lamp, both well below the vitreous transition temperature of the polylactic acid-PLA (melting point in the 150–160 °C range). We chose PLA for making the holder, a standard plastic used in 3D printing.

Fig. 1 **a** Lens profile with hyperbolic cosine approximation; **b** Cross section of the lens profile

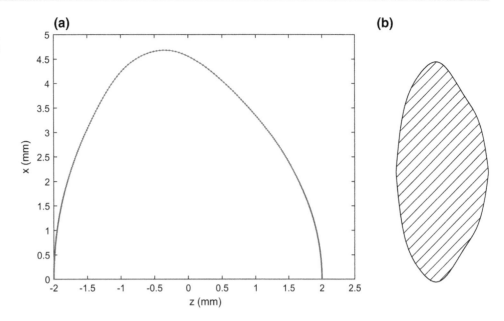

3 Results

Each piece of the lens holder is shown in the Fig. 2a and b.

The holder was created in two pieces, which were placed together with a free space inside to hold the lens (lens bed), Fig. 2a and b. The lens bed was designed based on the lens prototype. The lens bed has an extra offset of 0.5 mm due to the imprecision of the 3D printer. One piece is longer, to easily disassemble it and to remove it from the cuvette. The pieces have concentric holes for the optical path of the CARY slit. They also have four mechanical aligners, which hold them together and the lens stays firmly positioned inside the pieces.

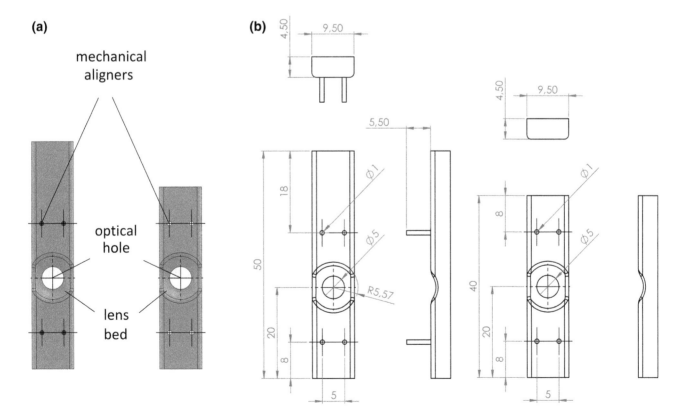

Fig. 2 **a** Front view with definitions; **b** Orthographic projection drawing with dimensions

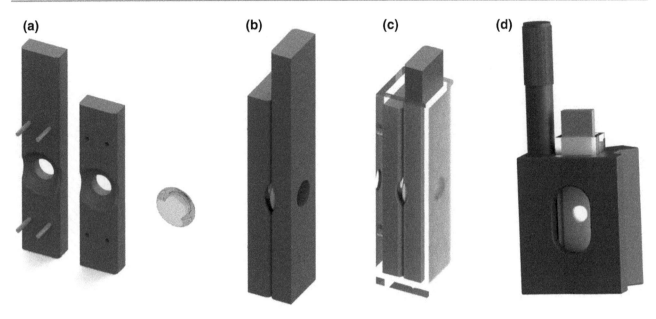

Fig. 3 **a** Lens prototype and disassembled holder; **b** lens holder assembled with lens prototype inside; **c** lens holder inside the cuvette; **d** lens holder in the spectrophotometer holder

Fig. 4 Diagram of the assembled system: the designed holder with the lens inside and placed in a cuvette for spectrophotometric analysis

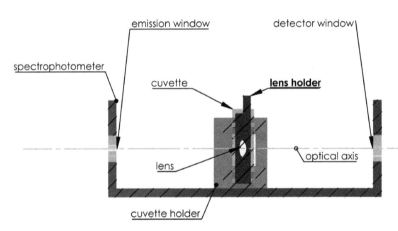

The pieces were printed in PLA in 20 min, with a ZMorph 3D printer, obtaining a very light and precise holder (Fig. 3) for fitting the lens, Fig. 3a and b and the cuvette, Fig. 3c. The lens holder was centered for the optical path of the spectrophotometer and the holder perfect fitted the cuvette, Fig. 3c and spectrophotometer holder, Fig. 3d. They were sanded and painted with black diffuse spray to reduce light inner reflection in the spectrophotometer.

The height of the lens holder hole was designed to match perfectly with the center of the spectrophotometer's slit window, schematically shown in Fig. 4.

4   Discussion

Due to the great difficulty that researchers have in relation to the handling and degradation of human lenses, there is a limitation in demonstrate the effects of ultraviolet radiation

on the human lens. In fact, in the scientific literature, studies suggest that ultraviolet rays cause damage to human lenses. However, these results are controversial and, most often, performed on lenses of animals rather than humans. Our intention when developing the lens holder is to decrease the difficulties of handling and conservation of the lens eye during experimental tests, permitting working with human lenses and avoiding controversial results.

Investigators suggest that the lens dimensions changes during ageing [12, 13]. Augusteyn [14], in an in vitro study, demonstrated that the lens equatorial diameter increases from around 8.8 mm at age 20–10.2 mm by age 99 while total sagittal thickness increases from around 4.2 mm at age 20–5.3 mm at age 99. From the previous information and using the mathematical model suggested by Kasprzak [10], we developed our lens prototype, which was used as a model for the lens holders made by us. It is important to note that the lens does not allow hooks, wires or any apparatus be attached to it without opacification of the tissue, which also makes manipulation difficult. Thus, with this in mind, the holder has been developed so that the lens fits perfectly inside it, without undergoing pressure and deformation, avoiding changes in the tissue. The advantage of this holder in relation to others is that the lens can be fitted in the vertical position (like a standing person) without undergoing pressure of its own weight, avoiding flattening of the lens at the bottom of the support. Besides that, the lens does not need to be manipulated during the experimental procedures, avoiding degradation and opacification of the tissue. Indeed, we find in literature lenses holders that is necessary a lot of manipulation [15], the use of hooks or wires attached to the lens or even the necessity of the maintaining other structures attached to lenses, limiting the study of different experimental protocols [16, 17]. These results reinforce the importance in lens preservation once the lens proteins are not turned over, characterizing the proteins in the center of the lens as the oldest proteins in the body [1–3].

There is an urgent need for studies that prioritize the human lens to understand the effects of ultraviolet radiation on this eye structure. As discussed previously, the difficulty of manipulation, the use of animals and, even experimental protocols that do not represent the natural environment, highlighted in researches conducted with the lens are limitations for results extrapolation to humans [9]. In this way, the holder for the lenses comes to increase the range of experimental options that can be performed with the human lens. In addition, it will be possible to mimic the natural environment, for example using a solar simulator, since the support stays inside a cuvette, allows the use of preservative substances for the lens.

In this context, to know with accuracy what the possible damages caused to lens due to chronic ultraviolet radiation exposure, will allow strategies to be developed aiming at eye protection. In addition, the use of sunglasses is an option to protect against ultraviolet rays. However, preliminary results from our research group indicate that ultraviolet protection degrades with the use and exposure of sunglasses to natural UVR. The results are extremely important because the inadequacy of sunglasses may not guarantee the safety of the eye structures. This holder will allow further studies of the human lenses and hence, the standards may be adapted accordingly. In fact, according to the Brazilian standard NBR ISO 12312-1:2015 [18] and international BS EN1836:2005; ANSI Z80.3-2009; AS/NZS 1067:2003 [19, 20], sunglasses must have protection filters against UVR according to the transmittance values allowed by wavelength. It is relevant to know that when the user wears sunglasses, the eye does not suffer mydriasis (pupil does not contract) and the eye stays without the natural protection of limiting the UVR that enters it. Thus, if the glasses do not have adequate protection, UV rays will reach the inner

structures of the eye in greater quantity, causing eye damage. In this sense, since studies demonstrating the effects of UVR in the lens are controversial, some UV protection standards for sunglasses do not consider necessary that lenses have full protection against ultraviolet radiation (219–400 nm). This reinforces the need for the studies with human lenses and exalts the relevance of the lens holder for future researches.

Therefore, we believe that from future results, we can know with certainty the effect of chronic exposure to ultraviolet solar radiation in the eyes, contributing for adequacy of standards and ISO regarding the use of sunglasses. Ensuring that sunglasses have adequate UV protection may be an important strategy to avoid the development of eye diseases.

In conclusion, the designed lens holder will permit that different experimental protocols can be conducted with human lenses because it facilitates the handling of lenses, avoiding excessive manipulation of the lens, preventing degradation and opacification of the tissue, generating reliable results.

Conflicts of Interest The authors declare that they have no conflict of interest.

References

1. Andley UP. The lens epithelium: focus on the expression and function of the alpha-crystallin chaperones. Int J Biochem Cell Biol. 2008;40(3):317–23.
2. Usha P. Andley. Crystallins in the eye: Function and pathology. Progress in Retinal and Eye Research 26 (2007) 78–98.
3. Hejtmancik JF, Riazuddin SA, McGreal R, Liu W, Cvekl A, Shiels A. Lens Biology and Biochemistry. Prog Mol Biol Transl Sci. 2015;134:169–201. https://doi.org/10.1016/bs.pmbts.2015.04.007.
4. Truscott RJ. Age-related nuclear cataract-oxidation is the key. Exp Eye Res. 2005 May;80(5):709–25.
5. Andley UP, Weber JG.Ultraviolet action spectra for photobiological effects in cultured human lens epithelial cells. Photochem Photobiol. 1995 Nov;62(5):840–6.
6. Wang F, Gao Q, Hu L, Gao N, Ge T, Yu J, Liu Y. Risk of eye damage from the wavelength-dependent biologically effective UVB spectrum irradiances. PLoSOne 2012;7(12):e52259. https://doi.org/10.1371/journal.pone.0052259.
7. Mercede Majdi, Behrad Y. Milani, Asadolah Movahedan, Lisa Wasielewski and Ali R. Djalilian. The Role of Ultraviolet Radiation in the Ocular System of Mammals. Photonics 2014, 1 (4), 347–368; https://doi.org/10.3390/photonics1040347.
8. Andley UP1, Malone JP, Townsend RR. Inhibition of lens photodamage by UV-absorbing contact lenses. Invest Ophthalmol Vis Sci. 2011 Oct 21;52(11):8330–41. https://doi.org/10.1167/iovs.11-7633.
9. Roger J.W. Truscott. Age-related nuclear cataract—oxidation is the key. Experimental Eye Research 80 (2005) 709–725.
10. Henryk T. Kasprzak. New approximation for the whole profile of the human crystalline lens. Ophthal. Physiol. Opt. Vol. 20, No. 1, pp. 31–43, 2000.

11. Masili M. and Ventura L. Equivalence between solar irradiance and solar simulators in aging tests of sunglasses; BioMed Eng OnLine (2016) 15:86 - https://doi.org/10.1186/s12938-016-0209-7 .

12. Robert C. Augusteyn. On the growth and internal structure of the human lens. Experimental Eye Research 90 (2010) 643e654. https://doi.org/10.1016/j.exer.2010.01.013.

13. Alexandre M. Rosen; David B. Denham; Viviana Fernandez; David Borja; Arthur Ho; Fabrice Manns; Jean-Marie Parel; Robert C. Augusteyn. In vitro dimensions and curvatures of human lenses. Vision Research 46 (2006) 1002–1009. https://doi.org/10.1016/j.visres.2005.10.019.

14. Augusteyn RC. On the contribution of the nucleus and cortex to human lens shape and size. Clin Exp Optom. 2018 Jan;101(1):64–68. https://doi.org/10.1111/cxo.12539.

15. Weale RA. Age and the transmittance of the human crystalline lens. J Physiol. 1988 Jan;395:577–87.

16. Trokel S. The physical basis for transparency of the crystalline lens. Invest Ophthalmol. 1962 Aug;1:493–501.

17. Artigas JM1, Felipe A, Navea A, Fandiño A, Artigas C. Spectral transmission of the human crystalline lens in adult and elderly persons: color and total transmis-sion of visible light. Invest Ophthalmol Vis Sci. 2012 Jun 26;53(7):4076–84. https://doi.org/10.1167/iovs.12-9471.

18. Associação Brasileira de Normas Técnicas. Proteção pessoal dos olhos—Óculos de sol e filtros de proteção contra raios solares para uso geral. Brasília: NBR ISO 12312-1:2015, 2015.

19. British Standards Institution, Europe. EN 1836, May. 2006.

20. Concil of Standards Australia, Australia. AS/NZS 1067, Apr. 2003.

Global CE Success Stories: Overview of 400 Submissions from 125 Countries

Yadin David and Tom Judd

Abstract

Health Technology (HT) is vital to global health care. The dependence of health, rehabilitation, and wellness programs on technology for the delivery of services has never been greater. It is essential that health technology be optimally managed. Clinical and biomedical engineers have been recognized by WHO as essential to providing this management. At the 1st International Clinical Engineering and HT Management Congress and Summit held in China in 2015, a resolution was adopted by the global Clinical Engineering (CE) country participants to identify and promote our unique qualifications, and to record the CE contributions to the improvement of world health status. A first group of CE Success Stories was captured, 150 from 90 countries—from the prior 10 years and presented to health leaders at the WHO World Health Assembly in 2016. In 2017, another 250 with a total of 125 countries were added from 2016–2017 presentations.

Keywords

Clinical engineering • Technology management
Outcomes • Success stories

1 Introduction

Health Technology (HT) is vital to health; the dependence of health, rehabilitation, and wellness programs on HT for the delivery of their services has never been greater. Therefore, it essential that competent and trained professionals manage in an optimal and safe way for better response to the burden of diseases and resources. Trained Clinical Engineers (CEs) are

academically-prepared and appropriately responsible for HT life-cycle management, fulfilling a critical role as members of the healthcare team focusing on availability and reliability of safe and effective technologies and outcomes.

Over the past fifty years growing concerns among CE professionals about lack of knowledge by government agencies and key stakeholders, coupled with the mute recognition for their vast contributions to the safe and effective creation and deployment of HT, led to programs that address these concerns. Knowledge about and recognition for the professionals of CE community who provide critical services will help recruit students and future practitioners into this needed field. Is CE practice important for health, rehabilitation, and wellness programs and are their contributions recognized? This paper shares the methodology and findings found following a three-year examination of published evidence.

Following the international congress on CE and HT management in Hangzhou, China in 2015, a Global CE Summit took place to determine whether regional issues are shared around the world and present common international challenges requiring global strategy for best dissemination. After order ranking the shared issues at the end of the Global CE Summit, the attending members voted that the (1) lack of understanding of and recognition for the CE contribution to improvements in healthcare delivery was a major concern for practitioners in the field, following by (2) the lack of sufficient education and training for both entering the field and for ongoing professional development. An action plan was devised to address these and other issues raised at the Summit. At the second global CE summit in Sao Paulo, Brazil, in 2017, these challenges were reviewed and confirmed with attendees adopting resolutions seeking to continue to address these concerns. The Summits' action plans included first, data collection identifying if CE contributions qualify as improvement to world health and wellness and can it be substantiated through evidence-based records. Addressing the second issue, an international survey of Body of Practice (BOP) and Body of Knowledge (BOK) was initiated and has been completed.

Y. David
Biomedical Engineering Consultants, LLC, Houston,
TX 77004, USA

T. Judd (✉)
IFMBE CED Secretary, Marietta, GA 30068, USA
e-mail: judd.tom@gmail.com

© Springer Nature Singapore Pte Ltd. 2019
L. Lhotska et al. (eds.), *World Congress on Medical Physics and Biomedical Engineering 2018*,
IFMBE Proceedings 68/3, https://doi.org/10.1007/978-981-10-9023-3_54

2 Methods

2.1 Rationale

A task force consisting of senior certified CEs from IFMBE/CED issued a global call for submissions of evidence-supported case studies of CE contributions to the improvement of delivery of healthcare services or of patient outcomes. In 2016, of the submitted studies, an aggregate volume of 150 responses from 90 countries was examined and found qualified as evidence-based contributions, see http://global.icehtmc.com/publication/healthteachnology.

Results were rated and tabulated into categories (Innovation, Improved Access, Health Systems, HT Management, Safety and Quality, and e-Technology) and incorporated into document that was submitted to WHO's World Health Assembly in May 2016, see http://global.icehtmc.com/publication/globalsuccess.

We expanded our review in 2017, as submissions and publications continued to collect, to include within our examination review of conference-accepted published data that was presented and published at IFMBE sponsored events. Our examination methodology identified 250 additional stories from 35 more countries—now raising the total volume over two years to 400 publications from 125 countries. These CE success stories point to improved outcomes with benefit from HT, and present overall demonstration of complex integrated systems that must be effectively managed for their optimal and safe clinical and business impact to be realized. Clinical outcomes included change in human life quality, care management decisions support, improving $365 \times 24 \times 7$ readiness, and reducing operational costs.

2.2 Definitions

For the present study, we classified the collected database into six categories with definitions:

- **Innovation**
  Through provision of new HT solutions, adaptation of existing, or a combination to address several issues.
- **Improved Access**
  Ease in reaching HT-related health services or facilities in terms of location, time, and ease of approach.
- **Health Systems**
  Positive impact from more efficient and effective deployment of HT at national or policy level.
- **Safety and Quality**
  HT's positive impact on health services safety or quality outcomes, or through HT human resource development.
- **Healthcare Technology Management (HTM)**
  Establishing or improving HTM methodology resulting in improved population health or wellness.

- **e-Technology**
  Improvements achieved due to deployment of internet-based health technology tools.

2.3 Measures

A successful project (or submission) was defined as satisfying two objective measures developed by the sponsors. These measures included timeliness, cost saving, deployment or adoption by care providers, impact on services, and overall projection for success. Each success metric was evaluated using three-point scale against a statement representing the success construct (1 = strongly disagree; 3 = strongly agree).

- Timeliness refer to whether the project/submission was implemented in timely manner. This was measure by the statement "The submission will impact outcomes in present time."
- The cost measure was evaluated whether the submission overall costs were within budget constraints and reasonable for the conditions in the region. This was assessed by statement of "The submission cost objectives can be met in the region".
- The next two combined into the statements "The submission will be deployed by its intended users" and "The submission will have positive impact on those who will adopt it."
- Finally, overall submission success expectations were assessed with the statement "All things considered, the submission will be a success."

Innovation is the beginning of the technology life cycle where new ideas offering solutions to current problems faced by healthcare providers or their patients. Clinical engineers are well positioned to understand the current problems and guide different or new approaches to resolve them. Innovation, in our category, meant to demonstrate the team approach to solving problems all the way from a concept and building a prototype and continuing with clinical trials and demonstration of compliance with standards, regulations, and intended outcomes. Improved Access to services follows the innovation stage same as Safety and Quality category, e-Technology category and HTM. Products and applications that are considered in successful deployment were rated high and included in category count of evidence-based category.

3 Results

Summaries from the six categories of submissions database are described below; from CED's 2016 Health Technologies Resources [1] document provided to the World Health Assembly, WHO's May 2017 Third Global Forum on Medical Devices [2]; (3), CED's September 2017 Sao Paulo II ICEHTMC [3] (S), and Others [4] from 2016–2017 IFMBE published sources (O):

3.1 Innovation

3.1.1 2016 = 17 Total; 13 + Countries

USA	India	Ethiopia	Australia	Uruguay
MG Hospital re Test for Superbugs [1]	Sharma, AMTZ—Local Production of Medical Devices [1] and MOH Cultivating Innovation [1]	Optimal design of O2 concentrators [1]	New Blood/Fluid Warner design [1]	Simini, CE driving Facility Design [1]
WHO: LRC Innovations [1], HT for Ebola Care [1]	**Bangladesh**: HT Policy and HTM [1]	**South Africa**: Local Production of HT in Africa [1]	**Italy**: Robotic Surgery [1]	**Colombia**: Business Opportunities in HT; Castaneda [1]
WHO, Technical specifications O2 concentrators [1]	**Malaysia**: Biomechanics for Amputee [1]	**Tanzania**: MCH (Maternal Child Health) rural HT [1]	**Peru**: Heavy Metals Detection [1]	**Canada**: Provincial Respiratory Outreach [1]

3.1.2 2017 = 61 More; 16 + Countries

WHO/Global	India	USA	UK	Brazil	Colombia
G-PATH (20 countries) MCH [2]	G-Prasant, GANDI—needs driven innov. [2]	G-MGH re LMIC Innovation [2]	G-Un.Oxford Child Pneu. Diag [2]	G-Orthostatic chair [2]	O-Cipro Pharm modeling and Circuits [4]
G-Priority devices, WHO HQ and EMRO [2]	G-Hypothermia alert for newborns-Bempu [2]	G-Early detection Brst Canc—Un WA [2]	G-Endo GI canc scrg Leeds Un [2].	G-Photometric test gestatational. age [2]	O-ECG signal modeling [4]
G-UNICEF: how we drive innovations [2]	G-Prevent Apnea prematurity-Bempu [2]	G-Field test neo phototherapy Little Spar [2]	**Norway, Denmark**	G-Prematurity light detection [2]	O-Mechanical knee modeling [4]
G-EPFL Digital X-ray [2]	G-Remote Monitor. critical infants-Bempu [2]	G-Test pre-eclampsia Un. OSU-Geneva [2]	O-Tiss engr impr. monitoring [4]	S-Travel ECG Telemedicine [3]	O-Mech ventilation from Pesticides [4]
G-IARC Thermal Coagulator [2]	G- AMTZ 2-HT Policy impr. Svc Del. [2]	G-MGH—Africa post-part hemo.. [2]		S-Flow Analyzer Blood Pump [3]	O-Parkinson EP Study [4]
G-e-stethoscope child Pneu diag, Un.Gen [2]	**Bangladesh**	**Senegal**	**Italy**	S-BME aid diagnose pathologies [3]	O-Respiratory system Simulation [4]
G-UNICEF devices for MCH, LaBarre [2]	G- Jaundice Screening, Harvard BWH [2]	G-O2 concentrators [2]	G-Phototherapy. and Transfusions [2]	S-Dig storage for surgcial video [3]	O-Permanent Magnet drug delivery [4]
G-WHO TB Diagnoses [2]	**China**	G-Inn. devices Inf Dieasess—FIND [2]	O-Current trends in HTA [4]	S-Hi Flow Nasal Therapy [3]	O-Parkinson EEG study [4]
G-WHO Digital Health Investmts [2]	S-Renal GRF Estimation		O-HTA Tablet for Dig Path. [4]	S-Locating HT via WiFi [3]	
G-WHO Assistive Devices - Gate [2]	**IFMBE**	**Australia**	**Croatia**	S-Remote Equip. Monitoring [3]	**Chile**
G-WHO Emergency Care [2]	G-BME and HTA Pecchia [2]	G-O2 storage [2]	O-Support Diabet. Pts remote mon. [4]	O-Resp control with exercise [4]	S-Clinical Simulation for Facility open [3].
G-HWO Emer NCD kit Refugees [2]	G-CED Innovation/Standards, David [2]			**Mexico**	
G-WHO Innovation in LRC 2 [2]	G-CE-IT Innovation, Castaneda, Judd [2]			G-Hand Orthosis [2]	

3.2 Access

3.2.1 2016 = 12 Total; 12 + Countries

WHO/Global	India	Africa	Australia	Argentina
Y David, the Rise of Telehealth [1]	Sharma, MOH Mobile Med Units [1]	Bartolo, Italy GHT: Mozam, Tanz., Malawi, Togo, DRC [1]	Sloane, Telehealth for Diabetes Care in Austral. and Canada [1]	Giles, HT for Provincial Access [1]
WHO Cancer Care initiative [1]	Sharma, MOH Free Diag. and Natl Dialysis [1]	**China**	**Albania**	**Cuba**
	Khambete, Telemedicine for Blindness [1]	Zhang Hanzhong, Ventilator access [1]	Picari, HTM for Diagnostic access [1]	Gonzalez, Telemed for Chronic Diseases [1]

3.2.2 2017 = 20 More; 9 + Countries

WHO/Global	India	Denmark	Syria	Brazil
G-WHO AFRO Kniazkov, HT Situation	G-Ameel, HTA and Nationall Innovation Portals	G-Andresen, Mobile Lab, IFBLS	S-Almonhamad Dialysis [3]	S-Lima, access to CT imaging [3]
G-WHO EMRO, Ismail, Strengthen. HT	**NGOs**: G-Cummings, PATH, Mkt Dynam and HT Dec. Making [1]	**Slovakia**: O-Lehocki, Telemedicine Diabetes [4]	**Kenya** G-Abbam, GE Kenya PPP	S-Mansur, access to mammograms [3]
G-WHO, Johnson, Emerg.. Surgery [1]	G-Prasad, IAEA, Radiation Therapy [1]	**Romania**: O-Sebesi, Tele-monitor elderly [4]	S-Galvin, Tele-diagnoses for coverage [3]	S-Valadares, Role of CE in Hosp. Network [3]
G-Butany, Lab sys, Un. Of Toronto [1]	G-HTTG Borras et al., Imaging Centers in Rural Areas [1]		**Mexico**: S-Ayala, CENETEC planning [3]	S-Martins, Linear Accelerator Study [3]

3.3 Health Systems

3.3.1 2016 = 37 Total; 22 + Countries

WHO/Global	India	Nigeria	Brazil
Worm, THET 1 partnerships [1]	MOH HTM via PPP, Sharma [1]	Esan, HT policy for HTM [1]	MOH RENEM invest., Conto [1]
Calil, WHO db CE natil. assns. [1]	**Singapore**: Toh, Global BME Ed. [1]	**Kenya**: Anyango, MOH HTM [1]	**Chile: Diaz**, Univ. of Valparaiso HT leadership [1]
YD, FH, Global HT Disast. Prep [1]	**China**: Gao, Inner Mongolia HTM [1]	**Ethiopia**: ideska, HTP and HTM [1]	**Colombia**: Garcia, HTP, HTR, HTM [1]
Hernandez, LA and C HT Trg [1]	**Taiwan**: Lin, Accred of BME/CE [1]	**Sierra Leone**: D Williams, Donations [1]	Molina, CES HT acquisition [1]
Worm, Lin. IFMBE dev coun [1]	**Japan**: CE Roles, Cert, Yoshioaka [1]	**Ghana**: Adjabu, CMBES Donations 1 [1]	**Mexico**: Ayala, National HTM, CENETEC [1]
Issakov, HTP Africa Health [1]	**Vietnam**: Tan, Survey of Roles for HTM [1]	**Cameroon**: Riha, HTP development [1]	Cardenas, HTA, HTR, HTM [1]
Voigt, global CE education [1]	**Indonesia**: Badri, BME Education development [1]	Ngaleu, HT investments [1]	**Peru**: Rivas, CENGETS, NIH, MOH for HTM [1]
THET, HTM in LRC 2013 [1]	**Italy**: AIIC Society Success Story 1 re HTM [1]	**Turkey**: Ugur, MOH track, price, HTM [1]	Rivas, Government Collaborative partnerships [1]
WHO MOH HT Indicators, Nagel [1]	**Albania**: Picari, MOH HT and HTR (HT regulation) [1]	Bilal, MOH HTM [1]	**Suriname**: Jie, MOH HTM [1]
WHO and ILO re BME/CE, Velaquez [1]			

3.3.2 2017 = 70 More; 30 + Countries

Global	India	Africa	Albania	Portugal	Brazil	Mexico	Colombia
G-A Lemgruber Med dev AMRO [1]	G-M. Gupta, SE Asia Reg. Perspective [1]	G-Worm, THET 2 HTM: Mal, Rwa, Benin, DRC [1]	S-Picari, HTR and survellance [3]	G-Maia, MRI purchase [1]	S-Ramos, CE curricula [3]	G-National HTP, Ayala [1]	G-Galeano, Univ. CE trg [1]
G-Velazquez WHO global HT Atlas [1]	G-Volun. HT Cert., Ameel [1]	S-Poluta, HTM in Africa [3]	Romania: G-Materio-vigilance [1]	Czech Rep. O-Kubatova, Hosp HTA [4]	S-Marciano, Selection process [3]	S-Bravo, CE Value Project Chain [3]	S-Inter Univ CE coop model [3]
G-WHO HT for Un. Hlth Cover., Kiely [1]	Taiwan: G-Lin, Intern BME Ed [1].	S-Mboule, Regional HM expert [3]	Moldova: O-Sontea—HTM [4]	Greece: G-Pallikar. HTM [1]	S-Fernandes, Intl MD standards [3]	S-Vernet, BME Society on CE [3]	S-Torres, replace planning [3]
G-HR book for BME, Velazquez [1]	S-Lin, CE MD Surveillance [3]	O-Worm, Potential of CE assns [4]	Italy: G-C Pettinelli, EURO HTR [1]	Bosnia, Her. G-Metrology [1]	S-Pires, CT Scan exam time [3]	Argentina S-Lencina, CE Present/future [3]	S-Garcia, CE Regional. nodes [3]
G-WHO Tech specs [2]	O-Lin, BME Ed in Asia [4]	Ghana—S-Adjabu, Donations 2 [3]	S-AIIC SS 2, Lago et al. [3]	Cuba: O-Trade barriers [4]	S-Melo, distribution of defibs [3]	IFMBE: G-CE SS, David, Judd [2]	O-Guerrero, Purchasing HT cycles [4]
G-Ismail, HTR in EMRO [2]	China: G-Bao Jiali, CE in China [2]	Bangladesh: G-Ashrafuzzaman—BME/CE develop [2]	O-Luschi, CIS/PACS Cardiology/ [4]	NGOs: G-Gutierrez, HTAi 2020 [2]	S-Calil, FDA standards International. [3]	G-Global CE Day, Iadanza [2]	O-Ortiz, Hosp Infrastructure. managemet [4]
S-AV, Role of HTM in WHO [3]	NGOs: G-HUMATEM, approp.. HT [2]	S-Ashrafuzzaman CE for MDR [3]	UK: O-Dan Clark, fund gap [4]	G-HUMATEM, BME trg [2]	S-Zaniboni, public acquisition [3]	G-CE Prof Soc. J Goh, E Iadanza [2]	O-Diaz, BME programs in LA and C [4]
S- AV, Role of HTM in UHC [3]	G-HUMATEM, Donations [2]	Japan: G-Igeta, Role of CE [2]	Peru: G-Emer Prep Rivas [1]	G-G Jiminez, MSF Med Eq Frmwk [2]	O-Fagundes, cost est procure [4]	G-Iadanza, CED Proj. 1[2]	G-Leandro P, HTA in LRC [2]
G- TIfenn H, EURO HT [2]	O-Judd, Hern. emerg ldrs help [4]	Kyrgyzstan - G-TIfenn H, EURO Med dev [2]	O-Luis, Comp HTM system [4]	S-Pecchia, HTAD fill HTA-HTM [3]	Canada: S-Ramirez, CE-HTM [3]		O-Iadanza CED Projects 2 [4]

3.4 Safety and Quality

3.4.1 2016 = 23 Total; 33 + Countries

WHO/Global	China	Jordan	Burkina Faso	Mexico
WHO BME HR book [1]; Easty, HFE book [1]	Shanghai, MD QC [1]	Six Sigma Case Study [1]	West Africa HC org: Cape Verde, Senegal, The Gambia, Guinea Bissau, Guinea [1]	Macias, Mex state Directorate [1]
Mullally, global trg partnerships [1]; HT Maint. Book, Wang [1]	Shenglin, Web based CMMS [1]	Kuwait: Al-awadhi, HT [1]	Sierra Leone, Liberia, Mali, Ivory Coast, Ghana, Togo, Benin, Burkina Faso, Nigeria, Niger [1]	Colombia: Garcia, MOH HTR [1]
Cheng, image of CEs and safety [1]	Japan: Mugitani,, HT policy [1]	New Guinea: QI and MCH [1]	Portugal/Mozambique–Secca, Training program in centra; Hosp. [1]	Brazil: Calil, CE Ed [1].
Worm, THET, managing the lifecycle [1]	Samoa-Fiji: Kapadia, HT user care [1]	Egypt/Italy: Saleh, PM prioritie [1] s	Germany—Tech Surveillance of. Infusion systems [1]	USA—Wear CE Cert. [1]
WHO HT indicators, Nagel 2009 [1]			Australia: Anne L-Smith, medical air misconnections [1]	Painter, CE Risk Mgmt [1]

3.4.2 2017 = 37 More; 22 + Countries

WHO/Global	China	UK	Kenya	Brazil
G-Abbam, GE, adopt HT in poor infrastructure [2]	S-Zingyi, Dyn Warning system consumables [3]	G-Young Pneumonia prevention [2]	G-Tsala, IVD product verification [2]	S-Vaz, False alarms in ICU [3]
G-Abbam, GE, skill devel. in emerging mkts [2]	S-Jing-ying Gao, Case StudyHT Mgmt Improv [3].	**Bosnia and Herg.**	**Dominican Rep.**	S-Oliveria, Waste mgmt [3]
G-WHO, HTR in the Americas, Lemgruber [2]	S-Zheng Kun, Vent alarms ICU [3]	O-Gurbeta, Tracking Inspection Processes [4]	S-Hernandez, Medical Gas Policy MOH [3]	S-Brito, Cold Chain HTA [3]
USA: S-Wang, CE RiskM, QI, Asset Mgmt [3]	**Taiwan**: G-Lin, Post market surveillance [1]	**Italy**: S-Marchesi, Digital Processes [3]	**IFMBE**: O-Wear, Credentialing [4]	S- Espinheira, Med Washer Disinfector [3]
S-Y David, Prevent Adverse Events [3]	**Saudi Arabia**	O-Iadanza, RM thermal ablation [4] [4]	G-David, CED global credentialing	S-Vincente, HT Surveillance
S-Y David, Hosp Integrated Networks RiskM	G-Nazeeh Alothmany, HT counterfeit [2]	**Greece**: O- Malataras, MD Vigilance	**Brazil**: O-Tsukahara, Calil, RCA HF [3]	S-DMR, Radiology report;
S-Easty, HT Risk Manage. using Hum Factors [3]	**Iran**: O-Ramezani, case study ESU [4]	S-Pallikarakis, HT Surveillance and HT Reg [3].	S-Santos, air compressor management [3]	S-Reatigui, Hosp Accred Radiology [2]

3.5 e-Technology

3.5.1 2016 = 22 Total; 16 + Countries

WHO/Global	China	Africa	Botswana	Kenya	Brazil
Wang et al., Global HTM, 2011 [1]	Zhou Dan, CE Cert. Impact [1]	THET and CEASA 2015 [1]	Tlhomelang - HTM [1]	Anyango— Opth maint. [1]	Santos, HTA applied to HTM [1]
Judd et al., Global HTM 2015 [1]	**Taiwan**: Chien, Intl. HTM model [1]	15 countries ↓ Burundi, DRC	**Uganda**: Mulepo, HTP and HTM for MOH [1]	Rugut— HydroC Refrig trg	**Cuba**: Castro, HTM
Clark, CED e-Course training [1]	**Bhutan**: Penjore, HTM and HTA [1]	Cameroon, Ethiopia; Ivory Coast, Nigeria [1]	**Sierra Leone**: Kabia— MOH HTM [1]	**Tanzania**: Mwizu, HTM [1]	**Chile**: Acevedo, Navy HTM [1]
Cordero—Orbis training [1]	**Laos**: Insal, HTM [1]	**South Africa**: Khalaf, HTM Math [1]	**Burkina Faso**: Emmanuel—MOH HTM [1]	Werlein, incr. mgmt. cap [1]	**Dominica**: Williams, HTM [1]
Hernandez et al., 2013 WHO [1]	**Kyrgyzstan**: Agibetov - HTM [1]	**Sudan**: Hassan, MOH HTM [1]	**Benin**: Soroheye - HTM [1]; **Gambia**: Nyassi - HTM [1]	**Ghana**: Adjabu HT Donations [1]	**El Salvador**: Juarez, HTM [1]
USA: Davis-Smith, KP CE Staff Best Practices [1]	**Australia**: Anne L-Smith, Endoscopy HTM [1]	**Rwanda**: Worm— HTM and training [1]	**Togo**: MCH Donations [1]	**Paraguay**: Galvin, HTM [1]	**Haiti**: Valliere, HTM [1]
Kosovo: Boshnjaku - HTM [1]	**Saudi Arabia**: Alkhallagi: HTM e-Library [1]	**Zambia**: Mullally —HTM and training [1]	**Senegal**: Sow, HTM Qual. Management [1]	**Puerto Rico**: Misla - HTM [1]	**Jamaica** Richards, HTM [1]

3.5.2　2017 = 26 More; 10 + Countries

Africa	China	Italy	Brazil	Colombia
G-Worm, Improved HTM Rwa, Mal, Ben, DRC [2]	S-Lu He-qing, PM fetal mon. [3]	S-Gemma, Italy WHO CC re HTM [3]	G-Avelar, CE impact in primary care [2]	S-Berrio, CE Finan. Mgmt in Hosp. [3]
Ethiopia: G-Desta, HTM [2]	S-Shaozhou Guang HTR [3]	O-Bibbo, Technical HTM [4]	S-Ferriera, Rain Forest HTM [3]	O-Cruz, Optimal Maint [4]
G-Desta, Warranty PM [2]	**Taiwan**: S-Lin, MD troubleshtg [3]	**Romania**: O-Corciova, HTM [4]	S-Petsgna, HTM based Inv. [3]	O-Torres, teach HTM [4]
Benin: G-Soroheye, Eval of devices [2]	O-Chen, in hosp HTM [4]	**Lebanon**: G-Farah- St George Hosp HTM [2]	S-Carneiro-HTM based on Man.; S-Bascani, HTM [3]	**Chile**: S-Avendano, HTM Un. Val. [3]
Nigeria: S-Esan, HTM challenges [3]	**Bangladesh**: S-Hossain, HTM in MIC [3]	S-Farah, MD Repair, Repl. [3]	S-Anderson, HTM Apheresis [3]	**Costa Rica**: S-Esquivel, HTM [3]
NGO: G-Smith, PATH—HT contracts [2]	**NGO**: G-Worm, Cordero, Role of BMETs [2]	**Bosnia and Herz**: O-Gurbeta-Dialysis HTM [4]	S-Silva, Sphyg. Manual vs Dig.; S-Neto Hosp. HTM [3]	**Ecuador**: S-Matamoros, HTM in Guayaquil [3]
	UK: G-Basit, CE apprentice [2]	**Peru**: S-Rivas, HTM enhance quality [3]	S-Davoglio, CT tube life;.; S–Pesregar, Mat Fet. HTM [3]	**Mexico**: S-Fernandez, HTM Priv Hosp [3]

3.6　e-Technology

3.6.1　2016 = 22 Total; 16 + Countries

WHO/Global	India	Spain/France	Italy
Clark, HTM on-line trg LA and C [1]; Gentles, Use of global CMMS [1]	Rausch, ICE data for HTM [1]	Quintero, EHR GUI and biosensors [1]	Tagliati, Int LIS in Lgiuria Region [1]
Hernandez, CE-IT Trends LA and C [1]	Sharma, CE IT in Developing Countries [1]	**Bulgaria, Greece**	Iadanza, CE from Developmt. to Systems book [1]
USA	**Japan**	Malataras, Bliznakov CMMS [1]	**Colombia**
Davis-Smith, Risk-KP CE-IT [1]	RTLS for eq location	**Portugal**	Quintero, Intro IHE in LA and C
Grimes HT Cybersecurity [1]	**IFMBE**	Abreu, Virtual Sensor Nodes [1]	**Venezuela, Ecuador**
Fraai, CE-ITI in EMR [1]	Sloane, CED Global CE-IT [1]	**Slovakia**: Jadud, MOH e-cat. MD [1]	Intell. Sys for PT ID, Silva [1]
Sloane, Judd, Dig hospitals, Saudi Arabia, Macedonia [1]	Knuma-Udah HT. Expert systems [1]	**Georgia**: Ubiquitous sensors [1]	**Haiti**: Judd, HTM strategies [1]

3.6.2　2017 = 26 More; 10 + Countries

WHO/Global	India	Greece	Brazil
G-Gentles et al., CMMS in LRC [2]	G-Rausch, ICE data for HTM [2]	G- Pallikarakis, Web CMMS [2]	G-Melo Telecomm mobile health [2]
G-Raab, CMMS in LRC [2]	**China**	**Italy**	G-Melo, Tele-radiology in Amazon [2]
G-Sloane, ICT Cybersecuirty [2]	G-Bao Jiali, Mobile NCDs [2]	S-Iadanza, Decision Support systems [3]	S-Osmam, BI in HT management [3]

(continued)

WHO/Global	India	Greece	Brazil
G-Noel, Medtronic Integrated Health [2]	S- Liu Jingxin, remote med imaging [3]	**Romania**	S-Lustosa, Geo-coding Dengue [3]
G-Olla, Leprosy mobile phones, Audacia [2]	**South Africa**	O-Fort, Wireless [4] trans and pt mon	S-Souza, Trans-op data with EMR [3]
S-Judd, Dev in Global CE-IT [3]	G-Khalaf, med IoT	**NGOs**	S-Garcia, dental chair HTM [3]
S-Sloane, Total Cost Ownership CE-IT [3]	S-Khalaf, Wireless body sensors [3]	G-Denjoy, COCIR, Interop and Digital Health	**Colombia**: G-Garcia, Networking among natl. CEs
O-Sloane, MD and ICT convergence [4]		**Uruguay**: O-Decia, HT location systems [4]	**Canada**: G-Ngoie, HTM mobile. apps [2]

4 Conclusions

Health Technology (HT) is vital to health; the dependence of health, rehabilitation, and wellness programs on HT for the delivery of their services has never been greater. Beyond the ongoing healthcare burdens of population growth, political and economic instability, disease management, disasters, millions of refugees, accidents, and terror attacks, world healthcare technological systems are facing enormous challenges to be innovative and optimally managed. The transition into health programs to the 21st century requires trained competent clinical engineering professionals. This paper describes the extensive study of published data on the vast contributions by CE that positively impact patient outcomes. This study shows that every region of the world including low resource regions face a challenge of improving health services while facing varied levels of infrastructure and human resources challenges. CEs play vital roles in all stages of healthcare technology life cycle management. From creation to planning, and from commissioning to utilization and integration; technology-based systems must and can be managed for optimal performance. In each of the life cycle stages requirements for trained and competent CE input makes critical difference as shown in the analyzed evidence reviewed here. It is our hope that government agencies and other interested parties will have better understanding of CEs role and thus will support their inclusion in the healthcare team of professionals.

References

1. IFMBE Clinical Engineering Division (CED), *Health Technologies Resource*, May 2016, http://cedglobal.org/global-ce-success-stories/
2. World Health Organization Third Global Forum on Medical Devices, May 2017, http://www.who.int/medical_devices/global_forum/3rd_gfmd/en/
3. IFMBE CED 2nd International Clinical Engineering and Health Technology Management Congress (II ICEHTMC) Proceedings, September 2017, http://cedglobal.org/icehtmc2017-proceedings/
4. Other IFMBE related CE Papers, http://cedglobal.org/global-ce-stories-other/

Links

1. WHO HQ http://www.who.int/medical_devices/en/ WHO EMRO http://www.emro.who.int
2. WHO AMRO http://www.who.int/about/regions/amro/en/
3. WHO Digital Health http://www.who.int/medical_devices/global_forum/Thedigitalhealthaltas.pdf
4. WHO Assistive Devices-GATE—https://mednet-communities.net/gate/
5. WHO Emergency www.who.int/medical_devices/global_forum/Essentialresourcesemergencycare.pdf
6. WHO NCD Kit Refugees http://www.who.int/medical_devices/global_forum/NCDkitrefugees.pdf
7. IFMBE, CED, HTA http://ifmbe.org/, http://cedglobal.org/ http://htad.ifmbe.org/
8. PATH https://www.path.org/ (Belgium, China, DRC, Ethiopia, Ghana, India, Kenya, Malawi, Mozambique, Myanmar, Peru, Senegal, RSA, Switzerland, Tanzania, Uganda, Ukraine, Vietnam, Zambia)
9. AWHP www.ahwp.info; Asian Harmonization Working Party - 30 countries, 3/17 Regulatory Authorities
10. HTAi https://www.htai.org/

Bed Management in Hospital Systems

E. Iadanza, A. Luschi, and A. Ancora

Abstract

The paper presents a design for a bed management web-application to efficiently provide for the allocation of beds inside hospitals to reduce the diversions (transfer of patient in other ward or hospital) and thus the number of outliers (patient admitted in not-right ward) which may cause a longer length of stay. Information system helps the role of Bed Manager to improve the performances of the hospital-care flow optimizing the clinical paths. The system itself analyzes the interaction between patients, admission status and personnel in order to reduce the length of stay and the cost of care for hospitals. The application is designed to be linked to an existing facility manager system to gather information about the number of beds and their physical location in each room.

Keywords

Bed • Management • Hospital • Care

1 Introduction

Nowadays the increasing care demand together with existing financial constraints force the hospitals to an efficient bed planning which aims to reduce the length of stay of patients without decreasing the quality of hospitalization and healthcare.

A patient can enter the hospital via an elective admission or via an emergency admission. The latter can imply to move the patient to a *proper bed*, i.e. a bed located in the right department according to his illness. The efficiency and speed

E. Iadanza · A. Luschi (✉)
Department of Information Engineering, University of Florence, Florence, Italy
e-mail: alessio.luschi@unifi.it

A. Ancora
Master's Degree in Biomedical Engineering, University of Florence, Florence, Italy

of this operation influences the further admission of other patients at the A&E (Accident and Emergency Department) with delays which can propagate through the workflow affecting the whole process.

Therefore, hospitals must face a dimensioning problem on the strategic level due to a limited budget and on the other hand a planning problem on the operating level. This can lead to patients admitted in non-proper ward because of a lack of available proper beds. This type of patients are called *outliers* and their treatment is harder than usual because of their spreading throughout the whole premises with great impact on physicians and nurses coordination.

Another issue which may negatively influence a patient's care-flow is the relocation to another hospital (*diversion*). That may introduce significant delays in the process, and therefore on patient's proper care, especially due to lack of correct procedures and communications between facilities [1] (Fig. 1).

1.1 Bed Cycle

United States' Agency for Health Care Research and Quality *AHRQ* subdivides hospital's beds in 11 classes:

- Intensive Care Unit
- Burn Intensive Care Unit
- Pediatric Intensive Care Unit
- Ward
- Pediatric Ward
- Psychiatric Ward
- Isolation Ward
- Operating Room
- Awakening Room
- Transition between ICU and ward
- Telemetry

A proper bed management aims to maximize the throughput for each one of these category, which obviously

© Springer Nature Singapore Pte Ltd. 2019
L. Lhotska et al. (eds.), *World Congress on Medical Physics and Biomedical Engineering 2018*,
IFMBE Proceedings 68/3, https://doi.org/10.1007/978-981-10-9023-3_55

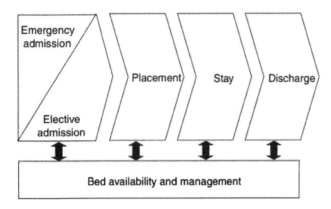

Fig. 1 Proper hospital care flow

have different bed cycles that must instead be minimized. Bed cycle is defined as the time which takes between two successive discharges of two different patients for the same bed [2].

All the operations and procedures which take place between a discharge request and its effective implementation must be minimized (notification, cleaning, assignment, transportation). This implies that every single actor (nurses, physicians, housekeeping staff) knows exactly what to do and how to activate the following step.

Bed Management Information Systems (BMIS) helps the Bed Manager to improve the process and to give information about criticalities and bed cycle minimization (Fig. 2).

2   Methods

The scope of this work is to implement a BMIS as an interface with a Workflow Management System (WMS) to rely on already existing information about spaces, rooms and technologies [3, 4, 5].

The application's main goal is to collect updated data from given stakeholders [6], aggregate and evaluate them, and then make the outputs available to the Bed Manager. Examples of possible outputs are the planned discharges, available beds grouped by category, number of outliers and available staff. The output must therefore be a dynamic representation of the actual status-quo.

Five user typologies are identified, each one with a dedicated panel of the application: nurse, housekeeping staff, ward physician, ED physician and bed manager.

Nurses usually have the most massive amount of information about the actual bed status and they must share them with the bed manager via the system. Housekeeping staff must be notified by nurses to begin the cleaning operation for a bed and then notify back its conclusion. Ward physicians have to feed the system with predicted length of stay of the patient according to Diagnosis Related Groups (DRG) standard. ED physician accesses the system to visualize the actual number of available beds so that he can easily decide if a patient could be moved to an inner ward or not.

The system relies on 12 main functions, shortly described in the following table, which are grouped in 3 categories: data entry, interface and evaluation/presentation (Table 1).

3   Results

The following application's screenshot (Fig. 3) shows the bed status for the Cardiology Ward.

The red cell identifies the occupied beds, the yellow ones are for a free but not yet cleaned beds, and the green rows represent available beds. The pie chart on the right summarizes the bed availability for the ward while on the top header there are shortcuts for the Elective Patients list section, the

Fig. 2 Bed management process

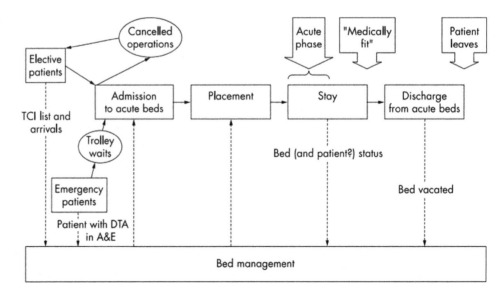

Table 1 List of BMIS's main functions

Function name	Description	User/Data	Category
LOS	Predicted length of stay	Ward physician	Data entry
Cleaning status	Bed cleaning progress	Housekeeping staff	Data entry
Bed status	Bed status in terms of availability	Nurse	Data entry
ED request	Request of emergency admission	ED physician	Data entry
Waiting list	List of emergency admission in order of priority	Ward and ED physicians. Bed manager	Data entry
Bed	Bed number for ward	WMS	Interface
Bed Availability	Actual and future bed availability	LOS, Bed status, ED request, Waiting list, Bed	Evaluation and presentation
Predicted bed availability	Admission management	Elective and emergency admissions	Evaluation and presentation
Bed typology	Incompatibility between bed typology and patient's illness	WMS, physician and nurse	Evaluation and presentation
Patient	A list of data about the patient: LOS, expected discharge, registration info, cleaning and transportation	Nurse	Evaluation and presentation
ED request response	A complete response to the ED Requests	Bed manager	Evaluation and presentation

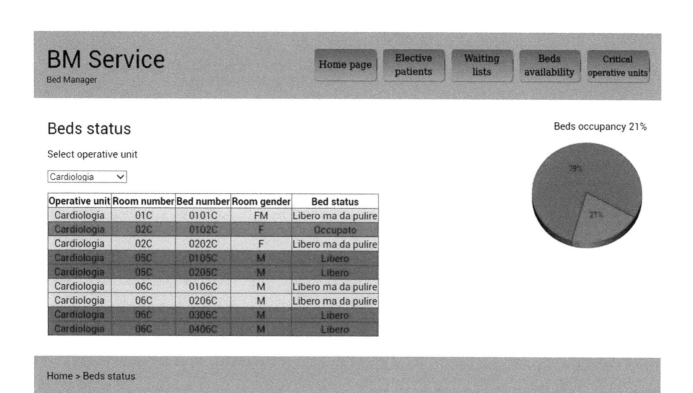

Fig. 3 Bed status for the Cardiology Ward

Waiting Lists, the Bed Availability and the Critical OU, i.e. the wards with more than 85% of occupied beds.

The following image shows the ED Requests. Here the Bed Manager views a list of all the admission requests made by the ED Physician and authorizes them to a specific bed of a specialized ward according to the information about the emergency admission [7], the bed status and localization retrieved from the WMS (Fig. 4).

Ed Requests

Beds occupancy 21%

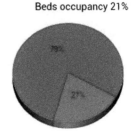

Request number	Name	Type of the requested bed	Priority	First diagnosis	Manage
3	Corrado Pascucci	Bariatrico	Alta	Emicrania e nausea	X
4	Giovanni Frasi	Ortopedico	Media	Fratture	X
5	Sonia Rossi	Cardiologico	Media	Aritmia	X
6	Luca Orlandi	Ortopedico	Bassa	Possibili fratture	X

Home > ED Requests

Fig. 4 ED requests

4 Conclusions

The designed BMIS aims to help the duties of the Bed Manager and let him make easier, faster and more accurate decisions. Even though the system has been developed together with nurse, physician and bed manager to best fit the actual users' requirements, it is still in a test phase in Le Scotte Hospital in Siena (Italy). Feedbacks and problematics are being recorded [8, 9] so that an official release could be developed in short time introducing the possible feature of decisional support algorithms with machine learning.

References

1. Landa, P., Sonnessa, M., Tànfani, E., Testi, A.: A Discrete Event Simulation Model to Support Bed Management. In: Proceedings of the 4th International Conference on Simulation and Modeling Methodologies, Technologies and Application, pp. 901–912 (2014).
2. Bachouch, R.B. et al.: An integer linear model for hospital bed planning. Int. J. Production Economics 140, 833–843 (2012).
3. Luschi, A., Marzi, L., Miniati, R., Iadanza, E.: A custom decision-support information system for structural and technological analysis in healthcare. In: IFMBE Proceedings, vol. 41, pp. 1350–1353 (2014).
4. Luschi, A., Miniati, R., Iadanza, E.: A Web Based Integrated Healthcare Facility Management System. In: IFMBE Proceedings, vol. 45, pp. 633–636 (2015).
5. Iadanza, E., Marzi, L., Dori, F., Gentili, G.B., Torricelli, M.C.: Hospital health care offer. A monitoring multidisciplinar approach. In: IFMBE Proceedings, 14 (1), pp. 3685–3688 (2007).
6. Guidi, G., Pettenati, M.C., Miniati, R., Iadanza, E.: Heart Failure analysis Dashboard for patient's remote monitoring combining multiple artificial intelligence technologies. In: Proceedings of the Annual International Conference of the IEEE Engineering in Medicine and Biology Society 2012, EMBS, pp. 2210–2213 (2012).
7. Bambi, F., Spitaleri, I., Verdolini, G., Gianassi, S., Perri, A., Dori, F., Iadanza, E.: Analysis and management of the risks related to the collection, processing and distribution of peripheral blood haematopoietic stem cells. Blood Transfusion, 7 (1), pp. 3–17 (2009).
8. Guidi, G., Pettenati, M.C., Melillo, P., Iadanza, E.: A machine learning system to improve heart failure patient assistance. IEEE Journal of Biomedical and Health Informatics, 18 (6), pp. 1750–1756 (2014).
9. Miniati, R., Dori, F., Iadanza, E., Fregonara, M.M., Gentili, G.B.: Health technology management: A database analysis as support of technology managers in hospitals. Technology and Health Care, 19 (6), pp. 445–454 (2011).

Navigation Algorithm for the Evacuation of Hospitalized Patients

E. Iadanza, A. Luschi, T. Merli, and F. Terzaghi

Abstract

The paper presents a model to support evacuation plans design for fire emergency management in healthcare facilities. It relies on existing path analysis algorithms such as Dijkstra and fire propagation simulation, also evaluating the level of criticality typical of healthcare facilities such as patients' speed based upon their illness and admission and architectural structure of wards. The algorithm automatically evaluates the safest evacuation path (which may not coincide with the shortest) for single typology of patient (ambulating, partially-ambulating, completely non-ambulating, auto-sufficient or not) and inpatient unit (ICU, ordinary ward, short-observation unit) in relation to the position of the fire trigger. The results of the algorithm are shown by using SVG-rendered graphic of existing hospital's layout.

Keywords

Fire • Evacuation • Hospital • Care • Navigation

1 Introduction

When it comes to approach the safety evaluation for facilities, there are a series of events which must be considered to prevent workers from accidents and injuries.

Prevention and protection in working environment is crucial and all the events which may cause damages to people (and to devices) must be identified, evaluated and eventually corrected. National and international regulations legislate on it and fire emergency is one of the highest and most dangerous aspect. Healthcare facilities like hospitals have a lot of technologies and procedures which involve combustible and oxidizing materials as well as ignition sources (MRI, CT, PET, bovies, thermos-ablation, etc.).

There are two main aspects of fire prevention: the first is about the technical and behavioral applications to prevent the ignition and to eventually detect and then extinguish the fire in its early stage (active protection); the second instead is toward the safe evacuation of the building once the fire has reached the flash-over and active protection has failed (passive protection). During this phase all the people inside the hospital must be safely evacuated, regardless of the reason they were in (physician, nurse, technicians, patients, visitors). However, while internal staff know the structure and how to move through, external people must be correctly led to the nearest emergency exit. Moreover, patients usually cannot walk properly and may need additional help during the evacuation according to their illness: this may imply the transportation of vital support systems and a subsequent decrease of evacuation speed. As a result, the interconnections among different users, speeds, spatial knowledge and spatial configuration lead to congestion issue.

This work aims to develop a consistent algorithm to evaluate the safest evacuation path in case of fire, according to the described factors: typology of user, speed, spatial position of ignition, fire and smoke propagation and congestion.

E. Iadanza · A. Luschi (✉)
Department of Information Engineering, University of Florence, Florence, Italy
e-mail: alessio.luschi@unifi.it

T. Merli
University of Florence, Florence, Italy

F. Terzaghi
Chief of Programming and Monitoring Unit, Hospital of Careggi, Florence, Italy

© Springer Nature Singapore Pte Ltd. 2019
L. Lhotska et al. (eds.), *World Congress on Medical Physics and Biomedical Engineering 2018*,
IFMBE Proceedings 68/3, https://doi.org/10.1007/978-981-10-9023-3_56

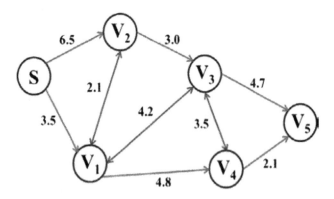

Fig. 1 Example of an interconnected weighted graph

2 Methods

2.1 Shortest and Safest Escaping Route

The first approach is to determine the shortest and safest route from a given point on the evaluated floor and the nearest emergency exit. This can be easily evaluated using a modified Dijkstra's algorithm [1]. An interconnected weighted graph is made on the CAD map of the floor: the weight represents the length of the path and the arrows between two different nodes indicate if the path can be travelled only toward or even backward. Then an adjacency matrix is implemented and used to evaluate the shortest path between a starting and an ending node (Fig. 1).

In case of multiple emergency exits the process is iterated among all the possible ending nodes (exits) and the shortest path among all the shortest paths to each exit is then chosen.

This approach does not take in consideration the spatial localization of the fire and how it is evolving. Therefore, it is crucial that a sensor system is installed on every node (each node represents a room o portion of an alley) and that it can communicate with all the other sensors along the possible escaping paths [2, 3]. The sensor must measure all the parameters that change during a fire such as temperature, humidity, air pressure and smoke density. If one of these measurements overcomes a threshold safety value, the node which the sensor is related to will be excluded from the graph. If the item was still in the group of the usable nodes, then the algorithm re-calculates to find another path which will now not include it. Therefore, the safest path might not be the same as the shortest, because it will result in the

shortest path among all the safe nodes, i.e. the ones with all the values below the threshold.

2.2 Speeds

As mentioned above, different users inside a hospital may have different movement speeds according to their illness and therapy they are undergoing. According to the destination of use of a room, together with the activity area of the department which the room belongs to, it is possible to determine the typology of patient and if it is plausible that he could be connected to a vital support system. All these factors affect the average speed of evacuation for that given room which is an input to the designed algorithm according to the following table (Table 1).

2.3 Congestion

The evacuation time increases with the number of persons that must be evacuated and can be assimilable to the flow of a fluid in a hydraulic model [4] according to the formula:

$$t = \frac{N}{D \cdot V \cdot W} \qquad (1)$$

where N is the number of people, D is the density of people, V is the speed evaluated at 2.2 and W is the width of the alleys crossed during the evacuation. The time t is in seconds. The total time of evacuation is given as a sum up between the escaping time of the first evacuated person and the rest of the people and must be less than the maximum admitted law values [5].

3 Results

The algorithm has been implemented on the San Luca Nuovo building of the Careggi Hospital in Florence (Italy) [6, 7]. Information about the destination of use and activity for single rooms, together with the expected number of persons (deduced from the number of beds, the time of the day and the scheduled visit program) is taken from the hospital WMS [8, 9, 10]. Pyrosim [11] has been used to simulate the propagation of a fire event and the evolution of the measurement parameters presented in 2.1 (Fig. 2).

Table 1 Different speeds for different types of patient

Subject group (number)	Mean (m/s)	Standard deviation range (m/s)	Range (m/s)	Interquartile range (m/s)
All disabled ($n = 107$)	1.00	0.42	0.10–1.77	0.71–1.28
With locomotion disability ($n = 101$)	0.80	0.37	0.10–1.68	0.57–1.02
No aid ($n = 52$)	0.95	0.32	0.24–1.68	0.70–1.02
Crutches ($n = 6$)	0.94	0.30	0.63–1.35	0.67–1.24
Walking stick ($n = 33$)	0.81	0.38	0.26–1.60	0.49–1.08
Walking frame or rollator ($n = 10$)	0.57	0.29	0.10–1.02	0.34–0.83
With out locomotion diability ($n = 6$)	1.25	0.32	0.82–1.77	1.05–1.34
Electric Wheelchair ($n = 2$)	0.89	–	0.85–0.93	–
manual wheelchair ($n = 12$)	0.69	0.35	0.13–1.35	0.38–0.94
Assisted manual wheelchair ($n = 16$)	1.30	0.34	0.84–1.98	1.02–1.59
Assisted ambuiant ($n = 18$)	0.78	0.34	0.21–1.40	0.58–0.92

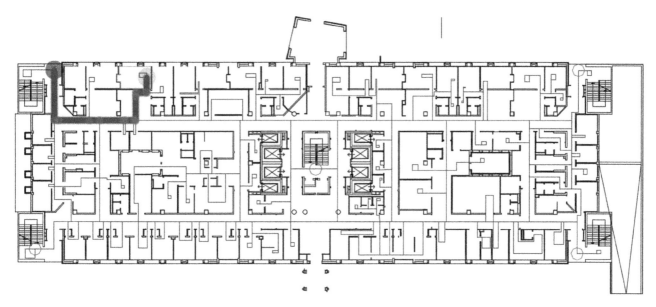

Fig. 2 Shortest route with no fire event in place

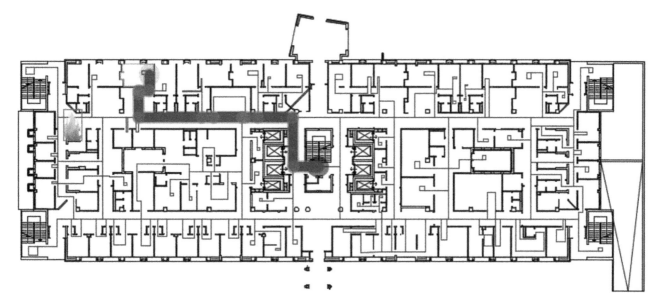

Fig. 3 Shortest safest route with a fire event in place

The starting node is a pulmonary outpatient. The WMS outputs a room surface of 19 sqm with a density of 0.2 person/sqm for the pulmonary ward. Thus, the number of people to evacuate is 4. The average speed is 1.2 m/s. The shortest route is about 30 meters long. According to (1) the average speed is reduced due to congestion to 0.98 m/s. The total escaping time is about 30 s (Fig. 3).

By introducing a fire event with an ignition in a nearby room, the algorithm re-calculates the escaping path because the fire would reach the nearest exiting mode before all the evacuated people (according to the example, all the 4 persons must safely reach the exit). Due to the longest and different path, the congestion parameter will be also recalculated. The new escaping time is about 47 s for a route of about 46 meters.

The escaping times can be compared to the maximum time admitted by the law [5] verifying their matching to the threshold (60 s for a path of 30 m and 300 s for a path of 46 m).

4 Conclusions

The algorithm finds its place of usability both for planning and verifying purposes. In fact, it can be used in development phases to design a correct layout for the emergency exits, and in later analysis to verify the accuracy and the compliance of the evacuation plans. However, the algorithm relies only on mathematical factors which by theirselves cannot fully describe all the human behaviors during a fire emergency. Therefore, the algorithm must be intended just as a decision support system because human validation of the obtained outputs is always needed.

References

1. Shastri, J.D.: Safe Navigation during Fire Hazards using Specknets. Master's thesis. University of Edinburgh (2006).
2. Guidi, G., Pettenati, M.C., Melillo, P., Iadanza, E.: A machine learning system to improve heart failure patient assistance. IEEE Journal of Biomedical and Health Informatics, 18 (6), pp. 1750–1756 (2014).
3. Miniati, R., Dori, F., Iadanza, E., Fregonara, M.M., Gentili, G.B.: Health technology management: A database analysis as support of technology managers in hospitals. Technology and Health Care, 19 (6), pp. 445–454 (2011).
4. Gwynne, S.M.V., Rosenbaum, E.R.: Employng the Hydraulic Model in Asseing Emergency Movement. SFPE Handb. Fire Prot. Eng., vol. 4, pp. 373–396 (2008).
5. D.M. 10 marzo 1998. Criteri generali di sicurezza antincendio e per la gestione dell'emergenza nei luoghi di lavoro. Italy (1998).
6. Guidi, G., Pettenati, M.C., Miniati, R., Iadanza, E.: Heart Failure analysis Dashboard for patient's remote monitoring combining multiple artificial intelligence technologies. In: Proceedings of the Annual International Conference of the IEEE Engineering in Medicine and Biology Society 2012, EMBS, pp. 2210–2213 (2012).
7. Bambi, F., Spitaleri, I., Verdolini, G., Gianassi, S., Perri, A., Dori, F., Iadanza, E.: Analysis and management of the risks related to the collection, processing and distribution of peripheral blood haematopoietic stem cells. Blood Transfusion, 7 (1), pp. 3–17 (2009).
8. Luschi, A., Marzi, L., Miniati, R., Iadanza, E.: A custom decision-support information system for structural and technological analysis in healthcare. In: IFMBE Proceedings, vol. 41, pp. 1350–1353 (2014).
9. Luschi, A., Miniati, R., Iadanza, E.: A Web Based Integrated Healthcare Facility Management System. In: IFMBE Proceedings, vol. 45, pp. 633–636 (2015).
10. Iadanza, E., Marzi, L., Dori, F., Gentili, G.B., Torricelli, M.C.: Hospital health care offer. A monitoring multidisciplinar approach. In: IFMBE Proceedings, 14 (1), pp. 3685–3688 (2007).
11. Thunderhead Engineering. PyroSim (https://www.thunderheadeng.com/pyrosim).

A Decision Support System for Chronic Obstructive Pulmonary Disease (COPD)

Ernesto Iadanza and Vlad Antoniu Mudura

Abstract

Obstructive chronic obstructive pulmonary disease (COPD) is a respiratory disease characterized by a chronic air flow limitation and associated with major economic and social problems. In an attempt to find a solution to these problems, numerous systems of clinical decision support for the management of patients with COPD have been developed in recent years. In particular, systems based on machine learning algorithms have been developed with the aim of monitoring the health status of patients and foreseeing and preventing exacerbations and hospital admissions. An in-depth research into scientific literature has shown that, in the state of the art, these goals have not yet been met and the performance of the current systems is not clinically acceptable. The aim of this work is the design and implementation of a new clinical decision support system that can at least partially fill the gaps present. The first step in the work was to try to replicate the performance of support systems for similar decisions, already present in scientific literature. Using the physiological parameters acquired by 414 patients using respiratory function tests, two predictive models were made using the same machine learning algorithms (neural network and support vector machine). The performance obtained was comparable to those of the scientific literature. The next step was to create a new predictive model, with superior performance to previous models. The machine learning algorithm chosen is C5.0. The performance obtained was significantly better than the two previous models. The new predictive model was implemented within a user interface, implemented in Java programming language, the COPD Management Tool. The software developed allows the evaluation and classification of the results of respiratory performance tests, with excellent performance, compared to the current state of the art and can therefore be used in many clinical applications.

Keywords

DSS • COPD • Machine learning

1 Introduction

Chronic Obstructive Pulmonary Disease (COPD) is a respiratory disease characterized by a chronic airflow limitation and associated with major economic and social problems. COPD is classified as the fourth leading cause of death in the world and in absence of countermeasures aimed to reduce risk factors it is expected to become the third leading cause of death by 2030 [1]. In 2015 there were 3.2 million deaths associated with COPD and the estimated global prevalence of COPD was about 175 million [2]. From an economic point of view, only in the United States, the annual estimated costs associated with COPD are about 50 billion dollars. In the next years, costs are expected to rise dramatically together with prevalence. Costs increase with increasing severity of the disease and most of them are linked with hospital admissions which in turn are mainly caused by exacerbation episodes [3].

In an attempt to find a solution to these problems, numerous clinical decision support systems (CDSSs) for the management of patients with COPD have been developed in recent years [4]. In particular, systems based on machine learning algorithms have been developed with the aim of monitoring the health status of patients and foreseeing and preventing exacerbations and hospital admissions [5]. An in-depth research into scientific literature has shown that, in the state of the art, these goals have not yet been met and the

E. Iadanza (✉) · V. A. Mudura
Department of Information Engineering, University of Florence, Florence, Italy
e-mail: ernesto.iadanza@unifi.it

© Springer Nature Singapore Pte Ltd. 2019
L. Lhotska et al. (eds.), *World Congress on Medical Physics and Biomedical Engineering 2018*,
IFMBE Proceedings 68/3, https://doi.org/10.1007/978-981-10-9023-3_57

performance of the current systems is not clinically accept-able. The aim of this work is the design and implementation of a new CDSS that can at least partially fill the current gaps.

2 Materials and Methods

2.1 Data

In order to train, validate and test the decision support system, data from 414 patients affected with COPD and obstructive ventilatory defect were acquired using pulmonary function tests. The following physiological parameters were acquired: Forced Expired Volume in one second (FEV1), Forced Vital Capacity (FVC), Slow Vital Capacity (SVC), FEV1/FVC ratio, FEV1/SVC ratio, Forced Expired Flow at 25–75% (FEF 25–75), Peak Expiratory Flow (PEF), Vital Capacity (VC), Total Lung Capacity (TLC), Residual Volume (RV), Functional Residual Capacity (FRC), Expiratory Reserve Volume, Diffusing Capacity (DLCO), Alveolar Volume (VA) and DLCO/VA ratio. All these

parameters were measured before and after bronchodilation. Other parameters were patients' age, height, bodyweight and sex. According to these parameters five expert pneumologists evaluated the severity of each patient's ventilatory defect and classified it as mild, moderate or severe.

2.2 Data Analysis and Predictive Model Training

Data were processed and analyzed using IBM SPSS Modeler 18.1 [6]. The aim of this phase was to develop a predictive model able to classify the patients' ventilatory defect in three categories (Mild, Moderate and Severe) according to the values of the physiological parameters previously described.

The first step was to try to replicate the performance of support systems for similar decisions, already present in scientific literature. Most of these systems and, in particular, the ones which reached better performances in terms of predictive accuracy, sensitivity and specificity were trained using Neural Networks and Support Vector Machines [7, 8].

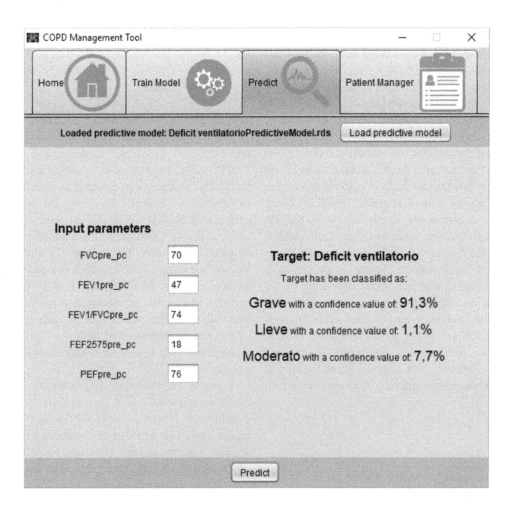

Fig. 1 COPD management tool user interface

Therefore, we used those machine learning techniques to train two different predictive models. We then calculated the performances of these predictive models in predicting the severity of patients' ventilatory defect.

Next step was training a new predictive model with better performances. In order to identify the most suitable machine learning technique for our data, IBM SPSS Modeler's auto classifier node was used. Performances of various machine learning techniques were compared: CART, Random Forest, QUEST, CHAID, Bayesian Networks, Logistic Regression, C5.0, KNN and others. The best performances were reached by the C5.0 algorithm. We therefore concluded that C5.0 was the most suited algorithm for our data. Finally, we trained a predictive model using the C5.0 algorithm and compared its performances with those reached by the predictive models trained with the Neural Network and the Support Vector Machine.

2.3 Predictive Model Implementation

The predictive model trained with the C5.0 algorithm was implemented within a user interface, implemented in Java programming language, the COPD Management Tool. A demonstrative view of the COPD Management Tool is shown in Fig. 1.

3 Results

Results and performances related to the three predictive models, respectively trained using Neural Network, Support Vector Machine and C5.0 algorithms, are reported in the tables below (Tables 1, 2 and 3).

Table 1 Support vector machine performances

Severity	Accuracy	Sensitivity	Specificity
Mild	0.94	0.81	0.97
Moderate	0.89	0.87	0.91
Severe	0.94	0.94	0.94

Table 2 Neural network performances

Severity	Accuracy	Sensitivity	Specificity
Mild	0.93	0.79	0.96
Moderate	0.90	0.86	0.93
Severe	0.97	0.98	0.96

Table 3 C5.0 performances

Severity	Accuracy	Sensitivity	Specificity
Mild	0.94	0.79	0.97
Moderate	0.92	0.91	0.94
Severe	0.97	0.99	0.96

4 Conclusions

Performances obtained with the Neural Network and the Support Vector Machine are comparable with those of the scientific literature. Performances obtained with the C5.0 algorithm are significantly better than those obtained with the two previous model.

The proposed approach, designed with the same systematic approach used in previous works from the authors for a Cardiac Heart Failure CDSS [9–13], allows the evaluation and classification of the results of pulmonary function tests, with excellent performance, compared to the current state of the art and can therefore be used in many clinical applications.

References

1. A. D. Lopez et al., "Chronic obstructive pulmonary disease: current burden and future projections," Eur. Respir. J., vol. 27, no. 2, pp. 397–412, Feb. 2006.
2. T. Vos et al., "Global, regional, and national incidence, prevalence, and years lived with disability for 310 diseases and injuries, 1990? 2015: a systematic analysis for the Global Burden of Disease Study 2015," Lancet, vol. 388, no. 10053, pp. 1545–1602, Oct. 2016.
3. S. Ray et al., "The clinical and economic burden of chronic obstructive pulmonary disease in the USA," Clin. Outcomes Res., vol. 5, p. 235, Jun. 2013.
4. D. Sanchez-Morillo et al., "Use of predictive algorithms in-home monitoring of chronic obstructive pulmonary disease and asthma: A systematic review," Chron. Respir. Dis., vol. 13, no. 3, pp. 264–283, 2016.
5. S. C. Peirce et al., "Designing and implementing telemonitoring for early detection of deterioration in chronic disease: Defining the requirements," Health Informatics J., vol. 17, no. 3, pp. 173–190, 2011.
6. "IBM SPSS Modeler." [Online]. Available: https://www.ibm.com/it-it/marketplace/spss-modeler. [Accessed: 31-Jan-2018].
7. M. Veezhinathan et al., "Detection of obstructive respiratory abnormality using flow-volume spirometry and radial basis function neural networks.," J. Med. Syst., vol. 31, no. 6, pp. 461–5, Dec. 2007.
8. R. Karakis, I. Guler, and A. H. Isik, "Feature selection in pulmonary function test data with machine learning methods," in 2013 21st Signal Processing and Communications Applications Conference (SIU), 2013, pp. 1–4.

9. Guidi, G., Iadanza, E., Pettenati, M.C., Milli, M., Pavone, F., Biffi Gentili, G., "Heart failure artificial intelligence-based computer aided diagnosis telecare system", Lecture Notes in Computer Science (including subseries Lecture Notes in Artificial Intelligence and Lecture Notes in Bioinformatics), 7251 LNCS, pp. 278–281, 2012.

10. Guidi, G., Pettenati, M.C., Miniati, R., Iadanza, E., "Random forest for automatic assessment of heart failure severity in a telemonitoring scenario" in Proceedings of the Annual International Conference of the IEEE Engineering in Medicine and Biology Society, EMBS, art. no. 6610229, 2013, pp. 3230–3233.

11. Guidi, G., Pettenati, M.C., Melillo, P., Iadanza, E., "A machine learning system to improve heart failure patient assistance", IEEE Journal of Biomedical and Health Informatics, 18 (6), art. no. 6851844, pp. 1750–1756, 2014.

12. Guidi, G., Pettenati, M.C., Miniati, R., Iadanza, E., "Heart Failure analysis Dashboard for patient's remote monitoring combining multiple artificial intelligence technologies" in Proceedings of the Annual International Conference of the IEEE Engineering in Medicine and Biology Society, EMBS, art. no. 6346401, 2012, pp. 2210–2213.

13. Guidi, G., Melillo, P., Pettenati, M.C., Milli, M., Iadanza, E., "Performance assessment of a Clinical Decision Support System for analysis of Heart Failure" in IFMBE Proceedings, 41, 2014, pp. 1354–1357.

Virtual Course for the Americas: Healthcare Technology Planning and Management Over the Life Cycle

Tobey Clark, Alexandre Lemgruber, Rossana Rivas, Francisco Caccavo, Tatiana Molina Velasquez, and Javier Comacho

Abstract

Clinical engineers are well suited as healthcare technology planning and management leaders working with stakeholders from clinical, administrative and other healthcare professions. However, education and training in this topic for healthcare professionals is limited outside the high income, developed countries. To provide accessible training and education to the low and middle income countries, a 100% virtual course, **Healthcare Technology Planning and Management**, was developed by the Technical Services Partnership—University of Vermont USA, a WHO Collaborating Center for Health Technology Management, to teach students on best practices to follow over the healthcare technology life cycle: *assessment, replacement, budgeting, acquisition, deployment, training, patient safety, compliance and maintenance.* The global state of medical devices, information systems and the convergence of technologies is part of the learning along with the setup and operation of a clinical engineering department in a healthcare system. The Healthcare Technology Planning and Management bi-lingual course was first taught on the Pan American Health Organization Virtual Campus for Public Health in 2015 to participants from the Caribbean and Latin America countries. The course was conducted a second time over a 15-week period in 2017. Overall, 96 students have successfully completed the course. The course showed significant interaction and engagement by participants in discussion boards and forums. For the next course session, it is hoped that the course will be translated to Portuguese with adaptation to Brazil to allow a three language offering. Also live workshops focused on solving real life healthcare technology challenges in the Americas are planned.

Keywords

Online course • Clinical engineering
Healthcare technology • Technology planning
Technology management

1 Introduction

1.1 Background

Healthcare technology is rapidly expanding in Latin America and the Caribbean countries. Health systems are understanding better the relevant influence of healthcare technology based on clinical, ethical, social, and economic health outcomes. Donations from developed countries of new and used equipment continue and vendors are eager to sell equipment and expand to this growing area.

In developing countries, reaching the full potential benefit of medical technology is difficult. Limited maintenance budgets and weak after sales support from manufacturers and distributors are all too common in addition to a deficient level of knowledge in the of the health staff and decision-makers.

To give developing nations the skills and tools to positively change this area, capacity building directed toward healthcare leaders including engineers, physicians, nursing, technical and administrative staff is necessary.

Online training is rapidly advancing worldwide due to the availability of communication networks and devices including computers, tablets and smartphones. The

T. Clark (✉)
University of Vermont, Burlington, VT, USA
e-mail: Tobey.Clark@its.uvm.edu

A. Lemgruber · F. Caccavo
Pan American Health Organization, Washington DC, WA, USA

R. Rivas
Pontifical Catholic University of Peru, Lima, Peru

T. M. Velasquez
Universidad CES, Medellin, Colombia

J. Comacho
Universidad EIA, Medellin, Colombia

© Springer Nature Singapore Pte Ltd. 2019
L. Lhotska et al. (eds.), *World Congress on Medical Physics and Biomedical Engineering 2018*,
IFMBE Proceedings 68/3, https://doi.org/10.1007/978-981-10-9023-3_58

educational resources of the World Wide Web are tremendous, there are no costs or lost time for travel to a classroom, and online training's asynchronous nature allows maximum flexibility for study and assignment completion time.

Virtual education was successfully applied by the University of Vermont in technical courses for the Americas beginning in 2007 [1]. Through grant funding (Pan American Health and Education Foundation), two online, interactive courses were developed for technical and clinical staff in hospitals—**Patient Care Equipment and Technology**, and **Advanced Medical Equipment Systems**. Faculty from University of Vermont, Pontificia Universidad Católica de Peru, and Universidad CES collaborated to produce and offer the courses in English and Spanish. Over 1000 students from 39 countries have taken these courses at the universities noted along with UTN Mendoza (Argentina) and on the Pan American Health Organization Virtual Campus for Public Health [2].

1.2 Healthcare Technology Planning and Management Course

A healthcare technology planning and management course was originally developed by Tobey Clark in 2008 and has been taught annually at the University of Vermont USA. In 2015, the course was placed on the Pan American Health Organization's Virtual Campus for Public Health. The Spanish version co-authored by Tobey Clark and Rossana Rivas using the content of the English course translated and adapted to Latin America with additional content created by Rossana Rivas. In 2015, Rivas and Clark taught the Spanish and English versions of the course respectively.

The Healthcare Technology Planning and Management online course provides students with a basic understanding of the principles of healthcare technology planning and management [3]—*assessment, budgeting, acquisition, deployment, education/training, patient safety, maintenance, and replacement/disposal.* Planning and management is focused on medical devices, clinical information systems, and converged technologies. Clinical engineering department setup, attributes and resources are presented along with the profession and global activities.

1.2.1 Learning Objectives

- To give students a basic understanding of the guiding principles of healthcare technology planning and management
- Provide a methodology for improving the quality of medical devices, clinical information systems and converged technology through planning and management

- Help students better communicate with technical staff, clinicians, regulators, administrators, and technology vendors.
- Develop the participant's interest, provide the tools for planning, managing and solving healthcare technology issues, and prepare them for further study and more advanced application of the principles.

1.2.2 Prerequisites
It is recommended that the student have completed university level courses in business, technology, engineering or management and have one-year experience in management or administration with responsibility for some aspect of healthcare technology.

1.2.3 Methods
The course utilizes web-based content including text, photos, diagrams, flow charts, other figures, video, audio, links to other websites, and other web attributes to deliver content to students.

1.2.4 Course Assessments
The assessments consist of five quizzes, two exams, three reports, and five interactive discussion question. The keys to course engagement are the communication between students, and student's interaction with instructors—*the discussion questions fostered this behavior.*

2 2017 Course Results

2.1 Background

The 2017 PAHO Virtual Campus for Public Health course began on September 11, 2017 and ended on December 10, 2017. The student selection process was coordinated from the PAHO Washington DC office and involved interaction with the PAHO regional and country offices in the Americas. Thirty-seven students registered for the English course from ten nations primarily in the Caribbean. Fifty-eight students from eighteen Latin America countries registered for the Spanish course. There was a course coordinator-professor for the Spanish course and two tutors—with one professor teaching the students in the English course while providing overall academic coordination for both courses. General coordination was provided by the PAHO Washington, DC office with platform services from PAHO Panama for the Virtual Campus for Public Health. The student's professions included engineers (39%), physicians (25%), biomedical and information technology technical staff (11%) with the remaining students from administration, pharmacy, economics, nursing, physics, dentistry, and architecture. (Fig. 1)

Student Background By Profession

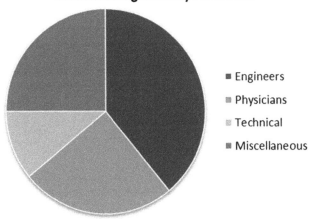

- ∎ Engineers
- ∎ Physicians
- ∎ Technical
- ∎ Miscellaneous

Fig. 1 Distribution of course participants by professional background

2.2 Course Completion

Of the original 95 students registered, 64 successfully completed the course. There were a number of students who after registering, never attended. Other students left the course due to work-course conflicts for reasons such as travel especially to remote areas, health problems, and disasters occurring during the course including hurricanes Irma and Maria affecting the Caribbean.

The final grading for the 67% of the students who successfully completed the course is below: (Table 1)

2.3 Course Evaluation

The course evaluation by sixty-two students provided the following results based on the thirteen question evaluation at the end of the course was:

- 95% or greater strongly agree/agree regarding questions on instructor knowledge, course well organized, course content valuable, assignments contributed to learning, clear objectives, grading understood, instructor contributes to online discussions; course valuable in improving career, met expectations, recommend the course; take another course with instructor. 95% or greater rated the course and instructor excellent or satisfactory

Table 1 Grade distribution

Grading	English	Spanish	Total
Passed with excellence (<90 points)	8	12	20
Passed with distinction (<80 points)	15	14	29
Passed (<70 points)	1	14	15

2.3.1 Course Highlights

- Discussion question answers and response to other participants at an exceptional level; Average of 2.5 quality posts by each student for each discussion question with constructive responds to other students showing outstanding interaction leading to high value.
- Report quality was very high for most students completing the course showing not only their strong backgrounds as working managers, but also engagement in the course.
- The class was dynamic, and the students kept in good communication with tutors and coordinators. The students affected by technical issues collaborated with the platform technical staff to solve problems.

2.4 Difficulties

- The quizzes and exams were not always completed in a satisfactory fashion due to participant's internet problems and virtual campus issues along with unfamiliarity with the content. All issues were rapidly resolved by the PAHO Virtual Campus team.

3 Comparison Between the 2015 and 2017 Courses

3.1 Background on the 2015 Course

The Healthcare Technology Planning and Management course was first offered on the PAHO Virtual Campus for Public Health over eleven weeks from September to November 2015. The initial registration of the course was 52 with 32 successfully completing the course. The grading was at a lower level with an overall pass rate of 59% with only 44% achieving Passed with Excellence or Distinction. The course and instructor rating were slightly lower than the 2017 course. The primary improvements suggested were to increase the course from ten weeks to fifteen weeks due to the heavy workload, for instructors to interact more with the students on the discussion boards, and improve the internet and platform reliability. All suggestions were implemented in the 2017 course.

4 Conclusion and Future Steps

Recommendations for improvement in the next course include:

1. Highlight more practical elements of HT planning and management by providing more real life case studies from Latin America and the Caribbean to complement principles.
2. The course intensity is difficult for some students. Due to the high degree of interaction and networking taking place on the discussion board, this high value area should be weighted more strongly.
3. Add additional webinars to the course as this promotes strong interaction and networking.
4. Correct platform problems especially those related to long exams.

The courses showed significant engagement by participants in a very relevant, timely and high impact area—healthcare technology planning and management. Students used real world examples of their healthcare technology planning successes, challenges and experiences in postings to the discussion board and forums. Reports provided flexibility in topics to allow students to present problems in their countries and discuss potential solutions. Feedback from course participants indicates that they wish to use the course learning to train others in their health system in the concepts presented, apply the knowledge for projects such as implementation of electronic health records, policies for acquisition requests, criteria for equipment replacement, and expanded resources for maintenance based on justification from the course.

Conflict of Interest The authors declare that they have no conflict of interest.

References

1. J. Tobey Clark, Online Courses in Medical Technology Application, Support and Management Improve Effectiveness and Patient Safety, World Congress on Medical Physics and Biomedical Engineering May 26–31, 2012, Beijing, China pp 715–717
2. Tobey Clark, Alexandre Lemgruber, Francisco Caccavo, Tatiana Molina, Federico Graciá, Rossana Rivas and Luis Vilcahuaman, Biomedical Technology Online Courses for the Americas, World Congress on Medical Physics & Biomedical Engineering, Toronto, CA, June 11, 2015
3. Tobey Clark, Are There Indicators to Determine Best Practice in HTM Programs? 1st International Clinical Engineering and Health Technology Management Congress, Hangzhou, China, October 21, 2015

Model HTM Application in Failure Analysis for Air Compressors in the Dental Service of Primary Health Care

Priscila Avelar and Renato Garcia

Abstract

This study presents a failure cause analysis in medical air compressors in dental services of primary health care in Santa Catarina, Brazil. The study classifies failures, from 2007 to 2016, associated with the three domains of the methodology developed for HTM: Human Resources (DHR), Infrastructure (DI) and Technology (DT). From these failures, 58% were associated to DT, 31% to the HRD and 11% to the DI. Data collection was from the HTM Information System of the IEB-UFSC CE and the application of HFMEA. This analysis identifies which failures are related to wear of parts due to their life cycle in DT. In the DI, the causes were the lack of electrical protection and inadequate electrical wirings with manufacturers and technical standards. In DHR, the lack of a manual purge procedure is a cause of failure. As a result, a checklist for functional equipment verification was implemented during the CE technical inspection, which identifies fault conditions and associated domains. For DT, a preventive maintenance program was implemented to replace oil and shorter its service life. In order to reduce the failures associated to the DI, adjustments were made in the compressor shelters with installation of electrical protection and resizing of the electrical system. For the failures associated with the DHR, didactic materials were developed for training with a proposal to improve operational routine, best practices and an installation program of automatic purger in units of greater demand for dental service. The results of these actions led to a reduction in the occurrence of failures and validated the application of HTM model developed for Primary Health Care is important contribution for add quality to primary health care system.

Keywords

Primary health care • Clinical engineering
HTM dental service

1 Introduction

The Brazilian MoH establishes in 2003 the National Health Policy, in which from the Brasil Sorridente Program seeks to guarantee promotion actions, prevention and recovery of oral health with expanded access to dental treatment, free of charge to Brazilians, through the Unified Health System (SUS) [1, 2].

In order to meet the epidemiological demands in dentistry, the primary health care system is structured by support units—health care centers—and reference centers—Dental Specialties Centers (CEA) and Immediate Care Units (UPA)—shown in Fig. 1. These structures are composed of different technological densities and function as a filter capable of organizing the flow of services in Primary Health Care System—HCS, from the simplest to the most complex.

The dental compressor is the main mean of supplying compressed air for the operation of the dental chair (pneumatic dental chair systems) and its components (dental and suction fittings). The technology management processes developed by Clinical Engineering seek to verify the conformity of the equipment from the analysis of causes involving failures, classified by the domains of human resources, infrastructure or by the technology itself.

This methodology, consolidated for the Clinical Engineering HTM model development by IEB-UFSC [3], establishes and implements actions in these domains that

P. Avelar · R. Garcia (✉)
Biomedical Engineering Institute, Federal University of Santa
Catarina, Florianópolis, Brazil
e-mail: renato@ieb.ufsc.br

P. Avelar
e-mail: priscila@ieb.ufsc.br

L. Lhotska et al. (eds.), *World Congress on Medical Physics and Biomedical Engineering 2018*,
IFMBE Proceedings 68/3, https://doi.org/10.1007/978-981-10-9023-3_59

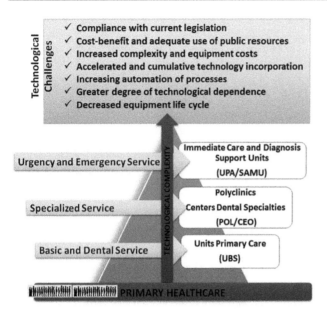

Fig. 1 Structure of primary health care system (PHCS) in Brazil

result in effectiveness, reliability and safety of the technological process and the fulfillment of health care demands.

The Health Technology Management (HTM) model, developed by IEB-UFSC Clinical Engineering, establishes and implements actions in the areas of human resources, infrastructure and technology. These actions seek as a result the effectiveness, reliability and safety of the technological process in health care.

In order for technology to have an impact on health care service quality, the technological process in health care is evaluated in order to obtain an infrastructure that provides the safe operation of medical equipment, the adequate and safe use of technologies by human resources, and the knowledge life cycle technologies to better plan and evaluate their cost-effectiveness.

The Clinical Engineering of the IEB-UFSC has validated and applied the HTM model for more than 15 years with the State Secretariat of Health of Santa Catarina [4–6] and in PHCS about 10 years in the primary health care centers of the Municipal Health Secretariat [7–9]. During this period, different peculiarities were identified of the HTM model applied in PHCS in relation to those practiced in Hospitals, such as logistics of technical visits to the sectors, complexity of the equipment, management and management processes, considering the three levels of attention among others.

The problems involve the inefficiency of the dental service by the dental compressor, are associated with several factors of human resources in the inadequate use due to little knowledge of the use of the technology and the lack of care regarding the water drainage procedure of the equipment; of infrastructure due to electrical installations without protection systems; and technology the frequency of failure due to

wear of the parts caused by improper dimensioning of equipment for a dental office and the unavailability of the technology.

This study aims to present a cause-of-failure analysis of dental compressor equipment associated with the context of primary health care dental services based on this HTM model, in the Florianópolis city PHCS, Santa Catarina, Brazil, is composed of 63 health care units distributed in 5 regional (north, south, east, center and mainland) and 4 sanitary districts (north, south, center and mainland) which manage the health care units.

2   Methodological

The methodology, based on a structural model, Fig. 2, is based on out HTM model [3] and in HFMEA tool application [10, 11].

For contextualization of the problem is necessary to know the technological process in which the dental compressor is inserted. This stage was based on bibliographic references that show studies on dental service and HTM in PHCS.

After the contextualization of the research and description of the technological process in the study, data were collected in the Clinical Engineering Information System for the period from 2007 to 2016. Data obtained from Service Orders were classified according to the domains of the HTM model and identified potential causes of failure that supported the application of the HFMEA tool.

The application on out HTM model made it possible to know the history of failures that have impacted the dentistry service unavailability.

In the application of the HFMEA, after the definition of the analyzed process, it followed with: (1) identification of the failure modes being written the problem and its non-conformities; (2) potential consequences of the defect by seeking a brief description of the consequence that may occur; (3) prioritizing the failures through the level of risk, classifying and assigning weights as to the occurrence of the cause and seeking to identify the frequency of appearance of the defects for a given sample and the severity of the effect with the severity weighting in terms of failure effect. Failure detection has been weighed against the ability to detect failure before it reaches the clinical body, the ability to detect the defect during preventive and corrective maintenance; (4) Preventive actions in search of solutions to minimize failures.

With the results of the preventive actions by the HFMEA, it was structured in the PDCA cycle the actions in our HTM model implemented in the health care units of the PHCS in Florianópolis city, Brazil.

Fig. 2 Structural model: methodology application in the study of dental compressor failures in primary health care systems

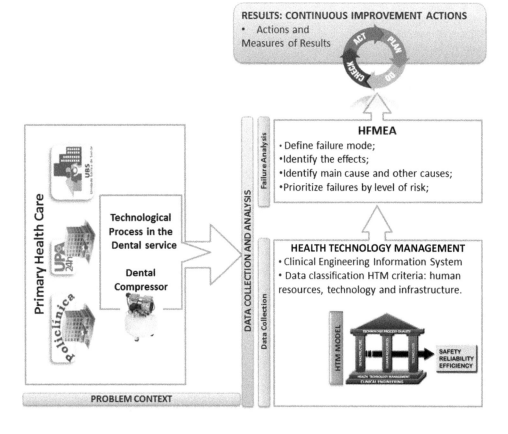

With the results of the preventive actions by the HFMEA, it was structure in the PDCA cycle the actions in our HTM implemented in the health care units of the PHCS in Florianopolis city, Brazil.

3 Results

In the analysis of the history of dental compressor management, from 2007 to 2016, 58% of clinical engineering activities involved technical problems of the technology itself, 31% of inappropriate use for human resources and 11% of problems with electrical network.

In addition to the management history classification in HTM domain, the HFMEA application made it possible to identify the incidence and repeatability of failure reasons. In Fig. 3, it presents the main problems and occurrence regarding the failures that caused repairs and corrective maintenance.

The CE aimed at achieving the reliability, effectiveness and safety of the technological process in dentistry, established short and medium term actions at the health care units of the PHCS. As shown in Fig. 4, actions in the HTM domains were structured in the PDCA Cycle and checked the results by indicators.

Fail mode	Occurrence	Detection			Repair Time (min)
		QUALY	PM	PC	
Motor locked	7	1	DE	DIP	120
External leakage	9	3	DE	DPP	30
Internal leakage	7	5	DO	DIP	120
Dead / blocked plug	5	3	DE	DIP	30
Connecting Rod	5	3	DP	DIP	60
Defective pressure switch	5	5	DE	DIS	60
Noise level	*	5	DO	DIP	60
Safety valve	*	5	DO	DIP	30

Domain HTM	Failure Analysis
Human Resources	Lack of water drainage
	Equipment is not switched off at the end of the day
Infrastructure (Electrical Grid)	Deregulated pressure switch
	motor locked
	Burnt capacitor
Technology	Leakage
	Lack of lubrication or Validity expired oil
	High noise level
	No pressure or Pressure regulation lack
	Reduced air production
	Piston Ring Wear
	Dirt on the filter

Fig. 3 Main faults that caused repairs and corrective maintenance of the dental compressor and results regarding the application of HFMEA

Fig. 4 Actions in the HTM domains were structured in the PDCA cycle and checked the results by indicators

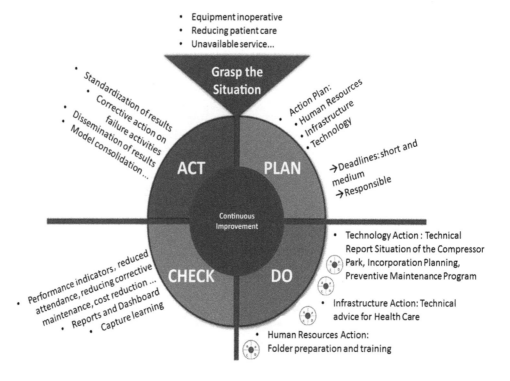

With failure modes and classification by the three domains of the HTM model, it was possible to identify that human resource failures were related to the periodic change of professionals in the clinical area.

Examples of the action implemented by the CE were the installation of automatic purger or automatic drainage of water from the equipment reservoir, in units of greater demand of dental care. Operational manual drainage of water was performed in the absence of automation of the process, and distribution of flyer of best practices of conservation of the equipment. The installation of the automatic purger provided an 80% reduction of water drainage failures.

The problems with infrastructure of the dental compressor shelters are considered the most challenging, because of the non-conformity of the buildings (shelter) with technical regulations. The recurring failures caused by overloading voltage, current and oscillation of the grid, resulting in the issuance of technical advice to adapt the electrical network equipment shelters. In the shelters were installed electrical protection system and resized the electrical system for better equipment performance.

As for technology, incorporation studies with issuance of technical report were presented to health care managers. The diagnosis of the current situation of the dental compressor park allowed establishing the planning of acquisition of new equipment to replace the equipment at the end of its useful life. It was also implemented a functional checklist by CE in order to minimize corrective maintenance and identify fault conditions and associated domains.

The implementation of the checklist allowed the standardization of the diagnosis of failures by the technical team of Clinical Engineering.

4 Conclusion

The process of identification and classification of dental compressor failures was performed by applying an analysis according to the criteria of the three domains is based on out HTM model and the HFMEA tool.

The application of the HTM model with the HFMEA tool provided the adequacy of the Clinical Engineering processes that sought: to reduce the failure to use the equipment improperly through operational training and installation of automatic purger; adequacy of infrastructure in the installation of electrical protection system and decision making in the renovation of the technology park.

The results of these actions led to a 30% reduction in the occurrence of failures and in the decision making process for the acquisition of new dental compressors. These actions allowed the validation of the HTM model and a contribution in the improvement of the dental services in PHCS.

This applied research, evidenced that there is not a single tool capable of answering all the problems of the CE in HTM activities. As observed, the management domains were associated to the HFMEA to support management decision making and specific model HTM for quality technological process improvement solutions over PDCA cycle.

Conflict of Interest The authors of this article declare that they have no conflict of interest.

References

1. BRASIL. Ministério da Saúde. Departamento de Atenção Básica. Brasília Homepage, http://dab.saude.gov.br/portaldab/ape_brasil_sorridente.php, last accessed 2017/12/21.

2. BRASIL. Ministério da Saúde. Passo a passo das Ações da Política Nacional de Saúde Bucal. Brasília: Ministério da Saúde, 2016.

3. Moraes, L.; Garcia, R. Proposta de um modelo de Gestão da Tecnologia Médico-Hospitalar. In: Anais do III Congresso Latino-Americano de Engenharia Biomédica CLAEB'2004. João Pessoa, SBEB, 2004. v. 5. p. 309–312.

4. Santos, Francisco de Assis Souza dos. Modelo multicritério para apoio no processo de incorporação de equipamento médico-assistencial. 2014. 151 p. Tese (Doutorado) - Universidade Federal de Santa Catarina, Centro Tecnológico, Programa de Pós-Graduação em Engenharia Elétrica, Florianópolis, 2014.

5. Rocco, E., Garcia, R. Definição de procedimentos para levantamento de produtividade e eficiência em serviços de manutenção de equipamentos eletromédicos - EEM. Florianópolis, 1998. xiii, 100f. Dissertação (Mestrado) - Universidade Federal de Santa Catarina, Centro Tecnológico. Disponível em: <http://www.bu.ufsc.br/teses/PEEL0516-D.pdf>. Acesso em: 17 maio 1999.

6. Garcia, R.; Santos, R.; Souza, R. E. H. Health care technology management applied to public hospitals in Santa Catarina – Brazil. In: Proceedings of First WHO Global Forum on Medical Devices, pages 9–11, 2010.

7. Avelar, P.; Silva, C. A. J. and Garcia, R. Clinical Engineering impact in Primary Health Care – Brazil. In: Proceedings of Third WHO Global Forum on Medical Devices, 2017.

8. Martins, J., Valderes, A. and Garcia, R. Use of the FMEA tool for decision-making in contract management of equipment maintenance and management in Municipal Health Secretariats. In XXI SABI, Córdoba, Argentina, 2017.

9. Rosa, F. and Garcia, R. Health Technology Management for Digital Medical Scales in Primary Healthcare. In VI Congreso Latino Americano de Ingeniería Biomédica – CLAIB2016, Bucaramanga, Colômbia.

10. DeRosier J, Stalhandske E, Bagian JP, Nudell T. Using health care Failure Mode and Effect Analysis: the VA National Center for Patient Safety's prospective risk analysis system. JtComm J QualImprov. 2002; 28(5):248–67.

11. Mcdermott, E. R.; Mikulak, J. R.; Beauregard, R. M. The Basics of FMEA. Taylor & F ed. New York: [s.n.].

Support in the Medical Equipment Incorporation Decision: Hyperbaric Oxygen Therapy Adjunct for Diabetic Foot Ulcers Therapy

Flávio Mauricio Garcia Pezzolla, Priscila Avelar, Jonas Maciel, and Renato Garcia

Abstract

This paper presents a study to assist of a Decision Support Systems and Clinical Engineering Health Technology Management. The methodology is based on methodological guideline for the evaluation of medical equipment addressing its main domains (Clinical, Admissibility, Technical, Operational and Economic) in order to verify the incorporation of Hyperbaric Chambers in comparison to the outsourcing service for the injuries treatment diabetes carriers in Santa Catarina, Brazil. The HBOT application is still very controversial, often generating doubts and making it difficult to make decisions about its incorporation for the public health services. As a result, the Systematic Review, Randomized Controlled Trials and clinical reports were selected in the clinical domain and the operational and technical domain, it is performed a comparative of equipment with its technological resources and service, seeking to analyze parameters that influence in its performance. In the economic domain, through the total cost of ownership, it was estimated it's direct and indirect costs related to the equipment's acquisition and inherent to the life cycle sustainability. HTA for medical equipment present several barriers due to the lack of evidence and quality information, it is expected that this work can generate scientific evidences of knowledge and instruments to enable new research involving hyperbaric chambers, as well as contribute to decision-making or other concomitant programs, due to the application of resources in a planned and adequate decision.

Keywords

Decision support • Clinical engineering
Health technology management • Incorporation decision
HBOT • Diabetic foot ulcers

1 Introduction

The decisions about incorporating, acquiring or covering new technologies and how to use them are among the most important decisions that a health system and its administrators must make [2, 12, 13].

The medical equipment incorporation decision-making process often needs to become systematized involving a multidisciplinary team of specialists [16].

A health technology management with this approach, including engineering and management knowledge applied to the technological process involved in health care reflects on a suitability for the use and continuous improvement of technology [9, 13].

Focusing on the technological process in health quality, it is sought to disseminate ATS actions as a contribution in the performance of clinical engineering, to a greater effectiveness in the process of incorporating medical-assistance equipment to a more appropriate choice of clinical needs [14, 15].

Hyperbaric oxygen therapy (HBOT) emerged under the hypothesis that various diseases and conditions benefit from increased oxygenation of tissues. Although it has been used in the treatment of chronic wounds for about 40 years, its application as routine therapy is still very controversial, in which it often raises doubts and makes a difficult decision regarding its use in public health services. Consequently, the health right judicialization and the limits of judicial action in

F. M. G. Pezzolla · P. Avelar · J. Maciel · R. Garcia (✉)
Instituto de Engenharia Biomédica—UFSC, Florianópolis, Brazil
e-mail: renato@ieb.ufsc.br; ieb-ufsc@ieb.ufsc.br

© Springer Nature Singapore Pte Ltd. 2019
L. Lhotska et al. (eds.), *World Congress on Medical Physics and Biomedical Engineering 2018*,
IFMBE Proceedings 68/3, https://doi.org/10.1007/978-981-10-9023-3_60

these cases sometimes end up influencing in decisions that overlap the incorporation cycle of this technology [8].

Therefore, the research objective, based on the application of HTA for equipment with Hyperbaric Chamber is to carry out a comparison of the outsourced treatment program with the incorporation of this technology in the state of Santa Catarina—Brazil public system.

2 Methodology

The methodology used to elaborate the research was based on the elaboration of studies methodological guideline for the evaluation of medical-assistance equipment developed by the Instituto de Engenharia Biomédica (IEB-UFSC) in partnership with the Brazilian Ministry of Health [5]. It recommends the collection of information in areas such as: Clinical, Admissibility, Technical, Operational and Economic (Fig. 1).

Initially the political question is imposed, that is, the need for information on the decision maker part [10, 14]. This step thus creates a bridge between the political issue and the HTA, since be considered as the starting point. After defining, it will be transformed into a series of HTA questions that will allow you to specify and filter the evidence gathered.

In the admissibility domain, the objective is to present legal and technical support to assess the relevance of a request, both in population and technical aspects. For this, parameters of health care coverage and sanitary control and marketing were evaluated.

For the Clinical Domain, it was necessary to define a specific key question with appropriate inclusion and exclusion criteria for a better research strategy in the scientific literature.

The search for evidence was carried through descriptors in Pubmed, CRD and Cochrane databases, allowing a systematized research, ensuring an overview of the best

Table 1 Pico Question

PICO	Equipment answers
Intervention	Hyperbaric oxygen therapy
Population	Diabetic patients with diabetic foot diagnosis
Comparison	Conventional treatment
Outcomes	Effectiveness and Safety

evidence available, according to the studies methodological quality. In this research we opted for the evidences of Systematic Reviews, Randomized Controlled Clinical Trials and clinical reports (Table 1).

For technical domain, it was performed a detailed analysis of technology, seeking to know their operating principle, its main applications, its configurations in order to draw a comparative between existing technologies.

In this process corresponds to the information generated by studies, field research and management tools applications and management of health technologies focused on equipment. These may come from regulations and should consider technical-operational aspects such as infrastructure, human resources and technology [6, 11].

For the economic domain, the Total Cost of Ownership survey was an important cost management technique used to financial estimate and evaluate the direct and indirect costs related to the health technology acquisition, as well as the costs inherent to maintaining its operation [5].

$$TCO = AC + OC + MC + TC + RC$$

where:

- Total Cost of Ownership—TCO
- Acquisition Cost—AC
- Operating Cost OC
- Maintenance Cost—MC
- Training Cost—TC
- Replacement Cost—RC

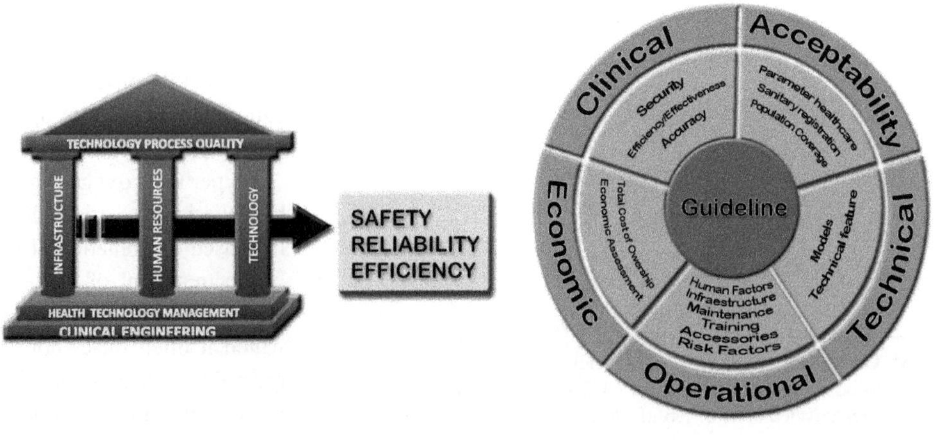

Fig. 1 Domains relevant to support the decision-making process to medical equipment incorporation

In this step an approach was taken of all the relevant costs of incorporation, the use as well as the service, the maintenance and the costs related to a specific supplier [7].

The HTA methodology for equipments is intended to assist in the health care decision-making process. Therefore, HTA can support Clinical Engineering to assist managers in decision making, incorporation or comparison of technologies.

3 Results

In the application of the methodology, the admissibility domain was performed through the verification in the Brazilian regulation system the equipment registry [1]. It was verified that only two devices were registered referring to Hyperbaric Chamber and according to ISO 14971, the Chamber is classified as a medium-risk medical device [4].

Following the survey, the incidence of Diabetes cases in the state of Santa Catarina between 1998 and 2013 was verified, showing the increasing number of cases registered, as shown in Fig. 2 [3].

In the literature review for the Clinical domain, the application of HBOT in the adjuvant treatment of diabetic foot ulcers was identified in four systematic reviews with five to eight controlled or non-randomized controlled trials of reduced samples. Moreover, were identified three not randomized clinical trials in which no presentedfinal conclusions.

The evidence of efficiency in reducing the risk of amputation was only greater in diabetic ulcers resistant to conventional treatment. However, it did not present benefits in randomized controlled trials of good methodological quality.

In the technological domain evaluation, the Hyperbaric Chamber operation was verified. The equipment consists of 01 Hyperbaric Chamber and 01 Set of Air Systems. The

Chamber can be Monoplace type or Multiplace for up to 12 people. The Air System Set for Multiplace comprises: compressor set, electric motor, hyperbaric air conditioning system and 01 oxygen cylinder.

Another technology domain aspect is maintenance related. It was observed that the system requires preventive maintenance and annual tests and a pressure test for hydraulic proof every 10 years, and this maintenance as well as the corrective should be performed by a specialized supplier. There is no protocol for different system layouts and their accessories.

The evaluation human resources domain revealed that a team of specialized professionals is needed for the proper use of the equipment. These professionals are Technicians and Nurses who prepare, monitor the equipment and accompany internally the patients during the treatment and the doctor who accompanies the patient. The training should be conducted directly with the manufacturer along with specialized courses on Hyperbaric Medicine in centers linked to the manufacturer and federal agencies.

For this equipment to be incorporated, it must comply with the infrastructure standards. Initially, a special room is needed to accommodate all elements of the system, whose size can vary from 4 to 6 m^2.

According to the available evidence related to economic evaluation, the total cost of ownership of the system was quantified for acquisition, with an estimate lifespan of 30 years [6].

In Table 2, it can be observed that the system capital and operating costs, with more than 30 years, exceed $8 million, having a total of 7.200 Sessions

The surveys compiled in Table 3 show the values of contracting the service in an outsourced standardized manner by value of session performed.

According to the estimated value of Hyperbaric Oxygen therapy session used in Santa Catarina, it is verified that the

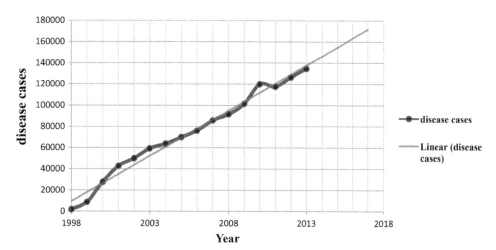

Fig. 2 Incidence of cases of diabetes people over 15 years in Santa Catarina

Table 2 Total cost of ownership of a 12-place hyperbaric chamber

Cost description	Estimation basis	Time (Year)	Estimated value USD $[a]
Acquisition	Search market	–	228.299
Planning	5% of the acquisition	–	11.409
Installation	Search market	–	15.736
Maintenance	$81.467/year	30	2.440.138
Operation	$188.727/year (Session, supplies)	30	5.664.991
Training	$2.202/year	30	66.092
Replacement	Value acquisition + 0,74% (IGP-DI)[b]	30	11.494
Total cost of ownership		30	8.438.159

[a]Quotation at R$3.18 in 2018-01-30
[b]General Price Index—Internal Availability in Brazil

Table 3 Estimated cost of annual session agreement

Description	Estimation basis	Time (Year)	Estimated value USD $*
State value	120 Sessions/10 a month	30	570.276
Municipal value	120 Sessions/10 a month	30	519.128

Table 4 Final recommendations in each domain

Domain	Recommendation	Considerations
Clinical	Low	The level of evidence is 2A and 2C [12]
Admissibility/ Patient	Medium	Another possibility of treatment: conventional treatment and the use of the technology is adjuvant
Operational/ Human Resources	Medium	Requirement of well-qualified human resources; The equipment is of medium complexity and constant training is required
Technical/ Infrastructure	Medium	Adequacy of infrastructure is complex
Technical/ Technology	Low	Maintenance needs are well delimited and should be performed by specialists
Economic	Low	For both treatment cases the equipment exceeds the estimates of inversion to the estimated session value of the technology

total cost of outsourcing within the equipment life span of 30 years totals a value of $519.128 for 2.160 sessions.

Considering the above, according to Table 4, the final recommendation was predominantly low for the incorporation of the Hyperbaric Chamber for the said case.

4 Conclusions

Through the study, it was possible to generate a recommendation to support the decision-making process for Hyperbaric Chamber incorporation into the Santa Catarina State—Brazil.

The efficiency of a technology incorporation is one of the primordial actions of clinical engineering in supporting the decision maker. Its main role in medical-assistance equipment is to ensure a selection and implementation of equipment more appropriate, avoiding possible insecure and inefficient technologies, with low growth in terms of cost-utility.

Currently, health technology assessments are still based on guidelines focused only on clinical assessments, such as those performed especially on pharmaceutical products. The Health Technologies Assessment for equipment [5], acts as a multidisciplinary field of scientific and technical knowledge, aiming at systematic studies of the clinical, social, ethical, legal and economic implications of a health technology, thus promoting subsidies for more reliable decision-making for health and practice policies.

This model allows clinical engineers to improve the health technology process quality, considering not only an analysis related to clinical evaluations and procurement, but also an analysis of the aspects related to maintenance, elimination, infrastructure and human resources necessary for the use of medical equipment. Thus, Clinical Engineering acts holistically and integrated with the needs of health systems.

HTA for medical equipment present several barriers due to the lack of evidence and quality information for a decision making. The performance of clinical engineering should include actions for a perspective of generating studies to improve scientific rigor related to equipment.

Conflict of Interest The authors of this article declare that they have no conflict of interest.

References

1. ASSOCIAÇÃO BRASILEIRA DE NORMAS TÉCNICAS. NBR ISO 14971 – Associação Brasileira de Normas Técnicas. Produtos para saúde – aplicação de gerenciamento de riscos a produtos para saúde. Rio de Janeiro, 2009, 86 p.

2. AUGUSTOVISKI, F.; PICHON-RIVIERE, A.; RUBINSTEIN, A. Critérios utilizados pelos sistemas de saúde para a incorporação de tecnologias. In: NITA, M. E. et al. Avaliação de tecnologias em saúde: evidência clínica, análise econômica e análise de decisão. Porto Alegre: Artmed, 2010.

3. Brasil, MINISTÉRIO DA SAÚDE, 2001b. Indicadores e Dados Básicos – IDB/SUS. Disponível em: <http://www.datasus.gov.br>. Accessin: Nov 2017.

4. BRASIL, Ministério da Saúde. Agência Nacional de Vigilância Sanitária.Tecnovigilância. 2001. Disponível em: <http://www.anvisa.gov.br/>. Access in: Sep 2017.

5. BRASIL. Ministério da Saúde. Departamento de Ciência e Tecnologia. Diretrizes metodológicas: elaboração de estudos para avaliação de equipamentos médicos assistenciais. Brasília: Ministério da Saúde, 2013b. 96 p.

6. BRASIL. Ministério da Saúde. Avaliação de tecnologias em saúde: ferramentas para a gestão do SUS/ Ministério da Saúde, Secretaria-Executiva, Área de Economia da Saúde e Desenvolvimento. – Brasília: Editora do Ministério da Saúde, 2009.

7. CALIL, S. J.; TEIXEIRA, M. S. Gerenciamento de Manutenção de Equipamentos Hospitalares, Série Saúde & Cidadania, v.11. São Paulo: Faculdade de Saúde Pública da Universidade de São Paulo, 1998.

8. FERNANDES, I. A. D. Judicialização Da Saúde: Estudos De Casos – Oxigenoterapia Hiperbárica Lesões Refratárias: lesões pé-diabético. Pouso Alegre – MG: FDSM, 2016.

9. GOODMAN. C S. HTA 101 Introduction to Health Technology Assessment. National Information Center on Health Services Research & Health Care Technology, USA, 2004.

10. MARGOTTI, A. E. Metodologia para incorporação de equipamento médico assistencial em hospitais utilizando a avaliação de tecnologias em saúde na engenharia clínica. Master's thesis, Universidade Federal de Santa Catarina, 2012.

11. PHILLIPS B, BALL C, SACKETT D, et al. Oxford Centre for Evidence-based Medicine - Levels of evidence. Grades of recommendation. Available from: http://www.cebm.net/index.aspx?o=1025.

12. SA'AID, H. B.; STEWART, D.; ENGLAND, I. Decision Making Processes for Introducing New Health Technology at Institutional Level: Decision Makers' Perspective. World Review of Business Research, v.1, n.2, p. 10–19, Mai. 2011.

13. SANTOS, F. A., GARCIA, R., Decision Process Model to the Health Technology Incorporation, In Proc. 32nd Annual International Conference of the IEEE EMBS, Buenos Aires - Argentina, pp. 414–417, 2010.

14. SÔNEGO, F. S. Estudo de Métodos de Avaliação de Tecnologias em Saúde aplicada a Equipamentos Eletromédicos. 2007. 92 f. Dissertação (Mestrado em Engenharia Elétrica) – Centro Tecnológico, Universidade Federal de Santa Catarina, Florianópolis, 2007.

15. WANG, B. Strategic Health Technology Incorporation. In: ENDERLE, J. D. Synthesis lectures on biomedical engineering. Morgan &Claypool, Princeton NJ, 2009.

16. WHO. World Health Organization. Medical Devices: managing the mismatch: an outcome of the priority devices project. Suiça, 2010.

Augmented Reality Technology as a Tool for Better Usability of Medical Equipment

Jonatas Magno Tavares Ribeiro, Juliano Martins, and Renato Garcia

Abstract

The use of medical equipment may be compromised by the lack of knowledge of users about important information and operational characteristics that may cause adverse events. Clinical engineering has as an important function to guide and qualify users in adapting to the use of technology to obtain safer and more reliable health technology processes. The new augmented reality tools, whose main objective is to overlap virtual information in reality through technology, are a good alternative to develop solutions focused on the orientation and qualification of users. This paper presents a proposal to develop a platform for support in the orientation and teaching of medical device users. With this augmented reality platform, through the use of mobile devices, the user can access in real-time information on procedures of adjustments, characteristics, ways of use and control of the medical equipment. This prototype developed for the pulmonary ventilator uses augmented reality in order to enhance its interactions with computer applications more naturally, it seeks to present information about the different ventilation modes, equipment initialization procedures and interactive contents to the user through links and videos. Modern pulmonary ventilators present challenges to users due to the need of knowledge, such as configurations and ways of using parameters; the patient's trigger in the assisted ventilation mode; use of assisted aspiration; pressure alarm setting. These actions are not usually carried out adequately generating possible adverse events, being this situation one of the main objectives of the use of the technology in the platform for the user support. The preliminary results obtained in this prototype characterize this solution as a support tool for activities developed by Clinical Engineering to improve processes in health care.

Keywords

Clinical engineering • Augmented reality
Health technologies management

1 Introduction

Modern medicine is increasingly dependent on technologies and consequently requires multidisciplinary in its processes, involving a large variety of professionals working together to achieve the proposed objectives.

The emergence of new clinical needs and, consequently, improvement of procedures through new techniques resulting from technological innovations implemented in the processes, seek to contribute to efficiency and effectiveness, impacting on improvements in the quality of the health care service. It is necessary to consider the term of usability when relating to the execution of some procedure through the technology. By definition, usability is the way we evaluate the use of a tool proposed to perform some task. Within the area of clinical engineering and healthcare processes, usability is defined as the user's experience when using technology and succeeding in the execution of tasks through it.

With all innovations in the medical field, the patient now becomes the user of hospital technology, increasing the need for qualification to use medical equipment appropriately. Therefore, with the expansion of the use of technology in the field of medicine, it is often necessary to have different types of professionals, such as the biomedical engineer. As the definition of [1], "...biomedical engineers apply the concepts, knowledge, and techniques of virtually all engineering

J. M. T. Ribeiro · J. Martins · R. Garcia (✉)
Biomedical Engineering Institute, Federal University of Santa Catarina, Florianópolis, Brazil
e-mail: renato@ieb.ufsc.br; ceged.tmh@ieb.ufsc.br

J. M. T. Ribeiro
e-mail: jonatas@prottaribeiro.com

© Springer Nature Singapore Pte Ltd. 2019
L. Lhotska et al. (eds.), *World Congress on Medical Physics and Biomedical Engineering 2018*,
IFMBE Proceedings 68/3, https://doi.org/10.1007/978-981-10-9023-3_61

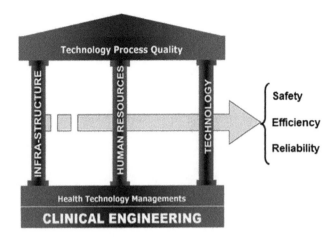

Fig. 1 IEB-UFSC HTM model, where technology process quality is based on 3 domains

disciplines to solve specific problems in the biosphere, i.e., the realm of biology and medicine".

The augmented reality technology is a new method to visualize an event, a process or an execution of an action, when applied in the management of medical technology, and it helps to increase the quality and effectiveness of how to teach the clinical staff and other users to use medical equipment properly.

The project developed at the first human factor engineering laboratory of Brazil [2], at the Biomedical Engineering Institute of the Federal University of Santa Catarina (IEB-UFSC), has one main objective to evaluate the impact of innovative technologies, when used as a tool decrease or avoid errors while using medical equipment, by applying usability technics, based on the 3 domains, as shown in Fig. 1. When applied to management model, it allows increasing health care safety, reliability and efficiency of technological processes. This is the main function for a clinical engineer, which seeks to merge technology applied at the medical field with medical equipment and the infrastructure.

2 Clinical Engineering and Augmented Reality Technology

The advance of technology enabled the development of several areas in our society and also affected directly the area of medicine and patient's health care, contributing with higher levels processes in performance and safety, resulting in a higher quality of life and health care.

The current context of the Clinical Engineer, involving processes and information technologies, is demanding from this professional application of new management practices and management of health technologies. Some of the main functions performed by the clinical engineering are [1]:

- Technology management: development, implementation and coordination of technology management programs. Specific activities related to the acceptance and installation of new equipment, incorporation of medical equipment in telemedicine/tele-health networks, preventive maintenance, and inventory management of equipment;
- Risk management: Appropriate assessments and actions related to adverse events, resulting from the failure or inappropriate use of medical equipment, through technical reports to health managers;
- Technology evaluation: Selection of new technologies and equipment for implementation, to verify and evaluate the impact of the results obtained from the establishment, producing reports containing comparisons, benefits and difficulties;
- Project management and facility planning: Direct assistance in the layout of new or existing clinical facilities, such as operating rooms, imaging centers and radiology centers, health technology simulation centers, among others;
- Training and capacity building: development of training modules for the clinical engineering team and clinical staff of the Health Care Establishment.

Among the functions listed above, there are several other areas that the Clinical Engineer can perform, either within a health care establishment or in the development processes of new healthcare products on digital platforms, such as the prototype developed and presented in this article, is using technology to monitor and evaluate the performance of medical equipment, like the mechanical ventilator [3].

Augmented Reality technology emerged from the development of Virtual Reality technology. Virtual reality is based on inserting the user into a virtual environment, following development's schedule to that program. Currently, virtual reality is used in different areas, such as: training and teaching; recreation; study; etc. [4].

Augmented Reality technology can be defined as "… the enrichment of the real environment with virtual objects using some technological device, working in parallel" [4] and its main objective is to overlap objects and virtual environments with the physical environment, through the device selected. Its development was facilitated due to the technological advance, mainly in the 90's, through the miniaturization of electronic components and the popularization of battery fed personal devices and the internet's expansion, increasing the capacity of connection between users transferring data.

The development of apps using augmented reality technology has reached a higher level of maturity, due to the internet, with technical support and specific programming libraries, available for download, resulting in a wide dissemination of knowledge and applicability in several areas, such as fixing bugs or searching innovative solutions.

Virtual reality and augmented reality can now be considered as the new generation of interface. This new type of technology differs from other generations, through user's new ways to do things and how quickly it can be done. These characteristics are highly desirable to follow technological trends emerging around the world.

3 Methodology

The prototype presented in this article is at stage of development, focusing on the evaluation of the technology and getting conclusions about its principles and utilization. The next step on the process of technology incorporation is to validate all results obtained with the prototype and compare them with all the traditional, inferring conclusions about how efficient they actually are.

The evaluation was made using an augmented reality technology application for devices running Android operational system. The application of the prototype was performed on a pulmonary ventilator, available at the human factor engineering laboratory, as referred on [5, 6].

The development of the training tool prototype was focused on studying the basics of pulmonary ventilator, to help users in intensive care units, to identify adverse events that were related to the equipment's lack of knowledge.

Studying the theme of augmented reality technology, a survey was made to identify all augmented reality technology apps available and its platforms. Then, apps were selected by checking the main characteristics for the application in medical equipment.

Then a study was realized to identify possible problems when developing an app using augmented reality technology. In this study, the criteria used for selection was to check the most widely used apps among members of the developers' community and select one compatible with the device used on the project (Samsung device with Android OS). For this type of OS, a popular tool was used, called Android Studio, which allows several types of programming languages interface development using graphic tools and XML. Java was selected as the main programming language for the app.

For the development of specific purposes apps special libraries are used which are basically preformatted with code sets, activating specific functions on devices. Usually these libraries are open-source and available to developers' community.

Based on studies mentioned earlier in this article, the prototype tool developed to increase usability of medical equipment was built using a specific augmented reality technology app, which controls the built in camera of the device, scanning images and then triggering the prototype start up, through digital data processing. An on-line tool was used to structure the prototype, by creating several projects, editing them and sharing them over the internet.

The content inserted (Fig. 2) on the prototype was selected according to their possible contributions to achieve the final objective (Fig. 3): user's instructions and contribution to the usability of the equipment. These data are divided as:

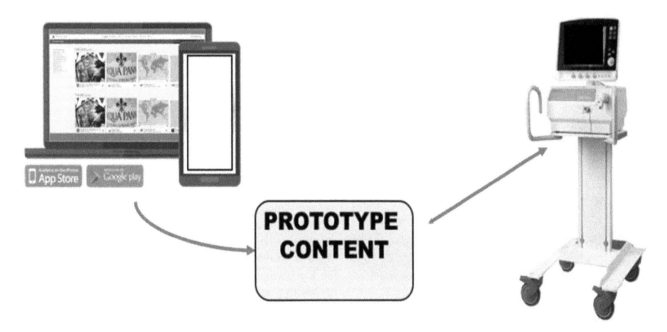

Fig. 2 Augmented reality prototype's diagram. Users can access information about the medical equipment through the prototype on the device

- Simplified and objective content presentation: Equipment ventilation modes available, with images from the data-sheet and identification keys, equipment structure and accessories;
- Secondary presentation of content: video content, allowing users to actually watch what they need to learn or how to perform on the medical equipment, websites containing information about the equipment or even online training tools developed by IEB-UFSC;
- Additional content presentation: video and audio format guideline content.

4 Results and Discussions

For validation purposes, test procedures were performed in three levels of equipment usage experience:

1 Beginner: Low need and experience in the use of medical equipment. Users usually perform some function that is not directly related to the equipment;

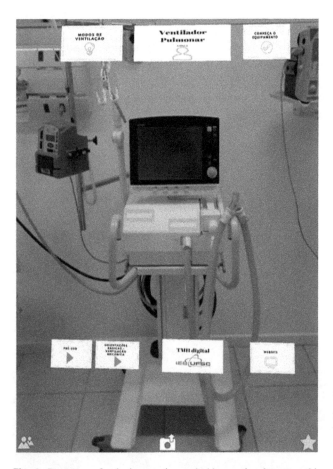

Fig. 3 Prototype of a device running android operational system; this is an example about how the user interact with the technology and how information are presented on the device's screen, when it's camera is pointing and focusing on the medical equipment

2 Intermediate: Average need and experience in the use of medical equipment. User who belong to the technical staff of the facility and have some technical (specific) knowledge about the equipment;

3 Advanced: High need and experience in the use of medical equipment. User belong to the clinical staff, such as doctors, physiotherapists and nursing professionals.

The validation method was based on evidence, analyzing results obtained by the application of two questionnaires. The first one elaborated was related to user's interview and their previous experience with this type of technology. The second one was related to user's experience during the validation process, with questions about personal opinions related to the prototype and technical questions about equipment's data inserted into the prototype.

According to the results obtained through the process of validation, it was possible to identify the impact that innovative technologies could cause in health care facilities. According to the answers obtained from the second questionnaire, it's possible to infer that the prototype is useful as an auxiliary tool for training purposes. This kind of technology allows important technology management model's updates and could be expanded to control other medical equipment, according to the facility's reality and necessities. With a greater control of the medical equipment, and consequently, greater control of the environment in which clinical engineer is inserted, it is possible to infer that the use of new technologies could also assist on management and quality control processes, increasing quality and safety on health care, and could also decrease adverse events due to the lack of knowledge when using any medical equipment.

Another important result verified was that the necessary time to learn how to use this technology in a consistent and efficient way was low, showing that even beginners could use it within a few minutes of practice.

5 Conclusion

As a part of the job of a clinical engineer, it is necessary to be aware for innovations and emerging technologies, in order to implement them on health care facilities successfully. Preparing the environment to receive new technologies is very important, leading to the development of better technology incorporation processes and training courses. As a result, it could be possible to avoid or reduce negative impacts on quality and performance of health care centers.

The results of the evaluation procedures realized on the technology solution prototype could be very useful for the clinical engineering, introducing an innovative methodology for teaching and training purposes on medical equipment. Augmented reality technology could be used as a usability tool in a big variety of situations on a health care facility, such as: user assistance tool; teaching tool; equipment identification and innovation processes.

Based on studies realized during this project, all results obtained and the discussions made, it is possible to verify that the use of new technologies could impact on a positive way health care processes and equipment maintenance, according to existing technology management models [7].

The next step on this project could be the development of a specific application system based on augmented reality technology to train and guide users of various medical equipment.

Then, it would be possible to define the structure and functionality of the app and the most appropriate interface, implementing different functions using augmented reality technology according to the specific need of the health care facility and based on the studies realized on this project.

So, augmented reality technology can be considered an useful alternative and innovative training tool to assist users in the use of medical equipment, and that this new type of technology actually represents a fast and intuitive new way to interact with situations involving modern technology.

Conflict of Interest The authors of this article declare that they have no conflict of interest.

References

1. Joseph D. Bronzino, F.: "The BiomedicalEngineering Handbook",first edn, CRC Press, 1995;

2. Rosa, F.; Garcia, R.; "Qualification in Medical Technology Using Clinical Engineering Laboratories". SABI 2017; 2017 Out 25-Out 27; Cordoba, Argentina;

3. Humberto Pereira da Silva, "Sistema Microcomputadorizado para a Avaliação de Desempenho de Ventiladores Pulmonares", Universidade Federal de Santa Catarina, 2001;

4. Cláudio Kirner, Robson Augusto Siscoutto: " Fundamentos da Realidade Virtual e Aumentada", IX Symposium on Virtual and Augmented Reality, Petrópolis, RJ, 2007;

5. Signori, M.; Garcia, R.,"Clinical Engineering Incorporating Human Factors Engineering into Risk Management", IFMBE Proceedings, v. 25, p. 449–452, 2009;

6. Reis, C. S.; Delgado, M. A.; V., J. A.; Garcia, Renato, "O Fator Humano nas Ocurrencias de Falhas com Tecnologias Médicas", In: XXIV Congresso Brasileiro de Engenharia Biomédica – CBEB2014, Uberlândia – MG. CBEB2014 – XXIV Congresso Brasileiro de Engenharia Biomédica, 2014. V.1. p. 867–870;

7. Carlos Gontarski Esperança, "Estudo de Metodologias para Gerenciamento de Ventiladores Pulmonares", Universidade Federal de Santa Catarina, 1996;

8. American College of Clinical Engineers (ACCE) Homepage, http://accenet.org/about/Pages/ClinicalEngineer.aspx, last accessed 2018/01/08;

9. Marcelo de Paiva Guimarães, Bruno BarberiGnecco, Rodrigo Damazio, "Ferramentas para Desenvolvimento de Aplicações de Realidade Virtual e Aumentada", IXSymposiumon Virtual andAugmented Reality, Petrópolis, RJ, 2007;

10. Dhiraj Amin,SharvariGovilkar, "ComparativeStudy ofAugmented Reality SDK′S", International Journal on ComputationalSciences&Applications, VOL. 5, 2015;

11. Vuforia Developer Homepage, https://developer.vuforia.com, last accessed 2018/01/10;

12. Aurasma App Homepage, https://www.aurasma.com/get-started, last accessed 2018/01/07;

13. Moraes, Luciano de; Garcia, R., "Proposta de um Modelo de Gestão da Tecnologia Médicohospitalar", IFMBEProceedings, v.5, n.2004, p. 309–312, 2004;

Hospitals With and Without Clinical Engineering Department: Comparative Analysis

Marcelo Horacio Lencina, Sergio Damián Ponce, Débora Rubio, Bruno Padulo, and Gustavo Ariel Schuemer

Abstract

In Buenos Aires province, first and second levels are provided by City government, whereas third and fourth levels are provided by 77 provincial Hospitals. The Public System is free of charge for all citizens. In these hospitals technical support of medical devices, preventive and corrective maintenance, has been and is carried out by the vendors´ technical services (manufacturers and distributors), or by companies that are dedicated to the technical service of medical equipment. These services do not include the permanent presence of technicians or engineers in the hospital. There is an exception to the procedure mentioned in 18 hospitals, which, by means of an agreement between the MoH and the National Technological University, have implemented clinical engineering departments. The department activities are carried out by graduates and undergraduate students, who perform tasks within the Hospital every weekday in an office assigned for that purpose, and with the support of the University structure. The object of this work is to establish qualitative and quantitative differences between two hospitals with the same characteristics of health services, one with department of clinical engineering and the other with contracted services, both with the modalities explained. For this purpose there were measurements of the time that medical devices have been out of use for damage, repair costs, customer satisfaction surveys of users, administration personnel and management.

Keywords

Clinical engineering • Management • Efficiency Effectiveness

M. H. Lencina (✉) · S. D. Ponce · D. Rubio
Facultad Regional, Universidad Tecnológica Nacional, Colón 332, San Nicolás, Argentina
e-mail: mhlencina@yahoo.com.ar

B. Padulo · G. A. Schuemer
Hospital Interzonal General de Agudos San Felipe, Moreno 31, San Nicolás, Argentina

L. Lhotska et al. (eds.), *World Congress on Medical Physics and Biomedical Engineering 2018*,
IFMBE Proceedings 68/3, https://doi.org/10.1007/978-981-10-9023-3_62

1 Introduction

The province of Buenos Aires is the one that has the largest area of the Argentina Republic, 307,571 Km^2, with a population of 15,625,084 inhabitants [1]. Within this territory is the Federal Capital or Autonomous City of Buenos Aires with a sur-face of 200 Km^2 and a population it is of 2,890,151 inhabitants [1]. Called Gran Buenos Aires at 19 Municipalities of the province surrounding the Federal Capital, these Municipalities has an area of 3,680 Km^2 and a population of 9,976,115 inhabitants [1].

This data suggests that the province is divided into two sectors (Fig. 1), the Gran Buenos Aires with a population density of 2,964.75 people per/Km^2 and the rest of the province with a population density of 18.56 people per km^2 [1].

The Ministry of Health has divided the province area in 11 Health Regions with 13,600 beds distributed in 77 hospitals, which are mostly located in the city of La Plata (provincial capital) and in the Gran Buenos Aires, 12 and 29 hospitals respectively [2].

The hospitals to be compare are the "San Roque" in Gonet town, near the city of La Plata, capital of the province; and the "San Felipe" of San Nicolas city, located 300 km North of the provincial capital (Fig. 1). Both hospitals are reference in their respective health regions and are third level according with the classification of the MoH [2].

The aim of this work is to demonstrate that hospitals that work with clinical Engineering Department are more efficient and effective in health technologies management.

2 Hospitals

2.1 Characteristics

Both hospitals have the same total amount of beds (160), but have some differences in critical care beds, as it can be seen in Table 1.

Fig. 1 Buenos Aires map with
Health division and GBA area
expanded map

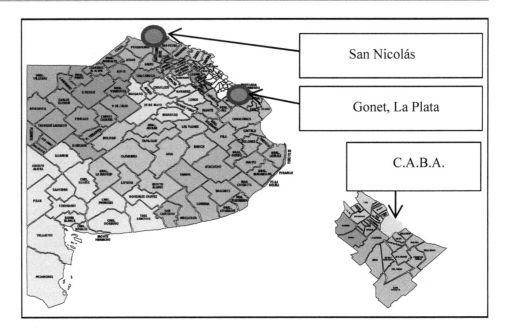

Table 1 Intensive care beds

Unit	Hospital San Roque	Hospital San Felipe
NICU	30	30
Adults ICU	14	7
CICU	6	4
PICU	0	8
TOTAL BEDS	50	49

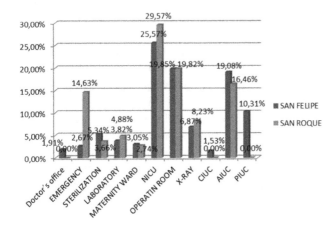

Fig. 2 Medical devices distribution per unit

Respective inventories, note that the "San Roque" has 328 medical devices, versus 262 of the "San Felipe". These are distributed in different units as shown in Fig. 2.

2.2 Operation

Hospital San Felipe, has in its staff a biomedical engineer and an electronics engineer with a specialty in clinical engineering, which, together with the Group of the National Technological University (U.T.N.), make the Clinical Engineering Department up. This group develops activities in the hospital through an agreement between the MoH and the U.T.N. since 1991.

The CED staff operates in an office which, in addition to repairs, calibrations of medical devices are carried out. Some of these calibrations are carried out in the laboratory of the University, since calibration instruments cannot be transported to the hospital.

CED develops its work from Monday to Saturday along with of the rest of the staff of the hospital, which provides the advantage of quickly attention to the solution of the problems that arise on a daily basis. The main activities carried out by the CED are:

- Planning and selecting new technologies.
- Developing specifications for procurement of acquisition and/or maintenance of medical devices.
- Inspecting initial and start-up of the new equipment.
- Monitoring and controlling the work carried out by technical service companies.
- Training users in the use of medical technologies.
- Managing supplies and spare parts.
- Preventive maintenance, according to the protocols of the manufacturers.
- Corrective maintenance: put in operating conditions, statistical data on failures.

Hospital San Roque has no engineers nor technicians on their staff for the management of medical devices. For some complex equipment, they have contracted maintenance service companies or the technical service of the manufacturer or Distributor. These are:

- X-ray equipment
- CT
- Processor for X-ray film
- Neonatology ventilator
- Incubators
- Sterilization equipment.

These represent 15.55% of its inventory.

Technical clauses of these contracted services are not made by the hospital, but by the medical technology of the MoH Office. In general, the methodology applied, is that companies must make a monthly visit to perform the preventive maintenance, without giving technical specification of these. As for corrective maintenance by failures, the company has 24 h to go and fix it.

Important or high price spare parts, as well as the inputs that are required for the operation of the equipment, are not included in the contract.

Compliance with the dates that are assigned by the clauses of the contract to go to the hospital is controlled by the heads of units, as responses of calling when equipment failure occurs. Therefore, it is a strictly administrative control.

3 Comparative Analysis

3.1 Cost Assessment

The agreement between the MoH and the U.T.N. operates with six students of the University, who, through a system of grants are trained and carry out activities of the CED. This activity takes place under the supervision and instruction of two clinical engineers. In 2017, the agreement, represented for Hospital San Felipe, a monthly cost of $85,683, approximately USD 4,284.

The CED in this period carried out 997 tasks on their equipment, in preventive and corrective maintenance. The total cost of supplies and spare parts for them was $42,276 (USD 2,114). This amount arises from the database of the CED. In this way, it can be appreciated that the higher percentage of tasks is performed in the following units: AICU (21.17%), Operating Room (17.79%), Sterilization (11.96%), PICU (10.84%), NICU (10.63%).

Repairs carried out by the official technical services for replacement of parts with respective calibration and certification, according to purchase orders issued by the hospital

amounts to $266,618.54 (USD 13,331). This amount corresponds the 11.5% for operating room equipment and the 88.5% for X- ray equipment and CT.

The total investment made by Hospital San Felipe in 2017 for the operation of theCED, was $1,337,090.54 (USD 66,855), as detailed in Table 2, and its evolution in Fig. 3.

In the same year Hospital San Roque carried out preventive and corrective maintenance with technical service companies. These included only part of the medical devices belonging to some units of the hospital, which are described

Table 2 Investments

Annual investments	Amount
Agreement	USD 51,410
Spare parts	USD 2,114
Purchases	USD 13,331
Total	USD 66,855

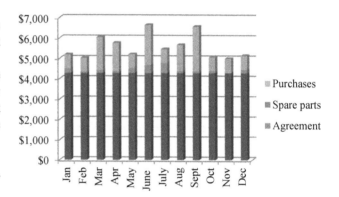

Fig. 3 Task cost evolution

Table 3 Maintenance investment

Unit	Annual amount
X-Ray	USD 104.488,80
NICU	USD 35.593,00
Sterilization	USD 12.954,00
Medical gases	USD 16.302,00

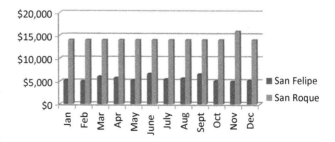

Fig. 4 Monthly cost comparative evolution

Fig. 5 Survey result

QUESTION	Hospital San Felipe				Hospital San Roque			
A	Inmediat	< 24 Hs	24 > x > 48	> 48 Hs	Inmediat	< 24 Hs	24 > x > 48	> 48 Hs
	23,3%	63,3%	6,7%	6,7%	0,0%	0,0%	6,7%	93,3%
B	< 24 Hs.	24 > x > 48	> 48 Hs		< 24 Hs	24 > x > 48	> 48 Hs.	
	15,0%	48,3%	36,7%		0,0%	0,0%	100,0%	
C	Yes	No	Sometimes		Yes	No	Sometimes	
	100,0%	0,0%	0,0%		72,4%	21,0%	6,7%	
D	Yes	No	Sometimes		Yes	No	Sometimes	
	85,0%	0,0%	15,0%		85,0%	6,7%	8,3%	
E	Yes	No			Yes	No		
	100,0%	0,0%			100,0%	0,0%		

in Table 3. The value of the contracted services was $3,367,116 (USD 168,356). The devices included in X-ray contract are: 1 CT (68.19%), 2 X-ray equipment (7.57%), 6 mobile X-ray (11.68%), 3 C arm fluoroscopy (4.95%), 1 mammography (2.33%), others (5.28%). In NICU are: 19 incubators (29.93%), 10 ventilators (70.07%). In sterilization are: 3 autoclave (100%). Medical gases distribution (100%).

Since inputs and spare parts are not included in these contracts, according to this mode of recruitment, an investment of $36,993.93 (USD 1,850) was made by the same.

Data show that the total investment in maintenance for four units of the hospital was $3,404,109.93 (USD 170,206).

Data obtained analysis from both hospitals for 2017, show that Hospital San Felipe for all units invested $1,337,090.54 (USD 66,855); while Hospital San Roque for four units, invested $3,404,109.93 (USD 170,206). Figure 4 show the comparative evolution.

Data emerges that the net investment made by Hospital San Roque is 154.56% higher than Hospital San Felipe.

It should also be considered that, while Hospital San Felipe includes all medical devices of its inventory, Hospital San Roque includes only 15.55% of them.

3.2 Surveys

Surveys with technical staff and nurses, heads of units and directors were done. The questions asked were: (A), What is the response time of the technical staff to assess the inconvenience of a medical device fails?, (B) How long does take since withdrawing a medical device of the Unit for repair and reinstatement?, (C) When you receive a medical device repaired, the verification of the correct operation is carried out?, (D) When you receive a new medical device, are made performance verifications?, (E) Do you consider important to have technical advice for decision-making?

The results of the responses obtained are shown in Fig. 5. It was also requested to respondents, to qualify from 1 to 10

for the received service. Average value described by the staff of Hospital San Roque was 4.38, while that of San Felipe was 7.25.

4 Conclusions

The results show that a hospital with CED is more efficient, since it achieved acceptable results with low investment of its budget, solving problems in the majority of medical devices and managing the solution of the most complex. The survey results show that the procedures adopted by the CED in Hospital San Felipe should be modified in order to achieve greater effectiveness.

In a Hospital without CED, the investment per maintenance contracts is very high and the works carried out have not technical control. Medical devices that are not under maintenance contract, to a fault, must wait too long for its solution, which is more expensive and at the same time decreases the ability to care for patients.

The survey shows, that the response time and medical devices reinstatement are better in Hospital with CED.

The University has agreements with other 17 provincial hospitals, and the results shown correspond to two chosen at random. For the purpose of simplifying the paper, it is that we chose to present only these two cases.

A report to the MoH with the results will be prepared, to incorporate personnel policy to be implemented to start the CED in hospitals that still does not have it.

References

1. I.N.D.E.C. Instituto Nacional de Estadísticas y Censos. Censo Nacional 2010.
2. Ministerio de Salud de la Provincia de Buenos Aires, http://www.ms.gba.gov.ar.

Evaluation of Downtime of Linear Accelerators Installed at Radiotherapy Departments in the Czech Republic

Vladimir Dufek and Ivana Horakova

Abstract

The National Radiation Protection Institute (NRPI) performed the study evaluating linear accelerator (linac) unscheduled downtime and other parameters related to linac failure (e.g. treatment cancellations, patients transferred to other linac, patients treated with modified dose fractionation) at radiotherapy departments in the Czech Republic. Thirteen radiotherapy departments with at least one linac (out of 21 departments in the Czech Republic) voluntarily participated in the study covering 29 out of 47 linacs. Downtime was evaluated for a one year period from July of 2016 to June of 2017. The methodology was as follows: NRPI designed the data entry form which was sent electronically to medical physicists at participating radiotherapy departments. Data related to linac failures were filled in. The completed forms were evaluated by NRPI. Unscheduled downtime was defined as time when linac cannot be operated during operating hours. Unscheduled downtime per linac per year ranged from 4 to 222 h (mean = 73 h, median = 61 h). Downtime percentage calculated as a ratio of downtime and total sum of operating hours per year ranged from 0.2 to 7.6% (mean = 2.8%, median = 2.2%). The number of treatment cancellations per linac per year ranged from 0 to 661. Unified methodology enabled objective comparison of linac downtime at particular radiotherapy departments in the Czech Republic. The study confirmed usefulness of having minimally two matched linacs at a department. The results of this study could help radiotherapy departments negotiate better service contract (e.g. agreement on maximum guaranteed downtime).

Keywords

Downtime • Linear accelerators • Radiotherapy

V. Dufek (✉) · I. Horakova
National Radiation Protection Institute, Bartoskova 28, 140 00
Prague, Czech Republic
e-mail: vladimir.dufek@suro.cz

© Springer Nature Singapore Pte Ltd. 2019
L. Lhotska et al. (eds.), *World Congress on Medical Physics and Biomedical Engineering 2018*, IFMBE Proceedings 68/3,
https://doi.org/10.1007/978-981-10-9023-3_63

1 Introduction

Linear accelerator (linac) failures complicate clinical operation at the radiotherapy departments. Moreover, these failures (particularly the prolonged ones) lead to the unscheduled interruptions in radiotherapy treatment and as a result complicate correct realization of the treatment from a radiobiology perspective.

Potential unreliability of the linacs increases the risk of unintended irradiation of patients, thus there is a link to radiation protection of patients. For the needs of the State Office for Nuclear Safety, which is national regulatory authority responsible for radiation protection and safety, the National Radiation Protection Institute (NRPI) performed the study evaluating linac downtime and its clinical impacts (e.g. treatment cancellations and patients transferred to other linac) at radiotherapy departments in the Czech Republic.

Thirteen radiotherapy departments out of 21 departments in the Czech Republic equipped with at least one linac voluntarily participated in the study covering 29 out of 47 linacs. Downtime and its clinical impacts were evaluated for a one year period from July of 2016 to June of 2017. Data were evaluated anonymously. Out of 29 linacs, 18 linacs of vendor No. 1 and 11 linacs of vendor No. 2 were analyzed within the study. Years of linac installation ranged from 2003 to 2016 (mean = 2010, median = 2009).

2 Materials and Methods

The parameters evaluated within the study were as follows: number of failures (number of interruptions) that led to linac downtime, downtime, downtime percentage, number of treatment cancellations, number of treatments when patients were transferred to other linac at the department, number of patients treated with modified dose fractionation and number of treatments when verification of patient positioning could not be performed.

The methodology used in the study was as follows: NRPI designed the data entry form in Excel which was sent electronically to medical physicists at participating radiotherapy departments. In this form, data related to linac failures were filled in (e.g. time of occurrence of linac failure, time of report linac failure to service organization, time of service engineer arrival, time of linac being repaired, time of interruption of clinical operation). Also clinical impacts, e.g. number of treatment cancellations and number of treatments when patients were transferred to other linac, were filled in. Completed forms were sent back to the NRPI, where forms from all departments were gathered, reviewed, corrected (if needed) and evaluated.

Unscheduled downtime was defined as time when linac cannot be operated during operating hours. Only interruptions of half an hour and longer were recorded. Preventive service maintenance work did not count as downtime.

Downtime percentage was calculated as a ratio of downtime and total sum of operating hours per year. The total sum of operating hours per year was calculated as a product of typical daily operating hours (e.g. 12 h) as reported by departments and the number of operating days per year. Operating hours relate to the hours when patients are treated. The hours for performing linac quality assurance are not counted.

Treatment cancellations are defined as instances when patients due to the linac failure could not be treated on the scheduled treatment day.

Apart from the downtime also the work time limitation was recorded. It is the time when linac is in clinical operation but there is a failure of one of its part, e.g. kilovoltage (kV) imaging system. Clinical impacts related to the work time limitation were evaluated.

3   Results

3.1   Downtime

The number of failures per year that led to linac downtime at 13 radiotherapy departments ranged from 2 to 44 (mean = 12, median = 9).

Unscheduled downtime per linac per year ranged from 4 to 222 h (mean = 73 h, median = 61 h). Downtime percentage per linac per year found out at 13 radiotherapy departments is shown in Fig. 1 and ranges from 0.2 to 7.6% (mean = 2.8%, median = 2.2%).

3.2   Clinical Impacts

The number of treatment cancellations per linac per year found out at 13 radiotherapy departments is shown in Fig. 2.

Two departments (No. 7 and 13) did not provide such data. The number of treatment cancellations per linac per year ranged from 0 to 661 (mean = 107, median = 56).

During linac downtime 9 out of 13 departments transferred patients to other matched linac at the department. The number of treatments when patients were transferred to other matched linac at the department ranged from 32 to 906 (mean = 216, median = 108).

Due to the linac downtime two departments modified dose fractionation for patients. At one department the dose fractionation was modified for five patients, the second department did not provide such data.

Due to the linac failures verification of patient positioning using kV imaging system could not be performed on seven linacs at five radiotherapy departments. At one department this verification could not be performed in approximately 210 treatments on one linac and in 266 treatments on the other linac. At the second department verification of patient positioning could not be performed in 16 treatments on one linac and in 5 treatments on the other linac. Three remaining departments did not provide such data.

4   Discussion

4.1   Downtime

As can be seen in Fig. 1, there is a variation in linac downtime percentage across the departments. The reason is that downtime percentage is influenced by many factors. Downtime percentage depends not only on reliability of the individual linac operation but also on other factors such as complexity of treatments (3D conformal radiotherapy (3DCRT), intensity modulated radiotherapy (IMRT), volumetric modulated arc therapy (VMAT) or stereotactic radiotherapy), skillfulness of individual service engineer, skillfulness of the personnel at the department (potential inexpert linac handling) and local conditions (e.g. service engineer is part of the personnel or his residence is very close to the department). There is also a variation in linac downtime percentage across the individual linacs at the departments.

Statistical analysis shows that the differences between the mean of downtime percentage for vendor No. 1 and 2 are not statistically significant at the alpha level of 0.05. Unscheduled downtime per linac of vendor No. 1 per year ranged from 0.2 to 5.4% (mean = 2.3%, median = 2.1%). Unscheduled downtime per linac of vendor No. 2 per year was higher and ranged from 1.0 to 7.6% (mean = 3.6%, median = 2.4%).

There is no relation between the age of the linacs and the downtime percentage and also there is no relation between the type of service contract (full service contract versus no service contract) and downtime percentage.

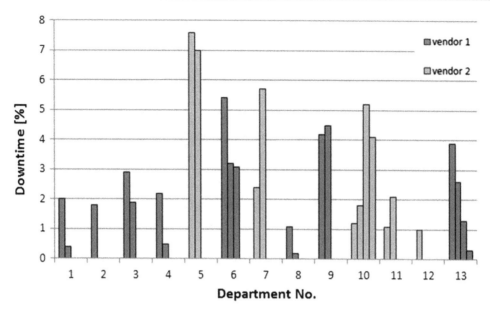

Fig. 1 Downtime percentage per linac per year found out at 13 radiotherapy departments. The columns represent individual linacs at radiotherapy departments. Blue columns represent linac vendor No. 1, green columns represent linac vendor No. 2 (Color figure online)

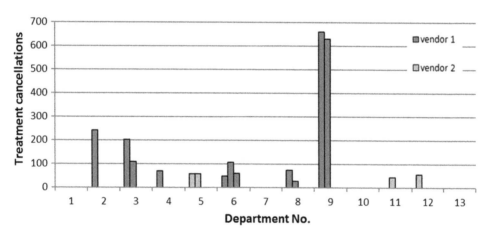

Fig. 2 Treatment cancellations per linac per year found out at 13 radiotherapy departments. The columns represent individual linacs at radiotherapy departments. Departments No. 7 and 13 did not provide such data. Department No. 1 and 10 had 0 treatment cancellations for all linacs. Department No. 4 had 0 treatment cancellations for second linac. Department No. 11 had 0 treatment cancellations for first linac. Blue columns represent linac vendor No. 1 and green columns represent linac vendor No. 2 (Color figure online)

On the basis of the detailed investigation of downtime, it can be mentioned that lower downtime was a little more often observed at linacs with a higher proportion of 3DCRT plans in relation to more complex techniques such as IMRT or VMAT. Conclusive findings related to downtime would require longer period of the study, e.g. 4 or 5 years.

For all failures leading to linac downtime we separated downtime into five items: time elapsed from occurrence of failure to sending a report on failure to service organization, time elapsed from sending a report to arrival of service engineer, time to repair, time of linac performance testing after the repair and time of the repairs performed by local physicists. We added up these partial times for all failures at all 13 departments and founded that 59% of the total downtime (calculated from all 13 departments) makes up the time to repair and 31% of the total downtime makes up the time to the arrival of service engineer. Contribution of the time of the repairs performed by local physicists to the total downtime is 6%, contribution of the time to the report on failure and contribution of the time of linac performance testing after the repairs is only 2%.

4.2 Clinical Impacts

The number of treatment cancellations depends on the fact whether the departments have another matched linac where in case of linac downtime patients could be transferred without the need of treatment plan recalculation. Two departments without treatment cancellations (department No. 1 and 10, see Fig. 2) have the matched linacs and transferred patients to the matched linacs. On the contrary, two departments with the highest number of treatment cancellations (department No. 9 and 2, see Fig. 2) do not have matched linacs, therefore could not transfer patients to the matched linacs.

For eliminating clinical impacts of downtime some departments had to add extra work shifts and at some departments performing linac quality assurance had to be rescheduled, e.g. on the weekend or on late in the evening.

Recording of clinical impacts related to downtime was too complicated and too time consuming at some departments, thus they did not provide these data.

5 Conclusion

Within the study, 29 out of 47 linacs were analysed. Hereby the overview of linac downtime in the Czech Republic was obtained. Unified methodology enabled objective comparison of linac downtime at particular radiotherapy departments.

The study confirmed usefulness of having minimally two matched linacs at a department. This, in case of linac downtime, enables transfer patients to another linac without the need of treatment plan recalculation and, as a result, decreases the number of treatment cancellations.

The results of this study could help radiotherapy departments negotiate better service contract. The specification of maximum guaranteed downtime together with the specification of linac operating hours in the service contract should be a good practice. Specified linac operating hours should be used for downtime evaluation. Each department should itself evaluate linac downtime and should require penalty in case of exceeding the guaranteed downtime.

The results of this study could also push service organizations forward to improve their service quality. Downtime should be one of the service quality indicators.

Recording of clinical impacts related to linac downtime was too complicated and too time consuming at some departments. Vendors should try to develop tools enabling simple evaluation of linac downtime and related clinical impacts.

Acknowledgements This work was done under a contract by the State Office for Nuclear Safety of the Czech Republic. The authors would like to thank the medical physicists and other personnel from radiotherapy departments who took part in this study for their cooperation.

Conflict of Interest The authors declare that they have no conflict of interest.

Application of Engineering Concepts in the Sterile Processing Department

Hermini Alexandre Henrique, Borges Ana Carolina, and Longo Priscilla

Abstract

This study has been performed in the Sterile Processing Department (SPD) at Women's Hospital "Professor Doctor José Aristodemo Pinotti in the State University of Campinas. The objective of this study is to evidence the importance of using engineering concepts to manage the processes of Sterile Processing Department. By applying the value stream mapping in the SPD it was possible to identify some problems in the process such as bottlenecks and nonconformities according to the Brazilian current sanitary legislation. Engineering tools were applied to understand how the processes were performed at CAISM. Measurements were collected from specific steps throughout the process to measure the cycle time of each step and then calculating the process lead time, the process cycle efficiency and the percentage of non-value-added time. The PDCA was also applied in order to implement and monitor the changes. Additionally, new methods and culture have been adopted to create a more collaborative, efficient and engaging environment.

Keywords

Process management • Value stream mapping
Process cycle efficiency • Percentage of non-value-added time • PDCA

1 Introduction

Sterile Processing Department (SPD) comprises a service within the healthcare facility in which healthcare products are decontaminated or sterilized in order to ensure high standard infection prevention and control and comply with the resolutions RDC n° 08/2009 [1] and RDC n° 15/2012 [2] of Brazil National Health Surveillance Agency (ANVISA) [3]. The SPD also maintains the flow of sterile supplies efficiently and effectively throughout the healthcare facility.

According to the World Health Organization (WHO) [4] in 2016 approximately 14% of hospitalized patients in Brazil acquired some hospital infection, and approximately 100,000 people die from these infections each year, demonstrating the impact of Hospital-Acquired Infection (HAI) in health care. Among all factors that contribute to the occurrence of HAI, it is possible to highlight the inadequate hand washing, the indiscriminate use of antibiotics and the use of contaminated items.

2 Methods

The value stream mapping (VSM) was drawn to understand how the processing of products was performed, in other words, it was meant to be a picture of the status quo. A tool of Human Factors Engineering called "shadowing" was used. This tool consists in following the employees routine, observing how they carried out each process activity and how they interacted with each other. The VSM can be seen in the Fig. 1.

Engineering tools were applied to define, measure, analyze, improve and control the process. Tools such as prioritization matrix, statistics software, PDCA cycle were used in this study as well.

3 Results

With prioritization matrix it was possible to identify the five main problems in the SPD of CAISM. Sixteen problems were found, however just the top five main problems were selected for this study.

The first process problem was the *lack of Standard Operating Procedures (SOP)*. The lack of SOP was present

H. A. Henrique (✉) · B. A. Carolina · L. Priscilla
Women's Hospital "Professor Dr. José Aristodemo Pinotti",
CAISM—UNICAMP, Campinas, São Paulo, Brazil
e-mail: hermini@g.unicamp.br

© Springer Nature Singapore Pte Ltd. 2019
L. Lhotska et al. (eds.), *World Congress on Medical Physics and Biomedical Engineering 2018*,
IFMBE Proceedings 68/3, https://doi.org/10.1007/978-981-10-9023-3_64

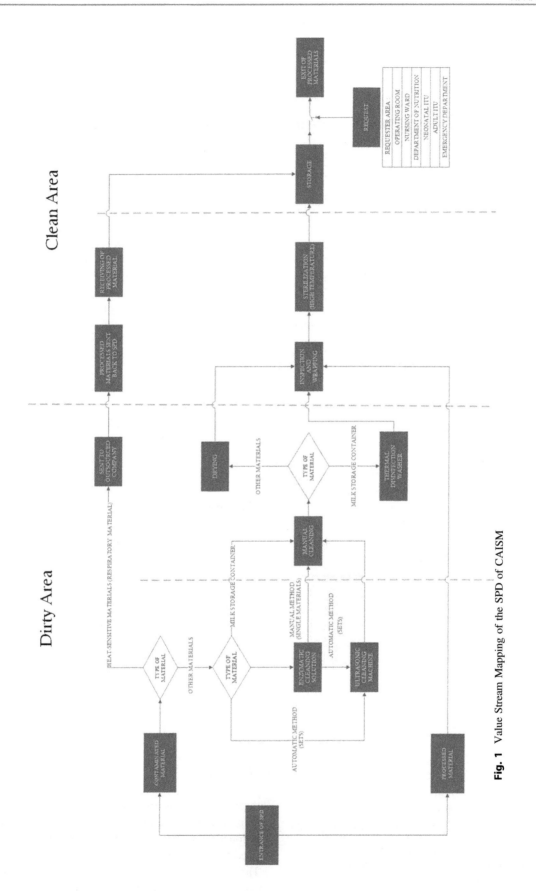

Fig. 1 Value Stream Mapping of the SPD of CAISM

Fig. 2 The top five main problems in the SPD

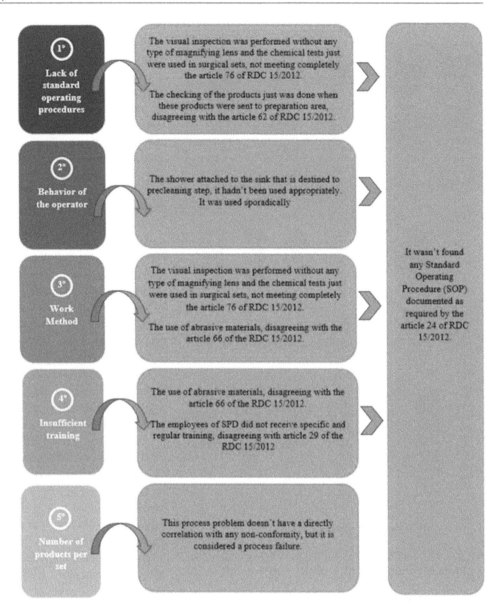

in all areas of the process from the dirty area to the storage, resulting in a systemic non-conformity. This problem not only affected the conformity with the RDC 15/2012, but also affected the KPI such as lead time and the variability of the process among others.

The second process problem was *behavior of the operator*. The employees behaved differently among distinct situations, depending on the group, the shift and what they believed to be the best for the process. Some operator preferred to work in dirty area while others in clean area, what impacted directly in the employee performance. The number of instruments also influenced employee performance in each activity. The ultrasonic cleaning cycle was interrupted several times by some employee once they were busy and needed to optimize their activities time,

therefore this kind of behavior also affected the cleaning efficiency.

The third process problem was *work method*. At the first process evaluation, the method used to perform each activity was defined by the employee. As consequence, the employees chose their own methods. The lack of standard allowed the employees to conduct their activities in different ways, not only affecting the attendance of the RDC 15/2012, but also impacting the time to carry out the activities.

The fourth process problem was *insufficient training*. The new employees learned their functions on the job from supervisors and senior employees. The training varied according to the senior employees. The rate of production was also low because the employees did not know how to

(a)

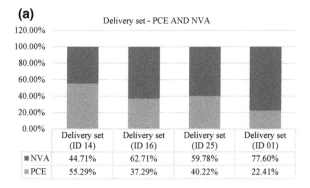

(b)

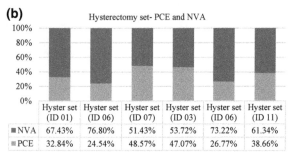

(c)

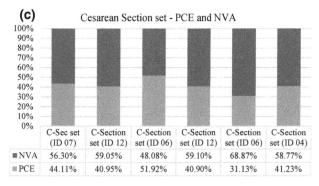

Fig. 3 **a** PCE and NVA time of delivery set. **b** PCE and NVA time of hysterectomy set. **c** PCE and NVA time of cesarean section set

perform the activities properly. Some employees lost considerable time to perform some activities because they did not know how to proceed according to the different steps of the process, as well as the name of the instruments, etc. When a change was applied, they did not receive a proper training, causing doubts, demotivation and low morale among other employees. Sometimes the changes were not accepted. Despite being aware of the new procedures, the employees did not respect, doing as they judged better for the process.

The fifth process problem was the *number of products per set*. Despite not affecting the conformity, it increased the cycle time to perform each activity.

The Fig. 2 shows a parallel between the top five main problems identified and the non-conformities found. Note

that one process problem can bring to one or more non-conformities in accordance with the Brazilian legislation.

This study also measured the process lead time for three different types of set. The sets were selected based on the number of instruments composed in each set in order to create a reliable sampling. The measurements were collected by two inspectors. They independently collected six times the lead time to process the cesarean section set and hysterectomy set. And the lead time to process the delivery set was collected four times by the same inspectors. The sampling was collected for six days.

When the lead time of the sets were collected and calculated, it was possible to conclude that the most of sets did not add any value along the supply chain. The work in progress was very long at that moment because the employees did not know how to optimize their time. Some wastes were inherent to the process, but others could be transferred to activities that added value or eliminated. The process took a long time to be completed, therefore an inventory was formed frequently.

After collecting information from the process, it was calculated the process cycle efficiency (PCE) and the percentage of non-value-added time (NVA). The Fig. 3a–c show the PCE and percentage of non-value-added time for each set.

All sets had a low process cycle efficiency which means that the process had several wastes. The main types of wastes were listed below.

The *waste of transport* was observed when the employees needed to collect products from the areas supported by the SPD. The collection was performed hourly in most of the areas what proved to have an inefficient distribution. Sometimes, the employees brought a low quantity of materials, but sometimes it came overwhelmed. Additionally, materials from specific areas such as materials of the department of nutrition were collected by the employees of this specific area, wasting time of these employees.

The *waste of motion* was observed when the employees needed to do excessive motion to fetch materials or to pass out materials to the next step. The assembly was not employee-friendly work environment because it was not in accordance with a natural work-flow, therefore the employees needed to do unnecessary motion. This type of waste could be seen in all areas. In the clean area, this effect was more severe because the employees had to walk up to take the sets that came from the dirty area as well as to take the wrapping materials, to seal the wrapped materials, to place the wrapped materials on the shelves and to prepare the materials for autoclave. The spaghetti diagram revealed an inefficient layout that caused stress on the employees and added more time to the process.

The *waste of waiting* was identified in all areas. In the dirty area the waste of waiting occurred when the employees waited for materials to be processed or waited the ultrasonic cleaning machine to finish its cycle. Other employees needed to wait the materials pass through the manual cleaning to start drying these materials. In the clean area, the same effect was identified when the employees waited for materials that came from the dirty area. During these waiting periods, the employees did other activities that had nothing to do with the SPD activities.

The *waste of defects* was identified when the materials needed to be reprocess because they were still dirty or stained. Other defects observed were the materials damaged because of the excess of enzymatic detergent or ill-use and rework at the sealing process because it was not well done.

The *waste of inventory* was observed when a batch was formed because a product was stopped in the assembly. The quantity of materials to be processed in coming days also caused waste of inventory and overwhelmed the SPD because there was not equipment enough to process all these materials. Because of the excessive number of materials composed in each set, the SPD had unbalanced production, causing waste as well. Another problem was at the storage because the SPD had a large quantity of materials in stock, therefore some materials got out of the expiration date and needed to be sterilized again.

To analyze if the process was under control and it was predictable, the Cp and Cpk indexes were used [5]. It was necessary to establish a specification limits for the process according to the client needs in order to calculate these indexes. The lower specification limit was 13497 s and the upper specification limit was 23000 s.

It was identified a Cp of 0,29 and a Cpk of 0,08. Therefore, this study concluded that the process was not capable to perform properly, it was not under control, it had a high variability and the process was not centered, therefore the processing of some sets was out of the specification limits. It was greater than the upper specification limit, taking more than 23000 s to be processed.

The capability and the sigma level of the process were also calculated. The study obtained a capability equal to 0,01 and a sigma level equal to 1,51. A sigma level equal to 1,51 revealed a low-quality process considering that the aim is to reach a six sigma level, besides 49,42% of the sets were out of the specification limits.

After some internal analyses, it was selected the PDCA to improve the process. The main changes implemented were listed below.

The *first change* applied was to abolish the enzymatic shower in the cleaning activity, but it is still mandatory to use a shower in order to comply with RDC 15/2012. The reason for this change is that the enzymatic solution can damage the instrument by causing stain or forming biofilm. Besides that, the enzymatic just works properly when the temperature is higher than 40 °C.

The *second change* was to implement regular training to all employees. The reason for applying this change was because it was possible to identify that some employees in SPD of CAISM had doubt in relation to the process, the name of the instruments, etc. In order to improve employee performance, satisfaction and behavior it was implemented a training program with the aim that all employees have the same consistent experience and background knowledge to perform their activities. The change was also necessary to comply with the RDC 15/2012.

The *third change* was to implement a medical washer-disinfector as one additional step in the process. The washer-disinfector was only used for milk storage container because the employees believed that the machine was damaging the instruments. After analyzing the reason why the washer-disinfector was staining the instrument, it identified the use of enzymatic in excess. In this regard, changes were implemented to make possible the correct use of the machine. The reason for this change was to optimize the efficiency of the process by using a washer-disinfector, eliminating some manual step such as the drying.

The *fourth change* was to oblige the use of magnifying lens to inspect the instruments during the preparation step. The reason for this change was to comply with RDC 15/2012 and to provide a more reliable processing of materials.

The *fifth change* will be implemented in further steps of the process optimization at the SPD. It is about a new method to trace the instruments along the entire process. The new method will not only allow a faster checking, but also improves the traceability.

4 Conclusion

In this study was described the use of engineering concepts to improve the sterilization process in the SPD and to obtain a redesign of the process. The basic concepts and tools of production engineering were covered as well as the legal requirements. A new design of the process was provided and, some changes were implemented. The improvement can already be seen in the hospital routine. Potential impediments, such as lack of time, organizational structure and culture issues were considered.

Further measurement will be collected in six months from the implementation date to evaluate how the changes have reflected in the process optimization and how the employees have dealt with the changes. Improvement in the work environment, the commitment of the employees and the reliability of the processing can already be noted. These changes have the aim of elevating the sigma level to three during the first six months, but for long term the intention is to achieve six sigma level. The expectation is also to comply with all requirement of the RDC 15/2012 after these six months and keeps the continuous improvement as part of the working routine.

Conflicts of Interest The research being reported in this full paper was supported by Unicamp. The authors of this full paper are member of Unicamp and act as master's students and teacher that are developing a new design of process in SPD. The terms of this agreement were reviewed and approved by Unicamp at Campinas in accordance with its policy of objectivity in research.

References

1. ANVISA: RESOLUTION—RDC N° 15, March 15 of 2012—Provides good practice requirements for the processing of health products and makes other arrangements. Brazilian Health Regulatory Agency. Brazilian Health Regulatory Agency (2012).
2. ANVISA: RESOLUTION—RDC N° 8, February 27 OF 2009—Provides measures to reduce the occurrence of infections by rapidly growing mycobacteria (RGM). Regulatory Agency. Brazilian Health Regulatory Agency (2009).
3. BRAZILIAN HEALTH REGULATORY AGENCY Homepage, http://www.anvisa.gov.br/servicosaude/controle/infectes%20 hospitalares_diagnostico.pdf, last accessed 2017/11/20.
4. HOSPITALAR TRADE FAIR Homepage: http://www.hospitalar.com/pt/portal-de-noticias/blog/82network-melhores-praticas/512-em-media-100-mil-pessoasmorrem-por-ano-por-causa-de-infeccao-hospitalar-no-brasil, last accessed 2017/05/21.
5. SIX SIGMA STUDY GUIDE.COM Homepage: http://sixsigmastudyguide.com/process-capability-cp-cpk/, last accessed 2017/10/23.

The Application of the Total Cost of Ownership Approach to Medical Equipment—Case Study in the Czech Republic

Petra Hospodková and Aneta Vochyánová

Abstract

Czech hospitals purchase the most expensive medical technology using public procurement process and they often make their purchasing decisions based on the lowest bid price. However, the management of hospital in most cases is unaware that the purchase price is only a minor part of the total cost of ownership of the equipment. This study intended to assess the nature of the Total Cost of Ownership (TCO) method in applications to medical equipment in the Czech Republic, and to carry out TCO analysis for selected medical equipment units. In order to evaluate the awareness and utilization of TCO in Czech hospitals, a quantitative research method (questionnaire) was used. The questions were addressed to the investment departments of the hospitals. To accomplish the research objectives, a TCO analysis for selected medical devices (X-ray machine, SPECT/CT and ultrasound scanner) was conducted. For each piece of equipment, four cost items were considered: acquisition cost; maintenance cost; operational cost and the cost of disposal. The results of questionnaire survey demonstrate that the TCO method was unknown to most of the respondents (67%) and the respondents usually make decisions based on the purchase price of medical equipment. All respondents who have already used the TCO method indicated that this method was useful to them. The results of the case study imply that the operating cost for the selected devices over five years are comparable to the purchase price. SPECT/CT scanner was the only one unit, where the acquisition cost was higher than the operational. In case of purchase of expensive medical technology, health care facilities often make decisions on the basis of the lowest bid price. According to the calculated TCO, this parameter is not the only important cost driver in the life cycle of medical equipment in hospital.

Keywords

Total cost of ownership • Medical equipment

1 Introduction

Healthcare providers founded by state or by regional government mostly use public resources for the acquisition of medical equipment in the Czech Republic. The formulations in the documentation of tenders, and mainly the purchase price, are often the controversial subject for the professional public. Due to the long lifespan of sophisticated medical equipment, the purchase price seems to be a misleading concept in the procurement process. This argument is supported by a number of studies based on the Total Cost of Ownership (TCO) method.

For healthcare providers, it is recommended to plan a purchase of medical equipment (including software updates) for approximately a five-year horizon. Device efficiency, safety criteria, utilization, ability for technological modernization and updates of medical equipment are the important factors to be taken into account during procurement [1].

The TCO method considers all cost items that need to be reflected before purchasing any technology. In addition to input costs, it also takes into account the costs that arise during the operation and disposal phases of the equipment lifecycle. TCO typically incorporates acquisition cost, energy, installation costs, regular maintenance, repairs, upgrades, personnel training, liquidation as well as personnel costs (which need to be assessed especially for new installation) [2, 3].

The total cost during the ownership of medical equipment often significantly exceeds the acquisition price and therefore the TCO method seems to be suitable for the assessment of devices in procurement process [2, 4, 5]. It was indicated that the purchase price might amount only to 20–25% of the total cost of ownership [6].

P. Hospodková (✉) · A. Vochyánová
Faculty of Biomedical Engineering, Czech Technical University in Prague, Nám. Sítná 3105, 272 01 Kladno, Czech Republic
e-mail: petra.hospodkova@fbmi.cvut.cz

© Springer Nature Singapore Pte Ltd. 2019
L. Lhotska et al. (eds.), *World Congress on Medical Physics and Biomedical Engineering 2018*,
IFMBE Proceedings 68/3, https://doi.org/10.1007/978-981-10-9023-3_65

The use of the TCO method is particularly useful in the assessment of large capital equipment. Its application is often in such areas as IT, construction, automotive, etc. [2, 7, 8]. In the healthcare sector, the TCO method is very rarely used for decision-making and is not standardized [1, 9].

Appropriate identification of total costs helps to make informed decisions, rationalizes performance requirements, reliability, sustainability, maintenance support, and other factors affecting life cycle costs, thus reducing the total cost of the asset [10, 11].

This study intended to assess the nature of the TCO method in applications to medical equipment in the Czech Republic, and to carry out TCO analysis for selected medical equipment units.

2 Methods (Methodology and Data Collection)

In order to evaluate the awareness and utilization of TCO in Czech hospitals, a quantitative research method (questionnaire) was used. All Czech bedside health facilities were included in the questionnaire, at the next step after-care facilities and day surgery, plastic surgery, elderly homes and hospices were excluded. The questionnaire was addressed to the investment department, competent decision-makers in two rounds. The concept of questions focuses on the overall knowledge of the TCO method and its application in practice.

To accomplish the research objectives, a TCO analysis for selected medical devices (X-ray machine, SPECT/CT and ultrasound scanner) was conducted. The TCO calculation process was based on the approach used in the studies of Morphonius [12] and Nierseen [13]. All costs are further discounted to the present value. The case study was also based on the Life-Cycle Costing Manual issued for the Federal Energy Management Program, which was used by Morfonios in his study [12]. The TCO process for the purpose of this study consisted of five basic steps [3, 14, 15]:

- Step 1: Evaluate the lifespan of the device.
- Step 2: Define the categories of costs for the evaluation.
- Step 3: Calculate the costs according to the defined categories.
- Step 4: Determine the relationship between costs.
- Step 5: Calculate the TCO.

TCO was composed of cost of acquiring, commissioning, operating, maintaining and disposing of medical equipment for a specified period of time.

$$TCO = Ca + Cc + Co + Cm + Cp + Cd \quad (1)$$

where Ca—Cost of Acquisition, Cc—Cost of Commissioning, Co—Cost of Operation, Cm—Cost of Maintenance, Cp—Cost of Production and Cd—Removal and Disposal cost minus any reclamation value [3].

X-ray, SPECT/CT and ultrasound scanner were selected for the purposes of the case study. Cost items for selected devices were identified with the help of medical technicians and managers at the department of medical technology. The list of cost items concerned for the selected equipment is presented in Table 1.

The acquisition cost included the cost of engineering, procurement, equipment cost, auxiliary equipment cost, inspections etc. Sales contracts, technical documentation and information from medical facilities were used as data sources.

Cost of Commissioning included the cost of construction, testing, training and technical support.

The operating cost was calculated from the active operating time of the medical device. The energy cost were calculated for ultrasonic devices using device parameters (length of active device operation, maximum device power and electricity price). The price of electricity was determined as the average value of electricity prices over the last 5 years. Direct measurement of energy consumed was used for SPECT/CT and X-ray devices. Personnel cost was calculated based on the typical number of operating personnel and examinations. Personal cost also included mandatory employer's contributions. We calculate the operating cost by the sum of all the above items.

Cost of maintenance was estimated on the basis of information in the equipment service documentation during the last 5 years. The cost of testing devices required by special legislation was calculated according to the records in the equipment documentation.

No cost of production was identified during the observed period. This usually includes production losses, quality cost, environmental cost and cost of redundancy.

Disposal cost included the following items: the uninstallation and ecological disposal of the device. In the case of X-ray devices, the disposal cost of the radiation source were concerned according to Czech legislation

3 Results

3.1 Questionnaire

The questionnaire was included in 155 healthcare facilities, 21 hospitals responded. In total, 13 questions were asked in the research. The following text offers the evaluation of some of them. Table 2

Table 1 Observed cost items for X-ray, SPECT/CT and ultrasound devices

	Cost item	SPECT/CT	X-Ray	Ultrasound scanner
Cost of acquisition	Acquisition price	X	X	X
	Workplace equipment cost	X	X	X
	Cost of mandatory protective aids		X	
	Inspections and documentation	X	X	X
Cost of commissioning	Construction work cost incl. shielding	X	X	X
	Cost of construction adjustment works of the application room and waiting area	X	X	
	IT cost (connection of the device with hospital information systems)	X	X	X
	Cost of mandatory testing under the radiation protection law	X	X	
	Personnel training cost	X	X	X
Operational cost	Energy	X	X	X
	Cost of consumed drugs (iodinated contrast agents, radiopharmaceuticals)	X		
	Cost of consumed material	X	X	X
	Cleaning and disposal cost	X	X	X
	Cost associated with the disposal of radioactive waste	X		
	Operating personnel	X	X	X
Cost of maintenance	Service contract cost	X	X	X
	Repair and maintenance cost outside the service contract (expensive parts, telecommunication support)	X	X	X
	Preventive maintenance cost			X
	SW update cost			X
Disposal costs	Cost of professional uninstalling of the device	X	X	
	Cost of ecological disposal according to the legislation	X	X	X
	Cost of ecological disposal of X-rays according to the legislation	X		

Table 2 The questionnaire results (n = 21)

Question	Response	Count	Percent (%)
Do you assess the operating cost of the medical device before purchasing it?	yes	17	81,0
	no	4	19,0
Do you assess the operating cost of a given medical device in a retrospective review after a certain period of use?	yes	13	61,9
	no	8	38,1
Do you know what Total Cost of Ownership (TCO) method consists in?	yes	7	33,3
	no	14	66,7
Do you have an overview of the operating cost of individual devices in your healthcare facility?	yes	15	71,4
	no	6	28,6
Do you have any idea how much it will cost to dispose of individual devices at the end of their life cycle?	yes	10	47,6
	no	11	52,4

Other inquired questions have shown that the governance of the hospital (21), medical technicians (19), doctors (13), economists (10), managers (7) and others (1) are the most involved in purchasing decisions. Another research question was "*Which parameter is crucial when deciding to purchase a medical device?*" First place is taken by device price, second was technical specification, next are: supplier's service, quality of the provided services and, last but not least, the brand of the device. The question that mapped the reasons for concluding a full-service agreement was "*If you*

Table 3 TCO of device Siemens Axiom Aristos MX (in thousands of CZK)

	2010	2011	2012	2013	2014	2015	Other
Number of examinations	n/a	43 930	47 383	49 055	49 039	50 295	n/a
Cost of acquisition and cost of commissioning*	9 601	–	–	–	–	–	n/a
Operational cost	–	8 697	8 538	8 547	8 968	9 201	n/a
Cost of maintenance	–	0	57	104	1 260	368	n/a
Disposal cost	–	–	–	–	–	–	250
TCO	9 601	8 697	8 595	8 861	10 228	9 569	250

Note *cost items was combined because they represent a one-off initial load

Table 4 Total Cost of Ownership of device Philips Precedence 6′ (in thousands of CZK)

	2010*	2011	2012	2013	2014	2015	Other
Number of examinations	n/a	633	1 046	878	882	752	–
Cost of acquisition and cost of commissioning	20 774	–	–	–	–	–	–
Operational cost	–	3 383	3 807	3 585	3 798	4 304	–
Cost of maintenance	–	668	581	1341	1351	1350	–
Disposal cost	–	–	–	–	–	–	350
TCO	20 774	4 051	4 388	4 926	5 149	5 654	350

Note * Medical device was acquired in 2007 and the acquisition price was converted to the 2010 net present value

have a full service contract with supplier companies, please state the main reason that helped you to negotiate this agreement." Eight facilities suggested that the full service contract has to be negotiated due to the guarantee of service workers' response and future carefree operation of equipment. Seven facilities stated that reason was a full service discount.

3.2 Case Study

X-ray system. The calculated total cost of acquiring Siemens Axiom Aristos MX and its five-year operation represented CZK 55,591,000. However, the purchase price was not the highest cost item. Purchase price amounted to CZK 9,601,000 (17.3%) and maintenance cost was CZK 1,789,000 (3.2%). Operating cost amounted to CZK 43,951,000 (79.1%) and the disposal cost was CZK 250,000 (0.4%). The highest TCO score was clearly the operation cost, with significant human resource cost. Among the disposal cost, the most significant item was the cost of spare parts (CZK 1,448,000). Table 3

SPECT/CT. The acquisition cost and commissioning cost in total was CZK 20,774,000 (45.9%) of the total cost of ownership evaluated over a five-year horizon. Maintenance cost reached CZK 5,291,000 (11.7%), of which the most significant was the cost of service contracts (change to full contract in 2012). Operating cost amounted to CZK

18,877,000 (41.7%) and disposal cost in total was CZK 350,000 (0.8%). It can be stated that the most significant cost category was operating cost. An important item within this category was the cost of consumables (CZK 10,750,000). The total cost of ownership in the monitored horizon was CZK 39,638,000 (Table 4).

Ultrasound system. The total cost of ownership over the five-year horizon for the GE Vivid 7 Dimension ultrasound system was CZK 11,156,000. Cost of acquisition and commissioning cost in total was CZK 3,889,000 (34.9%) according to the purchase contract. Maintenance cost amounted to CZK 565,000 (5.1%). The operating cost was calculated to CZK 6,652,000 (59.6%) and disposal cost to CZK 50,000 (0.4%). The cost of operation is the most significant cost item, as with previous devices. A high cost item was the personnel cost (CZK 6,480,000). In cost of maintenance, the highest cost over time was the cost of spare parts (CZK 525,000) (Table 5).

4 Discussion

The survey results show that the management of healthcare facilities, medical technicians, physicians and economists are the ones who are the most involved with the process of procurement of expensive medical equipment. This corresponds to the work of Konschak [16], but the representation of economists in the assessment of long-term investment

Table 5 Total Cost of Ownership of device GE Vivid 7 Dimension (in thousands of CZK)

	2010*	2011	2012	2013	2014	2015	Other
Number of examinations	n/a	2 872	3 094	2 968	2 838	3 294	–
Cost of acquisition and cost of commissioning	3 889	–	–	–	–	–	–
Operational costs	–	1 311	1 326	1 314	1 327	1 374	–
Cost of maintenance	–	3	98	3	3	458	–
Disposal costs	–	–	–	–	–	–	50
TCO	3 889	1 314	1 424	1 137	1 330	1 832	50

Note * Medical device was acquired in 2009 and the acquisition price was converted to the 2010 net present value

should be much higher. The purchasing price was the most frequent decision parameter among the healthcare facilities inquired, probably due to the long-established way of evaluating public procurement in the Czech Republic. At present, bids in public procurement process are assessed using principle of "most economically advantageous tender". The healthcare equipment can be evaluated by the lowest bid price, the quality of the offered solution, its reliability, the range of service, etc.

The frequency and the reasons, for selecting full service contracts, are similar to the ones that can be found in available literature. Sferrella states that the most common reasons for negotiating full service contracts from supply companies are as follows: carefree operation, guarantee of response, and discount on the service contract when signing the contract together with the purchase of the instrument [17]

TCO was calculated for selected SPECT/CT, X-ray, and ultrasound devices for a five-year horizon, which did not represent the total lifetime of the instrument. However, it has already been demonstrated at this time that operational cost outweigh the acquisition cost. The cost item that also significantly increased the TCO was the service and repair cost group. Therefore, it can be concluded that the purchasing organization should also consider the scope of the negotiated service contract. In addition, health care facilities should concentrate on choosing a supplier of consumables, as study results demonstrate that these items were among the third most important category in TCO. Personnel costs included in the TCO calculation are an item that is typically included into operational costs [14–16]. It should be noted that in the case of existing operation, this item should be considered irrelevant, and taken into account in the overall calculation only in terms of additional wage costs (incremental). The Inclusion of personnel cost into the calculation is suitable especially for new departments. During the device operation, it is also advisable to take into account the number of examinations and to track the average operating costs per examination. The method can also be supplemented by monitoring of marginal costs, i.e. costs incurred with each incremental unit of output (number of such units) [18].

5 Conclusion

The TCO method was unknown to most of the respondents and the respondents usually make decisions based on the purchase price of medical equipment.

The operating cost over five years for the devices selected for case study is comparable to the purchase price. SPECT/CT scanner was the only one unit, where the acquisition cost was higher than the operational. According to the calculated TCO, this parameter is not the only important cost driver in the life cycle of medical equipment in hospital.

Further research is recommended, especially in the field of operating costs and in the area of methods for estimating cost items.

Conflict of Interests The authors declare no conflict of interests.

References

1. Canadian Association of Radiologists: Lifecycle guidance for medical imaging equipment in Canada, https://car.ca/wp-content/uploads/car-lifecycleguidance-mainreport.pdf, last accessed 2017/12/20.
2. Hockel, D., Hamilton, T.: Understanding total cost of ownership. Healthcare Purchasing News. 35 (9), (2011).
3. Dabbs, T.: Optimizing Total Cost of Ownership (TCO), http://www.argointl.com/wp-content/uploads/2014/03/goulds-Optimizing-Total-Cost-of-Ownership-final2.pdf, last accessed 2017/12/20.
4. TRIMEDX: Total Cost of Ownership : The Influence of Clinical Engineering, http://www.hfma.org/brg/pdf/WhitePaperTCO_B5AAD6A0-E673-3BD0-FF9599EB16582879.pdf, last accessed 2017/12/20.
5. National Center for Education Statistics: Total Cost of Ownership: An Important Piece of Any Sustainability Plan., https://nces.ed.gov/programs/slds/pdf/TotalCostofOwnership.pdf, last accessed 2017/12/20.
6. BMET Wiki, Total Cost of Ownership, http://bmet.wikia.com/wiki/Total_Cost_of_Ownership, last accessed 2017/12/20.
7. Ferrin, B.G., Plank, R.E.: Total Cost of Ownership Models : Journal of Supply Chain Management. 38(2), 18–29 (2002).

8. Total Cost of Ownership: An introduction to whole-of-life costing, http://www.procurement.govt.nz/procurement/pdf-library/agencies/guides-and-tools/guide-total-cost-ownership.pdf, last accessed 2017/12/20.

9. Chakravarty, A., Debnath, J.: Life cycle costing as a decision making tool for technology acquisition in radio-diagnosis. Medical Journal Armed Forces India. 71(1), 38–42 (2015).

10. WERF: Overview: What is Life Cycle Costing? http://simple.werf.org/simple/media/tools/Life Cycle Cost Tool/index.html, last accessed 2017/12/20.

11. International Supply Chain Solutions: Total Cost of Ownership, http://www.iscsglobal.com/files/assets/Total%20cost%20of%20ownership.pdf, last accessed 2017/12/20.

12. Morfonios, A., Kaitelidou, D., Filntisis, G., Baltopoulos, G., Str, P., Myrianthefs, P.: Economic Evaluation of Multislice Computed Tomography Scanners Through a Life Cycle Cost Analysis. Indian Journal of Applied Research. 4, 158–161 (2014).

13. Nisreen, H.J., Salloom, A.J., Omer, N.M.: Medical Devices Service Life Cycle Cost Management in Al Karak Hospital as a Case Study. Journal of Accounting & Marketing. 4, 2–9 (2015).

14. Heilala, J., Helin, K., Montonen, J.: Total cost of ownership analysis for modular final assembly systems. International Journal of Production Research. 44(18–19), 3967–3988 (2006).

15. Weber, M., Hiete, M., Lauer, L., Rentz, O.: Low cost country sourcing and its effects on the total cost of ownership structure for a medical devices manufacturer. Journal of Purchasing and Supply Management. 16(1), 4–16 (2010).

16. Konschak, C.: Understanding the Total Cost of Ownership (TCO) analysis for IS in the healthcare setting, http://www.colinkonschak.com/images/TotalCostofOwnership.pdf, last accessed 2017/12/29.

17. Sferrella, S.: Equipment Service: Total Cost of Ownership. Radiology Business Journal (December 2012).

18. Sealey, C.W., Lindley, J.T.: Inputs, Outputs, and a Theory of Production and Cost at Depository Financial Institutions. The Journal of Finance. 32, 1251 (1977).

The Profile of Clinical Engineering in Espirito Santo, Brazil

L. A. Silva, F. L. Cunha, and K. L. Oliveira

Abstract

As a result of high rates of non-operating equipment due to the lack of maintenance and proper training, Clinical Engineering has been growing considerably in Brazil since the 1990s and, therefore, its demand. In the State of Espírito Santo (ES), the reality is different from found in large cities. Because of that, this article has the following purposes: to verify the presence of Clinical Engineers in ES; to analyze the team composition and the presence of predictive and preventive maintenance procedures; and to compare the proportion of CT and MRI equipment in ES with national and international data. For this, data gathering of 97 hospitals registered in the National Registry of Health Establishments was carried out to obtain a list of existing equipment in the places and the way that maintenance service of such equipment is performed, as well as the composition of the team. The results obtained shows most of the general hospitals have outsourced Equipment Maintenance Service and those responsible for Clinical Engineering do not have specialization in the area or a degree in engineering. In addition, there are 0.75 MRI equipment and 0.50 CT for each 500 thousand and 100 thousand inhabitants, respectively, in the Universal Healthcare Service, values below determined by ordinance MS 1101. Thus, these data corroborate the need of increasing the supply and training of professionals in Biomedical Engineering and Clinical Engineering areas to promote additional study in the technological management practices of such equipment.

Keywords

Clinical engineer • Biomedical engineer • Brazil

1 Introduction

Clinical Engineering has its origins in the 1940s in the United States with the implantation of a medical equipment course. However, it was not until the 1970s that the profession of clinical engineer was established, which is the professional responsible for the management, evaluation and transfer of hospital equipment [1, 2].

In Brazil, however, it was only in the 1990 s that the first courses of specialization in Clinical Engineering for electrical engineers were implemented: two of them in São Paulo; one in Paraíba; and one in Rio Grande do Sul. The trigger for the implementation of these courses was the high rate of non-operating equipment due to lack of maintenance and specific training [1]. Although there are currently several courses in Brazil, in the State of Espírito Santo (ES) there are still no specialized courses in the area, but since 2013 a degree in Biomedical Engineering has been offered by a private institution. It is important to point out the Brazilian Federal Council of Engineering and Agronomy (CONFEA) does not yet recognize Clinical Engineering as a specialty. It allows professionals such as technicians, managers, nurses, doctors, etc. to act in the position of clinical engineer [3, 4]. The Brazilian Association of Clinical Engineering (ABEClin) started an unsuccessful process in 2013 within CONFEA in order to obtain formal recognition by the professional council [5].

In this context of hospital technology management, the Information Technology Department of National Health System (Datasus) has made an online platform available to collect, process and disseminate data on health equipment: National Register of Health Establishments (CNES). On this basis it is possible to obtain information regarding health service providers, for instance: equipment; support services; professionals; among others [6, 7]. These data are extremely important to verify if there is a consonance between regulations applied by the federal government and what is seen in practice. As an example, it can be listed the one No. 1101/

L. A. Silva (✉) · F. L. Cunha · K. L. Oliveira
UCL—Faculdade Do Centro Leste, Serra, ES, Brazil
e-mail: 20092tcest0222@ucl.br

© Springer Nature Singapore Pte Ltd. 2019
L. Lhotska et al. (eds.), *World Congress on Medical Physics and Biomedical Engineering 2018*,
IFMBE Proceedings 68/3, https://doi.org/10.1007/978-981-10-9023-3_66

GM elaborated in 2002 by the Ministry of Health that establishes that for every 100 thousand inhabitants there must be 01 Computed Tomography (CT); and for every 500 thousand inhabitants 01 Magnetic Resonance (MRI) [8].

Therefore, the present study is aimed at: verifying the presence of Clinical Engineers in Hospitals in ES; analyzing the composition of the teams responsible for Clinical and Hospital Engineering service and the presence of corrective and preventive maintenance procedures in these hospitals, as well as analyzing the proportion of CT and MRI equipment comparing these data with national and international ones.

2 Methods

During the end of 2017, an investigation was carried out at the base of CNES (National Registry of Health establishments). This database provided by federal government lists various information about establishments provide health services in country, both those are part of private network as well as those compose public network [6, 7]. The mainly research criteria adopted is the place of sample space: all cities and establishments in the State of Espírito Santo with at least one General Hospital. With this, the following data were obtained from each hospital: business name; County; telephone; disabled or not; legal nature; equipment maintenance service (SME); diagnostic imaging equipment. For the latter, we sought only for the presence of CT and MRI devices available in Unified Health System (SUS), which is the public health system in Brazil.

With CNES data, a questionnaire was created with 12 questions that were used in individual research. The questions consisted of verifying: academic formation of the person responsible for Clinical Engineering or Equipment Maintenance; formation of the team of Clinical Engineering or Maintenance of Equipment; presence of procedures to perform corrective and preventive maintenance; existence of CT and MRI equipment. Such research was carried out by telephone, e-mail and on-site and had the purpose of verifying the consonance between actual data and those provided by CNES.

3 Results

According to Table 1, information was collected on 97 general hospitals. It is possible to observe 18.6% answered the questionnaire directly, 12.4% did not respond when sent by e-mail and 13.4% of hospitals were disabled.

It is also possible to note 36.1% reported (via phone call, e-mail or on-site) the SME is outsourced. Of these, 11 hospitals have equipment park managed by the same private company. The 24 remaining (24.7%) have a direct contract

Table 1 Result after collection at CNES database

Result	Number of hospitals	(%)
Answered questionnaire	18	18.6
Outsourced SME	35	36.1
Hospital disable	13	13.4
Unanswered e-mail	12	12.4
Others	19	19.6
Total	97	100.0

Table 2 Comparison between CNES data and those purchased individually for each hospital

Description at CNES	Number of hospitals	(%)
Own	7	29.2
Outsourced	12	50.0
Own and outsourced	3	12.5
No information	2	8.3
Total	24	100.0

with some technical assistance and, when necessary, a hospital employee triggers a service.

As mentioned in item 2, it is possible to obtain in CNES the form that SME is made: own; outsourced; and own and outsourced. In this way, a comparison was made between CNES data and those obtained individually from the 24 hospitals when contacted reported that service was outsourced.

It can be seen from Table 2 that half of these hospitals actually have outsourced services, while the rest go against what is registered with CNES.

As seen in Table 1, 18 hospitals responded directly to the questionnaire. Initially asked about academic training of the person in charge of the Clinical Engineering or Equipment Maintenance sector. It can be seen from Table 3 most of people in charge are not trained in specific area or even in Engineering.

When questioning about the form of SME, it is noticed there is no conformity between the data present in CNES's base and the reality of the hospital, according to Table 4. For example, in the sample of 5 hospitals that reported that SME was its own, only 2 had this information correctly at CNES.

Regarding the presence of a procedure for preventive and corrective maintenance, almost all (97.4%) hospitals reported there is a procedure, according to Table 5. It was also questioned about ease of finding equipment maintenance supplier in ES, 61.1% report there is facility, according to Table 6.

In addition to seeking information on SME, data were sought on the amount of CT and MRI equipment available at SUS. According to Table 7, it can be seen that for both cases

Table 3 Academic background of the responsible for the sector of clinical engineering or maintenance of equipment

Academic Background	Number of hospitals	(%)
Electrical engineering	5	27.8
Electronics technician	3	16.7
High school	3	16.7
Administration	3	16.7
Biomedical engineering	1	5.6
Mechanical engineering	1	5.6
Physiotherapy	1	5.6
Nursing	1	5.6
Total	18	100.0

Table 5 Presence of procedures for preventive and corrective maintenance

	Number of hospitals	(%)
Existing procedure	17	94.4
Non-existing procedure	1	5.6
Total	18	100.0

Table 6 Ease of finding equipment maintenance supplier in ES

	Number of hospitals	(%)
Easy	11	61.1
Regular	2	11.1
Difficult	5	27.8
Total	18	100.0

there is not even 1 equipment at SUS available for each general hospital in the registry.

Table 7 Number of CT and MRI equipment per 100 thousand and 500 thousand inhabitants, respectively

CT equipment/100 thousand inhabitants	0.50
MRI equipment/500 thousand inhabitants	0.75

4 Discussion

One cannot deny the importance of CNES in accessing information on health institutes. However, it still encounters several flaws. Among them, there is a lack of data update for simple information, such as telephone number to keep contact. In this context, there were several obstacles such as difficulty contact by telephone and e-mail, difficulty to contact the person in charge of SME sector and obtain correct data. In this case, it is also noticed there is no consensus between the type of SME base and the reality observed through the contact with each general hospital. For example, less than half of hospitals reported in the questionnaire that SME was actually compliant with the CNES, as seen in Table 4.

In relation to the people in charge of Clinical Engineering or Equipment Maintenance sector, it is noted that majority (61.1%) of those do not have a specialization in Clinical Engineering or, at least in Engineering. For example, 16.7% have a high school education level. This data was also noticed when we contacted the 24 hospitals in Table 2. Among those with specialization in Clinical Engineering or undergraduate degree in Biomedical Engineering, it is not

observed that such specialization or graduation was performed in ES. This information demonstrates the lack of specialized professional in this area in ES. In this context, the teams of hospitals that had an Engineer were also composed by at least 1 technician in electronics.

Besides the lack of consensus between the data of CNES and reality, it is also noticed that in other spheres it also occurs. As explained in item 1, Regulation No. 1101/GM provides minimum amounts of CT and MRI equipment to a certain number of people. In this research, it was observed, for both equipments, the legislation is not followed observed. In the case of the CT, this reality is not different from the State of São Paulo, for example, where there are 0.79 equipment per 100 thousand inhabitants [9]. When comparing these ES data to those available on World Health Organization (WHO) website, the values found for ES State are close to countries such as Costa Rica and Cuba for TC. As for as MRI, it is close to Jamaica and Albania [10].

Table 4 Comparison between CNES data and individually purchased for hospitals that answered the questionnaire

SME	Questionnaire	CNES			
		Own	Outsourced	Own and Outsourced	No Information
Own	5	2	1	1	1
Outsourced	7	5	1	1	0
Own and Outsourced	6	4	0	1	1
Total	18	11	2	3	2

5 Conclusion

Although Clinical Engineering has advanced in Brazil since the 1990 s, in ES the reality differs from that found in developed countries, such as United States, for example. It is still necessary to invest and promote the area, either through recognition of the profession by CONFEA or legislation that determines the presence of this professional inside hospitals. Thus, these data corroborate the need to increase supply and training of professionals in the area of Biomedical Engineering and Clinical Engineering in order to promote greater study in the technological management practices of such equipment.

Acknowledgements The authors would like to thank all hospitals for contributing their research and the possibility of contributing to the results of this work.

Conflict of Interest The authors declare that they have no conflict of interest.

References

1. RAMÍREZ, E. F. F.; CALIL, S. J.: Engenharia clinica: Parte I-Origens (1942–1996). Semina: Ciências Exatas e Tecnológicas, v. 21, n. 4, p. 27–33 (2004).
2. DE MORAES, Luciano et al. The multicriteria analysis for construction of benchmarkers to support the Clinical Engineering in the Healthcare Technology Management. European Journal of Operational Research, v. 200, n. 2, p. 607–615 (2010).
3. FRANÇA, A.S. D. A. Atribuição profissional na gestão de tecnologias em estabelecimentos de saúde no Brasil. Revista Organização Sistêmica, v. 7, n. 4, p. 130–141 (2016).
4. SOUZA, A. F.; MORE, R. F. O perfil do profissional atuante em engenharia clínica no Brasil. A Engenharia Biomédica como Propulsora de Desenvolvimento e Inovação Tecnológica em Saúde. Uberlândia-MG:[sn], p. 1086–1090 (2014).
5. DEL SOLAR, J. G. M.; SOARES, F. A.; MENDES, C. J. M. R. Brazilian clinical engineering regulations: health equipment management and conditions for professional exercise. Research on Biomedical Engineering, n. AHEAD, p. 0–0 (2017).
6. DE MATOS, C. A.; POMPEU, J. C. Onde estão os contratos? Análise da relação entre os prestadores privados de serviços de saúde e o SUS. Ciência & Saúde Coletiva, v. 8, n. 2 (2003).
7. FREITAS, M. B. D.; YOSHIMURA, E. M. Levantamento da distribuição de equipamentos de diagnóstico por imagem e freqüência de exames radiológicos no Estado de São Paulo. Radiol. bras, v. 38, n. 5, p. 347–354 (2005).
8. Portaria n° 1101/GM em 12 de junho de 2002, http://www.betim.mg.gov.br/ARQUIVOS_ANEXO/Portaria_1001;;20070606.pdf, last accessed 2018/01/21.
9. DE ARAÚJO, P. N. B.; COLENCI, R.; RODRIGUES, S. A. Mapeamento dos equipamentos e exames de diagnóstico por imagem no estado de São Paulo. Tekhne e Logos, v. 7, n. 2, p. 121–135 (2016).
10. Baseline country survey on medical devices, 2014 update: Medical Equipment, http://gamapserver.who.int/gho/interactive_charts/health_technologies/medical_equipment/atlas.html, last accessed 2018/01/25.

Reduced Cost Training Program in Minimally Invasive Surgery

A. H. Hermini, I. M. U. Monteiro, L. O. Z. Sarian, and J. C. C. Torres

Abstract

Minimally Invasive Surgery (MIS) procedures are now considered mainstay practice in the best hospitals worldwide. Despite its benefits, high costs and increased hazards appear as disadvantages for this technology. Suboptimal surgeon performance is a major hazard in many centers and can be overcome by an adequate training program. This work describes a reduced cost Training Program (TP) composed of three levels of theory and practice activities to improve the skills of the surgeons in MIS. The basic level is composed by 1 h of lessons on the videosurgery setup, optics and instruments, and a practice lessons composed of eight exercises in inanimate models. The intermediary level consists of 1 h of theory with 6 h of practice exercises in electrosurgery (chicken leg) and 4 h of laparoscopic suture in Neoderma simulation pieces. The third (advanced) level, consists of a hysterectomy in a living animal (pig). The investment in permanent assets for the basic and intermediary levels is roughly US$ 12,000,00; and for the advanced level, US$ 180,00 are needed per procedure. This TP has been applied in a multidisciplinary program involving medicine and clinical engineering areas in a University Woman's Hospital. Twenty-one medical residents started the program and 10 completed all levels.

Keywords

Training program • Surgeon skill • Laparoscopy

A. H. Hermini (✉) · I. M. U. Monteiro · L. O. Z. Sarian
J. C. C.Torres
State University of Campinas, CAISM, UNICAMP, Campinas,
São Paulo, Brazil
e-mail: hermini@unicamp.br

1 Introduction

Since the first laparoscopic hysterectomy (LH) performed by Reich [1], this procedure has grown in hospitals worldwide. LH has obvious advantages over open hysterectomy, e.g. reduced postoperative pain, reduced convalescence and small portholes leading to improved cosmetic outcomes (sometimes the great advantage in patient's perception). However, considerable adverse events have been reported [2]. One of the major factors associated with unsuccessful LH is suboptimal surgeon performance. The steep "learning curve" for laparoscopy procedures in general and LH in special has been pointed as chief among the disadvantages of laparoscopy [3]. Some methodologies have been developed to increase the surgeon skills in laparoscopy (very often named Fundamentals of Laparoscopic Surgery—FLS); most commonly, the trainee surgeon learns from performing surgery under the supervision of a more experienced fellow (supervised learning). Endotrainer box simulation and virtual simulation are often cited as adjuvant learning tools [4, 5]. However, it remains unknown which training model yields the best surgeon performance improvement, or are firmly defined the parameters used to define "surgeon performance". In addition, supervised learning poses risks, since the participation of untrained personnel in surgical procedures may be associated with extended surgical times, and additional complications. Virtual Reality Simulators, on the other hand, and currently beyond the financial possibilities of most public hospitals in Brazil, including our institution. Thus, we decided to develop a low-cost training problem to improve our professional's laparoscopy skills.

This paper describes a reduced cost training program to improve the surgeon skills according the University Hospital needs. Our training program starts addressing perception and fine motor control, through training in inanimate models, and culminates with actual surgical procedures in living animals.

2 Methods

Using the concepts of the MISTELS model [4, 5], we used endotrainer box constructed with opaque fiber, 10 flexible trocars ports, allowing to simulate different accesses for training and practice. The simulated cavity image is captured by a HD webcam and visualized by a computer video monitor, as shown in the Fig. 1

The training program encompasses three levels of progressive skill acquisition, simulating conditions faced by surgeons during real-life procedures. By the end of the program, students are expected to be able to perform complex activities, including perform surgery in a living animal, which in turn will make their further supervised training in human patients more efficient and safer.

The three levels of training are named: Basic, Intermediate and Advanced, and their contents are described below:

Basic: The basic level consists of theory classes and practice lessons. The objective is to give the trainees fine

(a)   (b)

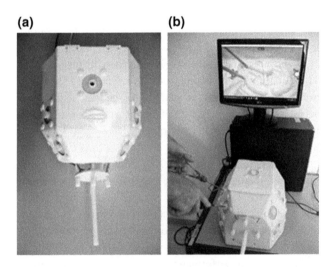

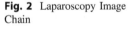
Fig. 1 **a** Endotrainer box; **b** Simulator system

motor coordination and 3D-in-2D perception. The theory classes deals with videosurgery setup and gaining familiarity with equipment. The classes address the components of the Image Chain, starting with the patient (Abdominal cavity is considered the first component of the system), presenting the Electromedical Equipments (camera, light source, insuflator, …), the optical devices (endoscope, light cable,…), the surgical instruments (scissors, grasps, trocars,…) and reaching the video monitor, where the patient cavity is visualized by the surgeon as shown in Fig. 2.

We consider important that the surgeon understands the image chain as a system and acquires fundamentals of operation of each component, becoming able to solve minor problems that could occur during the surgical procedure or to give more specific explanation for the clinical engineering team about the problems, optimizing its solution.

Next, the practical activity of this level is composed of 8 (eight) exercises, as presented below. Before starting with the exercises, a brief explanation is given. It is emphasized that all exercises should be performed using both hands ('two-hands surgery') and that all instruments must be kept in the field of vision. The 'two hands ability' is trained by forcing students into performing each exercise at least once with each hand.

The first exercise, named Pick and Place, was developed to give to the trainee the 2D x 3D perception and the haptic or kinesthetic sensation related to the endoscopic instruments, as grasps, scissors and needle holders, as presented in Fig. 3. Three rounds of each exercise, for each hand, are performed. During the entire session, the instrument held by the hand contralateral to that performing the exercise must be kept in the field of vision, as mentioned above.

Next, the trainee is exposed to perception and control skills, by being asked to put four training plastic pieces inside of a plastic vessel with 50 mm (2") diameter. The goal is to place the pieces inside the vessel without touching vessel walls. The simulated operating field is small and full of structures that cannot be touched, always keeping the

Fig. 2 Laparoscopy Image Chain

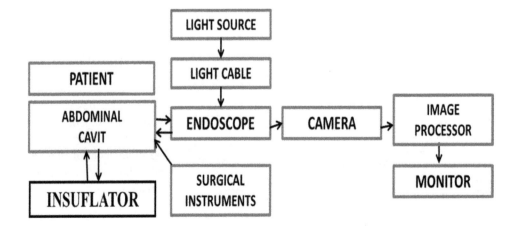

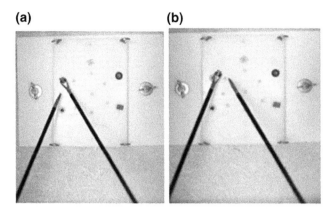

Fig. 3 Pick and place exercise **a** right; **b** left

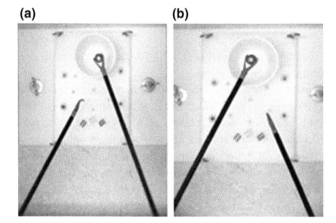

Fig. 4 Depth perception and control **a** right; **b** left

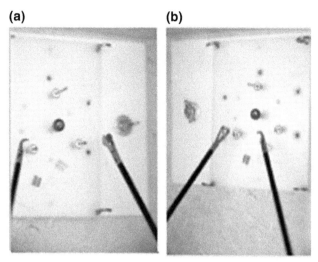

Fig. 5 3D fine perception and control **a** right; **b** left

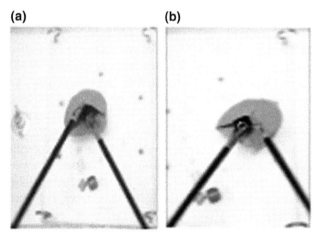

Fig. 6 Endobag simulation **a** right; **b** left

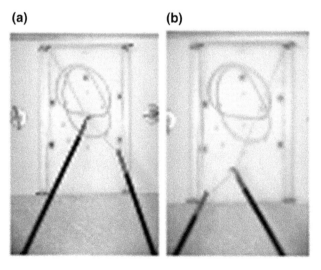

Fig. 7 Training traction and control **a** right; **b** left

instrument held by the opposite hand inside the field of vision, as presented below (Fig. 4).

After training depth perception and control, students are asked to place different shape rings with 4 mm internal diameter in a 2 mm pin, to train delicate movements and fine 3D control (Fig. 5).

Next, we simulate Endobag use. During this activity, both hands are used. The trainee is asked to place four plastic pieces inside the bag, similar to surgical procedures where it is necessary to place biopsy specimens inside the endobags (Fig. 6).

Traction and force control. During laparoscopic procedure, surgeons need to traction tissues in order to cut and move. Thus, they use force that must be controlled, that is, insufficient force may result in loss of traction and excessive forces may result in tissue damage. The fifth exercise in the series consists of pulling four office elastics, to train traction forces and control, as presented in Fig. 7.

Cutting tissue is inherent to the surgical procedure, meaning that damage to the tissue is a necessary part of the surgical intervention. However, the aphorism "don't cause greater damage than necessary" is valid, and students must

(a)   (b)

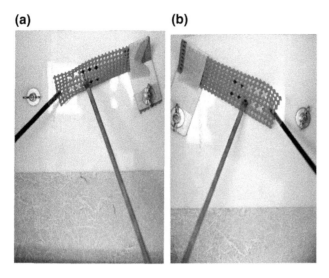

Fig. 8 Scissors cut training **a** right; **b** left

(a)   (b)

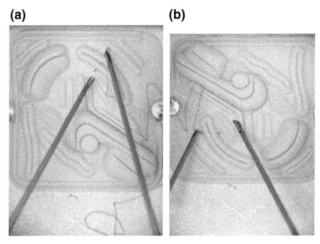

Fig. 10 Silicon model **a** right; **b** left

be trained to cut no longer or deeper than necessary. Due to optical magnification of the instruments, surgeons must learn to adjust range of motion. To train safe tissue dissection, we used an EVA 3 mm mash (piece and space) marked alternately, as shown in Fig. 8. Using laparoscopic scissors and clamps, the trainee was asked to hold and to cut inked tops without damaging the adjacent markings.

The final stage of the Basic Level is dedicated to preparing trainees to perform sutures. The first movement to be trained is 'pick and position' needles using the needle holder. We used foam to simulate and train deepness perception. The exercise consisted of the needle being passed through the peak of the foam piece, three times with each hand (Fig. 9).

When the students develop the ability to position and hold the needle, they repeat the exercise passing it through a silicon model that simulates the hardness of the tissue, as presented below (Fig. 10).

(a)   (b)

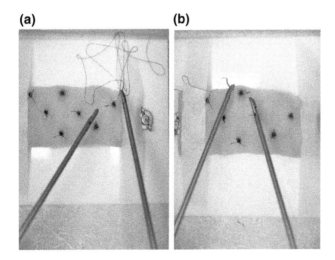

Fig. 9 Needle control foam model **a** right; **b** left

Students are scored according to the time spent to perform each exercise series.

Intermediary level: The intermediary level is aimed at developing students' electrosurgical and suture skills. This levels comprise theory classes (1 h) about the principles, risks and injury prevention, and 4 h of laparoscopic suture in Silicon model. After training using the model, the resident physician performs 6 h of practice in monopolar electrosurgical dissection and cut in chicken leg, as shown in Fig. 11a and b. Next, they will have the opportunity to rehearse bipolar hemostasis, as shown in Fig. 11c. After cutting the chicken leg, the incision is sutured, as presented in Fig. 11d. The score of the electrosurgical procedures is obtained evaluating if the tissue was dissected and if the incision was done. Points are subtracted if adjacent tissue was burned. Sutures are considered optimal if sufficient to tie the edges of the tissues together and hold them firmly together. When the trainees develop fine movements and perception in the 2D-in-3D view, they are deemed ready to advance to the next level.

Advanced level: The last, the advanced level, consists of a hysterectomy in a living animal (pig), using the ethical and animal protection procedure, approved by Internal Ethics Committee on the Use of Living Animals.

Trainees are supposed to excise anatomical structures indicated by the Fellow, and to log injuries. Importantly, approval depends on the animal being still alive by the end of the procedure. Figure 12 shows the external and internal views of the procedure.

Experimentally, we evaluated the trainee laparoscopic surgical skills prior to the TP measuring time and quality of three types suture: open suture (Fig. 13a), laparoscopic suture with direct view (Fig. 13b) and laparoscopic suture with camera view (Fig. 13c) comparing it with the values and levels obtained by a trained surgeon.

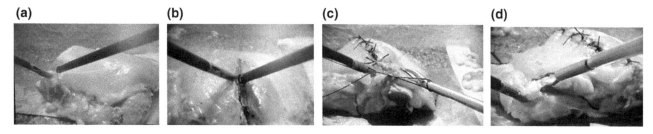

Fig. 11 Chicken leg exercises

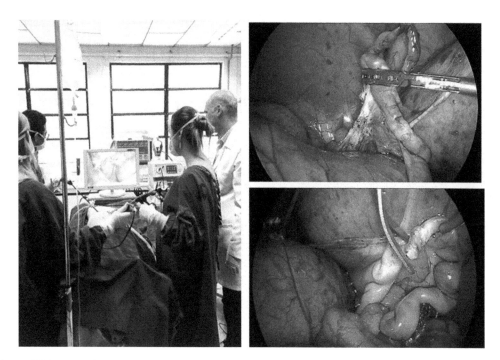

Fig. 12 Experimental living animal procedure

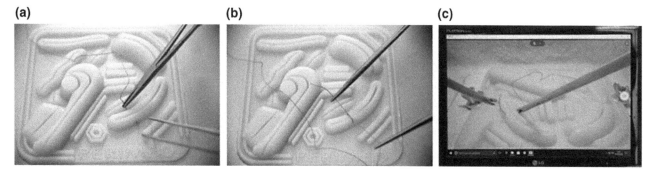

Fig. 13 Evaluation measuring

3   Results

During the two years of TP, a total of 21 residents partici-pated in the program, divided into two groups, one per year. Eleven finished only the Basic Level, stating as reasons to abandon: no time to continue or showing no further interest in performing laparoscopy during their careers. Ten sur-geons performed all the three levels (four residents of the first group and six of the second one), performing the living animal surgery. In the first year two surgeries were per-formed, each one with two residents. In the second year, two

Table 1 Activity measured time

Activity	Time	Realized
Novice open suture	00:26	3 knots OK
Novice lap suture direct view	04:44	3 knots OK
Novice laparoscopic suture camera view	05:00	Only needle pass
Trained surgeon laparoscopic suture camera view	02:50	2 Knots OK

animal surgeries were performed too, but with three residents forming each group. Four pigs were used for all procedures and all animals were still alive by the end of the procedures. In one surgery, the trainee cut a small blood vessel, which was sutured by the Fellow. After experimental surgery, the trained surgeons of the second group performed two hysterectomies, without any adverse events.

Table 1 shows the key achievements of the students and the time taken to perform.

4 Discussion and Conclusion

The initial investment was roughly U$ 8,000.00 to acquire the Endobox, the laparoscopic instruments, the models, the computer and furniture. At this moment the residents are performing some exercises of the Intermediate Level in more realistic simulators, however performed in the medical equipment manufactures plants. It was identified three new simulators and started the procurement expending U$ 4,000.00, thus enabling the University to realize full program in house.

Our low-cost laparoscopy TP resulted in adequate medical resident training in basic through advanced laparoscopy surgical skills, indicating its *cost-effectiveness viability*. The effectiveness can be assured by the successful surgery without patient injury. The cost can be figured when the expenses of our TP are compared to the expenses to make a similar training program in a private institution. In Brazil, the cost of a similar program is approximately U $ 2,500.00 per trainee. Calculating the total expenses of our TP, US$ 12,720.00 were spent (US$ 12,000.00 for investment in permanent assets [U$ 8,000.00 initial + U$ 4,000.00 new simulators] and US$ 720.00 to purchase four animals and anaesthetic drugs) to train 10 residents. Roughly US$ 25,000.00 would be spent to train the same 10 residents (US$ 2,500.00 × 10) in a similar private training program.

Conflicts of Interest The research being reported in this full paper was supported by Unicamp. The authors of this full paper are researchers and teaching member staff of Unicamp. All authors are involved in the development an innovative, low-cost Laparoscopy Training Program (FLS). The terms of this agreement were reviewed and approved by Unicamp at Campinas in accordance with its policy of objectivity in research.

References

1. Reich, H., Decaprio, J. and Fran Mcglynn: Minimally invasive surgery; Journal of Gynecologic Surgery, February 2009, 5(2): 213–216.
2. Brummer, T.H.I., Seppälä, T.T., Härkki, P.S.M.: National learning curve for laparoscopic hysterectomy and trends in hysterectomy in Finland 2000–2005; Human Reproduction, Volume 23, Issue 4, 1 April 2008, Pages 840–84.
3. Jaffray, B, Arch Dis Child; 90:537–542. https://doi.org/10.1136/adc.2004.062760, 2005.
4. Gerald M. Fried, G.M. at al: Proving the Value of Simulation in Laparoscopic Surgery, Annals of Surgery, Volume 240, Number 3, September 2004.
5. Feldman, L.S. et. Al: Relationship Between Objective Assessment of Technical Skills and Subjective In-Training Evaluations in Surgical Residents; Journal of American College of Surgeons, Vol. 198, No. 1, January 2004.

Evaluation of Benefit of Low Dose CT in the Diagnosis of Charcot Arthropathy

Markéta Štveráková, Michaela Steklá, Miroslav Selčan, and Petra Hospodková

Abstract

Diabetes mellitus (DM) is accompanied by many complications (nephropathy, neuropathy, ischemic heart disease etc.), while the foot ulcer falls among the most serious ones. The main problem for the right diagnosis of this syndrome is an appropriate choice of a diagnostic scenario. The aim of this study is to analyse the current possibilities of DM diagnostics using particular imaging modalities, and to compare them with a three-phase skeletal scintigraphy complemented by a low-dose CT. The principal method of this work was the cost effectiveness (C/E) analysis of individual diagnostic modalities in relation to a timely detection of the disease completed by a sensitivity analysis. The result of a multiple-criteria decision analysis calculated by means of the TOPSIS method was used as the effect for the C/E calculation. Nine basic criteria entered the analysis. The values of the final effects of individual modalities are in favour of the skeletal scintigraphy and the leucoscan plus the low-dose CT. Furthermore, total costs were calculated for individual diagnostic approaches from the perspectives of a medical facility and of a health insurance company. In both cases, the lowest costs were reached for the magnetic resonance, which has also significantly influenced the final value of the cost effectiveness. In the sensitivity analysis, variations in the skeletal scintigraphy and/or leukoscan sensitivity did not affect the final order of the diagnostic modalities provided unchanged weights of the criteria. The same conclusion was also in the case of the magnetic resonance. A change in the order was detected for the skeletal scintigraphy with the low-dose CT and for the leukoscan in the case of a decrease in specificity by 40 or more percent. The study opens a topical issue and forms a basis for a broader discussion concerning Charcot osteoarthropathy diagnostics within the professional community.

Keywords

Charcot arthropathy • Low dose CT • Cost effectiveness TOPSIS

1 Introduction

Charcot arthropathy is a serious diabetes mellitus complication. It is a serious, destructive affliction of the joints in a neuropathic leg. The aim of this study is to compare Charcot arthropathy diagnoses using a number of imaging modalities and focusing on three phase skeletal scintigraphy with low dose CT and evaluate its benefits.

Data collection was performed at a nuclear medicine hospital ward, which receives patients referred for skeletal scintigraphy from a podiatric out-patient service.

The standard procedure includes dynamic or early radiopharmaceutical detection, a full body scan with a two-hour interval SPECT/CT and a static image after 24 h.

If necessary, the procedure may also include an in vivo or in vitro marked leukocytosis examination. Based on the results of these tests, the doctor should decide whether the patient does indeed have Charcot arthropathy. An additional aim of this study is to define criteria that would help doctors unequivocally tell Charcot arthropathy from osteomyelitis based on the evaluation of radiopharaceutical detection in skeletal scintigraphy. Another goal is to develop a cost effectiveness analysis of each method in the context of early diagnosis of the disease. The study employs value engineering methods, multiple-criteria decision analysis and, last but not least, HTA methods to interpret the results.

M. Štveráková · M. Steklá · M. Selčan · P. Hospodková (✉)
Faculty of Biomedical Engineering, Czech Technical University in Prague, Nám. Sítná 3105, 272 01 Kladno, Czech Republic
e-mail: petra.hospodkova@fbmi.cvut.cz

© Springer Nature Singapore Pte Ltd. 2019
L. Lhotska et al. (eds.), *World Congress on Medical Physics and Biomedical Engineering 2018*,
IFMBE Proceedings 68/3, https://doi.org/10.1007/978-981-10-9023-3_68

377

1.1 Charcot Arthropathy

A total of 858,010 people were treated for diabetes in the Czech Republic in 2015 [1]. Diabetes mellitus is a group of chronic metabolic diseases that manifest primarily as hyperglycemia. There are many complications that may come with diabetes (nephropathy, neuropathy, ischemic heart disease, bone and joint pain, etc.) but certainly the most serious complication is the diabetic restless leg syndrome [2], a destructive condition, which damages leg tissue and is one of the most difficult complications to treat as it requires long-term hospitalisation and rehabilitation [3]. A more serious neuropathic leg condition is known as Charcot arthropathy. It is a progressive and highly destructive disease that affects one or more joints. Local osteoporosis occurs, accompanied by repeated micro injuries [4]. A suspected acute or subacute Charcot arthropathy is often diagnosed based on a physical leg oedema, elevated skin surface temperature (the difference between the affected and the non-affected leg must be greater than 2 °C -> acute phase, which might be complicated by a bilaterally elevated temperature), neuropathy, etc. Imaging methods used in Charcot arthropathy diagnosis include simple X-ray images, which, unfortunately, are only able to detect very advanced skeletal changes—the dislocation and fracture stage in the case of Charcot arthropathy. It is primarily useful in the localisation of changes. CT tests come with similar limitations but offer a more precise localisation [5]. A reliable examination method is the one using a four-phase dynamic skeletal scintigraphy. Magnetic resonance is a frequently debated modality in Charcot arthropathy diagnosis. It promises to detect even early stages of the disease and can also be used to monitor the progress of the disease. Edmonds et al. have shown that bone marrow oedema (in patients without changes detectable by X-ray) may lead to the development of subchondral fractures, erosion or cysts [6].

2 Methodology

2.1 Clinical Effectiveness

To ensure smooth cooperation with the podiatry team of the out-patient clinic, experts of the radiology ward created a scoring chart describing the results of CT and magnetic resonance examinations (Table 1). A group of specialists from the nuclear medicine department developed a set of evaluation criteria to assess skeletal scintigraphy and leukocyte scintigraphy/leukoscan results. Images from patients who had undergone the examination before the criteria were developed were re-examined any evaluated by relevant experts.

Table 1 CT and MR scoring table [7, 8, 9]

Attributes	Score
Signs of Charcot arthropathy	
Higher bone density (subchondral sclerosis)	
Bone fragments	
Bone dislocation	
Cartilage destruction	
Degeneration	
Deformation (metatarsus in the shape of a sharpened pencil)	
Bone marrow oedema under the joint space	
Total:	
Signs of osteomyelitis	
Regional osteopenia	
Periostal reaction	
Focal lesion of bone or loss of corticalis	
Endosteal scalloping	
Loss of trabecular architecture	
New bone apposition	
Bone marrow oedema around the location of structural changes	
Total:	
Final diagnosis	

Source own

The group of evaluating experts attributed values 0 or 1 to the above signs. 0 value means absence and 1 value means presence of the sign. The score for each condition is then added up and the high score thus confirms a diagnosis.

When evaluating bone scintigraphy (Table 2), the evaluating experts focused on the difference in accumulation calculated using the Ostnucline (or Xeleris) software. They recorded values at acquisition after 2 h after the application of radiopharmaceuticals for scintigraphy, the first reading for leukocyte scintigraphy/leukoscan and after 24 h. The resulting dataset points to a specific diagnosis.

The resulting value of clinical efficacy was then used in a multiple-criteria decision analysis.

2.2 Multiple-Criteria Decision Analysis

When choosing a suitable diagnostic method, one must take into account several criteria at once. Multiple-criteria decision analysis is a useful tool in this respect. The first step is to create a set of criteria and assign specific weight to each of them using Saaty's matrix, followed by value normalisation, partial version evaluation using the TOPSIS method and, finally, selection of the most suitable version or ranking of all possible versions [10]. In this study, specialists from the podiatry team evaluated the criteria using a questionnaire developed for these purposes.

2.3 Cost Effectiveness Analysis

The study also employs a cost effectiveness analysis (CEA) as it allows researchers to compare one or more versions of the diagnostic process and assess the magnitude of the benefits they bring in the context of the cost required [11].

$$CEA = \frac{E_1}{C_1} \geq \frac{E_2}{C_2} \qquad (1)$$

We considered the clinical effect calculated based on the multiple-criteria decision analysis, taking into account 9 criteria: radiation exposure, time requirements (from the patient's perspective), the cost of the examination [in CZK], sensitivity, specificity, clinical efficiency, required expertise on Charcot arthropathy, availability in time, organisational constraints. The criteria were defined by the expert team based on the Delphi method.

Multiple-criteria decision analysis evaluates several possible alternatives based on a set of criteria. The alternatives under consideration in this study are three different diagnostic methods used for Charcot arthropathy. The Multiple-criteria decision analysis is applied using TOPSIS a method widely used for medical technologies with a high precision of results [12, 13]. Due to the nature of this study, the researchers then performed a one-way sensitivity analysis.

3 Results

The final values were calculated using a clinical efficiency analysis in a retrospective study, a multiple-criteria decision analysis, a cost effectiveness study and a sensitivity study.

Data were collected retrospectively concerning patients who were referred for skeletal scintigraphy by the podiatry out-patient clinic due to a suspected Charcot arthropathy.

The results of all examinations performed on these patients were taken into account including their clinical state at their first visit to the podiatry clinic, the results of skeletal scintigraphy, the results of leukocyte scintigraphy/leukoscan (if performed), magnetic resonance results as well as the results of a clinical examination half a year later.

The control group consisted of 19 patients who underwent skeletal scintigraphy and low dose CT. There was also a group of 14 patients who did not undergo a low dose CT examination.

The cost figures were discussed with the head doctor of the relevant ward.

3.1 Multiple-Criteria Decision Analysis

The researchers and the podiatry team created a set of 9 criteria—radiation exposure, time requirements for the patient, the cost of the examination, sensitivity, specificity, clinical efficiency, expertise on Charcot arthropathy required from the examining specialist, waiting time, and organisational constraints.

The specialists were provided with a table listing the evaluation criteria for each method. They were also provided with a set of relative weights based on Saaty's method. The expert opinions where then processed using a TOPSIS multiple-criteria decision method.

Table 3 shows the ranking of the modalities under consideration. Skeletal scintigraphy combined with a leukoscan and low dose CT seems to be the most efficient. By comparison, the clinical effect of magnetic resonance is approximately half as important. Compared with the control group, there is a clear benefit of adding low dose CT.

3.2 Health Insurer's Perspective

The calculation of the cost the examination (in CZK) is based Decree No. 273/2016 Coll., on point value calculation, payments for healthcare services and regulation for 2017 [14]. The value of each variable in the equation was

Table 2 Scintigraphy evaluation table for suspected cases of Charcot arthropathy

	Skeletal scintigrafy	Static image after 24 h	Leukoscan/leukocyte scintigraphy	Static image after 24 h (leukocyte scintigraphy/leukoscan)
Degenerative changes	≤ 2	drop or stationary accumulation	minimal accumulation	minimal accumulation
Charcot arthropathy	>2	accumulation increases by at least 1,0	up to 1,5	no accumulation increase
Osteomyelitis	>2	accumulation increases by at least 1,1	over 2.0	small increase in accumulation

Source own

Table 3 Ranking based on TOPSIS

	Results	Ranking
Skeletal scintigrafy + leukoscan	0.6712	2
Skeletal scintigrafy + leukoscan + low dose CT	0.6724	1
Magnetic resonance	0.3287	3

determined based on the relevant document describing the calculation algorithm or based on the List of Healthcare Services. The payment includes a separate figure for the radiopharmaceutical (ZULP) used, which was added to the overall cost (Table 4).

3.3 Cost Effectiveness Analysis

Cost effectiveness was assessed based on the results of the multiple-criteria decision analysis and the cost calculated from the perspective of the healthcare provider and the health insurance company (Table 5).

From the health insurance company's perspective, magnetic resonance is the most cost effectiveness method (Table 6). Nuclear medicine methods are nearly 2.5× less cost efficient by comparison. However, they consist of two examinations that are performed in parallel. The cost of each examination is not calculated separately and only the sum total is considered here. As a result, this approach is much more expensive as it requires double application of radiopharmaceuticals, double manpower cost, etc.

3.4 Sensitivity Analysis

The sensitivity analysis was used to examine the effect of changes in sensitivity, specificity, clinical efficiency and their respective weights on the TOPSIS ranking. The analysis found that when an equal weight is assigned to the criteria, a change in skeletal scintigraphy and leukoscan sensitivity has no effect on the final ranking. The same conclusion was reached for magnetic resonance. In the case

of scintigrafie skeletu in combination with low dose CT and a leukoscan, the ranking changed—namely a drop in specificity by 40 or more percent, the ranking of nuclear medicine methods swapped. Changes in sensitivity or clinical efficiency in any of the versions at the same criteria weights did not result in changes in ranking. When the specificity weight changed by 180 or more percent, the ranking of the top two versions swapped.

4 Discussion

Our retrospective controlled observational study conducted at a partner healthcare provider found that the clinical efficiency of skeletal scintigraphy with leukoscan and low dose CT is only 2% lower than that of magnetic resonance, a method described by foreign studies as the best available method for diagnosing Charcot arthropathy. [15]. Compared with magnetic resonance, though, nuclear medicine methods are significantly more expensive both for the healthcare provider and for the health insurance company. This is due to the examination method where skeletal scintigraphy is used as the basic examination. A positive finding than means that the patient also needs to get a leukoscan (most typically) on top of the skeletal scintigraphy in order to rule out osteomyelitis. This procedure is reflected in the multiple-criteria decision analysis and the cost of the two procedures is added together because evaluating each procedure separately is meaningless. The high price is primarily due to the cost of radiopharmaceuticals [14] and represents the biggest disadvantage of this method. Although its clinical effect is nearly twice as high compared with magnetic resonance, the price is nearly 6.5 times higher. So in terms of cost effectiveness, magnetic resonance is the most efficient method. We also calculated ICER, which is CZK 60,445.84 per unit of incremental effect.

If we focus exclusively on the benefits of low dose CT in diagnosing Charcot arthropathy, we see a clear increase in effect when combined with skeletal scintigraphy. Adding this examination to the diagnostic protocol is, therefore, clearly meaningful. From the health insurer's perspective, the fact that it is not charged

Table 4 Health insurance payment

Service	Code	Time [in min.]	Point value	Value per point [CZK]	ZULP/ZUM [CZK]	Total payment [CZK]
Skeletal scintigraphy + 1× SPECT	47245 + 47269	125	1379.33	1.03	3016	6054
Skeletal scintigraphy + low dose CT (+ 1× SPECT)	47246 + 47269	126	1379.33		3016	6054
Leukoscan	47237 + 47269	170	1907.98		15720	19876
MR of legs without contrast agent	89713	120	797.76		0	5157

Table 5 Cost and effect calculation

Methods	Effect	Healthcare provider cost [CZK]	Health insur. company cost [CZK]
Skeletal scintigraphy + leukoscan	0.67121	21692.185	25930
Scintigrafie skeletu + leukoscan + low dose CT	0.67245	21692.185	25930
Magnetic resonance	0.32878	2903.765	5157

Source own

Table 6 CEA results

Methods	Insurance company CEA [CZK]
Skeletal scintigraphy + leukoscan	38631.68
Scintigrafie skeletu + leukoscan + low dose CT	38560.36
Magnetic resonance	15684.82

Source own

(the procedure has not been assigned a specific code) must be certainly of interest. Low dose CT is not considered a "thorough" CT diagnostic procedure. From the healthcare provider's perspective, the higher effect is "paid for" by a one-minute extension of the examination, i.e. a negligible addition to the staff cost [14]. In terms of equipment, the Czech Nuclear Medicine Society's latest version of the nuclear medicine concept of March 17, 2017 defines the required equipment of SPECT/CT and non-hybrid modalities are considered only an additional option [16]. This means that a healthcare provider purchasing new equipment for the nuclear medicine department should make sure that at least one of the devices is hybrid (SPECT/CT or a highly specialised PET/CT with PET/MR). Most nuclear medicine departments and wards already possess such devices or will be required to acquire one at the next update of their equipment. There is, therefore, no additional cost of buying low dose CT equipment specifically for the purposes of this examination. In the context of diagnosing Charcot arthropathy, the equipment is used to determine the precise location of the affliction and distinguish between Charcot arthropathy and osteomyelitis based on a scoring table available to radiologists.

The data from the Czech Republic used in this study offer a sample comparable to that used in the study described in "Role of magnetic resonance imaging in the diagnosis of osteomyelitis in diabetic foot infections" [15]. Authors of this article reached a similar conclusion—magnetic resonance is the most cost effectiveness method. The cost effectiveness analysis performed as part of our study concludes the same for Charcot arthropathy.

Conflict of Interests The authors declare no conflict of interests.

References

1. Prevention of diabetes mellitus, Institute of Health Information and Statistics of the Czech Republic, http://reporting.uzis.cz/cr/index.php?pg=statisticke-vystupy–morbidita–intervalova-prevalence-dle-diagnoz–prevalence-diabetu-mellitu&region=cr&year=2015, last accessed 2018/01/29.
2. Diabetes Care: Diagnosis and Classification of Diabetes Mellitus. Issue SUPPL, vol. 35, pp. 64–71, (2012).
3. Jirkovská, A.: Syndrome of diabetic foot: International consensus developed by the International Working Group for diabetic foot syndrome. Praha (2000).
4. Poch, T.: Diabetic foot - diagnostic, therapy, prevention. Practicus, 36–39, (2010).
5. Bém, R., Jirkovská, A., Fejfarová, V.: News in diagnosis and therapy of Charcot osteoarthropathy: Syndrome of diabetic foot. Bulletin HPB., vol. 14 (4), pp. 156–159. Praha (2006).
6. Edmonds, M., Petrova, N., Edmonds A. et al: What happens to the initial bone marrow oedema in the natural history of Charcot osteoarthropathy? Diabetologia, vol. 49 (1), pp. 684. (2006).
7. Schoots, IG, Slim, FJ, Busch-Westbroek, TE & Maas, M 2010,'Neuro-osteoarthropathy of the Foot-Radiologist: Friend or Foe?' Seminars in musculoskeletal radiology, vol. 14 (3), pp. 365–376, (2010).
8. Donovan, A., Schweitzer M.,: Use of MR Imaging in Diagnosing Diabetes-related Pedal Osteomyelitis. RadioGraphics, vol. 30, pp. 723–736, (2010).
9. Abdel, R., Samir, S.:Diagnostic performance of diffusion-weighted MR imaging in differentiation of diabetic osteoarthropathy and osteomyelitis in diabetic foot, vol. 89, pp. 221–225. (2017).
10. Korviny, P.: Theoretical foundations of multi-criteria decision, https://korviny.cz/Korviny/soubory/teorie_mca.pdf, last accessed 2018/01/30.
11. Goodman CS. Healthcare technology assessment: methods, framework, and role in policy making.American Journal of Managed Care, 1998:SP200–14; quiz SP215-6.
12. Chen, S., Cheng, S., Lan, T.: Multicriteria decision making based on the TOPSIS method and similarity measures between intuitionistic fuzzy values. Information Sciences., pp. 367–368, pp. 279-295, (2016).
13. Kubátová, I.: The use of value engineering and multi-criteria decision-making in health technology assessments. Praha (2015).
14. Government Regulation Num. 273/2016 Sb. Laws for people, https://www.zakonyprolidi.cz/cs/2016-273, last 2017/01/30.
15. Croll, S., Nicholas, G., Osborne, M., Wasser, T., Jones, S.: Role of magnetic resonance imaging in the diagnosis of osteomyelitis in diabetic foot infections. J Vasc Surg. vol. 24, pp. 266–70, (1996).
16. The Czech Nuclear Medicine Society. Concept of department of nuclear medicine, Praha (2017).

Development of Methodology of Evaluation for Medical Equipment Replacement for Developing Countries

Mario Andrés Alvarado and Sandra Luz Rocha

Abstract

Currently, decision-making in hospitals and health institutions in developing countries about the replacement or withdrawal of medical equipment mainly involves a series of subjective positions about the time at which these actions should be carried out. It is for the foregoing that not always have the technical, clinical or economic basis on the request for the withdrawal of any equipment. There are several quantitative methodologies in the world that evaluate a considerable amount of parameters about the equipment and with this determine the status in which it is, however, these methodologies are difficult to apply especially in the public healthcare system of countries in development due to the ranges of evaluation of the parameters are focused on countries that acquire the technology as it goes to market, or even are technology manufacturing countries. It is due to the above that in this project a methodology is designed to evaluate medical equipment so that it can be used in countries that do not have the degree of substitution of developed countries. It is hoped that with this methodology it will be possible to anticipate equipment obsolescence and make the corresponding substitution at the best time according to the possibilities of each institution, based on quantitative data. The methodology is based on the evaluation of eight parameters that include the technical, economic and clinical part of each equipment.

Keywords

Equipment replacement • Developing country
Equipment evaluation

1 Introduction

Medical technology management is nowadays an important tool available for a biomedical engineering department, because due to this it is possible to have a certain order in all the processes that are carried out within a hospital.

Medical equipment: medical devices requiring calibration, maintenance, repair, user training and decommissioning. Medical equipment is used for the specific purposes of diagnosis and treatment of disease or rehabilitation following disease or injury [1]. Once the equipment stops solving the above it's said that it is obsolete and must be submitted to the aforementioned evaluations to plan its replacement or withdrawal from the service.

1.1 Nowadays

The evaluation of technologies is a rare practice among health institutions in developing countries, especially in public sector. Nowadays in the world there are several methodologies that evaluate the status of medical equipment, from subjective methodologies that consider the experience of the personnel in the departments of biomedical engineering of hospitals, to quantitative methodologies that evaluate a considerable number of parameters and through a numerical analysis it is determined if for a medical equipment its replacement must be soon planed or it is still in good condition.

The main problem of any methodology designed in a developed country attends the realities that these countries live, for example, the AHA has an average life document of all medical equipment that serves as a reference in many methodologies, however, equipment such vital signs

M. A. Alvarado (✉) · S. L. Rocha
Instituto Nacional de Cancerología National Cancer Institute,
Mexico City, 08544, Mexico
e-mail: andres.alvaradon14@gmail.com

S. L. Rocha
e-mail: srochan@yahoo.com

M. A. Alvarado
Universidad Iberoamericana, Mexico City, Mexico

© Springer Nature Singapore Pte Ltd. 2019
L. Lhotska et al. (eds.), *World Congress on Medical Physics and Biomedical Engineering 2018*,
IFMBE Proceedings 68/3, https://doi.org/10.1007/978-981-10-9023-3_69

monitors, anesthesia machines, linear accelerators have an average age of 6, 8 and 10 years respectively [2]. For developed countries that acquire the technology as it goes to market, or even it is manufactured within the same country, it is easier to follow this standard of time, due to they have the level of planning and resources to be able to make a substitution within the suggested times, totally away from the reality of developing countries, since in most cases the equipment reach or even exceeds the dates of useful life that the manufacturer gives to the institutions. Another situation that could prevent these methodologies work in public institutions in developing countries is that they include the cost-benefit part of the analysis, this is because the public health sector of a developing country usually does not show economic benefits due to the using level of the equipment or the degree of attention provided.

The biggest problem of the above is not that the dates of life suggested by the AHA are exceeded, as the correct functioning of an equipment does not necessarily depend on the age, the problem is that the evaluation methodologies that currently exist can´t be used by developing countries, this has as consequences: equipment obsolescence, poor budgeting, purchase of non-priority equipment.

2 Methodology

A suggested form in which an evaluation model should be designed is in a personalized way, because each institution will have differences in terms of the information it has about its medical equipment, considering that is waited that the methodology be universal, the design was done by a different way, choosing parameters that can be obtained by any biomedical department, this allows to use the methodology in more than just one hospital. The main reference to make the present methodology was Fennigkoh and his model for replacement [3].

For the design of the methodology, the following objectives were considered:

- A methodology that can evaluate medical equipment of high and medium technology within any institution.
- A quantitative, objective and simple methodology to be used by the inventories staff.
- Evaluate the equipment from the technical (T), economic (E) and clinical (C) aspects, at the same time and not separately.

The methodology includes eight parameters to be evaluated, these cover the three aspects: T, E and C. These parameters were chosen from a large number of possibilities [4] because these eight can be obtained by any biomedical department.

To carry out the evaluation of the parameters it was decided not to group them according to their nature, they will not necessarily be grouped according to the aspect they belong (T, E, C), but to the specific weight they have at the time of evaluating the equipment. A main differentiator of this methodology against the rest of methodologies is that it has a maximum priority group made up of two parameters that depending on them scores, can determine the final result. The groups that were defined for the model are the "technical-economic" (TE), the "maximum priority" group (MP) and the "clinical" group (C). Each group includes, as previously mentioned, some of the eight parameters chosen for the methodology.

The methodology is based on a system of scores, medical equipment can obtain different values depending on the parameter that is being evaluated, at the end these values that the equipment obtained will be introduced in a mathematical model.

2.1 Parameters

2.1.1 Age
Time that a medical equipment can be in operation in some developed countries is often based on a period of seven years, where by official norm they must be replaced, however, as mentioned, due to in developing countries it is very complicated to give a life of seven years to an equipment that is expected even twenty years of use, it has been decided to use the average useful life that the manufacturer grants, if the age of the equipment exceeds that suggestion will get a 1 in the model, otherwise it will have a 0.

2.1.2 Maintenance Cost
An equipment that is part of a maintenance program in a hospital can extends its time of correct operation and avoid a period of long downtime. The maintenance program can be through external companies or by the biomedical department. The most important is to try that the cost of such maintenance (including costs of spare parts) does not exceed a considerable percentage of the value of the new equipment. Therefore, an equipment whose average maintenance cost of the last 3 years exceeds 15% of the total value of itself will get a 1 in the model, otherwise it will have a 0.

2.1.3 Downtime
Medical equipment must be available for use as much as possible, since any equipment that is not available when needed can detonate a dangerous situation and even threaten the lives of patients. It has been decided that an equipment must have at least an Uptime of 90%, the above is calculated as Eq. (1), if having a downtime greater than 10% obtains a 1, or a 0 otherwise for the model.

$$\frac{hours\ not\ available\ for\ clinical\ use}{number\ of\ clinical\ hours\ of\ use} \times 100 = Downtime \quad (1)$$

2.1.4 End of Support

An equipment can begin to present difficulties for a biomedical department when the manufacturing company stops supporting it, since from that moment any repairs that are required will have to be solved by the biomedical department, everything that implies spare parts will also leave of being manufactured and distributed by the manufacturer. If the equipment no longer has support get a 2, in case the equipment has support for a year or less it gets a 1, having support for 2–3 more years, gets 0.5, and if having more than 3 years of support the model gets a 0.

2.1.5 Availability of Consumables and/or Spare Parts

There is many equipment whose operation depends on the use of consumables, sometimes there is the possibility of obtaining consumables and spare parts with external suppliers, once the manufacturer has stopped supporting the equipment and it is even impossible to get them externally, the equipment despite working correctly will go to its obsolescence and will have to be considered an immediate replacement. If there are no consumables and/or spare parts available, the equipment will get a 1, otherwise it will be awarded 0. This parameter is closely related to the previous one, since obtaining a 1 the model immediately considered that in "end of support" it gets a 2.

$$RI = 0.4[age + cost + downtime] + 0.4[end\ of\ support \\ + availability\ of\ consumables] + 0.2[clinical\ efficiency \\ + user\ pref. + freq.\ of\ use]$$

$$(3)$$

2.1.6 Clinical Efficiency

The opinion of the professional responsible for operating an equipment has an important value in the evaluation, even if it is a subjective aspect, it can give us important information about how accepted an equipment is for its use within the institutions. Those equipment that according to the professionals give the best possible care to the patients, in matters of form, time, quality, will be considered as meeting a clinical efficiency and will get a 0 in the model, if professional thinks the device has a medium efficiency it gets 1, and 1.5 if it does not have clinical efficiency.

2.1.7 User Preference

An equipment that is liked by professionals is one that can be understood is fulfilling its purpose in the best way, therefore, if the preference is high the model gets a 0, if it is medium gets a 0.5, and if it is low it will be 1.

2.1.8 Frequency of Use

The equipment, as mentioned has a suggested lifespan to be in operation, and although, giving them continuous maintenance these times can be extended, it is necessary to evaluate the use that is given, because if it is intensive it could affect the time of life that it may have, the objective of this last parameter is to guarantee that an equipment with continuous use does not represent problems at the moment of its operation. If an equipment has a high use it will get a 1, medium use gets a 0.5, and low use gets a 0 to the model.

2.2 The Model

Every group of parameters has a percentage in the total result of 100% from the model. As follows: TE (40%), MP (40%) and C (20%).

The complete model is a linear sum (2) of each group and its possible scores of each parameter:

$$RI = 0.4 * TE + 0.4 * MP + 0.2 * C \quad (2)$$

where RI is the Replacement Index, confined to the range: $0.2 \leq RI \leq 3.0$

2.2.1 Replacement Thresholds

Replacement index represents the value of the equipment's status within a replacement plan, as follows:

$1.0 > RI$	Re-evaluate in two years
$1.0 \leq RI \leq 1.3$	Re-evaluate in one year
$1.4 \leq RI \leq 1.7$	Replace following year
$1.8 \leq RI$	Replace immediately

This model and its thresholds only represents a recommendation as a tool for a biomedical department, and does not represents a replacement mandate.

3 Conclusion and Results

The methodology was designed to be a support for health institutions in developing countries, especially those in the public sector. It is expected that this helps biomedical

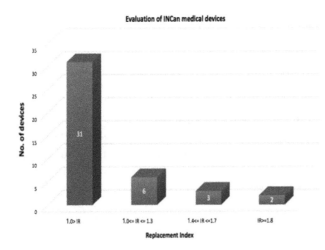

Fig. 1 Obtained results of the INCan equipment evaluation

departments to justify the request of budget to acquire new equipment when it is really necessary (having the quantitative fundament) and avoid purchases of non-priority equipment. The methodology offers objective information for making medical devices replacement decision, it is based on the economic, clinic and technical aspect, this offers the chance to know the complete status of every equipment. The methodology was tested on some of the medium and high technology equipment of the Instituto Nacional de Cancerología (INCan), in Mexico City, and the obtained results are shown below:

The results (Fig. 1) indicate that the most of the evaluated devices are in an acceptable status and should be re-evaluated in two years, 6 devices should be re-evaluated in one year, 3 should be replaced the following year at the latest and only 2 should be replaced immediately. Comparing the obtained results with Fennigkoh´s model (designed in a developed country) where results were: 25 should be re-evaluated in one year, 12 should be replaced next year and finally 5 must be urgent replaced, it is concluded that the methodology shown in this paper is less rigorous and can be applied to developing countries, since it adapts better to their reality.

References

1. WHO, http://www.who.int/medical_devices/definitions/en/, last accessed 2017/12/25.
2. American Hospital Association: Estimated Useful Lives of Depreciable Hospital Assets, American Hospital Association, Chicago (1998).
3. L. Fennigkoh: A Medical Equipment Replacement Model. Vol. 17 No. 1, 43–47 (1992).
4. J. Tobey Clark, Raymond D. Forsell: A Practicum for Biomedical Engineering & Technology Management Issues. 1st edn. Kendall Hunt, Pennsylvania (2008).
5. Joseph Dyro: The Clinical Engineering Handbook. 1st edn. Elsevier Academic Press, Pennsylvania (2004).
6. Y. David, W. von Maltzahn, N. Michael, B. Joseph: Clinical Engineering. 3rd edn. CRC Press, New York (2005).
7. WHO, http://apps.who.int/iris/bitstream/10665/44561/1/9789241501392_eng.pdf?ua=1, last accessed 2017/12/12.
8. Pacheco A., Pimentel AB., et al.: Metodología para evaluación de Equipo Médico.3(1), 22–26 (2002).

Toward a Novel Medical Device Based on Chromatic Pupillometry for Screening and Monitoring of Inherited Ocular Disease: A Pilot Study

Paolo Melillo, Antonella de Benedictis, Edoardo Villani,
Maria Concetta Ferraro, Ernesto Iadanza, Monica Gherardelli,
Francesco Testa, Sandro Banfi, Paolo Nucci,
and Francesca Simonelli

Abstract

Chromatic pupillometry is a relatively novel research tool for retinal function evaluation and may be an appropriate and easier way to diagnose and monitor inherited retinal diseases in paediatric population. Nevertheless, although the method is clinically feasible in paediatric populations, as shown by several non-ocular studies, only few studies, on a small size sample of paediatric subjects, are available. To the best of the authors' knowledge, no medical device based on chromatic pupillometry was CE-marked for diagnosis and/or monitoring of these conditions. Therefore, we designed a pilot study in order to evaluate clinical feasibility, reliability and utility of chromatic pupillometry. The study sample consists of sixty patients, affected by inherited ocular diseases. A pupillometric system, including definition of pupillometric protocols, have been set up. In the current paper, we present the comparison between the measurements obtained in one patient affected by Retinitis Pigmentosa and a healthy age-matched control in order to disclose differences in chromatic pupillometry parameters between case and control.

Keywords

Chromatic pupillometry • Inherited retinal diseases
Pilot study

1 Introduction

Inherited Retinal Diseases (IRDs) represent a significant, and often overlooked, cause of severe visual deficits in children. Both accurate diagnosis and proper clinical management and treatment are impaired by the high genetic heterogeneity of these conditions (with over 200 causative genes) and by the necessity to perform complex and sometime invasive clinical tests that are of difficult application in young children.

White light stimulus pupillometry has been developed to characterize lowest level of visual perception [1, 2] and then this test was successfully adopted as outcome measurement in gene therapy clinical trial for IRD [3–5]. More recently, chromatic pupillometry has been proposed as a highly sensitive and objective test to quantify the function of different light-sensitive retinal cells. Nevertheless, although the method is clinically feasible in pediatric populations and instrumentations are commercially available, only few cases of its application to pediatric ophthalmic patients are described in literature and to the best of the authors' knowledge, no medical device based on chromatic pupillometry was CE-marked for diagnosis and/or monitoring of IRDs.

P. Melillo (✉) · A. de Benedictis · F. Testa · F. Simonelli
Eye Clinic, Multidisciplinary Department of Medical, Surgical and Dental Sciences, University of Campania Luigi Vanvitelli, Naples, Italy
e-mail: paolo.melillo@unicampania.it

F. Simonelli
e-mail: francesca.simonelli@unicampania.it

E. Villani · M. C. Ferraro · P. Nucci
Department of Clinical Sciences and Community Health, University of Milan, Milan, Italy

E. Iadanza · M. Gherardelli
Department of Information Engineering, University of Florence, Florence, Italy

S. Banfi
Medical Genetics, Department of Precision Medicine, University of Campania Luigi Vanvitelli, Naples, Italy

S. Banfi
Telethon Institute of Genetics and Medicine, Pozzuoli, Italy

© Springer Nature Singapore Pte Ltd. 2019
L. Lhotska et al. (eds.), *World Congress on Medical Physics and Biomedical Engineering 2018*,
IFMBE Proceedings 68/3, https://doi.org/10.1007/978-981-10-9023-3_70

2 Primary Research Goals

The primary research goals of the pilot study are to:

1. evaluate the clinical feasibility of the chromatic pupillometric protocol in a paediatric population;
2. estimate the repeatability and reproducibility of chromatic PLR parameters in cases and controls;
3. assess the differences in chromatic PLR between ophthalmic paediatric patients with the target disease and healthy paediatric subjects.

3 Design of the Pilot Study

The study was designed as a multicentric case—control pilot study. The study was approved by the Ethics Committee of the University of Campania Luigi Vanvitelli, carried out in compliance with the Declaration of Helsinki, and an informed consent was obtained from all participants. The study schedule establishes a baseline visits, including informed consent, assessment of inclusion/exclusion criteria and pupillometry, and a 28 (±7) days follow-up visit to repeat pupillometry.

The study sample consists of sixty patients (cases), affected by inherited ocular diseases, and twenty healthy subjects (controls). Both cases and controls should be recruited provided they satisfy the following conditions: age between 8 and 16 years; willing to participate to this pilot study by subjects' assent and relatives' informed consent. Cases should have a clinical diagnosis of: non-syndromic inherited retinal dystrophies (i.e., Retinitis Pigmentosa; Leber Congenital Amaurosis) or inherited macular degeneration (i.e., Stargardt disease; Cone dystrophy), or optic disk pathologies (i.e., Leber hereditary optic neuropathy). Controls should have no known ocular diseases and a refraction error in absolute value lower than 5 dioptres.

The pupillmetric protocol is performed by using a multi-chromatic laboratory pupillometer with binocular/dual camera system that tracks both pupils and can stimulate right or left or both eyes. The video stream is captured at 30 Hz and the video frame is digitized into 640×480 pixels with 8-bit gray level resolution. Light is emitted through a diffusing screen (approx. $50° \times 35°$ of visual angle). Light intensities are defined in Lux and correspond to the scotopic CIE human luminosity sensitivity function.

The protocol for chromatic pupillometry was defined on the basis of existing literature data [2, 6–10]. The measurements started under scotopic conditions after 10 min of dark-adaption. The protocol was based on 1 s stimuli presented to both eyes in three different conditions as follows:

(a) 0 log Lux stimuli in the dark with a 15-s inter-stimulus interval, in order to evaluate rods; (b), 3 log Lux stimuli in the dark with a 30-s inter-stimulus interval, in order to evaluate cones, (c) after a 3-min adaption to blue background, 3 log Lux stimuli on the blue background, in order to evaluate intrinsically photosensitive retinal ganglion cells (ipRGCs). Each stimulus was repeated three times in order to assess short-term repeatability of the measures and at each condition, red stimuli were followed by matched blue stimuli (Table 1).

For each stimulus, the following parameters were computed:

- Maximum diameter (Max) = pupil diameter baseline before the constriction
- Minimum diameter (Min) = pupil diameter at the peak of the constriction
- Delta = Max–Min
- Maximum constriction = (Max–Min)/Max, given in percentage
- CLAT = latency of the constriction (delay of the onset of the pupil constriction)
- CV = mean constriction velocity
- MCV = maximum constriction velocity
- DV = mean dilation velocity (recovery after the constriction).

4 Preliminary Results

In the current paper, we present a comparison between the measurements obtained in one patient affected by a typical form of Retinitis Pigmentosa and a healthy age-matched control. The pupil responses were visualized as graphs and an example for each stimulus is reported in Fig. 1. As reported in Table 2, the case showed a range of lower value of CH compared to the control, particularly, in response to stimuli 1 and 2.

5 Discussions

In this paper, an objective measurement technique, i.e., chromatic pupillometry, was presented to evaluate the light-sensitive retinal cell populations in pediatric patients. A pilot study has been designed based on the previously published protocols and results. The preliminary results seems to confirm a clinically significant difference in the pupil response: for instance, a case affected by Retinitis Pigmentosa showed a lower pupil response, compared an age-matched control, particularly, to stimulus eliciting rods, consistently with the

Table 1 Summary of chromatic pupillometric protocol

#	Color	Intensity	Background
1	RED (622 nm)	LOW (0 log Lux)	Dark (after 10-min dark adaptation)
2	BLU (463 nm)	LOW (0 log Lux)	Dark
3	RED (622 nm)	HIGH (3 log Lux)	Dark
4	BLU (463 nm)	HIGH (3 log Lux)	Dark
5	RED (622 nm)	HIGH (3 log Lux)	Blue (after 3-min light adaptation)
6	BLU (463 nm)	HIGH (3 log Lux)	Blue

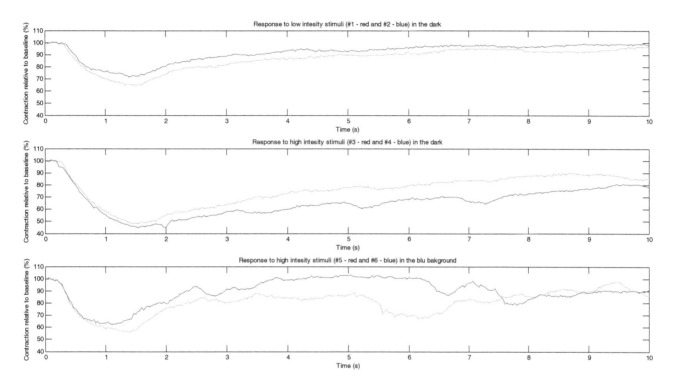

Fig. 1 Examples of pupil responses to each stimulus taken from the selected case

Table 2 Summary of comparison between the case and the age-matched control

Stimulus #	Maximum Diameter (mm^2)		Minimum Diameter (mm^2)		Maximum constriction (%)	
	Case	Control	Case	Control	Case	Control
1	6.97–7.23	6.6–6.79	4.58–4.95	3.45–3.63	29–35.3	45–48.4
2	6.85–7.06	6.18–6.64	4.86–5.11	3.55–3.9	27.6–29.1	36.9–46.1
3	6.72–6.8	4.48–6.4	1.77–3.28	2.45–2.91	51.6–74	45.3–61.7
4	6.42–6.87	6.47–7.6	3.02–3.13	2.71–3.01	53–54.4	54.4–64.2
5	3.27–3.84	3.94–4.46	1.93–2.17	2.12–2.29	37.9–46.2	46.2–49.6
6	3.44–3.51	4.1–5.01	2.1–2.21	2.22–2.51	37–39	41.5–55.8

unrecordable scotopic and markedly reduced photopic ERG. We selected the case here presented, since the patient showed a typical form of Retinitis Pigmentosa (i.e., classic fundus appearance, e.g. intraretinal pigment deposits, thinning and atrophy of the Retinal Pigment Epithelium in the mid- and far peripheral retina, with relative preservation in the macula, waxy pallor of the optic disc, attenuation of the retinal vessels; reduced and delayed ERG responses; visual field constriction), and consequently, would be a paradigmatic representative of the entire population affected by Retinitis Pigmentosa. Further developments of the current work include the analysis of cases with other inherited retinal diseases.

Since the pupil response to the lower stimulus was detectable also when the scotopic ERG was below the noise level, chromatic pupillometry could be adopted not only to support the diagnosis but also to monitor the disease progression and eventually to assess amelioration in visual function after an experimental therapy.

Acknowledgements The current study was supported by Italian Ministry of Education, Research and University under the project "Toward new methods for early diagnosis and screening of genetic ocular diseases in childhood" (Grant PRIN 20158Y77NT to Francesca Simonelli). The funders had no role in study design, data collection and analysis, decision to publish, or preparation of the manuscript. The authors declare that they have no conflict of interest.

References

1. Aleman, T.S., Jacobson, S.G., Chico, J.D., Scott, M.L., Cheung, A.Y., Windsor, E.A., Furushima, M., Redmond, T.M., Bennett, J., Palczewski, K., Cideciyan, A.V.: Impairment of the transient pupillary light reflex in Rpe65(-/-) mice and humans with leber congenital amaurosis. Invest Ophthalmol Vis Sci 45, 1259–1271 (2004).
2. Roman, A.J., Schwartz, S.B., Aleman, T.S., Cideciyan, A.V., Chico, J.D., Windsor, E.A., Gardner, L.M., Ying, G.S., Smilko, E. E., Maguire, M.G., Jacobson, S.G.: Quantifying rod photoreceptor-mediated vision in retinal degenerations: dark-adapted thresholds as outcome measures. Exp Eye Res 80, 259–272 (2005).
3. Simonelli, F., Maguire, A.M., Testa, F., Pierce, E.A., Mingozzi, F., Bennicelli, J.L., Rossi, S., Marshall, K., Banfi, S., Surace, E.M., Sun, J., Redmond, T.M., Zhu, X., Shindler, K.S., Ying, G.S., Ziviello, C., Acerra, C., Wright, J.F., McDonnell, J.W., High, K. A., Bennett, J., Auricchio, A.: Gene therapy for leber's congenital amaurosis is safe and effective through 1.5 years after vector administration. Mol Ther 18, 643–650 (2010).
4. Melillo, P., Pecchia, L., Testa, F., Rossi, S., Bennett, J., Simonelli, F.: Pupillometric analysis for assessment of gene therapy in Leber Congenital Amaurosis patients. Biomed Eng Online 11, 40 (2012).
5. Testa, F., Maguire, A.M., Rossi, S., Pierce, E.A., Melillo, P., Marshall, K., Banfi, S., Surace, E.M., Sun, J., Acerra, C., Wright, J.F., Wellman, J., High, K.A., Auricchio, A., Bennett, J., Simonelli, F.: Three-Year Follow-up after Unilateral Subretinal Delivery of Adeno-Associated Virus in Patients with Leber Congenital Amaurosis Type 2. Ophthalmology 120, 1283–1291 (2013).
6. Kardon, R., Anderson, S.C., Damarjian, T.G., Grace, E.M., Stone, E., Kawasaki, A.: Chromatic pupil responses: preferential activation of the melanopsin-mediated versus outer photoreceptor-mediated pupil light reflex. Ophthalmology 116, 1564–1573 (2009).
7. Kawasaki, A., Crippa, S.V., Kardon, R., Leon, L., Hamel, C.: Characterization of pupil responses to blue and red light stimuli in autosomal dominant retinitis pigmentosa due to NR2E3 mutation. Invest Ophthalmol Vis Sci 53, 5562–5569 (2012).
8. Kawasaki, A., Collomb, S., Leon, L., Munch, M.: Pupil responses derived from outer and inner retinal photoreception are normal in patients with hereditary optic neuropathy. Exp Eye Res 120, 161–166 (2014).
9. Kawasaki, A., Munier, F.L., Leon, L., Kardon, R.H.: Pupillometric quantification of residual rod and cone activity in leber congenital amaurosis. Arch Ophthalmol 130, 798–800 (2012).
10. Lorenz, B., Strohmayr, E., Zahn, S., Friedburg, C., Kramer, M., Preising, M., Stieger, K.: Chromatic pupillometry dissects function of the three different light-sensitive retinal cell populations in RPE65 deficiency. Invest Ophthalmol Vis Sci 53, 5641–5652 (2012).

Design Model for Risk Assessment for Home-Care Lung Ventilation

Ivana Kubátová, Martin Chlouba, and Ondřej Gajdoš

Abstract

Introduction: The study consists of two objectives. The first objective is to assess the risks associated with the whole process of treatment using home-care lung ventilation (HLV). The main source for describing such risks were foreign studies. The other objective is to compare risk assessment methods based on the current situation in the Czech Republic and in the world. Methods: The selected methods—Failure Mode and Effects Analysis (FMEA), Health Failure Mode and Effects Analysis (HFMEA), Fault Tree Analysis (FTA) and Root Cause Analysis (RCA)—were compared using comparative analysis, and subsequently examined, in the specific area of home-care lung ventilation. Results: The final objective is to design a model for risk assessment in the field of home mechanical ventilation based on the existing knowledge. This model includes both a prospective and a retrospective analysis. Their appropriateness of use is documented in this paper; according to their specificities they are used in a variety of areas. Discussion: Implementation of risk analyses in this area can contribute to increasing the safety and quality of provided care, and at the same time help attract more support to the home-care lung ventilation program.

Keywords

Risk analysis • FMEA • HFMEA • RCA
FTA • Model • Home-care lung ventilation

1 Introduction

At present, there is increasing emphasis on relocating patients to home care in those cases where immediate hospitalization is not required. This model of treatment in the home environment brings a number of benefits. One of these benefits is to reduce costs that are significantly higher when hospitalizing in a healthcare facility. Another indisputable advantage is the psychosocial effect on the patient and his family. In the home environment, the risk of hospitalization is reduced, self-sufficiency increases, and there is also a natural opportunity to engage in activities that are essential for the person. Due to a constant modernizing of health technologies, it is possible to move a wider range of people to the home care than in the past.

One of the state-of-the-art technologies of the present time, which allows moving patients from healthcare facilities to the home environment, is the home ventilator. By means of this technology it is possible to replace, partially or completely, spontaneous breathing in people with this need.

The aim of this study was to identify and describe the risks that are associated with the entire process of home pulmonary ventilation. On the basis of the findings, another aim was also designing a suitable model to identify and assess the risks in providing the treatment method. The core of the model is to support the process of moving patients to and providing home-care lung ventilation (HLV). To design this model, it was necessary to test individual risk analyzes and then identify their specific area of use. Verification of modulation, risk identification and appropriate use was carried out in cooperation with experts involved in home pulmonary ventilation issues.

I. Kubátová (✉) · M. Chlouba · O. Gajdoš
CzechHTA, Faculty of Biomedical Engineering, Czech Technical University in Prague, Prague, Czech Republic
e-mail: ivana.kubatova@fbmi.cvut.cz

2 Methods

In order to be able to design appropriate analyzes for the implementation of risk assessment for HLV, it is necessary to take into account the nature of the individual methods. Each method has its own specifics and is more or less suitable for each area of use. Figure 1 describes the characteristics of each analysis and its advantages and disadvantages.

Failure Mode and Effect Analysis (FMEA) is a systematic method whose purpose is to identify potential disorders, their causes and consequences. To initiate this analysis, it is necessary to define a specific system (process) and create a functional flowchart. This analysis should be the result of the work of a team of experts who are able to recognize and assess the degree and consequences of various kinds of potential deficiencies in the process. FMEA is considered to be a method of assessing the severity of potential failure modes and creating possibilities for their mitigation, which reduces risk. The principle of this analysis can be used for any work process that takes place in hospitals, healthcare facilities, and schooling systems and other areas [1].

Healthcare Failure Modes and Effects Analysis (HFMEA) is a prospective analysis, developed by the VA National Center for Patient Safety (NCPS), which combines elements of several different methods. In HFMEA an interdisciplinary team of experts is formed similarly as is the case of FMEA. The analysis uses a graphical representation identifying the failure and its cause, an array of values to determine risks and a decision tree algorithm to reveal the riskiest places in the process. This also includes an introduction of measures and their ex-post evaluation [2].

Fault Tree Analysis (FTA) serves to determining the cause of an adverse effect. In a deductive way, possible causes or failures are detected at successively lower functional levels of the system. A gradual deductive

identification, undesirable functions within the system, leads to finding the required level of the entity to which risk control measures may be applied. Such a procedure can reveal the continuity of the failures that most likely lead to the consequence [3].

Root Cause Analysis (RCA) is a method for finding causes for failures. This method, unlike proactive methods, is used when a specific failure has already occurred. RCA is therefore considered to be retrospective and reactive. In addition, the JCAHO requires this method to be performed in all accredited health facilities in the event an adverse event occurs [4].

These analyzes can serve as an effective tool for reducing risk factors in the field of healthcare processes, when implemented appropriately. Each of the individual analyzes uses a different strategy to achieve the desired results. However, the general purpose of these methods is the same. This goal is to minimize the risks arising in the sector and thereby improve the quality of care provided.

3 HLV Risk Assessment Model

To assess the health risks that may arise with HLV, HFMEA would be used by this proposal to prepare a nursing plan when a future HLV user is hospitalized. At this time, this person, together with an informal caregiver (in most cases a family member), could participate in the creation of the given analysis. The patient and the informal nurse, together with an expert team, could assess what risks are serious and likely for him. For these risks, appropriate measures could be put in place, even before being placed in home care. If any of the potential failures posed a subjectively excessive risk, an FTA would be used to identify more accurately the specific causes of the default. This method would also be used for those failures that offer a large number of primary causes.

Method	Time process	Type of Analysis	Determining the severity level	Identification of causes	Focus methods	Advantages	Disadvantages
FMEA	Prospective	Inductive	Yes	Yes	Process	A proven and often used method applicable in many areas	Ability to ignore the most serious failures through RPN (10/1/1); The possibility of choosing a too extensive process circle - congestion by quantity to the detriment of quality
HFMEA	Prospective	Inductive	Yes	Selected failures	Process	Possibility of double verification of the nature of the risk through the decision tree; A clear representation of the individual steps of the process using letters and numbers	Limited ability to evaluate the effect of the measure
RCA	Retrospectice	Deductive	No	Yes	One episode	Creating measures based on actual causes of failure; Recommendations of SAK and JCAHO	The necessary presence of all stakeholders; Insufficient analysis, if used alone - prevents only repetition of an already failed failure
FTA	Prospective/Retrospective	Deductive	No	Yes	One episode	Clear visualization of the issue	With a high degree of identified causes, it is difficult to correctly identify the priority of action

Fig. 1 Comparison of risk analysis

Using this method, it would be possible to clearly visualize the potential causes and then develop the necessary measures.

The FMEA, according to the proposed model, was used to identify the risks associated with the technical equipment needed for HLV. Using this analysis, it would be possible to analyze in detail the risks of each technical item the patient would use in the home care. In the same way, an FTA analysis could be used to identify the potential causes of individual failures. The essence of this analysis would be to create appropriate measures to prevent the consequences of technical equipment failure.

The latest risk analysis in this process is RCA. This analysis would be used in the case of rehospitalisation of the patient for unexpected reasons. It would then be possible to identify the root causes that led to the event. This part of the process is analogous to the treatment of undesirable events in accredited healthcare facilities where RCA is required to produce these events.

The whole process would be in charge of the same professional team that prepares the patient for the HLV in the given health facilities. With this team, the patient and the informal caregiver would work closely together. Analysis of technical risks would fall under the direction of companies that provide the equipment themselves. JCAHO recommends doing this method at least once a year (Fig. 2).

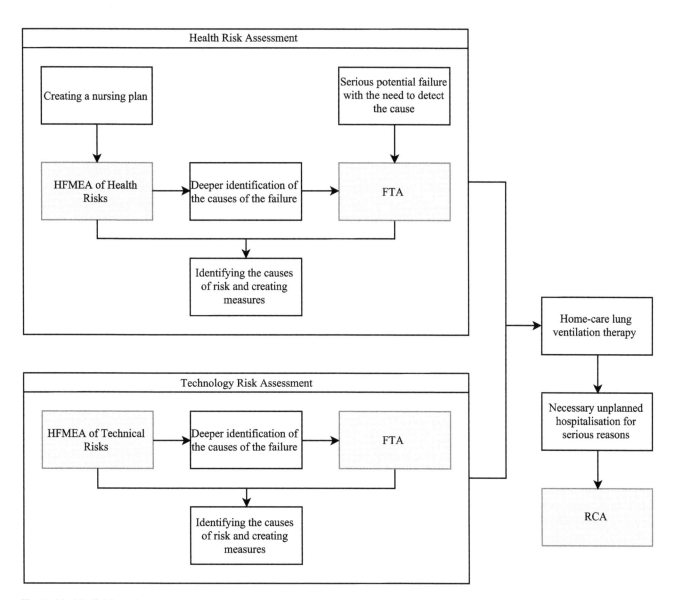

Fig. 2 Model of risk analysis

4   Discussion

In order to ensure an appropriate treatment process, it is necessary to identify as wide as possible range of potential risks and to create measures that would reduce the chance of these failures. Only in this way it would be possible that the quality and safety was increasing constantly and that this method fully utilized its undisputed potential.

At present, countless risk analyzes are being used worldwide to improve processes in a wide range of industries. The use of these methods is not left behind by the health sector. Organizations such as the World Health Organization or the Joint Commission on Accreditation of the Healthcare Organization are involved in implementing already existing methods from other areas, modifying these methods for healthcare needs, or developing new methods [5].

Individual analyzes are more or less suitable for implementation in a variety of areas of the healthcare sector. If the use of one analysis is not sufficient and therefore does not cover the broad range of risk identification in general, it is possible to use a combination of multiple analyzes. Incorporating a larger number of methods also adds to the potential for detection of possible risks, with a consequent possibility of action [6].

For this study it was necessary to describe the risk analyzes that are currently most commonly used in the healthcare sector. Subsequently, these analyzes were applied to the HLV field to verify their suitability for the above-stated treatment method.

One of the frequently used methods is the Failure Mode and Effect Analysis (FMEA). This method was developed by NASA, and has continued to expand into a variety of sectors over the years. It has been used in the health sector since 1999 [5]. Due to its potential in the field of technical risk identification, the method was also used for the purposes of this study.

The particular area studied was the identification of the risks associated with a technical failure of a self-contained fan for the home care. After consultations with service technicians having years of experience with technical troubles of fans, it was possible to create a schematic overview of the device. For specific components, the most common failures have been identified with the help of the above-mentioned experts. After identifying the potential causes of these failures and assessing them according to the magnitude of the risk, it was possible to propose remedial measures that would be given further attention in accordance with established priorities. The results of this analysis have shown that a frequent problem that leads to fan failures is the clogging of the components. As opposed to a sterile hospital environment, the device is exposed to a greater stress,

resulting in a faster wear and consequent defects. This factor was also described in a study by R. Gershon, where the purity and hygiene conditions of the home environment were assessed in patients with HLV. A surprisingly frequent phenomenon is the clogging of the fan turbine with cigarette smoke, which was confirmed not only by the service technicians, but also by the aforementioned study [7]. Frequent disturbances are then inherent mechanical damages that arise from the human factor. Such failures include breaking down some parts of the fan, falling over, exposing to excessive heat, or using non-original parts and consumables. Appropriate measures result primarily from sufficient awareness and caution of persons coming into contact with the instrument. It is also important to carry out functional tests and calibrations to verify a correct operation of the fan and detect possible malfunctions.

Another method used in this study is the Healthcare Failure Mode and Effect Analysis. This method, proposed in 2002 under the auspices of the National Center for Patient Safety, is a modification of the classic FMEA method. The advantage over conventional FMEA lies in its approach to assessing the degree of risk of individual failures. In the classic FMEA analysis, the main problem is the evaluation of the risk level for very serious but unlikely failures. Such situations may have fatal consequences for the patient but, due to their low probability of occurrence, they can be assessed by the FMEA as less risky. Such a failure would not be addressed in the FMEA case, despite being a life-threatening condition. For this reason, the implementation of the classic FMEA is not entirely appropriate for identifying and assessing health risks. This idea was proved by G. Faiella in her study carried out in Portugal in 2014 [5].

For the modified version, on the other hand, this situation is addressed by several tools, such as the decision tree, to verify the actual risk based on the expert opinion of those experienced in the field. The HFMEA method was therefore used in our study as a tool to analyze just the health risks. A specific process is the nursing plan that is tailored to each applicant for HLV. That is why each nursing plan is individual. Such a plan contains a number of different steps, varying according to the nature of the care itself. Some patients are dependent on continuous care and need to be aware of potential invasive inputs such as PEG. Therefore the level of action required is individual, and at the same time there may be different risks according to the state of health. To develop this method it was necessary to consult the whole issue with experts in the respective field. A consultation was conducted with the head of the Motol University Hospital Nursing Department concerning the follow-up intensive care and long-term intensive nursing care, a nurse from the same department, a professor at the Anaesthesiology and Resuscitation Clinic, and a co-ordinator of the Civic Association Life of Life, who also has practical experience in nursing care.

With the help of these people, a graphic design of the nursing process was created, which contains the generally valid steps in this treatment procedure. In individual steps, the identified potential failure was subsequently evaluated according to the degree of the risk posed to the patient. The above-mentioned fact that it is possible to verify and pay more attention to those failures that would not be considered risky due to a low probability of the risk, is confirmed by a failure associated with the tracheostomy cuff. Excessive pressure in the cuff may cause its damage and a subsequent rupture. This situation is very serious, however very unlikely. Hence, insufficient attention would be paid to the creation of measures for this failure when using the classic FMEA analysis. However, despite the low probability of using the tools modified by HFMEA, it is possible to assess this failure as a sufficient risk to continue to pay further attention to the design and implementation of the measures. A frequent cause that can result in a failure to compromise patient's health condition is a human factor again. In this case, there may be insufficient education of informal carers or a decrease in the quality of care provided due to their psychological and physical exhaustion. As a result of this analysis, it is possible to regularly monitor education of informal careers. Such a check could take place, for example, 6 and 12 months after the patient was moved to the home environment. The medical staff would then be able to verify the level of experience of these people and, if necessary, provide more information on the problem. The need for an adequate control of education is described in the 2006 study by A.K. Simonds. This study describes the frequent inadequate experience of informal carers with the operation of HLV techniques and the risks associated with this factor [8].

Another frequent risk is the lack of appropriate equipment to support smooth running of the whole home care process. Such a piece can be, for example, a pressure gauge which measures the pressure of the tracheostomic cuff. There may also be a shortage in supplies that HLV users receive from health insurance. This problem factor was confirmed by members of the expert team and also described by A. Gershon in his study [7].

The third method, the suitability of which was tested in this study for use in the HLV field, is the Fault Tree Analysis (FTA). This method was used to decipher the causes of failure in the HLV treatment process. One of the investigated failures was the rupture of the tracheostomy cuff, which was identified by the above-mentioned HFMEA method. Using the FTA, it was possible to identify the causes of the failure related to the defect of the cuff resulting from the manufacturing process or its

introduction, as well as the causes of its misuse, in this case by exposing the cuff to excessive pressure due to the impossibility of measuring it. Another failure that was subjected to this analysis was the failure to supply lungs with air from the fan. Here, a number of causes have been identified that are related both to the defect of the technical equipment and to the problem in the pace of the airway, for example obstruction. Using the FTA method, it was possible to identify the causes of potential failures in both cases and to create a clear graphical representation. Based on the above verification, it can be said that this method is very suitable for deeper identification of the causes of potential failures. A similar conclusion was reached by W. Hyman and E. Johnson of University of Texas, who tested this analysis in the field of medical alarms and subsequently recommended its use in this field [9].

The latter method, which has been applied in terms of suitability for use, is the retrospective Root Cause Analysis (RCA). The Ministry of Health published this method in their Journal 8/2012 as an appropriate method for reporting adverse events. For this reason, the RCA method is currently used in the Czech healthcare system, however, mostly only in accredited health facilities [10]. Its use, in the field of HLV risks, was described in this study in four situations. The nature of these situations consisted of a period of undesirable events. In these cases, either the patient died or was hospitalized from his/her home environment. In all these cases at least one cause was due to the absence of a secondary technical equipment that the HLV patient needs immediately.

After verifying the suitability of all the above-mentioned analyzes, it was possible to design a model for risk assessment in the HLV area. Individual analyzes come into this model with different goals, i.e. different perspectives for risk assessment. The designed model contains all of the above described analyzes. The use of several complementary analyzes was also described in the study by K. Shaqdan in 2014. Here, RCA and HFMEA were used, combined in a more effective model for risk assessment in the health sector [6]. The model designed in this work combines the same methods as the study and contains also two further analyzes that extend the area of risk assessment in the area under consideration.

In the proposed model, the FMEA method is a tool for identification and evaluation of technical risks. It has a potential to be used for all technical equipment associated with HLV treatment. The implementation of this analysis would fall under the direction of manufacturers and suppliers of technical equipment. Not only would this analysis serve to reduce the risks to patients, but at the same time it could contribute to a competitive advantage from the point of view of technical equipment innovations. The HFMEA method serves to assessing health risks that arise from the nursing plan. Creating this method would be done at a time when the patient is still hospitalized, and a nursing plan is being created for him.

With a larger number of people involved in the process of making the analysis, the range of opinions would increase. This view is described in the study by J. Derosir, who mentions the need to include as many experts as possible in the process of making this method [2]. The model proposed in this study, however, introduces an attempt not only to involve experts in the field when applying this method, but also patients and their families. If these persons were ready to carry out a detailed risk analysis, it would be possible to prevent potential failures even before the patient is released to his/her home environment, and thus before any chance of such risks to arise. It would be possible to focus on measures lowering those risks that are, according to a subjective opinion of the patient and informal caregivers, most likely and that the family is most concerned about. The benefits of such modifications to the composition of HFMEA team members are supported by R.G. van Kesteren in his 2001 study, where he presents the HLV risk assessment directly from patient´s and family member´s perspective, and discusses them on the basis of their recommended potential action [11].

The FTA method, in the developed model design, serves as a tool for deepening the analysis of root causes of the problem. This method could be used both on the basis of previous HFMEA and FMEA results, but also on any other revelation of significant potential risks. The latest validated method, the RCA, is conceived in the design as a means of retrospective assessing root causes of patient rehospitalization, a failure leading to a deterioration of patient's health condition, or an unexpected death unrelated to the progression of the disease. Using this analysis, it would be possible to avoid recurring faults that have already occurred and to increase the safety of the care provided.

As mentioned above, many risks arise from the fact that patients are not entitled to a contribution for a secondary equipment which they are often dependent on. This situation may result in a hospitalization, which would not be necessary if the devices were duplicated. The solution in the form of reimbursement of this equipment, which is standard in some European countries, is still only a speculative opinion in the Czech Republic [12]. The only way to contribute to the option of duplicate equipment from health insurers is to demonstrate indisputable benefits of this solution in all respects. One of the necessary parts, apart from demonstrating savings in the cost of potential hospitalization, is to reduce the risk of this therapy. In an effort to prove all the benefits that this option offers, we are faced with the problem of evaluating the necessary data that would clearly demonstrate the benefits. There is currently a problem in communication between the parties entering the HLV process. On the one hand, these are medical facilities that are responsible for the released patient. On the other hand, there are providers of technical equipment, especially of the fan. However, there is currently no authority to facilitate the necessary communication between these two parties, while at the same time obtaining valuable data on the course of this method. An example of such an authority are HLV Centers, which are established, for example, in the Netherlands, and which collect all necessary information about patients using home ventilators. This information on health, statistical data and, last but not least, risks can be used to further develop and support HLV [12].

5 Conclusion

The main goal of the study was to design a Risk Model for the HLV, applying all reasonable risk analyzes, in order to support the transfer and delivery process for the treatment options. This model includes both prospective and retrospective analyzes. The HFMEA method in the proposed modulation serves to evaluating health risks that arise from the nursing plan. The FMEA method includes the identification and assessment of risks associated with the technical equipment needed to provide HLV. The FTA method deepens the possibility of identifying causes of individual failures that can be detected in the previous two methods. The last method is the RCA, the use of which in is analogous to its current use in accredited healthcare facilities—a retrospective tool for creation of measures for undesirable events. These events may be defined as an unexpected hospitalization, serious deterioration of the state of health or death of the patient for reasons unconnected with the natural course of the illness.

Conflict of Interests The authors declare no conflict of interests.

References

1. ČSN EN ISO 60812. Analysis techniques for system reliability - Procedure for failure mode and effects analysis (FMEA), Praha, Czech Office for Standards, Metrology and Testing (2007).
2. Derosier, J., Stalhandske, E., Bagian, J. P., Nudell, T.: Using health care Failure Mode and Effect Analysis: the VA National Center for Patient Safety's prospective risk analysis system. The Joint Commission journal on quality improvement. Vol. 28, 248–267 (2002).
3. ČSN EN 61025. Fault Tree Analysis (FTA). Praha, Czech Office for Standards, Metrology and Testing (2007).

4. Škrla, P., Škrlová, M.: Řízení rizik ve zdravotnických zařízeních. Grada Publishing, ISBN 978-80-247-2616-5, (2008).
5. Faiella, G., Clemente, F., Rutoli, G., Romano, M., Bifulco, P., Cesarelli, M.: FMECA and HFMEA of indoor air quality management in home mechanical ventilation. In: IEEE MeMeA - IEEE International Symposium on Medical Measurements and Applications, Proceedings, (2014).
6. Shaqdan, K., Aran, S., Besheli, L. D., Abujudeh, H.: Root-Cause Analysis and Health Failure Mode and Effect Analysis: Two Leading Techniques in Health Care Quality Assessment. Journal of the American College of Radiology 11 572–579, (2014).
7. Gershon, R. R. M., Pogorzelska, M., Qureshi, K. A., Stone, P. W,, Canton, A. N., Samar, S. M., Westra, L. J., Damsky, M. R., Sherman., M.: Home Health Care Patients and Safety Hazards in the Home: Preliminary Findings (2017).
8. Simonds, A K. Risk management of the home ventilator dependent patient. Thorax 61, 369–371 (2006).
9. Hyman, W. A., Johnson, E.: Fault Tree Analysis of Clinical Alarms, [online]. http://thehtf.org/documents/FTA_of_Clinical_Alarms-Hyman_and_Johnson.pdf (2017).
10. Janssens, J. P., Penalosa, B., Degive, C., Rabeus, M., Rochat, T.: Quality of life of patients under home mechanical ventilation for restrictive lung diseases: a comparative evaluation with COPD patients. Monaldi archives for chest disease 51, 178–184 (1996).
11. van Kesteren, R. G., Velthuis, B., van Leyden, L. W.: Psychosocial problems arising from home ventilation. American journal of physical medicine & rehabilitation 80, 439–446 (2001).
12. Kampelmacher, M. J., Gaytant, M. A., Westermann, E. J. A.: Home mechanical ventilation in patients with neuromuscular diseases [online]. http://www.isno.nl/Neuromuscular_Info/Treatment/Treatment/Items/Home_mechanical_ventilation_in_patients_with_neuromuscular_diseases/Default.aspx (2010).
13. Chlouba, M.: Diploma Thesis Design model for risk assessment for homecare ventilation (supervisor: Kubátová, I.), Faculty of Biomedical Engineering, Czech Technical University in Prague (2017).

Rapid Manufacturing and Virtual Prototyping of Pre-surgery Aids

Magdalena Żukowska, Filip Górski, and Gabriel Bromiński

Abstract

Progressive development of rapid manufacturing and virtual prototyping have a significant influence not only in the industry and transport, but also in the medicine. Presurgical support and preparation of a surgeon with use of these technologies, especially in complex cases, can help prepare more precise plan of surgery and perform a simulated operation. The aim of these studies was to develop a methodology and manufacture an anatomical model of a kidney with a tumour, using rapid manufacturing technologies and virtual prototyping techniques. The model was a part of a presurgical support, allowing a doctor to become acquainted with an organ and a tumour and was also used for a simulative operation of partial nephrectomy. Due to the fact that model has two functions (preoperative planning and simulative operation), an important part during the production process was to consult procedures like cutting or suturing. Combination between technology of 3D printing and vacuum casting and silicon usage allowed to create a model, which imitates living tissues, especially the renal cortex and tumour. Transparency, which is a property of both models—physical and virtual—also plays a relevant role. Transparency helps surgeons in precise planning before operation. Doctors can familiarize themselves with arrangement of internal structures and pathologically altered areas. The collected information and tests performed with a cooperating hospital helped evaluation of created models, their usefulness and future implementation possibilities.

Keywords

Pre surgical support • Rapid prototyping • Virtual reality

1 Introduction

Rapid Prototyping and Addictive Manufacturing have become a big player in production and industry. Based on EY's Global 3D printing Report 2016, Pharma and Medical companies are on 3rd place in using 3D printing to create their own products and components. Authors of the report make a point of main benefits which are improved product quality (44%) and customized products as well (41%) [1]. It shows that Addictive Manufacturing is a technology which could help surgeons and patients. Rapid Prototyping in medicine can be used to create customized implants [2], prosthetics [3], surgical instruments [4] and personalized medical models as a pre-surgical support [5].

Virtual Reality and Augmented Reality are relatively new technologies, they have a significant influence on medicine and are part in creating new methods of treatment (Virtual Reality therapy) and can support surgeons work. Recent development in virtual prototyping allows to create applications, that also help doctors in many areas, like in psychological or occupational therapy, educational programs for future surgeons and digital models for pre-surgical planning [6, 7].

Presurgical support with use of these technologies, especially in complex cases, can help prepare more precise plan of surgery and perform a simulated operation.

The aim of the studies presented in the paper was to develop a methodology and manufacture an anatomical model of a kidney with a tumor, using rapid manufacturing technologies and virtual prototyping techniques.

M. Żukowska (✉) · F. Górski
Department of Management and Production Engineering, Poznan University of Technology, Poznan, Poland
e-mail: magdalena.k.zukowska@doctorate.put.poznan.pl

G. Bromiński
Department of Urology, Poznan University of Medical Science, Poznan, Poland

© Springer Nature Singapore Pte Ltd. 2019
L. Lhotska et al. (eds.), *World Congress on Medical Physics and Biomedical Engineering 2018*,
IFMBE Proceedings 68/3, https://doi.org/10.1007/978-981-10-9023-3_72

2 Methodology

Methodology of producing anatomical models is divided into two parts. First one is common for physical and virtual models. It includes stages of image segmentation and a spatial digital model creating. Depending on type of the model (physical or virtual), the second part of methodology is different. The diagram in Fig. 1 gives a summary of stages in methodology.

Process of model design begins with CT/MRI images of patient's abdomen (in case of kidney) or other body parts. Images are in the DICOM format (*Digital Imaging and Communications in Medicine*). In the next stage, images are imported to medical software like Mimics or InVesalius, that enables segmentation. Using automatic or manual segmentation, spatial digital model of a selected organ is created. Afterwards, model in STL format is imported to CAD software, where it is possible to remove compression artifacts and improve model quality. At this point, methodology is common for both models (physical and virtual) [8].

Finished digital model is then sent to Rapid Manufacturing systems and prepared for 3D printing. The model is sliced by software dedicated for 3D printers and saved in the G-Code format. Ready physical model needs post processing like smoothing and removal of support structures. Post processing is mostly contingent on technology of 3D printing [8].

In case of creating an application in Virtual Reality, finished digital model is import to software like Unity (game engine) or EON Studio. Then animation of model's parts, lights and visual effects are programmed [9].

The models prepared as described above can be part of a pre-surgical planning process.

3 Studies

Produced models were a part of presurgical support (as a test), allowing a doctor to become acquainted with an organ and a tumor. The physical model was also used for a simulative operation of partial nephrectomy. Due to the fact that model has two functions (preoperative planning and simulative operation), an important part during the production process was to consult procedures like cutting or suturing.

Transparency, which is a property of both models—physical and virtual—also plays a relevant role. Transparency helps surgeons in precise planning before the operation. Doctors can familiarize themselves with arrangement of internal structures and pathologically altered areas.

3.1 Physical Model

Production process of physical model was determined by its dual functionality. Combination between a technology of additive manufacturing and vacuum casting and silicone usage allowed to create a model, which imitates living tissues, especially the renal cortex and tumor.

First stage of building the model was 3D printing of internal parts: tumor, veins and arteries. All components were manufactured with use of FDM technology (*Fused Deposition Modeling*). Used materials was different for blood vessels (ABS in colors red and blue) and tumor (NinjaFlex in skin color). Decision about using other materials was made as an attempt to imitate appearance of living tissues for the tumor. The renal cortex was cast out of transparent silicone. Mold for the Vacuum Casting was 3D

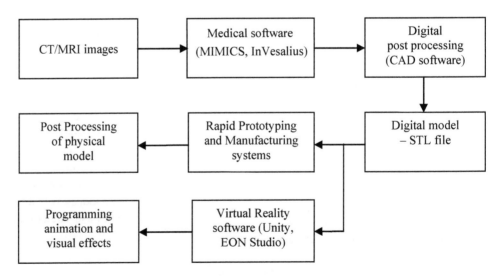

Fig. 1 Methodology of production anatomical models (physical and virtual) with the use of rapid manufacturing and virtual prototyping (based on [8, 9])

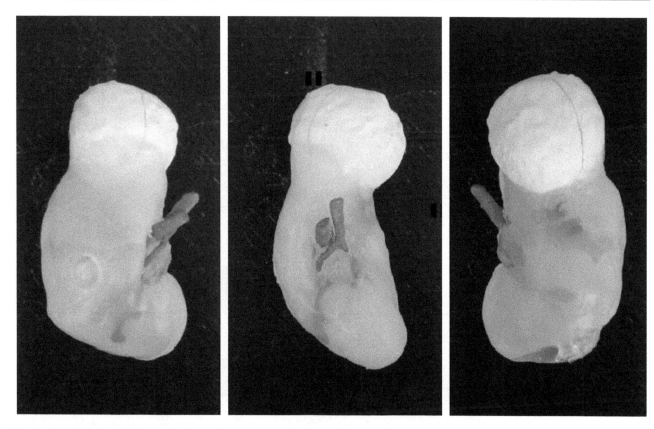

Fig. 2 The finished physical model—from left to the right: anterior, medial and inferior surfaces of kidney

printed. The mold was designed in CAD software, on the basis of STL Files. It was made by performing a Bool subtraction of the kidney with all components from a box shape. The mold was smoothed by covering its internal surface with a layer of polyurethane resin.

The Vacuum Casting process was crucial, because it was a moment when all the components were connected together by the silicone material filling the mold. Before casting, blood vessels and tumor had to be stably secure inside the mold cavity. Afterwards parts of the mold were closed. The mold contained a functioning gating and ventilation system. The silicone, which was used is a transparent material with ~45 Shore A toughness. The mold was poured with silicone and it was solidified after 24 h. Finished model needs slight post processing (smoothing). After this process, the anatomical model (Fig. 2) was ready to verify by the surgeons and test the planning and simulative operation.

3.2 Virtual Model

Digital model in the STL format, which was used in creating the physical model, has become base to produce a virtual model. The EON Studio 8 software was used to create a Virtual Reality application. Imported model was built from 4

separate elements: body of kidney (renal cortex), tumor, veins and arteries. All elements had the same zero point, so it was easy to place them together in a virtual scene. The first operation was creating a navigation system for the model. That helps moving virtual camera to view kidney under various angles, during planning operation. All the elements were textured for better realism and three light sources were added.

Second stage of creating the application was programming its interface. There were two interaction possibilities—transparency manipulation and exploded view animation of internal parts of the kidney (including tumor). After that, the application was slightly adjusted—a background image was added. Finished application (Fig. 3) was presented to surgeon for verification and testing.

4 Discussion

Based on tests and interview with a surgeon, the most important feature of both models (physical and virtual) was transparency. It increased precision in planning before operation. Doctors can familiarize themselves with arrangement of internal structures and pathologically altered areas. What is interesting, the virtual model was put by the

Fig. 3 Finished virtual model
with adjustable transparency

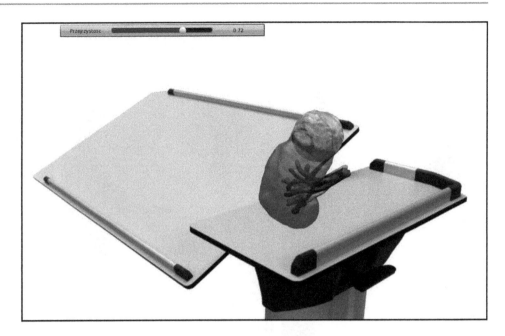

Fig. 4 Process of performing
simulative operation

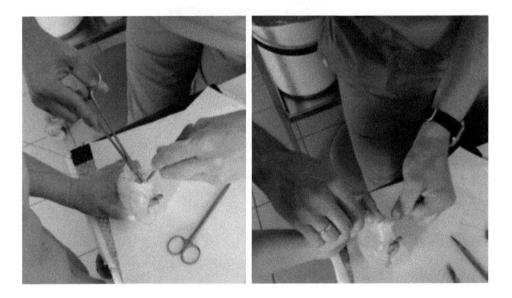

surgeon in first place, as a tool for planning before the operation. However, the physical model has an advantage of possibility of performing a simulated operation (Fig. 4).

Depending on the type of operation (open/laparoscopic) demand for the models is different. The presented virtual application has only basic function and was easy to produce, but it does not allow any type of surgery. However, development of the application would allow to perform virtual laparoscopic nephrectomy with use of haptic technology. Unfortunately, it creates huge costs. Physical models are cheaper but there is still need to improve level of imitation of living tissues and accuracy.

In terms of accuracy, it has been observed to be dependent on few things. First of all, the most important are source

CT/MRI images. When segmentation is realized on soft tissues, gap between images cannot be higher than 0.5–1 mm. Larger gap can deform view of organ and create mistakes in model, further resulting with problems in pre-surgical planning. In authors' opinion, this is a reason why anatomical physical models cannot be independently used in planning. Models accomplish a task only when they are used in planning together with CT/MRI scans, interview with patients and other medical procedures.

Produced mold was slightly too large and it caused problems with airtight closing of its halves. In consequence, the model had air bubbles and it was partially deformed. In the future, design changes should be introduced. The used silicone was also too hard. During simulative operation, one of sutures was broken, and two weeks after the procedure, all the sutures loosened and cut parts of the model. Next studies should involve testing other materials, like resins for 3D printing and other silicones to create better imitation of living tissues.

Programming applications in Virtual Reality enables to make more expanded models with structures smaller than 1 mm. It gives chance to precisely inspect an organ and its parts.

In general, it must be stated that procedures like segmentations and selection of rapid prototyping technology need good communications between doctors and engineers, to bring the doctors the most satisfying solution.

5 Conclusion

Presented studies show that rapid manufacturing and virtual prototyping are technologies worth using in medicine, especially in pre-surgery planning. Prepared anatomical models could help surgeons in preoperative planning and in performing simulated operations. It also could be part of education of students at medical studies and future specialists. Anatomical models are personalized and dedicated to a particular case. It increases patient's safety and reduces time necessary to perform an operation. Both virtual and physical models are useful, although urgency of operations rarely allows preparing both on time, therefore a decision must be made which type of model should be used to a given case. A simple virtual model is shorter and less costly than the physical one, but it does not allow touching and performing an actual operation activities.

Technologies like 3D printing and virtual reality are the future in biomedical engineering and that is why their medical applications should be developed continuously.

Conflict of Interest The authors declare that they have no conflict of interest.

References

1. Müller A., Karevska S.F.: How will 3D printing make your company the strongest link in the value chain? EY's Global 3D printing Report 2016. Ernst & Young GmbH (2016).
2. 3ders.org, Argentinan patient leads normal life with 3D printed cranial implant, http://www.3ders.org/articles/20150510-argentinan-patient-leads-normal-life-with-3d-printed-cranial-implant.html, accessed 2017/07.
3. S. Greene, D. Lipson, A. Mercado, A. Heain Soe, Design and manufacture of a scalable prosthetic hand through the utilization of additive manufacturing, SEMINAR, Worcester Polytechnic Institute.
4. BioFabris, 3D Printed Surgical Guides Make Their Malaysian Debut, http://biofabris.com.br/en/3d-printed-surgical-guides-make-their-malaysian-debut/, last accessed 2017/07.
5. European Association of Urology, Surgeons develop personalised 3D printed kidney to simulate surgery prior to cancer operation, http://www.alphagalileo.org/ViewItem.aspx?ItemId=140891&CultureCode=en, last accessed 2017/07.
6. Virtual and augmented reality in medical applications, https://www.esat.kuleuven.be/psi/research/virtual-and-augmented-reality-in-medical-applications, last accessed 2017/07.
7. IgnisVR, Arachnophobia, Virtual Reality Exposure Therapy, http://en.adquan.com/post-18-1034.html, last accessed 2017/07.
8. I. Gibson, L.K. Cheung, S.P. Chow, W.L. Cheung, S.L. Beh, M. Savalani, S.H. Lee, The use of Rapid Prototyping to assist medical applications, "Rapid Prototyping Journal", Nr 12, pp. 53–58 (2006).
9. Edward Pająk, Adam Dudziak, Filip Górski, Radosław Wichniarek, Techniki przyrostowe i wirtualne w procesach przygotowania produkcji, Poznań (2011).

Cobalt-60 Radiotherapy Units, Assessment of the Utilization or Disinvestment in Latin America

Daniel Martínez Aguilar, Alejandra Prieto-de la Rosa,
Esteban Hernández San Román, Arturo Becerril Vilchis,
Francisco Ramos Gómez, and Alessia Cabrera Yudiche

Abstract

Objectives: To assess whether the criteria for disinvestment in health technologies—clinical, technological advantage, safety, lifecycle, human factor and costs—are applicable for the withdrawal of cobalt-60 radiotherapy units (Co-60 RTUs) in Latin America, considering health outcomes, and that the economic context, replacing this technology with linear accelerators (LINACs) is not always feasible. **Methods**: A systematic review of articles published between 2003 and 2017 in PubMed, Cochrane Library and CRD on the current use of Co-60 RTUs and publications comparing them with LINACs has been made. With a manual search of the references of selected articles. **Results**: The clinical results indicate that Co-60 RTUs have a significant role in treatment of patients with tumors of head, neck, breast and some types of superficial soft tissue sarcomas of the extremities. The comparison between Co-60 RTUs and LINACs results in advantages for linear accelerators in: the variety of cancer type that can be treated, the delivery of treatment, lifecycle and safety. In terms of acquisition costs, although Co-60 RTUs are comparable to a low-energy LINACs. Considering the number of existing Co-60 RTUs in Latin America, their effectivity and safety in the treatment of some types of cancer and the shortage of skilled professionals, its use can still be beneficial. **Conclusions**: Whereas in Latin America more than 26% of radiotherapy equipment are Co-60 RTUs, available economic resources and staff are limited. A recommendation is to continue utilizing such equipment in some types of cancer where they can be used: head, neck, breast and superficial sarcomas extremity soft tissue and allocate the use of existing LINACs for other types of cancer and in special cases like pediatric patients.

Keywords

Cobalt-60 radiotherapy unit • Co-60 RTUs
Linear accelerator • LINAC • Latin America
Radiotherapy

1 Introduction

The Pan American Health Organization (PAHO) reports that cancer is the second cause of mortality in the Americas and that the capacity to respond to this public health problem is limited, especially in Latin America and the Caribbean (LAC) [1]. In 2012, in the Americas 2.8 million new cases and 1.3 million deaths were recorded as a result of cancer [2]. Cervical cancer mortality rates are 3 times higher in LAC than in Canada and the United States of America, reflecting large inequalities in health resources (health promotion, disease prevention, opportune diagnosis and treatment).

Radiotherapy, along with surgery and drug therapy, is one of the cornerstones of cancer treatment and it is considered a cost-effective treatment [3]. However, radiotherapy is the least considered treatment due to lack of access to the technology [4]. It is estimated that between 40 and 60% of patients with cancer will require treatment with radiotherapy for tumor control or as palliative therapy at some point

D. M. Aguilar (✉) · A. P. la Rosa · E. H. S. Román · F. R. Gómez
A. C. Yudiche
Centro Nacional de Excelencia Tecnológica en Salud
(CENETEC), Ciudad de México, Mexico
e-mail: daniel.martineza@salud.gob.mx; mimadapa@gmail.com

A. B. Vilchis
Servicios de Salud de Oaxaca/Facultad de Medicina, de la
Universidad Autónoma Benito Juárez de Oaxaca, Oaxaca, Mexico

F. R. Gómez
Facultad de Ciencias, Universidad Nacional Autónoma de
México, Mexico City, Mexico

© Springer Nature Singapore Pte Ltd. 2019
L. Lhotska et al. (eds.), *World Congress on Medical Physics and Biomedical Engineering 2018*,
IFMBE Proceedings 68/3, https://doi.org/10.1007/978-981-10-9023-3_73

during the course of the disease [3, 5, 6], either alone or in combination with surgery, drug therapy, hormonal therapy or immunotherapy [7].

In addition, it is estimated that the demand for radiotherapy services in developing countries will increase dramatically in the next 20 years [8].

Among the radiotherapy devices for cancer treatment are Cobalt-60 radiotherapy units (Co-60 RTUs, 1.2 meV) available since 1948, and the linear accelerators (LINACs, 4–25 MV) in use since 1952 [9]. In the decade of 1950 there were 1700 teletherapy devices worldwide, 90% were Co-60 RTUs installed and functioning. For 30 years, Co-60 RTUs were the main technology for radiotherapy [3]. As a result of diverse technological developments, in the 1980s, more than 90% of the radiotherapy units in the United States were LINACs and by 1990 Co-60 RTUs practically disappeared [10]. The situation in other developed countries is practically the same; in Japan and Korea Co-60 RTUs represent 1.4% of the cancer treatment teletherapy devices [11]. However, according to data from the International Atomic Energy Agency (IAEA) in 29 Latin American countries in 2017 there were 651 radiotherapy centers with 1,033 radiotherapy devices, of them 763 (73.9%) are LINACs and 270 (26%) are Co-60 RTUs [12]. In Mexico, there are currently 164 radiotherapy devices, 30 (18.2%) are Co-60 RTUs with an average age of the devices of 22.1 years and with an average remaining life of the cobalt sources of 4.1 years, considering a life time of two half-lives (10 years) of the source [13].

Although the use of Co-60 RTUs has decreased due to the development of safer and more efficient technologies, Co-60 RTUs are still used in Latin America due to its lower need of resources with respect to costs in training, treatment delivery, planning and maintenance [8, 14, 15], However, the production cost of Co-60 sources have increased in recent years and there is a growing concern about safety. Also, there has been a relative decrease in the cost of low energy LINACs (6 MV) [16], making these devices being more or less comparable by combining initial and subsequent costs.

The incorporation of new medical technologies into the market makes some equipment inefficient or changes its usage patterns; hence the need to assess whether the criteria for disinvestment in health technologies—clinical results, technological advantage, safety, durability, human factors and costs—are still effective in the use of Co-60 RTUs in Latin America considering health results. Due to economic reasons and the availability of trained human resources, the replacement of this technology is not always feasible [16, 17], although in some cases it will lead to the restriction of its use and the redefinition of its indications.

2 Materials and Methods

A systematic review of the literature was performed using PubMed, CRD and Cochrane Library of publications from January 2003 till September 2017 about the current use of Co-60 RTUs and publications comparing Co-60 RTUs with LINACs. A search in international health agencies and nuclear safety agencies was also made. The search returned 247 results excluding duplications across databases, all the abstracts were revised, from those, 35 articles were obtained in full text. The obtained articles were reviewed manually and consultations were made to experts to expand the selection criteria of the literature. A final filter was made where only 24 items that met the following criteria were taken into account: articles comparing results in patients treated with LINACs or Co-60 RTUs; articles that mention international trends in radiotherapy units; articles that detail the technical characteristics of LINACs and Co-60 RTUs and articles that describe the operating costs of radiotherapy units.

3 Results

3.1 Clinical Effectiveness Results

Radiotherapy is used in patients diagnosed with cancer in at least one stage of cancer [18]. The recommendation of use of a Co-60 RTU or a LINAC depends on the type, location of cancer and characteristics of the patient. Given the physical properties present in the radiation emitted by each device, a classification of cases can be made in which it is recommended to use one or another technology.

Co-60 RTUs still play a significant role in the treatment of patients with neck, head, breast tumors and some types of superficial soft tissue sarcomas in the extremities [6, 19].

Head and neck cancer. A retrospective analysis of the results of treatment in 1452 patients with head and neck cancer using Co-60 RTUs and low energy LINACs demonstrated that Co-60 RTUs has better local control in patients with neck lesions and with a high relapse tendency [20]; the dose distribution provided by LINACs is not sufficient in superficial lesions, an alternative to compensate this is with the use of beam-compensating materials for LINACs.

Breast cancer. The results in the treatment of breast cancer showed that Co-60 RTUs are a useful alternative since there are only little differences from those obtained with low energy LINACs as they present a slight increase in non-homogeneities without exceeding the limits established by the International Commission on Radiation Units and

Measurements (ICRU) [14]. Co-60 RTUs can generate an increase in the dose absorbed by the contralateral breast, a greater dose absorbed by the skin, or non-homogeneities in the treated breast, especially in the breast-conservation therapies. However, these disadvantages could be mitigated with an appropriate treatment plan and with simple accessories such as filters, wedges or blocks [14]. On the other hand, Co-60 RTUs are inadequate for procedures where esthetic results are sought, in these cases the use high energy X-rays (10–25 MV) is required.

Superficial sarcomas. The treatment of superficial soft tissue sarcomas in the extremities is recommended with Co-60 RTUs [6, 19], although it is not recommended in deeply located tumors since the maximum dose is reached at a depth of 0.5 cm for the average energy of cobalt (1.25 meV) [6]. In addition, the dose interval is directly proportional to the activity of the cobalt source contained in the device and it is not possible to regulate it, unlike LINACs. Because the penumbra of Co-60 RTUs is greater than 1 cm, damage to adjacent tissues is common. Low-energy LINACs can be used to treat superficial lesions and have a penumbra less than 0.5 cm, because of this property the damage to adjacent tissues is less.

3.2 Comparison of Technologies

LINACs allow more precise treatments and to treat deeper tumors, properties not shared by Co-60 RTUs; this is accomplished because the use of electron beams or X-rays with a wide range of energies: low energy 4–6 MV, medium energy 8–10 MV and high energy 15–25 MV, all of them of an order of magnitude higher than the one offered by Co-60 RTUs, which offers a single energy of 1.25 meV and is an inherent property of the source of Co-60 [21].

In LINACs the dose rate per minute can be modified, in Co-60 RTUs the dose rate (which decreases with time due to radioactive decay) is determined at the beginning of the treatment session and cannot be regulated. In addition, LINACs technology presents advantages regarding collimation of the beam and the diversity of treatment techniques such as conformal radiation therapy, intensity-modulated radiation therapy and other more sophisticated techniques, in the most recent treatment and planning modes [21].

Advantages of LINACs when compared to Co-60 RTUs. (a) the beam produced with Co-60 RTUs has a greater penumbra than the one produced by LINACs [19], which is one of the main arguments to consider this technology as obsolete; (b) treatment with cobalt produces higher superficial doses that can produce skin reactions (radiodermatitis) due to the contamination of low energy electrons; (c) the beam of Co-60 RTUs have a very low dose rate when compared with low energy LINACs, and is less penetrating than the higher energy beams of 10 and 20 MV [10].

The radiation beam generated by Co-60 RTUs is comparable to that produced by the low energy LINACs. However, the number of patients that can be treated in one day is three to four times lower in comparison to LINACs of medium and low energy, because LINACs have the ability to treat larger areas without the need to use multiple treatment fields [22].

Advantages of Co-60 RTUs when compared to LINACs. During the treatment, the dose rate of Co-60 RTUs remains constant and continues while the dose rate of LINACs presents variations that are adjusted electronically during the treatment [23, 24].

3.3 Normal Life-Expectancy of Devices

Co-60 RTUs have a life span of 20 years (can be up to 30 years) [12]. The Co-60 radiation source gradually loses activity; the dose rate drops about 1% per month [19]. Different authors say that the source should be replaced every half-life (between 5 and 7 years) and can reach up to 2 half-lives (10 years), after which it still emits radiation, although not clinically useful [9, 11]. For the purposes of this article, CENETEC "Centro Nacional de Excelencia Tecnológica en Salud" adopts the criterion that one half-life of the Co-60 radiation source is 5 years and that LINACs have a life span of 10 to 12 years [8].

3.4 Safety

The first safety issue of Co-60 RTUs is the ecological one, since the cobalt source presents a risk of accidental exposure during its transfer, use, storage and disposal until it ceases emitting radiation [19]. During its life span the risk of leak of radiation from the device increases in the event of earthquakes, acts of sabotage and in any scenario where the source is retracted. The disposal of the source of Co-60 is a major problem, due to the lack of trained personnel to manage the source at the end of its useful life, the misuse that can be given, its inadequate disposal as well as the risk of storing this type of materials in hospitals and centers with insufficient safety measures [6].

Regarding patient safety, special emphasis should be placed on the protection systems of the structures adjacent to the treatment area in both technologies. The lack of patient position verification, dose delivered verification, and beam collimation devices for irregular fields that allow complex therapies represents limitations in Co-60 RTUs and a potential risk for the patient due to the probable damage during treatment. Besides, the clinical staff could receive more indirect radiation doses during their activity in Co-60 RTU treatment room (accidental exposure) with respect to LINACs.

LINACs can be "turned off" when they are not in use, in contrast, the constant radiation emission from the Co-60 source represents a latent risk of a radiological accident due to a possible failure of the movement mechanism of the source during the displacement from a safe source storage position to an exposed position and vice versa [19].

3.5 Costs

Along with surgery, radiotherapy remains the most cost-effective treatment for cancer [6]. While high energy dual LINACs are expensive, low-energy LINACs compare favorably with Co-60 RTUs in terms of cost, as well as life span [6]. The costs of the Co-60 RTUs are raised by including the disposal and replacement of the source, so it presents a similar acquisition cost to the low energy LINAC, however the training, staff and maintenance costs and physical infrastructure are lower for Co-60 RTUs [6].

For Co-60 RTUs, in addition to its initial cost, the cost of replacing the source of Co-60 every 5 years (half-life) or up to 10 years (two half-lives) must be considered. For LINACs, the cost of capital is high and its implementation, operation, training, quality control programs and maintenance requirements are more complex and expensive.

The acquisition cost of Co-60 RTUs varies from $400,000 to $1,000,000 USD (2017) [7, 14]. LINACs are in a range of $3,500,000 USD (2017) with all the peripheral equipment needed. Among the international literature obtained from the systematic review there is no available information about the cost linked to the clinical result with both technologies and their diverse applications. There is only information about the combined cost components per fraction delivered (in USD) and range from $1.29 to $34.23 for Co-60 RTUs with a median of $4.87 and from $3.27 to $39.59 for LINACs with a median of $11.02 [25].

3.6 Human Factors

The IAEA recommends a staff of around 20 professionals (4–5 radiotherapeutic oncologists, 3–4 medical physicists, 7 radiotherapy technicians, 3 radiotherapy nurses, and a maintenance engineer) to operate a basic radiotherapy center to treat up to 1000 patients per year [7]. However, the global shortage of qualified radiotherapy professionals is recognized, and even more so in Latin America [8]. The learning curve in the use of technology is an important factor in the acquisition of LINACs since it is considered to be "easier to learn to use a Co-60 RTU" [24]. It is suggested that if there

is a substitution of technology (Co-60 RTU by LINAC), to maintain for a certain period of time the current technology, taking into account the time it will take to learn how to use the new technology [24].

3.7 Health Context

The technology used in Latin America is diverse and in general terms is unfavorable when compared with technology used in developed countries, nevertheless, the CONCORD 2 study concludes that inequity in access to timely diagnosis and treatment services is probably the cause of the wide dispersion in survival rate observed in countries regardless of whether they are high or low-income countries [25, 26]. The same study mentions that the lowest survival rate observed in Europe, in comparison with the United States, has been associated with GDP, health expenditure, investment in health technology and the suboptimal location of available resources [25].

4 Discussion

The efficacy and safety of Co-60 RTUs in the clinical domain are well defined, since this device delivers the prescribed dose limiting the associated radiotoxicity at acceptable levels in superficial tumors in the head and neck, breast, and soft tissue sarcomas in extremities. The energy provided by Co-60 RTUs is not adequate to treat tumors at deeper levels.

LINACs presents technological advantages over Co-60 RTUs, by having greater precision in the dose delivered, a greater range of depths for radiation delivery in tumor treatment, from superficial to deep in the thoracic or abdominal cavity tumor, moreover LINACs have a higher and variable dose rate.

Despite the technical disadvantages, Co-60 RTUs cannot be considered as an obsolete technology since they are used for some types of cancer with good results in well selected cases, being equivalent to a low energy LINAC [18]. In addition, modern Co-60 RTUs have multiple Co-60 sources that deliver higher dose rates, with penumbra reducers that decreases the dose to the critical structures adjacent to the tumor to be treated and some of them include a multileaf collimation systems. The development of more modern Co-60 RTUs is a research topic, these new devices try to minimize the disadvantages of Co-60 RTUs with regard to LINACs and at the same time try to maximize the advantages over them, such as the use of imaging systems that do not use ionizing radiation [27–32].

Considering that in Latin America more than 26% of radiotherapy equipment are Co-60 RTUs [12], and that financial and personnel resources are limited, the recommendation is to continue using these equipment in cancers of the head, neck, breast and soft tissues sarcomas in the extremities and allocate the use of LINAC for other types of cancer and for pediatric patients.

5 Conclusions

Considering the current infrastructure, costs and availability of specialized personnel in LAC, the actual need of radiotherapy units cannot be satisfied only with Co-60 RTUs or only with LINACs. To reduce the burden of disease both in cost of diagnosis and in treatment incurred by individuals and the health system of each country, it is convenient that the specialized radiotherapy units of the public system use both technologies to achieve the maximum benefit and the lowest cost, procuring the sense of equity in health care. Even in places where several public health institutions from different organizations are located, it is recommended to ensure that the investment is coordinated in terms of the acquisition and use of these technologies that are ultimately complementary in the treatment of pathologies associated with cancer.

Co-60 RTUs remains a viable and cost-effective option in well-selected clinical cases, a cost beneficial alternative for the treatment of at least 25% of cancer patients [3] requiring radiation therapy. In times of economic restriction, it is even more important to achieve a delicate balance between the use of LINACs and the Co-60 RTUs with the combined and complementary use of both technologies. Under ideal conditions of availability of resources, it is recommended the progressive replacement of Co-60 RTUs by LINACs, meanwhile Co-60 RTUs that are already installed must be kept operational until the end of their lifespan.

Conflict of Interest The authors declare that they have no conflict of interest.

References

1. Organización Panamericana de la Salud. Salud en las Américas Edición de 2012 Panorama regional y perfiles de país. Special Collection, Health in the Americas. 2012. http://iris.paho.org/xmlui/handle/123456789/3272.
2. Organización Panamericana de la Salud. El Cáncer en la Región de las Américas. 2012.
3. Thariat J, Hannoun-Levi JM, Myint S, Vuong Te, Gérard JP. Past, present, and future of radiotherapy for the benefit of patients. Nat. Rev. Clin. Oncol. 2013 (10): p. 52–60. https://doi.org/10.1038/nrclinonc.2012.203.
4. Atun R, Jaffray DA, Barton MB, Bray F, Baumann M, Vikram B, et al. Expanding global access to radiotherapy. The Lancet Oncology Commission. 2015; 16(10): p. 1153–1186. http://dx.doi.org/10.1016/S1470-2045(15)00222-3.
5. Symonds RP. Recent advances Radiotherapy. BMJ. 2001; 323 (7321): p. 1107–10. https://www.ncbi.nlm.nih.gov/pmc/articles/PMC1121599/.
6. Reddy KS. Choice of a Teletherapy Unit: Cobalt 60 Unit vs Linear Accelarator Services DGoH, editor. India: Government of India Ministry of Health & Family Welfare. p. 79–87.
7. International Atomic Energy Agency (IAEA). Setting up a Radiotherapy Programme: Clinical, Medical Physics, Radiation Protection and Safety Aspects. IAEA, Vienna (2008). http://www-pub.iaea.org/mtcd/publications/pdf/pub1296_web.pdf.
8. Salminen EK, Kiel, Ibbott GS, Joiner C, Rosenblatt E, Zubizarreta E, et al. International Conference on Advances in Radiation Oncology (ICARO): Outcomes of an IAEA Meeting. Radiation Oncology. 2011 Feb; 6(11). https://doi.org/10.1186/1748-717X-6-11.
9. Emergency Care Research Institute (ECRI). Health Care Product Comparison System (HPCS). https://hpcs.ecri.org/, last accessed 2018/01/21.
10. Hauri P, Hälg RA, Schneider U. Technical Note: Comparison of peripheral patient dose from MR-guided 60 Co therapy and 6MV linear accelerator IGRT. Med. Phys. 2017 Jul; 44(7): p. 3788–3793. https://doi.org/10.1002/mp.12293.
11. Fox C, Romeijn HE, Lynch B, Men, Aleman DM, Dempsey JF. Comparative analysis of Co intensity-modulated radiation therapy. Phys. Med. Biol. 2008 May; 53: p. 3175–3188. stacks.iop.org/PMB/53/3175.
12. Ji YH, Jung H, Yang K, Cho CK, Yoo S, Yoo HJ, et al. Trends for the Past 10 Years and International Comparisons of the Structure of Korean Radiation Oncology. Japanese Journal of Clinical Oncology. 2010 May; 40(5): p. 470–475. https://doi.org/10.1093/jjco/hyq001.
13. International Atomic Energy Agency (IAEA). DIRAC (DIrectory of RAdiotherapy Centres). https://dirac.iaea.org/.
14. Centro Nacional de Excelencia Tecnológica en Salud (CENETEC-Salud). Inventario Nacional EMAT. http://www.cenetec.salud.gob.mx/contenidos/biomedica/mapa.html.
15. Bese NS, Munshi A, Budrukkar A, Elzawawy A, Perez CA. Breast radiation therapy guideline implementation in low- and middle-income countries. Cancer. 2008 Oct; 113(8): p. 2305–2314. http://onlinelibrary.wiley.com/doi/10.1002/cncr.23838/epdf.
16. Massoud S. Challenges of Making Radiotherapy Accessible in Developing Countries. Cancer Control. 2013: p. 85–96 http://cancercontrol.info/wp-content/uploads/2014/08/cc2013_83-96-Samiei-varian-tpage-incld-T-page_2012.pdf.
17. Watt AM, Hiller JE, Braunack-Mayer J, Moss JR, Buchan H, Wale J, et al. The ASTUTE Health study protocol: Deliberative stakeholder engagements to inform implementation approaches to healthcare disinvestment. Implementation Science. 2012; 7(101): p. 1–12. http://dx.doi.org/10.1186/1748-5908-7-101.
18. Valentín B, Blasco JA. Identificación de oportunidades de desinversión en tecnologías sanitarias. Madrid (2012).
19. Ravichandran R. Has the time come for doing away with Cobalt-60 teletherapy for cancer treatments. J Med Phys [serial online]. 2009; 34(2): p. 63—65. http://www.jmp.org.in/text.asp?2009/34/2/63/51931.
20. Van Dyk J, Battista JJ. Cobalt60: An Old Modality, A Renewed Challenge. Current Oncology; 1995. No. 3. p. 1–31.
21. Fortin A, Allard J, Albert M, Roy J. Outcome of patients treated with cobalt and 6 MV in head and neck cancers. Head & Neck. 2001 Mar; 23(3): p. 181–188. http://dx.doi.org/10.1002/1097-0347(200103)23:33.0.CO;2-8.
22. IAEA. Radiation Oncology Physics: A Handbook for Teachers and Students. Vienna (2005). http://www-pub.iaea.org/MTCD/publications/PDF/Pub1196_web.pdf.

23. Adams EJ, Warrington P. A comparison between cobalt and linear accelerator-based treatment plans for conformal and intensity-modulated radiotherapy. Br J Radiol. 2008; 81(964): p. 304–10. https://doi.org/10.1259/bjr/77023750.

24. Healy BJ, van der Merwe D, Christaki KE, Meghzifene A. Cobalt-60 Machines and Medical Linear Accelerators: Competing Technologies for External Beam Radiotherapy. Clin Oncol (R Coll Radiol). 2017 Feb; 29(2): p. 110–115. https://doi.org/10.1016/j.clon.2016.11.002.

25. van der Giessen PH, Alert J, Badri C, Bistrovic M, Deshpande D, Kardamakis, et al. Multinational assessment of some operational costs of teletherapy. Radiother Oncol. 2004; 71(3): p. 347–55 https://doi.org/10.1016/j.radonc.2004.02.021.

26. Allemani C, Weir HK, Carreira H, Harewood R, Spika, Wang XS, et al. Global surveillance of cancer survival 1995–2009: analysis of individual data for 25 676 887 patients from 279 population-based registries in 67 countries (CONCORD-2). Lancet. 2005; 285(9972): p. 977–1010. http://dx.doi.org/10.1016/S0140-6736(14)62038-9.

27. Hauri P, Schneider P. Whole-body dose and energy measurements in radiotherapy by a combination of LiF:Mg,Cu,P and LiF:Mg,Ti. Z. Med. Phys. 2017; p. 15. http://dx.doi.org/10.1016/j.zemedi.2017.07.002.

28. Allen BJ. Workshop on Palliative Radiotherapy for Developing Countries, 1st ed: Comparative Cost Analyses for Co-60 and Linacs in Developing Countries. P. 124–127. https://www.researchgate.net/publication/285382473_Workshop_on_Palliative_Radiotherapy_for_Developing_Countries_1st_Edition.

29. Yang D, Wooten HO, Green O, Li HH, Liu S, Li X, et al. A software tool to automatically assure and report daily treatment deliveries by a cobalt-60 radiation therapy device. J Appl Clin Med Phys. 2016; 17(3): p. 492–501. https://doi.org/10.1120/jacmp.v17i3.6001.

30. Wojcieszynski AP, Hill PM, Rosenberg SA, Hullett CR, Labby ZE, Paliwal, et al. Dosimetric Comparison of Real-Time MRI-Guided Tri-Cobalt-60 Versus Linear Accelerator-Based Stereotactic Body Radiation Therapy Lung Cancer Plans. Technol Cancer Res Treat. 2017; 16(3): p. 366–372. https://doi.org/10.1177/1533034617691407.

31. Raghavan G, Kishan AU, Cao M, Chen AM. Anatomic and dosimetric changes in patients with head and neck cancer treated with an integrated MRI-tri-60Co teletherapy device. Br J Radiol. 2016; 89(1067). https://doi.org/10.1259/bjr.20160624.

32. Merna C, Rwigema JCM, Cao M, Wang PC, Kishan AU, Michailian A, et al. A treatment planning comparison between modulated tri-cobalt-60 teletherapy and linear accelerator-based stereotactic body radiotherapy for central early-stage non-small cell lung cancer. Med Dosim. 2016; 41(1): p. 87–91. https://doi.org/10.1016/j.meddos.2015.09.002.

Incorporating the Local Biological Effect of Dose Per Fraction in IMRT Inverse Optimization

Brígida da Costa Ferreira, Panayiotis Mavroidis, Joana Dias, and Humberto Rocha

Abstract

In intensity modulated radiation therapy (IMRT), the dose in each voxel of the organs at risk (OAR) can be strongly reduced compared to conformal radiation therapy (RT). Due to the sensitivity of late side-effects to fraction size, a smaller dose per fraction in the normal tissues represent an increased tolerance to RT. This expected reduction in biological effect may then be used as an additional degree of freedom during IMRT optimization. In this study, the comparison between plans optimized with and without a voxel-based fractionation correction was made. Four patients diagnosed with a head and neck (HN), a breast, a lung or a prostate tumor were used as test cases. Voxel-based fractionation corrections were incorporated into the optimization algorithm by converting the dose in each normal tissue voxel to EQD2 (equivalent dose delivered at 2 Gy per fraction). The maximum gain in the probability of tumor control (P_B), due to the incorporation of the correction for fractionation in each voxel, was 1.3% with a 0.1% increase in the probability of complications (P_I) for the HN tumor case. However, in plan optimization and evaluation, when tolerance doses were compared with the respective planned EQD2 (calculated from the 3-dimensional dose distribution), P_B increased by 19.3% in the HN, 12.5% in the lung, 6.2% in the breast and 2.7% in the prostate tumor case, respectively. The corresponding increases in P_I were 2.3%, 6.2%, 1.0% and 0.7%, respectively. Incorporating voxel-based fractionation corrections in plan optimization is important to be able to show the clinical quality of a given plan against established tolerance constraints. To properly compare different plans, their dose distributions should be converted to a common fractionation scheme (e.g. 2 Gy per fraction) for which the doses have been associated with clinical outcomes.

Keywords

Radiation therapy • IMRT optimization
Voxel-based fractionation corrections

B. da Costa Ferreira (✉)
School of Health, Polytechnic of Porto, Porto, Portugal
e-mail: bcf@ess.ipp.pt

B. da Costa Ferreira
I3N Department of Physics, Aveiro University, Aveiro, Portugal

P. Mavroidis
Department of Radiation Oncology, University of North Carolina at Chapel Hill, Chapel Hill, NC, USA

P. Mavroidis
Division of Medical Radiation Physics, Karolinska Institutet and Stockholm University, Solna, Stockholm, Sweden

B. da Costa Ferreira · J. Dias · H. Rocha
Institute for Systems Engineering and Computers at Coimbra, Coimbra University, Coimbra, Portugal

J. Dias · H. Rocha
CeBER and FEUC, Coimbra University, Coimbra, Portugal

1 Introduction

Historically, the dose tolerances to RT of the OAR were mostly derived from dose distributions irradiating tissues homogeneously with the conventional fractionation of 2 Gy per fraction. Tabulated dose tolerance values are closely followed during treatment plan optimization as if the organ would be irradiated homogeneously or with the same dose distribution as those who were used to derive such tolerances. However, with 3D Conformal RT and more recently with IMRT, most OAR are heterogeneously irradiated with large portions of their volumes receiving fractional doses lower than 2 Gy. Additionally, the tri-dimensional dose distributions vary greatly between patients and alternative fractionations are becoming increasingly used. Due to this variety of fractionation schedules and scarcity of patient clinical data, tolerance doses are derived from converting all delivered treatments to a 2 Gy fractionation schedule.

Late RT side-effects are radiosensitive to fraction size. Thus, normal tissue voxels irradiated with a dose per fraction

© Springer Nature Singapore Pte Ltd. 2019
L. Lhotska et al. (eds.), *World Congress on Medical Physics and Biomedical Engineering 2018*,
IFMBE Proceedings 68/3, https://doi.org/10.1007/978-981-10-9023-3_74

much lower than 2 Gy will have a larger tolerance to radiation than those regions irradiated with doses equal or higher than 2 Gy. With IMRT, high dose gradients are produced in the target volume borders and OARs located outside the planning target volume (PTV) will be irradiated with maximum fractional doses much smaller than 2 Gy. For instance, in a HN cancer patient prescribed to 70.2 Gy in 33 fractions, the tolerance dose of 45 Gy for the spinal cord was previously used [1]. In this case, the maximum dose per fraction in that organ will be 1.36 Gy and the tolerance dose would be 49.5 Gy (α/β of 3 Gy). For parallel organs the correction of the tolerance dose for fractionation is not as straightforward, as the biological effect depends on the 3D dose distribution irradiating the organ. During plan optimization and evaluation, the comparison between tabulated tolerance dose values and the planned dose for each structure should therefore be done using a common fractionation scheme.

In particularly difficult cases needing to improve the probability of tumor control, taking into account the dose per fraction at the voxel level may be an additional degree of freedom during inverse IMRT optimization to improve the quality of RT. Increased normal tissue tolerance to radiation due to the heterogeneity in the fractional dose map, associated with a reduced dose in the OAR obtained with typical IMRT dose distributions, may allow the improvement in target volume coverage and thus treatment outcome.

In this study, a comparison between treatment plans, which were optimized with and without voxel-based fractionation correction was made. In this context, the physical dose loses its meaning and plan evaluation should then be based on EQD2 values. The gain obtained with this approach was then also quantified.

2 Materials and Methods

Four test patients diagnosed with HN, breast, lung or prostate tumors were included in this study. All cases were planned with simultaneous integrated boost techniques using 7 or 9 equidistant beams, except for the breast tumor case where 5 beams placed around the target breast were used. In the HN tumor case, the prescription dose was 70.2 and 63.0 Gy delivered in 33 fractions to the primary tumor and adenopathies (PTV-T) and high risk lymph nodes (PTV-N), respectively. In the lung tumor case, the prescription dose was 70.0 and 56.0 Gy to gross disease and enlarged planning target volume, respectively delivered in 31 fractions. In the breast case, the prescription dose was 66.0 Gy to the tumor bed and 50.0 Gy to the whole breast delivered in 25 fractions; and in the prostate tumor case, the dose prescription was 74.2 and 56.0 Gy to gross disease and involved lymph nodes, respectively. The tolerance doses followed

Emami et al. 2013 [2] recommendations, except for the spinal cord where the tolerance dose of 45 Gy was used [1].

Corrections for fractionation were performed in each voxel of the normal tissues, for each plan obtained in each optimization iteration, using the Biologically Effective Dose (BED) concept [1]. α/β values of 10 were used for the HN and the lung tumors, 4 for the breast tumor and 3 for the prostate tumor and OARs. Repopulation was considered using a potential doubling time of 3 days and kick-off time of 28 days for the HN cancer, 5 days and 14 days for the lung cancer, and 4 days and 29 days for the prostate tumor, respectively. For the breast tumor case a potential doubling time of 15 days was used [3–6].

Plan comparison was made using conventional dose statistics and radiobiological metrics such as the uncomplicated tumor control probability (P_+), the probability of tumor control (P_B) and the probability of normal tissue complications (P_I). The relative seriality model and the linear-quadratic-Poisson model were used to determine the probability of response of the OARs and targets [7]. The model parameters used for each tumor volume are listed in Table 1. The dose-response parameters for the OARs are summarized in Mavroidis et al. [8].

The open access treatment planning system matRad (developed by DKFZ) was used for inverse IMRT optimization [9]. The voxel-based fractionation correction was additionally implemented into the code by identifying the target or OAR to which each voxel belongs. Thus, the evaluation of the objective function was based on the dose matrix converted to 2 Gy, instead of the physical dose matrix, to drive the optimization algorithm.

Three plans were simulated for each patient: (1) the initial plan, which was conventionally optimized based on physical dose (noVC); (2) a new plan which was re-optimized using the voxel-based fractionation correction and the same objective function, i.e., no adjustments were made on objectives and penalties (VC); and (3) a final plan, using the voxel-based fractionation correction, which was re-optimized adjusting objectives and penalties based on planned biological dose (VC+EQD2).

3 Results

By comparing VC with noVC, the maximum gain in the probability of tumor control for the patient with an HN tumor, was 1.3% with a 0.1% increase in probability of complications (Table 2). This was mostly due to an improvement in the irradiation of PTV-N. No significant differences in the dosimetry or in the response probabilities of the OARs were obtained except for a small increase in the minimum and mean dose in the oral cavity, which however did not result in an increase in the probability of injury.

Table 1 Dose-response parameters for the tumor cases used in this study

ROI	D_{50}/Gy	γ	ROI	D_{50}/Gy	γ
Head and neck			*Prostate*		
GTV-T	55.0	8.0	Prostate	74.0	7.5
PTV-T	54.0	7.5	PTV-T	73.0	6.0
PTV-N	46.0	4.8	PTV-N	42.0	3.5
Lung			*Breast*		
GTV	52.0	7.5	Tumor bed	50.0	4.0
CTV	50.0	5.0	CTV	35.0	2.0
PTV	40.0	4.0	PTV	30.0	1.0

*GTV is gross tumor volume and CTV is the clinical target volume

Table 2 Results for the HN cancer case. The first group of columns report the difference between the results obtained with plan VC and plan noVC. The second group of columns report the differences between plan VC+EQD2 and noVC. Dosimetry reports differences in biological the dose, i.e., corrected to a dose per fraction of 2 Gy

ROI	VC − noVC				VC+EQD2 − noVC			
	ΔP_+/%	ΔP_B/%	ΔP_I/%		ΔP_+/%	ΔP_B/%	ΔP_I/%	
	1.1	1.3	0.1		17.0	19.3	2.3	
	$\Delta P_{B/I}$/%	ΔD_{98}/Gy	ΔD_{mean}/Gy	ΔD_2/Gy	$\Delta P_{B/I}$/%	ΔD_{98}/Gy	ΔD_{mean}/Gy	ΔD_2/Gy
GTV-T	0.1	0.2	0.1	−1.7	1.4	3.1	1.6	1.9
PTV-T	0.0	0.2	0.0	−0.1	1.1	3.8	1.5	1.8
PTV-N	1.3	1.1	0.2	−0.1	18.6	9.5	3.6	3.0
Spinal cord	0.0	0.0	0.4	−0.3	0.0	0.0	4.0	5.6
Ips. Parotid	0.0	0.0	−0.1	0.4	0.6	0.3	5.3	9.4
Ctr.Parotid	0.0	−0.4	0.0	1.5	0.3	0.6	4.3	7.7
Oral cavity	0.0	2.4	3.0	1.0	0.1	0.4	3.5	3.1
RVR	0.1	0.0	0.2	0.0	1.2	0.0	0.5	1.9

*D_{98} and D_2 is the near-minimum and near-maximum dose, respectively and D_{mean} is the mean dose

When plan optimization and evaluation were based on planned EQD2 values (VC+EQD2), compared to conventional optimization, the probability of tumor control increased by 19.3% in the HN tumor case. This was mostly due an 9.5 Gy increase in the near-minimum biological dose in PTV-N and about 3 Gy in PTV-T. An increase of less than 0.6% in the probability of xerostomia and mucositis was obtained.

The most critical OAR in HN cancer RT is the spinal cord. For the plan noVC, the spinal cord had a maximum physical dose of 44.5 Gy and an EQD2 value of 38.7 Gy given some freedom to improve tumor coverage. Thus, in plan VC+EQD2 the probability of tumor control increased from 61.7% to 81.0% for a final EQD2 value to spinal cord of 44.5 Gy (physical dose of 49.5 Gy).

When applying the voxel-based fractionation correction by itself negligible benefits were obtained for the other tumor cases. For the lung, breast and prostate tumor case, the differences in P_B between the plan VC and noVC were

0.3%, 0.02% and -0.3%. However, when the optimization was based on EQD2 planned values, gains in P_+ of 6.4%, 5.2% and 1.9% for the lung, breast and prostate tumor, respectively, were obtained due to increased P_B by 12.5%, 6.2% and 2.7% and in P_I by 6.2%, 1.0% and 0.7%, respectively.

The improvements obtained in the lung cancer case with the plan VC+EQD2, compared to noVC plan, were mostly due to improvements in the probability of tumor control of the external PTV and CTV by 9.5% and 5.0%, respectively, due to an increase of about 6 Gy in the near-minimum dose in these volumes. Simultaneously, an increase in the probability of injury in the ipsilateral lung by 5% was obtained. The difference in the probability of RT side-effects for the heart, liver, spinal cord and esophagus between VC+EQD2 and noVC was below 1%. Similarly, for the breast cancer case, the largest improvements were obtained by the better CTV and PTV irradiation increasing the probability of tumor control from 58.9% with noVC to 65.1% with VC+EQD2 while probability of complications in heart, lungs, and contralateral breast remained almost the same.

For the prostate cancer case, the gains by using these two new approaches were small. The gain in using EQD2 values for plan evaluation, compared to conventional optimization, was about 1.8% for both PTVs due to an increase in the minimum dose in these volumes while the difference in the probability of complications in the rectum, bladder and femoral heads between the two methods remained below 0.4%.

4 Discussion

The heterogeneity obtained with IMRT dose distributions, reducing not only the total dose but also the dose per fraction delivered to the OARs, suggests that there might be room for improving target volume coverage. The numerical correction of tabulated tolerance doses to different fractionation schedules is not sufficient during IMRT planning as the dose distribution changes in each optimization iteration. Furthermore, incorporating voxel-based fractionation corrections allows a compatible comparison between planned doses and established tolerances doses using a common fractionation scheme. In this study, when this correction was incorporated into the optimization algorithm, the gain, obtained by just re-optimizing the plan without adjusting any of the objectives and penalties, was small. This comparison demonstrated the shortcomings of evaluating plan quality based on physical dose objectives and constraints. This emulates clinical procedures that do not take into account the full potential of RT fractionation. This can be though as a reverse double-trouble effect that is being completely overlooked.

When plans, optimized based on fractionation corrected biological dose, were compared against plans produced based on physical tolerance doses, the gain in probability of uncomplicated tumor control was substantial especially for the most difficult tumors cases. When dose per fraction in each voxel of the normal tissues was considered, and it was smaller than 2 Gy, the values of the biological dose in the OAR allowed to raise the minimum dose in the (outer) target volume significantly improving the probability of tumor control. Inevitably this also resulted in an increase in the probability of complications. However, the gain obtained in

the probability of tumor control outweighed the loss in the probability of RT side-effects.

For tumor cases with OARs partially located in the PTV, irradiated with simultaneous integrated boost techniques, the effect is the opposite as these OARs will be irradiated with a dose per fraction larger than 2 Gy. In this case, voxel-based optimization algorithms will take this into account investing its efforts in trying to reduce the dose in those voxels.

This work followed the assumption that each OAR is characterized by a homogenous radiosensitivity. The ability to determine the internal organization of the OARs in a more precise and quantitative manner would bring additional benefits for treatment planning. Normal tissue response to RT needs to be thoroughly investigated to maximize treatment individualization and the success of delivered therapies.

Conflicts of Interest None.

References

1. Ferreira, B.C.; Lopes, M.C.; Mateus, J.; et al. Radiobiological evaluation of forward and inverse IMRT using different fractionations for head and neck tumours. Radiat Oncol. 5:57 (2010).
2. Emami, B.; Lyman, J.; Brown, A.; et al. Tolerance of normal tissue to therapeutic irradiation. Int J Radiat Oncol Biol Phys. 21:109–122 (1991).
3. Wyatt, R.M.; Beddoe A.H.; Dale R.G. The effects of delays in radiotherapy treatment on tumour control. Phys Med Biol. 48 (2):139–55 (2003).
4. Partridge, M.; Ramos, M.; Sardaro, A.; Brada, M. Dose escalation for non-small cell lung cancer: analysis and modelling of published literature. Radiother Oncol. 99(1):6–11 (2011).
5. Pedicini, P.; Strigari, L.; Benassi M. Estimation of a self-consistent set of radiobiological parameters from hypofractionated versus standard radiation therapy of prostate cancer. Int J Radiat Oncol Biol Phys. 85(5):e231–7 (2013).
6. Fowler J.F. Optimum overall times II: extended Modelling for head and neck radiotherapy. Clinical Oncol. 20:113–126 (2008).
7. Källman, P.; Ågren, A.; Brahme, A. Tumour and normal tissue responses to fractionated non-uniform dose delivery. Int J Radiat Biol. 62(2):249–62 (1992).
8. Mavroidis, P.; Ferreira, B.C.; Papanikolaou, N.; Lopes, M.C. Analysis of fractionation correction methodologies for multiple phase treatment plans in radiation therapy. Med Phys. 40(3):031715 (2013).
9. Wieser, H.P.; Cisternas, E.; Wahl, N. et al. Development of the open-source dose calculation and optimization toolkit matRad. Med Phys. 44(6):2556–2568 (2017).

Accurately Evaluating Settling Responses of Ionization Chambers Used in Radiation Therapy Depend on the Accelerating Tube

Tetsunori Shimono, Yasuyuki Kawaji, Tatsuhiro Gotanda, Rumi Gotanda, and Hiroshi Okuda

Abstract

The growing use of multiple small fields in radiotherapy treatments such as intensity-modulated radiation therapy has increased the importance of small-field dosimetry. In this study, we investigate the settling response of a set of ionization chambers exposed to 4, 6, and 10 MV stereotactic radiotherapy X-ray fields. Previous studies had reported that lack of pre-irradiation could result in settling response errors of up to several percentage points. While the use of ionization chambers does contribute to this behaviour, the most obvious factors affecting the settling response appear to be the area of the insulator, the material used for the central collector electrode, and the accelerating tube. The results of this study show that Farmer-style ionization chambers with electrode connections that are guarded up to an active air volume settle quickly (within five minutes); moreover, the changes in their responses are small. On the other hand, small ionization chambers exhibit settling times of 11–24 min. In this study, the settling times of small ionization chambers were found to be dependent on the accelerating tube.

Keywords

Small ionization chamber • Response • Accelerating tube

T. Shimono (✉) · Y. Kawaji
Faculty of Health Sciences, Junshin Gakuen University, Fukuoka, Japan
e-mail: shimono.t@junshin-u.ac.jp

T. Gotanda · R. Gotanda
Kawasaki University of Medical Welfare, Okayama, Japan

H. Okuda
Department of Radiology, Hoshigaoka Medical Center, Osaka, Japan

© Springer Nature Singapore Pte Ltd. 2019
L. Lhotska et al. (eds.), *World Congress on Medical Physics and Biomedical Engineering 2018*,
IFMBE Proceedings 68/3, https://doi.org/10.1007/978-981-10-9023-3_75

1 Introduction

TG-51 and IAEA TRS-398 recommend that ionization chamber measurements should be performed repeatedly until a stable reading is obtained [1, 2]. Ionization chambers exhibit a settling response that depends on the time elapsed since the previous irradiation [3–5]. This pre-irradiation depends on whether the settling is dose-rate dependent or time dependent and whether irradiation is required if the instrument is biased for an extended period or not. Previous studies have reported on the required total dosage and period of this pre-irradiation [6].

It has been observed that an NE2571 ionization chamber exposed to a 60Co beam settles after approximately 15 min of irradiation [7]. This suggests that the settling behaviour mechanisms of ionization chambers vary depending on the material and construction. During pre-irradiation, large changes in the response appear to arise from several mechanisms that are dominated by the effects of irradiation on the unguarded portions of the insulated stem. Ionization chambers designed by different manufacturers exhibit settling curves of different shapes. Small chambers exhibit longer settling times. Because access to radiation facilities in a radiotherapy clinic is often limited, the additional time required for ionization chamber stabilization is starting to become a concern. For settling to occur, an ionization chamber must be exposed to radiation. The settling times observed are specific to the ionization chambers and the accelerating tube. This study investigates the effects of pre-irradiation on a variety of typically small radiotherapy chambers and attempts to relate the observed behaviour to the design of the small ionization chambers and to the accelerating tubes.

417

2 Materials and Methods

In this study, commercially available similar ionization chambers were used for evaluating the pre-irradiation effect. A wide variety of commonly used ionization chambers was irradiated using various X-ray beams, and their responses were monitored until stabilization occurred. We used two commercial Farmer-type ionization chambers and three small ionization chambers, to compare the magnitudes of the pre-irradiation effects in the different types of ionization chambers, as well as in individual chambers. In the TG-51 protocol of AAPM, $\%dd(10)\times$ was adopted, which was defined as the photon component of the percentage depth dose at a depth of 10 cm for a field size of 10×10 cm^2 on the surface of a water phantom at a source–surface distance of 100 cm. It was also used as the beam quality index for high-energy photon radiation. Measurements were performed using a microtron electron accelerator (HTM2210; Hitachi), an Elekta Synergy medical linear accelerator, and a Varian's Clinac iX system linear accelerator. Table 1 shows the physical characteristics of the following ionization chambers, which were used in this study: PTW30001, PTW31015, PTW31016, PTW31006 (PTW Co., Freiburg, Germany), and NE2571 (Nuclear Enterprises, Reading, United Kingdom). The chambers were connected to RAM-TEC 1000D and RAMTEC 1000plus (Toyo Medic, Tokyo, Japan) electrometers. All measurements were defined by a Microtron HTM2210 on the beam central axis at a depth of 10 cm in water, obtained at an SCD of 100 cm. Ion chamber readings were obtained by maintaining the bias voltage constant at -300 V. The ion chambers were exposed to photon beams for just the 100 MU measurement. The dark current was measured with and without the ionization chamber in the charge mode for 15 min. In this study, the dark current constituted approximately $\pm0.01\%$ of the measured current.

3 Results

Figure 1 shows the relative responses of the NE2571 and PTW30001 ionization chambers to a constant dose of 10 MV photon beam measured at a dose rate of 300 MU/min. A voltage of -300 V was applied to the ionization chambers. The two chambers were placed at a depth of 10 cm in a field of size 10×10 cm^2. The ionization chamber that contained the chamber stem and connector was exposed to the radiation beam and was shielded with lead blocks of thickness 20 cm and area 10×10 cm^2. The scattering from the lead shield was measured and was found to be negligible. The settling times of the NE2571 ionization chamber for microtron, Synergy, and Clinac iX were 10 min, 8 min, and 7 min, respectively (Fig. 1). The PTW30001 ionization chamber demonstrated a settling time of 5 min for all of microtron, Synergy, and Clinac iX (Fig. 2). The PTW31015 ionization chamber demonstrated settling times of 13 min, 12 min, and 11 min for microtron, Synergy, and Clinac iX, respectively (Fig. 3). The settling times of the PTW31016 ionization chamber for microtron, Synergy, and Clinac iX were 16 min, 14 min, and 13 min, respectively (Fig. 4), and that of the PTW31006 ionization chamber were 24 min, 22 min, and 20 min, respectively (Fig. 5). The percentages of decrease in the settled readings of the NE2571 and PTW30001 ionization chambers, from the first reading, were 0.50% to 0.60%, while that of the PTW31015, PTW31016, and PTW31006 ionization chambers were 0.90% to 1.80%.

Figure 6 shows the response of the PTW31016 chamber to a 10 MV photon beam for various dose rates. Depending on the dose rate, the response changes just after the measurement begins. It is different, and the measurement back show, in particular, a large change in the value at 50 MU, when compared to the stable measurements in the case of other dose rates.

Table 1 Characteristics of the ionization chambers used for the study

Chamber	Cavity volume (cm^3)	Thimble	Electrode	Insulator	Unguarded distance (mm)
NE 2571	0.68	Graphite	Aluminum	PCTFE(KEL-F)[a]	26
PTW 30001	0.6	PMMA	Aluminum	PE/PF[b]	0
PTW 31015	0.03	PMMA	Aluminum	PE/PF[b]	0
PTW 31016	0.016	PMMA	Aluminum	PE/PF[b]	0
PTW 31006	0.015	PMMA	Steel	PE/PF[b]	0

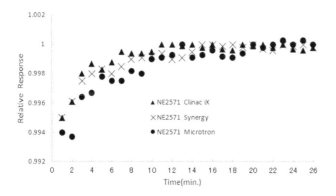

Fig. 1 Response of a NE2571 ionization chamber (Farmer-type) used to a constant dose of 10 MV photon beam from a Microtron, Synergy, and Clinac iX

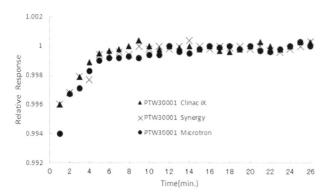

Fig. 2 Response of a PTW 30001 ionization chamber (Farmer-type) used to a constant dose of 10 MV photon beam from a Microtron, Synergy, and Clinac iX

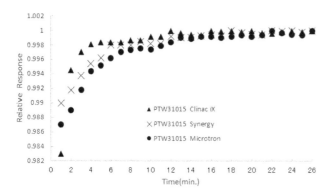

Fig. 3 Response of a PTW 31015 small ionization chamber used to a constant dose of 10 MV photon beam from a Microtron, Synergy, and Clinac iX

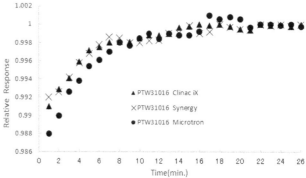

Fig. 4 Response of a PTW 31016 small ionization chamber used to a constant dose of 10 MV photon beam from a Microtron, Synergy, and Clinac iX

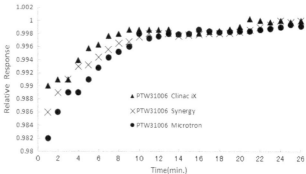

Fig. 5 Response of a PTW 31006 small ionization chamber used to a constant dose of 10 MV photon beam from a Microtron, Synergy, and Clinac iX

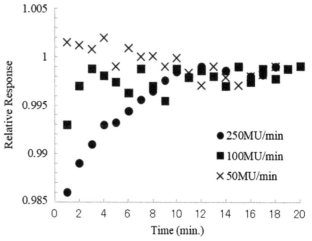

Fig. 6 Response of the PTW31016 chamber to a 10 MV photon beam for variousdose rates

The central electrodes in the stem of the PTW31015 and PTW31016 ionization chambers were aluminium and that of the PTW31006 ionization chamber was steel. Small air-volume thimble chambers require a longer initial settling time.

4 Discussion

The settling behaviours in five ionization chambers were observed. In this study, the magnitude of the average settling effect was a typical settling time of 11–24 min in the three small ionization chambers. The 0.6 cm^3 NE 2571 and PTW30001 chambers demonstrated short average settling times but displayed a wide variety of settling behaviours depending on the individual ionization chamber. The change in the measurements of the small ionization chamber volume was specific, unlike that of the other ionization chambers. Figures 3, 4 and 5 illustrate the wide variety in the settling times and magnitudes of the responses for different chamber models. When an ionization chamber is exposed to a radiation beam, the instrument requires a significant period to settle. The settling time required can vary from 5 to 24 min, and the magnitude of the change can vary from <0.1% to 1.8%. Individual chambers must be measured to determine the magnitude of the change in the response. Variations in the material of the ionization chamber produce variations in the ionization chamber response. The accelerator of Synergy is a traveling-wave acceleration tube, whereas Clinac's accelerator is a standing-wave electron linear accelerator. The microwave power sources of Synergy and Clinac are magnetron and klystron, respectively. The electron beam energy W at the accelerator exit is e (electronic charge) × E (electric field strength) × S (distance). The length S of the standing-wave linear acceleration tube is shorter than that of the traveling-wave linear acceleration tube. In the case of the same energy, the electric-field intensity is large when the acceleration tube is short. In this study, the settling time of the standing-wave linear accelerating tube was short.

5 Conclusion

Our study of ionization chamber settling behaviours was time consuming, and each chamber required a significant amount of time for configuration. The settling behaviours varied among different chambers of the same model, making the identification of the causes difficult. Small ionization chambers that were fully guarded typically exhibited settling times of 11–24 min and the associated changes in response of up to 1%. Most models of small ionization chambers showed time-dependent settling behaviours that were independent of the beam quality or dose rate. Large changes in the response appeared to arise from different mechanisms, and were dominated by the linear accelerating tubes. Small ionization chambers exhibited large pre-irradiation effects. In addition, large pre-irradiation effects were observed when steel was used as a central electrode, instead of polymer. Therefore, in order to eliminate inadvertent errors, pre-irradiation of ionization chambers is necessary.

Conflict of Interest The authors hereby declare that they have no conflicts of interest.

References

1. Almond PR, Biggs PJ, Coursey BM, et al. AAPMs TG-51 protocol for clinical reference dosimetry of high-energy photon and electron beams. Med. Phys. 26:1847–70 (1999).
2. Andreo P, Burns DT, Hohlfeld K, et al. Absorbed dose determination in external beam radiotherapy. IAEA Technical Report Series; No. 398 (2000).
3. Cacak RK, Hendee WR. Chapter 3. Ionization chamber dosimetry, techniques of radiation dosimetry. pp. 104–105, ed. Mahesh K.; Vij DR. Wiley Eastern Limited, India; (1985).
4. Liversage WE. The effects of X-rays on the insulating properties of polytetrafluoroethylene (P. T. F. E.). Brit J Radiol. 296: 434–436 (1952).
5. Fowler JF, Farmer FT. Conductivity induced in insulating materials by X-rays. Nature. 13: 317–318 (1954).
6. DeBlois F, Zankowski C, Podgorsak EB. Saturation current and collection efficiency for ionization chambers in pulsed beams. Med. Phys. 27:1146–55 (2000).
7. McCaffrey JP, Downton B, Shen H, et al. Pre-irradiation effects on ionization chambers used in radiation therapy. Phys Med Biol. 50:121–133 (2005).

Monte Carlo Dosimetry of Organ Doses from a Sweeping-Beam Total Body Irradiation Technique: Feasibility and First Results

Levi Burns, Tony Teke, I. Antoniu Popescu, and Cheryl Duzenli

Abstract

Total body irradiation (TBI) is a radiation treatment often purposed to suppress the immune system prior to a bone marrow transplant. Several toxicities can arise in TBI, and high-quality dose volume data for organs at risk are required if one is considering any change from a well-established technique. We present a novel Monte Carlo (MC) dosimetry technique to acquire this data based on our current TBI technique that accounts for a sweeping-beam Cobalt-60 delivery, a stationary flattening filter, patient-specific lung compensators, and two patient treatment positions. For each patient, a virtual MC phantom is created including the planning CT image in each treatment position (supine and prone). The results from dose simulations on each phantom were summed together geometrically with a deformable registration tool. Dose volume statistics for lungs, liver, thyroid, and kidneys are obtained. The preliminary results of a retrospective study using this technique on patients who have received TBI at our clinic indicate that, for a total body prescription dose of 12 Gy ± 10%, the mean body dose ranged from 11.19 to 12.15 Gy with smaller patients receiving lower mean body doses than larger patients. The mean dose delivered to the thyroid was the highest of the contoured organs receiving up to 12.84 Gy, and the lung doses were the most heterogeneous, with standard deviations up to 0.73 Gy in individual patients. This high-quality dose data shows promise for use in both routine quality assurance of our current technique, and to provide baseline data for development of a new technique. The technique could also be adapted to TBI techniques at other clinics that include compensators, flattening filters, moving beams, and/or multiple treatment positions.

Keywords

Total body irradiation • Monte Carlo • Dosimetry Cobalt • Sweeping beam • Organ dose

1 Introduction

Total body irradiation (TBI) entails the delivery of a uniform radiation dose to the entire patient. Its use is often part of a conditioning regimen to prevent the onset of graft-versus-host disease (GvHD) in patients who will receive a bone marrow transplant (BMT) [1, 2]. A recent survey across Canadian cancer centres estimates that TBI is received by roughly 400 people nation-wide per year [3]. This study, in conjunction with an international survey of a similar nature [4], illustrates a large amount of diversity in how TBI is delivered, with approaches including translating treatment couches, parallel-opposed pair treatments, and volumetric modulated arc therapy (VMAT). Several toxicities are reported from TBI, the most important of which is interstitial pneumonitis which can be fatal. It is important to acknowledge the difficulty in discerning whether toxicities following TBI are due to the radiation treatment or other aspects of the patient's care, such as concurrent chemotherapy or the BMT itself [5], although precautions may still be taken during TBI to reduce the incidence of these effects.

L. Burns (✉) · I. A. Popescu · C. Duzenli
Department of Physics & Astronomy, University of British Columbia, Vancouver, BC, Canada
e-mail: levi.burns@bccancer.bc.ca

L. Burns · I. A. Popescu · C. Duzenli
Medical Physics, BC Cancer (Vancouver Centre), Vancouver, BC, Canada

T. Teke
Department of Computer Science, Mathematics, Physics and Statistics, University of British Columbia (Okanagan), Kelowna, BC, Canada

T. Teke
Medical Physics, BC Cancer (Southern Interior), Kelowna, BC, Canada

© Springer Nature Singapore Pte Ltd. 2019
L. Lhotska et al. (eds.), *World Congress on Medical Physics and Biomedical Engineering 2018*,
IFMBE Proceedings 68/3, https://doi.org/10.1007/978-981-10-9023-3_76

Our clinic currently uses a sweeping-beam Cobalt-60 technique to deliver TBI. While our treatment is effective, we are considering an upgrade to a linac-based method that will improve the patient experience and open the door to possible customization of organ doses through intensity modulation. To ensure continuity of clinical care, it is desirable to establish dosimetric equivalence of any new technique with the current technique, prior to implementation. We thus require accurate, high-quality organ dose data from the current treatment which has, until recently, been unavailable for individual patients. The gold standard for obtaining this data would be Monte Carlo (MC) dosimetry. Some efforts at implementing MC procedures into the TBI treatment planning pipeline have been described in the literature, primarily to evaluate the homogeneity of the dose distribution and the effectiveness of flattening filters, but organ dose data from TBI has yet to be reported [6–9].

In this paper, we present a technique for TBI dosimetry using Monte Carlo simulations based on our sweeping-beam technique, including a deformable registration to add together doses from two different treatment positions. We demonstrate proof of principle with initial findings for a small retrospective series of patients who have received TBI at our centre.

2 Methods

2.1 Description of TBI Technique

Patients in our clinic receive TBI from a repurposed Cobalt-60 unit (Theratron 780C, Atomic Energy of Canada Ltd., Ottawa, Canada). An early version of our technique is described by Sherali and El-Khatib [10]. A schematic of the treatment is given in Fig. 1. The patient is positioned at an extended SSD, typically around 160 cm, depending on the size of the patient. The field (70×70 cm^2 at 160 cm from the source) covers a section of the patient at a given time, and the gantry repeatedly sweeps out a 90-degree angle during treatment to irradiate the entire body. Each fraction is delivered in two parts, with the patient lying in supine and prone positions. Before each treatment, CT image volumes are collected from the patient in both positions, from the level of the pelvis to the level of the chin. These are used to design patient-specific lead lung compensators for each treatment position for each patient to bring the lung dose down to prescription level. The positions of the lung compensators are verified by a physicist at the first treatment with a digital imaging panel placed underneath the treatment bed. A stationary polystyrene flattening filter is centered beneath the Cobalt source to ensure a uniform dose is delivered to the whole patient.

2.2 Monte Carlo Phantom Production

To run a Monte Carlo simulation, a phantom corresponding to the modelled system must first be created in the simulation software. We have designed a method to produce these phantoms using Python, Matlab, and C++ codes. The phantoms are designed for use with the EGSnrc Monte Carlo simulation software [11, 12]. A complete phantom for a patient is shown in Fig. 2.

The patient CT image is converted to a MC phantom using the software CTcreate [13] in the EGSnrc platform. The lung blocks for the patient treatment are retrieved and converted to a phantom with an in-house Matlab script. A phantom for the polystyrene flattening filter was created by taking a CT scan of it and using CTcreate. The phantoms of the lung blocks, flattening filter, and the plastic sheet on which the lung blocks rest are appended as new slices above the original patient phantom at a height and position corresponding to the treatment specifications. Variable resolutions are used in the beam axis direction to model the large air gaps between phantom components as single air voxels. This reduces the computational time that would be required to simulate histories through large areas of air voxels. The resolution in the other directions is 0.25×0.25 cm^2 for all phantom components.

2.3 Patient Selection and Ethics Approval

Patient cases selected for this retrospective study had been prescribed a TBI treatment of 1200 cGy in 6 fractions since August 2015. Cases were excluded from the study if the CT images revealed resected organs. For this initial study, both male and female patients with a variety of body sizes were included, including pediatric patients. The study has been approved by our institutional Research Ethics Board.

2.4 Monte Carlo Simulations and Deformable Registration

A model of the Cobalt source and housing was produced by Mora et al. [14] and is used in our simulations. Simulations of this source while the beam is sweeping are made possible by Source 21 of DOSXYZnrc [15]. Three billion particle histories are simulated for each treatment position with maximum dose uncertainties below 2% in the body and below 3% in the lungs. The simulations are run in parallel in up to 256 jobs simultaneously (8 GB RAM, 2.00 GHz processors) and take between 60 and 90 min to complete depending on the size of the phantom.

After the simulations are performed, the flattening filter, lung blocks, and plastic compensator tray are removed from

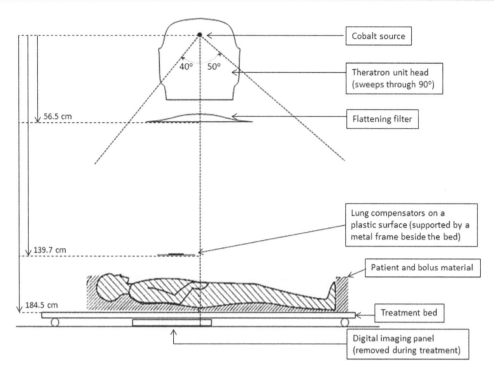

Fig. 1 Schematic of our sweeping-beam total body irradiation technique

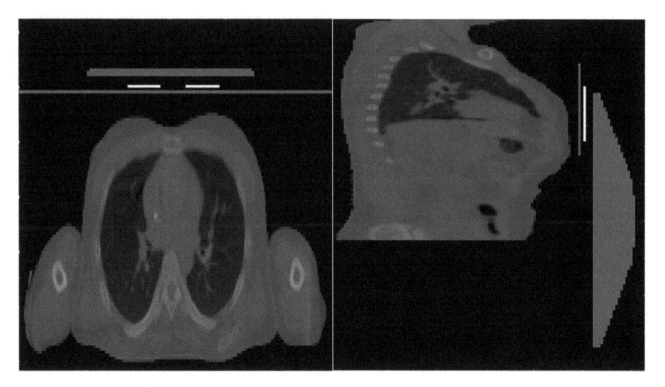

Fig. 2 Axial (left) and sagittal (right) slices of a Monte Carlo phantom for a TBI patient in the supine position. The three structures above the patient are the flattening filter, the lung compensators (white), and the plastic tray on which the lung compensators rest. The visualization software used is unable to display variable resolutions, although this depiction remains useful to conceptualize the geometry of the MC phantoms used. In the simulations, these structures are positioned at appropriate distances as indicated in Fig. 1

the phantom to restore a uniform resolution along the beam direction for analysis. Dose distributions are de-noised with a 3D Savitzky-Golay filter [16].

The doses given by an EGSnrc MC simulation are given in units of Gy per particle incident on a linear accelerator head target [17], which is not the circumstance for our purposes. To scale the relative doses from the simulation to real doses, a calibration measurement was undertaken by irradiating a solid water phantom at TG-51 conditions [18], with an ion chamber dosimeter at the standard reference point. An identical geometry was modelled as an EGSnrc phantom and simulated with two billion particle histories. The computed dose and the measured dose at the calibration point from this experiment were used to scale patient doses to Gy. Our sweeping Monte Carlo source has been validated in water tank measurements in previous work [19] to be described in an upcoming report.

To add the doses from each treatment position in an appropriate manner, the simulated dose files from supine and prone treatments are transferred to MIM Maestro™ v6.6.8 (MIM Software Inc., Cleveland, OH) where a deformable registration tool is available. The deformable registration is intensity-based and fully automatic.

2.5 Analysis

Organ dose data is analyzed in MIM Maestro™ following the deformable registration. The contoured organs are the kidneys, lungs, liver, and thyroid. Additionally, the whole body dose was considered for all patients. In the body contour, because the clinical protocol for CT simulations only covers part of the patient, three axial slices were removed from each of the superior and inferior edges of the body contour, as these doses are missing a scatter dose contribution from the rest of the body not included in the CT image volume. For a similar reason, the arms are removed manually from each body contour because the simulations do not incorporate bolus material that is used in patient positioning, and thus the arms are missing a scatter dose contribution in the simulations in comparison to the actual treatment. Moreover, the different positions of the arms between the supine and prone positions render this a challenging area for deformable registration of dose distributions.

3 Results

The TBI treatments of 11 patients were simulated and the dose results for organs and the whole body were collected. The supine and prone dose distributions for a patient are shown in Fig. 3 along with the final distribution after deformable registration.

In Fig. 4, the mean body dose is plotted against the size of the patient, as measured by the patient mid-separation depth at the level of the umbilicus. All mean body doses fall within the 12 Gy ± 10% prescription (range: 11.19–12.15 Gy), but within this constraint, the data suggest that smaller patients receive lower mean body doses.

Preliminary organ dose statistics are shown in Table 1, representing the results across the 11 simulated treatments. In addition to this data across the patient population, individual organ doses were examined on a case-by-case basis. For each patient, the doses to the thyroid and kidneys were consistently above the mean body dose, while the doses to the liver and lungs were consistently below, and the lung doses appear to be the most heterogeneous with standard deviations up to 0.73 Gy in one case. A further discussion of these results is being prepared in an upcoming publication.

4 Discussion

First and foremost, we have demonstrated the feasibility of employing Monte Carlo dosimetry for a sweeping-beam Cobalt-60 TBI technique incorporating lung blocks, a flattening filter, a moving beam, and two treatment positions. This data will be invaluable for setting clinical dose constraints in the design of a new TBI technique. The Monte Carlo technique has potential to be implemented as a routine quality assurance tool. Further studies are underway to optimize and streamline simulations, and to simulate a diverse patient population and obtain further information about organ doses.

In this study, three billion histories per phantom simulated were used. This number was chosen based on the authors' previous experience with MC simulations. Experiments are currently underway to determine if simulations with a lower number of particle histories per phantom could be used to obtain results of similar quality, in the interest of reducing simulation times.

The technique to produce these phantoms currently requires about 75 min of manual work per patient by an expert user. For clinical implementation, this would be an unacceptable amount of time for quality assurance of more common treatments, but it is not unfathomable given that the average number of TBI cases in our clinic is less than one patient per week. There is potential to further streamline the phantom production workflow and this is a current area of focus.

Validating the algorithm for a prone-to-supine deformable registration is a challenging task and is beyond the scope of this work. However, the RegReveal tool in MIM Maestro™ allows for qualitative inspection of local performance of the registration algorithm, and the CT images contain a large amount of bony anatomy to guide the

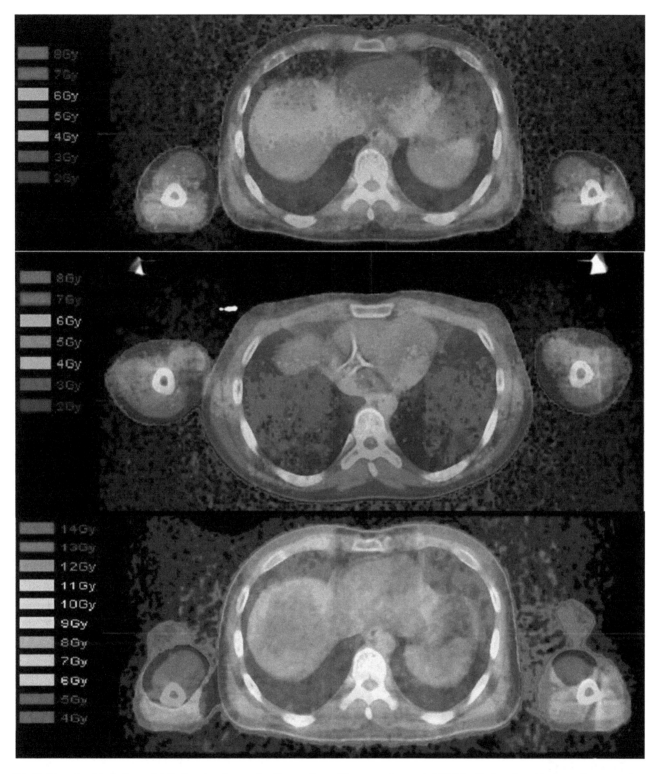

Fig. 3 The simulated dose distribution is shown for the supine (top) and prone (middle) simulations separately, and the final distribution after deformation is projected onto the supine image (bottom). The prone image (middle) is displayed upside-down for comparison purposes with the supine images

registration, including the entire spine. It should also be noted that the quality of the registration can be improved further by the use of "locks" by adept users of MIM Maestro which allow for iterative deformations of challenging areas while preserving existing deformations in successful areas. Because the placement of locks would vary by user, we have

Fig. 4 Mean body dose plotted against patient mid-separation depth at the level of the umbilicus. The error bars represent the standard deviation of dose to voxels within the body contour

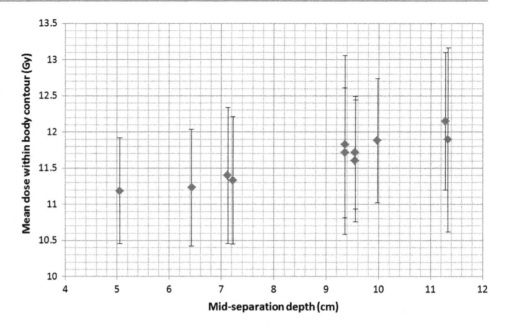

Table 1 A summary of organ dose statistics for 11 TBI patients. Each mean dose is the average across the patients. The maximum and minimum doses across the patients are also given

Organ	Mean dose (Gy)	Minimum dose (Gy)	Maximum dose (Gy)
Lung (left)	11.32	10.96	11.89
Lung (right)	11.25	10.73	11.75
Kidney (left)	11.86	11.22	12.52
Kidney (right)	11.86	11.15	12.34
Liver	11.37	10.98	11.73
Thyroid	12.07	11.10	12.84
Body contour	11.63	11.19	12.15

not made any attempt to implement them at this stage in the interest of the reproducibility of our results.

Finally, it can be noted that our technique in producing MC phantoms, where material slices such as plastic filters and lead compensators are appended slice-by-slice at appropriate heights above a patient CT scan, could be applied to TBI techniques at other clinics, or to other treatments where physical objects lie in the path of the beam, with minor adjustments.

5 Conclusions

We have created a method to obtain Monte Carlo organ dose information of our sweeping-beam TBI technique, and it could feasibly be implemented as a routine quality assurance tool for our treatments in the future. Preliminary retrospective results indicate that patient doses are consistent with

their prescriptions with some patterns of variation within the clinically acceptable dose limits for each organ.

Acknowledgements This work was supported in part by an Affiliated Fellowship from the UBC Faculty of Graduate and Postdoctoral Studies.

Conflict of Interest The authors declare that they have no conflict of interest.

References

1. Hill-Kayser CE, Plastaras JP, Tochner Z, Glatstein E: TBI during BM and SCT: review of the past, discussion of the present and consideration of future directions. Bone Marrow Transplantation 46(4), 475–484 (2011).
2. Wills C, Cherian S, Yousef J, Wang K, Mackley HB: Total body irradiation: a practical review. Appl Rad Oncol 5(2), 11–17 (2016).
3. Studinski RCN, Fraser DJ, Samant RS, MacPherson MS: Current practice in total-body irradiation: results of a Canada-wide survey. Current Oncology 24(3) (2017).
4. Geibel S, Miszczyk L, Slosarek K, Moukhtari L, Ciceri F, Esteve J, et al: Extreme heterogeneity of myeloablative total body irradiation techniques in clinical practice. Cancer 120(17), 2760–2765 (2014).
5. Kelsey CR, Horwitz ME, Chino JP, Craciunescu O, Steffey B, Folz RJ, et al: Severe pulmonary toxicity after myeloablative conditioning using total body irradiation: an assessment of risk factors. Int J Radiat Oncol Biol Phys 81(3), 812–828 (2011).
6. Charakova R, Muntzing K, Krantz M, Hedin E, Hertzman S: Monte Carlo optimization of total body irradiation in a phantom and patient geometry. Phys Med Biol 58(8), 2461–2469 (2013).
7. Charakova R, Krantz M: A Monte Carlo evaluation of beam characteristics for total body irradiation at extended treatment distances. J Appl Clin Med Phys 15(3), 182–189 (2014).
8. Serban M, Seuntjens J, Roussin E, Alexander A, Tremblay JR, Wierzbicki W: Patient-specific compensation for Co-60 TBI

treatments based on Monte Carlo design: A feasibility study. Phys Med 32(1), 67–75 (2016).

9. Liu X, Lack D, Rakowski JT, Knill C, Snyder M: Fast Monte Carlo simulation for total body irradiation using a 60Co teletherapy unit. J Appl Clin Med Phys 14(3), 133–149 (2013).

10. Sherali H, El-Khatib E: Total body irradiation with a sweeping 60-Cobalt beam. Int J Radiat Oncol Biol Phys 33(2), 493–497 (1995).

11. Rogers DWO, Faddegon BA, Ding GX, Ma CM, We J, Mackie TR: BEAM: A Monte Carlo code to simulate radiotherapy treatment units. Med Phys 22(5), 503–524 (1995).

12. Kawrakow I: Accurate condensed history Monte Carlo simulation of electron transport. I. EGSnrc, the new EGS4 version. Med Phys 27(3), 485–498 (2000).

13. Walters B, Kawrakow I, Rogers DWO: DOSXYZnrc users manual Report PIRS-794. National Research Council of Canada (2017).

14. Mora GM, Maio A, Rogers DWO: Monte Carlo simulation of a typical 60Co therapy source. Med Phys 26(11), 2494–2502 (1999).

15. Lobo J, Popescu IA: Two new DOSXYZnrc sources for 4D Monte Carlo simulations of continuously variable beam configurations, with applications to RapidArc, VMAT, TomoTherapy and CyberKnife. Phys Med Biol 55(16), 4431–4443 (2010).

16. Kawrakow I: On the de-noising of Monte Carlo calculated dose distributions. Phys Med Biol 47(17), 3087–3103 (2002).

17. Popescu IA, Shaw CP, Zavgorodni SF, Beckham WA. Absolute dose calculations for Monte Carlo simulations of radiotherapy beams. Phys Med Biol 50(14), 3375–3392 (2005).

18. Almond PR, Biggs PJ, Coursey BM, Hanson WF, Huq MS, Nath R, Rogers DWO: AAPM's TG-51 protocol for clinical reference dosimetry of high-energy photon and electron beams. Med Phys 26(9), 1847–1870 (1999).

19. Teke T: Monte Carlo techniques for patient specific verification of complex radiation therapy treatments including TBI, VMAT and SBRT lung. Ph.D thesis (not published), University of British Columbia (2012).

Rapid Prototyping, Design and Early Testing of a Novel Device for Supine Positioning of Large Volume or Pendulous Breasts in Radiotherapy

Levi Burns, Scott Young, Joel Beaudry, Bradford Gill, Robin Coope, and Cheryl Duzenli

Abstract

Here we describe the development of a novel device for breast positioning in supine radiotherapy that reduces breast sag and skin folds for patients with large or pendulous breasts. The overall aim of this work is to provide a practical and robust means of reducing high grade skin toxicity (moist desquamation) which tends to occur in skin folds. Participants with breast cup size D or greater were recruited to this ethics board approved prototype design study. Brassiere size, cup size, breast diameter, body mass index, height, weight, skin folds and torso dimensions were measured. Participants were positioned in treatment position on a breast board, with arms above the head and skin folds were identified and measured. 3D optical surface imaging provided initial design ideas and a rapid prototyping process using 3D printing was employed to arrive at a suitable design. The final clinical device consists of a curved carbon fibre breast support scoop suspended from a rigid frame that is compatible with commercially available breast boards. In addition to reducing skin folds, the device better positions the breast on the chest wall to help minimize the volume of normal tissue being irradiated and facilitates rapid setup. We present results of preliminary testing of the device, including dose buildup incurred by the carbon fibre scoop, skin fold reduction data and treatment planning data from CT simulations with and without the device. Surface dose with the device in place remains less than 80% of the prescription dose to the breast. Skin folds were reduced and reductions in irradiated volumes of lung and body were achieved compared with clinical plans without the supportive device. The novel breast support shows great potential to address a long-standing problem for a significant population of patients undergoing radiotherapy for breast cancer.

Keywords

Breast radiotherapy · Supine positioning
Rapid prototyping · Carbon fibre · Skin toxicity

1 Introduction

Acute radio-dermatitis is a frequent complication of whole breast radiation therapy. Previous studies [1–8] have shown that approximately 60–80% of all breast patients undergoing whole breast radiation therapy experience grade 2 or higher skin toxicity according to Radiation Therapy Oncology Group (RTOG) toxicity criteria [9] and 15–40% experience moist desquamation depending on the patient population and radiotherapy technique. For patients with large, pendulous or ptotic breasts, skin toxicities in the infra-mammary skin fold and axilla regions are of particular concern. If skin folds can be minimized and skin sparing restored to these areas, a reduction in moist desquamation should be achievable.

IMRT works well to improve the homogeneity of dose within the breast, but does not address the problem of high dose in skin folds and only a marginal improvement in skin toxicity has been attributed to IMRT [10, 11]. Prone patient positioning is another technique that has recently received significant attention. While prone positioning may eliminate skin folds for some patients, it is known to introduce additional problems, and is not a viable option for the majority of patients [12–14].

The variation in the size and shape of women's breasts creates a challenge when designing accessories to achieve good breast position for RT. Suggested supine breast

L. Burns (✉) · S. Young · J. Beaudry · B. Gill · C. Duzenli
Medical Physics, BC Cancer—Vancouver Centre, Vancouver, BC, Canada
e-mail: levi.burns@bccancer.bc.ca

C. Duzenli
e-mail: cduzenli@bccancer.bc.ca

L. Burns · C. Duzenli
Physics and Astronomy, University of British Columbia, Vancouver, BC, Canada

R. Coope
Genome Sciences Centre, BC Cancer—Vancouver Centre, Vancouver, BC, Canada

© Springer Nature Singapore Pte Ltd. 2019
L. Lhotska et al. (eds.), *World Congress on Medical Physics and Biomedical Engineering 2018*,
IFMBE Proceedings 68/3, https://doi.org/10.1007/978-981-10-9023-3_77

positioning devices include brassieres, plastic breast cups, plastic rings, and thermoformed plastic supports. Despite many attempts to solve this problem, there is a lack of evidence to indicate that any of the current devices makes a clinical difference [13] in skin toxicity.

Here we present a novel device for breast positioning in supine radiotherapy that supports the breast and reduces skin folds for patients with large or pendulous breasts. A curved carbon-fiber breast support is suspended from a rigid frame that is compatible with commercially available breast boards. The device also re-positions the breast to help minimize the volume of normal tissue, heart and lung being irradiated. A rapid prototyping process using human models and 3D printing was used to arrive at this design. We present the design and results of preliminary testing of the device, including a dosimetric study of dose buildup incurred by the positioning device and early treatment planning data.

2   Materials and Methods

2.1   Study Participants

Participants (both patients and non-patients) with breast cup size D or greater, or having skin folds greater than 1 cm depth, were recruited under ethics board approval to participate in an initial design study (9 participants), followed by an early testing study (4 patients) of the prototype breast support. The range of body size and shape represented in this sample was adequate to cover $\geq 80\%$ of the population expected to benefit from improved breast positioning. Participants were placed in supine treatment position on a breast board, with one or both arms above the head. In the initial design phase, body and skin fold measurements, photos and optical 3D scans were taken. In the early testing phase, patients undergoing whole breast adjuvant radiotherapy underwent a second CT simulation with the breast support in place, in addition to their standard CTsim for treatment planning. Treatments were performed without the support, pending results of this planning study. The current standard clinical breast support at our institution consists of either a small foam wedge placed in the infra-mammary fold, a thermoplastic shell, or no support.

2.2   Prototyping

A multi-disciplinary team of oncologists, radiation therapists and medical physicists developed the following design requirements:

1. Maintain a skin dose to $\leq 80\%$ of the prescribed dose to the planning target volume

2. Minimize skin folds while fitting the largest possible range of breast sizes with one device each for left and right breasts
3. Clear the bore of a GE Lightspeed RTTM 16 CT scanner (60 cm diameter)
4. Assembled support device to weigh <1 kg and be convenient to affix and remove in 5 min or less
5. Support device to be reusable, cleanable and indexed for individual patient settings

Computer aided design (CAD) was done in Solidworks 2016TM (Dassault Systèmes, Waltham, MA). Breast support and frame connector initial rapid prototypes were 3D Printed (Stratasys uPrint SE plus) from ABS plastic. The clinical version of the breast support was made from layers of 0.4 mm heat cured carbon fiber sheet (Prepreg 3 K, Fibre Glast Developments Corporation, Brookville, Ohio). A support frame and indexed brackets were made from 30 mm square carbon fiber tubes (Dragon Plate Inc., Elbridge, NY) with machined Delrin™ connectors.

2.3   Surface Dose Measurement

The carbon fibre sheet pre-preg material is 37% resin (resin density 1.2 g/cm^3) with the remainder being carbon fibre, for an overall density of 0.634 g/cm^3. Preliminary studies were performed using a Markus^{TM} (PTW Freiburg GmbH) parallel plate ionization chamber and Solid Water^{TM} (CSP Medical, Sarnia, Ont) to measure dose buildup properties of the carbon fibre sheet. This data guided decision making on the number of carbon fibre sheets to use in construction of the support to meet design criteria. Dosimetric measurements on the final carbon fiber support were performed using EBT3 GafChromic™ (Ashland) film with and without the support in contact with a silicone model of the breast.

2.4   Early Clinical Testing

Subsequent to the initial support prototyping, the fully assembled carbon fibre support was tested on 4 patients. In this phase of testing, patients undergoing breast radiotherapy participated in an evaluation study to compare treatment plans with and without the support, to assess patient setup reproducibility from treatment planning to treatment unit, to measure the time required to set up the breast support on patients and to confirm that skin folds could be reduced. We report here on two left breast and two right breast cancer patients undergoing adjuvant whole breast radiotherapy. CT scans were performed with and without the support in place with the patient in treatment position. Treatment plans consisted of tangents alone (2 patients) or wide tangents plus supraclavicular and axilla fields (2 patients) to include nodal regions. In the case of breast plus nodal irradiation, a wide tangent arrangement was used to include the internal mammary nodes. Skin fold depth and dose volume statistics for ipsilateral lung, heart and body were measured. Field size, beam energy and monitor units were also compared for treatment plans with and without the support. Setup time was recorded at both CTsim and treatment unit where portal images with the support in place were acquired. The new breast support was not used in treatment delivery at this phase of the study. Day to day setup reproducibility will be the subject of future testing.

3 Results

3.1 Design

The final design prototype for the supine breast positioning support is shown in Fig. 1. The breast support "scoop" follows the curve of the body laterally across the chest and in the superior-inferior direction. The shape across and along

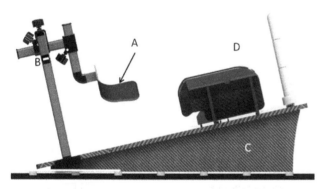

Fig. 1 Lateral and angled views showing: **a** The carbon fiber breast support "scoop"; **b** The carbon fibre support frame with Delrin™ joints and locking devices; **c** Existing supine breast board back support; and **d** existing head rest, wing-board and hand grip

the chest wall was modelled from the 3D surface imaging and CT data. The fit is snug to the skin along the infra-mammary and lateral breast-chest wall. The scoop is angled laterally such that the wall of the scoop is approximately perpendicular to the treatment beam and is constructed of three layers of carbon fibre sheet, heat curved over an aluminum mold.

The lateral, superior-inferior and vertical position of the scoop is indexed and adjustable for each patient. The frame fits within the CT simulator bore and does not collide with the linear accelerator. The support fit all participants in the study and visibly eliminated all inferior and lateral skin folds. Radiation therapists' experience when testing the support has been positive. The current assembled support prototype weighs 870 g. The frame clears the abdomen of the largest participant in the study. The setup with and without the support in place on a patient is shown in Fig. 2.

3.2 Surface Dose Measurement

Dose buildup data for the carbon fibre sheet for 6 and 10 MV X-ray beams, is shown in Fig. 3. Buildup in the carbon fibre sheet occurs less rapidly than buildup under the same depth in solid water. Each data point on the carbon fibre curve corresponds to an additional sheet of the material. 3 sheets of carbon fibre increases the surface dose from 26 to 48.6% in the 6 MV beam and from 18.2 to 38.4% in the 10 MV beam.

Surface dose maps with and without the carbon fibre scoop in contact with a silicon beast phantom are shown in Fig. 4, for 6 MV photons. A single lateral tangential field at 100 cm SAD delivered dose normalized to 100% at a depth of 5 cm. The curvature of the scoop results in a longer radiation path length through the device as it curves around the inferior portion of the breast. To assist with interpretation of this data, the experimental setup is shown in Fig. 5. Maximum dose buildup is observed at the inferior edge of the breast due to the curve of the scoop and tangential beam incidence. The maximum increase in surface dose due to the support is 30%, compared with no support.

3.3 Early Clinical Testing

Skin folds were successfully removed on all patients using the carbon fibre breast support, as demonstrated for a patient shown in Fig. 6. Note that the plastic elbow holding the scoop of the prototype breast support is visible in the right image inferior to the breast. The carbon fibre scoop is not visible in this image. All components of this support assembly will be ultimately be constructed from carbon fibre material.

(a)   **(b)**

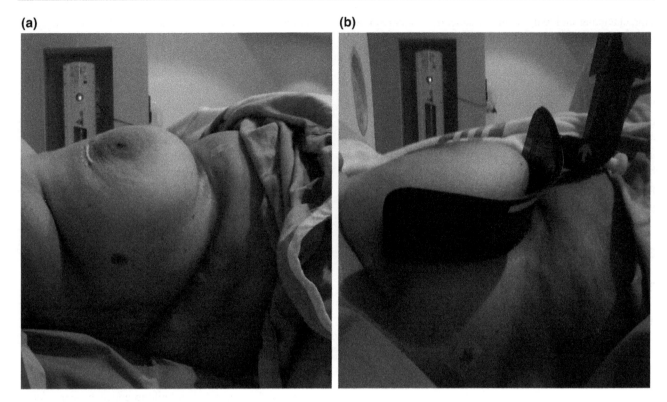

Fig. 2 Demonstration of the carbon fibre support device used on a patient. Left: without the support; Right: with the support

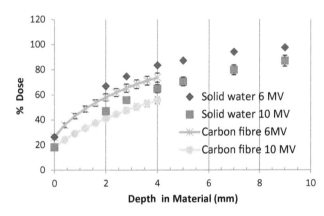

Fig. 3 Dose buildup data in carbon-fibre sheet versus solid water for 6 and 10 MV X-rays, measured with the Markus ionization chamber. All data is normalized to dmax in solid water

The treatment plans for the same patient with and without the support are shown in Fig. 7. Dose volume statistics and skin fold depth for the first 4 study patients are shown in Table 1.

4 Discussion

The carbon fibre material is less dense than water and produces less buildup than water equivalent material of the same thickness. This has distinct advantages over thermoplastic materials that have been suggested as solutions to the problem of breast positioning. For normal beam incidence, the carbon fibre scoop increases the entrance dose by approximately 20% compared with the bare surface for a 6 MV beam. For tangential incidence at the inferior edge of the breast, the carbon fibre support increases the entrance dose for the lateral beam by up to 30%. Despite the increase in entrance dose for the lateral beam, the total combined skin dose for a lateral and medial tangential pair of 6 MV beams with the support in place meets the design criteria of $\leq 80\%$ of the prescribed target dose, since exit dose from the medial beam is not significantly impacted by the support and accounts for more than 40% of the lateral surface dose.

Skin folds in all patients in this study were effectively reduced below what could be measured. Tangential field

Fig. 4 2D surface dose distribution **a** without the "scoop" in contact with the breast phantom surface and **b** with the "scoop". The colour scale and number labels indicate the measured dose as a percentage of the dose prescribed at depth 5 cm in phantom. The irradiation was performed using a 6 MV tangential beam. The values on the horizontal axis correspond to the angular position of the film corresponding to the angles shown in Fig. 5

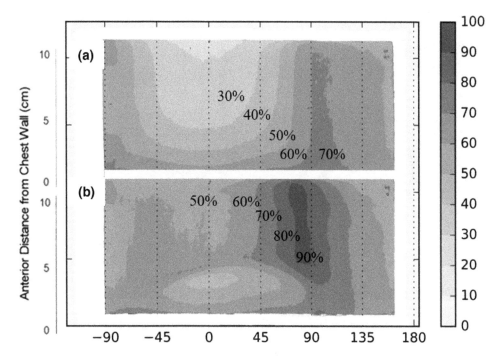

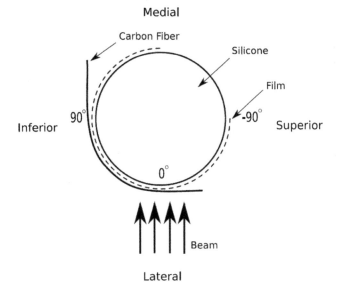

Fig. 5 Schematic showing setup for the EBT3 GafChromicTM film measurements corresponding to the data in Fig. 4

lengths were reduced by amounts corresponding to the infra-mammary skin fold depth. Tangential field widths increased by up to 1.5 cm with the breast supported, however this had no negative impact. While the clinical utility is currently being tested on a larger patient population, our hypothesis is that radiation tangent field borders are extended inferiorly and/or posteriorly to include all of the breast tissue when the breast sags. This can result in more lung, heart and body in the field than would be necessary had the breast been supported. Monitor units did not change significantly in this study, however with the support in place, acceptable plans could be achieved using more 6 MV beam weighting compared with 10 MV.

Many left breast patients are currently treated using the deep inspiration breath hold (DIBH) technique in order to reduce dose to the heart. The carbon fibre support is compatible with the DIBH technique, however, in order to keep the number of repeat CT scans to a minimum in this study, DIBH scans were not obtained with the support in place. Without DIBH, absolute irradiated lung volumes and body volumes were reduced using the support but heart dose was not. An ongoing study is evaluating the impact of DIBH combined with the carbon fibre support for left breast cases.

The device designed in this study supports the breast, removes skin folds and can be expected to reduce the volume of normal tissue irradiated. It remains to be seen how much reduction in moist desquamation and dose to organs at risk is achievable for a larger patient population. Future studies are planned to determine day to day setup reproducibility and clinical outcomes. Compatibility with MRI will also be assessed.

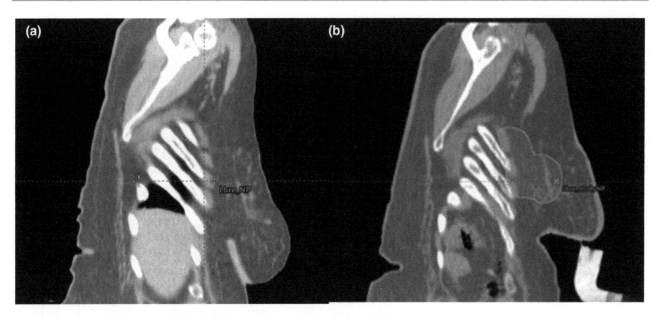

Fig. 6 Sagittal CT images demonstrating the removal of a 5 cm infra-mammary skin fold. The contralateral breast without support is shown in (**a**), and in (**b**) the treated breast is supported using the device. The patient was symmetric left to right

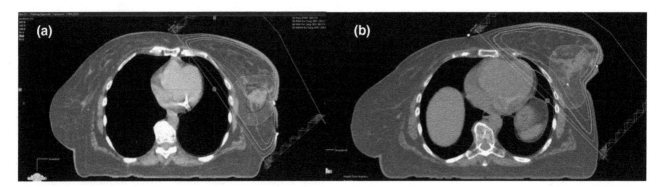

Fig. 7 Treatment planning images for the same left breast patient shown in Fig. 6, without support in (**a**) and with the support in (**b**). Note that the image on the left was taken during deep inspiration breath

hold, resulting in larger lung volume and a more inferior position of the heart visible on this CT slice

Table 1 Treatment planning parameters for the 4 study patients with and without the breast support. The numbers in the table represent the difference in each parameter with the support compared to without support. Thus, negative numbers indicate reductions in irradiated volumes and skin fold depths achieved with the support

	With support—without support	
	Left breast	*Left breast + nodes*
	Tangents only	*Wide tangents*
V50% body	−9.6 cc	−218.3 cc
V20 Gy ipsilateral lung	−58.9 cc	−90.3 cc
Skin fold depth	−5.1 cm	−2.0 cm
	Right breast	*Right breast + nodes*
	Tangents only	*Wide tangents*
V50% body	−39.7 cc	−645.2 cc
V20 Gy ipsilateral lung	−113.3 cc	−55.5 cc
Skin fold depth	−0.7 cm	−1.5 cm

5   Conclusions

A novel supine breast support and positioning device has been designed, constructed and preliminarily tested. This device has the potential to be universally adopted to reduce skin folds and to better position the breast for women with large volume, pendulous or ptotic breasts who are undergoing radiation therapy. Testing of the device in the clinical setting is currently ongoing.

Acknowledgements The authors wish to acknowledge Tania Arora, Keri Smith, Christina Cumayas, Terra Menna, Drs. P. Lim, A. Nichol, E. Chan and PA Ingledew for their dedicated efforts toward meeting the goals of this project.Funding for the work has been generously provided by the Canadian Cancer Society (Grant # 319456).Conflict of Interest: US Provision Patent Number 6246494 filed.

References

1. Porock D, Kristjanson L, Nikoletti S, Cameron F, Pedler P: Predicting the severity of radiation skin reactions in women with breast cancer. Oncol Nurs Forum 25, 1019–1029 (1998).
2. Bentel G, Marks L, Whiddon C, Prosnitz L: Acute and late morbidity of using a breast positioning ring in women with large/pendulous breasts. Radiother Oncol 50, 277–281 (1999).
3. Arenas M, Hernandez V, Farrus B, Muller K, Gascon M, Pardo A, et al. Do breast cups improve breast cancer dosimetry? A comparative study for patients with large or pendulous breasts. Acta Oncol 53, 795–801 (2014).
4. Cohen R, Freedman G, Li T, Li L, Brennan C, Anderson P, et al. Effect of bra use during radiotherapy for large breasted women: acute toxicity and treated heart and lung volume. Int J Radiat Oncol Biol Phys 72, S182–S183 (2008).
5. Gray J, McCormick B, Cox L, Yahalom J. Primary breast irradiation in large-breasted or heavy women: Analysis of cosmetic outcome. Int J Radiat Oncol Biol Phys. 21, 347–354 (1991).
6. Harper J, Franklin L, Jenrette J, Aguero E. Skin toxicity during breast irradiation: pathophysiology and management. South Med J. 97, 989–993 (2004).
7. Hymes S, Strom E, Fife C. Radiation dermatitis: clinical presentation, pathophysiology and treatment. J Am Acad Dermatol. 54, 28–46 (2006).
8. Wright J, Takita C, Reis I, Zhao W, Lee E, Nelson O, Hu J. Prospective evaluation of radiation-induced skin toxicity in a race/ethnically diverse breast cancer population. Cancer Medicine 5, 454–464 (2016).
9. Cox J, Stetz J, Pajak T. Toxicity criteria of the Radiation Therapy Oncology Group (RTOG) and the European Organization for Research and Treatment of Cancer (EORTC). Int J Radiat Oncol Biol Phys. 72, 1341–1346 (1995).
10. Michalski A, Atyeo J, Cox J, Rinks M, Morgia M, Lamoury G. A dosimetric comparison of 3D-CRT, IMRT, and static tomotherapy with an SIB for large and small breast volumes. Med Dosim. 39, 163–168 (2014).
11. Pignol J, Olivotto I, Rakovitch E, Gardner S, Sixel K, Beckham W, et al. A multicenter randomized trial of breast intensity-modulated radiation therapy to reduce acute radiation dermatitis. J Clin Oncol. 108, 2085–2092 (2008).
12. Mulliez T, Veldeman L, van Greveling A, Speleers B, Sadeghi S, Berwouts D, et al. Hypofractionated whole breast irradiation for patients with large breasts: A randomized trial comparing prone and supine positions. Radiother Oncol. 108, 203–208 (2013).
13. Probst H, Bragg C, Dodwell D, Green D, Hart J. A systematic review of methods to immobilise breast tissue during adjuvant breast irradiation. Radiography 20, 70–81 (2014).
14. Varga Z, Hideghety K, Mezo T, Nikolenyi A, Thurzo L, Kahan Z. Individual positioning: a comparative study of adjuvant breast radiotherapy in the prone versus supine position. Int J Radiat Oncol Biol Phys. 75, 94–100 (2009).

Radiotherapy Quality Assurance Using Statistical Process Control

Diana Binny, Craig M. Lancaster, Tanya Kairn, Jamie V. Trapp, and Scott B. Crowe

Abstract

Statistical process control (SPC) is an analytical decision-making tool that employs statistics to measure and monitor a system process. The fundamental concept of SPC is to compare current statistics in a process with its previous corresponding statistic for a given period. Using SPC, a control chart is obtained to identify random and systematic variations based on the mean of the process and trends are observed to see how data can vary in each evaluated period. An upper and lower control limit in an SPC derived control chart indicate the range of the process calculated based on the standard deviations from the mean, thereby points that are outside these limits indicate the process to be out of control. Metrics such as: process capability and acceptability ratios were employed to assess whether an applied tolerance is applicable to the existing process. SPC has been applied in this study to assess and recommend quality assurance tolerances in the radiotherapy practice for helical tomotherapy. Various machine parameters such as beam output, energy, couch travel as well as treatment planning parameters such as minimum percentage of open multileaf collimators (MLC) during treatment, planned pitch (couch travel per gantry rotation) and modulation factor (beam intensity) were verified against their delivery quality assurance tolerances to produce SPC based tolerances. Results obtained were an indication of the current processes and mechanical capabilities in the department rather than a vendor recommended or a prescriptive approach based on machine technicalities. In this study, we have provided a simple yet effective method and analysis results to recommend tolerances for a radiotherapy practice. This can help improve treatment efficiency and reduce inaccuracies in dose delivery using an assessment tool that can identify systematic and random variations in a process and hence avoid potential hazardous outcomes.

Keywords

Radiation therapy • Statistical process control
Quality assurance • Tomotherapy

1 Introduction

Recent advances in radiotherapy have sparked the need to reform quality check processes and tighten specifications set on treatment delivery systems [1]. Quality assurance (QA) of radiotherapy systems is a process to identify Type A and Type B errors against baseline or tolerance levels and is generally performed according to published guidelines and recommendations to ensure that the process quality conforms with existing standards [2–4]. The increased complexity of the treatment planning and delivery process requires thorough evaluation of QA procedures and subsequent dosimetric measurements to make informed decisions in the practice of radiation oncology [5, 6]. However, it is also necessary to determine the treatment system capabilities prior to imposing tolerances recommended by local or national standards [2, 7, 8]. With evolving treatment techniques, QA programs that ensure treatment dose is within a clinical tolerance are no longer sufficient and the process of quality assurance should include identifying and improving underlying uncertainties in the process to minimise variations [9].

Computational methods [10–12] have been applied in radiotherapy to verify if systems operate within their specifications. Statistical process control (SPC) [13] is one such tool that converts data into information to document, correct and improve system performance by testing if the mean and

D. Binny (✉)
Radiation Oncology Centres, Redlands, Australia
e-mail: diana.binny@roc.team

D. Binny · T. Kairn · J. V. Trapp · S. B. Crowe
Queensland University of Technology, Brisbane, Australia

C. M. Lancaster · T. Kairn · S. B. Crowe
Cancer Care Services, Royal Brisbane and Women's Hospital, Brisbane, Australia

© Springer Nature Singapore Pte Ltd. 2019
L. Lhotska et al. (eds.), *World Congress on Medical Physics and Biomedical Engineering 2018*,
IFMBE Proceedings 68/3, https://doi.org/10.1007/978-981-10-9023-3_78

the dispersion of the measured data is stable over the period of analysis.

TomoTherapy Hi-Art II system (Accuray, Inc., Sunnyvale, CA) is a hybrid between a 6 MV linear accelerator and a helical megavoltage CT (MVCT) scanner capable of helically delivering intensity modulated radiation therapy (IMRT) combined with the advancing translational motion of a treatment couch [14]. The ratio of maximum to average of non-zero leaf open times is restricted to a particular value (also known as the modulation factor) between unity and five to enable optimised treatment delivery [14, 15].

Planning studies have demonstrated dosimetric advantages of using helical tomotherapy for sites such as breast, prostate, brain and head and neck over non-rotational treatment techniques [16–19]. However, delivery aspects of the treatment plan rely heavily on the user's TPS input parameters, such as modulation factor (MF), field width (FW) and pitch [14, 20]. These parameters are modified in the TPS to produce an optimal plan.

Despite the numerous amount of tomotherapy planning studies [14, 21–25] to recommend optimal parameters, no consensus was observed in recommendations to derive optimal plan parameters based on SPC methods. Additionally, no published data were found to recommend optimal dosimetric and mechanical tolerances in regard to machine output and tomotherapy couch position accuracy respectively. Knowledge of such a relationship would benefit quality assurance to improve treatment deliverability and detect possible flaws in the machine behaviour before an unforeseeable event.

In this work, we demonstrate SPC utilisation for (i) dosimetric, (iii) mechanical and (iii) patient-specific QA for tomotherapy.

2 Method and Materials

During an SPC analysis, a control chart is obtained that shows how a process varies over time [1, 26]. A bold center line (CL) in this control chart corresponds to the mean of the process which is also the reference for data point dispersion. The upper control limit (UCL) and lower control limit (LCL) indicate the range of the process. When the data fall within the UCL and LCL, the process is said to be within control (with only random or Type A causes affecting the process) and out of control (due to Type B or non-random causes) when the points fall outside the range. The control limits are calculated from Eqs. (1–3) [27].

$$UCL = \overline{X} + 3\frac{\overline{mR}}{d_2\sqrt{n}} \tag{1}$$

$$CL = \overline{X} \tag{2}$$

$$LCL = \overline{X} - 3\frac{\overline{mR}}{d_2\sqrt{n}} \tag{3}$$

where R is the range of the group, d_2 is a constant and depends on the continuous set of n measurements. For all cases considered in this study, n is 1 and d_2 is 1.128 [28]. $\overline{mR}$ is the average of the moving range or the absolute values of the difference between two consecutive measurements ($mR_i = |x_i - x_{i-1}|$) and $\overline{X}$ is the mean of the dataset. As a pre-requisite to use control charts the data was tested for normal distribution. The Ander-Darling [29, 30] (AD) statistic was used to test the hypothesized distribution F(x) for normality according to the below Eq. (4):

$$A_n^2 = -n - \sum_{i=1}^{n} \frac{2i-1}{n}[\ln(FX_i)) + \ln(1 - F(X_{n+1-i}))] \tag{4}$$

where $(X_1 < \ldots < X_n)$ are the ordered sample data points and n is the number of data points in the data distribution. In the AD test, the decision to reject a null hypothesis (H_0) is based on comparing the p-value [31] for the hypothesis (h) test with the specified significance level of 5% such that a h value of 0 would indicate that the distribution is normal and 1 otherwise. The process capability, c_p is used to compare the variation process of the data with respect to the upper and lower specified limits relative to the dispersion of process data and is calculated from Eq. (5) [12].

$$c_p = \frac{USL - LSL}{6\sigma} \tag{5}$$

where, USL and LSL are upper and lower user specified limits and σ is the standard deviation of the data distribution. A c_p value of 1 would indicate that the process is within action limits and a $c_p > 1$ would mean that the process is well within specification limits. A c_p value less than 1 indicates the process is outside a permissible range for a given action limit. However, in some cases a high c_p process can still perform poorly [1, 12, 28], therefore process acceptability index c_{pk} is also used to assess if the process center is relative to the user specified limit and is calculated from Eq. (6) [12]

$$c_{pk} = \min\left(\frac{USL - \overline{X}}{3\sigma}, \frac{\overline{X} - LSL}{3\sigma}\right) \tag{6}$$

SPC was used to assess tomotherapy rotational and static beam output and energy. Rotational output difference for two tomotherapy machines T1 and T2 were plotted against time for a period of three years. Out of process control points were investigated and post correction control charts were obtained again to verify process stability.

TomoTherapy translational couch movement was also tested using SPC and variations (offsets) in its IEC X, Y and Z directions were assessed using SPC over a four-year period. Baseline comparisons for IEC offset measurements using the Step-wedge Helical module were set at user specified action levels of ±2 mm and process indices for action levels of ±1 mm. Since there is no current protocol to adhere to for these limits, process indices c_p and c_{pk} were employed to quantify the process behaviour.

SPC was also used on set of 28 head and neck, 19 pelvic and 23 brain pre-treatment plans verified using 3D diode array ArcCHECK (Sun Nuclear Corporation (SNC), Melbourne, FL) and an Exradin A1SL ionisation chamber (Standard Imaging, Middleton, WI) placed at the center of the ArcCHECK in a polymethyl methacrylate (PMMA) cylinder for relative fluence and absolute dose measurements respectively. Gamma [32] and point dose variations (planned versus measured dose) were compared against parameters such as %LOT, sinogram segments, gantry period, modulation factor actual, etc.

The normality, capability and acceptability values with their corresponding probability were calculated from the measurement data using the MATLAB 2016a program (The MathWorks, Natick, NA, USA).

3 Results and Discussion

3.1 Dosimetric QA

Figure 1 shows tomotherapy output variations for a six-week period in which a magnetron was changed. A Type B or systematic uncertainty was observed and was

subsequently investigated. Analysis and machine log book recordings found this to be a malfunctioning magnetron which was replaced following which the system appeared to be back in control.

3.2 Mechanical QA

Figure 2 shows T2 IECZ axis offset measurements were retrospectively assessed using SPC for the first 180 observations. Machine logbooks indicated that the z-axis encoder was calibrated when the system reported an out of tolerance measurement at the end of first 90 observations. Calculated c_p and c_{pk} indices showed that ±2 mm was the most appropriate tolerance for the given dataset.

SPC analysis performed on the dataset showed that systematic variations (as indicated by the data points greater than the calculated LCL of −1.6 mm in Fig. 2a) were present prior to the out of tolerance behaviour (>2 mm) and could have been repaired prior to couch failure. This further demonstrates the need for a statistical process control QA.

Table 1 shows calculated LCL and UCL limits derived using SPC for the existing couches for T1 and T2 which are lower than the pre-set tolerance of 2 mm on the system.

3.3 Patient-Specific QA

Based on SPC analysis on planning parameters, it was concluded that different treatment sites have varied MLC distribution patterns and hence different parameters to aid in achieving optimal dose distribution. Recommendations were provided based on machine performance (assessed using gamma delivery and point dose variations from plan) for

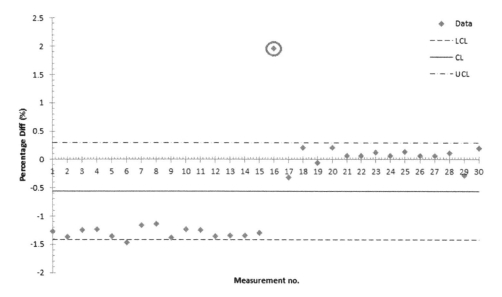

Fig. 1 T2:SPC analysis for static output measurement variation during a six-week period before and after magnetron replacement. Red circle indicates a higher output during the period which the magnetron was replaced. (Adapted from Binny et al. Investigating output and energy variations and their relationship to delivery QA results using Statistical Process Control. Physica Medica 38 (2017): 105–110.)

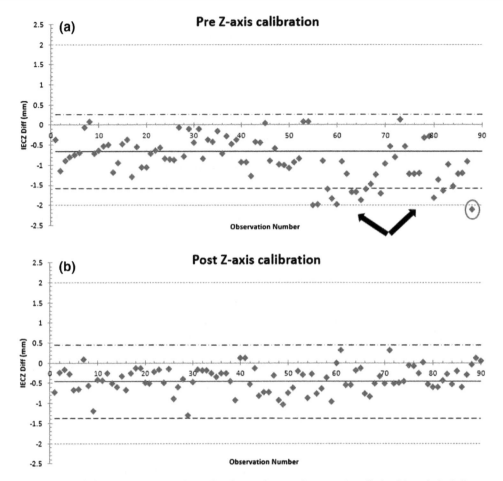

Fig. 2 Retrospective (**a**) pre-and (**b**) post-z-axis encoder calibration measurements assessed using SPC for unit T2 for the first 180 observations. Black arrows indicate out of control points below the user-specified limit of ±2 mm. Red circle indicates out of control point above ±2 mm action limit. Blue dashed lines represent the user specified limit of ±2 mm. (Adapted from Binny et al. Statistical process control and verifying positional accuracy of a cobra motion couch. Journal of applied clinical medical physics 18.5 (2017): 70–79.)

Table 1 Control chart based parameters for a three-monthly couch analysis period for units T1 and T2. (Adapted from Binny et al. Statistical process control and verifying positional accuracy of a cobra motion couch. Journal of applied clinical medical physics 18.5 (2017): 70–79.)

SPC parameters	IEC offsets					
	T1			T2		
	X	Y	Z	X	Y	Z
UCL (mm)	0.65	0.342	0.566	0.818	0.539	0.236
LCL (mm)	−0.775	−0.421	−1.298	−0.725	−0.192	−1.996
CL (mm)	−0.062	−0.039	−0.366	0.046	0.174	−0.88
σ	0.3436	0.165	0.377	0.295	0.171	0.542
AD*	Not normal	Normal	Not normal	Normal	Normal	Normal
No. of observations	90					

each patient specific site analysed in this study. An example of the difference in a plan parameter (Modulation factor (actual)) specific to treatment site is shown in Figs. 3 and 4.

Therefore, a single recommendation for gamma % pass rate or % point dose tolerance in addition to various plan parameters may not be sufficient.

Fig. 3 Brain SPC analysis: modulation factor (actual)

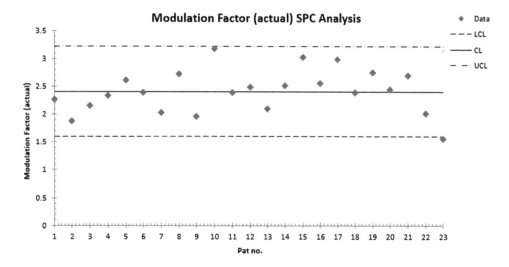

Fig. 4 Head and neck analysis: modulation factor (actual)

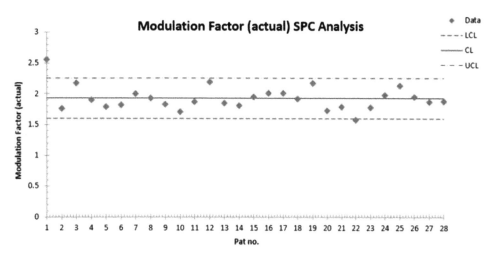

4 Conclusion

This study investigated the usefulness of SPC methods for dosimetric, mechanical and patient-specific QA parameters for tomotherapy. Our results have highlighted that treatment and machine specific tolerances should be used in addition to conforming to various clinical standards. SPC analysis can help in identifying special cause of variations thereby optimising tomotherapy system maintenance and aiding in improved treatment delivery outcomes.

Conflict of Interest This research did not receive any grant from funding agencies in the public, commercial, or not-for-profit sectors. The authors have no conflict of interest to declare.

References

1. Breen SL, Moseley DJ, Zhang B, Sharpe MB. Statistical process control for IMRT dosimetric verification. Med Phys. 2008;35:4417–4425.

2. Mezzenga E, D'Errico V, Sarnelli A, Strigari L, Menghi E, Marcocci F, et al. Preliminary Retrospective Analysis of Daily Tomotherapy Output Constancy Checks Using Statistical Process Control. PLoS One. 2016;11:e0147936.

3. Kutcher GJ, Coia L, Gillin M, Hanson WF, Leibel S, Morton RJ, et al. Comprehensive QA for radiation oncology: report of AAPM radiation therapy committee task group 40. Med Phys. 1994;21:581–618.

4. Millar M, Cramb J, Das R. ACPSEM Supplement. ACPSEM Position Paper: Recommendations for the Safe Use of External Beams and Sealed Brachytherapy Sources in Radiation Oncology. 1997.

5. Baskar R, Lee KA, Yeo R, Yeoh K-W. Cancer and radiation therapy: current advances and future directions. Int J Med Sci. 2012;9:193–199.

6. Urruticoechea A, Alemany R, Balart J, Villanueva A, Vinals F, Capella G. Recent advances in cancer therapy: an overview. Curr Pharm Des. 2010;16:3–10.

7. Binny D, Lancaster CM, Kairn T, Trapp JV, Crowe SB. Monitoring Daily QA 3 constancy for routine quality assurance on linear accelerators. Phys Medica. 2016;32:1479–1487.

8. Binny D, Lancaster CM, Trapp JV, Crowe SB. Statistical process control and verifying positional accuracy of a cobra motion couch using step wedge quality assurance tool. Journal of Applied Clinical Medical Physics. 2017;18:70–79.

9. Pawlicki T, Chera B, Ning T, Marks LB. The systematic application of quality measures and process control in clinical radiation oncology. Semin Radiat Oncol. 2012;22:70–76.

10. Binny D, Lancaster CM, Kairn T, Trapp JV, Back P, Cheuk R, et al. Investigating the use of image thresholding in brachytherapy catheter reconstruction. Australas Phys Eng Sci Med. 2016:1–7.

11. Langen KM, Papanikolaou N, Balog J, Crilly R, Followill D, Goddu SM, et al. QA for helical tomotherapy: Report of the AAPM Task Group 148a. Med Phys. 2010;37:4817–4853.

12. Sanghangthum T, Suriyapee S, Srisatit S, Pawlicki T. Retrospective analysis of linear accelerator output constancy checks using process control techniques. J Appl Clin Med Phys. 2013;14.

13. Ziegel ER. Understanding statistical process control. Technometrics. 1993;35:101–102.

14. Binny D, Lancaster CM, Harris S, Sylvander SR. Effects of changing modulation and pitch parameters on tomotherapy delivery quality assurance plans. J Appl Clin Med Phys. 2015;16:85–105.

15. Ryczkowski A, Piotrowski T. Influence of the modulation factor on the treatment plan quality and execution time in Tomotherapy in head and neck cancer: In-phantom study. J Cancer Res Ther 2013;9:618.

16. Lee T-F, Fang F-M, Chao P-J, Su T-J, Wang LK, Leung SW. Dosimetric comparisons of helical tomotherapy and step-and-shoot intensity-modulated radiotherapy in nasopharyngeal carcinoma. Radiother Oncol. 2008;89:89–96.

17. Mavroidis P, Ferreira BC, Shi C, Lind BK, Papanikolaou N. Treatment plan comparison between helical tomotherapy and MLC-based IMRT using radiobiological measures. Phys Med Biol. 2007;52:3817.

18. Schubert LK, Gondi V, Sengbusch E, Westerly DC, Soisson ET, Paliwal BR, et al. Dosimetric comparison of left-sided whole breast irradiation with 3DCRT, forward-planned IMRT, inverse-planned IMRT, helical tomotherapy, and topotherapy. Radiother Oncol. 2011;100:241–246.

19. Sheng K, Molloy JA, Larner JM, Read PW. A dosimetric comparison of non-coplanar IMRT versus Helical Tomotherapy for nasal cavity and paranasal sinus cancer. Radiother Oncol. 2007;82:174–178.

20. Yawichai K, Chitapanarux I, Wanwilairat S. Helical tomotherapy optimized planning parameters for nasopharyngeal cancer. Journal of Physics: Conference Series: IOP Publishing; 2016. p. 012002.

21. Bresciani S, Miranti A, Di Dia A, Maggio A, Bracco C, Poli M, et al. A pre-treatment quality assurance survey on 384 patients treated with helical intensity-modulated radiotherapy. Radiother Oncol. 2016;118:574–576.

22. Kapatoes J, Olivera G, Reckwerdt P, Fitchard E, Schloesser E, Mackie T. Delivery verification in sequential and helical tomotherapy. Physics in Medicine and Biology. 1999;44:1815.

23. Webb S. Intensity-modulated radiation therapy: dynamic MLC (DMLC) therapy, multisegment therapy and tomotherapy. An example of QA in DMLC therapy. Strahlentherapie und Onkologie: Organ der Deutschen Rontgengesellschaft[et al]. 1998;174:8–12.

24. Westerly DC, Soisson E, Chen Q, Woch K, Schubert L, Olivera G, et al. Treatment planning to improve delivery accuracy and patient throughput in helical tomotherapy. International Journal of Radiation Oncology Biology Physics. 2009;74:1290–1297.

25. Hooper H, Fallone B. Sinogram merging to compensate for truncation of projection data in tomotherapy imaging. Med Phys. 2002;29:2548–2551.

26. Noyez L. Control charts, Cusum techniques and funnel plots. A review of methods for monitoring performance in healthcare. Interact Cardiovasc Thorac Surg. 2009;9:494–499.

27. Buttrey SE. An excel add-in for statistical process control charts. J Stat Softw. 2009;30:1–12.

28. Montgomery DC. Statistical quality control: Wiley New York; 2009.

29. Nelson LS. The Anderson-Darling test for normality. J Qual Technol. 1998;30:298.

30. Pettitt A. Testing the normality of several independent samples using the Anderson-Darling statistic. J Appl Stat. 1977;26:156–161.

31. Bland JM, Altman DG. Multiple significance tests: the Bonferroni method. BMJ. 1995;310:170.

32. Low DA, Harms WB, Mutic S, Purdy JA. A technique for the quantitative evaluation of dose distributions. Med Phys. 1998;25:656–661.

Stereotactic Radiosurgery for Multiple Brain Metastases: A Dose-Volume Study

Tanya Kairn, Somayeh Zolfaghari, Daniel Papworth, Mark West, David Schlect, and Scott Crowe

Abstract

A substantial number of cancer patients develop brain metastases, which often present as multiple lesions. Stereotactic radiosurgery (SRS) can be used to treat brain metastases, with some incidental dose to the healthy brain. This study evaluated the effect of the number and combined volume of metastatic lesions on the dosimetric quality and the deliverability of a small sample of SRS test treatments. Five simulated static conformal arc treatments of 4–12 brain metastases were planned for linac-based multi-isocentre delivery to a head phantom. Film measurements were used to verify dose calculation and treatment delivery accuracy. Several of the treatment plans were considered clinically acceptable when local dose prescriptions (14–18 Gy) were used, but when the prescription dose to all metastases was increased to match the RTOG 0320 recommended value of 24 Gy, no plans resulted in a V12 less than 10 cm^3. Agreement between planned and measured dose was poorest for the treatments of 10 and 12 metastases, due to increased disagreement in out-of-field regions. Using the multi-isocentre static conformal arc method, it is possible to deliver treatments to relatively large numbers (at least 12) and total volumes (at least 8 cm^3) of brain metastases without excessive radiation doses being delivered to the healthy brain, provided that reduced prescription doses are acceptable.

Keywords

Radiation therapy · Dosimetry · Stereotactic

T. Kairn (✉) · S. Zolfaghari · S. Crowe
Queensland University of Technology, Brisbane, QLD, Australia
e-mail: t.kairn@gmail.com

T. Kairn · D. Papworth · M. West · D. Schlect
Genesis Cancer Care Queensland, Genesis Care, Brisbane, QLD, Australia

T. Kairn · S. Crowe
Royal Brisbane and Women's Hospital, Brisbane, QLD, Australia

1 Introduction

A substantial number of cancer patients develop brain metastases, 50–60% of which present as multiple lesions. Stereotactic radiosurgery (SRS) can be used to treat brain metastases with tightly conformal radiation beams, with the aim of sparng the healthy brain tissue and potentially avoiding many of the neurocognitive effects of surgery or whole brain radiotherapy [1].

Generally, the more metastases a patient has, the higher the dose delivered to the healthy brain by the radiosurgery treatment. For patients with very large numbers of metastases, whole brain radiotherapy may be the only acceptable option. However, the threshold at which linac-based stereotactic techniques become untenable remains debatable: treatments of 10–14 metastases within a single course of radiosurgery have been reported in the literature [2, 3], while many centres limit the number of radiosurgically treated metastases to 3 or 4 [1, 4].

The volume of healthy brain receiving a dose of 12 Gy is widely recognised as a useful predictor of radionecrosis and consequent neurocognitive decline [5, 6]. QUANTEC reported a significant risk of radiation injury from radiosurgery when the volume of brain tissue receiving 12 Gy exceeds 5–10 cm^3 [7] and Blonigen et al. recommend a specific limit of 7.9 cm^3 [5].

In this study, a small number of test treatment plans were used to investigate the effects of the number and combined volume of metastatic lesions on the dosimetric quality and the deliverability of linac-based multi-isocentre SRS treatments, with particular emphasis on the volume of healthy brain planned to receive 12 Gy.

2 Method

Five simulated cranial metastases cases were created by contouring planning target volumes (PTVs) on a CT scan of a head phantom (model 605, CIRS Inc, Norfolk, USA). The

© Springer Nature Singapore Pte Ltd. 2019
L. Lhotska et al. (eds.), *World Congress on Medical Physics and Biomedical Engineering 2018*,
IFMBE Proceedings 68/3, https://doi.org/10.1007/978-981-10-9023-3_79

Table 1 Diameters (numbers, in cm) and volumes (numbers in parentheses, in cm^3) of PTVs used to create test treatment plans

Structure	12 met plan	10 met plan	8 met plan	6 met plan	4 met plan
PTV1	2.0 (4.19)	2.0 (4.19)	2.0 (4.19)	2.0 (4.19)	2.0 (4.19)
PTV2	1.0 (0.52)	1.0 (0.52)	1.0 (0.52)	1.0 (0.52)	0.5 (0.07)
PTV3	1.0 (0.52)	1.0 (0.52)	1.0 (0.52)	0.5 (0.07)	0.5 (0.07)
PTV4	1.0 (0.52)	1.0 (0.52)	1.0 (0.52)	0.5 (0.07)	0.5 (0.07)
PTV5	1.0 (0.52)	1.0 (0.52)	0.5 (0.07)	0.5 (0.07)	
PTV6	1.0 (0.52)	1.0 (0.52)	0.5 (0.07)	0.5 (0.07)	
PTV7	1.0 (0.52)	0.5 (0.07)	0.5 (0.07)		
PTV8	1.0 (0.52)	0.5 (0.07)	0.5 (0.07)		
PTV9	0.5 (0.07)	0.5 (0.07)			
PTV10	0.5 (0.07)	0.5 (0.07)			
PTV11	0.5 (0.07)				
PTV12	0.5 (0.07)				
Total	(8.12)	(7.07)	(6.02)	(4.97)	(4.39)

number of metastases per case ranged from 4 to 12, with minimum and maximum diameters of 0.5 and 2.0 cm and total volumes ranging from 4.4 to 8.1 cm^3 (see Table 1). All of these volumes were subtracted from the brain contour (plus a 1 mm dose fall-off margin) to produce a "healthy brain" contour.

Five corresponding SRS treatments were planned for delivery as non-coplanar static conformal arcs, with one isocentre for each PTV, using a Varian iX linear accelerator (Varian Medical Systems, Palo Alto, USA) with a Brainlab m3 micro-multileaf collimator (Brainlab AG, Feldkirchen, Germany). The Brainlab iPlan RT Dose (v. 5.4) treatment planning system (TPS) was used, with reference to published small field beam data [8].

PTV doses of 14–18 Gy were prescribed to a 90% covering isodose, according to local practise. After the volumes of healthy brain receiving 12 Gy from these treatment plans were found to be within acceptable limits (less than 7.9 cm^3 [5]) for four out of the five treatment plans, the prescriptions for all PTVs in all treatment plans were increased 24 Gy. The 24 Gy prescription was taken from the protocol for RTOG 0320, which was a clinical trial designed for cases with 1–3 brain metastases [4].

Treatment delivery was verified using Gafchromic EBT3 film (Ashland Inc., Covington, USA) placed in a transverse plane through the head phantom, which was immobilised, imaged and treated as though it was a cranial SRS patient. Film analysis and comparisons with planned dose planes were completed using established techniques [9, 10].

3 Results and Discussion

Dose profile comparisons shown in Fig. 1 indicate that the dose calculation provided by the TPS accurately represented the delivered dose when the number of metastases was less than or equal to eight. When the number of metastases was increased to ten or twelve, disagreement between the film measurement and the TPS dose calculation increased, with the TPS generally predicting higher out-of-field doses than the measurements showed. These results suggest that the TPS dose calculation can be used as a worst-case-scenario estimate of the healthy brain dose.

The treatment plan information listed in Table 2 shows the increasing numbers of beams and monitor units that were needed to treat the increasing numbers of metastases. These results highlight the challenges encountered when planning multi-isocentre linac-based SRS treatments for increasing numbers of metastases. More treatment targets means more challenging geometric and dosimetric constraints, as beams need to enter and exit each PTV without intersecting with the beams used to treat the other PTVs.

Dose-volume data listed in Table 2 indicate that reductions in prescription dose that are needed (compared to a prescription suited to the treatment of just 1–3 metastases [4]) in order to limit the volume of healthy brain treated to 12 Gy, as the number of metastases increases. Table 2 shows that several of the test plans examined in this study resulted in clinically acceptable 12 Gy coverage of the healthy brain (less than 7.9 cm^3 [5]), when local dose prescriptions (14–18 Gy) were used, but when the prescription dose to all metastases was increased to match the RTOG 0320 recommended value of 24 Gy, no plans resulted in a V12 less than 10 cm^3.

As the number of metastases per case increased, so did the amount of time required to plan and deliver the treatments. The four metastases treatment plan was completed in less than half a day, whereas devising the complex beam arrangements for the ten and twelve metastases cases required one to two days of treatment planning time (no inverse planning was used). Delivering the treatments to the

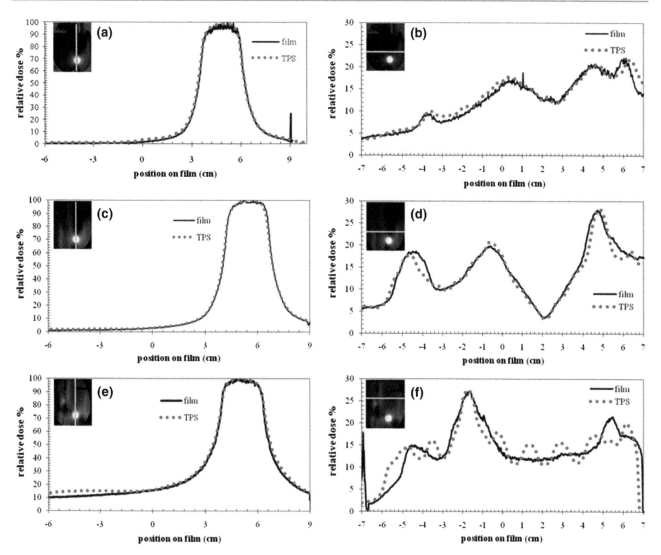

Fig. 1 Dose profiles from TPS (dotted lines) and film measurements (solid lines), from **a** and **b** the treatment plan for four metastases, **c** and **d** the treatment plan for eight metastases, and **e** and **f** the treatment plan for twelve metastases. Insets show film dose planes with lines indicating profile locations

Table 2 Treatment plan parameters: Number of PTVs per treatment (N_{PTV}), total volume of PTVs per treatment (V_{PTV}), number of arc beams per treatment ("Arcs"), local and RTOG prescription doses (P, in Gy), total numbers of MU per treatment using local and RTOG prescriptions, volumes of healthy brain irradiated to 12 Gy when using local prescriptions and when using the RTOG prescription ($V(12)$, in cm^3)

N_{PTV}	V_{PTV}	Arcs	P_{local}	MU_{local}	$V(12)_{local}$	P_{RTOG}	MU_{RTOG}	$V(12)_{RTOG}$
4	4.39	21	16–18	11,343	5.26	24	15,538	12.45
6	4.97	34	16–18	16,517	6.38	24	22,823	16.03
8	6.02	44	16	19,992	7.26	24	30,004	21.85
10	7.07	58	16	25,217	9.58	24	37,686	28.60
12	8.12	69	14	26,361	5.22	24	45,295	35.44

phantom (including phantom setup and repeated imaging) increased from three to eight hours of linac time, as the number of metastases increased from four to twelve. This treatment time would need to be divided over several days, to minimise physical discomfort and short-term side effects, if real human patients were involved.

4 Conclusion

Using a multi-isocentre static conformal arc method, it is possible to deliver treatments to relatively large numbers and total volumes of brain metastases without excessive radiation doses being delivered to the healthy brain, provided it is possible to compromise the prescription dose. If prescription doses above 18 Gy are required for such cases, the decision to use SRS and the particular SRS method selected for use may both need to be reconsidered.

Compliance with Ethical Standards The authors declare that they have no conflict of interest. This article does not contain any studies with human participants or animals performed by any of the authors.

References

1. Tsao, M.N., Rades, D., Wirth, A., et al.: Radiotherapeutic and surgical management for newly diagnosed brain metastases: An American Society of Radiation Oncology evidence-based guidelines. Pract. Radiat. Oncol. 2, 210–225 (2012). https://doi.org/10.1016/j.prro.2011.12.004.
2. Hunter, G.K., Suh, J.H., Reuther, A.M., et al.: Treatment of five or more brain metastases with stereotactic radiosurgery. Int. J. Radiat. Oncol. Biol. Phys. 83(5), 1394–1398 (2012). https://doi.org/10.1016/j.ijrobp.2011.10.026.
3. Schultz, D., Modlin, L., Jayachandran, P., et al.: Repeat courses of stereotactic radiosurgery (SRS), Deferring whole-brain irradiation, for new brain metastases after initial SRS. Int. J. Radiat. Oncol. Biol. Phys. 92(5), 993–999 (2015). https://doi.org/10.1016/j.ijrobp.2015.04.036.
4. Sperduto, P.W., Wang, M., Robins, H.I., et al.: A phase 3 trial of whole brain radiation therapy and stereotactic radiosurgery alone versus WBRT and SRS with temozolomide or erlotinib for non-small cell lung cancer and 1 to 3 brain metastases: Radiation Therapy Oncology Group 0320. Int. J. Radiat. Oncol. Biol. Phys. 85(5), 1312–1318 (2013). https://doi.org/10.1016/j.ijrobp.2012.11.042.
5. Blonigen, B.J., Steinmetz, R.D., Levin, L., et al.: Irradiated volume as a predictor of brain radionecrosis after linear accelerator stereotactic radiosurgery. Int. J. Radiat. Oncol. Biol. Phys. 77(4), 996–1001 (2010). https://doi.org/10.1016/j.ijrobp.2009.06.006.
6. Sahgal, A., Barani, I., Novotny, J., et al.: Prescription dose guideline based on physical criterion for multiple metastatic brain tumors treated with stereotactic radiosurgery. Int. J. Radiat. Oncol. Biol. Phys. 78(2), 605–608 (2010). https://doi.org/10.1016/j.ijrobp.2009.11.055.
7. Lawrence, Y.R., Li, X.A., el Naqa, I., et al.: Radiation Dose-Volume effects in the brain. Int. J. Radiat. Oncol. Biol. Phys. 76(3), S20–S27 (2010). https://doi.org/10.1016/j.ijrobp.2009.02.091.
8. Kairn, T., Charles, P.H., Cranmer-Sargison, G., et al.: Clinical use of diodes and micro-chambers to obtain accurate small field output factor measurements. Australas. Phys. Eng. Sci. Med. 38(2), 357–367 (2015). https://doi.org/10.1007/s13246-015-0334-9.
9. Kairn, T., Hardcastle, N., Kenny, J., et al.: EBT2 radiochromic film for quality assurance of complex IMRT treatments of the prostate: Micro-collimated IMRT, RapidArc, and TomoTherapy. Australas. Phys. Eng. Sci. Med. 34(3), 333–343 (2011). https://doi.org/10.1007/s13246-011-0087-z.
10. Kairn, T., West, M.: Six isocentres and a piece of film: Comprehensive end-to-end testing of a cranial stereotactic radiosurgery system. Australas. Phys. Eng. Sci. Med. 39(1), 341–342 (2016). https://doi.org/10.1007/s13246-015-0410-1.

Retrospective Audit of Patient Specific Quality Assurance Results Obtained Using Helical Diode Arrays

Liting Yu, Tanya Kairn, and Scott B. Crowe

Abstract

A retrospective audit was performed for existing patient specific quality assurance (PSQA) results measured on two ArcCheck helical diode arrays (Sun Nuclear Corporation). Twenty-five volumetric modulated arc therapy (VMAT) and thirty-two helical tomotherapy (HT) treatment plans were re-analysed using SNC patient software (version 6.2) and in-house gamma analysis code (developed in Python). Global gamma analyses were performed on the measured and calculated data (2%/2 mm) to identify the registration shift which provided the greatest gamma agreement index (GAI). Audit results indicated that when the ArcCheck devices were used for VMAT and HT, 1 mm longitudinal (Y) registration shifts frequently provided better GAI results than no shift. Specifically, the SNC and Python codes both identified a significant trend for longitudinal shifts for both ArcCheck devices. No significant trend was observed for roll (X) registration shifts. Measurements performed with physical shifts of the ArcCheck device improved the GAI results with no shift applied, suggesting that unless there is a co-incidental offset in the position of the radiation isocentre on both the TomoTherapy unit and the linac, there may be an actual displacement of the centre of the diode array and the marking lines on the two ArcCheck devices. This behaviour is dependent on gamma evaluation criteria used. The results of this study confirm the necessity of undertaking regular audits of QA results, as well as the need to consider sources of geometric uncertainty when selecting gamma evaluation criteria
and when applying automatic geometric shifts to measured data.

Keywords

ArcCheck • Shift • PSQA • VMAT • Tomotherapy

1 Introduction

Routine patient specific quality assurance (PSQA) testing of modulated radiotherapy treatment plans can produce large numbers of test results and reports. These results should be periodically reviewed and audited as part of continuous quality improvement in radiotherapy treatment delivery. Regular auditing of QA results is an integral component of a QA program. Published retrospective audits of PSQA results have described the relationship between QA pass rates, treatment parameters [1, 2] and evaluation criteria [3, 4]. The characterisation of QA trends allows refinement of QA processes.

ArcCheck (AC) helical diode arrays (Sun Nuclear Corporation) are used at our clinic for routine PSQA of modulated radiotherapy treatments, including volumetric modulated arc therapy (VMAT) and helical tomotherapy (HT). It was noticed that the auto-shift function in the SNC patient software (version 6.2.2; Sun Nuclear, Melbourne, FL) was sometimes used in order to improve the gamma agreement index (GAI), however no certain protocol is yet in place.

In the manufacturer specification, it was stated that the AC placement accuracy is within 0.5 mm. Several studies in the literature have investigated the sensitivity of AC to translational and rotational offsets [5–7]. Studies by Yang et al. [8] and Fan et al. [9] both investigated AC performance when introducing a 1 mm translational shifts in major axes, and concluded that AC is sensitive to this shift magnitude especially in longitudinal directions.

L. Yu · T. Kairn · S. B. Crowe (✉)
Royal Brisbane and Women's Hospital, Brisbane, QLD, Australia
e-mail: sb.crowe@gmail.com

L. Yu
e-mail: nancy.yu@health.qld.gov.au

L. Yu · T. Kairn · S. B. Crowe
Queensland University of Technology, Brisbane, QLD, Australia

© Springer Nature Singapore Pte Ltd. 2019
L. Lhotska et al. (eds.), *World Congress on Medical Physics and Biomedical Engineering 2018*,
IFMBE Proceedings 68/3, https://doi.org/10.1007/978-981-10-9023-3_80

It is believed that it could be beneficial to conduct a retrospective audit on a random sample of the recent PSQA results. The aim of this study is to re-evaluate the previous QA results, perform auto-shifts, and identify whether there is a trend in the proposed shift values. The findings of this study could be useful as a guide for improving clinical practice in the future. This study also exemplifies the important and actionable information that can be obtained from such auditing.

2 Methods

A retrospective audit was performed for 85 existing PSQA results measured on two AC devices. The AC, model 1220, is a cylindrical water-equivalent phantom with 1386 diode detectors arranged in a spiral pattern. The detector array has its diameter and length of 21 cm. Measurements were taken when a solid PMMA insert is placed in the central cavity. The AC was setup against the external lasers for VMAT PSQA and by image registration for HT PSQA. Dose distribution was calculated by Eclipse treatment planning system (TPS) (version 13.7) and TomoTherapy TPS (version 4.2.5) on a virtual AC dataset provided by the manufacturer using 2 and 2.5 mm resolution dose grid respectively.

Twenty-five VMAT treatment plans, including 53 individual arcs, and 32 HT treatment plans were re-analysed using the SNC patient software (version 6.2.2) and validated against the in-house gamma analysis code (developed in Python 3.5.2). All VMAT and HT plans were previously delivered on one of the three beam-matched Varian iX linear accelerators (Varian Medical Systems, Palo Alto, CA) and two Hi-Art TomoTherapy units (Accuray Incorporated, Sunnyvale, CA, USA). 2D global gamma analyses were performed on the AC measured dataset and the SNC extracted calculated dataset using gamma criteria of 2%/2 mm with a 5% lower dose threshold (LDT). Absolute dose comparisons were performed for VMAT and relative dose comparisons were performed for HT.

In the SNC patient software, roll (X) and longitudinal (Y) alignment shifts were calculated to identify the optimal alignment shift between the measured and calculated datasets, that is, the shift which provided the greatest gamma agreement index (GAI). The optimal alignment was identified using the 1 mm search interval provided in the SNC auto-shift function. In the in-house gamma code, search intervals of 1 and 0.1 mm were used to identify the optimal shift within a ±2 mm window. GAI was calculated between each shifted and the measured dataset until the optimal shift (producing the highest GAI, with minimum shift magnitude) was found. Where more than one shift produced the highest GAI value, the average of values with the minimum shift

magnitude was taken. Regression analyses were performed on the GAI and calculated Y shift values of all the beams.

As suggested by the analysis results of most VMAT plans from the SNC software, measurements were repeated with a 1 mm longitudinal shift applied to the AC device, achieved with a couch shift following conventional laser positioning, in order to identify whether the calculated shift values were actual physical shifts or merely software bugs. Same shift analyses were performed on the repeated measurements.

3 Results

Table 1 summarised the GAI values calculated using both the SNC and the in-house software. From the original measured datasets of 25 VMAT plans (53 beams) in this study, a −1 mm Y shift was suggested for 30 beams (56.6%) by the SNC software and 34 beams (64.2%) by the in-house gamma code.

Where a −1 mm optimal Y shift was identified, the mean GAI was improved by 1.3% (from $97.8 \pm 1.5\%$ to $99.1 \pm 0.9\%$) in SNC and by 1.2% (from $97.7 \pm 1.4\%$ to $98.9 \pm 0.8\%$) in the in-house gamma code. GAI values calculated before and after −1 mm Y shifts were significantly different ($p < 0.001$). The mean optimal Y shift for all data was -0.57 ± 0.50 mm using the SNC, -0.63 ± 0.48 mm using the in-house software with a 1 mm search interval. These shifts were not significantly non-zero.

Of the 32 HT plans, SNC software showed 21 plans (65.6%) having a −1 mm Y shift and the average Y shift was -0.84 ± 0.57 mm, whereas the gamma code calculated 24 plans (75%) having a −1 mm Y shift and the average Y shift was -0.91 ± 0.59 mm.

Figure 1 (left) presents the optimal X and Y shifts calculated using the in-house code with a 0.1 mm search interval. The mean optimal Y shift calculated with a 0.1 mm

Table 1 Mean and standard deviation (SD) of GAI, best GAI from SNC and calculated shifts from SNC and gamma code for 53 VMAT arcs and 32 HT plans

	VMAT		HT	
	Mean	SD	Mean	SD
GAI %	98.18	1.48	93.58	4.45
Best GAI %	99.02	0.94	95.76	3.54
SNC mean shift x	0.47	0.80	0.16	0.63
SNC mean shift y	−0.57	0.50	−0.84	0.57
Gamma code mean shift x (1 mm resolution)	0.38	0.86	0.31	0.93
Gamma code mean shift y (1 mm resolution)	−0.64	0.48	−0.91	0.59

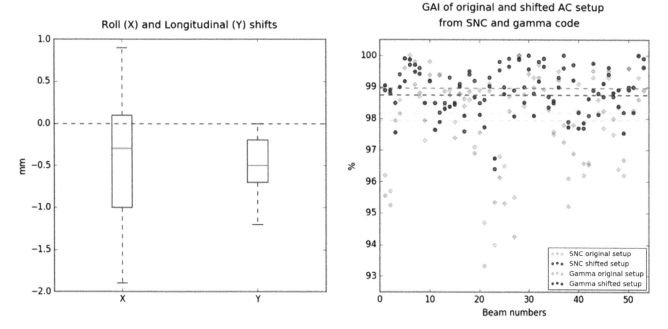

Fig. 1 Left: Optimal shift values for 25 VMAT plans (53 arcs) calculated using the in-house gamma code with a 0.1 mm resolution. Right: GAI individual and mean values (dashed lines) from 53 VMAT arcs when AC was at original and offset positions

search interval was significant $(p < 0.005)$ at -0.49 ± 0.33 mm. No significant X shift was observed.

From the repeated measurements where the AC has been offset by 1 mm in the Y direction, as suggested for most plans by both SNC and gamma code, analyses performed by SNC showed that the average Y shift increased by 0.65 mm (from -0.57 ± 0.50 to 0.08 ± 0.27 mm). The gamma code gave comparable results where the average Y shift increased by 0.78 mm (from -0.63 ± 0.48 to 0.15 ± 0.46 mm) for 1 mm search interval and by 0.59 mm (from -0.49 ± 0.33 to 0.10 ± 0.37) for 0.1 mm search interval). As shown by Fig. 1 (right), the overall GAI improved when the AC is offset by -1 mm in the longitudinal Y direction.

4 Discussion

According to our results from both SNC and the in-house gamma code, applying shifts may significantly increase GAI. This trend might not have been recognised without an audit as most plans in this study are previous clinical passed plans which have initial high GAI. In these cases usually physicists will not attempt to perform a shift to look for better GAI since they have already passed. It is worth mentioning that this trend could impact local practices especially for borderline QA results where a calculated shift may improve the GAI from below to just above the action level. In these cases care must be taken to decide whether the calculated shift should be accepted, taking into account potential setup differences and alignment tolerances. For example, if a calculated optimal shift is in the opposite direction of the observed trend, perhaps this would less likely to be accepted.

Templeton et al. [10] studied HT PSQA deliveries and followed the same setup procedure as our clinical practice. They reported that with the help of imaging guided setup, auto-shifting should not significantly improve GAI for clinical plans, though may result in substantial improvements for plans introduced with manufactured errors, where auto-shifting could mask positional errors. However they did not describe what improvements might have been produced with auto-shifting for those clinical plans. We found that applying a 1 mm shift in the Y direction, as suggested by the software, has made the mean GAI increase by approximately 1–2% for most of the already clinically passed plans, which given the action level (95%), is quite high. And this is true for both VMAT plans (where the AC was setup using the lasers) and HT plans (where the AC was setup using the MVCT image registration). It is worth mentioning that the arbitrariness of the selection of the PSQA plans in this study has excluded the possibility of any biased results on any specific linac. Our repeated measurements on VMAT have confirmed the calculated shifts are real, although the cause is still unknown.

The -1 mm Y shift was observed for 2 modalities, using 2 positioning techniques (lasers and MV imaging), and was corrected by physically shifting the AC, which implied a systematic behavior. Ahmed et al. [11] have reported a new AC model replacing the old model in 2014. They mentioned electronics, interfaces and material changes in the new model. No literature has been found comparing differences between the new and old models. Unless there is a co-incidental offset in the position of the radiation isocentre on both the TomoTherapy unit and the linac, it is possible that there may be a physical offset of approximately 0.5 mm between the centre of the diode array and the surface markers, for our two older-model AC devices, which happens to be the accuracy of the detector locations after attachment to the cylinder that stated in the manufacturer specification.

It was tested that the AC was sensitive to interpolation. By interpolating the resolution of SNC extracted TPS calculated dataset from 1 to 0.1 mm resolution, the mean GAI of VMAT plans increased by 2%. Our observations were consistent with Templeton et al. that auto-shifting should not be conducted blindly as part of the analysis process just to find a better GAI, especially in cases where the phantom has been setup by the aid of image registration, which on HT generally achieves the tolerance of 1 mm accuracy to the radiation isocentre.

Our findings are not solely dependent on SNC software analysis tool as it is like a black box. Manufacturers of commercial analysis software generally do not disclose detailed information how the gamma calculation is computed in their package [12]. The in-house gamma code acted as an important additional tool and independent check which provided us with confidence of the results acquired from SNC.

5 Conclusion

Audit results indicated that when the AC devices were used for VMAT and HT, 1 mm longitudinal (Y) registration shifts frequently provided better GAI results than no shift. This trend might not have otherwise been identified without an audit. The SNC and Python in-house gamma code both identified a significant trend for longitudinal shifts for both AC devices (mean values of −0.57 and −0.64 mm for VMAT, and −0.84 and −0.91 mm for HT). No significant trend was observed for roll (X) registration shifts. The results indicated that the actual shift is 0.5 mm in the SI Y direction.

Measurements performed with physical shift of the AC device improved the overall GAI results, suggesting the displacement to be systematic. The results of this study confirmed the necessity of undertaking regular audits of QA results, as well as the need to consider sources of geometric uncertainty when selecting gamma evaluation criteria and when applying automatic geometric shifts to measured data.

References

1. Childress NL, White RA, Bloch C, Salehpour M, Dong L, Rosen II (2005) Retrospective analysis of 2D patient-specific IMRT verifications. Medical physics 32 (4):838–850
2. Crowe S, Kairn T, Middlebrook N, Sutherland B, Hill B, Kenny J, Langton CM, Trapp J (2015) Examination of the properties of IMRT and VMAT beams and evaluation against pre-treatment quality assurance results. Physics in Medicine & Biology 60 (6):2587
3. Crowe SB, Sutherland B, Wilks R, Seshadri V, Sylvander S, Trapp JV, Kairn T (2016) Relationships between gamma criteria and action levels: Results of a multicenter audit of gamma agreement index results. Medical physics 43 (3):1501–1506
4. Steers JM, Fraass BA (2016) IMRT QA: Selecting gamma criteria based on error detection sensitivity. Medical physics 43 (4):1982–1994
5. Koren S, Ma C (2011) Gamma Analysis Criteria for Planar and Cylindrical Dose Arrays. International Journal of Radiation Oncology Biology Physics 81 (2):S834
6. Lin MH, Koren S, Veltchev I, Li J, Wang L, Price RA, Ma CM (2013) Measurement comparison and Monte Carlo analysis for volumetric-modulated arc therapy (VMAT) delivery verification using the ArcCHECK dosimetry system. Journal of applied clinical medical physics 14 (2):220–233
7. Vieillevigne L, Molinier J, Brun T, Ferrand R (2015) Gamma index comparison of three VMAT QA systems and evaluation of their sensitivity to delivery errors. Physica Medica 31 (7):720–725
8. Yang W, Wallace R, Huang K, Cook R, Fraass B (2012) SU-E-T-363: Sensitivity of ArcCheck to Delivery Errors in IMRT/VMAT Treatment. Medical Physics 39 (6):3787–3787
9. Fan Q, Park C, Lu B, Barraclough B, Lebron S, Li J, Liu C, Yan G (2015) SU-E-T-584: Optical Tracking Guided Patient-Specific VMAT QA with ArcCHECK. Medical physics 42 (6):3470–3470
10. Templeton AK, Chu JC, Turian JV (2015) The sensitivity of ArcCHECK - based gamma analysis to manufactured errors in helical tomotherapy radiation delivery. Journal of applied clinical medical physics 16 (1):32–39
11. Ahmed S, Nelms B, Kozelka J, Zhang G, Moros E, Feygelman V (2016) Validation of an improved helical diode array and dose reconstruction software using TG-244 datasets and stringent dose comparison criteria. Journal of applied clinical medical physics 17 (6):163–178
12. Hussein M, Clark C, Nisbet A (2017) Challenges in calculation of the gamma index in radiotherapy–towards good practice. Physica Medica: European Journal of Medical Physics 36:1–11

Optimising a Radiotherapy Optical Surface Monitoring System to Account for the Effects of Patient Skin Contour and Skin Colour

Candice Milewski, Samuel Peet, Steven Sylvander, Scott Crowe, and Tanya Kairn

Abstract

Optical surface monitoring systems (OSMSs) are designed to assist patient setup and patient motion management during radiotherapy treatments. Systems use projected and reflected patterns of coloured light on the patients surface and therefore depend on the skin's optical absorbance and reflectance properties, which can vary with surface shape and colour. This study aimed to identify optimal operating parameters for the Catalyst HD OSMS (C-rad, Uppsala, Sweden) when used to monitor the surfaces of 3D-printed objects with various convex and concave surfaces, one of which was painted in six different colours with various levels of red and black saturation (from light pink to dark grey). The degree of surface detection was assessed via the Catalyst HD interface, with different levels of gain (100–600%) and signal integration time (1–7 ms). The OSMS was able to detect horizontal and convex shapes more consistently than vertical or steeply angled surfaces. The OSMS was not able to detect the darkest surface at all, even with the highest gain and the longest integration times. Mid-grey surfaces were detectable only when the integration time was increased to 2 s. All pink surfaces were easily detectable at the shortest integration time, with the OSMS performing best when red saturation was highest. Further work is recommended, as the red undertone of all human skin may lead to improved results for real patients.

C. Milewski · S. Peet · S. Sylvander · S. Crowe (✉) · T. Kairn
Royal Brisbane and Women's Hospital, Brisbane, QLD, Australia
e-mail: sb.crowe@gmail.com

T. Kairn
e-mail: t.kairn@gmail.com

C. Milewski
Université Paris-Sud, Orsay, France

C. Milewski
Polytech Marseille, Marseille, France

S. Peet · S. Crowe · T. Kairn
Queensland University of Technology, Brisbane, QLD, Australia

However, these preliminary results indicate that careful commissioning and optimisation of OSMS systems may be required before they can be used in radiotherapy treatments for a broad patient cohort.

Keywords

Radiation therapy • Surface monitoring • Quality assurance

1 Introduction

Optical surface modelling systems (OSMSs) have been under development for radiotherapy applications for more than a decade [1], but have only begun to be widely adopted in recent years [2–8]. Today, OSMSs that identify and track the position of the patient's external surface are being investigated and used for treatment setup and motion monitoring for breast radiotherapy with and without breath holds [1, 6], stereotactic body radiotherapy [4] and cranial stereotactic radiosurgery [2, 3, 5] as well as for motion phantom development [8].

OSMSs operate by projecting patterns of coloured light onto the patient's surface and detecting the reflected signal optically. Early systems used only one camera [7] or two closely-spaced cameras [1], but contemporary systems generally use three cameras spaced around the treatment room [3, 5] to read the optical information and provide comprehensive stereo photogrammetry. Displacements or distortions of the resulting two-dimensional images are analysed, to provide a real-time indication of patient positioning accuracy [3, 5, 6].

Because the operation of these systems depends on the optical absorbance and reflectance properties of the patient's surface, the results and reliability of using an OSMS can be affected by visible features of the patient's skin, such as contour shape and skin colour. Mancosu et al. found that the geometric uncertainty of the Varian OSMS (Varian Medical

Fig. 1 Test palette used to assess the OSMS's response to different colours (Color figure online)

Systems, Palo Alto, USA) was increased by 0.5 mm when one of the three cameras used by the system was blocked [3]; steeply varying contours may have a similar affect, degrading the performance of the OSMS by limiting the surface information that is visible to all of the OSMS cameras. Similarly, the optical information detectable by the cameras may be affected by the colour of the patient's skin, as the optical reflectance of darker skin is dramatically reduced compared to lighter skin [9]. This effect is visible in a photograph published by Mancosu et al., which showed that the red speckle pattern produced by the Varian OSMS system was reflected cleanly by the light-coloured regions of the head phantom used in the study, as well as the white surface on which it was placed, but was largely absorbed by the dark-coloured region at the base of the phantom [3].

While these effects have not been specifically investigated using a three-camera (stereo-photogrammetric) OSMS, Stieler et al. provided a detailed examination of the effects of using test objects with varying geometries and colours with the Catalyst single-camera OSMS (C-Rad, Uppsala, Sweden) [7]. Stieler et al. found that the single-camera system was able to reliably detect lighter-coloured objects with horizontal surfaces more reliably than darker-coloured objects or surfaces oriented away from the camera [7].

This study used the Catalyst HD three-camera system (C-Rad, Uppsala, Sweden) to repeat and expand upon some of the phantom tests completed by Stieler et al. [7], to provide an assessment of the sensitivity with which a contemporary OSMS system is able to detect surfaces with various contours and colours, with the goal of identifying optimal operating parameters for monitoring surfaces with different optical properties.

2 Method

Catalyst HD is an OSMS designed to assist radiotherapy patient setup, monitor intra-fraction motion, and to hold the radiation beam in response to out-of-tolerance patient movements or during respiratory-gated radiotherapy treatments. Catalyst HD uses three ceiling-mounted projector/camera units that project/detect a pattern of blue light to/from the patients surface. Each unit employs three high-power LEDs to project a pattern of blue light (450 nm) which is reflected from the patient and detected at a rate of

202 frames/second using monochrome charge-coupled device (CDD) cameras. This optical information is used to reconstruct the surface of the patient and compare with a reference image.

Six differently-shaped test objects were created from white polylactic acid (PLA), using a 3D printing technique [10], for use in evaluating the systems sensitivity to skin contour. White was chosen as the PLA colour due to its high reflectance (being expected to give the best results in terms of spatial accuracy and visibility) and due to the frequent local use of white thermoplastics, tape, pillows and sheets, during patient setup. The chosen shapes were curved and rectilinear, concave and convex, covering the range of shapes used by Stieler et al. [7] plus two additional clinically-likely contours (a wedge and a half-cylinder).

After the white half-cylinder test object was found to be easily detectable by the OSMS at the shortest integration time, it was selected for use in the skin colour study and painted six different colours with various levels of red and black saturation (from light pink to dark grey) shown in Fig. 1. These particular colours were selected to provide some consistency and comparability with the work of Stieler et al. [7].

The ability of the OSMS to detect the test objects was evaluated using a visual indicator of exposure built into the system's software, which allowed a subjective assessment of the accuracy of the detection of the object to be completed simultaneously with an objective assessment of the suitability of the level of exposure detected by the OSMS (see Fig. 2). The degree of surface detection achieved by the OSMS was assessed and optimised by adjusting the imaging gain (100–600%) and signal integration time (1000–7000 ms). Optimal settings of these parameters must be determined by the user; gain and integration time cannot be set by the system automatically.

3 Results and Discussion

Results presented Fig. 3 show that the three-camera-based OSMS achieved full detection of all white test objects with the lowest integration time (1000 ms). As the integration time was increased the reliability of the system decreased, especially for the higher gain settings, due to over-exposure of the test objects. Over-exposure causes insensitivity to surface

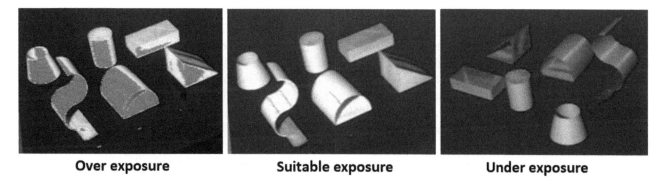

Fig. 2 Screen-shots of OSMS interface showing 3D printed test objects with different levels of detected exposure. Regions of optical over-exposure are shown as red and regions of under-exposure are shown as grey. Objects that are not detected at all (e.g. the black surface of the treatment couch) are shown as blue (Color figure online)

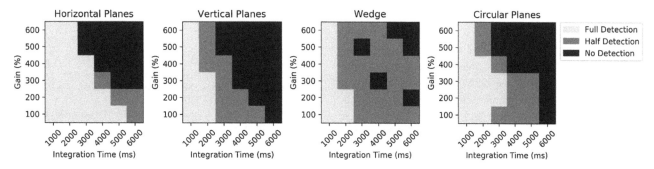

Fig. 3 Imaging parameter test results for six white test objects, stratified by surface type (Color figure online)

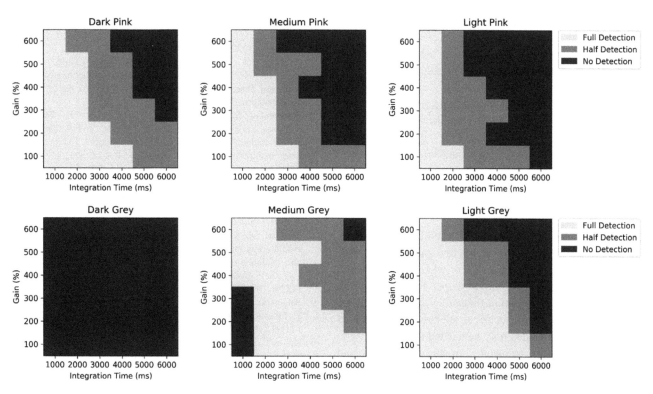

Fig. 4 Imaging parameter test results for coloured half-cylinder test object, stratified by surface colour (Color figure online)

contours (over-exposed regions appear flat) and therefore compromises the positioning accuracy of the system. Evidently, when using this system to image white objects such as sheets, thermoplastics and marker-tapes, the use of minimal integration times and low gain settings is advisable.

Generally, the OSMS was able to detect horizontal and convex shapes more consistently than vertical or steeply angled surfaces. However, comparing the result shown in Fig. 3 with the results reported by Stieler et al. [7] indicates that the use of Catalyst HD, with three cameras spread around the ceiling of the treatment room, instead of Catalyst, with one camera attached to the ceiling at the end of the treatment couch, leads to an obvious improvement in the system's ability to detect vertical or slanted surfaces that are not directly facing the end of the couch.

Figure 4 shows that the OSMS performed well when the half-cylinder was painted pink. For all three shades of pink the test object was easily detected with the shortest image integration time. When the test object was painted with the two lighter shades of pink, the OSMS performed similarly to when it was used to monitor the white test objects; increasing the integration time and the gain quickly led to over-exposure. When the half-cylinder test object was painted a darker shade of pink, with increased red saturation, it remained fully detectable by the OSMS even with moderate increases in gain and integration time.

Figure 4 shows that after the half-cylinder object was painted dark grey, the OSMS was not able to detect it at all, even with the highest gain and the longest integration times. A similar result was obtained by Stieler et al. [7] when they attempted to detect a horizontal, black surface with their one-camera system. Figure 4 also indicates that when the test object was painted medium-grey it was detectable only when either the integration time was increased to at least 2000 ms or the gain was increased to at least 400%, whereas when the object was painted light-grey it was detectable across a broad range of imaging parameters.

4   Conclusion

During clinical use of the Catalyst HD OSMS, it is advisable to select short integration times and low gain settings when monitoring patients with light skin or white immobilisation equipment, regardless of contour shape. For patients with darker skin, longer integration times may be needed. If the skin surface cannot be detected at all, then light coloured (high reflectance) markers may be needed.

Further work is recommended in this area, as the red undertone of all human skin may lead to improved results for real patients. However, these preliminary results indicate that careful commissioning and optimisation of OSMS systems may be required before they can be used in radiotherapy treatments for a broad cohort of patients.

There may be patients for whom the use of an OSMS is unsuitable, due to the geometry of the targeted anatomy or low skin reflectance at the optical imaging wavelength of the chosen system. For such patients, more-conventional image guidance techniques, based on matching of internal (not external) anatomy, would be preferable.

Compliance with Ethical Standards   The authors declare that they have no conflict of interest. This article does not contain any studies with human participants or animals performed by any of the authors.

References

1. Bert, C., Metheany, K.G., Doppke, K., Chen, G.T.: A phantom evaluation of a stereo-vision surface imaging system for radiotherapy patient setup. Med. Phys. 32(9), 2753–2762 (2005). https://doi.org/10.1118/1.1984263
2. Li, G., Ballangrud, A., Kuo, L.C., et al.: Motion monitoring for cranial frameless stereotactic radiosurgery using video-based three-dimensional optical surface imaging. Med. Phys. 38(7), 3981–3994 (2011). https://doi.org/10.1118/1.3596526
3. Mancosu, P., Fogliata, A., Stravato, A., et al.: Accuracy evaluation of the optical surface monitoring system on EDGE linear accelerator in a phantom study. Med. Dos. 41(2), 173–179 (2016). https://doi.org/10.1016/j.meddos.2015.12.003
4. Wen, N., Li, H., Song, K., et al.: Characteristics of a novel treatment system for linear acceleratorbased stereotactic radiosurgery. J. Appl. Clin. Med. Phys. 16(4), 125–148 (2015). https://doi.org/10.1120/jacmp.v16i4.5313
5. Cerviño, L.I., Detorie, N., Taylor, M., et al.: Initial clinical experience with a frameless and maskless stereotactic radiosurgery treatment. Pract. Radiat. Oncol. 2(1), 54–62 (2012). https://doi.org/10.1016/j.prro.2011.04.005
6. Schönecker, S., Walter, F., Freislederer, P., et al.: Treatment planning and evaluation of gated radiotherapy in left-sided breast cancer patients using the Catalyst TM/Sentinel TM system for deep inspiration breath-hold (DIBH). Radiat. Oncol. 11, 143 (2016). https://doi.org/10.1186/s13014–016-0716-5
7. Stieler, F., Wenz, F., Shi, M., Lohr, F.: A novel surface imaging system for patient positioning and surveillance during radiotherapy. Strahlenther. Onkol. 189(11), 938–944 (2013). https://doi.org/10.1007/s00066-013-0441-z
8. Lempart, M., Kügele, M., Snäll, J., et al.: Development of a novel radiotherapy motion phantom using a stepper motor driver circuit and evaluation using optical surface scanning. Australas. Phys. Eng. Sci. Med. 40(3), 717–727 (2017). https://doi.org/10.1007/s13246-017-0556-0
9. Anderson, R.R., Parrish, J.A.: The optics of human skin. J. Investig. Dermatol. 77(1), 13–19 (1981). https://doi.org/10.1111/1523-1747.ep12479191
10. Kairn, T., Crowe, S.B., Markwell, T.: Use of 3D printed materials as tissue-equivalent phantoms. IFMBE Proc. 51, 728–731 (2015). https://doi.org/10.1007/978-3-319-19387-8_179

Absolute Calibration of the Elekta Unity MR Linac Using the UK Code of Practice for High-Energy Photon Dosimetry

G. Budgell, P. Gohil, J. Agnew, J. Berresford, I. Billas, and S. Duane

Abstract

Absolute dosimetry for MR Linacs is complicated by non-standard reference conditions, the non-suitability of the UK secondary standard chamber for use in water and the effects of the magnetic field on the response of Farmer field chambers. Measurements were made on a pre-clinical 7 MV Elekta Unity MR Linac. Reference conditions were chosen as isocentre (143.5 cm SAD), 10 cm deep, 10×10 cm^2 field and gantry angle 90° to avoid output variation from liquid helium levels dropping in the surrounding annulus. TPR 20/10 was measured as 0.698 at isocentre. Measurements on a conventional linac demonstrated that TPR 20/10 does not vary between SAD 100 cm and 143.5 cm. The UK National Physical Laboratory (NPL) provides calibration factors, N_D, for the secondary standard NE 2611A thimble chamber in terms of absorbed dose to water. N_D was taken for the value of the measured TPR 20/10. Independent intercomparisons were made at 6 MV on a conventional linac between the NE2611A and two PTW waterproof Farmer 30013 chambers. The response of these chambers varies very slowly with energy hence it is reasonable to assume this introduces minimal uncertainty. Ion recombination and polarity correction factors measured on the MR Linac were unchanged by the magnetic field. An additional correction factor to account for the 1.5 T magnetic field on the Farmer chambers was measured as 0.986 taking the ratio of TP corrected readings before and after ramp-up of the magnetic field. An orientation parallel to the magnetic field (along the bore of the MR Linac) was chosen to minimise the magnitude of this factor. An independent audit was performed by the NPL using alanine pellets and Farmer chambers calibrated via a water calorimeter from the Netherlands primary standards laboratory which operates within the MR Linac. This gave agreement with our calibration to within 1.0%.

Keywords

MR linac • Absolute dosimetry • Reference dosimetry

1 Introduction

Absolute dosimetry for MR Linacs is complicated by a number of factors. These include:

- The use of non-standard reference conditions compared with international dosimetry protocols such as the UK Code of Practice normally used in our centre [1].
- Not being able to make direct inter-comparison measurements in the clinical beam between field Farmer chambers and secondary standard chambers. This is a particular issue in the UK since the only recognized secondary standard chamber is the NE 2611A thimble chamber which is not waterproof and air gaps around ion chambers in magnetic fields are known to cause errors in dose measurement.
- The effects of the magnetic field on the response of Farmer field chambers, which are dependent on the orientation of the chambers relative to the magnetic field and beam.
- The effects of the magnetic field on the photon beam, which need to be understood in order to correctly determine magnetic correction factors.
- The practical difficulties of setting up ion chambers in water phantoms to measure in an MR Linac bore without lasers or cross-wires or physical pointers to assist alignment.

The purpose of this work was to establish absolute dosimetry for the Elekta Unity 1.5 T MR Linac installed in

G. Budgell (✉) · P. Gohil · J. Agnew · J. Berresford
The Christie NHS Foundation Trust, Manchester, M20 4BX, UK
e-mail: geoff.budgell@christie.nhs.uk

I. Billas · S. Duane
National Physical Laboratory, Hampton Rd, Teddington,
Middlesex, TW11 0LW, UK

our centre via the UK Code of Practice for high-energy photon dosimetry.

2 Methods

A pre-clinical 7 MV Elekta Unity 1.5 T MR Linac has been installed in our clinic—a variety of measurements as specified below have been made on this machine to determine and verify the chain for absolute dosimetry based on the UK National Physical Laboratory's (NPL) absorbed dose calibration service. The field chambers chosen were two PTW waterproof Farmer 30013 chambers used with a voltage of +250 V, the secondary standard chamber used was the NE 2611A thimble chamber recommended in the UK Code of Practice.

2.1 Reference Conditions

The isocentre of the Unity MR Linac is at a larger distance than on a standard linac due to the linac wave-guide having to be being positioned outside the envelope containing the MR coils. Reference conditions were chosen as isocentre (143.5 cm SAD), 10 cm deep, 10×10 cm^2 field and gantry angle 90°. This gantry angle is chosen to avoid output variation from liquid helium levels dropping in the surrounding annulus through which the beam passes during delivery. This would first affect beams at gantry angle zero and could potentially create a dose change of around 0.7% compared with a full annulus.

The 10 cm depth is deeper than the recommended depth for this energy of 5 cm in the Code of Practice, which is also the depth at which Intercomparison measurements were made on a standard linac. However, the difference in depth is expected to have minimal effect on the Intercomparison ratio and 10 cm is the preferred standard depth for planning systems.

In addition, the Farmer chamber alignment has been chosen to align parallel with the magnetic field (along the bore of the magnet in the in-plane direction). This orientation has been shown to minimize the size of the magnetic field correction [2] and must be consistently used in order to avoid introducing errors into output measurements.

2.2 Equipment and Measurement Set-Up

A water phantom has been custom made to carry out output checks under the stated reference conditions (Fig. 1). A QA platform has also been custom made which attaches to the treatment couch using the positioning bar at a specified position. The phantom is positioned on a block of solid

water on the QA platform and driven to a predetermined couch position. This method provides a quick and consistent way of setting up the phantom with the ion chamber at isocentre. The MR Linac MV imager is then used to acquire portal images at gantry angles 0° and 90°—the x and y pixel values at the centre of the chamber are visually identified in each image by placing the cursor at the centre of the visible air volume and checked to be within 3 pixels (0.7 mm) of the known isocentre pixel values. This checks that the phantom and chamber set-up is correct. Output measurements can then be acquired.

2.3 Energy Specifier

The energy specifier used in the Code of Practice is the Quality Index or TPR 20/10, the ratio of corrected instrument readings with the chamber at 20 and 10 cm depths in a water phantom with a 10×10 cm field at the chamber and a constant source-chamber distance (SCD). The Quality Index can be directly measured in the MR Linac since the SCD is not specified and the Quality Index is assumed to be independent of SCD. However, normal clinical practice would be to measure TPR 20/10 at 100 cm SCD—the MR Linac SCD of 143.5 cm is significantly different from this. To check that the assumption still holds TPR 20/10 measurements were made at SCDs of 100 and 143.5 cm on a conventional Elekta linac at energies of 6 and 10 MV and compared against each other.

TPR 20/10 measurements were made on the MR linac using the 143.5 cm SCD in the water phantom shown in Fig. 1 but with the chamber orientation changed to point straight down (Fig. 2), aligned perpendicular with the magnetic field and beam directions. The same QA platform, solid water and MV imaging technique were used to set up the phantom—the gantry angle of 90° is still used but the

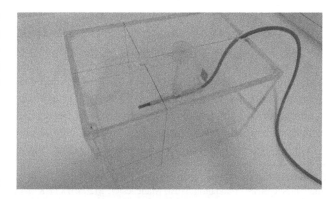

Fig. 1 30 cm $\times$ 20 cm $\times$ 20 cm water phantom used for quality index and output measurements. The Farmer chamber is oriented correctly for output measurements in this picture; the 30 cm dimension would be positioned along the bore of the MR Linac

phantom is positioned across the bore and then rotated through 180° to achieve the two required depths of 10 cm and 20 cm.

2.4 Calibration Factors

The NPL provides energy specific calibration factors, N_D, for the secondary standard NE 2611A thimble chamber in terms of absorbed dose to water. N_D was taken from the current valid calibration certificate for the value of the measured TPR 20/10 on the MR Linac. This assumes that the spectrum of the 7 MV FFF beam of the MR Linac is sufficiently close to the spectrum of the conventional Elekta SL linac used by NPL for providing absorbed dose calibrations to make no significant difference, This is a reasonable assumption given that N_D varies slowly with energy and differences of less than 0.6% are seen between conventional and FFF beams in Monte Carlo calculations [3]. The validity of this assumption will be proved by the results of the independent audit (Sect. 2.6) which did not rely on the same assumption.

UK guidance on FFF beams suggests application of a 0.997 correction factor for FFF beams [3] due to the lighter filtration in an FFF beam. However, in the MR Linac the beam travels through additional material—the annulus containing helium and structural walls of the MR system. Hence it is expected that the beam will be harder than a standard FFF beam. This correction factor has therefore not been applied.

Two sets of independent Intercomparisons were made at 6 MV on a conventional linac between the NE2611A secondary standard and the two PTW waterproof Farmer 30013 chambers and these values were used to transfer the calibration to the field chambers. Again, this assumes that the Intercomparison ratio is the same between the 6 MV clinical beam and the MR Linac beam. Given that the response of these chambers is known to vary very slowly with energy

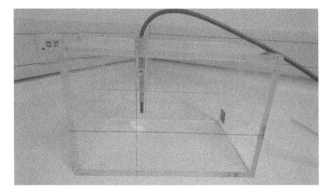

Fig. 2 30 cm × 20 cm × 20 cm water phantom with Farmer chamber oriented correctly for TPR 20/10 measurements; in this case the 30 cm dimension would be positioned across the bore of the MR Linac

(essentially no change between 6 and 10 MV) it is reasonable to assume this introduces minimal uncertainty.

2.5 Correction Factors

Ion recombination and polarity correction factors were measured for the Farmer chambers under the reference conditions in the water phantom both during installation of the MR Linac before the magnetic field was turned on and after the magnetic field was switched on.

A magnetic field correction factor, k_B, was derived for each field Farmer chamber by measuring the temperature pressure corrected readings from the chambers under the reference conditions both before (during installation) and after the magnetic field was turned on.

However, the simple ratio of these two values cannot be taken directly and used as the magnetic field correction factor because the beam itself is shifted when the magnetic field is turned on. There is a lateral shift in the beam in the cross-plane direction. However, at 10 cm deep this shift has negligible effect on the output at the isocentre. There is also a shift in the percentage depth dose curve, resulting in a dose reduction of around 0.5% at the 10 cm reference depth. This value has been calculated using Monte Carlo simulations by O'Brien et al. [2]. We independently checked this value by modelling reference conditions in the Elekta Monaco TPS which incorporates a magnetic field calculation in its Monte Carlo dose calculations.

The magnetic field correction factor, k_B is therefore calculated as the ratio of the Farmer chamber corrected readings corrected for the change in dose at isocentre when the magnetic field is on.

2.6 Independent Audit

An independent audit was performed by the NPL using both alanine pellets and Farmer chambers calibrated via a water calorimeter from the Netherlands primary standards laboratory which operates within the MR Linac. Both these methods avoid using transfer of correction factors via a conventional linac and allow the assumptions used in our method to be validated by an independent dosimetry chain.

3 Results

3.1 Energy Specifier

The TPR 20/10 measured for our Unity MR Linac was 0.698; this value remained constant at both zero magnetic field and the clinical 1.5 T magnetic field.

TPR 20/10 measurements at SCDs of 100 and 143.5 cm on the conventional Elekta linac were within 0.5% at both 6 and 10 MV, confirming the validity of using this Quality Index at extended SCDs.

3.2 Correction Factors

Ion recombination was measured as 1.005 for the Farmer chambers at +250 V at both 0 and 1.5 T hence the magnetic field did not affect the ion recombination correction. Polarity was measured as 1.001 at 0 T and 1.000 at 1.5 T, which again shows no significant difference with magnetic field.

The ratio of temperature/pressure corrected readings for the Farmer chambers at +250 V was 0.986 (0 T/1.5 T readings) for the chosen reference conditions. When corrected for the shift in depth dose this gives a value for k_B of 0.981.

Absorbed dose to water, D_w, under the reference conditions measured with a Farmer chamber can therefore be expressed as

$$D_w = Reading \times N_D \times k_B \times f_{elec} \times TPC \times ICR_{raw}$$
$$\times \frac{f_{ion,sec}(6\,\text{MV}) \times f_{pol,sec}(6\,\text{MV})}{f_{ion,field}(6\,\text{MV}) \times f_{pol,field}(6\,\text{MV})}$$
$$\times f_{ion,field}(7\,\text{MV}) \times f_{pol,field}(7\,\text{MV})$$

Where:

N_D	is the secondary standard calibration factor for the measured TPR 20/10
k_B	is the magnetic correction factor, 0.981 for these chambers at +250 V
f_{elec}	is the electrometer correction factor
TPC	is the standard temperature and pressure correction
ICR_{raw}	is the ratio of raw readings made between the secondary standard and field chamber at 6 MV on a conventional linac
f_{ion}	(6 MV) are the ion recombination factors for the secondary standard and field chambers on the conventional linac at 6 MV
f_{pol}	(6 MV) is the polarity factors for the secondary standard and field chambers on the conventional linac at 6 MV
$f_{ion,field}$	(7 MV) is the ion recombination factor for the field chamber on the MR Linac at 7 MV
$f_{pol,field}$	(7 MV) is the polarity factor for the field chamber on the MR Linac at 7 MV.

3.3 Independent Audit

This gave agreement with our calibration to within 1.0%, providing reassurance that the assumptions made in this calibration process are valid.

4 Conclusion

It is possible to establish absolute dosimetry for the Elekta Unity 1.5 T MR Linac via the UK Code of Practice for high-energy photon dosimetry using standard waterproof field Farmer chambers given careful attention to the measurement of the magnetic field correction for the chambers.

Conflict of Interest The authors declare that they have no conflict of interest.

References

1. Lillicrap, S. *et al:* Code of Practice for high-energy photon therapy dosimetry based on the NPL absorbed dose calibration service. Phys. Med. Biol. 35(10), 1355–1360 (1990).
2. O'Brien, D. *et al:* Reference dosimetry in magnetic fields: formalism and ionization chamber correction factors. Medical Physics 43, 4915–27 (2016).
3. Budgell, G. *et al:* IPEM Topical Report 1: Guidance on implementing Flattening Filter Free (FFF) radiotherapy in the UK. Phys. Med. Biol. 61 8360–8394 (2016).

Calibration Seed Sampling for Iodine-125 Prostate Brachytherapy

Scott Crowe and Tanya Kairn

Abstract

Use of iodine-125 seeds for intraoperative planning and delivery of low dose rate brachytherapy treatments for prostate cancer requires that the air kerma strength of the seeds be checked against the vendor-supplied calibration certificate at the beginning of the surgical procedure. In practise, activity checks of multiple sources are difficult to achieve in surgery. This study therefore investigated the reliability of a calibration method that sampled only one seed, or a small number of seeds, by evaluating the consistency the air kerma strengths of all seeds in three small batches of 10, 20 and 30 iodine-125 seeds and calculating the probability of achieving results representative of each batch, within different levels of uncertainty. For the cartridges containing 10, 20 and 30 seeds, the mean differences between the source strengths identified by physical measurement and their decay-corrected calibration certificate values were respectively $5.0\% \pm 4.4\%$, $-2.7\% \pm 3.7\%$, and $0.4\% \pm 2.8\%$. Assays of 5 randomly sampled seeds were shown to produce results within 3% of the mean air kerma strength of each batch, with larger assays producing less uncertainty. For these seeds, there was a greater than 30% chance that a randomly selected seed would have an activity that differed by more than 3% from the mean activity of all seeds in the cartridge. Although attractive as an efficiency measure, the testing of just one seed from a cartridge of iodine-125 seeds has a significant probability of

producing an activity measurement that is not representative of the activity of the other seeds in the cartridge, potentially leading to substantial inaccuracies in implant dosimetry.

Keywords

Radiation therapy • Brachytherapy • Quality assurance

1 Introduction

Implanted iodine-125 seeds can be effectively used for low dose rate (LDR) brachytherapy treatments for localised adenocarcinoma of the prostate, using intraoperative pre-planning with trans-rectal ultrasound. This intraoperative technique requires that the air-kerma strength of the seeds be checked against the vendor-supplied calibration certificate prior to implantation. In practise, air-kerma strength checks of multiple sources are difficult to achieve when seeds are packaged in a sterile cartridge, as test seeds must be extracted at the beginning of the surgical procedure. The number of seeds checked, i.e. the assay size, should be optimised to minimise checking time and seed expense, while ensuring a satisfactory level of uncertainty is met.

Detailed statistical studies of source strength homogeneity, assaying methods and clinical impact of source strength inhomogeneity have been presented by various authors [1–5]. Ramos and Monge [2] produced recommendations on the assay size required to achieve uncertainties (two standard deviations) of 1–0.05%, for batch sizes up to 160 seeds; based on historical data consisting of 2030 normally-distributed in-air kerma measurements. Yue et al. [3] described a method by which the standard deviation in the assay source strength mean could be used to estimate potential deviation from the batch source strength mean, for normally distributed source strengths.

However, the AAPM's Low Energy Brachytherapy Source Calibration Working Group have specifically

S. Crowe (✉) · T. Kairn
Royal Brisbane and Women's Hospital, Brisbane, QLD, Australia
e-mail: sb.crowe@gmail.com

S. Crowe · T. Kairn
Queensland University of Technology, Brisbane, QLD, Australia

© Springer Nature Singapore Pte Ltd. 2019
L. Lhotska et al. (eds.), *World Congress on Medical Physics and Biomedical Engineering 2018*,
IFMBE Proceedings 68/3, https://doi.org/10.1007/978-981-10-9023-3_83

recommended that the results of statistical modelling not be used to identify the required assay size, because detailed information about the distribution of source strength as a function of radionuclide and manufacturer is lacking [6]. A normal distribution of source strengths cannot be assumed without substantial knowledge of the source production process and its potential fallibilities. Rather, AAPM report 098 [6] recommends the testing of 10% of the seeds or 10 seeds, whichever number is larger, if sources are loose or non-sterile. For sterile source assemblies, the report recommends testing of 5% of the seeds or 5 seeds, whichever number is smaller [6].

In an effort to comply with the AAPM's published recommendations while producing results efficiently enough to avoid surgical delays, the Nucletron SeedSelectron afterloader (Elekta, Stockholm, Sweden) contains and array of diodes that assay all seeds as the implant is delivered, as a backup system, allowing the manufacturer to claim that only one seed from each batch needs to be assayed before the implantation procedure commences [4]. This ambitious system has been criticised for providing results that are too imprecise to satisfy AAPM recommendations, due to the inconsistency of the afterloader's diode response [4]. Specifically, Perez-Calatayud et al. [4] devised an alternative system for calibrating ^{125}I seeds from the Nucletron SeedSelectron system, using an attachment for the PTW SourceCheck flat ionisation chamber (PTW Freiburg GmbH, Freiburg, Germany) that allowed ten seeds to be assayed simultaneously and averaged.

This study used experimental (rather than statistical) measurements to investigate the consistency of the strength of all seeds in three small batches of 10, 20 and 30 ^{125}I seeds, with the aim of identifying the optimal assay sizes required to achieve the AAPM recommended tolerance of 3% uncertainty [6] and exemplifying the uncertainty that may be introduced if only one seed is assayed.

2   Method

Three cartridges of 10, 20 and 30 seeds from three different batches of ^{125}I seeds were obtained from the vendor. The seeds were manually extracted from the cartridges, and their air-kerma strengths (S_K) were measured using a calibrated PTW SourceCheck ionisation chamber mounted in a PMMA block, using a method similar to that described by Perez-Calatayud et al. [4] except that the strength of each seed was measured separately.

Results were analysed by calculating

(a) the difference between the measured S_K for each seed and the decay-corrected air kerma strength calibration certificate value for the batch ($S_K^{C_i}$),
(b) the difference between the measured S_K for each seed and the mean of all S_K values for the batch ($\bar{S}_K$),
(c) the mean of all the differences (a), calculated over each batch and over all three batches combined, and
(d) the mean of all the differences (b), calculated over each batch and over all three batches combined.

For each set of measurements, statistical sampling methods were used to characterise the 2σ standard deviation (δ) in the mean deviation between two values: the mean of S_K measurements for assays of size n between 1 and $N-1$ ($\bar{S}_{K,assay}$) and the mean of corresponding batch-specific data ($\bar{S}_K$). That is, the 2σ uncertainty for assay size n (δ_n) can be defined as:

$$\delta(n) = \sqrt{\frac{\sum_{i \in S}\left(\bar{S}_{K,i} - \bar{S}_K\right)^2}{n-1}} \times 2 \qquad (1)$$

where i is a possible assay of size n, of all possible combinations S.

For the 10 seed and 20 seed measurement data, these values were calculated by iterative evaluation of all possible assay combinations (1,022 and 524,286 combinations, respectively). For the 30 seed and combined measurement data, random sampling (Monte Carlo) methods were used to simulate the selection of 10^7 random-sampled assays of random size n between 1 and $N-1$.

The 2σ standard deviation was selected for evaluation because it would correspond to a confidence interval of approximately 95%, if the seed strengths were normally distributed, or a confidence interval of at least 75% (by Chebyshev's inequality), if the seeds strengths were not normally distributed. That is, regardless of the true distribution of the source strengths, the mean air kerma strength of a sample assay should be within 2σ of the cartridge (or batch) mean air kerma strength at least 75% of the time.

3   Results and Discussion

The method of extracting seeds one at a time and measuring them separately was relatively slow and inefficient, especially since three 60 s readings per measurement were found to be necessary, to average out small variations in electrometer response. Only 19 measurements were obtained for

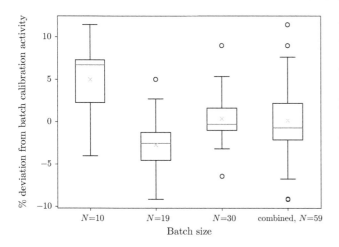

Fig. 1 Percent deviation between activity measurements and decay-corrected calibration activity values

the cartridge containing 20 seeds, due to difficulty extracting one of the seeds from the cartridge.

Figure 1 presents differences between measurements (S_K) and decay-corrected calibration certificate data $(S_K^{C_i})$, for each batch.

For the cartridges containing 10, 19 and 30 seeds, the mean differences $(\pm 1\sigma)$ between S_K and $S_K^{C_i}$ were respectively $5.0\% \pm 4.4\%$, $-2.7\% \pm 3.7\%$, and $0.4\% \pm 2.8\%$, with 80, 53 and 17% of seeds disagreeing with $S_K^{C_i}$ by more than 3%. If

a normal distribution had been assumed, these results would suggest there was a greater than 30% chance that a randomly selected seed would have an activity that differed by more than 3% from the mean strength of all seeds in the cartridge.

These results are consistent with those reported by Perez-Calatayud et al. [4], using the same dosimetry system and sources from the same vendor; where a systematic deviation of $-2.1\% \pm 0.7\%$ (mean $\pm 1\sigma$) was observed when analysing single-seed calibration data from over 48 implants.

Figure 2 presents 2σ standard deviations $(\delta(n))$ for each set of source strength measurements. These results show that when a random sample of 5 seeds from each batch is assayed, the resulting air kerma strength measurement $(\bar{S}_{K,assay})$ can be expected to be within $\leq 3\%$ of the batch $\bar{S}_K$, supporting the AAPM recommendations [6]. Values of $\bar{S}_{K,assay}$ within 1% of the batch $\bar{S}_K$ are likely to be achieved with an assay of between 7 and 9 sources, with uncertainty decreasing with the number of seeds assayed.

The assay size required to meet 1% uncertainty (7–9 seeds from a population of 10–59 seeds) was found to be greater when evaluated using these physical measurements of clinical (vendor-supplied) ^{125}I seeds than the assay sizes (e.g. 3 seeds from a population of 30 seeds) reported to achieve 1% uncertainty when a statistical sampling method assuming a normal distribution was used [2].

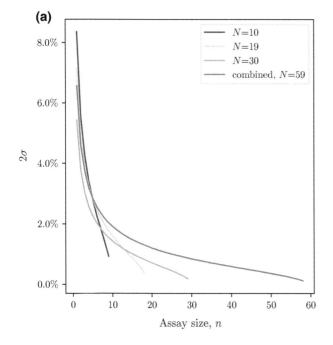

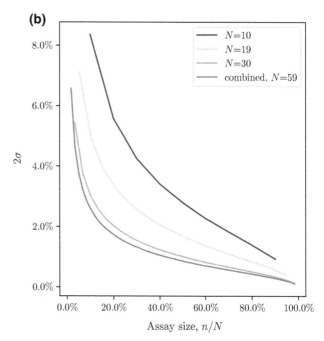

Fig. 2 2σ for relative deviation between mean assay source strength $(\bar{S}_{K,assay}$, for size n) and mean cartridge source strength $(\bar{S}_K$, for size N), presented against **a** assay sample size and **b** assay sample size as a percentage of batch

4 Conclusion

The results of this experimental study broadly support the AAPM recommendation that when checking the air kerma strength of LDR brachytherapy sources prior to implantation, an assay size of at least 5 seeds is needed to reduce uncertainty to within 3%. A larger assay size of 7–9 seeds may be advisable, to reduce uncertainty due to assay size (and source strength variation) to within 1%, given the numerous other uncertainties affecting measurements of air kerma strength in a surgical setting.

While the measurement results produced in this study suggest that a normal distribution of ^{125}I air kerma strengths cannot be assumed, especially if the number of sources in the batch is small, it is nonetheless possible to use an analysis of standard deviations to conclude that there may be a greater than 30% chance of one randomly selected seed having an activity that differs by more than 3% from the mean strength of all seeds in the cartridge. The use of one seed per LDR procedure as an indicator of the activity of all implanted seeds is inadvisable and may lead to substantial errors in implant dosimetry.

Acknowledgements This work was made possible by the thoughtful ordering and preservation of seed cartridges that was provided by Candice Deans, Mark West and George Warr.

Compliance with Ethical Standards The authors declare that they have no conflict of interest. This article does not contain any studies with human participants or animals performed by any of the authors.

References

1. Rosenzweig, D.P., Schell, M.C., Yu, Y.: Toward a statistically relevant calibration end point for prostate seed implants. Med. Phys. 27(1), 144–150 (2000). https://doi.org/10.1118/1.598877
2. Ramos, L.I., Monge, R.M.: Sampling size in the verication of manufactured-supplied air kerma strengths. Med. Phys. 32(11), 3375–3378 (2005). https://doi.org/10.1118/1.2089627
3. Yu, N.J., Haffty, B.G., Yue, J.: On the assay of brachytherapy sources. Med. Phys. 34(6), 1975–1982 (2007). https://doi.org/10.1118/1.2734723
4. Perez-Calatayud, J., Richart, J., Guirado, D., et al.: I-125 seed calibration using the SeedSelectron® afterloader: a practical solution to fulfill AAPM-ESTRO recommendations. J. Cont. Brachyther. 4(1), 21–28 (2012). https://doi.org/10.5114/jcb.2012.27948
5. Nuñez-Cumplido, E., Perez-Calatayud, J., Casares-Magaz, O., Hernandez-Armas, J.: Influence of source batch SK dispersion on dosimetry for prostate cancer treatment with permanent implants. Med. Phys. 42(8), 4933–4940 (2015). https://doi.org/10.1118/1.4926848
6. Butler, W.M., Bice Jr., W.S., DeWerd, L.A., et al.: Third-party brachytherapy source calibrations and physicist responsibilities: Report of the AAPM Low Energy Brachytherapy Source Calibration Working Group. Med. Phys. 35(9), 3860–3865 (2008). https://doi.org/10.1118/1.2959723

Automated VMAT Treatment Planning for Complex Cancer Cases: A Feasibility Study

Savino Cilla, Anna Ianiro, Gabriella Macchia, Alessio G. Morganti, Vincenzo Valentini, and Francesco Deodato

Abstract

Treatment plans for high-risk prostate and endometrial cancer are highly complex due to large irregular-shaped pelvic target volumes, multiple dose prescription levels and several organs at risk (OARs) close to the targets. The quality of these plans is highly inter-planner dependent. We aimed to assess the performance of the Auto-Planning module present in the Pinnacle treatment planning system (version 16.0), comparing automatically generated plans (AP) with the historically clinically accepted manually-generated ones (MP). Twenty consecutive patients (10 for high-risk prostate and 10 for endometrial cancer) were re-planned with the Auto-Planning engine. Planning and optimization workflow was developed to automatically generate "dual-arc" VMAT plans with simultaneously integrated boost. Primary target (PTV1) included the prostate and seminal vesicles or the upper two thirds of vagina; PTV2 included the lymph-nodal drainage. PTVs were simultaneously irradiated over 25 daily fractions at 45 Gy for the PTV2 and 65/55 Gy to the prostate/endometrial PTV1. For AP plans, a progressive optimization algorithm is used to continually adjust initial targets/OARs objectives. Tuning structures and objectives are automatically added during optimization to increase the dose fall-off outside targets and improve the dose conformity. Various dose and dose-volume metrics, as well as conformity indexes and healthy-tissue integral dose were evaluated. A Wilcoxon paired-test was performed for plan comparison ($p < 0.05$ as statistical significance). All AP plans fulfilled the clinical dose criteria for OARs and PTV coverage. Dose coverage metrics for both PTVs were very similar with AP showing slight better results for PTV1. For both anatomical sites, differences in DVHs were no significant in overall dose range for rectum, bladder and small bowel. However, AP plans provided significant better conformity and an average decrease in Integral Dose of 6–10%. The Pinnacle Auto-Planning module is capable of efficiently generating highly consistent treatment plans, meeting our institutional clinical constraints.

Keywords

Automatic planning • VMAT
Simultaneous integrated boost

1 Introduction

The advanced developments in external beam radiation therapy over the past few decades have greatly improved plan quality in terms of dose conformity to the target while minimizing dose to the surrounding healthy tissues. However, complex treatment planning with intensity-modulated radiotherapy (IMRT) or volumetric modulated arc therapy (VMAT) can be challenging in order to achieve clinically acceptable plans and several trial-and-error optimization processes are usually required.

Despite the obvious benefits of IMRT and VMAT, the planning and quality assurance processes required for these advanced techniques are more and more complex and time-consuming and can have a significant impact on departmental resources.

S. Cilla (✉) · A. Ianiro
Medical Physics Unit, Foundation of Research and Cure "John Paul II", Catholic University of Sacred Heart, Campobasso, Italy
e-mail: savino.cilla@fgps.it

G. Macchia · V. Valentini · F. Deodato
Radiation Oncology Unit, Foundation of Research and Cure "John Paul II", Catholic University of Sacred Heart, Campobasso, Italy

A. G. Morganti
Radiation Oncology Unit, Department of Experimental, Diagnostic and Specialty Medicine, S.Orsola-Malpighi Hospital, University of Bologna, Bologna, Italy

V. Valentini
Radiation Oncology Department, Foundation University Hospital "Agostino Gemelli", Catholic University of Sacred Heart, Rome, Italy

© Springer Nature Singapore Pte Ltd. 2019
L. Lhotska et al. (eds.), *World Congress on Medical Physics and Biomedical Engineering 2018*,
IFMBE Proceedings 68/3, https://doi.org/10.1007/978-981-10-9023-3_84

Previous studies [1] have also shown that the planner experience has a large impact on plan quality, prompting the need to automate the planning process in order to improve best practice [2–4].

In this study we hypothesized that automated radiotherapy treatment planning have the potential to increase consistency, improve plan quality and reduce workload. The current study aims to validate the performance of the Auto-Planning module implemented for clinical use in the clinical version of Pinnacle 16.0 treatment planning system (Philips Healthcare, Fitchburg, WI). This module is designed to be "one button planning", able to create quality treatment plans using a single optimization preset (including beam set-up, dose prescription, objectives and priorities for target volumes and organs at risk). In order to investigate the potential of Auto-Planning engine, this strategy was applied to complex high-risk prostate and endometrial cancer, where large irregular-shaped pelvic target volumes, multiple dose prescription levels and several organs-at-risk close to the targets represent a challenge for the generation of high quality plans. The treatment plans generated by Auto-Planning were then compared with historical, clinically accepted VMAT treatment plans generated by experienced physicists.

2 Materials and Methods

The Auto-Planning software was evaluated by replanning twenty previously delivered clinical plans for high-risk prostate (10 patients, group 1) and endometrial cancer (10 patients, group 2). All patients were treated with an accelerated SIB approach. Primary targets (PTV1) included the prostate and seminal vesicles (group 1) or the upper two thirds of vagina (group 2); nodal targets (PTV2) included the lymph-nodal drainage. Target volumes were simultaneously irradiated over 25 daily fractions at 45 Gy for the PTV2 and 65 Gy to the prostate and 55 Gy to the endometrial PTV1 s. This resulted in daily fractions of 1.8 Gy for the PTV2 and in 2.6 Gy and 2.2 Gy for the prostate and endometrial PTV1, respectively. The main organs-at-risk (OARs) were the small bowel, the rectum, the bladder and the femoral heads. Small bowel was contoured as individual loops.

The original clinical manually optimized plans were created using the Oncentra MasterPlan TPS for 6-MV beams from an Elekta Precise linac (Elekta Ltd., Crawley, UK). SIB-VMAT plans were created by means of the "dual-arc" feature and generated using the optimization process described by Cilla et al. [5]. The entire gantry rotation is described in the optimization process by a sequence of 86 control point, i.e., every 4°. Collimator was set at 10° to minimize the tongue-and-groove cumulative effect. Since SIB-VMAT planning is challenging when highly

concave-shaped target volumes are present [5], some non-anatomic dummy volumes were also defined to guide the optimization process and to prevent dose dumping in undefined areas.

Each of the 20 treatment plans were transferred to Pinnacle 16.0 TPS and re-planned with the Auto-Planning engine without knowledge of the clinically delivered treatment plans. Auto-Planning is fully integrated into Pinnacle v.16.0 TPS to automate the inverse planning process [4]. The core Auto-Planning algorithm is based on the regional optimization concept introduced by Cotrutz and Xing [6]. In the Auto-Planning engine, a template of configurable parameters (the so-called "Technique") was defined for each treatment protocol and tumour site. The Techniques include the definition of all beam parameters and planning goals. The Auto-Planning module uses the Technique definition to iteratively adjust VMAT planning parameters to best meet the planning goals. During optimization, Auto-Planning uses a progressive optimization algorithm to continually adjust targets/OARs optimization objectives based on the initial values set by the user to meet or further decrease OARs doses with minimal compromise to PTV coverage. Additionally, planning structures and objectives are automatically added during optimization to manage targets uniformity and conformity by reducing cold/hot spots and controlling dose fall-off outside targets. Each Technique was defined according to local standards, including prioritization between target coverage and dose to organs at risk. Each Technique was tuned on the basis of previous five pilot patients for each anatomical sites, independent of the 20 study patients. In each Techniques, the definition of beam parameters and planning goals was the same used for manual optimized plans. All plans were calculated using the Pinnacle3 collapsed cone algorithm with a dose grid resolution of 3 mm.

The quality of automated plans was evaluated comparing the target coverage, homogeneity, conformity, and OARs sparing with respect to manual plans. DVHs were calculated for all involved structures. Target dose distribution was evaluated using mean dose (Dmean), near-maximum dose (D2%), near-minimum dose (D98%), dose irradiated to 95% of target volumes (D95%) and conformity indexes (CI).

For both PTVs, the CI is defined as the volume encompassed by the 95% isodose cloud of each dose prescription divided by the PTV volumes.

$$\text{CI} = \frac{V95\%}{PTV}$$

To quantify the ability to deliver highly heterogeneous doses as requested for SIB, a metric called Dose Contrast Index (%DCI) was calculated [5]. The ideal DCI (iDCI) was defined as the ratio between the prescription doses to the PTV1 and PTV2. An actual DCI was defined as the mean dose to the PTV1 divided by the mean dose to the PTV2. The ratio of DCI and iDCI multiplied by 100 defines the percentage DCI (%DCI) quantifying the deviation of the actual DCI from the ideal iDCI. A %DCI value closer to 100% indicates a better dose contrast.

Last, the integral dose (ID) received by non-tumor tissue (NTT = whole body minus PTV2) was calculated as:

$$ID = average\ dose * volume(Gy * cc)$$

For the OARs, metrics included the mean dose for each structure and the values for the percentage of the OARs volumes irradiated at various dose levels (Vx), converted to equivalent dose at 2 Gy/fraction. The differences between manual and automated plans were quantified using Wilcoxon matched-pair signed rank with a statistical significance at p < 0.05.

3 Results

Figure 1 shows the dose distributions for the original clinical plan and automated plan for a representative high-risk prostate cancer patient. Tables 1 and 2 present the results of the comparison of all patient plans for PTVs and OARs. All AP plans fulfilled the clinical dose criteria for OARs and PTV coverage. Dose coverage metrics for both PTVs were very similar with AP showing slight better results for PTV1. For both anatomical sites, differences in DVHs were no significant in overall dose range for rectum, bladder and small bowel. However, AP plans provided significant better conformity for nodal targets (CI2) and, although not statistically significant, an average decrease in Integral Dose of about 6–11%. A slight increased number of monitor units was found in the AP plans that probably reflect a larger degree of MLC modulation.

4 Discussion

Comparison of automated plans and previous clinical manual plans showed only small dosimetric differences in target coverage and dose irradiation to OAR. In particular, we showed that in complex treatments as high-risk prostate or endometrial cancer, the AutoPlanning optimization is suitable for clinical practice. While the small differences in tumor coverage will certainly have no clinical impact, the reduced integral dose to normal tissue could be of clinical impact for the patients. Integral dose is an important parameter to evaluate the delivery efficiency of a planning system. In the present study, AP plan reported a mean decrease of integral dose of 6–10% indicating that the AutoPlanning module represent a treatment platform with an

optimal physical efficiency. AP plans provided also a better conformity to concave-shaped target volumes with respect to MP plans. Manual SIB-VMAT planning is challenging for complex tumor cases and it's necessary to define some irregular dummy volumes in anterior and posterior areas of axial slices to prevent areas of high doses in healthy tissues. This way, not only the volume definition process itself can be significantly more time consuming, but also the overall optimization and calculation time may be longer, as dose-volume constraints and penalty factors to dummy volumes are interactively determined during the optimization process. On the contrary, the Auto-planning engine is able to automate, during the optimization process, the implementation of several tuning structures (up to 25–30 structures in pelvic tumor cases). These include expanded structures of PTVs and OARs, residual OAR structures where overlaps between targets are removed, residual targets structures where overlaps between non compromised OARs are removed, body structures used to control body dose, ring structures to control dose falloff and structures to control target uniformity by reducing the areas of hot and/or cold spots. The integration of all these tuning structures in the optimization process has the potential to greatly improve the overall dose distribution.

It must be emphasized that original clinical plans were performed by experienced medical physicists with a very high level of planning skills, gained in almost ten years of experience with VMAT. As a result, also the inter-planner variation become very limited over time. Despite this aspect, our results show a slight improvement in plan quality for Auto-Planning versus the clinical plans. Other studies reported different results showing a more substantial improvement [2, 3]. In addition, automated plans also reported a lower variance, thus reducing the inter- and intra-planner variability and achieving higher plan consistency. Although clinically acceptable automated plans were already obtained in a "one-button click" procedure without planner intervention, all automated plans underwent a second optimization to further increase dose conformity because of the complex clinical anatomical sites. Thus, in our perception, in challenging clinical cases, the automated plans should be considered a high quality starting point for further optimization. In other words, our initial experience show that the impact of Auto-Planning seems likely to be a tool to increase the overall quality of dose planning, rather than a tool that could completely remove the need of manual optimization.

It must be also underlined that the Autoplanning engine requires a training set before its clinical implementation. The Technique, including all the beam geometries, settings, optimization options and goals, must be accurately implemented by experienced medical physicists or dosimetrists. Then, the optimization of the Technique undergoes an

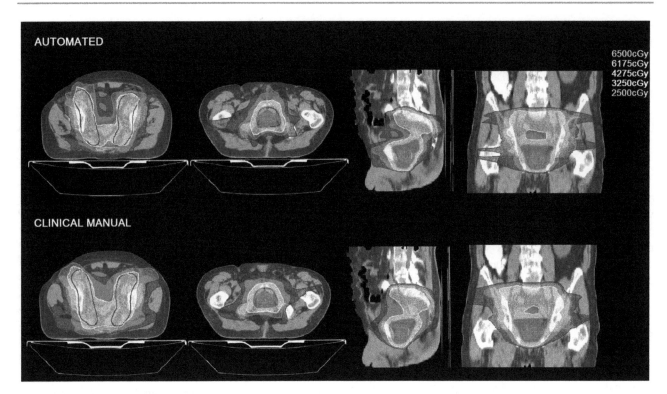

Fig. 1 Dose distributions in axial, sagittal and coronal plans of an automatic plan and a clinical manual plan for a representative high-risk prostate patient

Table 1 Summary of results for PTVs

High-risk prostate (n = 10)				Endometrial (n = 10)		
	Manual	Autoplan	p	Manual	Autoplan	p
PTV1						
D98% (%)	96.0 ± 1.0	97.3 ± 1.5	**0,023**	98.6 ± 1.6	98.7 ± 0.7	0,636
D95% (%)	97.5 ± 1.2	98.6 ± 1.2	**0,034**	99.5 ± 1.5	99.6 ± 0.6	0,713
D50% (%)	102.3 ± 1.1	102.3 ± 0.5	0,821	104.0 ± 1.5	102.6 ± 0.6	**0,027**
D2% (%)	104.5 ± 1.3	105.2 ± 0.4	0,290	106.5 ± 1.0	105.4 ± 0.9	0,074
CI1	1.20 ± 0.09	1.27 ± 0.10	0,111	1.71 ± 0.24	1.56 ± 0.15	0,600
PTV2						
D98% (%)	94.7 ± 0.7	94.1 ± 1.0	0,212	94.6 ± 3.9	93.8 ± 3.1	0,401
D95% (%)	96.7 ± 0.4	96.5 ± 0.9	0,406	97.6 ± 1.4	97.1 ± 1.1	0,318
D50% (%)	102.5 ± 1.1	104.6 ± 0.9	**0,002**	104.2 ± 1.0	104.8 ± 1.1	0,208
CI2	1.59 ± 0.09	1.44 ± 0.08	**0,002**	1.63 ± 0.13	1.35 ± 0.06	**0,001**
%DCI	96.2 ± 3.3	96.5 ± 1.1	0,450	90.1 ± 0.8	89.0 ± 0.7	**0,021**
ID (*10^3)	263 ± 38	248 ± 38	0,131	320 ± 69	288 ± 62	0,248
MUs	528 ± 69	534 ± 43	0,050	540 ± 122	599 ± 98	0,141

iterative process in which the autoplanning engine is performed on test patients for each specific anatomical site. The optimization goals in the Technique are eventually adjusted until the overall results reached the expected outcomes.

Finally, the overall treatment planning time with Auto-Planning was reduced to less than 1 h, with only few interventions from the planner for the second round optimization. This strategy reduces the need for multiple plan reviews and frees the planner from the long manual trial-and-error planning process.

In conclusion, automated treatment planning can be adopted as the current standard workflow in a radiation

Table 2 Summary of results for OARs

High-risk prostate (n = 10)				Endometrial (n = 10)			
	Manual	Autoplan	p	Manual	Autoplan	p
Rectum						
Dmean (Gy)	40.9 ± 2.0	40.0 ± 1.9	0,226	42.0 ± 1.8	39.5 ± 2.0	**0,046**
V50 Gy (%)	31.9 ± 8.5	30.7 ± 7.5	0,406	24.6 ± 7.3	24.4 ± 6.1	1,000
V60 Gy (%)	21.2 ± 6.4	20.8 ± 5.2	0,705	12.4 ± 5.3	13.3 ± 4.2	1,000
V65 Gy (%)	15.5 ± 5.3	15.9 ± 4.4	1,000	4.9 ± 3.7	6.4 ± 2.8	0,462
Bladder						
Dmean (Gy)	44.2 ± 5.9	43.2 ± 5.7	0,496	44.4 ± 4.0	42.7 ± 3.0	0,529
V65 Gy (%)	19.2 ± 12.7	19.7 ± 11.6	0,762	9.9 ± 6.7	9.7 ± 6.4	0,916
V70 Gy (%)	14.8 ± 10.0	15.9 ± 9.9	0,650	0.7 ± 0.8	0.6 ± 1.1	0,224
Small bowel						
Dmean (Gy)	12.9 ± 5.2	12.5 ± 4.5	0,762	17.9 ± 8.3	17.2 ± 7.1	0,753
V15 Gy (cc)	115 ± 42	120 ± 58	0,174	119 ± 44	128 ± 51	0,123
V50 Gy (cc)	0.1 ± 0.2	0.1 ± 0.2	0,256	0.4 ± 0.8	1.1 ± 3.9	0,168

oncology clinic to generate high quality treatment plans for prostate and endometrial cancer. Future studies are needed to expand automated treatment planning to other tumour sites, such as head-and-neck, and to other treatment techniques such as extracranial stereotactic radiotherapy.

References

1. Nelms BE, et al. Variation in external beam treatment plan quality: an inter-institutional study of planners and planning systems. Pract Radiat Oncol 2(4),296–305 (2012).

2. Gintz D, et al. Initial evaluation of automated treatment planning software. J Appl Clin Med Phys 17(3):331–346 (2016).

3. Krayenbuehl J, et al. Evaluation of an automated knowledge based treatment planning system for head and neck. Radiat Oncol 10:226 (2015).

4. Hazzell I, et al. Automatic planning of head and neck treatment plans. J Appl Clin Med Phys 17(1):272–282 (2016).

5. Cilla S, et al. Assessing the feasibility of volumetric-modulated arc therapy using simultaneous integrated boost (SIB-VMAT): An analysis for complex head-neck, high-risk prostate and rectal cancer cases. Med Dosim 39(1),108–16 (2014).

6. Cotrutz C, et al. IMRT dose shaping with regionally variable penalty scheme. Med Phys 30(4), 544–551 (2003).

Monte Carlo Validation of Output Factors Measurements in Stereotactic Radiosurgery with Cones

Nicolas Garnier, Régis Amblard, Rémy Villeneuve, Rodolphe Haykal, Cécile Ortholan, Philippe Colin, Joël Herault, Sarah Belhomme, Mourad Benabdesselam, and Benjamin Serrano

Abstract

The purpose of this work is to assess 8 detectors variation performance for output factor (OF) determination, in a clinical 6 MV Varian Clinac 2100C stereotactic radiosurgery mode for cone irradiation using Monte Carlo simulation as reference. We study 10 cones with diameters between 30 and 4 mm. The evaluated detectors were ionization chambers: pinpoint and pinpoint 3D (PTW), diodes: SRS, P and E (PTW), Edge (Sun Nuclear), microdiamond (PTW) and EBT3 radiochromic films (Ashland). The OF were normalized with the 30 mm cone and compared to Monte Carlo (MC). For the 8 detectors, the OF measurements reproducibility was very high: r = 0.99 and p < 0.0001. The uncertainty of the MC calculation was lower than 0.8% (type A). The results show: pinpoints (axial position) underestimate OF until −2.3% for cone diameter $\geq$ 10 mm and down to −12% for smaller cones. Non-shielded (SRS and E) and shielded (P and Edge) diodes overestimate OF respectively up to 3.3% and 5.2% for cone diameter $\geq$ 10 mm but in both case >7% for smaller cones. Microdiamond slightly overestimates OF, 3.7% for all the cones and EBT3 film is the closest to MC with maximum difference ±1% whatever the cone size. Film is the more accurate detector for OF determination of stereotactic cones but it is restrictive to use. Pinpoints and diodes, respectively due to inappropriate size of sensitive volume and composition do not seem appropriate without corrective factors below 10 mm diameter cone. Microdiamond appears the best detector for all cones despite its sensitive volume size.

Keywords

Output factor • Stereotactic cone • Monte Carlo Detectors

1 Introduction

Effectiveness of linac-based in stereotactic radiosurgery (SRS) with small cone sizes (few millimeters) bring to more and more frequent use, especially for treatment of brain (metastases, trigeminal neuralgia, AVM and other brain localizations) [1–3]. To ensure the quality of these treatments with small field sizes, measurements of percentage depth-dose curves, tissue-phantom ratios, profiles and output factor should be well achieved in spite the size and composition of the detectors [4–7]. In this study we will focus on a high precision of the dosimetry measurements of output factors (OF) for application of small photon fields in SRS cone irradiation with diameters between 30 and 4 mm.

The requiring determination of OF will not target on correction factors of the OF as study in several research groups [4, 8]. Our purpose is to assess the variation in performance of 8 detectors for OF calculation in a clinical 6 MV linear accelerator photon beam using the Monte Carlo (MC) Penelope [9] simulation as reference.

N. Garnier (✉) · R. Amblard · R. Villeneuve · R. Haykal · B. Serrano
Medical Physics Department, Princess Grace Hospital Center, 98000 Monte Carlo, Monaco
e-mail: nicolas.garnier@chpg.mc

C. Ortholan · P. Colin
Radiotherapy Department, Princess Grace Hospital Center, 98000 Monte Carlo, Monaco

J. Herault
Medical Physics Department, Centre Antoine Lacassagne, 06200 Nice, France

S. Belhomme
Medical Physics Department, Institut Bergonié, 33076 Bordeaux, France

N. Garnier · M. Benabdesselam
Institut de Physique de Nice UMR 7010, University Nice Sophia Antipolis, 06108 Nice, France

L. Lhotska et al. (eds.), *World Congress on Medical Physics and Biomedical Engineering 2018*,
IFMBE Proceedings 68/3, https://doi.org/10.1007/978-981-10-9023-3_85

2 Materials and Methods

2.1 Conventional LINAC-Based Device

The measurements performed for this study were conducted by means of a linear accelerator Clinac 2100C (Varian Medical System) at 6 MV photon beam with an energy index ($TPR_{20,10}$) of 0.669. Linac is equipped with the accessory slot mounted cone system developed by BrainLAB. The cone set consists of 10 cones with diameters of 30, 25, 20, 17.5, 15, 7.5, 5 and 4 mm at the isocenter. The field size defined by the jaws behind the cones was set to 4×4 cm^2. The Linac nominal dose rate was fixed at 600 MU/min.

Output factor was measured at source to surface distance (SSD) of 90 and 10 cm depth. In this work, the output factor (OF_{coll}) was defined by (1):

$$OF_{coll}^{det} = \frac{D_{coll}^{det}}{D_{30mm}^{det}} \qquad (1)$$

Where D_{coll} represents the measured dose for each collimator and D_{30mm} the reference dose measured with the 30 mm diameter collimator. The latter was used for reference instead of the standard 10×10 cm^2 field for two reasons: (i) it is closer to the small fields whilst there is still sufficient electronic equilibrium and good agreement between measurements made with various types of detectors [10]. (ii) in this way, we do not move jaw's size and therefore it is not necessary to take into account the jaw back scattered in Monte Carlo simulation.

2.2 List of Used Detectors

Seven active detectors and a passive detector (Radiochromic film EBT3) were used for output factor measurements in this study (Table 1).

Diode and microdiamond detectors were used in axial orientation. Ionization chambers were used in both axial and radial orientations.

2.3 Setup and Measurements

Active detectors. Measurements with active detectors were made using a PTW MP3 scanning water phantom controlled by Mephysto software. "True Fix" system was used to position detectors. This system allows precise positioning of effective points of measurement of various detectors on the surface of the water phantoms. After each detector or cone change, In-plane and Cross-plane profiles were made to center the detector. Dose measurement was performed with PTW UNIDOS Webline electrometer.

All measurements were achieved with 100 MU and averaged over a series of at least three repeated runs.

Passive detector. Measurements with GAFCHROMIC EBT3 were made in a solid water equivalent phantom with dimensions of 30 cm $\times$ 30 cm $\times$ 30 cm. Films were cut into 5×5 cm^2 24 h before irradiation. The upper right corner was marked at the time the film was cut to define orientation. Since the relation between pixel value and absorbed dose is non-linear, a calibration dose is necessary. The calibration of EBT3 films is performed by doing eight exposures for the dose range varying from 0 to 500 cGy with 6 MV beam. For films response stabilization, 48 h standby times are observed and scanned with Epson 11000XL flatbed scanner in transmission mode with a resolution of 150 dpi and 48-bit RGB format. Exploitation of the films was performed with Film QA Pro software (Ashland) including multichannel correction [11]. The red color channel was used to calculate absorbed dose on EBT3 films. In order to find the beam center of the film and to place automatically a region of interest (ROI), a computer code was written on MATLAB software. The code search along the in-plane and cross plane directions to find out the beam center from full width at half maximums (FWHM) of profiles. The ROI size was 0.6 mm for the 4 mm cone diameter and 1 mm for the other cone. The film absolute dose was evaluated by taking the average of the voxels dose on the ROI. Films measurements were averaged over a series of thirteen times.

UM number issued for each cone was calculated in order to obtain an absorbed dose in the film of about 4 Gy whatever the cone size is (Table 2).

This method allows to work in ideal dose range for the film and to obtain the same signal to noise ratio and thus the same uncertainty whatever the cone size is. For films, the output factor (OF_{coll}) was defined by (2):

$$OF_{coll}^{EBT3} = \frac{D_{coll}^{EBT3}}{D_{30mm}^{EBT3}} \times \frac{UM_{30mm}}{UM_{coll}} \qquad (2)$$

Where D_{coll} represents the EBT3 measured dose for studied collimator, D_{30mm} the reference EBT3 dose measured with the 30 mm collimator. UM_{30mm} is the UM number used with the 30 mm collimator and UM_{coll} the UM number used for studied collimator.

2.4 Monte Carlo Simulation

The PENELOPE code [9] is one of the several general-purpose MC packages available intended for simulation of particle transport in radiation therapy. This code is reliable mostly due to the advanced physics and algorithms for their electron transport component. Here, the user-code

Table 1 Summary of detectors characteristics. (water: Zeff = 7.42)

Label	Type	Active volume dimensions	Material	Zeff
PinPoint 31014 PTW	Air filled ionization chamber	Ø 2 mm 5 mm height 15 mm^3	Wall: 0.57 mm PMMA 0.09 mm graphite Electrode: Ø 0.3 mm Al	7.64
PinPoint 3D 31016 PTW	Air filled ionization chamber	Ø 2.9 mm 2.9 mm height 16 mm^3	Wall: 0.57 mm PMMA 0.09 mm graphite Electrode: Ø 0.3 mm Al	7.64
Diode SRS 60018 PTW	Unshielded diode	Disk, Ø 1.13 mm 250 µm thick 0,3 mm^3	Silicon	14
Diode P 60008 PTW	Shielded diode	Disk, Ø 1.13 mm 2.5 µm thick 0,0025 mm^3	Silicon	14
Diode E 60017 PTW	Unshielded diode	Disk, Ø 1.13 mm 30 µm thick 0,03 mm^3	Silicon	14
Diode Edge Sun Nuclear	Shielded diode	Square, 0.8 × 0.8 mm^2 30 µm thick 0,019 mm^3	Silicon	14
MicroDiamond 60019 PTW	Synthetic diamond	Disk, Ø 2.2 mm 1 µm thick 0,004 mm^3	Diamond	6
EBT3 Film GAFCHROMIC Ashland	Radiochromic film	278 µm thick	H(56.8), Li(0.6), C(27.6), O(13.3), Al (1.6) (% of each atom)	7.26

Table 2 Number UM delivered according to the cone diameter

Cone diameter (mm)	30	25	20	17.5	15	12.5	10	7.5	5	4
UM number	576	586	599	607	619	639	668	721	830	925

penEasy [12] was used. PenEasy is a modular, general-purpose main program for the PENELOPE Monte Carlo system including various source models, tallies and variance-reduction techniques (VRT). The code includes a new geometry model for performing quadratic and voxelized geometries.

The treatment heads of the Clinac 2100C were simulated according to manufacturer specifications. The geometry of the accelerator is composed of: target, primary collimator, flattening filter, monitor chambers, mirror, beryllium plate, jaws and collimator cone. Source characteristic for the 6 MV photon beam was determined iteratively by varying the energy of the primary electron beam, its energetic dispersion and its shape [13–16]. Parameters used for the primary electron beam was based on monoenergetic 5.95 meV impinging on the target with a Gaussian spatial distribution and a full width at half maximum (FWHM) of 1 mm. With these parameters, PDD and profile comparison between simulation and measurement does not exceed ±1% in homogenous water phantom for field sizes comprised between 2×2 cm^2 and 20×20 cm^2.

Interaction forcing variance reduction and Phase Space File (PSF) techniques were used in the simulation of the treatment head. Bremsstrahlung event is forced in target with factor equal to 20. The phase space was realized just before collimator cone because geometry is not modified upstream. PSF is read several times (between 2 and 5 times) in order to obtain the desired statistical uncertainty. All the variance reduction techniques applied were tested in order to prove that they do not change the physics of the calculation and they provide an unbiased estimate of any scored quantity

The transport energy cut-off of photon and charged particle were respectively 10 and 100 keV. The threshold energies for charged radiative particle and inelastic collisions were set equal to 10 keV. The parameters C1 and C2, modulating the limit between detailed and condensed charged particle simulation, were set to 0.05.

The small volumes of water used for the calculation of D_{coll} in Eq. (1) were taken to be a rectangular parallelepiped with 1 mm side and 1 mm height centered in the beam axis.

3 Results

Table 3 presents the results of the OF measurements for cone diameters between 30 and 4 mm, performed with the passive detector, the active detectors and the Monte Carlo simulations on a Clinac 2100C linear accelerator. Figure 1

shows the comparison between the OF measured by all the detectors and the OF simulated by Monte Carlo considered as the reference.

We can observe large variations: −35 to +10% (Fig. 1) as described in considerable numbers of publications [4–8].

3.1 Statistical and Reproducibility Aspects of the OF Determination

For the OF determination with the Monte Carlo simulations the statistical uncertainties (type-A) were lower than 0.8%.

In the case of active detectors (diodes, ionization chambers and microdiamond) we repeat 3 times the measurements

Table 3 OF measured with all the active detectors, the passive detector and simulated with Penelope for different cone size diameters on a Clinac 2100C linear accelerator. All the OF were calculated with the 30 mm cone diameter as normalization value

Detector	Cone diameter									
	30	25	20	17.5	15	12.5	10	7.5	5	4
PinPoint //	1.000	0.982	0.958	0.941	0.915	0.875	0.824	0.741	0.605	0.512
PinPoint ⊥	1.000	0.982	0.954	0.933	0.905	0.860	0.801	0.699	0.509	0.379
PinPoint3D //	1.000	0.982	0.956	0.938	0.910	0.870	0.816	0.727	0.580	0.474
PinPoint3D ⊥	1.000	0.985	0.955	0.933	0.903	0.854	0.793	0.693	0.529	0.415
Diode SRS	1.000	0.983	0.963	0.949	0.931	0.901	0.863	0.799	0.694	0.623
Diode P	1.000	0.984	0.967	0.955	0.940	0.914	0.878	0.822	0.710	0.615
Diode E	1.000	0.983	0.963	0.950	0.930	0.900	0.863	0.798	0.694	0.626
Edge	1.000	0.985	0.965	0.954	0.936	0.910	0.872	0.808	0.698	0.622
Microdiamond	1.000	0.983	0.962	0.948	0.928	0.896	0.855	0.788	0.672	0.597
EBT3	1.000	0.986	0.961	0.945	0.924	0.884	0.838	0.758	0.652	0.579
MC simulation	1.000	0.983	0.959	0.944	0.920	0.885	0.835	0.764	0.648	0.581

Fig. 1 Comparison of all the OF measured for the passive and active detectors with the gold standard Monte Carlo simulation OF. The OF measurements of the pinpoint and pinpoint 3D detectors were done by positioning them parallel and perpendicular to the axis of the photon beam

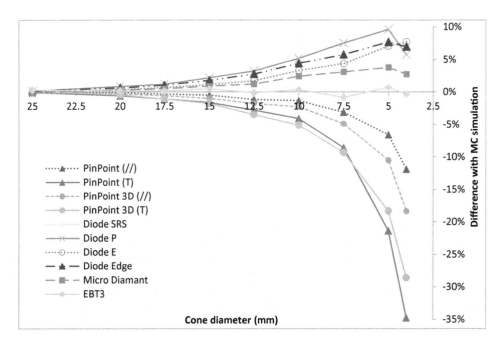

in a water tanker at 3 different days. The uncertainty based on the TRS-398 report uncertainties [17] were respectively 0.1% for the pinpoint chambers, 0.2% for the diodes (SRS, P, E and Edge) and less than 0.3% for the microdiamond.

EBT3 radiochromic film measurements for OF estimation were averaged over a series of thirteen irradiations. This passive detector is known to have noise uncertainty [11, 18] but in accordance with film dosimetry multichannel correction [11, 19, 20] we obtained a relative error uncertainty less than 1.5%.

3.2 Ionization Chambers: Pinpoint and Pinpoint 3D

We used two different orientation of the pinpoint chambers, perpendicular: PinPoint ($\perp$), and parallel: PinPoint (//) of the beam irradiation. The pinpoint and the pinpoint 3D are quite similar in their characteristics with a very similar sensitive volume. The difference is on the isotropic size of the pinpoint 3D which must induce for radial ($\perp$) and axial (//) positions a same answer. In fact Fig. 1 shows better results of the OF measurements in parallel (axial) position for the both detectors. The difference between PinPoint ($\perp$) and PinPoint (//) on the OF measurements are, for cone less than 10 mm diameter, 3% (cone 10 mm) up to 23% (cone 4 mm) and for pinpoint 3D 2.8% (cone 10 mm) up to 10% (cone 4 mm). One can observe a smallest gap for the pinpoint 3D.

The detectors, in radial position, show in comparison with the MC simulation an underestimation of the OF until −2.3% for cone diameter $\geq$ 10 mm and down to −12% for smaller cones only for the pinpoint.

3.3 Diodes: E, P, SRS and Edge

The non-shielded diodes (SRS and E) and the shielded ones (P and Edge) overestimate OF measurements in comparison with MC simulation respectively up to 3.3% and 5.2% for cone diameter $\geq$ 10 mm but in both case >7% for smaller cones. In Fig. 1 one can see that the shielded diodes further overestimate the OF than the non-shielded ones. The shielded component that is interesting for large fields (>10 cm 10 cm) is rather a handicap for our study with small cones [21].

3.4 Microdiamond

The output factor measured with the microdiamond slightly overestimates the value, in comparison with the MC simulation, up to 3.7% for all the cones (Fig. 1). This over-response as observed by Ralston and et al. [5] is in this study less important. This active detector for the smallest cones (4 and 5 mm) has promising results.

3.5 Radiochromic EBT3 Film

Using radiochromic EBT3 film to determine OF is commonly accepted and validated in the literature [6, 8, 22]. Then, as expected, the OF measurement with this passive detector, compared to active ones, is closest to MC simulations. The maximum difference with MC is ±1% whatever the cone size is.

4 Discussion

Figure 2 shows a 1D perpendicular dose profile of a 4 mm diameter cone for a 6 MV photon beam, and the size of the different active detectors: microdiamond, pinpoint, diodes SRS, P, E and Edge. For a smallest field size one can easily consider that first of all we are in a non-equilibrium conditions in SRS treatment inducing a lack of charged particle equilibrium on detectors and secondly the crucial importance of the size of the detector used for OF estimation. At the same time the variation of the electron spectrum induces changes in the stopping power ratios [23, 24]. This effect will be directly in relation with the detectors composition (Table 1) and the non-water equivalent. In Fig. 1, one can class the detectors as those which overestimate the OF (diodes and microdiamond), those who underestimate the OF (ionization chambers) and the film whose answer is equivalent of the MC as reference values. Whereas as explained by Charles et al. [25], the definition of the small field size, as for cone in SRS treatment, output factor measurements should be considered. So, for cone diameter size > of 12.5 mm all the detectors used in this study are able at ±2% to estimate OF. For smaller cones the size of the detectors is the main reason of the underestimation of OF: it is the case of the pinpoint ionization chambers due to a largest air cavity volume and the partial volume effects induced [7, 21].

Diodes in Fig. 2 are the smallest detectors and are commonly used for dosimetry because of their high spatial resolution and small sensitive volume. The overestimation answers of these detectors for OF determinations results from the energy, angular, dose rate dependences and the high density of silicon (Table 1) in comparison with water [4, 8, 26, 27]. For the shielded diodes (Edge, P) the deviation with the MC simulations on the OF measurements is greater than that obtained for the unshielded ones (SRS, E) due to the poor estimation of the scattered radiation in the small field cones [8]. Finally, one can see that the major studies

Fig. 2 1D Dose profile
perpendicular of a 4 mm diameter
cone photon field. The sizes of the
different detectors used are
indicated by colored arrows

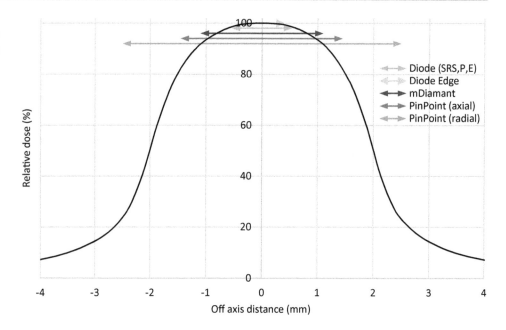

using diode for SRS treatment with circular cone definitely introduces correction factors for the OF determination [4, 8, 27, 28].

The microdiamond detector which is a synthetic diamond material overestimates a little bit the OF. This detector has a good signal to noise ratio and a much better water equivalence than diode detectors studied here, its size (Fig. 2) is small and supplied it a good spatial resolution. Morales and et al. [29] show, in a 6 MV SRS Novalis Trilogy linear accelerator equipped with BrainLAB circular cone (30–4 mm diameters) that microdiamond possesses good dosimetric properties. Chalkley and et al. [30] made the same study on a CyberKnife system and concluded that the microdiamond is an excellent detector. These findings are in agreement with the present work.

The very good results of the film EBT3, confirmed in many publications [6, 8, 22], is close to being a perfect detector: dosimetrically water equivalent, high spatial resolution, minimal energy dependence [31]. However, it is complicated to use and is not a real-time dosimeter and can have uncertainties linked to film polarization, scanner non-uniformity and handling techniques [26]. But for the first determination of OF for a new machine (Cyberknife, linear accelerator) it is an unavoidable detector with recognized accuracy.

5 Conclusions

Monte Carlo simulation, as our gold standard, help us to determine on the wide range of detectors we used which are the most appropriate for measuring the OF. It is known that the radiochromic film, especially EBT3 film is the more

accurate detector for OF determination of stereotactic cones but it is restrictive to use. Pinpoints and diodes, respectively due to inappropriate size of sensitive volume and composition does not seem appropriate without corrective factors below 10 mm diameter cone. MicroDiamond seems to be the best detector for all cones despite its sensitive volume size.

Acknowledgements We thank the Scientific Center of Monaco (CSM) for its collaboration and financial support in this study.

Ethical Requirements The authors declare that they have no conflict of interest.

References

1. Chin, L.S., Regine, W.F.: Principle and practice of stereotactic radiosurgery. 2nd edn. Springer, London (2015).
2. Hayat, M. A.: Brain metastases from primary tumors. Epidemiology, biology and therapy. Vol1. Elsevier, San Diego (2014).
3. Pokhrel, D., Sumit, S., McClinton, C., Saleh, H., Badkul, R., Jiang, H., Stepp, T., Camarata, P., Wang, F.: Linac-based stereotactic radiosurgery (SRS) in the treatment of refractory trigeminal neuralgia: detailed description of SRS procedure and reported clinical outcomes. J. Appl. Clin. Med. Phys. 18(2), 136–143 (2017).
4. O'Brien, D. J., Leon-Vintro, L., McClean, B.: Small field detector correction factors for silicon-diode and diamond detectors with circular 6 MV fields derived using both empirical and numerical methods. Med. Phys. 43(1), 411–423 (2016).
5. Ralston, A., Tyler, M., Liu, P., McKenzie, D., Suchowerska, N.: Over-response of synthetic microdiamond detectors in small radiation fields. Phys. Med. Biol. 59(19), 5873–5881 (2014).
6. Garcia-Garduno, O., Larraga-Gutierrez, J. M., Rodriguez-Villafuerte, M., Martinez-Davalos, A., Celis, M. A.: Small photon beam measurements using radiochromic film and

Monte Carlo simulations in a water phantom. Radiother. Oncol. 96 (2), 250–253 (2010).

7. Lechner, W., Palmans, H., Solkner, L., Grochowska, P., Georg, D.: Detector comparison for small field output factor measurements in flattening filter free photon beams. Radiother. Oncol. 109 (3), 356–360 (2013).

8. Bassinet,C., Huet, C., Derreumaux, S., Brunet, G., Chéa, M., Baumann, M., Lacornerie, T., Gaudaire-Josset, S., Trompier, F., Roch, P., Boisserie, G., Clairand, I.: Small fields output factors measurements and correction factors determination for several detectors for a CyberKnife and linear accelerators equipped with microMLC and circular cones. Med. Phys. 40(7), 071725 (2013).

9. Salvat, F., Fernández-Varea, J. M., Sempau, J.: PENELOPE, A Code System for Monte Carlo Simulation of Electron and Photon Transport. OECD Nuclear Energy Agency, Issy-les-moulineaux (2010).

10. Scott, A. J. D., Nahum, A. E., Fenwick, J. D.: Using a Monte Carlo model to predict dosimetric properties of small radiotherapy photon field. Medical Physics 35, 4671–4684 (2008).

11. Micke, A., Lewis D. F., Yu, X.: Multichannel film dosimetry with nonuniformity correction. Med. Phys. 38(5), 2523–2534 (2011).

12. Sempau, J., Beadal, A., Brualla, L.: A PENELOPE-based system for the automated Monte Carlo simulation of clinacs and voxelized geometries-application to far-from-axis fields. Med. Phys. 38, 5887–5895 (2011).

13. Ding, G. X.: Energy spectra, angular spread, fluence profiles and dose distributions of 6 and 18 MV photon beams: results of Monte Carlo simulations for a Varian 2100EX accelerator. Phys. Med. Biol. 47(7), 1025–1046 (2002).

14. Antolak, J. A., Bieda, M. R., Hogstrom, K. R.: Using Monte Carlo methods to commission electron beams: a feasibility study. Med. Phys. 29(5), 771–786 (2002).

15. Deng, J., Jiang, S. B., Kapur, A., Li, J., Pawlicki, T., Ma, C. M.: Photon beam characterization and modelling for Monte Carlo treatment planning. Phys. Med. Biol. 45(2), 411–427 (2000).

16. Chaney, E. L., Cullip, T. J., Gabriel, T. A.: A Monte Carlo study of accelerator head scatter. Med. Phys. 21(9), 1383–1390 (1994).

17. Andreo, P., Burns, D. T., Hohlfeld, K., Saiful Huq, M., Kanai, T., Laitano, F., Smyth, V., Vynckier, S.: Absorbed dose determinations in external beam radiotherapy. An international code of practice for dosimetry based on standards dose to water. IAEA technique Report Series TRS-398, (2000).

18. Vera-Sanchez, J. A., Ruiz-Morales, C., Gonzalez-Lopez, A.: Characterization of noise and digitizer response variability in radiochromic film dosimetry. Impact on treatment verification. Eur. J. Med. Phy. 32(9), 1167–1174 (2016).

19. Mayer, R. R., Ma, F., Chen, Y., Belard, A., McDonough, J., O'Connell, J. J.: Enhanced dosimetry procedures and assessment for EBT2 radiochromic film. Med. Phys. 39(4), 2147–2155 (2012).

20. Mendez, I.: Model selection for radiochromic film dosimetry. Phys. Med. Biol. 60(10), 4089–4104 (2015).

21. Benmakhlouf, H., Sempau, J., Andreo, P.: Output correction factors for nine small field detectors in 6MV radiation therapy photon beams: a Penelope Monte Carlo study. Med. Phys. 41(4), 041711 (2014).

22. Fan, J., Paskalev, K., Wang, L., Jin, L., Li, J., Eldeeb, A., Ma, C.: Determination of output factors for stereotactic radiosurgery beams. Med. Phys. 36(11), 5292–5300 (2009).

23. Das, I. J., Ding, G. X., Ahnesjo, A.: Small fields: None-equilibrium radiation dosimetry. Med. Phys. 30(1), 206–215 (2008).

24. Verhaegen, F., Das, I. J., Palmans, H.: Monte Carlo dosimetry study of 6 MV stereotactic radiosurgery unit. Phys. Med. Biol. 43 (10), 2755–2768 (1998).

25. Charles, P. H., Cranmer-Sargison, G., Thwaites, D. I., Crowe, S. B., Kairn, T., Knight, R. T., Kenny, J., Langton, C. M., Trapp, J. V.: A practical and theoretical definition of very small field size for radiotherapy output factor measurements. Med. Phys. 41(4), 041707 (2014).

26. Tyler, M., Liu, P. Z. Y., Chan, K. W., Ralston, A., McKenzie, D. R., Downes, S., Suchowerska, N.: Characterization of small-field stereotactic radiosurgery beams with modern detectors. Phys. Med. Biol. 58(21), 7595–7608 (2013).

27. Dieterich, S., Sherouse, G. W.: Experimental comparison of seven commercial dosimetry diodes for measurement of stereotactic radiosurgery cone factors. Med. Phys. 38(7), 4166–4173 (2011).

28. Khelashvili, G., Chu, J., Diaz, A., Turian, J.: Dosimetric characteristics of the small diameter BrainLAB cones used for stereotactic radiosurgery. J. Appl. Clin. Med. Phys. 13(1), 4–13 (2012).

29. Morales, J. E., Crowe, S. B., Hill, R., Freeman, N., Trapp, J. V.: Dosimetry of cone-defined stereotactic radiosurgery fields with a commercial synthetic diamond detector. Med. Phys. 41(11), 111702 (2014).

30. Chalkley, A., Heyes, G.: Evaluation of a synthetic single-crystal diamond detector for relative dosimetry measurements on a CyberKnife. Br. J. Radiol. 87, 1035 (2014).

31. Butson, M. J., Cheung, T., Yu, P. K. N.: Weak energy dependence of EBT gafchromic film dose response in the 50 kVp–10 MVp x-ray range. Appl. Radiat. Isot. 64, 60–62 (2006).

Sensitivity of Electronic Portal Imaging Device (EPID) Based Transit Dosimetry to Detect Inter-fraction Patient Variations

Omemh Bawazeer, Sivananthan Sarasanandarajah, Sisira Herath, Tomas Kron, and Pradip Deb

Abstract

The sensitivity of EPID-based transit dosimetry to detect patient variations between treatment fractions is examined using gamma analysis and a structural similarity (SSIM) index. EPID images were acquired for 3-dimensional conformal (3DCRT) and dynamic intensity modulated (dIMRT) radiation therapy fields in multiple fractions. Transit images were converted to doses, transit dose in the first fraction considered the reference dose. Variations in patient position or weight were then introduced in the subsequent fractions. Positional variations were examined using a lung and a head and neck phantoms. Anatomical variations were examined using a slab phantom in three scenarios, with solid water simulating tissue, medium-density fiberboard simulating fat, and Styrofoam simulating lung. The dose difference between the first and subsequent fractions was computed using various gamma criteria and the SSIM index. Using a criterion of 3%/ 3 mm, EPID can detect positional variations $\geq$ 4 mm, and tissue and fat variations $\geq$ 1 cm, whereas it cannot detect lung variations up to 4 cm. The sensitivity for 3DCRT is higher than for dIMRT. EPID can detect the most variations when using 3%/1 mm. With the SSIM index, EPID can detect a 2 mm positional variation and 1 cm of lung variation. The factor that optimized the sensitivity of EPID was a reduction in the distance to the agreement criteria. Our study introduces the SSIM as an alternative analysis with high sensitivity for minimal variations.

Keywords

Transit EPID dosimetry • Patient variation • SSIM index

O. Bawazeer · S. Sarasanandarajah (✉) · P. Deb
Discipline of Medical Radiations, RMIT University, Melbourne, VIC, Australia
e-mail: S.Sarasanandarajah@iaea.org

O. Bawazeer
Discipline of Sciences, Umm Al-Qura University, Mecca, Saudi Arabia

S. Sarasanandarajah · S. Herath · T. Kron
Department of Physical Sciences, Peter MacCallum Cancer Centre, Melbourne, VIC, Australia

S. Sarasanandarajah
Division of Human Health, International Atomic Energy Agency, Vienna, Austria

T. Kron
The Sir Peter MacCallum Department of Oncology, University of Melbourne, Melbourne, VIC, Australia

1 Introduction

Inter-fraction patient variations, particularly positional and anatomical variations, have been reported as a relevant type of error in radiotherapy [1, 2]. One study conducted over a 10-year period showed that a 40.4% of all the radiotherapy errors were caused by patient setup errors [3]. The movement of the isocenter point by 1 cm was reported to yield a dose difference in organ-at-risk (OAR) of up to 84.2% [4]. Discrepancies of 20% in dose were reported due to morphological changes [5].

The electronic portal imaging device (EPID) has demonstrated its value in transit dosimetry [2], however, studies are limited in the literature regarding the threshold of detectability using EPID for patient variations. Most studies have adopted a common gamma criterion of 3%/ 3 mm [6, 7]. The aim of the present study was to quantify the sensitivity of EPID transit dosimetry for detecting patient variations using various gamma criteria. Variations that cannot be detected by gamma analysis were examined using a structural similarity index (SIMM).

© Springer Nature Singapore Pte Ltd. 2019
L. Lhotska et al. (eds.), *World Congress on Medical Physics and Biomedical Engineering 2018*,
IFMBE Proceedings 68/3, https://doi.org/10.1007/978-981-10-9023-3_86

2 Methods

An a-Si-500 EPID mounted to a Varian linear accelerator 21iX (USA) was used. The EPID was irradiated with 6 MV using a dose rate of 600 and 400 MU/min, at a source-to-detector distance of 150 cm, and with the gantry and collimator set to zero. The cine acquisition mode in the IAS3 software (version 8.2.03) was used to acquire images, and multiple cine image results from each delivery were summed to give one image. The reproducibility of EPID signal were measured and it was less than $\pm 0.2\%$ (1 SD) in two months' time intervals.

Transit images were converted to dose based on the method introduced by Sabet [8], with modification. Raw EPID images were multiplied pixel by pixel with a backscatter correction matrix, and followed by a 2D in-field area correction. The resulting images have been subtracted by an out-field correction factor. Finally, the relative dose images were multiplied by an absolute dose calibration factor. These corrections were applied using an in-house program written in MATLAB (VR2012). A total of two hundred transit EPID images were converted to doses.

Positional variation was examined using head and neck (Alderson RANDO, USA) and stationary lung (CIRS, USA) phantoms. The position of the phantom was moved 2, 3, 4, or 5 mm to the right of the isocenter using the couch position indicators. For anatomical variations (e.g. reduced weight), a phantom consisting of twenty slabs of solid water, each 1 cm in thickness, was examined for three scenarios. Twenty slabs of solid water mimicked tissue. Five slabs of medium-density fiberboard embedded in fifteen slabs of solid water mimicked fat. Five slabs of Styrofoam embedded in fifteen slabs of solid water mimicked lung. Variations were introduced by removing slabs of 1, 2, 3, or 4 cm in each scenario. For each variation, four 3-dimensional conformal (3DCRT) and four clinical dynamic intensity modulated (dIMRT) radiotherapy fields were delivered.

The dose difference between the first and subsequent fractions were computed using gamma analysis. The SNC Patient software (Sun Nuclear Corporation, USA) was used to run the gamma analysis in absolute dose mode, using a global analysis with a 10% threshold and a criterion of 3%/3 mm. The additional criteria of 3%/2 mm, 3%/1 mm, 3%/2 mm, 2%/2 mm, 2%/1 mm, and 1%/1 mm were subsequently applied. A pass rate less than 98% was considered as a detection threshold. Variations that could not be detected by 3%/3 mm were evaluated using the SSIM index for comparisons between reference and variations doses using MATLAB. SSIM index is a method for measuring the similarity between two images. It provides a decimal value from 1 (high similarity) and less if there is dissimilarity. The SSIM index is defined as a product of three comparison elements: luminance(L), contrast(c), and structural image quality (s), and described by the following equation [9].

$$\text{SSIM}_{(x,y)} = \left[1_{(x,y)}\right]^{\alpha} \cdot \left[c_{(x,y)}\right]^{\beta} \cdot \left[s_{(x,y)}\right]^{\gamma}$$

Where $\alpha > 0$, $\beta > 0$, and $\gamma > 0$ are parameters used to adjust relative importance of three elements, a default setting in MATLAB was used for these parameters.

3 Results

3.1 Sensitivity of EPID Using Gamma Analysis

The sensitivity of EPID for positional variations using a criterion of 3%/3 mm is shown in Fig. 1a, b. The smallest positional variation that can be detected is 4 mm, with pass rates of $97.7\% \pm 2.6\%$ (1SD) for the H&N site and $97.2\% \pm 2.9\%$ (1SD) for the lung site. EPID shows a slightly higher sensitivity for the lung than for the H&N treatment site, with pass rates of $92.19\% \pm 3\%$ (1SD) and $97.7\% \pm 1.46\%$ (1SD), respectively, when a 5 mm positional variation was introduced. In addition, EPID exhibits higher sensitivity for 3DCRT than for dIMRT fields, as pass rates for a positional variation of 5 mm were $93.97\% \pm 5.12\%$ (1SD) and $95.9\% \pm 4.05\%$ (1SD), respectively. Figure 1c, d shows the sensitivity of EPID using various gamma criteria. Result demonstrated that the pass rates depend on the tolerance of the distance to an agreement more than the tolerance of dose difference. Criteria of 3%/3 mm and 2%/3 mm yield pass rates higher than criteria of 3%/2 mm and 2%/2 mm. The lowest gamma pass rates correspond to the lowest distance tolerances, which are 3%/1 mm, 2%/1 mm, and 1%/1 mm.

The sensitivity of EPID using 3%/3 mm for anatomical variations is illustrated in Fig. 2a EPID can detect tissue and fat variations ≥ 1 cm, with higher sensitivity for 3DCRT than for dIMRT fields. In contrast, it cannot detect lung variations up to 4 cm in either type of field. Using various criteria in Fig. 2b, c and d, EPID was unable to detect lung variations unless a criterion of 1%/1 mm was applied. The results were similar for positional variations; the lowest gamma pass rates were linked to the lowest distance tolerances.

3.2 Sensitivity of EPID Using the SSIM Index

Figure 3 shows that EPID could detect the smallest positional variation (2 mm), and the SSIM indexes were 0.92 ± 0.01(1SD) and 0.95 ± 0.02(1SD) for H&N and lung sites, respectively. EPID shows a higher sensitivity for the

Fig. 1 Average gamma pass rates for positional variations **a** and **b** with a criterion of 3%/3 mm. **c** with various gamma criteria for a H&N treatment site. **d** with varied gamma criteria for a lung treatment site. Error bar represents the standard deviation of the gamma pass rates of the fields

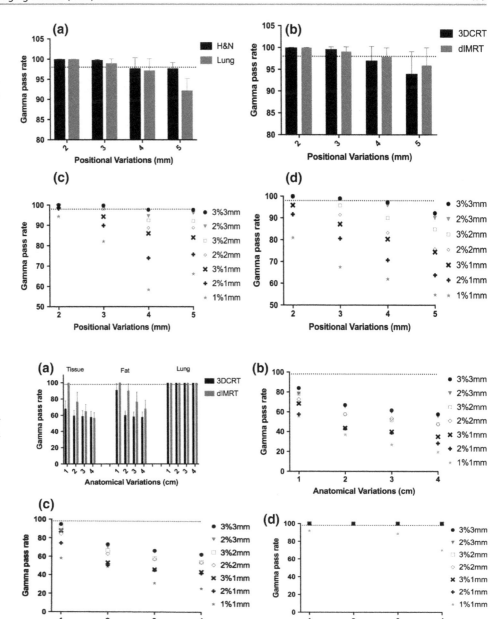

Fig. 2 Average gamma pass rates for anatomical variations **a** with a criterion of 3%/3 mm. **b** with varied criteria for tissue variation, **c** with varied criteria for fat variation **d** with varied criteria for lung variation. Error bar represents the standard deviation of the gamma pass rates of the fields

H&N treatment site than for the lung site. For anatomical variations (lung), EPID can detect minimal lung variations (1 cm), with SSIM indexes of $0.97 \pm 0.001(1SD)$ obtained for 3DCRT and $0.94 \pm 0.01(1SD)$ for dIMRT fields, with higher sensitivity observed for dIMRT than for 3DCRT fields.

4 Discussion

The analysis method described here could increase or decrease the sensitivity of EPID for detecting dosimetry errors. Study found there is difference between an image-based evaluation method (SSIM) and a typical

dose-based evaluation method (gamma analysis) using the EPID. The use of gamma analysis, when introducing positional variations, confirmed that EPID has a slightly higher sensitivity for lung than for H&N treatment sites. One explanation might be that the H&N site consists mainly of bony structures, whereas the lung phantom consists mainly of air cavity structures. A shift in an air cavity will change the transmission dose to a much larger extent than will a shift in a bony structure [7]. Interestingly, EPID had a higher sensitivity for 3DCRT than for dIMRT. This is associated with the uniformity of 3DCRT versus the non-uniformity of dIMRT, and gamma analysis normalized the points to a maximum dose. Investigation of the beam profiles shows a flat profile for 3DCRD, and most points in the field area is

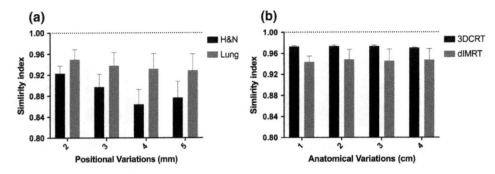

Fig. 3 Average SSIM indexes for, **a** positional variations, **b** anatomical variations (lung). Error bar represents the standard deviation of the SSIM Index of the fields

similar to the maximum dose. Application of gamma analysis gave all these points similar passing or failing criteria. By contrast, with dIMRT, considerable variation in dose profile is encountered across the field area, so the application of gamma analysis yields some points with passing and others with failing criteria. A decrease in the distance to agreement tolerance increases the sensitivity of detecting errors with EPID, and a criterion of 3%/1 mm detected the most variations.

With the SSIM index, the sensitivity of EPID increases to enable detection of small variations, such a 2 mm variation in position or a 1 cm anatomical variation. Unlike gamma analysis, EPID using the SSIM index has a higher sensitivity for lung than for H&N treatment sites, and for dIMRT compared to 3DCRT fields. This could be because the SSIM index computes the contrast between images. The contrast will be more dominant in bony structures in the H&N than in the air cavities in lung sites, and the contrast is greater in the dIMRT field than in the 3DCRT field. However, interpretation of error outcomes from SSIM may require further work to optimize the SSIM algorithm parameters.

5 Conclusion

When using gamma analysis, the sensitivity of EPID transit dosimetry for patient variations depends on the treatment site, the type of delivered technique, and the tissue heterogeneities. Using a gamma analysis tool, the main factor that optimized the sensitivity of EPID is reducing the tolerance of distance to agreement. However, gamma analysis has limitations with respect to detecting a minimal positional variation of 2 mm and anatomical lung variations. Our study offers baseline information about the use of the structural similarity index as an alternative analysis method that has higher sensitivity for these variations.

Acknowledgement The first author acknowledges the financial support given by Saudi Arabian Cultural Mission for this research.

Conflict of Interest Declaration The authors declare that they have no conflict of interest.

References

1. Mijnheer, B.J., et al., *Overview of 3-year experience with large-scale electronic portal imaging device–based 3-dimensional transit dosimetry.* Practical radiation oncology, 2015. **5**(6): p. e679–e687.
2. Bojechko, C., et al., *A quantification of the effectiveness of EPID dosimetry and software-based plan verification systems in detecting incidents in radiotherapy.* Medical physics, 2015. **42**(9): p. 5363–5369.
3. Yeung, T.K., et al., *Quality assurance in radiotherapy: evaluation of errors and incidents recorded over a 10 year period.* Radiotherapy and Oncology, 2005. **74**(3): p. 283–291.
4. Passarge, M., et al., *A Swiss cheese error detection method for real-time EPID-based quality assurance and error prevention.* Medical Physics, 2017. **44**(4): p. 1212–1223.
5. Fidanzio, A., et al., *Routine EPID in-vivo dosimetry in a reference point for conformal radiotherapy treatments.* Physics in medicine and biology, 2015. **60**(8): p. N141.
6. Bojechko, C. and E. Ford, *Quantifying the performance of in vivo portal dosimetry in detecting four types of treatment parameter variations.* Medical physics, 2015. **42**(12): p. 6912–6918.
7. Hsieh, E.S., et al., *Can a commercially available EPID dosimetry system detect small daily patient setup errors for cranial IMRT/SRS?* Practical Radiation Oncology, 2017. **7**(4): p. e283-e290.
8. Sabet, M., et al., *Transit dosimetry in dynamic IMRT with an a-Si EPID.* Medical & Biological Engineering & Computing, 2014. **52**(7): p. 579–588.
9. Wang, Z., et al., *Image quality assessment: from error visibility to structural similarity.* IEEE transactions on image processing, 2004. **13**(4): p. 600–612.

A Study of Single-Isocenter for Three Intracranial Lesions with VMAT-Stereotactic Radiosurgery: Treatment Planning Techniques and Plan Quality Determination

Wisawa Phongprapun, Janjira Petsuksiri,
Puangpen Tangboonduangjit, and Chumpot Kakanaporn

Abstract

Objective: To compare a modified single-isocenter technique between (1) 6 MV and 6FFF and (2) fixed collimator angles and adjusted collimator angles for three intracranial lesions by using VMAT-SRS. Materials and methods: Twenty patterns of three intracranial lesions varying in size and location were generated. The VMAT plans using Eclipse version 13.6 were initially generated according to the University of Alabama, Birmingham's (UAB's) guidelines. Planning parameters including 6 MV, 6FFF, and collimator angles were further modified. All plans were normalized to achieve a 99% dose coverage with 20 and 24 Gy to 5 mm and 10 mm lesions, respectively. Dosimetric parameters, including CI_{RTOG}, $CI_{Paddick}$, GI, HI, mean dose to the normal brain, and V_{5Gy} and V_{12Gy}, were analyzed using Wilcoxon or paired t-test. Results: The 6 MV plans with adjusted collimator angle provided better CI_{RTOG} (1.217 vs. 1.266, p = 0.007) and $CI_{Paddick}$ (8.30 vs. 8.13, p = 0.007), while the 6FFF plans were not statistically different. For both energies, the adjusted collimator angles were less than V_{5Gy} (p < 0.01), V_{12Gy} (p < 0.01) and GI (p < 0.001) compared to the fixed collimator angles of UAB protocol, while the HI index was similar. The plans with 6FFF offered superior plan quality than 6 MV for target coverage (CI_{RTOG} 1.222 vs. 1.266, p = 0.005 and $CI_{Paddick}$ 0.832 vs. 0.813, p = 0.002), dose fall off (GI 7.246 vs. 8.264, p < 0.001) and normal brain sparing (V_{12Gy} 3.802 vs. 4.224, p < 0.001 and V_{5Gy} 22.092 vs. 24.966, p < 0.001). Conclusion: The optimization of collimator angles show an improvement in dose fall-off and normal brain sparing relative to the fixed collimator angles. Plans with 6 FFF provide a better plan quality than 6 MV.

Keywords

Stereotactic radiosurgery • Single isocenter VMAT Brain metastases

1 Introduction

Brain metastasis is the most common brain tumor. Most patients with brain metastases present with multiple lesions. In general, treatment of brain metastasis involves whole brain radiation therapy (WBRT), which leads to acute and late toxicities. Thus, stereotactic radiosurgery (SRS) has been implemented to reduce complications. Gamma Knife SRS is a preferred treatment for multiple intracranial lesions due to its dose characteristic of achieving rapid dose fall off in the adjacent normal tissues. However, the major limitation of Gamma Knife is it is only for use in treating the head and upper neck regions. Also, the invasive rigid fixation system is undesirable for most patients. Therefore, linear accelerator (Linac) has been implemented for SRS treatment to treat not only brain lesions but also other disease sites, utilizing stereotactic body radiation therapy (SBRT).

Many studies have compared the plan qualities of Gamma Knife and the single iso-center VMAT technique by using a LINAC machine for multiple brain metastases treatment [1, 2]. The results showed that the conformity index (CI) of the single isocenter technique was better than that of the Gamma Knife technique However, the gradient index (GI) and low dose area were inferior to the Gamma Knife technique. In 2012, Clark et al. [3] demonstrated quality plans and provided treatment planning guidelines for single isocenter VMAT-SRS for multiple brain lesions. However, there was no beam energy determination for the techniques.

W. Phongprapun · P. Tangboonduangjit
Master of Science Program in Medical Physics, Faculty of Medicine Ramathibodi Hospital, Mahidol University, Bangkok, Thailand

W. Phongprapun (✉) · J. Petsuksiri · C. Kakanaporn
Division of Radiation Oncology, Faculty of Medicine Siriraj Hospital, Mahidol University, Bangkok, Thailand
e-mail: boatwisawa@hotmail.com

© Springer Nature Singapore Pte Ltd. 2019
L. Lhotska et al. (eds.), *World Congress on Medical Physics and Biomedical Engineering 2018*,
IFMBE Proceedings 68/3, https://doi.org/10.1007/978-981-10-9023-3_87

In addition, the effects of collimator angles to achieve quality plans were studied. Wu et al. [4] studied the optimization of beam geometry for multiple lesions, using single isocenter VMAT. The results showed that optimal beam geometry selection would be able to improve rapid dose fall off by reducing the amount of island blocking problems [5].

Therefore, the major objective of our study was to determine the optimal collimator angle for single isocenter VMAT-SRS by using the dynamic conformal arc (DCA) technique. The secondary objective was to compare the plan qualities of 6 megavoltage flat beam (6 MV) and 6 megavoltage flattening filter free (6FFF).

2 Methods and Materials

2.1 Simulated Lesions on CT Image Set

Eight spherical targets with a diameter of 0.5 cm and eight spherical targets with a 1.0 cm diameter were delineated and prepared for treatment planning on an anonymous patient CT data set using Eclipse treatment planning system version 13.6 (Varian medical system, Palo Alto, USA). All targets were equally distributed inside the brain and separated into four quadrants. The lesions were synchronously divided into 2 spherical layers, of which the inner layer was within 5 cm of the center of the brain. The lesions outside the inner layer were classified as the outer layer lesions. We randomly and manually divided these lesions into 20 subsets of three lesions.

2.2 Treatment Planning

Twenty plans were randomly generated, using single isocenter non-coplanar VMAT, with 3 lesions per plan. All treatment plans were performed with the Eclipse treatment planning system. The machine was a Varian TrueBeam STx linear accelerator with 120 high definition multileafs collimator (MLC). The anisotropic analytical algorithm (AAA) was used for dose calculations with 6 MV and 6FFF beams, at a 600 MU/min and 1400 MU/min maximum dose rate, respectively.

For the isocenter placement, we used the inverse volume centroid (presented by Stanhope et al. [6]) for our study. We created a spherical contour of 1.4 cm diameter around each lesion. Then, the spherical contour was subtracted with target lesions. After that, we combined three subtracted spherical contours of target lesions for isocenter placement and treatment planning. All plans used jaw tracking, an optimization grid of 1.25 mm minimum grid, and a 1 mm calculated grid for dose calculation. The dose prescriptions were 20 and 24 Gy for irradiated targets with a 1.0 and 0.5 cm diameter, respectively. The objective was to achieve at least a 99% dose coverage for each lesion.

Initially, we used the UAB's guidelines [3], which recommended using 4 arcs comprised of 1 coplanar full arc and 3 non-coplanar partial arcs, as shown at Table 1.

Subsequently, we modified the specific planning parameters, including the collimator angle and beams, with and without flattening filter, to achieve quality plans. Thus, there were four planning techniques in this study, namely, 6 MV photon with fixed collimator angle (6 MV-Fixed coll.) following UAB's guidelines; 6FFF photons and fixed collimator angles (6FFF-Fixed coll.); 6 MV photons with adjusted collimator angle (6 MV-Adjusted coll.); and 6FFF photons with adjusted collimator angle (6FFF-Adjusted coll.).

For each arc, the collimator angle was adjusted to minimize the "island blocking problem" by the MLCs between the targets. The adjusted collimator angles were generated by DCA for every arc. The adjusted collimator angle was defined as minimal island blocking compared to other collimator angles with the DCA technique.

2.3 Plan Comparisons

We quantitatively assessed plan qualities [7] with CI_{RTOG}, $CI_{Paddick}$, GI, HI (heterogeneity index), mean dose to the normal brain, V_{5Gy} and V_{12Gy}. Doses to the critical structures were evaluated, including the brainstem, optic nerves, optic chiasm and bilateral hippocampi. In addition, the total monitor unit (MU) and beam-on time were evaluated.

The RTOG and Paddick conformal index, GI and HI were calculated for each lesion for all plans. CI_{RTOG} was defined as the ratio of prescription isodose volume (V_{PI}) to tumor volume (TV; $CI_{RTOG} = V_{PI}/TV$). $CI_{Paddick}$ was defined as the ratio of tumor volume within the prescription isodose volume squared to the tumor volume multiplied by the prescription isodose volume [$CI_{Paddick} = (TV$ in $V_{PI})^2/(TV \times V_{PI})$].

GI was defined as the ratio of half of the dose prescription volume ($V_{50\%}$) to V_{PI} (GI = $V_{50\%}/V_{PI}$). HI was the ratio of the maximum dose to the dose prescription. We compared the plan qualities of 6 MV-Fixed coll. versus 6FFF-Fixed coll., and 6 MV-Fixed coll. versus 6 MV-Adjusted coll. The Wilcoxon signed test was used to compare CI_{RTOG}, $CI_{Paddick}$, GI, and HI. Moreover, other indices were compared by paired sample t-test.

3 Results

A summary of the plan quality evaluation for all plans is at Table 2. Our study showed that the conformity indices were superior with a 1 cm lesion than a 0.5 cm lesion. The dose

Table 1 Field geometry following the UAB guidelines for single-isocenter VMAT

Field	Arc length (degree)	Couch angles (degree)	Collimator angle (degree)	Arc direction	Start angle (degree)	Stop angle (degree)
1	360	0	45	CW	181	179
2	180	315	45	CCW	179	359
3	180	45	315	CCW	1.0	181
4	180	90	45	CW	181	1.0

Table 2 The plan evaluation for single isocenter VMAT multi-lesions according to lesion sizes

Lesions size	Planning techniques	Mean CI_{RTOG}	Mean $CI_{Paddick}$	Mean GI	Mean HI
0.5 cm diameter lesions	6 MV-Fixed coll	1.484 ± 0.403	0.697 ± 0.126	12.652 ± 2.474	1.219 ± 0.062
	6FFF-Fixed coll	1.402 ± 0.319	0.727 ± 0.112	10.373 ± 1.887	1.275 ± 0.070
	6 MV-Adjusted	1.397 ± 0.229	0.719 ± 0.097	10.522 ± 1.944	1.253 ± 0.087
	6FFF-Adjusted	1.367 ± 0.170	0.727 ± 0.112	10.737 ± 1.887	1.307 ± 0.085
1.0 cm diameter lesions	6 MV-Fixed coll	1.088 ± 0.037	0.907 ± 0.028	4.674 ± 0.333	1.501 ± 0.065
	6FFF-Fixed coll	1.073 ± 0.027	0.918 ± 0.022	4.390 ± 0.387	1.573 ± 0.123
	6 MV-Adjusted	1.069 ± 0.031	0.920 ± 0.024	4.390 ± 0.370	1.460 ± 0.151
	6FFF-Adjusted	1.075 ± 0.029	0.915 ± 0.023	4.254 ± 0.352	1.573 ± 0.123
All lesions	6 MV-Fixed coll	1.266 ± 0.335	0.813 ± 0.136	8.264 ± 4.333	1.374 ± 0.155
	6FFF-Fixed coll	1.222 ± 0.269	0.832 ± 0.123	7.246 ± 3.434	1.439 ± 0.181
	6 MV-Adjusted	1.217 ± 0.225	0.830 ± 0.121	7.149 ± 3.347	1.367 ± 0.163
	6FFF-Adjusted	1.207 ± 0.186	0.831 ± 0.108	6.524 ± 2.774	1.445 ± 0.155

comparison to normal organs is illustrated at Fig. 1. The total MU and beam-on time is at Table 3.

3.1 Fixed Coll. Versus Adjusted Coll.

The adjusted collimator angles of each arc were obtained from the DCA planning technique by manual adjustment. Table 2 shows that the CI_{RTOG} and $CI_{Paddick}$ of the 6 MV plans with an adjusted collimator angle were significantly better than those of the plans with a fixed collimator (CI_{RTOG} 1.217 vs. 1.266, $p = 0.007$; $CI_{Paddick}$ 8.30 vs. 8.13, $p = 0.007$), while the 6FFF plans were not statistically different. As for the GI, the 6 MV and 6FFF plans with adjusted collimator angle were significantly superior to the plans with a fixed collimator ($p < 0.001$). Nonetheless, HI was not statistically different.

In the case of normal organ sparing, the adjusted collimator angle plans provided a lower mean brain dose, brainstem dose, $D_{40\%}$ of the hippocampus, V_{12Gy}, and V_{5Gy} than the fixed collimator angle plans ($p \leq 0.015$; Fig. 1 and Table 3). The percentage reduction of the V_{12Gy}, V_{5Gy}, and mean dose of the normal brain were 10.8%, 14.8%, and 6.7% for the 6 MV plans, and 6.4%, 9.6%, and 4.5% for the 6FFF plans, respectively. In addition, the total MU and beam-on time of the plans were not statistically different ($p \geq 0.173$).

3.2 6 MV-Fixed Coll. Versus 6FFF-Fixed Coll.

The dose conformity for the 6FFF-Fixed coll. plans ($CI_{RTOG} = 1.222$, $CI_{Paddick} = 0.832$) were significantly better than for the 6 MV-Fixed coll. plans ($CI_{RTOG} = 1.266$, $CI_{Paddick} = 0.813$), $p = 0.005$ and $p = 0.002$, respectively. Furthermore, the HI and GI of 6FFF-Fixed coll. plans ($GI = 7.246$, $HI = 1.439$) were significantly superior to the 6 MV-Fixed coll. plans ($GI = 8.264$, $HI = 1.374$), $p < 0.001$ (Table 2). In terms of normal organ sparing, 6FFF-Fixed coll. allowed lower mean brain doses, D_{40} of bilateral hippocampi, V_{12Gy}, and V_{5Gy} than 6 MV-Fixed coll. ($p \leq 0.032$). However, the doses to the other organs were not statistically different ($p > 0.05$), as shown at Fig. 1 and Table 3. The beam-on time with 6FFF was less than 6 MV, while the mean total MU increased.

3.3 6 MV-Adjusted Coll. Versus 6FFF-Adjusted Coll.

The CI_{RTOG} and $CI_{Paddick}$ were not statistically different ($p \geq 0.707$), while the HI and GI of 6FFF-Adjusted coll. plans were significantly better than the 6 MV-Adjusted coll., $p < 0.001$. Moreover, the 6FFF-Adjusted coll. plans were able to reduce the dose to the normal organ (Fig. 1).

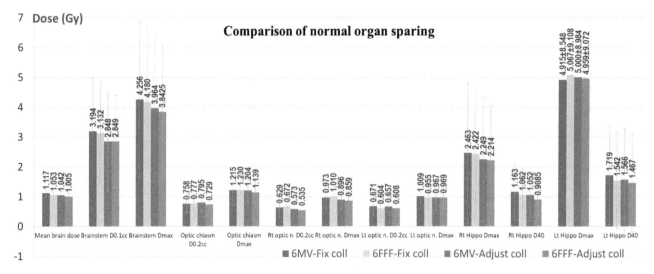

Fig. 1 Comparison of doses to normal organs

Table 3 Plan evaluation of planning parameter for each technique

Plans	V_{12Gy} (cc)	V_{5Gy} (cc)	Total MU	Beam-on time (min)
6 MV-Fixed coll.	4.224 ± 0.766	24.966 ± 3.869	6339.7 ± 759	10.550 ± 1.290
6 MV-Adjusted coll.	3.769 ± 0.915	21.281 ± 4.325	6677.6 ± 1410	11.179 ± 2.316
6FFF-Fixed coll.	3.802 ± 0.835	22.092 ± 4.374	7314.2 ± 1213	5.203 ± 0.862
6FFF-Adjusted coll.	3.561 ± 0.965	19.971 ± 4.458	7253.2 ± 1757	5.188 ± 1.256

4 Discussion

This study shows that single isocenter VMAT plans can produce high quality plans with an optimized collimator angle and beam energy as follows:

4.1 Effect of Collimator Angle on Plan Quality

Wu et al. showed that optimization geometry of treatment in a single isocenter VMAT-SRS could decrease the mean and volume dose to normal brain tissues [4]. They reported that V_4, V_6, V_{12}, and the mean dose of normal brain were reduced by 17.6%, 13.7%, 3.1%, and 8.4%, respectively with the optimization geometry. Our study confirmed that the plans with an adjusted collimator angle presented superior dose conformity V_{12Gy}, and V_{5Gy} than plans with a fixed collimator angle. In contrast, the total MU and beam-on time were increased with an adjusted collimator angle by the effect of field size reduction with tumor-shaped MLC. Another difficulty with MLC is the island blocking in between the lesions. The adjusted collimator angle was able to reduce island blocking problems. Moreover, the planning time was increased from manually adjust collimator angle by the DCA technique by approximately 30–45 min.

4.2 Effect of Photon Beam Energy to Plan Quality

As for the beam energy, the maximum dose rate of 6FFF (1400 MU/min) was more than that of 6 MV (600 MU/min). Thus, the range of dose rate modulations for VMAT optimization of 6FFF was superior to 6 MV. Based on those characteristics, our results showed better dose conformity, dose fall off, V_{12Gy}, and V_{5Gy} with 6FFF than 6 MV. In addition, the beam-on time with 6FFF was less than with 6 MV.

Dzierma et al. [8] studied the dose comparison of intracranial lesions for 6 MV and 7FFF beams. Their results showed no significant difference in dose conformity. However, the 7FFF plans provided better dose fall-off, normal organ sparing, volume of 40 and 5% prescription doses than 6 MV plans. Those results were similar to the findings in our study.

Nonetheless, the beam profile of the 6 MV and 6FFF beams showed no difference in the small field size [9].

4.3 Effect of Lesion Size and Location to Plan Quality

The other factors to affect the dose conformity were the size and location of the lesions. Audet et al. revealed a worse conformity index for smaller lesions than larger lesions [7].

Also, the conformity index was impaired for lesions that were close to normal structures. Our results confirmed those findings by having a wide variation of CI and doses to normal organs. The plan qualities of our study showed similar results with the study by Clark et al. [3].

5 Conclusion

The results of this study show that a single-isocenter technique for 3 intracranial lesions with adjusted collimator angles by DCA provided improved dose conformity, dose fall-off and normal brain sparing than fixed collimator angles. 6FFF also offered better plan qualities than 6 MV beams. Additionally, V_{5Gy}, V_{12Gy}, and doses to normal organs were improved with 6FFF.

References

1. Thomas, E.M., et al., Comparison of plan quality and delivery time between volumetric arc therapy (RapidArc) and Gamma Knife radiosurgery for multiple cranial metastases. Neurosurgery, 2014. **75**(4): p. 409–17; discussion 417–8.

2. McDonald, D., et al., Comparison of radiation dose spillage from the Gamma Knife Perfexion with that from volumetric modulated arc radiosurgery during treatment of multiple brain metastases in a single fraction. J Neurosurg, 2014. **121 Suppl**: p. 51–9.

3. Clark, G.M., et al., Plan quality and treatment planning technique for single isocenter cranial radiosurgery with volumetric modulated arc therapy. Pract Radiat Oncol, 2012. **2**(4): p. 306–13.

4. Wu, Q., et al., Optimization of Treatment Geometry to Reduce Normal Brain Dose in Radiosurgery of Multiple Brain Metastases with Single-Isocenter Volumetric Modulated Arc Therapy. Sci Rep, 2016. **6**: p. 34511.

5. Kang, J., et al., A method for optimizing LINAC treatment geometry for volumetric modulated arc therapy of multiple brain metastases. Med Phys, 2010. **37**(8): p. 4146–54.

6. Stanhope, C., et al., Physics considerations for single-isocenter, volumetric modulated arc radiosurgery for treatment of multiple intracranial targets. Pract Radiat Oncol, 2016. **6**(3): p. 207–13.

7. Audet, C., et al., Evaluation of volumetric modulated arc therapy for cranial radiosurgery using multiple noncoplanar arcs. Med Phys, 2011. **38**(11): p. 5863–72.

8. Dzierma, Y., et al., Planning study and dose measurements of intracranial stereotactic radiation surgery with a flattening filter-free linac. Pract Radiat Oncol, 2014. **4**(2): p. e109–16.

9. Sangeetha, S. and C.S. Sureka, Comparison of Flattening Filter (FF) and Flattening-Filter-Free (FFF) 6 MV photon beam characteristics for small field dosimetry using EGSnrc Monte Carlo code. Radiation Physics and Chemistry, 2017. **135**: p. 63–75.

Evaluation of Time Delay and Fluoroscopic Dose in a New Real-Time Tumor-Tracking Radiotherapy System

Masayasu Kitagawa, Ayaka Hirosawa, and Akihiro Takemura

Abstract

A combined system comprising the LINIAC TrueBeam (Varian Medical Systems, Palo Alt, CA) and a new real-time tumor-tracking radiotherapy system, SyncTraX FX4® (Shimadzu Co., Kyoto, Japan), was installed in our institution. It consists of four pairs of an X-ray tube and a flat panel detector. The system was assessed on beam-on time delay between TrueBeam and SyncTraX FX4 and fluoroscopic dose during a real-time tracking. Delay time was measured by using a tumor-tracking radiotherapy phantom (CALIB PHANTOM ASSY, Shimadzu Co., Kyoto, Japan), in the cases of flattening filtered (FF) 6 MV photon beam, flattening filter-free (FFF) 6 MV photon beam, FF-10 MV photon beam, and FFF-10 MV photon beam with the LINAC True-Beam. Half-value layer (HVL) in mm Al, effective kVp, and air-kerma rate during fluoroscopy were measured using a solid-state detector for the tube voltage (70–110 kV) and the current (50–100 mA). The LINAC delayed from the real-time tumor tracking system on beam-on by 140.9 ± 8.5, 119.8 ± 3.8, 126.1 ± 3.2 and 116.8 ± 9.7 ms for FF-6 MV, FFF-6 MV, FF-10 MV and FFF-10 MV, respectively. The HVL, effective kVp and air-kerma rates from X-ray tube #1 (X-ray tube #1 and #2 were embedded in the patient's head side floor) were 4.98 ± 0.00 mm, 111.2 ± 0.1 kV and 9.14 ± 0.04 mGy/min for 110 kV X-ray at 100 mA. The HVL, effective kVp and air-kerma rates from X-ray tube #3 (X-ray tube #3 and #4 were embedded in the patient's feet side floor), were 5.20 ± 0.00 mm, 110.0 ± 0.1 kV and 11.87 ± 0.06 mGy/min for 110 kV X-ray at 100 mA. These tube voltage and current are the maximum conditions of this study. The time delay of the real-time system is longer than the old system that used image intensifiers. The air-kerma rate from X-ray tube #3 was higher than that from X-ray tube #1.

Keywords

Real-time tumor tracking • Time delay
Fluoroscopic dose

1 Introduction

In radiation therapy, tumor motion during respiration results in significant geometric and dosimetric uncertainties and these lead varying dose of a target in thorax and abdomen. Conventionally, large internal margins (IMs) are required to fully cover the varied tumor positions in free breathing [1]. Techniques for managing respiratory-induced tumor motion, breath-holding [2], tumor-tracking radiotherapy [3], and dynamic tumor tracking delivery techniques [4] are effective in reducing the IMs to reduce normal tissue dose.

A combined system comprising the TrueBeam (Varian Medical Systems, Palo Alt, CA) linear accelerator and a new real-time tumor-tracking radiotherapy system, SyncTraX FX4® (Shimadzu Co., Kyoto, Japan), was installed in our institution. The concepts of SyncTraX FX4 are almost the same as those of the previous Mitsubishi-developed RTRT system (Mitsubishi Electronics Co., Ltd., Tokyo, Japan) [5]. This system consists of four pairs of an X-ray tube and a flat panel detector (Fig. 1). The system perform fluoroscopy using two of four pairs for tumor tracking therapy. The combination of two pairs is selected from the presets which are preconfigured four combinations of these two pairs. The combinations of the X-ray tube #1 and #3 (see Fig. 1) and the X-ray tube #2 and #4 (see Fig. 1) is not preconfigured as

M. Kitagawa (✉) · A. Hirosawa
Department of Medical Technology, Toyama Prefectural Central Hospital, Toyama, Japan
e-mail: cassis030333@gmail.com

M. Kitagawa · A. Hirosawa
Division of Health Sciences, Graduate School of Medical Sciences, Kanazawa University, Kanazawa, Japan

A. Takemura
Faculty of Health Sciences, Institute of Medical, Pharmaceutical and Health Sciences, Kanazawa University, Kanazawa, Japan

© Springer Nature Singapore Pte Ltd. 2019
L. Lhotska et al. (eds.), *World Congress on Medical Physics and Biomedical Engineering 2018*,
IFMBE Proceedings 68/3, https://doi.org/10.1007/978-981-10-9023-3_88

a preset. Some studies have been conducted on real-time tumor-tracking radiotherapy systems, however those systems used image intensifiers [1, 5] and no study have investigated this system.

The aims of this study were to evaluate time delay and fluoroscopic dose of this new real-time tumor-tracking radiotherapy system.

2 Materials and Methods

2.1 Evaluation of Time Delay Using Oscilloscope

Time delay measurement was performed using the oscilloscope (DSO6054A, Agilent Technologies Inc., Santa Clara, CA). The oscilloscope was used to record the timing of beam-on signal from the TrueBeam and gate signal from the SynTraX FX4 together during the tumor-tracking radiotherapy. The calibration phantom CALIB PHANTOM ASSY (Shimadzu Co., Kyoto, Japan) in which eight fiducial markers were embedded was used in the tumor tracking examination. The TrueBeam irradiated radiation at flattening filtered 6 MV photon beam (FF-6 MV), flattening filter-free 6 MV photon beam (FFF-6 MV), flattening filtered 10 MV photon beam(FF-10 MV) and flattening filter-free 10 MV photon beam(FFF-10 MV). The dose rates were set to 300 MU/min for FF-6 MV and FF-10 MV, 600 MU/min for FFF-6 MV, and 1200 MU/min for FFF-10 MV, which are the maximum dose rates in tumor-tracking radiotherapy.

Figure 2 shows a schematic of the oscilloscope image during tumor-tracking radiotherapy using SyncTraX FX4. The yellow line corresponds to the fluoroscopic X-ray beam-on signal from SyncTraX FX4, the green line corresponds to the gate signal from SyncTraX FX4, and the purple line corresponds to the beam-on signal from TrueBeam.

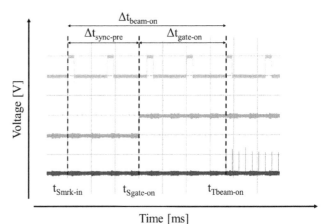

Fig. 2 Output from the oscilloscope. The yellow line corresponds to the fluoroscopic X-ray beam-on signal from SyncTraX FX4 (a), the green line corresponds to the gate signal from SyncTraX FX4 (b), and the purple line corresponds to the beam-on signal from TrueBeam (c)

The beam-on delay ($\Delta t_{beam-on}$) was defined as;

$$\Delta t_{beam-on} = \Delta t_{sync-pre} + \Delta t_{gate-on} \# \tag{1}$$

$$\Delta t_{sync-pre} = t_{Sgate-on} - t_{Smrk-in} \# \tag{2}$$

$$\Delta t_{gate-on} = t_{Tbeam-on} - t_{Sgate-on} \# \tag{3}$$

where $\Delta t_{sync-pre}$ is time duration that SyncTraX FX4 prepares the gate-on signal; from the time that a tracked fiducial marker go into the gating area ($t_{Smrk-in}$) to the time at the gate-on signal ($t_{Sgate-on}$). $\Delta t_{gate-on}$ is time duration from gate-on signal to TrueBean generated beam-on signal ($t_{Tbeam-on}$). The beam-on signal is displayed on the oscilloscope when TrueBeam is permitted to irradiates by SyncTraX and actually irradiates photon beams in tumor-tracking radiotherapy. Those time and time delay were measured three time.

2.2 Measurements of Half-Value Layer (HVL) in mm AL, Effective kVp, and Air Kerma Rate

Figure 1 shows arrangement of X-ray tubes and flat panel detectors. The source-to-isocenter distance (SID) of the X-ray tube #1 is the same as X-ray tube #2 and the SID is equal to 2353 mm. Similarly, that of X-ray tube #3 and #4 was 2081 mm. Thus, all the measurements were performed with X-ray tube #1 and #3. The solid-state detector (AGMS-DM, Radcal Accu-Gold, Monrovia, CA), which can analyze HVL, effective kVp, and air kerma rate of X-ray at the same time, was set at the isocenter. The measurement duration of HVL, effective kVp, and air kerma rate was 60 s. The tube voltage was set to 70, 80, 90, 100, or 110 kV, and the tube current was set to 50, 63, 80, 90, or 100 mA. The frame rate of fluoroscopy was set to

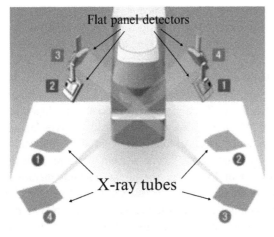

Fig. 1 Positions of X-ray tubes and flat panel detectors

30 frames per second, and the pulse duration was 4.0 ms. The HVL, effective kVp, and air kerma rate were measured three times and the mean ± SD for three measurements was calculated at each combination of the X-ray tube voltage and current.

3 Results

3.1 Evaluation of Time Delay

Table 1 shows the $\Delta t_{sync-pre}$, $\Delta t_{gate-on}$, and $\Delta t_{beam-on}$ for each beams. The $\Delta t_{sync-pre}$ is stable among the beams and the time was about 59 ms. However, the mean values of the $\Delta t_{gate-on}$ for the four beams were varied from 81.9 ms to 57.9 ms. The $\Delta t_{beam-on}$ was also varied depending upon the variation of the $\Delta t_{gate-on}$. The $\Delta t_{gate-on}$ and $\Delta t_{beam-on}$ of the FF beams were longer than that of the FFF beams.

3.2 HVL, Effective kVp, and Air Kerma Rate Measured Using Solid-State Detector for Various Fluoroscopic Conditions

Tables 2 and 3 show the results of the HLVs and effective kVp measurement. The HVLs and effective kVp were consistent with the nominal voltage. However, the HVL values at the tube current of 50 mA seemed to have lower than these at other tube currents each tube voltage.

Figures 3 and 4 show relationship between the X-ray tube current and air-kerma rate (mGy/min) from the X-ray tube #1 and #3. The relationship of the air-kerma rate and tube current had linear relation. However, the air-karma rate for the Tube #3 at the tube current of 50 mA seemed to have lower air-karma rate than the regression line.

4 Discussion

Shiinoki et al. [1] reported the $\Delta t_{sync-pre}$ of SyncTraX were 45.3 ± 0.5 ms, 45.6 ± 0.5 ms, 45.2 ± 0.4 ms, and 45.7 ± 0.4 ms in the case of FF-6 MV, FFF-6 MV, FF-10 MV, and FFF-10 MV photon beams, respectively. The

Table 1 The time delay of the combined system TrueBeam and SyncTraX FX4

Energy of photon	$\Delta t_{sync-pre}$ [ms]	$\Delta t_{gate-on}$ [ms]	$\Delta t_{beam-on}$ [ms]
FF-6 MV	59.1 ± 0.6	81.9 ± 9.0	140.9 ± 8.5
FFF-6 MV	59.0 ± 0.3	60.8 ± 3.6	119.8 ± 3.8
FF-10 MV	59.0 ± 0.5	67.1 ± 2.9	126.1 ± 3.2
FFF-10 MV	59.0 ± 0.5	57.9 ± 9.4	116.8 ± 9.7

mean values of $\Delta t_{sync-pre}$ of SyncTraX were shorter than these of SyncTraX FX4. This was due to the difference of the imaging processing between image intensifiers and flat panel detectors.

The $\Delta t_{beam-on}$ includes $\Delta t_{sync-pre}$, so the variation of the $\Delta t_{beam-on}$ among the beams was similar to that of the $\Delta t_{beam-on}$. These of FF beams (low dose rate) were longer than FFF beams (high dose rate). TrueBeam keeps outputting beam-on signal pulse during tumor-tracking radiotherapy, however, the beam is held by SyncTraX FX4 if the tracked fiducial marker is out of the gating area. When the tracked fiducial marker return in the gating area, SyncTraX FX4 permits TrueBeam beam-on. TrueBeam starts irradiation at the moment of the next beam-on pulse timing. In addition, the oscilloscope begun to display the beam-on signal. So, the $\Delta t_{gate-on}$ were influenced by the timing of the tracked marker into gating area and the beam-on pulse frequency. The beam-on pulse frequency at the high dose rate is shorter than that at the low dose rate. Thus, the $\Delta t_{gate-on}$ and $\Delta t_{beam-on}$ are influenced by the dose rates of the beams. Smitn et al. [6] reported increase of target velocity and time delay lead to increase in IM in tumor-tracking radiotherapy. In tumor-tracking radiotherapy with SyncTrax FX4, using the FFF beams could employ thinner IM than the FF beams.

HVL depends on the target material and filters of the X-ray tube, and the tube voltage. In this study, only tube voltage has changed in the range of 70–110 kV. The HVLs and the effective X-ray energy increased related to the tube voltages. The HVLs at the current 50 mA were lower than these at other tube currents each tube voltage. There is some possibility X-ray energy was changed at the current 50 mA. Japanese industrial standards 4702 recommends that the difference between a nominal tube voltage and effective tube voltage peak should be less than 10% [7]. The kVp values of SyncTraX FX4 meet the requirement.

The solid-state detector can measure a kVp value with the error of 2% or less and a HVL with the error of 10% [8]. The HVL values errors between the X-ray tube #1 and #3 were range of 1.1–4.4%, these were within the solid-state detector's maximum possible percent errors. The kVp values errors between the X-ray tube #1 and #3 were range of 1.1–5.9%, some of values were out of the solid-state detector's error. We think that the difference of the effective kVp between X-ray tube #1 and #3 were caused by manufacturing variation of the X-ray tubes.

The air-kerma rates increased corresponding to the increment of the tube current. These were consistent with previous study [1]. The air-kerma rate from X-ray tube #3 was higher than that from X-ray tube #1 because the SID of the X-ray tube #1 was longer than that of the X-ray tube #3. Almost of all differences between the air-karma rates of the X-ray tube #1 and #3 followed the inverse-square law.

The maximum air-kerma rate values in this study was 11.87 ± 0.06 mGy/min for 110 kV at 100 mA of X-ray

Table 2 The HVLs of each nominal tube voltage and current

Voltage [kV]	X-ray tube	Current [mA]				
		50	63	80	90	100
70	Tube #1	3.13 ± 0.00	3.22 ± 0.01	3.19 ± 0.00	3.19 ± 0.00	3.18 ± 0.00
	Tube #3	3.21 ± 0.00	3.27 ± 0.00	3.28 ± 0.00	3.28 ± 0.00	3.28 ± 0.00
80	Tube #1	3.55 ± 0.00	3.70 ± 0.00	3.66 ± 0.00	3.67 ± 0.00	3.68 ± 0.00
	Tube #3	3.64 ± 0.01	3.74 ± 0.00	3.77 ± 0.00	3.79 ± 0.00	3.79 ± 0.00
90	Tube #1	3.89 ± 0.00	4.09 ± 0.00	4.10 ± 0.00	4.11 ± 0.00	4.11 ± 0.00
	Tube #3	3.98 ± 0.00	4.21 ± 0.00	4.25 ± 0.00	4.26 ± 0.00	4.27 ± 0.00
100	Tube #1	4.27 ± 0.00	4.55 ± 0.00	4.55 ± 0.00	4.57 ± 0.00	4.57 ± 0.00
	Tube #3	4.39 ± 0.00	4.67 ± 0.00	4.71 ± 0.00	4.72 ± 0.00	4.74 ± 0.00
110	Tube #1	4.55 ± 0.01	4.93 ± 0.00	4.96 ± 0.00	4.98 ± 0.00	4.98 ± 0.00
	Tube #3	4.68 ± 0.00	5.10 ± 0.00	5.16 ± 0.00	5.18 ± 0.00	5.20 ± 0.00

Unit [mm]

Table 3 The effective kVp values of each nominal tube voltage and current

Voltage [kV]	X-ray tube	Current [mA]				
		50	63	80	90	100
70	Tube #1	69.6 ± 0.1	70.1 ± 0.0	69.2 ± 0.1	69.3 ± 0.1	69.0 ± 0.1
	Tube #3	67.2 ± 0.1	67.3 ± 0.1	67.3 ± 0.1	67.0 ± 0.1	66.9 ± 0.1
80	Tube #1	80.1 ± 0.1	82.1 ± 0.1	80.0 ± 0.1	79.8 ± 0.1	80.0 ± 0.1
	Tube #3	77.3 ± 0.1	77.4 ± 0.1	77.6 ± 0.1	77.7 ± 0.1	77.6 ± 0.1
90	Tube #1	91.3 ± 0.1	93.6 ± 0.1	91.2 ± 0.1	91.1 ± 0.1	90.9 ± 0.1
	Tube #3	88.7 ± 0.1	88.1 ± 0.0	88.8 ± 0.1	88.7 ± 0.1	88.9 ± 0.3
100	Tube #1	102.1 ± 0.1	103.8 ± 0.1	101.7 ± 0.1	101.5 ± 0.1	101.3 ± 0.3
	Tube #3	99.9 ± 0.1	99.8 ± 0.1	99.7 ± 0.2	99.4 ± 0.1	99.7 ± 0.1
110	Tube #1	112.9 ± 0.2	114.2 ± 0.1	111.8 ± 0.1	111.8 ± 0.1	111.2 ± 0.1
	Tube #3	110.7 ± 0.2	110.4 ± 0.1	110.4 ± 0.2	109.9 ± 0.1	110.0 ± 0.1

Unit [kV]

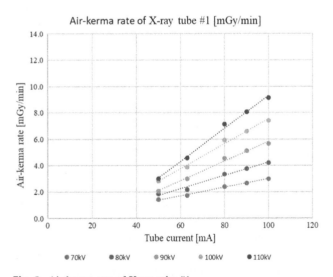

Fig. 3 Air-kerma rate of X-ray tube #1

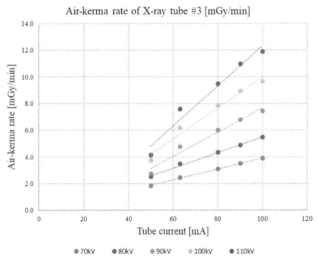

Fig. 4 Air-kerma rate of X-ray tube #3

tube #3. International atomic energy agency (IAEA) recommends fluoroscopic dose should be 25 mGy/min or less [9]. All combinations of tube voltage and tube current examined in this study were under the recommended dose.

5 Conclusions

We evaluated time delay and fluoroscopic dose in real-time system using the TrueBeam linear accelerator and the Sync-TraX FX4 system. The time delay of the SyncTraX FX4 system were 140.9 ± 8.5 ms, 119.8 ± 3.8 ms, 126.1 ± 3.2 ms, and 116.8 ± 9.7 ms in the case of FF-6 MV, FFF-6 MV, FF-10 MV, and FFF-10 MV photon beams. The time delay of FF beams longer than that of FFF beams. The air-kerma rate from the X-ray tube #1 and #3 of SyncTraX FX4 was 9.14 ± 0.04 and 11.87 ± 0.06 mGy/min. The air-kerma rate at the isocenter from X-ray tube #3 is higher than that from X-ray tube #1. The fluoroscopic dose of SyncTraX FX4 were lower than IAEA guidance level.

References

1. T. Shiinoki, S. Kawamura, T. Uehara, et al: Evaluation of a combined respiratory-gating system comprising the TrueBeam linear accelerator and a new real-time tumor-tracking radiotherapy system: a preliminary study. Journal of Applied Clinical Medical Physics 17(4), 202–213 (2016).
2. M. Nakamura, K. Shibuya, T. Shiinoki, et al: Positional Reproducibility of Pancreatic Tumors under End-exhalation Breath-hold Conditions Using a Visual Feedback Technique. Int. J. Radiation Oncology Biol. Phys. 79(5), 1565–1571 (2011).
3. Berson A. M., Emery R., Rodriguez L., et al: Clinical experience using respiratory gated radiation therapy: comparison of free-breathing and breath-hold techniques. Int. J. Radiation Oncology Biol. Phys. 60(2), 419–426 (2004).
4. Poels K., Dhont J., Verellen D., et al: A comparison of two clinical correlation models used for real-time tumor tracking of semi-periodic motion: a focus on geometrical accuracy in lung and liver cancer patients. Radiother Oncol. 115(3), 419–424 (2015).
5. H. Shirato, S. Shimizu, T. Kunieda, et al: Physical aspects of a real-time tumor-tracking system for gated radiotherapy. Int. J. Radiation Oncology Biol. Phys. 48(4), 1187–1195 (2000).
6. Smith W. L., Becker N.: Time delays in gated radiotherapy. Journal of Applied Clinical Medical Physics 10(3), 140–154 (2009).
7. JIS 4702: General requirements for high-voltage generators of medical X-ray apparatus. (1999).
8. Brateman L.F., Heintz P.H.: Solid-state dosimeters: a new approach for mammography measurements. MedPhys 42(2), 542–557(2015).
9. IAEA: International Basic Safety Standards for Protection against Ionizing Radiation and for the Safety of Radiation Sources Safety Series No.115. (1996).

Stereotactic Radiotherapy for Choroidal Melanoma: Analysis of Eye Movement During Treatment, Eye Simulator Design and Automated Monitoring System Development

F. Souza, J. Valani, O. D. Gonçalves, D. V. S. Batista, S. C. Cardoso, and D. D. Pereira

Abstract

Malignant choroidal melanoma is a choroidal tumor arising in the layer of blood vessels located beneath the retina. It is a rare occurrence, affecting five to every million inhabitants and rapidly evolving into metastasis. The stereotactic radiotherapy is one of the possible treatments for the disease, it consists in the application of small photon beams directed to the PTV, the prescribed dose is delivered in multiple sessions. The success of treatment depends directly, among others biomedical factors, on the precision of the application and how the treatment method can hold still the patient's eye. This study aims: (a) to analyze the eye's movement during the treatment for choroidal melanoma and the non-invasive method for fixating the patient's eye; (b) to develop a monitoring software for patient eye gaze; (c) design a mechanical eye to verify the monitoring system. As a result, it is expected that the system will be improved, raising the precision and accuracy of application and reducing the damage to healthy tissues and side effects. After analyzing the positions of the iris center and calculating its displacements relative to the orthophoria point the standard deviation values of 0.368 mm and 0.364 mm were found for the X-axis and Y-axis, respectively. The LED lamp is enough for a noninvasive fixation of the patient's eye during treatment. The mechanical eye now is able to verify the monitoring system software, as both work with resolutions within the limit given by the eye movement analyzes.

Keywords

Stereotactic radiotherapy • Choroidal melanoma
Eye simulator

1 Introduction

1.1 Malignant Choroidal Melanoma

Malignant choroidal melanoma is a tumor arising in the layer of blood vessels, named choroid, located beneath the retina. It is a type of uvea melanoma and rapidly evolves into metastasis [1, 2].

The incidence of uvea melanoma is estimated to be 6–7 occurrences per million inhabitants per year in Europe [3]. Choroidal melanoma accounts for 68–91% of melanomas occurring in the uvea [4].

There are several types of treatments for this pathology, the choice of which one will be performed is related, among other factors, to the location and size of the tumor.

Radiotherapy has become the most common treatment used when it is intended to preserve the eye with the melanoma and its vision. This technique uses ionizing radiation to interact with cellular environment. It modifies the intracellular elements, causing harmful biological effects to the cell. By delivering the right amount of radiation in the tumor tissues, it can destroy malignant cancer cells as well as helps to preserve healthy tissues surrounding the tumor [5].

External beam radiotherapy has been applied for decades in order to treat tumors that cannot be treated by brachytherapy [6]. Dunavoelgyi R. et al. showed that the tumor control rate of the EBRT after a 20 months period did not have significant difference when compared to results obtained using other types of treatments such as gamma knife, brachytherapy and proton therapy [7].

There are some obstacles that prevent the treatment from achieving the maximum precision in case of choroidal melanoma by EBRT. The doses are delivered in multiple

F. Souza (✉)
Polytechnich School, Federal University of Rio de Janeiro,
Rio de Janeiro, RJ 21941-901, Brazil
e-mail: felipemlucas@poli.ufrj.br

J. Valani · O. D. Gonçalves · S. C. Cardoso · D. D. Pereira
Physics Institute, UFRJ, 68528 Rio de Janeiro, RJ 21941-909,
Brazil

D. V. S. Batista
Institute of Radioprotection and Dosimetry, National Commission
of Nuclear Energy, Rio de Janeiro, RJ 22780-160, Brazil

© Springer Nature Singapore Pte Ltd. 2019
L. Lhotska et al. (eds.), *World Congress on Medical Physics and Biomedical Engineering 2018*,
IFMBE Proceedings 68/3, https://doi.org/10.1007/978-981-10-9023-3_89

sessions, so that the patient needs to be accurately repositioned between them. Another problem is that the eye moves during treatment and without a system to fix the eye, the PTV needs to be larger, increasing the damage to healthy tissues and consequently, to the vision.

Hence, the success of this treatment depends directly, among other biomedical factors, on the precision of the application and how the treatment method can hold still the patient's eye.

Therefore, this work aims:

(a) To analyze the movement of patient's eye during treatment of choroidal melanoma using a led lamp as a non-invasive method for immobilize the patient's gaze;
(b) To develop a software to monitoring the patient's gaze;
(c) Design a mechanical eye to calibrate the monitoring system.

2 Materials and Methods

2.1 Radiotherapy Equipments

The data to develop the prototype of the monitoring system was obtained from a radiotherapy section performed on a Novalis® equipment from Brainlab® with the patient's skull immobilized using a stereotactic mask made of thermoplastic material (a) and positioned under a structure called array (b), both manufactured by Brainlab®.

In order to analyze the iris displacement during the radiotherapy treatment a webcam (c), a ten meters HDMI

cable (d) and a laptop located on the outside of the treatment room (see Fig. 1) were adapted to the standard setup.

In addition, during the treatment the patient was instructed to look with the dis-eased eye at the LED light (e) located on the same vertical axis of the isocenter and positioned above the webcam.

2.2 Motion Analysis Method

The video of the patient's eye was recorded and later the displacements were analyzed. The videos were transferred to the software Tracker [8], a motion analysis software, and separated as their lighting quality.

In the analysis software, the video was digitally processed to increase the contrast between the iris and the conjunctiva. As a result, it was possible to locate the center of the iris by calculating the centroid of an ellipse and then following the movement of the iris center in an automated way, returning the values of the position in the x and y axis (horizontal and vertical, respectively) in each frame of the video.

Displacement Analysis. With the position of the iris center in each frame x_i and y_i where 'i' represents the frame of the video, it was possible to calculate its displacement relative to the first frame of the video. In other words, the displacement between each frame and the beginning of treatment. As the subject was instructed to look at the LED at the beginning of treatment, the eye on this frame is considered the orthophoria point, the condition of normal equilibrium of the eye muscles or muscular balance.

2.3 Automated Monitoring Software

To automatically monitoring the eye movement during the choroidal treatment, a software was developed using Python programming language. The software detects the center of the iris calculating the centroid of an ellipse coincident with the iris.

The algorithm for this procedure converts red-green-blue (RGB) video to grayscale, so each pixel is an element of an $m \times n$ matrix (where $m \times n$ is the resolution of the video) with an intensity ranging from 0 (totally black) to 255 (totally white). In each frame, this value is read, and if it is a given threshold, that element is stored. After a scan through each element of the array, the centroid position for the elements that have been stored is calculated (see Fig. 2).

2.4 Mechanical Eye Simulator

To calibrate the monitoring system, a mechanical eye model was developed. The mechanical structure of the model was

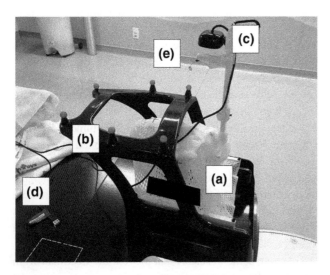

Fig. 1 Equipment setup for stereotactic radiotherapy for choroidal melanoma

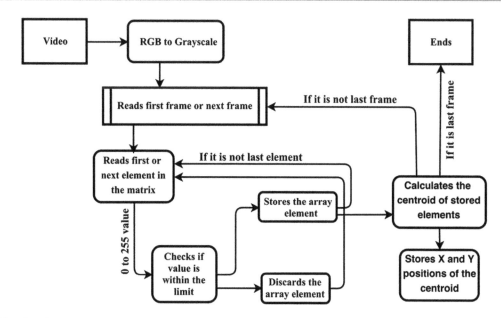

Fig. 2 Algorithm flowchart

printed in a 3D—cube 3D 3rd generation printer. Cylindrical bearing were used to perform the coupling between the moving parts, thereby reducing friction and possible misalignments. The eye itself was doll eye with similar dimensions to those of a human eye, according to Gullstrand Schematic Eye [9].

The movement of the structure is performed by two 5 V 28BYJ-48 step motors (vertical and horizontal movements), both in a half-step mode, controlled by a Raspberry Pi 3 Model B that sends control information to an ULN200 chip on the controller board.

A home button was made using gold foiled copper plates attached to the structure base, linked to the GPIO ports of the

Raspberry Pi 3 Model B and a metal rod connected to the drive shaft and connected to the ground.

3 Results

3.1 Displacement of the Iris Center Relative to the Orthophoria Point

After analyzing the positions of the center of the iris in the software and calculating its displacements, the cumulative displacement distribution relative to the orthophoria point and displacement histogram were plotted as shown in

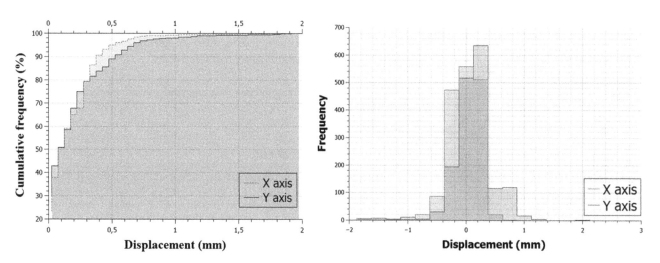

Fig. 3 Cumulative distribution of displacements and displacement histogram relative to the orthophoria point

Fig. 3. The standard deviation values of 0.368 mm and 0.364 mm were found for the X-axis and Y-axis, respectively.

3.2 Monitoring Software

In preliminary tests, the program proved to be stable being able to detect the centroid of a circumference, store the values and return them to the user. It was also possible to detect variation of the position of the centroid over 1 unit of pixel for both the vertical axis and the horizontal axis.

With this precision, it was possible to calculate the required distance between lens and object equivalent to 104.2 mm which allows 1 pixel in the recording to have a dimension of 0.10×0.10 mm.

3.3 Mechanical Eye

Two pitch motors were coupled to the independent turning system controlled by a Raspberry Pi 3 Model B (see Fig. 4a). The PLA is a lightweight material and was used to print the simulator [10]. The advantage of having a light system is lower resistance from friction (see Fig. 1b).

Because of the gyroscope-based design, the simulator can independently move on both horizontal and vertical axis, performing movements similar to that of a human eye (see Fig. 4b, c) [11].

The motors (see Fig. 4e) are used in a half-step mode, which increases the resolution of the mechanism, reaching approximately 0.1 degree in each half-step.

A home button was planned in order to remove accumulated errors. Besides that, it is used to ensure that the system has a pre-established origin (see Fig. 4d).

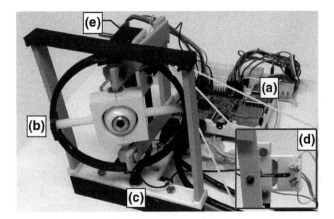

Fig. 4 Mechanical eye simulator and its components

4 Conclusion

A person, when oriented to look at the LED, moves his eye within a certain limit. The data showed that this variation is compatible with other similar studies [12] and that it is within the safety margin established by the radiotherapists, which is 3.5 mm for all sides of GTV.

The LED lamp is enough for a noninvasive fixation of the patient's eye during treatment. Besides that, it is known that during the treatment the visual acuity of the patient decays as a side effect [13], consequently decreasing his ability to locate the LED lamp as the therapy progresses. An alternative to the current system would be necessary due to the loss of visual acuity of the eye affected by the disease.

The eye tracking software is able to read centroids with a resolution of 0.10×0.10 mm (1 pixel) when using a camera with resolution of 720×480 pixels, positioned 104.2 mm away from the center of the eye. The recording speed was set to 60 fps, the maximum allowed due to software and hardware limitations. Future studies should be performed to determine a minimum interval between frames.

The mechanical eye moves in both axes, which can perform movements comparable to those of the patient's eye during treatment. The resolution of 0.2 degrees of rotational motion is equivalent to a 0.04 mm translation of the iris center into a projection in a plane parallel to the plane of the camera (considering the eyeball as a sphere of 24.00 mm of diameter).

The resolution of the mechanical eye is compatible to the eye movement of the patient treatment allowing the verification of results and the traceability of the measurements.

In case of successful verification of the software, it will be possible to develop a new algorithm causing the radiation to be stopped giving an unintended movement of the eye. In other words, stop the radiation when the patient moves the eye away from the treatment position for a given time. The displacement time needs further studies in order to ignore the blink of the eye, for example. This new algorithm will allow the reduction of damage to healthy tissues present in the eyeball, reducing the side effects.

Conflict of Interest The authors declare that they have no conflict of interest.

References

1. Chauvel P, Sauerwein W, Bornfeld N et al. Clinical and technical requirements for proton treatment planning of ocular diseases. In: Wiegel T, Bornfeld N, Foerster MH, et al. Editors. Radiotherapy of ocular disease. Basel: Karger; 1997, Pp. 133–142.
2. Dieckmann K, Georg D, Zehetmayer M et al. LINAC based stereotactic radiotherapy of uveal melanoma: Four years clinical experience. Radiother Oncol 2003;67:199–206.

3. Cruz, D. P. & lopes, J. M. Características clínico-laboratoriais e sobrevida em doentes com melanoma. Arquivos de medicina vol. 23, nº 2, pp. 45–57.

4. Bicas, H. E. A. Oculomotricidade e seus fundamentos, Arq. Bras. Oftalmol. 2003;66:687–700.

5. Bernhard Petersch, K. D. Automatic real-time surveillance of eye position and gating for stereotactic radiotherapy of uveal melanoma. *Medical Physics*, pp. 3521–3527.

6. Freire, J.E., Brady, L.W., DePotter, P. et al. Malignant melanoma of the posterior uvea. in: C.A. Perez, L.W. Brady et al, (Eds.) Principles and practice of radiation oncology. 3rd ed. Lippincott-Raven, Philadelphia, PA; 1997:871–874.

7. Dunavoelgyi R., Dieckmann K., Gleiss A., Sacu S., Kircher K., Georgopoulos M. et al. (2011). Local tumor control, visual acuity, and survival after hypofractionated stereotactic photon radiotherapy of choroidal melanoma in 212 patients treated between 1997 and 2007. I. J. Radiation Oncology Biology Physics, pp. 199–205.

8. Loo W, Tze L. Video Analysis and Modeling Performance Task to Promote Becoming Like Scientists in Classrooms. American Journal of Educational Research. 2015; 3(2):197–207.

9. D. A. Robinson the mechanics of human saccadic eye movement, j. Phy8iol. (1964), 174, pp. 245–264.

10. Bose S.,Vahabzadeh S. and Bandyopadhyay A. Bone tissue engineering using 3D printing 2013;16;496–504.

11. Robinson, D. A., Gordon, J. L. & Gordon, a. S. (16 de Abril de 1986). A Model of the Smooth Pursuit Eye Movement System. *Biological Cybernetics Springer-Verlag*, p. 15.

12. Joachim B, Bernhard P, Dietmar G et al. A noninvasive eye fixation and computer-aided eye monitoring system for linear accelerator-based stereotactic radiotherapy of uveal melanoma. Int J Radiation Oncology Biol Phys 2003;56:1128–1136.

13. Dunavoelgyi R, Dieckmann K, Gleiss A et al. Radiogenic side effects after hypofractionated stereotactic photon radiotherapy of choroidal melanoma in 212 patients treated between 1997 and 2007. Int J Radiation Oncology Biol Phys 2012;83:121–128.

Configuration of Volumetric Arc Radiotherapy Simulations Using PRIMO Software: A Feasibility Study

Jorge Oliveira, Alessandro Esposito, and João Santos

Abstract

Volumetric Modulated Arc Therapy (VMAT) uses non-uniform intensity fields allowing volumetric complex dose distributions. The simultaneous MultiLeaf Collimator (MLC) motion and Gantry rotation pose difficulties in the dose distribution calculation by Treatment Planning Systems (TPS). Furthermore, a dedicated Quality Assurance (QA) program and patient-specific dose verifications are requested. Monte Carlo dose calculation in Radiotherapy (RT) is a gold standard due to its most detailed description of radiation-matter interaction. Recently, the PRIMO software was proposed, providing several built-in RT units models, including TrueBeam. Nevertheless, VMAT is not implemented yet. In this work, TrueBeam was simulated in PRIMO using 6 and 10 MV in Flatness Filter Free (FFF) mode and at 15 MV with Flatness Filter inserted. The results were validated by Gamma Function (2%, 2 mm) based on reference measurements in water tank. The VMAT complex dynamic delivery is divided into a customizable number of probabilistically sampled static configurations of jaws, leaves and gantry angles. In-house algorithms were developed to interpolate the LINAC geometrical information along the process once the planned information is retrieved from the TPS output DICOM file. A Graphical User Interface (GUI) was developed to assist the user to configure PRIMO to simulate complex deliveries. Static simulations in reference conditions showed always >97% of Gamma points <1 for PDD and profiles at various depths and fields sizes for the 6, 10 and 15 MV primary beam respectively. The GUI properly read, manipulated and wrote the configuration data in a "ppj" format, which was accepted by PRIMO. The dynamic jaws, MLC and gantry motion were positively assessed by visual inspection of the static beam configuration in PRIMO. Dynamic irradiations were simulated and the gamma function tests against reference dose distributions showed good agreement with typical QA criteria. A GUI to configure PRIMO for VMAT irradiations allowed to enable a flexible workflow for simulating a general dynamic treatment.

Keywords

External beam radiotherapy • VMAT
Monte Carlo simulation • PENELOPE • PRIMO

1 Introduction

In External Radiotherapy (RT), the idea of modulating the radiation field through a dynamic collimation system, developed into highly complex treatment modalities such as Intensity Modulated Radiation Therapy (IMRT), that makes use of collimation system movement to modulate the radiation field in intensity, and Arc Radiotherapy (ART) or Volumetric Modulated Arc Therapy (VMAT) technique, which extends the IMRT concept introducing the rotation of the gantry synchronized with Mulitleaf Collimator (MLC) motion during the treatment [1]. The introduction of continuous motion of the beam modifiers components and treatment delivery, introduces uncertainties and presents complexity in the dose distribution calculation by Treatment Planning Systems (TPS). In order to deal with the introduced uncertainties, a dedicated Quality Assurance (QA) program and patient-specific dose verifications [2] are mandatory. The patient specific QA and the necessary occasional

J. Oliveira (✉) · J. Santos
Medical Physics, Radiobiology and Radiation Protection Group IPO Porto, Research Center (CI-IPOP), Portuguese Oncology Institute of Porto (IPOPorto), Porto, Portugal
e-mail: Jorgeduardo@gmail.com

A. Esposito
Radiation Oncology Department, Princess Alexandra Hospital, Brisbane, QLD, Australia

© Springer Nature Singapore Pte Ltd. 2019
L. Lhotska et al. (eds.), *World Congress on Medical Physics and Biomedical Engineering 2018*,
IFMBE Proceedings 68/3, https://doi.org/10.1007/978-981-10-9023-3_90

corrections are time consuming procedures The use of Monte Carlo (MC) methods [3] can be a useful tool to provide an independent dose calculation to understand the reasons of occasional fail of the QA verifications. Several codes have been available for simulation of linacs, such as GEANT4, EGSnrc/BEAMnrc, PENELOPE, FLUKA and MCNP [4] but the not so trivial set-up of the RT model and treatment added to the generally long calculation time, have prevented its routinely clinical use. Recently, a new software simulation system named PRIMO, based on PENELOPE features, was developed [5]. It is provided with several built-in linac models, including a TrueBeam model, named FakeBeam [6], used along this work since a new unit has been recently commissioned at our institute. This software has a user-friendly approach, which is a suitable and competitive characteristic for clinical activity. Although advanced features such as IMRT/ART are not introduced yet, PRIMO can simulate a maximum of 180 static fields per simulation. This feature can be used to implement dynamic procedures simulations according to the Static Component Simulation (SCS) [7] or Position Probability Sampling (PPS) [8] strategies for dynamic treatment simulation. In this regard, previously promising results to test the feasibility of study of PRIMO to simulate an IMRT [9] provided the motivation to extend the feasibility study of PRIMO usage to simulate a VMAT treatment.

In order to close the gaps in a possible workflow that integrates the TPS and PRIMO to simulate the dynamic treatments, tools were developed and the code encapsulated in a Graphical User Interface (GUI) application.

2   Methods

2.1   Monte Carlo PRIMO Simulation

PRIMO makes use of the PENELOPE code calculation features. It follows a user-friendly approach and provides different linac heads and MLC models. In addition, PRIMO allows multi-beam simulations, with different geometric setup for each beam. The complete simulation is organized in a pipeline consisting in three stages, named s1, s2 and s3. Stage s1, is the tuning of the primary beam parameters in order to get agreement with the experimental measurements set and outputs a Phase Space file (PHSP), phsp1, calculated at a position immediately above the movable jaws. Stage s2, simulates the transport of the particles in phsp1 through the collimation system, including jaws and MLC, and outputs a phps2 calculated at the entrance of the phantom. Stage s3, tracks the particles stored in phsp2 into the phantom. While the s1 and s2 outputs PHSP files that follow the IAEA standard [10], the s3 outputs a 3D dose distribution in a specific PRIMO output.

All the simulations in this study were performed with PRIMO version 0.1.5.1307 installed on a machine with an Intel®Xeon® CPU E5-2660 V3@2,60 GHz with 16 Gb of RAM and with 32 CPU cores available.

2.2   PRIMO Models Validation

TrueBeam Linac head. To get a working model of the TrueBeam linac head and taking advantage of the PHSP importation functionality of PRIMO, a set of PHSP files pre-calculated with GEANT4, available from Varian, were imported into a PRIMO project based on a Varian 2100 geometry model as recommended in the software user manual [11]. The final s1 output phsp1 consisted on the composition of 10 Varian files to represent a number of primary histories in the order of 10^9. In order to validate the stages s2 and s3 simulations, they were modeled to match the commissioning conditions. The evaluation of the simulations quality was performed for the PDD and dose profiles of 6FFF, 10FFF and 15X MV energies using the intrinsic PRIMO Gamma analysis tool. The pre-established passing criterion was >95% point with gamma <1 and tolerances of 2%–2 mm as the Gamma parameters.

Collimator. For validation purposes, a particular arrow shape pattern, Fig. 1, was defined using a 120HD MLC and a setup with the geometry and materials usually used in QA measurements was simulated (i.e. a 15 cm height RW3 multi-slab phantom with Source Surface Distance—SSD— of 95 cm). Both simulations and acquired experimental data was compared. In the experimental case, a Gafchromic film was inserted between two adjacent slabs at 5 cm depth in the phantom. This process was performed for the 6FFF and 10FFF MV energies.

2.3   Dynamic Simulation Workflow Validation

Movement sampling. In a typical VMAT procedure the beam is always on, while the linac gantry, jaws and MLC components perform a continuous synchronized motion. At the planning phase, the TPS defines a set of control points relating the position of every component with the cumulative Monitor Units (MUs) delivered. From this, it is created a corresponding normalized MU_{index} set that spans between 0.0 and 1.0. Following a Probability Position Approach (PPS) this cumulative distribution function is randomly inversely sampled. The end result is a custom number of static beams configurations, that represent the dynamic procedure. Each beam contributes with a different weight, proportional to the sampled MU interval.

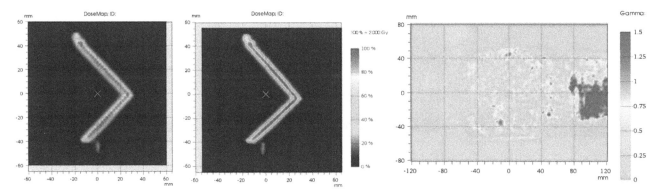

Fig. 1 From left, Gafchromic measurement, PRIMO calculation, gamma evaluation

2.4 Graphical User Interface Application (GUI)

Building on previous results [9], in order to expand the procedure to VMAT simulations, some open gaps and missing links in the workflow between TPS and PRIMO were approached with new MATLAB® code. The main application features were: to import of DICOM RT-Plan or the TrajectoryLogFile, to create a PRIMO input file by randomizing static configurations to reproduce the dynamic delivery and to export 2D dose distributions.

VMAT like plan. For the validation of the proposed workflow, of the sampling algorithms and of the developed functionalities, a 6FFF TPS pre-planned VMAT treatment, calculated with Accuros XB dose algorithm in a Varian Eclipse TPS, with a resulting 216 control points, was exported from the TPS in DICOM format and reimported into the GUI application where a sampling of 180 randomly interpolated fields was generated. The sampled configuration was exported to the corresponding "ppj" PRIMO project file created for this simulation. The simulation was taken in a 2 steps way. First, the validated PHSP for the 6FFF energy at s1 was used as the radiation source for the s2 stage. The simulation of the s2 was carried out followed by the s3 stage of dose deposition calculation in a phantom OCTA-VIUS4D®. A CT scan of the phantom was imported by PRIMO and treated as a homogeneous RW3 material. The same plan in the same set-up was irradiated with a film inserted at the position corresponding to the middle of the phantom, in vertical direction, and sandwiched between RW3 slabs.

Calculated dose evaluation. Each irradiated Gafchromic film was scanned with an Expression 10000XL scanner (Seiko Epson Corp.) and processed with DoseLab from PTW where the corresponding dose calibration curve was applied in order to obtain the 2D dose distributions. The resulting output.tiff file from process was subsequently converted to DICOM format. On the other hand, with the aid of the GUI, a slice corresponding to the film location was extracted from the simulation integrated dose distribution

and exported as a DICOM file. The DICOM files were then imported by the Verisoft PTW software for a gamma 2D evaluation. All the dose distributions originated from the simulation, film measurement and TPS calculation were compared with each other using Verisoft software.

3 Results

3.1 Models Validation

Truebeam Linac head. The PDD and dose profiles were evaluated with the Gamma analysis tool available in PRIMO that in all cases reported more than 95% of gamma points within the mentioned accepted criteria. In general, small size fields report a lower number of accepted gamma points on the beam profiles, and the percentage of accepted points for the energies of 10 and 15 MV is higher than for the energy of 6 MV. Hereafter, the validated PHSP of the PRIMO linac model can be confidently used for all the MC in future simulations related to this unit and using these validated energies.

Collimator. The evaluation of the dose distributions, Fig. 1, of the MLC conformation simulated with MC and calculated with the TPS are resumed in Table 1.

3.2 Dynamic Simulation

After the s1 phase, the set-up with the GUI and the simulation of the collimation and dose deposition stages for the present VMAT like case took about 12 h. The splitting

Table 1 Collimator model evaluation with Gamma criteria 2%–2 mm

Energy	MC versus Gaf. film (%)	TPS versus Gaf. film (%)
6FFF	94,9	89,8
10FFF	98,0	93,7

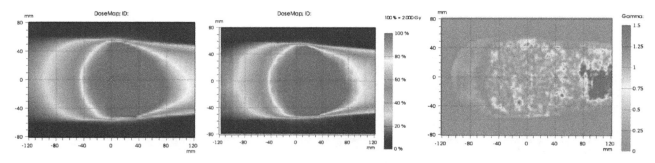

Fig. 2 From left, Gafchromic measurement, PRIMO calculation, gamma evaluation

factor in phantom was set at 64, and 180 sampled static fields were configured to reproduce the dynamic procedure. The final average uncertainty on dose was 0,78%. The Gamma evaluation of the MC simulation and TPS calculation, Fig. 2, of the VMAT like plan used in the validation test showed an agreement of 92,2% with a Gamma criteria of 2%–2 mm between the MC simulation and the Gafchromic. The same analysis between the TPS calculation and the Gafchromic measurement showed 96,1% of accepted Gamma points.

4 Discussion

For a feasible usage of PRIMO in a workflow capable of simulate dynamic procedures, some essential questions had to be addressed: can PRIMO properly simulate the linac model? Can the dynamic motion be implemented in PRIMO and simulated with sufficient accuracy? Can the overall PRIMO Monte Carlo simulation be fast and easily configured?

For the present study, the physical system consisted of a Varian TrueBeam with a 120HD MLC operating with energies of 6FFF and 10FFF MV. The PRIMO model of the linac was based on the Varian 2100 geometry model but pre-calculated PHSP were imported and used as the radiation source. The Gamma evaluation of the models against the commissioning data resulted in their validation since for all tested situations, more than 95% of Gamma points were <1 in PDD and dose profiles in water tank. As for the PRIMO 120HD MLC model the results show that it is reliable with 94,9 and 98,0% 2D Gamma points for the static simulation with energies 6FFF and 10FFF respectively.

The algorithm used to implement the PPS approach to allow PRIMO to simulate a dynamic procedure when tested with a VMAT like case, resulted in a 92,2% of accepted Gamma points. Although less than the usual 95% criteria used in the QA validation, the passing gamma rate is within an acceptable range of comparison and gives reliability to the used sampling algorithm used, reinforced by the results obtained in parallel works with less modulated and IMRT

cases. However, inferior results like the one shown here may point to the need that, in some cases, more than the 180 number of fields per simulation should be used. This is still not allowed by PRIMO itself, which imposes a maximum of 180 fields per project, but may be overcome in the near future.

Finally, although PRIMO does not have intrinsically this possibility, it could be successfully integrated in a dynamic treatment simulation workflow. The workflow was encapsulated in a MATLAB GUI application with the aim of facilitate a future clinical implementation.

5 Conclusion

With the aid of PRIMO, complete and reliable models of the Varian TrueBeam linac with a MLC 120HD were validated. The PRIMO possibility to simulate multiple fields in the same project is a suitable characteristic that allowed implementing dynamic deliveries simulations. This was possible by implementing a visual application that bypassed the necessity of any code generation by the final user. As a result, unlike what is generally offered by the common MC codes, an easy, visual and unified workflow was developed to drive the PRIMO to simulate dynamic procedures. However, some open issues remain, as it is the case of the possible limitations that the 180 maximum fields can impose, which affects the best optimized simulation results achievable. These questions will be addressed in future on-going work, including detailed description of the GUI developed and used in this work. Meanwhile, the GUI, which visually helps automatic and fast retrieval of TPS information and PRIMO configuration, can play the role of implement the workflow, providing the link between the TPS, user and the PRIMO. This is a precious characteristic that perfectly matches the clinical demanding and that contributes to push forward the Monte Carlo use in the daily routine.

Acknowledgements This article is a result of the project NORTE-01-0145-FEDER-000027, supported by Norte Portugal

Regional Operational Programme (NORTE 2020), under the POR-TUGAL 2020 Partnership Agreement, through the European Regional Development Fund (ERDF).The author J. Oliveira acknowledges the research grant CI-IPOP-BIFMRPR2018/UID/DTP/00776/POCI-01-0145-FDER-006868.

Conflict of Interest The authors declare that they have no conflict of interest.

References

1. K. Otto, "Volumetric modulated arc therapy: Imrt in a single gantry arc," *Med. Phys.*, no. 35, pp. 310–317, 2008.
2. P. Andersson, M. Krantz, R. Chakarova, and R. Cronholm, "Monte carlo patient specific pre- treatment qa system for volumetric modulated arc therapy," Swedish Radiation Safety Authority, 2017.
3. F. Verhaegen and J. Seco, *Monte Carlo Techniques in Radiation Therapy*. CRC Press.
4. A. Lallena, L. Brualla, and M. Rodriguez, "Monte carlo systems used for treatment planning and dose verification," *Strahlenther. Onkol.*, 2016.
5. J. Sempau, M. Rodriguez, and L. Brualla, "Primo: A graphical environment for the monte carlo simulation of varian and elekta linacs," *Strahlenther. Onkol.*, no. 1889, pp. 881–886, 2013.
6. M. F. Belosi et. al, "Monte carlo simulation of truebeam flattening-filter-free beams using varian phase-space files: Comparison with experimental data," *Med. Phys.*, no. 41, 2014.
7. Shih and R.e.a., "Dynamic wedge versus physical wedges: A monte carlo study," *Med. Phys.*, no. 28, pp. 612–619, 2001.
8. F. Verhaegen et al., "A method of simulating dynamic multileaf collimators using monte carlo techniques for intensity-modulated radiation therapy.," *Phys. Med. Biol.*, no. 46, pp. 2283–2298, 2001.
9. A. Esposito et al., "PRIMO Software as a tool for Monte Carlo Simulations of Intensity Modulated Radiotherapy: A Feasibility study," *Phys. Med.*, vol. 32, no. 3, p. 205, 2016.
10. C.-M. Ma *et al.*, "Phase-space database for external beam radiotherapy," IAEA - International Atomic Energy Agency, Technical report, 2005.
11. J. Sempau, L. Brualla, and M. Rodriguez, *PRIMO User's Manual - SOFTWARE VERSION 0.1.5.1202.*

Evaluation of Deformable Image Registration Between High Dose Rate Brachytherapy and Intensity Modulated Radiation Therapy for Prostate Cancer

Noriomi Yokoyama, Akihiro Takemura, Hironori Kojima, Kousuke Tsukamoto, Shinichi Ueda, and Kimiya Noto

Abstract

High risk prostate cancer is treated with a combination of intensity-modulated radiation therapy (IMRT) and high dose rate brachytherapy (HDR-BT). Deformable image registration (DIR) techniques used for dose accumulation sums dose distributions. The accuracy of DIR would get worse when density of an organ in a pair of registering two images differs greatly each other. Needles and contrast medium are used in HDR-BT. In this study, the effect of needles and contrast medium for DIR accuracy was evaluated. Six patients with prostate cancer were enrolled, who were treated with the combination of HDR-BT and IMRT. In the HDR-BT plan, needleless image (NI) and needleless and no-contrast medium image (NCI) were created from the original HDR-BT plan image (OI) to investigate the influence of needles and contrast medium. Both DIR and rigid registration (RR) were performed on the OI, NIs and NCIs by using MIM Maestro ver. 6.7.6 (MIM software Inc, Cleveland, USA) and after that the dose distribution of HDR-BT (used as the reference image) and IMRT were accumulated. The Dice Similarity coefficient (DSC) between DIR and RR were analyzed and compared each other. The mean DSC values of the prostate with DIR on OI, NI and NCI were 0.51, 0.57 and 0.73, respectively. The DSC with DIR on NCI was higher than DSC with DIR on OI and NI. The DSC values improved by removing the contrast medium.

Keywords

Radiotherapy • Brachytherapy
Intensity-modulated radiation therapy
Deformable image registration

1 Introduction

Adaptive radiotherapy technique (ART) which requires several radiotherapy planning, is used to reduce the side effects in many institution. Deformable image registration (DIR) became important technology in ART because it helps delineation of organs and targets in re-planning, which is a time-consuming task. DIR techniques can be used for registration between different planning CT images, and dose accumulation which sums several dose distributions. Registration and dose accumulation with DIR were applied for various clinical sites, such as head and neck, prostate. For prostate cancer, the DIR techniques register and sums a dose distribution of intensity-modulated radiation therapy (IMRT) and a dose distribution of high dose rate brachytherapy (HDR-BT).

Brock KK et al. evaluated the accuracy, reproducibility, and computational performance of deformable image registration algorithms at multiple institutions on common datasets and reported that organ position change decreased the DIR accuracy [1]. In addition, Ikeda et al. evaluated the accuracy of DIR with a digital phantom of the pelvic regarding changes in contrast using the commercial software MIM Maestro ver. 6.7.6 (MIM software Inc, Cleveland, USA) and reported that the DIR accuracy got worse related to the HU difference of the prostate between two images [2].

High risk prostate cancer is treated with a combination of IMRT and HDR-BT. In the HDR-BT, metal needles place into a prostate and contrast medium filled in a bladder. This has the potential to make DIR accuracy worse because large difference of CT value occurred between a reference image and a source image.

N. Yokoyama (✉) · K. Tsukamoto
Divisions of Health Sciences, Graduate of School of Medical
Sciences, Kanazawa University, Kanazawa, Japan
e-mail: shougun621@gmail.com

A. Takemura
Faculty of Health Sciences, Institute of Medical, Pharmaceutical
and Health Sciences, Kanazawa University, Kanazawa, Japan

H. Kojima · S. Ueda · K. Noto
Department of Radiological Technology, Kanazawa University
Hospital, Kanazawa, Japan

© Springer Nature Singapore Pte Ltd. 2019
L. Lhotska et al. (eds.), *World Congress on Medical Physics and Biomedical Engineering 2018*,
IFMBE Proceedings 68/3, https://doi.org/10.1007/978-981-10-9023-3_91

In this study, we evaluated the effect of needles and contrast medium to DIR accuracy.

2 Materials and Methods

2.1 Patients and Treatment Plans

Pairs of an IMRT plan and a HDR-BT plan from six patients with prostate cancer were collected for this study. The average age of six patients was about 68 years (ranging from 63 to 74 years). All patients were classified as high risk prostate cancer. Five patients were treated with IMRT for the whole pelvis in which the total dose was 46 Gy and daily fractions was 2 Gy. One patient was treated with IMRT for the prostate in which the total dose was 40 Gy and the daily fractions was 2 Gy. In addition, all patients treated with ^{192}Ir HDR-BT to the prostate. These prescribed doses were varied depending on patients. Three patients were received 13 Gy once. Three patients were received 19 Gy in twice. Even though the prescribed dose of ^{192}Ir HDR-BT was different, there were no differences in dose constraints of rectum and bladder. The dose constraint of the rectum for the HDR-BT was that $V_{75\%}$ was less than 1 cc and $V_{100\%}$ was zero. To the bladder, $V_{75\%}$ was less than 1 cc. To the prostate, $V_{100\%}$ was more than 95% volume of the prostate, $V_{150\%}$ was less than 40% and D_{90} of the prostate received more than 90–95% of the prescription dose.

The treatment planning system Monaco (Elekta AB, Sweden) was used for IMRT planning and Oncentra Brachy (Elekta AB, Sweden) for HDR-BT planning. The Monaco applies Monte-Carlo algorithm as the dose calculation algorithm and Oncentra Brachy applies the collapsed cone algorithm.

2.2 Image Processing for Needles and Contrast Medium

The needles in the original CT images (OI) for HDR-BT were painted by the density of the prostate and were erased the high density area to create needleless images (NI). Addition to the needles, the density of bladder wall and the density of contrast medium in a bladder (about 400 HU) were also replaced to 30 and 10 HU to create needleless and no-contrast medium images (NCI).

2.3 Dose Accumulation

A process workflow for creating the DIR-based dose accumulation is shown in Fig. 1. The planning CT images for the HDR-BT, as source images, were deformed to match the

planning CT images for the IMRT, as reference images in the RR and DIR process. The DIR software MIM Maestro ver. 6.7.6 (MIM software Inc, Cleveland, USA) executed both the RR and DIR procedures.

In the RR procedure, the planning CT images for the HDR-BT were roughly aligned by shifting and rotation according to bone structures by hand, then the automatic RR was performed. The RR procedure was performed on the pair of the CT images for IMRT and the OI, NI or NCI, and after that the dose distributions of the HDR-BT plan were moved according to the RR results and each of these dose distribution was summed with the dose distribution of the IMRT plan.

In the DIR procedure, once the RR was finished, then DIR was performed. The pairs of images were the same as the RR. Deformation vector field (DVF), which is one of the DIR results indicates corresponding voxels between the IMRT planning CT images and the OI, NI or NCI. Based on the DVF, the delineate organs and the dose distributions of the HDR-BT plan were deformed, and then accumulated to obtain a total dose distribution.

2.4 Evaluation

The D_{98}, D_{50}, D_2 and D_{mean} for the prostate, rectum and bladder were calculated from the accumulated dose distribution.

The Dice similarity coefficient (DSC) [3] of the bladder, prostate and rectum was used to evaluate the RR and DIR accuracy. The DSC results were tested by Tukey's honestly significant difference test. Statistical analysis was performed with IBM SPSS Statistics Version 24 (IBM COMPANY, Chicago, USA).

3 Result

3.1 Evaluation of Effect of Needles and Contrast Medium for DIR Accuracy

The results of the DIR using the OI as the HDR-BT plan image were denoted as DIR-OI, and the results of the DIR using the NI and the NCI were denoted as DIR-NI and DIR-NCI as well. The mean DSC values of bladder for the RR, DIR-OI, DIR-NI and DIR-NCI were 0.63, 0.42, 0.41 and 0.73, respectively, these of prostate were 0.75, 0.51, 0.57 and 0.73, respectively, and these of rectum were 0.46, 0.48, 0.48, and 0.55, respectively (Fig. 2). Overall, for all structures, the difference in DSC value were seen among DIR-OI, DIR-NI and DIR-NCI. This results showed that needle hardly affect DIR accuracy. The DSC of bladder in

Fig. 1 A process workflow for creating the DIR-based dose accumulation

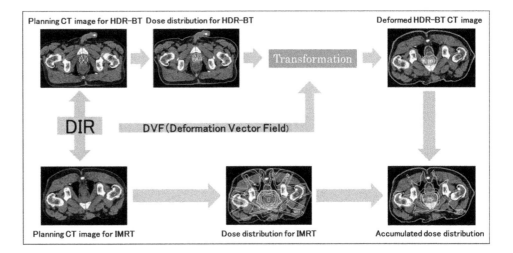

the DIR-NCI was better than that in DIR-NI, and the results for the prostate and rectum were similar to those for bladder.

Comparing RR and DIR-NCI, the DSC of the bladder and rectum in the DIR-NCI was better than these in the RR. In contrast, for the prostate, the RR was slightly better than the DIR-NCI.

For the prostate, there was significant difference between RR and DIR-OI (p < 0.05). Also, it was between DIR-NCI and DIR-OI (p < 0.05).

3.2 Evaluation of DVH Parameters

The D_2 of rectum for the RR, DIR-OI, DIR-NI and DIR-NCI were 54.1, 47.1, 48.1 and 49.3 Gy, respectively, and these of bladder were 55.9, 58.6, 61.2, and 51.7 Gy, respectively (Fig. 3). For the D_2 of the rectum, the accumulated dose by the RR produced higher D_2 value than the DIR-based dose accumulation (p < 0.05). For the D_2 of the bladder, the DIR-NI has the highest D_2 values of 61.2 Gy and the DIR-NCI has the lowest value of 51.7 Gy. There was significant difference in bladder D_2 between DIR-NI and DIR-NCI (p < 0.01).

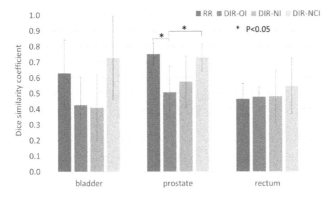

Fig. 2 Dice similarity coefficients for the bladder, prostate and rectum

4 Discussion

We evaluated the impact of needles and contrast medium in the planning CT image for the HDR-BT for DIR accuracy as the DIR was performed with the CT images of IMRT and HDR-BT. From Fig. 2, the DSC values was greatly decreased when the contrast medium existed in the planning CT image for HDR-BT. There were two reasons for this; first, large difference (>400 HU) occurred between the CT images for IMRT and the CT images for HDR-BT because of the contrast medium in bladder. Ikeda et al. reported that the DSC decreased by approximately 0.1 when the difference in CT value of the prostate between the reference image and the source image was 40 HU [4]. The result in this study was consistent with the results of that study. In addition, the bladder volume in the resulting deformed images shrunk because of the contrast medium in bladder. Tanner, C et al. evaluated the volume and shape preservation of enhancing lesions when applying non-rigid registration to a time series of contrast enhancing MR breast images and reported that volume reduction of the contrast medium part in the image after DIR occurred because of free form deformation of elements constituting DIR algorithm [4]. In the study, MIM Maestro employs a Demons based free form deformation algorithm. Thus, the DIR accuracy decreased due to volume reduction.

Comparing RR and DIR-NCI, the DSC of the prostate with the RR was slightly better than that with the DIR-NCI. The reason would be that artifacts generated by the needles could not be completely removed from the planning CT image for the HDR-BT, thus, the artifact made the DIR accuracy worse.

Regarding D_2 of the rectum, the RR resulted in higher D_2 value in the accumulated dose distribution than the DIR-OI, DIR-NI and DIR-NCI. The reason might be that the rectum was deformed by the DIR and was moved to the low dose region.

Fig. 3 D_{98}, D_{50}, D_2 and D_{mean} for prostate, rectum and bladder

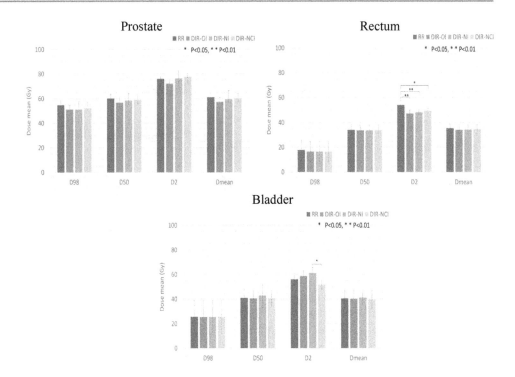

Regarding D_2 of the bladder, the D_2 of the bladder for RR, DIR-OI, DIR-NI and DIR-NCI were 55.9, 58.6, 61.2 and 51.7 Gy, respectively. The DIR-NI has the highest D_2 values of 61.2 Gy and the DIR-NCI has the lowest value of 51.7 Gy. DSCs of the bladder for RR, DIR-OI, DIR-NI and DIR-NCI were 0.63, 0.42, 0.41 and 0.73, respectively. The DIR-NI has the highest DSC value of 0.73 and the DIR-NCI has the lowest value of 0.41. The lower D_2 of the bladder was related to the DSC value. These results suggested that DIR accuracy affected the accuracy of the accumulated dose. Thus, we should pay attention to the DIR accuracy for dose accumulation.

5 Conclusion

We investigated the effect of needles and contrast medium for DIR accuracy. We found that contrast medium affected DIR accuracy more than needles. The DIR accuracy affected the accuracy of the accumulated dose. Thus, the DIR accuracy should be checked carefully for dose accumulation.

References

1. Kristy K. Brock. Results of a multi-institution deformable registration accuracy study (MIDRAS). Radiation Oncology Biol. Phys (2010); 76(2): pp. 583–596.
2. T. Ikeda, A. Takemura, S. Ueda, H. Kojima, K. Noto et al. Characterization of deformable image registration for the pelvic region regarding changes in contrast, noise, and prostate shifting. AAPM 58th annual meeting & exhibition (2016).
3. L.R. Dice. Measures of the Amount of Ecologic Association Between Species. Ecology (1945), 26(3): pp. 297–302.
4. Tanner, C., Schnabel, J.A., Chung, D., Clarkson, M.J., Rueckert, D., Hill, D.L.G., Hawkes, D.J et al. Volume and shape preservation of enhancing lesions when applying non-rigid registration to a time series of contrast enhancing MR breast images. Medical Image Computing and Computer Assisted Intervention (2000); Vol 1935: pp. 327–337.
5. N. Kadoya et al. Evaluation of rectum and bladder dose accumulation from external beam radiotherapy and brachytherapy for cervical cancer using two different deformable image registration techniques. Journal of Radiation Research (2017), pp. 1–9.
6. Laura E. van Heerden et al. structure-based deformable image registration: Added value for dose accumulation of external beam radiotherapy and brachytherapy in cervical cancer. Radiotherapy and Oncology (2017).

Intensity Modulated Radiotherapy (IMRT) Phantom Fabrication Using Fused Deposition Modeling (FDM) 3D Printing Technique

John Paul Bustillo, Roy Tumlos, and Randal Zandro Remoto

Abstract

Design and fabrication of patient-specific radiotherapy phantom is now more accessible and cost-effective using 3D printing technology. This study fabricates a 3D printed radiotherapy phantom for quality assurance of Intensity Modulated Radiotherapy (IMRT). Using an IMRT Thorax anthropomorphic phantom (CIRS) as a substitute for an actual patient, a 3D printed radiotherapy phantom was designed based on a patient computed tomography (CT) scan during treatment planning. Before printing the phantom, the tissue equivalence of Acrylonitrile Butadiene Styrene (ABS) and Polylactic Acid (PLA) polymers used in 3D printing was characterized by quantifying its CT number and relative electron density to water. In the 3D printed phantom fabricated, it was shown that soft tissue and lungs can be simulated using PLA 100% infill $\left(\rho_{e,w}^{130kV} = 0.99 \right)$ and 20% infill plastic $\left(\rho_{e,w}^{130kV} = 0.20 \right)$.

Keywords

IMRT • 3D printing • Radiotherapy phantom

J. P. Bustillo (✉) · R. Tumlos
Department of Physical Sciences and Mathematics, College of Arts and Sciences, University of the Philippines Manila, Manila, Philippines
e-mail: jobustillo@up.edu.ph

J. P. Bustillo · R. Z. Remoto
The Graduate School, University of Santo Tomas, Manila, Philippines

R. Z. Remoto
Section of Radiation Oncology, National Kidney and Transplant Institute, Quezon City, Philippines

1 Introduction

The success of a complex radiotherapy treatment lies in the accuracy of the treatment delivery [1]. Radiotherapy modalities such as Intensity Modulated Radiotherapy (IMRT) are validated before the actual treatment. A pre-treatment verification in phantom is done to check the treatment delivery accuracy. In-phantom dosimetry is an indirect method of measuring the radiation dose that will be delivered to the patient during the actual radiotherapy. It uses a patient simulator (phantom) with good tissue equivalence that can act as the patient [2]. This is part of the clinical quality assurance to validate the dose given to the target tumor and to the organs at risk near it [3, 4].

Due to the complex nature of IMRT delivery, a precise pre-treatment phantom dose measurement should be done [5]. However, anthropomorphic phantoms are expensive and usually not readily available in the facility. Thus, water equivalent slabs, which are not accurate to patient's anatomy, are used as substitutes to anthropomorphic phantoms. In addition, standard homogeneous phantom does not simulate accurately the actual heterogeneity in patient anatomy [6].

This study fabricates a 3D printed patient-specific phantom as a tool in radiotherapy quality assurance. The 3D printing material characteristics important in radiation interaction and dosimetry were assessed to check the coherence of the phantom material with the tissue being simulated. The criticality of correct phantom geometry and the accuracy of tissue equivalence of the materials are significant to represent the actual radiotherapy patient well. This will help professionals in radiation oncology department in checking the accuracy of their dose delivery with the planned treatment. Moreover, this can give us a cost-effective alternative to assess the planned treatment other than using standard commercial phantoms.

The Fused Deposition Modeling (FDM) 3D printing technique was utilized in simulating soft tissue and lung

© Springer Nature Singapore Pte Ltd. 2019
L. Lhotska et al. (eds.), *World Congress on Medical Physics and Biomedical Engineering 2018*,
IFMBE Proceedings 68/3, https://doi.org/10.1007/978-981-10-9023-3_92

while bone was not simulated in this study. All the current limitations of FDM 3D printing technology (long printing time, presence of shells and holes, thermal warping) were considered during the experiment.

2 Materials and Methods

2.1 Characterization of 3D Printing Polymers

Two 3D printing polymers were characterized prior to the fabrication of the 3D printed phantom. Acrylonitrile Butadiene Styrene (ABS) and Polylactic acid (PLA) are the two commonly used materials in 3D printing. ABS plastic have been investigated by various studies [7–9] as a material for 3D printed phantom. On the other hand, PLA plastic has been investigated as a building material of bolus in conforming radiation beams [10]. The infill percentages (ratio of thermoplastic volume to air volume) are varied during fabrication. This study uses 20, 40, 60, 80, and 100% infill percentages. Thus, this parameter changes the physical density of the printed object.

ABS and PLA blocks of $5 \times 5 \times 5$ cm^3 was designed using Autodesk 123D (Autodesk, Inc., California, USA). Once the 3D design is done, it was exported into a Stereolithography (STL) file format. This STL file was then transferred to slicer software. The said file was then converted into a G-code that the 3D printer directly understands.

After the actual printing of the blocks, it was scanned in CT using the standard protocol of the hospital for IMRT radiotherapy simulation. The cubes were scanned in three different orientations relative to its printing position. This was done to check for possible orientation dependence especially for printed objects with infill percentages less than 100%.

The following physical properties of the samples were identified in the CT scan images: CT number (HU), physical density (ρ), and relative electron density ($\rho_{e,w}$) relative to water. These material properties were identified by creating a 30×30 cm^2 region of interest (ROI) inside the set of images of sample blocks. In addition, CT number profiles were acquired at the middle of each samples for all the orientations scanned to check the uniformity of printing.

Using the image analysis software ImageJ (National Institute of Health, Bethesda, Maryland, USA), the CT number in Hounsfield units (minimum, maximum, mean and standard deviation) was acquired in the created ROI on each axial image of the sample. By checking the CT number at different slices, the uniformity of the printed object can be assessed. Also, the relationship of the average CT number and infill percentage of the block was obtained [11].

2.2 Fabrication of 3D Printed Phantom

This study used an anthropomorphic phantom with various tissue equivalent inserts (CIRS IMRT Thorax Phantom) as the 'patient'. This methodology, shown in Fig. 1, is adopted from a previous study [8]. It was done because it is not possible to directly measure the dose in vivo for a real patient. Moreover, it avoids ethical issues due to the privacy of patient data.

The CT scan images acquired during the treatment planning were imported into 3D Slicer software (www.slicer.org). Using the 'volume rendering' module of the software, a volumetric model was created as shown in Fig. 2. The ROI for the images was adjusted to remove the phantom holder from the 3D design generated using the 'crop volume' module of the program.

Moreover, the 'model maker' module of the program was utilized to create a stereolithography (STL) file. A marching cube algorithm was used in forming a polygonal mesh (includes vertices, edges, and faces) from a three dimensional array of medical data such as the voxels of CT. Then, the smoothing and decimation were also defined during the process.

The model was modified further using available 3D graphics software (Autodesk Netfabb) to separate the model into sub-sections. In addition, the holder of the detectors was designed by creating a plain cylinder and a 3D model of the two ionization chambers (FC65-G and CC01). The plain cylinder and the chamber STL files were then subtracted to each other using Boolean Operation algorithm of Autodesk Netfabb.

3 Results and Discussion

3.1 CT Number Tissue Equivalence Characterization

Proper calibration of the CT number using a material with known relative electron density values is essential in treatment planning system commissioning. Accurate mapping of electron density distribution increases the accuracy for dose

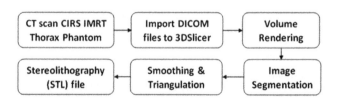

Fig. 1 Flowchart in creating a STL file (3D CAD Model)

Fig. 2 Volume rendering of the CT images using 3D Slicer: **a** Axial View. **b** Reconstructed 3D model. **c** Sagittal View. **d** Coronal View

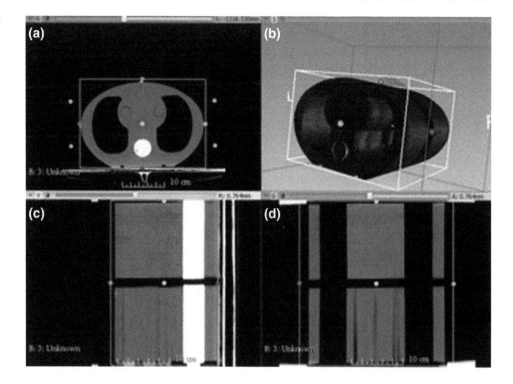

calculation. In addition, it enables determination of electron density in vivo.

In this study, the Siemens Somatom Emotion 16-slice CT is used in scanning the 3D printed cubes and phantom for characterization. Its CT number is calibrated for 130 kV using the electron density reference plugs (lung, bone, muscle adipose) of CIRS IMRT Thorax phantom. The CT calibration curve of the relative electron density is shown in Fig. 3. The calibration was done by creating a circular ROI with 1 cm diameter to record the HU measured with the known value of relative electron density and mass density.

3.2 Electron and Mass Density of 3D Phantom Material

The 3D printed cubes with varying infill percentages were scanned first in a 16-slice CT Somatom Emotion using RT_Thorax CT protocol (exposure of 90mAs, peak voltage 130 kV, 1 mm slice thickness and convolution kernel type "B41 s"). These factors are being used for radiotherapy patients of NKTI.

CT images were imported into ImageJ image analysis software (National Institute of Health, Bethesda, Maryland, USA). Using this program, the dimension, profile and the CT number statistics of the images can be acquired. A 30×30 mm^2 region of interest (ROI) was made at the middle portion of every slice to acquire CT number readings. The pixels at the boundary of the cube were avoided due to possible partial volume averaging CT artifact that can affect the average CT number [12].

There are 45 square ROIs acquired from each sample cubes. To ensure that the acquired means are representative to the sample, the recorded mean CT number per slice was averaged over the volume of the material. The summary of the data gathered is shown in Table 1.

The mean CT number measured for the solid ABS plastic (100% infill) gives an HU value of −99.96 HU. This value is approximately the same with the values acquired by Bibb et al. which are −110.38 and −98.31 HU for two ABS plastic samples. ABS material has already been used in fabricating a tissue equivalent IMRT phantom [7, 8]. The effect of high standard deviation (SD) was investigated

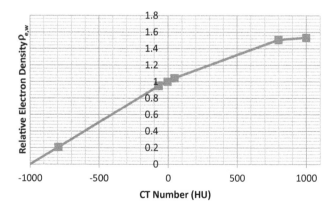

Fig. 3 Graph of relative electron density for Siemens Somatom Emotion CT (130 kV)

Table 1 Mean CT number and SD for each cube scanned using 130 kV

Sample	130 kV	
	Mean	SD
ABS 100%	−99.96	3.20
ABS 80%	−287.58	8.79
ABS 60%	−402.68	23.09
ABS 40%	−641.27	1.44
ABS 20%	−804.98	2.59
PLA 100%	−18.88	12.55
PLA 80%	−236.06	6.01
PLA 60%	−346.11	8.39
PLA 40%	−615.75	6.19
PLA 20%	−798.03	2.43

Table 2 Mean relative electron density and mass density (g/cm^3) of samples

Sample	130 kV		Tissue
	$\rho_{e,w}$	ρ	Equivalence
ABS 100%	0.915	0.926	Adipose
ABS 80%	0.724	0.733	
ABS 60%	0.606	0.614	
ABS 40%	0.363	0.368	Lung
ABS 20%	0.196	0.197	Lung
PLA 100%	0.988	0.99	Solid water
PLA 80%	0.777	0.786	
PLA 60%	0.664	0.672	
PLA 40%	0.389	0.394	Lung
PLA 20%	0.203	0.205	Lung

further by doing a t-test using the scanned cubes in different orientations. It was identified that the there is no significant difference ($p > 0.05$) between the mean CT number for different scanning orientation. The presence of high SD is expected given that the printed objects have pores.

The equivalent relative electron densities of the mean CT number per sample were interpolated using the lookup table of the CT machine used (CIRS). In doing so, the tissue equivalence of the 3D printed cubes was then identified. Figure 4 shows the experimental relative electron density in reference to water for each infill percentages. In addition, PLA 100% polymer $\rho_{e,w}^{130kV} = 0.988$ is more water equivalent in terms of relative electron density than ABS 100% polymer $\rho_{e,w}^{130kV} = 0.915$.

Moreover, the graph of relative electron density of the cubes against the infill percentages is observed to have a linear correlation. A Pearson product-moment correlation

coefficient was calculated to assess the relationship of relative electron density with the infill percentages of the samples. There is a positive linear correlation between the infill percentages and $\rho_{e,w}$ for ABS 130 kV ($r = 0.996$, $p = 0.000$), and PLA 130 kV ($r = 0.994$, $p = 0.001$). This linear relationship is consistent with the result of a previous study for high impact polystyrene 3D printing polymer that used the same methodology for characterization [11]. The tissue equivalence of each cube is further shown in Table 2.

All the samples investigated have a relative electron density less than 1 (ideal water). It shows that tissues with higher electron density relative to water such as bone, liver and brain cannot be simulated using ABS and PLA material. In this study, lung and solid water equivalent materials were used in fabricating a radiotherapy phantom due to this limitation. Thus, denser 3D printing materials such as cyano-acrylate and polyphenylsulfone should be considered in future studies [12].

3.3 3D Printed Phantom Fabrication

3.3.1 Construction of 3D Surface Model of Phantom

The soft tissue and the lung part of the CIRS phantom were segmented using the editor module of the program. The spine part of the phantom was not simulated because denser 3D printing filament is not readily available. The threshold technique was utilized during contouring by defining a range of CT number values.

After segmentation, a marching cubes algorithm was run for the two segmented parts (soft tissue and lungs) of the CT images to create a triangulated 3D mesh. A smoothing iteration of 30 using Laplacian filter and 0.25 decimation

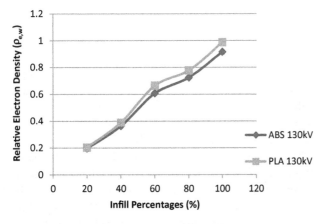

Fig. 4 The mean relative electron density for both ABS and PLA cubes

(a) **(b)** **(c)**

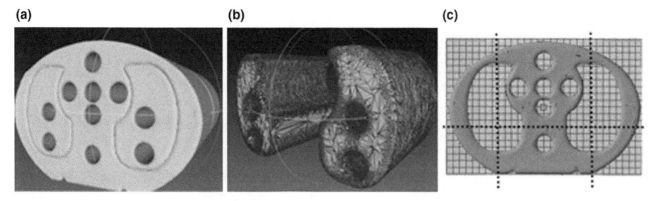

Fig. 5 **a** STL file generated in 3DSlicer program **b** Polygonal Mesh of Lung STL file (0.5 decimation applied) **c** Pattern used in cutting the phantom

Fig. 6 3D printed phantom together with the FC65-G and CC01 chamber inserts

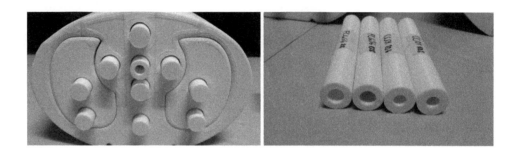

factor (reduction in polygons) were applied to the stereolithography file generated to remove the CT slicing effect [13]. In addition, a joint smoothing algorithm was applied to make sure that all the models would fit together. The STL file was modified further as shown in Fig. 5c using the AutoDesk Netfabb to divide the CAD model into several segments to conform to the maximum printing capacity of the 3D printer. The inserts that will fill the holes of the phantom were created using AutoDesk 123D software by designing a cylinder with 160 mm length and 25 mm diameter.

3.3.2 Printing of Phantom

As seen in the tissue equivalence characterization done to two 3D printing polymer materials, PLA has a CT number and relative electron density near to soft tissue (for 100% infill) and to lung (for 20% infill). In addition, PLA is known to be biocompatible because it is derived from corn starch and it is more capable to be printed into large object without thermal warping unlike ABS material [8]. The 3D printed phantom used in this study was made entirely from PLA polymer due to its more accurate CT number for soft tissue and lung. In addition, PLA does not easily warp (shrinking upon cooling) which guarantees the geometrical accuracy during fabrication. The printing was done using the 3D printers of 3D2gO Philippines Inc.

A printing temperature of 215 °C was applied on the PLA extruder and a heated bed of 36 °C. After printing the phantom into segments, it was assembled by making sure that the gap between segments is minimized. Sanding was also done to smoothen the surface and to remove excess plastic. Figure 6 shows the finished product of the printing.

The 3D printed phantom was then scanned in the Somatom Emotion 16-slice CT to investigate the contour of the phantom and to compare it with the CIRS phantom. Figure 7 shows the CT slice in different views together with the reconstructed volume using 3DSlicer program. It can be seen that the contrast between the lung and soft tissue is evident in the scan. The division created in the lung part and insert of the phantom can be seen clearly in the coronal view. This division is one limitation of the 3D printing method used in printing large objects.

Although the thickness of the printed phantom is limited to 21.8 cm, the dosimeter inserts were adjusted to make sure that the sensitive volume is located at the middle portion of the phantom. This is done to make sure that there is enough scatter during measurement. Moreover, the external geometry of the 3D printed phantom and the CIRS IMRT thorax phantom (patient) was assessed using the Dice Similarity Coefficient (DSC) defined at (−900 to 1000 HU). Using the 3DSlicer algorithm, the DSC was identified to be 0.805 (where 1 refers to a perfect geometry match). This is a good

Fig. 7 DICOM images of the 3D printed phantom: **a** Axial View. **b** Reconstructed 3D Volume, **c** Sagittal View. and **d** Coronal View

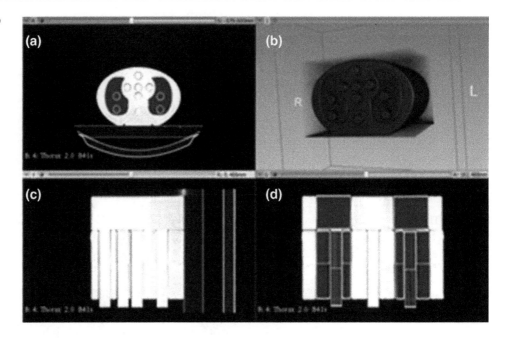

geometry match given that the whole thickness of the CIRS phantom was not fabricated and air gap is present at the middle portion.

Two 3D printers were utilized during the process and it shows that the CT number is also dependent on the printer used during fabrication. Thus, it is recommended that future researchers use only a single printer in printing radiotherapy phantom parts to avoid this variation in CT number. On the other hand, the junction of each segments is not visible in the CT line profile.

4 Conclusion and Recommendation

In the 3D printed phantom fabricated, soft tissue and lungs were simulated using PLA plastic. It was shown that PLA 100% and PLA 20% have almost the same relative electron densities for soft tissue and lung, respectively. Moreover, this material could be printed into large segments without too much thermal warping. It was observed that the geometry of the 3D printed phantom is in good match with the commercial CIRS IMRT phantom. Thus, it is feasible to design and create a patient specific phantom using open source software and a desktop 3D printer.

Acknowledgements The authors would like to thank the Accelerated Science and Technology Human Resource Development Program (ASTHRDP) scholarship of the Department of Science and Technology–Science Education Institute (DOST-SEI) for funding this research.

Conflict of Interest The authors declare that they have no conflict of interest.

References

1. Nakamura, M., Minemura, T., Ishikura, S., Nishio, T., Narita, Y., & Nishimura, Y. An on-site audit system for dosimetry credentialing of intensity modulated radiotherapy in Japanese Clinical Oncology Group (JCOG) clinical trials. Physical Medical, 32, 987–991. (2016).
2. Leary, M., Kron, T. Keller, C., Franich, R. Lonski, P. Subic, A. & Brandt, M. Additive manufacture of custom radiation dosimetry phantoms: An automated method compatible with commercial polymer 3D printers. *Materials and Design*. 86. 487–499. (2015).
3. Rehman, J. Tailor, R. Isa, M. Afzal, M. Chow, J., & Ibbot, G. Evaluation of secondary cancer risk in spine radiotherapy using 3DCRT, IMRT, and VMAT: A phantom study. *Medical Dosimetry*. 40. 70–75. (2014).
4. Sini, C., Broggi, S., Fiorino, C. Cattaneo, G. & Calandrino, R. (2015). Accuracy of dose calculation algorithm for static and rotational IMRT of lung cancer: A phantom study. *Physica Medica*, 31, 382–390.
5. Sumida, I., Yamaguchi, H., Das, I., Kizaki, H., Aboshi, K., Tsujii, M. Yamada, Y., Suzuki, O., Seo, Y., Isohashi, F. & Ogawa, K. (2016). Intensity-modulated radiation therapy dose verification using fluence and portal imaging device. *Journal of Applied Clinical Medical Physics*. Volume 17. 259–271.
6. International Atomic Energy Agency (IAEA). (2008). Commissioning of Radiotherapy Treatment Planning Systems: Testing for Typical External Beam Treatment Techniques. IAEA-TECDOC-1583, IAEA, Vienna.
7. Kumar, R., Sharma, S. D., Despande, S., Ghadi, Y., Shaiju, V.S., Amols, H.I. & Mayya, Y.S. (2009). Acrylonitrile Butadiene Styrene (ABS) plastic-based low cost tissue equivalent phantom for verification dosimetry in IMRT. Journal of Applied Clinical Medical Physics, Volume 11, Number 1.
8. Ehler, E.D., Barney, B.M., et al. (2014). Patient specific 3D printed phantom for IMRT quality assurance. *Phys. Med. Biol.* 59, 5763–5773.

9. Kairn, T., Crowe, S.B. & Markwell, T. (2015). Use of 3D Printed Materials as Tissue-Equivalent Phantoms. World Congress on Medical Physics and Biomedical Engineering, June 7–12, 2015.

10. Zou, W., Fisher, T., Zhang, M., Kim, L., Chen, T., Narra, V., Swann, B., Singh, R., Siderit, R., Yin, L., Teo, B.K., Mckenna, M., McDonough, J. & Ning, Y.J. (2015). Potential of 3D printing technologies for fabrication of electron bolus and proton compensators. *Journal of Applied Clinical Medical Physics*. Volume 16. 90–98.

11. Madamesila, J., McGeachy, P., Barajas, J. & Khan, R. (2016). Characterizing 3D printing in the fabrication of variable density phantoms for quality assurance of radiotherapy. *Physica Medica*, 32, 242–247. https://doi.org/10.1016/j.ejmp.2015.09.013.

12. Bibb, R., Thompson, D. & Winder, J. (2011). Computed tomography characterization of additive manufacturing materials. *Medical Engineering & Physics* 33, 590–596.

13. Canters, R.A., Lips, I.M., Wendling, M., Kusters, M., Zeeland, M., Gerritsen, R.M., Poortmans, P. & Verhoef, C.G. (2016). Clinical implementation of 3D printing in the construction of patient specific bolus for electron beam radiotherapy for non-melanoma skin cancer. Radiotherapy and Oncology, 121, 148–153. https://doi.org/10.1016/j.radonc.2016.07.011.

CT Spectrometry with a Portable Compton Spectrometer with Stationary and Moving Tube

Ricardo Terini, Vincent Morice, Denise Nerssissian, and Elisabeth Yoshimura

Abstract

The knowledge of energy spectra of CT X-ray beams is essential to completely characterize beam quality and equipment performance. However, CT photon fluxes are too high to be directly measured with most of photon counting detectors. This work describes a Compton spectrometer designed at LRDMP, based on a CdTe detector with proper Al–Pb-Al collimators and shields, to obtain spectra of CT beams, from the measured spectra of 90°-scattered beams. A MatLab® computer program, including the Waller-Hartree formalism, was developed, to correct measured data, and then reconstruct the spectrum of the beam incident on the scatterer. Tests at LRDMP with direct and scattered standard CT beams showed, after data processing and normalization, similarity between correspondent spectra of reconstructed and directly measured beams. Shielding and scatterer thickness influence were carefully investigated. The system was tested in clinical measurements in a GE690 CT scanner, using CT lasers and scout radiographies for alignment. HVL values obtained from the reconstructed spectra, with the stationary tube, agree within 3% with those measured in QC tests. We also double-checked, with good accuracy, the actual scattering angle and the kVp values, through the energy shift of K lines and spectra end point, respectively. Although several exposures might be necessary to acquire each spectrum with good statistics, the total acquisition time was no longer than two minutes for each one. Furthermore, measurements with rotating tube were made, showing that accumulated spectra shape are like those obtained with the stationary tube.

Keywords

Compton spectrometry • Computed tomography CdTe detector

1 Introduction

In the 1970s, the development of computed tomography (CT) revolutionized diagnostic radiology. Afterwards, continuous technological developments introduced helical and multislice scanning, improving the speed and quality of the obtained images.

Despite the benefits, however, there is a growing concern about the radiation absorbed doses to patients and the risks associated with CT examinations [1].

The typical organ doses from a common radiography projection range from 1–20 mGy to the highest irradiated organ. CT procedures, however, result in organ doses in the range of 10–100 mGy! In order to optimize the balance of image quality and patient dose, the knowledge of the beam spectra is of great value. The complete characterization of the X-ray beam before and after its penetration through the patient's body might promote a more accurate assessment of the absorbed dose due to each kind of diagnostic procedure. Also, the spectrum knowledge makes it possible to obtain parameters that characterize beam quality and equipment performance.

Usually, X-ray beams are well characterized when the electron acceleration voltage (kVp), 1st and 2nd half-value layers (HVL), or even the homogeneity coefficient (1stHVL/2ndHVL) are known [2]. However, the most complete specification of X-ray beam characteristics is given by the spectral distribution. This measurement is not always feasible in clinical practice and even in certain laboratories due to

R. Terini (✉) · D. Nerssissian · E. Yoshimura
Laboratory of Radiation Dosimetry and Medical Physics (LRDMP), Institute of Physics, University of São Paulo, São Paulo, SP, Brazil
e-mail: ricardoaterini@gmail.com

V. Morice
Grenoble INP—ENSIMAG, École Nationale Supérieure d'Informatique et de Mathématiques Appliqués, Grenoble, France

© Springer Nature Singapore Pte Ltd. 2019
L. Lhotska et al. (eds.), *World Congress on Medical Physics and Biomedical Engineering 2018*,
IFMBE Proceedings 68/3, https://doi.org/10.1007/978-981-10-9023-3_93

the lack of proper equipment or staff knowledge and practice and due to the long measurement time.

X-ray beam spectra can be obtained from measurements made directly of the incident beams or indirectly of the scattered beams. As the primary beam of a clinical CT scanner is very intense, its direct measurement can damage the spectrometer or produce a distorted spectrum due to high dead time and pulse pile-up.

Several researchers [3, 4], have measured the spectrum of X-ray beams of clinical equipment by means of Compton spectrometry using Ge detectors and Lucite® or carbon scatterers. Although their good energy resolution, these detectors need nitrogen cooling and pinhole collimation, so being more feasible in laboratory. More recently, Maeda et al. [5] have used a compact Schottky CdTe detector for this kind of measurement in the range 70–100 kVp.

Vieira et al. [6] described the principles of a portable Compton spectrometer using a small CdTe detector and a PMMA (polymethyl methacrylate) rod scatterer, describing its preliminary applications in lab to mammography and radiology beams. Duisterwinkel et al. [7] have built a prototype of a portable Compton spectrometer, and have presented results of clinical CT X-ray spectra.

This work describes a portable Compton spectrometer with a $3 \times 3 \times 1$ mm^3 CdTe detector and its application to characterize different X-ray beams. At LRDMP, a set of RQT standard beams [2] had their spectra assessed with measurements of direct and scattered beams (90°). Clinical CT scanner beams were also measured after 90° scattering. Improvements in previous projects [6, 7] introduced the possibility of double-checking the actual scattering angle from the measured spectrum, as well as to reconstruct primary beam spectra from the measured ones utilizing the Waller-Hartree approximation [8]. Influence of shielding, scatterer thickness and positioning, both in lab and in the CT gantry, were investigated aiming to reduce spurious counts and to improve accuracy in incident spectra reconstruction.

2 Materials and Methods

2.1 Compton Spectrometer Design and Calibration

The Compton spectrometer was developed using an Amptek XR-100T-CdTe detector as the central element. The entire assembly was built on a 62×41 cm^2 aluminum base provided with holes to fix the other parts. Four conical collimators (5 mm diameter central hole in 10×10 cm^2 Pb (4 mm thick) covered with Al (2 mm thick) plates) are part of the device, two on the main beam before the scatterer, and two for the 90° scattered beam that will reach the detector. The Pb thickness is sufficient to attenuate diagnostic X-rays

Fig. 1 Compton spectrometer positioned for Philips X-ray tube beam measurements. The incident beam (blue arrow) is collimated, scattered at 90° by the PMMA rod, and collimated again, and then reaches the detector (within the shielded box) (Color figure online)

and the Al coating absorbs Pb characteristic X-rays. Collimators are attached to rails fixed in the base, so that their positions may be varied. The scatterer is a PMMA $(C_5H_8O_2)_n$ cylinder with 10, 8, 6 or 4 mm diameter, attached to an Al socket (Fig. 1).

A set of extra plates and a lid, all made with the same Al–Pb-Al material structure complete the detector shielding. In almost all measurements, a tungsten (W) collimator with a 2 mm diameter hole (2 mm thick) was attached few mm near the detector window in a hollow stainless-steel cylinder adapted to the detector.

The detector was energy calibrated just before or after the X-ray beam measurements, using the same amplifier gain, by measuring X- and γ-ray spectra from calibration radioactive sources (Am-241, Ba-133 and Eu-152). Gaussian curves were then fitted by the least squares method to the main peaks of such spectra, to get the data to construct the Energy versus Centroid curve for the chosen amplification.

2.2 Determination of Scattering Angle and Reconstruction of Primary Spectra

In each measurement session, both in lab and in clinic, the device was mounted and the scatterer properly aligned with the tube window. The required voltage and filter were set in the emission system, and tube current was selected to a value low enough to avoid pulse pile-up and excess dead time on the detector.

Each measured spectrum was corrected for detector efficiency, photoelectron escape, and absorption in materials between source and detector [9]. Then, the effective scattering angle θ, close to 90°, was double checked by determining the position of the W characteristic lines in the scattered spectra. A weighted Gaussian least square fitting to the $K_{\alpha1}$ and $K_{\alpha2}$ characteristic lines of the scattered X-ray spectrum was made. The average energy shift (E–E') due to the Compton effect can be used to evaluate the angle θ Eq. (1) with better accuracy than visual evaluation based on the experimental set-up.

$$\therefore \quad \theta = \cos^{-1}\left[1 - 511\left(\frac{E-E'}{EE'}\right)\right] \qquad (1)$$

where $\alpha = E/mc^2 = E(keV)/511$.

As most of the detected photons are product of incoherent (Compton) scattering of beam hitting the PMMA rod, the entire spectrum is distorted compared to the primary.

The reconstruction of the actual spectrum incident on the scatterer from the measured one was made considering the energy of the scattered photons, E', according to the Compton displacement law Eq. (1), so depending on E and θ. In addition, it was assumed that the scattered beam intensity follows Klein-Nishina differential cross section, corrected for the PMMA incoherent scattering function. This formalism is known as the Waller-Hartree approximation, which adopts the Independent Atomic Model, in which electrons are not free but bound to atoms [8, 10].

We implemented, in MatLab® computing environment, all the calculation process from the acquired spectra (of the X-ray beam and calibration sources) to the final reconstructed spectrum. The program, built in our lab, includes subroutines to perform all the tasks described before: the energy calibration of the detector, the correction of the measured X-ray spectrum, the assessment of the effective scattering angle, and the reconstruction of the primary spectrum from the measured one. After the calculations, the program provides the primary beam spectrum in terms of fluence rate and air kerma rate. Some other calculations, such as the determination of beam half-value layer, were implemented in the same routine.

2.3 Compton Spectrometry of Standard X-Ray Beams

In LDRFM, the Compton spectrometer was used to measure the spectrum of standard RQT beams for CT [2], characterized on an industrial constant potential Philips equipment. Alignment was achieved with lasers and an optical bench.

Measurements were made, for comparison, with beams 90° scattered by the PMMA rod (Fig. 1), as well as with direct RQT beams (without scatterer), 1 and 5 m away from tube focal spot, respectively. Tests were also conducted to verify the influence of scatterer diameter and the effectiveness of shields.

2.4 Compton Spectrometry of Clinical CT X-Ray Beams

The Compton spectrometer here described was used at a hospital, measuring radiation beams from a GE Discovery 690 CT scanner, after its scattering by the PMMA cylinder. In this case, the scanner tube was stopped at one side of the gantry. Adjustable feet allowed levelling of the device on the exam table. Lasers and radiographs were used to align the system (Fig. 2). Better results were obtained by improving the shielding under the detector to protect it from the radiation scattered in the gantry and from the direct radiation.

CT beam spectra were obtained setting current values of 10 and 50 mA so that the dead time was <10%. Between 35 and 50 consecutive 2 s exposures were needed in order to accumulate enough counts for each spectrum. Using beam width of 8 mm, X-ray beam spectra of 80, 100, 120 and 140 kV, have been acquired with the spectrometer, with or without a bow-tie filter.

After the measurement of each spectrum, a 2 mm thick W absorber was inserted, replacing the W collimator, and the measurement was repeated. These gross spectra, before corrections, were subtracted from those measured with collimator. This way it was possible to reduce the contribution of spurious counts due to transmission through the tungsten shielding (transmission penumbra) or to backscattering by surrounding materials.

3 Results and Discussion

3.1 Spectra of Standard X-Ray Beams in Lab

The variation of the diameter of the scattering PMMA rod did not influence, in general, the relative shape and the energy resolution of the scattered beam spectrum. On the

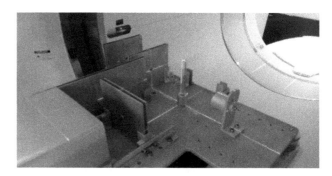

Fig. 2 Compton spectrometer positioned on the exam table for measurements on the GE PET-CT 690, showing the alignment of the scatterer by means of the system positioning Laser

other hand, as expected, it was observed that the measured intensity increases with the rod diameter.

Figure 3 shows that, after processing and normalizing adequately the measured data, the reconstructed beam spectra are very similar to the directly measured ones. This occurs except on the characteristic peak region. It is noticeable that in the reconstructed spectra the peaks are wider than in directly measured ones. A 20% enlargement was observed at the $K_{\alpha 1}$ peak −1.0 keV FWHM for the reconstructed and 0.8 keV for the direct one. This energy broadening is caused by factors like: the energy resolution of the detector (0.7 keV for the 59.54 keV γ-ray peak of Am-241 spectrum), the finite size of tube focal spot, as well as the irradiation geometry (the scatterer irradiated region is also finite) (~ 0.3 keV), and the Doppler shift, due to the distribution of electron velocities in scatterer atoms ("Doppler broadening"), which is one of the main contributions to the Compton profile of the measured lines [6].

Evaluated values of 1st HVL were: 8.5 mmAl (reconstructed beam, from the scattered at 1 m from the focal spot) and 9.6 mmAl (direct beam, 5 m from the focus).

3.2 X-Ray Compton Spectrometry at a Clinical CT Scanner

Figure 4 compares two 140 kV spectra of CT beams filtered or not by a small inner bow-tie filter, in the z-axis, at the center of the gantry, after correction and reconstruction using the described routine. Differences between both spectra are small, but significant. In the upper inset, it is also possible to observe the end part of the reconstructed spectra, which can be used to verify the actual tube voltage in each measurement [11].

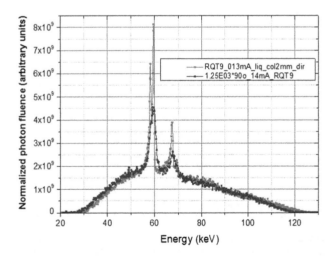

Fig. 3 Comparison between X-ray spectra of RQT 9 (120 kV) beams, directly measured (0.13 mA, 5 m focus-detector distance) and from 90° scattered beam (14 mA, 1 m focus-scatterer distance and reconstructed as described before), all measured with the CdTe spectrometer. Both spectra appear normalized to the same area

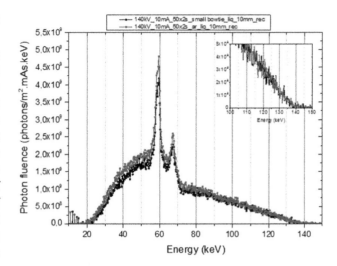

Fig. 4 Reconstructed 140 kV spectra (10 mA, 50 rotations), measured with the Compton spectrometer in the GE CT scanner, without (blue) or with (black) a small bow tie filter. In the inset, the end part of the reconstructed spectra (Color figure online)

Comparisons between HVL values obtained from standard QC tests and from Compton spectral measurements of the whole set of beams have shown differences <3%, all covered by the uncertainty interval.

Furthermore, measurements with rotating tube were made with a horizontal centered 5 mm in diameter scatterer fixed at the boundary of the Al base, showing that accumulated spectra shape are similar to those obtained with the stationary tube.

4 Conclusions

A portable Compton spectrometer, with a PMMA scatterer and a CdTe detector to receive 90°-scattered photons, was constructed at LRDMP and applied to measure laboratory and clinical X-ray beams. Tests confirmed the effectiveness of its shielding, in both setups.

Although several exposures might be necessary to acquire each clinical spectrum with good statistics, the total acquisition time was no longer than two minutes for each one. The alignment using radiography and the evaluation of the actual scattering angle by the measured spectrum increase accuracy and decrease setup time. Thus, Compton spectrometer could be applied to periodic checks of CT scanners, together with other common QC tests.

Acknowledgements This work was partially supported by CNPq Brazilian agency through DTI-1 grant number 380362/2016-3. INCT Radiation Metrology in Medicine (CNPq and FAPESP) also supported this work. R. A. Terini thanks São Paulo Research Foundation (FAPESP) for supporting the participation in 2018 WCMPBE, through grant 2018/00009-1.

References

1. Smith-Bindman, R., Lipson, J., Marcus, R., Kim, K.P., Mahesh, M., Gould, R., González, A.B., Miglioretti, D.L.: Radiation dose associated with common computed tomography examinations and the associated lifetime attributable risk of cancer. Arch. Intern. Med., 169(22), 2078–2086 (2009).

2. International Atomic Energy Agency (IAEA): Dosimetry in diagnostic radiology: an international code of practice. Technical Reports Series n° 457. IAEA, Vienna (2007).

3. Yaffe, M., Taylor, K.W., Johns, H.E.: Spectroscopy of diagnostic x-rays by a Compton-scatter method. Med Phys 3, 328–34 (1976).

4. Matscheko, G., Carlsson, G.: Measurement of absolute energy spectra from a clinical CT machine under working conditions using a Compton spectrometer. Phys Med Biol 34, 209–22 (1989).

5. Maeda, K., Matsumoto, M., Taniguchi, A.: Compton-scattering measurement of diagnostic x-ray spectrum using high-resolution Schottky CdTe detector. Med Phys 32, 1542–7 (2005).

6. Vieira, A.A., Linke, A., Yoshimura, E.M., Terini, R.A., Herdade, S.B.: A portable Compton spectrometer for clinical X-ray beams in the energy range 20–150 keV. Appl Radiat Isot 69, 350–7 (2011).

7. Duisterwinkel, H.A., van Abbema, J.K., van Goethem, M.J., Kawachimaru, R., Paganini, L., van der Graaf, E.R., Brandenburg, S.: Spectra of clinical CT scanners using a portable Compton spectrometer. Med Phys 42, 1884–94 (2015).

8. Ribberfors, R., Berggren, K.F.: Incoherent x-ray scattering functions and cross sections (dd)incoh by means of a pocket calculator. Phys Rev A 26, 3325–33 (1982).

9. Tomal, A., Santos, J.C., Costa, P.R., Lopez Gonzales, A.H., Poletti, M.E.: Monte Carlo simulation of the response functions of CdTe detectors to be applied in x-ray spectroscopy. Appl. Radiat. Isot. 100, 32–7 (2015).

10. Hubbell, J.H., Veigele, W.J., Briggs, E.A., Brown, R.T., Cromer, D.T., Howerton, R.J.: Atomic form factors, incoherent scattering functions, and photon scattering cross sections. J Phys Chem Ref Data 4, 471–538 (1975).

11. Terini, R.A., Pereira, M.A.G., Künzel, R., Costa, P.R., Herdade, S. B.: Comprehensive analysis of the spectrometric determination of voltage applied to X-ray tubes in the radiography and mammography energy ranges using a silicon PIN photodiode. Br J Radiol 77, 395–404 (2004).

Evaluation of Effective Energy Distribution of 320-Multidetector CT Using GAFCHROMIC EBT3

Tatsuhiro Gotanda, Toshizo Katsuda, Rumi Gotanda, Takuya Akagawa, Tadao Kuwano, Nobuyoshi Tanki, Hidetoshi Yatake, Yasuyuki Kawaji, Takashi Amano, Shinichi Arao, Atsushi Ono, and Akihiko Tabuchi

Abstract

Knowledge of the effective energy of 320-multidetector computed tomography (CT) is important for quality assurance and quality control. Evaluation in two dimensions is necessary because the effective energy varies depending on the shape of the wedge filter located in the CT device. The purpose of this study was to measure the two-dimensional effective energy distribution of the CT using GAFCHROMIC EBT3 (EBT3), which has a weak energy dependence. The exposure parameters of the 320-multidetector CT were 120 kV, 500 mA, and 5.0 s, and the X-ray tube was stopped at the 0 o'clock position. To avoid scattered radiation, the distance between the EBT3 and other scatterers was set to 200 mm or more. The Al filter thickness was increased from 2 to 20 mm. The irradiated area was divided into 54 compartments, and the density attenuation ratio was measured. The half-value layers (HVLs) were determined using the density attenuation ratios. The effective energies were obtained from the HVLs, and the two-dimensional effective energy distribution was evaluated. Because the thickness of the wedge filter in the longitudinal direction (parallel to the bed) remained unchanged, the variation in the effective energy was negligible in this direction. On the other hand, in the lateral direction (perpendicular to the bed), because the wedge filter gradually thickened from the center to the side, the effective energy from the center to the side increased. The two-dimensional effective energy distribution of the CT could thus be measured using EBT3.

Keywords

Computed tomography • Effective energy
GAFCHROMIC film

1 Introduction

The effective energy of an X-ray beam is one of the standard quality assurance (QA) and quality control (QC) tests for various radiological systems. In the radiation-quality management of complex X-ray generators such as X-ray computed tomography (CT), the effective energy is important. In general, the half-value layer (HVL), which is used to calculate the effective energy, is measured by means of an ionization-chamber (IC) dosimeter. However, there are two major problems. The first problem is the distribution of the effective energy by a wide X-ray beam. The effective energy of the CT varies depending on the shape of the wedge filter located in the CT device. It is necessary to measure at each position, because the area of the actually used X-ray beam is wide. Therefore, a method using an IC dosimeter for measuring the X-ray beam center only cannot be applied. The second problem is the influence of the scattered radiation caused by the geometric layout of the X-ray CT device, because the X-ray tube, bed, and detector are closer to the dosimeter as compared to an X-ray apparatus in general. In the HVL measurement using the IC, the sensitivity to the scattered radiation is high and measurement error is considered to have occurred. To solve these problems, attention has been focused on the characteristics of the radiochromic film.

T. Gotanda (✉) · R. Gotanda · T. Amano · S. Arao A. Ono
A. Tabuchi
Kawasaki University of Medical Welfare, Okayama, Japan
e-mail: tat.gotanda@mw.kawasaki-m.ac.jp

T. Katsuda · N. Tanki
Butsuryo College of Osaka, Osaka, Japan

T. Akagawa
Tokushima Red Cross Hospital, Tokushima, Japan

T. Kuwano · N. Tanki
Okayama University, Okayama, Japan

H. Yatake
Kaizuka City Hospital, Osaka, Japan

Y. Kawaji
Junshin Gakuen University, Fukuoka, Japan

© Springer Nature Singapore Pte Ltd. 2019
L. Lhotska et al. (eds.), *World Congress on Medical Physics and Biomedical Engineering 2018*,
IFMBE Proceedings 68/3, https://doi.org/10.1007/978-981-10-9023-3_94

Radiochromic films are easily used to measure absorbed doses, because they do not require development processing and exhibit a density change that depends on the absorbed dose [1, 2]. The main characteristics of radiochromic film that distinguish it from an IC dosimeter are that it can be used to measure two-dimensional dose distributions and can serve as a general X-ray film. As a result, it is used in dose measurement and effective energy measurement techniques [3–5]. In this study, GAFCHROMIC EBT3 (EBT3: International Specialty Products, Wayne, NJ, USA) dosimetry film was used as the radiochromic film because it exhibits only slight energy dependency errors in comparison with other radiochromic films. In a previous experiments, the density-absorbed dose calibration curve was linearly correlated in the low dose range (100 mGy or less), and the energy dependence error of 30 to 60 keV was about 0.2% [6]. In addition, it has been proven that the influence of scattered radiation on EBT3 is small at 50 mm or beyond [7]. The purpose of this study was to measure the two-dimensional effective energy distribution of the 320-multidetector CT using the EBT3 with a weak energy dependence.

2 Materials and Methods

2.1 GAFCHROMIC EBT3

EBT3 is yellow and rectangular, measuring 205 mm × 255 mm, with a specified thickness of 280 µm. Although the scan mode for image acquisition is typically the transmission mode, several studies suggested that reflection mode improved the measurement precision for the low absorbed dose range [8]. Therefore, this study was performed using reflection mode. When the active component is exposed to radiation, it changes color to dark blue. EBT3 is suitable for absorbed dose measurement in the diagnostic range, because it is recommended for dosimetry in a wide dose range (0.01–40 Gy) [1]. The disadvantage is that there are non-uniformity errors of ± 3% reported on the EBT3 data sheet [1]. The EBT3 was kept at room temperature (20–25 °C) in a shaded bag.

2.2 HVL Measurement of the 320-Multidetector CT

Exposure methods. In this study, a 320-multidetector CT (Aquilion one, Toshiba Medical Co., Ltd., Tochigi) was used. The exposure parameters were 120 kV, 500 mA, and 5.0 s, and the X-ray tube was stopped at the 0 o'clock position. The geometric arrangement of the experimental setup of the exposure method for measuring the CT HVL is

shown in Fig. 1. The aluminum filter was set up near the X-ray tube, and the EBT3 was set up 50 mm above the isocenter of the CT. To avoid scattered radiation, the distance from the aluminum filter to the EBT3 was set at 200 mm. To avoid back-scattered radiation from the bed, the distance between the bed and the EBT3 was set at 200 mm. To avoid back-scattered radiation from the bed, the distance between the bed and the EBT3 was set at 200 mm. The aluminum filter has a square shape form with dimensions of 200 mm × 200 mm, and its thickness increases from 2 to 20 mm (the specific thicknesses are 2, 4, 6, 8, 10, 15 and 20 mm).

Scanning and Analysis of the EBT3 Film. EBT3 was scanned using an A3 flat-bed scanner in RGB (48-bit) mode, 100 dpi, with the protection of a film of liquid crystal for removal of the Moire artifact. The EBT3 was scanned in reflection mode. To scan the EBT3 film using this setting, regular white paper with a uniform density was attached to the back of the EBT3 film. In addition, the EBT3 film was scanned both before and after exposure to eliminate the non-uniformity error of the film layer. The EBT3 film was scanned 24 h after exposure. The EBT3 was kept at room temperature (20−25 °C) in a shaded bag.

The image data from the EBT3 film was divided into R, G, and B modes (16 bits each), and the R mode was used for high-density contrast. It was converted to greyscale and analyzed using ImageJ version 1.48v image analysis software (National Institutes of Health, Maryland, USA). To measure the increase in the density of the EBT3 film, the image data before exposure were subtracted from the after-exposure data in terms of pixel units in two dimensions.

The region of interest (ROI) was set on the irradiated area of the EBT3 at each thickness of the aluminum filter and was sized at 600 pixels (152.4 mm) × 900 pixels (228.6 mm). Then, the ROI was divided into 54 compartments (6 × 9), and an average density pixel value was calculated in each compartment. The attenuation curves for HVL measurement

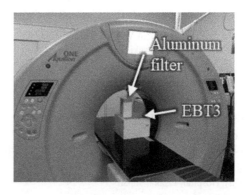

Fig. 1 Geometric arrangement of the exposure method for measuring the CT HVL

were obtained using the density pixel values of each compartment. The HVLs were then calculated using the attenuation curves of each compartment. In addition, the effective energies were obtained from the HVLs, and the two-dimensional effective energy distribution was evaluated. The conversion from the HVL to the effective energy was calculated using the data of Seltzer and Hubbell at the National Institute of Standards and Technology [9].

HVL Measurement Using the Multi-Semiconductor Detector. HVL measurement using the multi semiconductor detector (NOMEX, PTW, Freiburg) was performed in order to compare it with the effective energy distribution using the EBT3. The HVL of only the center of the X-ray beam of the 320-multidetector CT was measured three times, and then averaged. The exposure parameters were 120 kV, 50 mA, 1.0 s. The choice of the arrangement of the exposure apparatus was the same as for the EBT3. The results for the HVL of only the center of the X-ray beam on the EBT3 were compared with the HVL using a multi-semiconductor detector, in order to evaluate the applicability of EBT3.

3 Results

3.1 Distribution of the Effective Energy

Figure 2 shows the distribution of the effective energy using the EBT3. Because the thickness of the wedge filter in the longitudinal direction (parallel to the bed) remained unchanged, the variation in the effective energy was negligible in this direction. On the other hand, in the lateral direction (perpendicular to the bed), because the wedge filter gradually thickened from the center to the side, the mean of the effective energy distribution from the center to the side (nine compartments along the lateral direction) were 53.0, 53.6, 54.3, 56.8, 61.9, 63.2, 75.1, 99.6, and 220.3 keV.

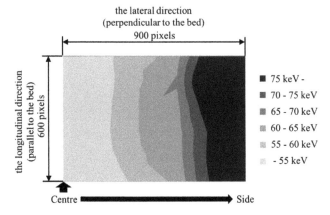

Fig. 2 Distribution of the effective energy using the EBT3

3.2 Comparison of the EBT3 and Multi-semiconductor Detector

Table 1 shows each value measured with the multi-semiconductor detector. The HVL of the 320-multidetector CT using the multi-semiconductor detector was 7.25 mm. Effective energies of the EBT3 and multi-semiconductor detector were 53.0 keV and 51.0 keV, respectively. There was an error of approximately 3.9%.

4 Discussion

In this study, the effective energy distribution of CT was measured using three main characteristics of the EBT3. First, the effective energies were evaluated in all areas of the CT X-ray beam, because the EBT3 can measure the dose distribution in two dimensions. Second, the HVL measurement using EBT3 is valid in the energy range of the CT because the energy-dependent error of the EBT3 is less than 1% from 30 to 60 keV. Third, the problems of the geometric layout of the X-ray CT apparatus can be eliminated, because the influence of scattered radiation on EBT3 is small at 50 mm or beyond. As a result, the effective energy, similar to that of a multi-semiconductor detector, was obtained at the center of the X-ray beam. However, it was observed that the error increased at the side of 190.5 mm or more from the center. As a reason, it is considered that the absorbed dose incident on the EBT 3 was too low. The pixel values at the center of the X-ray beam without filter and with a 20-mm Al filter were −6291.70 and −1585.08, respectively. On the other hand, the pixel values at the side of the X-ray beam were −335.00 and −94.25, respectively. Considering that the non-uniformity error of EBT 3 is approximately 400 pixel values, it is suggested that the difference in pixel value at the side of the X-ray beam is low and the error increases. Therefore, when the distance from the X-ray beam center is less than 190.5 mm, the EBT3's sensitivity is similar to that of the multi-semiconductor detector and can measure the CT's effective energy distribution.

Table 1 Values measured with a multi-semiconductor detector

	Average
Dose [Gy]	13.657E − 03
Dose rate [Gy/s]	12.660E − 03
Exposure time [s]	1.079E + 00
Practical Peak Voltage [kV]	120.9
Average kVp [kV]	123.8
Max kVp [kV]	123.8
HVL [mm Al]	7.25

5 Conclusion

There is no standard method to measure the effective energy of the 320-multidetector CT. This study succeeded in measuring the effective energy with a precision of less than 5% in the CT X-ray beam center. In addition, the effective energy could obtain two-dimensional distributions when the distance from the X-ray beam center was less than 190.5 mm. Therefore, it is considered that the method provides a precise estimate of the CT effective energy for QA and QC.

Acknowledgements The authors would like to thank Mr. Kouji Ochi for providing various equipment used in this investigation. This study was supported by JSPS KAKENHI, Grant Number 16K21553.

Conflict of Interest The authors hereby declare that they have no conflicts of interest.

References

1. Ashland Homepage. Available at: http://www.gafchromic.com/documents/BallCube_II_EBT2_3_box_insert_201404.pdf, last accessed 2017/10/31.
2. Gotanda R, Katsuda T, Gotanda T et al.: Computed Tomography Phantom for Radiochromic Film Dosimetry, Australas. Phys. Eng. Sci. Med., 30(3), 194–199 (2007).
3. Gotanda T, Katsuda T, Gotanda R et al.: Evaluation of effective energy for QA and QC: measurement of half-value layer using radiochromic film density, Australas. Phys. Eng. Sci. Med. 32(1), 26–29 (2009).
4. Gotanda T, Katsuda T, Gotanda R et al.: Half-value layer measurement: simple process method using radiochromic film, Australas. Phys. Eng. Sci. Med. 32(3), 150–158 (2009).
5. Gotanda T, Katsuda T, Gotanda R et al.: Evaluation of effective energy using radiochromic film and a step-shaped aluminum filter, Australas. Phys. Eng. Sci. Med. 34(2), 213–222 (2011).
6. Gotanda T, Katsuda T, Gotanda R et al.: Energy response of the GAFCHROMIC EBT3 in diagnosis range. IFMBE Proceedings 51, 752–755 (2015).
7. Gotanda T, Katsuda T, Kawasaki A et al.: Influence of scattered radiation on Gafchromic EBT3. IFMBE Proceedings 65, 1037–1040 (2017).
8. Gotanda T, Katsuda T, Akagawa T et al.: Evaluation of GAFCHROMIC EBT2 dosimetry for the low dose range using a flat-bed scanner with the reflection mode, Australas. Phys. Eng. Sci. Med. 36(1), 59–63 (2013).
9. Seltzer SM and Hubbell JH.: Table and Graphs of Photon Mass Attenuation Coefficients and Mass Energy-Absorption Coefficients for Photon Energies 1 keV to 20 Mev for Elements Z = 1 To 92 And Some Dosimetric Materials, National Institute of Standards and Technology (1995).

Monte Carlo Calculations of Skyshine Neutron Doses from an 18 MeV Medical Linear Accelerator

Nobuteru Nariyama

Abstract

Neutron skyshine doses were calculated using a Monte Carlo code, FLUKA, by changing the heights and openings of shield walls. In the calculation geometry, the distance from the floor of the room to the roof was varied from 3 to 10 m. The target into which electrons were injected was located at a height of 2.3 m. The distance between the shield wall and target was varied from 1.5 to 5.79 m. Further, the height of the atmosphere was varied from 10 to 40 m. Consequently, the dose outside the shield wall increased with the height of the atmosphere and became saturated at 30 m; the neutrons reaching the ground were observed to be scattered mostly below 30 m. For neutron skyshine, NCRP Report No. 51 provided an expression stating that the neutron dose at ground level was proportional to the solid angle of shield walls and inversely proportional to the square of the distance from the target to the ceiling plus 2 m. In the Monte Carlo results, the doses conformed to the expression below a height of 5 m and above a distance of 2.6 m from the target to the shield wall. Under other conditions, the doses became higher than those predicted by the expression, which indicated the large contribution of neutrons emitted immediately above the target to the skyshine doses.

Keywords

Skyshine • Monte carlo • Linear accelerator

1 Introduction

Skyshine is a phenomenon wherein radiation scattered by the atmosphere above a ceiling reaches the ground level outside the shield. The radiation is crucial when the ceiling is thin. Previously, simple equations for accelerator facilities have been proposed [1]; however, their consistency with experimental data was not satisfactory [2]. With an increase in the accelerating energy, the contribution of neutrons to the skyshine dose increases. The neutron doses from an 18 MeV accelerator are higher than photon doses for skyshine [2]. In the equation for neutron skyshine, the dose is proportional to the solid angle of shield walls and inversely proportional to the square of the distance from the target to the ceiling plus 2 m. By using a Monte Carlo code, the doses can be obtained with variation of these parameters. In this study, neutron skyshine doses were calculated using a Monte Carlo code by changing the ceiling heights and openings of shield walls.

2 Calculation Method

2.1 A Monte Carlo Method

Figure 1 shows the vertical section view of the Monte Carlo calculation model. An electron of 18 MeV was impinged on the lead cylindrical target horizontally. The target was surrounded by a cylindrical shield. No roof shield was assumed.

A Monte Carlo transport code FLUKA (Ver. 2011.2c.6) [3] was used with the flair interface (Ver. 2.3-0) for the calculations. A multicore CPU was used by applying different random number seeds on each one-day run. In the calculation geometry of Fig. 1, the distance from the floor of the room to the top of the shield was varied from 3 m to 10 m. The cylindrical target of radius 3.3 cm was located at a height of 2.3 m from the ground. A large dependence of neutron doses on the target size was not observed compared

N. Nariyama (✉)
Japan Synchrotron Radiation Research Institute,
Sayo, Hyogo 679-5198, Japan
e-mail: nariyama@spring8.or.jp

© Springer Nature Singapore Pte Ltd. 2019
L. Lhotska et al. (eds.), *World Congress on Medical Physics and Biomedical Engineering 2018*,
IFMBE Proceedings 68/3, https://doi.org/10.1007/978-981-10-9023-3_95

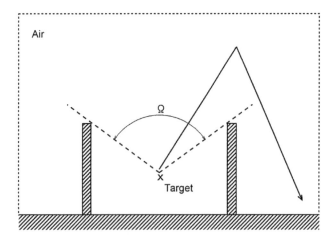

Fig. 1 Calculation model for neutron skyshine. The shaded region was assumed to absorb the radiation completely. The solid line with the arrow indicates the path of a neutron

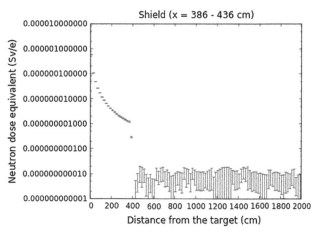

Fig. 2 Neutron dose equivalent per incident electron calculated using the Monte Carlo code for the model of Fig. 1 as a function of the distance from the target. The doses at the points more than 4.36 m from the target, which are almost constant, are only due to the skyshine. The location of the shield was 386 cm

with photon doses [4]. The distance between the cylindrical shield wall and target was varied from 1.5 to 9.5 m. Further, the height of the atmosphere was varied from 10 to 40 m. As the shield and ground were assigned as "black hole" in the input, all the particles impinging on the materials were absorbed immediately.

2.2 An Equation for Neutron Skyshine

NCRP Report No. 51 [1] provided the equation for neutron skyshine as follows:

$$H = 0.84 \times 10^5 B_{ns} \Phi_O \Omega / d_i^2, \qquad (1)$$

where H is the neutron dose equivalent rate, B_{ns} is the roof shielding transmission ratio for neutrons, Φ_O is the neutron fluence rate at 1 m from the target, Ω is the solid angle of the shield walls, and d_i is the distance from the target to ceiling plus 2 m.

3 Monte Carlo Calculation Results

3.1 Atmosphere Height

Figure 2 shows the calculated neutron dose equivalent distribution with respect to the distance from the target. In the Monte Carlo calculation model, the shield was located in the range of 3.86 to 4.36 m from the target, and the height of the shield was 4.97 m. The top height of atmosphere was 20 m. Inside the shield wall, the doses steeply decreased with the distance from the target. Outside the wall, the doses, which were due to the skyshine, were approximately constant up to 20 m. The average values in the flat region are plotted in

Figs. 3, 4, 5, and their errors were approximately 100% as shown in Fig. 2.

Figure 3 shows the neutron skyshine dose equivalent with variation of the top height of atmosphere, which was normalized at 30 m. The top height of atmosphere corresponds to the height of the dotted rectangle area of Fig. 1. From 10 m, the doses increased with the top height of the air layer, and gradually approached to the saturation value around 30 m. Thus, the contribution at 15 m was half of that at 30 m in the saturation condition, and the contribution at the height above 30 m was observed to be small. This is attributable to the attenuation in the air and the increase of visual angle from the scattering points with the increase in the distance.

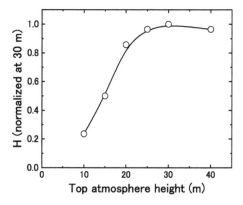

Fig. 3 Neutron skyshine dose equivalent normalized at the distance from the target of 30 m with variation of the height of the air region, i.e., the dotted rectangle area of Fig. 1

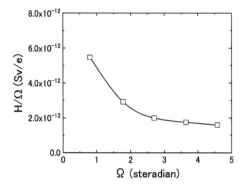

Fig. 4 Neutron skyshine dose equivalent divided by Ω as a function of Ω. The distance from the target to the shield wall was varied

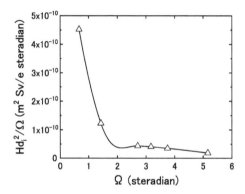

Fig. 5 Neutron skyshine dose equivalent multiplied by d_i^2/Ω as a function of the solid angle. The height of the shield wall was varied

3.2 Solid Angle of Shield Walls with Variation of Room Size

Figure 4 shows the neutron skyshine dose equivalent normalized with the solid angle Ω of the shield wall when Ω was changed by changing the distance from the target to the shield. The height of the shield wall was 4.97 m. Thus, when the distance from the target to the inner surface of the shield was 3.86 m, the value of Ω was 2.7 steradian. The distance was varied as 1.5, 2.6, 3.86, 5.79, and 9.5 m. According to Eq. (1), the dose H is proportional to Ω. In Fig. 4, the ordinate values H/Ω decreased with Ω at first, and approached a constant value at approximately $\Omega = 2.7$ steradian. Thus, the dose was almost proportional to the solid angle at the distance above 2.7 m. However, a room with dimensions smaller than 2.7 m is not realistic.

3.3 Solid Angle of Shield Walls with Variation of Shield Wall Height

The change of the shield wall height influences both Ω and d_i in Eq. (1). Figure 5 shows the values of Hd_i^2/Ω with respect to Ω. In the calculation model, the shield heights were varied as 3, 4, 4.5, 4.97, 7, and 10 m, i.e., $d_i = 2.7, 3.7, 4.2, 4.67, 6.7,$ and 9.7 m, and $\Omega = 5.16, 3.75, 3.17, 2.71, 1.43,$ and 0.666, respectively. The location of the shield was varied from 3.86 to 4.36 m from the target. Similar to Fig. 4, the ordinate values decreased with Ω steeply in the small Ω region, and became almost constant, i.e., Eq. (1) was still satisfied. The flat region above $\Omega = 2.5$ became almost the same as Fig. 4. When the shield was high, Ω became small, and H became higher than that predicted using Eq. (1). The heights of 7 and 10 m are practically too high.

4 Conclusions

Neutron skyshine dose equivalent from an 18 MeV electron linear accelerator was calculated using a Monte Carlo code FLUKA. Most of the neutrons reaching the ground outside the shield were scattered in the air below 30 m in height. The doses were proportional to the solid angle Ω above 2.5 steradian. Even when the wall height was changed, the doses were proportional to Ω/d_i^2 at $\Omega > 2.5$. The behavior of neutrons emitted immediately above the target was considered to deviate from Eq. (1), and the contribution was larger than predicted. In practical facilities, as Ω is assumed to be approximately 2.6, the relation among H, Ω, and d_i in Eq. (1) is maintained.

Conflict of Interest The authors have no conflict of interest to report.

References

1. National Council on Radiation Protection and Measurements: Radiation protection design guidelines for 0.1–100 MeV particle accelerator facilities. NCRP Report No. 51, Washington, DC (1977).
2. McGinley, P.H.: Shielding Techniques for Radiation Oncology Facilities. 2nd edn. Medical Physics Publishing, Madison (2002).
3. Ferrari, A., Sala, P.R., Fassò, A., Ranft, J.: FLUKA: a multi-particle transport code. CERN-2005–10 (2005).
4. Nariyama, N.: Evaluation of angular distribution of radiation doses to upgrade RF-gun test facility in SPring-8. In: Proceedings of the 12th annual Meeting of Particle Accelerator Society of Japan, pp. 874–877 (2015).

Evaluation in the Use of Bismuth Shielding on Cervical CT Scan Using a Female Phantom

Fernanda Stephanie Santos, Priscila do Carmo Santana,
Thessa Cristina Alonso, and Arnaldo Prata Mourão

Abstract

Computed Tomography (CT) has become an important tool to diagnose cancer and to obtain additional information for different clinical questions. Today, it is a very fast, painless and noninvasive test that can be performed high quality images. However, CT scans usually require a higher radiation exposure than a conventional radiography examination. The aim of this study is to determine the dose variation deposited in thyroid and in nearby radiosensitive organs, such as: lenses, pharynx, hypophysis, salivary gland, spinal cord and breasts with and without the use of bismuth shielding. A cervical CT scan was performed on anthropomorphic female phantom model Alderson Rando, from the occipital to the first thoracic vertebra, using a GE CT scanner, Discovery model with 64 channels. Dose measurements have been performed by using radiochromic film strips to recorder the individual doses in the organs of interest. After the phantom cervical CT scan, the radiochromic film strips were processed for obtaining digital images. Digital images were worked to obtain the dose variation profiles for each film. With the data obtained, it was found the organ dose variation. The results show us that the thyroid received the highest dose, 24.59 mGy, in the phantom, according to the incidence of the primary X-ray beam.

Keywords

Computed tomography • Dosimetry • Bismuth shielding

1 Introduction

The Computed tomography (CT) is the most common technique used for diagnostic purpose. It is a very fast test that can be performed high quality images. However, the increasing demand for CT had a considerable impact on doses provided to patients and on the exposure of the population as whole [1]. In fact, the worldwide average annual per-capita effective dose from medical procedures has approximately doubled in the past 10–15 years [2]. The dose evaluation in CT is one of many steps that can contribute for reducing patient doses. The cervical CT scans are commonly used for diagnosis of soft tissue, vascular changes, fractures, extent of injuries, dysplasia and other diseases with instability, so it can be associated with a high radiation dose to organs such as thyroid, lenses, salivary gland, pharynx, breast, parotid gland, spinal cord and hypophysis, when compared with conventional radiology. The main objective of this study was to analyze the absorbed doses in a cervical CT scan with and without the use of bismuth shield in the neck.

2 Materials and Methods

The experiment to observe the reduction dose due to bismuth shield was conducted using a GE CT scanner, Discovery model of 64 channels. An Alderson Rando female anthropomorphic phantom was used to perform head CT scans from the occipital to the first thoracic vertebra (Fig. 1).

This phantom is composed of a human skeleton surrounded by a material, physically and chemically similar to the soft tissues of an adult human body [3]. The body and head are structured in transected-horizontally into 2.5 cm

F. S. Santos (✉) · A. P. Mourão
Department of Nuclear Engineering, Federal University
of Minas Gerais, Belo Horizonte, Brazil
e-mail: fernanda.stephaniebh@yahoo.com.br

P. do Carmo Santana
Department of Image and Anatomy, Federal University
of Minas Gerais, Belo Horizonte, Brazil

A. P. Mourão
Biomedical Engineering Center, Federal Center for Technological
Education of Minas Gerais, Belo Horizonte, Brazil

T. C. Alonso
Nuclear Technology Development Center, Belo Horizonte,
Minas Gerais, Brazil

© Springer Nature Singapore Pte Ltd. 2019
L. Lhotska et al. (eds.), *World Congress on Medical Physics and Biomedical Engineering 2018*,
IFMBE Proceedings 68/3, https://doi.org/10.1007/978-981-10-9023-3_96

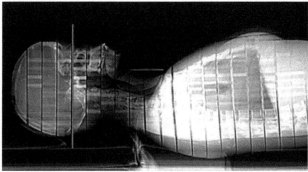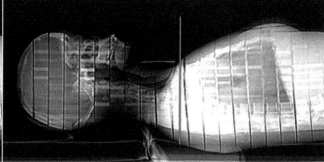

Fig. 1 Lateral scout of head and chest with thyroid bismuth shielding, with limits of region scanning yellow line (Color figure online)

thick slices. The slices that make up the body phantom have holes that allow placing dosimeters within the phantom [4, 5]. Figure 2 shows the positioning of the female phantom in the gantry isocenter. The parameters of the protocol used in this study are shown in Table 1.

Dose measurements have been performed by using GAFCHROMIC XR-CT radiochromic film strips to register the individual doses in the organs of interest such as lenses, thyroid, pharynx, breast, hypophysis, spinal cord, parotid gland and salivary gland, with and without bismuth shield on the neck area (Fig. 3). The Fig. 4 shows the axial cervical CT images with and without a 1 mm thick piece of bismuth shielding.

The radiochromic films are self-developing dosimetry films, insensitive to visible light making it easy to work with during analysis and provide greater spatial resolution in the sub millimeter range. They have been used extensively in combination with flat bed document scanners to measure patient doses [6, 7]. Metrological reliability of the radiochromic films was demonstrated through homogeneity and repeatability tests and by calibrating it in a reference

Table 1 CT scan parameters

Voltage (kV)	Electric current(mA)	Tube time(s)	Pitch	Distance (mm)	Thickness beam(mm)
120	175	0.8	0.984	150	40

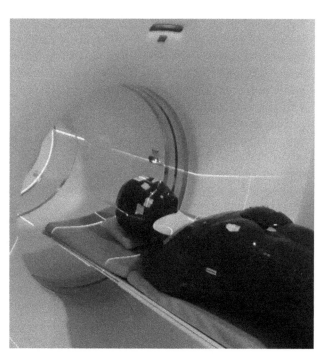

Fig. 3 The phantom Alderson female with the bismuth shielding on the neck spot

radiation for CT (RQT9) that were reproduced in the Calibration Laboratory of the Development Center of Nuclear Technology (CDTN/CNEN) [7, 8].

3 Results

Absorbed doses in the organs positions such as: thyroid, lenses, pharynx, hypophysis, breast, spinal cord in cervical area, parotid and salivary glands are shown in Table 2.

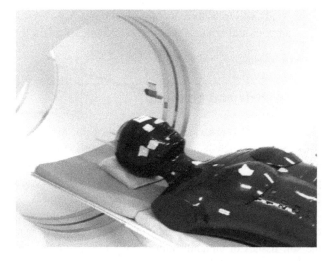

Fig. 2 Positioning of the Alderson female phantom in the gantry

Fig. 4 Axial cervical CT images: **a** without and **b** with bismuth shielding over the neck

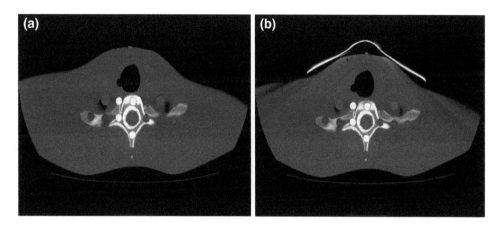

Table 2 Mean absorbed dose in some organ positions in the phantom during cervical CT scans with and without bismuth shielding on neck

Organ position	Mean absorbed dose (mGy)		Dose reduction (%)
	Without bismuth shielding	With bismuth shielding	
Lenses	11.53 ± 0.58^{a}	12.69 ± 0.80	–
Pharynx	16.52 ± 0.75	13.31 ± 0.65	20
Thyroid	24.59 ± 0.69	15.51 ± 0.33	37
Salivary gland	18.87 ± 0.55	19.34 ± 0.85	–
Hypophysis	7.33 ± 0.79	5.58 ± 0.25	24
Breast	0.27 ± 0.13	0.25 ± 0.12	1
Spinal cord	16.55 ± 0.47	15.67 ± 0.55	6
Parotid gland	18.15 ± 0.66	16.0 ± 0.62	12

[a]Standard deviation

These results allow us to observe that the use of bismuth shield led to a decrease in radiation dose deposited in the neck and all organs studied.

The highest recorded dose was 24.59 mGy, occurred in the thyroid position that stressed the situation of unnecessary radiation exposure. However, with the use of bismuth shield had a dose reduction of 37%. The recorded doses due to scans with and without bismuth shielding showed significant differences. The decrease dose in pharynx, hypophysis and parotid gland was 20%, 24% and 12% respectively, that was desired due to the use of bismuth shielding. It is expected that thyroid shielding would degrade image quality and would increase the image noise, however the results of this work suggests that it might be an acceptable procedure to be used for dose reduction mainly during CT examinations that would provide high doses to radiosensitive organs. In organs like salivary gland and lenses had a little increase of the dose due to the scatter radiation. The use of bismuth shielding is simple and efficient to

Fig. 5 Influence of bismuth shielding on absorbed doses for some organ positions of the phantom

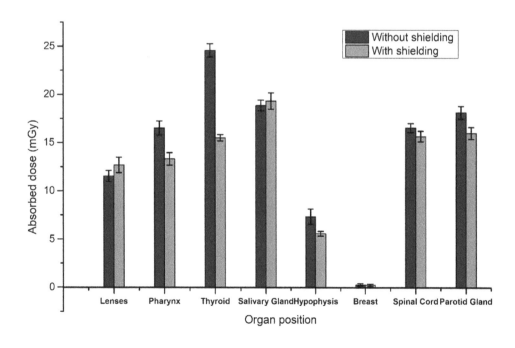

reduce absorbed doses to the thyroid and nearby organs. The graphic showed in the Fig. 5 allows observe the influence of bismuth shielding on absorbed doses in the organs studied.

4 Conclusions

The absorbed doses were determined during cervical CT scans with and without bismuth shielding on thyroid of an Alderson Rando female anthropomorphic phantom. Dose values were significantly reduced and they suggested that the use of bismuth would be, in some cases, a proper procedure for protection as the conditions used for both scans were the same.

Acknowledgements The authors are grateful to CAPES, CNPq and FAPEMIG for the support. Also, the Technology Center in Nuclear Medicine of UFMG is acknowledged for producing the images and CDTN/CNEN for the use of the phantom.

Conflict of Interest The authors declare that they have no conflict of interest.

References

1. Alonso, T., Mourão, A., Santana, P.: Assessment of breast absorbed doses during thoracic computed tomography scan to evaluate the effectiveness of bismuth shielding. Applied Radiation and Isotopes (117), 55–57 (2016).
2. Goo, H.: CT radiation dose optimization and estimation: an update for Radiologists. Korean J Radiol (13), 1–11 (2012).
3. Aleme, C., Lyra, M., Mourão, A.: Evaluation in the use of Bismuth Shielding on cervical spine CT scan using a Male Phantom. In: INTERNATIONAL SYMPOSIUM ON SOLID STATE DOSIMETRY 2014, pp. 664–669. Cusco, Peru (2014).
4. Gbelcova, L., Nikodemova, D, Horvathova, M.: Dose reduction using Bismuth shielding during paediatric CT examinations in Slovakia. Radiation Protection Dosimetry (147), 160–163 (2011).
5. Colletti, P., Micheli, O., Lee, K.: To shield or not to shield: application of bismuth breast shields. AJR (200), 503–507 (2013).
6. Giaddui, T., Cui, Y., Galvin, W., Chen, Y., Xiao, Y.: Characteristics of Gafchromic XRQA2 films for kV image dose measurement. Medical Physics (39), 842–850 (2012).
7. Mourão, A., Alonso, T., Silva, T.: Dose profile variation with voltage in head CT scans using radiochromic films. Radiation Physics and Chemistry (95), 254–257 (2014).
8. Costa, K., Gomez, A., Alonso, T., Mourão, A.: Radiochromic film calibration for the RQT9 quality beam. Radiantion Physics and Chemistry (140), 370–372 (2017).

Temporal Characterization of the Flat-Bed Scanner Influencing Dosimetry Using Radiochromic Film

Rumi Gotanda, Toshizo Katsuda, Tatsuhiro Gotanda, Nobuyoshi Tanki, Hidetoshi Yatake, Yasuyuki Kawaji, Tadao Kuwano, Takuya Akagawa, Akihiko Tabuchi, Atsushi Ono, and Shinichi Arao

Abstract

Radiochromic films (RFs) have been developed for the measurement of the absorbed dose of low-energy photons. RFs are self-developing and radiation sensitive, and the amount of darkening is proportional to the absorbed dose. RFs are easy to handle due to their insensitivity to interior room light. However, the precision of the measurement has been questioned because of the change in density caused by the scan timing of the image acquisition using a flat-bed scanner. In this study, the density change of a flat-bed scanner was investigated using the temporal and the repetition scans. To obtain the image density, Gafchromic XR-QA2 films (XR-QA2s) were irradiated at 0 and 20 mGy (air-kerma) using 75 kVp (30 keV). The XR-QA2s were scanned every hour (0–6 h) from power activation to investigate the temporal light source change of a flat-bed scanner (EPSON ES-10000G). In addition, ten consecutive scans were performed every hour. The scan parameters were RGB (48-bit) mode, 100 dpi, and reflection mode. Image data of the XR-QA2s were divided into R, G, and B modes, and the R (16-bit) mode was used. The temporal light source change after power activation was small. However, in ten consecutive scans, the density of the first scan was the highest. The densities decreased with more scans.

This result indicated that the precision of the dose measurement has about a 3% error due to the repeated scans. To obtain an accurate dose measurement, the image data obtained under the same conditions, such as the same time from power activation or same number of consecutive scans, must be used.

Keywords

Radiochromic film • Temporal characteristics
Flat-bed scanner • Radiation dosimetry

1 Introduction

Recently, radiochromic films (RFs) have been developed for measurement of the absorbed dose of low-energy photons. RFs are self-developing and radiation sensitive, and the amount of darkening is proportional to the absorbed dose. RFs are easy to handle due to their insensitivity to interior room light [1].

In this study, Gafchromic XR-QA2 film (ASHLAND) was used. The measured dose and energy range of the XR-QA2 are designed to be 1–200 mGy and 20–200 kVp, respectively. XR-QA2 was designed as a QA tool for radiological diagnosis, but it is also used for dose measurement in CT examinations and radiography [2–6]. To measure the absorbed dose, a calibration curve is necessary. The calibration curves vary according to the tube voltage. The energy response is the best at around 50 kVp (25 keV) and decreases with the increase in tube voltage within the range of 50–150 kVp. Additionally, the energy response decreased rapidly below 40 kVp [7]. In terms of the directional characteristics, the sensitivity decreased by 76% at 90°, where the incident angle became parallel to the film plane. To obtain an accurate measurement of the absorbed dose, there are various key parameters (e.g., energy response, dose characteristics, and directional characteristics) that need to be considered in the measurement method [7–9].

R. Gotanda (✉) · T. Gotanda · A. Tabuchi · A. Ono · S. Arao
Kawasaki University of Medical Welfare, Okayama, Japan
e-mail: gotandar@mw.kawasaki-m.ac.jp

T. Katsuda · N. Tanki
Butsuryo College of Osaka, Osaka, Japan

N. Tanki · T. Kuwano · T. Akagawa
Okayama University, Okayama, Japan

H. Yatake
Kaizuka City Hospital, Osaka, Japan

Y. Kawaji
Junshin Gakuen University, Fukuoka, Japan

T. Akagawa
Tokushima Red Cross Hospital, Tokushima, Japan

© Springer Nature Singapore Pte Ltd. 2019
L. Lhotska et al. (eds.), *World Congress on Medical Physics and Biomedical Engineering 2018*,
IFMBE Proceedings 68/3, https://doi.org/10.1007/978-981-10-9023-3_97

XR-QA2 is a reflective-type film, and a flatbed scanner is used for the measurement of the image density [1]. To measure the increase in density, the XR-QA2 is scanned before and after irradiation using a flatbed scanner. An ES-10000G (EPSON) light source used in this study is a high-brightness xenon fluorescent lamp [10]. The xenon fluorescent lamp takes time to stabilize the luminance. Therefore, the precision of the measurement is questionable because of the change in density caused by the scan timing of the image acquisition using the ES-10000G. To obtain an accurate measurement of the absorbed dose, the influences of the light source properties have to be determined.

In this study, the density change from a flat-bed scanner was investigated using the temporal and the repetition scans.

2 Materials and Methods

2.1 Irradiation Method

The XR-QA2 was irradiated at 0 and 20 mGy (air-kerma) using 75 kVp (30 keV, 2.5 mmAleq; TOSHIBA DRX-1603B). Figure 1 shows the geometric arrangement of the irradiation method. The XR-QA2 was cut to a width of 100 mm and a length of 254 mm. The distances between the focus of the x-ray tube and the XR-QA2 or semiconductor detector are 45 or 90 cm, respectively. The x-ray irradiation area was 100 mm 40 mm at the film center. The irradiation dose was monitored with a semiconductor detector placed behind the film.

2.2 Analysis of Temporal Characterization

The XR-QA2 is a four-layer laminate: substrate (yellow polyester), adhesive layer, active layer, and substrate (white polyester). Substrate layers are dyed yellow or white to protect the active layer from ultraviolet rays or to create reflective-type film, respectively [1]. To adjust the color shading in the yellow polyester layer of the XR-QA2 and to evaluate the practical density increase of each film, the XR-QA2 was scanned before and after irradiation. The scan times after irradiation were decided to be after a period of 48 h to stabilize the darkening of the XR-QA2. In the scans after irradiation, the XR-QA2 was scanned every hour (0– 6 h) from power activation to investigate the temporal light source change of the flat-bed scanner (EPSON ES-10000G). In addition, ten consecutive scans were performed. The scan parameters were RGB (48-bit) mode, 100 dpi, and reflection mode.

Image data of the XR-QA2 were divided into R, G, and B modes, and the R (16-bit) mode was used. The R (16-bit) mode data of the films were inverted and changed into grayscale and were analyzed with Image J 1.51 k (National Institutes of Health (NIH)). In the image data before and after exposure, a ROI for analysis was chosen at the same position on each film (Fig. 2), and the pixel values of the ROI were measured. To adjust the color shading in the yellow polyester layer of the XR-QA2 and to evaluate the practical density increase in each film, the pixel values of the ROIs on the pre-irradiation films were subtracted from the pixel values of the ROIs on the post-irradiation films. The pixel values after this subtraction were defined as the net pixel value (NPV). The NPV of each film was analyzed for determining the influences of the temporal light source changes after power activation and the repeated scans.

3 Results

Figure 3 shows the NPVs of each temporal and repetition scan. The temporal light source change after activation was small. However, of the ten consecutive scans, the density of the first scan was highest. The densities decreased with the increasing number of scans. Therefore, in the first scan of each hour, the luminance of the light source was low, and the luminance increased as the number of scans increased.

Figure 4 shows the pixel value ratios of each temporal and repetition scan. This result indicates that the time it takes for the luminance to stabilize is longest immediately after activation.

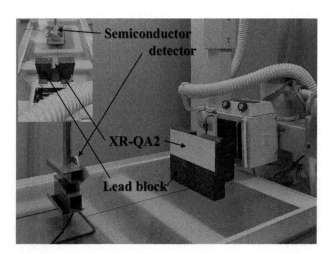

Fig. 1 Geometric arrangement of the irradiation method

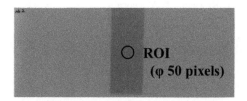

Fig. 2 Setting position of the region of interest (ROI)

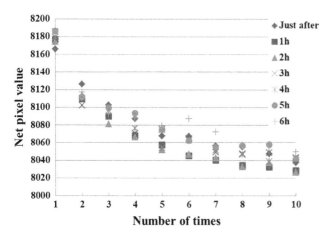

Fig. 3 NPVs of each temporal and repetition scan

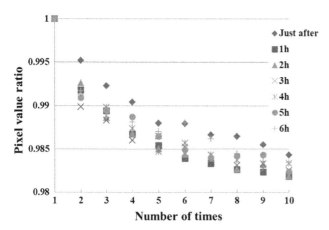

Fig. 4 Pixel value ratios for each temporal and repetition scan

4 Discussion

In dosimetry using a radiochromic film, the film density is proportional to the absorbed dose. In addition, using a flat-bed scanner with a high luminance xenon lamp as a light source, it takes time to stabilize the luminance of the light source just after power activation. Therefore, a roughly 3% error due to the temporal characterization of the scanner is contained in the dose measurement. To obtain an accurate dose measurement, the scan must be performed 1 h after power activation. In addition, the image data obtained after the seventh consecutive scan must be used.

5 Conclusion

These results indicate that the precision of the dose measurement contains a roughly 3% error due to the temporal characterization of the flat-bed scanner with a high luminance xenon lamp as a light source. To obtain an accurate dose measurement, image data obtained under the same conditions, such as the same time from power activation or same number of consecutive scans, must be used.

Conflict of Interest The authors hereby declare that they have no conflict of interests.

References

1. ASHLAND website. Gafchromic™ radiology films. http://www.gafchromic.com/gafchromic-film/index.asp
2. Gotanda R, Katsuda T, Gotanda T et al.: Computed Tomography Phantom for Radiochromic Film Dosimetry, Australas. Phys. Eng. Sci. Med., 30(3), 194–199 (2007)
3. Gotanda R, Katsuda T, Gotanda T et al.: Dose Distribution in Pediatric CT Head Examination using a New Phantom with Radiochromic Film, Australas. Phys. Eng. Sci. Med., 31(4), 339–344 (2008)
4. Gotanda T, Katsuda T, Gotanda R et al.: Evaluation of effective energy for QA and QC: measurement of half-value layer using radiochromic film density, Australas. Phys. Eng. Sci. Med. 32(1), 26–29 (2009)
5. Gotanda T, Katsuda T, Gotanda R et al.: Half-value layer measurement: simple process method using radiochromic film, Australas. Phys. Eng. Sci. Med. 32(3), 150–158 (2009)
6. Gotanda T, Katsuda T, Gotanda R et al.: Evaluation of effective energy using radiochromic film and a step-shaped aluminum filter, Australas. Phys. Eng. Sci. Med. 34(2), 213–222 (2011)
7. Gotanda R, Sato H, Nakajima E et al.: Energy response characteristics of radiochromic film at CT radiation quality. European Journal of Medical Physics 32(3), 293 (2016)
8. Gotanda T, Katsuda T, Gotanda R et al.: Energy response of the GAFCHROMIC EBT3 in diagnosis range. IFMBE Proceedings 51, 752–755 (2015)
9. Gotanda T, Katsuda T, Akagawa T et al.: Evaluation of GAFCHROMIC EBT2 dosimetry for the low dose range using a flat-bed scanner with the reflection mode, Australas. Phys. Eng. Sci. Med. 36(1), 59–63 (2013)
10. EPSON web site. http://www.epson.jp/products/back/hyou/scanner/es10000g.htm

The Relationship Between Eye Lens Doses and Occupational Doses for Different Centres of Interventional Cardiology

Lucie Sukupova and Zina Cemusova

Abstract

This study shows the level of eye lens doses and the relationship between eye lens doses $H_p(3)$ and doses above the apron $H_p(10)$ for 14 interventional cardiologists in 3 different centres of interventional cardiology. The doses were measured repeatedly with TLDs simultaneously placed above the apron and close to the left eye. For each patient's procedure, the kerma-area product value (P_{KA}) was known; values of $H_p(3)$ and $H_p(10)$ above the apron were normalized to total patients' doses for procedures performed by a given cardiologist in the measuring period. The $H_p(10)/P_{KA}$ values for IC 1: IC 2 : IC 3 were as follows: 0.19 : 0.69 : 0.35 μSv/Gy*cm^2 respectively. The $H_p(3)/P_{KA}$ values for IC 1 : IC 2 : IC 3 were as follows: 0.15 : 0.34 : 0.29 μSv/Gy*cm^2 respectively. The values in IC 1 were 2–4 times lower than in IC 2 and IC 3. The expected annual $H_p(3)$ doses were estimated; the highest was almost 16 mSv, the lowest 2 mSv. The results show different approaches to occupational radiation protection by cardiologists in the three centres (the use of an automatic contrast medium injector, shielding placed on patients to absorb scatter radiation from patients) as well as different approaches by cardiologists within the same centre. The local practice in IC 1 was more dose-saving than in the other centres. The values of $H_p(10)/H_p(3)$ were reproducible for each cardiologist, so the $H_p(10)$ value could be used for a retrospective estimation of the eye lens dose when needed, even when the eye lens dosimeter was not worn.

Keywords

Eye lens dosimetry • Occupational dose
Interventional cardiology

1 Introduction

In the last decade, procedures in interventional cardiology have become ever more complex. Cardiologists performing these procedures usually wear personal protective shieldings (apron, thyroid shield), so as to protect their torso and neck against radiation, but other body parts usually remain unshielded, i.e. the head, arms and legs. Some cardiologists even wear protective goggles.

The sensitivity of the eye lens has been a point of discussion for a few years. Finally, the eye lens was proclaimed to be more radiosensitive than thought previously, mainly as regards the cataract. A threshold for the cataract was set to 0.5 Sv for a single exposure and 5.0 Sv for a chronic exposure. As a consequence, the limit for the eye lens was lowered, and the former equivalent dose limit of 150 mSv/year was replaced by the limit of 20 mSv [1], as recommended by the IAEA and ICRP.

This substantial legislative change brings forth the challenge of assessing the eye lens dose with sufficient accuracy. For this purpose, special eye dosemeters measuring the quantity $H_p(3)$ (personal dose equivalent at the depth of 3 mm), which is the operational quantity for the eye lens dose, have been developed. Another discussed possibility is based on the measurement with ordinarily used personal dosemeters [2, 3].

In compliance with ICRP recommendations [4], physicians performing interventional procedures are equipped with a whole–body dosemeter, which can be placed on the torso or neck. The usually provided quantities are $H_p(10)$ and $H_p(0.07)$, i.e. the personal dose equivalent at the depth of 10 and 0.07 mm, respectively. With a help of an

L. Sukupova (✉)
Institute for Clinical and Experimental Medicine, Videnska 1958/9, 140 21 Prague 4, Prague, Czech Republic
e-mail: lucie.sukupova@gmail.com

Z. Cemusova
National Radiation Protection Institute, Prague, Bartoskova 1450/28, 140 00 Prague 4, Czech Republic

© Springer Nature Singapore Pte Ltd. 2019
L. Lhotska et al. (eds.), *World Congress on Medical Physics and Biomedical Engineering 2018*,
IFMBE Proceedings 68/3, https://doi.org/10.1007/978-981-10-9023-3_98

appropriate algorithm, these quantities may be employed for the assessment of the dose to the eye lens [5–7].

Many studies dealing with eye lens dosimetry [4, 8–13] have already been published, but unfortunately no study has, as yet, been carried out in the Czech Republic. Therefore we were motivated to carry out a study dealing with eye lens doses for interventional cardiology centres to find out what the typical eye lens doses are and whether they are comparable to those in other countries. Our secondary goal was to assess the approach of cardiologists, as regards their occupational doses and eye lens doses. Our last, but not least important task was to recommend a way to perform dosimetry of the eye lens in clinical practice.

2 Materials and Methods

Three out of 23 interventional cardiology centres in the Czech Republic agreed to take part in the study about doses in interventional cardiology. The first centre (IC 1) employs 5 cardiologists, the second one (IC 2) employs 5 cardiologists too and the third one (IC 3) employs 4 cardiologists. Cardiologists in all three cardiology centres perform the same range of interventional procedures, mainly coronary angiography, sometimes accompanied by additional measurements, and percutaneous transluminal coronary angioplasty. All the procedures were performed in their standard manner, with the cardiologists wearing a protective apron with a thyroid shield and usually using ceiling–mounted radiation shielding as well as lower body protection shielding attached to the patient table.

The doses of interventional cardiologists were measured with dosemeters placed in three positions, but only (b) and (c) are assessed and discussed below:

(a) The left front side of the chest below the apron
(b) The left front side of the chest above the apron
(c) The left side of the head as close to the eye as possible.

When the procedures are performed, an access from the patient's right radial or femoral artery is used; therefore the cardiologist is usually standing on the patient's right hand side, with his left side oriented closer to the source of scatter radiation (the patient) than his right side. This is the reason why dosemeters were worn on the left hand body side of the cardiologist.

For all measurements, thermoluminescent dosimeters (TLDs) of LiF:Mg,Cu,P were used. The $H_p(10)$ quantity was measured in measuring positions (a) and (b). The eye lens dose, position (c), was measured with the TLD placed in the appropriate cover (EYE-D, Radcard) in the $H_p(3)$ quantity. For $H_p(10)$, TLDs were calibrated in ^{137}Cs and read with the automatic TLD reader Harshaw 6600. For eye lens dose

measurements, the TLDs were calibrated in RQR6 and with the TLD reader Harshaw 3500. The reading process of TLDs and the dose calculation are described in [14]. The TLDs, their reading and dose calculations (the detection threshold of TLDs is 0.017 mSv) were provided by the National Radiation Protection Institute in accordance with the method approved by the State Office for Nuclear Safety [15].

The cardiologists were required to wear the sets of TLDs for a three week period, repeated three times. During some periods, there was a lack of procedures for certain cardiologists due to their absence (holidays), so these measurements were not taken into account. The cardiologists were asked to wear all the TLDs at once or none at all, not just some of them, due to the better correlation among measured doses. During the measuring periods, the kerma–area product (P_{KA}) values from the performed procedures were collected for each cardiologist.

3 Results and Discussion

For each cardiologist and for each measuring period, the number of procedures, total P_{KA}, $H_p(3)$ and $H_p(10)$ values were collected. The dependence of the total $H_p(3)$ on the total P_{KA} value for centres IC 1, IC 2 and IC 3 is shown in Fig. 1. $H_p(3)$ are values measured on the eye without considering the wearing of goggles, because some cardiologists only wear the goggles occasionally and some of them never. A similar figure for the dependence of $H_p(10)$ above the apron on the total P_{KA} is illustrated in Fig. 2.

The average values of $H_p(10)/P_{KA}$ for IC 1 : IC 2 : IC 3 are as follows: 0.19 : 0.69 : 0.35 µSv/Gy*cm². As regards $H_p(3)/P_{KA}$, the values for IC 1 : IC 2 : IC 3 are as follows: 0.15 : 0.34 : 0.29 µSv/Gy*cm². In general, the values in IC 1 are 2–4x lower than in IC 2 and 2x lower than in IC 3. Local practice in IC 1 is much more dose-saving than in the other centres. For IC 2, normalized doses are the highest on average for the whole centre.

It is evident from Figs. 1–4 that $H_p(3)$, $H_p(10)$, $H_p(3)/P_{KA}$ and $H_p(10)/P_{KA}$ are highest in IC 2, as compared with IC 1

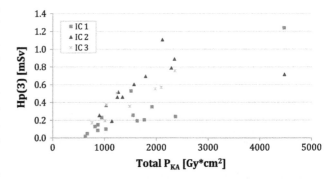

Fig. 1 Dependence of $H_p(3)$ on the total P_{KA} for IC 1, IC 2 and IC 3

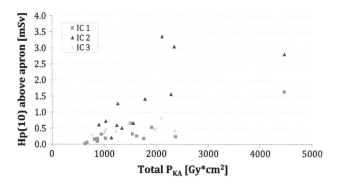

Fig. 2 Dependence of $H_p(10)$ on the total P_{KA} for IC 1, IC 2 and IC 3

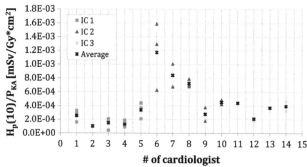

Fig. 4 $H_p(10)$ normalized to total P_{KA} for each cardiologist

and IC 3. In IC 2, neither an automatic contrast injector, nor any additional shielding on the patients was used. In IC 1, the cardiologists used the automatic contrast injector and they also shielded the scatter from patients using an apron placed on the patients' abdomen. This approach was recommended in the study [16], where the authors proved that using shielding against the scatter radiation from the patients can lower the cardiologists' doses by 30–50%. But the excellent result from IC 1 was not caused solely by the injector and the shielding, but also by the fact that the head of this centre is a cardiologist well aware of the detrimental effects of radiation. Cardiologist #4 is wary of radiation, and usually leaves the room when performing the cine acquisition. His average normalized dose $H_p(3)/P_{KA}$ is the lowest of all the cardiologists included in this study. Usually, however, only one step back from the patient is recommended to cardiologists when the contrast medium is applied and the cine acquisition is performed. It is not necessary to leave the room.

Cardiologist #6 had three quite different normalized values of $H_p(10)/P_{KA}$. He was confronted with his dose after the first measuring period (the highest value in Figs. 3 and 4) and he tried to improve his practice. The improvement was obvious in the following measuring periods. The main problem was that this cardiologist suffers from back pain, probably due to wearing the apron, so that when he applies

contrast medium (a contrast injector is not used), he is usually bent over the patient.

All the measured $H_p(3)$ doses were normalized to the number of days when the cardiologists were wearing the set of dosimeters. Afterwards, the expected annual eye lens doses were estimated. The estimated annual value of $H_p(3)$ for each cardiologist is shown in Table 1.

As one may see from Table 1, the estimated annual $H_p(3)$ values are below the new limit of 20 mSv. The highest estimated annual $H_p(3)$ value is around 16 mSv. Most $H_p(3)$ doses are even below 10 mSv. After the correction for the attenuation of the goggles (the eye lens dose is roughly 4x lower [5]). At least for IC 1, the eye lens doses would be around 1 mSv. In Table 1, the estimated $H_p(3)$ of cardiologist #4 is the lowest. This can be caused just by the approach of this cardiologist, who leaves the room when the cine acquisition is performed. When other cardiologists were

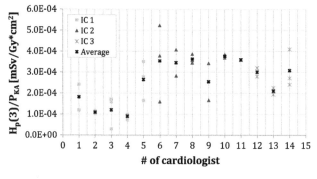

Fig. 3 $H_p(3)$ normalized to total P_{KA} for each cardiologist

Table 1 Estimated annual eye lens dose for each cardiologist

# of cardiologist	Days of wearing dosimeters	$H_p(3)$ in all 3 measuring periods [mSv]	Estimated annual $H_p(3)$ [mSv]
1	54	0.77	5.2
2	33	0.45	4.9
3	54	0.30	2.0
4	54	0.24	1.6
5	54	2.02	13.7
6	63	2.71	15.7
7	43	0.77	6.5
8	43	1.48	12.6
9	43	0.65	5.5
10	43	1.06	9.0
11	17	0.37	8.0
12	38	1.31	12.6
13	38	0.37	3.5
14	53	1.43	9.9

Table 2 Average values of $H_p(10)/H_p(3)$ with standard deviation for each cardiologist

# of cardiologist	$H_p(10)/H_p(3) \pm SD$
1	1.41 ± 0.09
2	0.93 ± 0.07
3	1.27 ± 0.09
4	1.43 ± 0.39
5	1.28 ± 0.04
6	3.47 ± 0.45
7	2.43 ± 0.05
8	2.00 ± 0.03
9	1.09 ± 0.01
10	1.20 ± 0.14
11	1.23 (only 1 meas.)
12	0.70 ± 0.19
13	1.78 ± 0.19
14	1.38 ± 0.56

asked why they are not doing the same, they replied that they prefer to stay close to the patient and to the catheters.

Average values of $H_p(10)/H_p(3)$ are shown in Table 2.

All the cardiologists behave in a reproducible way, the ratio $H_p(10)/H_p(3)$ above the apron (excluding cardiologists #4, #6, #14) is like a personal footprint of each cardiologist. The eye lens dosimetry is not mandatory, but the eye lens dose can be estimated from known $H_p(10)$ above the apron value to estimate $H_p(3)$.

4 Conclusion

Each of the three centres of interventional cardiology has its own local practice and each cardiologist has his own approach. In IC 2, a new automatic contrast injector was bought a few months ago and shielding against the scatter from patients was recommended (apron, disposable pads). The annual $H_p(3)$ dose could be estimated from performed measurements and it was below the limit of 20 mSv. Eye lens doses of each cardiologist can be estimated retrospectively from known $H_p(10)$ value using the ratio determined for each individual cardiologist.

Conflict of Interest The authors declare that they have no conflict of interest.

References

1. Pinak M. ICRP 2013: 2nd International Symposium on the System of Radiological Protection. Dose limits to the lens of the eyes: New limit for the lens of the eye–International Basic Safety Standards and related guidance. RSM/NRRW. IAEA, 2013.
2. Farah, J., Struelens, L., Dabin, J., et al. A correlation study of eye lens dose and personal dose equivalent for interventional cardiologists. Radiat. Prot. Dosim. 157(4), 561–569 (2013).
3. Clerinx, P., Buls, N., Bosmans, H., et al. Double-dosimetry algorithm for workers in interventional radiology. Radiat. Prot. Dosim. 129, 321–327 (2008).
4. O'Connor, U., Walsh, C., Gallagher, A., et al. Occupational radiation dose to eyes from interventional radiology procedures in light of the new lens dose limit from the International Commission on Radiological Protection. Br J Radiol. 2015; 88(1049): 20140627.
5. Geber, T., Gunnarsson, M., Mattsson, S. Eye lens dosimetry for interventional procedures–Relation between the absorbed dose to the lens and dose at measurement positions. Radiation Measurements 2011; 46: 1248–1251.
6. Covens, P., Berus, D., Buls, N., et al, F. Personal dose monitoring in hospitals: Blobal assessment, critical applications and future needs. Radiat Prot Dosimetry 2007; 124, 250–259.
7. Cemusova, Z., Ekendahl, D., Judas, L. Assessment of eye lens doses in interventional radiology: a simulation in laboratory conditions. Radiat. Prot. Dosim. 170, 256–260 (2016).
8. Vanhavere F, Carinou E, Domienik J, et al. Measurements of eye lens doses in interventional radiology and cardiology: Final results of the ORAMED project. Radiation Measurements 2011; 46(11): 1243–1247.
9. Sanchez, RM., Vano, E., Fernandez, JM., et al. Occupational eye lens doses in interventional cardiology. A multicentric study. J Radiol Prot. 2016; 36(1): 133–143.
10. Haga, Y., Chida, K., Kaga, Y., et al. Occupational eye dose in interventional cardiology procedures. Sci Rep. 2017; 7(1): 569.
11. Matsubara, K., Lertsuwunseri, V., Srimahachota, S., et al. Eye lens dosimetry and the study on radiation cataract in interventional cardiologists. Phys Med. 2017; 44: 232–235.
12. Jupp, T., Kamali-Zonouzi, P. Eye lens dosimetry within the cardiac catheterization laboratory–Are ancillary staff being forgotten? Radiat Prot Dosimetry 2018; 178(2): 185–192.
13. Thrapsanioti, Z., Askounis, P., Datseris, I., et al. Eye lens radiation exposure in Greek Interventional Cardiology Article. Radiat Prot Dosimetry 2017; 175(3): 344–356.
14. Cassata, J. R., Moscovitch, M., Rotunda, J. E., Velbeck, K. J. A new paradigm in personal dosimetry using LiF:Mg,Cu,P. Radiat Prot Dosimetry. 2012; 101(1–4): 27–42.
15. Ekendahl, D. The determination of the personal doses from external irradiation by the use of the system TLD Harshaw 6600. National Radiation Protection Institute, 2012.
16. Musallam, A., Volis, I., Dadaev, S., et al. A randomized study comparing the use of a pelvic lead shield during trans-radial interventions: Threefold decrease in radiation to the operator but double exposure to the patient. Catheter Cardiovasc Interv. 2015; 85(7): 1164–1170.

Radiotherapy Dose Measurements Using a Fluorescing Quinine Solution

Scott Crowe, Steven Sylvander, and Tanya Kairn

Abstract

Quinine solutions fluoresce when exposed to ionising radiation, through the production and absorption of Čerenkov radiation. This study evaluated the feasibility of using 'household' tonic water as a radiotherapy dosimeter. Tonic water samples were irradiated with static beams for a variety of energies and dose rates: 6 and 10 MV photons at 600 MU/min; 6 MV flattening-filter-free photons at 1400 MU/min; 6, 9, 12, 15 and 18 meV electrons at 400 MU/min and 6 meV electrons at approximately 2474 MU/min (used for total skin electron irradiation). A sliding window IMRT field was delivered using a 6 MV photon beam at 600 MU/min, to assess dynamic response. Fluorescence was successfully recorded using a monochrome low light CCD camera placed on the treatment couch as well as the treatment room visual monitoring system in the linear accelerator control area. Energy dependence and dose rate independence were observed. While limitations in the bit-depth and focal length of the camera prevented precise quantitative analysis of depth dose profiles for conventional dose rates (≤ 600 MU/min), performance for higher dose rates (in terms of signal-to-noise in depth dose profiles) was comparable to radiochromic film. Potential use includes measurement of dose in the build-up region, efficient checks of beam energy and tomographic reconstruction of 4D dose delivery, though further optimisation of fluorescent signal acquisition is required.

Keywords

Radiation therapy • Dosimetry • Fluorescence

S. Crowe (✉) · S. Sylvander · T. Kairn
Royal Brisbane and Women's Hospital, Brisbane, Australia
e-mail: sb.crowe@gmail.com

S. Crowe · T. Kairn
Queensland University of Technology, Brisbane, Australia

1 Introduction

The ability to achieve real-time multi-dimensional measurements of water equivalent dose distributions in radiation therapy is becoming increasingly desirable due to changes in treatment complexity. Dosimetry solutions based on the acquisition of a radiation-induced light signal (e.g. by scintillation [1, 2] or Čerenkov effect [3–9]) offer an attractive solution.

The imaging of Čerenkov radiation, in particular, has allowed the 2D and 3D measurement of dose in water phantoms [3–7] and visualisation of superficial dose in vivo [7–9]. Glaser et al. [3] demonstrated the feasibility of optical imaging of photon beam dose in quinine-doped water. The quinine (1.0 g/L) was introduced to act as a fluorophore, producing isotropic fluorescence after absorption of Čerenkov radiation. This approach was later used by the authors to acquire images of modulated radiotherapy treatment plans [4–6].

Whereas previous studies have used water that was deliberately doped with quinine, in-house, the current study instead investigated the feasibility of producing measurable radio-fluorescence using a commercial tonic water product. The use of a 'household' beverage for this purpose is advantageous because the quinine solution is likely to be very consistent [10] and very widely available, allowing easy replication of this study and potential broad adoption of the methods proposed in this work.

2 Method

The tonic water used in this study was Schweppes Indian Tonic Water (Schweppes, Geneva, Switzerland), reported to contain a quinine concentration of 68 ppm (or mg/L) [10]. Measurements were performed with the tonic water poured into two transparent containers: a square drinking glass (approximately 200 mL) and a cylindrical PET jar (approximately 1 L).

© Springer Nature Singapore Pte Ltd. 2019
L. Lhotska et al. (eds.), *World Congress on Medical Physics and Biomedical Engineering 2018*,
IFMBE Proceedings 68/3, https://doi.org/10.1007/978-981-10-9023-3_99

An initial proof-of-concept experiment was performed by irradiating tonic water in the drinking glass using a True-Beam linear accelerator (Varian Medical Systems, Palo Alto, United States) operating at a beam energy of 6FFF at 600 and 1400 MU/min. Sources of ambient light within the bunker were covered with used radiographic film packets. Two consumer-grade point-and-shoot cameras (Olympus Corporation, Tokyo, Japan) were unable to detect fluorescence, though it could be seen on the treatment room monitor screens. Collimation and dose-rate effects were observed.

Based on these results, a scientific low light CCD camera was used for the feasibility study. Limitations in focal length required the camera to be placed at approximately 1 m from the tonic water on the treatment couch. The experimental setup was isolated from ambient light sources by a thick black cloth draped over a purpose-built frame that was placed on the treatment couch. Images were captured using 2 s exposures with the TSView software (V7.1.1.7, Tucsen Photnoics, Fuzhou, China).

Measurements were performed for the delivery of the following fields using a Varian iX linear accelerator (Varian Medical Systems, Palo Alto, USA):

1. square field electron beams of variable dose rates of 400 MU/min and approximately 2474 MU/min (total skin electron therapy beam [11]), at a constant nominal accelerating potential of 6 meV;
2. square field electron beams of variable nominal accelerating potentials of 6, 9, 12, 15 and 18 meV, at a constant dose rate of 400 MU/min;
3. square field photon beams of variable nominal accelerating potentials of 6 and 10 MV, at a constant dose rate of 600 MU/min; and
4. intensity-modulated radiation therapy (IMRT) beam, "M2Mill120.DOS", provided by Varian Medical Systems for customer acceptance testing of the linear accelerator.

Images were processed and analysed using ImageJ (V1.50a, National Institutes of Health, Bethesda, USA). A "dark image" (acquired without beam delivery) was subtracted from acquired images, where quantitative analysis was required. Percentage depth dose (PDD) profiles were extracted and smoothed using a mean filter with a window size of 9 pixels (approximately 1 mm), unless otherwise stated.

3 Results and Discussion

Figure 1 shows fluorescence images obtained using the 6 meV electron beam, delivered at dose rates of 400 and 2474 MU/min. Clearly, the image obtained using the higher dose

Fig. 1 Fluorescent response at dose rates of 400 MU/min (left) and 2474 MU/min (right), for a 6 meV electron beam, presented with identical window settings

rate is more intense than the image obtained using the lower dose rate. The ratio of fluorescence intensities in the same region of interest in both images was equal to the ratio of the dose rates, within 0.2%, indicating that the dose-rate-response of the tonic water is linear over at least the range 400–2474 MU/min, for a 6 meV electron beam. This is consistent with linearity reported for photon beams by Glaser et al. [3]. The signal-to-noise ratio of the high-dose-rate electron beam image was similar to that of radiochromic film.

Figure 2 presents the florescence response achieved using electron beams across a range of nominal energies. For the each of the beams with nominal energies from 6 to 12 meV, the increase in the depth of maximum dose (R_{100}) with increasing beam energy is obvious in the fluorescence image, as is the steep dose fall-off beyond R_{100}. For the higher energy beams, R_{100} is close to the bottom of the glass and therefore the dose fall-off at greater depths is not visible. PDD profiles for the five electron energies, as produced by tonic water in the 1 L PET jar, are presented in Fig. 3.

Comparison of the results shown in Fig. 2, which all have the same window settings as each other, indicate that the signal intensity produced by the 15 and 18 meV beams is noticeably less than the signal intensity produced by the lower-energy beams. Analysis of the fluorescence images suggested that the dose at R_{100} produced by 18 meV beam was approximately 16% lower than the dose at R_{100} produced by the 9 meV beam. Although the higher energy beams (which have longer practical range of electrons in water) are expected to be more affected by the loss of lateral electron scatter (due to the small size of the glass of tonic water) than the lower energy beams, this substantial variation of measured signal with beam energy is nonetheless suggestive of a possible energy dependence in the fluorescence response.

An energy-dependent fluorescence response was similarly suggested by the results produced by the static photon beams. PDDs derived from the florescence images suggested that the maximum dose from the 10 MV photon beam was approximately 30% higher than the maximum dose from the 6 MV photon beam, despite both beams being calibrated to give the same maximum dose at the same dose rate.

Fig. 2 Fluorescent response for electron beams with nominal energies of (left to right) 6, 9, 12, 15 and 18 meV, presented with identical window settings

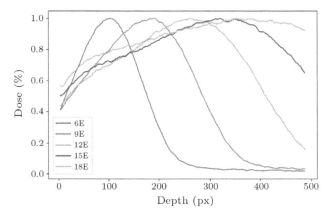

Fig. 3 Electron PDDs for 6, 9, 12, 15 and 18 meV beams

The PDDs obtained from fluorescence images of the static electron and photon beams (e.g. Fig. 2) also differ from PDDs measured using conventional means (with ionisation chambers in water) during routine quality assurance testing of the linear accelerator.

Despite these differences, the ease with which an indication of dose variation with depth can be obtained via fluorescence imaging with household tonic water means that

if a consistent experimental setup is designed and used periodically, fluorescence measurements have the potential to provide rapid and reliable evaluations of beam energy consistency. Alternatively, fluorescence measurements may be used to provide information about the features of radiotherapy beams that are difficult to measure accurately using conventional dosimeters, such as in the electron build-up region and in small apertures from modulated radiotherapy treatments.

Figure 4 exemplifies the results that can be achieved when tonic water is used to provide fluorescence images of modulated radiotherapy beams. The images in Fig. 4 were extracted from a video recording of the tonic water response during the deliver of an IMRT beam. This illustration shows both the variation in dose with depth (in the vertical direction) and the modulation of the beam delivery (in the horizintal direction—leaves moved from left to right). More-sophisticated analysis of results such as this may be achieved through tomographic reconstruction of dose delivery, to potentially to enable time-dependent (four-dimensional) verification of modulated treatment delivery.

Fig. 4 Fluorescent response sampled at 10 s intervals (left) and integrated response (right) for intensity modulated radiation therapy beam

4 Conclusion

This feasibility study showed that commercial tonic water can be used as a simple substitute for in-house quinine-water mixtures, to obtain useful fluorescence measurements of radiotherapy beams. Potential applications for this technique include measurement of dose in the build-up region, efficient checks of beam energy and tomographic reconstruction of 4D dose delivery. Careful selection of imaging technology and imaging parameters is advisable, to optimise the efficiency, accuracy and consistency of tonic water fluorescence results.

Compliance with Ethical Standards The authors declare that they have no conflict of interest. This article does not contain any studies with human participants or animals performed by any of the authors.

References

1. Pönisch, F., Archambault, L., Briere, T.M., et al.: Liquid scintillator for 2D dosimetry for high-energy photon beams. Med. Phys. 36(5), 1478–1485 (2009). https://doi.org/10.1118/1.3106390.
2. Goulet, M., Rilling, M., Gingras, L., et al.: Novel, full 3D scintillation dosimetry using a static plenoptic camera. Med. Phys. 41(8), 082101 (2014). https://doi.org/10.1118/1.4884036.
3. Glaser, A.K., Davis, S.C., Voigt, W.H.A., et al.: Projection imaging of photon beams using Čerenkov-excited fluorescence. Phys. Med. Biol. 58(3), 601–609 (2013). https://doi.org/10.1088/0031-9155/58/3/601.
4. Glaser, A.K., Andreozzi, J.M., Davis, S.C., et al.: Video-rate optical dosimetry and dynamic visualization of IMRT and VMAT treatment plans in water using Cherenkov radiation. Med. Phys. 41(6), 062102 (2014). https://doi.org/10.1118/1.4875704.
5. Pogue, B.W., Glaser, A.K., Zhang, R., Gladstone, D.J.: Cherenkov radiation dosimetry in water tanks video rate imaging, tomography and IMRT & VMAT plan verification. J. Phys. Conf. Ser. 573, 012013 (2015) https://doi.org/10.1088/1742-6596/573/1/012013.
6. Bruza, P., Andreozzi, J.M., Gladstone, D.J., et al.: Real-time 3D dose imaging in water phantoms: reconstruction from simultaneous EPID-Cherenkov 3D imaging (EC3D). J. Phys. Conf. Ser. 847, 012034 (2017). https://doi.org/10.1088/1742-6596/847/1/012034.
7. Pogue, B.W., Zhang, R., Glaser, A., et al.: Cherenkov imaging in the potential roles of radiotherapy QA and delivery. J. Phys. Conf. Ser. 847, 012046 (2017). https://doi.org/10.1088/1742-6596/847/1/012046.
8. Andreozzi, J.M., Zhang, R., Glaser, A. K., et al.: Camera selection for real-time in vivo radiation treatment verification systems using Cherenkov imaging. Med. Phys. 42(2), 994–1004 (2015). https://doi.org/10.1118/1.4906249.
9. Andreozzi, J.M., Zhang, R., Gladstone, D.J., et al.: Cherenkov imaging method for rapid optimization of clinical treatment geometry in total skin electron beam therapy. Med. Phys. 43(2), 993–1002 (2016). https://doi.org/10.1118/1.4939880.
10. Streller, S., Roth, K.: Eine Rinde erobert die Welt: Von der Apotheke an die Bar. Chemie. in. unserer. Zeit. 46(4), 228–247 (2012). https://doi.org/10.1002/ciuz.201200593.
11. Fitchew, R.S., Nitschke, K.N., Christiansen, P.G.: Total skin electron beam therapy using a Dynaray 18 linear accelerator. Australas. Phys. Eng. Sci. Med. 8(4), 182–187 (1985).

Linac Leakage Dose Received by Patients Treated Using Non-coplanar Radiotherapy Beams

Tanya Kairn, Scott Crowe, and Samuel Peet

Abstract

As non-coplanar beams are increasingly used to deliver cranial and extra-cranial stereotactic radiotherapy treatments, radiation leakage from the accelerating waveguide, bending magnet and other components of the medical linear accelerator (linac) that are not conventionally brought into proximity to the patient becomes increasingly concerning. In this study, the leakage dose in the patient plane was measured using optically stimulated luminescence dosimeters placed along the treatment couch, at 10 cm intervals. A Varian iX linac was operated in 6 and 10 MV photon mode, with all jaws and multi-leaf collimators closed. Dose measurements were made (a) using a "standard" setup, with couch and gantry at zero degrees and (b) using a worst-case non-coplanar setup, with the couch at 90° and the gantry at 30° (rotated over the couch). Results indicated that the leakage dose in the patient plane was uniformly low (less than 2 cGy/10,000MU) at all measurement positions for both energies, using the standard setup. However, when the gantry and couch were rotated, there was a systematic increase in dose to 4.2 cGy/10,000MU at a point 80 cm from isocentre (below the bending magnet). While these doses are within recommended out-of-field dose limits for static beam treatments, if IMRT/VMAT factors are applied then the leakage dose to the patient from non-coplanar treatments may become unacceptable. Specific checks of out-of-field dose from non-coplanar beam directions are advisable prior to acceptance of new or modified linacs.

Keywords

Radiation therapy • Radiation protection • Dosimetry

1 Introduction

The minimisation of patient exposure to out-of-field radiation is an important aspect of radiation protection for radiotherapy, helping to reduce the risk of radio-carcinogenesis. Vendors and users of clinical radiotherapy treatment machines take care to confirm that new medical linear accelerators (linacs) are adequately shielded, in compliance with national and international standards. For example, the International Electrotechnical Commission (IEC) standard 60601-2-1 requires that out-of-field radiation leakage reaching the patient plane from the radiation sources within each medical linac be limited to less than 0.2% (with an average of 0.1%) of the central axis radiation dose in a reference field [1]. This limit increases to 0.5% at other points at 1 m distance from the electron accelerating waveguide and the photon source, and the IEC requires the recording and reporting of all dose maxima at points "which can come into close proximity to the patient" [1].

Measurements of radiation leakage completed during linac acceptance and commissioning often identify regions of increased leakage at the front and top of the linac, where radiation is scattered from the flattening filter, target, bending magnet and accelerating waveguide. However, if treatments are only performed using a conventional radiotherapy setup, with coplanar beams, without rotations of the treatment couch, it can be legitimately assumed that this multi-directional leakage makes a negligible contribution to the patient dose because its sources cannot be brought into "close proximity to the patient".

As new treatment methods are investigated and implemented, it is important to review the assumptions behind leakage dose assessments. Recently, there has been a growing interest in leakage doses delivered to patients

T. Kairn (✉) · S. Crowe · S. Peet
Royal Brisbane and Women's Hospital, Brisbane, QLD, Australia
e-mail: t.kairn@gmail.com

T. Kairn · S. Crowe · S. Peet
Queensland University of Technology, Brisbane, QLD, Australia

© Springer Nature Singapore Pte Ltd. 2019
L. Lhotska et al. (eds.), *World Congress on Medical Physics and Biomedical Engineering 2018*,
IFMBE Proceedings 68/3, https://doi.org/10.1007/978-981-10-9023-3_100

during intensity modulated and volumetric modulated arc radiotherapy (IMRT and VMAT) treatments [2, 3], which require substantially increased numbers of monitor units to deliver each cGy of dose (MU/cGy) and therefore produce correspondingly increased leakage doses, compared to static conformal radiotherapy treatments [4–6]. However, previous studies of leakage dose to the patient have been largely focussed on leakage from the collimator side of the linac gantry [2, 3, 7].

As couch rotations are increasingly used to deliver non-coplanar beams during cranial and extra-cranial stereotactic radiotherapy treatments, radiation leakage from previously ignored sources (e.g. the flattening filter, the bending magnet, the top of the waveguide) becomes increasingly concerning. This study therefore investigated dose in the patient plane when both the gantry and the couch were rotated, to simulate part of a possible non-coplanar beam arrangement and provide an indication of the importance of re-evaluating leakage doses under non-conventional treatment conditions.

2 Method

Radiation leakage was measured using a Varian iX linear accelerator with an Exact couch (Varian Medical Systems, Palo Alto, USA) operating in 6 and 10 MV photon modes (abbreviated to "6X" and "10X"). Optically stimulated luminescence dosimeters (OSLDs) were placed along the treatment couch at 10 cm intervals, from 10 cm superior to 190 cm inferior of the isocentre, covering the length that could potentially be occupied by a patient's body, during a cranial treatment.

OSLDs were used in this study due to their reusability, reproducibility, sensitivity and linearity in the low-dose range [8], despite the need to account for their potential over-response to out-of-field radiation [9] and to correct for the varying compositions and responses of OSLDs within each batch [10]. The specific OSLDs used were Landauer nanoDots (Landauer, Glenwood, USA), which were calibrated by exposure to a known dose from the linac's primary beam, annealed using a 3 h exposure to fluorescent light in a Gammasonics Manual OSL Annealing Lightbox (Gammasonics Institute for Medical Research Pty Ltd, Lane Cove, Australia) and read out using a Landauer Microstar OSLD reader (Landauer, Glenwood, USA).

Dose measurements were made using (a) a conventional setup, with couch and gantry at zero degrees (head-up position, referred to as the "zero" configuration) and (b) a non-coplanar setup, with the couch at 90° and the gantry at 30° (gantry rotated over couch, referred to as the "rotated" configuration), as shown in Fig. 1. The specific "rotated" position of the couch and gantry were selected to give a

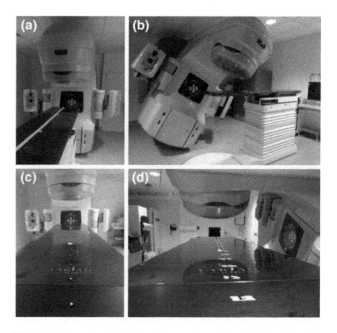

Fig. 1 Photographs indicating linac and couch orientation from the point of view of the bunker entrance (**a** and **b**) and the end of the couch (**c** and **d**), when the gantry and couch are in the zero configuration (**a** and **c**) and when the gantry and couch are in the rotated configuration (**b** and **d**)

plausible worst-case-scenario measurement; couch-gantry arrangements such as this are occasionally used during radiosurgical treatments of lesions located near the orbits, pituitary or frontal lobes of the brain, or in cases where large numbers of differently oriented beams are required to treat multiple brain metastases while sparing the healthy brain.

For each measurement, the linac delivered 20,000 MU, with all jaws and multi-leaf collimators closed. The 6X and 10X beams were evaluated separately, due to Lonski et al. observation that photon beams with different energies can produce increased leakage at different points on the linac's housing [11].

3 Results and Discussion

Figure 2 summarises the results of this study. In regions primarily affected by transmission through collimators (within 30 cm of the isocentre), Fig. 2 shows that the measured dose from the 10X beam exceeded the measured dose from the 6X beam, due to reduced attenuation of the higher-energy (more penetrating) primary beam. Further from isocentre, where the effects of laterally scattered radiation were more prevalent, differences between the 6X results and the 10X results were diminished, except in the region beyond 100 cm from isocentre where the "rotated" result showed the effects of increased scatter from the 6X beam. Apparently, the increased scatter from lower-energy

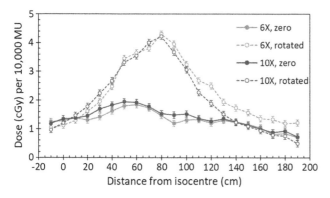

Fig. 2 Doses from the 6X and 10X beams, measured along the treatment couch when the gantry and couch were in the zero configuration and when the gantry and couch were both in the rotated configuration (as specified in legend)

photon beams that has been identified on the waveguide side of the exterior housing of the linac head [11] is also detectable in the patient plane, if the couch and gantry are both rotated.

Generally, data shown in Fig. 2 indicate that the leakage dose measured in the patient plane was uniformly low (less than 2 cGy/10,000MU) at all measurement positions for both energies, using the standard "zero" setup. However, when the gantry and couch were both rotated, there was a systematic increase in dose towards 4.2 cGy/10,000MU at a point 80 cm from isocentre (directly below the end of the travelling waveguide and the side of the bending magnet).

The maximum leakage dose measured in the patient plane (4.2 cGy/10,000 MU) is equivalent to 0.042% of the maximum central axis dose in a 10×10 cm^2 reference field, which is well within the 0.2% limit set by the relevant IEC standard [1]. However, if the linac is used to deliver IMRT or VMAT treatments, then the leakage results should be multiplied by a modulation factor, to account for the increased MU/cGy and therefore the increased leakage delivered during modulated radiotherapy treatments. Modulation factors can range from 2 to 10 [4]. For example Saleh et al. retrospective evaluation of the workload of ten radiotherapy bunkers over a ten year period produced mean modulation factors of 5.0 for IMRT and 4.6 for VMAT [5]. If the measurements obtained in this study are adjusted using a clinically realistic [5] modulation factor of 5, then the maximum leakage measurement is increases to be slightly beyond the IEC's limit.

Elekta Synergy and Siemens Primus linacs have been found to produce more leakage than the Varian iX model used in this work by factors of 2 and 2.5, respectively [11]. It is therefore advisable to repeat this study using different treatment systems, to establish the possible effects of leakage dose on patients treated using non-coplanar beam arrangements using different types of linear accelerator.

4 Conclusion

Checks of out-of-field dose from non-coplanar beam directions are advisable prior to acceptance of new or modified linacs. Vendors should consider applying additional shielding to linacs that are capable of delivering non-coplanar treatments, especially linacs that are designed and promoted specifically for use in delivering stereotactic radiotherapy treatments, where non-coplanar beams are common.

Compliance with Ethical Standards The authors declare that they have no conflict of interest. This article does not contain any studies with human participants or animals performed by any of the authors.

References

1. International Electrotechnical Commission: Medical electrical equipment - Part 2–1: Particular requirements for the basic safety and essential performance of electron accelerators in the range 1–50 MeV. 3rd edition. International standard, IEC 60601-2-1 (2009).
2. Wiezorek, T., Georg, D., Schwedas, M., et al.: Experimental determination of peripheral photon dose components for different IMRT techniques and linear accelerators. Z. Med. Phys. 19(2), 120–128 (2009). https://doi.org/10.1016/j.zemedi.2009.01.008.
3. Peet, S.C., Wilks, R., Kairn, T., Crowe, S.B.: Measuring dose from radiotherapy treatments in the vicinity of a cardiac pacemaker. Phys. Medica. 32(12), 1529–1536 (2016). https://doi.org/10.1016/j.ejmp.2016.11.010.
4. International Atomic Energy Agency: Radiation protection in the design of radiotherapy facilities. Safety reports series No. 47. International Atomic Energy Agency, Vienna (2006).
5. Saleh, Z.H., Jeong, J., Quinn, B., et al.: Results of a 10 year survey of workload for 10 treatment vaults at a high throughput comprehensive cancer center. J. Appl. Clin. Med. Phys. 18(3), 207–214 (2017). https://doi.org/10.1002/acm2.12076.
6. Kairn, T., Crowe, S.B., Trapp, J.V.: Correcting radiation survey data to account for increased leakage during intensity modulated radiotherapy treatments. Med. Phys. 40(11) 111708 (2013). https://doi.org/10.1118/1.4823776.
7. Bordy, J.M., Bessieres, I., d'Agostino, E., et al.: Radiotherapy out-of-field dosimetry: Experimental and computational results for photons in a water tank. Radiat. Meas. 57, 29–34 (2013). https://doi.org/10.1016/j.radmeas.2013.06.010.
8. Jursinic, P.A.: Characterization of optically stimulated luminescent dosimeters, OSLDs, for clinical dosimetric measurements. Med. Phys. 34(12), 4594–4604 (2007). https://doi.org/10.1118/1.2804555.
9. Scarboro, S.B., Followill, D.S., Kerns, J.R., et al.: Energy response of optically stimulated luminescent dosimeters for non-reference measurement locations in a 6 MV photon beam. Phys. Med. Biol. 57(9), 2505–2515 (2012). https://doi.org/10.1088/0031-9155/57/9/2505.
10. Asena, A., Crowe, S.B., Kairn, T., et al.: Response variation of optically stimulated luminescence dosimeters. Radiat. Meas. 61, 21–24 (2014). https://doi.org/10.1016/j.radmeas.2013.12.004.
11. Lonski, P., Taylor, M.L., Franich, R.D., et al.: Assessment of leakage doses around the treatment heads of different linear accelerators. Radiat. Prot. Dosim. 152(4), 304–312 (2012). https://doi.org/10.1093/rpd/ncs049.

Optically Stimulated Luminescence Dosimeters as an Alternative to Radiographic Film for Performing "Head-Wrap" Linac Leakage Measurements

Tanya Kairn, Holly Stephens, Scott Crowe, and Samuel Peet

Abstract

The linac "head-wrap", where a new or modified linac is covered with radiographic film as a means to identify regions of increased radiation leakage, is an important part of the linac acceptance/commissioning process. However, as radiographic film and developing equipment decrease in availability and increase in cost, a simple, reusable, non-chemical solution becomes increasingly desirable. This study investigated whether discrete dose points measured using optically stimulated luminescence dosimeters (OSLDs) could be used to detect regions of increased radiation, as a substitute for radiographic film. After establishing the ability of the OSLDs to detect leakage and differentiate between high and low leakage doses, via a set of proof-of-concept measurements made in known high and low leakage regions on a Varian iX linac, a systematic evaluation of leakage at the surface of the linac head was undertaken. 60 OSLDs were positioned at regular intervals over the linac head by a member of the research team who was unfamiliar with the expected patterns of linac leakage. The OSLD measurements were able to detect linac head leakage and quantify high and low doses (from 0.6 to 44.7 cGy per 10,000 MU) with sufficient geometric precision to guide the use of an ionisation chamber to measure leakage doses in the patient plane. Reusable point dosimeters such as OSLDs are a promising solution to the problem of diminishing availability of film stock for linac head-wrap tests.

Keywords

Radiation therapy • Solid state dosimetry • Radiation protection

1 Introduction

The linac head-wrap measurement is an important part of the process of accepting and commissioning each new (or modified) medical linear accelerator (linac). Generally, the measurement involves running the most penetrating photon beam with radio-sensitive film covering the radiation emitting parts of a the linac, in order to produce film darkening in regions of elevated radiation leakage. Qualitative results of the head-wrap allow ionisation chamber measurements to be localised in regions of maximum leakage, to verify the linac's compliance with international limits [1] and ensure minimal out-of-field dose is delivered to radiotherapy patients.

The use of radiographic (silver-halide) film has long been specifically recommended for this purpose [1], as it is sensitive enough to provide an obvious visible response to relatively low leakage doses after just ten or twenty thousand dose monitor units (MU) have been delivered [2]. However, in recent years, radiographic film production has been scaled back as radiotherapy and radiography departments have moved towards electronic patient imaging techniques [3]. This change in imaging practice has also led to a decrease in the availability (and increase in the expense) of film developing chemicals and film processors [4]. It is becoming increasingly difficult for radiotherapy departments to source the equipment needed to perform linac head-wraps using radiographic film.

It may be desirable to replace the use of radiographic film with newer types of self-developing radiochromic film, which have a growing role in radiotherapy departments [3]. However, while radiochromic film is sensitive enough to detect and measure low doses of radiation after appropriate scanning and analysis [5], up to five times as many MU need

T. Kairn (✉) · S. Crowe · S. Peet
Royal Brisbane and Women's Hospital, Brisbane, QLD, Australia
e-mail: t.kairn@gmail.com

T. Kairn · S. Crowe · S. Peet
Queensland University of Technology, Brisbane, QLD, Australia

H. Stephens
Genesis Cancer Care Queensland, Genesis Care, Brisbane, QLD, Australia

H. Stephens
University of Adelaide, Adelaide, SA, Australia

© Springer Nature Singapore Pte Ltd. 2019
L. Lhotska et al. (eds.), *World Congress on Medical Physics and Biomedical Engineering 2018*,
IFMBE Proceedings 68/3, https://doi.org/10.1007/978-981-10-9023-3_101

to be delivered to produce a response visible to the human eye, when using radiochromic film rather than radiographic film [6]. Additionally, the relatively small size and large expense of currently available radiochromic film sheets mean they are better suited to performing spot checks in regions of concern, than to measuring dose around the whole linac head.

Given the current limitations of radiochromic and radiographic film, a simple, reusable, non-chemical system for evaluating linac head leakage is highly desirable. This study therefore investigated whether discrete dose points measured using optically stimulated luminescence dosimeters (OSLDs) could be used to detect regions of increased radiation during a modified head-wrap measurement.

2 Method

Initially, a proof-of-concept study was completed, where 30 OSLDs were taped to the exterior housing of a Varian iX linac (Varian Medical Systems, Palo Alto, USA) in regions of known high and low leakage (based on previous radiographic film measurements), to establish the ability of the OSLDs to detect leakage and differentiate between high and low leakage doses. The OSLDs were irradiated with a leakage dose by closing the linac's jaws and multi-leaf collimators and delivering 20,000 MU, the same number of monitor units as have been successfully used locally for head wraps using radiographic film.

After the pilot study produced positive results, a blind experiment was undertaken, using 60 OSLDs that were positioned at regular intervals over the linac head by a member of the research team who was unfamiliar with the usual patterns of linac leakage as well as the specific leakage pattern of the linac under investigation.

3 Results and Discussion

Doses measured on the linac housing during the proof-of-concept study ranged from 0.9 cGy/10,000 MU in known low-leakage regions to 47.6 cGy/10,000 MU in known high-leakage regions. The blind measurements of linac head leakage at points systematically spread around the linac head produced similar results to the proof-of-concept study, ranging from 0.6 to 44.7 cGy/10,000 MU, suggesting that OSLD point dose measurements could be used to identify regions of maximum linac leakage (without prior knowledge of the expected patterns of linac head leakage) with sufficient geometric precision to guide the use of an ionisation chamber to measure leakage doses in the patient plane.

Figure 1 provides a summary of the results from the blind OSLD leakage measurements. The highest leakage doses were found in regions around the sides of the collimator, as well as regions lateral to the bending magnet and the flattening filter. The OSLD measurements of leakage dose were generally low in well-shielded regions that would be directly facing the patient during coplanar treatments (without couch rotation) at the front face of the collimator and the underside of the gantry.

Figure 1 shows several regions where the film is dark but the doses measured with OSLDs are low. Although the pre-existing film irradiation result shown in Fig. 1 was never quantified, identification of which OSLDs were located at which regions of film darkening provided a surrogate OSLD-based measurement of the film response, which suggested that the radiographic film produced substantial darkening at doses as low as 1.6 cGy/10,000 MU. Evidently, even if radiographic film is used to provide a continuous indication of the leakage patterns around the linac head, the quantitative data provided by performing simultaneous

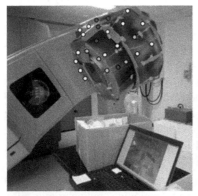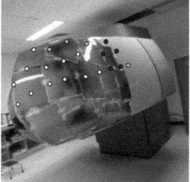

Fig. 1 OSLD results, overlaid on a photograph of a previous head-wrap film result. Coloured circles indicate OSLD measurement points: white shading indicates doses less than 5 cGy/10,000 MU; light shading (yellow) indicates doses greater than 5 but less than 10 cGy/10,000 MU; dark shading (orange) indicates doses greater than 10 cGy/10,000 MU (Colour figure online)

OSLD irradiations has the potential to improve the efficiency of eventual ionisation chamber measurements, minimising the time taken to set up and complete measurements in regions where the film is dark but the leakage is low.

Film measurements (rather than discrete point dose measurements such as those used in this work) are recommended for head wraps [1], to eliminate the possibility of missing or misplaced internal shielding or gross errors in electron beam steering [2], because "discrete measurements at selected points will not necessarily detect a small area of high leakage" [2]. The results reported in this work suggest that OSLD point dose measurements are able to detect even the small regions of elevated leakage that exist around a well-shielded and appropriately-steered linac. However, the risk of failing to detect a small but intense source of leakage radiation during a head wrap conducted using OSLDs alone should be minimised, by using the largest possible number of OSLDs, positioned as closely together as is practical, or by repeating the test using several different sets of measurement points.

4 Conclusion

Reusable point dosimeters such as OSLDs are a promising solution to the problem of diminishing availability (and increasing expense) of film stock for linac head-wrap tests. The results of this study suggest that in circumstances where the use of a continuous dosimetry medium is not practical or not possible, sets of systematic point dose measurements performed using sufficient numbers of regularly spaced OSLDs may provide suitable information to guide the positioning of ionisation chamber measurement points, to assess linac head leakage.

Compliance with Ethical Standards The authors declare that they have no conflict of interest. This article does not contain any studies with human participants or animals performed by any of the authors.

References

1. International Electrotechnical Commission: Medical electrical equipment - Part 2–1: Particular requirements for the basic safety and essential performance of electron accelerators in the range 1–50 MeV. 3rd edition. International standard, IEC 60601-2-1 (2009)
2. Cramb, J.: Radiation Protection in External Beam Radiotherapy. In: Trapp, J., Kron, T. (eds.) An Introduction to Radiation Protection in Medicine. pp. 171–203. Taylor & Francis, New York (2008)
3. Das, I.J.: Introduction. In: Das, I.J. (ed.) Radiochromic Film: Role and Applications in Radiation Dosimetry. pp. 22–25. CRC Press, New York (2017)
4. Kairn, T.: Intensity-modulated radiotherapy and volumetric-modulated arc therapy. In: Das, I.J. (ed.) Radiochromic Film: Role and Applications in Radiation Dosimetry. pp. 194–227. CRC Press, New York (2017)
5. Aland, T., Moylan, R., Kairn, T., Trapp, J.: Effect of verification imaging on in vivo dosimetry results using Gafchromic EBT3 film. Phys. Medica. 32(11), 1461–1465 (2016). https://doi.org/10.1016/j.ejmp.2016.10.020
6. Butson, M.J., Peter, K.N., Cheung, T. and Metcalfe, P.: High sensitivity radiochromic film dose comparisons. Phys. Med. Biol. 47(22), N291-N195 (2002). https://doi.org/10.1088/0031-9155/47/22/402

Measurement of Percentage Depth-Dose Profiles in Very Small Fields

Shadi Khoei, Mark West, and Tanya Kairn

Abstract

This study aimed to develop a method for accurately measuring small field percentage depth-dose (PDD) profiles. An SNC 3D Scanner cylindrical water tank (Sun Nuclear Corp, Melbourne, USA) was used in combination with an unshielded PTW diode 60017 (PTW Freiburg GmbH, Freiburg, Germany) to measure PDDs in very small radiation beams from a Varian Truebeam STx linac (Varian Medical Systems, Palo Alto, USA). Two PDD measurement methods were investigated; (a) a ray-tracing technique, where a complex automated process was used to continuously drive the diode into the centre of the field while varying the measurement depth, and (b) a conventional technique, where the centre of each field was identified using orthogonal profile scanning and then the tank platform was iteratively shifted to place the diode at the centre of the field before PDD scanning. For both methods, the effects of the low measurement signal were mitigated by scanning very slowly (2 s integration times). The field size was reproducible within 0.1 mm even after collimator and carousel repositioning. Differences of up to 3% were identified between the PDDs measured using the conventional and ray tracing methods. Comparison of these results with PDDs from larger fields suggested that the ray-tracing method was over-measuring the dose (or under-measuring depth) in the very small fields. Although time consuming, the use of a PDD measurement technique where the centre of the field is identified manually is advisable, unless the reliability more sophisticated ray-tracing techniques can be convincingly established.

Keywords

Small field dosimetry • Very small field • Diode

1 Introduction

Stereotactic radiosurgery (SRS) involves the delivery of tumoricidal doses of radiation to cranial lesions while sparing the surrounding brain tissue, using highly-conformal and very small radiation beams [1]. In order for SRS and SRT treatment doses to be predicted accurately by a radiotherapy treatment planning system (TPS), a set of accurate beam configuration data must be measured and input into the TPS [1, 2]. This data includes absolute dose calibration information as well as relative dose data, including point dose factors, lateral dose profiles and percentage depth-dose (PDD) profiles, measured in liquid water.

Accurate measurements of beam characterisation and TPS configuration data are especially difficult for small fields (≤ 3 cm across) [3] and very small fields (≤ 1.5 cm across) [4] due to: lack of lateral charged particle equilibrium since the size of the field becomes smaller compared to the range of the dose-depositing electrons [3, 4]; the partial obstruction of the photon source by closely-spaced beam collimators [3]; and volume averaging due to the relatively large size of dosimeter active volumes compared to field areas [3].

While improved small field dosimetry techniques have been developed over the last decade, research has largely focussed on measuring small field scatter factors [5–9] and lateral profiles [7, 10]. Recent international small field dosimetry guidelines provide methods for evaluating scatter factors and measuring lateral profiles, with minimal guidance on PDD measurements [1, 7].

PDD measurements are especially challenging when radiation fields are small due to the difficulty of keeping the dosimeter centred on the beam axis throughout the measurement (the beam may be slightly and unavoidably non-vertical due to gantry sag) and the dosimetric consequence of drifting

S. Khoei · M. West
Cancer Care Queensland, Genesis Care, Brisbane, Australia

T. Kairn (✉)
Royal Brisbane and Women's Hospital, Brisbane, Australia
e-mail: t.kairn@gmail.com

T. Kairn
Queensland University of Technology, Brisbane, Australia

© Springer Nature Singapore Pte Ltd. 2019
L. Lhotska et al. (eds.), *World Congress on Medical Physics and Biomedical Engineering 2018*,
IFMBE Proceedings 68/3, https://doi.org/10.1007/978-981-10-9023-3_102

off axis (the substantial effect of slightly shifting the measurement point away from the central axis is one of the defining features of very small fields [4]). The need to verify the size and reproducibility of the radiation field, which is an established requirement of accurate small field dosimetry [4, 10, 11], is also important when measuring small field PDDs, as the effects of field size inconsistency have not been evaluated over the large range of depths scanned by PDD measurements.

This study aimed to develop a method for accurately measuring small field PDD profiles, while exploring some of the capabilities of a contemporary water-tank-based dosimetry system.

2 Method

This study used a nominal 6 MV flattened photon beam, produced by a Varian Truebeam STx linac (Varian Medical Systems, Palo Alto, USA), which was collimated using a HD-MLC to produce a nominal 0.5×0.5 cm^2 field.

All measurements were performed using an SNC 3D Scanner cylindrical water tank (Sun Nuclear Corp, Melbourne, USA) in combination with an unshielded PTW diode 60017 (PTW Freiburg GmbH, Freiburg, Germany). The unshielded diode was chosen for this work due to the relatively small size of its collecting volume (≤ 0.0001 cm^3) and its established suitability for relative dosimetry in small fields [5, 8, 9]. The possibility of the diode over-responding at depth, due to increasing low-energy scatter from the water, was investigated by comparing a conventionally-scanned PDD measured using the unshielded diode in a 4×4 cm^2 field against a conventionally-scanned PDD measured using a CC13 spherical ionisation chamber (IBA Dosimetry GmbH, Schwarzenbruck, Germany) in the same field. Additional conventionally-scanned PDDs measured with the unshielded diode in 1×1 and 2×2 cm^2 fields were used to evaluate whether the PDD scans from the 0.5×0.5 cm^2 field were physically realistic.

Two different methods were used to acquire PDD profiles along the central axis of the 0.5×0.5 cm^2 field, from a depth of 30 cm to the water surface, with the source-to-surface distance set to 100 cm.

The first method of PDD measurement involved using the "ray-trace" technique provided by the SNC 3D Scanner. This technique was specifically designed to provide a straightforward means to measure PDDs in small fields, via a largely-automated process. An auto-setup step determined the centre of the radiation field (the midpoint between the 50% relative dose points) by scanning at a shallow depth and a deep depth [12], and then linearly interpolating and extrapolating to determine the beam centre as a function of depth [12]. Since the detector movement in the cylindrical water tank is restricted to cylindrical directions (vertical, rotation and diameter, see Fig. 1), the automatic positioning

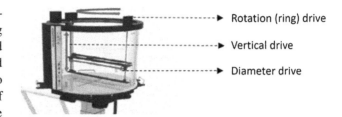

Fig. 1 Photograph of cylindrical water tank showing dosimeter drive directions

of the diode at the centre of the field required all three drives to operate simultaneously throughout the PDD scan, in order to apply the offsets identified at the setup step. At each measurement point, the vertical drive adjusted the measurement depth while the ring drive rotated to the required offset angle and the diameter drive shifted the diode laterally, to place it in the required position.

The second method used a conventional vertical-scanning process, after first completing additional small-field-specific setup and validation steps. The radiation field centre (the midpoint between 50% relative dose points) was identified by repeatedly scanning at 0.1 mm resolution in the in- and cross-plane directions at two depths (5 and 20 cm) iteratively shifting the tank platform to place the diode at the centre of the field before PDD scanning.

Field size reproducibility was evaluated by obtaining additional lateral scans after repositioning (retracting/extending) the MLC leaves, changing the beam energy (rotating the carousel to check the effect of any slight changes in flattening filter position) and turning on and off the field light (to check the effect of monitor chamber repositioning, a feature of the Truebeam linac). Scan parameters were optimised to maximise the signal, which would otherwise have been very low due to the small size of both the diode's active volume and the radiation field.

3 Results and Discussion

Data in Tables 1 and 2 respectively show the precision with which the diode was positioned in the centre of the field for the conventional PDD scans and the reliability with which

Table 1 Location of measured field centre relative to tank centre position, before and after tank position adjustment. Scan directions are defined relative to the plane of the linac's bending magnet (BMAG, in-plane or cross-plane) and relative to the direction of motion of the MLC leaves (parallel or perpendicular). All measurements are in cm

Depth	Direction, BMAG	Direction, MLC	Centre before adj.	Centre after adj.
5	Cross-plane	Parallel	0.025 ± 0.002	-0.0002 ± 0.0019
5	In-plane	Perpendicular	-0.0033 ± 0.0006	Not adjusted
20	Cross-plane	Parallel	0.013 ± 0.001	0.001 ± 0.006
20	In-plane	Perpendicular	0.0040 ± 0.0002	Not adjusted

Table 2 Measured radiation field size in cm, in the direction of MLC motion (measured 105 cm from the photon source)

Measurement conditions	Field width
Initial setup	0.602 ± 0.002
Leaves retracted (opened) and driven back to field position	0.605 ± 0.006
Leaves extended (closed) and driven back to field position	0.564 ± 0.003
Leaves retracted (opened) and driven back to field position	0.600 ± 0.003
Tank shifted to improve alignment with centre of field	0.603 ± 0.002
Leaves retracted (opened) and driven back to field position	0.603 ± 0.002
Carousel rotated, monitor chamber repositioned	0.603 ± 0.002

the linac reproduced the same very small field throughout the study. The PDDs shown in Fig. 2a establish the reliability of the unshielded diode for measuring relative dose at

depth despite changing phantom scatter (despite the difference between the effective atomic number (Z_{eff}) of the silicon diode and the Z_{eff} of water); the diode and ionisation chamber measurements agree within 0.4% at all points beyond the electron build-up region. The optimal small-field scanning method identified in this study used step-by-step scanning with a signal integration time of 2 s per point, to maximise signal-to-noise and achieve the smooth small field PDDs shown in Fig. 2b–d. Together, these results (Tables 1 and 2 and Fig. 2a–d) validate the preparation and measurement steps used to measure small field PDDs in this study.

Figure 2b compares the PDDs obtained using ray-tracing and conventional techniques, for the 0.5×0.5 cm^2 field, and shows that the two curves differ by up to 3% at depth, with the PDD measured using the ray-tracing method decreasing less steeply than the PDD measured using the conventional method.

Figure 2c and d show the additional PDDs that were measured to investigate the difference between the results in the 0.5×0.5 cm^2. (The use of a conventional scanning

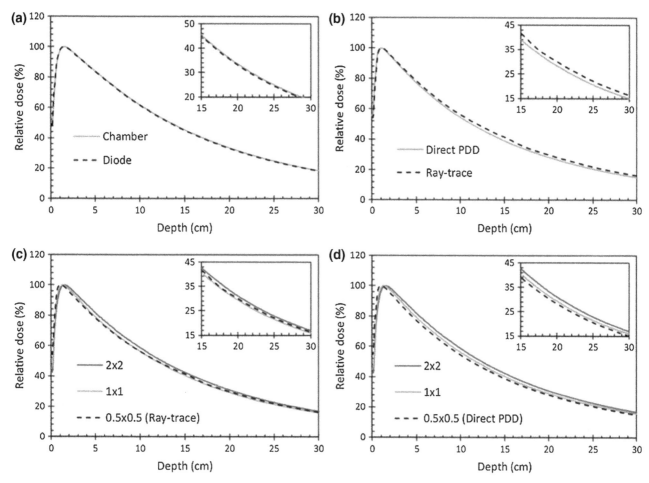

Fig. 2 PDD measurement results. **a** Diode and ionisation chamber measurements in the 4×4 cm^2 field. **b** PDDs measured conventionally and using the ray-tracing method, with the unshielded diode in the 0.5×0.5 cm^2 field. **c** PDD measured using the ray-tracing method in the 0.5×0.5 cm^2 field, compared with PDDs measured in 1×1 and 2×2 cm^2 fields. **d** PDD measured conventionally in the 0.5×0.5 cm^2 field, compared with PDDs measured in 1×1 and 2×2 cm^2 fields. Insets show detail at depth

method was assumed to produce reliable results in the 1×1 and 2×2 cm^2 fields, which are large enough to be unaffected by small gantry sag effects.) It is expected that at any depth well beyond the build-up region, there should be an increase in relative dose with increasing field size, due to an increase in phantom scatter at that depth. Figure 2c and d both show the overall increase in relative dose at depth following the change of field size from 1×1 to 2×2 cm^2. However, the PDD from the 0.5×0.5 cm^2 field that was obtained using the ray-trace method (shown in Fig. 2c) crosses the PDD from the 1×1 cm^2 field at approximately 10 cm depth and shows larger dose per depth compared to the results taken for the 1×1 cm^2 field.

This unphysical behaviour, which suggests that, for example, relative dose at 15 cm depth from a 0.5×0.5 cm^2 field is greater than relative dose at the same point from a 1×1 cm^2 field, does not occur when the 0.5×0.5 cm^2 PDD is measured using the conventional method (Fig. 2d). These results suggest that use of the mechanically-demanding ray-tracing method may result in over-measurement of the dose (or under-measuring of the depth) in very small fields.

4 Conclusion

When using a conventional PDD scanning technique to acquire small field beam data for use in treatment planning, three additional setup and verification steps should be completed. Firstly, while a suitable dosimeter for small field measurements must be used, the response of that dosimeter should be benchmarked against an ionisation chamber in a larger field, to establish its reliability. Secondly, PDD scanning parameters should be selected to maximise signal-to-noise in the small field; slow step-by-step scanning may be needed. Thirdly, repeated profile scanning should be used to verify the reproducibility of the size of the small field under relevant conditions, and to identify the centre of the field at shallow and deep depths so that the dosimeter can be positioned appropriately to scan along the central axis.

Although it is time consuming and requiring careful consideration, a conventional PDD scanning technique, including these additional small-field-specific steps, is an efficient and reliable means to collect PDD data for very small fields and should remain the recommended method for

doing so, until the reliability more sophisticated ray-tracing techniques can be convincingly established.

Compliance with ethical standards The authors declare that they have no conflict of interest. This article does not contain any studies with human participants or animals.

References

1. International Commission on Radiation Units & Measurements.: ICRU Report 91, Prescribing, Recording, and Reporting of Stereotactic Treatments with Small Photon Beams. J. ICRU. 14 (2), 1–145 (2017). https://doi.org/10.1093/jicru/nd013.
2. Kairn, T., Charles, P.H., Crowe, S.B., Trapp, J.V.: Effects of inaccurate small field dose measurements on calculated treatment doses. Australas. Phys. Eng. Sci. Med. 39(3), 747–753 (2016). https://doi.org/10.1007/s13246-016-0461-y.
3. Das, I.J., Ding, G.X., Ahnesjö, A.: Small fields: Nonequilibrium radiation dosimetry. Med. Phys. 35, 206–215 (2008). https://doi.org/10.1118/1.2815356.
4. Charles, P.H., Cranmer-Sargison, G., Thwaites, D.I., et al.: A practical and theoretical definition of very small field size for radiotherapy output factor measurements. Med. Phys. 41(4), 041707 (2014). https://doi.org/10.1118/1.4868461.
5. Cranmer-Sargison, G., Weston, S., Sidhu, N.P., Thwaites, D.I.: Experimental small field 6MV output ratio analysis for various diode detector and accelerator combinations. Radiother. Oncol. 100(3), 429–435 (2011). https://doi.org/10.1016/j.radonc.2011.09.002.
6. Charles, P.H., Crowe, S.B., Kairn, T.: Monte Carlo-based diode design for correction-less small field dosimetry. Phys. Med. Biol. 58, 4501–4512 (2013). https://doi.org/10.1088/0031-9155/58/13/4501.
7. International Atomic Energy Agency.: Dosimetry of Small Static Fields Used in External Beam Radiotherapy, an International Code of Practice for Reference and Relative Dose Determination. IAEA Technical Reports Series No. 483. Vienna (2017).
8. Dieterich, S., Sherouse, G.W.: Experimental comparison of seven commercial dosimetry diodes for measurement of stereotactic radiosurgery cone factors. Med. Phys. 38(7), 4166–4173 (2011). https://doi.org/10.1118/1.3592647.
9. Kairn, T., Charles, P.H., Cranmer-Sargison, G., et al.: Clinical use of diodes and micro-chamber to obtain accurate small field output factor measurements. Australas. Phys. Eng. Sci. Med. 38, 357–367 (2015). https://doi.org/10.1007/s13246-015-0334-9.
10. Cranmer-Sargison, G., Charles, P.H., Trapp, J.V., Thwaites, D.I.: A methodological approach to reporting corrected small field relative outputs. Radiother. Oncol. 109(3), 350–355 (2013). https://doi.org/10.1016/j.radonc.2013.10.002.
11. Kairn, T., Asena, A., Charles, P.H., et al: Field size consistency of nominally matched linacs. Australas. Phys. Eng. Sci. Med. 38(2), 289–297 (2015). https://doi.org/10.1007/s13246-015-0349-2.
12. Sun Nuclear Corporation, SNC Dosimetry Software Reference Guide, (2017).

Long-Term Reliability of Optically Stimulated Luminescence Dosimeters

Tanya Kairn, Samuel Peet, Liting Yu, and Scott Crowe

Abstract

Optically stimulated luminescence dosimeters (OSLDs) can be used as accurate and re-usable dosimeters for radiotherapy applications. OSLDs have been observed to decline in sensitivity with repeated use and it is important to determine whether this decline in sensitivity is associated with a decline in reliability. This study used three batches of OSLDs (purchased in 2012, 2014 and 2016) that had been repeatedly re-used in a mature in vivo dosimetry programme over a period of up to five years and evaluated the consistency of their response over repeated irradiation-readout-bleaching cycles. Each irradiation delivered 105 cGy to all OSLDs, using a 12 meV electron beam from a Varian iX linear accelerator. The five- and three-year-old OSLDs respectively displayed 86% and 89% of the sensitivity of the one year old OSLDs, but when a correction factor for each OSLD was derived based on the first measurement result and applied to each subsequent reading, all OSLDs were able to measure the 105 cGy test dose accurately, within standard deviations of 2.0% for the OSLDs from 2012 and 1.3% for the OSLDs from 2014 and 2016. If a mean calibration value was applied to the readings from each batch of OSLDs, instead of applying a measurement-derived sensitivity correction factor to each individual OSLD reading, the standard deviations increased to an unacceptable 6.1, 5.6 and 2.9%. Well-used three- and five-year-old OSLDs were shown to be capable of providing measurements with similar accuracy to a more recently-purchased batch of OSLDs, when measurement-derived sensitivity correction factors were applied to each result. If this extra step is included in the OSLD measurement process, then the same OSLDs may be reliably used for years without needing to be retired and replaced.

Keywords

Radiation therapy • Solid state dosimetry Semiconductors

1 Introduction

Optically stimulated luminescence dosimeters (OSLDs) are an attractive solution for radiotherapy dosimetry due to their small size, ease of use, sensitivity and reusability [1, 2]. OSLDs have been investigated and used for a variety of radiotherapy applications, including quality assurance and end-to-end testing [3, 4], dose measurements in non-reference and out-of-field conditions [5–7], in vivo dosimetry [8–10] and dosimetry auditing [2, 11].

OSLDs function similarly to thermoluminescent dosimeters (TLDs); incident radiation frees electrons which are then trapped between the valence and conduction bands of a doped semiconductor, until released by the application of additional energy and allowed to fall back to the valence band by the emitting a quantifiable light signal. Whereas the trapped electrons in TLDs are read out and cleared (annealed) using controlled heating, the energy needed to read out and clear (bleach) trapped electrons from OSLDs is applied using optical light.

Like TLDs and other solid state dosimeters, OSLDs have been found to exhibit a sensitivity variation with total accumulated dose [1, 12]. Different rates of decline in OSLD response have been observed after OSLD chips have been exposed to total doses above 10–20 Gy [1, 4, 12]. For example, Opp et al. measured a decrease in response of almost 7% after OSLDs were repeatedly irradiated and bleached until they were exposed to a total accumulated dose of 16 Gy [4], and Jursinic observed that "above 20 Gy, the

T. Kairn (✉) · S. Peet · L. Yu · S. Crowe
Royal Brisbane and Women's Hospital, Brisbane, Australia
e-mail: t.kairn@gmail.com

T. Kairn · S. Peet · L. Yu · S. Crowe
Queensland University of Technology, Brisbane, Australia

© Springer Nature Singapore Pte Ltd. 2019
L. Lhotska et al. (eds.), *World Congress on Medical Physics and Biomedical Engineering 2018*,
IFMBE Proceedings 68/3, https://doi.org/10.1007/978-981-10-9023-3_103

OSLD sensitivity begins to drop by about 4% per 10 Gy of additional accumulated dose" [12].

As OSLD systems are increasingly implemented for use in routine and specialised radiotherapy dosimetry services, it is important to understand the long-term stability of OSLD response. This study therefore aimed to re-evaluate the sensitivity of thoroughly used batches of OSLDs, to provide an indication of whether decreases in OSLD sensitivity are associated with decreases in OSLD reliability and thereby provide guidance on the frequency with which batches of OSLD chips should be retired and replaced.

2 Method

This study used three batches of Landauer nanoDot OSLDs (Landauer, Glenwood, USA), which had been purchased in 2012, 2014 and 2016. The OSLDs were reported by the manufacturer (and found at inital commissioning, immediately after purchase) to produce a consistent response within ±2%. The OSLDs were then repeatedly re-used in a mature in vivo dosimetry programme over a period of up to five years. The sensitivity and consistency of the response of all available OSLDs was evaluated in 2017, over three repetitions of the bleaching-irradiation-readout cycle.

Bleaching was performed using a Gammasonics Manual OSL Annealing Lightbox (Gammasonics Institute for Medical Research Pty Ltd, Lane Cove, Australia) which contained an array of fluorescent tubes, beneath a light-diffusing screen (see Fig. 1a). A bleaching time of 3 h was selected, to reduce the mean residual signal to 0.26 ± 0.01 cGy and to correspond to established fluorescent-light bleaching conditions [12].

A Varian iX linac (Varian Medical Systems, Palo Alto, USA) was used to deliver 105 cGy to all OSLDs at each irradiation, using a 25×25 cm^2 12 meV electron beam, which produced a large region of relatively flat dose, sufficient for irradiating up to 100 OSLDs at once, in a

purpose-fabricated array (see Fig. 1b) at the depth of maximum dose in a water-equivalent plastic phantom (see Fig. 1c).

All OSLDs were read out after each irradiation, using a Landauer Microstar OSLD reader containing an array of 38 light emitting diodes (LED) which provided a 532 nm (green) light source for stimulating the luminescence [3]. Individual OSLDs were identified and tracked using numbers and barcodes on stickers that were attached to the OSLD housings by the manufacturer. Four OSLDs were removed from the sample due to loss (detachment) or fading of their identifying numbers and barcodes (see Fig. 1d).

3 Results and Discussion

The results of irradiating all OSLDs to the same dose three times, during three repetitions of the bleaching-irradiation-readout cycle are shown in Fig. 2a–c.

Examination of the number of raw counts per cGy from each measurement (shown in Fig. 2a) indicates that the OSLDs from 2012 and 2014 respectively displayed 86% and 89% of the sensitivity of the newer OSLDs. This result provides a useful long-term verification of the decrease in OSLD sensitivity with total accumulated dose that was predicted via short-term OSLD response studies [1, 4, 12].

When OSLD response was corrected using a batch calibration factor (a correction calculated from the mean over/under response of all OSLDs in each batch), the OSLD measurements appeared unacceptably variable, especially for the older batches (see Fig. 2b), due to the response variations of the individual OSLD chips [13]. The mean standard deviations calculated over the three repetitions of the bleaching-irradiation-readout cycles were 6.1, 5.6 and 2.9% for the batches of OSLDs from 2012, 2014 and 2016, respectively.

After a locally-derived correction factor for each OSLD was determined based on the first reading of the first

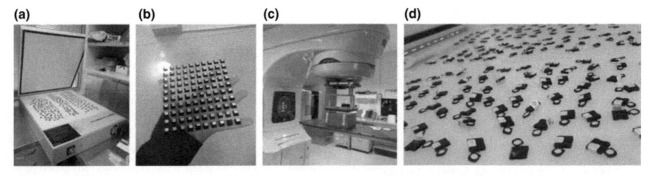

(a)   **(b)**   **(c)**   **(d)**

Fig. 1 Photographs of **a** the lightbox used for OSLD bleaching, **b** the OSLD irradiation array, **c** the OSLD irradiation setup and **d** OSLDs in bleaching configuration, with chips exposed, showing peeling identification stickers on the oldest OSLDs

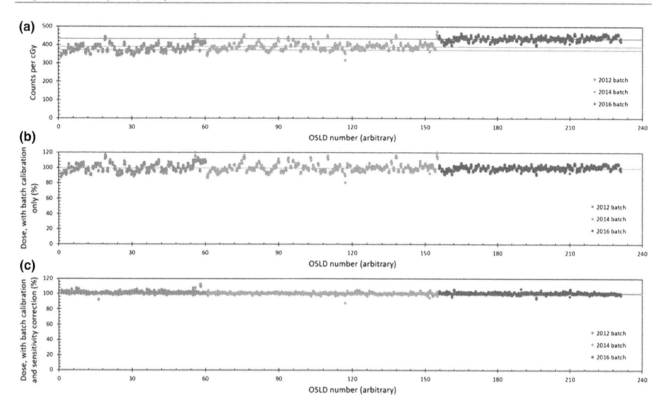

Fig. 2 OSLD dose response, plotted as **a** number of counts per cGy of delivered dose, **b** measured dose as a percentage of delivered dose, where measurements were corrected using a batch calibration factor only, and **c** measured dose as a percentage of delivered dose, where measurements were corrected using a locally-measured sensitivity factor for each OSLD. Horizontal lines indicate mean values from each batch of OSLDs

measurement (without reference to the vendor-supplied sensitivity factor) and applied to each subsequent reading, all OSLDs were able to measure the 105 cGy test dose accurately, within standard deviations of 2.0% for the OSLDs from 2012 and 1.3% for the other two batches of OSLDs (see Fig. 2c). Using this method, approximately 90% of the OSLDs from 2014 and 2016 were able to provide measurements within ±2% of the known delivered dose, while 75% of the OSLDs from 2012 were able to provide the same level of accuracy.

Data shown in Fig. 2c indicates that several outliers exist, where measured doses differ from the known delivered doses by 5% or more. This result suggests that even when local sensitivity correction factors are measured, used and frequently updated, it may also be advisable to use more than one OSLD to perform each dose measurement, to minimise the likelihood of results being confounded by anomalous OSLD over/under-response.

4 Conclusion

Three batches of OSLDs that had been repeatedly re-used in a mature in vivo dosimetry programme have been found to be capable of providing radiotherapy dose measurements with similar levels of accuracy, despite the batches' varying ages and levels of previous accumulated dose. To achieve this result, it was necessary to measure and apply a local sensitivity correction factor to each OSLD reading. If this extra step is included in the OSLD measurement process, then it may be possible to use the same OSLDs repeatedly and reliably for years, without needing them to be retired and replaced.

Compliance with Ethical Standards The authors declare that they have no conflict of interest. This article does not contain any studies with human participants or animals performed by any of the authors.

References

1. Jursinic, P.A.: Characterization of optically stimulated luminescent dosimeters, OSLDs, for clinical dosimetric measurements. Med. Phys. 34(12), 4594–4604 (2007). https://doi.org/10.1118/1.2804555.

2. Dunn, L., Lye, J., Kenny, J., et al.: Commissioning of optically stimulated luminescence dosimeters for use in radiotherapy. Radiat. Meas. 51, 31–39 (2013) https://doi.org/10.1016/j.radmeas.2013.01.012.

3. Villani, D., Mancini, A., Haddad, C.M., Campos, L.L.: Application of optically stimulated luminescence 'nanoDot' dosimeters for dose verification of VMAT treatment planning using an anthropomorphic stereotactic end-to-end verification phantom. Radiat. Meas. 106, 321–325 (2017). https://doi.org/10.1016/j.radmeas.2017.03.027.

4. Opp, D., Nelms, B.E., Zhang, G., Stevens, C. and Feygelman, V., 2013. Validation of measurement guided 3D VMAT dose reconstruction on a heterogeneous anthropomorphic phantom. Journal of Applied Clinical Medical Physics, 14(4), pp. 70–84. https://doi.org/10.1120/jacmp.v14i4.4154.

5. Charles, P.H., Crowe, S.B., Kairn, T., et al.: The effect of very small air gaps on small field dosimetry. Phys. Med. Biol. 57, 6947–6960 (2012). https://doi.org/10.1088/0031-9155/57/21/6947.

6. Scarboro, S.B., Followill, D.S., Kerns, J.R., et al.: Energy response of optically stimulated luminescent dosimeters for non-reference measurement locations in a 6 MV photon beam. Phys. Med. Biol. 57(9), 2505–2515 (2012). https://doi.org/10.1088/0031-9155/57/9/2505.

7. Peet, S.C., Wilks, R., Kairn, T., Crowe, S.B.: Measuring dose from radiotherapy treatments in the vicinity of a cardiac pacemaker. Phys. Medica. 32(12), 1529–1536 (2016). https://doi.org/10.1016/j.ejmp.2016.11.010.

8. Austin, M.J., Bergstrand, E.S., Bokulic, T., et al.: Development of Procedures for In VivoDosimetry in Radiotherapy. IAEA (International Atomic Energy Agency) Human Health Report 8 (2013).

9. Butson, M., Haque, M., Smith, L., et al.: Practical time considerations for optically stimulated luminescent dosimetry (OSLD) in total body irradiation. Australas. Phys. Eng. Sci. Med. 40(1), 167–171 (2017). https://doi.org/10.1007/s13246-016-0504-4.

10. Nabankema, S.K., Jafari, S.M., Peet, S.C., et al.: Wearable glass beads for in vivo dosimetry of total skin electron irradiation treatments. Radiat.Phys. Chem. 140, 314–318 (2017). https://doi.org/10.1016/j.radphyschem.2016.12.013.

11. Lye, J., Dunn, L., Kenny, J., et al.: Remote auditing of radiotherapy facilities using optically stimulated luminescence dosimeters. Med. Phys. 41(3), 032102 (2014). https://doi.org/10.1118/1.4865786.

12. Jursinic, P.A.: Changes in optically stimulated luminescent dosimeter (OSLD) dosimetric characteristics with accumulated dose. Med. Phys. 37(1), 132–140 (2010). https://doi.org/10.1118/1.3267489.

13. Asena, A., Crowe, S.B., Kairn, T., et al.: Response variation of optically stimulated luminescence dosimeters. Radiat. Meas. 61, 21–24 (2014). https://doi.org/10.1016/j.radmeas.2013.12.004.

Feasibility Study of Alanine Dosimeter for Carbon-Beam Dosimetry

H. Yamaguchi, M. Shimizu, Y. Morishita, K. Hirayama, Y. Satou, M. Kato,
T. Kurosawa, T. Tanaka, N. Saito, M. Sakama, and A. Fukumura

Abstract

Alanine dosimeters are useful tools for measuring both kilo-gray level doses and radiation therapy level doses. The National Metrology Institute of Japan is planning to develop a dose measurement service for carbon-beam therapy. The purpose of this study was to verify that the alanine dosimeter could be used for carbon-beam dosimetry. We irradiated the alanine dosimeter with ^{60}Co-gamma rays and a carbon beam using a range of 10–30 Gy. For the gamma-ray irradiation, the alanine dosimeter was placed in a 5 g cm^{-2} water phantom. For the carbon-beam irradiation, the spread out Bragg peak of the 290 meV/u carbon ion was used, and the dosimeter was located at the centre of the peak region. The alanine spectra obtained under the carbon-beam irradiation were almost the same as the spectra under the gamma-ray irradiation. The alanine dosimeter irradiated with the carbon beam show a smaller signal than that obtained under the gamma-ray irradiation for the same dose. The alanine signal increased as the dose increased over the entire dosage range for both the gamma-ray and carbon-beam irradiation. This result suggests that the alanine dosimeter has potential for carbon-beam dosimetry.

Keywords

Alanine • Carbon beam • Dosimetry

1 Introduction

The number of radicals generated from L-α-alanine by irradiation can be measured using an electron spin resonance (ESR) spectrometer. The ESR signal, which is defined as the peak-to-peak height of the centre of the alanine spectrum, increases with the absorbed dose up to approximately 100 kGy [1]. Thus, alanine dosimeters are widely used for calibrating the dosimeters routinely used in high-dose irradiation facilities. The advantages of an alanine dosimeter include a density close to that of water, an ESR signal with long-term stability, the ability to use it as a mailed dosimeter, and the lack of a significant change in the signal of the dosimeter under normal temperature and humidity conditions.

Because of these advantages, alanine dosimeters are also applied to radiation therapy level dose measurement, a reference dosimetry service, and international comparisons [2–4]. There is a different dose measurement situation for carbon-beam therapy because the physical characteristics of an ion beam are different from those of photons and electrons. Nakagawa et al. [5] reported that the ESR signal of the alanine increased as the absorbed dose increased to 60 Gy by carbon-beam irradiation. They also showed that the ESR signal decreased as the linear energy transfer (LET) increased. Herrmann et al. [6] calculated and measured the depth dose curves for a mono-energy beam and spread out Bragg peak (SOBP) beam using stacked alanine pellets for a carbon beam. The measured alanine responses agreed with the calculated alanine responses for both the mono-energy beam and SOBP beam. The alanine response for the carbon beam relative to the ^{60}Co-gamma rays was also calculated and measured. They reported that the relative response (effectiveness) had a value of approximately 0.80–0.95 depending on the carbon-beam energy. However, the

H. Yamaguchi (✉) · M. Shimizu · Y. Morishita · K. Hirayama ·
Y. Satou · M. Kato · T. Kurosawa · T. Tanaka · N. Saito
National Metrology Institute of Japan, AIST, Tsukuba, Ibaraki
305-8568, Japan
e-mail: hidetoshi.yamaguchi@aist.go.jp

K. Hirayama · Y. Satou
Graduate School of Medical Health and Sciences, Komazawa
University, Setagaya-Ku, Tokyo, 154-8525, Japan

M. Sakama · A. Fukumura
National Institutes for Quantum and Radiological Science and
Technology, National Institute of Radiological Sciences,
Inage-Ku, Chiba-Shi, Chiba, 263-8555, Japan

© Springer Nature Singapore Pte Ltd. 2019
L. Lhotska et al. (eds.), *World Congress on Medical Physics and Biomedical Engineering 2018*,
IFMBE Proceedings 68/3, https://doi.org/10.1007/978-981-10-9023-3_104

relative response for the SOBP of the carbon beam is not available.

In Japan, the National Metrology Institute of Japan (NMIJ) is developing an alanine dosimetry system for carbon-beam therapy to provide an independent peer review service. In this study, we compared the signals of the alanine dosimeter using ^{60}Co-gamma rays and the SOBP, and measured the relative response of the alanine dosimeter.

2 Materials and Methods

2.1 Alanine Dosimeter

Commercial alanine pellets (Harwell Dosimeters Ltd, Lot number: AX584) were used for the experiments. The weight ratio of the composition was 90.9% L-α-alanine and 9.1% paraffin wax. The nominal pellet diameter was 4.8 mm, and the nominal height was 2.7 mm. The dosimeter consisted of four alanine pellets and a case made of conductive poly ether ether ketone (PEEK). The case was 13.2 mm in diameter and 22 mm in height.

2.2 Irradiation with ^{60}Co-Gamma Rays

Irradiation with ^{60}Co-gamma rays was performed at the NMIJ using a 46 TBq source (this was the value in December, 2017) [7]. Figure 1a shows the experimental setup for the ^{60}Co-gamma-ray irradiation. The source-chamber distance (SCD) was 100 cm, and the diameter of the irradiation field was 11 cm. The beam entered a water phantom horizontally. An ionization chamber (PTW, TN31013) and electrometer (EMF Japan, EMF520) were used to determine the absorbed dose rate at a reference depth of 5 g cm^{-2} in the water phantom. The chamber was calibrated beforehand using another water phantom. The determined absorbed dose rate was used to calculate the delivered dose to the alanine dosimeter. The dose rate was 0.24 Gy/min in December, 2017. The doses delivered to the alanine dosimeter were approximately 10, 20, and 30 Gy, and three dosimeters were used for each dose point.

The alanine dosimeter was located with its centre in the centre of the ionization chamber. To prevent the alanine dosimeter from contacting the water, the dosimeter was inserted into a sleeve with a rod. The sleeve and rod were made of conductive PEEK. The wall thickness of the sleeve was 1.3 mm, and the rod had the same diameter as the alanine dosimeter.

2.3 Irradiation with Carbon Beam

Irradiation with a carbon beam was performed at the Heavy Ion Medical Accelerator in Chiba at the National Institute of Radiological Sciences (QST/NIRS-HIMAC). The nominal accelerator energy of the carbon beam was 290 meV/u. An SOBP with a width of 6 cm was formed by a ridge filter. The experimental setup for carbon-beam irradiation is shown in Fig. 1b. Since the purpose of this study was that the alanine dosimeter was feasible for carbon-beam dosimetry, the irradiation was conducted in air, and the SOBP was formed by a rage shifter. The centre of the SOBP was located at a water depth of 11 g cm^{-2}. The ionization chamber and alanine dosimeter were irradiated in air. The size of the irradiation field was set to 10×10 cm^2. The ionization chamber and alanine dosimeter were held in place by holders made of polyoxymethylene and fixed to the centre of the irradiation field. The ionization chamber was positioned at a $0.75r_{cyl}$ deeper depth than the centre of the SOBP, where r_{cyl} is the inner radius of the ionization chamber [8]. The alanine dosimeter was positioned at a depth of 11 g cm^{-2} considering the density, the radius of the pellet and the wall thickness of the case.

An ionization chamber (PTW, TN30013) and electrometer (EMF Japan, EMF522) were used to measure the dose at the reference point, which was the centre of the SOBP, before the irradiation of the alanine dosimeter. The ionization chamber was calibrated using ^{60}Co-gamma rays in the

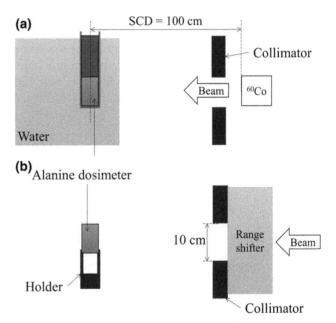

Fig. 1 Experimental setup: **a** setup for gamma-ray irradiation and **b** setup for carbon-beam irradiation

NMIJ. The absorbed dose to water was determined using the following equation described in TRS-398 [8].

$$D_{w,Q} = M_Q N_{D,w,Q_0} k_{Q,Q_0}. \tag{1}$$

Here, M_Q is the reading of the ionization chamber corrected by the ion recombination, polarity, temperature, and pressure, and N_{D,w,Q_0} is a calibration factor for gamma rays. k_{Q,Q_0} is a radiation quality factor that depends on the chamber model. The value of k_{Q,Q_0} is listed in the reference [8]. The determined dose was used to calibrate a monitor unit of the carbon-beam irradiation system. The dose delivered to the alanine dosimeter was almost the same as that used for the gamma-ray irradiation. Three dosimeters were used for each dose point.

2.4 ESR Measurement

The alanine signals were measured using a spectrometer (Bruker Biospin, EMX micro) in the NMIJ. A schematic view of the ESR system is shown in Fig. 2. The alanine pellet was inserted from the top of the outer tube. The inner tube was evacuated by a vacuum pump, and the pellet was fixed on the inner tube. The Cr^{3+} reference sample was inserted to approximately the centre of the cavity using a marker accessory provided by Bruker Biospin. Because it is known that the alanine signal has angular dependence [9, 10], the alanine pellet was rotated in 72° steps (0°, 72°, 144°, 216°, 288°) by the actuator connected to the inner tube. The signals of the alanine and Cr^{3+} were measured at these angles. The measured ESR signal of one dosimeter, S, was calculated from

$$S = \frac{1}{nm} \sum_{i=1}^{n=4} \sum_{j=1}^{m=5} \left(\frac{S_{i,j}^a}{S_{i,j}^r M_i} - \bar{S}_{u.a.} \right), \tag{2}$$

where n and m are the number of pellets in one dosimeter and number of rotations, respectively. $S_{i,j}^a$ and $S_{i,j}^r$ are the measured ESR signals of the alanine and Cr^{3+}, respectively. Figure 3a shows the typical spectrum of the alanine pellet. The peak-to-peak height of the alanine signal at approximately 0.34548 T (central peak) was used for $S_{i,j}^a$, and that of the Cr^{3+} signal at approximately 0.34969 T was used for $S_{i,j}^r$. The mass of an alanine pellet, M_i, was measured using an electronic balance (SHIMADZU, AUW120D). $\bar{S}_{u.a.}$ is the averaged ESR signals for four alanine pellets without irradiation.

The following parameters were used for the ESR measurement. The microwave power was 0.3 mW, modulation amplitude was 0.26 mT, time constant was 1310 ms, sweep time was 30.72 s, and sweep width was 1 mT. A total of 1024 data points were used. Before and after irradiation, the alanine pellets were stored in a desiccator. Wieser et al. [11] reported that no fading occurred when alanine pellets were stored in an environment of 22 °C and 50% RH. On the other hand, Anton [12] reported that the ESR signal of irradiated Harwell alanine pellets showed an increase of 0.5–1.5% within one day of irradiation, and became stable after a week under laboratory conditions (temperature: 21–22 °C, relative humidity: 25–60%). We stored the alanine pellets in

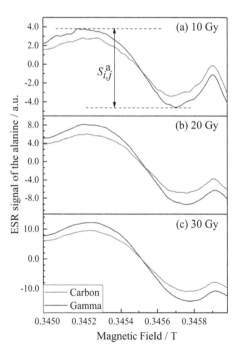

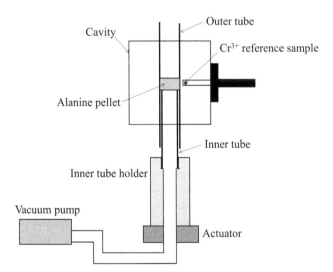

Fig. 2 Schematic view of ESR measurement system. The alanine pellet is inserted from the outer tube fixed to the cavity. The pellet is fixed on the inner tube by the vacuum pump. The inner tube is indirectly connected to the actuator and can be rotated

Fig. 3 ESR spectrum of alanine. The red line shows the alanine spectrum with carbon-beam irradiation, and the blue line shows that with gamma ray irradiation. The absorbed doses are 10, 20, and 30 Gy for (**a**), (**b**), and (**c**), respectively (Colour figure online)

a desiccator at a temperature of 18.9 ± 2.1 °C and a relative humidity of 39.6% ± 1.3% using saturated salt of potassium carbonate. The ESR measurements began at least two weeks after the irradiation.

3 Results and Discussion

3.1 ESR Spectrum

The blue and red lines in Fig. 3 show the ESR spectra of the alanine pellets irradiated by gamma rays and carbon beams. Figures 3a, b and c show the results for the delivered doses of 10, 20, and 30 Gy, respectively. The alanine spectra obtained with irradiation by the carbon beam did not show remarkable differences compared to those obtained with the irradiation by gamma rays, except for the peak-to-peak height. Apparent background noise is seen in the spectrum for 10 Gy. Thus, the lower limit for dose measurement with the present method is approximately 10 Gy for a carbon beam.

3.2 Response of Alanine Dosimeter

Figure 4 shows the ESR signals for the carbon beam and ^{60}Co-gamma rays. Both ESR signals increase linearly with the delivered dose. The calibration curves fitted by a linear function are also shown. The gradient of the calibration line of the carbon beam is obviously smaller than that of the gamma rays. This difference resulted because the LET of the carbon ion was larger than that of the gamma rays. A portion of the alanine molecules may have been broken into pieces by the carbon beam without generating radicals. The gradient of each calibration line and the uncertainty of the gradient are listed in Table 1. The uncertainty of the gradient of

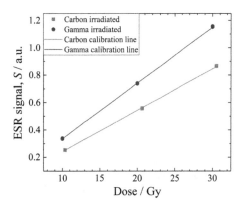

Fig. 4 Experimental results of ESR signals for carbon-beam and gamma-ray irradiation. The filled squares are the results for the carbon beam, and the filled circles are the results for the gamma rays. Each line represents a calibration line fitted by a linear function

Table 1 Gradients of calibration lines and uncertainties

	Gradient [Gy^{-1}]	Uncertainty ($k = 1$) [%]
Carbon beam	0.03021	0.65
Gamma ray	0.04066	0.31

the calibration line for the carbon beam is larger than that for the gamma rays.

The relative response η of the alanine dosimeter is defined as

$$\eta = \frac{g_c}{g_\gamma}, \tag{3}$$

where g_c and g_γ are the gradients of the calibration lines for the carbon beam and gamma rays, respectively. The present relative response is 0.7431 ± 0.0054 ($k = 1$). Herrmann et al. performed a similar experiment at a plateau region for a 300 meV/u carbon beam. They used another definition for the relative response, which was evaluated using the dose ratio for the same signal level (isoresponse) [6]. They reported that the relative response was 0.95. Although we also analyzed the response using this definition, the response was 0.7559 ± 0.0085 ($k = 1$).

The reason for this difference between the present result and Herrmann's result is that we irradiated the alanine dosimeter at the centre of the SOBP, while they irradiated the dosimeter at the plateau region of the carbon beam. The smaller relative response may suggest that more alanine molecules are destroyed in the SOBP region than in the plateau region.

4 Conclusion

In this study, we tested an alanine dosimeter designed for dose measurement in carbon-beam therapy. An SOBP beam of 290 meV/u carbon ions was used to irradiate the alanine dosimeter in a range of 10–30 Gy, and the calibration line was compared with that of ^{60}Co-gamma-ray irradiation. There were no apparent differences in the ESR spectra between the carbon-beam irradiation and gamma-ray irradiation. Because the measured ESR signals increased with the dose, a third-party evaluation for carbon-beam therapy could be performed using the alanine dosimeter at least within the present dose. The gradient of the calibration line for the carbon beam was smaller than that for the gamma-ray beam. The relative response derived in the SOBP region was approximately 22% smaller than that obtained at the plateau region by Herrmann et al. [6]. These results suggest that the response of the alanine dosimeter depends on the LET of the beam. In the future, we will irradiate the alanine dosimeter using different beams, e.g., in the plateau region. We will also extend the range of the dose to determine whether the alanine dosimeter could be used for carbon-beam dosimetry. Since the results of this study confirmed that the alanine dosimeter was feasible for carbon-beam dosimetry, we planned the experiment in water for measuring absorbed dose to water.

Acknowledgements This study was carried out partly under the Research Project with Heavy Ions at QST/NIRS-HIMAC.

Conflict of Interest The authors have no conflict of interest to declare.

References

1. Regulla, D.F., Deffner, U.: Dosimetry by ESR spectroscopy of alanine. Int. J. Appl. Radiat. Isot. 33(11), 1101–1114 (1982).
2. Sharpe, P.H.G., Sephton, J.P.: Alanine dosimetry at NPL - the development of a mailed reference dosimetry service at radiotherapy dose levels. In: IAEA-SM-356/R6 Proc. 183–189 (1998).
3. Garcia, T., Anton, M., Sharpe, P.: EURAMET.RI(I)-S7 comparison of alanine dosimetry systems for absorbed dose to water measurements in gamma- and x-radiation at radiotherapy levels. Metrologia. 49(1A), 06004 (2012).
4. McEwen, M., Sharpe, P., Vörös, S.: Evaluation of alanine as a reference dosimeter for therapy level dose comparisons in megavoltage electron beams. Metrologia. 52(2), 272–279 (2015).
5. Nakagawa, K., Ikota, N., Sato, Y.: Heavy-ion-induced sucrose and L-α-alanine radicals investigated by EPR. Appl. Magn. Reson. 33 (1–2), 111–116 (2008).
6. Herrmann, R., Jäkel, O., Palmans, H., Sharpe, P., Bassler, N.: Dose response of alanine detectors irradiated with carbon ion beams. Med. Phys. 38(4), 1859–1866 (2011).
7. Morishita, Y., Kato, M., Takata, N., Kurosawa, T., Tanaka, T., Saito, N.: A standard for absorbed dose rate to water in a Co-60 field using a graphite calorimeter at the national metrology institute of Japan. Radiat. Prot. Dosim. 154, 331–339 (2013).
8. Andreo, P., Burns, D.T., Hohlfeld, K., Huq, M.S., Kanai, T., Laitano, F.: IAEA Technical Report Series No. 398 (2000).
9. Dolo, J.M., Garcia, T.: Angular response of alanine samples: From powder to pellet. Radiat. Meas. 42(6–7), 1201–1206 (2007).
10. Kojima, T., Kashiwazaki, S., Tachibana, H., Tanaka, R., Desrosiers, M.F., McLaughlin, W.L.: Orientation effects of ESR analysis of alanine-polymer dosimeters. Appl. Radiat. Isot. 46(12), 1407–1411 (1995).
11. Wieser, A., Lettau, C., Fill, U., Regulla, D.F.: The influence of non-radiation induced ESR background signal from paraffin-alanine probes for dosimetry in the radiotherapy dose range. Appl. Radiat. Isot. 44(1–2), 59–65 (1993).
12. Anton, M.: Postirradiation effects in alanine dosimeter probes of two different suppliers. Phys. Med. Biol. 53(5), 1241–1258 (2008).

Local Dose Survey on Paediatric Multi-detector CT: A Preliminary Result

L. E. Lubis, A. F. Jundi, A. Susilo, A. Evianti, and D. S. Soejoko

Abstract

For paediatric patients, where there exists a wide range of patient size, and where radiation risk is two to three times more prevalent than on adults, the need of a more stringent care on dose optimization becomes more crucial and demanding, particularly on computed tomography (CT) procedures. To first investigate the need of optimization, a preliminary dose survey was carried out by recording dose data from head, chest, and abdomen MDCT procedures for paediatric patients. The survey from a total of 343 paediatric CT patients was conducted on March-September 2017 at Harapan Kita Maternal and Children's Hospital as Indonesia's primary referral hospital for paediatric patients. Results per age group were compared with available works from other regions or countries to first indicate the need for optimization. The preliminary result indicated that head scans for 10–15 years old patients require optimization. The evaluation of protocol selections is proposed as an appropriate action.

Keywords

Computed tomography • Paediatric • Dose

1 Introduction

The world-wide use of multidetector computed tomography (MDCT) for both adult and paediatric patients has increased over time since its first introduction in the 1990's [1]. In Indonesia, the increasing number of CT scanners, 706 units as in 2016, reflects this trend as well. In accordance to the trend, there is an increased concern about the biological risks of CT scan-induced radiation to the patients, particularly when the ALARA principle is applied in general.

With stochastic risk as high as two to three times that of adult patients [2–4] and the existence of late onset during longer life expectancy [5, 6], paediatric patients requires more attention regarding the radiation dose. The requirement of optimization to ensure the appropriateness of selections of scan parameters has thus become more demanding. To first indicate the need for optimization, and to identify the possible change of practice required in further duly optimization, a dose survey is considered appropriate. Several studies have been performed regarding dose in paediatric CT as either local [1, 7–9], multi-center [10, 11], nation-wide [12–18], even international [19] surveys. Some of these studies either compared their results to existing national Diagnostic Reference Level (DRL) to draw conclusions for optimization or even led to the establishment of their DRLs. In Indonesia, national DRL is currently unavailable.

As an Indonesian national reference hospital for paediatric patient with new, dedicated MDCT for paediatric use, Harapan Kita Maternal and Childern's hospital took the initiative to perform a preliminary survey on dose metric for paediatric MDCT. Standard metrics of $CTDI_{vol}$ and DLP [3, 20–23] were recorded and preliminary analysis were made. As there is currently no DRL present in Indonesia to compare with, we compare results with other studies to indicate our need of optimization and provide guidance.

2 Materials and Methods

The survey on $CTDI_{vol}$ and DLP for head, abdomen, and chest scans was performed during the period of March to September 2017 from a newly-installed Siemens SOMATOM Perspective 128 slice MDCT. Values were recorded from patient dose report with the device having less than 10% discrepancy with measured metrics according to compliance test result using

L. E. Lubis (✉) · A. F. Jundi · D. S. Soejoko
Department of Physics, Faculty of Mathematics and Natural Sciences, Universitas Indonesia, Depok, 16424, Indonesia
e-mail: lukmanda.evan@sci.ui.ac.id

A. Susilo · A. Evianti
Department of Radiology, Harapan Kita Maternal and Children's Hospital, Jakarta, 11420, Indonesia

© Springer Nature Singapore Pte Ltd. 2019
L. Lhotska et al. (eds.), *World Congress on Medical Physics and Biomedical Engineering 2018*,
IFMBE Proceedings 68/3, https://doi.org/10.1007/978-981-10-9023-3_105

Table 1 Number of patients for each age groups and studies

Age groups	Head	Abdomen-pelvis	Chest
0–1 years	53	33	24
1–5 years	60	41	9
5–10 years	20	37	4
10–15 years	11	46	5

Table 2 Recorded dose metric for paediatric CT compared with several previous works

Works	$CTDI_{vol}$ (mGy)				DLP (mGy.cm)			
	0–1 y	1–5 y	5–10 y	10–15 y	0–1 y	1–5 y	5–10 y	10–15 y
Head								
Shrimpton et al. 2015 [17]	25	40	60	–	350	650	860	–
Järvinen et al. 2015 [16]	23	25	29	35	330	370	460	560
Galanski et al. 2006 [15]	33	40	50	50	390	520	710	920
Buls et al. 2010 [11]	35	43	49	50	280	473	637	650
Veit et al. 2010 [14]	33	40	50	60	400	500	650	850
Roch and Aubert 2012 [13]	30	40	50	–	420	600	900	–
Watson and Coakley 2010[a] [8]	–	7	16	–	–	68.7	154.8	–
Fukushima et al. 2012[b] [12]	–	–	–	–	820	1000	1040	1120
Kritsaneepaiboon et al. 2012 [10]	26	29	39	45	402	570	613	801
Vassileva et al. 2015 [19]	26	36	43	53	440	540	690	840
This work	36	42	45	87	635	885	1024	2194
Abdomen-pelvis								
Galanski et al. 2006 [15]	–	–	–	–	70	125	240	500
Buls et al. 2010 [11]	–	–	5	7.5	–	110	220	330
Veit et al. 2010 [14]	4	6	8	13	85	165	250	500
Ruiz-Cruces 2015 [18]	–	–	–	–	95	150	190	340
Roch and Aubert 2012 [13]	–	4	5	7	–	80	120	245
McCollough et al. 2011[c] [27]	20 (no age group)				–	–	–	–
Watson and Coakley 2010[a] [8]	3	5	–	8	67	153	–	502
Kritsaneepaiboon et al. 2012 [10]	7.7	8.9	13.8	16.8	222	276	561	764
Vassileva et al. 2015[d] [19]	5.2	7	7.8	9.8	130	250	310	460
This work	2.9	2.4	2.5	5.2	98	90	112	261
Chest								
Galanski et al. 2006 [15]	–	–	–	–	28	55	105	205
Buls et al. 2010 [11]	–	1.5	2	3.5	–	35	55	130
Veit et al. 2010 [14]	2	3.5	5	8	30	65	115	230
Ruiz-Cruces 2015 [18]	–	–	–	–	46	82	125	200
Roch and Aubert 2012 [13]	–	3	4	5	–	30	65	140
Watson and Coakley 2010[a] [8]	3	3	5	–	55	82.6	152	–
Kritsaneepaiboon et al. 2012 [10]	4.5	5.7	10	15.6	80	140	305	472
Vassileva et al. 2015[d] [19]	5.2	6	6.8	7.3	130	140	170	300
This work	1.5	2.9	2.4	4.5	42	91	58	137

[a] reporting mean values instead of 3rd quartile
[b] different age grouping (<1, 1–7, 8–12, and 13–19 years)
[c] reporting $CTDI_W$ measured in standard phantom
[d] using 32 cm standard phantom as reference

Table 3 Technical exposure parameters ranges recorded for all scans of interest

Exposure factor (min-max)	Patient age groups			
	0–1 y	1–5 y	5–10 y	10–15 y
Head				
Voltage (kV)	(80–130)	(110–130)	(80–130)	(110–130)
Effective mAs	(130–210)	(66–350)	(150–215)	(213–350)
Scan length (mm)	(152.3–282.1)	(114.3–264.3)	(191.3–272.8)	(213.8–280.7)
DLP (mGy.cm)	(242.41–1021.87)	(216.35–1841.62)	(318.71–1054.15)	(788.15–2367.92)
CTDI (mGy)	(12.54–50.14)	(11.52–86.99)	(13.95–48.07)	(36.86–86.99)
FOV (mm)	(130–201)	(148–221)	(186–229)	(200–242)
Increment (mm)	(0.5–1.0)	(0.5–0.7)	(0.6–0.7)	(0.5–0.7)
Rotation time (s)	(1.0–1.0)	(1.0–1.5)	(1.0–1.0)	(1.0–1.5)
Ref mA	(10.99–24.40)	(9.19–35.21)	(12.57–19.09)	(8.85–17.37)
Abdomen-pelvis				
Voltage (kV)	(80–110)	(80–110)	(80–130)	(80–110)
Effective mAs	(15–60)	(35–69)	(43–235)	(47–119)
Scan length (mm)	(90.2–469.2)	(233.7–407.7)	(264.3–610.5)	(245.8–729.2)
DLP (mGy.cm)	(10.28–78.23)	(30.05–202.87)	(50.61–1544.26)	(55.44–353.01)
CTDI (mGy)	(0.97–2.53)	(1.02–5.09)	(1.21–58.43)	(2.01–6.43)
FOV (mm)	(133–214)	(182–246)	(168–410)	(180–489)
Increment (mm)	(0.5–5.0)	(0.5–5.0)	(0.5–5.0)	(0.5–5.0)
Rotation time (s)	(0.6–0.6)	(0.6–0.6)	(0.6–1.0)	(0.6–0.6)
Ref mA	(37.48–188.12)	(41.76–72.80)	(8.27–63.33)	(9.98–69.50)
Chest				
Voltage (kV)	(80–110)	(80–110)	(110–110)	(80–130)
Effective mAs	(11–65)	(19–84)	(22–32)	(40–59)
Scan length (mm)	(168.7–298.6)	(229.7–304.2)	(243.7–305.7)	(307.5–348.7)
DLP (mGy.cm)	(12.74–49.07)	(22.46–72.22)	(46.35–58.54)	(57.70–135.87)
CTDI (mGy)	(0.70–1.9)	(0.97–2.37)	(1.68–2.4)	(1.65–4.52)
FOV (mm)	(131–217)	(186–223)	(209–219)	(243–310)
Increment (mm)	(1.3–1.4)	(1.4–1.5)	(1.4–1.4)	(1.3–1.4)
Rotation time (s)	(0.6–0.6)	(0.6–0.6)	(0.6–0.6)	(0.6–0.6)
Ref mA	(27.95–101.06)	(35.16–70.11)	(26.39–29.13)	(18.92–48.84)

16 cm standard phantom. A total of 343 paediatric patients are separated in four age groups of 0–1, 1–5, 5–10, and 10–15 years, following IAEA studies [24, 25] and PiDRL 2016 final complete draft [26].

Along with the dose metric, technical scan parameters (nominal tube voltage, effective tube current, scan length, FOV, pitch/increment, rotation time, reference current, mean $CTDI_{vol}$ and mean DLP) were also recorded for assessment. All technical exposure parameters were averaged to represent each trend, while the 3rd quartile (75th percentile) for the dose metrics were calculated and compared with results from various studies. The 3rd quartile values were chosen as a comparative measure to simulate practice against available DRLs. A subsequent analysis is performed on each technical

exposure parameters when values are found to exceed respective results from other studies as this would indicate the need for practice change regarding exposure setting.

3 Results and Discussion

Number of patients for respective age groups for each study are enlisted in Table 1. Very low number of patient is due the device being newly-installed (February 2017) and retrospective studies were not possible.

$CTDI_{vol}$ and DLP values are presented in Table 2 for head, abdomen-pelvis, and chest, respectively. For each anatomy, a set of other works are presented as a comparison with several notes. The work of Watson and Coakley (2010) is presented as mean values [8], while Fukushima et al. (2012) applied different age grouping in the head data [12]. On abdomen-pelvis and chest examination data, the international study of Vassileva et al. (2015) uses 32 cm CTDI phantom as reference [19]. There were no age group separation on the work of McCollough et al. (2011) [27].

From Table 2, it can be deducted that our abdomen-pelvis and body scans delivered lower typical exposure index compared with other published works (Galanski et al. 2006; Buls et al. 2010; Veit et al. 2010; Ruiz-Cruces 2015; Roch and Aubert 2012; McCollough et al. 2011; Watson and Coakley 2010; Kritsaneepaiboon et al. 2012; and Vassileva et al. 2015). This can further imply that for optimization, the next appropriate step will be image quality assessment by radiologists to gather information not on reducing dose trend but on possible image quality improvement.

On the other hand, head scan tends to deliver slightly higher $CTDI_{vol}$ on younger patients (0–1 years) and pre-adolescent patients (10–15 years). From Table 3, the use of high tube voltage (130 kVp) was prevalent, and it was observed to have been used by 67% of all head scans for the age group. It is more interesting to observe that 130 kVp was being used on abdomen-pelvis and chest scan of the same patient age group, despite the probability of larger object. This has led to slightly higher $CTDI_{vol}$ and DLP. Head scans for pre-adolescent patients yields on typically higher $CTDI_{vol}$ and DLP (45% and 96% higher than the highest comparable study, respectively).

Additionally, it has also been identified by observing Table 3 that higher effective mAs and kVp was applied. Unlike other projection in which pre-adolescent patients are scanned with effective mAs only slightly higher than adjacent, younger age groups (15% on abdomen-pelvis), head scans apply 66% higher effective mAs. Subsequent investigation has revealed that patients under this age group underwent head scans using adult protocols—indicating the need of a simple change in practice.

4   Conclusion

Based on the recorded $CTDI_{vol}$ and DLP, head scan procedure for patients aging 10–15 years old gave typically higher dose than compared works, while other procedures delivered lower dose. This preliminary work, thus, leads to a conclusion that a change in practice can first be proposed on head scans for pre-adolescent patients (10–15 years of age). Other than the specific situation, the other conditions need investigation on the image quality. More thorough studies are required on the clinical and technical image quality produced by the low dose to further indicate the possibility on improving the image quality as a mean of optimization.

Acknowledgements This research is supported by the International Atomic Energy Agency (IAEA) through Coordinated Research Project E2.40.20 entitled "Evaluation and Optimization of Paediatric Imaging" (contract number 19108).

Conflicts of Interest The authors declare that they have no conflict of interest.

References

1. Hwang J-Y, Do K-H, Yang DH, et al (2015) A Survey of Pediatric CT Protocols and Radiation Doses in South Korean Hospitals to Optimize the Radiation Dose for Pediatric CT Scanning. Medicine (Baltimore) 94:e2146. https://doi.org/10.1097/md.0000000000002146
2. Committee to Assess Health Risks from Exposure to Low Levels of Ionizing Radiation (2006) Health Effects of Exposure to Low Levels of Ionizing Radiation (BEIR VII Phase 2). National Academy Press, Washington DC
3. International Atomic Energy Agency (2013) IAEA Human Health Series No. 24: Dosimetry in Diagnostic Radiology for Paediatric Patients. International Atomic Energy Agency, Vienna
4. Lubis LE, Bayuadi I, Pawiro SA, et al (2015) Optimization of dose and image quality of paediatric cardiac catheterization procedure. Phys Medica 31:659–68. https://doi.org/10.1016/j.ejmp.2015.05.011
5. Brenner DJ, Hall EJ (2007) Computed Tomography—An Increasing Source of Radiation Exposure. N Engl J Med 357:2277–2284. https://doi.org/10.1056/nejmra072149
6. Strauss KJ, Goske MJ, Kaste SC, et al (2010) Image Gently: Ten Steps You Can Take to Optimize Image Quality and Lower CT Dose for Pediatric Patients. Am J Roentgenol 194:868–873. https://doi.org/10.2214/ajr.09.4091
7. Brady Z, Ramanauskas F, Cain TM, Johnston PN (2012) Assessment of paediatric CT dose indicators for the purpose of optimisation. Br J Radiol 85:1488–98. https://doi.org/10.1259/bjr/28015185
8. Watson DJ, Coakley KS (2010) Paediatric CT reference doses based on weight and CT dosimetry phantom size: local experience using a 64-slice CT scanner. Pediatr Radiol 40:693–703. https://doi.org/10.1007/s00247-009-1469-1
9. Mokhtar A, Elawdy M, El-Hamid MA, et al (2017) Radiation dose associated with common computed tomography examination. Egypt J Radiol Nucl Med 48:701–705. https://doi.org/10.1016/j.ejrnm.2017.03.005
10. Kritsaneepaiboon S, Trinavarat P, Visrutaratna P (2012) Survey of pediatric MDCT radiation dose from university hospitals in Thailand: a preliminary for national dose survey. Acta radiol 53:820–826. https://doi.org/10.1258/ar.2012.110641
11. Buls N, Bosmans H, Mommaert C, et al (2010) CT paediatric doses in Belgium: a multi-centre study
12. Fukushima Y, Tsushima Y, Takei H, et al (2012) Diagnostic reference level of computed tomography (CT) in japan. Radiat Prot Dosimetry 151:51–57. https://doi.org/10.1093/rpd/ncr441
13. Roch P, Aubert B (2013) French diagnostic reference levels in diagnostic radiology, computed tomography and nuclear medicine: 2004–2008 review. Radiat Prot Dosimetry 154:52–75. https://doi.org/10.1093/rpd/ncs152
14. Veit R, Guggenberger R, Noßke D, Brix G (2010) Diagnostische Referenzwerte für Röntgenuntersuchungen. Radiologe 50:907–912. https://doi.org/10.1007/s00117-010-2066-x
15. Galanski M, Nagel HD, Stamm G (2006) Pädiatrische CT-Expositionspraxis in der Bundesrepublik Deutschland Dienstanschriften der Autoren

16. Järvinen H, Seuri R, Kortesniemi M, et al (2015) Indication-based national diagnostic reference levels for paediatric CT: a new approach with proposed values. Radiat Prot Dosimetry 165:86–90. https://doi.org/10.1093/rpd/ncv044

17. Shrimpton PC, Hillier MC, Meeson S, Golding SJ (2014) Doses from Computed Tomography (CT) Examinations in the UK – 2011 Review

18. Ruiz-Cruces R, Cañete S, Perez Martínez M (2015) Estimación de las dosis a las poblaciones en España como consecuencia del radiodiagnóstico médico

19. Vassileva J, Rehani M, Kostova-Lefterova D, et al (2015) A study to establish international diagnostic reference levels for paediatric computed tomography. Radiat Prot Dosimetry 165:70–80. https://doi.org/10.1093/rpd/ncv116

20. International Atomic Energy Agency (2007) IAEA Technical Report Series No. 457: Dosimetry in Diagnostic Radiology: an International Code of Practice. International Atomic Energy Agency, Vienna

21. The International Commission on Radiation Unit and Measurement (2005) ICRU Report 74: Patient Dosimetry for X Rays Used in Medical Imaging. Oxford University Press

22. International Atomic Energy Agency (2014) IAEA Diagnostic Radiology Physics: A Handbook for Teachers and Students. IAEA, Vienna

23. Bauhs JA, Vrieze TJ, Primak AN, et al (2008) CT Dosimetry: Comparison of Measurement Techniques and Devices. RadioGraphics 28:245–253. https://doi.org/10.1148/rg.281075024

24. Vassileva J, Rehani MM, Applegate K, et al (2013) IAEA survey of paediatric computed tomography practice in 40 countries in Asia, Europe, Latin America and Africa: procedures and protocols. Eur Radiol 23:623–631. https://doi.org/10.1007/s00330-012-2639-3

25. Vassileva J, Rehani MM, Al-Dhuhli H, et al (2012) IAEA Survey of Pediatric CT Practice in 40 Countries in Asia, Europe, Latin America, and Africa: Part 1, Frequency and Appropriateness. Am J Roentgenol 198:1021–1031. https://doi.org/10.2214/ajr.11.7273

26. (2016) European Guidelines on DRLs for Paediatric Imaging (Final complete draft)

27. McCollough C, Branham T, Herlihy V, et al (2011) Diagnostic Reference Levels From the ACR CT Accreditation Program. J Am Coll Radiol 8:795–803. https://doi.org/10.1016/j.jacr.2011.03.014

Advantages in the Application of Conductive Shielding for AC Magnetic Field in MRI Exam Rooms

Rafael Navet de Souza and Sergio Santos Muhlen

Abstract

Due to the increasing demand for installation of Magnetic Resonance Imaging (MRI) equipment in clinics and hospitals, and the difficulty to choose the location for installation of these equipment because of their high sensitivity to 60 Hz magnetic field sources, this study was carried out to present a practical solution for shielding 60 Hz magnetic field in the hospital environment. MRI is a technique that produces 3-D (volumetric) tomographic images of high resolution without using ionizing radiations. The quality of images is greatly influenced by magnetic fields of the environment, especially of 50/60 Hz, which results in the need of shielding the space where the equipment are installed. This study proposes the use of aluminum in the construction of the MRI room shield for the many advantages presented by this material when compared to ferromagnetic materials: it is lighter, easier to handle, bend, rivet or weld, it does not rust and dispenses sturdy supports for fixing on walls, resulting in the best cost-benefits ratio in short and long terms. We performed computational simulation and experiments with shields of rectangular geometries assembled with aluminum and ferromagnetic materials. The aluminum shield has proved to be advantageous, since it presents shielding effectiveness to 60 Hz magnetic field similar to those of Fe–Si GNO under certain conditions, and radio frequencies shielding also, with lower cost in the installation and maintenance of shielding in MRI rooms.

Keywords

Magnetic shielding • Electromagnetic interference
Magnetic resonance imaging • Hospital environment

1 Introduction

Magnetic Resonance Imaging (MRI) systems use magnetic fields and radio frequency (RF) waves. The technique is based on the phenomenon of Nuclear Magnetic Resonance (NMR), discovered in 1938 and used since then in the analysis of chemical compounds. NMR is a physical phenomenon in which the nucleus of an atom of a given substance, in the presence of an external magnetic field, absorbs and emits energy in the form of RF. It is possible to determine properties of the substance by correlating the energy absorbed at each frequency of the magnetic spectrum (in the MHz range), such as spectroscopy.

MRI systems are well-accepted powerful diagnostic tools, but are also amongst technologies that are very susceptible to Electromagnetic Interference (EMI) extremely low frequency (AC and DC magnetic field) and RF. The planning for the installation of an MRI system often represents a major technical and economic challenge, open study opportunities for innovative and creative solutions, provided they prove to be efficient. The installation needed of each system are changing quickly, following the fast evolution of MRI equipment, the complexity and sensibility of the tests available, as well as the increased physical restrictions and levels of EMI at proposed installation sites.

Choosing the place and preparing for a clinical installation of MRI equipment requires special considerations that are not previously encountered in a clinical environment. Most current clinical and hospital environments have not been designed for the needs and constraints that are necessary for the operation of an MRI system, requiring them to be suitably adapted with appropriate technical solutions. The factors involved in defining a place for the installation of

R. N. de Souza · S. S. Muhlen (✉)
Department of Biomedical Engineering, University of Campinas, Campinas, Brazil
e-mail: Smuhlen@g.unicamp.br

R. N. de Souza
e-mail: navet_r@yahoo.com.br

R. N. de Souza
Siemens Healthineers Brazil, Rio de Janeiro, Brazil

© Springer Nature Singapore Pte Ltd. 2019
L. Lhotska et al. (eds.), *World Congress on Medical Physics and Biomedical Engineering 2018*,
IFMBE Proceedings 68/3, https://doi.org/10.1007/978-981-10-9023-3_106

MRI equipment in a diagnostic center are much more complex than for other imaging equipment. In addition to the usual requirements for suitable foundation and structure, their effects on the magnetic field and the effect of the static magnetic field on other devices present in the site [1] should be considered.

The RF operating frequencies in commercial MRI equipment range from 12 to 298 MHz, which means that it is very important to prevent RF waves dispersed in the environment from affecting MRI equipment. Therefore, RF shielding is required in all projects, since its absence results in the impoverishment of the signal-to-noise ratio of the images, which can limit or compromise the diagnostics [2]. Normally, conductive materials are used for RF shielding because of their attenuation effectiveness to electromagnetic fields and easy handling for fixing and fold according to the room design.

In addition to RF, there are two other types of disturbances caused by external magnetic fields in MRI equipment. The first is often referred as Direct Current (DC) or Quasi-DC. This type of disturbance is due to the proximity between MRI equipment and devices that use DC at a great intensity for their operation, e.g. subway, electric trains, trams and similar equipment that operate with high intensity DC currents. Common sources of Quasi-DC interference include automobiles, trucks and other ferromagnetic objects moving close to the location proposed for the MRI equipment. The solution to this type of interference is the containment of the field lines of the magnet in order to reduce the interaction of these field lines with the disturbing source. Only ferromagnetic materials are effective to contain the static magnetic fields.

The second type of interference is created by electrical devices operating with Alternating Currents (AC), such as transformers, transmission lines, power cables, equipment switching and other rapid variations in the intensity or orientation of the electric current. High intensity 50/60 Hz magnetic fields can affect image quality in MRI equipment [3].

Ferromagnetic materials have high magnetic permeability and produce effective shielding to contain the static, DC and AC magnetic field. However, when adopting this type of material for shielding AC magnetic field, the magnetic saturation must be taking into account, since it substantially decreases the shielding capacity [4]. Considering that, in our comparative study the shielding will be applied to MRI equipment working with high intensity of magnetic field (from 0.2 to 7.0 T), which increases the probability of saturation of metal sheets.

The objective of this study is to evaluate the advantages of using aluminum in 60 Hz magnetic field shields in MRI exam rooms, motivated by the high demand for installation of MRI equipment in Brazil. Although this shielding

technology is not new for other applications, such as substations, underground power cables and transformers, it still has apace for improvement and development for cost reduction. We performed experiments and computational simulations using ferromagnetic and conductive materials to compare their Shielding Effectiveness (SE).

The SE is calculated by ratio of magnetic field intensity inside the shield H_i, to that outside the shield H_o, H_o/H_i [5].

2 Materials and Methods

The experiments using ferromagnetic and conductive materials for shielding AC magnetic fields and the computational simulation considered "aluminum alloy 1200" as conductive material and "Fe–Si alloy E185 NGO" (non-grain-oriented) as ferromagnetic material. These alloys are commonly used on shielding of MRI exam rooms because of their affordability and availability in Brazil. Their main properties are shown in Table 1.

2.1 Experiments

The experiments aimed to compare the SE of the aluminum and Fe–Si NGO shields in a controlled environment, and to compare them with the computational model in order to validate our model. The experiments consisted in performing 60 Hz magnetic field measurements by varying the distance between the measuring point and the generating source, the thickness and the material of the shielding.

The shield is a metallic cubic structure (1 m side) in which the sheets of shielding materials (aluminum or Fe–Si NGO) are fastened, so as to make it easier to change the number of sheets (total thickness) and types of materials. The shield is shown in Fig. 1, with aluminum sheets.

Measurements of shielding efficiency were performed using a triaxial magnetic field sensor (magnetometer) (STL, model DM-050) mounted on a tripod inside the cubic shield and connected to a computer via a coaxial cable. The values obtained by the magnetometer of density of magnetic flux were expressed in nanotesla (nT).

The 60 Hz magnetic field was generated by a coil of a single rectangular turn (19 × 12 m loop), implemented with

Table 1 Material properties

Material	Electrical conductivity [10^6 S/m]	Relative permeability [H/m] @ 1.5 T
Aluminum	34.5	1
Fe–Si NGO	2.6	1.530

Fig. 1 Cubic structure with aluminum sheets for shielding used for experiments, showing the coaxial cable of the magnetometer

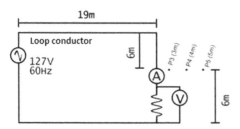

Fig. 2 Circuit used for computational simulation and experimental setup. A rectangular loop conductor (19 × 12 m) was used to generate the magnetic field, and P3–P5 are the measuring points where the sensor is placed, inside the shield

a flexible cable (φ 4 mm²) suspended 60 cm from the ground. This cable supplied a resistive load (2 heaters) to generate a current of 18.4 A_{rms}. Figure 2 shows the electric circuit used on the computational and experimental setup. The points P3–P5 are the measuring points representing the position of the center of the shield and vary from 3 to 5 m with measurements steps of 1 m.

These distances were defined because in practical MRI rooms they are the minimum distances between a source of magnetic field (e.g. transformer or power cables), and the isocenter position of the MRI magnet.

2.2 Computational Simulation

The computational simulation was performed with the software Comsol Multiphysics®, for a configuration with a conductor with a known current, and a cubic metallic box, as performed on the experiments.

Table 2 Parameters used for simulation

Description	Value
Frequency f	60 Hz
Temperature T	20 °C
Current I	18.4 A_{rms}
Thickness t	of 0.5–5 mm
Distance d	of 3–5 m
Electrical conductivity σ	see Table 1
Relative permeability μ_r	see Table 1

The Ampère law was used on the basis of calculation. Due the low density of magnetic flux $\vec{B}$ [nT], we are not considering the non-linearity of Fe–Si NGO assuming that the relative permeability will be constant. The parameters used for simulation are shown in Table 2.

3 Results

The results presented in the Fig. 3 were obtained in the experiments and computational simulation. The maximum error between computational simulation and experiments was smaller than 10% when compared the values obtained to density of magnetic flux in nT.

4 Discussion

According to the literature [6], the results in Fig. 3 show that the aluminum shielding is more effective for AC magnetic field than the shielding of Fe–Si NGO.

As foreseen, the non-linearity of Fe–Si NGO was not affected by the low density of magnetic flux $\vec{B}$ used for experiments and simulation, the maximum error obtained between simulation and experiments was smaller than 10% which is acceptable, and are due to some approximations

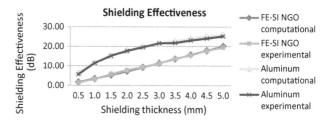

Fig. 3 Graphic obtained in the experiments and computational simulation for the SE comparing the results for aluminum and Fe–Si NGO

used in the simulator and the possibility of measurement errors.

Conductive materials (copper, aluminum) have higher SE in RF shielding and are already used due to the easy handling for fixing and fold according to the design of the room. They are lighter, do not suffer from saturation effect due the strong static magnetic field of the magnets (0.2 T a 7.0 T) or even at high dynamic field (AC 50/60 Hz), unlike Fe–Si NGO who has their properties directly affected from the high magnetic field.

5 Conclusion

As aluminum is already used for RF shielding (together with Fe–Si NGO for AC magnetic fields shielding), it is easier to just increase the thickness of aluminum and eliminate the need of the ferromagnetic shield, reducing the current constrains of shielding construction and getting a better attenuation factor.

With the use of conductive materials for AC and RF shielding it is possible to reduce time of shielding installation and consequently reduce costs of structure construction to support the ferromagnetic materials, representing better results to short and long terms.

Due to Fe–Si GNO be used to contain the static magnetic field and the aluminum to be a paramagnetic material and not effective to this type of shielding and to be little diffused

for AC magnetic shielding for MRI exam rooms, the first option was the use of Fe–Si GNO. In some cases the use of Fe–Si GNO will be better than aluminum.

The experimental results and the simulation of Shielding Effectiveness, together with the constructive advantages discussed above, allow us to conclude that the aluminum shields present better cost/benefits ratio than ferromagnetic shields for MRI exam rooms.

References

1. American Association of Physicists in Medicine: Site planning for magnetic resonance imaging systems. In: AAPM report, n. 20 (1986).
2. MRI suite shielding requirements, http://www.controlledpwr.com/whitepapers/mri_room_shielding.pdf, last accessed 2013/10/12.
3. Kellogg, J.: Electromagnetic interference (EMI) and structural vibration effects on MRI site construction and installation requirements, http://www.ets-lindgren.com/pdf/ETSL_0508_kellogg.pdf, last accessed 2013/10/10.
4. Yichao, Z., Cheng, G., Lihua, S., Bihua, Z.: Analysis and test of EM shielding for low-frequency magnetic field. In: 2007 International Symposium on electromagnetic compatibility, pp. 345–349, Qingdao, China (2007).
5. Cooley, W.W.: Low-frequency shielding of nonuniform enclosures: In: IEEE Transaction on Electromagnetic Compatibility, v. EMC-10, n. 1, March 1968.
6. Machado, V.M.: Magnetic field mitigation shielding of underground power cables. In: IEEE Transactions on Magnetics, vol. 48, n. 2, February 2012.

Development of Optical Computed Tomography for Evaluation of the Absorbed Dose of the Dyed Gel Dosimeter

Takuya Wada, Kazuya Nakayama, Akihiro Takemura,
Hiroaki Yamamoto, Hironori Kojima, Naoki Isomura, and Kimiya Noto

Abstract

Optical computed tomography (optical CT) is a reading device of the dyed gel dosimeters. We are developing the optical CT for the evaluation of three dimensional radiation absorbed dose distribution in the dyed gel dosimeters. We made dyed gel with leuco crystal violet and the dyed gel will be contained in vials. The dyed gel dosimeters were irradiated with 10 MV X-ray beam at 100–2000 MU. The optical CT we developed was consists of a liquid crystal monitor VL-176SE (FUJITSU, Japan) as a light source and a camera uEye XS (iDS, Germany). The dyed gel dosimeter was rotated by a step of every 0.9° with the stepper motor ST-42BYH 1004-5013 (MERCURY MOTOR, China) in a water tank and be taken 400 projections per rotation. Volume data was reconstructed from the projection images with the image processing software Plastimatch. The correlation between the absorbed dose and signal values of the dyed gel dosimeters in the reconstructed image was analyzed. The developed optical CT could reconstructed the images of the dyed gel dosimeters and the signal values of the dyed gel dosimeters in the reconstructed images had linear response related to the dose up to 20 Gy.

T. Wada
Division of Health Sciences, Graduate School of Medical
Sciences, Kanazawa University, Kanazawa, Japan
e-mail: wawawa.wa.3lta@gmail.com

K. Nakayama (✉) · A. Takemura
Faculty of Health Sciences, Institute of Medical, Pharmaceutical
and Health Sciences, Kanazawa University, Kanazawa, Japan
e-mail: knaka@kenroku.kanazawa-u.ac.jp

H. Yamamoto
School of Health Sciences, College of Medical, Pharmaceutical
and Health Sciences, Kanazawa University, Kanazawa, Japan

H. Kojima · N. Isomura · K. Noto
Deparment of Radiological Technology, Kanazawa University
Hospital, Kanazawa, Japan

Keywords

Optical computed tomography (optical CT)
Dyed gel dosimeter • Lueco crystal violet

1 Introduction

Dyed gel dosimeters are expected to measure three dimensional (3D) dose distribution for quality control (QC) of a radiotherapy plan. They are water-equivalent radiation dosimeters and change their color as oxidation reactions. Fricke dosimeter, PRESAGE, micel gel dosimeter, etc. were reported as the dyed dosimeters [1]. The optical computed tomography (optical CT) was used as a reading device of the dyed gel dosimeters [2]. The Vista (Modus Medical Devices Inc., ON, Canada) is known as a practical optical CT [3]. Wolodzko et al. investigated about the feasibility of employing a new, simple and inexpensive technique for optical tomographic imaging of radiation gel dosimeters [4]. Dekker et al. have researched the suitable reconstruction algorithm for gel dosimetry by using optical CT [5].

Developing an optical CT allows us to investigate the imaging methods and image reconstruction methods, and it is cost-effective. We are also developing an optical CT to read 3D absorbed dose distribution in dyed gel dosimeters [6].

The purpose of this paper was to confirm the dose response of dyed gel dosimeters obtained with the optical CT system.

2 Materials and Methods

2.1 Overview of the Developed Optical CT

Figure 1 shows the optical CT components and layout. The optical CT consisted of a liquid crystal monitor VL-164E (FUJITSU, Tokyo, Japan) as a light source and a camera

© Springer Nature Singapore Pte Ltd. 2019
L. Lhotska et al. (eds.), *World Congress on Medical Physics and Biomedical Engineering 2018*,
IFMBE Proceedings 68/3, https://doi.org/10.1007/978-981-10-9023-3_107

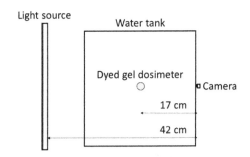

Fig. 1 Top view of the optical CT

(uEye XS, iDS, Obersulm, Germany). The camera built in a CMOS sensor of 5 million pixel, its matrix size was 2592 × 1944 pixels and maximum resolution was 0.19 mm/pixel. However, in this study we acquired 8-bit gray scale images of 640 × 480 pixels to reduce the data volume. Images of 400 projections per a rotation were obtained and stored. The data volume for one acquisition was 60 MB in total. The monitor was used as a light source because the light color can be controlled. The monitor was connected to a laptop computer to change light color. The distance between the camera and a dyed gel dosimeter was 17 cm and the distance between the camera and the light source was 42 cm. The dyed gel dosimeter was placed in water. Placing a dyed gel dosimeter in water during image acquisition with the optical CT can reduce the effect of refraction and reflection of the dyed gel dosimeter. During one rotation, 200 projection images of a dyed gel dosimeter were acquired. The rotation step was 0.9°. A single board computer (Raspberry Pi 2 Model B, Raspberry Pi Foundation, Cambridge, UK) controlled the camera and the motor and the computer can be controlled via a local area network. The camera control program using the OpenCV 3.0 library (Open Source Computer Vision Library) were compiled by using GCC 4.9.2. The optical CT preserved the captured images. The stored images were transferred to another computer and a 3D volume data was reconstructed from the projection images by Feldkamp reconstruction method using a free software, Plastimatch 1.6.4 (Plastimatch development team 2010) without filter. The matrix size of the reconstructed cross-sectional image was 200 × 200 pixels.

The dyed gel dosimeter was fixed directly to the motor shaft. A vial lid was attached to the end of the motor to make the replacement of samples easy. A stepper motor (ST-42BYH 1004-5013, MERCURY MOTOR, Shenzhen, China) was used as the motor. It was a unipolar type motor, and the static torque was 4.4 kgf cm. The step angle was 0.9° ± 5%. It took 400 images per rotation in about 10 min. The exposure time of the camera was set to 66 ms. In every step, rotation was hold in about 1 s before acquiring a projection image to suppress the influence of vial vibration. It takes 26 s for imaging time and 400 s for holding time in one

acquisition. By optimizing the hold time, it might be possible to shorten the total time.

2.2 Manufacturing of Dyed Gel Dosimeters

Dyed gel dosimeters based on lueco crystal violet (LCV) as a pigment was prepared [7, 8]. The dyed gel dosimeters was composed of five components: 4.0 wt% gelatin (TypeA, Sigma-Aldrich, Oakville, ON, Canada), 25.0 mM trichloroacetic acid (Sigma-Aldrich, Oakville, ON, Canada), 4.0 mM Triton x-100 (Sigma-Aldrich, Oakville, ON, Canada), 1.0 mM LCV (Sigma-Aldrich, Oakville, ON, Canada), and 96.0 wt% ultra-pure water. The dyed gel was poured into glass cylindrical vial (φ19 × 70 mm, NICHIDEN-RIKA GLASS CO.LTD., Hyogo, Japan) and 4.5 ml acrylic cuvettes (AZ ONE, Osaka, Japan) with 10 mm light path.

The dyed gel dosimeters were cooled in a refrigerator for a day and then put out of the refrigerator to return them to room temperature before irradiation. They were irradiated with 10 MV X-ray beam delivered by the Elekta Synergy (Elekta, Stockholm, Sweden) at the depth of 5 cm in water and the dose rate was 570 MU/min. The irradiation dose range was 100–2000 MU (0.988–19.7 Gy). The field size was 15 × 15 cm^2. The dyed gel dosimeters laid on the bottom of the water tank. Water equivalent phantom, three slices of the tough water phantoms of 5 cm thickness (457–350, Gammex, Nederland, Netherlands) was stacked under the water tank in consideration of back scattering.

2.3 Measurement of Dyed Gel

The projection images of the dyed gel dosimeters were obtained with the optical CT and reconstructed images were calculated from the projection images. The region of interest (ROI) was placed at the center of the dyed gel in the reconstructed cross sectional image and mean signal values in the ROI was measured. The measurement was repeated in 10 slices around the center of the gel dosimeter. Transmittance of all cuvettes were measured with the spectrophotometer (S-1000, SHIMADZU, Tokyo, Japan) with the light of the wavelength 600 nm. The ultrapure water was taken for the reference transmittance. Each cuvette was measured 10 times and average transmittance was calculated.

3 Results

The reconstructed cross sectional images of the dyed gel dosimeters obtained with the original optical CT system were in Fig. 2. Figure 2a was a cross sectional image of the

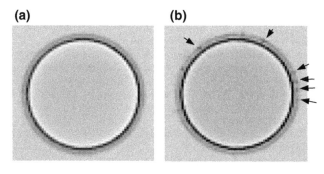

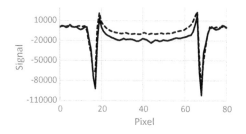

Fig. 2 The reconstructed cross sectional images of the un-irradiated dyed gel dosimeter (**a**) and the irradiated dyed gel dosimeter (**b**)

Fig. 3 The signal profiles at the center of the reconstructed cross sectional image: the dashed line represents for the un-irradiated dyed gel dosimeter; the solid line represents for the irradiated dyed gel dosimeter

un-irradiated dyed gel dosimeter, and Fig. 2b was a cross sectional image of the dyed gel dosimeter which was delivered the absorbed dose of 19.7 Gy. The dyed gel in Fig. 2b was darker than it in Fig. 2a. Black dots were observed in Fig. 2b because air bubbles have stuck to the vial surface during imaging.

Comparison of the signal profiles of the dyed gel was shown in Fig. 3. In the positional range of 20–65 pixel, the signal values of the irradiated dyed gel dosimeter was lower than that of un-irradiated dyed gel dosimeter. The signal drop-off was observed in the signal profiles. That was because the light was refracted and/or reflected at the boundaries of the glass vial, water and gel.

The dose response of dyed gel dosimeter obtained with the optical CT system was shown in Fig. 4. Linear dose response was observed in the dose range of 0–20 Gy. However, the signal values of 1 Gy was higher than 0 Gy. There was no difference between the signal values of 8 and 10 Gy.

The measured results of the cuvettes were depicted as dashed line and triangle markers in Fig. 4. The dose-transmittance response showed the same trend of the obtained results by optical CT.

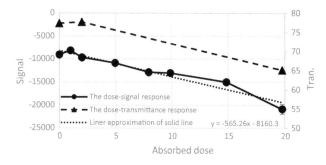

Fig. 4 The transmittance and dose-signal response of dyed gel dosimeter

accuracy of dose measured with the optical CT. Using water left for a night will reduce bubbles on surface of containers. Third, dyed gel original signal value were decreased by reflection and refraction at the regions close to the glass vial. It is necessary to change the glass vials to other containers that are transparent and its refractive index is close to that of water. Also, the dose resolution of 1.4 Gy was calculated from twice the standard deviation of the gel signal values at 0 Gy (2SD = 812) and the equation of liner approximation.

The reasons why that the signal values of the dyed gel dosimeter irradiated by 1 Gy was higher than that un-irradiated and that there was no signal difference between 8 and 10 Gy were unknown at this time. One of reasons might be the unstable production of dyed gel. However, the dose-transmittance response of the dyed gel dosimeters was similar to the dose-intensity response, so the optical CT was able to correctly reconstruct the dyed gel dosimeters.

4 Discussions

There was three problems remained in the imaging with the optical CT. First, the flicker in which brightness changed due to blinking backlight did not appear, however the moire appeared. It was considered the interference between the pattern of the liquid crystal monitor and the image sensor matrix of the camera. The moire appear in the image as artifacts, it might produce inaccuracies in the dosimetry with dyed gel dosimeter. A solution is to adjust the position of the monitor and camera or to install a scatterer at the front of the monitor. Second, bubbles were attached to the container of the dyed gel dosimeters. Bubbles might affect in the

5 Conclusions

We developed the optical CT system and scanned dyed gel dosimeters with the optical CT system. There was a linear correlation between the signal values of the dyed gel dosimeter in the reconstructed cross sectional image and the dose. The result of the dose response was similar to the result in the other paper [4] and the result of the transmittance with the spectrophotometer, so that the original optical CT system was able to measure the dose in dyed gel dosimeters correctly. Further investigations about unevenly irradiated dyed gel and reproducibility of both preparation for dyed gel and reading it should be conducted to show the original optical CT system is more reliability.

References

1. Shunji Usui. Radiochromic Hydrogel Dosimeters. Jpn. J. Med.Phys. vol. 37 (2):95–98(2017).
2. Munehiro Yonehara, Ryo Wakana, Kazuhisa Sakakibara, Jun'ichi Kotoku, Takanori Kobayashi, Hiroaki Gotoh. Jpm. J. Med. Phys. vol. 37 No.2:117–121(2017).
3. Tim Olding, Oliver Holmes, L John Schreiner. Cone beam optical computed tomography for gel dosimetry I: scanner characterization. Phys. Med. Biol. 55 2819–2840 (2010).
4. John G Wolodzko, Craig Marsden, Alan Appleby. CCD imaging for optical tomography of gel radiation dosimeters. Med. Phys. 26 2508–2513 (1999).
5. Kuris H Dekker, Jerry J Battista, Kevin J Jordan. Technical Note: Evaluation of an iterative reconstruction algorithm for optical CT radiation dosimetry.
6. Kazuya Nakayama, Takenori Kobayashi, Takuya Wada, Akihiro Takemura. Optical CT for polymer gel dosimetry: Three-dimensional measurement of absorbed dose using a polymer gel dosimeter. Journal of Wellness and Health Care vol.41 (1) 137–142 (2017).
7. Kevin J Jordan, Nikita Avvakumov. Radiochromic leuco dye micelle hydrogels: I. Initial investigation. Phys. Med. Bio. 54 6773–6789 (2009).
8. A T Nasr, K Alexander, L J Schrelner, K B McAuley. Leuco-crystal-violet micelle gel dosimeters: I. Influence of recipe components and potential sensitizers. Phys. Med. Biol. 60 4665–4683 (2015).

Assessment of Neonatal Entrance Surface Doses in Chest Radiographic Examinations at East Avenue Medical Center

Franklyn Naldo, Bayani San Juan, and Melanie Marquez

Abstract

A significant development in the initial diagnosis and evaluation of illnesses of neonates is the use of a chest x-ray. The technology is essentially useful to hospitalized and prematurely-born neonates suffering from respiratory and cardiovascular complications. Neonates are known to be more radiosensitive than adults because of the high mitotic rate of neonatal cells. They are also at a higher risk of inducing stochastic effects due to their long life expectancy. Despite such risk, physicians still require neonates to undergo radiographic examinations to monitor treatment progress while in the Neonatal Intensive Care Unit (NICU). Therefore, the radiation doses they receive during a radiographic examination should be kept at a minimum without compromising the diagnostic image quality. In this study, the entrance surface dose (ESD) for neonates undergoing diagnostic chest radiography in the NICU at East Avenue Medical Center was measured. The ESD for chest anteroposterior (AP) and lateral (LAT) projections ranges from 0.022 to 0.080 mGy and 0.023–0.080 mGy, respectively. Reference levels set by international organizations were used for benchmark comparisons with the results of the present study.

Keywords

Entrance surface dose • Neonate • Radiosensitive

F. Naldo (✉) · B. San Juan
The Graduate School, University of Santo Tomas, Manila, Philippines
e-mail: naldo.franklyn@gmail.com

B. San Juan
e-mail: sanjuanbayani@yahoo.com

M. Marquez
Department of Radiological Sciences, East Avenue Medical Center, Quezon City, Philippines
e-mail: marquez_melanie@yahoo.com

1 Introduction

The United Nations Scientific Committee on the Effects of Atomic Radiation (UNSCEAR) released a report in 2008 estimating that about 3.6 billion diagnostic examinations are being performed annually in the world [1]. The ICRP, on the other hand, reported in 2012 that diagnostic x-rays comprise more than 90% of the total medical exposures. With these figures, there is a growing concern over the exposure of patients to ionizing radiation especially on its use in medical diagnosis [2–5].

The World Health Organization (WHO) refers to a neonate as a newborn infant less than 28 days since its birth, thus, requiring appropriate care during this critical period [6]. Prematurely-born neonates often suffer from medical complications due to diseases in the respiratory and cardiovascular systems [4, 7–9]. These neonates require special attention and must be under long periods of hospitalization for medical attendants to monitor the neonates' treatment progress.

During treatment, a diagnostic x-ray is a fundamental tool in assessing most pediatric pathologies [10]. Multiple chest radiographs are required in the Neonatal Intensive Care Unit (NICU) within a short period of time for the neonate's diagnosis, follow-up, treatment and treatment progress [4, 7].

Neonates are more radiosensitive than adults due to the former's many actively dividing cells [9]. During radiographic examinations, more tissues of these neonates may be exposed to the primary beam of radiation than those of adult tissues because of the neonates' smaller sizes. They also have relatively longer life expectancy than adults after radiation exposure, hence, there is a greater chance for the manifestation of stochastic effects, such as cancer [3–5, 7]. In addition, neonates are said to be vulnerable to the cumulative effects of radiation exposure over their lifetime [11].

A more accurate understanding of radiation doses, as well as the factors that affect them, is essential in assessing patient doses. According to the IAEA, dose monitoring is one way

© Springer Nature Singapore Pte Ltd. 2019
L. Lhotska et al. (eds.), *World Congress on Medical Physics and Biomedical Engineering 2018*,
IFMBE Proceedings 68/3, https://doi.org/10.1007/978-981-10-9023-3_108

of assessing patient doses to ensure optimal protection of patients [12]. It is then important to make sure that radiation doses received by neonates during a radiographic examination are kept at a minimum while maintaining adequate image quality [7, 8, 13]. At present, there is not enough information available in the Philippines regarding the doses received by neonates in the NICU i.e., whether or not radiation doses administered to them are within international standards.

2 Materials and Methods

2.1 Patient Samples

A total of 107 neonates of both genders admitted and treated in the NICU at East Avenue Medical Center (EAMC) were examined. The ethics and review board approved the study and a waiver of informed consent was given. The authors declare that they have no conflict of interest. The procedures were performed in neonates with chest anteroposterior (AP) and lateral (LAT) projections. For each neonate, the following parameters were recorded: date of birth, weight, sex, patient thickness and date and time of radiographic examination. The exposure factors were selected manually by the radiographers. Radiographic data for each exposure such as projection, tube voltage, mAs settings, focus to detector distance were also recorded. All the images included in the study from the two projections passed the quality standard set by the department.

2.2 X-Ray Machine

The x-ray machine used was a Shimadzu Mobile ArtEco MUX10 model with 2.5 mm Al total filtration and 0.7 mm focal spot size. This unit was exclusively used in the NICU. Chest x-rays were acquired using computed radiography (CR) digital radiographic receptors. A quality control test was done prior to data gathering to ensure the performance and reproducibility of the exposure parameters of the machine. The RaySafe Platinum Xi R/F detector and base unit were used in performing the quality control test of the x-ray machine.

2.3 Dose Calculation

The doses were evaluated using the indirect assessment of entrance surface doses (ESD) using the equation [12]:

$$K_i = Y(d)P_{lt}\left(\frac{d}{d_{FTD} - t_P}\right)^2 \quad (1)$$

where K_i is the incident air kerma obtained by multiplying the x-ray tube output $Y(d)$ measured at a distance d from the tube focus, tube loading P_{lt} and the correction for the effect of varying distances.

$$K_e = K_i B \quad (2)$$

where K_e, is the entrance surface air kerma. B is the appropriate backscatter factor for water based on the selected field size and tube voltage given in Appendix VIII of Technical Reports Series 457 [12].

3 Results and Discussion

The results were based on 107 neonates with diagnostic procedures admitted in the NICU at EAMC. Summary statistics were presented in tables and graphs as mean ± standard deviation or median (interquartile range) for quantitative characteristics. The summary of the patient demographic data and exposure parameters used for neonates with chest AP and LAT projections was presented in Table 1.

The minimum age of the neonate was 0 which means having an x-ray examination few hours after birth. The mean values of the parameters were the same for both projections except for the tube voltage. The slightly higher tube voltage value could be due to the thicker body part being radiographed in the lateral projection. The mean and median values for both projections for the parameters weight, patient thickness, tube voltage and tube loading were almost equal which means that these values were normally distributed.

The exposure factors used for both projections were lower compared with Commission on European Communities (CEC) with values ranging from 60 to 65 kVp [9]. The tube voltage determines the penetrating ability of the x-ray beam to pass through deeper tissues [14]. Large patient thickness requires high tube voltage setting to ensure sufficient penetration of the x-ray beam to the portion of the body being radiographed. There was a trend in the use of high kVp technique compensated by the use of low mAs value in producing a good quality radiograph while reducing the patient dose. However, an increase in the tube voltage decreases the contrast on the image. So, the tube voltage should provide a balance between contrast and patient dose.

There was a wide variation in the exposure factors set by the radiographers within an age group due to lack of standardization in the radiographic procedures. The deviation in

Table 1 Patient demographic data and exposure parameters

Projection	Age (days) n = 107	Weight (kg) n = 107	Thickness (cm) n = 107	Tube voltage (kVp) n = 107	Tube loading (mAs) n = 107
Chest AP					
Mean ± SE	5.1 ± 8.0	2.2 ± 0.8	10.8 ± 3.2	50.9 ± 2.1	1.9 ± 0.19
Median (IQR)	1 (7)	2.2 (1.4)	10.2 (5.1)	50 (0)	2.0 (0.4)
Min–Max	0–27	0.97–4	2.5–17.8	50–60	1.2–2.2
Chest LAT					
Mean ± SD	5.1 ± 8.0	2.2 ± 0.8	10.9 ± 3.1	53.2 ± 2.8	1.9 ± 0.2
Median (IQR)	1 (7)	2.2 (1.4)	10.2 (4.7)	55 (5)	2.0 (0.2)
Min–Max	0–27	0.97–4	2.5–17.8	50–60	1.6–2.2

the exposure factors used, even with the same patient thickness, indicates that the examinations were taken by different radiographers. No exposure table was followed for uniformity.

The patients considered in this study were neonates who were considerably smaller compared with an average-sized 5-year old and typical adult patients. Thus, the use of lower tube voltage and high mAs setting technique was employed by the radiographers in the neonates during radiographic examinations. One reason that could be attributed to this was the fact that neonates in the Philippines were comparatively smaller than the neonates in western countries.

Scatterplot was used to graph the occurrence of one variable with respect to another. This scatterplot shows the distribution of weight and age of the neonates (see Fig. 1). Most of the neonates were newborn with varying weights.

The calculated ESD values for chest AP ranged from 0.022 to 0.080 mGy with a mean of 0.043 and median of 0.042 mGy. The mean and median values were almost equal which means that the ESD values were normally distributed.

The obtained ESD values for LAT ranged from 0.023 to 0.080 mGy with a mean of 0.051 and median of 0.052 mGy. The mean and median values for this projection were almost equal which signifies that the ESD values are normally distributed as well.

The present study showed a wide variation of the ESD values. This was evident from the range of ESD values obtained in both AP and LAT projections. These variations indicate that patient doses could be reduced by paying more attention to the exposure factors without the loss of image quality, highlighting the importance of quality control programs. (Figure 2)

The mean ESD value obtained in this study was lower than the diagnostic reference levels (DRL) set by both the CEC and the National Radiological Protection Board (NRPB). Although the obtained mean ESD value satisfies the DRLs set by CEC and NRPB, the large variation of ESD values for each neonate still indicates that there was a lack of standardization in the selection of exposure factors by the radiographers in the NICU at EAMC. The obtained ESD

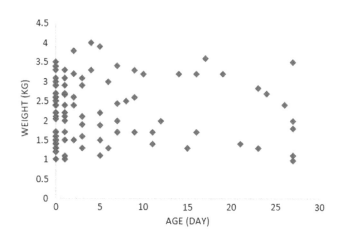

Fig. 1 Scatterplot showing the age and the corresponding weight of the 107 neonates

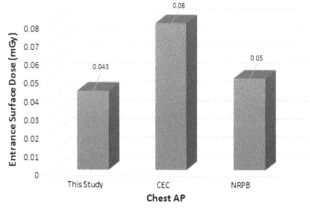

Fig. 2 Comparison of the obtained mean ESD value with CEC and NRPB

value for the lateral projection was not compared because of the lack of available data for this projection.

4 Conclusion

The relatively low ESD values measured in the study can be due to the following reasons: (1) the mobile x-ray machine used was brand new; hence, the outputs for various settings were optimized; and (2) the patient size in the Philippines is relatively smaller than others. However, the results also revealed that the doses imparted to neonates showed wide variations. This could be attributed to the differences in: (1) the exposure factors set by the different radiographers even within the same patient thickness; and (2) the distance between the x-ray tube and the detector.

The survey carried out in this study may serve as a baseline data of neonatal doses in the Neonatal Intensive Care Unit at East Avenue Medical Center and hopefully, will contribute to the establishment of the local diagnostic reference level of the hospital and the Philippine Diagnostic Reference Levels (DRLs) in the future.

References

1. UNSCEAR, 2008. Medical exposure to ionizing radiation. United Nations, New York, NY (Sources and effects of ionizing radiation. United Nations Scientific Committee on the Effects of Atomic Radiation. Report. Vol. I).
2. ICRP, 2013. Radiological protection in paediatric diagnostic and interventional radiology. ICRP Publication 121. Ann. ICRP 42(2).
3. Porto, L., et al., Evaluation of entrance surface air kerma in pediatric chest radiography. Radiat. Phys. Chem. (2014), http://dx.doi.org/10.1016/j.radphyschem.2014.02.014.
4. Alzimami, K., et al., Evaluation of radiation dose to neonates in a special care baby unit. Radiat. Phys. Chem. (2013), http://dx.doi.org/10.1016/j.radphyschem.2013.11.035.
5. Frayre, A., et al.: Radiation dose reduction in a neonatal intensive care unit. Applied Radiation and Isotopes 71, 57–60 (2012).
6. World Health Organization, http://www.who.int/topics/infant_newborn/en/, last accessed 2016/11/11.
7. Bouaoun A., Ben-Omrane, L., Hammou, A. Radiation doses and risk to neonates undergoing radiographic examinations in intensive care units in Tunisia. Int J Cancer Ther Oncol 2015; 3(4):342.
8. Conradie, A., Herbst, CP. Evaluating the effect of reduced entrance surface dose on neonatal chest imaging using subjective image quality evaluation. Phys. Med. (2016), http://dx.doi.org/10.1016/j.ejmp.2016.07.005.
9. Toossi, M., Malekzadeh, M.: Radiation Dose to Newborns in Neonatal Intensive Care Units. Iranian Journal of Radiology 9(3), 145–149 (2012).
10. Matthews, K., et al. An evaluation of paediatric projection radiography in Ireland, Radiography (2013), http://dx.doi.org/10.1016/j.radi.2013.10.001.
11. Yu, C.: Radiation Safety in the Neonatal Intensive Care Unit: Too Little or Too Much Concern? Pediatr Neonatol 51(6), 311–319 (2010).
12. International Atomic Energy Agency.: Dosimetry in Diagnostic Radiology: An International Code of Practice (IAEA Technical Report Series No. 457). IAEA, Austria (2007).
13. Oloowookere, C., Obed, R., Babalola, I., Bello, T. Patent dosimetry during chest, abdomen, skull and neck radiography in SW Nigeria. Radiography 17, 245–249 (2011).
14. Bushong, S.:Radiologic Science for Technologists: Physics, Biology, and Protection. 10th edn. Mosby, Inc., USA (2013).

The Metrological Electron Accelerator Facility (MELAF) for Research in Dosimetry for Radiotherapy

Andreas Schüller, Stefan Pojtinger, Markus Meier, Christoph Makowski, and Ralf-Peter Kapsch

Abstract

The Metrological Electron Accelerator Facility (MELAF) of the Physikalisch-Technische Bundesanstalt (PTB) offers access to well characterized high-energy (0.5–50 MeV) electron and photon radiation fields also for external researchers. This work outlines the capabilities of the facility to foster new collaborations. As example, the experimental determination of ionization chamber typical correction factors for Magnetic-Resonance guided Radiotherapy (MRgRT) is presented.

Keywords

High-energy electron radiation • High-energy photon radiation • Dosimetry • MRgRT • Magnetic field Ionization chamber • LINAC

1 Introduction

The PTB, Germany's national primary standard laboratory, operates the Metrological Electron Accelerator Facility (MELAF) for service and research in the field of dosimetry for external beam radiotherapy [1]. The PTB also offers access to its metrologically well characterized radiation fields for external researchers with other research projects beyond dosimetry.

The purpose of this work is to outline the capabilities and the properties of our facility in order to foster new collaborations. Our facility is equipped with two medical electron linear accelerators (LINAC) for the generation of high-energy photon and electron radiation, as well as a research LINAC, which operates on the same principle as medical LINACs, with adjustable energy up to 50 MeV. The properties of the research LINAC (e.g. spectral electron fluence, beam current, etc.) can be measured with small uncertainties. Therefore, radiation effects can be studied as a function of the fundamental physical quantities.

An electromagnet with magnetic flux density up to 1.4 T with sufficient space between the pole shoes to place a water phantom can be positioned in front of each accelerator for tests of dosimetry procedures for Magnetic-Resonance guided Radiotherapy (MRgRT) [2, 3].

In addition, a reference field of ^{60}Co γ radiation from a 130 TBq source, monoenergetic neutron fields up to 20 MeV and an ion microbeam are available on site. Furthermore, the PTB provides an S1 laboratory for cell culture and microbiological preparations with qualification for genetically modified cells.

Access to the facility is granted without fees on the basis of joint research collaborations or otherwise charged according to "Regulations Governing Charges for Services Supplied by PTB" (application anytime, decision by PTB).

2 Capabilities of the Facility

The MELAF is placed in a dedicated building with four irradiation rooms. All irradiation rooms are equipped with lasers to align the object under irradiation and air conditioning. Motorized precise XYZ positioning systems are available for the fixation and controlled movement of the object under irradiation as e.g. for the measurement of a depth dose distribution in a water phantom by means of an ionization chamber.

A. Schüller (✉) · S. Pojtinger · M. Meier · C. Makowski R.-P. Kapsch
Physikalisch-Technische Bundesanstalt, Bundesallee 100, 38116 Brunswick, Germany
e-mail: andreas.schueller@ptb.de
URL: http://www.ptb.de/MELAF

© Springer Nature Singapore Pte Ltd. 2019
L. Lhotska et al. (eds.), *World Congress on Medical Physics and Biomedical Engineering 2018*,
IFMBE Proceedings 68/3, https://doi.org/10.1007/978-981-10-9023-3_109

2.1 Medical LINACs

There are two irradiation rooms equipped with Elekta Precise medical LINACs (Elekta Instrument AB, Stockholm, Sweden). Both can be operated independently. In total, 9 electron beam qualities (nominal energy 4, 6, 8, 10, 12, 15, 18, 20, or 22 MeV), and 6 photon beam qualities (nominal accelerating voltage 4, 6, 8, 10, 15, or 25 MV) can be generated. Both medical LINACs are equipped with a multileaf collimator for investigations in small and irregularly shaped fields, which allows field sizes of any shape up to 40×40 cm at 1 m distance to the source. Typical dose rates are 0.1–5 Gy/min.

2.2 Research LINAC

A photo of the research LINAC is shown in Fig. 1. It consists of two sections: A low-energy Sect. (0.5–10 MeV) and a high-energy Sect. (6–50 MeV). At both accelerator sections the electron beam can be deflected into the dedicated beam line in the respective separate irradiation hall. Both beam lines are equipped with devices for a precise characterization of the beam. Figure 2 shows a photo of the 6–50 MeV beam line. The research LINAC provides a pulsed beam with about 2.5 μs pulse duration at a variable pulse repetition rate from 1 to 100 Hz. The kinetic energy of the electrons can be varied continuously from 0.5 to 50 MeV. For the accurate determination of the energy a magnetic spectrometer is used [4]. The energy width of the beam is less than 1% of the mean energy and the relative measurement uncertainty of the mean energy amounts to 0.125%. The charge of each beam pulse, i.e. the number of radiated electrons, can be measured non-destructively with a measurement uncertainty of about 0.1% [5]. The pulse charge can be varied continuously (typical range: 0.1–200 nC). The drift of the mean pulse charge (after warming-up phase) is less than 0.1%/h. For the analysis of the beam profile and the beam divergence, several wire scanners and removable YAG:Ce screens are installed at the beamline. The profile of a typical optimized beam has a Gaussian shape with a FWHM of about 4 mm [5]. At the end of the beam line the electrons either pass through a beam exit window for electron irradiations or impinges on a bremsstrahlung target for the generation of high-energy photons with therapeutically relevant dose rates. The resulting photon field corresponds to a Flattening Filter Free (FFF) radiation field from a medical LINAC. Dose rates are up to several Gy/s.

2.3 Dosimetry

Numerous different kinds of ionization chambers, calibrated traceably to the PTB primary standards, are available for dose measurements at the highest accuracy level. For instant verication of the shape of the radiation field a 27×27 cm detector matrix (PTW OCTAVIUS Detector 1500) and a computed radiography system based on 24×30 cm storage plates (Kodak ACR-2000i) are available which can be used even for absolute dosimetry [6]. Different water and PMMA phantoms as well as an anthropomorphic Alderson phantom are available. Furthermore, it is possible to create IMRT fields by means of a Monaco radiotherapy treatment plan system and to measure the 3D dose distribution by means of a OCTAVIUS 4D dosimetry system.

2.4 Magnet for MRgRT Dosimetry

MRgRT is a new technique providing real-time Magnetic Resonance Imaging (MRI) and irradiation of the tumour volume simultaneously by means of a MRI scanner integrated with a LINAC [7] (MR-LINAC). Due to the strong magnetic field of the MRI scanner, the reading of the commonly used ionization chambers may deviate from the value at standard conditions, i.e. without a magnetic field, by up to several percent [2, 3]. For experiments regarding the development of a reference dosimetry protocol a transportable electromagnet (BRUKER ER073) is available at the MELAF.

Figure 3a shows the magnet in front of one of the medical LINACs, while the isocenter of the LINAC is between the pole shoes. The maximal magnetic flux density amounts to 1.4 T, which corresponds to the magnitude at the existing MR-LINACs [2, 7].

Figure 3b shows the magnetic flux density as measured by means of a Hall probe as a function of the position in the plane between the pole shoes at nominal 1.4 T. The uniform area with <0.5% drop is 110 mm in diameter. The space between the pole shoes is 100 mm (for up to 1.0 T) or 70 mm (for up to 1.4 T), respectively. For both configurations a dedicated water phantom with corresponding width is available. Figure 3c shows an ionization chamber in the water phantom between the pole shoes at the isocenter of the medical LINAC marked by laser lines. The ionization chamber is mounted at a precise four axis remote positioning system installed atop of the magnet (see Fig. 3a).

Fig. 1 Photo of the research LINAC

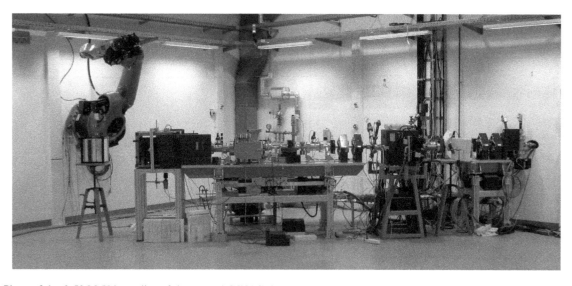

Fig. 2 Photo of the 6–50 MeV beam line of the research LINAC (beam enters from right)

An example for an application is the experimental verification of a MC calculation of the response of an ionization chamber in a magnetic field as shown in Fig. 4 for a farmer type ionization chamber (PTW 30013, Freiburg, Germany) irradiated by photons at 6 MV accelerating voltage. For simulation and measurement, the magnetic field vector, the LINAC beam direction and the chamber rotational axis were each pairwise perpendicular. The ionization chamber was placed at 10 cm depth inside the water phantom. Simulations were done using EGSnrc with the recently published algorithm for enhanced electron transport in electromagnetic fields [8].

(a)

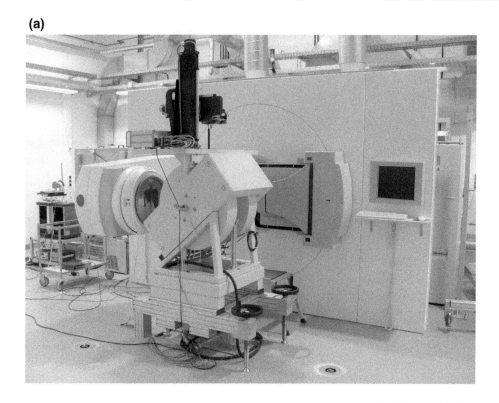

(b) 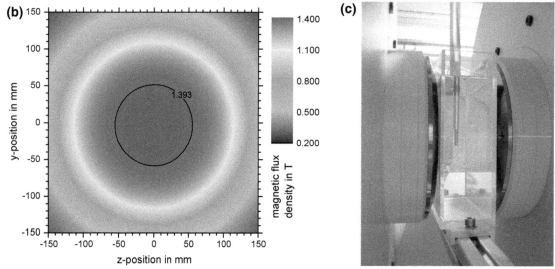 **(c)**

Fig. 3 **a** Electromagnet in front of a medical LINAC. **b** Measured magnetic flux density as a function of the position in the plane in the middle between the pole shoes at nominal 1.4 T. **c** An ionization chamber in a water phantom between the pole shoes of the electromagnet at the isocenter of the medical LINAC (marked by laser lines)

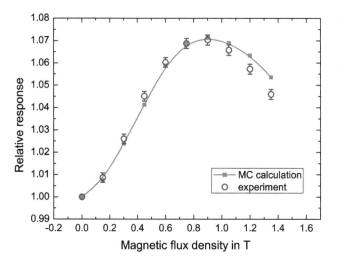

Fig. 4 Relative response of the ionization chamber with respect to the response without magnetic field as a function of the magnetic flux density for a PTW 30013 chamber at 10 cm water depth irradiated by photons of 6 MV accelerating voltage

3 Conclusion

The PTB offers access to its metrologically well characterized radiation fields and radiation dosimetry equipment for external researchers in order to exploit the potential of the existing infrastructure. MELAF provides capabilities way beyond standard conditions of medical LINACs as e.g. in energy range, in electron fluxes (dose rates) and in the traceable measurement of those parameters. Furthermore, experiments can be carried out for calibration of dosimetry equipment for MR-LINACs as well as for investigation of magnetic field effects on dose deposition and detectors in general.

Author's Statement The authors declare that they have no conflict of interest.

References

1. Derikum, K.: A dedicated irradiation facility for radiotherapy dosimetry. IFMBE Proceedings 25/1, 53–55 (2009). https://doi.org/10.1007/978-3-642-03474-9_14
2. Meijsing, I., Raaymakers, B.W., et al.: Dosimetry for the MRI accelerator: the impact of a magnetic eld on the response of a Farmer NE2571 ionization chamber. Phys. Med. Biol. 54, 29933002 (2009). https://doi.org/10.1088/0031-9155/54/10/002
3. Reynolds, M., Fallone, B. G., Rathee, S.: Dose response of selected ion chambers in applied homogeneous transverse and longitudinal magnetic fields. Medical Physics 40, 042102 (2013). https://doi.org/10.1118/1.4794496
4. Renner, F., Schwab, A., Kapsch, R.-P., Makowski, C., Jannek, D.: An approach to an accurate determination of the energy spectrum of high-energy electron beams using magnetic spectrometry. Journal of Instrumentation 9, P03004 (2014). https://doi.org/10.1088/1748-0221/9/03/P03004
5. Schüller, A., Illemann, J., Renner, F., Makowski, C., Kapsch, R.-P.: Traceable charge measurement of the pulses of a 27 MeV electron beam from a linear accelerator. Journal of Instrumentation 12, P03003 (2017). https://doi.org/10.1088/1748-0221/12/03/P03003
6. Aberle, C., Kapsch, R.-P.: Characterization of a computed radiography system for external radiotherapy beam dosimetry. Phys. Med. Biol. 61, 40194035 (2016). https://doi.org/10.1088/0031-9155/61/11/4019
7. Raaymakers, B.W., et al., Integrating a 1.5 T MRI scanner with a 6 MV accelerator: Proof of concept. Phys. Med. Biol. 54(12), N229N237 (2009). https://doi.org/10.1088/0031-9155/54/12/N01
8. Malkov, V.N., Rogers, D.W.O.: Charged particle transport in magnetic fields in EGSnrc. Medical Physics 43, 4447 (2016). https://doi.org/10.1118/1.4954318

Incidence of Burnout Among Medical Dosimetrists in Portugal

Dina Gonçalves and Ana Sucena

Abstract

Burnout is a pathologic response to chronic occupational stress resulting from the lack of coping strategies that assist the individual to comply with the working demands. Burnout is a cause of serious consequences for both the individual and the organization. Given the lack of studies on the incidence of this disease in medical dosimetrists in Portugal, this study aims to compensate for this gap. The method used for data collection was a survey divided into three parts. The first part was designed to evaluate the sociodemographic and professional conditions of the sample. The second part of the survey was based on the Copenhagen Burnout Inventory (CBI) and the third part was based on the Maslach Burnout Inventory (MBI). The survey was distributed to all medical dosimetrists working in public and private institutions in Portugal which gave the authorization to participate in this study. We evaluated 17 medical dosimetrists (physicists and radiation therapists), engaged in six Portuguese public and private institutions. Our results reveal an incidence of burnout between 35.3% (CBI) and 88.2% (MBI), manifested by worrying results on the sub-scales of personal burnout, work-related burnout, emotional exhaustion and personal accomplishment. For this study, the questionnaires were adapted and tested, having been considered valid for the Portuguese population, however, more detailed formal validation should be carried out in future studies.

Keywords

Burnout • Medical dosimetrists • Copenhagen burnout inventory • Maslach burnout inventory

D. Gonçalves (✉) · A. Sucena
Escola Superior de Saúde do Porto, IPP, Porto, Portugal
e-mail: dinapereira_goncalves@hotmail.com

A. Sucena
e-mail: asucena@ess.ipp.pt

1 Introduction

The term burnout was used for the first time in 1969 by Bradley, however, it only gained popularity when it was used, in the 1970s (1974), by the one who is generally considered its founder, Hebert Freudenberg [1–3].

Burnout syndrome arises in response to chronic occupational stress associated with a lack of coping strategies [4]. According to Maslach & Jackson [5], the burnout syndrome can be described as emotional exhaustion, depersonalization, and reduction of personal achievement.

Burnout has been the object of studies in several countries, revealing significant values for the incidence of this syndrome among health professionals in all dimensions [1, 4, 6–11]. Concerning the general population, there is almost no data on the prevalence of burnout [6].

The present study aims to determine the incidence of burnout syndrome in medical dosimetrists in Portugal and to verify the association of burnout characterizing components with personal and/or professional characteristics of the medical dosimetrists under study.

1.1 Research Hypothesis

We anticipate a significant incidence of burnout syndrome among medical dosimetrists. Since medical dosimetrists have less contact with patients than professionals in other areas of radiation therapy, it is expected that the incidence of burnout among medical dosimetrists will be lower than among the other professionals in the oncology area.

2 Materials and Methods

This study is observational, cross-sectional. The sampling method was a simple random type.

© Springer Nature Singapore Pte Ltd. 2019
L. Lhotska et al. (eds.), *World Congress on Medical Physics and Biomedical Engineering 2018*,
IFMBE Proceedings 68/3, https://doi.org/10.1007/978-981-10-9023-3_110

The population of this study covers all medical dosimetrists that are practicing in a public or private institution in Portugal.

The questionnaire developed for this study was based on Fonte's questionnaire [1] and adapted for the study population. The questionnaire is divided into three parts. The first part consists of the sociodemographic and professional evaluation of participants. The second part is based on the Copenhagen Burnout Inventory (CBI) and the third part is based on the Maslach Burnout Inventory (MBI).

The study was evaluated by the ethics committee of the institution Escola Superior de Tecnologia da Saúde do Porto, as well as by three other institutions that participated in the study.

The data collection was done through the Google Forms platform, which included a brief introductory explanation of the study as well as informed consent and the questionnaire. The results of the collection were compiled using Microsoft Office Excel® software.

The study included a total of 17 medical dosimetrists in public or private institution in Portugal.

Statistical analysis of the variables was supported by the Statistical Package for Social Science (SPSS version 20), applying the significance level of $p < 0.05$, with a 95% confidence interval.

3 Results

3.1 Evaluation of Burnout Syndrome According to CBI

About 35.3% of participants revealed high levels of burnout. The level of burnout was defined based on three dimensions, alternatively or cumulatively: personal, work-related, and patient-related. Among those individuals with a high level of burnout, the distribution of the three dimensions was more representative in relation to the dimensions "Personal burnout" and "Burnout related to work".

The analysis of the data revealed the existence of an association between the variables "Age" and "Weekly workload" with the Personal Burnout dimension.

3.2 Evaluation of Burnout Syndrome According to MBI

In the Depersonalization subscale, there was no register of elevated burnout levels. The majority (94.1%) of the professionals surveyed reported low levels of burnout. Regarding the Emotional Exhaustion subscale, about 76.5% of respondents presented low levels of burnout. It is relevant to mention that 17.6% of the sample exhibited a high level of

burnout in this subscale. In the Personal Accomplishment subscale, the existence of a high percentage of the sample with low and medium levels of personal accomplishment was observed. These levels of personal accomplishment translate into high and medium levels of burnout, respectively, 47.1% and 35.3%.

A data analysis reveals an association between the variables "Educational qualifications", "Basic training", "Practice in another institution", "Weekly workload" and "Work schedule" with the Emotional Exhaustion dimension. We also found an association between the variable "Time of exercise in the employer institution" and the dimension Depersonalization.

4 Discussion and Conclusion

Burnout syndrome is a growing public health problem. Burnout has serious consequences for both the individual (physical and psychological) and the organization (monetary —absenteeism and reduction of productivity) [10].

In this study, we found an incidence of 35.3% (CBI) and 88.2% (MBI) of high and medium levels of burnout among Portuguese medical dosimetrists. The low incidence of Burnout related to the patient may be a direct result of the nature of the profession, in the sense that medical dosimetrists are not in direct contact with the patient and therefore are not thus subject to the (negative) influence exerted by the patient.

The high incidence of Personal Burnout alerts to the importance of the variable "personal life", along with professional requirements, in the perception of physical and emotional demands by the medical dosimetrist.

The low incidence of high levels of burnout in the Depersonalization dimension can be justified by the low interpersonal interaction of the medical dosimetrists.

The high incidence of high levels of burnout in the Personal Accomplishment dimension can be justified by the monotony of work, lack of correspondence between expectations and the reality of work, lack of career progression, lack of autonomy and feedback for the work performed.

The association between some of the variables and the different dimensions of burnout suggest that those variables are risk factors for the development of the burnout syndrome among medical dosimetrists.

Although according to literature, age, sex, and marital status are risk factors for the development of burnout, the present study does not corroborate these data. These variables were not found to have a statistically significant association with the development of the syndrome.

For this study, the questionnaires were adapted and a pilot study was run with a small sample of participants. Results from this pilot study indicated the questionnaires were valid for the Portuguese population, however more detailed formal validation should be carried out in future studies.

For future studies, it is recommended to include an assessment of the personality type since there are studies that prove the relationship between this component and the risk of developing burnout [12–14].

Compliance with Ethical Standards The authors declare that the study is in compliance with ethical standards. **Conflicts of Interest** The authors declare that they do not have a conflict of interest. **Informed Consent** Consent was obtained from the subjects during the data collection.

References

1. Fonte C (2011) Adaptação e validação para português do questionário de Copenhagen Burnout Inventory (CBI). Repositório Digit da Univ Coimbra
2. Kristensen TS, Borritz M, Villadsen E, Christensen KB (2005) The Copenhagen Burnout Inventory: A new tool for the assessment of burnout. Work Stress 19:192–207
3. Schaufeli W, Buunk B (2003) Burnout: An overview of 25 years of research and theorizing. Handb Work Heal Psychol 2:282–424
4. Cumbe V (2010) Síndrome de burnout em médicos e enfermeiros cuidadores de pacientes com doenças neoplásicas em serviços de oncologia. Repositório da Fac Med Univ do Porto
5. Maslach C, Jackson SE (1981) The measurement of experienced burnout. J Organ Behav 2:99–113. https://doi.org/10.1002/job. 4030020205
6. Trigo TR, Teng CT, Hallak JEC (2007) Síndrome de burnout ou estafa profissional e os transtornos psiquiátricos. Rev Psiquiatr Clínica 34:223–233
7. Silva M, Queirós C, Cameira M, et al (2015) Burnout e engagement em profissionais de saúde do interior: norte de Portugal / Burnout and engagement among health professionals from interior: north of Portugal. Psicol Saúde Doenças 16:286–299
8. Shanafelt TD, Gradishar Wj Fau - Kosty M, Kosty M Fau - Satele D, et al (2014) Burnout and career satisfaction among US oncologists. J Clin Oncol TA - Oncol, J Clin 22:7. https://doi.org/ 10.1200/jco.2013.51.8480
9. Howard M (2013) The incidence of burnout or compassion fatigue in medical dosimetrists as a function of various stress and psychologic factors. Med Dosim 38:88–94. https://doi.org/10. 1016/j.meddos.2012.07.006
10. Akroyd D, Caison A, Adams RD (2002) Clinical investigation: Radiation oncology practice: Burnout in radiation therapists: the predictive value of selected stressors. Int J Radiat Oncol Biol Phys 52:816–821. https://doi.org/10.1016/s0360-3016(01)02688-8
11. Sousa F (2012) Síndrome de burnout em terapeutas de radioterapia: Um estudo no maior centro ibérico de radioterapia. Non Publ Rep
12. Alarcon G, Eschleman KJ, Bowling NA (2009) Relationships between personality variables and burnout: A meta-analysis. Work Stress 23:244–263
13. Bakker AB, Van der Zee KI, Lewig KA, Dollard MF (2006) The relationship between the big five personality factors and burnout: A study among volunteer counselors. J Soc Psychol 146:31–50
14. Glazer S, Stetz TA, Izso L (2004) Effects of personality on subjective job stress: a cultural analysis. Pers Individ Dif 37:645–658. https://doi.org/10.1016/j.paid.2003.10.012

CT Extended Hounsfield Unit Range in Radiotherapy Treatment Planning for Patients with Implantable Medical Devices

Zehra Ese, Sima Qamhiyeh, Jakob Kreutner, Gregor Schaefers, Daniel Erni, and Waldemar Zylka

Abstract

Radiotherapy (RT) treatment planning is based on computed tomography (CT) images and traditionally uses the conventional Hounsfield unit (CHU) range. This HU range is suited for human tissue but inappropriate for metallic materials. To guarantee safety of patient carrying implants precise HU quantification is beneficial for accurate dose calculations in planning software. Some modern CT systems offer an extended HU range (EHU). This study focuses the suitability of these two HU ranges for the quantification of metallic components of active implantable medical devices (AIMD). CT acquisitions of various metallic and non-metallic materials aligned in a water phantom were investigated. From our acquisitions we calculated that materials with mass-density $\rho > 3.0$ g/cm^3 cannot be represented in the CHU range. For these materials the EHU range could be used for accurate HU quantification. Since the EHU range does not effect the HU values for materials $\rho < 3.0$ g/cm^3, it can be used as a standard for RT treatment planning for patient with and without implants.

Z. Ese (✉) · D. Erni
Laboratory for General and Theoretical Electrical Engineering (ATE), Faculty of Engineering, CENIDE Center of Nanointegration Duisburg-Essen, University of Duisburg-Essen, Duisburg, Germany
e-mail: zehra-ese@outlook.de

Z. Ese · W. Zylka
Faculty of Electrical Engineering and Applied Natural Sciences, Westphalian University Gelsenkirchen, Gelsenkirchen, Germany

Z. Ese · J. Kreutner · G. Schaefers
MR:comp GmbH, Gelsenkirchen, Germany

J. Kreutner · G. Schaefers
MRI-STaR—Magnetic Resonance Institute for Safety, Techchnology and Research GmbH, Gelsenkirchen, Germany

S. Qamhiyeh
Department of Radiotherapy, University Hospital Essen, Essen, Germany

Keywords

Extended hounsfield unit • Radiotherapy • Medical implants

1 Introduction

The number of patients with implantable medical devices receiving radiation therapy is increasing [1–3]. However, the interactions of ionizing radiation and AIMDs are not well established yet. There are many concerns regarding the beam energy, dose and dose-rate limitations to the AIMD with respect to malfunction that may occur in these devices and impact patient safety. Many institutions still refer to the report of the American Association of Physicists in Medicine (AAPM) task group no. 34 [4] from 1994, even though this report only focuses on pacemakers and does not consider other AIMDs. The report recommends to avoid positioning the pacemaker in the direct beam and to limit the absorbed dose to the pacemaker to 2 Gy. However, case reports show that malfunctions can occur even below this threshold [2]. Unfortunately, a precise determination of the dose the AIMD is exposed to during radiotherapy is not possible, both because it is in general not located in the direct beam and local active measurements during radiation are not possible yet [1].

Current RT treatment planning systems (TPS) are not able to model the out-of-field doses beyond the beam penumbra [5, 6]. The accuracy of dose calculations is known to decrease with increasing distance from the field edge. However, peripheral doses depend on beam energy, field size and distance from the field edge and are usually esti-mated based on previous publications [7]. RT TPS are mainly based on CT data which provides tissue dependent radiodensity values, which are the base for accurate dose calculation in TPS. High-density materials cause image artefacts on CT images which can lead to undefinable geometries of both the object and anatomical structures due

L. Lhotska et al. (eds.), *World Congress on Medical Physics and Biomedical Engineering 2018*,
IFMBE Proceedings 68/3, https://doi.org/10.1007/978-981-10-9023-3_111

to overlapped artefacts [8–10]. To achieve an appropriate dose calculation, the HU value of implant materials have to be determined. However, the majority of CT scanners used in radiology provide only a conventional HU range, which suffices to properly represent human body tissues. HU values of dense materials considerably exceed the maximum of CHU, because of which these materials are usually mapped to CHU's maximum [9].

A qualitative quantification of Hounsfield units of high density materials have been reported in [11]. In this study we examine HU values of specific implant materials with high purity and known mass density at the CHU range and compare them to those at the EHU range. We analyse the suitability of the CHU and EHU ranges for the representation of high-density materials which would allow to draw conclusions regarding the superiority of EHU in accurate dose calculations in order to increase patient safety in RT.

2 Material and Method

The Hounsfield scale is defined based on the X-ray linear attenuation coefficient of water $\mu(water)$. The HU value (also known as CT number) of a particular material with X-ray linear attenuation coefficient μ is calculated by

$$HU = 1000\,\text{HU}\left[\frac{\mu}{\mu(water)} - 1\right]. \qquad (1)$$

Note that this definition does not impose a minimum or maximum on the Hounsfield scale.

The CT images for this study were acquired with two different ranges of the Hounsfield scale: the CHU and the EHU range, respectively. Both ranges are defined by imposing restrictions to the Hounsfield scale (1), i.e. by limiting it to particular upper and lower HU values. A CT image acquired at the CHU range uses 12 bit per pixel to represent an HU interval from $HU_{min} = -1024$ HU to $HU_{max} = 3071$ HU. Consequently, at CHU all values above the maximum will be in saturation and are represented as a constant value of $HU_{max} = 3071$ HU. In contrast, the EHU range reflects a 10 times broader HU interval of -10240 HU to 30710 HU. Choosing this interval allows for a differentiation between the high-density metal and its surroundings. However, metals may cause artefacts on CT images in both HU ranges which complicates a proper geometric delineation of the metallic objects and the structures nearby.

A Siemens Somatom Force dual energy CT scanner providing the mentioned HU ranges was used. CT images samples of two material categories were acquired: (i) metallic objects: titanium, chromium, aluminium, silicon, copper and (ii) non-metallic objects: carbon, epoxy, ceramic and plastic. All metallic objects were coin shaped with a

Table 1 HU values (mean ± standard deviation) of different metallic and non-metallic materials quantified in the conventional and extended HU range on CT images. The purity P and mas density ρ are shown for the metallic objects. (*samples offered by umicore GmbH, Essen, Germany*)

Material	P [%]	ρ [g/cm^3]	CHU [HU]	EHU [HU]
Metallic				
Silicon (Si)	99.999	2.34	1959 ± 71	1965 ± 19
Aluminium (Al)	99.500	2.67	2132 ± 83	2148 ± 172
Titanium (Ti)	99.900	4.50	3070 ± 0.65	6543 ± 173
Chromium (Cr)	99.600	7.14	3070 ± 0.77	8722 ± 128
Copper (Cu)	99.990	8.92	3071 ± 0.32	11552 ± 452
Non-metallic				
Carbon			278 ± 20	270 ± 10
Epoxy			1314 ± 31	1310 ± 3
Ceramic			3023 ± 56	3160 ± 112
Plastic			50 ± 16	50 ± 15

diameter $D = 21$ mm, a thickness of $d = 5$ mm and a purity above $P = 99.5\%$ (see Table 1). The physical mass densities of the metallic objects are summarized in Table 1.

A water equivalent solid state slab phantom, made of a white polystyrene material (RW3), was positioned in a water filled acrylic tank. The test objects were embedded and fixed between two RW3 slabs, which were kept apart by two thin separators positioned at the edges of the slab phantom. Several RW3 slabs were positioned under the object to account for any backscattering by the patient table. With the intention to achieve the highest possible image quality for dense materials, we investigated several CT protocols. Following protocol was then chosen for the CT acquisitions: tube energies with $E_1 = 100$ kVp and $E_2 = 150$ kVp, slice thickness $\Delta d = 0.6$ mm, pitch-factor $P = 0.6$ mm and a rotation time $t_R = 0.5$ s. All metallic materials in the phantom were aligned in scanning direction and scanned in one pass. The same procedure was performed for non-metallic materials. Both setups were acquired in CHU and EHU scales, without any computed image corrections (e.g. metal-artefact-reduction). For the analysis of HU values the commercial radiation therapy treatment planning software Varian Eclipse ^{TM} (edition 13.5) was used.

The metallic samples were identified on the CT images according to their physical geometry. The central volume of the samples was manually selected (40% of the total sample volume) (see Fig. 1), which were used for HU quantification. In order to obtain data from identical volumes the same region of interest (ROI) was assigned on both CT images reconstructed in CHU and EHU range. The average HU value and its standard deviation σ of the ROI was quantified with the software Varian Eclipse ^{TM} (edition 13.5).

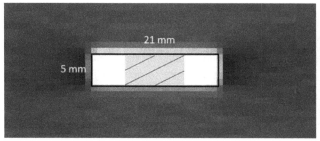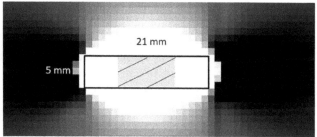

Fig. 1 Cropped and magnified CT images of silicon (left) and chromium (right) samples in EHU acquisition. The position of the object is indicated by the rectangle and the ROI (5 mm × 8.5 mm) for HU quantification is represented by hatched lines

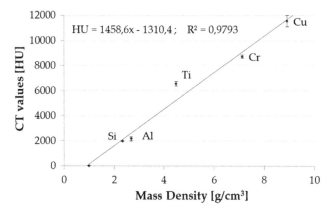

Fig. 2 HU values versus mass density of the metallic materials. A linear regression based on these values were added

3 Results

Table 1 compares the HU values of various metallic and non-metallic materials, quantified in the conventional and extended HU range on CT images. The HU values of non-metallic materials, except ceramic show up almost identical results in both HU ranges. This effect is also seen for two metallic samples: silicon and aluminium. The HU value of the metallic objects titanium, chromium and copper is $HU_{max} = (3071 \pm 0.58)$ HU in the conventional HU range. In comparison, the HU values in the extended HU range differs strongly for each of these materials (see Table 1).

Figure 2 displays the quantified HU values of the metallic samples as function of their mass densities (Table 1 non-metallic samples were not added due to unknown mass-density). The relation between Hounsfield value and mass density was approximated by a linear regression [12] to determine an approximate maximum mass-density to be quantified by CHU. Based on the unweighted linear fit with a correlation coefficient of $R^2 = 0.9793$ materials with mass density below $\rho = 3.0$ g/cm^3 can be well quantified within CHU range. Due to the limited number of materials used in

this study this value should be considered as an approximation. It should be noted, that this linear fit is not suitable for CT calibration in TPS, which actually relies on individual points instead of linear regression.

Figure 3 shows cross-sectional profiles through the metallic objects on the CT images, reconstructed at the CHU and EHU range, respectively. In Fig. 3a the profiles of silicon and aluminium are shown, while in Fig. 3b the profiles of titanium, chromium and copper are illustrated. For silicon and aluminium the profiles in the CHU and EHU range are equal in value. No differences in HU value quantification in both ranges was obtained. Slightly cupping artefacts are

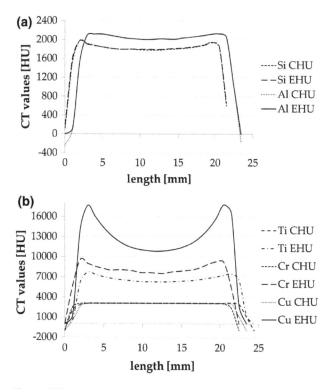

Fig. 3 HU value distribution shown as cross-sectional profiles through the CT images of the metallic objects. For each metal, profiles at the CHU and EHU range are drawn. The ROI used for HU quantification is between 6.3 and 14.7 mm

observed for both metals in both HU ranges. The profiles at the CHU range of titanium, chromium and copper are almost similar, diverges slightly at the edges of the samples. The HU values at the center of the metals are constant, no cupping artefacts are observed. In contrast, the profiles at the EHU range for these metals proceeds quite different from each other. The higher the mass density of the metal, the higher the HU value along the cross-section of the samples. In EHU range all metals show cupping artefacts. The higher the mass density, the more sever the cupping artefact.

4 Discussion and Conclusion

The HU values of the non-metallic materials carbon, epoxy, plastic and the metallic materials silicon and aluminium show identical HU value distribution at the CHU and EHU range, respectively. All of these materials have a relative low mass density (see Table 1). According to Fig. 2, all materials with an approximate value of $\rho < 3.0$ g/cm^3 can be properly represented in the CHU range with slightly cupping artefacts. However, for these materials the EHU range has no effects on the HU values and could be also used for radiation treatment planning. Since the mass density of titanium, chromium and copper is $\rho > 3.0$ g/cm^3, the CHU range is not capable for representing the corresponding HU value. Therefore, the HU value of these metals is quantified as the maximum HU value at the CHU range ($HU_{max} = 3071$ HU). The cross-sectional profiles through the metallic objects show cupping artefacts. So the HU values at the edges are higher compared to the center of the object. The cupping artefact is a physical based artefact, which is caused by the beam-hardening effect [13]. The X-ray beam used in computed tomography systems is a polyenergetic beam. Once the beam passes through the metal the low energy photons get absorbed. The total beam energy increases and the attenuation coefficient of the metal passed decreases [14]. For the materials with $\rho > 3.0$ g/cm^3 no cupping artefacts were observed in the CHU range (see Fig. 3b). A constant cross-sectional profile through the metal samples was measured. In this case, beam-hardening effects are still present but cannot be distinguished at the CHU range, since HU values above 3071 HU cannot be represented in the CHU range. We consider for ceramic to be an CHU inaccuracy, thus the EHU value appears different although not significant within statistical margins.

In conclusion, the EHU range better reproduces HU values of high-density materials thus rendering itself appropriate for the purpose of dose calculations in radiotherapy treatments. However, the quantified HU value strongly depend on the selected ROI size and position. The calculated HU values for the metallic and non-metallic samples used in this study could be used with restrictions to expand the CT calibration curves used in radiotherapy treatment planning software to account for more accurate dose calculations for patients with metallic implants. In order to enable exact in vivo dose monitoring of implants, further investigations will be proceed.

Funding and Conflict of Interest This study is supported by the Federal Ministry for Economic Affairs and Energy on the basis of a decision by the German Bundestag, grant no. ZF4205702AW6. The authors declare that they have no conflict of interest.

References

1. Gauter-Fleckenstein, B., Israel, C.W., Dorenkamp, M., Dunst, J., Roser, M., Schimpf, R., Steil, V., Schaefer, J., Hoeller, U., Wenz, F.: DEGRO/DGK guideline for radiotherapy in patients with cardiac implantable electronic devices. Strahlentherapie Onkologie 191, 393–404 (2015)
2. Zaremba, T.: Radiotherapy in Patients with Pacemakers and Implantable Cardioverter-Defibrillators. PhD Thesis, Aalborg University (2015)
3. Prisciandro, J.I., Makkar, A., Fox, C.J., Hayman, J.A., Horwood, F., Pelosi, L., Moran, J.M.: Dosimetric review of cardiac implantable electronic device patients receiving radiotherapy. Journal of Applied Clinical Medical Physics 16, 1–8 (2014)
4. Marbach, J.R., Sontag, M.R., Van Dyk, J., Wolbarst, A.B.: Management of Radiation Oncology Patients with Implanted Cardiac Pacemakers: Report of AAPM Task Group No.34. Medical Physics 21(1), 85–90 (1994)
5. Bourgouin, A., Varfalvy, N., Archambault, L.: Estimating and reducing dose received by cardiac devices for patients undergoing radiotherapy. Journal of Applied Clinical Medical Physics 16(6), 411–420 (2015)
6. Huang, J.Y., Followill, D.S., Wang, X.A., Kry, S.f.: Accuracy and sources of error out-of-field dose calculations by commercial treatment planning system for intensity-modulated radiation therapy treatments. Journal of Applied Clinical Medical Physics 14(2), 4139 (2015)
7. Howell, R.M., Scarboto, S.B., Kry, S.F., Yaldo, D.Z.: Accuracy of out-of-field dose calculations by a commercial treatment planning system. Physics in Medicine and Biology 55(23), 6999–7008 (2010)
8. Kairn, T., Crowe, S.B., Kenny, J., Mitchell, J., Burke, M., Schlect, D., Trapp, J.V.: Dosimetric effects of a high-density spinal implant. Journal of Physics: Conference Series, 7th International Conference on 3D Radiation Dosimetry 444, 1–4 (2008)
9. Mullins, J.P., Grams, M.P., Herman, M.G., Brinkmann, D.H., Antolak, J.A.: Treatment planning for metals using an extended CT number scale. Journal of Applied Clinical Medical Physics 17 (6), 179–188 (2016)
10. Gossman, M.S., Graves-Calhoun, A.R., Wilkinson, J.D.: Establishing radiation therapy treatment planning effects involving implantable pacemakers and implantable cardioverter-defibrillator. Journal of Applied Clinical Medical Physics 11(1), 33–45 (2009)

11. Ese, Z., Kressmann, M., Kreutner, J., Schaefers, G., Erni, D., Zylka, W.: Influence of conventional and extended CT scale range on quantification of Hounsfield units of medical implants and metallic objects. tm-Technisches Messen 85(5), 343-350 (2018)

12. Hubbell, J.H.: Photon Cross Sections, Attenuation Coefficients, and Energy Absorption Coefficients From 10 keV to 100 GeV. Nat. Stand. Ref. Data. Ser., Nat. Bur. Stand. (U.S.), 29, 85 (1969)

13. Barrett, J.F., Keal, N.: Artifacts in CT: Recognition and Avoidance. RadioGraphics 24(6), 1679–1691 (2004)

14. Xin-ye, N., Liugang, G., Mingming, F., Tao, L.: Application of Metal Implant 16-Bit Imaging: New Technique in Radiotherapy. Technology in Cancer Research and Treatment 16(2), 188–194 (2017)

3D Absorbed Dose Reconstructed in the Patient from EPID Images for IMRT and VMAT Treatments

Fouad Younan, Jocelyne Mazurier, Frederic Chatrie, Ana Rita Barbeiro, Isabelle Berry, Denis Franck, and Xavier Franceries

Abstract

A back-projection method has been used in this study to reconstruct the 3D absorbed dose matrix in the patient from EPID images for IMRT and VMAT fields. Images were acquired with the Clinac 23iX aS-1000 imager (Varian) and a 6 MV beam. Then a calibration step was performed to transform the grey levels of the pixels into absorbed dose in water via a response function. Correction kernels were also used to correct for the scatter within the EPID. The dose was then back-projected into the patient for all EPID parallel planes for each gantry angle. Finally the total dose was obtained by summing the 3D dose associated for each gantry angulation. First, the 3D dose reconstruction algorithm was tested for 20 IMRT and VMAT prostate and head and neck treatments in a homogeneous cylindrical phantom. Then the algorithm was used for IMRT brain cancer plans on 20 real patients. The EPID reconstructed 3D dose distributions were then compared to the planned dose from TPS (Treatment Planning System, Eclipse Varian) with a 3D global gamma index of 3% and 3 mm criteria. The percentage of points of which the gamma index was larger than unity was greater than 97% for all IMRT treatments both in the phantom and in the patients and over 96% for all VMAT treatments checked in the cylindrical phantom. Our 3D reconstruction algorithm, validated for homogeneous medium, can be used to verify the dose distribution for IMRT and VMAT fields using in vivo EPID images.

Keywords

Epid • Back-projection algorithm • 3D absorbed dose
Dose verification • In vivo dosimetry

F. Younan (✉) · J. Mazurier · D. Franck
Groupe Oncorad-Garonne, Service de Radiothérapie,
Clinique Pasteur, L'Atrium, 1, rue de la Petite-Vitesse,
31300 Toulouse, France
e-mail: fouad.younan@inserm.fr

F. Younan · I. Berry
INSERM UMR 1214, TONIC (Toulouse NeuroImaging Centre),
31059 Toulouse, France

F. Younan · I. Berry
Université Toulouse III Paul Sabatier, UMR 1214,
31059 Toulouse, France

F. Chatrie · A. R. Barbeiro · X. Franceries
Inserm, UMR 1037 CRCT, 31000 Toulouse, France

F. Chatrie · A. R. Barbeiro · X. Franceries
Université Toulouse III-Paul Sabatier, UMR 1037 CRCT,
31000 Toulouse, France

F. Chatrie
LAAS-CNRS, 31000 Toulouse, France

I. Berry
Department of Nuclear Medicine, CHU Rangueil,
1 Avenue du Professeur Jean Poulhès, 31400 Toulouse, France

1 Introduction

With the increasing complexity of external radiation therapy techniques such as volumetric-modulated arc therapy (VMAT), a lot of commercial tool has been created to perform patient quality assurance by using the electronic portal imaging device (EPID). However, only few of them were dedicated to in vivo dosimetry. The purpose of this work was to develop a back-projection algorithm that uses in vivo EPID images in order to reconstruct the 3D absorbed dose within the patient or phantom during IMRT and VMAT treatments.

2 Materials and Methods

2.1 Image Acquisition

Images were acquired using a Varian Clinac 23iX accelerator for 6 MV treatments at 600 MU/sec dose rate. The amorphous silicon EPID had an active area of 40×30 cm^2 and was mounted on Exact-arm. Technical details can be found in the

© Springer Nature Singapore Pte Ltd. 2019
L. Lhotska et al. (eds.), *World Congress on Medical Physics and Biomedical Engineering 2018*,
IFMBE Proceedings 68/3, https://doi.org/10.1007/978-981-10-9023-3_112

literature [1]. All EPID images were acquired at 150 cm from the source. The raw images were automatically dark field (DF) and flood field (FF) corrected by the Varian IAS3 software before being multiplied by the number of frames. The algorithm described below was implemented on Matlab R2015b (The MathWorks, Inc.) with a 2.5 GHz quad core processor.

IMRT Versus VMAT

During IMRT treatments, the Varian integrated mode was used. In this mode many frames (between 100 and 200 per image) were acquired and averaged, thus leading to one image per gantry angle. For VMAT treatments, images were rather acquired in cine mode for which the averaged number of frames per image was set to 6 in the IAS3 software (1 image acquired for each 4 degrees of rotation). In addition, the image acquisition was synchronized with beam pulse to avoid strip artefact [2].

2.2   Measurement Uncertainty

Uncertainty related to the detector measurements is the result of propagation of uncertainties related to the repeatability and the reproducibility of the imager. The repeatability was evaluated by making 10 successive irradiations without attenuator placed between the source and the detector, for a reference field size of 10×10, a fixed dose rate of 600 UM/min at the energy of 6 MV. This was done for an interval period of 30 s. The reproducibility was evaluated by irradiating every day the EPID under the same conditions over a period of 4 months. The repeatability and the reproducibility of the detector were then estimated by dividing the standard deviation by the average of the measurements, (see discussion for further details).

2.3   Grey Levels to Dose in Water Calibration

This step consists in calculating the dose in water at the distance of 150 cm using EPID images without patient [3]. The model includes a first pre-calibrated image called S_{EPID} that accounts for the non-uniformity of the detector response and also of the influence of the backscatter radiation.

This pre-calibrated image S_{EPID} is then transformed into absorbed dose to water $D_{EPID \rightarrow Water}$ through a conversion function associated with a dose redistribution kernel as follows:

$$D_{EPID \rightarrow Water} = \left[(f_{Dose} \times S_{EPID}) \otimes^{-1} K_{material} \right] \otimes K_{penumbra}.$$

$$(1)$$

where f_{Dose} is the dose conversion function, $K_{material}$ is a kernel that accounts for the difference between water and the materials within the EPID. $\otimes$ and $\otimes^{-1}$ denote respectively the convolution and deconvolution operator.

$K_{penumbra}$ is a kernel used to correct the shape of the penumbra. The EPID dose image obtained is then used in a back-projection algorithm to reconstruct the 3D dose distribution in the patient.

2.4   Reconstruction Within Phantom or Patient

Our algorithm uses a back-projection method described by Wendling et al. [4–6]. From an EPID image obtained behind the patient (or phantom), the 2D dose distribution is reconstructed inside the volume in every parallel plane separated by 1 mm thick, leading to a 3D dose distribution for each beam angle. Then the entire absorbed dose of the IMRT or VMAT treatment is obtained by summing every 3D dose calculated for each EPID image.

For each plane, the reconstructed 2D dose D_{plane} consists of the combination of the primary dose Pr_{plane} and the scattered dose Sc_{plane} at every pixel.

$$D_{plane} = Pr_{plane} + Sc_{plane}.$$

$$(2)$$

Primary dose, Pr_{plane}

The primary dose distribution in a plane is calculated, pixel by pixel, from the EPID dose image $D_{EPID \rightarrow Water}$ accounting for the inverse square law ISL and an attenuation correction factor ACF as follows:

$$Pr_{plane}(d) = D_{EPID \rightarrow Water} \cdot ISL(d) \cdot ACF(d) \qquad (3)$$

$$= D_{EPID \rightarrow Water} \cdot \left(\frac{DSE + d}{SDD} \right)^{-2} \cdot \left(\frac{T(d)}{T_{tot}} \right), \qquad (4)$$

where $Pr_{plane}(d)$ is the primary dose in a given plane at a depth, d.

DSE is the distance from the source to the entrance of the phantom, and SDD is the source detector distance. T_{tot} represents the total transmission calculated by dividing the EPID portal dose image Eq. (1) measured with and without the patient or phantom in the beam. $T(d)$ is referred as the transmission at the depth d, calculated by taking into account the radiological thickness through the phantom d'.

$$T(d) = T_{tot}^{\frac{d}{d'}} \cdot exp \left\{ \beta \cdot d'^2 \times \left[-\frac{d}{d'} + \frac{d^2}{d'} + \frac{d^3}{d'} \right] \right\}. \qquad (5)$$

In the above equation, the exponential function corrects the transmission for the beam hardening effect. The coefficient β was fitted by comparing the depth dose curve obtained on axis in a homogeneous slab phantom of 20 cm thick, with that obtained with an ionization chamber after 100 MU irradiation at a field size of 10×10 cm^2.

Scattered dose, Sc_{plane}

To calculate the scattered component into the reconstruction plane, the primary dose obtained before is multiplied by a third-order polynomial ω, function of the transmission calculated at the given plane d, Eq. (5). To determine ω, it is first obtained experimentally and then fit as a function of $T(d)$ [4]. 100 MU were delivered in isocentrically aligned slab phantom of several thicknesses for a 10×10 cm^2 field size. The primary dose calculated using Eq. (4) at the isocenter was then compared to the dose measured with an ionization chamber, $D_{chamber}$ in the same conditions. Thus, for every thickness of the phantom, the scattered-to-primary ratio is calculated as

$$\omega = \frac{D_{chamber} - Pr_{plane}}{Pr_{plane}}, \qquad (6)$$

and then fit to a third-order polynomial function of the transmission.

After multiplying $Pr_{plane}(d)$ by the corresponding ω, the result is then convolved with the kernel K_m accounting for the field-size dependence of the scattered dose distribution in the reconstruction plane [4].

In-phantom and In-vivo dosimetry

First, the 3D dose reconstruction algorithm was tested for 10 prostate and head and neck plans both for IMRT and VMAT treatments. All plans were checked using a homogeneous cylindrical phantom of 25 cm diameter.

Then the algorithm was tested on 20 patients treated for a brain cancer with IMRT plans.

2.5 Gamma Evaluation

Planned and in vivo reconstructed dose distributions were compared using a 3D global gamma analysis [7], implemented in our algorithm. The criteria were 3% of the maximum dose as dose-difference criterion and 3 mm as distance-to-agreement criterion, with a 10% threshold.

3 Results

3.1 Measurement Uncertainty

Uncertainties related to the repeatability and the reproducibility of the detector were successively 0.16 and 0.41%. The global uncertainty associated with the imager was therefor about 0.44%.

3.2 Reconstruction Within Phantom or Patient

A result of 3D EPID dosimetry for a VMAT plan of prostate cancer checked on the phantom is shown in Fig. 1. Three orthogonal planes intersecting the isocentre are presented EPID-reconstructed dose and the resulting γ analysis. In this case, 3D gamma evaluation (3% maximum dose, 3 mm) revealed a mean $\gamma \langle \gamma_{mean} \rangle = 0.34$ and a percentage of points with $\gamma \leq 1 \langle \mathbb{P}_{\gamma \leq 1} \rangle = 99\%$. Detailed results of all IMRT and VMAT plans checked in phantom are presented in Table 1.

For all IMRT plans of the 20 patients treated for a brain cancer, the $\%\gamma \leq 1$ was in average greater than 97% between planned and in vivo reconstructed dose. The average values for all patients of the mean γ and the $\%\gamma \leq 1$ were successively 0.31 ± 0.14 and 98.16 ± 0.35.

4 Discussion

The 3D back-projection algorithm was implemented in our institution to reconstruct the 3D absorbed dose in homogeneous medium (patient or phantom). Results obtained for the IMRT plans are sufficiently high ($\gamma 3\% - 3$ mm on average greater than 97%) to be used on patients for little inhomogeneity localization as for example prostate, pelvis or brain. Concerning VMAT plans, the reconstruction was done from 1 EPID image each 4 degrees rotation, thus leading to some loss of information. The only way to reduce this loss was to reduce the acquisition frame parameter for cine mode on IAS3 software, for example 3 instead of 6. Unfortunately this changing caused undesirable effects such as stopping the gantry during irradiation. Moreover, results obtained after reconstructing the 3D dose from the acquisition of 1 image each 4 degrees angulation, are sufficiently appreciable as shown in Fig. 1. This back-projection algorithm will be tested with the aS-1200 imager on TrueBeam accelerator (Varian) for which the image acquisition system named XI is designed to record every frame acquired during the treatment. Thereby, better results are expected on TrueBeam accelerator, especially for VMAT treatments.

The algorithm will be also improved to include patient inhomogeneity for the 3D absorbed dose calculation.

The method used to reconstruct EPID raw images into absolute dose is purely analytical in the way that it uses empirical kernels, functions and radiological lengths to convert the grey level pixels of the detector into absolute dose inside the irradiated volume. This means that the result will always be the same if the input grey scale of the EPID remains unchanged. In other words, the only way to quantify the uncertainty related to the algorithm is to quantify the uncertainty of the detector.

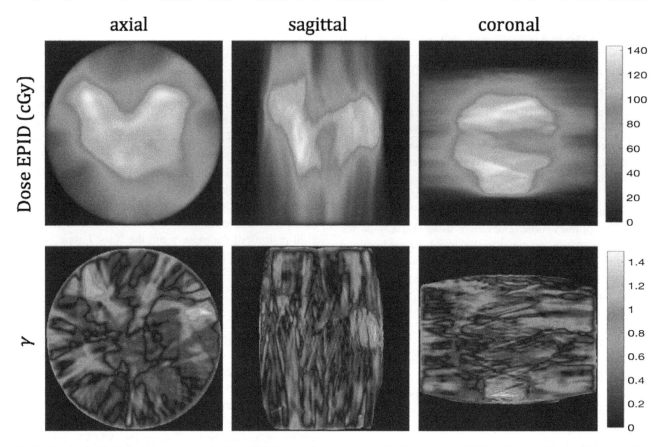

Fig. 1 Result of a 3D dosimetric reconstruction for a VMAT plan of prostate cancer checked on the cylindrical phantom. Above are presented the EPID-reconstructed dose in 3 orthogonal planes. The corresponding γ verification for each plane is shown below

Table 1 Average values of the mean γ and the $\%\gamma \leq 1$ obtained for all IMRT and VMA plans after comparing the 3D EPID-reconstructed dose with the corresponding TPS planned dose

Localization	IMRT		VMAT	
	$\%\gamma \leq 1$	Mean γ	$\%\gamma \leq 1$	Mean γ
Prostate	99.40 ± 0.34	0.35 ± 0.18	98.18 ± 1.00	0.38 ± 0.11
Head and neck	98.74 ± 1.01	0.40 ± 0.21	97.30 ± 0.30	0.41 ± 0.29

Many studies are currently underway in our department to evaluate in the one hand the robustness and the precision of the algorithm and, in the other hand the capability of the algorithm to detect sources of errors during the treatment such as the air gap, the gantry rotation (during IMRT) and the position of the patient on the couch.

5 Conclusion

The back-projection algorithm proposed in this study is sufficiently precise to reconstruct the 3D dose distribution in homogeneous medium (patient or phantom) for IMRT and VMAT plans using Clinac aS-1000 imager (Varian). The next step is to perform the algorithm from EPID TrueBeam images (Varian) on patients with large heterogeneity treated with VMAT plans.

References

1. J. V. Siebers, J. O. Kim, L. Ko, P. J. Keall, and R. Mohan, "Monte Carlo computation of dosimetric amorphous silicon electronic portal images," Med. Phys., vol. 31, no. 7, pp. 2135–2146, 2004.
2. B. Liu, J. Adamson, A. Rodrigues, F. Zhou, F. Yin, and Q. wu, "A Novel Technique for VMAT QA with EPID in Cine Mode On Varian TrueBeam," in Medical Physics, 2013, vol. 40, no. 6, p. 502.
3. J. Camilleri, J. Mazurier, D. Franck, P. Dudouet, I. Latorzeff, and X. Franceries, "2D EPID dose calibration for pretreatment quality control of conformal and IMRT fields: A simple and fast convolution approach," Phys. Medica, vol. 32, no. 1, pp. 133–140, 2016.

4. M. Wendling, R. J. W. Louwe, L. N. McDermott, J. J. Sonke, M. Van Herk, and B. J. Mijnheer, "Accurate two-dimensional IMRT verification using a back-projection EPID dosimetry method," Med. Phys., vol. 33, no. 2, pp. 259–273, 2006.

5. M. Wendling, L. N. McDermott, A. Mans, J. J. Sonke, M. Van Herk, and B. J. Mijnheer, "A simple backprojection algorithm for 3D in vivo EPID dosimetry of IMRT treatments," Med. Phys., vol. 36, no. 7, pp. 3310–3321, 2009.

6. R. Pecharromán-Gallego et al., "Simplifying EPID dosimetry for IMRT treatment verification," Med. Phys., vol. 38, no. 2, pp. 983–992, 2011.

7. D. Low, W. B. Harms, S. Mutic, and J. A. Purdy, "A technique for the quantitative evaluation of dose distributions.," Med. Phys., vol. 25, no. 5, pp. 656–61, 1998.

Implementation of a Novel Uncertainty Budget Determination Methodology for Small Field Dosimetry

David Eduardo Tolabin, Rodolfo Alfonso Laguardia, and Sebastian Bianchini

Abstract

This paper presents the implementation of a novel methodology for the determination of the uncertainty associated to the process of beam measurement in small fields. Field output factors were measured for different fields sizes in a 6 MV photon beam. The uncertainties related to the detector's positioning were taken into account, as well as its characteristics. Contributions of the beam limiting device, scanning system, instrumental and correction factors introduced for field output factor calculations were also analyzed. A broad range of detectors, including ionization chambers with different active volumes and one stereotactic diode for radiosurgery were used. Expanded uncertainty levels under 3% for $k = 3$ were obtained with the stereotactic diode for all the studied field sizes. Ionization chambers' uncertainty levels were substantially larger in most cases.

Keywords

Uncertainty budget determination • Small field dosimetry
Field output factor measurements
Positioning uncertainty • Field correction factor
uncertainty

1 Introduction

The use of small fields in radiotherapy and radiosurgery has substantially increased with the outcome of new technologies and the development of complex techniques [1].

D. E. Tolabin (✉) · S. Bianchini
Fundación INTECNUS, 8400 San Carlos de Bariloche, Argentina
e-mail: david.tolabin@gmail.com; david.tolabin@intecnus.or6.ar

S. Bianchini
e-mail: sebastianbianchini@gmail.com

R. A. Laguardia
Instituto Superior de Tecnologías y Ciencias Aplicadas, 11300 La Habana, Cuba
e-mail: rodocub@yahoo.com

However, precise and accurate measurements involve more complex procedures, leading to higher levels of uncertainty and differences of up to 12% in field factor measurements reported by several institutions have been found [2].

The development of a concise, straight-forward and reproducible methodology has the potential of being useful for the determination of uncertainties in a radiotherapy service, as well as serving as a means of data comparison between different institutions.

2 Materials

All measurements were performed in the radiotherapy service of "Fundación INTECNUS", of San Carlos de Bariloche. The dosimetry equipment either belongs to the institution or was provided by IBA Dosimetry GmbH.

2.1 Equipment and Planning Tools

An Elekta Synergy Platform® linear accelerator was used, with a nominal accelerating potential of 6 MV and a beam quality $Q = 0.682$ [3]. The Linac's beam limiting device (BLD) is the MLC Elekta Agility®. An IBA BluePhantom$^{2®}$ water phantom was used to carry out the measurements, together with a calibrated IBA Dose^2 electrometer [4].

2.2 Detectors

Four field detectors were used, the stereotactic diode IBA Razor Detector and the ionization chambers IBA CC13, CC01 and Razor NanoChamber®, being the latter loaned for testing by the manufacturer. Their characteristics are summarized in Table 1.

© Springer Nature Singapore Pte Ltd. 2019
L. Lhotska et al. (eds.), *World Congress on Medical Physics and Biomedical Engineering 2018*,
IFMBE Proceedings 68/3, https://doi.org/10.1007/978-981-10-9023-3_113

Table 1 Main technical characteristics of the field detectors used in this work

	CC13	CC01	NanoChamber	Razor detector
Sensitive volume [cm^3]	0.13	0.01	0.003	0.6 mm; 0.02 mm[a]
Sensitivity [nC/Gy]	3.6	0.4	0.11	4.6
Bias voltage [V]	±300	±300	±300	0
Beam quality [MV]	^{60}Co-25	^{60}Co-25	^{60}Co-25	^{60}Co-15
Electrode	C-552	Steel	Graphite-EDM3	p-Silicon[b]
Wall	C-552	C-552	C-552	ABS/Epoxy[c]
Wall thickness [mm]	0.4	0.5	0.5	N/A
Leakage current [fA]	3	3	3	N/A

[a]Diameter and thickness, respectively
[b]Sensitive material
[c]Enclosure material

3 Methods

Field output factors for clinical square fields of a nominal side, L(c), of 40, 30, 20, 15, 10 and 5 mm were measured in water using an isocentric technic with a depth, d, of 10 cm. The Razor Detector was positioned parallel to the beam axis, while the ionization chambers where positioned perpendicular to it. The measurement uncertainties were estimated with the mentioned methodology.

3.1 Field Output Factor Measurements

To determine the field output factors, $\Omega_{Q_{clin},Q_{msr}}^{f_{clin},f_{msr}}$, the formalism adopted in [5], originally proposed by Alfonso et al. [6] was followed, being:

$$\Omega_{Q_{clin},Q_{msr}}^{f_{clin},f_{msr}} = \frac{M_{Q_{clin}}^{f_{clin}}}{M_{Q_{msr}}^{f_{msr}}} \cdot k_{Q_{clin},Q_{msr}}^{f_{clin},f_{msr}} \quad (1)$$

where $M_{Q_{clin}}^{f_{clin}}$ and $M_{Q_{msr}}^{f_{msr}}$ are the readings of the clinical field and machine specific reference field, respectively and $k_{Q_{clin},Q_{msr}}^{f_{clin},f_{msr}}$ is a field correction factor, which takes into account the difference between the detector's response in each field [6].

Daisy-Chain Correction Method. The Daisy-Chain (DC) correction method [7, 8] was used to normalize the

reference detector and an intermediate field, f_{int} of 40×40 mm^2. As a consequence, $\Omega_{Q_{clin},Q_{msr}}^{f_{clin},f_{msr}}$ is calculated as:

$$\Omega_{Q_{clin},Q_{msr}}^{f_{clin},f_{msr}} = \left[\frac{M_{Q_{int}}^{f_{int}}}{M_{Q_{msr}}^{f_{msr}}} \cdot k_{Q_{int},Q_{msr}}^{f_{int},f_{msr}} \right]_{CC13} \cdot \left[\frac{M_{Q_{clin}}^{f_{clin}}}{M_{Q_{int}}^{f_{int}}} \cdot k_{Q_{clin},Q_{int}}^{f_{clin},f_{int}} \right]_{Det} \quad (2)$$

The choice of the CC13 chamber as the reference detector and f_{int} allows to consider $\left[k_{Q_{int},Q_{msr}}^{f_{int},f_{msr}} \right]_{CC13} \approx 1$.

Irradiation Field Size Determination. The irradiation field sizes were determined according to the procedures specified in [5]. Table 2 presents the nominal square field sides, L(c) together with the measured irradiation square field sides, S(c) [4].

3.2 Uncertainty Determination Budget Methodology

To determine the typical uncertainty [9] of $\Omega_{Q_{clin},Q_{msr}}^{f_{clin},f_{msr}}$, Eq. 4 is considered. Each factor of the equation has a typical uncertainty, which is the result of several contributions and are summed in quadrature [9–11]. Since the uncertainty of $\left[k_{Q_{int},Q_{msr}}^{f_{int},f_{msr}} \right]_{CC13}$ can be disregarded, the typical combined uncertainty of $\Omega_{Q_{clin},Q_{msr}}^{f_{clin},f_{msr}}$ can be expressed as:

$$u_{\Omega_{clin,msr}} = \sqrt{\left[u_{M_{msr}}^2 \right]_{CC13} + \left[u_{M_{int}}^2 \right]_{CC13} + \left[u_{M_{int}}^2 \right]_{Det} + \left[u_{M_{clin}}^2 \right]_{Det} + \left[u_{k_{clin,int}}^2 \right]_{Det}} \quad (3)$$

response of the detectors, using a reference field of 100×100 mm^2, a CC13 ionization chamber as the

Definition and Estimation of u_M. The uncertainty of the readings "M" of each detector, for any given field size

Table 2 Nominal and irradiation field sides of the Linac used

L(c) [mm]	100	40	30	20	15	10	5
S(c) [mm]	100	41.2	31.4	21.2	16.2	11.2	6.9

$(u_{M_{msr}}, u_{M_{int}}, u_{M_{clin}})$, is mainly related to the detector's positioning and monitor system of the linear accelerator. The positioning of the detector can be affected by the collimating system, u_{coll}, and scanning system, u_{scan}. Regarding the uncertainties of monitor system, u_{mon}, the uncertainties related to the measuring instrumental, such as electrometer and cable, are included in one term. Thus, the typical uncertainty of the detector's reading is given by:

$$u_M = \sqrt{u_{coll}^2 + u_{scan}^2 + u_{mon}^2} \qquad (4)$$

Estimation of u_{coll}. As stated above, this contribution is related to the detector's reading, depending on the accuracy of the beam collimating system. This uncertainty does not depend on the detector and in this case measurements were made only with the stereotactic diode [4].

Estimation of u_{scan}. This source of positioning uncertainties is related to the scanning system used to measure, in this case an automatic water phantom, and is influenced by the detector used. Hence, all the detectors were used to estimate this contribution [4].

Estimation of u_{mon}. The contribution of the beam monitor system is also related with the stability of the detector used and the electrometer used [4].

Definition and Estimation of $u_{k_{clin,int}}$. The field correction factors' uncertainties were extracted from the available literature [5]. Since they are normalized to the reference field, with the exception of the CC13, a ratio of two correction factors was calculated due to the application of the Daisy Chain Method [4], as a consequence:

$$[u_{k_{clin,int}}]_{Det} = \sqrt{[u_{k_{clin,msr}}^2]_{Det} + [u_{k_{int,msr}}^2]_{Det}} \qquad (5)$$

3.3 Estimation of $u_{\Omega_{Q_{clin},Q_{msr}}^{f_{clin},f_{msr}}}$

The typical uncertainty of $u_{\Omega_{Q_{clin},Q_{msr}}^{f_{clin},f_{msr}}}$ is calculated according to Eq. 5. Since a 99% confidence level is requested by the radiotherapy service, an expansion factor $k = 3$ has to be applied. In consequence, the measurement results of $\Omega_{Q_{clin},Q_{msr}}^{f_{clin},f_{msr}}$ are expressed as:

$$\Omega_{Q_{clin},Q_{msr}}^{f_{clin},f_{msr}} = [\Omega_{Q_{clin},Q_{msr}}^{f_{clin},f_{msr}}]_{measured} \pm U_{\Omega_{Q_{clin},Q_{msr}}^{f_{clin},f_{msr}}} \qquad (6)$$

where $U_{\Omega_{Q_{clin},Q_{msr}}^{f_{clin},f_{msr}}}$ is the expanded uncertainty:

$$U_{\Omega_{Q_{clin},Q_{msr}}^{f_{clin},f_{msr}}} = k \cdot u_{\Omega_{Q_{clin},Q_{msr}}^{f_{clin},f_{msr}}} \qquad (7)$$

4 Results

4.1 Uncertainty Budget Determination

The uncertainties related to the determination of small field output factors were estimated for the four detectors used after applying Eq. 5.

Estimation of u_M. To estimate this uncertainty, several measurements were carried out, since there are three present components in this term, u_{coll}, u_{scan} and u_{mon}.

Estimation of u_{coll}. The relative shift between the inline profiles, determined by the diaphragms, is approximately constant for all field sizes, presenting low values. In the crossline direction, determined by the MLC's leaves, there is a more significant shift for the smaller fields. The collimating system uncertainty estimation made, presented in Fig. 1, shows a decreasing behavior as a function of the field size, strongly influenced by the crossline positioning uncertainties.

Estimation of u_{scan}. This source of uncertainties was estimated for all the field detectors. The Razor Detector was used with all the clinical field sizes, while the ionization chambers were used for L(c) = 5, 20 and 40 mm. The uncertainties of the remaining fields were interpolated linearly in order to obtain a representative value.

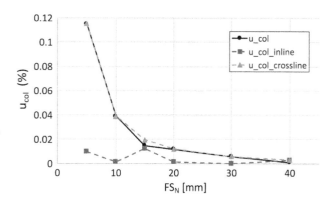

Fig. 1 Typical uncertainty of the beam collimating system (continuous curve)

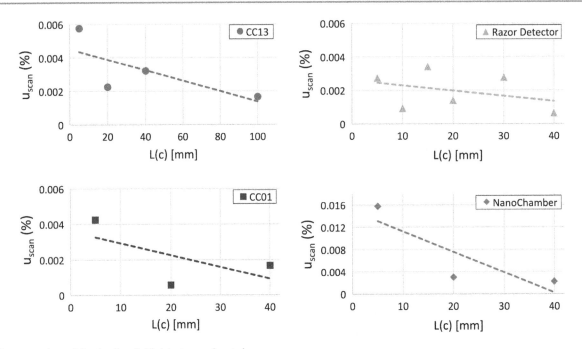

Fig. 2 u_{scan} estimated for the four field detectors under study

The stereotactic diode presents the lowest levels of uncertainty, similar to the ones of the CC13 and CC01 chambers, lower than 0.01%. On the other hand, the NanoChamber presents levels of uncertainty up to three times greater, as seen in Fig. 2.

Estimation of u_{mon}. The monitor system and instrumental uncertainties were estimated together with the detectors' stability uncertainties. The results are illustrated in Fig. 3.

The Razor Detector and CC13 present similar uncertainty levels, while the CC01 and NanoChamber uncertainties are substantially higher, especially in the case of the latter.

Estimation of u_{kclin}. The uncertainty values, presented in Fig. 4, were extracted from the available literature [5]. The values selected for the NanoChamber were the ones of the CC01. There is an exponential increasing behavior in the

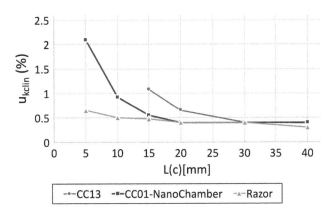

Fig. 4 Field correction factors' uncertainties used in this work

case of the ionization chambers, while the diode's uncertainty is lower than 1% for all the fields under study.

Typical uncertainty values. Figure 5 shows the contributions to the uncertainty budget for each detector. The constant behavior of some of these contributions is due to the Daisy-Chain method application.

For the CC13 and Razor Detector, the most significant source of uncertainty is related to the correction factors. With the exception of the NanoChamber, the correction factors and clinical fields' contributions decrease as the field size increases.

The typical uncertainty values, calculated from this data, are summarized in Table 3. In general terms, the best results were obtained with the Razor Detector. Although the CC13 presents similar and even lower levels of uncertainties in

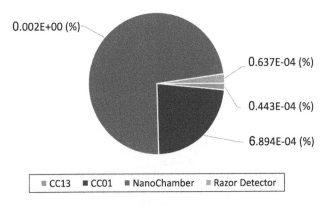

Fig. 3 Monitor system uncertainties. The estimations strongly depend on the detector's choice

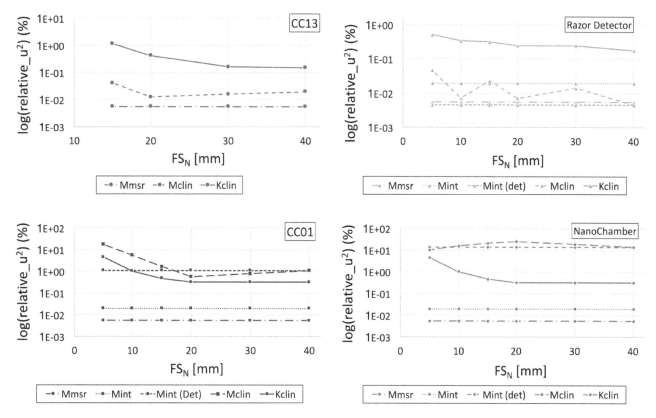

Fig. 5 Contributions to the typical uncertainty for each field detector used

Table 3 Typical relative uncertainty values of the field detectors used in the present work

L(c)(S(c)) [mm]	CC13 (%)	CC01 (%)	NanoChamber (%)	Razor detector (%)
40 (41.2)	0.42	1.58	5.30	0.46
30 (31.4)	0.43	1.48	5.73	0.54
20 (21.2)	0.67	1.41	6.26	0.53
15 (16,2)	1.10	1.80	5.93	0.61
10 (11.2)	–	2.77	5.54	0.61
5 (6.9)	–	4.76	5.36	0.77

some cases, this detector is not suitable for field sizes smaller than 20×20 mm^2.

4.2 Field Output Factors and Their Related Expanded Uncertainties

After applying the methodology, the results, together with the measured output factors, are presented in Fig. 6. The level of confidence is approximately 99% ($k = 3$).

The Razor Detector presents levels of expanded uncertainty of less than 3% for all the fields under study, being in agreement with the required standards of the service.

The CC13 ionization chamber fulfills this requirement only for fields of 20×20 mm^2 and larger and is not considered as appropriate for measurements in smaller fields.

The CC01 ionization chamber levels of expanded uncertainty were higher than expected, being above 3% in all cases. If a 95% confidence level is considered as acceptable ($k = 2$), its use would be recommended for fields up to 15×15 mm^2.

The NanoChamber presents levels of expanded uncertainty in the order of 15% for all the fields under study. A greater number of measurements might be necessary when implementing the methodology to lower these uncertainties.

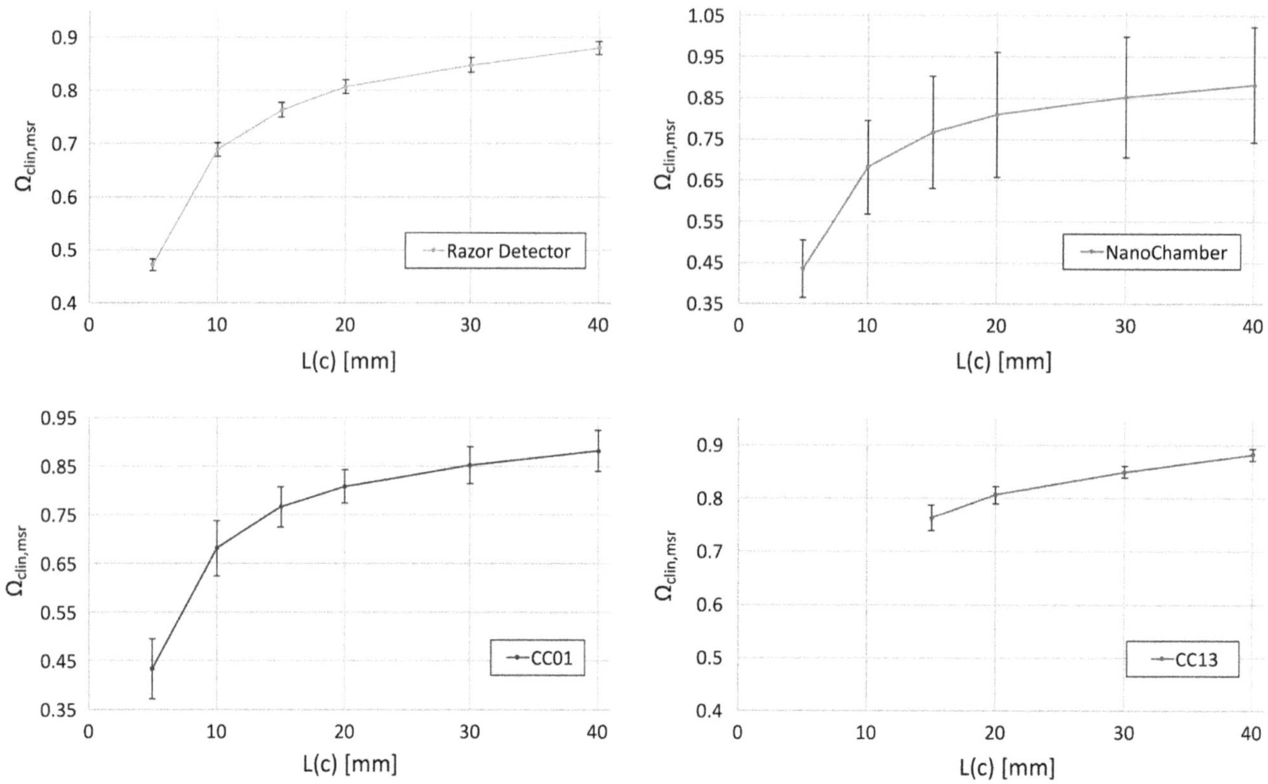

Fig. 6 Field output factors measured with each detector and their expanded uncertainties (k = 3)

5 Conclusions

A novel measurement uncertainty budget methodology for small fields was implemented after measuring field output factors, leading to the following conclusions:

- The methodology is straightforward. simple, and flexible, including the most relevant sources of uncertainty in the beam measurement process. Additionally, it can be used as a tool to assess the performance of field detectors and helping in the selection of the most suitable one for small field dosimetry.
- In the case of the Razor Detector and the CC13 ionization chamber, the highest uncertainty contribution is associated to $k_{clin,int}$. Uncertainties related to the detector's readings are relevant in the case of the CC01 and NanoChamber.
- Uncertainties under 3% with a 99% confidence level were obtained with the Razor Detector for all the field sizes under study.
- According to our results, the ionization chambers under study are suitable only for measurements in fields larger than 15 × 15 mm^2 and further measurements should be made with the NanoChamber.

Acknowledgements We would like to thank "Fundación INTECNUS" and "Instituto Balseiro" for allowing us to work at their radiotherapy service, as well as IBA Dosimetry GmbH, for providing the necessary instrumental and software for testing.

Conflicts of Interest The authors declare they have no conflicts of interest.

References

1. Podgorsak, E.: Radiation Oncology Physics: A Handbook for Teachers and Students. 1st edn. International Atomic Energy Agency, Vienna (2005).
2. Das, I. et al.: Choice of radiation detector in dosimetry stereotactic radiosurgery-radiotherapy. J Radiosurgery 3(4), 177–186 (2000).
3. IAEA: Technical Report Series no. 398: Absorbed dose determination in external beam radiotherapy. International Atomic Energy Agency, Vienna (2000).
4. Tolabin, D.: Medidas de dosimetría relativa y análisis de incertidumbres en campos pequeños. Master's degree in medical physics thesis. Instituto Balseiro, S.C. de Bariloche (2017).
5. IAEA: Technical Report Series no. 483: Dosimetry of small static fields used in external beam radiotherapy. International Atomic Energy Agency, Vienna (2017).
6. Alfonso, R. et al.: A new formalism for reference dosimetry of small and nonstandard fields. Am. Assoc. Phys. Med. 35(11), 5179–5186 (2008).

7. Lárraga-Gutiérrez, J.: Experimental determination of field factors (Ωfclin,fmsr;Qclin,Qmsr) for small radiotherapy beams using the daisy chain correction method. Phys. Med. Biol. 60(15) 5813–5831 (2015).

8. Spretz, T.: Dosimetría de campos pequeños de fotones en radioterapia. Intercomparación entre distintos detectores. Master's degree in medical physics thesis. Instituto Balseiro, S.C. de Bariloche (2016).

9. Granados, C. et al.: Incertidumbres y tolerancias de la dosimetría en radioterapia. Hospital Central de Asturias—Hospital General, Oviedo (1997).

10. IAEA: Measurement Uncertainty. A Practical Guide for Secondary Standards Dosimetry Laboratories. International Atomic Energy Agency, Vienna (2008).

11. BIPM: Evaluation of measurement data—Guide to expression of uncertainty in measurement. JCGM, Paris (2008).

Fundamental Research and Experimental Work on Properties of Tungsten Micro- and Nanoparticle Structured Composite Material

G. Boka, Y. Dekhtyar, S. Bikova, Y. Bauman, P. Eizentals, A. Svarca, and M. Kuzminskis

Abstract

Tungsten micro- and nanoparticle structured composite have been demonstrated recently as a promising material for protection of the radiation therapy patient against radiation. The shielding properties of the composite to a great extent depends on the homogeneity. The present research concentrates on the experimental investigation of the homogeneity of the synthesized composite in dependence on the material mixing methods, particle size, and concentration. The material radiation attenuation properties were explored as well. It was observed that the tendency to form agglomerates becomes greater if tungsten particle size decreases. The best particle distribution uniformity in the composite was obtained with ultrasound disperser. Most effective radiation absorption was observed for the samples with a particle size of 500 nm and 50 nm.

Keywords

Tungsten nanoparticles • Radiation therapy
Tungsten wax composite

1 Introduction

The incidence of malignant tumors remains the major health problem worldwide. One of the most typical cancer treatment procedures is radiation therapy. The main struggle in radiation therapy is to provide a necessary dose to the cancer target volume while protecting the surrounding healthy tissue. If tumor volume is close to the eyes, ears or lips, individually applicable metal alloy (Bi–Sn–Pb) masks are used to protect these organs. The manufacturing process of these masks is time-consuming and complex. In addition, the lead is known to be a toxic material. Another serious problem is electron backscatter from the mask, especially from the material edges. This effect leads to skin dose increase by 30–70% [1].

In recent years, polymer materials with metal nanoparticles became widely used for protection of electronics from radiation [2]. This idea was adopted to simplify protective mask manufacturing process, hereby new polymer-based tungsten micro/nanoparticle structured composite material was developed [3]. The material is plastic after warm—up in hands, non-allergic and non-toxic. The non-toxicity of the tungsten is a great advantage, compared to the lead, being in use typically. The transmission factor of the developed material in the range of beam energies 6–12 meV was close to one of the Bi–Sn–Pb alloy and does not exceed 2.7%, that makes the material suitable for manufacturing of radiation therapy masks.

The absorption properties of the developed polymer wax —tungsten composite strongly depend on its homogeneity. The present paper explores the homogeneity of the synthesized composite in dependence on the material mixing methods, particle size, and concentration. In addition, the radiation attenuation of the obtained composite materials was studied.

2 Materials and Methods

The composite was made by mixing melted polymer wax with tungsten powder of three different fraction size: 50 nm, 500 nm, and 5 μm [4] (Fig. 1). The proportion of composite compound mass was 2:1 (tungsten particles:micro wax). For material mixing, two different methods were used. One of them is ultrasonic processor UP50H and the other—ultrasonic dispergator [4]. The samples were molded in a 1 cm thick plates sized 4×4 cm^2 [5].

G. Boka · S. Bikova · Y. Bauman · A. Svarca
Riga East University Hospital, Clinic of Therapeutic Radiology and Medical Physics, Riga, Latvia

G. Boka · Y. Dekhtyar · S. Bikova · P. Eizentals (✉) · M. Kuzminskis
Riga Technical University, Riga, Latvia
e-mail: peteris.eizentals@gmail.com

© Springer Nature Singapore Pte Ltd. 2019
L. Lhotska et al. (eds.), *World Congress on Medical Physics and Biomedical Engineering 2018*,
IFMBE Proceedings 68/3, https://doi.org/10.1007/978-981-10-9023-3_114

Fig. 1 Micro wax (left) and tungsten particle powder (right)

The homogeneity of the material was studied using scanning electron microscope Hitachi TM 3000 with resolution 30 nm. The preparation of substrate was difficult due to the plasticity of the material there was not possible to cut a thin material slice. Instead, thin material layers were smeared [4]. The composite which showed the best homogeneity was used for further radiation absorption property research.

Composite's absorption was characterized by transmission factor. The measurements were performed for electron radiation supplied from a linear accelerator Clinac 2100C/D (Varian Medical Systems, USA). The sample was irradiated with electron energy 9 meV which is usually used to treat tumors close to skin surface [4]. Delivered dose was 3 Gy to the depth of the dose maximum. An incident radiation beam was perpendicular to the surface of a solid water phantom, on which the specimens were located. The used radiation detector was Scanditronix Wellhöfer plane parallel ion-chamber type PPC05 with its active volume 0.05 cm^3. Measurements were repeated five times for each test sample. The transmission factor was calculated using the equation

$$TF(\%) = \frac{Q(pC)}{Q_{of}(pC)} \cdot 100 \qquad (1)$$

where Q(pC) is chamber registered charge with the composite specimen in the radiation field, Q_{of} (pC)—chamber registered a charge for open field without a specimen.

The planar dose distributions perpendicular to the central axis were measured by using Gafchromic EBT2 self-developing films in solid water phantom [3]. RGB Canon ScanLide100 scanner was used to scan exposed films. For the film, the optical density reconstruction only red color channel was used because for doses up to 3 Gy scanner

provides the most sensitive response for this color channel. The scanner output resolution was set to 75 dpi. Acquisition and analysis of images were performed using PTW (Freiburg, Germany) software. Dose distributions were calculated using the obtained optical density—dose calibration curves. The scattering effect for metal alloy and the composite material was compared by subtracting one dose distribution from another, using the equation

$$Scattering\ effect = \frac{D_{metal\ block} - D_{composite\ block}}{D_{metal\ block}} \qquad (2)$$

3 Results and Discussion

Eight different material structures were made: three of them were made with every fraction size (5 μm, 50 nm, and 500 nm) and micro wax mixed with the ultrasonic processor; three mixed with ultrasonic dispergator; two samples were the compound of all three fractions particles and micro wax mixed either with the ultrasonic processor or ultrasonic dispergator.

Scanning electron microscope (SEM) images were analyzed in different magnifications (200, 500 and 1500 times) to get precise homogeneity evaluation. For better image comparison, just images with 500-time magnification were analyzed in this paper.

For the material with 5 μm tungsten particles, more inhomogeneity was observed in a sample mixed with US processor. No significant difference was observed between particle distributions in a material with different mixing methods. It is described as fact that particles with big sizes

Fig. 2 SEM images of the prepared samples. 50 nm particles (top row), 500 nm particles (middle row) and 5 μm particles (bottom row), particles mixed with US processor (left column) and particles mixed with US dispergator (right column)

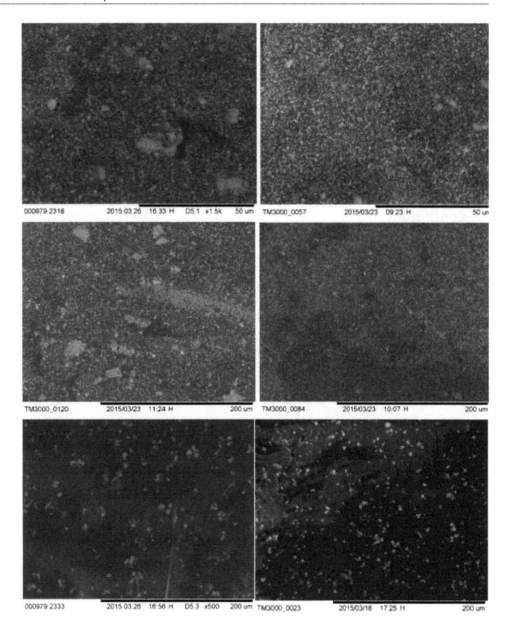

do not tend to make agglomerates, so relatively good homogeneity can be obtained with both mixing methods [4] (Fig. 2).

For the composite with 500 nm sized particles, the difference in homogeneity of the samples, mixed with different methods are well expressed. The material mixed with US processor had a considerable amount of agglomerates with size up to 60 μm. The inhomogeneity change material absorption properties. For material mixed with US dispergator, particles distribution was with better homogeneity.

The homogeneity differences between two material samples with 50 nm particles mixed with different methods were observed, as well. Sample mixed with US processor was more non-homogenous, comparing to one mixed with US dispergator.

For samples with the best homogeneity, radiation absorption properties were analyzed. As it follows the image analysis, the better homogeneity was obtained by means of US dispergator. Three different material structures were made: three of them were made with every fraction size of particles (5 μm, 50 nm, and 500 nm) mixed with micro wax. Two of samples was the compound of all three fraction size particles mixed with micro wax. All materials were mixed with US dispergator. Transmission factor for all samples is summarized in Table 1.

The highest transmission factor was observed for composite made of 5 μm particles. Surprisingly, use of the mixture of all three fractions did not reduce transmission factor, it remains higher for material with smaller particles. All transmission values are below tolerance value of 5%.

Table 1 Transmission factors for samples

Tungsten fraction size in sample	Transmission factor (%)
5 μm	3.6
500 nm	2.1
50 nm	2.1
5 μm + 500 nm + 50 nm first	2.6
5 μm + 500 nm + 50 nm second	2.3
First research result	1.5

4 Conclusions

This paper introduced a novel material for radiation protection mask production for cancer treatment. The new material consists of tungsten micro and nanoparticles and micro wax base.

The best homogeneity of the material could be achieved by mixing of components with US dispergator. US processor could be used for mixing only for particles, bigger than 5 μm.

Absorption properties are better for materials manufactured with smaller particles 500 and 50 nm (Transmission factor 2.1%). It means, that this material can be used in electron radiation therapy protection mask manufacturing.

Conflict of Interest The authors declare that there is no conflict of interest.

References

1. Krumeich F. Properties of Electrons, their Interactions with Matter and Applications in Electron Microscopy. Switzerland: ETH Zurich, 2010. 24 p.
2. Marcos J., Jurado M., Carapelle A., Orava R. Radiation Shielding of Composite Space Enclosures. Spain: TECNALIA, 2013. 10 p.
3. G. Boka, E. Reine, A. Svarca, M. Kuzminskis, S. Bikova, Y. Bauman, Y. Dekhtyar, "Absorption and scatter properties of tungsten structured composite material", ENCY2015, Budapest, May 28–30, 2015.
4. Svarca A., "Synthesized composite with tungsten nanoparticles for radiation protection in radiotherapy", Master's thesis, Riga Technical University, 2014.
5. Reine E., "Absorption and scattering properties of tungsten composite material", Master's thesis, Riga Technical University, 2013.

Assessment of Low Doses During Diagnostic Procedures Using BeO Detectors and OSL Technique

Anna Luiza M. C. Malthez, Ana Clara Camargo, Ana Paula Bunick, Danielle Filipov, Celina Furquim, Elisabeth Yoshimura, and Nancy Umisedo

Abstract

Advances in imaging techniques with the use of radiation increased the quality and power of medical diagnostic. At the same time, concerns about the doses to patients arose in the radiation protection community. According to UNSCEAR (United Nations Scientific Committee on the Effects of Atomic Radiation), the second main contribution to the population annual dose is from exposure in medical procedures. Although the doses are relatively low in diagnostic, the number of exams for the same patient is growing, bringing especial attention to doses to children or to radiosensitive regions in the body. International organizations recommend the establishment of reference levels in diagnostic procedures and the tracking of the patient exposures. Considering this, we studied the use of optically stimulated luminescence (OSL) and thermoluminescent (TL) techniques together for assessment of skin doses in simulated (pediatric phantom) and real patients during diagnostic procedures. BeO and LiF detectors were used in mammography and radiographic procedures, and in CT examinations. The results of TL and OSL were compared to each other and to international reference levels, when it was possible. Both detectors were able to evaluate doses in the range of 15 µGy to 100 mGy. BeO detectors presented compatible results with LiF detectors for doses in the range of few mGy and low uncertainties in the range of µGy, for both, adults and pediatric patients. We can conclude that the advantages of OSL technique combined to intrinsic characteristics of BeO (tissue equivalence and flat energy response for photons up to 100 keV) can be explored for assessment of patient doses in diagnostic procedures.

Keywords

Beryllium oxide detectors • Optically stimulated luminescence • Patient doses

1 Introduction

According to international organizations like IAEA (International Atomic Energy Agency) and UNSCEAR (United Nations Scientific Committee on the Effects of Atomic Radiation), exposures of patients undergoing medical procedures represent the most significant source of artificial exposure to ionizing radiation. Furthermore, the collective doses due to medical exposures are growing very fast in the last decades [1, 2].

Several recent works discuss how to estimate the risk and quantify the patient exposures to low-dose levels [3–6]. It is noteworthy that the adverse effects of low-dose exposure in medical imaging are not well established and more efforts are required in the evaluation and assessment of risk [7]. Particularly, there is a concern about the repetition of examinations of the same patient, especially in children, as the risk of cancer induction is higher than in adults, and about examinations with high exposure and exposure of more radiosensitive regions of the body, such as mammography and CT [2, 6, 8].

In this sense, IAEA, EC (European Commission) and UNSCEAR have reinforced the importance of optimization of protection, mainly of patients submitted to radiodiagnostic procedures, establishing of the Diagnostic Reference Levels (DRL), regional or local, for several X rays exams [1, 8]. Countries or regions, where national DRL are not established could use local dosimetry or practical data, or the published values that are appropriate for the local circumstances [1].

A. L. M. C. Malthez (✉) · A. C. Camargo · D. Filipov
C. Furquim
Federal University of Technology - Parana, Curitiba, Brazil
e-mail: annaluizacruz@gmail.com

A. P. Bunick
Pelé Pequeno Príncipe Research Institute, Curitiba, Brazil

E. Yoshimura · N. Umisedo
Institute of Physics, University of Sao Paulo, Sao Paulo, Brazil

L. Lhotska et al. (eds.), *World Congress on Medical Physics and Biomedical Engineering 2018*,
IFMBE Proceedings 68/3, https://doi.org/10.1007/978-981-10-9023-3_115

The results of local dosimetry and proposed regional or national DRL are based on indirect methods, using exposure parameters (tube current, exposure time and tube voltage in X rays exams) or direct methods using ionization chamber and passive detectors, like thermoluminescent (TL) and optically stimulated luminescence (OSL) detectors. In both methods, evaluated index (dose-length product) or quantities (entrance skin dose) can be associated or correspondent to a DRL. Some passive detectors (LiF and BeO) present some advantages over active detectors, such as: small size, relative low cost, tissue equivalence, relatively low energy dependence, a broad range of linearity of dose response, high sensitivity to several types and energy of ionizing radiation and availability in various sizes and shapes [9].

Considering the growing of medical exposure, the goal of this work was to assess doses in simulated (phantom) and real patients during diagnostic procedures like mammography (in adults) and CT examinations and radiographies in pediatric patients, using BeO detectors with OSL technique. When it was possible, the results from BeO detectors were compared with dose values obtained from LiF:Mg,Ti detector with TL technique and reference levels.

2 Materials and Methods

2.1 Detectors and Readouts

The detectors used in this work were BeO (Thermalox 995 square chips with 4.5 mm side) and LiF:Mg, Ti (TLD100—Bicron square chips with 3.1 mm side) with the OSL and TL techniques, respectively. LiF detectors were used, when it was possible, as standard detector for comparison.

Those were previously selected (similar sensitivities within the same batch) and subjected to appropriate annealing prior to use, erasing residual signals stored in the detectors. The thermal annealing of BeO detectors consisted in 30 min at 400 °C, and of LiF detectors, in 1 h at 400 °C followed by 2 h at 100 °C.

The OSL readouts of BeO were performed in a Risø TL/OSL DA reader (DTU—*Nutech*) equipped with blue light stimulation. The settings for optimal OSL readouts of BeO were stimulation during 40 s with 90% of power of blue stimulation LED and detection of OSL signal using a transmission filter Hoya U340 in front of PMT feed with 1125 V. Integrating the curve from 1 to 20 s and from 21 to 40 s and subtracting those OSL intensities we have the corresponding signal to be associated to exposure. The TL readouts of LiF detectors also were performed in Risø TL/OSL reader, with appropriated detection window, and in a RA 04 TLD READER-ANALYSER (RadPro Gmbh) using a heating rate of 5 and 10 °C/s, respectively, and final temperature of 400 °C for stimulation.

2.2 Diagnostic Procedures

The assessment of patient doses, with both BeO and LiF detectors, was done using sets with at least 2 detectors of each type per point or position in the phantom or patient. Preventing light exposure, the detectors were packed in black plastic bags (radiotransparent).

This study was carried out at the Regional University Hospital of Campos Gerais in the city of Ponta Grossa (Brazil) and at Pequeno Príncipe Children's Hospital in the city of Curitiba (Brazil). The Ethics Committees in Research of both hospitals approved previously all procedures and use of data involving patients (registered as 51155915.0.00 00.5580 and 2.290.098 at Plataforma Brasil national database).

Pediatric CT exams. In order to simulate real exposure conditions in diagnostic procedures with newborn patients, a plastic pediatric phantom with a length of 50 cm filled with water was used. The exams were performed in two different CT (Philips MX16 and Philips Neusoft, both 16 channels). Nine tests with pediatric and adult exam protocols (Table 1) were run. In order to approach the actual skin dose received by the patients submitted to this type of examination, the detectors were positioned in the region corresponding to the frontal skull of the phantom. The protocols used during the exams are described in Table 1, as well as the evaluated dose values.

Pediatric radiography exams. Eight patients (ages ranging from 1 month up to 13 years old), 4 males and 4 females, who underwent chest X-ray examinations in the anteroposterior projection were arbitrarily chosen, while hospitalized in a Pediatric Intensive Care Unit. The examination was performed with mobile X-ray equipment, positioned near the patient bed, with a central beam perpendicular to the region examined. The detectors were positioned on the patient skin in the region of incidence of the central X-ray beam. The evaluated entrance skin doses and parameters used during the exams are described in Table 2.

Mammography exams. Five female patients undergoing routine mammography examination were selected according to the size of the breast (mean breast). In these tests the automatic protocol was selected in the equipment and the detectors were positioned outside the radiation field. This part of the study aimed to evaluate the scattered dose received in the regions of the iris, thyroid, gonads and the opposite breast during mammography exams.

2.3 Approaches to Assessment of Dose

The OSL and TL intensities from BeO and LiF detectors, corrected for background radiation contribution, were converted in absorbed dose in air using calibration factors previously evaluated to ^{60}Co photon energy, in the case of patient doses during pediatric radiography exams. For mammography and CT exams, calibration factor for a specific beam energies were evaluated for both detectors. Calibration factors were evaluated through a calibration curve using a calibrated ionization chamber for ^{60}Co photon energy in air. In the case of pediatric radiography exams, appropriate energy dependence correction factors, obtained for each tube voltage were applied to get the patient dose [10].

In the case of TL technique, it is necessary a non-irradiated subset of detectors to evaluate the TL intensity corresponding to the background. However, in the case of OSL, the background signal of detector can be evaluated using the same OSL intensity used to evaluate the dose, without other detectors, if measurements are performed immediately after the annealing treatments.

3 Results and Discussion

Table 1 presents the results of skin doses evaluated from OSL and TL intensities from BeO and LiF detectors during CT procedure in pediatric phantom. Although it is not usual and the reference levels or local dosimetry data are not based on the entrance skin dose for CT, in this test, the aim was to compare dose evaluations with BeO and LiF detectors for assessment of doses in surface of phantom during the procedure.

The doses evaluated with BeO detectors presented a lower uncertainty when compared with the doses evaluated with LiF detectors, being the absolute values similar in almost all protocols. Also, the results obtained with both detectors showed that specific protocols for pediatric patients resulted in a lower skin dose in the tests with pediatric phantom.

Table 1 Skin dose (SD) evaluated with LiF (TL) and BeO (OSL) detectors in CT examinations of pediatric phantom using several protocols

CT equipment	Protocol	Tube voltage (kV)	Current × Time (mAs)	Time (s)	Pitch	SD (mGy) OSL	SD (mGy) TL
Philips	Pediatric skull	120	250	9.07	0.6713	24.6 ± 0.2	34 ± 2
	Adult skull	120	369	12.06	0.5035	35.7 ± 0.3	51 ± 4
	Skull trauma	120	399	22.02	0.5035	38.6 ± 0.6	61 ± 3
	Skull trauma	120	399	17.02	0.6713	37.4 ± 0.5	53 ± 7
	Pediatric skull	90	338	7.90	0.5035	9.3 ± 0.5	11 ± 2
Neusoft	Skull routine	140	381	17.04	1	60 ± 3	79 ± 5
	Skull routine	120	347	8.03	0.8631	33 ± 4	39 ± 1
	Skull trauma	140	299	21.09	0.6713	54 ± 1	58 ± 6
	Skull trauma	140	299	21.09	0.5035	55 ± 1	58 ± 6
	Skull less than one year	120	120	19.08	1.009	17.6 ± 0.4	21 ± 1

Table 2 Entrance skin dose (ESD) evaluated from BeO (OSL) detectors in chest X ray examinations of pediatric patients hospitalized in an intensive care unit

Patient	Age[a]	Weight (kg)	Tube voltage (kV)	Current–time product (mAs)	Focus skin distance (cm)	ESD (μGy)
1	3 m + 5 d	3.17	50	2	78	73 ± 5
2	13 y + 5 m + 25 d	72	55	5	94	122 ± 28
3	2 y + 2 m + 28 d	11.2	50	2	99	68 ± 11
4	1 m + 30 d	3.9	50	2	73	81 ± 7
5	1 m + 30 d	8.83	50	2	90	63 ± 4
6	1 m + 13 d	4.6	62	2	62	122 ± 6
7	3 y + 7 m + 27 d	14.7	65	2	65	153 ± 7
8	1 y + 7 m + 29 d	12	63	2	67	131 ± 25

[a]years (y) + months (m) + days (d)

Table 3 Dose evaluated with LiF (TL) and BeO (OSL) detectors during mammography examination

Patient	Dose (µGy)							
	Left breast		Thyroid		Iris		Gonads	
	OSL	TL	OSL	TL	OSL	TL	OSL	TL
1	17 ± 3	94 ± 19	52.7 ± 0.3	112 ± 8	51 ± 5	113 ± 6	97 ± 20	97 ± 20
2	22 ± 1	45 ± 16	29 ± 4	46 ± 18	43 ± 12	22 ± 22		
3	39 ± 8	65 ± 3	50 ± 14	100 ± 24	26 ± 7	248 ± 158	22 ± 3	74 ± 100
4	26 ± 2	38 ± 6	44 ± 1	71 ± 28	21 ± 4	56 ± 4	20 ± 15	33 ± 20
5	25 ± 3	94 ± 20	42 ± 3	32 ± 16	25 ± 2	37 ± 37		
6	17 ± 3	94 ± 19	52.7 ± 0.3	112 ± 8	51 ± 5	113 ± 6	97 ± 20	97 ± 20
7	22 ± 1	45 ± 16	29 ± 4	46 ± 18	43 ± 12	22 ± 22		
8	39 ± 8	65 ± 3	50 ± 14	100 ± 24	26 ± 7	248 ± 158	22 ± 3	74 ± 100

Table 2 presents the results of entrance skin doses in pediatric patients submitted to chest X rays exams. For this type of examination, previous studies recommended reference levels lower than 100 µGy, but they can change mainly due differences in adopted parameters and place of exams, besides the patients characteristics (size and age) [1, 2]. As previously, BeO detectors presented low uncertainties for the eight patients with several sizes and ages.

For the doses evaluated in peripheral regions during a mammography examination, the associated uncertainties are higher (Table 3); as the doses evaluated are in the range of tens of µGy, and are due to scattered radiation. Comparing the results of the two detectors, the uncertainties and the obtained values are usually lower with the BeO detectors. Moreover, in some cases, the assessment of doses was possible just with BeO detectors as uncertainties were higher than absolute value, as dose in the area of gonads in patients 3 and 8, for example.

4 Conclusions

For the assessment of patient dose during diagnostic procedures, in the case of CT and mammography exams, BeO detectors used with OSL technique present similar results to those obtained with LiF detectors, which are largely applied in situ dosimetry. Also, the dose levels estimated with BeO detectors were in the established range of international DRL, in the case of pediatric patients submitted to chest X rays exams.

Although more studies are required, as scattered radiation distribution and the protocols adopted, in addition to the anatomical characteristics of the patient, may influence the evaluation of the received doses, the results obtained with BeO detectors indicate that the optimized reading process using the OSL technique provide good dose estimates in diagnostic procedures, even for doses of tens of µGy.

Conflicts of Interest The authors declare that they have no conflict of interest.

References

1. IAEA Homepage, https://www.iaea.org/topics/patients#news, last accessed 2018/01/10.
2. UNSCEAR. Sources, effects and risk of ionizing radiation. In Report to the General Assembly with Scientific Annexes, vol. II, Scientific Annex B (2013).
3. Järvinen, H., Vassileva, J., Samei, E., Wallace, A., Vano, E., Rehani, M.: Patient dose monitoring and the use of diagnostic reference levels for the optimization of protection in medical imaging: current status and challenges worldwide. J. Medical Imaging, 4(3), 031214 (2017). https://doi.org/10.1117/1.jmi.4.3.031214.
4. Brenner D., Elliston C., Hall E., Berdon W. Estimated risks of radiation-induced fatal cancer from pediatric CT. AJR Am J Roentgenol. 2001 Feb; 176(2):289–96. [PubMed]
5. Pearce MS, Salotti JA, Little MP, McHugh K, Lee C, Kim KP, et al. Radiation exposures from CT scans in childhood and subsequent risk of leukemia and brain tumors, an historical cohort study. Lancet (2012).
6. Mathematical modelling of radiation-induced cancer risk from breast screening by mammography
7. Hobbs, J.B., Goldstein, N., Lind, K.E., Elder, D., Dodd III, G.D., Borgstede, J.P.: Knowledge of Radiation Exposure and Risk in Medical Imaging. J. Am. Coll. Radiol.,15, 34–43 (2018).
8. European Commission. Radiation protection N. 180: Medical radiation exposure of the European population. Part 1/2. Luxembourg: Publications Office of the European Union (2015).
9. Malthez, A. L.M.C., Freitas, M.B., Yoshimura, E.M., Umisedo, N. K., Button, V.L.S.N. OSL and TL techniques combined in a beryllium oxide detector to evaluate simultaneously accumulated and single doses. Applied Radiation and Isotopes, 110, 155–159 (2016).
10. Malthez, A. L.M.C., Freitas, M.B., Yoshimura, E.M., Button, V.L.S.N. Experimental photon energy response of different dosimetric materials for a dual detector system combining thermoluminescence and optically stimulated luminescence. Radiation Measurements 71, 133–138 (2014).

Development of a Patient Dosimetry Record System in an Oncological Hospital in Mexico

Reyna O. M. González, Gerardo A. D. Enríquez, Sandra N. L. Rocha,
Rafael S. Samra, Jorge L. P. Castillo, Yolanda N. Villaseñor,
and Hector E. Galván

Abstract

Medical imaging studies represent an important tool for patient diagnosis. However some may be harmful due to the amount of ionizing radiation used in them, this is the main reason why it is necessary to keep track of the dose received by each patient. There are many countries in which, as a norm, a dosimetry record of each patient who has undergone imaging studies using ionizing radiation is taken. Mexico is trying to implement this kind of registration, however, commercial systems are still not available. The aim of this article is to describe the development and implementation of a system that allows doctors to assess and monitor the total effective dose of each patient. This system is aimed to work under the consideration of certain requirements and as long as the right protocols are taken.

Keywords

Medical imaging • Ionizing radiation • Dosimetry record
Effective dose • DICOM • Equivalent dose
Image metadata

1 Introduction

Diagnostic Imaging and Nuclear Medicine studies are an effective tool especially in an oncology facility where are needed for the diagnostic and also to verify the treatment response.

Radiation and radiation dose reduction are at the top of the list of controversial topics in medical imaging today [1].

R. O. M. González (✉) · G. A. D. Enríquez · S. N. L. Rocha · R.
S. Samra · J. L. P. Castillo · Y. N. Villaseñor · H. E. Galván
Instituto Nacional de Cancerología, Cuidad de México, Mexico
e-mail: rmarianag@hotmail.com

R. O. M. González · G. A. D. Enríquez · S. N. L. Rocha · R. S.
Samra
Universidad La Salle, Cuidad de México, Mexico

To our knowledge, commercial systems to automatize the dose collection data are still not available in Mexico; however, there are certain free open-source software designed to perform this kind of task, specifically for CT (Computed Tomography) Scans.

1.1 Medical Imaging and Ionizing Radiation

Medical imaging involves performing diagnostic tests in which medical equipment are used to obtain a detailed image of structures inside the body. They help getting a quickly and safely detection of many diseases; we know many of these studies use ionizing radiation to obtain these images [2, 3].

The damage caused by radiation in organs depends on the study type, radiation type, tissue radiosensitivity and absorbed dose (Table 1). Absorbed dose is expressed in a unit called gray (Gy). Our work deals with radiation risk due to partial body irradiation using effective dose. This magnitude considers how dose imparted to each body part contributes to whole body risk of developing cancer or hereditary effects. Effective dose measures ionizing radiation in terms of its potential to cause damage, it is reported in sievert (Sv) [4].

2 Dosimetry Record

The developed system provides information of the dosimetry history of any patient undergoing medical imaging studies and it is oriented towards the calculation of the effective dose in two image modalities: Computed Tomography (CT) and Mammography (MG). This system automatically collects and analyzes data of the radiation received by the patient on both modalities, regardless of the equipment manufacturer. This system is able to work even considering the different protocol issues found in the Institute and allows for an effective and accurate dose calculation.

© Springer Nature Singapore Pte Ltd. 2019
L. Lhotska et al. (eds.), *World Congress on Medical Physics and Biomedical Engineering 2018*,
IFMBE Proceedings 68/3, https://doi.org/10.1007/978-981-10-9023-3_116

Table 1 Tissue weighting factors [7]

Tissue	Weighting factors	Sum of weighting factors
Bone marrow, colon, lung, stomach, breast, rest of the tissues	0.12	0.72
Gonad	0.08	0.08
Bladder, esophagus, liver, thyroid	0.04	0.16
Bone surface, brain, salivary glands, skin	0.01	0.04
	Total	1

The system is made up by two main parts. The first one is represented in Fig. 1, it runs on a server where the data that comes from the PACS is received and processed to perform the effective dose calculations; it also collects basic information about the patient and their imaging studies. The system automatically downloads the patient's daily studies, allowing the doctor to work with updated information. Finally everything is stored in a database for future reference.

The second part of the system (Fig. 1) provides information about the patient's effective dose through a graphical interface. When entering the program a list of all patients who have imaging studies is deployed, patients may be searched by patient ID, name, age or gender. Double clicking on the patient's name shows specific radiological information divided in the following three tabs.

- First tab "General Information", this tab shows general dose information of the patient, it is possible to see all the studies that have been done to the patient and how much dose each study brings. The total dose displayed in the circle is proportional to the sum of doses in all studies (Fig. 2).

- Second tab "Dose", the second tab is comprised of a series of graphs. The system displays the patient's total effective dose within a period, it has the ability to select an specified time period (Fig. 3).
- Third tab "Body Diagram"; an outline of a human body is shown. Adopting the International Commission on Radiological Protection (ICRP) recommendations of 20 mSv per year for staff, this scheme will display three colors: red, yellow and green, being this last one an acceptable dose and the red one being a dose greater than 20 mSv. Finally the pie chart shows how effective dose is divided among the image modalities (Fig. 4).

3 Server-PACS Communication

To obtain the information data from the PACS it was necessary to incorporate a DICOM image server, through the DCM4CHEE system. DCM4CHEE is a free software program which enables communication, collection and storage of DICOM images [5, 6]. The query and retrieve images from the PACS were done through a supplement of the DCM4CHEE called DCM4CHE2 DICOM Toolkit, this add-on also contains many tools for the control and management of DICOM files. We worked particularly with the tool dcmrcv, which allowed receiving studies from the PACS.

4 Obtaining and Managing Data for Effective Dose Calculation

Most of the medical imaging equipment report issued dose, this data is indispensable for the calculation of the effective dose, its location varies depending on the modality of study;

Fig. 1 Diagram of the system's first part: General work process; and second part: Graphical Interface

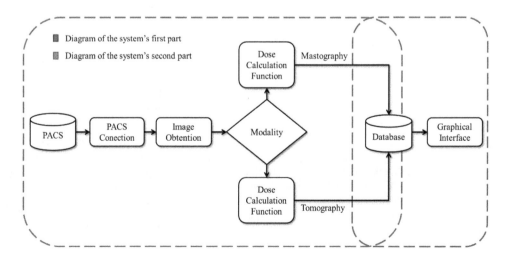

Fig. 2 Overview

Fig. 3 Dose data

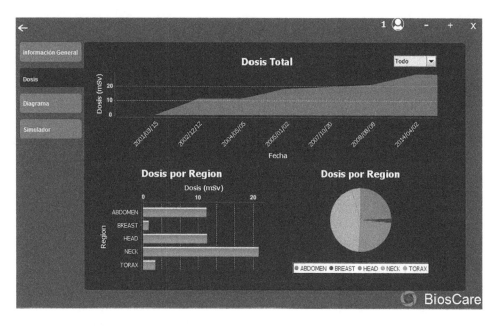

that is why a separation between these calculations and the method of data collection was done.

4.1 Mammography Effective Dose Calculation

The digital Mammography equipment report average glandular dose in the DICOM image metadata. This is the key point for the effective dose calculation. The main metadata of interest is "Organ Dose" that indicates the equivalent dose reported in cGy, note this is not the same as effective dose, since this last one is defined as the sum of weighted tissue of the equivalent doses in all the specific tissues and organs of the body [7].

$$E = \sum_T w_T H_T \qquad (1)$$

The equivalent dose in the tissue or the organ and its weighting factor for the tissue, is shown in Table 1, which as its name implies, ponders the irradiation effect for each organ or tissue depending on its mass.

Considering the formula for effective dose and taking into account that the weighting factor for mammography is 0.12 for both breasts, we can take a factor of 0.06 for each breast, the following expression is obtained for the final effective dose calculation.

$$E[mSv] = (H\_T[cGy])(0.06)(100) \qquad (2)$$

Fig. 4 Dose per region scheme

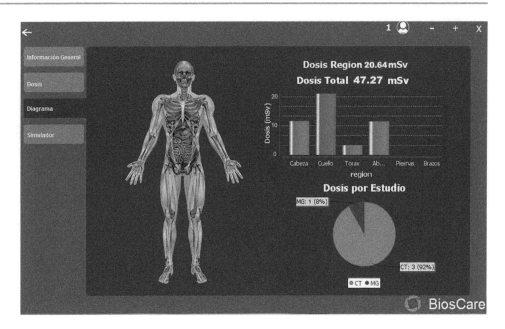

The "Organ Dose", reported by the equipment, is multiplied by 100 to obtain the final effective dose in mSv.

4.2 Tomography Effective Dose Calculations

The DICOM Standard has defined a set of Radiation Dose Structured Report (RDSR) for recording and storing dose details in a DICOM study; however, this function requires the payment of a license. Having this type of licenses would facilitate dose calculations significantly [8].

Although DICOM RDSR is standard for new CT scans there are still many CT scans that report the DLP (Dose Length Product) and additional dose data in a radiation dose report.

The key fields to extract in the report are the total DLP and the DLP per regions. All this fields are of great importance to get an accurate effective dose.

To obtain these data, in legacy CT modalities, an algorithm was developed to convert the image data, inside the metadata, to a binary system of 512×512 which then was converted into a JPG image. Once the JPG image was obtained an OCR (Optical Character Recognition) was used to obtain the data.

To calculate the effective dose we used the "Monte Carlo conversion factors" (Ta-ble 2) represented in the following formula as the value "k" related to the body part scanned by the CT [9].

$$E[mSv] = k(DLP) \qquad (3)$$

The Monte Carlo technique is used to calculate dose conversion coefficients for dose estimations in different parts of the body.

Each report of absorbed dose contains the body regions where the study was conducted. The system gets the region as well as the DLP for each of them, the effective dose of each region is calculated and added to get a total effective dose of the study (Table 2).

5 Results and Validation

To validate the data delivered by the software, a team of medical physicists of the National Cancer Institute of México (INCan) verified that the dose data as well as the effective dose calculation were correct for both modalities. Our sample was 18 random patients; 8 mammography and 10 CT Scans.

For the first stage validation a test was done to ensure that the DLP and AGD (Average Glandular Dose) values reported by the equipment were within a range of 10% for Mammography and 20% for Tomography. The physicists then took DLP values and regions of each patient study and proceed to make manual calculations using the formula for effective dose for both Tomography and Mammography.

Table 2 Montecarlo coefficients [9]

Tissue	Monte Carlo coefficient
Head and neck	0.0031
Head	0.0021
Neck	0.0059
Chest	0.014
Abdomen y Pelvis	0.015
Trunk	0.015

Table 3 Comparison of patient's effective dose between the system and manual calculations

Mammography		Tomography	
Manually calculated effective dose [mSv]	System's effective dose [mSv]	Manually calculated effective dose [mSv]	System's effective dose [mSv]
0.1512	0.1512	24.969	24.960
0.2082	0.2082	32.49	32.490
0.3252	0.3252	42.513	42.510
0.2346	0.2346	30.0705	30.070
1.2882	1.4886	31.2345	31.230
0.2196	0.2196	9.027	28.039
0.2538	0.2538	11.884963	12.180
0.7368	0.7368	42.498	43.610
		39.474	39.580
		35.385	34.980

The dose data readings for mammography were correct; in tomography small differences were found, however the system was able to identify and correct most of those differences.

The next step was to verify, in the same studies, the correct calculation of effective doses. For this the effective dose values thrown by the system against the effective dose data calculated manually by the physicists, in each modality, were compared. The results were as follows:

There was a case in tomography where the difference was significant, however, it was due to the fact that in the manual calculation an incorrect K factor was taken (Table 3).

6 Conclusions, Currents and Future Work

Having a system that provides information about how much radiated a patient is, could be considered as an extra tool of great importance to provide the patient better medical quality care because it could prevent unnecessary or repeated studies.

The system is developed to work with the majority of medical imaging equipment for these modalities available in the market. In tomography it is necessary to make some adjustments depending on the institution where it would be installed, this is because for the calculation of the effective dose as we know, we need to know the specific region where the study was conducted and we realize that hospitals in Mexico do not always follow the same protocols so it would be needed to make some changes for its adaptation.

Currently it is featured the effective dose calculations for mammography and tomography but is planned to expand to planar X-ray and angiography which are also widely used in the medical field and contribute significantly to the total dose of the patient. In addition we plan to incorporate into the system a dose simulator to indicate how much dose the patient could receive, given a certain study, and which would be the organs and regions most affected.

Conflict of Interest The authors of this paper declare that they have no conflict of interest.

References

1. Consumer Reports Magazine.: LOS PELIGROS DE REALIZARSE DEMASIADAS TOMOGRAFÍAS COMPUTARIZADAS (CT SCANS). Consumer Reports, (2014).
2. Macovski, A.: Medical imaging systems. Prentice Hall, (1983).
3. Prince, J. L., & Links, J. M.: Medical imaging signals and systems. Upper Saddle River, NJ: Pearson Prentice Hall, (2006).
4. United Nations Scientific Committee on the Effects of Atomic Radiation.: Sources and effects of ionizing radiation, UNSCEAR 1996 report to the General Assembly, with scientific annex, (1996).
5. Mildenberger, P., Eichelberg, M., & Martin, E.: Introduction to the DICOM standard. European radiology, (2002).
6. Parisot, C.: The DICOM standard. The International Journal of Cardiac Imaging, (1995).
7. ICRP.: The 2007 Recommendations of the International Commission on Radiological Protection. ICRP Publication 103. Ann. ICRP Vol. 37 (2–4), (2007).
8. Clunie, D.: Extracting, Managing and Rendering DICOM Radiation Dose Information from Legacy & Contemporary CT Modalities. Corelab Partners, (2010).
9. McCollough, C., Cody, D., Edyvean, S., Geise, R., Gould, B., Keat, N., & Morin, R.: The measurement, reporting, and management of radiation dose in CT Report of AAPM Task Group 23. American Association of Physicists in Medicine, (2008).

Electronic Portal Imaging Devices Using Artificial Neural Networks

Frédéric Chatrie, Fouad Younan, Jocelyne Mazurier, Luc Simon,
Laure Vieillevigne, Régis Ferrand, Ana Rita Barbeiro,
Marie-Véronique Le Lann, and Xavier Franceries

Abstract

The aim of this work was to use the Artificial Neural Network (ANN) in External Beam Radiation Therapy (EBRT), especially for pre-treatment patient-specific quality assurance of Conformational Radiation Therapy (CRT) and Intensity-Modulated Radiation Therapy (IMRT) using Electronic Portal Imaging Device (EPID). The EPIDs need frequent calibration and complex setting in order to be used with dedicated dosimetry software. The idea was to create a model with ANN algorithms allowing the reconstruction of the 2D dose distribution comparable with a corresponding Treatment Planning System (TPS) solution. The supervised ANN algorithms work with two phases—learning and recognition. Learning was performed using data sets regarding CRT and IMRT composed of 8 and 11 input/output respectively. To compare ANN predicted and planned results the global gamma index was used, obtaining a $\gamma_{(2\%, 2\,mm)} = 99.78\%$ and $\gamma_{(2\%, 2\,mm)} = 99.7\%$, respectively. This first work showed the capability of ANN to reconstruct the absorbed dose distribution based on EPID signals.

Keywords

Quality assurance • EPID • External radiotherapy
Artificial neural network

F. Chatrie (✉) · J. Mazurier · L. Simon · L. Vieillevigne
R. Ferrand · A. R. Barbeiro · X. Franceries
Inserm, UMR1037 CRCT, 31000 Toulouse, France
e-mail: frederic.chatrie@laas.fr

F. Chatrie · X. Franceries
Université Toulouse 3-Paul Sabatier, 31000 Toulouse, France

F. Chatrie · M.-V. Le Lann
LAAS-CNRS, 31000 Toulouse, France

L. Simon · L. Vieillevigne · R. Ferrand
Departement de Physique Medicale, IUCT Oncopole, 31000 Toulouse, France

F. Younan · J. Mazurier
Service de Radiothérapie, Groupe Oncorad-Garonne, Clinique Pasteur, 31000 Toulouse, France

1 Introduction

ANN have received a considerable boost of attention in recent years due to the increased computing capacity and the employment of graphics processing units (GPUs). In this study, neural networks are used in the domain of EBRT for the CRT and IMRT pre-traitment verification and in a future for in vivo dosimetry. Other works, have implemented ANN algorithms for treatment verification [1, 2]. Kalantzis et al. have developed a similar approach, using different data training.

During the last years, the EPID has been used for dosimetric purpose [3] unlike its initial usefulness, patient positioning. Several studies have shown the possibility of the EPID for dosimetry based on analytical [4, 5] or Monte Carlo [6, 7] methods. In this work, the capability of using an artificial intelligence way is proposed.

2 Materials and Methods

2.1 The ANN Principle

Artificial neurons have been inspired from the biological neurons but they are only an abstract mathematical representation which are far from the complexity of the real neurons. ANN is composed by a set of artificial neurons organised in several layers according their determined architecture.

Like humans, artificial neural networks work via two phases—learning and recognition. Learning can be done via several ways. One is unsupervised learning, which consists to organize the only provided inputs data and making a link between them, in order to create a pattern depending on the architecture used. Another one is supervised learning (the type of learning that is used throughout this work) which

© Springer Nature Singapore Pte Ltd. 2019
L. Lhotska et al. (eds.), *World Congress on Medical Physics and Biomedical Engineering 2018*,
IFMBE Proceedings 68/3, https://doi.org/10.1007/978-981-10-9023-3_117

entails the creation of an organised architecture of neurons by linking input and output data. In the latter, the output data brings an additionnal information like a "teacher" to help the learning.

The supervised ANN learning phase consists on setting different weights associated to the level of importance of signals coming from the underlying layer. The predicted single neural result is calculated by the following expression:

$$y_{neural} = f\left(\left(\sum_{i \in \mathbb{N}} w_i * x_i\right) + b\right)$$

where w_i are the weights corresponding to the signals x_i, b a bias and f an activation function.

In order to set the weights, several optimisation algorithms can be used (e.g. gradient backpropagation, quasi-Newton, Levenberg-Marquardt) minimising, iteratively, the error between the predicted result and the provided output for all data set.

The learning phase can be time consuming but once the ANN weight has been established, the use of the neural network during the recognition phase will be instantaneous.

The output could be either a label for the classification or a float value for the regression mode. The regression mode, the one used, permits to modelise complex or non-linear functions.

For both, unsupervised and supervised technics, the recognition phase consists of using new input data which has not been trained to predict output data according to the learning done.

2.2 ANN Application for EBRT Dosimetry

The Cancer University Institute of Toulouse (IUCT) and Pasteur clinic dispose of a wide range of techniques: CRT, IMRT and Volumetric Modulated Arc-Therapy (VMAT).

Regarding the EPID measurements, all EPID images were taken during 6MV treatment delivery, with a dose rate of 600 MU/min, from a-Si1000 EPID (Varian) mounted with Exact-arm on a Clinac 23iX equipped with a multi-leaf collimator (120 leaves). The EPID has an active area of 30×40 cm^2 and the images were acquired using the half-resolution mode (EPID spatial resolution is reduced to 384×512 pixels from 768×1024). All measurements were done to 150 cm Source Detector Distance (SDD). The quasi-raw EPID signals were used as input datas by the algorithms, since only dark-field and flood-field corrections were included. The dark field image correction corresponds to an average of several frames without any radiation to remove the eletronic noise. The flood field image is acquired by an uniform irradiation larger than the area of detection in order to correct for the pixel sensitivity differences. The

image acquisition software IAS3 has cine and integrated mode. This latter was used, where one image is reconstructed from an average of frame acquired during radiation delivery. The maximum frame rate for this equipment is 9.574 frames/sec.

Regarding the TPS absorbed dose distributions, all 2D plane images were calculated in Eclipse^{TM} at the maximum depth dose in a water phantom.

A selected region of interest, after applying a threshold of a 10% (CRT) and 5% (IMRT) of the TPS dose, was considered. The created mask was applied for both data sets, EPID and TPS images, in order to only keep and evaluate the pixels containing relevant treatment information.

During the learning phase, corrected EPIDs and desired absorbed dose distributions from TPS are used as the neural network's input and output information, respectively. Once the learning is completed—a process that requires many images—future EPIDs can be used to predict the delivered absorbed dose distribution, allowing its comparison with the planned treatment. Non-negligeable difference between the desired and delivered distributions can indicate potential problems during the treatment delivery.

2.3 The ANN Parameters and Architecture

The algorithms have been developped with MATLAB® (Matlab R2015b, MathWorks), before moving onto the Tensorflow Google tools Development. The work presented here only concerns the Matlab Development using the Neural Network toolbox version 8.4.

The simplest ANN model have been chosen, a non-recurrent feed-forward multi-layer ANN is used with one input, one hidden and one output layer. Each node (called neuron) composing the input layer corresponds to each input signals. All imput nodes are linked to all hidden nodes which are also linked to all output nodes. The number of input and output nodes is equal to respectively, the number of inputs and outputs.

The number of hidden nodes is defined by the user and has an importance. There is no theoretical limits for them but this parameter is correlated with the generalisation of the model [8, 9]. It must be correctly calibrated for the application, else there is a risk of overtraining. Overtraining appears when the network learn very well the training data sets including a noise with complicated model, therefore, the recognition phase with a new data set will give wrong results. In this case, the model loses its generalization capability.

The optimisation algorithm used for setting the weights was Levenberg-Marquardt and the activation function, both output and hidden layers was sigmoïd and linear respectively.

3 Results

The learning phase was performed using 8 and 11 input/output data sets, respectively from CRT and IMRT treatments. All of the used data sets (both input corrected EPIDs and output absorbed dose distributions) consisted of 384×512 pixels before applying the threshold, as described in the previous section. The EPID that was used during the recognition phase was not one of those used during the learning phase. Figures 1, 2, 3, 4 show the absorbed dose distributions calculated by the neural network and the distributions originally planned, for CRT and IMRT respectively.

The global gamma index, γ, was used to evaluate the difference between the calculated and planned distributions. γ gives the number of pixels (as a percentage) that respect a given objective. In these cases, the objective was that pixels should have a difference $\leq 2\%$ at a distance ≤ 2 mm. In case of Figs. 1 and 2 concerning all brain instance, the $\gamma_{2\%,2\,mm}$ was found to be 99.78%. In case of Figs. 3 and 4, regarding Head and Neck instance, $\gamma_{2\%,2\,mm}$ value is 99.7%, both, highlighting the neural network's capability to predict the absorbed dose distribution based on EPID images. The global gamma index has only been calculated within the selected region of interest.

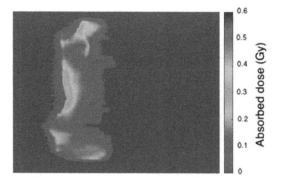

Fig. 3 Calculated absorbed dose distribution (ANN)

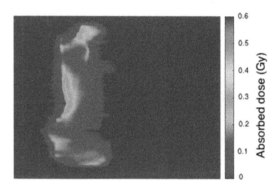

Fig. 4 Planned absorbed dose distribution (TPS)

4 Discussion

These initial results showed that machine learning could be used to reconstruct the delivered absorbed dose distributions. However, there is a lot of parameters to consider extending these algorithms for in vivo dosimetry. Indeed, modelling the heterogeneity geometry with ANN method can be complex. For this type of methods the provided data are obviously very important and serve to established the modelisation.

Currently, it is still challenging to know the absorbed dose that is truly received by the patient during treatment sessions. The EPID imager can provide information relative to the photon beam passing through the patient, which can potentially be used to produce real-time in vivo dosimetry via artificial intelligence methods.

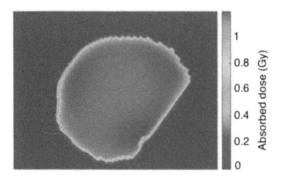

Fig. 1 Calculated absorbed dose distribution (ANN)

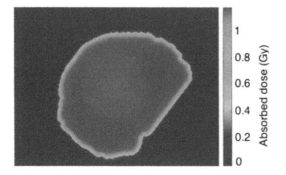

Fig. 2 Planned absorbed dose distribution (TPS)

5 Conclusion

It was shown in this study, that patient-specific quality assurance of CRT and IMRT based on EPID can be performed with neural network algorithms.

Next works would be focused on generalising the neural network by increasing the number of data sets used during the learning phase, whilst maintaining elevated performance. Additionally, these algorithms would be extended for reconstructing the 2D or 3D in vivo absorbed dose distribution for CRT, IMRT and VMAT technics.

Lastly, it would be interesting to compare results between neural networks and deep learning.

References

1. Tang, X. and Lin, T. and Jiang, S.: A feasibility study of treatment verification using EPID cine images for hypofractionated lung radiotherapy. In: Phys Med Biol, vol. 54, pp. 1–8. (2009).
2. Kalantzis, G. and Vasquez-Quino, L. A. and Zalman, T. and Pratx, G. and Lei, Y.: Toward IMRT 2D dose modeling using artificial neural networks: a feasibility study. In: Med Phys, vol. 38, pp. 5807–5817. (2011).
3. van Elmpt, W. and McDermott, L. and Nijsten, S. and Wendling, M. and Lambin, P. and Mijnheer, B.: A literature review of electronic portal imaging for radiotherapy dosimetry. In: Radiother Oncol, vol. 88, pp. 289–309. (2008).
4. Camilleri, J. and Mazurier, J. and Franck, D. and Dudouet, P. and Latorzeff, I. and Franceries, X.: Clinical results of an EPID-based in-vivo dosimetry method for pelvic cancers treated by intensity-modulated radiation therapy. In: Phys Med, vol. 30, pp. 690–695. (2014).
5. Wendling, M. and Louwe, R. J. and McDermott, L. N. and Sonke, J. J. and van Herk, M. and Mijnheer, B. J.: Accurate two-dimensional IMRT verification using a back-projection EPID dosimetry method. In: Med Phys, vol. 33, pp. 259–273. (2006).
6. Siebers, J. V. and Kim, J. O. and Ko, L. and Keall, P. J. and Mohan, R.: Monte Carlo computation of dosimetric amorphous silicon electronic portal images. In: Med Phys, vol. 31, pp. 2135–2146. (2004).
7. van Elmpt, W. J. and Nijsten, S. M. and Schiffeleers, R. F. and Dekker, A. L. and Mijnheer, B. J. and Lambin, P. and Minken, A. W.: A Monte Carlo based three-dimensional dose reconstruction method derived from portal dose images. In: Med Phys, vol. 33, pp. 2426–2434. (2006).
8. Hinton, G.: Learning Translation Invariant Recognition in a Massively Parallel Network. Lecture Notes Computer Science, Springer-Verlag, New York, 1987.
9. Weigend, A. and Huberman, A. and Rumelhart, D.: Predicting the future: A connectionist approach. In: Int. J. Neural Syst., vol. 1, pp. 193209. (1990).

Study of the X-Ray Attenuation as a Function of the Density and Thickness of the Absorbent: Cortical Bone and BaSO4

Andrea Vargas-Castillo and Angel M. Ardila

Abstract

In the present paper a non-metallic material to attenuate X-rays is proposed, mainly for medical imaging applications in order to reduce the radiation dose received by patients due to dispersion. For this purpose, the filing of cortical bone and barium sulfate (BaSO4) were characterized using X-ray diffraction, energy-dispersive X-ray analysis and Raman spectroscopy techniques. Attenuation capacity of the X-rays was determined using an X-rays equipment (10–30 kV) and a Geiger-Müller detector, bearing in mind that the intensity of the transmitted radiation depends on the thickness and density of the material, having a 1 mm lead sheet as reference. In addition, a radiation attenuation comparison dose emmited by a dental X-ray generator using TLD-100 thermoluminescent crystals and periapical radiographic plates is presented, identifying that BaSO4 is the material that attenuates the better this type of radiation compared to cortical bone.

Keywords

X-rays • Cortical bone • Barium sulfate (BaSO4) • Attenuation of radiation • XRD • EDX • Raman spectroscopy

1 Introduction

Most of the adverse health effects produced by ionizing radiation are classified as deterministic or stochastic effects. Deterministic effects are produced from a threshold dose at which the severity of the consequences increases with the amount of the received dose, and they are mainly related with the malfunctioning of cells after receiving high doses, including cell death. Stochastic effects are probabilistic events and can change between individuals; its probability of occurrence increases with the dose received and among these are cancer and inheritable effects [1].

Radiation protection is responsible for ensuring the safety of people against ionizing radiation harmful effects. To prevent the deterministic effects, the protection consists about not exceeding the thresholds defined in each situation, while for stochastic effects the doses must be maintained as low as possible, since for any given dose an effect can be produced, so that if one desired to obtain a null effect the received dose must be zero. In fact, some interventional procedures that require prolonged exposure times and the acquisition of multiple diagnostic images could produce deterministic effects on personnel and patients [2]. To characterize the photon beam penetration into an absorbing medium, suppose that a monoenergetic and collimated X-rays beam is emitted towards a material of thickness t that is placed in front of a detector. The result should be an exponential attenuation of the X-rays, according to the equation [3]:

$$\frac{I}{I_0} = e^{-\mu t}, \tag{1}$$

where $\mu = \tau(\text{photoelectric}) + \sigma(\text{Compton}) + \kappa(\text{pair})$ is the *linear attenuation coefficient*, defined as the probability per unit length path that an X-ray photon be removed from the beam, I is the number of transmitted photons and I_0 is the incident intensity. In order to eliminate the dependence of the linear attenuation coefficient with the density of the absorber, the *mass attenuation coefficient* is defined as [3],

$$\text{mass attenuation coeficient} = \frac{\mu}{\rho}. \tag{2}$$

It is known that the exposition to low doses of radiation can also result in severe health effects, although they are less frequent and only detectable through complex

A. Vargas-Castillo (✉)
Medical Physics Group, Physics Department, Universidad Nacional de Colombia, Bogotá, Colombia
e-mail: anvargasc@unal.edu.co

A. M. Ardila
Applied Physics Group, Physics Department, Universidad Nacional de Colombia, Bogotá, Colombia

© Springer Nature Singapore Pte Ltd. 2019
L. Lhotska et al. (eds.), *World Congress on Medical Physics and Biomedical Engineering 2018*,
IFMBE Proceedings 68/3, https://doi.org/10.1007/978-981-10-9023-3_118

epidemiological studies of large populations [4]. Following the Linear No-Threshold (LNT) model, the aim of this study was to find a material capable of attenuate the doses received by dispersion in patients submitted to diagnostic images (mainly dental images) making a comparison between two highly viable materials for this purpose.

2 Materials and Methods

Bones can easily absorb X-rays and contain calcium, whose atomic number is greater than the ones of the chemical elements that constitute the tissues, producing a high contrast in a radiography. Also, the density of the cortical bone is 1.9 g/cm^3 [5]. On the other hand, the barium sulfate BaSO4, constituted by the elemental barium of atomic number 56 and density 3.62 g/cm^3, is often used as a contrast agent in X-ray images and Computed Axial Tomography [6]. Since μ depends, in general, on the photons energy hν and on the nature of the absorber (its density and atomic number) and for the photoelectric effect (μ/ρ) is proportional to $(Z/h\nu)^3$, while for the Compton scattering it is independent of Z [3], the cortical bone and barium sulfate are proposed as viable materials for the desired objectives; besides they are easy to obtain and cheap.

For samples preparation, a hydraulic press was used to compress the pulverized material of the porcine femur cortical bone and the barium sulfate. A total of 10 samples were made with a pressure of 156 MPa and with the same mass to ensure that the pellets have the same height for both cortical bone and barium sulfate. Small variations of the density were obtained changing the mass from 1.5 to 3.0 g, leaving constant pressure. The characterization techniques used to obtain information about the physical, chemical and structural properties of the materials were X-ray diffraction, Energy-Dispersive X-ray spectrometry and Raman spectroscopy, using a PANalytical Difframeter, a Tescan Vega 3 microscope with a Bruker Quantax EDS and an EnSpectr spectrometer R532, respectively.

2.1 Cortical Bone

Cortical bone showed to have a high coincidence with hydroxyapatite (HAp or HA) apparently amorphous, according to the reference code 96-431-7044 [7]. The diffractogram at Fig. 1a shows the spectrum obtained for the material compared to the main peaks of HAp pattern. Also it was identified that the material has an hexagonal structure with a spatial group P63/m with network parameters a = b = 9.419 Å and c = 6.881 Å, very similar to the pure phase HAp, which varies according to the Ca:P ratio. In Fig. 1b we present the comparison between diffractograms corresponding to the tablets of different densities: 1.54, 1.55, 1.58 and 1.59 g/cm^3, showing that apparently the compression do not modify the chemical composition neither the mechanical properties of the material, resulting that there is no shift of the main peaks and do not appear new peaks.

The EDX analysis indicates the presence of calcium, phosphorus and oxygen, corresponding to the HAp chemical formula $Ca_5(PO_4)_3(OH)$. Due to the amount of fat present in the material, since it is composed by organic substances, it was possible to compact the cortical bone into tablets allowed an easy manipulation of it.

2.2 Barium Sulfate BaSO4

Barium sulfate pellets were obtained through a process that had two steps: first the barium sulfate was pressed at 156 MPa at room temperature and then it was sintered at 1000 °C for 12 h [8]. After the first step we obtained a tablets set made of barium sulfate, while after the later one we made them more compact, which was required because the grain size is so small (1–3 μm) that the compressed tablets are not compact enough to be easily manipulated. With this sintering, the material showed a density of 4.0 g/cm^3. To verify that the chemical composition and the cristalline structure of the material do not change after heating, XRD analysis was performed and the results are

Fig. 1 Cortical bone analysis by XRD

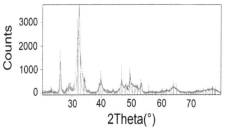

(a) Cortical bone diffractogram. Remarkable peaks at 2θ = 25.9°, 29.01°, 31.83°, 32.369°, 39.85°, 46.73° and 53.25°.

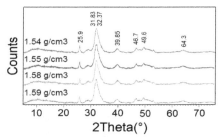

(b) Diffractograms comparison for samples of cortical bone with different densities.

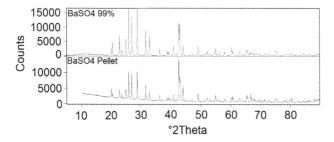

Fig. 2 Barium sulfate diffractogram before (up) and after (down) sintering, with remarzkable peaks at $2\theta = 20.0526°$, $20.5192°$, $22.8631°$, $24.9321°$, $25.9260°$, $26.9099°$, $28.8140°$, $31.5886°$, $32.8599°$, $42.6402°$, $42.9920°$, $49.0688°$, $54.8609°$, $60.3009°$, $65.3756°$ and $75.2764°$

Table 1 Elemental composition of barium sulfate by EDX

El	AN	Series	unn.C [wt%]	norm. C [wt%]	Atom. C [at.%]	Error (1 Sigma) [wt%]
Ba	56	L-series	52.29	59.95	17.81	1.53
O	8	K-series	21.31	24.43	62.30	3.96
S	16	K-series	13.63	15.63	19.89	0.59

shown in Fig. 2. According to the diffractograms it can be seen that the parameters obtained with the HighScore Plus database program coincide with the letter 96-900-4486 [9]. The material has an orthorhombic structure and space group Pnma, with parameters a = 8.8790 Å, b = 5.4540 Å and c = 7.1540 Å.

Similarly, EDX analysis was performed for barium sulfate, sintered and non-sintered, obtaining that it is composed of barium, sulfur and oxigen, as expected in both cases. The proportions of these elemental components are shown in Table 1. In addition, analysis was performed by Raman spectroscopy, finding that its characteristic peaks coincide with those of the barite, as is presented in Fig. 3.

3 Results and Discussion

To perform the radiation study of attenuation and intensity as a function of thickness and density, an experimental arrangement was used in which samples of thicknesses 0.4, 0.8, 1.2, 1.6 and 2.0 cm for cortical bone and 0.26, 0.52, 1.1, 1.6 and 2.1 cm for barite were located between the source (Gendex 770, 70 kVp and 7 mA) and the detectors: periapical photographic plates and thermoluminescent crystals TLD-100. Additionally, crystals were placed on the samples, to determine the incident intensity on the material. Determining the dependence of the transmitted intensity as a function of the thickness, varying the energy from 10 to

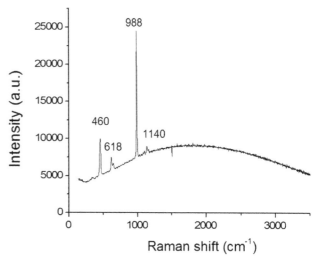

Fig. 3 Raman spectrum of barium sulfate

30 keV, the LD Didactic GmbH X ray equipment with 1 mA beam current was used.

3.1 X-Rays Attenuation

Using the program ImageJ, the images obtained from the photographic plates were analyzed for the two materials, wich allowed to make a histogram of the gray scale and compare it with the reference lead sheet (1 mm), as shown in Fig. 4.

Barium sulfate gives a greater attenuation than the cortical bone, since for approximately 2.0 cm of thickness there is a radiation attenuation of 98%, while for the same thickness the cortical bone attenuates 83%. This means that it is

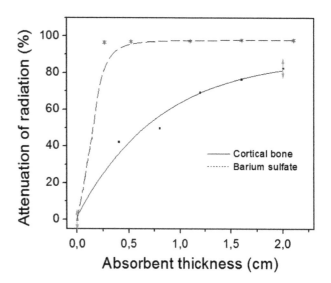

Fig. 4 Percentages of attenuation of the X-rays by barium sulfate and cortical bone depending on sample thickness

Fig. 5 Dependence between the transmitted intensity and the absorbent thickness for the barium sulfate and the cortical bone

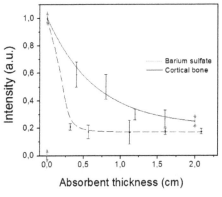
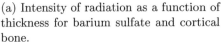

(a) Intensity of radiation as a function of thickness for barium sulfate and cortical bone.

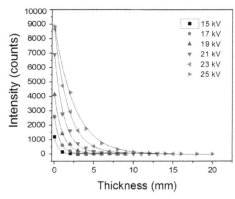

(b) Intensity of radiation as a function of thickness and beam energy (keV) for the cortical bone.

not necessary to use a large barium sulfate thickness to attenuate the X-rays, as is the case for cortical bone, because the exponential increase of the barite is much more abrupt for small values of the thickness.

On the other hand, cortical bone density was varied with values of: 1.54, 1.55, 1.58 and 1.59 g/cm³ and for barium sulfate: 3.61, 3.92, 4.02 and 4.44 g/cm³. It was observed that there are no remarkable changes in the percentage of attenuation for small variations in the material density, compared with the strong dependence of it with the absorbent thickness.

3.2 Intensity as a Function of Thickness

Thermoluminescent crystals gave information about the incident and transmitted intensity through the material for different thicknesses. In Fig. 5a we can see that for the samples of thickness 2.0 cm the barium sulfate attenuates the radiation up to 0.17, while the cortical bone attenuates only up to 0.25. The dependence of the transmitted intensity as a function of the thickness for the cortical bone tablets is presented in Fig. 5b.

For barium sulfate, measurements to determine the attenuation capacity of the X-rays were made for different exposure times fixing the thickness at 2.6 mm, since the tests carried out previously it could be concluded that some

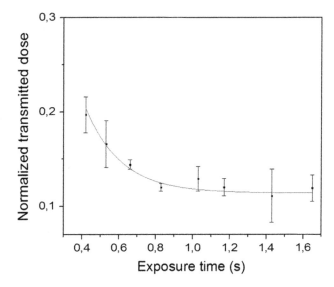

Fig. 6 Normalized transmitted dose for different exposure times: 0.42, 0.53, 0.66, 0.83, 1.03, 1.17, 1.43 and 1.65 s

millimeters of the material are sufficient to attenuate large amounts of radiation. The exposure time was increased from 0.42 to 1.65 s in the Gendex 770 with 70 kV and the results are shown in the Fig. 6.

This way it is possible to observe that the material attenuates between the 80 and 90% of the incident radiation for the exposure times used, using barium sulfate with

Table 2 Comparative table for barium sulfate and cortical bone pellets

	μ (cm^{-1})	μ/ρ (cm^2/g)	HVL (cm)	ρ g/(cm^3)
Barium sulfate	8.52 ± 1.37	2.12 ± 0.50	0.13 ± 0.02	4.00 ± 0.34
Cortical bone	1.44 ± 0.30	1.04 ± 0.22	0.72 ± 0.21	1.38 ± 0.01

2.6 mm thickness. Finally, Table 2 shows values of interest for the proposed materials that allows a comparison between the attenuation capacity of them.

4 Conclusions

Barium sulfate shows better results in the measurements made for radiation attenuation, in comparison with the cortical bone. However, the cortical bone is a promising material, since the raw material is easy to obtain, allowing the preparation of the samples anywhere in the world and offering better physical qualities for handling than for the BaSO4.

Acknowledgements Authors want to thank everyone who made this work possible, particularly professors Julio Evelio Rodriguez and Ricardo Parra from the Universidad Nacional de Colombia and the New Materials Physics Group doctoral student Jorge Ignacio Villa.

References

1. ICRP Publication 103: The 2007 Recommendations of the International Commission on Radiological Protection, Ann. ICRP 37 (2–4), 2007 (2007).
2. Zakeri, F. et al: Biological efects of low-dose ionizing radiation exposure on interventional cardiologists. Occupational Medicine, Vol 60, Issue 6, 1 September 2010, pages 464–469. https://doi.org/10.1093/occmed/kqq062.
3. Knoll, G.: Radiation Detection and Measurement. Third edition (2000).
4. Gonzalez, A.: Los efectos biolgicos de las dosis bajas de radiacin ionizante: Una visin ms compleja. Informes especiales UNSCEAR. Boletin OIEA 4/1994 (1994).
5. Mayles, P. et al: Handbook of Radiotherapy Physics - Theory and Practice. Taylor & Francis, p 40, (2007).
6. Aiken, T.: Legal and Ethical Issues in Health Occupations. Second edition. Saunders Elsevier, page 136 (2009).
7. Lyudmyla, I. et al: Isomorphous Substitutions of Rare Earth Elements for Calcium in Synthetic Hydroxyapatites. Inorg. Chem., 2010, 49 (22), pp 10687–93. https://doi.org/10.1021/ic1015127 (2010).
8. Felder, E. et al: Process for the preparation of barium sulfate of increased owability and density, suitable as a radio-opaque component in radiographic contrast agents, product obtained according to this process and the radiographic contrast agent produced therefrom. United States Patent, (1986).
9. Jacobsen, S. et al: Rigid-body character of the SO4 groups in celestine, anglesite and barite. The Canadian Mineralogist, pp 1053–60, (1998).

Development of 3D Printed Phantom for Dose Verification in Radiotherapy for the Patient with Metal Artefacts Inside

Diana Adlienė, Evelina Jaselskė, Benas Gabrielis Urbonavičius, Jurgita Laurikaitienė, Viktoras Rudžianskas, and Tadas Didvalis

Abstract

One of the problems performing IMRT dose planning or plan verification using anthropomorphic phantoms for head and neck cancer patients is the presence of possible artefacts (dental crowns, metal dental implants, dental restoration materials.) inside the oral cavity. In many cases these artefacts are not accounted but may cause deviations from patient treatment plans due to enhanced scattering dose from the metal artefacts. Exploiting 3D printing technologies 3D dosimetry phantom corresponding to patient-specific anatomic structures with precisely positioned artefacts can be produced. Application of such 3D phantom in radiation therapy may contribute to more accurate dose planning and thus more efficient patient treatment. In this work we propose newly developed patient-specific 3D printed phantom of lower jaw with teeth which can be used for patient specific QA in IMRT. Developing this phantom DICOM image of the real patient was used for 3D reconstruction of patient's lower jaw with teeth. 3D shape of this anatomic structure was printed out using Zortrax M300 3D printer. High Impact Polystyrene (HIPS) which is characterized as having satisfactory bone tissue equivalency was used as printing material. Keeping in mind a real patient, possibility of covering of corresponding teeth with a metallic crown in the 3D printed jaw construction was foreseen. Prepared construction was fixed in dose gel matrix, thus forming dosimetry phantom for the evaluation of possible radiation treatment errors caused by artefacts located in this anatomic region of the patient. Investigation of 3D printed lower jaw phantom has shown its feasibility for the assessment of dose distortions related to the presence of metal artefacts in the mouth of the patient. Also potential for the application of 3D printed phantoms in radiotherapy quality assurance has been shown.

Keywords

Dose verification • IMRT • Metal artefacts
3D printed patient specific phantom

1 Introduction

Intensity modulate radiation therapy (IMRT) has been widely adopted as advanced treatment technique for head and neck cancers. Compared with conventional RT, IMRT has better survival outcomes and less treatment-related side effects [1]. IMRT treatment requires high precision patient specific quality assurance (QA) [2, 3]. However usual pretreatment QA in IMRT is performed using standard rectangular or cylindrical phantoms thus not accounting patient specific anatomy or possible artificial artefacts inside the body. The limits and deficiencies of the algorithms used in the treatment planning systems can lead to large errors in dose calculation. These errors can be crucial for the treatment of head and neck cancer patients having artefacts (metallic dental crowns and/or metallic dental implants for dentures) [4], since high atomic number of dental crowns/implants leads to major problems at providing an accurate dose distribution in radiotherapy and contouring tumors and organs caused by the artefact in head and neck tumors [5].

In order to reduce the likelihood of treatment dose errors, validation of patient doses calculated using treatment planning system of the irradiation unit (LINAC) is needed comparing them with experimentally measured

D. Adlienė (✉) · E. Jaselskė · B. G. Urbonavičius
J. Laurikaitienė · V. Rudžianskas
Kaunas University of Technology, Studentų G. 50, 51368 Kaunas, Lithuania
e-mail: diana.adliene@ktu.lt

V. Rudžianskas
Oncology Institute of Lithuanian University of Health Sciences, Eivenių G. 2, 50161 Kaunas, Lithuania

T. Didvalis
Hospital of Lithuanian University of Health Sciences Kaunas Clinics, Eivenių G. 2, 50161 Kaunas, Lithuania

© Springer Nature Singapore Pte Ltd. 2019
L. Lhotska et al. (eds.), *World Congress on Medical Physics and Biomedical Engineering 2018*,
IFMBE Proceedings 68/3, https://doi.org/10.1007/978-981-10-9023-3_119

doses obtained applying the same radiation treatment parameters. Application of patient specific phantoms for dose measurements may contribute significantly to more effective treatment procedure avoiding unnecessary dose deviations.

3D printing technologies and growing variety of 3D printing materials provide excellent possibilities for the fabrication of patient specific 3D anatomic structures and development of patient specific 3D phantoms for verification of calculated patient treatment doses [6]. Recent investigations [7–12] demonstrate successful application of multifunctional printed/fabricated structures for individualized radiotherapy: bolus, supporting means for patient's immobilization or dosimetry phantoms. Development of the conceptually new patient anatomic region/tumor specific 3D printed phantoms is also progressing very fast [13, 14–17]. Along with the usual 3D printing procedures some new concepts, e.g. ionizing irradiation based 3D printing approach using special polymer dose gels for fabrication of free standing tumor specific shapes (dose phantoms) has been reported recently [18, 19].

A lot of attempts have been done investigating experimentally [5, 20, 21] and theoretically [22, 23] scattering dose effects caused by dental crowns, dental implants and different teeth restoration materials. It was clearly shown that such artefacts are responsible for the dose distribution/dose changes observed during radiotherapy treatment procedure as compared to doses calculated using algorithms provided by treatment planning system and may cause additional health problems for treated cancer patients. However almost all investigations have been performed treating the artefact as a separate object located at the certain position in tissue equivalent surrounding (phantom construction) but not accounting patient-specific teeth related shapes and locations of organs, especially jaws.

In this paper we discuss newly developed patient-specific 3D printed phantom of lower jaw with teeth which can be used for patient specific QA in IMRT.

2 Materials and Methods

2.1 Tissue Equivalent 3D Printing Materials

Patient specific 3D printed phantoms can be produced using various technologies and various types of different nearly tissue equivalent printing materials. Physical and radiation attenuating properties of commercially available 3D printing materials thought for fabrication of patient specific 3D printed phantoms were investigated by many authors [6, 21, 24–29]. 3D printing materials recording similar properties to those of human anatomic structures/tissues located in the head region are provided in the Table 1.

It should be noted, that equivalency of 3D printing material to a certain tissue was extrapolated taking into account also radiation attenuation properties of 3D printed materials in the treatment energy range of 1.0–10.0 meV [27, 28]. It is also very important, that recently patented Polytetrafluoro-ethylene (PTFE) based 3D printing material [29] will provide more realistic results for bone and especially teeth investigations using 3D printed phantoms. At the moment ABS with 60% W infill is used to represent teeth tissue [6].

Due to the fact, that bone/tooth equivalent 3D printed materials are not commercially available yet, pre-selection of eligible for phantom printing materials (HIPS, ABS and PLA) has been performed. For this purpose X-ray attenuation properties were evaluated using XCOM database. Considering differences in mass densities of materials (Table 1), HIPS showed the best agreement of mass attenuation data with the data of bone in the energy region of 1.0–10.0 meV. Observed mismatch of the results at lower energies can be explained by the fact, that high atomic number of bone's calcium component causes an increase in the interaction coefficients for low energy photons. This material was chosen for the printing of patient specific phantom.

2.2 3D Printing Parameters

The main criteria choosing the 3D printing technology for patient-specific 3D phantom was the diversity of the available 3D printing materials and availability of sufficient large print volume. Zortrax M300 3D printer having one of the largest print volumes (300 × 300 × 300 mm) was selected and Fused Deposition Modeling (FDM) technology was applied for printing of patient specific 3D phantom.

It was decided not to produce the 3D phantom of the head, but limit printing to the anatomic region of interest and fabricate patient specific lower jaw 3D phantom, including teeth, which can be placed into the volume, filled with near tissue equivalent dose gel, and used for dosimetry purposes. Also the option to cover several teeth with a metallic crowns was foreseen.

3 Development of Patient Specific Lower Jaw 3D Phantom

In order to develop realistic patient specific 3D shape of the lower jaw, the same CT images of the patient were used for preparing of IMRT treatment plan and for 3D reconstruction of the anatomic structure of interest. Clinical case with laryngeal squamous cell carcinoma was selected for evaluation, taking into account that CT images of the lower jaw of

Table 1 Mass density of 3D printed materials and corresponding biological tissues

Body tissue	ρ, g/cm³	3D printing material	ρ, g/cm3
Muscle	1.04 [27]	High impact polystyrene (HIPS)	1.04–1.05 [28]
Brain	1.05 [28]	Acrylonitrile Butadiene Styrene (ABS)	1.06–1.08 [28]
Bone	1.92 [27]	Polytetrafluoro-ethylene (PTFE)	2.12–2.19 [29]
Thyroid gland	1.05 [28]	Polylactic acid (PLA)	1.06–1.43 [28]
Teeth	2.20 [21]	Polytetrafluoro-ethylene (PTFE)	2.12–2.19 [29]
Skin	1.09 [28]	Polyactic acid (PLA)	1.06–1.46 [28]

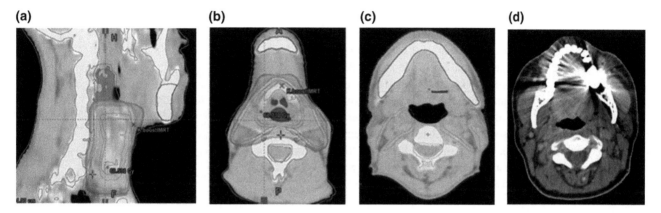

(a) (b) (c) (d)

Fig. 1 Case: laryngeal squamous cell carcinoma: **a**—dose distribution in sagittal view; **b**—dose distribution in transversal view; **c**—transversal view with marked parotid glands; **d**—CT image indicating scattering effects due to the presence of metal tooth

this patient indicated presence of metal tooth. Also the related scattering effects were clearly seen in the CT image (Fig. 1d). Treatment planning in the presence of metal artefacts can be handled by contouring these scattering objects and manually assigning HU to the corresponding regions thus adjusting treatment plans. However scattering effects still remain and contribute to the uncertainties in delivered doses. Cancer treatment plan using IMRT mode of Clinac iX linear accelerator (Varian medical systems) applying 6 meV energy for each irradiation field are provided in Fig. 1. The mean doses to particular organs were as follows: Parotic gland (R)—17.91 Gy, Parotid gland (L)—17.58 Gy, Spinal cord—34.64 Gy and Mandible—18.04 Gy (Dmax = 43.59 Gy). The total treatment dose of 50 Gy, 2 Gy/fr was planned for delivery to the tumor.

In order to reconstruct anatomic structure in 3D, a set of original patient's CT scans (CT scanner Siemens Light Speed RT16) in DICOM format recording anatomic area of interest was selected. Several software packages for this type of 3D reconstruction are available. We have used two freely available software packages: 3D Slicer [30] and InVesalius3 [31]. Both of them were similarly effective to 3D reconstruct selected anatomic structure, converting DICOM files to STL format. Extracted STL file was converted to the command

set of ZORTRAX M300 printer using special Z-Suite code and corresponding 3D structure was printed. Printed 3D model of the lower jaw allowed covering of the tooth with 0.2 mm thick steel thus simulating dental metal crown as it is shown in Fig. 2b.

In order to create possibility for dosimetric evaluation of metal artefact related distortions of the initially preplanned patient's treatment dose distributions, 3D printed model was fixed in the nearly tissue equivalent dose gel volume [32], keeping the margins around the 3D construction as it is shown in Fig. 2c.

Two experiments have been performed. In the first experiment gel filled construction with suspended 3D printed lower jaw model, including teeth, was irradiated in linear accelerator (one fraction) using irradiation parameters taken from IMRT plan. In the second experiment the same lower jaw model was used, however one tooth was covered with thin metal foil thus imitating presence of possible dental metal artefact. Irradiation conditions were the same as in the previous case. Dose distributions in irradiated dose gels were assessed using MRI technique and following usual dose evaluation procedure of irradiated dose gels. Experimental results were compared with TPS calculated dose values using gamma analysis. Comparison of doses revealed, that

Fig. 2 **a**—CT image of lower jaw; **b**—image of the 3D printed lower jaw; **c**—image of dose gel filled PP container with inmersed lower jaw 3D printed model in it. The arrays indicate position of tooth covered with metal crown

(a)
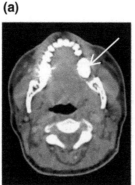

(b)
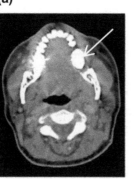

(c)
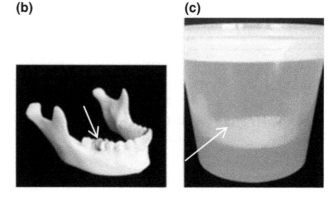

the fraction of points of interest, for which $\gamma < 1$ was evaluated, was 98% in the first case (lower jaw without artefacts). The figure was changed when the lower jaw model with metal artefact was used. In this case the fraction of points corresponding to gamma factor $\gamma < 1$ was 94.5% only. Obtained high dose discrepancies between experimental and calculated doses might be related to dosimetry method uncertainties (6–8%), but also might indicate that the Varian Eclipse anisotropic algorithm (AAA), version 8.6 was not sensitive enough to account the presence of metal artefacts in the anatomic region where lower jaw is located, even if these artefacts were indicated during the dose planning procedure. Taking into account that scattered dose from these artefacts may contribute to the development of different deterministic effects and even enhance the risk of radiation induced necrosis of lower jaw (if maximum dose > 60 Gy will be achieved) further investigations are needed.

4 Conclusions

Patient specific 3D model of lower jaw, including teeth, was proposed and fabricated using 3D printing technique. The possibility to insert metal artefacts (metal dental crowns, implants and dental restoration materials was also foreseen. It was shown that dose gel filled construction in which 3D model of the lower jaw is fixed may serve as a dosimetry phantom. This phantom can be used for the assessment of discrepancies between TPS calculated and measured doses caused by artefacts located in the region of interest, thus implementing patient specific QA in radiation therapy. Further investigations are needed prior to adapt this method for verification of the cancer patient treatment.

Acknowledgements This work was partly supported by the research grant of Lithuanian Research Council No. S-MIP-17-104.

Conflict of Interest The authors of this article declare that they have no conflict of interest.

References

1. Beadle, BM., et al. Improved survival using intensity-modulated radiation therapy in head and neck cancers: a SEER-Medicare analysis. Cancer 120, 702–10 (2014).
2. Report 83: Prescribing, Recording, and Reporting Photon-Beam Intensity-Modulated Radiation Therapy (IMRT). Journal of the ICRU 10(1) 1–106 (2010).
3. ESTRO Booklet 9: Guidelines for the Verification of IMRT. Brussels: ESTRO; 2008.
4. AAPM Report No. 85: tissue inhomogeneity corrections for megavoltage photon beams. Medical Physics Publishing, Madison, WI 53705-4964 (2004).
5. Kamomae, T., et al. Dosimetric impact of dental metallic crown on intensity-modulated radiotherapy and volumetric-modulated arc therapy for head and neck cancer. J Appl Clin Med Phys. 17, 234–245 (2016).
6. Jaselské, E., et al. Dosimetric characteristics of 3D printed materials. In: Medical physics in the Baltic States. Proceedings of the 13th International Conference on Medical Physics, pp. 52–55, Kaunas University of Technology, Kaunas (2017). ISSN: 1822-5721.
7. Sung-Woo, K., et al. Clinical Implementation of 3D Printing in the Construction of Patient Specific Bolus for Photon Beam Radiotherapy for Mycosis Fungoides. Prog Med Phys. 28(1) 33–38 (2017).
8. Michiels, S., et al. Towards 3D printed multifunctional immobilization for proton therapy: Initial materials characterization. Med Phys. 43(10) 5392 (2016).
9. Unterhinninghofen, R., et al. 3D printing of individual immobilization devices based on imaging ñ analysis of positioning accuracy. Radiotherapy and Oncology 115 (1) S199–S200 (2015).
10. McCowan, P. et al. On The Physical and Dosimetric Properties of 3D Printed Electron Bolus Fabrcated Using Polylactic Acid. Radiotherapy and Ocology. 120 (1) S47 (2016).
11. Burleson, S., et al. Use of 3D printers to create a patient-specific 3D bolus for external beam therapy. J. Appl. Clin. Med. Phys. 16 (3) 5247 (2015).
12. Holtzer, N.A., et al. 3D printing of tissue equivalent boluses and molds for external beam radiotherapy. Radiotherapy and Oncology. 111(1) S279 (2014).
13. Mayer, R., et al. 3D printer generated thorax phantom with mobile tumor for radiation dosimetry. Rev Sci. Instrum. (2015). 86p.
14. Alssabbagh, M., et al. Evaluation of 3D printing materials for fabrication of a novel multifunctional 3D thyroid phantom for medical dosimetry and image quality. Rad. Phys. and Chem. 135, 106–112 (2017).

15. Craft, D. and Howell, R. Preparation and fabrication of a full-scale, sagittal-sliced, 3D-printed, patient-specific radiotherapy phantom. J Appl Clin Med Phys. 5, 285–292 (2015).

16. Kamomaea, T., et al. Three-dimensional printer-generated patient-specific phantom for artificial in vivo dosimetry in radiotherapy quality assurance. Med. Phys. 44, 205–211 (2017).

17. Woon Yea, J., et al. Feasibility of a 3D-printed anthropomorphic patient-specific head phantom for patient specific quality assurance of intensity modulated radiotherapy. PLoS ONE. 12(7), e0181560 (2017).

18. Adlienė, D., et al. Application of dose gels in HDR brachytherapy. IFMBE Proc. 2015, V51, 724–727.

19. Adlienė, D., et al. First approach to ionizing radiation based 3D printing: fabrication of free standing dose gels using high energy gamma photons. NIMB62907 (2018). Accepted for publication.

20. Shimamoto, H., Sumida,I. and Kakimoto,N. Evaluation of the scatter doses in the direction of the buccal mucosa from dental metals. J. Appl. Clin. Med. Phys.16 (3) 233–243 (2015).

21. Azizi, M., et al. Dosimetric evaluation of scattered and attenuated radiation due to dental restorations in head and neck radiotherapy. J. Rad. Research&Appl. Sci. (2017), https://doi.org/10.1016/j.jrras.2017.10.004.

22. Spirydovicha, S., Papieza, L., Langera, M. High density dental materials and radiotherapy planning: Comparison of the dose predictions using superposition algorithm and fluence map Monte Carlo method with radiochromic film measurements. Radiotherapy and Oncology 81, 309–314 (2006).

23. Çatli, S. High-density dental implants and radiotherapy planning: evaluation of effects on dose distribution using pencil beam convolution algorithm and Monte Carlo method. J. Appl. Clin. Med. Phys. 16 (5) 46–52 (2015).

24. Hersch, C.A. Applying 3-Dimensional Printing for Radiation Exposure Assessment. National Cancer Institute. Cancer.gov. (2017).

25. Savi, M., et al. Density comparison of 3D printing materials and the human body. IJC Radiology, pp 1–3 (2017).

26. Craft, D., Burgett, E., Howell, R. Evaluation of single material and multimaterial patient-specific, 3D-printed radiotherapy phantoms. Radiotherapy and Oncology. EP-1435(2017).

27. Jeong, H., et al. Development and evaluation of a phantom for multipurpose in Intensity Modulated Radiation Therapy. Nucl. Eng. Tech. 4(43) 399–404 (2011).

28. Alssabbagh, M., et al. Evaluation of nine 3D printing materials as tissue equivalent materials in terms of mass attenuation coefficient and mass density. Int. J. Advanced and Applied Sciences. 4(9) 168–173 (2017).

29. https://www.3m.co.uk/3M/en_GB/design-and-specialty-materials-uk/products/.

30. www.slicer.org.

31. https://invesalius.github.io/.

32. Adlienė, D., et al. Application of optical methods for dose evaluation in normoxic polyacrylamide gels irradiated at two different geometries. NIM-A. 741, 88–94 (2014).

Magnetic Resonance Cancer Nanotheranostics

V. E. Orel, M. Tselepi, T. Mitrelias, A. D. Shevchenko, O. Y. Rykhalskiy, T. S. Golovko, O. V. Ganich, A. V. Romanov, V. B. Orel, A. P. Burlaka, S. N. Lukin, and C. H. W. Barnes

Abstract

It is well known that the magnetic spin effects during nanotherapy can cause tumor cell apoptosis and necrosis based on redox reactions. The current study was carried out on C57Bl/6 mice with Lewis lung carcinoma. The magnetic nanocomplex administration combined with the impact of magnetic resonance system (1.5 T) showed maximal antitumor and antimetastatic effects. The electron spin resonance spectra have been used as diagnostic markers and recorded a change in the tumor redox state based on chemical species such as NO-FeS-proteins and ubisemiquinone. The technology of magnetic resonance nanotheranostics could potentially allow to improve the antitumor effect of chemotherapeutic agents in disseminated cancer therapy.

Keywords

Magnetic resonance • Magnetic nanocomplex Doxorubicin • Tumor

V. E. Orel (✉) · O. Y. Rykhalskiy · A. V. Romanov
Medical Physics and Bioengineering Research Laboratory,
National Cancer Institute, Kiev, Ukraine
e-mail: valeriiorel@gmail.com

V. E. Orel · O. Y. Rykhalskiy
Biomedical Engineering Department, Igor Sikorsky Kyiv
Polytechnic Institute, Kiev, Ukraine

M. Tselepi · T. Mitrelias · C. H. W. Barnes
Cavendish Laboratory, University of Cambridge, J.J. Thomson
Avenue, Cambridge, UK

A. D. Shevchenko
G.V. Kurdyumov Institute for Metal Physics, Kiev, Ukraine

T. S. Golovko · O. V. Ganich
Research Department of Radiodiagnostics, National Cancer
Institute, Kiev, Ukraine

V. B. Orel
Bogomolets National Medical University, Kiev, Ukraine

A. P. Burlaka · S. N. Lukin
R.E. Kavetsky Institute of Experimental Pathology, Oncology and
Radiobiology of the NAS of Ukraine, Kiev, Ukraine

M. Tselepi
Department of Physics, University of Ioannina, Ioannina, Greece

1 Introduction

The magnetic spin effects as a result of the hyperfine spin-coupling between nuclear spins and electron spins can play an important role in redox reactions during nanotherapy causing tumor cell apoptosis and/or necrosis. Magnetic resonance imaging can potentially induce double-strand breaks in DNA through induction of reactive oxygen and nitrogen species by an electromagnetic field [1, 2].

Therefore, it is reasonable to exploit MRI and magnetic resonance nanotheranostics simultaneously in total body diagnostics and treatment similar to whole body hyperthermia and radiotherapy that are applied for treatment of cancer with extensive metastases. Based on our early studies of cell cultures [3], one can assume that the administration of the magnetic nanocomplex (MNC) in animals with a transplanted tumor and following nuclear magnetic resonance effect could have an impact on the kinetics of tumor growth and metastasis.

This pilot study aims to investigate the possible use of magnetic resonance nanotheranostics for Lewis lung carcinoma with hematogenous dissemination.

2 Methods

2.1 Magneto-Mechano-Chemical Synthesis

The MNC comprised the iron oxide (II, III) commercial nanoparticles Fe_3O_4 (Sigma-Aldrich) in diameters of 30–50 nm and antitumor drug DOXO (Pfizer, Italy). The MNC synthesis was performed in a magneto-mechanical reactor (NCI) based on the magnetic resonance phenomena [4].

2.2 Magnetic Studies

"Vibrating Magnetometer 7404 VSM" (Lake Shore Cryotronics Inc., USA) was utilized to determine magnetic properties of the MNC in magnetic fields up to 13 kOe.

2.3 Electron Spin Resonance Spectroscopy

Electron spin resonance (ESR) spectra have been applied for the study of reactive oxygen species (ROS) in the tumor. g-factors in samples were studied with the spectrometer RE1307. 500 mg of soft tissue tumors was placed in a special mold and frozen in liquid nitrogen. Qualitative and quantitative changes in functioning of the mitochondrial electron transport chains (ETC) were evaluated by the shape and location of ESR signal (Landé g-factor of spectroscopic splitting) and amplitude of ESR signal in the control and experimental samples of the tissue. The following parameters were investigated: state of FeS-proteins N2 in the ETC of mitochondria (g = 1.94), level of ubisemiquinone in the mitochondrial ETC (g = 2.0023), formation dynamics of NO–FeS-proteins complexes of N-type in the ETC (g = 2.03) and dynamics of structural changes in the mitochondrial ETC with generation of triplet signal in the ESR spectra (g = 2.007) [5].

2.4 Magnetic Resonance System

The Intera 1.5T (Philips Medical Systems) magnetic resonance system with a synergy body coil was used in this study. The T1-weighted images were produced with an TE (echo time) 20 ms and TR (repetition time) 100 ms.

Simultaneous tumor treatment and visualization were performed on the backs of animals immobilized in horizontal volume coil for 15 min. The temperature inside the tumor reached up to 37 °C after a 15 min electromagnetic irradiation (ER) by the magnetic resonance system. The temperature was measured with fiber optics thermometers TM-4 (Radmir). A tomogram was taken in three orthogonal projections.

2.5 Laboratory Animals, Tumor Transplantation and Treatment of Lewis Lung Carcinoma

All animal procedures were carried out with the humane care of the animals according to the Law of Ukraine N 3447–IV on the protection of animals from cruelty and European Directive 2010/63/EU on the protection of animals for scientific purposes.

The study was carried out on male C57Bl/6 mice (n = 40, 18–20 g, vivarium of National Cancer Institute). Lewis lung

carcinoma was transplanted to animals as described in the paper [6]. The animals were divided into 4 groups, 10 per group: control (intact mice), conventional DOXO, magnetic complex comprised of the ferromagnetic nanoparticles and paramagnetic DOXO without magnetic resonance tomography (MRT) treatment, magnetic complex comprised of the ferromagnetic nanoparticles and paramagnetic DOXO with MRT. DOXO (3 mg/kg) or MNC (3 mg/kg DOXO and 3 mg/kg Fe_3O_4, 0.2 ml in sodium chloride solution) was intraperitoneally injected on the 3rd day after tumor transplantation. The treatment was given four times every 2 days. The following whole body electromagnetic irradiation of the animals was carried out over 15 min on the Intera 1.5T. The size of animal tumors was measured with a caliper.

The nonlinear kinetics of the animal tumor growth was evaluated by the growth factor φ according to the autocatalytic equation and braking ratio κ [4]. The effectiveness of anticancer magnetic nanotherapy was evaluated through statistical comparisons of data performed with Statistica 13.0 (© StatSoft, Inc., 2015) software by using the Student's t-test when the data complied with the conditions of normality.

3 Results

3.1 Magnetic Properties of Nanocomplex

Table 1 presents the magnetic parameters of magneto-mechano-chemically synthesized MNC from obtained measurements of the magnetic hysteresis loop and ESR spectrum. The MNC had properties of a soft ferromagnetic material.

3.2 Magnetic Resonance Imaging

The typical whole-body MRI scans with Lewis lung carcinoma after the conventional DOXO and MNC with ER by MRT treatment on 25th day after tumor transplantation are compared in Fig. 1. The tumor volume was reduced by 1.9 times after administration of MNC and ER compared to the conventional DOXO. A red line indicates the tumor.

Table 1 Magnetic properties of MNC comprising ferromagnetic iron oxide nanoparticles and doxorubicin

The magnetic hysteresis loop	
Saturation magnetic moment m_s, emu/g	8.17
Coercive field Hc, Oe	19.14
Area of the hysteresis loop, erg/g	610.12
Electron spin resonance spectrum	
g-factor	2.50

Fig. 1 The whole-body MRI scan of mice bearing Lewis lung carcinoma after the conventional DOXO (**a**) and MNC with ER by MRT (**b**) treatment on 25th day after tumor transplantation

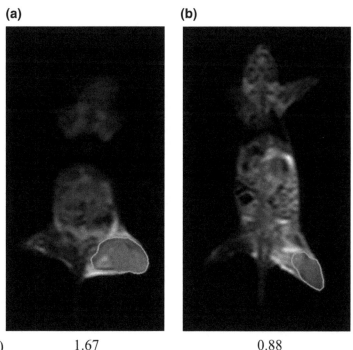

(a) **(b)**

Volume tumor (cm³)   1.67   0.88

3.3 Nonlinear Kinetics of Tumor Volume

The conventional DOXO treatment (group 2) resulted in minimal tumor response (Table 2). The combined therapy of MNC and ER by MRT (group 4) showed maximal antitumor effect. The treatment by MNC alone (group 3) resulted in an equal antitumor effect compared with group 2. The average number of lung metastatic foci per mouse in groups 3 and 4 compared to group 1 (control) and group 2 (conventional DOXO) had a tendency to decrease up to 25 days after tumor transplantation.

3.4 Electron Spin Resonance Spectroscopy

ESR directly detects paramagnetic species (stable radicals) that can induce modifications in redox reactions within the tumor. A triplet signal characterizing the ETC state was recorded near $g = 2.007$ in the ESR spectra for the control and investigated samples. The bar charts in Fig. 2 were obtained from tumor ESR spectra.

After DOXO administration and the introduction of MNC or particularly MNC + ER by MRT, the content of NO-FeS-protein complex of N-type in the respiratory chain of the mitochondria decreased by 49%, 70% and 65%, respectively. NO-FeS-protein complexes of N-type in the ETC could inhibit the ATP synthesis in the mitochondria of cancer cells and lead to cell death. The latter provides further support for the working hypothesis of our early paper [4] that magneto-mechano-chemically synthesized MNC and ER increase the oxidative assault on cancerous cells being already under the ROS overproduction and consequently render them vulnerable to further oxidative stress. According to the presented data, administration of magneto-mechano-chemically

Table 2 The growth kinetics and metastasis of Lewis lung carcinoma for 25 days after tumor implantation

N	The group of animals	Growth factor φ, day^{-1}	Braking ratio κ, relative units	Average number of lung metastatic foci per mouse
1	Control (without treatment)	0.31 ± 0.01	1.00	15.4 ± 2.8
2	Conventional DOXO	0.30 ± 0.01	1.04	18.5 ± 3.5
3	MNC	0.30 ± 0.01	1.05	11.3 ± 2.6
4	MNC + ER by MRT	0.25 ± 0.01^{abc}	1.23	10.7 ± 5.9

[a]Statistically significant difference from group 1 (control), $p < 0.05$; [b]Statistically significant difference from group 2 (conventional DOXO), $p < 0.05$; [c]Statistically significant difference from group 3 (MNC), $p < 0.05$

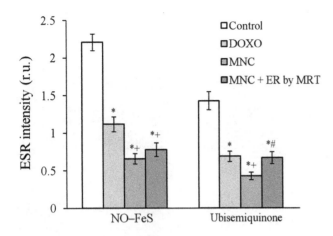

Fig. 2 Redox status of Lewis lung carcinoma on 25th day after tumor transplantation. *Statistically significant difference from the control group, $p < 0.05$; +Statistically significant difference from conventional DOXO, $p < 0.05$; #Statistically significant difference from MNC, $p < 0.05$

synthesized MNC and ER more effectively involves ROS mechanisms as compared to the conventional DOXO.

Ubisemiquinone is a free radical resulting from the removal of one hydrogen atom with its electron during the process of dehydrogenation of a hydroquinone. Such radical can potentially donate its unpaired electron to O_2, thereby generating superoxide. During hypoxia in tumors, the process of ROS generation can be amplified [7]. The level of ubisemiquinone in the mitochondrial ETC of tumor cells in groups 2, 3 and 4 respectively lowered by 52%, 70%, 53% compared to the control group. This could indicate deregulation of an electron transfer in the mito-chondrial ETC of tumor cells under the influence of applied antitumor factors.

4   Conclusions

The technology of magnetic resonance nanotheranostics allows to improve antitumor effect of the drug and monitor the treatment effectiveness. The current research is only the first step, however it could open new prospects for clinical use of nanotechnology in personalized cancer treatment and individually optimized treatment protocols for cancer patients with disseminated tumor cells in the future.

Conflict of Interest The authors declare that they have no conflict of interest.

References

1. F.S. Barnes, B. Greenebaum, The effects of weak magnetic fields on radical pairs, Bioelectromag. 36 (2015) 45–54, https://doi.org/10.1002/bem.21883.
2. H. Jaffer, K.J. Murphy, Magnetic resonance imaging-induced DNA damage, Canadian Assoc. Radiol. J. 68 (2017) 2–3.
3. N. Bezdenezhnykh, V. Orel, N.I. Semesyuk et al. The character-ization of viability and expression of adhesion proteins and cellular cytoskeleton for A-549 lung cancer and T47D breast cancer cell lines when using magnetic nanotherapy technology. Clinical Oncology. 24 (2016) 72–77.
4. V. Orel, A. Shevchenko, A. Romanov et al. Magnetic properties and antitumor effect of nanocomplexes of iron oxide and doxorubicin, Nanomed. Nanotech. Biol. Med. 11 (2015) 47–55, https://doi.org/10.1016/j.nano.2014.07.007.
5. L. Blumenfeld, Problems of biological physics. Springer-Verlag, Berlin, 1981.
6. T. Matsuzaki, T. Yokokura, Inhibition of tumor metastasis of Lewis lung carcinoma in C57BL/6 mice by intrapleural administration of Lactobacillus casei, Cancer. Immunol. Immun. 25 (1987) 100–104, doi:https://doi.org/10.1007/BF00199948.
7. A.J. Giaccia, M.C. Simon, R. Johnson, The biology of hypoxia: the role of oxygen sensing in development, normal function, and disease, Genes Dev. 18 (2004) 2183–2194, https://doi.org/10.1101/gad.1243304.

Development of a Spherical Ultrasound Transducer for Transcranial Low-Dose Ultrasound Hyperthermia Used in Brain Tumor Nanodrug Delivery

Chun-Chiang Shen, Gin-Shin Chen, Yung-Yaw Chen, and Win-Li Lin

Abstract

Previous studies showed that low-dose focused ultrasound hyperthermia (UH) could enhance the delivery and therapeutic efficacy of nanodrug for brain metastasis of breast cancer. In this study, our purpose is to design an ultrasound transducer that can be used for clinical application for brain tumor nanodrug delivery. Computer simulation has been used to calculate the pressure field, temperature distribution and thermal dose in the brain for different parameters of a spherical array ultrasound transducer. An ultrasound transducer with 112 disc elements (1 MHz, 2 cm in diameter) was constructed and its heating ability was characterized using a skull phantom filled with hydrogel. The simulation results showed that the position of peak pressure was slightly offset when the transducer was mechanically moved 3 cm away from the center point in all directions. There was no significant side-lobe developed when the transducer was moved away from the center 2 cm. Temperature distribution showed that the heating zone was an ellipsoid with temperature higher than 42 °C for a maximum intensity of 121 W/cm^2. Phantom experiment showed that the 42 °C color-changeable hydrogel would turn to be white at the focal point when the transducer was moved 2 cm in all directions with an electrical power 55 W. A spherical ultrasound phased array was developed and characterized, and the results showed that this ultrasound transducer has the potential used in transcranial short-time ultrasound hyperthermia for brain tumor nanodrug delivery.

Keywords

Spherical ultrasound transducer • Hyperthermia
Brain tumor • Nanodrug delivery

1 Introduction

In past decades, focused ultrasound has become a popular and important technology in non-invasive surgery with more than 30 clinical indications now in trials (FUS Foundation 2017) [1]. One of the potential applications is for brain therapy and surgery. Several researches showed positive results, such as ultrasound hyperthermia, BBB opening for drug delivery, histotripsy, and tumor ablation [2–5]. However, the characteristic of skull bone that causes significant attenuation and defocusing (aberration effects) to the acoustic wave is still a critical obstacle for transcranial ultrasound therapy. High acoustic absorption of skull bone is the main reason that overheats skull surface during HIFU treatment. Current HIFU system equipped cooling system to cool down surface temperature to prevent thermal damage of skin, skull bone or normal brain tissue. There are two ways to calibrate defocusing issue: measuring phase distortion and acoustic time reversal. Ideally, measuring phase distortion could compensate the distortion by using CT images before treatment. However, the result is limited because of the difficulty for patients to keep still during the entire treatment [6–9]. Acoustic time reversal can compensate the phase distortion by receiving the signal from inside the brain. A recent study showed that receiving the signals emitted by the individual bubble could generate only coarse aberration correction [10].

C.-C. Shen (✉) · W.-L. Lin
Institute of Biomedical Engineering, National Taiwan University, No. 1, Sec.1, Jen-Ai Road, Taipei, 100, Taiwan
e-mail: abc1234567704@gmail.com

W.-L. Lin
e-mail: winli@ntu.edu.tw

G.-S. Chen · W.-L. Lin
Institute of Biomedical Engineering and Nanomedicine, National Health Research Institutes, No. 35, Keyan Road, Zhunan, Miaoli, Taiwan
e-mail: gschen@nhri.org.tw

Y.-Y. Chen
Department of Electrical Engineering, National Taiwan University, No. 1, Sec. 4, Roosevelt Road, Taipei, Taiwan
e-mail: yychen@ntu.edu.tw

© Springer Nature Singapore Pte Ltd. 2019
L. Lhotska et al. (eds.), *World Congress on Medical Physics and Biomedical Engineering 2018*,
IFMBE Proceedings 68/3, https://doi.org/10.1007/978-981-10-9023-3_121

Previous studies showed that low-dose focused ultrasound hyperthermia (UH) could enhance the delivery and therapeutic efficacy of nanodrug for brain metastasis of breast cancer without damaging the surrounding normal tissues. Ultrasound was used to deliver proper energy non-invasively at the target region to raise its temperature to a proper value to achieve the thermal dose that could increase perfusion, vascular permeability and interstitial micro-convection. Due to the above mechanism, low-dose ultrasound hyperthermia could enhance the delivery of nanodrug to improve the cancer treatment [11, 12].

In this study, our purpose is to develop a spherical transducer for transcranial low-dose ultrasound hyperthermia used in brain tumor nanodrug delivery. Numerical simulation has been used to calculate pressure field, temperature and thermal dose distributions in the brain. An ultrasound transducer with 112 disc elements (1 MHz, 2 cm in diameter) was constructed and its heating ability was characterized using a skull phantom filled with hydrogel.

2   Materials and Methods

2.1   Skull Model

Figure 1 shows the skull model which is obtained from Ten24's online store. This model is the high-res capture of a male human skull with a completely free anatomically correct 3D scan which uses the company's 30-camera photogrammetry rig. The dimension of anterior-to-posterior is 17.1 cm, the width is 13.5 cm and the height is 8.29 cm. Uniform skull thickness is constructed by 0.71 cm which is the average thickness of women. In order to set the brain base as the reference plane for numerical simulation, we rotated the orbitomeatal line to line up to the y-axis. Furthermore, we removed points of head below the reference plane to increase simulation efficiency, and set the middle of the reference plane as the center of brain.

2.2   Numerical Simulation

2.2.1   Acoustic Pressure Field
Acoustic pressure field for the simulation was calculated by each piston element with a radius of 1 cm circular transducer. The ultrasound beam propagated inside the brain was generated by a phased-array piston transducer using Rayleigh-Somerfeld integral [13]. The Rayleigh-Somerfeld integral is defined as

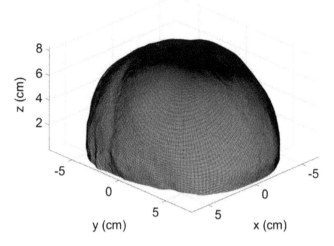

Fig. 1 Skull model after reconstruction. The dimension of anterior-to-posterior is 17.1 cm, the width is 13.5 cm and the height is 8.29 cm

$$p(x, y, z, t) = i \frac{\rho c k}{2\pi} u \int_{surface} \frac{e^{-i(\omega t - kr)}}{r} dS, \qquad (1)$$

where p is the pressure, t is the time, x, y and z, represent the position where the pressure is calculated, i is the imaginary unit, ρ is the density, c is the sound speed, u is the surface velocity of the piston transducer, k is the wave number, ω is the angular frequency, r is the distance between calculated position and transducer.

Acoustic wave propagated through the skull model into the brain by using the described method to solve a two-layer propagation problem [14]. The skull surface was divided into small patches as the second source. The normal velocity of the second source would be considered to deliver ultrasound wave into the brain as a piston transducer. The particle velocity transmission coefficient for the source on the outer layer of the skull is defined here as

$$\mathrm{T} = \frac{2[\rho_1 c_1 \cos(\theta_{i_2}^i)]}{\rho_2 c_2 \cos(\theta_{i_2}^t) + \rho_1 c_1 \cos(\theta_{i_2}^i)} \frac{\cos(\theta_{i_2}^i)}{\cos(\theta_{i_2}^t)}, \qquad (2)$$

where the incidence and transmission angles $\theta_{i_2}^i$, $\theta_{i_2}^t$ satisfy the Snell's law, i.e., $\sin \theta_{i_2}^t / \sin \theta_{i_2}^i = c_2 / c_1$.

In this study, a radius of 16 cm spherical ultrasound transducer with 121 elements (Fig. 2) was designed. The transducer was moved 3 cm away from the center point in all directions to calculate the acoustic signals in the water between the transducer and the skull model and signals propagated into the brain. The (average) intensity was

Fig. 2 **a** 3D and **b** 2D (top view) schematics of a spherical ultrasound transducer (a radius of 16 cm) with 121 elements (1 cm radius for each) arranged circularly

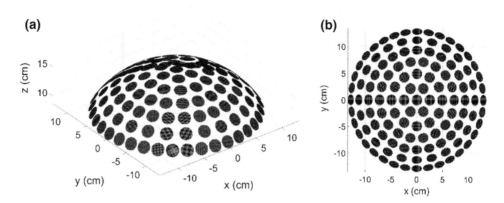

Fig. 3 **a** Picture of the spherical ultrasound transducer consisting of 112 disc elements, acrylic housing, and 112 coaxial and 2 ZIF connectors, and **b** experiment setup for characterizing the heating ability of the transducer

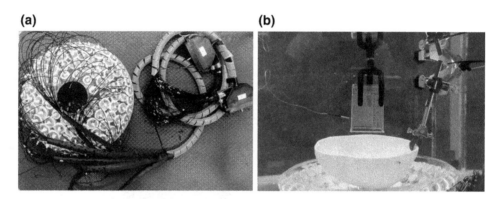

converted from summing the contribution of all the elements of the transducer at the interested region. The converted relationship between pressure and intensity is defined as here

$$I = \frac{|P|^2}{2\rho c} \qquad (3)$$

where I is the intensity, p is the pressure, ρ is the density and c is the sound speed. All simulations were performed using MATLAB 2017Ra.

2.2.2 Temperature and Thermal Dose Calculation

Temperature estimation is an important parameter for modeling several therapies like ultrasound hyperthermia (radiofrequency, HIFU ablation or laser). Bio-heat transfer equation (BHTE) is a typical way to estimate temperature distribution. When solving the equation by using finite element method, time step is less than 0.1 s to prevent coarse results [15]. Diffusion and tissue perfusion are considered in BHTE, the tissue temperature over time is defined here as

$$\rho c_t \frac{\partial T}{\partial t} = c\nabla^2 T - \omega_b c_b (T - T_{ar}) + Q \qquad (4)$$

where ρ and c_t is the tissue density and specific heat, $k\nabla^2 T$ is the model of diffusion with thermal conductivity k, $\omega_b c_b (T - T_{ar})$ is modeled the effect of perfusion, where ω_b, c_b and T_{ar} is the perfusion rate, specific heat of blood and

arterial blood temperature. The parameter above is based on values presented in Duck [16], Cooper and Tezek [17], Connor and Hynynen [18] and Connor and Hynynen [19].

The resulting temperature of bio-heat transfer equation was used to compute the thermal dose of the tissues. The thermal dose is defined as

$$\text{Thermal dose} = \int_{t=0}^{t_{end}} R^{(T-43)} dt, \begin{cases} R = 2 & for T \geq 43\,^\circ C \\ R = 4 & for 37^\circ C < T < 43\,^\circ C \end{cases}$$

$$(5)$$

Tissue will be damaged when the delivered thermal dose is greater than the specific threshold value during treatment.

2.3 Transducer

In order to treat tumors deep inside the brain, Fig. 3a shows a therapeutic transducer with an observation window is fabricated as the spherical shell due to its strong focusing geometry and filled with 112 disc elements (1 MHz, 2 cm in diameter). A 256-channel power amplifier was used in the study. The driving frequency of the power amplifier is between 0.8 and 1.2 MHz, so 1 MHz is chosen to characterize the heating ability through the skull phantom filled with color-changeable hydrogel in water, Fig. 3b.

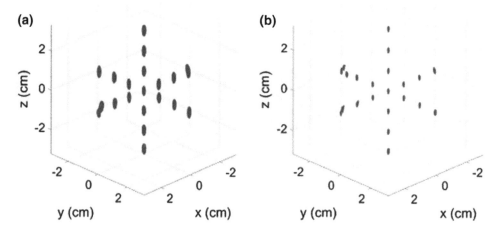

Fig. 4 Mechanically move the focus 3 cm away the center in each direction for **a** 0.5 MHz and **b** 1 MHz

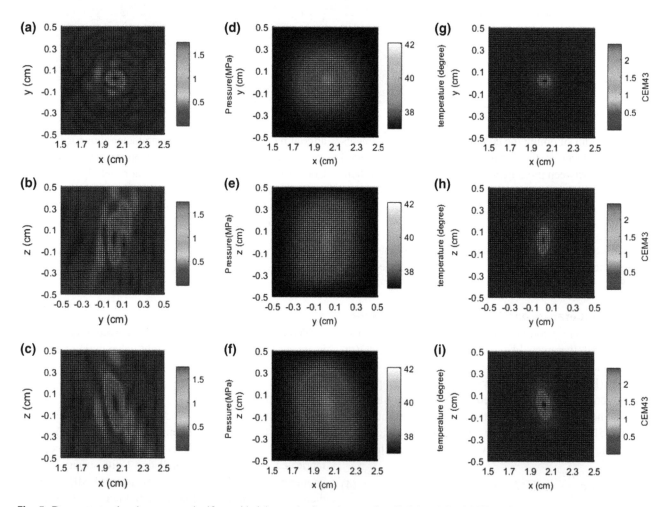

Fig. 5 Demonstrates that there are no significant side lobes as the focus is moved to (2, 0,0 cm) for 1 MHz. **a**, **b**, **c** are the pressure fields, **d**, **e**, **f** temperature distributions, **g**, **h**, **i** thermal dose distributions shown on the x-y plane, y-z plane and x-z plane, respectively

Fig. 6 Experimentally measured the focal dimension when moved 1, 2 and 3 cm away from the center along x-axis (**a** and **d**), y-axis (**b** and **e**), and z-axes (**c** and **f**)

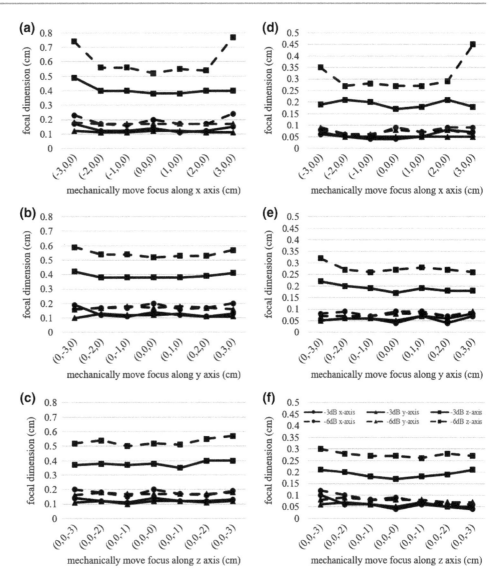

3   Results and Discussion

As the results shown in Fig. 4, there are no significant side lobes developed in all directions when mechanically moved the focus 2 cm away the center in each direction for 0.5 or 1 MHz.

The simulation results for the acoustic pressure field, the temperature distribution and the thermal dose distribution in the x-y plane, the x-z plane and the y-z plane are shown in the left, middle and right columns of Fig. 5, respectively, when the transducer is mechanically moved to locate the focus at (2, 0, 0 cm) with a driving frequency of 1 MHz. As shown in Fig. 5c, the pressure field is slightly tilt to the center, and the side lobes lower than −15 dB is observed surrounding the focal point. The heating temperature is controlled to 42 °C (brain temperature: 37 °C, perfusion rate: 0.00833 1/s) in a 600 s sonication, and the temperature

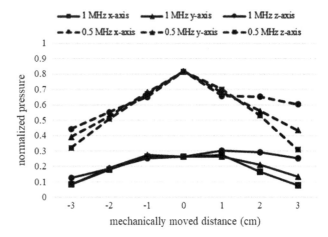

Fig. 7 Relationship between moved distance and pressure normalized to the pressure without skullcap

Fig. 8 **a** Temperature responses with an electrical power of 55 W for a duration of sonication 540 s **b** the size of target lesion (red arrow) for a 540 s sonication

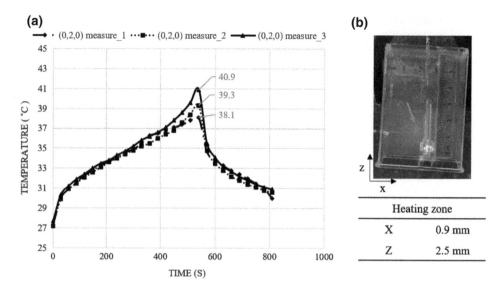

distribution indicates the heating zone is an ellipsoid with temperature higher than 42 °C for a maximum intensity of 121 W/cm². Thermal dose is controlled below 3 min of CEM₄₃ which is harmless to normal brain tissue.

For well-controlled treatments, the variation of focal dimension and pressure loss due to the skull bone should be well calculated. In order to demonstrate the variation, the −3 and −6 dB focal dimensions along x-axis (lines with squares), y-axis (lines with triangle) and z-axis (lines with dots) are shown in Fig. 6 [1]. The focal dimension changed less than 1 mm when moving 2 cm away from the center in all directions. Furthermore, the position of peak pressure was slightly offset when the transducer was mechanically moved 3 cm away from the center point in all directions. However, when moving 3 cm away from the center along x-axis, side lobes which is lower than −3 dB to the main lobe were observed, and the shape of the focal zone slightly tilt towards to the moving direction. Figure 7 shows that the pressure normalized to the pressure without skull bone decreases when the focus is moved away from the center. Ultrasound signal loss is about 18.4–69% for 0.5 MHz, and 73.8–92.4% for 1 MHz. The signal loss is getting smaller when moved along positive z-axis because the focal point is getting closer to brain surface. The loss variation is mainly due to the angle between the directions of ultrasound signal and the normal vector of the skull surface. It gets smaller when the focus moves away from the center. In numerical simulation, we only take the shape of the skull bone and uniform skull thickness into account. In clinical application, the aberration effect caused by the discontinuity of skull thickness is still an important issue to be discussed.

Phantom experiments showed that the 42 °C color-changeable hydrogel would turn to be white at the focal point when the transducer was moved 2 cm away the

center in all directions as an electrical power of 55 W was on for a duration of 540 s. Figure 8 shows the heating results at (0, 2, 0 cm) measured by thermal couples. The focal dimension measured by image J shows a lateral dimension of 0.9 mm and an axial dimension of 2.5 mm, which close to the simulation results. The heating zone at (2, 0, 0 cm), (−2, 0, 0 cm) and (0, 0, −2 cm) produces a small white lesion which is hard to be measured when the same electrical power 55 W is used. The 42 °C color-changeable hydrogel changed its color when the peak temperature recorded by the thermal couples is not higher than 42 °C and this may be due to the position limitation of thermal couples.

4 Conclusion

A multi-element spherical ultrasound transducer was developed and characterized. The results showed that this ultrasound transducer has the potential for using in transcranial short-time ultrasound hyperthermia to enhance brain tumor nanodrug delivery.

References

1. K. W. Ferrara, J. Liu, O. Le Baron and D. N. Stephens (2016) Development of a spherically focused phased array transducer for ultrasonic image-guided hyperthermia. Physics In Medicine And Biology, vol. 61, no. 14, p. 5275.
2. J. G. Lynn and T. J. Putnam (1944) Histology of cerebral lesions produced by focused ultrasound. Amer. J. Pathol., vol. 20, no. 3, p. 637.
3. W. J. Fry, W. Mosberg Jr, J. Barnard, and F. Fry (1954) Production of focal destructive lesions in the central nervous system with ultrasound*. J. Neurosurg., vol. 11, no. 5, pp. 471–478.

4. J. Barnard, W. Fry, F. Fry, and J. Brennan (1956) Small localized ultrasonic lesions in the white and gray matter of the cat brain. AMA Arch. Neurol. Psychiatry, vol. 75, no. 1, pp. 15–35.
5. G. Young and P. Lele (1964) Focal lesions in the brain of growing rabbits produced by focused ultrasound. Exp. Neurol., vol. 9, no. 6 pp. 502–511.
6. K. Hynynen and F. A. Jolesz (1998) Demonstration of potential noninvasive ultrasound brain therapy through an intact skull. Ultrasound Med. Biol., vol. 24, no. 2, pp. 275–283.
7. F. Marquet, M. Pernot, JF. Abury, G. Montaldo, L. Marsac, M. Tanter and M. Fink (2009) Non-invasive transcranial ultrasound therapy based on a 3D CT scan: Protocol validation and in vitro results. Phys. Med. Biol., vol. 54, no. 9, p. 2597.
8. N. McDannold, G. Clement, P. Black, F. Jolesz, and K. Hynynen (2010) Transcranial MRI-guided focused ultrasound surgery of brain tumors: Initial findings in three patients. Neurosurgery, vol. 66, no. 2, p. 323.
9. Sukovich. J, Xu. Z, Kim. Y, Cao. H, Nguyen. T.S, Pandey. A, Hall. T, Cain. C (2016) Targeted lesion generation through the skull without aberration correction using histotripsy. IEEE Trans. Ultrason. Ferroelectr. Freq. Control. vol. 63, no. 5, pp. 671–682.
10. J. Gateau, L. Marsac, M. Pernot, J.F. Aubry, M. Tanter, and M. Fink (2010) Transcranial ultrasonic therapy based on time reversal of acoustically induced cavitation bubble signature. IEEE Trans. Biomed. Eng., vol. 57, no. 1, pp. 134–144.
11. Wu SK, Chiang CF, Hsu YH, Lin TH, Liou HC, Fu WM and Lin WL (2014) Short-time focused ultrasound hyperthermia enhances liposomal doxorubicin delivery and antitumor efficacy for brain metastasis of breast cancer. Int J Nanomedicine, vol. 2014:9(1), pp. 4485–4494.
12. Wu SK, Chiang CF, Hsu YH, Liou HC, Fu WM and Lin WL (2017) Pulsed-wave low-dose ultrasound hyperthermia selectively enhances nanodrug delivery and improves antitumor efficacy for brain metastasis of breast cancer. Ultrason Sonochem, vol. 36, pp. 198–205.
13. O'Neil H. T. (1949) Theory of focusing radiators. J. Acoust. Soc. Am., vol. 21, p. 516.
14. Sun J and Hynynen K (1999) The potential of transskull ultrasound therapy and surgery using the maximum available skull surface area. J Acoust. Soc. Am., vol. 105, pp. 2519–2527.
15. Dillenseger J, Esneault S (2010) Fast FFT-based bioheat transfer equation computation," Comput Biol Med., vol. 40, no. 2, pp. 119–123.
16. Duck (1990) Physical Properties of Tissue: A Comprehensive Reference, London: Academic.
17. Cooper T E and Tezek G J (1972) A probe technique for determining the thermal conductivity of tissue. J. Heat Transfer, vol. 94, no. 2, pp. 133–140.
18. Connor C W and Hynynen K (2002) Bio-acoustic thermal lensing and nonlinear propagation in focused ultrasound surgery using large focal spots: a parametric study. Phys. Med. Biol. vol. 47, no. 11, pp. 1911–1928.
19. Connor CW, Hynynen K (2004) Patterns of thermal deposition in the skull during transcranial focused ultrasound surgery. IEEE Trans Biomed Eng., vol. 51, pp. 1693–1706.

An in Vitro Phantom Study to Quantify the Efficacy of Multi-tine Electrode in Attaining Large Size Coagulation Volume During RFA

Sundeep Singh and Ramjee Repaka

Abstract

The present study aims at evaluating the efficacy of commercially available RITA's StarBurst® XL multi-tine electrode in attaining large size coagulation volumes (≥ 3 cm in diameter) during radiofrequency ablation (RFA) application. In vitro studies have been conducted on the cylindrical shaped polyacrylamide based tissue-mimicking phantom gels utilizing different active lengths of the multi-tine electrode, viz., 2 cm, 3 cm, 4 cm and 5 cm. A temperature-controlled RFA has been performed at a target tip temperature of 95 °C for 5 min. The variations in the power supply, the target tip temperature and the size of coagulation volume have been reported for different active lengths of the multi-tine electrode. The study revealed that the increase in active length of the multi-tine electrode results in more energy deposition and consequent rise in the coagulation volume during RFA procedure. Further, a simplified novel statistical correlation between the coagulation volume and active length of the multi-tine electrode has been proposed.

Keywords

Radiofrequency ablation • Temperature-controlled ablation • Multi-tine electrode • Coagulation volume Tissue-mimicking phantom

1 Introduction

Over the past few decades, percutaneous thermal ablation has become the most promising and widely applied technique for treatment of focal tumors in liver, lung, kidney, breast, adrenal glands and, head and neck [1]. The goal of thermal ablation is quite similar to that of surgery. However, instead of physical excision during surgery the target tissue is treated in situ by the application of fatal high and low temperatures to a focal zone in and around the tumor [2]. The thermal ablation methods can be classified into different categories on the basis of temperature level and energy source. Notably, radiofrequency ablation (RFA), microwave ablation and laser ablation utilizes electromagnetic energy to induce heat within the biological tissue. During high intensity focused ultrasound (HIFU) ablation heat is generated by the application of ultrasound waves. The cryoablation is the thermal ablative modality that uses extreme cold temperature to freeze and destroy the tumor. Among the different minimally invasive thermal treatment modalities, RFA is the most extensively studied and widely applied technique in clinical practice for treatment and palliation of various types of primary and secondary tumors in different organs. The potential advantages of RFA include low cost, minimal scarring, less pain, increased preservation of surrounding tissue, improved cosmesis, shorter hospitalization time and lower morbidity [3].

RFA is a minimally invasive thermal ablative technique that utilizes the percutaneous insertion of radiofrequency electrode into the target tissue with the aid of image guidance techniques, viz., computed tomography (CT), ultrasound (US), or magnetic resonance imaging (MRI) [4]. Usually, ultrasound is preferred during RFA due to its low cost, capability of real-time visualization and absence of ionizing radiation hazard [5]. Once positioned, high-frequency alternate current (450–550 kHz) is delivered by the radiofrequency power generator that flows from the electrode to the ground pads placed at patient's back or thigh, forming a closed electric circuit. As the current flows through the biological tissue, frictional (resistive or joule) heat is induced due to the agitation of free ions (Na^+, K^+, Cl^- etc.) when they attempt to follow change in direction (polarity) of high-frequency alternate current. As a result, two zones of heating are induced: direct heating zone and indirect heating zone. The direct heating zone is localized

S. Singh (✉) · R. Repaka
Department of Mechanical Engineering, Indian Institute of
Technology Ropar, Rupnagar, Punjab, India
e-mail: sundeep.singh@iitrpr.ac.in

© Springer Nature Singapore Pte Ltd. 2019
L. Lhotska et al. (eds.), *World Congress on Medical Physics and Biomedical Engineering 2018*,
IFMBE Proceedings 68/3, https://doi.org/10.1007/978-981-10-9023-3_122

around the active electrode and has high current density. While, the indirect heating zone is a consequence of heat transfer (conduction) from direct heating zone to more peripheral areas around the electrode [6, 7]. The goal of RFA is to achieve temperature between 60 and 100 °C that basically leads to the destruction of tumor cells by the induction of protein coagulation [8]. Importantly, the higher temperatures during RFA procedure should be strictly below 100 °C to avoid tissue boiling, vaporization and carbonization that usually leads to drastic decline in electrical and thermal conductivities along with poor image resolution [9]. Further, RFA planning is hampered if the ablated tumor is near the large blood vessels that cause a heat sink effect, and thus decreases the size of coagulation volume.

RFA has already emerged as an effective alternative treatment modality for treating early and small tumors in those patients for whom surgery is not a viable option [10]. However, till date, the major limitation of RFA is the poor efficacy in treating large size tumors, i.e., ≥ 3 cm in diameter. To overcome this deficiency, present in vitro study is focused on evaluating the efficacy of commercially available RITA's StarBurst® XL multi-tine electrode in achieving large size coagulation volumes. Importantly, the considered mono-polar multi-tine electrode consists of nine curved tines that are deployed in the shape of Christmas tree configuration from a central cannula of single needle. The power is supplied to the nine-tine electrode using temperature-controlled mode that mitigates any chances of tissue charring close to the active tip of the electrode during RFA procedure. The study has been conducted utilizing different active lengths of nine tine electrode, viz., 2 cm, 3 cm, 4 cm and 5 cm. The variations in the power supply, the target tip temperature and the size of coagulation volume has been reported for different active lengths of the electrode. Furthermore, simplified statistical correlations between the obtained coagulation volumes and active lengths of the electrode have been developed to provide a priori prediction of coagulation volume for different active lengths of the multi-tine electrode during RFA application.

2 Materials and Methods

A commercial electrosurgical radiofrequency generator (RITA 1500X, AngioDynamics Inc., Latham, NY) has been used for all experimental in vitro studies conducted on polyacrylamide based tissue-mimicking phantom gel. The generator has the capability of delivering radiofrequency power up to 250 W $\pm$ 20% at a frequency of 460 kHz $\pm$ 5%. The RFA procedure has been performed utilizing Automatic Temperature Control (ATC) mode of

radiofrequency generator that is the most reliable and frequently used technique in clinical practices. In ATC mode the power delivery is automatically controlled based on the average of the temperature readings of all selected thermocouples (with accuracy of $\pm$ 3 °C) embedded at the tips of multi-tine electrode. The power supplied initially increases monotonically till the target temperature is reached in the ATC mode and declines afterwards to limit the pre-set target temperature. However, the supplied power may still rise if the target temperature falls. In the present study, the pre-set target temperature has been considered to be below 100 °C (i.e. 95 °C) to eradicate any adverse effects of charring and overheating of the biological tissue during RFA application. In the present study, maximum power has been considered to be 150 W. The RFA procedure has been performed for 5 min post the target tip temperature has been reached plus an additional cool down cycle of 30 s. Figure 1 illustrates the experimental set-up of RFA used in the present in vitro study.

A cylindrical shaped mould of 10 cm height and 8 cm diameter has been used for the fabrication of polyacrylamide based tissue-mimicking phantom gel that uses bovine serum albumin protein [11]. All the refrigerated phantom gels were allowed to attain the ambient room temperature before the start of RFA procedure (Fig. 2a). A multi-tine RITA Starburst XL electrode (AngioDynamics Inc., Latham, NY) has been inserted into the phantom gel (Fig. 2b) and then deployed with a gentle push (Fig. 2c) to different active lengths. The fabricated phantom gels are transparent and turns into ivory white at a coagulation temperature above 50 °C (Fig. 2d). After the completion of RFA procedure, coagulated phantom gel has been dissected along the longitudinal plane (passing through the electrode insertion axis) to measure the dimensions of ellipsoid-shaped ivory white regions by visual examinations. The maximal longitudinal (L) and maximal transverse dimensions (W) of the dissected phantom gels have been measured macroscopically. Further, the coagulation volume (V) has been calculated using the ellipsoid formula $(\pi/6)LW^2$ [12].

3 Results and Discussion

The variation of power supplied during temperature-controlled RFA for different active lengths of the multi-tine electrode, viz., 2 cm, 3 cm, 4 cm and 5 cm, have been depicted in Fig. 3. It is evident from Fig. 3 that, the output power (or energy deposition) increases as the deployment of the active part of electrode increases. This can be attributed to the fact that, the surface area of the active part of electrode increases with the increase in deployment

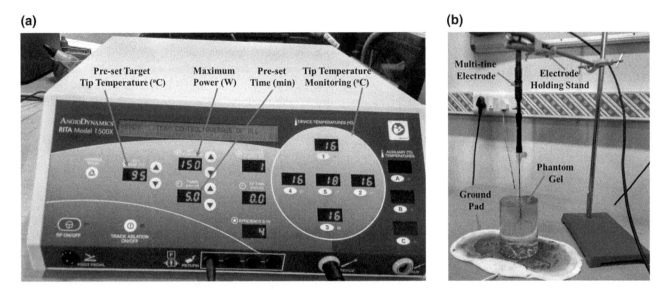

Fig. 1 Experimental set-up for performing in vitro RFA on tissue-mimicking phantom gel, **a** RITA 1500X® radiofrequency generator, **b** RITA StarBurst® XL electrode inserted into the tissue-mimicking phantom gel placed on ground pad

Fig. 2 The picture shows different stages of tissue-mimicking phantom gel, **a** fresh transparent phantom, **b** insertion of electrode into the phantom, **c** deployment of tines of electrode by gentle push of piston down the shaft of the electrode, and **d** coagulated phantom gel turning into ivory white with well-defined boundaries post RFA procedure

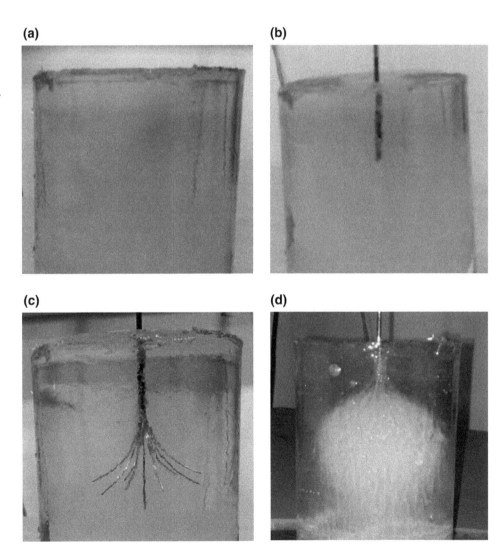

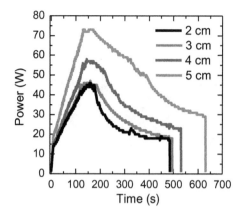

Fig. 3 Variation of power supplied with time for different active lengths of multi-tine electrode

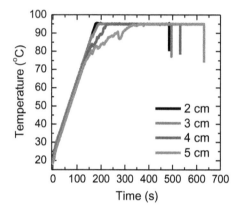

Fig. 4 Variation of target tip temperature with time for different active lengths of electrode

of multi-tine electrode and that allows more dispersion in the current density and subsequent enhancement in the energy deposition during RFA. Further, the maximum power applied at the active part of the multi-tine electrode deployed to 2 cm, 3 cm, 4 cm and 5 cm has been found to be 46 W, 47 W, 58 W and 73 W, respectively.

The variations of the average value of target tip temperature of the multi-tine electrode deployed to different active

lengths have been shown in Fig. 4. It can be clearly observed from Fig. 4 that, the time required to reach the pre-set target temperature of 95 °C increases as the deployment of electrode increases. This can be attributed to the fact that more the active length of the multi-tine electrode, more will be the power expended to reach the target tip temperature. Moreover, the abrupt decline in the average temperature at the end of RFA procedure (see Fig. 4) represents the cool down cycle of 30 s. Further, the average temperature of ablation for each deployment of the multi-tine electrode is over 55 °C that signifies the attainment of complete ablation during RFA.

Table 1 shows the comparison of coagulation volume obtained from the in vitro studies on polyacrylamide based tissue-mimicking phantom gel during RFA. It is clearly apparent from Table 1 that, there is a considerable rise in the coagulation volume with increase in deployment of the active part of the multi-tine electrode. Further, the time required to attain the target tip temperature along with the equivalent spherical diameter (D), maximal longitudinal (L) and transverse dimensions (W) of coagulation zone for different deployment lengths has also been reported in Table 1. It is noteworthy to mention that, the equivalent spherical diameter (D) has been computed by equating the obtained coagulation volume (V) with $(\pi/6)D^3$.

Based on the data obtained from the in vitro studies of temperature-controlled RFA on tissue-mimicking phantom gels, a simplified correlations have been developed (see Fig. 5) by using curve fitting toolbox of MATLAB®. The proposed correlations will aid in the prediction of longitudinal dimension (Fig. 5a), transverse dimension (Fig. 5b), coagulation volume (Fig. 5c) and equivalent spherical diameter (Fig. 5d) from the known values of deployment length of active part of multi-tine electrode. It can be clearly seen from Fig. 5 that, for all developed correlations the goodness of fit (R^2) is greater than 0.95 that signifies the accuracy of proposed correlations. It is noteworthy to mention that, the coagulation volume attained during RFA is significantly dependent on the blood perfusion rate of the

Table 1 Comparison of coagulation volume attained for different values of active length of multi-tine electrode during temperature-controlled RFA

Active length of electrode (cm)	Time to reach target tip temperature (min)	Maximal longitudinal dimension L (cm)	Maximal transverse dimension W (cm)	Coagulation volume V (cm³)	Equivalent spherical diameter D (cm)
2	3.1	4	3	18.85	3.30
3	3.3	4.3	3.5	27.58	3.75
4	3.9	5	4	41.89	4.31
5	5.5	5.3	4.7	61.30	4.89

Fig. 5 Proposed correlation as a function of active length of electrode for **a** longitudinal dimension, **b** transverse dimension, **c** coagulation volume, and **d** equivalent spherical diameter

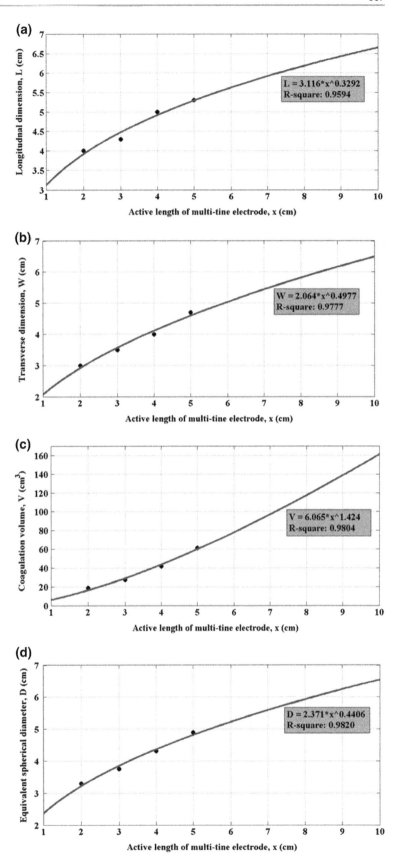

tumor and surrounding healthy tissue. Since, tissue-mimicking phantom gels does not possess any blood perfusion, thus it has been neglected in the present in vitro study. Future studies will be focused on addressing the effect of blood perfusion along with an embedded tumor in the phantom gels.

4 Conclusions

In vitro experimental studies have been conducted to address the problem of attaining ≥ 3 cm diameter tumor ablation during RFA application. A temperature-controlled RFA has been performed on polyacrylamide based tissue-mimicking phantom gels utilizing commercially available mono-polar multi-tine electrode deployed in the shape of Christmas tree configuration to different active lengths. The results revealed that, there is a considerable rise in the energy deposition within the phantom gel with the increase in the deployment lengths of multi-tine electrode during RFA. Moreover, it has been found that the deployment lengths of 2 cm, 3 cm, 4 cm and 5 cm can successfully attain the ablation diameter of 3.3 cm, 3.75 cm, 4.31 cm and 4.89 cm, respectively. Further, a simplified correlation $[D \ (\text{cm}) = 2.371*(\text{Deployment length in cm})^{0.4406}]$ has been proposed for the prediction of ablation diameter at different deployment lengths of multi-tine electrode during temperature-controlled RFA.

Acknowledgements Authors would like to acknowledge Indian Institute of Technology Ropar for providing essential infrastructure and necessary support to carry out the present research.

Conflict of Interest The authors declare that they have no conflict of interest.

References

1. Chu, K.F., Dupuy, D.E.: Thermal ablation of tumours: biological mechanisms and advances in therapy. Nature Reviews Cancer 14 (3), 199–208 (2014).
2. Brace, C.: Thermal tumor ablation in clinical use. IEEE Pulse 2(5) 28–38 (2011).
3. Singh, S., Repaka, R.: Parametric sensitivity analysis of critical factors affecting the thermal damage during RFA of breast tumor. International Journal of Thermal Sciences 124, 366–374 (2018).
4. Gazelle, G.S., Goldberg, S.N., Solbiati, L., Livraghi, T.: Tumor ablation with radio-frequency energy 1. Radiology 217(3), 633–646 (2000).
5. Singh, S., Repaka, R.: Quantification of thermal injury to the healthy tissue due to imperfect electrode placements during radiofrequency ablation of breast tumor. ASME Journal of Engineering and Science in Medical Diagnostics and Therapy 1(1), 011002(1–10) (2018).
6. Ahmed, M., Brace, C.L., Lee, F.T., Goldberg, S.N.: Principles of and Advances in Percutaneous Ablation. Radiology 258(2), 351–369 (2011).
7. Zhang, B., Moser, M.A., Zhang, E.M., Luo, Y., Liu, C., Zhang, W.: A review of radiofrequency ablation: Large target tissue necrosis and mathematical modelling. Physica Medica 32(8), 961–971 (2016).
8. Miller, M.W., Ziskin, M.C.: Biological consequences of hyperthermia. Ultrasound in Medicine and Biology 15(8), 707–722 (1989).
9. Singh, S., Repaka, R.: Temperature-controlled radiofrequency ablation of different tissues using two-compartment models. International Journal of Hyperthermia 33(2), 122–134 (2017).
10. Singh, S., Repaka, R.: Effect of different breast density compositions on thermal damage of breast tumor during radiofrequency ablation. Applied Thermal Engineering 125, 443–451 (2017).
11. Bu-Lin, Z., Bing, H., Sheng-Li, K., Huang, Y., Rong, W., Jia, L.: A polyacrylamide gel phantom for radiofrequency ablation. International Journal of Hyperthermia 24(7), 568–576 (2008).
12. Singh, S., Repaka, R.: Numerical study to establish relationship between coagulation volume and target tip temperature during temperature-controlled radiofrequency ablation. Electromagnetic Biology and Medicine, 37(1), 13–22 (2018).

Temperature-Induced Modulation of Voltage-Gated Ion Channels in Human Lung Cancer Cell Line A549 Using Automated Patch Clamp Technology

Sonja Langthaler, Katharina Bergmoser, Alexander Lassnig, and Christian Baumgartner

Abstract

In cancer cells specific ion channels exhibit altered channel expression, which can drive malignant and metastatic cell behavior. Hence, therapeutic strategies modulating ion channels prove to be promising in cancer therapeutics. Alterations in temperature, even small deviations from normothermia, may cause changes in electrophysiological processes, since activation and conductivity of various ion channels are temperature-dependent. In this pilot study, we focused on a basic understanding of the effects of temperature-alterations on voltage-gated ion channels of A549 cells using an automated patch-clamp system. The measurements were carried out in whole-cell voltage-clamped configuration applying test pulses between −60 and +60 mV. For positive voltages the ion-current curves showed an instantaneously increased conductance, followed by a slow current increase provoked by later activating voltage-gated ion channels, indicating the time-delayed response of additional channels. To investigate the temperature-dependent electrophysiological behavior, six cells (passages 7–10, n = 34) were examined at room temperature and normal body temperature. Compared to normal body temperature, reduced temperatures revealed a higher whole-cell current at negative voltages (63.4% (±18.5%), −60 mV) and lower currents (52.6% (±27.3%), +60 mV) at positive voltages, indicating a hypothermia-induced modulation of voltage-gated channels in the lung cancer cell line A549.

Keywords

Ion channels • Non-small cell lung cancer Hypothermia

1 Introduction

Lung cancer is one of the most prevalent forms of tumour worldwide [1] and often advanced at diagnosis [2]. Despite therapy survival is still poor, hence, medical research constantly tries to find new drugs and innovative treatment approaches to improve the outlook for patients with lung cancer. Substantial advances in the understanding of cancer biology as well as electrophysiological aspects of cancer afford novel therapies to effectively fight and cure this disease [3].

Ion channels play a fundamental role in almost all basic cellular processes like proliferation, differentiation and apoptosis [4, 5]. Indeed, specific ion channels have been recognized to influence malignant cancer cell behavior and thus have an immense potential for the development of therapeutic targets for cancer therapy [6].

Temperature is an important factor for the survival of cells influencing the permeability of the cell membrane and offering the possibility to improve drug metabolism rate. As a result, hyperthermia is used supportive to established treatment modalities such as chemo- and radiation therapy [7, 8]. Several studies already investigated further positive effects of hyperthermal conditions in cancer treatment by evaluating cell death and survival rates [9–11], heat shock proteins [10] or alterations of cancer blood flow (see e.g. in a survey of genotoxic effects of Speit et al. [11]). Likewise hypothermia has already shown negative effects on the cell growth, however, a stagnation of cell division or apoptosis can only be induced in profound hypothermia (body temperature <24 °C) [10, 12, 13]. Nevertheless, temperature alterations, even small deviations from normal body temperature, might provide a key factor in terms of

S. Langthaler (✉) · K. Bergmoser · A. Lassnig · C. Baumgartner
Institute of Health Care Engineering with European Testing Center
of Medical Devices, Graz University of Technology, Graz, Austria
e-mail: s.langthaler@tugraz.at

K. Bergmoser
CBmed—Center for Biomarker Research in Medicine, Graz,
Austria

© Springer Nature Singapore Pte Ltd. 2019
L. Lhotska et al. (eds.), *World Congress on Medical Physics and Biomedical Engineering 2018*,
IFMBE Proceedings 68/3, https://doi.org/10.1007/978-981-10-9023-3_123

electrophysiological processes since activation of various ion channels as well as the conductivity show temperature dependency [14–18].

In this pilot study, we focus on a more profound understanding of temperature-induced changes in the functionality of voltage-gated ion channels in human lung cancer cell line A549 using an in-house adapted automated patch-clamp system. Several preliminary tests were performed to establish a suitable test protocol and quality parameters. In a first experimental run, measurements at hypo- and normothermia were conducted to investigate the temperature dependent electrophysiological behavior of A549 cells.

2 Methods

Cells used in this study. Cells from the human lung cancer cell line A549 were purchased from the Center for Medical Research (ZMF) (Medical University Graz, Austria) obtained from the American Type Culture Collection (ATCC) and cultivated in Dulbecco's Modified Eagle Medium (Gibco® DMEM) supplemented with 10% fetal bovine serum (FBS) and 1% penicillin-streptomycin. Cells were maintained at standard conditions, 37 °C, 5% CO_2 in humidified atmosphere. The cell condition is particularly important with regard to automated patch clamp experiments, since cells cannot be collected individually. Hence, cells were harvested using Trypsin/EDTA every other day at confluence levels 50–70%.

Patch clamp recordings. Patch clamp recordings were performed using an automated patch clamp system with an integrated heating element and perfusion system (port-a-patch, Nanion, Munich, Germany) and NPC®-1 chips (Nanion, Munich) with chip resistance of 3–5 MΩ.

The external bath solution consisted of (in mM) 4 KCl, 140 NaCl, 1 $MgCl_2$, 2 $CaCl_2$, 5 D-Glucose, 10 HEPES/NaOH, pH 7.4, the intracellular solution contained (in mM) 50 KCl, 10 NaCl, 60 K-Fluoride and 10 HEPES/KOH. The cells were sealed in a solution containing (in mM) 80 NaCl, 3 KCl, 10 $MgCl_2$, 35 $CaCl_2$, 10 HEPES/NaOH, pH 7.4.

The measurements were carried out as voltage clamp and membrane currents measured in whole-cell configuration. Hence, we defined a pulse protocol consisting of an initial and re-pulse of −80 mV for 100 and 800 ms test pulses between −60 and +60 mV (increment 10 mV).

Pre-study experiments were carried out solely under hypothermal conditions at room temperature (22–26 °C). The subsequent experiments were performed at temperature levels of severe hypothermia (21.2–24.3 °C) [19] and normal body temperature (37.0±1 °C). The temperature at the cell

site was controlled using an external calibrated temperature measurement device (IR thermometer optris® CS LT).

Study protocol. In the pre-study, experiments with several cells from different cultures and number of passages were performed. It is important to note that the duration of measurements for a single cell experiment is limited. In the course of multiple preliminary test runs, a significant increase of the ion currents was observed after about 10 min, which may occur by a change in osmolarity of the solutions or indicates that the cell membrane or seal becomes instable. Similarly, each alignment and measurement can affect the cell membrane condition, whereby the number of measurements is also limited.

To ensure stable cell conditions and reliable results, experiments considered for data analysis were limited to a test time of 10 min and 10 measurements per cell.

Computational approach and data analysis. The recorded data comprise information on the membrane currents over time at the different voltage steps (sampling rate 20 kHz), seal resistance, series resistance and the cell capacity for assessing the quality of measurements. Main quality criteria for reliable measurements are met showing an internal resistance (R_s) not greater than 50 MΩ and a seal resistance (R_{memb}) at least 4 times higher than the internal resistance to avoid a high leakage current.

Analysis of the measured ion current curves as well as data visualization and processing was carried out using the software MATLAB. For determining the corresponding current-voltage relations the currents at the end of the voltage pulse were selected. Therefore, to avoid errors as result of large leakage currents at the end of the voltage pulse, we selected the maximum current at 99.2% of the voltage pulse length.

3 Results

3.1 Time Dependency

Figure 1a shows the response of a sample run in A549 cells at room temperature (22.4 °C). It is known from literature that the voltage-gated outward current of A549 cells is mainly carried by K^+ [20]. Note that measurements in whole-cell mode represent the total ion current across the cell membrane (whole-cell current), also including currents driven by other voltage-gated ion channels like Ca^{2+}, Na^+ and Cl^-.

The reversal potential Rp resides in the range between −20 and 10 mV. Ion currents activated by positive voltages (10–60 mV) can be separated into two parts: an instantaneously activating current, mainly driven by hIK channels

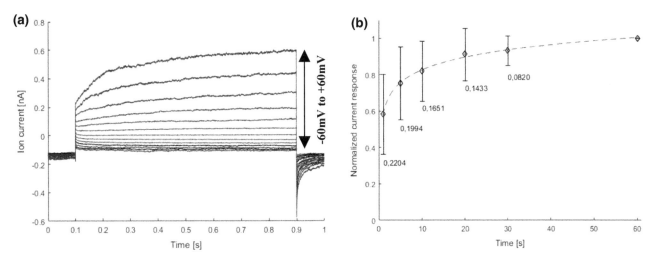

Fig. 1 **a** Behavior of voltage-gated ion channels in A549 at different voltages at room temperature (22.4 °C). **b** Mean standard current response over a time period of 60 s at +60 mV (13 cells, n = 55 single measurements)

[20, 21], followed by a slowly activating current provoked by later activating voltage-gated ion channels (e.g. Kv1.3 [20, 21]). A further increase in current could be observed with longer pulse durations, indicating a time-delayed response of additional channels. It was found that the registered current amplitudes at the end of an 800 ms test pulse correspond to about 60% (58.3% (±22.04%), +60 mV) of the maximum current measured at a pulse length of 60 s (13 cells, n = 55 single measurements). Figure 1b shows the averaged, normalized current values and the corresponding standard deviations at a voltage of +60 mV.

3.2 Effects of Hypothermia on Current Response

For the investigation of temperature-induced effects on the electrophysiological behavior, current amplitudes of a total of six cells (passages 7–10, $n_{hypothermia} = 16$, $n_{normothermia} = 18$) were measured at hypothermal condition (21.2–24.3 °C) and normal body temperature (37.0 ±1 °C). Figure 2 illustrates the mean percentage change and standard deviations of the current response in hypothermia referred to normal body temperature at the applied voltage steps (−60 to +60 mV). Compared to normothermia, reduced temperatures showed significantly higher whole-cell currents of about 60% and thus an increased ion channel activity at voltages below −20 mV. In contrast, constantly lower whole-cell currents were revealed at positive voltages and therewith decreased ion channel activity under hypothermal conditions. Due to different zero crossings of the individual measurement curves registered between −30 and 0 mV, higher standard deviations at voltage step −20 mV (16.5% (±79.1%)) and −10 mV (−32.3% (±86.1%)) are evident.

4 Conclusion

Potassium channels are known to play an important role in cell apoptosis as well as proliferation and migration of malignant cells. As a consequence, therapeutic strategies modulating ion channels could prove to be promising in cancer therapeutics. However, a profound understanding of ion channel function under different environmental conditions is crucial in order to set up new treatment modalities.

In this pilot study, we have demonstrated that temperature obviously modulates the size of registered whole-cell currents and thus the electrophysiological characteristics of investigated cancer cells. As indicated by the voltage-dependent increase and decrease of whole-cell currents, voltage-gated ion channels can be assumed to be affected by temperature alterations. In addition, the individual ion-current curves show significant changes in the instant conductance, so that primarily a high temperature-sensitivity of calcium-activated potassium channels (hIK), which also belong to the voltage-gated superfamily, can be expected. Therefore, the results support our hypothesis of a temperature-induced modulation of voltage-gated ion channels under hypothermal conditions in lung cancer cells. However, due to the small sample size (n = 6 cells) it should be noted that more comprehensive experiments are needed to confirm these first promising results in terms of interpretation and validation of findings.

In a next step we will also investigate possible effects of hyperthermia on ion channel activity and its impact on the resting membrane potential. These findings will serve as an input for a model-description of temperature-induced alterations of ion-channel activity in non-excitable cancer cells. Eventually this may help to gain a more comprehensive

Fig. 2 Mean percentage deviation of whole-cell currents at hypothermia referred to normothermia (6 cells, n = 34 single measurements)

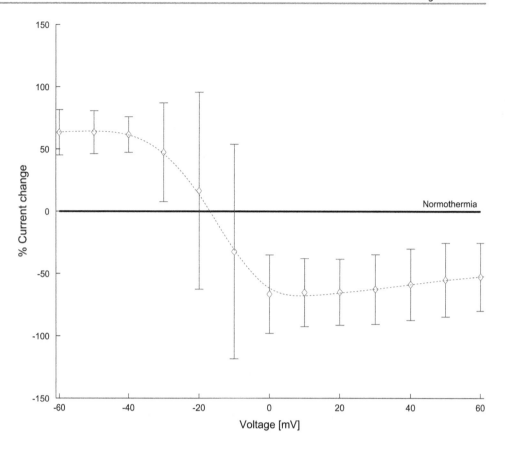

knowledge of ion channel characteristics in cancer to further develop and assess thermal therapies in cancer treatment.

Author's Statement The authors declare that they have no conflict of interest.

References

1. Ma X., Yu H.: Global Burden of Cancer. Yale J Biol Med 79(3–4), 8 –94 (2006).
2. Ramalingam S.S., Owonikoko T.K., Khuri F.R.: Lung cancer: New biological insights and recent therapeutic advances. CA Cancer J Clin 61(2), 91–112 (2011).
3. Manegold C., Thatcher N.: Survival improvement in thoracic cancer: progress from the last decade and beyond. Lung Cancer 57, Suppl 2:S3–S5 (2007).
4. Yang M., Brackenbury J.W.: Membrane potential and cancer progression. Front Physiol 4:185 (2013).
5. Rao V.R., Perez-Neut M., Kaja S., Gentile S.: Voltage-Gated Ion Channels in Cancer Cell Proliferation. Cancers (Basel) 7(2), 849–875 (2015).
6. Huang X., Jan L.Y.: Targeting Potassium channels in cancer. J Cell Biol 206(2), 151–62 (2014).
7. Pang C.L.K.: Hyperthermia in Oncology. CRC Press Tylor Francis Group, Boca Raton (2016).
8. Jha S., Sharma P.K., Malviya R.: Hyperthermia: Role and Risk Factor for Cancer Treatment. Achievements in the Life Sciences 10, 161–167 (2016).
9. Remani R., Ostapenko V.V., Akagi K., Bhattathiri V.N., Nair M. K., Tanaka Y.: Relation of transmembrane potential to cell survival following hyperthermia in HeLa cells. Cancer Letters 144, 117–123 (1999).
10. Kalamida D., Karagounis I.V., Mitrakas A., Kalamida S., Giatromanolaki A., Koukourakis M.I.: Fever-Range Hyperthermia vs. Hypothermia Effect on Cancer Cell Viability, Proliferation and HSP90 Expression. PLoS One 10(1), e0116021 (2015).
11. Speit G., Schütz P.: Hyperthermia-induced genotoxic effects in human A549 cells. Mutat Res 747–748:1–5 (2013).
12. Urban N., Beyersdorf F.: Hypothermia and its effect on tumor growth. Z Herz- Thorax- Gefäßchir 31, 222–227 (2017).
13. Sano M.E., Smith L.W.: The Behavior of Tumor Cells in Tissue Culture Subjected to Reduced Temperatures. Cancer Res 2, 32–39 (1942).
14. Peloquin J.B., Doering C.J., Rehak R., McRory J.E.: Temperature dependence of Cav1.4 calcium channel gating. Neuroscience 151 (4), 1066–83 (2008).
15. Vandenberg J.I., Varghese A., Lu Y., Bursill J.A., Mahaut-Smith M.P., Huang C.L.: Temperature dependence of human ether-a-go-go-related gene K + currents. Am J Physiol Cell Physiol 291(1), C165–75 (2006).
16. Mauerhöfer M., Bauer C.K.: Effects of Temperature on Heteromeric Kv11.1a/1b and Kv11.3 Channels. Biophys J 111(3), 504–523 (2016).

17. Wawrzkiewicz-Jalowiecka A., Dworakowska B., Grzywna Z.J.: The temperature dependence of the BK channel activity – kinetics, thermodynamics, and long-range correlations. Biochim Biophys Acta 1859(10), 1805–1814 (2017).

18. Yang F., Zheng J.: High temperature sensitivity is intrinsic to voltage-gated potassium channels. Elife 3, e03255 (2014).

19. Khasawneh F.A., Thomas A., Thomas S.: Accidental Hypothermia. Hospital Physician 16–21 (2006).

20. Roth B.: Exposure to sparsely and densely ionizing irradiation results in an immediate activation of K+ channels in A549 cells and in human peripheral blood lymphocytes. (Doctoral Thesis). Technische Universität Darmstadt, Darmstadt (2014).

21. Gibhardt C.: Radiation induced activation of potassium-channels: The role of ROS and calcium. (Doctoral Thesis). Technische Universität Darmstadt, Darmstadt (2014).

A Volumetric Delta TCP Tool to Quantify Treatment Outcome Effectiveness Based on Biological Parameters and Different Dose Distributions

Daniella Fabri, Araceli Gago-Arias, Teresa Guerrero-Urbano, Antonio Lopez-Medina, and Beatriz Sanchez-Nieto

Abstract

Intra-tumor variability of oxygenation and clonogenic cell density causes tumor non-uniform spatial response to radiation. Strategies like dose redistribution/boosting, whose impact should be quantified in terms of tumor control probability (TCP), have been proposed to improve treatment outcome. In 1999, Sánchez-Nieto et al. developed a tool to evaluate the impact of dose distribution inhomogeneities, compared to a reference homogeneous dose distribution, in terms of TCP. DVH data were used to calculate the so-called ΔTCP, defined as the difference in TCP arising from dose variations in individual DVH-bins. In this work, we develop an open source tool to calculate volumetric ΔTCP and evaluate the impact on TCP of: (i) Spatial dose distribution variations with respect to a reference dose; (ii) Spatial radiosensitivity variations with respect to a reference radiosensitivity; (iii) Simultaneous variation in dose distribution and radiosensitivity. ΔTCP calculations can be evaluated voxel-by-voxel, or in a user defined subvolume basis. The tool capabilities are shown with 2 examples of H&N RT treatments and subvolume contours data providing information about tumor oxygenation status. ΔTCP values are computed for a homogeneous dose to a well oxygenated tumor volume (with a homogeneous 5% vascular fraction), as reference condition, with respect to the same dose now considering 3 oxygenation levels and 3 cell density values (10^4, 10^6 and 10^7 cells/mm^3, respectively). ΔTCP values are also computed for the comparison of a homogenous dose distribution vs a redistributed dose distribution delivered to the non-homogeneous tumor.

Keywords

Tumor control probability • Tumor heterogeneity
Response to radiation

1 Introduction

Tumor control is the principal aim of curative Radiotherapy. It has been shown that tumors have a non-uniform spatial response to radiation [1]. Besides intrinsic radiosensitivity, two main factors induce radiation response variabilities within the volume: tumor cell density and oxygenation status spatial variations. There is clinical evidence that hypoxia plays an important role in malignant progression and treatment outcome, especially for some types of cancer, at the radiation dose levels that can be safely delivered using uniform dose distributions [2–6]. Other observations support the non-uniform distribution of cells in tumors [7].

Nowadays it is possible to deliver highly conformal non-homogeneous dose distributions to the tumors, and research fields are exploring strategies of dose escalation/redistribution strategies, commonly referred to as dose painting, based on spatial information from functional imaging. In order to adequately develop strategies for the redistribution (or boosting) of dose, the predicted effectiveness of a dose distribution can be quantified in terms of tumor control probability (TCP).

In 1999 Sánchez-Nieto and Nahum introduced the concept of the ΔTCP DVH-bin-based model [8]. They developed a tool to evaluate the impact on TCP of the

D. Fabri · A. Gago-Arias (✉) · B. Sanchez-Nieto
Instituto de Física, Pontificia Universidad Católica de Chile, Avda Vicuña Mackenna, 4860 Santiago, Chile
e-mail: agago@fis.uc.cl

T. Guerrero-Urbano
Department of Clinical Oncology, Guy's and St Thomas Hospitals. NHS Foundation Trust, Great Maze Pond, London, SE1 9RT, UK

A. Lopez-Medina
Medical Physics and Radiological Protection Department, Galaria-Hospital do Meixoeiro-Complexo Hospitalario Universitario de Vigo, 36214 Vigo, Pontevedra, Spain

D. Fabri
Unidad de imágenes cuantitativas avanzadas, Departamento de imágenes, Clinica Alemana de Santiago, Vitacura, 5951 Santiago, Chile

© Springer Nature Singapore Pte Ltd. 2019
L. Lhotska et al. (eds.), *World Congress on Medical Physics and Biomedical Engineering 2018*,
IFMBE Proceedings 68/3, https://doi.org/10.1007/978-981-10-9023-3_124

inhomogeneities found in a dose distribution, compared to a reference homogeneous dose distribution. DVH data was used to calculate the so-called ΔTCP, defined as the contribution to TCP differences arising from dose variations in individual DVH-bins.

Based on this concept, we propose in this work a ΔTCP subvolume-wise model using three dimensional patient information and the medical imaging/treatment planning computational tools currently available. The developed tool evaluates the impact on TCP of different dose distributions with the possibility of incorporating information about the spatial distribution of oxygen and predicted clonogenic cell density in the target.

2 Methods

2.1 Volumetric ΔTCP Tool

An opensource software tool was developed in C++ language using the Insight Toolkit ITK. The following information is loaded into the program as input data: patient anatomical information (planning CT), tumor contours (CTV), possible subvolumes within the tumor (BTV) and the reference and tested dose distributions.

The program allows to evaluate the impact on TCP of: (i) Spatial variations in a test dose distribution with respect to the dose prescription, from now on referred as reference dose distribution (i.e. boosted or redistributed vs homogenous); (ii) Spatial radiosensitivity and tumor cell density variations observed from functional imaging studies (PET, MRI); (iii) Simultaneous variation in dose distribution and radiosensitivity.

In order to address this, the tumor volume is divided in voxels or subvolumes and a voxel-control-probability (VCP_{ijk}) is defined as the probability of controlling all the tumor cells inside a voxel ijk, being calculated as:

$$VCP_{ijk} = -N_{ijk} \cdot e^{-\alpha\left(1+\frac{\alpha}{\beta}D_{ijk}\right)\cdot D_{ijk}} \quad (1)$$

with N_{ijk} being the number of clonogenic cells inside voxel ijk, D_{ijk} the total dose deposited on the voxel and α and β the intrinsic radiosensitivity parameters of the tumor cells. From this, TCP can be calculated in terms of the VCPs as:

$$TCP = \sum_x g_x(\sigma_\alpha) \cdot \prod_{ijk} VCP_{ijk}(\alpha_x) \quad (2)$$

where $g_x(\sigma_\alpha)$ is a Gaussian distribution introduced to consider interpatient intrinsic radiosensitivity variations (changes in α_x).

In order to include the effect of hypoxia on the VCP calculation, oxygen pressure levels are assigned to different regions of the tumor, using previously calculated oxygen

histograms corresponding to different vascular fractions (VF) [9]. Oxygen-corrected radiosensitivity parameters, α (p), are then calculated using published values of oxygen enhancement ratios:

$$\alpha(p) = \frac{\alpha_{ox}}{OER\alpha_m}\left(\frac{p \cdot OER\alpha_m + k_m}{p_{ijk} + k_m}\right). \quad (3)$$

Where p correspond to the oxygen pressure distribution associated to each histogram, $OER\alpha_m$ is maximum oxygen enhancement ratio, α_{ox} the radiosensitivity parameter on normoxic conditions and k_m is the oxygen partial pressure at which half-maximum sensitization is achieved [10].

Sánchez-Nieto and Nahum showed that the VCP distribution was not a useful indicator of impact on tumor control by itself, due to its lack of sensitivity to dose changes. (e.g., if 64 Gy are delivered to 5000 voxels with 8×10^4 cells/voxel, VCP and TCP values of 99.99 and 92.8% can be obtained using average radiosensitivity values of H&N. If dose is decreased to 54 Gy in all voxel, VCP would minimally change to 99.95%, while TCP would decrease to 8.4%).

Using all the tumor VCPs, a ΔTCP can be associated to each voxel to show the impact on the global TCP of having the tested conditions instead of the reference ones (e.g., $(D_r, \alpha(p_{ox})_x)$ versus $(D_t, \alpha(p)_x)$. This ΔTCP is computed as:

$$\Delta TCP_{ijk} = \sum_x g_x(\sigma_\alpha) \cdot TCP(\alpha(p)_x)\left[1 - \frac{VCP_{ijk}(\alpha(p_{ox})_x, D_r)}{VCP_{ijk}(\alpha(p)_x, D_t)}\right] \quad (4)$$

Since voxel contributions are sometimes too small, volume wise calculations (performed for groups of voxels selected by contouring or by individual selection attending to their characteristics) can be also evaluated by the tool. The same equations presented above can be used, with subindex ijk referring to the different subvolumes in this case.

2.2 Tool Capabilities Exemplification

In order to show the capabilities of the tool, two examples are presented here. Both analysis are associated to the same H&N patient, using treatment data available from the FiGaro project [11].

Three predefined subregions are defined on this patient (see Fig. 1): (i) CTV, defined as the clinical target volume on the planning CT; (ii) GTV, active volume identified on a pre-chemotherapy 18F-FDG-PET/CT study; and (iii) BTV, active volume identified on a 18F-FDG-PET/CT study 3 weeks after administration of neoadjuvant cisplatin %FU chemotherapy. The radiotherapy treatment prescription dose is 65 Gy to the CTV and 71.5 Gy to the BTV (dose boost), all delivered in 30 fractions.

Fig. 1 Subvolumes defined on the patient CT. Top left, CTV (yellow), with a volume of 112.7 cm^3, and an assigned VF of 5% and 10^4 mm^{-3} clonogenic cell density. Top right and bottom panels, contours of GTV (blue), with a volume of 10.9 cm^3, 3% VF and 10^6 mm^{-3} clonogenic cell density values assigned and BTV (barn red) with a volume of 4.4 cm^3 and assigned VF of 2.5% and 10^7 mm^{-3} clonogenic cell density (Color figure online)

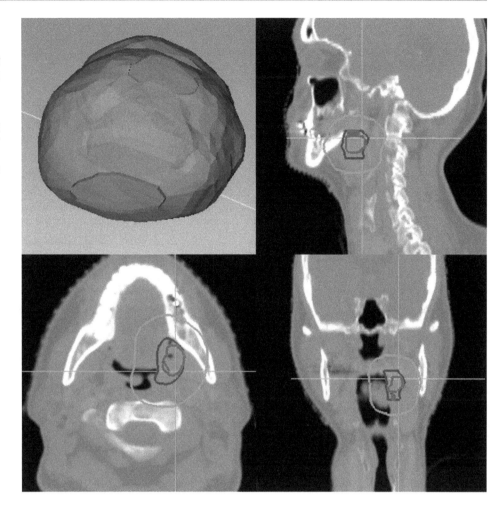

2.3 ΔTCP Analysis Examples

Two scenarios are analyzed, with four ΔTCP calculations associated: two voxel wise and two subvolume wise, respectively.

On the first scenario (referred to as Case 1), we compare the effect of delivering a homogeneous dose to a well oxygenated tumor volume (with a homogeneous 5% VF), as reference condition, with respect to the effect of having VF variations on the sub regions above described (see caption of Fig. 1), as tested condition. This comparison is performed by applying voxel wise and sub volume wise ΔTCP calculations.

On the second scenario (Case 2), we evaluate the impact on TCP of the boosted dose distribution versus a homogeneous dose distribution, considering the heterogeneous oxygenation of the tumor described above. Again, voxel wise and subvolume wise calculations are presented.

An intrinsic well oxygenated alpha radiosensitivity parameter value was taken from literature, $\alpha_{ox} = 0.35\,\mathrm{Gy}^{-1}$, using a fixed $\frac{\alpha}{\beta} = 10\,\mathrm{Gy}$ and a 20% interpatient radiosensitivity variation, $\sigma_\alpha = 0.06\,\mathrm{Gy}^{-1}$.

Additionally, a complementary parametric sensitivity study was performed to analyze the impact on TCP of the values assigned to different parameters, namely α_{ox}, and the VF and cell density values assigned to the three tumor sub regions. Reference values of VF with associated oxygenation histograms corresponding to well oxygenated and moderately and extreme hypoxic tissues were used [9, 12], as well as cell density values ranging within experimentally observed limits [13, 14].

3 Results

Results for the four examples are shown on Table 1. TCP$_t$ (tumor control associated to the tested condition) and TCP$_r$ (tumor control associated to the reference condition) values are presented, showing quite small variations. The largest change in TCP corresponds to the Case 1 voxel wise calculation, where, for the tested condition, a homogeneous dose distribution is delivered to the tumor, and TCP is calculated considering oxygenation status variations,

Table 1 TCP and ΔTCP values associated to the two scenarios investigated under voxel and subvolume wise calculations. For the voxel wise calculation, TCP values associated to the tested and reference conditions, TCP$_t$ and TCP$_r$ respectively, as well as average, minimum and maximum ΔTCP values are presented. For the subvolume wise calculation, global ΔTCP values associated to each region of the tumor are presented

		CASE 1	CASE 2
Voxel wise	TCP$_t$	0.99985	0.999997
	TCP$_r$	1	0.999977
	Average ΔTCP	-5.68×10^{-11}	1.1×10^{-11}
	ΔTCP$_{max}$	0	5.3×10^{-7}
	ΔTCP$_{min}$	-3.21×10^{-6}	-1.8×10^{-8}
	sd ΔTCP	7.83×10^{-9}	1.4×10^{-9}
Subvolume wise	TCP$_t$	0.999922	1
	TCP$_r$	1	0.999952
	ΔTCP CTV	-0.000034	6.21×10^{-6}
	ΔTCP GTV	-0.000433	0.00026
	ΔTCP BTV	-0.0042	0.00256

versus a reference TCP calculated for a homogeneous oxygenation. In this tested condition TCP decreases in 0.015%.

Individual voxel contributions to ΔTCP are extremely small, taking values of the order of 10^{-11}. The subvolume analysis leads to larger ΔTCP values. For Case 1, all ΔTCP values corresponding to the subvolumes described in Fig. 1 are negative. This is expected as a non-homogeneous oxygenation status through the tumor would decrease the probability of controlling it when it is treated with a homogeneous dose distribution.

For Case 2, ΔTCP associated to CTV is very small even though the number of pixels on this area represent the 92% of the volume, on the other hand, the ΔTCP values associated to GTV and BTV regions are positive, leading to an improvement of the total TCP.

Voxel ΔTCP values are very useful to analyze the relevance of local changes through the analysis of map distributions. Figure 2 shows axial, sagittal and coronal views of the ΔTCP distributions associated to the voxel wise calculations, where positive values indicate TCP improvement under the tested condition, compared to the reference one.

These maps may help to spatially asses the clinical relevance, in terms of TCP, of having hypoxic regions or density variations within a tumor, as well as the effect of applying dose resditribution/boosts in a treatment.

Table 2 shows the results of the parametric sensitivity study, where very small variations were observed for the parameter values investigated.

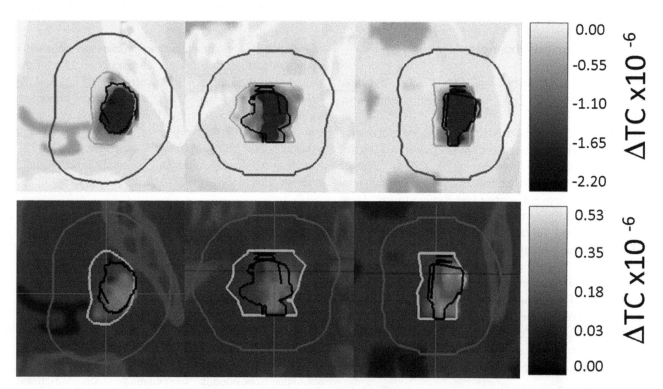

Fig. 2 ΔTCP map distributions associated to the voxel by voxel calculation. Axial, sagittal and coronal views of the distributions corresponding to Case 1 and Case 2, in top and bottom images respectively (Color figure online)

Table 2 Percentage of TCP variation (in last column) due to changes in radiosensitivity, VF and tumor cell density parameter values

α_{ox}	Vascular Fraction associated to the subvolume			Number of clonogenic cells per mm^3 associated to the subvolume			% variation of TCP (%)
	CTV (%)	GTV (%)	BTV (%)	CTV	GTV	BTV	
0.3	5	5	5	10^6	10^6	10^6	REF
0.35	5	5	5	10^6	10^6	10^6	0.00010
0.25	3	3	3	10^6	10^6	10^6	0.07770
0.3	3	3	3	10^6	10^6	10^6	0.00030
0.35	3	3	3	10^6	10^6	10^6	0.00010
0.25	2.5	2.5	2.5	10^6	10^6	10^6	0.08810
0.3	2.5	2.5	2.5	10^6	10^6	10^6	0.00030
0.3	5	3	2.5	10^6	10^6	10^6	0.00000
0.3	5	3	2.5	10^6	10^8	10^6	0.00090
0.3	5	3	2.5	10^6	10^8	10^8	0.00160
0.3	5	3	2.5	10^4	10^6	10^8	0.00060
0.3	5	3	2.5	10^2	10^6	10^{1o}	0.06810

4 Discussion and Conclusion

An open source tool that allows to identify the impact on TCP of delivering inhomogeneous dose distributions to non-homogeneous responding tumors has been developed and tested. Voxel wise ΔTCP calculations give very small quantitative results, but allow a useful qualitative analysis when represented as ΔTCP map distributions. Subvolume analysis lead to larger values, making easier to interpret, on a macroscopic scale, the effects associated to the tested conditions.

Acknowledgements D. F. and A. G-A. and B. S-N acknowledge the support of FONDECYT 2015 Postdoctoral Grant 3150422, FONDE-CYT 2017 Iniciación Grant 11170575 and Fondo de apoyo a la organización de reuniones científicas (VRI UC 2015). The authors thank the help of Christopher Thomas concerning the treatment planning and transfer of anonymized patient data between institutions.

Conflict of Interest The authors declare that they have no conflict of interest.

References

1. Bentzen, SM. Theragnostic imaging for radiation oncology: Dose-painting by numbers. Lancet Oncol. 6, 112–117 (2005).
2. Lee, N. Y., and Le, Q-T.: New developments in radiation therapy for head and neck cancer: Intensity modulated radiation therapy and hypoxia targeting. Seminars in Oncology 35(3): 236–250 (2008).
3. Okunieff, P., de Bie, J., Dunphy, EP., Terris, DJ., Höckel, M.: Oxygen distributions partly explain the radiation response of human squamous cell carcinomas. British Journal of Cancer. 74,185–190 (1996).
4. Briztel, DM. Sibley, GS. Prosnitz, LR. Scher, RL. and Dewhirst, MW.: Tumor hypoxia adversely affects the prognosis of carcinoma of the head and neck. Int.J.Radiat.Oncol.Biol.Phys. 38, 285–289 (1997).
5. Brizel, DM.: Targeting the future in head and neck cancer. The lancet oncology;10, 204–205 (2009).
6. Nordsmark, M., Bentzen, SM., Rudat, V. et al: Prognosis value of tumor oxygenation in 397 head and neck tumors after primary radiation therapy. An international multicenter study. Radiother. Oncol.77,18–24 (2005).
7. Nutting, C M. Corbishley, C M. Sanchez-Nieto, B. Cosgrove, V P. Webb, S. Dearnaley, D P: Potential im-provements in the therapeutic ratio of prostate cancer irradiation: dose escalation of pathologically identified tumour nodules using intensity modulated radio-therapy. The British Journal of Radiology, 75, 151–161 (2002).
8. Sanchez-Nieto, B. Nahum, A. E.: The delta-TCP concept: a clinically useful measure of tumor control probability. Int. J. Radiation Oncology Biol. Phys., Vol. 44, No. 2, pp. 369–380, (1999).
9. Espinoza, I. Peschke, P. Karger, CP: A model to simulate the oxygen distribution in hypoxic tumors for different vascular architectures. Med Phys. 40(8):081703(2013).
10. Wouters B G and Brown J M: Cells at intermediate oxygen levels can be more important than the 'hypoxic fraction' in determining tumor response to fractionated radiotherapy. Radiat. Res.147, 541–50 (1997).
11. Michaelidou, A., et al.: PO-0609: 18F-FDG-PET in Guiding Dose-painting with IMRT in Oropharyngeal Tumours (FiGaRO)-Early Results. Radiother. and Oncol. 123 S318 (2017).
12. Mönnich, D. Troost, E. G. Kaanders, J. H. Oyen, W. J. Alber, M. Thorwarth, D: Modelling and simulation of [18F]fluoromisonida-zole dynamics based on histology-derived microvessel maps. Phys. Med. Biol. 56, 2045–2057(2011).
13. Webb, S. Nahum, AE: A model for calculating tumour control probability in radiotherapy including the effects of inhomogeneous distributions of dose and clonogenic cell density. Phys Med Biol. Jun;38(6):653–66 (1993).
14. Del Monte, U. Does the cell number 10^9 still really fit one gram of tumor tissue? Cell Cycle 8(3):505–6 (2009).

Part XI

Nuclear Medicine and Molecular Imaging

Establishment of an Analysis Tool for Preclinical Evaluation of PET Radiotracers for In-Vivo Imaging in Neurological Diseases

Fabian Schadt, Samuel Samnick, and Ina Israel

Abstract

Aim: Analysis of in vivo acquired data remains a challenging issue in preclinical studies using positron emission tomography (PET). The aim of this study is the implementation of a tool which should allow a semi-automated analysis of PET data independently from the administrated radiotracer and the imaging modality for preclinical investigations. By registering anatomical data sets, it additionally shall offer a more detailed data analysis allowing a statistical analysis also for smaller sub-regions. *Materials and methods*: Data used for primarily implementation of the tool were acquired on a Siemens µPET scanner (Inveon®) and were based on the investigation of the glucose metabolism in a subarachnoid hemorrhage (SAH) model in Sprague Dawley rats in vivo using FDG-PET. We used the software programs Matlab (version 2016a) and Fiji for data analysis and visualization. In addition, statistical tests were performed in order to determine regions with trending/significant differences in the SUV of sham and SAH animals. *Results*: Following data import, data were separated into predefined time periods and artefacts were eliminated. Afterwards, a volume of interest (VOI) was defined by the threshold of the Standardized-Uptake-Value (SUV). Before masking each data set with its segmented VOI, all data sets were intensity-normalized, eliminating the full body intensity differences caused by the different amount of injected activity. After masking, data sets of the sham operated animals were registered on the best orientated sham data set, reduced on the VOI and shifted into the center of the 3D space. By averaging aligned data sets based on all data sets from seven sham rats, we generated a FDG-template of a Sprague-Dawley rat brain. This PET data template was the basis for the evaluation of registered data sets. Afterwards, an anatomical MR-based atlas of the brain of Sprague-Dawley rat was co-registered on the template for a better sub-classification of the acquired data. *Conclusion*: These preliminary data show that the described method represents a very promising tool for data analysis in the preclinical evaluations of PET radiotracers for neurological applications.

Keywords

Preclinical imaging • Positron emission tomography
Data analysis • Matlab • Subarachnoid hemorrhage
PET • SAH • FDG

List of Abbreviations

ANOVA	Analysis of variance
EEG	Electroencephalography
FDG	2-[^{18}F]Fluoro-2-deoxy-glucose
fMRI	Functional magnetic resonance imaging
MEG	Magnetoencephalography
MIRT	Medical Image Registration Toolbox
MRI	Magnetic resonance imaging
OSEM2D	2D-ordered subsets expectation maximization
PET	Positron emission tomography
SAH	Subarachnoid hemorrhage
SPM	Statistical Parametric Mapping
SUV	Standardized-Uptake-Value
VOI	Volume of Interest

1 Introduction

Positron emission tomography (PET) is a non-invasive imaging modality, which allows analysis of functional, physiological and biochemical processes in vivo by use of radiolabeled tracers. PET is clinically established in oncology and cardiology. In neurology, magnetic resonance

F. Schadt (✉) · S. Samnick · I. Israel
Department of Nuclear Medicine, University Hospital Würzburg,
Oberdürrbacher Str. 6, 97080 Würzburg, Germany
e-mail: e_schadt_f@ukw.de

© Springer Nature Singapore Pte Ltd. 2019
L. Lhotska et al. (eds.), *World Congress on Medical Physics and Biomedical Engineering 2018*,
IFMBE Proceedings 68/3, https://doi.org/10.1007/978-981-10-9023-3_125

imaging remains the method of choice. Despite strong competition from other functional imaging methods like functional magnetic resonance imaging (fMRI), or electroencephalography/magnetoencephalography (EEG/MEG), PET still remains a very promising tool in studying neurotransmission and neuromodulation studies, e.g. in neurostimulation [1] or neuroinflammation [2] non-invasively in vivo. Radiotracers for PET imaging are the prototype of imaging agents for functional brain imaging [3] and additionally they offer a wide-range variety of examination possibilities, like atherosclerosis [4, 5], myocardial inflammation [6] or cardiac wound healing [7].

Because of these extensive applications, molecular imaging has been continuously improved during the last decades. Shorter acquisition times [8], lower radiation exposures [9] and higher image resolution [10] are just some examples in which the developments of PET imaging significantly improved. Besides these developments, preclinical image data analysis and interpretation remains very challenging.

The aim of this work was to develop an improved analysis tool which should allow a semi-automated analysis of PET data. The tool was planned to be useable for preclinical data analysis in vivo, independently from the PET radiotracer and the imaging modality.

2 Materials and Methods

2.1 Radiochemistry

2-[^{18}F]Fluoro-2-deoxy-glucose (FDG) used in the present study was produced in-house by a standard synthetic method using the GE-PETtrace cyclotron and the GE-Fastlab® synthesis unit (GE Medical Systems, Uppsala, Sweden) as described previously [2, 11]. Before use, the FDG injection solution in physiological saline was analyzed by HPLC and TLC for radiochemical purity.

2.2 Animals and Animal Experiments

Data used for the primarily implementation of the tool are based on the investigation of the glucose metabolism in a subarachnoid hemorrhage (SAH) model in Sprague Dawley rats in vivo using FDG-PET. The SAH model in rat has been established in the department of neurology of the university medical school Würzburg and has been used routinely for preclinical research.

All animal experiments were carried out according to the Guide for the Care and Use of Laboratory Animals

published by the US National Institutes of Health (NIH Publication No. 85-23, revised 1996) and in compliance with the German animal protection law. The experiments were approved by the district government of Lower Franconia (Regierung von Unterfranken AZ: 55.2-2531.01-40/12). For the experiments, 14 male Sprague-Dawley rats (310–350 g) were purchased from Harlan Winkelmann GmbH (Borchen, Germany), and maintained in the animal facility of the University of Würzburg, Department of Neurosurgery. The rats were randomly assigned to one of the two experimental groups: (1) Subarachnoid hemorrhage (SAH) induced by the endovascular filament model as described previously [12, 13] (n = 7) and (2) sham operated controls (n = 7). The animals were anesthetized and maintained with 2% isoflurane anesthesia in 100% oxygen during the whole procedure.

2.3 PET Measurements

The FDG-PET scans were performed in rats at 3 h and on days 1, 4 and 7 after induction of SAH or sham operated animals. The food was removed 6 h before tracer injection, water was available anytime. Animals were kept under a 2% isoflurane anesthesia in 100% oxygen and warmed during the tracer injection and PET scan continuously. 31,44 ± 3, 41 MBq FDG were injected via the tail vain. 40 min after injection, animals were placed in prone position at the animal bed of the PET scanner and PET acquisition was started. Emission scan duration was 20 min, splitted into two time frames each 10 min. Subsequently, a transmission scan was started immediately.

Data acquisition was performed on a Siemens Inveon microPET scanner (Siemens Medical Solutions, Knoxville, USA). The measured data sets were reconstructed with an OSEM2D algorithm provided by the embedded software package Inveon Acquisition Workplace (Version 1.5.0.28). Spatial resolution of the images was about 0.16 cm * 0.16 cm, while the voxel size of the reconstructed image was (0.776 mm * 0.776 mm * 0.796 mm) $\approx$ 0.480 mm^3.

2.4 Software Programs

For data analysis and visualization, the software programs MATLAB (version 2016a, MathWorks, United States) and Fiji [14, 15] were used. Registration algorithm for aligning the images is based on the Medical Image Registration Toolbox (MIRT) of Myronenko [16]. The MRI atlas of the Sprague Dawley rat brain used for a better sub-classification of the VOI was developed by Papp et al. [17–19].

2.5 Standardized Uptake Value

Data analysis was evaluated on the basis of the standardized uptake value (SUV). The standardized uptake value is a semi-quantitative index for PET quantification in clinical practice [20] and is defined as radiotracer concentration $[MBq/cm^3 \triangleq g]$ divided by the division of injected activity [MBq] to body weight [g].

$$SUV = \frac{radiotracer\ concentration}{\dfrac{injected\ activity}{body\ weight}} \quad (1)$$

3 Results and Discussion

3.1 Creation of a Template and Data Registration

After data import, each reconstructed data set was separated in its individual time frames (40–60 min post-injection, separated in 10 min time intervals). Additionally, in a first pre-processing step, artefacts at border regions as well as single annihilations were eliminated by the tool. Before data analysis, the volume of interest (VOI) for each data set was defined by threshold using Fiji [14, 15]. After defining the VOI, the VOI were centered and the orientations of the individual VOI were determined within the 3D space in order to figure out the best orientated VOI. This process was first performed for the different measurement days of a single sham animal, then for the averaged, individual sham animals. The data set with the best oriented VOI was considered as reference data set for the registration. Before registration, all data sets were intensity-normalized on the data set with the highest sum of intensities. This process is necessary for eliminating the full body intensity differences caused by the different amount of injected activity.

3.2 Template and Data Registration

Result of the registration process is an averaged data set (template) based on 28 measurements of the healthy Sprague Dawley rats (seven rats, four measurements each) (Fig. 1a). The created template was used as reference data set to align the individual sham and SAH data sets. Next, all data sets (template, sham and SAH) were summed with their gradient images, normalized and then, the sham and SAH data sets were registered on the template. Based on the knowledge that the radiotracer will be processed in a similar way within one experimental group, the added gradient images provide a better classification along the border crossings. Since the original data sets were modified

for a better registration, only the shifting matrices of each registration process were extracted and then applied on the original data sets.

Next, in order to perform a more detailed data analysis, the data sets should be co-registered with an MRI data atlas of a Sprague Dawley rat brain (Fig. 1b, [17–19]), sub-classifying the VOI of the PET data within specific sub-regions. But before a co-registration could be carried out, some preparations still had to be made.

First, the aligned data sets were roughly reduced on the VOI while retaining the dimensions of the 3D space as a multiple of the size of the MRI data set and then shifted into the center of the downsized 3D space. Second, since the MRI data atlas of the Sprague Dawley rat brain (Fig. 1b, [17–19]) contained no harderian glands, these glands also needed to be excluded within the PET data. Therefore, the templates were imported into Fiji [14] and their harderian glands, which have a strong FDG uptake but were out of interest, were manually excluded (Fig. 1a). The edited templates were then stored as binary masks for multiplying the binary masks with all data sets (template, sham and SAH), excluding unimportant information while keeping the structures of interest (Fig. 1a). Third, since the MRI resolution is much higher than the PET resolution (~1:4044), the multiple slices of the MRI data set were averaged over 16 slices each in order to attain the same resolution and dimensions of the PET data. The result of the co-registration is visualized in Fig. 1c, d.

3.3 Statistical Analysis

The tool offers several possibilities to perform data analysis. In order to find outliers, the SUVs for the whole brain, the gray and white matter of the individual animals are box plotted for all time frames. The lower and upper quartile includes 50% of all data. Lower and upper whiskers are defined at the 1.5 fold of the interquartile range. As an example, Fig. 2 shows the SUVs of the sham and SAH animals for the whole brain for all four measurements (3 h and on days 1, 4 and 7 after induction of SAH or sham operation).

An outlier is defined as an observation that is well outside of the expected range of values in a study or experiment. In this tool we defined outliners as data points outside the lower and upper whiskers. Therefore, the following diagrams should show whether an animal has several outliers within the measurements. If so, this could be considered an exclusion criterion from the study for that specific animal. For easier allocation the SUVs are color-coded for each animal. Furthermore, as additional information, the VOI currently being investigated on outliers and its largest regions are shown in 3D (template).

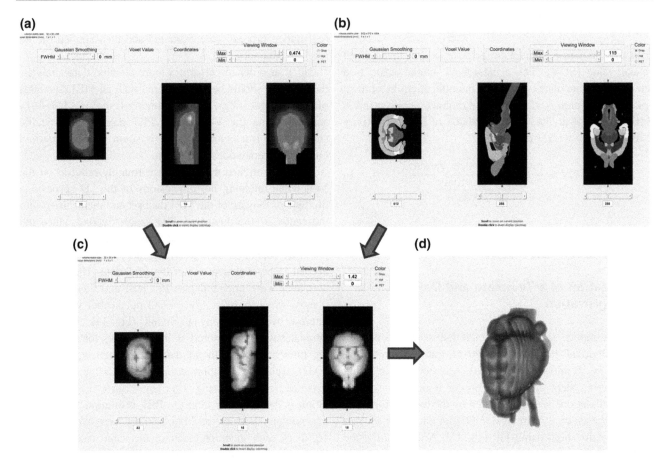

Fig. 1 **a** FDG Sprague Dawley rat brain template based on 28 different data sets of seven healthy control animals. Harderian glands were manually excluded after registration. **b** Anatomical MRI data atlas of a Sprague Dawley rat brain [17–19]. **c** Co-registered image of the functional FDG template and the anatomical MRI atlas data **d** 3D visualization of the (co-registered) FDG template (purple) and the MRI atlas (cyan) in Fiji [14, 15]. Graphics (**a**)–(**c**) were visualized with VolumeViewer3D (Medical Image Reader and Viewer by Schaefferkoetter, version 3.3), **d** with ImageJ 3D Viewer [21] (Color figure online)

In addition, QQ-plots give visually information whether the SUVs of the sham and the SAH data are normal distributed. The data can be regarded as normally distributed if the curves of the individual measurements run best along the norm line (Fig. 3, red line). Based on the registered MRI atlas, the number of the QQ-plots may vary depending on the interest and sub-classification of the atlas. Here, for the Sprague Dawley rat brain, the 20 largest regions of the first time period (40–50 min after injection) are plotted for normal distribution (Fig. 3). Measurement days one to four correspond to the measurements at 3 h, 1 day, 4 days and 7 days after induction of SAH or sham operation. Beside the visual representations, two numerical tests for normal distribution are computed, the Anderson-Darling test and the Shapiro-Wilk test.

Similarity measurements such as correlation coefficient (Pearson, Spearman), Joint Histogram, Joint Entropy and Mutual Information are calculated as well.

Based on the number of animals, t-tests (Σ(Animals) $\geq$ 30) or repeated-measures ANOVA are performed

in-between the SAH groups and the sham operated control group for each measurement day and over the time course of the measurement series for each pathology (Fig. 4). Additionally, since the amount of statistical results can become very large, the statistical results of all time periods are automatically checked for trends and significant effects. For the statistical test in-between the experimental groups, the trending or significant results are also visualized, respectively to Fig. 1c, d.

Furthermore, the statistical results over the time course of the measurement series are plotted, as seen in the example of Fig. 4. Again, measurements one to four correspond to the measurements at 3 h, 1 day, 4 days and 7 days after induction of SAH or sham operation. For the significance of the results, it is assumed, that there is an almost constant line for the sham animals, while for the diseased animals it may vary. However, for the data analysis between the sham and the SAH experimental groups, no significant effects were observed.

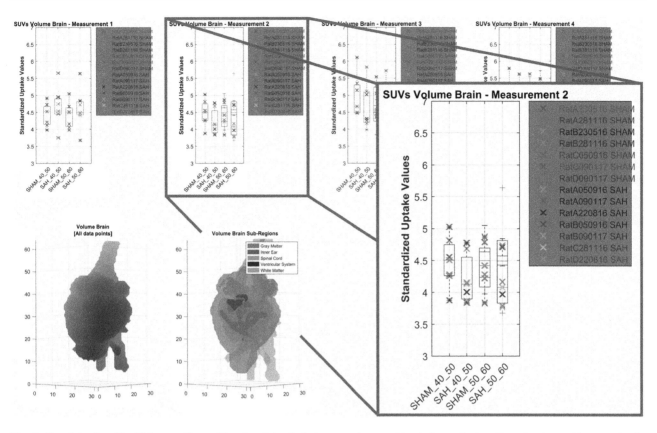

Fig. 2 Boxplots for identifying outliers: The boxplots include color-coded SUVs for sham and SAH animals (here: with the whole brain as VOI). Animals, showing several SUVs lying outside the boxplots for multiple measurements (3 h, 1d, 4d and 7d after injection) can be considered for exclusion. In order to avoid errors in the assignment, the VOI to be examined and its largest structures are additionally displayed in 3D (template) (Color figure online)

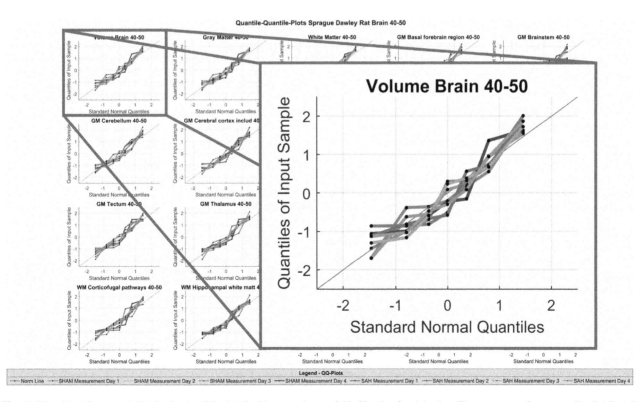

Fig. 3 Visual test for normal distribution by QQ-plot: In this example, the curves of the 20 largest regions of the Sprague Dawley rat brain are presented for all measurement days (0, 1, 4 and 7), for the time interval of 40–50 min after injection. The curves are then normally distributed, when they correspond to the norm line (red) in the best possible way (Color figure online)

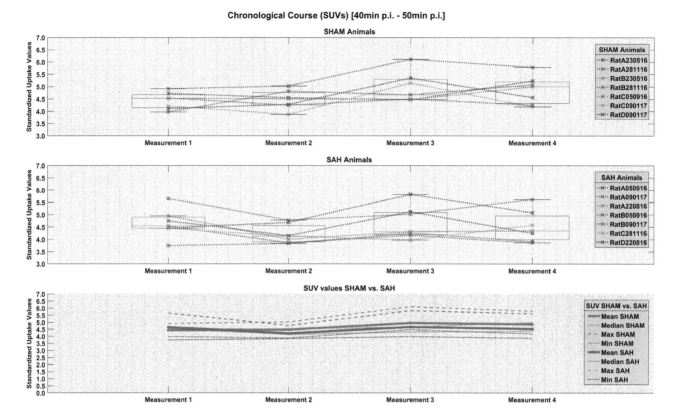

Fig. 4 Time course of the measurement series: Between the sham and the SAH Sprague Dawley rats, no significant effects were observed over the four measurements (3 h, 1 day, 4 days and 7 days after induction of SAH or sham operation)

4 Conclusion

In conclusion, data of the present analysis show that the presented tool could be a suitable alternative for analyzing PET data, independently from the PET tracer and the VOI itself. It offers the possibility to create a new PET template based on the own measured data and therefore allows a better alignment of the acquired data than registering on a given template. Even though, the added MRI atlas does not originate from the measured animals, it still gives a considerable gain of information compared to PET data alone allowing a detailed sub-classification. Nevertheless, it still should be seen critical in very small regions as well as in peripheral regions.

Data used for implementation offered the advantage, that FDG has a good homogeneous uptake in the brain. Even though the number of animals is sufficient for the implementation of the tool, for statistical analysis, the number of animals should be further increased in further studies.

Unfortunately, the SAH lesions were too small for an accurate delineation by FDG-PET. In the upcoming steps, the proposed tool needs to be tested using data showing no homogeneous uptake, e.g. when using specific radiotracer or for disease patterns such as metastases or lesions to confirm

its potential as image analysis tool in PET imaging in neurological diseases.

Acknowledgements This project was funded by a grant from the Deutsche Forschungsgemeinschaft (SFB688, Z02 project). The authors would like to thank Dr. N. Lilla (Department of Neurosurgery, University Hospital Würzburg, Germany) for her support with the animal model and Dr. J. Tran-Gia (Department of Nuclear Medicine, University Hospital Würzburg, Germany) for his technical advices.

Conflicts of Interest The authors declare that they have no conflict of interest.

References

1. Jech R. (2008) Functional Imaging of Deep Brain Stimulation: fMRI, SPECT, and PET. In: Tarsy D., Vitek J.L., Starr P.A., Okun M.S. (eds) Deep Brain Stimulation in Neurological and Psychiatric Disorders. Current Clinical Neurology. Humana Press.
2. Israel I, Ohsiek A, Al-Momani E, Albert-Weissenberger C, Stetter C, Mencl S, Buck AK, Kleinschnitz C, Samnick S, Siren AL (2016) Combined [(18)F]DPA-714 micro-positron emission tomography and autoradiography imaging of microglia activation after closed head injury in mice. Journal of neuroinflammation 13:140.

3. Mier W, Mier D (2015) Advantages in functional imaging of the brain. Frontiers in human neuroscience 9:249.

4. Li X, Bauer W, Kreissl MC, Weirather J, Bauer E, Israel I, Richter D, Riehl G, Buck A, Samnick S (2013) Specific somatostatin receptor II expression in arterial plaque: (68)Ga-DOTATATE autoradiographic, immunohistochemical and flow cytometric studies in apoE-deficient mice. Atherosclerosis 230:33–39.

5. Li X, Bauer W, Israel I, Kreissl MC, Weirather J, Richter D, Bauer E, Herold V, Jakob P, Buck A, Frantz S, Samnick S (2014) Targeting P-selectin by gallium-68-labeled fucoidan positron emission tomography for noninvasive characterization of vulnerable plaques: correlation with in vivo 17.6T MRI. Arteriosclerosis, thrombosis, and vascular biology 34:1661–1667.

6. Lapa C, Reiter T, Li X, Werner RA, Samnick S, Jahns R, Buck AK, Ertl G, Bauer WR (2015) Imaging of myocardial inflammation with somatostatin receptor based PET/CT - A comparison to cardiac MRI. International journal of cardiology 194:44–49.

7. Tillmanns J, Schneider M, Fraccarollo D, Schmitto JD, Langer F, Richter D, Bauersachs J, Samnick S (2015) PET imaging of cardiac wound healing using a novel [68 Ga]-labeled NGR probe in rat myocardial infarction. Molecular imaging and biology: MIB: the official publication of the Academy of Molecular Imaging 17:76–86.

8. Alessio, A. M., M. Sammer, et al. (2011). "Evaluation of optimal acquisition duration or injected activity for pediatric 18F-FDG PET/CT." J Nucl Med 52(7): 1028–1034.

9. Roberts, F. O., D. H. Gunawardana, et al. (2005). "Radiation dose to PET technologists and strategies to lower occupational exposure." J Nucl Med Technol 33(1): 44–47.

10. Peng, H. and C. S. Levin (2010). "Recent Developments in PET Instrumentation." Current pharmaceutical biotechnology 11(6): 555–571.

11. Lapa C, Reiter T, Kircher M, Schirbel A, Werner RA, Pelzer T, Pizarro C, Skowasch D, Thomas L, Schlesinger-Irsch U, Thomas D, Bundschuh RA, Bauer WR, Gartner FC (2016) Somatostatin receptor based PET/CT in patients with the suspicion of cardiac sarcoidosis: an initial comparison to cardiac MRI. Oncotarget 7:77807–77814.

12. Bederson JB, Germano IM, Guarino L (1995) Cortical blood flow and cerebral perfusion pressure in a new noncraniotomy model of subarachnoid hemorrhage in the rat. Stroke 26:1086–1091; discussion 1091-1082.

13. Lilla N, Fullgraf H, Stetter C, Kohler S, Ernestus RI, Westermaier T (2017) First Description of Reduced Pyruvate Dehydrogenase Enzyme Activity Following Subarachnoid Hemorrhage (SAH). Frontiers in neuroscience 11:37.

14. Rasband, W.S., ImageJ, U. S. National Institutes of Health, Bethesda, Maryland, USA, https://imagej.nih.gov/ij/, 1997–2016.

15. Ollion J, Cochennec J, Loll F, Escude C, Boudier T (2013) TANGO: a generic tool for high-throughput 3D image analysis for studying nuclear organization. Bioinformatics 29:1840–1841.

16. Myronenko, A. (2010). Non-rigid image registration regularization, algorithms and applications. Department of Science & Engineering, Oregon Health & Science University. Doctor of Philosophy in Electrical Engineering: 177.

17. Papp, E. A., T. B. Leergaard, et al. (2014). "Waxholm Space atlas of the Sprague Dawley rat brain." Neuroimage 97: 374–386.

18. Kjonigsen, L. J., S. Lillehaug, et al. (2015). "Waxholm Space atlas of the rat brain hippocampal region: three-dimensional delineations based on magnetic resonance and diffusion tensor imaging." Neuroimage 108: 441–449.

19. Sergejeva, M., E. A. Papp, et al. (2015). "Anatomical landmarks for registration of experimental image data to volumetric rodent brain atlasing templates." J Neurosci Methods 240: 161–169.

20. Boellaard R (2009) Standards for PET image acquisition and quantitative data analysis. Journal of nuclear medicine: official publication, Society of Nuclear Medicine 50 Suppl 1:11S–20S.

21. Schmid B, Schindelin J, Cardona A, Longair M, Heisenberg M (2010) A high-level 3D visualization API for Java and ImageJ. BMC bioinformatics 11:274.

Radio-Guided Surgery with β^- Radiation: Tests on Ex-Vivo Specimens

C. Mancini-Terracciano, V. Bocci, M. Colandrea, F. Collamati,
M. Cremonesi, R. Faccini, M. E. Ferrari, P. Ferroli, F. Ghielmetti,
C. M. Grana, M. Marafini, S. Morganti, S. Papi, M. Patané, G. Pedroli,
B. Pollo, A. Russomando, M. Schiariti, G. Traini, and E. Solfaroli
Camillocci

Abstract

Radio-Guided Surgery (RGS) is a surgical technique
aimed at assisting the surgeon to reach as complete a
resection of the tumoural lesion as possible. Established
methods to date make use of γ-emitting tracers to
radio-mark the neoplastic tissue. However, in case of
uptake from healthy organs around the lesion the large
penetration of photons yields a non-negligible background that can limit the RGS application. The adoption
of β^- radiation has been proposed to overcome this limit.
To validate the entire RGS procedure, from the evaluation
of the tracer uptake of the tumor, to the assumptions on
the bio-distribution and the signal detection, tests on
ex vivo specimens of meningioma brain tumour were
performed. Meningioma was selected due to the well
known high receptivity to a β^- emitting radio-tracer
already in use in the clinical practice: ^{90}Y-labelled
DOTATOC. Patients were enrolled according to the
tumour Standard Uptake Value (SUV $>$ 2) and the
expected Tumour to Non-tumour Ratio (TNR $>$ 10)
estimated from ^{68}Ga-DOTATOC PET images. After
injecting the patients with 93–167 MBq of
$^{90}Y - DOTATOC$, 26 samples excised during surgery
were examined with a dedicated β^- detecting probe to
assess the sensitivity of millimetre-sized tumour remnants
in case of administration of low activity value compatible
with those injected for diagnostic exams. Even injecting
as low as 1.4 MBq/kg of radio-tracer, tumour remnants
greater than 0.06 ml would be discriminated by the
healthy tissue in few seconds.

Keywords

Radioguided surgery • Meningioma brain tumor
β^- radiation • Intraoperative imaging

C. Mancini-Terracciano (✉) · V. Bocci · F. Collamati · R. Faccini
· M. Marafini · S. Morganti · G. Traini · E. Solfaroli Camillocci
Istituto Nazionale di Fisica Nucleare, Sezione di Roma, Rome,
Italy
e-mail: carlo.mancini.terracciano@roma1.infn.it

M. Colandrea · M. Cremonesi · M. E. Ferrari ·
C. M. Grana · S. Papi · G. Pedroli
Istituto Europeo di Oncologia, Milan, Italy

P. Ferroli · F. Ghielmetti · M. Patané · B. Pollo · M. Schiariti
Fondazione Istituto Neurologico Carlo Besta, Milan, Italy

M. Marafini
Museo Storico della Fisica e Centro Studi e Ricerche E. Fermi,
Rome, Italy

A. Russomando
Universidad Tcnica Federico Santa Mara, Valparaiso, Chile

R. Faccini · G. Traini · E. Solfaroli Camillocci
Dip. Fisica, Sapienza University of Rome, Rome, Italy

E. Solfaroli Camillocci
Dip. Scienze di Base e Applicate per l'Ingegneria, Sapienza
University of Rome, Rome, Italy

E. Solfaroli Camillocci
Scuola di Specializzazione in Fisica Medica, Sapienza University
of Rome, Rome, Italy

1 Introduction

Radio-guided surgery (RGS) is a surgical technique that
enables the surgeon to evaluate, in real time, the complete-
ness of the tumour lesion resection [1]. It could also help to
minimise the amount of healthy tissue removed, in all the
cases where such attention could be of utmost importance for
the impact on the patient. It represents a significant surgical
adjunct to intra-operative detection of millimetre-sized
tumour residues, providing the surgeon with vital and
real-time information regarding the location and the extent
of the lesion, as well as assessing surgical resection margins.
RGS is crucial for those tumours where surgical resection is
the only possible therapy, since it reduces the probability of
tumour recurrence. It is based on the availability of a
radiopharmaceutical with an uptake larger on the lesion than
on the healthy tissue, such a quantity is called Tumour to
Non-tumour Ratio (TNR). The radiopharmaceutical is
administered to the patient prior to the surgery.

© Springer Nature Singapore Pte Ltd. 2019
L. Lhotska et al. (eds.), *World Congress on Medical Physics and Biomedical Engineering 2018*,
IFMBE Proceedings 68/3, https://doi.org/10.1007/978-981-10-9023-3_126

Traditional RGS approaches use a γ (photons) emitting tracers with a detection probe optimised to detect them [2, 3]. Since γ radiation can traverse large amounts of tissue, any uptake of the tracer in the nearby healthy tissues represents a non-negligible background, often preventing the practical usage of this technique.

To mitigate this effect it was suggested in literature the use of β^+ decaying tracers [4]. The emitted positrons in fact have a limited penetration and their detection is local. Nonetheless, positrons annihilate with electrons in the body and produce γs with an energy of 511 keV: the background persists and actually increases in energy. The improvement with respect to the use of pure emitters is that a dual system can be devised where the background can be measured separately and subtracted from the observed signal. This approach has been studied in preclinical tests [5] but it is not yet in use in the clinical practice. The largest limitations range from the time needed to identify a residual, the dimensions of the probes and the dose absorbed by the medical personnel. A better solution to the current limits of RGS would be to eliminate the background from radiation.

A better solution to the current limits of RGS would be to eliminate the background from γ radiation using pure β^- (i.e. electrons) emission [6].

Indeed, electrons with an energy of few MeV have a penetration in water and human tissues of the order of millimetres and the γ contamination can be neglected, as the *Bremsstrahlung* contribution has an emission probability above 100 keV of 0.1%. In addition, we measured the sensitivity of our probe to *Bremsstrahlung* photons to be lower than 1% [7].

Using β^- radiation allows the development of a probe more compact than the one used in traditional RGS. Moreover, detecting electrons and operating with low radiation background provides a clearer delineation of the margins of lesioned tissues. For all these reasons, this novel approach requires a smaller injected activity to detect tumor residuals with respect to the one using γ radiation. This reduces also the medical personnel radiation exposure at a level which can be considered negligible [6].

We performed several studies with both Monte Carlo simulations (MC) [7] and laboratory tests on phantoms [8] to predict the applicability of this technique. We started with brain tumours, namely meningioma and glioma [9], and abdominal neuro-endocrine tumors [10] that are known to express receptors to ^{90}Y-labeled [1,4,7,10-tetraazacyclododecane-N,N',N'', N'''-tetraacetic acid0-D-Phe1,Tyr3]octreotide (DOTATOC), a radio-pharmaceutical already in clinical use for therapy. ^{90}Y-DOTATOC is a suitable radio-tracer for β^- RGS since ^{90}Y is a pure β^- emitting radionuclide. The feasibility studies showed that for those tumours the receptivity to the tracer and the TNRs were high enough to detect 0.1 ml lesioned

residuals with a dedicated detecting probe within few seconds. These studies used the PET images with ^{68}Ga-labeled DOTATOC assuming that the bio-distribution of the tracer did not change when labeled with ^{90}Y. The patient ^{68}Ga-DOTATOC activity distribution was used to sample the distribution probability of ^{90}Y in a MC simulation program (FLUKA [11]) to estimate the counting rate on the probe.

To strengthen the feasibility studies and test the entire RGS procedure, from the evaluation of the tracer uptake of the tumour to the assumptions on the bio-distribution and the signal detection, tests on ex vivo specimens of meningioma brain tumour were performed. A description of the test on a first patient as a proof-of-principle was published in Ref. [12]. In this article, the results of tests on further three patients are reported. The patients were injected with ^{90}Y-labeled DOTATOC prior to surgery and a prototype of the intraoperative detecting probe was exposed to the radiation of the excised samples. Decreasing activity values (93–167 MBq) were administered in order to evaluate the minimum detectable activity in meningioma. Moreover, tumor samples of different sizes were measured with the probe to estimate the device sensitivity to small meningioma residuals (~ 0.1 ml).

2 Methods

2.1 Patient Treatment Protocol

Three patients with radiological diagnosis of meningioma were selected for the present study. They gave written informed consent to participate in the clinical trial (EUDRACT 2013-004033-32) approved by IEO Ethic Committee (institutional review board).

To assess the in vivo presence of somatostatin receptors, a PET scan with ^{68}Ga-DOTATOC was performed about two weeks prior to the surgical intervention to ensure a sufficient uptake to test the technique. As in the preliminary studies, the biodistribution was assumed to be dependent on the drug delivery mechanism and not affected by the labelling radionuclide. The patients were enrolled if the PET scans, performed one hour after the injection of 4 MBq/kg, revealed that the bulk tumor had an average SUV > 2 and a TNR > 10. These threshold values were estimated in a previous study [9].

Twenty-four hours before surgical intervention, the patients were injected with ^{90}Y-DOTATOC. The treatment protocol was the same as for the patient of the proof-of-principle test [12], except for the injected activity. Indeed, the first patient was injected with 300 MBq (4.7 MBq/kg) while for these tests the activity was lowered case by case according to the individual radiotracer uptake (SUV, TNR) and the patient weight (167, 111 and 93 MBq resulting in 1.6 MBq/kg for the first one and 1.4 MBq/kg for the others), aiming at evaluating the minimum value that makes a 0.1 ml tumour volume detectable.

Before surgery the probe was placed in proximity to the skin of the patient in several spots to estimate the level of background. After surgery, that was performed following the clinical routine, the extracted tumor and the attached dura were sectioned in samples.

The specimens were placed in calibrated boxes designed and built to host them and estimate their volumes (see the next section). Then, the counting rate of each sample was measured with the intraoperative probe prototype to detect the activity in them.

Finally, the specimens underwent histology to evaluate their actual tumoral nature.

2.2 Samples Characterization

The volume of each sample was measured estimating height (h) and area (A_s) using boxes built on purpose: a set of rectangular calibrated vessels with height of 1, 2, 3 and 4 mm and a reference grid with a millimetre step placed on the floor of each box.

Each sample was placed in the vessel that best fitted its thickness to measure h. Using this vessel, during the measurement the probe was nearly in contact with the sample without applying pressure on it and avoiding shape distortion.

The area was computed acquiring photos of the specimens with the grid behind and redrawing its shape off line. The specimen profile was drawn on each picture as shown in Fig. 1 (the yellow contour line) and its area was measured converting the area in pixel to millimetre using the reference grid. The figure also shows the area covered by the probe tip (the blue circle): we had to consider the percentage of the sample covered by the probe, given the short penetration of the electrons.

Assuming an uniform distribution for h and A_s in a range of 0.5 mm, and an uncertainty of 5% for both, h and A_s, then the volume uncertainty is: $\sigma_h = 0.5/\sqrt{12} = 0.14$ mm.

2.3 The Detection System

The β^- detection system used in this test was a prototype of the intraoperative probe developed for RGS of brain tumor (see Fig. 2) [13].

The radiation sensitive element was a plastic scintillator made of mono-crystalline para-terphenyl doped to 0.1% in mass with diphenylbutadiene. The high light yield of this material (larger than typical organic scintillators), its non-hygroscopic property, and the low density resulting in

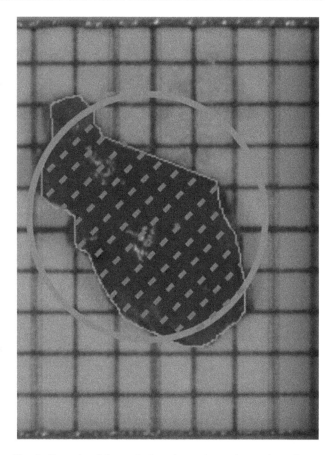

Fig. 1 Example of the method used to estimate the specimens' area. The overlaid contour is used to estimate the sample size as explained in the text; the blue circle ($d = 6$ mm) shows the projection of the active area of the probe (Color figure online)

low sensitivity to photons [14] make the para-terphenyl an optimal candidate to detect the β^- decays.

The scintillator tip was 5 mm in diameter and 3 mm in height and was enclosed by a black ABS (Acrilonitrile Butadiene Stirene) ring with external diameter of 12 mm. A 15 μm-thick aluminum front-end sheet covered the detector window ensuring the light sealing. The scintillator light was read by a silicon photomultiplier (SiPM from SensL). This assembly was encapsulated in an easy-to-handle aluminum cylindrical body (diameter 12 mm and length 14 cm).

A portable electronics based on Arduino Due, with wireless connection to PC or tablet was used for the read out [15]. The electronic noise measured with this probe prototype and its electronics is less than 0.5 counts/s.

3 Results and Discussion

The Tables 1, 2 and 3 summary the results obtained measuring the specimens excised from the three patients. In them the geometrical properties of each sample

Fig. 2 The β^- probe prototype

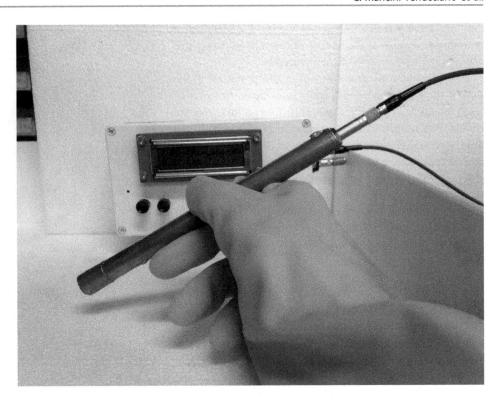

Table 1 Patient 1. Label of the sample, surface (A_s) and volume (V), counting rate acquired and category of the sample according to the histological analysis

Name	A_s [mm^2]	V [ml]	Counts [cps]	Medical report
1-A	47 ± 2	0.047 ± 0.007	1.80 ± 0.05	Dura mater
1-B	19 ± 1	0.038 ± 0.003	2.66 ± 0.16	Dura mater
1-C	101 ± 5	0.404 ± 0.025	50.1 ± 1.1	Meningioma
1-E $_1$	65 ± 3	0.259 ± 0.016	44.6 ± 1.5	Tumour margin
1-E $_2$	44 ± 2	0.176 ± 0.011	43.7 ± 0.7	Tumour margin
1-E $_3$	16 ± 1	0.064 ± 0.004	13.7 ± 0.5	Tumour margin
1-F $_1$	26 ± 1	0.102 ± 0.006	15.0 ± 0.3	Meningioma
1-F $_2$	43 ± 2	0.171 ± 0.011	20.2 ± 0.3	Meningioma
1-F $_3$	23 ± 1	0.092 ± 0.006	45.9 ± 1.5	Meningioma

Table 2 Patient 2. Label of the sample, surface (A_s) and volume (V), acquired counting rate and category of the sample according to the histological analysis

Name	A_s [mm^2]	V [ml]	Counts [cps]	Medical report
2-A	60 ± 1	0.060 ± 0.009	1.3 ± 0.1	Dura mater
2-B $_1$	108 ± 2	0.22 ± 0.02	3.4 ± 0.2	Dura mater infiltrated
2-B $_2$	115 ± 2	0.23 ± 0.02	4.6 ± 0.3	Dura mater infiltrated
2-D	56 ± 1	0.056 ± 0.008	13.4 ± 0.5	Meningioma
2-E	31 ± 1	0.031 ± 0.005	10.4 ± 0.3	Meningioma

(surface and volume), the counting rate measured with the probe and the report from the histological analysis are listed. To test the probe sensitivity on small meningioma residuals, the specimens were cut in sub-samples and measured with the probe. The number subscript to the sample name keeps track of the specimen split from the same sample.

These tests showed that meningioma specimens > 0.06 ml are detected with a counting rate of at least 10 cps. These values have to be compared with the expected value of about 1 cps for the healthy brain (see Ref. [12]). Therefore for a tumour specimen, a volume of 0.06 ml is enough to yield a signal that could be clearly discriminated from the healthy tissue administering to the patient at least 1.4 MBq/kg in a few seconds, as the first one produces roughly 10 times the counts than the second one. The procedure we applied did not allows to discriminate if the dura mater was infiltrated, however it was not in the purposes of this work.

Table 3 Patient 3. Label of the sample, surface (A_s) and volume (V), acquired counting rate and category of the sample according to the histological analysis

Name	A_s [mm^2]	V [ml]	Counts [cps]	Medical report
3-A	73 ± 2	0.146 ± 0.013	1.3 ± 0.2	Dura mater
3-B	144 ± 4	0.576 ± 0.035	4.6 ± 0.2	Dura mater infiltrated
3-D $_1$	147 ± 4	0.588 ± 0.036	68.9 ± 2.1	Tumour margin
3-D $_2$	67 ± 4	0.268 ± 0.017	42.5 ± 0.6	Tumour margin
3-D $_3$	44 ± 4	0.177 ± 0.011	35.6 ± 0.2	Tumour margin
3-D $_4$	23 ± 4	0.092 ± 0.006	22.3 ± 0.6	Tumour margin
3-D $_5$	13 ± 3	0.039 ± 0.003	8.9 ± 0.4	Tumour margin
3-E $_1$	45 ± 3	0.136 ± 0.009	10.8 ± 0.4	Meningioma
3-E $_2$	18 ± 3	0.055 ± 0.004	4.0 ± 0.3	Meningioma
3-F $_1$	19 ± 4	0.075 ± 0.005	10.6 ± 0.3	Tumour margin
3-F $_2$	17 ± 4	0.070 ± 0.004	13.3 ± 0.4	Tumour margin
3-F $_3$	11 ± 3	0.034 ± 0.002	4.1 ± 0.1	Tumour margin

References

1. Mariani G., Giuliano AE, Strauss HW. *Radioguided Surgery: A Comprehensive Team Approach.* New York: Springer; 2006.
2. Tsuchimochi M and Hayamaand K. Intraoperative gamma cameras for radioguided surgery: Technical characteristics, performance parameters, and clinical application, *Phys. Med.* 2013; 29:126–38.
3. Schneebaum S, Even-Sapir E, Cohen M, et al. Clinical applications of gamma-detection probes - radioguided surgery. *Eur J Nucl Med.* 1999; 26: S26–35.
4. Daghighian, F. et al. Intraoperative beta probe: A device for detecting tissue labeled with positron or electron emitting isotopes during surgery. *Med. Phys. 21, 153 (1994).*
5. Bogalhas, F. et al. Development of a positron probe for localization and excision of brain tumours during surgery. *Phys. Med. Biol. 54, 4439 (2009).*
6. Solfaroli Camillocci E, Baroni G, Bellini F, et al. A novel radioguided surgery technique exploiting β^- decays. *Sci Rep.* 2014; 4: 4401.
7. Mancini-Terracciano C., Donnarumma R., Bencivenga G., et al. Feasibility of the β^- Radio-Guided Surgery with a Variety of Radio-Nuclides of Interest to Nuclear Medicine. *Physica Medica 43, 127–133 (2017).*
8. Russomando A, Bellini F, Bocci V, et al. An Intraoperative β^- Detecting Probe For Radio-Guided Surgery in Tumour Resection. *IEEE Trans. Nucl. Sci.* 2016; 63(5):2533.
9. Collamati F, Pepe A, Bellini F, et al. Toward radioguided surgery with β^- decays: uptake of a somatostatin analogue, DOTATOC, in meningioma and high-grade glioma *J Nucl Med.* 2015; 56:3–8.
10. Collamati F, Bellini F, Bocci V, et al. Time evolution of DOTATOC uptake in Neuroendocrine Tumors in view of a possible application of Radio-guided Surgery with beta- Decays *J Nucl Med.* 2015; 56:1501–6.
11. Battistoni G, Cerutti F, Fassò A, et. al. The FLUKA code: Description and benchmarking. *AIP Conf. Proc.* 2006; 896:31–49.
12. Solfaroli-Camillocci E, Schiariti M, Bocci V et al. First ex-vivo validation of a radioguided surgery technique with β^- radiation. *Physica Medica* 2016; 32:1139–1144.
13. Solfaroli Camillocci E, Bocci V, Chiodi G, et al. Intraoperative probe detecting β- decays in brain tumour radio-guided surgery *Nucl. Inst. and Met. in Ph. Res., A* 2016; 1:4.
14. Angelone M, Battistoni G, Bellini F, et al., Properties of para-terphenyl as detector for alpha, beta and gamma radiation. *IEEE Trans. Nucl. Sci.* 2014; 61(3):1483.
15. Bocci V, Chiodi G, Iacoangeli F, et al. The ArduSiPM a compact trasportable Software/Hardware Data Acquisition system for SiPM detector. *IEEE NSS/MIC* 2014.
16. Hoffman EJ, Tornai MP, Janecek M, Patt BE and Iwanczyk JS. Intraoperative probes and imaging probes. *Eur. J. Nucl. Med.* 1999; 26:913–935.
17. Riccardi L, Gabusi M, Bignotto M, et al. Assessing good operating conditions for intraoperative imaging of melanoma sentinel nodes by a portable gamma camera. *Ph. Med.* 2015; 31(1):92–97.
18. Kaviani S, Zeraatkar N, Sajedi S, et al. Design and development of a dedicated portable gamma camera system for intra-operative imaging. *Ph. Med.* 2016; 32(7):889–897.

Investigation of Time-Activity Curve Behavior in Dynamic [11C]-(R)-PK11195 PET in Cortical Brain Regions: Preliminary Results

Giordana Salvi de Souza and Ana Maria Marques da Silva

Abstract

Dynamic Positron Emission Tomography (PET) allows quantification of underlying physiological processes in a tissue or organ of interest by modeling the radioactivity concentration measured in time (time-activity curve, i.e. TAC). Although the [11C]-(R)-PK11195 PET radiotracer binds to activated microglia and therefore images neuroinflammation, its quantification is still challenging. The aim of this study was to investigate a novel method for analyzing [11C]-(R)-PK11195 TAC behavior from dynamic PET. Seven healthy subjects underwent dynamic 60 min [11C]-(R)-PK11195 PET scans, and TACs were generated for 30 brain cortical regions using the AAL-Merged atlas. The proposed method supposes healthy subjects have similar TACs, which allows the construction of a "healthy template". Then, TACs of patients can be compared to the healthy template to determine the likelihood that their behavior is abnormal. To evaluate the differences between a healthy region and an inflamed region, a cut-off value for abnormality was created. The method was then tested on a multiple sclerosis (MS) patient. The proposed method was able to identify a number of brain cortical regions with distinct behavior in one MS patient as compared to the healthy template. Further studies are required to evaluate the applicability of the proposed method with additional MS patients, and with data acquired in different scanners and reconstructed with other algorithms.

Keywords

Dynamic images • PET • [11C]-(R)-PK11195

G. Salvi de Souza · A. M. Marques da Silva (✉)
Laboratory of Medical Imaging, Science School, PUCRS, Porto Alegre, Brazil
e-mail: ana.marques@pucrs.br

G. Salvi de Souza
e-mail: giordana.souza@acad.pucrs.br

1 Introduction

Neurodegeneration mediated by an inflammatory response involves the activation of macrophages resident in brain, called microglia, and the release of neurotoxic and pro-inflammatory factors, including cytokines, free radicals, nitric oxide and eicosanoids that can damage neurons and glial cells. Microglia are constantly sweeping brain tissue to detect any signs of damage and infections, and when these last occur, microglia launches a defensive response generally known as microglial activation [1]. The microglial activation increases the expression of the translocator protein 18 kDa (TSPO), which can be considered as a sensitive biomarker of microglial activation and therefore of inflammation. In contrast, in the central nervous system (CNS) of a healthy adult, microglia appear as small branched monocytic cells with little apparent activity [2].

With the use of Positron Emission Tomography (PET), it is possible to image the expression of TSPO with the aid of the radiotracer [11C]-(R)-PK11195. This radiotracer is used to study several conditions, including Alzheimer's disease [3], multiple sclerosis [4–6], normal aging, dementia [7] and depression [8]. One of the advantages of PET is that it allows for the quantification of underlying physiological processes. For that purpose, dynamic images are acquired, time-activity curves (TAC) are generated and analyzed through pharmacokinetic modeling.

Although [11C]-(R)-PK11195 PET enables visualization of neuroinflammation or of microglial activation in brain diseases. However, dynamic imaging quantification is still challenging, due to the high expression of TSPO in peripheral organs where the radiotracers are bound specifically, the large contribution of non-specific binding in the brain and the absence of a brain reference region devoid of specific binding [9].

The aim of this study was to propose a new method for analyzing TAC behavior from dynamic [11C]-(R)-PK11195

L. Lhotska et al. (eds.), *World Congress on Medical Physics and Biomedical Engineering 2018*,
IFMBE Proceedings 68/3, https://doi.org/10.1007/978-981-10-9023-3_127

PET in a simple straightforward way, without the need of a plasma input function or a reference region.

2   Methodology

To investigate the dynamic behavior of $[^{11}C]$-(R)-PK11195 in neuroinflammation, PET images we analyzed from volunteer subjects and multiple sclerosis patients scanned at the Brain Institute of Rio Grande do Sul (InsCer). The Ethical Committee (CAAE 23949813.7.0000.5336, opinion 1.094.228) approved the study.

Dynamic PET acquisitions (60 min) were performed on a Discovery 600 (GE Healthcare). The data was acquired in list mode and then rebinned into 23 time frames: 1×15 s, 3×5 s, 3×10 s, 2×30 s, 3×60 s, 4×150 s, 5×300 s and 2×600 s. The temporal rebinned in the second frame was chosen to sample in detail the uptake in this moment.

The study group comprised seven (7) healthy subjects (mean age 29 years, age range 21–36 years; 4 men, 3 women), injected with (372 ± 73) MBq $[^{11}C]$-(R)-PK11195 activity. The method was validated with one multiple sclerosis (MS) patient with high score (5) in Kurtzke Expanded Disability Status Scale (EDSS), age of 41 years and injected activity was 362 MBq.

All images were analyzed using PMOD (3.5, PMOD Technologies, Zurich, Switzerland). Statistical analysis was performed using SPSS software (SPSS Statistics, version 17; IBM).

2.1   Method Rationale

The method proposed in this study assumes the overall behavior of TACs differ between healthy and affected regions. Therefore, the goal was to construct a regional TAC template representing the behavior seen in healthy subjects. However, TACs are dependent on injected activity, and therefore require normalization before inter-subject comparisons.

The first step of the model was to transform the TACs into normalized SUV-based TACs. For that purpose, each subject's regional TAC values were normalized by injected dose and weight. In addition, for each subject, the area under the TAC curve (AUC) for each brain region was calculated, and a global brain AUC from the healthy population. At a second step, mean healthy TACs were constructed for each region by averaging the normalized TACs across healthy subjects. Then, each individual regional TAC was compared to the healthy template by Spearman correlation to understand the variability between healthy subjects in their TAC

behavior. Finally, a cut-off for abnormality was determined using Spearman correlation values bellow two standard deviations between the normalized TAC and the healthy template.

2.2   Method Validation

SUV (Standard Uptake Value) values were extracted from each cortical brain region TAC, using the AAL-Merged atlas (3.5, PMOD Technologies, Zurich, Switzerland).

For each healthy subject (i), the AUC was calculated for each brain region (k), composed by r regions, in the time interval (j), as shown in Eq. 1.

$$AUC_i = \sum_{k=1}^{r} \sum_{j=0}^{\infty} (t_{j+1} - t_j) \left(\frac{SUV(t_{j+1}) + SUV(t_j)}{2} \right) \tag{1}$$

The AUC mean of m healthy subjects was calculated, and then, for each subject, each region (k) was normalized by value of its own AUC and the AUC mean of the healthy group. The healthy template $hTAC_k$ is calculated by the mean of normalized TACs ($nTAC_{ik}$) for each brain region, for a healthy population of m individuals.

A cut-off for abnormality was determined using the mean and standard deviation (SD) of the Spearman correlations (R) between the healthy template $hTAC_k$ and $nTAC_{ik}$ by region. The cut-off value (v_k) for each brain region (k) was defined as shown in Eq. 2.

$$v_k = \overline{R_{controls}} - 2 \cdot SD \tag{2}$$

Correlation values below the cut-off value (v_k) would indicate a significant difference in brain uptake in this region, relative to the same brain region in the healthy TAC. To validate the method, TAC of one MS patient with high score in Kurtzke EDSS was compared with the healthy template $hTAC_k$.

3   Results

3.1   Healthy Time-Activity Curve

The Shapiro-Wilk test showed that AUC values of the healthy group obeyed a normal distribution, within a significance level of 5%. Figure 1 shows the healthy time-activity curve, $hTAC_k$, by brain region, obtained from normalized TACs.

Figure 2 (left) shows whole brain TACs for each healthy subject and the whole brain $hTAC$ (black line) and the effect of the data normalization (right).

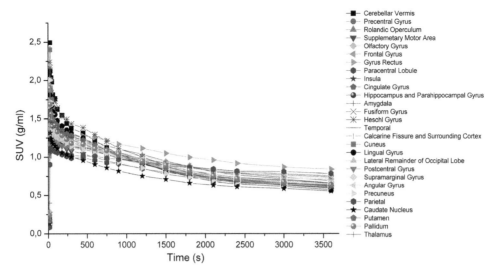

Fig. 1 Healthy TAC, $hTAC_k$, by brain region

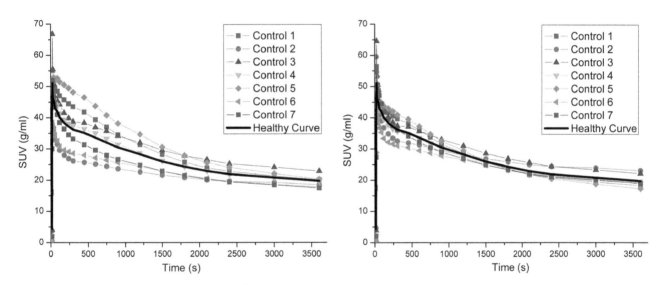

Fig. 2 Left. Whole brain TACS for each healthy subject and $hTAC_k$. Right. Normalized whole brain TACs and $hTAC_k$

3.2 Cut-off Values and MS Patient Data

The method was tested on a MS patient with high score (5) in Kurtzke EDSS. We identified lower correlation values, bellow the cut-off, in the following brain regions: cerebellar vermis, rolandic operculum, supplementary superior area, paracentral lobule, insula, cingulate gyrus, hippocampus and parahypocampal gyrus, amygdala, fusiform gyrus, helsch gyrus, temporal lobe, calcarine fissure and surrounding cortex, cuneus, lingual gyrus, lateral remainder of occipital lobe, supramarginal gyrus, precuneus, putamen, cerebellum and cerebellum crus. Precentral, olfactory gyrus, frontal gyrus, gyrus rectus, postcentral gyrus, angular gyrus, parietal

lobe, caudate nucleus, pallidum, thalamus showed high correlation with the healthy population-based TAC.

Table 1 shows the correlation mean and SD values between $hTAC_k$ and $nTAC_{ik}$, the cut-off value (v_k) for exemplary brain regions and the correlation values between $hTAC_k$ and the MS patient TAC.

4 Discussion

This study investigated the kinetic behavior of $[^{11}C]$-(R)-PK11195 PET scans in 30 brain regions in healthy subjects. Cut-off values for each brain region were created, from the

Table 1 Mean and SD of Spearman correlation, cut-off values v_k of some brain regions, and the Spearman Correlation of one MS patient

Brain region	$\overline{R}_{control}$	$SD(\overline{R}_{control})$	v_k	MS
Cerebellar vermis	0,971	0,034	0,902	0,543
Precentral gyrus	0,862	0,099	0,664	0,854
Olfactory gyrus	0,776	0,499	−0,221	0,636
Frontal gyrus	0,878	0,138	0,602	0,912
Hippocamp/ Parahippocam gyrus	0,978	0,030	0,918	0,636
Amygdala	0,928	0,087	0,755	0,532
Fusiform gyrus	0,933	0,101	0,730	0,593
Heschl gyrus	0,990	0,009	0,973	0,602
Lingual gyrus	0,966	0,036	0,895	0,849
Lateral remain occipital lobe	0,906	0,130	0,647	0,14
Postcentral gyrus	0,857	0,105	0,647	0,789
Supramarginal gyrus	0,946	0,044	0,858	0,75
Angular gyrus	0,729	0,331	0,068	0,78
Caudate nucleus	0,892	0,138	0,617	0,679
Putamen	0,891	0,126	0,640	0,636
Thalamus	0,927	0,080	0,767	0,856
Cerebellum	0,964	0,046	0,871	0,296

mean and SD of Spearman correlation between $hTAC_k$ and $nTAC_{ik}$. Some brain regions show higher variability (angular and olfactory gyrus), even within healthy subjects, resulting in a small cut-off, does not discriminating healthy or abnormal regions.

Although the method was tested with only one MS patient, correlation values bellow the cut-off, were identified in several cortical brain regions. This preliminary result should be validated with more MS patient data, evaluating the correlation between affected brain regions and clinical symptoms.

Although AUC has been used to analyze dynamic PET scans [10, 11], none use a healthy template. Kreisl et al. (2010) used AUC to differentiate [^{11}C]-(R)-PK11195 and [^{11}C]PBR28 uptake in different organs [10]. Shiiba et al. (2014) studied the efficiency of AUC in dynamic imaging with ^{123}I-MIBG to distinguish Lewy body disease and Parkinson's syndrome, achieving 93% sensitivity and 100% specificity [11].

Further studies are required to investigate the effect on healthy TAC for images acquired in different scanners and using other reconstruction algorithms.

5 Conclusions

The proposed method is able to identify brain cortical regions with a distinct behavior as compared to a healthy template, indicating higher chances of it being affected by neuroinflammation in [^{11}C]-(R)-PK11195 PET scans using a healthy population-based TAC. The method was tested preliminarily with a MS patient and cortical regions affected by neuroinflamation were found.

Compliance with Ethical Standards Conflicts of Interest The authors declare that they have no conflict of interest.

References

1. Peterson PK and Toborek M. *Neuroinflammation and Neurodegeneration*. New York, NY: Springer New York (2014).
2. Winkeler A, Boisgard R, Martin A, et al. Radioisotopic Imaging of Neuroinflammation. *J Nucl Med*, 51(1):1–4 (2010).
3. Ma L, Zhang H, Liu N, et al. TSPO ligand PK11195 alleviates neuroinflammation and beta-amyloid generation induced by systemic LPS administration. *Brain Res Bull*, 121:192–200 (2016).
4. Banati RB, Newcombe J, Gunn RN et al. The peripheral benzodiazepine binding site in the brain in multiple sclerosis Quantitative in vivo imaging of microglia as a measure of disease activity, *Brain*,123: 2321–2337 (2000).
5. Narciso LDL, Schuck PN, Dartora CM, et al., Semiquantification Study of [11C](R)PK11195 PET Brain Images in Multiple Sclerosis, *Rev Bras Fís Méd*, 10(2):39–43 (2016).
6. Giannetti P, Politis M, Su P et al. Microglia activation in multiple sclerosis black holes predicts outcome in progressive patients: An in vivo [11C](R)-PK11195-PET pilot study. *Neurobiol Dis*, 65: 203–210 (2014).
7. Yokokura M, Terada, T, Bunai, T. and et al. Depiction of microglial activation in aging and dementia: Positron emission tomography with [^{11}C]DPA713 versus [^{11}C](R)PK11195, *J Cereb Blood Flow Metab* 37(3):877–889 (2017).
8. Kopschina Feltes P, de Vries EF, Juarez-Orozco LE, et. al. Repeated social defeat induces transient glial activation and brain hypometabolism: A positron emission tomography imaging study. *J Cereb Blood Flow Metab*(2017).
9. Hinz, R., Boellaard, R. Challenges of quantification of TSPO in the human brain. *Clin Transl Imaging*, 3(5):403–16 (2015).
10. Kreisl, W.C., Fujita, M., Fujimura, et al. Comparison of [^{11}C]-(R)-PK 11195 and [^{11}C]PBR28, two radioligands for translocator protein (18 kDa) in human and monkey: Implications for positron emission tomographic imaging of this inflammation biomarker, *Neuroimage*, 49(4):2924–2932 (2010).
11. Shiiba, T., Nishii, R., Sasaki, M., et al. Assessment of the efficacy of early phase parameters by ^{123}I-MIBG dynamic imaging for distinguishing Lewy body-related diseases from Parkinson's syndrome. *Ann Nucl Med*, 29(2):149–156(2014).

Image Quality Performance of a Dedicated Cardiac Discovery NM 530c SPECT: Impact of Time Acquisition and Reconstruction Parameters

Mariana Saibt Favero and Ana Maria Marques da Silva

Abstract

The aim of this study is to evaluate the impact of acquisition time and reconstruction parameters in image quality in a dedicated cardiac Discovery NM 530c CZT SPECT scanner. Anthropomorphic torso phantom with a cardiac insert and a cold lesion starting with 84 kBq/ml, 62 kBq/ml and 3 kBq/ml in myocardium, liver and background, respectively, simulated high-dose stress ^{99m}Tc acquisition. During two half-lives, images were acquired hourly to simulate lower doses, using both 3 and 5 min time acquisition. Reconstruction parameters were used separately per type of acquisition (low-dose and high-dose), according to manufacturer recommendation. Image quality and cold lesion visibility were evaluated using normalized standard deviation (NSD), ventricle-wall cavity contrast (VCC), contrast-to-noise ratio (CNR), and lesion effective contrast (EC). Variations in acquisition time did not affect quality parameters and lesion visibility for the appropriate reconstruction parameters, mainly in higher activities. However, image quality indicators are highly sensitive to the reconstruction parameters, producing significant differences ($p < 0.05$) in NSD, VCC, CNR and EC. Concluding, the reconstruction parameters need to be chosen carefully, considering patient's characteristics and administered activity, in order to produce the best relation image quality/lesion visibility and dose.

Keywords

SPECT • Cardiology • CZT • Image quality Discovery NM 530c

M. S. Favero
Nucleorad, Porto Alegre, RS, Brazil

A. M. Marques da Silva (✉)
Laboratory of Medical Imaging, Science School, PUCRS, Porto Alegre, Brazil
e-mail: ana.marques@pucrs.br

1 Introduction

Instrumentation in nuclear cardiology field has evolved significantly in recent years. Concerns about radiation dose and acquisition time have driven the evolution of dedicated SPECT scanners for cardiology [1]. Designs with new collimators, such as multi-pinhole or focused collimators arranged in optimized geometries for cardiac exams, were implemented to increase the gamma photon detection sensitivity.

Some of the new SPECT cameras use solid state detectors and photodiode arrays, instead of scintillation crystals and photomultipliers, making it possible to reduce camera dimensions and reduce up to 7 times acquisition time, with similar reduction in radiation exposure [2]. SPECT cameras with cadmium-zinc-telluride (CZT) detectors have several advantages compared to conventional Anger SPECT cameras. While guides recommend a 20–30 min acquisition time for Anger SPECT systems, cardiac SPECT camera with CZT detectors can acquire images in 3–6 min [3].

Discovery NM 530c, a GE Healthcare ultra-fast cardiac SPECT camera, features CZT detectors and a multi-pinhole system [4]. D-SPECT, produced by Spectrum Dynamics, utilizes the same CZT detectors, with dynamic parallel-hole collimators, which focuses on the myocardium [5]. They were created to meet the demands of nuclear cardiology, possessing high sensitivity, as well as improvement in spatial, temporal and energetic resolutions, allowing the acquisition time reduction.

Previous studies [6–9] have shown that myocardial perfusion exams in patients with CZT SPECT, has a significant dose reduction and reduced acquisition time with image quality comparable to conventional Anger SPECT scanner.

Imbert et al. (2012) compared Discovery NM 530c and D-SPECT CZT cameras with Anger cameras, and reported better physical performance of CZT cameras [10]. However, both CZT cameras are inherently different, with better spatial resolution and contrast-to-noise ratio in Discovery NM 530c,

© Springer Nature Singapore Pte Ltd. 2019
L. Lhotska et al. (eds.), *World Congress on Medical Physics and Biomedical Engineering 2018*,
IFMBE Proceedings 68/3, https://doi.org/10.1007/978-981-10-9023-3_128

whereas detection sensitivity is higher in D-SPECT. Dartora et al. (2015) showed better cold lesion detectability of images acquired in a cardiac phantom in the Discovery NM 530c when compared to the conventional Anger camera [11].

However, studies found pitfalls and artifacts in obese patients [12], and the influence of small changes in patient positioning on the images acquired in a CZT GE Discovery NM 530c [13]. In the literature, there are no studies showing the impacts generated by acquisition parameters modifications, such as acquisition time and image processing parameters in the visibility of cardiac lesions.

The objective of this work is to evaluate the impact of changes in acquisition time and reconstruction parameters in image quality and cold lesions visibility in a multi-pinhole GE Discovery NM 530c scanner.

2 Materials and Methods

A commercially anthropomorphic torso phantom with cardiac insertion (Data Spectrum Corporation, Hillsborough, NC, USA) was used in this study. The phantom compartments were initially filled with ^{99m}Tc solutions at different concentrations: 83.8 kBq/ml for the left ventricle wall (cardiac insert), 46.2 kBq/ml for the liver, and 2.8 kBq/ml for the thorax and left ventricle cavity, approximately in the ratio 30:15:1. A solid cold lesion was positioned in the infero-basal region. Phantom was placed at Discovery NM 530c (GE Healthcare) scanner field of view center. Prior to this study all quality control tests recommended by manufacturer were performed.

Acquisitions were carried out over 12 h (two half-lives), and images were acquired hourly, with acquisition times of 3.0 and 5.0 min.

The manufacturer offers only two reconstruction options: High dose (HD) for stress studies or Low Dose (LD) for rest studies. The dedicated reconstruction algorithm is based on a 3-D iterative Bayesian reconstruction combining accurate scanner modelling. Reconstruction algorithm parameters are optimized separately per type of acquisition (LD, HD), based on the optimal trade-off between reconstructed uniformity of healthy myocardium and contrast of lesions, using phantom studies [14].

Four quality indicators were calculated. Normalized standard deviation, NSD [14] and contrast between the ventricle wall and the cavity, VCC [15] were evaluated in the phantom region without myocardial lesion. Phantom region with the insertion of the inferobasal cold lesion were evaluated by contrast-noise ratio, CNR and the effective contrast in the lesion, EC [14].

Mean and standard deviation (SD) of quality indicators were performed in three subsequent slices. Differences between different acquisition times and reconstruction parameters were evaluated using ANOVA single factor analysis, with $p < 0.05$. All statistical analyses were performed with the software OriginPro® 2017 (OriginLab Corporation, Northampton, MA, USA).

3 Results and Discussion

Table 1 shows representative results of NSD, VCC, CNR and EC when the acquisition time is 3 and 5 min, for the same reconstruction parameters.

Acquisition time variation of 3 and 5 min does not present significant difference in image quality NSD, VCC and CNR, for a higher activity concentration values. However, for lower activities, acquisition time affect NSD ($p = 0.06$) and EC ($p = 0.20$), producing a 20% decrease in effective contrast [15].

Table 2 summarizes the results of NSD, VCC, CNR and EC for a fixed acquisition time of 3 min. Images were reconstructed with the manufacturer's indicated reconstruction parameters, using HD (>48 kBq/ml) and LD for lower activities. Mean and SD of the quality indicators were calculated in three subsequent slices.

Figure 1 shows the results for a acquisition time of 3 min using box-plot graphs. HD reconstruction indicates grouped data with myocardium > 48 kBq/ml and LD for lower. Normality for quality indicators using Shapiro-Wilk test was rejected.

When reconstruction parameters are chosen according to the manufacturer's indications (HD for myocardial

Table 1 NSD, VCC, CNR and EC for representative activities in the myocardium, for 3 and 5 min acquisition time and same reconstruction parameters

Myocardium activity (kBq/ml)	Acq time (min)	Recon param	NSD (%)	VCC	CNR	EC
51.0	3	LD	29.4 ± 0.3	83.3 ± 0.2	4.51 ± 0.01	55.2 ± 0.1
51.0	5	LD	30.5 ± 0.3	87.4 ± 0.3	4.62 ± 0.02	56.1 ± 0.1
28.0	3	LD	31.2 ± 0.1	85.8 ± 0.2	4.86 ± 0.04	46.4 ± 0.2
28.0	5	LD	31.4 ± 0.1	85.8 ± 0.2	4.90 ± 0.01	55.6 ± 0.3

Table 2 NSD, VCC, CNR and EC means and standard deviations, for different concentrations of activities in the myocardium, for a fixed 3 min acquisition time, and reconstruction parameters indicated by manufacturer. Data are expressed as mean ± SD

Myocardium activity (kBq/ml)	Acq time (min)	Recon param	NSD (%)	VCC	CNR	EC
83.8	3	HD	18.3 ± 0.1	73.3 ± 0.1	3.98 ± 0.01	56.3 ± 0.1
72.7	3	HD	21.9 ± 0.3	77.3 ± 0.2	4.02 ± 0.01	56.2 ± 0.1
64.5	3	HD	15.5 ± 0.1	87.3 ± 0.1	4.52 ± 0.02	55.3 ± 0.3
56.1	3	HD	18.3 ± 0,1	89.2 ± 0.2	4.53 ± 0.02	55.4 ± 0.2
51.0	3	HD	31.9 ± 0.1	88.2 ± 0.3	4.54 ± 0.02	55.2 ± 0.1
46.2	3	LD	21.9 ± 0.1	93.5 ± 0.4	4.49 ± 0.06	50.2 ± 0.1
40.8	3	LD	23.3 ± 0.2	90.4 ± 0.4	4.46 ± 0.01	40.5 ± 0.2
35.2	3	LD	23.5 ± 0.2	90.2 ± 0.3	4.44 ± 0.01	40.4 ± 0.3
31.9	3	LD	24.5 ± 0.1	90.2 ± 0.3	4.86 ± 0.02	40.3 ± 0.1
28.0	3	LD	22.1 ± 0.2	90.0 ± 0.1	4.84 ± 0.04	46.6 ± 0.2
25.7	3	LD	22.6 ± 0.1	89.9 ± 0.1	4.84 ± 0.04	48.8 ± 0.1
22.6	3	LD	22.3 ± 0.1	83.2 ± 0.1	4.87 ± 0.02	48.5 ± 0.4

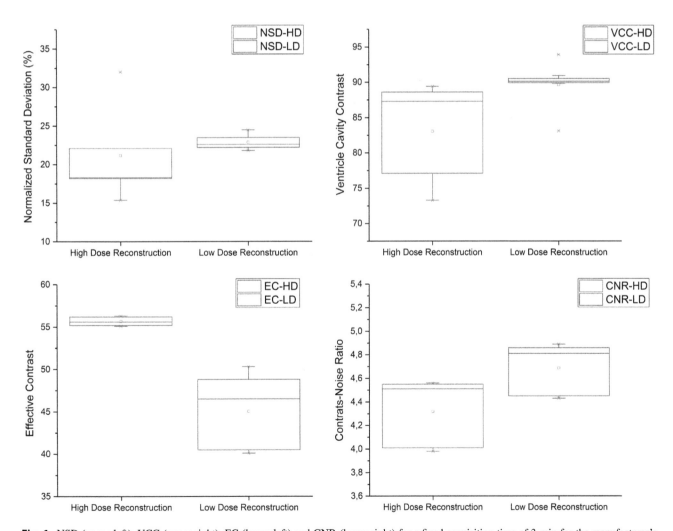

Fig. 1 NSD (upper left), VCC (upper right), EC (lower left) and CNR (lower right) for a fixed acquisition time of 3 min for the manufacturer's indicated reconstruction parameters, using HD reconstruction for myocardium activities higher than 48 kBq/ml and LD for lower activities

Table 3 Results of NSD, VCC, CNR and EC mean and standard deviation, for some concentrations of activities in the myocardium, for a fixed 3 min acquisition time and with both reconstruction parameters (HD and LD). Data are expressed as mean ± SD

Myocardium activity (kBq/ml)	Acq time (min)	Recon param	NSD (%)	VCC	CNR	EC
64.5	3	HD	15.5 ± 0.1	87.3 ± 0.1	4.52 ± 0.02	55.3 ± 0.3
64.5	3	LD	13.6 ± 0.1	77.4 ± 0.2	4.99 ± 0.01	40.4 ± 0.2
51.0	3	HD	31.9 ± 0.1	88.2 ± 0.3	4.54 ± 0.02	55.2 ± 0.1
51.0	3	LD	29.4 ± 0.3	83.3 ± 0.2	4.51 ± 0.01	55.2 ± 0.1
46.2	3	HD	32.2 ± 0.2	88.4 ± 0.3	4.54 ± 0.02	55.2 ± 0.1
46.2	3	LD	21.9 ± 0.1	93.5 ± 0.4	4.49 ± 0.06	50.2 ± 0.1

concentration > 48 kBq/ml and LD for <48 kBq/ml), the analysis show statistically significant differences between image indicators when activities decrease until two half-lives.

As the activity decreases, image quality NSD decrease, with higher variability in HD than in LD group data. NSD characterize uniformity in the healthy cardiac tissue. A large NSD variability near the limit for using recommended HD parameter (48 kBq/ml) was observed. NSD results in 31.9%–51 kBq/ml, very different from the posterior value (18.3%), for a slightly higher activity (56.1 kBq/ml).

CNR also decreases with activity reduction with high variability in both HD and HD groups. VCC increase and then decrease close to 48 kBq/ml, where the reconstruction parameter changes and then increases again. EC shows values varying in opposite direction regarding VCC, but with the same dependence with the activity values close to 48 kBq/ml. These results are consistent with the separated optimization in the reconstruction per type of acquisition (LD, HD) in this scanner, based on the optimal trade-off between uniformity of healthy myocardium and contrast of perfusion defects, cited by Volokh et al. (2018) [14].

Table 3 summarizes some results for NSD, VCC, CNR and EC for a fixed acquisition time of 3 min and images reconstructed using HD and LD parameters.

Huge variability is shown changing the reconstruction parameters without considering the actual activity concentration in the heart. These results reveal the importance of careful selection of the reconstruction parameters, taking in consideration patient weight and size and uptake.

4 Conclusion

We evaluated the impact of acquisition time and reconstruction parameters in the image quality and cold lesions visibility of a cardiac anthropomorphic phantom scanned by a CZT SPECT Discovery 530c. Acquisition time does not affect quality and lesion visibility indicators for the appropriate reconstruction parameters, mainly at higher activities.

This means a cost-saving reduction in radiopharmaceutical dosage with similar image quality indicators and a corresponding patient and personnel radiation exposure reduction, contributing to cost-effectiveness in the nuclear cardiology clinic. However, image quality indicators are sensitive to the reconstruction parameters choice, considering patient´s characteristics and administered activity, in order to produce the best relation image quality/lesion visibility and dose.

Compliance with Ethical Standards Conflicts of interest The authors declare that they have no conflict of interest.

References

1. Acampa, W., Buechel, R.R., and Gimelli, A.: Low dose in nuclear cardiology: state of the art in the era of new cadmium–zinc–telluride cameras. European Heart Journal – Cardiovascular Imaging 17, 591–595, (2016).
2. Slomka, P. J., Berman, D. S., Germano, G.: New Cardiac Cameras: Single-Photon Emission CT and PET. Seminary Nuclear Medicine, 44:232–251 (2014).
3. Takahashi Y, Miyagawa M, Nishiyama Y, Ishimura H, Mochizuki T.: Performance of a semiconductor SPECT system: comparison with a conventional Anger-type SPECT instrument. Ann Nucl Med, 27, 11–16 (2013).
4. Patton J, Slomka P, Germano G, Berman D.: Recent technologic advances in nuclear cardiology. J Nucl Cardiol 14, 501–513 (2007).
5. Bocher M, Blevis IM, Tsukerman L, Shrem Y, Kovalski G, Volokh L.: A fast cardiac gamma camera with dynamic SPECT capabilities: design, system validation and future potential. Eur J Nucl Med Mol Imaging, 37, 1887–1902 (2010).
6. Oddstig J, Hedeer F, Jogi J, Carlsson M, Hindorf C, Engblom H. Reduced administered activity, reduced acquisition time, and preserved image quality for the new CZT camera. J Nucl Cardiol, 20, 38–44 (2013).
7. Duvall WL, Wijetunga MN, Klein TM, et al.: Stress-only Tc-99 m myocardial perfusion imaging in an Emergency Department Chest Pain Unit. J Emerg Med, 42, 642–650 (2011).
8. Perrin, M., et al.: Stress-first protocol for myocardial perfusion SPECT imaging with semiconductor cameras: high diagnostic performances with significant reduction in patient radiation doses. EJNMMI, 42: 1004–1011 (2015).
9. Oldan, J. D., et al.: Prognostic value of the cadmium-zinc-telluride camera: A comparison with a conventional (Anger) camera. J Nucl Cardiology, https://doi.org/10.1007/s12350-015-0181-9 (2015).

10. Imbert L, Poussier S, Franken PR, et al.: Compared performance of high-sensitivity cameras dedicated to myocardial perfusion SPECT: a comprehensive analysis of phantom and human images. J Nucl Med 53, 1897–1903 (2012).

11. Dartora C.M., Favero M.S., da Silva A.M.M.: Detectability in SPECT Myocardial Perfusion Imaging: Comparison between a Conventional and a Semiconductor Detector System. IFMBE Proceedings, 51:169–172. Springer, Cham https://doi.org/10.1007/978-3-319-19387-8_41 (2015).

12. Fiechter, M. et al.: Cadmium-zinc-telluride myocardial perfusion imaging in obese patients. Journal of Nuclear Medicine, 53(9): 1401–1406, (2012).

13. Hindorf C, Oddstig J, Hedeer F, Hansson MJ, Jögi J, Engblom H.: Importance of correct patient positioning in myocardial perfusion SPECT when using a CZT camera. J Nucl Cardiol, 21, 695–702 (2014).

14. Volokh L., Lahat C., Binyamin E., Blevis I.: Myocardial perfusion imaging with an ultra-fast cardiac SPECT camera—a phantom study. In: Nuclear Science Symposium Conference Record, 2008. NSS '08. IEEE, Dresden, Germany. 4636–4639 (2008) https://doi.org/10.1109/nssmic.2008.4774437.

15. Caobelli, F., Ren Kaiser, S., Thackeray, JamJ. et al.: IQ SPECT Allows a Significant Reduction in Administered Dose and Acquisition Time for Myocardial Perfusion Imaging: Evidence from a Phantom Study. Journal of Nuclear Medicine. 55. 2064–2070 (2014).

A Surgical Robotic System for Transurethral Resection

Junchen Wang, Jiangdi Zhao, Xuquan Ji, Xuebin Zhang, and Hanzhong Li

Abstract

Transurethral surgery is a noninvasive interventional procedure that delivers a tubular surgical instrument via the urethra to access surgical sites in the prostate or bladder for abnormal tissue resection. In the clinical practice, a rigid resectoscope is inserted via the urethra to access the surgical site. The resectoscope is manually operated by the surgeon. Due to the limited field of the view (FOV) of the resectoscope, difficult hand-eye coordination, hand tremor, and lack of depth perception, the procedure is laborious, having a risk of perforation and damage to healthy tissue. We develop a master-slave robotic system for transurethral resection. The system is composed by a user console and a slave robot. The user console provides surgical vision and intuitive human-robot interaction interface, and the slave robot performs the surgery accordingly. The slave robot further consists of a 6 DOFs (degree of freedom) serial robot arm and a 1 DOF end-effector. The robot arm is used to accurately position and orient the resectoscope inside the body. The end-effector is designed to hold the resectoscope and precisely implement the linear motion of the cutting loop by reproducing the user's operation at the console side. Preliminary experiments were performed to evaluate the proposed system and the results have confirmed its effectiveness.

Keywords

Transurethral resection • Surgical robot • End-effector

1 Introduction

Transurethral surgery is an interventional procedure that accesses the surgical site via the urethra. Typical clinical indication includes transurethral resection of the prostate (TURP) and bladder tumor (TURBT). Due to the non-invasiveness of the procedure, it has become the golden standard treatment of benign prostate hyperplasia and non-muscle-invasive bladder cancer. A resectoscope is the most widely used surgical instrument that performs transurethral surgery [1]. It is a rigid endoscope with a monopolar electrical cutting loop at the distal. The resectoscope is inserted into the patient's urethra and positioned inside the prostate or bladder. Surgeons manually orient the resectoscope and operate a handle to control the cutting loop forward and backward to excise tissue under the vision guidance of the endoscope. However, manual manipulation of the resectoscope is laborious and non-intuitive. Due to hand tremor and lack of depth perception, it also has a risk of perforation and damage to healthy tissue [2]. Furthermore, limited by the narrow field of view and motion dexterity of the endoscope, surgeons have difficulties in understanding the global surgical environment and reaching all suspect areas. These make the surgical outcomes variable and highly dependent on surgeons' experience and ability.

Robotic systems have been proposed to assist transurethral surgery. The advantages of robotic assistance include access dexterity, precise motion and targeting, intuitive interaction, high repeatability, improved visualization. Yoon et al. [3] reported a shape memory alloy (SMA) based automated steering mechanism for bladder surveillance. The system only controls the imaging fiber and does not provide resection functionality. Sarli et al. [4] developed a customized resectoscope prototype. This resectoscope has a

J. Wang (✉) · J. Zhao · X. Ji
School of Mechanical Engineering and Automation, Beihang University, Beijing, China
e-mail: wangjunchen@buaa.edu.cn

X. Zhang · H. Li
Peking Union Medical College Hospital, Beijing, China

J. Wang
Beijing Advanced Innovation Center for Biomedical Engineering, Beihang University, Beijing, China

stem at the distal that provides working channels for flexible surgical tools. However the control of the surgical tools was not considered. Goldman et al. [5, 6] proposed a minimally invasive telerobotic platform for transurethral surveillance and intervention. The main body of the robot consists of an outer sheath and an inside dexterous continuum arm with access channels for the parallel deployment of multiple visualization and surgical instruments. The flexible continuum arm can be driven to bend so that it can reach the place where the traditional resectoscope has difficulty to access. A laser cautery fiber is inserted into one of the access channels to cauterize abnormal tissue. The shortcoming of the system is the inefficiency in excising tissue due to the limited heat effect of laser cautery. Hendrick et al. [7] presented a multi-arm hand-held robotic system for transurethral laser prostate surgery. The system provides surgeons with two concentric tube manipulators that can aim the laser and manipulate tissue simultaneously. Russo et al. [8] also developed a robotic platform for laser-assisted transurethral surgery of the prostate. A multi-lumen catheter is inserted into the working element of the resectoscope in which a laser fiber is integrated. The catheter is driven by three cables to change the direction. Additionally, three fiber Bragg grating (FBG) sensors are attached to the catheter to measure the force. However, compared with the electrical cutting loop, laser ablation is low efficient and is not a mainstream method for transurethral resection. For the consideration of fast clinical acceptance, this paper presents a commercial resectoscope compatible master-slave robotic system for transurethral surgery.

2 System Overview

Figure 1a illustrates the proposed master-slave robotic system. The slave robot is set up at the surgical site, which consists of a commercial resectoscope, a 6R robot arm, and an end-effector to hold and control the resectoscope. The master site consists of a computer workstation, a monitor to display the endoscopic view, a joystick as the interactive device to control the robot arm and end-effector. The mechanical design of the end-effector is shown in Fig. 1b. Its main function is holding the resectoscope and controlling the linear motion of the cutting loop. The resectoscope is fixed on the end-effector's supporting frame which can be attached to the robot arm. The central axis of the resectoscope coincides with the z axis of the robot's tool coordinate system (TCS). A DC motor is amounted at the frame to drive the cutting loop handle forward and backward by a cogged belt. The joystick shown in Fig. 1c is employed to control the motion of the robot arm and implement the cutting operation remotely. The x-y-z information from the joystick is

processed by the workstation to obtain the desired pose of the resectoscope, and the pose instruction is further sent to the robot arm to execute. The position of the cutting loop handle is also obtained from the joystick, and is sent to an embedded system to drive the motor. The control scheme is shown in Fig. 2.

3 Resectoscope Orientation

The robot arm is used to orient the resectoscope inside the body so that the cutting loop can reach the target site. To mimic the surgeon's operation, the robot arm is controlled to implement two types of movement. One is moving along the z axis of the TCS. This is achieved by directly controlling the robot arm. The other is fixed point rotation. The rotation center is set to be on the TCS's z axis, and can be shifted along the axis so that the rotation is always performed with respect to the patient's pubis. The rotation axis is determined from the joystick controlled by the surgeon. To achieve intuitive manipulation, the rotation axis in the TCS should be correlated with the endoscopic camera's axis, since the surgeon operates the joystick at the master site by monitoring the real-time endoscopic image. As shown in Fig. 3a, let Ox and Oy denote the camera's horizontal and vertical axes where O is the camera's center. If the user wants to pivot the resectoscope so that the camera's center moves towards $\mathbf{P}$ in the image, he/she may push the joystick towards the same direction shown in Fig. 3b. The rotation axis $\mathbf{r}_{TCP}$ in the TCS can be calculated as $\mathbf{r}_{TCP} = \mathbf{R} \cdot (-\sin\theta, \cos\theta, 0)^T$ where $\mathbf{R}$ is the rotation transformation from TCS to the camera's coordinate system (CCS); $\theta = \text{atan}(y, x)$ and (x, y) is the output of the joystick. $\mathbf{R}$ can be obtained by a so-called robot hand-eye calibration procedure. After obtaining the rotation axis in TCS, the robot arm can be programmed to execute the desired pivot.

4 Cutting Loop Control

The linear motion of the cutting loop is driven by the DC motor with an encoder. The DC motor is controlled in PWM (pulse width modulation) mode using an STM32 micro-controller. The basic idea is to reproduce the surgeon's cutting operation (push or pull the joystick) at the end-effector side, which means the position of the cutting loop should follow the joystick's output. A classical position PID control model is applied to achieve this goal, as shown in Fig. 4. The input position is from the joystick's output represented by the encoder's pulse count. The output position is measured by the encoder. Position error is fed into the PID controller to determine the duty ratio of the PWM wave,

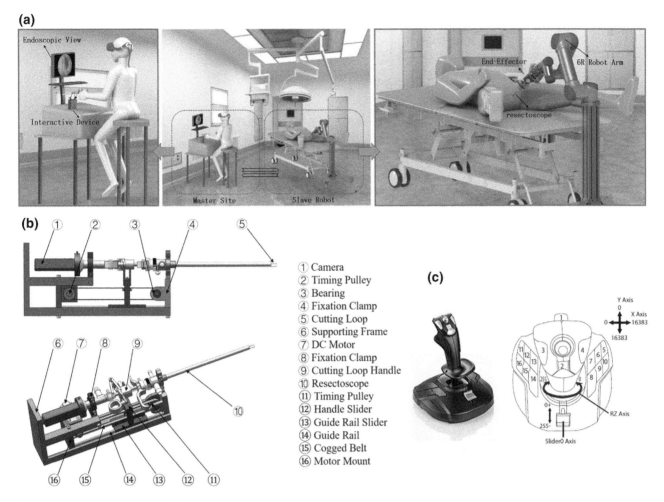

(a)

Endoscopic View

Interactive Device

Master Site   Slave Robot

End-Effector   6R Robot Arm

resectoscope

(b)

① ② ③ ④ ⑤

⑥ ⑦ ⑧ ⑨

⑩

⑯ ⑮ ⑭ ⑬ ⑫ ⑪

① Camera
② Timing Pulley
③ Bearing
④ Fixation Clamp
⑤ Cutting Loop
⑥ Supporting Frame
⑦ DC Motor
⑧ Fixation Clamp
⑨ Cutting Loop Handle
⑩ Resectoscope
⑪ Timing Pulley
⑫ Handle Slider
⑬ Guide Rail Slider
⑭ Guide Rail
⑮ Cogged Belt
⑯ Motor Mount

(c)

Y Axis
0
X Axis
0   16383
16383

RZ Axis

255-

Slider0 Axis

Fig. 1 **a** System overview. **b** End-Effector. **c** Joystick

which is used to control the speed of the motor. When the motor is rotating, the encoder records the total pulse count which is considered as the current real position.

5 Experiments

Experiments were performed to evaluate the system. The experimental scene is shown in Fig. 5a. The robot arm was UR5 from Universal Robots. The end-effector was fabricated using 3D printing, with a total weight of 2 kg. The encoder had a resolution of 2000 pulses per round. The linear motion range of the cutting loop was approximately 21 mm, yielding the total pulse count of 9250.

We first evaluated the accuracy of resectoscope orientation by simulating the TURP procedure. A circular marker with the radius of 2.5 cm shown in Fig. 5b was employed to evaluate the accuracy. The black/white dots in each black/white sector were set to be targets. An operator was asked to manipulate the joystick at the remote site only by watching the endoscopic view. For quantitative evaluation,

the cutting loop was replaced by a stainless steel pin shown in Fig. 5c. The operator was asked to orient the resectoscope so that the pin can touch each target dot on the marker as precisely as possible. Note that once the pinpoint have touched the marker area for the first time, it cannot be adjusted and the error is immediately measured. The error was calculated by measuring the distance between the pinpoint and the dot center, as shown in Fig. 5d. The results are shown in Table 1. The mean error was 0.76 mm.

Then we evaluated the time delay of executing cutting operation. The joystick was changed to the cutting loop control mode. Two IMU (inertial measurement unit) sensors (JY901) were attached to the joystick lever and the cutting loop handle, respectively. The operator was asked to push the joystick which would eventually drive the cutting loop. The acceleration signals from both IMU sensors (Fig. 6a) were collected with time stamps. The time difference where the signals abruptly changed was used to estimate the time delay. Ten trials were performed and the time delays are shown in Fig. 6b. The average time delay was 150 ms.

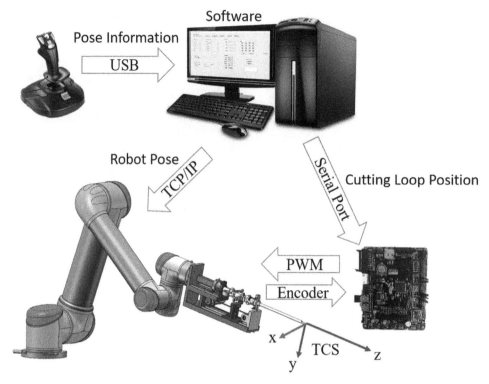

Fig. 2 Control scheme

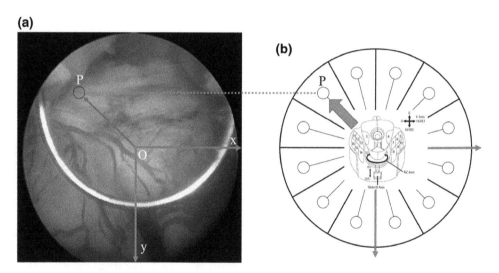

Fig. 3 Intuitive orientation control. **a** Pivoting resectoscope so that the principal point O moves towards P. **b** Corresponding manipulation of joystick

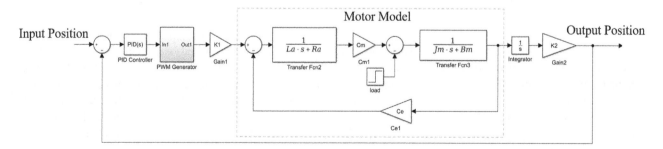

Fig. 4 Control model

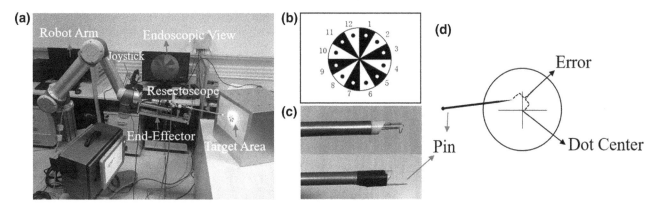

Fig. 5 Experiments. **a** Experimental scene. **b** Circular marker. **c** Pin substituting cutting loop. **d** Error measurement

Table 1 Accuracy evaluation results

Dot index	1	2	3	4	5	6	7	8	9	10	11	12
Error (mm)	0.57	0.71	0.53	0.74	0.32	1.05	0.42	1.20	0.80	0.88	1.29	0.60

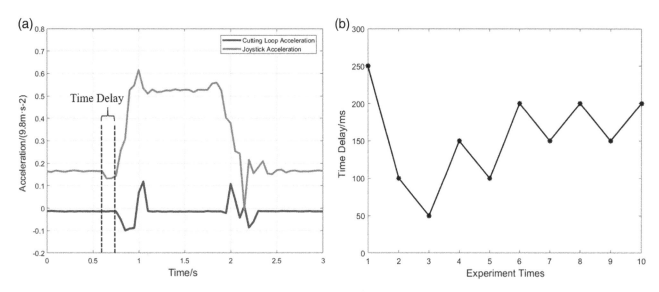

Fig. 6 **a** Time delay measurement. **b** Time delay results

6 Conclusion

This paper presented a robotic system for transurethral resection. System configuration, end-effector design, resectoscope orientation, cutting loop motion control were elucidated. Experimental results have shown the initial effectiveness of the system. Further research includes image guidance and force sensing.

Acknowledgements This work was partially supported by National Science Foundation of China (Grant No. 61701014).

Conflicts of Interest The authors declare that they have no conflict of interest.

References

1. Tasci, A.I., Ilbey, Y., Tugcu, V., Cicekler, O., Cevik, C., Zoroglu, F.: Transurethral resection of the prostate with monopolar resectoscope: Single-surgeon experience and long-term results of after 3589 procedures. Urology **78**(5), 1151–1155 (2011).
2. Rassweiler, J., Teber, D., Kuntz, R., Hofmann, R.: Complications of transurethral resection of the prostate (turp)-incidence, management, and prevention. European Urology **50**(5), 969–980 (2006).

3. Yoon, W.J., Park, S., Reinhall, P.G., Seibel, E.J.: Development of an automated steering mechanism for bladder urothelium surveillance. Journal of Medical Devices 3(1), 11,004 (2009).

4. Sarli, N., Giudice, G.D., Herrell, D.S., Simaan, N.: A resectoscope for robot-assisted transurethral surgery. ASME. J. Med. Devices (2016).

5. Goldman, R.E., Bajo, A., MacLachlan, L.S., Pickens, R., Herrell, S. D., Simaan, N.: Design and performance evaluation of a minimally invasive telerobotic platform for transurethral surveillance and intervention. IEEE Transactions on Biomedical Engineering 60(4), 918–925 (2013).

6. Pickens, R.B., Bajo, A., Simaan, N., Herrell, D.: A pilot ex vivo evaluation of a telerobotic system for transurethral intervention and surveillance. Journal of endourology 29(2), 231–234 (2015).

7. Hendrick, R.J., Herrell, S.D., Webster, R.J.: A multi-arm hand-held robotic system for transurethral laser prostate surgery. In: 2014 IEEE International Conference on Robotics and Automation (ICRA), pp. 2850–2855 (2014).

8. Russo, S., Dario, P., Menciassi, A.: A novel robotic platform for laser-assisted transurethral surgery of the prostate. IEEE Transactions on Biomedical Engineering 62(2), 489–500 (2015).

Sophisticated Hydrodynamic Simulation of Pulmonary Circulation for the Preclinical Examination of Right Heart Circulatory Assist Device

Yusuke Tsuboko, Yasuyuki Shiraishi, Akihiro Yamada, Kiyotaka Iwasaki, Mitsuo Umezu, and Tomoyuki Yambe

Abstract

To evaluate systemic circulatory support devices such as left ventricular assist system, surgical heart valve prosthesis, and transcatheter aortic valve, various in vitro hydrodynamic tests have been performed. As these devices are being applied to the pulmonary circulatory support in recent years, novel evaluation platform for right heart support is increasingly demanded. This study aims to develop a pulmonary mechanical circulatory simulation system to assess the hydrodynamic performance of newly designed artificial cardiovascular devices. For the construction of the system, we developed the pneumatically-driven polymer right atrial and ventricular models with the pulmonary arterial valve chamber, silicone-made peripheral pulmonary artery model, and a venous reservoir. A woven polyester vascular graft and commercially available mechanical bileaflet valve were installed into the valve chamber. Then, the right ventricular pressure and pulmonary arterial pressure were regulated by the peripheral resistive unit. As a result, we successfully obtained the standard conditions of our mechanical circulatory system to be 28/3 (systolic/diastolic) mmHg of right ventricular pressure, 29/7 mmHg of pulmonary arterial pressure, 6 mmHg of mean right atrial pressure, and 3.0 L/min of pulmonary flow rate. To carry out the sophisticated assessment for the support of the pulmonary surgical and percutaneous treatments, we are preparing the next step with the reproduction of respiratory changes in pulmonary peripheral resistance, and the patient-specific shape vascular model including catheter access vessels. Under the highly simulated both pulmonary anatomical morphology and hemodynamic function conditions, effective preclinical examination of newly designed surgical or percutaneous pulmonary circulatory support devices can be performed.

Keywords

Pulmonary mechanical circulatory simulator
Right heart support device • Preclinical evaluation

Y. Tsuboko (✉) · K. Iwasaki · M. Umezu
Institute for Medical Regulatory Science, Organization for University Research Initiative, Waseda University, 2-2 Wakamatsu-cho, Shinjuku-ku, Tokyo, 1628480, Japan
e-mail: tsuboko@aoni.waseda.jp

Y. Shiraishi · A. Yamada · T. Yambe
PreClinical Research Center, Institute of Development, Aging and Cancer, Tohoku University, 4-1 Seiryo-machi, Aoba-ku, Sendai, 9808575, Japan

K. Iwasaki
Cooperative Major in Advanced Biomedical Sciences, Graduate School of Advanced Science and Engineering, Waseda University, 2-2 Wakamatsu-cho, Shinjuku-ku, Tokyo, 1628480, Japan

K. Iwasaki · M. Umezu
Department of Modern Mechanical Engineering, Graduate School of Creative Science and Engineering, Waseda University, 2-2 Wakamatsu-cho, Shinjuku-ku, Tokyo, 1628480, Japan

1 Introduction

To evaluate systemic circulatory support devices such as left ventricular assist system, surgical heart valve prosthesis, and transcatheter aortic valve, various in vitro hydrodynamic tests have been performed. As these devices are being applied to the pulmonary circulatory support in recent years, novel evaluation platform for right heart support is increasingly demanded. In pulmonary circulation which has one-fifth of pressure in comparison with systemic circulation, the effect of kinetic energy is relatively high (Fig. 1) [1]. The authors have been developing a mechanical mock circulatory system for the evaluation of pediatric pulmonary artery reconstruction [2, 3]. This study aims to develop a sophisticated pulmonary mechanical circulatory simulation system as a platform for evaluation of hydrodynamic performance in newly designed pulmonary circulatory support devices.

© Springer Nature Singapore Pte Ltd. 2019
L. Lhotska et al. (eds.), *World Congress on Medical Physics and Biomedical Engineering 2018*,
IFMBE Proceedings 68/3, https://doi.org/10.1007/978-981-10-9023-3_130

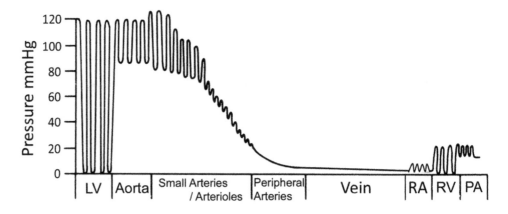

Fig. 1 Blood pressure distribution through the circulation system (created based on data from [1])

2 Materials and Methods

2.1 Construction of Pulmonary Mechanical Circulatory System

A mechanical circulatory system was designed to simulate a natural pulmonary circulation. The system consists of a pneumatic-driven right ventricle (RV) and a right atrium (RA) with a bileaflet polymer valve, a pulmonary valve chamber with a visualization port, a pulmonary arterial compliance tube, a pulmonary peripheral resistance unit, and a venous reservoir (Fig. 2). The peripheral resistance of the PA model was adjusted at the resistive unit attached to the PA model. The RA and the RV could be driven synchronously under the different timing patterns with compressed air through the air ports. We also developed a pneumatic driver for the interactive contraction between the RA and the RV. The pressure in the upper air chamber was supplied and controlled by an air pressure regulator. The

system achieved the synchronous motion with ventricular driving signals in the atrium model, as well as its pumping rate and systolic fraction.

2.2 Hydrodynamic Performance Test

Considering the reproducibility of experiments, a room temperature saline was used as a circulatory medium. Commercially available woven polyester vascular graft (Maquet, Hemashield, Japan) and mechanical bileaflet valve (St. Jude Medical, Regent, USA) were installed into the valve chamber. Then, the pressure of the RV and the PA were regulated by the resistive unit for maintaining those values within physiological ranges [4, 5] as shown in Table 1. The measurements were performed at 60 beats/min with diastole of 600 ms. During systolic periods, the RV was contracted to 400 ms with 25 mmHg. The RA was contracted for 150 ms with 5 mmHg prior to the ejection of the RV. The atrioventricular interval (50 ms) was chosen to be

Fig. 2 Whole view of pulmonary mechanical circulatory system

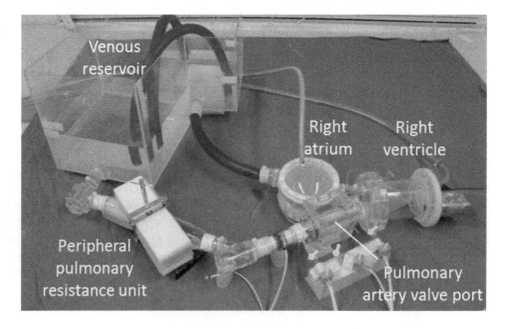

Table 1 Physiological pressure ranges of pulmonary circulation in the human

Parameters mmHg	Physiological range		
	Systole	Diastole	Mean
Right atrium	–	–	2–6
Right ventricle	20–30	0–5	–
Pulmonary artery	15–30	2–8	10–15

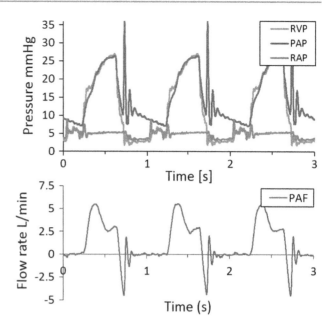

Fig. 3 Hemodynamic waveforms obtained from pulmonary mechanical circulatory system. RVP; right ventricular pressure, PAP; pulmonary arterial pressure, RAP; right atrial pressure, PAF; pulmonary arterial flow

similar to physiological conditions of the atrial contraction during sinus rhythm. RA and RV systolic fraction were 15, 40%, respectively. We measured the RV pressure, PA pressure waveforms and RA pressure by the pressure transducers (Nihon Kohden, DX-300, Japan). The pulmonary flow was obtained at RV outflow portion using the electromagnetic blood flow probe (Nihon Kohden, FF-200T, Japan).

2.3 Calculation of Pulmonary Artery Input Impedance

Vascular input impedance represents the afterload of the heart and depends on the resistance and elastance of the vasculature and the inertia of the blood [6–9]. The pulmonary input impedance of system was acquired using pulmonary pressure and flow measurements using Fourier analysis algorithms. The modulus Z and phase angles ϕ of impedance for 14 frequencies (harmonics) were calculated by

$$Z(\omega) = |PAP(\omega)|/|PAF(\omega)| \tag{1}$$

$$\phi(\omega) = \theta(\omega) - \varphi(\omega) \tag{2}$$

where ω is the frequency, $|PAP(\omega)|$ and $|PAF(\omega)|$ are the spectra of the average PAP and PAF. The length of the average PAP and PAF was dependent on the heart rate. θ is the PAP phase angle and φ is the flow phase angle. The impedance at zero harmonic (Z_0) was obtained from the impedance modulus at zero harmonic. The characteristic impedance (Z_c) was calculated as the average impedance modulus from harmonics 1–14.

3 Results

3.1 Hemodynamic Simulation

As a result, we successfully obtained the standard conditions of the mechanical circulatory system to be 28/3 mmHg (RVP), 29/7 mmHg (PAP), 6 mmHg (mean RAP), and 3.0 L/min (mean PAF). In this study, we could simulate the

physiological waveforms in natural pulmonary circulation by our mechanical circulatory simulator (Fig. 3).

3.2 Pulmonary Arterial Input Impedance

We calculated pulmonary input impedance from pulmonary arterial pressure and flow in our mechanical circulatory simulation condition (Fig. 4). The fluctuation could be obtained in high harmonics. Longitudinal impedance at 0 Hz and characteristic impedance were 524.2, 153.2 dyne sec cm^{-5}, respectively.

4 Discussions

In this study, we achieved hemodynamic simulation that could be derived in the natural pulmonary artery and right ventricle with atrial contraction. To achieve the sophisticated evaluation of the pulmonary support devices, the optimal driving contractile conditions including physiological atrioventricular balances should be considered. In our pulmonary mechanical circulatory simulator, we employed room temperature saline as the circulating medium. Fluid viscosity effects on the circulatory condition followed by the pressure and flow changes should be examined in the future study that could simulate the natural blood viscosity. We calculated the pulmonary input impedance from simultaneous pulmonary arterial pressure and flow measurement.

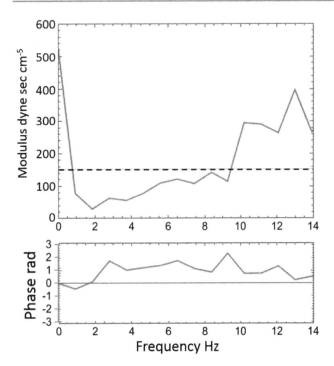

Fig. 4 Modulus and phase of pulmonary input impedance calculated by pulmonary arterial pressure and flow. Solid line; input impedance, Dashed line; characteristic impedance (Z_c)

Natural pulmonary arterial input impedance characteristics [6, 9] were well reproduced in our mechanical simulator. The variation of these pulmonary resistance could be changed by the ventilation pressure difference, and this pulmonary impedance changes would also be implanted into the mock circulatory examination for regulatory purposes. Therefore, pulmonary impedance variation caused by respiratory phase and the negative pressure condition in the thoracic cavity should be considered. Under the highly simulated both pulmonary anatomical morphology and hemodynamic function conditions, effective preclinical examination of newly designed surgical or percutaneous pulmonary circulatory support devices can be performed.

5   Conclusion

The purpose of this study was to develop the evaluation platform for pulmonary circulatory assist devices. We successfully constructed of mechanical pulmonary circulatory simulator which was capable of pressure and flow condition in healthy or diseased right heart hemodynamics.

Acknowledgements This work was partly supported by the Cooperative Research Project Program of Joint Usage/Research Center at the Institute of Development, Aging and Cancer, Tohoku University.

Conflicts of Interest The authors have no relationships that could be construed as a conflict of interest.

References

1. Smith, J.J., Kampine, J.P.: Circulatory Physiology: The Essentials. 2nd Japanese edn. Igaku-Shoin, Tokyo (1989).
2. Suzuki, I., Shiraishi, Y., Yabe, S., Tsuboko, Y., Sugai, T.K., Matsue, K., Kameyama, T., Saijo, Y., Tanaka, T., Okamoto, Y., Feng, Z., Miyazaki, T., Yamagishi, M., Yoshizawa, M., Umezu, M., Yambe, T.: Engineering analysis of the effects of bulging sinuses in a newly designed pediatric pulmonary heart valve on hemodynamic function. Journal of Artificial Organs 15(1), 49–56 (2012).
3. Tsuboko, T., Shiraishi, Y., Matsuo, S., Yamada, A., Miura, H., Shiga, T., Hashem, M.O., Yambe, T.: Effect of right atrial contraction on prosthetic valve function in a mechanical pulmonary circulatory system. Journal of Biomechanical Science and Engineering 11(6) 15-00356 (2016).
4. Badesch, D.B., Champion, H.C., Sanchez, M.A.G., Hoeper, M.M., Loyd, J.E., Manes, A., McGoon, M., Naeije, R., Olschewski, H., Oudiz, R.J., Torbicki, A.: Diagnosis and assessment of pulmonary arterial hypertension. Journal of American College of Cardiology 54 (1), 55–66 (2009).
5. Nichols, W.W., O'Rourke, M.F., Vlachopoulos, C.: McDonald's Blood Flow in Arteries: Theoretical, Experimental and Clinical Principles. 6th edn. CRC Press, Boca Raton/FL (2011).
6. Murgo, J.P. and Westerhof, N.: Input impedance of the pulmonary arterial system in normal man. Effect of respiration and comparison to systemic impedance. Circulation. 54, 666–73 (1984).
7. Lucas, C.L., Radke, N.F., Wilcox, B.R., Henry, G.W., Keagy B.A.: Maturation of Pulmonary Input Impedance Spectrum in Infants and Children with Ventricular Septal Defect. American Journal of Cardiology 57, 821–827 (1986).
8. Weinberg, C.E., Hertzberg, J.R., Ivy, D.D., Kirby, K.S., Chan, K. C., Valdes-Cruz, L., Shandas, R.: Extraction of Pulmonary Vascular Compliance, Pulmonary Vascular Resistance, and Right Ventricular Work from Single-Pressure and Doppler Flow Measurements in Children with Pulmonary Hypertension: a New Method for Evaluating Reactivity. Circulation 110, 2609–2617 (2004).
9. Hunter, K.S., Lee, P.F., Lanning, C.J., Ivy, D.D., Kirby, K.S., Claussen, L.R., Chan, K.C., Shandas, R.: Pulmonary vascular input impedance is a combined measure of pulmonary vascular resistance and stiffness and predicts clinical outcomes better than pulmonary vascular resistance alone in pediatric patients with pulmonary hypertension. American Heart Journal 155(1), 166–174 (2008).

Education Successes Applied to Disaster Preparedness, Meeting Infectious Diseases and Malnutrition Challenges

R. Rivas, Y. David, and T. Clark

Abstract

Hurricanes, floods and earthquakes have devastated part of the Caribbean and Latin America Region: "Maria" hurricane in Puerto Rico, flooding disaster in Peru, the several serious earthquakes in Mexico, Colombia, Chile, Argentina, occurred recently; it is required to ensure the effective functioning of the health systems at the potential situation of an emergency. Natural disasters and other emergencies around the world put the populations at risk; they may cause diseases and/or the disruption of health systems, facilities and services. The health risks of a disaster can be mitigated by building capacities of individuals, community and the country with developed or developing economies to protect health, Preparedness should address all the health disciplines. Responding to the effects of climate change on 2017 the National Institute of Health of Peru trained health workers from the 6 regions seriously affected by dengue disease, the result improved the effectiveness of the regional laboratories through the on time distribution and correct use of supplies to respond to the emergency. On 2014, the Philippines trained to prepare health professionals working in hospitals to detect and safely manage Ebola virus disease (EVD): public, private and local government hospitals were engaged. The confidence in managing EVD increased significantly (P = 0.018) with 96% of participants feeling more prepared to safely manage EVD cases. In other country like Ethiopia the adequate preparedness in emergencies and disaster response improved the capacities and better understanding of context specific causes of acute malnutrition and contributed to prevent the increase of severe acute malnutrition in the Horn of Africa in 2011 and other places in 2013 and 2014.

Keywords

Natural disasters • Preparedness • Education Health systems • Resilience

1 Introduction

Hurricanes, floods and earthquakes have devastated part of the Caribbean and Latin America Region: "Maria" hurricane in Puerto Rico, flooding disaster in Peru, the several serious earthquakes in Mexico, Colombia, Chile, Argentina, occurred recently [1].

The effective functioning of the health systems during the situation of an emergency is critically needed and depending on multidiscipline preparation. Natural disasters and other emergencies around the world put the populations at risk; they may cause diseases and/or the disruption of health systems, facilities and services.

Emergency Preparedness-EP follows an iterative cycle [2]:

1. Coordinating: mechanisms that include multisector and partners participation.
2. Financing: available financial and resources from local, national or international sources.
3. Assessing risk and capacity: EP program and plans should be based on all-hazards assessments of risk, and of the available capacity to manage the priority risks.
4. Planning: aimed at developing consensus and agreement not only on content, but also with regard to roles in implementation and financing.
5. Implementing: includes the definition of roles for stakeholders to oversee and monitor progress; responsibilities and accountabilities identified; sufficient resources

R. Rivas (✉)
Pontifical Catholic University of Peru, Lima, Peru
e-mail: privas@pucp.edu.pe

Y. David
IUPESM-HTTG, Houston, TX, USA

T. Clark
University of Vermont, Burlington, VT, USA

© Springer Nature Singapore Pte Ltd. 2019
L. Lhotska et al. (eds.), *World Congress on Medical Physics and Biomedical Engineering 2018*,
IFMBE Proceedings 68/3, https://doi.org/10.1007/978-981-10-9023-3_131

Fig. 1 Operationalizing emergency preparedness, a strategic framework for emergency preparedness, WHO 2016

available; and an adequate time lag between development and implementation of the plan.

6. Evaluating and Taking corrective actions: using pre-defined indicators and standardized tools and processes, this should be reported accordingly.
7. Exercising: co-operatively test and evaluate emergency policies, plans and procedures (Fig. 1).

EP is defined as: "…the knowledge and capacities and organizational systems developed by governments, response and recovery organizations, communities and individuals to effectively anticipate, respond to, and recover from the impacts of likely, imminent, emerging, or current emergencies" WHO, Framework for a Public Health Emergency

Operations Centre, 2015. EP is aimed to events which include infectious diseases and others caused by natural, technological and societal hazards [2].

The aim of this paper is to contribute to the achievement of health systems resilience through the training of a health emergency workforce.

2 Infectious Diseases and Malnutrition Following Natural Disasters

Infectious diseases are reported following natural disasters in developing countries. Disasters increase the risk factors for infectious diseases transmission by affecting preexisting poor water, sanitation and sewage systems [3]. See Table 1:

In other side malnutrition could increase following natural disasters; they cause higher morbidity and mortality in developing countries compared to middle-income and high-income countries due to higher vulnerabilities of the population, weaker healthcare system, and limited surge capacity [4].

3 Disaster Preparedness-Philippines, Ethiopia, Peru

3.1 Hospital Preparedness in Philippines

A training workshop to detect and safely manage Ebola virus disease-EVD was organized by the Philippines Department of Health and WHO. Teams with 5 members of health professionals from public, private and local government hospitals participated. Each workshop extended over 3 days with 18 lectures and 10 practical or small group sessions, including three practical sessions to don (put on) and doff (take off) personal protective equipment-PPE [5]. See Table 2:

Table 1 Risk factors and communicable diseases following natural disasters, Kouadio et al. [3

Mayor risk factors following natural disasters	Communicable disease
Population displacement from no endemic to endemic areas	Malaria, Dengue fever
Overcrowding (close and multiple contacts)	Diarrhea, Acute Res-piratory Infection-Pneumonial Influenza, Measles, Meningococcal, Meningitis, TB
Stagnant water after flood and heavy rains	Diarrhea, Leptospirosis, Malaria, Dengue fever
Insufficient/contaminated water and poor sanitation conditions	Diarrhea, Hepatitis
High exposure and proliferation to disease vectors	Leptospirosis, Malaria, Dengue fever
Insufficient nutrient intake/ Malnutrition	Diarrhea, Acute Respiratory Infection-Pneumonial Influenza, Measles, TB
Low vaccination coverage	Measles
Injuries	Tetanus, Cutaneous mucormycosis

Table 2 Structure of workshop on hospital management of EVD, Carlos et al. 2015

Carlos et al   Hospital preparedness training for Ebola virus disease, Philippines

Table 1. Structure of workshop on hospital management of EVD*

Session	Type of activity	Materials used
Day 1		
Opening	Formal opening with support from WHO country office and DOH	
Introduction	Lecture	
Ebola – basics, natural history and epidemiology of the West African outbreak; Reston Ebolavirus in the Philippines	Lectures	
Screening and triage	Lecture; small group work on six cases	Participants' sheet of cases Facilitators' guide Appendices B.2 and B.3 of the DOH Interim Guidelines
Treatment and discharge	Lecture	
Laboratory support and biosafety	Lecture	
Laboratory confirmation of EVD	Lecture	
Infection control for EVD	Lecture Practical session on removing gloves	Individual gloves and alcohol-based hand rub Glow-powder and UV lights
Ethical issues about clinical activities in EVD patients	Group discussion	
PPE for EVD: donning and doffing	Demonstration Practical sessions	Donning and doffing schedules Facilitators' guide Individual PPE
Day 2		
Isolation and patient flow	Lecture	
Designing isolation units for your hospital	Practical session	Plans of each participant's hospital
Management of sharps and post-exposure management for EVD	Lecture Role plays	
Environmental cleaning and waste management for EVD	Lecture	
Transport of EVD patients	Lecture	
PPE for EVD: donning and doffing	Practical sessions	Donning and doffing schedules Facilitators' guide Individual PPE
Day 3		
Comment on plans for isolation units	Commentary of each group's plans for an isolation unit in their hospital	Photographs of individual hospital isolation plans in PowerPoint presentation
Community health and support	Lecture	
Staff health and support in EVD	Lecture	
Safe and dignified burial for EVD	Lecture	
Epidemic management and surveillance	Lecture	
Contact tracing	Lecture	
Role of subnational laboratories	Lecture	
Question and answer session relevant to hospital management of EVD	Interactive session with DOH representative	
PPE for EVD: donning and doffing	Practical session with addition of red water-based paint used to contaminate PPE	Donning and doffing schedules Facilitators' guide Individual PPE Water-based paint
Additional specialized workshop for medical technologists		
EVD risk assessment and biosafety for laboratory personnel	Lecture	
Referral system, transport and storage of EVD specimens	Lecture	
Laboratory waste management, decontamination and laboratory emergencies for EVD	Lecture	
Laboratory procedures with PPE for EVD	Practical sessions: blood collection, specimen processing; packaging for transport	Individual PPE Blood collection equipment Safety cabinet Packaging and transportation materials
Closing ceremony	Speeches by WHO country office and DOH; presentation of certificates	Certificates of participation

* Modules and guide are available at http://www.wpro.who.int/philippines/mediacentre/features/ebolatraining_materials/and

DOH, Department of Health; EVD, Ebola virus disease; PPE, personal protective equipment; UV, ultraviolet rays; WHO, World Health Organization.

Confidence in managing EVD increased significantly (P = 0.018) with 96% of participants feeling more prepared to safely manage EVD cases. It was effective at increasing the level of knowledge about EVD and the level of confidence in managing EVD safely was effective at increasing the level of knowledge about EVD and the level of confidence in managing EVD safely.

3.2 Preparedness for Nutrition in Ethiopia

Recurrent droughts in the last 30 years contributed to continuous loss of assets and depleted communities especially poorest. As result, households in six drought regions are chronically food insecure have had associated higher prevalence of acute malnutrition that evolves to emergency levels even with mild rainfall performance. This case illustrates the importance of accurate, timely and reliable nutrition information to improve: planning, implementation and monitoring of emergency nutrition responses [6].

Global Nutrition Cluster-GNC concluded that adequate preparedness especially expansion and training Health Extension Workers-HEW on Community Management of Acute Malnutrition-CMAM in health posts as well as annual procurement of Therapeutic Feeding Programme-TFP supplies' timely distribution mitigates the impact of drought crisis and prevents unprecedented increase in acute malnutrition among under five children in drought affected woredas.[1]

On 2014 the preparedness for nutrition cluster according to GNC included: (A) biannual nutrition surveys conducted in 21–25 selected sites in the country. (B) Trend analysis regarding the nutrition situation and trends in TFP admissions are conducted and compared with previous years and (C) Training to HEW/HWs on Severe Acute Malnourished-SAM management.

Long term resilience programming with emergency response capacities and better understanding of context specific causes of acute malnutrition were identified as effective solutions among others to prevent the increase of severe acute malnutrition. The cluster strategic plan coordinated 30 partners: government institutions, UN agencies, NGOs and donors.

"El Niño" effects Peru: A warming of the central to eastern tropical Pacific Ocean, "El Niño" affects in eastern and southern Africa, the Horn of Africa, Latin America and the Caribbean, and the Asia-Pacific region since 2015.

Severe drought and associated food insecurity, flooding, rains, and temperature rises due to "El Niño" are causing disease outbreaks, malnutrition, and other diseases [7].

The number of cases of dengue increased in Peru. See Fig. 2.

Piura, La Libertad, Tumbes, Ica, Ancash, Ayacucho, Lambayeque, Ucayali, Loreto have the higher number of cases of dengue in the country. See Fig. 3:

National Institute of Health's assessment applied to the laboratories of Piura determined the following urgent requirements [8]:

1. Improvement of infrastructure.
2. Human resources trained.
3. Acquisition of Equipment for Laboratories: serological and molecular tests.
4. Improvement of capacities to increase the number of tests and supplies for diagnostic.

Some of the actions implemented by NIH on 2017 were:

- Staff from NIH-Lima was transferred to the laboratories of Piura for 3 months.
- Acquisition of equipment for the laboratories of Piura: serological and molecular diagnosis for dengue.

NIH remarked the relevance of (a) improving the training of human resources and (b) the need of technology acquisition for diagnostic to guarantee the effective response of the laboratories in Piura in short term.

Fig. 2 Dengue disease: 2000–2017, MoH, Peru, 2017

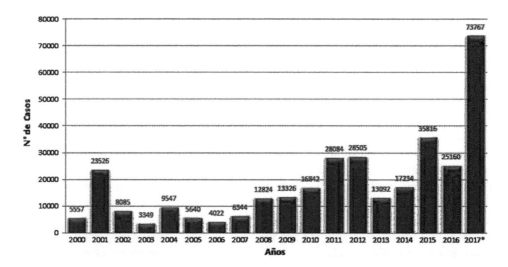

---

[1]Administrative structure similar to a district.

DEPARTAMENTOS	AÑOS																	
	2000	2001	2002	2003	2004	2005	2006	2007	2008	2009	2010	2011	2012	2013	2014	2015	2016	2017*
PIURA	2620	11713	101	1726	37	51	865	282	1702	4029	8393	183	1181	1979	2675	20043	7610	48300
LA LIBERTAD	1496	5718	3	0	263	259	10	1482	267	134	728	17	104	23	63	2072	4650	6632
TUMBES	192	1803	13	50	1552	183	243	79	51	830	1177	104	592	250	1700	7418	1089	5010
ICA	0	0	0	0	0	0	0	0	0	0	0	0	0	0	0	3	323	4436
ANCASH	0	4	824	1	8	4	1	8	77	224	50	0	1068	454	0	118	454	1889
AYACUCHO	0	0	0	0	0	0	0	0	0	0	0	0	0	1	0	268	2637	1826
LAMBAYEQUE	0	813	45	79	1868	804	77	656	718	674	291	10	491	25	147	1103	1662	1621
UCAYALI	97	682	2977	182	1413	69	174	182	931	1069	121	1770	1056	1059	1493	350	1007	682
LORETO	518	510	2499	784	2580	1772	1995	1720	7232	3723	1322	21245	14382	4479	7049	1630	1686	674
SAN MARTIN	218	179	42	46	577	172	170	677	541	448	307	1437	2322	1208	1574	220	335	557
CUSCO	0	0	2	0	0	2	0	0	0	0	0	57	0	2	227	248	1100	550
CAJAMARCA	18	1100	1176	114	383	1127	123	125	464	473	784	688	3208	85	295	218	281	429
LIMA	0	2	0	0	0	443	10	91	0	235	90	0	314	102	4	9	58	368
MADRE DE DIOS	21	103	12	0	0	85	2	314	45	798	2952	1956	2047	2272	1117	966	469	359
JUNIN	7	48	207	116	192	114	189	378	8	245	140	87	736	781	508	774	931	250
HUANUCO	29	159	132	107	356	143	128	28	110	257	214	136	336	67	129	307	728	113
AMAZONAS	341	692	30	143	312	409	35	320	648	158	273	305	587	247	207	37	90	49
CALLAO	0	0	0	0	0	0	0	0	0	0	0	0	0	0	0	0	0	15
PASCO	0	0	22	1	6	3	0	2	30	29	0	87	80	56	33	32	50	6
HUANCAVELICA	0	0	0	0	0	0	0	0	0	0	0	0	0	0	0	0	0	1
APURIMAC	0	0	0	0	0	0	0	0	0	0	0	0	0	0	0	0	0	0
TACNA	0	0	0	0	0	0	0	0	0	0	0	0	0	0	0	0	0	0
MOQUEGUA	0	0	0	0	0	0	0	0	0	0	0	0	0	0	0	0	0	0
AREQUIPA	0	0	0	0	0	0	0	0	0	0	0	0	0	0	0	0	0	0
PUNO	0	0	0	0	0	0	0	0	0	0	0	2	1	2	13	0	0	0
Total general	5557	23526	8085	3349	9547	5640	4022	6344	12824	13326	16842	28084	28505	13092	17234	35816	25160	73767

Fig. 3 Dengue disease per region: 2000–2017, MoH, Peru, 2017

4 Actions in Peru

4.1 Laboratory Technology: On-Line Course on Technology Planning and Management

Clinical engineers play an important role in emergency planning by applying healthcare technology management principles to determine technology needs, supporting logistics, providing asset inventories and applying proven educational methods throughout the technology life-cycle.

To maximize the value of EP education and training especially, where learners are at a distance from classrooms hybrid interactive EP programs can be valuable. Hybrid training is achieved by combining strategic classroom sessions with ongoing webinars and online courses. This strategy prepares personnel to effectively respond to disasters and post-disaster challenges without an extreme expenditure of scarce resources. This training approach to public health in Peru can be valuable with pertinent evidence for other health sectors around the world that face similar health challenges.

To support the Peruvian National Institutes of Health, a hybrid training program has been developed to improve public health laboratories in Lima and in the regions of Peru. The course provides personnel with a basic knowledge of the principles of Health Technology Planning and Management focusing on improvement in operations quality assurance, maintenance, and emergency preparedness. After an initial live workshop to start the course, ten one hour webinars will be held mixed with a midterm live session. This allows NIH laboratory staff in the regions to learn remotely without the expense and disruption of travel. The course will end with student team presentations in a workshop format. During the course, students will have access to resources via an online platform: text, pictures, diagrams, flow charts, figures, and links to the resources of the World Wide Web. Assessments will include Tests and Exams, Discussion Questions along with the final report/project.

5 Conclusions and Future Steps

Emergency Preparedness should address all the health disciplines; Clinical Engineering has an essential role in this regard by applying: healthcare technology management principles, technology life-cycle and others to contribute to the achievement of health systems resilience.

In other side:

1. Health services during and post disasters are more effective when EP program has been planned well, supported, and practiced. Clinical Engineering is an important resource applied to this activity.
2. Building capacities on Emergency Preparedness-EP have in incorporating Hybrid interactive EP Programs a valuable support.
3. Timing and level of effective response regarding public health diseases are two strong positive aspects to promote Hybrid Interactive EP Programs specially in developing countries.
4. Contents of the program in Peru should include the elaboration of improvements related to the topics and applied to NIH public health laboratories.

The health risks, especially infectious diseases and malnutrition, of a disaster can be mitigated by building capacities of individuals, community and the country with developed or developing economies to protect health. Hybrid EP Programs are effective to provide the training and information required. EP Preparedness program should address all the health disciplines.

Conflict of Interest The authors declare that they have no conflict of interest

References

1. EU Homepage, https://ec.europa.eu/echo/news/new-eu-aid-natural-disasters-latin-america-and-caribbean, last accessed 2018/01/28.
2. World Health Organization: A strategic framework for emergency preparedness. WHO, Geneva, (2016).
3. Kouadio I, Aljunid S, Kamigaki T, Hammad K, Oshitani H, (2012) Infectious diseases following natural disasters: prevention and control measures. Expert Review of Anti-infective Therapy 10(1): 95–104.
4. Pradhan PMS., Dhital R., Subhani H. Nutrition interventions for children aged less than 5 years following natural disasters: a systematic review protocol. BMJ Open, (2015).
5. Carlos C., Capistrano R., Fay Ch., de los Reyes M., Lupisan S., Cor puz A., Aumentado Ch., Lee L., Hall J., Donald J., Counahan M., Cureless M., Rhymer W., Gavin M., Lynch Ch., Black M., Anduyon A., Buttner P., Speare R.: Hospital Preparedness for Evola virus disease: a training course in the Philippines. WPSAR, Vol 6, No 1, (2015).
6. Global Nutrition Cluster Net Homepage, http://nutritioncluster.net/wp-content/up-loads/sites/4/2016/03/Ethiopia-Country-Lessons-Learned-progress-update-Final.pdf, last accessed 2018/01/28.
7. WHO Homepage: http://www.who.int/features/2016/el-nino-health-impacts/en/, last accessed 2018/01/28.
8. García M.: Rol del INS en la preparación y respuesta de Emergencias y Desastres. VI Reunión Nacional de la Red de Laboratorios en Salud Pública. Instituto Nacional de Salud-INS, Perú, (2017).

Part XIV

The Valley of Death in Medical Technology Transfer
Why It Exists and How to Cross It

Experience from Industrial Graduate (PhD) Schools

Maria Lindén and Mats Björkman

Abstract

Traditionally, research education is performed within the universities, and the PhD students are working within a research group. However, technical development and also research is performed within companies, and the need to keep up with the latest findings in research and to strengthen the competence within the private business sector is increasing. At Mälardalen University, we have experience from working in several Industrial Graduate Schools. The collaboration with the companies gets intensified and deepened trough such programmes, and the university tends to keep the good contact with previous PhD students and their companies also many years after their graduation. The Graduate Schools also give the companies good insight in the university world. Presently, we are involved in two Graduate Schools, and several of the PhD projects are focusing within Biomedical Engineering. Further, one of the graduate schools is linked to the research profile Embedded Sensor Systems for Health, which is supported from the same financier. Companies are involved also in the research profile, and through these activities, the Industrial PhD students form a critical mass and can exchange both experience and knowledge with other companies and with university researchers.

Keywords

Industrial graduate school • Collaboration with industry

M. Lindén (✉) · M. Björkman
Mälardalen University, Västerås, Sweden
e-mail: maria.linden@mdh.se

M. Björkman
e-mail: mats.bjorkman@mdh.se

1 Collaboration Between Academia and Industry

The importance of collaboration between academia and industry is recognized by many research financiers, among these the European Commission, but also at national levels. Academia and industry working together will give benefits to both sides, more specifically industry will gain knowledge in their development and the universities will work with challenges of large relevance and importance for innovation development. By this, the research environments of the universities will develop as well as the environments at the participating industries.

There is an increasing need of knowledge development within industry today. The development in technology is very rapid, and a close collaboration with universities helps industry to follow the research development. Also, there is a need to further educate the industry staff, and providing research education for persons working in industry will give the companies inside expertize.

2 The Concept of Industrial PhD

The main idea behind industrial graduate schools (industrial PhD education) is to enroll a person working in industry as a PhD student at a university. The person is already working with complex development and/or analysis at the company, and by collaboration with a university, the problem solving can be elevated to a higher level. The research problem that the PhD student works with is of relevance both to the company, which wants to solve a certain issue, and to the university, since it is so complex that it cannot be solved by conventional methods. It is also of major interest to the company to increase the internal knowledge level.

In addition to work on the research project, there is a requirement to take relevant courses for the PhD student. The amount of courses required varies depending on the

L. Lhotska et al. (eds.), *World Congress on Medical Physics and Biomedical Engineering 2018*,
IFMBE Proceedings 68/3, https://doi.org/10.1007/978-981-10-9023-3_132

subject, but normally equaling 1–1.5 years of study. One part of the courses is general research education, the other part aims at deepening the knowledge of the research problem, and can also consist of self-studies on a specific subject.

3 The Concept of Industrial Graduate (PhD) Schools

Long-term objectives from a doctoral student perspective are to enable higher career opportunities through industrially relevant education and research, as well as awareness of opportunities in industry and academia. Further, to provide opportunities for graduate students to work in both academic and industrial environments, gaining knowledge and experience valuable for a future career, and to establish long-term international contacts and networks with leading universities or research institutes, with other companies, and with other PhD students.

From a collaboration perspective, the objectives for pursuing an industrial graduate school are to establish a long-term platform for collaboration between universities and the participating companies based on co-production, giving results beneficial to both parties and also leading to a wider network—both academically and industrially, with increased possibilities for personal mobility. Further, it will provide opportunities for graduate students to work in both academic and industrial environments with knowledge and experience valuable for a future career, and with increased personal mobility, and to assure industrial relevance of the research and education at the universities. Additionally, it might increase industrial involvement in education at the Master level, and also to some extent at the Bachelor level.

Objectives from a university environment perspective are to contribute to research of highest international standard regarding methods and tools, e.g. in areas such as reliable embedded sensor systems, contributing to state-of-the-art competence for future product development. It will also help to obtain deep knowledge, insight and understanding of the area of reliable embedded sensor systems in the participating companies, which in turn leads to academic awareness and knowledge of industrial competence and industrial needs in the area of reliable embedded sensor systems and to assure industrial relevance of the research and education. Also, it will increase the industrial involvement in education at the Master level, and also to some extent at the Bachelor level. It will also enable higher career opportunities for both the industrial PhD students and other students through industrially relevant education and research, as well as increased awareness of opportunities in industry and potential of engaging students. It will also be instrumental in deployment

of the latest research results, primarily to the industrial PhD school partners, secondly to industry in general.

Objectives from a business sector perspective are to obtain strategic competence by research education of existing industrial specialists and/or new PhD students, to establish a long-term platform for collaboration with universities based on co-production, giving results beneficial to both parties, and to develop the profession and function that will contribute to new competitive products, including technical development. Additionally, a national and international network of leading scientists and industrial practitioners is developed. The status and knowledge is increased in industry by cooperation with other industrial partners in and new and improved methods supporting technology development is provided.

4 Experiences from Several Industrial Graduate Schools

The research environment of Mälardalen University (MDH) has experience of managing several successful Graduate Schools in co-production with industry. The Swedish Knowledge Foundation funding agency (KKS) [1] has funded most of these. The current KKS-funded graduate schools ITS ESS-H, ITS-EASY and Innofacture, as well as the former KKS-funded graduate school SAVE-IT, are hosted by the School of Innovation, Design and Engineering, at Mälardalen University, and ITS ESS-H, SAVE-IT and ITS-EASY belong(ed) to the same strategic research area, Embedded Systems (ES). Within the graduate school ITS-EASY, there have been four industrial PhD students active in the area of Biomedical Engineering. Within the graduate school ITS ESS-H, there are also four industrial PhD students active in the area of Biomedical Engineering.

ITS-EASY mainly focuses on Embedded Software and Systems, and ITS ESS-H is an important complement to ITS-EASY by focusing on reliable embedded sensor systems. The main industrial domains considered in ITS-EASY are automation, telecommunication and vehicles and the primary topics covered by ITS-EASY are related to system and software engineering of embedded systems. Within ITS ESS-H, reliable embedded sensor systems it the main topic, including the core competence areas of the research profile Embedded Sensor Systems for Health (ESS-H); sensor systems, signal processing, intelligent decision support, and reliable communication, together with area of reliable hardware systems.

With the industrial graduate schools, we have been collaborating with both large and small companies. The collaboration typically works very well, with a big interest from the companies to take part in meetings and joint activities of the schools. Since one of the objectives of these graduate schools is to give the industrial PhD students a research context bigger than what is available at the companies, these joint meetings and activities are important to create a relationship between PhD students, and between participating companies, as well as providing training for the PhD students. The joint activities usually take the form of either 1–2 days meetings, or travel to conferences of joint interest to the participating PhD students.

Another form of activities that are important to the wider knowledge of the PhD students is visits to the participating companies. By getting to know the participating companies better, new ideas for collaboration topics can results from this. This is one of the unexpected benefits of these industrial graduate schools; they have established new collaborations between companies that have not previously collaborated.

There are of course also challenges with industrial graduate schools. For an industrial PhD student, the company is the employer, and critical work tasks may interrupt the PhD work for shorter or longer periods of time. On a higher business level, long-term interests of the company may change, forcing the PhD student to re-focus the research efforts of the PhD studies. However, these problems have only been occurring in a few cases.

Our experience is that the quality of the PhD work of industrial PhD students is high. We have this far, within the Biomedical Engineering area at MDH, three PhD theses at MDH as results from industrial PhD students in industrial PhD schools: Kaisdotter Andersson [2], Kade [3], and Ljungblad [4]. In addition, we have two Licentiate (a degree halfway to PhD) theses from industrial PhD students: Gerdtman [5], and Du [6].

To conclude, the concept of industrial PhD students is beneficial to companies and universities as well, and enables a mutual exchange of knowledge that are good for both parties. The concept of industrial graduate schools enables a critical mass of industrial PhD students that enhances the situation of the PhD students and enables collaboration and knowledge exchange also between participating companies. According to our experience, the specific challenges with industrial PhD students are small compared to the benefits.

References

1. KKS About, in English, http://www.kks.se/om-oss/in-english/, last accessed 2018/01/31.
2. Kaisdotter Andersson, A., Improved breath alcohol analysis with use of carbon dioxide as the tracer gas, Mälardalen University Press Dissertations No. 83 (2010).
3. Kade, D., Head-mounted Projection Display to Support and Improve Motion Capture Acting Mälardalen University Press Dissertations No. 203 (2016).
4. Ljungblad, J., High Performance Breath Alcohol Analysis, Mälardalen University Doctoral Dissertation No. 240 (2017).
5. Gerdtman, C., Avancerade alternativa inmatningsenheter till datorer för funktionshindrade, Mälardalen University Press Licentiate Theses No. 133 (2011).
6. Du, J., Signal processing for MEMS sensor based motion analysis system, Mälardalen University Press Licentiate Theses No. 228 (2016).

Part XV

Women in Medical Physics and Biomedical Engineering

Gender Balance in Medical Physics—Lost in Transition?

Loredana G. Marcu and David Marcu

Abstract

Before the toppling of the communist regime in Romania in 1989, the gender distribution in physics college students was bent towards males with an approximate two-thirds to one-third ratio. A recent statistical analysis made among researchers that have applied for research grants in the area of health science (including medical physics) shows that the number of male applicants is around double the number of female applicants. However, a look at the current gender balance among clinical medical physicists shows a clear bend towards females. Theoretically, this could result in an increased number of female researchers in medical physics, though practically this does not happen due to the large clinical workload in the Romanian hospitals which limits the time and effort necessary for scientific explorations. This fact leads us to the actual category of skilled people who undertake research: the academic staff. Among them, the gender balance is off, as the male to female ratio is nearing 3. Given these facts, it is interesting to think about the choices made by males and females after graduation, entering the workforce. Despite the general decline in science students over the last few decades, medical physics keeps being an attractive educational offer presented by several Romanian universities. An interesting shift is that among current medical physics students, the large majority are female both at undergraduate and postgraduate levels. Does this mean that in 20 years we will see a dominance of females in higher academic positions and research grant applications? Can we afford to just let nature take its course, or do we have to actively intervene to encourage women to choose an academic and research career after graduation?

Keywords

Science education • Gender inequity • Science career

1 Introduction

According to the 2015 report of the UNESCO Institute for Statistics on women in science, employees in STEM (science, technology, engineering, and mathematics) are predominantly men, totaling over 70% as a worldwide average [1]. While this gender imbalance has diminished over the last decades in various scientific fields, overall, they are still dominated by men [2].

Statistics show that women outnumber men in overall undergraduate enrolments, however men are more likely to major in science or mathematics and also to pursue a career in these fields [3]. While these statistics are based on American college graduates, the trend is found in European countries as well. The scientific literature points out several aspects that lead to different preferences towards curricular subjects between men and women and implicitly, to different career choices. Psychologists consider that male predilection towards sciences are based on (i) evolutionary perspective, (ii) differences in brain structure and function between sexes, (iii) biopsychosocial aspects related to environmental factors, (iv) sociocultural influences including family and school [4].

The goal of this paper is to present the gender distribution among Romanian students, academic professionals and researchers (both young and mature) in the field of physical sciences, and to illustrate the effects of the last decades on gender balance in this area.

L. G. Marcu (✉) · D. Marcu
Faculty of Science, University of Oradea, 410087 Oradea, Romania
e-mail: loredana@marcunet.com

L. G. Marcu
Division of Health Sciences, University of South Australia, Adelaide, SA 5005, Australia

© Springer Nature Singapore Pte Ltd. 2019
L. Lhotska et al. (eds.), *World Congress on Medical Physics and Biomedical Engineering 2018*,
IFMBE Proceedings 68/3, https://doi.org/10.1007/978-981-10-9023-3_133

2 From School to Workplace

Until 1989, the gender distribution among Romanian physics college students was bent towards males with an approximate two-thirds to one-third ratio. Over the last couple of decades this distribution has balanced out, and today is even shifted towards female students. Several factors contributed to this change, including new carrier opportunities through interdisciplinary education, increased number of female teachers playing a role model, and continuous global female emancipation.

This shift in gender distribution among medical physics students is already observed among clinical medical physicists. As the profession of medical physicist is rather new in our country, the relative proportion of female medical physicists is expected to correspond to the relative proportion of female medical physics students. According to the Romanian College of Medical Physicists, the percentage of active female members versus male members in 2018 is 70.6% versus 29.4%, while ten years ago it was 45% versus 55%. Theoretically, this female dominance should lead to a preponderance of female researchers in medical physics, though reality paints a different picture. Due to the heavy clinical workload of the Romanian hospitals, the time and effort necessary for scientific explorations is severely limited. As a result, the majority of scientific research is undertaken by the higher education academic staff. An analysis of current members of physics departments among Romanian universities shows that the gender balance of male to female academics is close to 3 (see Fig. 1 for the physics faculties and departments in the country).

It is interesting to notice the difference in gender balance between the two career choices. The 'medical physicist' profession has been officially introduced in Romanian legislation in 2005, leaving a short 13-year period for the formation of this professional community. The gender ratio within this group closely follows the ratio seen within current physics students. An academic career takes long years to build, explaining why the gender balance within faculty members still reflects the student ratio from a few decades ago.

3 Gender Distribution in Research Related to Physical Sciences

In order to observe gender distribution among grant applications, first we evaluated the types of available calls as a function of their eligibility criteria to determine the level of seniority of likely applicants.

For a better overview and trend prediction, two types of research grants are assessed below regarding the gender of the applicants: (i) complex research projects (aimed at mature, established researchers) and (ii) postdoctoral research grants (aimed at young researchers). Based on the aforementioned science graduate student statistics that was dominated in the past by males, it is expected that in the first grant category, the same gender imbalance be found (Fig. 2).

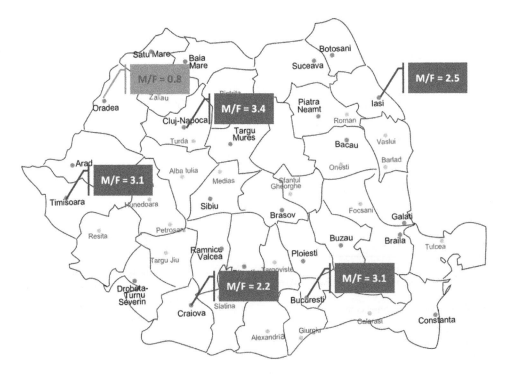

Fig. 1 Male to female ratio among academic staff within physics faculties and departments

Fig. 2 *Mature researchers:*
Male versus female candidates
that addressed the call for
complex research projects within
a national competition (data
processed from [5])

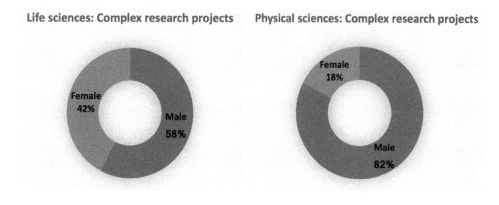

Fig. 3 *Young researchers:* Male
versus female candidates that
addressed the call for postdoctoral
research projects within a national
competition (data processed from
[5])

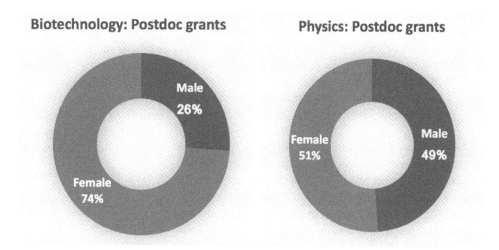

Figure 2 presents the male versus female candidates that addressed the call for complex research projects within a national competition. Two research areas that are of interest for the current study are presented: life sciences and physical sciences. The charts in Fig. 2 clearly illustrate the male preponderance in both fields (58% males in life sciences and 82% males in physical sciences).

In contrast with the results illustrated in Fig. 2 are the statistics from the postgraduate research competition (Fig. 3), where the applicants were young researchers (postdoctoral). The new gender distribution among science students is reflected already in the group of young researchers, with 51% female applicants in the field of physics and an impressive 74% female applicants in the field of biotechnology and applied health sciences.

4 Where Are We Headed?

Over the last few decades there was a noticeable decline in science students, a trend that is not only local, but universal. Nevertheless, medical physics keeps being an educational offer that attracts local students and is therefore, presented by several Romanian universities. A remarkable observation is that among current medical physics students the large majority are female both at undergraduate and postgraduate levels. It will be interesting to see if this trend will be further translated into a dominance of females in higher academic positions and complex research grant applications in 20 years' time.

One of the reasons for the increase in female students undertaking medical physics studies might be the strong interdisciplinary aspect of medical physics, which beside physics and medicine incorporates several other scientific disciplines, ranging from chemistry to bioinformatics. Since gender distribution differs in various science fields [2], the interdisciplinary aspect of medical physics education can stimulate a more homogeneous group of interest, decreasing the gender imbalance. Another reason determining women to choose careers that historically were stereotyped as male-specific is the global emancipation of women. This has been determined by a strong cultural shift and accelerated by the high level of connectivity enabled by modern technology.

The traditional trend of women becoming science educators rather than science researchers has led to the situation where current students have an abundance of female role models playing an important part in their science upbringing, which might also determine more young women to choose a career in science research.

As a subjective reason for women's presence in the field we could mention the fortunate aspect of medical physics that allows the specialist to be directly involved in other people's wellbeing, which creates a strong professional fulfilment.

It is very pleasing to see that women are entering this field of science in great numbers. As shown within the body of this work, the current trend has not only reached gender balance, but crossed over into a larger presence of young female professionals. If the trend continues, medical physics in Romania will be clearly dominated by women for the foreseeable future. We must ask ourselves if this situation will serve the best interest of the scientific community. We believe that a well-balanced community of professionals is best suited towards the advancement of science.

Gender balance in research and workforce has several advantages. It has been shown by Ingalhalikar et al. [6] that the structure of the male brain facilitates connectivity between perception and coordinated action, while the female brain enables communication between the analytical and intuitive processing modes. Also, Stoet et al. [7] has shown that women tend to outperform men at assignments that require multitasking, while men tend to more easily focus on a specific task. The ideal research team has members that can provide a variety of cognitive abilities, attention to details, ability to easily transition between highly focused tasks and last, but not least, collaborate effectively with empathy.

It is a well-known fact in science that the very act of measuring a system introduces changes to said system. While sometimes the changes are instant (the introduction of a measuring device in an electrical circuit), other times the changes happen over time as a 'snowball effect'. The fact that today we are acknowledging and measuring the gender imbalance in our scientific community, will hopefully raise awareness and future actions aimed at achieving gender balance.

5 Final Note

As mentioned above, gender distribution can be influenced by role models that play inspirational roles in someone's life. I would like to finish by exemplifying a recent personal experience that involved a large number of medical physics students asked to actively partake in the organisation of an important event. For the last 5 years I have been organising at the University of Oradea the International Day of Medical Physics (IDMP), alongside IOMP (International Organization for Medical Physics). Since in November 2017 the scientific world celebrated 150 years from the birth of Marie Sklodowska-Curie, the 'mother' of medical physics, the topic chosen for this occasion was "Medical Physics: Providing a Holistic Approach to Women Patients and Women Staff Safety in Radiation Medicine". Whether the topic, or the name of Marie Curie stimulated the creativity and willingness to contribute to this event, fact is that an exceptionally high female interest was shown in activity involvement, to such an extent that 100% of the presenters were women and 100% of the high-school teachers that involved the students in these activities were female. Interestingly enough, none of the four previous events (IDMP in 2013–2016) gathered presenters that represented one gender in unanimity.

Conflict of interest statement The authors declare no conflict of interest.

References

1. UNESCO Institute for Statistics. Women in Science. 2015. http://www.uis.unesco.org/ScienceTechnology/Documents/fs34-2015-women%20in%20science-en.pdf.
2. Kerkhoven, A.H., Russo, P., Land-Zandstra, A.M., et al.: Gender stereotypes in science education resources: a visual content analysis. PLoS One 11(11), e0165037 (2016).
3. Reuben, E., Sapienza, P., Zingales, L.: How stereotypes impair women's careers in science. Proc Natl Acad Sci U S A. 111(2), 4403–4408 (2014).
4. Halpern, D.F., Benbow, C.P., Geary, D.C., et al.: The science of sex differences in science and mathematics. Psychol Sci Public Interest. 8(1), 1–51 (2007).
5. Executive Agency for Higher Education, Research, Development and Innovation Funding, uefiscdi.ro, accessed on 14/01/2018.
6. Ingalhalikar, M., Smith, A., Parker, D., et al.: Sex differences in the structural connectome of the human brain. Proc Natl Acad Sci U S A. 111(2), 823–828 (2014).
7. Stoet, G., O'Connor, D.B., Conner, M., et al.: Are women better than men at multi-tasking? BMC Psychology. 1:18 (2013).

Crude Oil in Drinking Water: Chitosan Intervention

Eileen E. C. Agoha

1 Introduction

Water pollution affects drinking water from rivers, streams, lakes and oceans all over the world. In many developing countries, it is usually a leading cause of death by people drinking from polluted water.

Oil exploration in Nigeria has been characterized by crude oil spills, illegal oil refining and pipeline vandalism with attendant pollution of water sources and massive environmental degradation. An estimated 240,000 barrels of crude oil are spilled in the Niger Delta every year Anonymous [5].

Owaza is one of the villages situated at the banks of the Imo River in Ukwa West Local Government Area, Abia State, Nigeria, and obtains its drinking water from Owaza Imo River. For more than five decades Owaza has become important because of petroleum operations in the area commonly known as Owaza Flow Stations. Thus crude oil spills are not uncommon in this community and the river often overflows its banks during such oil spills. As observed by Amnesty International, hundreds of crude oil spills continue to blight the Niger Delta and some of the major operators have lost control of their operations Anonymous [4].

Like most villages in the developing countries where municipal water supplies are non-existent, the villagers depend on the crude oil polluted river for their daily water needs. Reports have indicated that the principal risk to human health associated with the consumption of polluted water is microbiological in nature, although the dangers of chemical contamination should not be underestimated (WHO 2014; Ziemer 2007). However water treatment chemicals Tayor [15]; Guibal et al. [9] play an important role in maintaining public and environmental quality, and also improve the quality of drinking water.

Chitosan, a linear polysaccharide consisting of β-(1→4)-linked 2-amino-2-deoxy D-glucose residues has been successfully used in the removal of a great variety of water pullutants No and Meyers [14], Guibal et al. [9], Agoha and Mazi [2], Agoha et al. [3].

The inhabitants of Owaza have been drinking from the Owaza Imo River for more than five decades and with no alternative source of drinking water. It is envisaged in this study to investigate the quality of water from Owaza Imo River using chitosan as a treatment chemical information form this study could help medical experts in understanding the cause of some diseases prevalent among the inhabitants of Owaza Community.

The Objectives of this Study Were:

1. To produce chitosan from snail shells waste.
2. To determine the physical, chemical and microbiological qualities of water samples from Owaza Imo River.
3. To determine the physical, chemical and microbiological properties of chitosan treated water samples from Owaza Imo River.

2 Materials and Methods

2.1 Materials

Water samples were collected from Owaza Imo River in Ukwa West Local Government Area, Abia State, Nigeria and stored in closed glass jars. Chitosan was obtained from the shell wastes of the land snail (*Achatina fulica*) as described by Agoha and Mazi [2].

E. E. C. Agoha (✉)
Department of Food Science and Technology, Abia State
University, Umuahia Campus, Umuahia, PMB 7010, Nigeria
e-mail: eecagoha@yahoo.com

© Springer Nature Singapore Pte Ltd. 2019
L. Lhotska et al. (eds.), *World Congress on Medical Physics and Biomedical Engineering 2018*,
IFMBE Proceedings 68/3, https://doi.org/10.1007/978-981-10-9023-3_134

2.2 Water Treatment

Water samples (1 L each) were separately pre-filtered into coagulation beakers and treated with chitosan at different concentrations (0.1, 0.2 and 0.3 mg/L) and allowed to stand for 1 h for formation of flocs and sludge and then filtered. Another 1 L of the water sample was pre-filtered into a coagulation beaker, but without chitosan addition and no further filteration (i.e. untreated water).

2.3 Water Analysis

Physical properties—odour, taste and turbidity of the water samples were determined following the ASTM [6] method. p^H was determined using the Surgifield SM601A membrane p^H meter.

Chemical properties—total hardness, calcium and magnesium were determined following the method of James [12]. Total alkalinity was determined as described by Udoh and Ogunwale [16]. The heavy metals lead, zinc, iron, chromium, copper and nickel were determined by the Atomic Absorption Spectrophotometer (AAS, Perkin Elmer 60011AAS) according to James [12].

2.4 Microbiological Analysis

Microbiological examination was carried out following the ICMSF [11] specifications.

3 Results and Discussion

3.1 Physical/Chemical Properties

The water sample from Owaza Imo River had a brown colour, but with a turbidity of 0.009 units Hazen. It also had a hydrocarbon odour and taste with a p^H value of 5.76 which indicated that the Owaza Imo River was highly acidic suggesting the presence of free carbon dioxide from crude oil pollution. Thus the water did not meet the WHO standard for drinking water (WHO 2011). When the water samples were treated with chitosan at concentrations of 0.2 and 0.3 mg/L, the p^H values were 6.50 respectively and 6.75 respectively which were similar to WHO recommended values of 6.50 to 8.50 for drinking water (WHO).

Alkalinity, total hardness, calcium and magnesium ions in the untreated water samples were 118.4, 29.36, 24.92 and 3.77 mg/L respectively and the above values were within the WHO acceptable standards for drinking water. On treatment with chitosan at the various concentrations, the values were still acceptable (WHO 2004).

Table 1 showed that the untreated water samples from Owaza Imo River had high concentrations of the heavy metals lead 0.84 mg/L, iron 3.27 mg/L, chromium 0.47 mg/L, nickel 0.41 mg/L and these values were above the maximum allowable concentrations for surface potable water (WHO 2011), while zinc 2.38 mg/L and copper 0.92 mg/L met the WHO standard of 5.0 mg/L and 1.00 mg/L respectively. Water sample treated with chitosan concentration of 0.3 mg/L reduced the lead content to 0.39 mg/L representing 56.25% reduction. Lead is toxic and accounts for most of the causes of paediatric heavy metal poisoning which affects the bones, kidney and thyroid glands (Neil [13]; WHO 2011; Harvey et al. [10]). At the same chitosan concentration of 0.3 mg/L reductions in other heavy metals were zinc 52.10%, iron 29.97%, copper 36.96%, chromium 34.04% and nickel 51.22%. The above findings indicated that the water did not meet the set standards in the water criteria (WHO 2011). Copper and zinc are trace elements essential to maintain the body metabolism, but in high doses copper can cause anaemia, liver and kidney damage and stomach and intestinal irritation. Also nickel is needed in small amounts in the human body to produce red blood cells. However, long term exposure like in drinking water can cause decreased body weight, heart and liver damage and skin irritation (Banflavi [7]; Vallero and Letcher [17]; Aggrawal [1]). On the other hand, chrominium can cause cancer, erythyma and exudative eczema and dermatitis.

3.2 Microbiological Properties

The results indicated a high total plate count of 252.67×10^4 cfu/mL in the raw water while chitosan treatment at 0.1, 0.2 and 0.3 mg/L concentrations reduced the total plate counts to 176.0×10^4, 151.67×10^4 and 135.0×10^4 cfu/mL respectively. This showed a reduction in the bacteria load suggesting the antibacterial activity of chitosan against the bacteria in the water (Friedman and Juneja [8]). Similarly, the coliform count was reduced to a most probable number (MPN) of 8.67/100 mL in the untreated water to 2.00/100 mL water after treatment with

Table 1 Heavy metals content of raw and chitosan treated water samples (mg/L)

Heavy metals	WHO std.	Raw water	Water + 0.1 mg chitosan	Water + 0.2 mg chitosan	Water + 0.3 mg chitosan
Pb	0.05	0.84	0.76	0.55	0.39
Zn	5.00	2.38	1.59	1.25	1.14
Fe	0.30	3.27	3.17	2.77	2.29
Cu	1.00	0.92	0.82	0.76	0.58
Cr	0.05	0.47	0.36	0.33	0.31
Ni	0.02	0.41	0.28	0.25	0.20

Values are means of triplicate determinations

0.3 mg/L chitosan concentration representing 76.93% reduction in the coliform count. Bacteria isolates from the water samples included *Micrococcus*, *Pseudomonas*, *Bacillus*, *Proteus*, *Staphylococcus* and *E. coli*, confirming the poor sanitary quality of the water.

4 Conclusion

The study has demonstrated that Owaza Imo River was highly contaminated with crude oil and heavy metals, and hygienically unfit for human consumption. There is an urgent need to create awareness about the health implications of drinking water polluted with crude oil particularly in the Niger Delta area of Nigeria, and the need for the provision of safe water for the inhabitants. Lastly, chitosan has both chelating and bactericidal properties and should be further investigated as a cheap and alternative water treatment chemical for the production of safe drinking water.

Conflict of Interest The authors declare that they have no conflict of interest.

References

1. Aggrawal, A. 2014. Textbook of Forensic Medicine and Toxicology. New Dehli; Avichal Publishing Company. ISBN978-81-7739-419-1.
2. Agoha, E.E.C., and Mazi, E.A. 2009. Biopolymers from African Giants Snail Shells Waste: Isolation and characterization. IFMBE Proceedings 25/x, pp. 249–251.
3. Agoha, E.E.C, Atowa, C., Okafor, F., and Ozodigwe, C.A. 2015. Chitinous Biopolymers for the treatment of Crude Oil Polluted Water.
4. Anonymous, 2014. Amnesty International's Global Issues.
5. Anonymous, 2017. The Guardian Newspaper. 6 Nov. 2017.
6. ASTM, 2006. American Society for Testing Materials. Protocol of accepting drinking water testing methods. International Library Service, Provo, Utah.
7. Banflavi, G. 2011. Heavy Metals, Trace Elements and their Cellular Effects. In Banflavi, G. Cellular Effects of Heavy Metals. Springer. pp. 3–28 ISBN 9784007.
8. Friedman, M., Juneja, V.K. 2010. Review of Antimicrobial and Antioxidative Activities of Chitosans in Food. J Food Project. 73:1737–1761.
9. Guibal, E., Van Vooren, M., Dempsey, B.A., Roussey, J. 2006. A review of the use of Chitosan for the Removal of Particulate and Dissolved Contaminants. Sep. Science Technology. 41(II):2487–2514.
10. Harvey, P.J., Handley, H.K., Taylor, M.P. 2015. Identification of the sources of metal(lead) contamination in drinking waters in north-eastern Tasmania using lead isotopic components. Environmental Science and Pollution Research. 22:12276–12288.
11. ICMSF. 1978. International Commission on Microbiological Specification for Food in Microbial Ecology of Food Commodities. Academic Press, New York, USA. pp. 522–532.
12. James, C.S., 1995. Experimental Procedures: In Analytical Chemistry of Food. Chapman and Hall. New York.
13. Neil, N., 2003. Food Science. The biochemistry for food and nutrition. 3rd Ed. pp. 71–75.
14. No, H.K., and Meyers, S.P. 2000. Application of Chitosan for treatment of waste waters. Review on Environmental Contamination Toxicology. 163:1–28.
15. Taylor, U.R. 2000. Chemical analysis manual for food and water. 5th Ed. 1:20–26.
16. Udoh, A. and Ogunwale, C., 1986. Standard Test Methods for water. pp. 20–22.
17. Vallero, D.A. and Letcher, T.M. 2013. Unraveling Environmental disaster. Elsevier. ISBN 9780123970268.

Women in Biomedical Engineering and Medical Physics in the Czech Republic

Lenka Lhotska

Abstract

In last decades the effort to increase the number of women applying for studies at technical universities, and consequently for a job in the field of technology, can be seen in European countries. We analyzed the situation in the Czech Republic and found differences in ratio of females in individual engineering disciplines. We can see a positive trend in biomedical engineering and medical physics where the percentage of female students almost reached 50% in bachelor and master study. However, in PhD study and in research in general the number of females is lower. Successively, it is also lower in jobs in the health care sector. Although the jobs are on the border between engineering and life sciences, they are considered more as engineering jobs. And employment of women in any field of engineering is still considered nontraditional. We discuss the position of females as employees and present their potential advantage in biomedical engineering and medical physics jobs. As a potential support to professional development we propose mentoring and show its positive impact on personal and professional growth.

Keywords

Biomedical engineering • Medical physics
Gender • Employment • Mentoring

L. Lhotska (✉)
Czech Institute of Informatics, Robotics and Cybernetics,
Czech Technical University in Prague, Prague, Czech Republic
e-mail: lenka.lhotska@cvut.cz; lhotska@cvut.cz

L. Lhotska
Faculty of Biomedical Engineering, Czech Technical University
in Prague, Prague, Czech Republic

1 Introduction

In last decades the effort to increase the number of women applying for studies at technical universities, and consequently for a job in the field of technology, can be seen in European countries. We analyzed the situation in the Czech Republic and identified differences in ratio of females in individual engineering disciplines. There is a positive trend in biomedical engineering and medical physics where the percentage of female students almost reached 50% in bachelor and master study. However, in PhD study and in research in general the number of females is lower. Successively, it is also lower in jobs in the health care sector. They are mostly employed as ordinary employees since only the largest hospitals have separate biomedical engineering departments. Moreover, we still have the so-called vertical segregation in employment. When it comes to comparison of positions in the companies or institutions, males are more frequently present on hierarchically higher positions. Less than 10% of the top positions and about 20% of high managerial positions are occupied by females. Recently we have analyzed and discussed the perception of women in engineering in general. We have also asked our students for their opinion. Regarding the prejudices about women in engineering, students are aware of their existence, but they personally do not make any differences between females and males. That is a very positive conclusion bringing hope that the situation in engineering fields will change, finally resulting in gender balance.

The fact that there is a low percentage of women in the field of technology, engineering and science in the Czech Republic can be explained based on the results of two studies conducted by Public Opinion Research Centre and Ministry of Labour and Social Affairs of the Czech Republic in 2006 and 2007. According to them, Czech society is conservative considering the division of the male and female roles in the society. In addition, employment of a woman in any field of engineering is considered nontraditional [1].

© Springer Nature Singapore Pte Ltd. 2019
L. Lhotska et al. (eds.), *World Congress on Medical Physics and Biomedical Engineering 2018*,
IFMBE Proceedings 68/3, https://doi.org/10.1007/978-981-10-9023-3_135

This clearly points out to social stereotypes in this country, considering only the close connection of men and technology to be natural.

Nowadays, various organizations both on global/European and local level present role models—women who overcome gender stereotypes and become successful in the area of engineering. These organizations, as for example IEEE Women in Engineering [2], Women's Engineering Society [3] or Zkus IT (= "Try IT") [4] and Zeny a veda (= "Women and Science") [5] in the Czech Republic are dedicated to promoting and supporting women in engineering, as well as inspiring young women to achieve their potential as engineers, scientists and technical experts. Also, numerous programs within the Sixth and Seventh Framework Programmes of the European Union were dedicated to gender issues, e.g. PROMETEA [6] or DIVERSITY [7]. In general all of them have a common goal—reaching gender balance in engineering.

The aim of the paper is to present briefly current situation and gender disbalance at universities in the Czech Republic in different study fields. Then we discuss female employment in engineering jobs and their professional career. The ratio of men and women in higher positions is commented in context of horizontal and vertical segragation. Finally mentoring as a means of guidance in personal and professional growth is mentioned.

2 Education

Technical education has been traditionally understood as a male domain. Engineering requires abilities that have been habitually associated with men. Similarly, electrical engineer or computer expert almost automatically imply male gender while female representatives of these areas could be perceived as an oddity and sometimes approached with distrust though we are living in the 21st century. That suggests that various gender patterns or even prejudices resonate in society as well as technical schools themselves and thus influence women's decision to enter engineering programs [8].

In the Czech Republic, most of the undergraduate study programs are structured into Bachelor and Master levels. There are only few exceptions, as for example medicine. It can be observed that at Bachelor and Master programs there is a greater percentage of women. At the PhD level, percentage of women among all students is increasing, but it is still lower than the percentage of men.

Different situation is observed at engineering faculties at all educational levels. Technical disciplines are studied more by men: overall percentage of women is about 11% and slowly decreasing, while this percentage in male population is around 40%. However, if we analyze the numbers in interdisciplinary areas, such as biomedical engineering, biomedical informatics, eHealth and telemedicine, we see different ratio. There are usually up to 30% of female students. The discussions with students confirm our assumptions, why they decided to apply for these particular interdisciplinary programs. The main motivation to enter interdisciplinary study programs is their variability and interaction in different settings than just "be confined in a laboratory setting, design room or a manufacturing floor".

Generally, more women are among university graduates at the first two levels, but at the PhD level men make approximately two-thirds of the total number. Women do not finish their doctoral studies during/after their maternity leave relatively frequently and if they get the degree they very often leave the institution or remain at lower academic positions. This obviously supports the idea that it is difficult to combine professional career with family care in the male-dominated area.

Several Czech technical universities have joined the campaigns promoting technical education among girls and women. The largest institutions—Czech Technical University in Prague and Brno University of Technology—have both supported the web portal www.zkusit.cz trying to attract women to ICT sector. The CTU also runs a much more vibrant web portal www.holkypozor.cz ("attention girls") that informs about interesting events happening at the CTU while often including perception of the current female students. The initiative also tries to build on the reputation of excellent female scientists and highly successful and recognized alumnae.

3 Employment

There were many discussions about status of women in technology and engineering and also in research in last decades. Ratio of male and female university students corresponds to gender ratio in population in most European countries. However we can see many differences across the disciplines and in jobs—the phenomena are called horizontal and vertical segregation.

Recently the effort to increase the number of women applying for studies at technical universities, and consequently for a job in the field of technology, can be seen in European countries. There exist already many positive examples in several countries. However the situation in the Czech Republic is different. Czech society is still conservative considering the division of the male and female roles in the society. Therefore, employment of a woman in any field of engineering is not yet perceived as natural. We cannot be surprised if we find out that most of the higher positions in engineering jobs at universities, research institutes, hospitals, companies, and governmental sector are occupied by men.

3.1 Horizontal and Vertical Segregation

Horizontal and vertical segregations are terms expressing differences in representation of females and males in disciplines and in hierarchy. Horizontal segregation means different representation of men and women in individual disciplines and sectors. Women are more frequently active in the so-called soft disciplines (humanities and social sciences) and employed mostly in governmental (37%) and non-profit (38%) sectors. In entrepreneurial sector there are only 15% of females in R&D. There is great contrast between soft disciplines (43% of female researchers) and engineering (only 12% of female researchers).

Vertical segregation expresses concentration of men and women on different levels of hierarchy. Women are more frequently represented on lower positions in the hierarchy, while men have majority on decision making positions. Almost the same situation is across disciplines and sectors. We explored it in more detail in academic sector and we can illustrate it by numbers from the universities in the Czech Republic: among 25 rectors of public universities there are only 2 females. On the position of a director of a research institute of the Academy of Sciences there are 12 females out of total number of 62 directors. Similar situation exists in other sectors.

In general, the number of university educated women in the Czech Republic represents 56% of all graduates, in PhD study the ratio is lower—about 43%. This is still a high number that changes rapidly later. When we observe the numbers in research and development area the females constitute only 26%. They are concentrated in specific scientific areas and in lower positions in the hierarchy. Definitely one of the reasons is care for children and family. If a woman does not find support in her family, it is usually difficult to continue the career in particular in areas where, for example, frequent and whole-day presence in laboratories is required. These facts show that without systematic work and support from the side of the institutions it is almost impossible to reach more satisfactory results.

3.2 Change in Scientific Culture and Institutions

Since 2009 EU has been supported approaches and projects aimed at cultural and institutional (structural) changes at academic and research institutions. Acceptance of responsibility at the institution is a key assumption for the activities that should lead to systemic changes and their sustainability. Support of gender equality is not a partial or marginal goal. Strengthening gender equality is an important value and inseparable part of the strategic development of an institution. The main aim is change of culture of academic environment that should become more open, free of bias and traditional stereotypes. Such environment may offer equal opportunities for all gifted and qualified researchers of both sexes to participate actively in high quality research. One of the results of these activities is introduction and support of the mentoring programs (see Sect. 4).

3.3 The Female Advantage

Women in biomedical engineering and medical physics jobs can advantageously use their mental abilities, such as empathy, communication, etc. Women are seen as less intimidating when reviewing complex technology with nurses. Women who pick this career are recognized and valued in some circumstances and create a tremendous positive presence. There are some distinct advantages to sending a woman technician with strong communication skills to clinical environment. They are seen as less invasive when they have to communicate and translate with clinical staff that is not technologically trained. Additionally, as service providers, female clinical engineers have an advantage when working in certain clinical areas. For example, it is easier to send a female into a sensitive department, such as mammography. Obstetrics is another area in which women might be better suited than men. Women possess other unique characteristics that can significantly benefit the industry: the ability to listen, to understand someone else's point of view, teamwork, and the ability to communicate. Women have the innate ability to multitask and adapt to their environment better than men.

4 Mentoring

Mentoring is a process of continuous and dynamic feedback between two individuals to establish a relationship through which one person shares knowledge, skills, information, and perspective to foster the personal and professional growth of the other. It is a different relationship than supervision which is usually pre-established and does not necessarily lead to the personal growth of the individual. Mentoring has already proved to be an efficient instrument that helps people to progress in their career. In recent years it has become increasingly popular. We present here basic characteristics of mentoring and few examples of existing programs.

4.1 Mentoring and Models of Mentoring Programs

Mentoring is a partnership between mentor and mentee normally working in a similar field or sharing experiences. It is a helpful relationship based upon mutual trust and respect.

The main aim of a mentoring program is support of professional development of postdocs and PhD students and help to start their scientific career. This complex aim consists of several partial goals: help young researchers to identify their career goals and steps to reach them; ease orientation in scientific career system; mediate them important contacts with the peers and more experienced researchers and extend their professional network; strengthen mutual exchange of experience among program participants; support self-confidence of young researchers.

Mentoring is quite frequently used within an institution to help new members of staff. It is interesting that active and successful mentoring programs can be found in life sciences but not so frequently in engineering or strongly interdisciplinary areas, in particular on the edge of engineering and medicine or biology. Mentoring programs can be based on different models of interaction. Some forms could be better adjusted for newcomers, some for more experienced employees. A group peer, collaborative mentoring model founded on principles of adult education is a good example of the latter mentoring form.

With fast development of interdisciplinary R&D, interdisciplinary mentoring has become more important and prevalent over the recent years. Interdisciplinary mentorship is the tool for scientists to help produce synergy in group, and to generate multifocal ideas and complex solutions to complex challenges. We should mention that it is frequently more useful and enriching if the mentor and mentees are coming from different disciplines and even from different institutions because they can view all issues from slightly different points, bring new opinion and perspectives, and are not bound by processes and relations in the institution of the mentees.

The areas covered by mentoring are extensive and diverse: networking (professional, educational, supporting); career support; role model; communication skills; research progress; supervision; scientific writing; presentation skills; combination of professional and private life. However, we have to note that the coverage of these areas in a single mentor—mentee relation need not be exhaustive. The content must be individualized based on the situation and previous experience of the particular mentee.

4.2 Relation Between Mentor and Mentee

A mentor is a guide who can help the mentee to find the right direction and who can help him/her to develop solutions to career issues. The mentor relies upon having had similar experiences to gain an empathy with the mentee and an understanding of his/her issues. Mentors are usually experienced researchers or university teachers from the same or similar scientific discipline. However, it is recommended

that they come from a different institution than the mentee. As mentioned above, in interdisciplinary areas it is welcome when mentor and mentee come from different disciplines. Mentoring provides the mentee with an opportunity to think about career options and progress. A mentor should help the mentee to believe in himself/herself and boost his/her confidence. A mentor should ask questions and formulate challenges, while providing guidance and encouragement. Mentoring allows the mentee to explore new ideas in confidence. It is a chance to look more closely at oneself, one's own issues, opportunities and what he/she wants in life. Mentoring is about becoming more self-aware, taking responsibility for one's own life and directing the life in the direction he/she decides, rather than leaving it to chance.

We realize that women, particularly working mothers, often need to balance their working lives with responsibilities at home. We recognize that issues outside the workplace may be hampering progress at work. Mentors and mentees should match using their own criteria-career considerations or aspects of their personal circumstances. Most of the mentoring programs provide guidelines and training for both mentors and mentees, but the issues discussed vary depending upon the issues being faced by the mentee.

4.3 Examples of Mentoring Programs

Recently several professional engineering societies, e.g. [9–11], and many American universities have started to organize mentoring programs. They have different forms with relations to type of mentees addressed. Many American universities organize mentoring programs for their students. They offer special programs for women in engineering and various programs for students of different years of study. Mentors and mentees is a peer mentoring program where freshmen are matched with juniors, and sophomores with seniors, in the same major. Grad mentoring program is a program providing peer mentoring for incoming graduate students. Professional mentoring matches undergraduates with professional female engineers from industry.

Recently, a mentoring program has been initiated in the Czech Republic [12]. It is organized by the National Contact Centre Gender & Science. Till now more than 50 mentees have participated, both females and males coming from research—postdocs and PhD students. The mentors are experienced researchers at higher than postdoctoral level working at public universities or research institutions, interested in sharing their experience with younger colleagues. Based on the evaluation of feedback from both mentors and mentees it has been decided to continue because all participants were very satisfied and it helped the mentees in their career development, they learned how to deal with new experience, tackle problems, and manage time better.

Both parties agreed that their relationship was based on mutual trust, respect, mentor's experience and mentee's commitment. It was important that the mentor demonstrated proper professional behavior, shared what he/she knows, and developed a shared connection with the mentee.

5   Conclusion

Although technologies are perceived as prospective students still do not prefer technical schools and universities. There are still stereotypes that women do not understand engineering and technologies. There are too few positive examples. It is not exceptional that women after maternity leave stay at the same position and do not proceed in their career. However, the experts claim that technical disciplines are attractive for females. Women employed in these disciplines earn more than in other fields, they are less endangered by unemployment and their work is usually more flexible, which is an advantage regarding care for children and family.

In the Czech Republic, there has been recently started an initiative that focuses on issues connected with professional development of females, in particular in research and engineering disciplines. The initiative has been transformed later to a project. The implementation of the project will be performed through realization of the so-called plan of gender equality. Its goals and activities correspond to three strategic areas. They are focused on professional placement and career development of females and young researchers of both sexes. An example in this area is a mentoring program for PhD students and postdocs, educational activities and training, career consultancy, gender sensitive setup of evaluation processes and career advancement or working conditions allowing coordination of work and family life fixed in official documents of the institution. Second area is increasing representation of women in decision-making and leading positions. The aim is setup of transparent and open rules for career advancement, elimination of gender bias and strengthening motivation of women themselves (for example training increasing competences of females aspiring to leader positions or to membership in decision-making boards or advancement in academic hierarchy). Third area is inclusion of gender perspective into knowledge development. That means inclusion of gender dimension into all phases of research cycle, starting from research intention and hypotheses over composition of the research team, selection of methodology, data analysis up to result interpretation and publication. Activities in this area are focused on education and increase of competences how systematically consider gender perspective and work with gender category as analytical variable in different scientific disciplines. And last but least, the positive examples of successful individuals should be more presented publicly.

Acknowledgements The research has been supported by the CVUT institutional resources.

Conflict of Interest The author declares that she has no conflict of interest.

References

1. Women in IT. Online: http://www.zkusit.cz/zeny-v-it.pdf (in Czech). Last accessed 2018/2/4.
2. IEEE Women in Engineering. Online: http://wie.ieee.org/. Last accessed 2018/2/4.
3. Women's Engineering Society. Online: http://www.wes.org.uk. Last accessed 2018/2/4.
4. Women in IT. Online: http://www.zkusit.cz (in Czech). Last accessed 2018/2/4.
5. Gender and Science. Online: http://www.zenyaveda.cz. Last accessed 2018/2/4.
6. PROMETEA Project. Online: http://www.prometea.info. Last accessed 2018/2/4.
7. Stefankova, J., Caganova, D., Moravcik, O.: "Promoting gender diversity research in Slovakia within 7th Framework Programme," Electronics and Electrical Engineering, No. 6 (102), 59–62 (2010).
8. Lhotská, L., V. Radisavljevič Djordjevič, V.: Gender Ratio in Engineering Disciplines: Why Are There Differences. In *Global Telemedicine and eHealth Updates: Knowledge Resources*. Grimbergen: International Society for Telemedicine and eHealth (ISfTeH), vol. 6, 268–272 (2013).
9. IEEE EMBS: http://embs.chronus.com/about. Last accessed 2018/2/4.
10. IET: http://www.theiet.org/membership/career/mentoring/. Last accessed 2018/2/4.
11. Am. Society of Mechanical Engineers: http://www.asme.org/career-education/mentoring. Last accessed 2018/2/4.
12. Gender and Science: http://genderaveda.cz/en/mentoring-en/. Last accessed 2018/2/4.

Part XVI

Emerging Technology in Diabetes Care Advances Real-Time Diabetes Monitoring Systems

A Combined-Predictor Approach to Glycaemia Prediction for Type 1 Diabetes

Kyriaki Saiti, Martin Macaš, Kateřina Štechová, Pavlína Pit'hová, and Lenka Lhotská

Abstract

Glycaemia prediction plays a vital role in preventing complications related to diabetes mellitus type 1, supporting physicians in their clinical decisions and motivating diabetics to improve their everyday life. Several algorithms, such as mathematical models or neural networks, have been proposed for blood glucose prediction. An approach of combining several glycaemia prediction models is proposed. The main idea of this framework is that the outcome of each prediction model becomes a new feature for a simple regressive model. This approach can be applied to combine any blood glycaemia prediction algorithms. As an example, the proposed method was used to combine an Autoregressive model with exogenous inputs, a Support Vector Regression model and an Extreme Learning Machine for regression model. The multiple-predictor was compared to these three prediction algorithms on the continuous glucose monitoring system and insulin pump readings of one type 1 diabetic patient for one month. The algorithms were evaluated in terms of root-mean-square error and Clarke error-grid analysis for 30, 45 and 60 min prediction horizons.

Keywords

Diabetes • Glycaemia prediction • Combined-predictor
Insulin pump • Diabetes care

K. Saiti (✉)
Department of Cybernetics, Czech Technical University in Prague, Prague, Czech Republic
e-mail: saitikyr@fel.cvut.cz

M. Macaš · L. Lhotská
Czech Institute of Informatics, Robotics and Cybernetics, Czech Technical University in Prague, Prague, Czech Republic

K. Štechová · P. Pit'hová
FN Motol University Hospital, Prague, Czech Republic

1 Introduction

Diabetes is defined as a group of metabolic diseases in which a patient has high blood-sugar, either due to pancreas inability to produce enough insulin, or because cells do not respond to insulin as expected. Two of the most important (short term) complications of diabetes are hypoglycaemia and hyperglycaemia, both of which are life-threatening [2]. Supporting patients to manage diabetes has gained global interest with recent efforts aiming at controlling the short-term complications [6, 9]. Type 1 diabetes and type 2 are chronic diseases and their conventional therapies are mainly dependent on diet management, physical exercise, exogenous insulin infusion and drug administration.

There are several developed diabetes management systems aiming to assist diabetics in self-managing disease, such as CareLink personal software (Medtronic, Northridge, CA) [8]. One of the most crucial components of any diabetes management system is the blood glucose level prediction method. Accurate short and long term predictions could enable diabetics to take control actions, such as adjusting the insulin dosage, taking extra food or choosing meals low in carbohydrates. Several specific data-driven methods have been proposed for blood glucose level prediction such as regression prediction, artificial neural networks and support vector machines. Finan et al. [1] implemented a family of linear models included Autoregressive with exogenous inputs model, autoregressive moving average with exogenous inputs model and a Box-Jenkins Model. Georga et al. [5] implemented a Support Vector Machine model in order to provide individualized glucose predictions. Zarkogianni et al. [11] presented a Feedforward neural network, a Self-Organized Map and a Neuro-Fuzzy network.

Results from each prediction algorithm could vary between patients or even for the same patient. Dassau et al. [4] and Buckingham et al. [3] proposed a hypoglycemia prediction algorithm combination framework. Wang et al. [10] proposed an adaptive-weighted-average framework for

© Springer Nature Singapore Pte Ltd. 2019
L. Lhotska et al. (eds.), *World Congress on Medical Physics and Biomedical Engineering 2018*,
IFMBE Proceedings 68/3, https://doi.org/10.1007/978-981-10-9023-3_136

combining different models by giving a weight to each prediction output using the sum of squared prediction error.

In this study, a stacking regression framework was suggested to use any blood glucose level prediction method and calculate the final prediction outcome by using the output of each individual model as predictor for a simple regressive model. Moreover, we examined the abilities of this framework by combining an Autoregressive, a Support Vector Regression and an Extreme Learning Machine for regression model.

2 Materials and Methods

Patients provided continuous glucose monitoring by using insulin pumps which were wirelessly connected to small subcutaneous glucose sensors. It represents 288 glucose concentration measurements per day. Patients used this system at least 30 days (one sensor can be used for 6 days). Insulin dose measurements were obtained by downloading insulin pump memory as well as documentation of all meals by using smart phone camera and recording meals and activity in a detailed logbook including food list and activities. Currently 16 patients are involved in this study. Using data from one female type I diabetic patient, autoregressive with exogenous inputs (ARX), Support Vector Regression (SVR), Extreme Learning Machine for regression (ELM) models were created in order to implement the proposed framework, predict blood glucose levels and compare the results.

It is generally accepted that various patients have different blood glucose characteristics, have different reaction even after taking the same insulin dosages and of course different metabolic rate. Nevertheless, even with data from the same patient each prediction algorithm could work differently; therefore, a stacking regression framework is proposed to combine different models by applying a standard regression on individual predictions. As an example, the above mentioned framework was used to combine ARX, SVR and ELM as these models have been used extensively in blood glucose level prediction. To make this study self-contained, some brief model descriptions are given in the following sections. All the derivations were done in MATLAB 2016b (The MathWorks, Inc., 1 Apple Hill Drive, Natick MA, USA).

ARX model prediction For implementing the ARX model we used the information about long-acting (basal) and fast-acting insulin (bolus) dosages as well as the blood glucose levels from a given training set. The ARX model can be stated algebraically as [7]:

$$\hat{y}_{ph}^{ARX} = \mathbf{X} \times \theta \qquad (1)$$

where matrix $\mathbf{X}$ contains the predictors, parameters θ are estimated using the ordinary least squares method and ph is the prediction horizon.

SVR model prediction For developing SVR model blood glucose levels were used. SVR estimates a function by minimizing an upper bound of the generalization error. Given a training set $(x(k), y(k+ph))$ where x are the predictors and $y(k+ph)$ are the measured values at a specific prediction horizon, the basic idea of SVR is to map the data x into a higher-dimensional feature space via a nonlinear mapping and perform a linear regression in this feature space. The regression function can be defined as [10]:

$$\hat{y}_{ph}^{SVR} = \sum_{i=1}^{n} (a_i - a_i^*) K(\mathbf{x}_i, \mathbf{x}) + b \qquad (2)$$

ELM model prediction Using only the information about blood glucose levels w implemented ELM algorithm with ten hidden layers, we chose the hyperbolic sigmoid function as the activation function and the predicted output can be calculated be the following equation [10]:

$$\hat{y}_{ph}^{ELM} = \mathbf{H}^T \times \beta \qquad (3)$$

where $\mathbf{H}$ is the hidden layer output matrix and β is the output weight vector.

Stacking regression framework We defined $\mathbf{R}$ as the matrix which includes outputs of individual models in order to calculate the coefficients Φ according to the following formula:

$$\Phi = (\mathbf{R}^T \mathbf{R})^{-1} \mathbf{R}^T \, y(ph) \qquad (4)$$

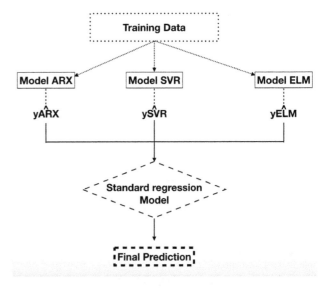

Fig. 1 Stacking regression framework

Table 1 Results for ph = 6 steps ahead (30 min), ph = 9 steps ahead (45 min) and ph = 12 steps ahead (60 min)

Training test							Testing set					
	RMSE			$CEG_{(zoneA)}$			RMSE			$CEG_{(zoneA)}$		
	30	45	60	30	45	60	30	45	60	30	45	60
ARX	23.24	29.54	35.13	87.32	80.25	74.23	21.62	27.02	29.72	89.00	83.55	78.78
SVR	22.34	28.64	34.23	88.32	81.86	76.23	22.58	26.48	31.17	90.46	84.16	78.05
ELM	23.96	32.07	35.49	86.32	77.61	74.32	21.08	27.20	30.63	89.14	81.67	78.34
Gray Combined	20.18	26.48	33.33	88.88	82.36	77.45	20.36	25.58	26.48	91.20	84.99	79.32

where $\mathbf{R} = [\hat{y}_{ph}^{ARX} \hat{y}_{ph}^{SVR} \hat{y}_{ph}^{ELM}]$ and $y(ph)$ is a vector contains the measured blood glucose values at each prediction horizon. The final prediction is given by the following equation:

$$\hat{y}(ph) = \mathbf{R} \times \Phi \qquad (5)$$

Figure 1 shows the basic principals of the stacking regression model implementing for combining results from three models; ARX, SVR and ELM. Moreover, this framework can be applied with any other model combination.

3 Experimental Results and Discussion

To quantify the prediction performance we measured the Root-Mean-Square-Error (RMSE) for each model and we used the Clarke Error grid analysis (CEG). In brief, data that fall into Zone A are considered clinically accurate so a larger percentage in zone A means better prediction performance [1].

A comparison of the prediction performance between the proposed framework and the three individual algorithms shows that stacking regression could give more reliable outputs. Examining the results, we conclude that a model can give better results to a particular prediction horizon than another one on training data as well as the result for the same case could be different on testing data.

In detail, according to Table 1 for ph = 60 min, SVR model performed better than the other two individual algorithms ($RMSE_{ARX} = 35.13$, $RMSE_{SVR} = 34.23$, $RMSE_{ELM} = 35.49 \, [mg/dl]$) on the training set while on testing set the ARX model had the best performance ($RMSE_{ARX} = 29.72$, $RMSE_{SVR} = 31.17$, $RMSE_{ELM} = 30.63 \, [mg/dl]$). The proposed framework gave the most satisfactory results at each prediction horizon for both training and testing set.

4 Conclusion

To sum up, the role of an accurate prediction model in self-diabetes management devices has proved to be crucial and could be able to improve patients' every day life. It can been seen that combining different algorithms could give more sufficient predictions and could overcome the

weaknesses of each model although the results for long-term predictions ($ph \geq 60$ min) are not yet reliable enough. Future work would be focused on improving results for long-term predictions by implementing and combining more mathematical models and neural networks.

Acknowledgements Research has been supported by the AZV MZ CR project No. 15-25710A "Individual dynamics of glycaemia excursions identification in diabetic patients to improve self managing procedures influencing insulin dosage" and by CVUT institutional resources (SGS grant application No. SGS17/216/OHK4/3T/37).

Conflict of Interest The authors declare that they have no conflict of interest.
Statement of Informed Consent The study protocol and patient informed consent have been approved by the University Hospital Motol ethical committee.
Protection of Human Subjects and Animals in Research The procedures followed were in compliance with the ethical standards of the responsible committee on human experimentation (institutional and national) and with the World Medical Association Declaration of Helsinki on Ethical Principles for Medical Research Involving Human Subjects.

References

1. A.Finan, D., C.Palerm, C., III, F.J.D., Seborg, D.E.: Effect of input excitation on the quality of empirical dynamic models for type 1 diabetes. AIChE (2009)
2. Alberti, K.G.M.M., Zimmet, P.f.: Definition, diagnosis and classification of diabetes mellitus and its complications. part 1: diagnosis and classification of diabetes mellitus. provisional report of a who consultation. Diabetic medicine 15(7), 539–553 (1998)
3. Buckingham, B., Chase, H.P., Dassau, E., Cobry, E., Clinton, P., Gage, V., Caswell, K., Wilkinson, J., Cameron, F., Lee, H., et al.: Prevention of nocturnal hypoglycemia using predictive alarm algorithms and insulin pump suspension. Diabetes care 33(5), 1013–1017 (2010)
4. Dassau, E., Cameron, F., Lee, H., Bequette, B.W., Zisser, H., Jovanovič, L., Chase, H.P., Wilson, D.M., Buckingham, B.A., Doyle, F.J.: Real-time hypoglycemia prediction suite using continuous glucose monitoring: a safety net for the artificial pancreas. Diabetes care 33(6), 1249–1254 (2010)
5. Georga, E., Protopappas, V., Ardigò, D., Marina, M., Zavaroni, I., Polyzos, D., Fotiadis, D.I.: Multivariate prediction of subcutaneous glucose concentration in type 1 diabetes patients based on support vector regression. IEEE Journal of biomedical and health informatics 17(1) (2013)

6. Knobbe, E.J., Buckingham, B.: The extended kalman filter for continuous glucose monitoring. Diabetes technology & therapeutics 7(1), 15–27 (2005)

7. Macaš., M., Lhotská, L., Štechová, K., Pit'hová, P., Saiti, K.: Particle swarm optimization based adaptable predictor of glycemia values. In Cybernetics (CYBCONF) pp. 1–6 (2017)

8. Medtronic: Carelink personal software, http://www.professional.medtronicdiabetes.com/para/carelink-personal-software-init

9. Palerm, C.C., Bequette, B.W.: Hypoglycemia detection and prediction using continuous glucose monitoring—a study on hypoglycemic clamp data. J Diabetes Sci Technol 5(5), 624–629 (2007)

10. Wang, Y., Wu, X., Mo, X.: A novel adaptive-weighted-average framework for blood glucose prediction. Diabetes technology & therapeutics 15(10), 792–801 (2013)

11. Zarkogianni, K., Mitsis, K., Litsa, E., Arredondo, M., Fico, G., A. Fioravanti, Nikita, K.: Comparative assesment of glucose prediction models for patients with type 1 diabetes mellitus applying sensors for glucose and plysical activity. International Federation for Medical and Biological Engineering 2015 (2015)

Improving Prediction of Glycaemia Course After Different Meals—New Individualized Approach

Lenka Lhotska⬤, Katerina Stechova, and Jan Hlúbik

Abstract

Motivation and objectives: Diabetes is one of the biggest medical problems nowadays, having different forms and different mechanism of development but the ultimate result is the same—hyperglycaemia. Hyperglycaemia leads to development of chronic diabetic complications, which are the most frequent cause of worsening patient's life quality and often shortening life expectancy. All diabetes mellitus (DM) type 1 patients and some of DM type 2 patients require full insulin substitution. It is not simple to adjust insulin dose to different meals and different daily activities. To help patients with this challenge we started to develop an application for smart phones having new features in comparison with existing applications. We concentrated on individual response to different types of meals (division based on glycaemic index), to physical activity and individually different basal metabolic rate. *Material and methods*: So far 24 patients, mostly insulin pump users, were enrolled. Patients used during the study at least 4 weeks RT-CGM and during this period they were asked to document all food and drinks containing carbohydrates by smart phone camera. Patients wrote during this time a detailed logbook as well. The detailed nutritional analysis of patient's food was done as well as evaluation of other condition (level of depression, measurement of basal metabolic rate). *Results*: The quality of photos was problematic but the biggest problem was to analyze mixed meal from the photography. It was not possible without at least short patient's description. Patient's diet was unhealthy (high fat content etc.) and patients despite remedial nutritional reeducation made mistakes in carbohydrates counting which was reflected in their glycaemia profiles. *Conclusion*: It seems that using photos with brief notes is an acceptable solution and adding a personalized database of favourite meals with correct nutritional data (which we are developing now) may be very helpful. Then the patient only confirms selected meal and does not need to insert all data again. Based on data from the insulin pump and the glucose sensor and inserted information about the planned meal from the patient, the application can recommend the prandial bolus to be injected before meal.

Keywords

Diabetes • Glycaemia course • Individualization Metabolism

1 Introduction

Diabetes is defined as a metabolic disease which is characterized by the disruption of glucose homeostasis leading to hyperglycaemia [1].

Glucose homeostasis. Glucose is the main energetic source for a human body. Glucose is acquired from food or can be produced (gluconeogenesis) by human liver and kidneys from alternative sources (lipids and proteins). Blood glucose concentration is maintained in very narrow interval (fasting glycemia <5.6 mmol/L, glycaemia after meal or 2 h after glucose administration in glucose tolerance test is <7.8 mmol/L). The main hormone acting in glucose homeostasis is insulin which is produced by pancreatic beta cells. Production of insulin is basal (approximately 50% of total daily production) and stimulated when glycaemia is increasing after a meal. This stimulated (prandial) response

L. Lhotska (✉) · J. Hlúbik
Faculty of Biomedical Engineering, Czech Institute of Informatics, Robotics and Cybernetics, Czech Technical University in Prague, Prague, Czech Republic
e-mail: lenka.lhotska@cvut.cz; lhotska@cvut.cz

J. Hlúbik
e-mail: hlubikjan@post.cz

K. Stechova
University Hospital Motol, Prague, Czech Republic
e-mail: katerina.stechova@lfmotol.cuni.cz

is quick and very effective so under the usual circumstances maximal glycaemia increment after meal is 2 mmol/L [2].

Types of diabetes. There are different types of diabetes which differ in the pathogenesis of glucose homeostasis disruption. The most frequent is type 2 diabetes mellitus (DM2). The crucial moment in DM2 pathogenesis is the increase of insulin resistance. In current DM2 treatment schema insulin is considered as a part of second level treatment options when glycaemia is not well controlled by other drugs.

The second most common is type 1 diabetes mellitus (DM1) which pathogenesis is totally different from DM2 but the result is the same—hyperglycaemia. Leading cause of this disease is a reaction of patient's immune system towards beta cells so DM1 belongs to autoimmune diseases. DM1 is more frequent in young individuals but can developed also in adults. Prior insulin discovery DM1 was the fatal disease and treatment by insulin injections is an only way how to save patient's life [1, 3].

There are other types of diabetes, namely gestational diabetes and "other specific forms of diabetes" which are not relevant to our topic so they are mentioned here only for complexity and will not be discussed further.

Hyperglycaemia consequences. Lasting hyperglycaemia leads to the development of chronic diabetic complications. These are classified as microvascular resp. macrovascular. Macrovascular complications are the result of accelerated atherosclerosis and include ischemic disease of the heart (for example myocardial infarction), brain (stroke) and lower limbs. Macrovascular complications are leading cause of premature death in diabetic patients. Microvascular complications are specific for diabetes and are represented by neuropathy, nephropathy and retinopathy. They lead to important worsening of life quality and to shortening of life expectancy too. Complication termed "diabetic foot" has a combined etiology (it develops due to microvacular as well as macrovascular changes) [4].

Insulin therapy. It is well known that since there is currently no possibility to restore a self-insulin production when beta cells are destroyed, an only way of treatment is via insulin injections to mimic physiological situation as much as possible. DCCT (Diabetes Control and Complication) trial clearly showed that to minimize the risk of chronic diabetic complications development full insulin substitution must be done in the form of intensified insulin regime (IIR). There are basically two types of IIR—multiple daily insulin injections (MDI) regime and insulin pump therapy. MDI regime consists of 1–2 injections of basal insulin (long acting insulin) and of prandial injections of rapidly acting insulin (usually 3 injections daily). In insulin pumps only rapidly acting insulins are used and the pump doses automatically basal dose according to the preset program (for example 1 insulin unit (IU) per hour) in a form of microboluses and patients give themselves (manually by pushing pump buttons) prandial boluses prior meal. In last two decades two innovations brought intensified insulin regimes closer to the ultimate therapeutic goal—to reach near normoglycaemia without increasing the risk of serious hypoglycaemia development. Hypoglycaemia is an acute diabetic complication which can develop in patients treated by insulin/insulin secretatogues and can be life threating [5].

Smarter insulin–insulin analogues. The first innovation was an introduction of insulin analogues (chemically altered and enhanced a molecule of human insulin). Basal insulin analogues are designed to produce long and stable level in the organism and rapidly acting analogues act quickly and shortly to minimize postprandial glycaemia increment as well as the risk of overcorrection and hypoglycaemia development prior the next meal [6, 7].

Diet and life style. DM1 patients are educated in carbohydrate counting and healthy life style (including healthy diet and exercise) is recommended. DM1 patients having appropriate body weight can eat even sweet food (in reasonable amount) when their glycaemia is within normal range. Carbohydrate counting still represents a big challenge for patients and educators [1].

Continuous glucose monitoring. The second crucial moment in the optimizing diabetes therapy was the introducing of continuous glucose monitoring (CGM). It is not possible to tailor insulin dose precisely without knowing actual glycaemia and the direction of its change. At the time of DCCT trial (90th of the last century) blood glucose monitoring by using personal glucometers started to spread. But still, even by frequent glucose monitoring (prior each meal, 2 h after each meal and at least once during the night) many hypoglycaemia as well as hypoglycaemia episodes remained unrecognized. At the time of introducing first insulin analogues continuous glucose monitoring started to be implemented into diabetic patients' glycaemia selfmonitoring. The principle of the most frequently used glucose sensors nowadays is based on glucose oxidase reaction—gentle small electrode (covered by the enzyme glucose-oxidase) is inserted into subcutaneous tissue where the reaction of glucose and oxygen is catalyzed by this enzyme. In the cascade of chemical reactions electron current is generated and is converted to electric signal which amplitude reflects glucose concentration. Glucose sensor is attached to the transmitter and the signal is sent wirelessly into a receiver. As a receiver a special device, insulin pump or smart phone can be used. This CGM type is called real time CGM (RT-CGM), displayed glucose concentration is actualized every 5 min and the trend of glucose concentration change is shown by trend arrows as well. RT-CGM system is equipped with hypo-resp. hyperglycemia alarms which threshold can be individualized. When a smart phone is used as RT-CGM receiver—a caretaker can be informed for example about hypoglycemia by sending warning sms. Blinded CGM exists too and is used as glucose concentration holter. By analyzing its record, a doctor can reveal problems in diabetes compensation when a patient for some reason is not suitable for RT-CGM.

Another system represents "Flash Glucose Monitoring" system which can be described as a glucose scanner. The system registers glucose concentration values all time but glucose concentration value and trend is displayed only when scanning device is activated so no warning is possible [8–10].

2 Materials and Methods

Study group. So far 24 patients suffering from type 1 diabetes were enrolled. All patients signed a written consent approved by the local ethical committee and all patient data were analyzed anonymously. Patient's further characteristics are as follows: 15 females, 9 males; median of age 40 years (range 20–59 years), median of diabetes duration 23 years (range 9–39 years). In insulin pump users (19/24) median of this type treatment was 7 years (range 1 month to 31 years). Seven out of 24 patients already have at least one chronic diabetic complication present (mostly retinopathy). To evaluate the efficacy of insulin therapy treatment several parameters must be taken into account. The first one is glycosylated haemoglobin (HbA1c) which reflects a mean glycaemia during last three months. Physiological values are ≤ 42 mmol/mol. Patient's HbA1c should be ≤ 53 mmol/mol. Otherwise a change of therapy is recommended. This condition only 5/24 patients fulfilled. Median of HbA1c in the study group was 60 mmol/mol, range 39–96 mmol/mol, SD 12.3 mmol/mol. Other criteria for compensation evaluation represent glycaemia profiles which will be discussed separately. Other parameters are important too. One of them is BMI (Body Mass Index) which should be in a range 19–25 [11]. Optimal BMI was present in only 10/24 patients (42% of the study group). Overall median of BMI in the tested group was 25.5 (range 20.4–34). Another parameter reflecting diabetes stabilization is total daily insulin dose per kg which should be <0.6 IU/kg [12]. Eight patients from the study had a higher total daily insulin dose per kilogram than this recommended threshold.

Methodology. All patients underwent prior the beginning of the study standardized reeducation with a special focus on diet and carbohydrates counting. Patients used during the study at least 4 weeks RT-CGM and during this period they were asked to document all food and drinks containing carbohydrates (meaning milk, juice etc.) by smart phone camera with time index recording. Patients wrote during this time a detailed logbook as well. The logbook was focused on meal description and patient's estimation of carbohydrates

and glycaemic index of the meal. Physical activity and eventual health problems or special circumstances were recorded too. Patients were asked at least during the first and last week of the study to use NutriData which is an on-line application for diet recording. The data from this application were analyzed by NutriPro software and the intake of energy, carbohydrates, proteins and lipids was evaluated by an experienced nutritional specialist. RT-CGM data and insulin data from insulin pump were downloaded and analyzed by using provider's software (CareLinkPro Medtronic, Dexcom Studio and by DiaSend). The RT-CGM analysis was done by an experienced diabetologist. Energy expenditure was analyzed by using vivo smart fitness wrist band and basal metabolic rate was measured by BIA (Bioimpedance Analysis; Bodystat Quadscan 4000 device). Patients also filled Beck inventory which is used routinely for the screening of depression.

3 Results

The quality of photos taken by patients. During the study 4464 photos were analyzed. Patients documented their meals without important gaps (meaning minimum missing meals photo documentation) but the problem was the quality of photos which was not caused by the quality of camera. Only 3/24 patients were able to provide perfect photo documentation (100% photos in a very good quality).

Identification of a meal from the pictures. The problem with photos quality can be handled by repeated patients instructing and practical training but the crucial problem was the identification of mixed meals (meaning for example pasta with meat and vegetables)—Fig. 1a. It is possible to estimate a food size with a good accuracy by comparing it to the object with standardized size (a pen, a teaspoon etc.) but to identify fully automatically meal content is now technically impossible. Automated identification of the food is applicable for one-component food (for example a banana—Fig. 1b) or for commercially prepared food having EAN code [13].

Willingness to use diet recording application. Patient's main complaint was that additional time is required to do this and they do not want to be concerned with their food analysis and diabetes so long. One patient summarized the problem with low motivation clearly: "We simply want eat what we want without additional thinking. As other people do."

Psychological problems. The presence of depression according to standardized questionnaire (Beck inventory) was high in the study group (13/24 patients), 5/12 patients use antidepressive drugs regularly. Higher depressivity is of course connected to worse co-operation and lower motivation, on the other hand the situation is well understandable as

Fig. 1 **a** Example of mixed meal, **b** example of one component food

many patients suffer important daily difficulties due to diabetes, its complications and present comorbidities.

Nutritional analysis results. Results (median, range) in % of recommended daily value: Energy intake: 117% (94–180%), total carbohydrates 109% (80–120%), mono + oligosaccharides 133% (93–180%), total fat 144% (120–210%), saturates 133% (105–196%), cholesterol 97% (93–110%), proteins 98% (83–132%), fiber 72% (63–110%). Only one patient had absolutely appropriate diet as well as energy expenditure. According to results on carbohydrates and lipid intake patients were categorised into 3 groups with different compliance to diabetic diet (Group 1—all results within 100% + 1SD of recommended values, Group 2—none parameter >100% + 2SD of recommended value, Group 3—at least one parameter >100% + 2SD of recommended value). Twelve patients (50% of the study group) scored 3 (the worst category, 4 of these 12 patients are obese and 4 of them are overweight). We observed the correlation between the compliance category and sex (women scored better, p = 0.049, r = 0.63) as well as with HbA1c level when patients scoring 3 had higher HbA1c at baseline (p = 0.019, r = 0.69) as well as after diet re-education combined with 4 weeks of continuous glucose monitoring (p = 0.037, r = 0.57). Level of depression was highly correlated to the compliance category when depressive patients scored worse and they had high mainly mono + oligosaccharides intake (p = 0.01, r = 0.74). Patients did not evaluate their diet as unhealthy and they considered only total carbohydrate intake as important.

Accuracy of patient's carbohydrates content estimation. Patients underestimated carbohydrate content importantly but mistakes in GI (glycaemic index of food) evaluation were insignificant. Sometimes overestimation of carbohydrates occurred as well with subsequent hypoglycaemia due to insulin overdosing. Median of "carbohydrate mistake" was 15 g (range 0–45 g).

Energy expenditure and the reaction to physical activity. Only 8/34 patients have a regular physical activity, one patient has hard manual work with high energy expenditure, 7/8 patients are active sportsmen. The most frequent reported sport activity was jogging and cycling, one patient goes in for martial arts. These trained patients did not react by important glycaemia dropping when they were introduced to submaximal physical activity (submaximal according to heart rate). Physical activity of other patients was low and after introducing mentioned physical activity for 30 min their glycaemia dropped quickly (Fig. 2). Basal metabolic rate (median) in the study group was 23 kcal/kg (range 17–29 kcal/kg). Women energy expenditure was more frequently lower than recommended in comparison with men (p = 0.028).

CGM results. Median of time spent in normoglycaemia was 47% (33–63%) and no serious hypoglycaemia as well as hyperglycaemia occurred. Hypoglycaemia events (mild hypoglycaemias) were connected in 40% to previous physical activity and in 36% to hyperglycaemia overcorrection. Glycaemia over target range occurred in 63% after meal and was strongly influenced by meal carbohydrate content underestimation (p < 0.001, Fig. 3). Patients insufficiently used advanced insulin pump features (bolus calculators, different bolus types etc.) and/or online dietary advisors (online nutritional databases) but they were satisfied with using smart phones as the quick and convenient meal logbook.

4 Discussion

We do not have another option for acquiring the information about meals then manual input provided by the patient. Since it is so strongly dependent on his/her perception of the health problem and motivation to cope with it we need to

Fig. 2 Hypoglycaemia at 2 pm
due to prior physical activity

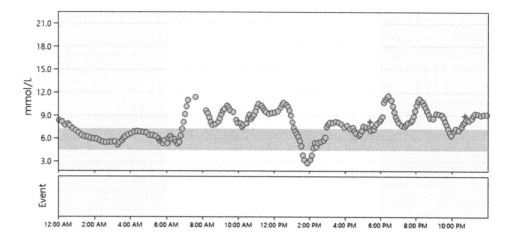

Fig. 3 Hyperglycaemia after
sweet breakfast eaten at 8 am

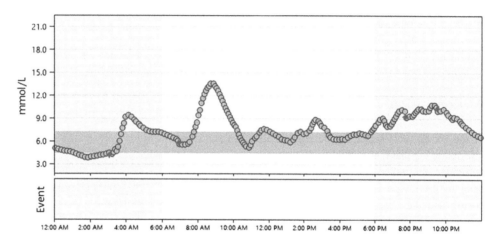

find a way that is least obtrusive on one side but informative on the other side. It seems that using photos with brief notes is an acceptable solution. Additional option we are working on is to create a personalized database of favourite meals which can be finally completed with more detailed information about nutrition values, including glycaemic index. Then the patient only confirms selected meal and does not need to insert all data again. Based on data from the insulin pump and the glucose sensor and inserted information about the planned meal from the patient, the application can recommend the prandial bolus to be injected before meal.

The case study shows that it is not a simple task to design and develop an application that can help a patient even when we have data input from several devices. If there is another input required from the user the final operation and reliability of the application is dependent on the precision of manually inserted data.

The presented results and data analysis indicate that the input data and information are key components for good quality decision making on insulin dose. In principle the mechanism can be represented by a control system with feedback.

In our case, the control system can be described in the following way: input data are represented by the information about meals and physical activity, the system (body with its metabolism) must be described by a model respecting the individual characteristics of the given person, measuring element is the glucometer (RT-CGM), controller and effector represent the insulin pump. It is obvious that there might be disturbances we cannot predict and usually we cannot measure them, as for example psychical stress. If we can get as precise information as possible about composition and quantity of meals and physical activity we can use it for better control of the system. We have to keep in mind that the values acquired from RT-CGM are delayed, the time delay is known. Thus this information must be included in the model.

The ultimate goal in the treatment using the insulin pump technology is the closed loop concept when patient has an insulin pump, glucose sensor and insulin dosage is driven by an algorithm integrated into computer.

There are several groups working on CL concept. All groups report a reasonable progress to maintain normoglycaemia (or near normoglycaemia) during nights. A great problem is how to fit insulin dose to very variable day activities, different meals, stress etc. The main "technical" problem is that currently available rapidly acting insulin analogues are not still "rapid enough" because insulin kinetics after its artificial parenteral application is different from the physiological situation.

In one of the most recent papers where OL (open loop which means not fully automatized system) was compared to two CL algorithms there was reported that time spent in the target range was quite similar in CL and OL. While mean glucose level was significantly lower in OL, percentage of time spent in hypoglycaemia was almost threefold reduced during CL. It means that nowadays CL is superior in prevention of hypoglycaemia but unfortunately is not now so superior to OL in the achievement of normoglycaemia [14]. There was presented a day and night closed-loop control in adults with type 1 diabetes as a comparison of two closed-loop algorithms driving continuous subcutaneous insulin infusion versus patient self-management.

5   Conclusion

In the paper we focused on detailed analysis of data and information content and quality in outpatient treatment. In particular, the patient study group has the diagnosis of diabetes mellitus type 1. The patients are dependent on insulin dosing and they have to estimate rather precisely the food composition in relation to insulin units. The analysis showed that even experienced patients sometimes make mistakes because it is not always easy to estimate carbohydrates content in mixed meal, especially in ready meals. Based on the analysis and long-term experience of one of the authors (KS) we started to design an application that can serve as a recommendation system for the patients both for meal composition estimation and insulin bolus.

Acknowledgements Research has been supported by the AZV MZ CR project No. 15-25710A "Individual dynamics of glycaemia excursions identification in diabetic patients to improve self managing procedures influencing insulin dosage."

Conflict of Interest The authors declare that they have no conflict of interest.
Statement of Informed Consent Informed consent was obtained from all individual participants included in the study.

Protection of Human Subjects and Animals in Research All procedures performed in studies involving human participants were in accordance with the ethical standards of the institutional and/or national research committee and with the 1964 Helsinki declaration and its later amendments or comparable ethical standards.

References

1. Pelikánová T, Bartoš V (eds). Praktická diabetologie. Maxdorf s.r. o. Prague, 5th edition, 2011.
2. Röder PV, Wu B, Liu Y, Han W. Pancreatic regulation of glucose homeostasis. Exp Mol Med. 2016 Mar 11;48:e219. https://doi.org/10.1038/emm.2016.6.
3. Copenhaver M, Hoffman RP. Type 1 diabetes: where are we in 2017? Transl Pediatr. 2017 Oct;6(4):359–364.
4. Olansky L. Advances in diabetes for the millennium: chronic micro-vascular complications of diabetes. MedGenMed. 2004 Aug 9;6(3 Suppl):14.
5. Donner T. Insulin—Pharmacology, Therapeutic Regimens and Principles of Intensive Insulin Therapy. In: De Groot LJ, Chrousos G, Dungan K, Feingold KR, Grossman A, Hershman JM, Koch C, Korbonits M, McLachlan R, New M, Purnell J, Rebar R, Singer F, Vinik A, editors. Endotext [Internet]. South Dartmouth (MA): MDText.com, Inc.; 2000–2015 Oct 12. Accessed 18.1.2018.
6. Cichocka E, Wietchy A, Nabrdalik K, Gumprecht J. Insulin therapy—new directions of research. Endokrynol Pol. 2016;67 (3):314–24.
7. Ziegler R, Freckmann G, Heinemann L. Boluses in Insulin Therapy. J Diabetes Sci Technol. 2017 Jan; 11(1):165–171.
8. Institute for Quality and Efficiency in Health Care. Continuous Interstitial Glucose Monitoring (CGM) with Real-Time Measurement Devices in Insulin-Dependent Diabetes Mellitus [Internet]. Cologne, Germany: Institute for Quality and Efficiency in Health Care (IQWiG); 2015 Mar 25. Accessed 18.1.2018.
9. Gómez AM, Henao Carrillo DC, Muñoz Velandia OM. Devices for continuous monitoring of glucose: update in technology. Med Devices (Auckl). 2017 Sep 12;10:215–224.
10. van Beers CA, DeVries JH. Continuous Glucose Monitoring: Impact on Hypoglycemia. J Diabetes Sci Technol. 2016 Nov 1; 10 (6):1251–1258.
11. Polsky S, Ellis SL. Obesity, insulin resistance, and type 1 diabetes mellitus. Curr Opin Endocrinol Diabetes Obes. 2015 Aug; 22 (4):277–82.
12. http://www.diab.cz/dokumenty/standard_DM_I.pdf   Accessed 18.1.2018.
13. Saiti K, Macaš M, Lhotská L. Importance and Quality of Eating Related Photos in Diabetics. M.E. Renda et al. (Eds.): ITBAM 2016, Springer LNCS 9832, pp. 173–185, 2016.
14. Luijf YM, et al. Day and Night Closed-Loop Control in Adults With Type 1 Diabetes. Diabetes Care. 2013 Dec; 36(12):3882–7.

Part XVII

The status of Bioengineering, Biomedical Engineering and Clinical Engineering Education in Latin America. "Curriculum, Accreditation, Innovation and Research"

Innovative Biosensor of Circulating Breast Cancer Cells; *a Potential Tool in Latin America Oncology Rooms*

César A. González and Herberth Bravo

Abstract

The detection of circulating tumoral cells represents the possibility for therapy monitoring as well as prevent metastasis in oncology patients. Portable and economical technologies are required in most of the oncology rooms of third level Latin America hospitals. In this work an innovative biosensor of circulating breast cancer cells on the basis of bioimpedance spectroscopy measurements assisted with magnetic nanoparticles is presented. The technical proposal involves a microfluidic system for cancer cells separation by immunomagnetic technique and its detection by multifrequency bioimpedance measurements. An experimental proof of concept to detect a typical breast cancer cell line was developed to evaluate the sensitivity of the system. The results shown the technical proposal feasibility as portable, inexpensive and non-invasive biosensor of circulating tumoral cells, as well as its technical feasibility for implementation in oncology rooms of Latin America third level hospitals.

Keywords

Biosensor • Bioimpedance • Cancer • Cell

1 Introduction

According to the World Health Organization (WHO) Breast Cancer (BC) is one of the largest health problems in the world. BC is the most frequently diagnosed and the leading cause of cancer death among women, accounting for 23% of total cancer cases and 14% of cancer deaths. BC is also the leading cause of death among women in economically developing countries in Latin America. Metastasis is the main cause of morbidity and mortality related to BC. In addition, some studies show that 30% of women with primary BC harbor micro metastasis in their bone marrow. Almost half of the patients with localized BC treated with surgery have a high possibility of recurrence, which attempts to be reduced by subjecting the patient to oncological treatment (chemotherapy, radiotherapy and others) that is not often effective because it does not consider the amount of cancer cells in systemic circulation. [1–4].

Circulating Tumor Cells (CTC) are cancer cells that originate from primary/metastatic solid tumors and are in transit through the circulatory system. The detection and characterization of CTC allows us to have a control of the oncological treatment of the patient with BC and to define if it is effective. The CTC count allows a window of possibility for general survival in operable and advanced breast cancer. However, the isolation of CTC is a challenge due to its extreme rarity, there are approximately between 1 and 10 CTC for every 10E9 of the total blood cells. [5–7].

Currently several methods and technologies for the detection of CTC that are based on their physical and biological properties are under development. This is a very complicated process since the ideal method must be highly sensitive, reproducible and easy to implement in a clinical setting, in addition to the very low concentration of CTC in the blood flow. Nowadays, better technologies are sought to avoid the disadvantages of these methods, such as the use of expensive and specialized equipment, expert and qualified personnel, the lack of accessibility that the population has to these, among others. Due to the above, it is necessary to search for new technological methods that allow us to dissuade these inconveniences and be able to have high quality technologies at a low cost and with easy access [4, 7, 8]. In most existing methods, a first enrichment step is carried out to increase the sensitivity of the assay. Then we continue with a detection step that will ideally protect the integrity of CTC [8].

C. A. González (✉)
Instituto Politécnico Nacional-Escuela Superior de Medicina, 11320 CDMX, Mexico
e-mail: cgonzalezd@ipn.mx

H. Bravo
Sociedad Mexicana de Ingeniería Biomédica, 11300 CDMX, Mexico

Our research group has proposed the detection of CTC through magnetic bioimpedance-assisted spectroscopy measurements with magnetic nanoparticles coupled with antibodies that recognize surface proteins [9]. The central idea is to attach magnetic nanoprobes made of iron oxide coated with a polysaccharide to which an antibody is adhered, on the surface of CTC, separate and anchor them by immunomagnetic methods on the surface of an array of micro-electrodes and perform bioimpedance measurements as detection technique. As a first approach to the proposed technique, in this research work we propose to evaluate the technical feasibility of using electrical bioimpedance measurements to detect breast cancer cells with the use of a microfluidic device.

2 Methodology

2.1 Biosensor Description System

The block design of the system consists of four modules: (1) Microfluidic infusion pump; (2) Electrical-Ionic Interface; (3) SciospecTM Module; and (4) The PC. The infusion pump mircofluidic has adapted an insulin syringe, which performs the controlled infusion of the analytes into the chamber of the Electrical-Ionic Interface module by means of a capillary tube. The Electrical-Ionic interface represents the main point of the experiment, since it is where the isolation of cancer cells and measurement of bioimpedance through gold electrodes is made. The SciospecTM module is the instrument that allows us to interact with the electrodes of the Electrical-Ionic Interface, that is, the one that injects the potential difference and at the same time measures the current to estimate the impedance of the system. The PC module allows us to program the SciospecTM instrument and data storage (Fig. 1).

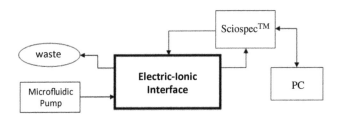

Fig. 1 Block diagram of the system

2.2 Bioimpedance Measurements in Microfluidic Device

The impedance reading module brings with it the SciospecTM software, which is provided by the same company that manufactured the product, allowing communication with the PC through a USB port, which provided the option of making a measurement protocol, achieving with this a mapping of 126 logarithmically spaced steps in a frequency range of 100 Hz to 1 MHz. The measurements were made in triplicate (Fig. 2).

2.3 Immunomagnetic Isolation of Cancerous Breast Cells in Microfluidic Device

We used the breast cancer cell line SK-3, such cell line represents an advanced and very aggressive phase of BC. The bioconjugate was elaborated on the basis of the A-10 CHEMICELL INC protocol, which is a coupling procedure with carbodiimides for magnetic particles. The technique demands the identification of molecular markers to promote an antigen-antibody reaction, which causes the cancer cells to adhere to the nanoparticles within the bioconjugate and thereby achieve immunomagnetic isolation, that is, to be able to trap the cells anchored to the nanoparticles with a magnet in the chamber of the microfluidic device and thus measure its impedance [10, 11].

Fig. 2 Physical appearance of the system modules

2.4 Design of the Evaluation Experiment

In order to demonstrate changes in the electrical bioimpedance as a function of the isolation and anchoring of cancer cells in the electrical-ionic interface, an experiment was designed in two independent tests in order to generate the following conditions: (A) SK-3—infusion and magnetic anchoring of cancer cells of the SK-3 line incubated with nanoprobes, and (B) Negative Control—infusion and magnetic anchoring of nanoprobes without the presence of cancer cells. A concentration of 50 cells/500 L, was evaluated in order to estimate the sensitivity of the system, each test was performed in triplicate (see Fig. 3).

3 Results

Figures 4 shows the bioimpedance spectra in magnitude and phase, for a concentration of 50 cells/500 μL. The magnitude spectrum shows significant sensitivity at low frequencies to discriminate the presence of cancer cells with respect to the condition in which only bioconjugate was infused.

4 Discussion

The relevant observation is that the highest sensitivity to discriminate the presence of cancer cells was observed in magnitude at low frequencies, basically below 10 kHz, and not sensitive in phase was observed, such observations indicate in principle the influence of structural changes in the evaluated biological sample, those changes must be associated to the complex and size of the cells in the whole volume of the case sample and not present in the control. Evaluating the potential utility of bioimpedance measurements to detect breast cancer cells by this proposed method, we determined very well a positive result, which, it should be noted that it is only possible to validate the sensitivity under the

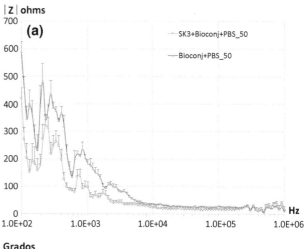

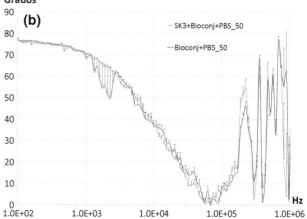

Fig. 4 Bioimpedance measurements at multiple frequencies, concentration of 50 cancer cells/500 μL in contrast to control without cancer cells. **a** Magnitude (ohms) and **b** phase (degrees)

experimental conditions that were defined for the experiment, not so for a clinical validation, which leads us to the need to investigate even more in this procedure, for instance; characteristical bioimpedance spectra for the presence of different CTC and even blood cells are required studied, in addition, noise sensitivity studies as a function of differents CTC concentration in the sample are warranted, in order to the renewal or improvement of the used device, being possible to do it with a smaller budget and with this to be able to take to the communities most in need in Latin America.

5 Conclusion

The implementation of a small system as a biosensor for the isolation and detection of breast cancer cells through the use of a microfluidic device and measurements of bio impedance assisted with magnetic nanoparticles was feasible at low cell concentrations and magnetic nanoparticles. Measurements of the magnitude of bioimpedance at low frequencies seem to

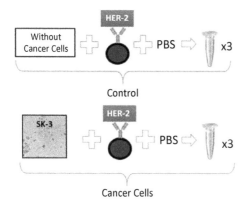

Fig. 3 Experimental evaluation design

offer the best sensitivity of the system, which indicates that this device together with its measurement procedure is viable as a new CTC detection technique.

Conflict of Interest Authors declare that there is not conflict of interest with the content in this paper.

References

1. Ahmedin Jemal, DVM, PhD et.al., "Global Cancer Statistics". CA: A Cancer Journal Clinicians; 61(2): 69–90, 2011.
2. "Cancertoday", IARC, GLOBOCAN. France, 2012. [Fecha de acceso 20 de Junio del 2017]. URL disponible en: https://gco.iarc.fr/today/home.
3. Felicia Marie Knaul, PhD, et al. "Utilización correcta de las técnicas de detección de cáncer de mama en las mujeres mexicanas". Salud Pública de México; 56(5): 538–546, Septiembre–Octubre 2014.
4. Luz F. Sua, Nhora M. Silva, Marta Vidaurreta, María L. Maestro Sara R. Fernández, Silvia Veganzones, Virginia de la Orden, José M. Román. "Detección inmunomagnética de células tumorales circulantes en cáncer de mama metastásico: nuevas tecnologías". Revista Colombiana de Cancerología; 15(2): 104–109, 2011.
5. Athina Markou, Areti Strati, Nikos Malamos, Vassilis Georgoulias and Evi S. Lianidou. "Molecular Characterization of Circulating Tumor Cells in Breast Cancer by a Liquid Bead Array Hybridization Assay". Clinical Chemistry; 57 (3): 421–430, 2011.
6. Luz Fernanda Sua Villegas, Nhora María Silva Pérez, Marta Vidaurreta Lázaro, María Luisa Maestro de las Casas, Sara Rafael Fernández y Silvia Veganzones de Castro. "Actualidad y futuro en las técnicas de cuantificación de células tumorales circulantes: su importancia en tumores sólidos epiteliales." Revista del Laboratorio Clínico; 4(3): 163–169, 2011.
7. Daniel L. Adams, Steingrimur Stefansson, Christian Haudenschild, Stuart S. Martin, Monica Charpentier, Saranya Chumsri, Massimo Cristofanilli, Cha-Mei Tang, R. Katherine Alpaugh. "Cytometric Characterization of Circulating Tumor Cells Captured by Microfiltration and Their Correlation to the Cell Search VR CTC Test." Cytometry, Part A (2015); 87A: 137–144, 2015.
8. Noh Gerges, Janusz, and Nada Jabado. "New technologies for the detection of circulation tumor cells". British Medical Bulletin; 94: 49–64, 2010.
9. Jesús G Silva, Rey A Cárdenas, Alan Quiróz, Virginia Sánchez, Lucila M. Lozano, Nadia M. Pérez, Jaime López, Cleva Villanueva y César González. "Impedance Spectroscopy assisted by Magnetic Nanoparticles as Potential Biosensor Principle of Breast Cancer Cells in Suspension". Physiological Measurement, Mexico, 2014.
10. L. F. E. Huerta-Núñez. G. Cleva Villanueva-Lopez. A. Morales-Guadarrama. S. Soto. J. López. J. G. Silva. N. Perez-Vielma. E. Sacristán Marco E. Gudiño-Zayas. C. A. González. "Assessment of the systemic distribution of a bioconjugated anti-Her2 magnetic nanoparticle in a breast cancer model by means of magnetic resonance imaging". Journal of Nanoparticle Research; 18:284, 2016.
11. Chemicell, "Covalent Coupling Procedure on fluid MAG-ARA by Carbodiimide Method". Chemicell GmbH • Coupling Protocol • fluidMAG-ARA 1.1.

Part XVIII
Application of EM Field Medical Diagnostics

Prototype of Simplified Microwave Imaging System for Brain Stroke Follow Up

Jan Tesarik, Luis F. Diaz Rondon, and Ondrej Fiser

Abstract

Stroke could be detected by Microwave Imaging method (MWI) in the future. At ELEDIA@CTU laboratory was designed prototype of MWI system for stroke follow up/detection. The system was able to detect five from six strokes which were then reconstructed by existing algorithm for difference Microwave Imaging. These results were very valuable for next research where for example new antenna elements need to be design as well as realistic phantom of human head.

Keywords

Microwave imaging • Stroke detection Dielectric properties

1 Introduction

Stroke is the second most common non-traumatic cause of mortality in the world [1] and can be distinguish on ischemic or haemorrhagic in general. Time plays key role in diagnosis of stroke and its shortening is main goal for successful treatment. Every minute dies 1.9 million neurons if stroke is not treated [2]. One way how to short a time between onset of stroke and its detection/classification is to use Microwave Imaging (MWI) method. Microwave Imaging is a new perspective non-harmful method which could be used for diagnosis of tissues state in the future similarly as CT or MRI. Potential of MWI hardware is at its fast diagnosis in the order of minutes, use of non-ionizing radiation, can be compact, light and relatively low cost. Such equipment could be placed directly in emergency cars as already presented by company Medfield Diagnostics [3] for example, so a time between onset of stroke and treatment would be shortened. Principle of MWI is based on interaction of human tissues with microwaves and non-invasive measurements of so called dielectric properties [4]. If the difference in dielectric parameters for normal brain tissue and ischemic/haemorrhagic brain tissue exists as well as for healthy and disease tissues, MWI can be used.

The aim of this contribution is to present some results of stroke detection provided by first prototype of MWI system designed at ELEDIA@CTU laboratory as well as describe methods and progresses leading to create MWI system and its experimental setup.

2 Methods

2.1 Design of Microwave Imaging System

Simplified Microwave Imaging system was designed in COMSOL Multiphysics as octagonal container respected an average size of human head with eight Bow Tie antennas placed in one plain around as shown in Fig. 1a. Width of system was set on 16 cm, length on 20 cm and deep on 20 cm too. Inside the system was possible to placed cylinders with different diameters d_1 = 20 mm and d_2 = 30 mm on three different positions p_1, p_2 and p_3 as shown in Fig. 1b. Cylinders could be filled with liquid so that way stroke could be simulated. System as well as cylinders was printed from PLA material on 3D printer. Bow Tie antennas was designed by numerical parametric study in COMSOL Multiphysics where parameters of antenna a, b, c (mm) for resonance frequency 1 GHz were found. Eight Bow Tie antennas were fabricated by chemical etching method. Example of fabricated Bow Tie antenna is shown in Fig. 1c.

2.2 Phantom of Human Head and Strokes

To provide perfect contact between antennas and head phantom without any gaps a liquid phantom of human head adopted from IEEE standard [5] was chosen and prepared.

J. Tesarik (✉) · L. F. Diaz Rondon · O. Fiser
Faculty of Biomedical Engineering, Czech Technical University in Prague, Prague, Czech Republic
e-mail: jan.tesarik@fbmi.cvut.cz

© Springer Nature Singapore Pte Ltd. 2019
L. Lhotska et al. (eds.), *World Congress on Medical Physics and Biomedical Engineering 2018*,
IFMBE Proceedings 68/3, https://doi.org/10.1007/978-981-10-9023-3_139

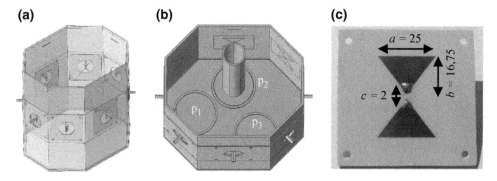

Fig. 1 Simplified model of MWI system (**a**), positions p$_1$, p$_2$ and p$_3$ of cylinders and upper view on MWI system (**b**) and example of fabricated Bow Tie antenna with dimensions of its parameters (**c**)

Table 1 Weight percentage of used substances for preparing of human head phantom and target dielectric parameters of this phantom for frequency 900 MHz

	Propylene glycol	NaCl	Water[a]
wt%	64.81	0.79	34.40
ε_r (—)		41.80	
$\sigma(S \cdot m^{-1})$		0.97	

[a]Deionized and demineralized water

Table 2 Measured dielectric parameters of prepared phantom of human head (Phantom) and stroke phantoms as well as absolute difference $\Delta\varepsilon_r$ (—) and $\Delta\sigma(S \cdot m^{-1})$ between stroke phantoms and Phantom

	Phantom	Hem1	Hem2	ISCH1	ISCH2	ISCH3
ε_r (—)	40.65	48.63	55.47	25.06	19.28	32.04
$\sigma(S \cdot m^{-1})$	0.95	1.28	1.52	0.83	0.74	0.85
$\Delta\varepsilon_r$ (—)		7.98	14.82	−15.59	−21.37	−8.61
$\Delta\sigma(S \cdot m^{-1})$		0.33	0.57	−0.12	−0.21	−0.1

Weight percentage of used substances and target dielectric parameters for frequency 900 MHz are summarized in Table 1. Stroke phantoms were prepared according to same recipe, based on assumption that ischemic tissue cause local decrease of dielectric parameters, haemorrhagic tissue cause local increase of dielectric parameters respectively. This assumption comes from [6]. It was prepared two hemorrhagic phantoms marked as Hem1 and Hem2 and three ischemic phantoms marked as ISCH1, 2, 3 as summarized in Table 2. Dielectric parameters of this stroke phantoms were measured by dielectric probe SPEAG DAK-12 and handheld VNA Keysight as well as prepared phantom of human head.

2.3 Experimental Setup

Printed model of Microwave Imaging system was fitted with Bow Tie antennas and through switching matrix R&S® ZN-Z84 was connected to VNA R&S® ZNB 4 as shown in Fig. 2a. MWI system was filled with liquid phantom of human head and on given positions was placed cylinders with stroke phantoms Fig. 2b. The power input from VNA was set to maximal value 13 dBm, width of inter-frequency filter was set to 100 kHz and operating frequency to 1 GHz. For that configuration all stroke phantoms to position p$_3$ and to the cylinders with diameter d_1 and d_2 were placed. Into position p$_1$ stroke phantoms Hem2 and ISCH2 with diameter

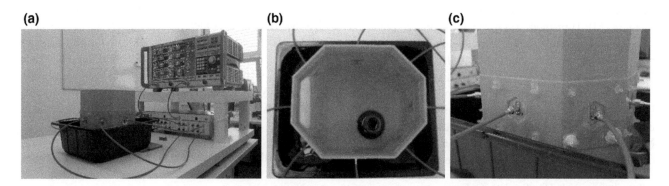

Fig. 2 Experimental setup of MWI system, switching matrix and VNA (**a**), detailed view on liquid phantom of human head and position of stroke phantom (**b**) and detailed view on connection of SMA ports and coaxial cables (**c**)

Table 3 Comparison of measured changes in dielectric parameters Δ and reconstructed changes δ for position p_3 and diameter d_2, green colour marks successful reconstruction and linker marks that was not possible to determine which type of stroke was placed in MWI system

	Hem1	Hem2	ISCH1	ISCH2	ISCH3
$\Delta\varepsilon_r$ (—)	8	15	−16	−21	−9
$\Delta\sigma(S \cdot m^{-1})$	0.3	0.6	−0.1	−0.2	−0.1
$\delta\varepsilon_r$ (—)	–	–	−11	−15	−10
$\delta\sigma(S \cdot m^{-1})$	–	0.5	–	–	–

d_2 were placed. The measured output was S-matrix (8×8) where its values (S-parameters) are changing according to specify combination of stroke phantom its position and diameter. Each combination was measured five times.

Measurements were simplified by the symmetricity of MWI system, it was used principle of reciprocity.

2.4 Data Processing

Reconstruction algorithm for differential microwave imaging was used and is described for example in [7]. Necessary input for this algorithm is intensity of the electric field each of eight antennas in x, y, z coordinates which was calculated numerically for model of MWI system in configuration with as well as without stroke phantom. Another input was difference between measured "stroke" S-matrix and measured "background" S-matrix. The goal was to find out the values

Fig. 3 Reconstructed change of relative permittivity $\delta\varepsilon_r$ and electrical conductivity $\delta\sigma$ for stroke phantoms Hem2 (**a**) and (**b**), ISCH2 (**c**) and (**d**) on position p_3 for diameter d_2 and for Hem2 (**e**) and (**f**) on position p_1 for diameter d_2

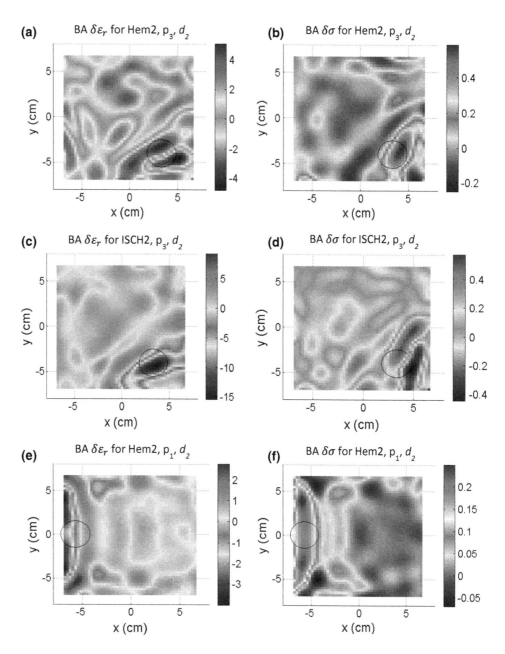

of δO vector, where each value is change in relative permittivity and conductivity, respectively.

3 Results

3.1 Measured Dielectric Parameters

In the Table 2 are summarized measured dielectric parameters of prepared phantom oh human head and stroke phantoms together with absolute difference in dielectric parameters of stroke phantoms and human head phantom $\Delta \varepsilon_r$ (—) and $\Delta \sigma (\mathrm{S} \cdot \mathrm{m}^{-1})$.

3.2 Reconstructed Strokes

On each of figures a–f is shown reconstructed change of relative permittivity $\delta \varepsilon_r$ (—) and electrical conductivity $\delta \sigma (\mathrm{S} \cdot \mathrm{m}^{-1})$. Black circle marks actual position of stroke phantom in MWI system. Values of reconstructed dielectric parameters are summarized in Table 3.

4 Discussion

By designed MWI system was possible to differentiate and reconstruct some of simulated strokes in phantom of human head. As shown in Fig. 3a–d and in Table 3 haemorrhagic stroke Hem2 was distinguished from background (phantom of human head) based on increase of electrical conductivity while ischemic strokes ISCH1, 2, 3 were distinguished based on decrease of relative permittivity. Hem1 was not possible to reconstruct probably due to so low $\Delta \sigma$. MWI system was able to measure change of dielectric parameters only for diameter of stroke phantoms $d_2 = 30$ mm. Results for diameter d_1 was not reached probably due to fact that diameter of stroke phantom is lower than half of wavelength of propagating EM wave which is 2.3 cm. Reconstruction of strokes placed on position p_1 was distorted as shown in Fig. 3e and f because of symmetricity of stroke phantoms towards MWI system, the number of independent values in S-matrix was decreased. Measurements were influenced by unstable environment in laboratory and used Bow Tie antennas which radiated to all directions, so microwave absorbers built around MWI system was used. Reconstructed changes of dielectric parameters could be affected by time instability of prepared head and stroke phantoms.

5 Conclusion

First prototype of MWI system in Czech Republic was designed and measurements of changes in dielectric parameters between phantom of human head and stroke phantoms following by reconstruction of simulated strokes was implemented. MWI system was able to differentiate five from six prepared strokes as summarize in Table 3. Sensitivity of system was influenced by position of strokes its diameters and difference in dielectric parameters of stroke phantoms and human head phantom. Nowadays new antenna elements with ground plane are testing and in near future new heterogeneous phantom of human head will be designed. Results provided by first prototype of MWI system are very valuable for research of next generation of MWI systems and testing of MWI method for stroke detection.

Acknowledgements This research has been supported by the research program of the Czech Science Foundation, Project no. 17-00477Y, Physical nature of interactions of EM fields generated by MTM structures with human body and study of their prospective use in medicine. The authors declare that they have no conflict of interest.

References

1. "WHO | The top 10 causes of death," *WHO*, 2017.
2. J. L. Saver, "Time is brain—quantified," *Stroke*, vol. 37, no. 1, pp. 263–6, 2006.
3. "Medfield Diagnostics| Safer diagnostics through microwave technology," [Online]. Available: http://www.medfielddiagnostics.com/.
4. J. Vrba and D. Vrba, "Temperature and Frequency Dependent Empirical Models of Dielectric Properties of Sunflower and Olive Oil," vol. 22, no. 4, 2013.
5. "IEEE 1528-2013: IEEE Recommended Practice for Determining the Peak Spatial-Average Specific Absorption Rate (SAR) in the Human Head from Wireless Communications Devices: Measurement Techniques," [Online]. Available: https://standards.ieee.org/findstds/standard/1528-2013.html.
6. S. Semenov, R. Svenson, G. Simonova, A. Bulyshev, E. Souvorov, Y. Sizov, A. Nazarov, V. Borisov, A. Pavlovsky, M. Taran, G. Tatsis and A. Starostin, "Dielectric properties of canine acute and chronic myocardial infarction at a cell relaxation spectrum. I. Experiments," in *Proceedings of the 19th Annual International Conference of the IEEE Engineering in Medicine and Biology Society*, 1997.
7. R. Scapaticci, O. M. Bucci, I. Catapano and L. Crocco, "Differential Microwave Imaging for Brain Stroke Followup," *International Journal of Antennas and Propagation*, vol. 2014, pp. 1–11, 2014.

Samples of Dry Head Tissues Phantoms for Brain Stroke Classification

Jan Tesarik, Tomas Pokorny, and Lukas Holek

Abstract

Phantoms are necessarily for testing of MWI method. At ELEDIA@CTU laboratory phantom for testing of brain stroke classification are designing. For that purpose, new materials/substances were tested. It was prepared a few series of phantom samples with different weight percentages of given materials. Fabricated samples were measured by dielectric probe. Measured dielectric parameters shown that some of samples can mimic head tissues like skin, skull and CSF. Results of this contribution will be used for fabrication of dry heterogeneous human head phantom.

Keywords

Microwave imaging • Phantoms of head tissues
Dielectric properties

1 Introduction

Microwave Imaging (MWI) is new perspective non-harmful method for diagnosis of human tissues state. In comparison with current conventional imaging techniques, MWI systems could be relatively cheap, fast and portable devices which use non-ionizing microwave radiation. Research in applications of microwaves is focused mainly on two areas, timely breast cancer detection [1] and brain stroke follow up, its classification respectively [2]. For testing and validating of MWI method phantoms of human tissues are used. Phantoms could be distinguished on homogeneous and heterogeneous in general. Another differentiating is possible based on its structure to liquid, semi-dry (agars) or dry. A large summary of phantom types and its application including MWI is available for example in [3]. At ELEDIA@CTU

J. Tesarik (✉) · T. Pokorny · L. Holek
Faculty of Biomedical Engineering, Czech Technical University in Prague, Prague, Czech Republic
e-mail: jan.tesarik@fbmi.cvut.cz

laboratory a simplified prototype of MWI system for brain stroke follow up was developed [4]. Homogeneous liquid phantom of human head for testing of MWI system and stroke detection was used. Liquid phantoms are not time stable [5]. It is necessary to developed dry heterogeneous phantom of human head with realistic shapes and layers for more detailed testing of MWI systems.

The aim of the contribution is to present measured dielectric parameters of fabricated phantom samples of head tissues and compare it with dielectric parameters of head tissues obtained by fourth order Cole–Cole model. Results should be base for future design and fabrication of dry heterogeneous human head phantom.

2 Methods

2.1 Materials and Phantom Samples Fabrication

According to studies [6, 7] a new material suitable for mixing with graphite powder and carbon black powder as shown in Table 1 was chosen. The combination of first three listed materials in Table 1 should provide time-stable, flexible and low viscosity mixture which preparation should be relatively easy. To decrease a viscosity of mixture was possible to add an acetone. Acetone (L) was added independently on total weight of each sample. Viscosity of mixture is very important for future fabrication of heterogeneous head phantom where different mixtures will create different layers of phantom.

In total 20 samples was fabricated. There were five basic series. Graphite series (GX), carbon black series (CBX), graphite 20% series (G20CBX), graphite 25% series (G25CBX) and carbon black 4% series (GXCB4) as summarized in Table 2. Weight percentages of our materials was mixed in the kitchen mixer, put to the printed bowls and vacuumed (approximately -0,8 atm) to eliminate air bubbles. Examples of fabricated samples are shown in Fig. 1.

© Springer Nature Singapore Pte Ltd. 2019
L. Lhotska et al. (eds.), *World Congress on Medical Physics and Biomedical Engineering 2018*,
IFMBE Proceedings 68/3, https://doi.org/10.1007/978-981-10-9023-3_140

Table 1 Materials and its specification used for fabrication of phantom samples

Material	Specification	Producer
Urethan rubber	PMC®-121 30/Wet	Smooth-On, Inc
Graphite powder	282863-graphite	Sigma-Aldrich
Carbon black powder	45527 carbon black	Alfa Aesar
Acetone	99.9% acetone	Lach-Ner, s.r.o.

Table 2 Weight percentages of given materials for phantom samples fabrication

Name	Urethan rubber (wt %)	Graphite powder (wt %)	CB powder (wt%)	Acetone (L)
G10	90	10	–	–
G20	80	20	–	–
G30	70	30	–	–
G40	60	40	–	0.3
G50	50	50	–	0.3
CB1	99	–	1	–
CB5	95	–	5	0.1
CB10	90	–	10	0.1
G25CB2	73	25	2	0.3
G25CB2.5	72.5	25	2.5	0.3
G25CB4	71	25	4	0.3
G25CB6	69	25	6	0.3
G10CB4	86	10	4	–
G15CB4	81	15	4	–
G20CB4	76	20	4	0.3
G25CB4	71	25	4	0.3
G30CB4	66	30	4	0.3
G20CB4	76	20	4	0.3
G20CB2	78	20	2	–
G20CB0.5	79.5	20	0.5	–

Fig. 2 Experimental setup for measurements of dielectric parameters of fabricated samples, *1*—dielectric probe, *2*—VNA

2.2 Measurements and Data Processing

Dielectric parameters of each of fabricated samples were measured in ten repetitions by open-ended coaxial reflection probe method. The experimental setup, shown in Fig. 2, was consisted of handheld VNA Keysight N9913A FieldFox, PC with corresponding software and dielectric probe SPEAG DAK-12 which can perform measurements in frequency range from 10 MHz to 3 GHz [8]. Dielectric probe was calibrated by OSL calibration method. Measured data was processed in MATLAB®, statistically evaluated using the uncertainty of type C and compared with dielectric parameters of skin, skull and CSF represented by fourth order Cole–Cole model also implemented in MATLAB®.

3 Results

Measured trends of dielectric parameters of some of fabricated phantom samples are shown in this chapter. In each figure is displayed the interval given by uncertainty of type C (error bars) which with a certain probability (95%) includes actual value of the measured relative permittivity and conductivity, respectively.

4 Discussion

Because of large number of phantom samples only relevant results are presented. Relevant means that results which were the most similar to Cole–Cole model of dielectric parameters of head tissues (skull, skin, CSF). From Figs. 3, 4 and 5 is evident that each of three human head tissues can be mimicked by mixing of urethan rubber, graphite powder and carbon black powder in specific weight percentages. Relative errors for each of presented phantom samples with respect to Cole–Cole model of given head tissues for frequency 1 GHz are summarized in Table 3. The little higher

Fig. 1 Some of fabricated samples of human head tissue phantoms

Fig. 3 Measured relative permittivity (**a**) and conductivity (**b**) of fabricated phantom sample G40 compare with Cole–Cole model of skin, skull and CSF

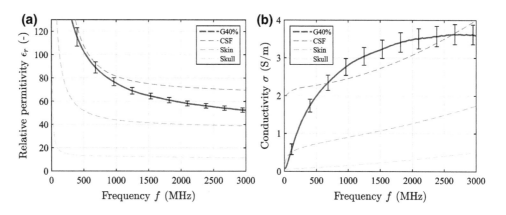

Fig. 4 Measured relative permittivity (**a**) and conductivity (**b**) of fabricated phantom sample G15CB4 compare with Cole–Cole model of skin, skull and CSF

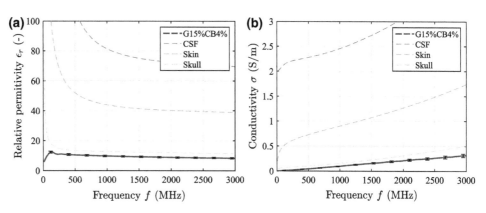

Fig. 5 Measured relative permittivity (**a**) and conductivity (**b**) of fabricated phantom sample G25CB2.5 compare with Cole–Cole model of skin, skull and CSF

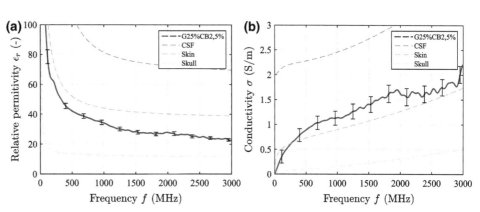

Table 3 Relative errors $\delta(\%)$ for each of presented phantom samples with respect to Cole–Cole model of given head tissues for frequency 1 GHz

	G40/CSF	G15CB4/Skull	G25CB2.5/Skin
$\delta\varepsilon_r$ (%)	7.34	22.90	22.70
$\delta\sigma$ (%)	14.55	37.32	26.04

relative errors could be caused by inaccuracy of dielectric probe and by fact that surface of fabricated phantom samples was not absolute smooth. The rest of fabricated phantom samples mimicked given head tissues only in one of two dielectric parameters or did not at all. Some of samples mimicked brain tissues like white matter or grey matter which is going to be very valuable for next research in phantom design and fabrication area.

5 Conclusion

The aim of this contribution was to present some of results of phantom samples fabrication suitable for brain stroke classification. It was found out that mixture of urethan rubber, graphite powder, carbon black powder and acetone in specific weight percentages can mimic head tissues like skin, skull and CSF. This weight percentages will be very valuable for fabrication of dry heterogeneous phantom of human head in the future which needs to be fabricated at ELEDIA@CTU laboratory for testing of brain stroke classification/detection by MWI system.

Acknowledgements This research has been supported by the research program of the Czech Science Foundation, Project no. 17-00477Y, Physical nature of interactions of EM fields generated by MTM structures with human body and study of their prospective use in medicine. The authors declare that they have no conflict of interest.

References

1. S. Kwon and S. Lee, "Recent Advances in Microwave Imaging for Breast Cancer Detection," *International Journal of Biomedical Imaging*, vol. 2016, pp. 1–25, 21 12 2016.
2. M. Persson, A. Fhager, et al, "Microwave-Based Stroke Diagnosis Making Global Prehospital Thrombolytic Treatment Possible," *IEEE Transactions on Biomedical Engineering*, vol. 61, no. 11, pp. 2806–2817, 11 2014.
3. A. T. Mobashsher and A. M. Abbosh, "Artificial Human Phantoms: Human Proxy in Testing Microwave Apparatuses That Have Electromagnetic Interaction with the Human Body," *IEEE Microwave Magazine*, vol. 16, no. 6, pp. 42–62, 7 2015.
4. J. Tesařík, "Bow Tie Antenna and Simplified Model of System for Brain Stroke Follow Up," 2016.
5. J. Vrba, J. Karch and D. Vrba, "Phantoms for Development of Microwave Sensors for Noninvasive Blood Glucose Monitoring," *International Journal of Antennas and Propagation*, vol. 2015, pp. 1–5, 1 3 2015.
6. J. Garrett and E. Fear, "Stable and Flexible Materials to Mimic the Dielectric Properties of Human Soft Tissues," *IEEE Antennas and Wireless Propagation Letters*, vol. 13, pp. 599–602, 2014.
7. A. Santorelli, O. Laforest, E. Porter and M. Popović, "Image Classification for a Time-Domain Microwave Radar System: Experiments with Stable Modular Breast Phantoms".
8. "DAK 4 MHz to 3 GHz, SPEAG," [Online]. Available: https://www.speag.com/products/dak/dak-dielectric-probe-systems/dak-4-mhz-3-ghz/.

High-Water Content Phantom for Microwave Imaging and Microwave Hyperthermia

Michaela Kantova, Ondrej Fiser, Ilja Merunka, Jan Vrba, and Jan Tesarik

Abstract

The main goal of this contribution is the investigation, design, and evaluation of an agar-based high-water content phantom for microwave imaging and hyperthermia. The contribution attempts to specify preparing procedure to get the most uniform results and describes the problems during the fabrication of the phantom, solves the problem with air bubbles. The effect of the size of PE powder particles to dielectric properties (40–48 and 150 μm) was evaluated. Dielectric properties of agar phantoms with different compositions of grape sugar and polyethylene powder were compared for lowering the relative part of the complex permittivity over the frequency range of 10–2995 MHz. Proposed phantom also consists of agar powder, distilled water, sodium chloride and TX-151. The relative permittivity was decreased to the value under 60 at frequency 434 MHz for muscle phantom with available equipment and fabrication (with 13.23% PE powder, 40–48 μm).

Keywords

Hyperthermia • Tissue-mimicking phantom
Agar phantom

M. Kantova (✉) · I. Merunka · J. Vrba
Faculty of Electrical Engineering, Department of Electromagnetic Field, Czech Technical University Prague, Prague 6, Czech Republic
e-mail: kantomi1@fel.cvut.cz

O. Fiser · J. Tesarik
Faculty of Biomedical Engineering, Department of Biomedical Technology, Czech Technical University in Prague, Kladno, Czech Republic

1 Introduction

For the design of microwave hyperthermia and microwave imaging systems, is an essential step the possibility of testing. Ideally, a phantom of biological tissue should accurately represent all the properties of the biological tissue [1]. For microwave applications, the necessary parameters of the phantom are electrical conductivity and complex relative permittivity. In recent years, great effort has been devoted to designing phantoms and hyperthermia applicators. One of the new approaches are applicators based on zero-order mode resonator metamaterial structures [2, 3].

One of the first gel phantoms was developed by Guy [4]. The base ingredient for simulating high-water-content tissue is TX-150, the other substances are a saline solution and polyethylene powder. Another alternative is a phantom based on gelatin, water, kerosene and sunflower oil described by Lazebnik et al. [5] for the broadband frequency range 500 MHz–20 GHz. One of the first compositions of agar phantoms designed for hyperthermia comes from Kato and Ishida [6]. The first phantom was designed to test RF hyperthermia at 13.56 MHz. It consists of 3 components—agar powder (2%), NaCl (0.43%) and water (97.57%). Ito et al. created an agar phantom that represents the electrical parameters of a high-water tissue suitable for the frequencies 300–2450 MHz [7]. Phantom consists of agar powder, deionized water, TX-151, PE powder, NaCl and preservative. The polyethylene powder is used to modify relative permittivity, while the electrical conductivity is adapted in particular by the proportion of NaCl (but also by PE powder). The agar solution cannot be mixed with PE powder directly, so the TX-151 is used to increase viscosity. Due to the high melting point (approx. 80 °C), agar phantoms are suitable for applications using higher temperatures (ablation, hyperthermia) [1].

2 Phantom Design

The agar type of phantom was selected for designing our phantom for its easy preparation, availability and high melting point. The proposed phantom is based on a recipe [7], which consists of deionized water, agar, sodium chloride, TX-151, polyethylene powder, and preservative. Our measurement of the parameters of this muscle tissue model did not achieve the required values with available equipment. It was necessary to modify the phantom composition to obtain appropriate properties. The proportion of PE powder adjusts the relative permittivity; the electrical conductivity is regulated mainly by the amount of sodium chloride, less by PE powder. Various recipe variants were prepared, based on a change of the proportion of PE powder and sodium chloride in the resulting mass to get lower permittivity and conductivity. For reducing the permittivity was necessary to increase the percentage of PE powder in the mixture, to reduce the conductivity, decrease sodium chloride. We used 300 g of distilled water and proportions of other substances. Figure 1 shows the results of different versions when changing the ratio of PE powder and sodium chloride along with the reference model from IT'IS Foundation [8].

During the preparation, many air bubbles get into the sample, which in case of non-removal, would devalue the measured results. For removing the bubbles was used a vacuum system. The consistency of the final mixture changes with the percentage of PE powder, with 49 g of PE powder the mixture was more solid than with using 30 g of PE powder.

Due to our observation, with the increasing proportion of PE powder permittivity decreases. Conductivity is significantly affected by the presence of sodium chloride, less with the proportion of PE powder. With decreasing sodium chloride, conductivity decreases. The following Fig. 2 compares the measured permittivity with a PE particle size of 40–48 μm and a powder with particle's size 150 μm. Almost three times bigger particles have just a little effect on the measured permittivity.

The possibility of substitution PE powder with the more available grape sugar has been inspected. However, measured data shows that replacing cannot be fully achieved. Figure 3 shows samples with various percentage of grape sugar. Problematic is the consistency of the phantom, which is very adhesive.

3 Proposed Phantom

The best results were achieved with the phantom with 13.23% of the polyethylene powder. The detailed composition of the phantom is in Table 1. Comparison of this phantom with muscle tissue is shown in the following Fig. 4 along with the measurement uncertainty.

The following preparation procedure ensures the best stable results with available equipment. Distilled water with sodium chloride is placed in the pot. The mixture of TX-151

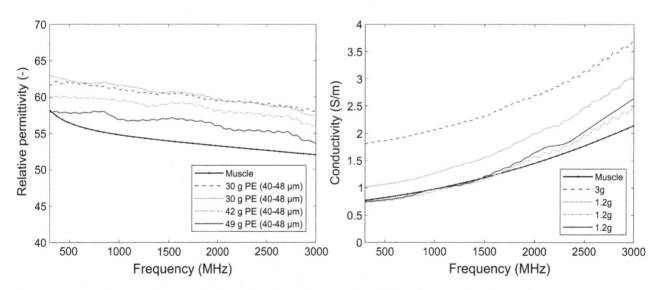

Fig. 1 Relative permittivity and conductivity of samples with varying proportion of PE powder and sodium chloride along with the muscle tissue properties by IT'IS Foundation [8]

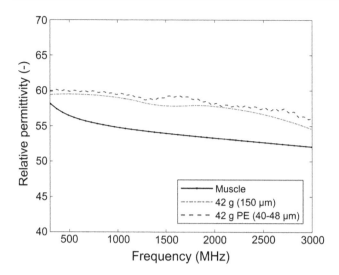

Fig. 2 Relative permittivity with varying particle's size along with the muscle tissue properties by IT'IS Foundation [8]

and agar powder is sprinkled several times by using mesh into the liquid and quickly mixed. The mixture is brought above to 85 °C (still mixed), which leads to the agar being solid. Polyethylene powder is sprinkled and quickly mixed into the liquid. The liquid is poured into the beaker and put into the vacuum system. Then vacuum system (approx. 0.6 bars due to the solidness of phantom) is used several times to remove air bubbles from the mixture. When the phantom is solid, it is covered and kept at room temperature. The mask, gloves and safety glasses should be worn during the whole fabrication.

Relative permeability and specific conductivity of all phantoms were measured using the open-ended probe kit SPEAG DAK-12 in conjunction with VNA Keysight FieldFox N9923A. The measurement was performed at room temperature. It was necessary to avoid too much pressure on the sample to avoid distortion of the measurement. Multiple measurements were taken to obtain

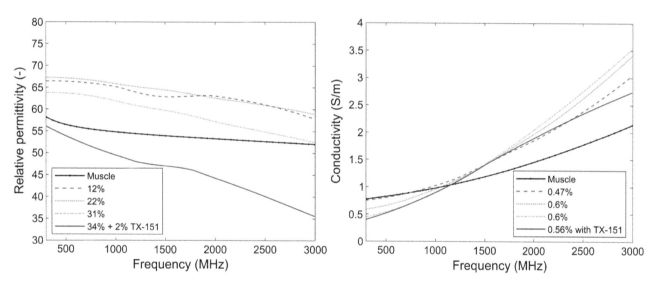

Fig. 3 Relative permittivity and conductivity of samples with varying proportion of grape sugar and sodium chloride along with the muscle tissue properties by IT'IS Foundation [8]

Table 1 The composition of proposed muscle phantom, substances in grams and percentages by weight

Ingredients	Percentage by weight (%)	Mass (g)
Distilled water	80.99	300
PE powder	13.23	49
TX-151	2.97	11
Agar powder	2.48	9.2
Sodium chloride	0.32	1.2

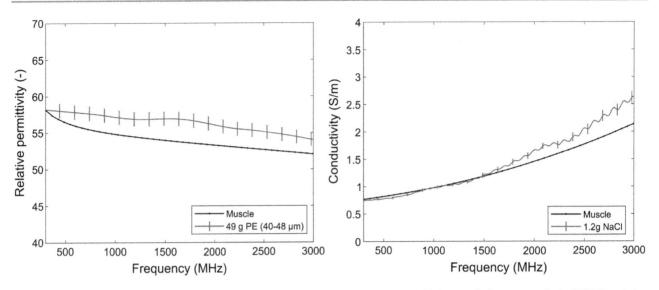

Fig. 4 Relative permittivity and conductivity of proposed muscle equivalent phantom along with the muscle tissue properties by IT'IS Foundation [8]

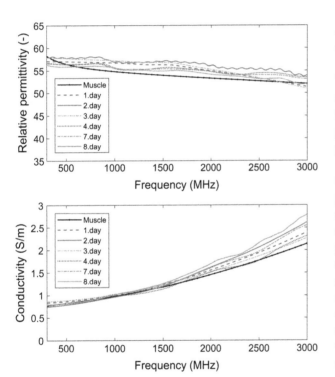

Fig. 5 Relative permittivity and conductivity of proposed muscle equivalent phantom measured for eight days along with the muscle tissue properties by IT'IS Foundation [8]

measurement uncertainty. Long-term stability of parameters was examined as shown in Fig. 5. The parameters did not change drastically during eight days.

4 Conclusion

The appropriate type of phantom was chosen for our application, prepared and measured. Due to the required properties, it was necessary to adapt the phantom. Proposed phantom is an alternative to the mentioned phantoms proposed in [4–7].

The recipe is suitable for hyperthermia applications over the frequency range from 434 MHz to 2.45 GHz. Appropriate values of relative permittivity and conductivity were achieved by gradual modification of the composition; consideration was also given to the consistency of the final phantom. The procedure of preparation requires vacuum system for removing any remaining air bubbles in the sample. Proposed phantom consists of approximately 81% distilled water, 13% polyethylene powder (particle size of 40–48 µm), 3% TX-151, 2.5% agar powder and 0.3% sodium chloride. The properties of the proposed phantom were measured over eight days, and no major changes were recorded. The possibility of replacing PE powder with grape sugar has also been investigated. Measured data shows that this substitution is not appropriate.

Acknowledgements This work has been supported by a grant from the Czech Science Foundation, number 17-20498 J: "Non-invasive temperature estimation inside of human body based on physical aspects of ultra-wideband microwave channel." The authors declare that they have no conflict of interest.

References

1. A. Dabbagh, B. J. J. Abdullah, C. Ramasindarum, and N. H. A. Kasim, "Tissue-Mimicking Gel Phantoms for Thermal Therapy Studies," *Ultrasonic Imaging*, vol. 36, no. 4, pp. 291–316, 2014.
2. D. Vrba, J. Vrba, D. B. Rodrigues, and P. Stauffer, "Numerical investigation of novel microwave applicators based on zero-order mode resonance for hyperthermia treatment of cancer," *J. Franklin Inst.*, vol. 354, no. 18, pp. 8734–8746, Dec. 2017.
3. D. Vrba, and J. Vrba, "Novel Applicators for Local Microwave Hyperthermia Based on Zeroth-Order Mode Resonator Metamaterial," *International Journal of Antennas and Propagation,* vol. 2014, pp. 1–7, 2014.
4. A. W. Guy, "Analyses of Electromagnetic Fields Induced in Biological Tissues by Thermographic Studies on Equivalent Phantom Models," in *IEEE Transactions on Microwave Theory and Techniques*, vol. 19, no. 2, pp. 205–214, February 1971.
5. M. Lazebnik, E. Madsen, G. Frank and S. Hagness, "Tissue-mimicking phantom materials for narrowband and ultrawideband microwave applications," *Physics in Medicine and Biology*, vol. 50, no. 18, pp. 4245–4258, 2005.
6. T. Ishida and H. Kato, "Muscle Equivalent Agar Phantom for 13.56 MHz RF-Induced Hyperthermia," *Shimane J. Med. Sci.*, vol. 4, no. 2, pp. 134–140, 1980.
7. K. Ito, K. Furuya, Y. Okano, and L. Hamada, "Development and characteristics of a biological tissue-equivalent phantom for microwaves," *Electron. Commun. Japan (Part I Commun.*, vol. 84, no. 4, pp. 67–77, 2001.
8. P.Hasgall, E.Neufeld, M.Gosselin, A.Klingenböck, and N.Kuster, "IT'IS database for thermal and electromagnetic parameters of biological tissues.," *IT'IS Foundation*, 2017. [Online]. Available: www.itis.ethz.ch/database. [Accessed: 28-Jan-2018].

Processing of Standard MR Images Prior Execution of the MR-Based Electrical Properties Tomography (MREPT) Method

Luis F. Díaz Rondón and Jan Tesarik

Abstract

Magnetic resonance-based electrical properties tomography (MREPT), uses information of the B_1^+ field distribution in MRI, and computes electrical properties relative permittivity and electrical conductivity via the Helmholtz equation. The method can be done using standard MRI sequences, boosting the method into a more realistic clinical environment. MRI images need to undergo certain image pre-processing to correct ailments like noise and phase shifting. Recent publications on the topic report encouraging results and give a detailed explanation on the theory in which the process is based. However, little or none explanation is given to the processing to which the standard MRI images need to undergo in order for the method to be correctly implemented. Emphasis will be put on processing and corrections that needed to be applied to the retrieved MRI images to arrive to the results that are here reported.

Keywords

MREPT • Electrical properties • Relative permittivity
Electrical conductivity • In vivo • MRI

1 Introduction

Electrical properties for biological tissues can be computed in vivo using MR imaging if the B_1^+ field distribution in the scanned sample is determined. MREPT is a technique that is able to map relative permittivity (ε_r) and electrical conductivity (σ) via the Helmholtz equation [2]. Standard MRI sequences, Gradient Recalled Echo (GRE) low-flip-angle and Spin Echo (SE), can be used to derive a complex field distribution approximation [5]:

$$\sqrt{B_1^+ B_1^-} \rightarrow \sqrt{|I_{GRE}|} \cdot exp(\frac{1}{2} \angle I_{SE}). \tag{1}$$

$$\varepsilon_r = -\frac{1}{\omega^2 \mu_0 \varepsilon_0} Re\left\{ \frac{\nabla^2 \sqrt{B_1^+ B_1^-}}{\sqrt{B_1^+ B_1^-}} \right\}, \tag{2}$$

$$\sigma = \frac{1}{\omega \mu_0} Im\left\{ \frac{\nabla^2 \sqrt{B_1^+ B_1^-}}{\sqrt{B_1^+ B_1^-}} \right\}, \tag{3}$$

In Eq. (1), $|I_{GRE}|$ stands for image intensity of the magnitude image from a GRE sequence with low flip angle; $\angle I_{SE}$ stands for image intensity of the phase image of a SE sequence; and ω is the Larmor frequency. Equations (2) and (3) solve for relative permittivity and electrical conductivity respectively [5].

An experiment was done in which three liquid samples, enclosed in 250 ml plastic containers, with varying electrical properties, underwent MREPT following procedures previously published by other research groups. The results of this experiment were published in [3]. During this, some difficulties were encountered that were not covered in much detail in other publications and thus the reason for this document. Here will be covered the processing steps on the GRE and SE resulting images that took place in order achieve the satisfactory results reported on [3].

2 Pre-processing Prior MREPT

The scans were made on a 3T SIEMENS MRI scanner. From the scanner, all images were exported as DICOMS and imported into, and processed using MATLAB (MathWorks, USA).

L. F. Díaz Rondón (✉) · J. Tesarik
Faculty of Biomedical Engineering, Czech Technical University in Prague, Prague, Czech Republic
e-mail: diazluis@fbmi.cvut.cz

© Springer Nature Singapore Pte Ltd. 2019
L. Lhotska et al. (eds.), *World Congress on Medical Physics and Biomedical Engineering 2018*,
IFMBE Proceedings 68/3, https://doi.org/10.1007/978-981-10-9023-3_142

2.1 GRE Low-Flip-Angle Magnitude

Square root The magnitude image coming from the low-flip-angle GRE sequence undergoes a simple square root operation yielding the magnitude term of the complex number seen in Eq. (1). A filtering stage is necessary to proceed. Getting rid of the noise in the image gave a considerable amount of trouble, as different filtering options, normally used in image processing, were applied with no satisfactory results. Noise needs to be filtered out in such subtle way as to conserve the data which in this case, is supposed to represent the magnitude values in space of the RF magnetic field distribution. Figure 1a shows the original magnitude image coming from the GRE sequence, while the square-rooted GRE can be seen on Fig. 1c.

De-noise The taken approach was the wavelet transform. By using a basic built-in de-noising function in Matlab (MathWorks, USA) [6], it was possible to reduce the noise in the image at such point that real results matched the expectations. Figure 2d and e show the surface of one

scanned sample before and after the wavelet de-noising function.

2.2 SE Phase

Re-scaling SIEMENS displays phase images using intensity values from a scale of 0 to 4096, with 2048 being the zero phase. The image needs to be operated in such way that it could be re-scaled from (π) to $(-\pi)$. Figure 1e shows original phase image from SE, and Fig. 1f, the image after re-scaling.

Phase correction SE phase images, may come with offset and spatially varying errors, known as zero-order and first-order phase errors. In [1], a very simple and effective method for first- and zero-order phase correction is very well explained, and this was the one successfully applied during the experiments. This method starts by stating that the image intensity from SE image is expressed by

Fig. 1 Resulting images from the different stages of their processing. **a** Original GRE low-flip-angle magnitude image. **b** GRE low-flip-angle image with background noise removed and edges extracted. **c** GRE magnitude square-rooted. **d** Square-root of GRE low-flip-angle surface fitted. **e** Original SE phase image. **f** SE phase image, re-scaled from $(\pi, -\pi)$, with background noise removed and edges extracted. **g** SE phase-corrected. **h** Half of SE phase-corrected surface-fitted. **i** ε_r map (numerical simulations). **j** ε_r map (MREPT). **k** σ map (numerical simulations). **l** σ map (MREPT)

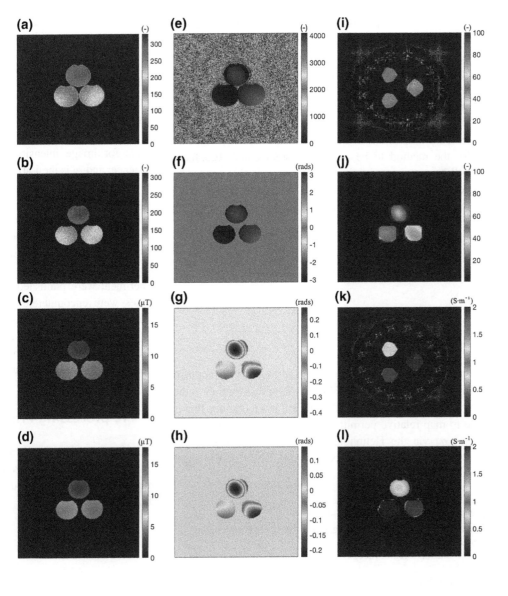

Fig. 2 **a** Histogram: SE phase values distribution from original re-scaled SE phase image. **b** Histogram: SE phase values distribution after first-order phase correction. **c** Histogram: SE phase values distribution after first-order and zero-order phase correction. **d** Noisy surface from a sample in square-rooted GRE magnitude image. **e** Sample's surface from square-rooted GRE magnitude image after wavelet transform de-noising. **f** Sample's surface from de-noised square-rooted GRE magnitude image after fitting

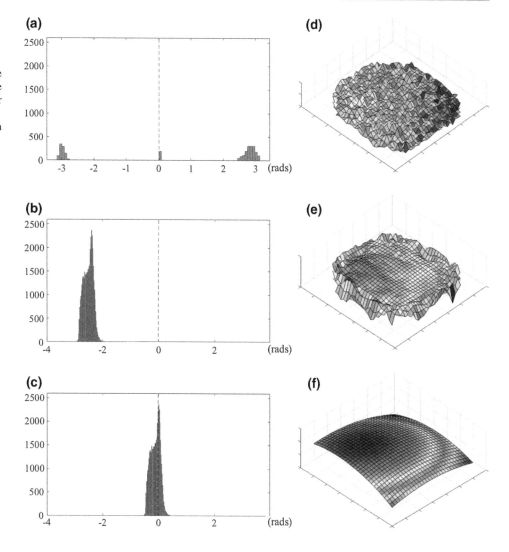

$$\hat{f}(x, y) = f(x, y) exp^{i\phi_0} exp^{i\epsilon_1 x},$$

where ϕ_0 is the zero-order phase error, ϵ_1 the first-order phase error, and $f(x, y)$ is the true image intensity. These two error terms are determined using an autocorrelation function and histogram operations, and then by inverse multiplication the error terms are cancelled. In Fig. 2a–c can be seen how the phase values are distributed in the image, at the beginning, after first-order error is corrected, and after zero-order error is corrected. Figure 1g shows SE phase image after successful phase correction. After the correction, the phase image can then be multiplied by 0.5 (Fig. 1h), following Eq. (1).

2.3 Surface Fitting

Computation of the Laplacian in a 2D or 3D image can be easily done by convolution using the appropriate kernel. Problem resides in noisy images and also in kernel position, as edges may be inside a different tissue type, increasing noise and/or

giving out false information. In order to deal with this issue, mainly the noise, a surface fitting took place. Every surface of the samples were extracted from the original image, and individually underwent a second-order polynomial surface-fit (Fig. 2f), following a parabola-fitting approach suggested and applied in [4]. Once all surfaces, from the magnitude and phase images were fitted, they were inserted back into their original format. End results can be seen on Fig. 1d and h. Done this, all needed to satisfy the equations and solve for relative permittivity and electrical conductivity, was to join square-rooted de-noised-surface-fitted GRE magnitude image, and halved phase-corrected surface-fitted SE phase image, into a complex-valued matrix, following Eq. (1).

2.4 MREPT and Resulting Electrical Properties

For every pixel, electrical properties were computed using Eqs. (2) and (3). Resulting images can be seen on Fig. 1j and l, for relative permittivity map and electrical

Table 1 Mean values of electrical properties of liquid samples: real values, results from numerical simulations, and results from MREPT

Sample #	Real		Sims.		MREPT	
	$\varepsilon_r(-)$	$\sigma(S \cdot m^{-1})$	$\varepsilon_r(-)$	$\sigma(S \cdot m^{-1})$	$\varepsilon_r(-)$	$\sigma(S \cdot m^{-1})$
1	73.9	1.24	76.9	1.22	75.5	1.27
2	48.1	0.17	44.9	0.17	48.3	0.11
3	53.7	0.34	52.2	0.33	56.6	0.34

conductivity map respectively. Figure 1i and k are correspondent electrical properties maps resulting from numerical simulations [3] (Table 1).

3 Conclusion

Magnetic Resonance-based Electrical Properties Tomography manages to provide a closer look into mapping of electrical properties of biological tissues in vivo. For those that are interested in following in this line of research it is important to understand that processing of delivered MR images is necessary, and its complexity will depend on the complexity of the imaged tissue.

Acknowledgements This research has been supported by the research program of the Czech Science Foundation, Project no. 17-00477Y, Physical nature of interactions of EM fields generated by MTM structures with human body and study of their prospective use in medicine.

References

1. Ahn, C.B., Cho, Z.H.: A New Phase Correction Method in NMR Imaging Based on Autocorrelation and Histogram Analysis. IEEE Transactions on Medical Imaging **6**(1), 32–36 (1987). https://doi.org/10.1109/tmi.1987.4307795
2. Bulumulla, S.B., Lee, S.K., Yeo, D.T.B.: Conductivity and permittivity imaging at 3.0T. Concepts in magnetic resonance. Part B, Magnetic resonance engineering **41B**(1), 13–21 (2012). https://doi.org/10.1002/cmr.b.21204
3. Diaz, L.F., Vrba, J., Vrba, D.: Extraction of electrical properties of strokes from magnetic resonance scans - testing on simplified head phantoms. PIERS - Progress in Electromagnetics Research Symposium, Singapoor (2017)
4. Katscher, U., Djamshidi, K., Voigt, T., Ivancevic, M., Abe, H., Newstead, G., Keupp, J.: Estimation of Breast Tumor Conductivity Using Parabolic Phase Fitting. In: ISMRM 20th Annual Meeting, Melbourne, vol. 20, p. 3482. ISMRM (2012)
5. Lee, S.K., Bulumulla, S., Lamb, P., Hancu, I.: Measurement of electrical properties of biological tissue at radio frequencies using magnetic resonance imaging. In: 2015 9th European Conference on Antennas and Propagation (EuCAP), pp. 1–4 (2015)
6. MathWorks: Understanding waveletes. URL https://www.mathworks.com/videos/series/understanding-wavelets-121287.html

Metamaterial Sensor for Microwave Non-invasive Blood Glucose Monitoring

Jan Vrba, David Vrba, Luis Díaz, and Ondřej Fišer

Abstract

In our previous paper, a metamaterial and microstrip transmission line sensors for non-invasive blood glucose level monitoring were designed. In this paper two different models of dielectric properties of blood glucose solutions are used to evaluate sensors' sensitivity by means of numerical simulations. Model A is adopted from professional literature and the model B was created by our group using information about dielectric properties of blood plasma-glucose solutions, recently published dielectric properties of red blood cell cytoplasm-glucose solutions and an electromagnetic mixing formula. Both sensors shows smaller sensitivity for the model B than for the model A. Due to non-linear dependency of dielectric properties on glucose concentration predicted by model B a lower sensitivity for high glucose concentrations was observed. The metamaterial sensor shows approximately 10-times higher sensitivity than the microstrip sensor of the same length.

Keywords

Microwave sensor • Wearable sensor • Glucose Monitoring

1 Introduction

The amount of glucose in the blood directly affects blood's dielectric properties in microwave frequency range, which makes it possible to detect changes in glucose concentrations (GC) using microwave sensors [1–4].

Several different sensors were designed to measure blood GC based on microwave resonators in past. Some have already been used for in vivo measurements and it has been shown [5–7] that the blood glucose level can be estimated in real-time. At the same time the need for individual calibration of each subject being measured by an invasive glucose meter was emphasized.

In this paper, numerical models of the metamaterial and microstrip TL sensors in virtual contact with blood samples which dielectric properties are defined by models A and B (glucose and frequency dependent models of dielectric properties of blood-glucose solutions) were created and sensors' S-parameters were computed in COMSOL Multiphysics [8]. Numerical results were used for evaluation of sensors' sensitivity.

2 Methods

The main aim of this paper is to evaluate the sensitivity of both sensors to GC in the blood. The two main components necessary for sensitivity evaluation are the mathematical model of the dielectric properties of the blood-glucose solution and numerical models of the sensors. The first mentioned is briefly described in Sect. 2.1 and the other in Sect. 2.2. The sensitivity of a microwave TL sensor is given in Sect. 2.3.

2.1 Models of Dielectric Properties

As mentioned above there are two models considered in this paper. The model A was adopted form [9]. In [4], it was observed that the model A significantly overestimate dependency of the dielectric properties on GC. Furthermore, non-linear behavior was observed in [9] (was not taken into account when creating the model) and [10]. From those reasons a new model (here denoted as model B) was created by our group (details about the model B go beyond this article and will be published elsewhere) as follows. In general a mixing formula, volumetric ratio between RBC and blood plasma and models of dielectric properties of blood plasma [11] and of RBC cytoplasm [10] were used. Both

J. Vrba (✉) · D. Vrba · L. Díaz · O. Fišer
Czech Technical University in Prague, Zikova 1903/4, 166 36
Prague 6, Czech Republic
e-mail: jan.vrba@fbmi.cvut.cz

© Springer Nature Singapore Pte Ltd. 2019
L. Lhotska et al. (eds.), *World Congress on Medical Physics and Biomedical Engineering 2018*,
IFMBE Proceedings 68/3, https://doi.org/10.1007/978-981-10-9023-3_143

permittivity as well as conductivity are more dependent on the GC in model A. Correspondingly, higher sensor sensitivity is expected for model A.

2.2 Numerical Models of Metamaterial and Microstrip Sensors

The geometry of the metamaterial (MTM) transmission-line based sensor as well as substrate dielectric properties were adopted from [3]. The second sensor is based on microstrip (MS) TL and has the same length and substrate as the MTM sensor. Both sensors are microwave 2-ports. Numerical models were created in full-wave numerical simulation tool COMSOL Multiphysics and they corresponds to sensors' prototypes inserted into bottom wall of the container for blood sample (see Figs. 1 and 2). The two models of dielectric properties of blood-glucose solutions were implemented in MATLAB [13] and imported into numerical models.

2.3 Sensor Sensitivity

In our case phase shift $\Delta\varphi$ of transmission coefficient S_{21} is used for evaluation of GC. Sensors' sensitivity is defined as

$$\Phi = \frac{\Delta\varphi}{\Delta c_{bg}}, \qquad (1)$$

where c_{bg} is glucose in blood concentration (mmol/l). The evaluation of the sensor sensitivity was performed as follows. First the transmission coefficients as functions of frequency were obtained from numerical simulations for two different c_{bg}, namely for 3 and 15 mmol/l. By subtracting the phases of the two S_{21} for different c_{bg} and dividing the resulted curve by the difference of the two considered c_{bg} we obtained an estimate of sensor sensitivity Φ in $(°\cdot\ell\cdot\text{mmol}^{-1})$.

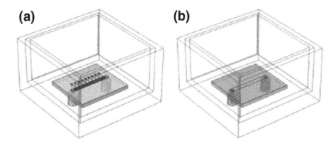

(a)   **(b)**

Fig. 1 Geometries of the numerical models of MTM (**a**) and MS (**b**) TL sensors

Fig. 2 Photos of the MTM sensor inside a container for blood sample. Length of the sensor (distance between the axes of the SMA connectors) is 2 cm

3 Results

For identifying two frequencies with high sensitivity and not too low module of S_{21} (> -30 dB) a set of numerical simulations of transmission coefficient were performed for c_{bg} ranging from 1 to 20 mmol/l. Phases of S_{21} as a function of c_{bg} at the two frequencies and for both mathematical models (A and B) are depicted in Figs. 3 and 4 for MTM and MS TL sensors, respectively.

Fig. 3 Numerically simulated dependence of phase of parameter S_{21} on glucose in blood concentration c_{bg} for the MTM sensor. For model A at 1,757 GHz (**a**) and for model B at 1,757 GHz (**b**)

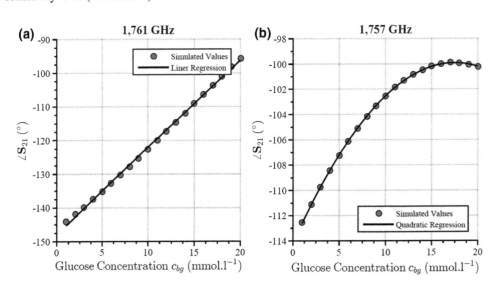

Fig. 4 Numerically simulated dependence of phase of parameter S_{21} on glucose in blood concentration c_{bg} for the MS TL sensor. For model A at 1,757 GHz (**a**) and for model B at 1,757 GHz (**b**)

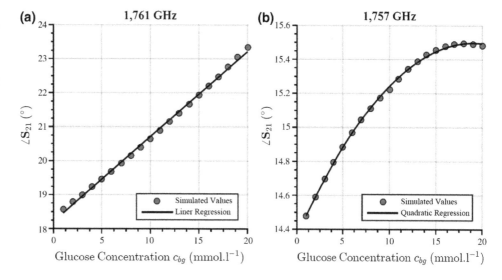

4 Conclusions and Outlook

In this work two different models of dielectric properties of blood glucose solutions were used to numerically evaluate sensitivity of the MTM and MS TL sensors. Sensors shows smaller sensitivity for the new model B than for the model A. For usual GCs c_{bg} < 10 mmol/l the phase of transmission coefficient has for both models almost linear dependency on GC. The linear dependency is shown in case of model A even for high GCs. In case of model B the non-linearity starts to be significant. The lower sensitivity for high GCs does not have to represent a problem. In case of detection of high GC in blood the measurement have to be repeated by other method. On the other hand the current invasive glucometers also show lower accuracy for higher glucose levels.

Although the sensitivity of the MTM sensor is considerably higher than that of the MS LT sensor, it is relatively small. Sensitivity of the MTM sensor, however, due to the high flexibility in the implementation of the sensor unit cells should not be difficult to further increase. Currently experimental validation of the mathematical and numerical models presented here is being prepared. Based on the results we plan to modify sensor geometry to set its sensitivity to desired value.

The sensitivity of the both sensors can be compared based on results shown in Figs. 3 and 4. The metamaterial sensor shows approximately 10-times higher sensitivity than the microstrip sensor of the same length.

Acknowledgements This research has been supported by the research program of the Czech Science Foundation, Project no. 17-00477Y.

References

1. Y. Hayashi, L. Livshits, A. Caduff, and Y. Feldman, "Dielectric spectroscopy study of specific glucose influence on human erythrocyte membranes," *Journal of Physics D: Applied Physics*, vol. 36, no. 4, pp. 369–374, Feb. 2003.
2. B. Freer and J. Venkataraman, "Feasibility study for non-invasive blood glucose monitoring," in *2010 IEEE Antennas and Propagation Society International Symposium (APSURSI)*, Jul. 2010, pp. 1–4.
3. J. Vrba and D. Vrba, "A Microwave Metamaterial Inspired Sensor for Non-Invasive Blood Glucose Monitoring," *Radioengineering*, vol. 24, no. 4, pp. 877–884, Dec. 2015.
4. J. Vrba, J. Karch, and D. Vrba, "Phantoms for Development of Microwave Sensors for Noninvasive Blood Glucose Monitoring, Phantoms for Development of Microwave Sensors for Noninvasive Blood Glucose Monitoring," *International Journal of Antennas and Propagation*, vol. 2015, 2015, p. 570870, Mar. 2015.
5. B. Jean, E. Green, and M. McClung, "A microwave frequency sensor for non-invasive bloodglucose measurement," in *IEEE Sensors Applications Symposium, 2008. SAS 2008*, 2008, pp. 4–7.
6. M. Sidley and J. Venkataraman, "Non-invasive estimation of blood glucose a feasibility study," in *2013 IEEE Applied Electromagnetics Conference (AEMC)*, Dec. 2013, pp. 1–2.
7. H. Choi, J. Naylon, S. Luzio, J. Beutler, J. Birchall, C. Martin, and A. Porch, "Design and In Vitro Interference Test of Microwave Noninvasive Blood Glucose Monitoring Sensor," *IEEE Transactions on Microwave Theory and Techniques*, vol. PP, no. 99, pp. 1–10, 2015.
8. COMSOL AB, COMSOL Multiphysics User's Guide, Version 5.1, 2015.
9. Adhyapak, M. Sidley, and J. Venkataraman, "Analytical model for real time, noninvasive estimation of blood glucose level," in *2014 36th Annual International Conference of the IEEE Engineering in Medicine and Biology Society (EMBC)*, Aug. 2014, pp. 5020–5023.

10. E. Levy, G. Barshtein, L. Livshits, P. B. Ishai, and Y. Feldman, "Dielectric Response of Cytoplasmic Water and Its Connection to the Vitality of Human Red Blood Cells: I. Glucose Concentration Influence," *The Journal of Physical Chemistry B*, vol. 120, no. 39, pp. 10214–10220, Oct. 2016.

11. T. Karacolak, E. C. Moreland, and E. Topsakal, "Cole-cole model for glucose dependent dielectric properties of blood plasma for continuous glucose monitoring," *Microwave and Optical Technology Letters*, vol. 55, no. 5, pp. 1160–1164, May 2013.

12. P. Hasgall, E. Neufeld, M. Gosselin, A. Klingenboeck, and N. Kuster. IT'IS database for thermal and electromagnetic parameters of biological tissues. Version 3.0, September 1, 2015, https://doi.org/10.13099/vip21000-03-0.

13. The Mathworks, Inc., MATLAB R2015a, Program and User's Manual, 2015.

A New Numerical Model of the Intra-aortic Balloon Pump as a Tool for Clinical Simulation and Outcome Prediction

Silvia Marconi, Carla Cappelli, Massimo Capoccia,
Domenico M. Pisanelli, Igino Genuini, and Claudio De Lazzari

Abstract

Counterpulsation is the form of circulatory support provided by the intra-aortic balloon pump (IABP), a device widely used in the clinical setting to assist the left ventricle by mechanical control of blood volume displacement within the aorta. The principles underlying the IABP function are well established but the interactions with the cardiovascular system make its performance rather complicated and still not completely understood. Here we propose a novel IABP numerical model as a tool with potential for clinical application in terms of simulation and outcome prediction in critical patients. The analysis of IABP assistance is presented in terms of the effects on stroke volume, mean aortic diastolic pressure, aortic end-systolic pressure, aortic end-diastolic pressure, endocardial viability ratio and pressure-volume loop shift. Balloon inflation/deflation rate and timing are also considered. The new model was implemented in CARDIOSIM© software simulator of the cardiovascular system.

Keywords

Cardiovascular system • IABP • Numerical model
Haemodynamics • Software simulation

1 Introduction

The intra-aortic balloon pump (IABP) has been the subject of significant research from an experimental, theoretical and clinical point of view [1, 2]. Despite the controversy generated by the shock trial [3], the IABP maintains a role to play in the clinical setting. Alternative percutaneous assist devices such as Impella and TandemHeart show a superior haemodynamic profile but require more technical expertise and no real survival benefit is observed. Central or peripheral veno-arterial extracorporeal membrane oxygenation (VA-ECMO) is a more advanced approach with variable survival although its timing and benefit remain controversial [4]. Despite the criticism and its limitations, the IABP remains widely used because of its immediate availability and ease of insertion. A combined IABP and VA-ECMO strategy is advocated but controversy remains [5].

The aim of this study was the development of a numerical model that could give appropriate information about the main haemodynamic and energetic parameters affected by the use of counterpulsation in order to monitor its efficacy in a simulation setting. A previous IABP module [6, 7] already existing in CARDIOSIM© software simulator, developed in the Cardiovascular Modelling Lab of the Institute of Clinical Physiology (Rome) [4–7], was upgraded with a new lumped parameter numerical model to reproduce the behaviour of

S. Marconi · C. De Lazzari (✉)
National Research Council, Institute of Clinical Physiology,
Rome, Italy
e-mail: claudio.delazzari@ifc.cnr.it

C. Cappelli
Faculty of Civil and Industrial Engineering, Sapienza University,
Rome, Italy

M. Capoccia
Cardiac Surgery Unit, University Hospitals of Leicester NHS
Trust, Leicester, UK

M. Capoccia
Department of Biomedical Engineering, University of Strathclyde,
Glasgow, UK

D. M. Pisanelli
National Research Council, Institute of Cognitive Sciences and
Technologies, Rome, Italy

I. Genuini
Department of Cardiovascular, Respiratory, Anaesthetic and
Geriatric Sciences, Sapienza University, Rome, Italy

I. Genuini · C. De Lazzari
National Institute for Cardiovascular Research (I.N.R.C.),
Bologna, Italy

L. Lhotska et al. (eds.), *World Congress on Medical Physics and Biomedical Engineering 2018*,
IFMBE Proceedings 68/3, https://doi.org/10.1007/978-981-10-9023-3_144

the systemic arterial circulation and the IABP combined with specific features for electrocardiogram (ECG) timing intervention.

The proposed new model allows real time changes of the IABP drive in relation to ECG timing and pressure. Also, appropriate synchronization with the dicrotic notch that precedes the inflation point of the balloon can be carried out leading to a reduction in peak systolic pressure and coronary perfusion pressure.

Preliminary simulation results showed that the new lumped parameter numerical model could reproduce the effects induced by the IABP on diastolic, systolic and mean

2 Methods

CARDIOSIM$^©$ consists of seven modules: left and right heart, systemic and pulmonary arterial sections, systemic and pulmonary venous sections and finally coronary circulatory network (Fig. 1). All the available modules can be assembled using relatively complex numerical representations. Several devices for mechanical circulatory and ventilation assistance are implemented in our software [7, 9]. Atrial and ventricular behaviour is modelled with a modified time-varying elastance approach [6, 7]. The following equations allow the simulation of ventricular interaction:

$$P_{lv}(t) = \left[\frac{e_{sp}(t) \cdot e_{lv}(t)}{e_{lv}(t) + e_{sp}(t)}\right] \cdot \left[V_{lv}(t) - V_{lv,0}\right] + \left[\frac{e_{lv}(t)}{e_{lv}(t) + e_{sp}(t)}\right] \cdot P_{rv}(t) + \left[\frac{e_{sp}(t)}{e_{lv}(t) + e_{sp}(t)}\right] \cdot P_{lv,0}$$

$$P_{rv}(t) = \left[\frac{e_{sp}(t) \cdot e_{rv}(t)}{e_{sp}(t) + e_{rv}(t)}\right] \cdot \left[V_{rv}(t) - V_{rv,0}\right] + \left[\frac{e_{rv}(t)}{e_{sp}(t) + e_{rv}(t)}\right] \cdot P_{lv}(t) + \left[\frac{e_{sp}(t)}{e_{sp}(t) + e_{rv}(t)}\right] \cdot P_{rv,0}$$

systemic arterial pressures. The activated IABP model increases cardiac output, coronary blood flow and decreases left atrial pressure. The new IABP model can also be used for educational purposes to train medical personnel, paramedics, students in medicine and bioengineering in a controlled clinical simulation setting [6–9].

P_{lv} (P_{rv}) is the left (right) ventricular pressure; V_{lv} (V_{rv}) is the left (right) ventricular volume; $V_{lv,0}$ ($V_{rv,0}$) is the resting left (right) ventricular volume; e_{sp} is the septal elastance; e_{lv} (e_{rv}) is the left (right) ventricular elastance.

The following equations allow the simulation of atrial interaction [6, 7]:

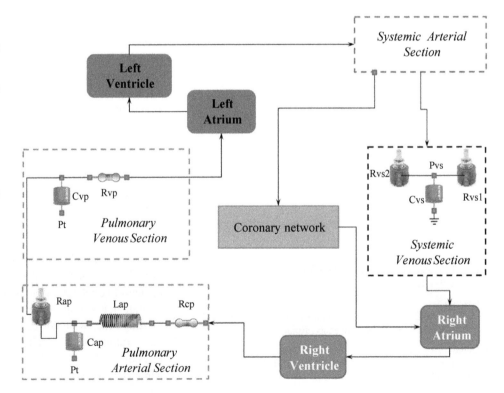

Fig. 1 General electrical analogue representation of the cardiovascular network. Systemic and pulmonary venous sections are modelled using simple representation. Rvs1 and Rvs2 are two variable resistances. Cvs (Pvs) is the systemic venous compliance (pressure). Pulmonary arterial section is modelled with a modified windkessel and variable peripheral pulmonary arterial resistance (Rap). Lap (Cap) is the pulmonary arterial inertance (compliance). Cvp (Rvp) is the pulmonary venous compliance (resistance). Pt is the mean intra-thoracic pressure

$$P_{la}(t) = \left[\frac{e_{Asp}(t) \cdot e_{la}(t)}{e_{la}(t) + e_{Asp}(t)}\right] \cdot \left[V_{la}(t) - V_{la,0}\right] + \left[\frac{e_{la}(t)}{e_{la}(t) + e_{Asp}(t)}\right] \cdot P_{ra}(t) + \left[\frac{e_{Asp}(t)}{e_{la}(t) + e_{Asp}(t)}\right] \cdot P_{la,0}$$

$$P_{ra}(t) = \left[\frac{e_{Asp}(t) \cdot e_{ra}(t)}{e_{Asp}(t) + e_{ra}(t)}\right] \cdot \left[V_{ra}(t) - V_{ra,0}\right] + \left[\frac{e_{ra}(t)}{e_{Asp}(t) + e_{ra}(t)}\right] \cdot P_{la}(t) + \left[\frac{e_{Asp}(t)}{e_{Asp}(t) + e_{ra}(t)}\right] \cdot P_{ra,0}$$

P_{la} (P_{ra}) is the left (right) atrial pressure; V_{la} (V_{ra}) is the left (right) atrial volume; $V_{la,0}$ ($V_{ra,0}$) is the resting left (right) atrial volume; e_{la} (e_{ra}) is the left (right) atrial elastance. The coronary section was modelled as described in [4, 7].

The scheme described in Fig. 1 was used to test the new IABP model. Figure 2 shows the electrical analogue of the new IABP numerical model inserted in the systemic arterial tree. The systemic arterial section includes four modified windkessel models reproducing the aortic (R_{AT}, L_{AT} and C_{AT}), thoracic (R_{TT}, L_{TT} and C_{TT}) and abdominal (R_{ABT1},

L_{ABT1}, C_{ABT1} and R_{ABT2}, L_{ABT2}, C_{ABT2}) tracts respectively. The IABP behaviour was emulated by the compliances C_{ABP1}, C_{ABP2} and C_{ABP3} activated in parallel with the systemic compliances, the resistances R_{ABP1}, R_{ABP2} and R_{ABP3} and a generator reproducing the balloon driving pressure.

The IABP resistances are volume-dependent. The balloon drive can be controlled to make changes in the driving, vacuum and plateau pressures. Also, appropriate ECG synchronisation is achieved by modifying the balloon cycle duration and making further changes to the rapid inflation

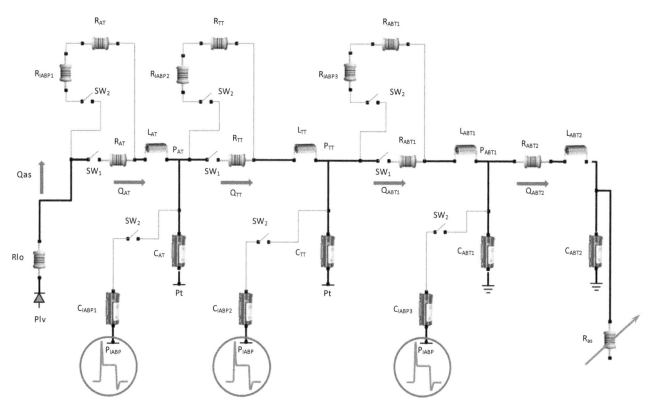

Fig. 2 Electrical analogue representation of the systemic arterial tree and IABP. P_{IABP} is the driving IABP pressure. R_{as} represents the variable peripheral arterial systemic resistance. P_{AT}, P_{TT} and P_{ABT1} represent the aortic, thoracic and abdominal pressures respectively. When all SW_1 are ON and all SW_2 are OFF the IABP is not activated, when all SW_1 are OFF and all SW_2 are ON the IABP is activated. The abdominal section is divided in two different tracts (R_{ABT1}, L_{ABT1}, C_{ABT1} and R_{ABT2}, L_{ABT2}, C_{ABT2})

Fig. 3 Screen output with
simulated diseased states obtained
using CARDIOSIM© software
simulator

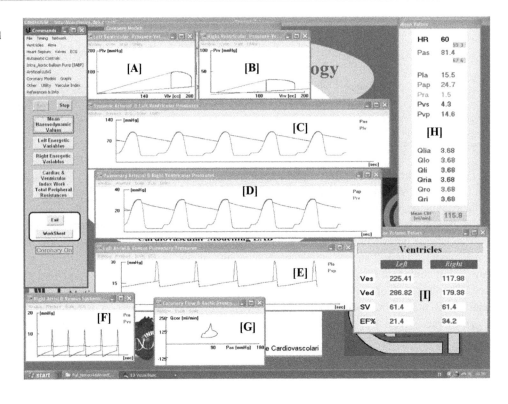

Fig. 4 Screen output with
simulated diseased and assisted
(with IABP) conditions obtained
using CARDIOSIM© software
simulator

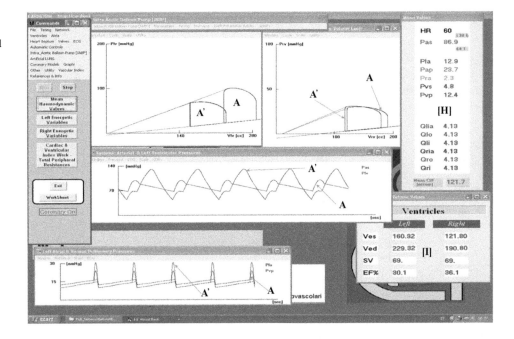

time, peak inflation artefacts, plateau pressure, rapid deflation, deflation artefact and baseline return. The IABP is ON
when all SW_1 are OFF and all SW_2 are ON. The effects of
IABP assistance were analysed on diseased states reproduced from literature data following IABP activation and
ECG synchronisation.

3 Results

Figure 3 shows CARDIOSIM© screen output when diseased
states were reproduced from literature data. Window A
(B) in Fig. 3 shows the left (right) ventricular

pressure-volume loop. The instantaneous left ventricular and systemic arterial pressures (right ventricular and pulmonary arterial pressures) are plotted in window C (D). The instantaneous left atrial and pulmonary venous pressures (right atrial and systemic venous pressures) are plotted in window E (F). Window G shows the coronary blood flow-aortic pressure loop.

The mean values of pressures and flows calculated during the cardiac cycle are reported in panel H with particular reference to the mean (P_{as}), systolic and diastolic aortic pressures and the mean coronary blood flow (CBF). Panel I shows the end-systolic (V_{es}) and end-diastolic (V_{ed}) ventricular volumes, the stroke volume (SV) and the ejection fraction (EF%) for both ventricles. Figure 4 shows the effects induced by IABP activation. The heart rate (HR) remains unchanged at 60 beat/min during the assistance. A rightward shift of the left ventricular loop is observed with reduction of end-diastolic and end-systolic volumes. The instantaneous aortic pressure before (A) and after (A') IABP activation is listed in the middle window. The results obtained with the proposed new IABP numerical model are consistent with current literature data. In fact, the analysis of Figs. 3 and 4 shows:

- an increase in "augmented diastolic aortic blood pressure" up to 130.6 [mmHg];
- a decrease in "assisted aortic end-diastolic blood pressure" up to 44.1 [mmHg];
- a decrease in mean left atrial pressure (P_{la}) from 15.5 [mmHg] to 12.9 [mmHg];
- a decrease in mean venous pulmonary pressure (P_{vp});
- an increase in cardiac output (CO) from 3.68 [l/min] to 4.13 [l/min];
- an increase in CBF from 115.8 [ml/min] to 121.7 [ml/min].

4 Conclusions

Although preliminary simulation results are quite promising, further data evaluation from patients treated with IABP is required to validate and improve our model.

Acknowledgements This work was supported by the Italian Ministry of Education, University and Research (M.I.U.R.) Flagship InterOmics Project (cod. PB05).

References

1. He P, Dubin SE, Moore TW, Jaron D. Guidelines for Optimal Control of Intraaortic Balloon Pumping: Experimental Studies. ASAIO J. 7(4), 172–179 (1984).
2. Jaron D, Moore TW, He P. Control of Intraaortic Balloon Pumping: Theory and Guidelines for Clinical Applications. Ann. Biomed. Eng. 13, 155–175 (1985).
3. Thiele H, Zeymer U, Neumann FJ, et al. Intraaortic Balloon Support for Myocardial Infarction with Cardiogenic Shock. N. Engl. J. Med. 367, 1287–1296 (2012).
4. Ergle K, Parto P, Krim SR. Percutaneous Ventricular Assist Devices: A Novel Approach in the Management of Patients with Acute Cardiogenic Shock. Ochsner J. 16, 243–249 (2016).
5. Nuding S, Werdan K. IABP plus ECMO – Is one and one more than two? J. Thorac. Dis. 9(4), 961–964 (2017).
6. De Lazzari C. Interaction between the septum and the left (right) ventricular free wall in order to evaluate the effects on coronary blood flow: numerical simulation. Comput. Methods Biomech. Biomed. Engin. 15(12), 1359–1368 (2012).
7. De Lazzari C, Genuini I, Pisanelli D, et al. Interactive simulator for e-Learning environments: a teaching software for health care professionals. Biomed. Eng. Online 13(1), 172 (2014).
8. De Lazzari C, Darowski M, et al. Ventricular energetics during mechanical ventilation and intraaortic balloon pumping – Computer simulation. J. Med. Eng. & Tech., 25(3), 103–111 (2001).
9. De Lazzari C, Stalteri D. CARDIOSIM© Software Simulator: http://cardiosim.dsb.cnr.it/.

Proposal of Electrode for Measuring Glucose Concentration in Blood

Klara Fiedorova, Martin Augustynek, and Tomas Klinkovsky

Abstract

The article deals with the design of the electrode system and the appropriate measuring circuit for the measurement of blood glucose concentration, since the determination of the level of blood glucose is an integral part of many medical procedures for determining the state of the human organism. The aim of the concept is to create two functional devices with different design of the electrode system and to perform a series of measurements to verify the functionality and the data statistically process. There was used the principle of resistive sensing of non-electric quantities, so the designed systems using gold electrodes working as electrolytic sensors. The problem relating to the measuring circuit, which is the same for both electrode systems, has been solved by a microampere meter which can convert the generated signal to a suitable measured electrical signal. The signal is formed by the reaction of glucose and enzyme, the resulting value being directly proportional to the glucose concentration. At this work, created systems fully meet specified conditions and requirements. The research confirmed the functionality of the designed units. On the basis of obtained data is possible to compare and evaluate the constructed electrode systems and to make the basis for further development in the subject matter.

K. Fiedorova (✉) · M. Augustynek · T. Klinkovsky
VSB–Technical University of Ostrava FEECS K450, 17. listopadu 15, 708 33 Ostrava–Poruba, Czech Republic
e-mail: klara.fiedorova@vsb.cz

M. Augustynek
e-mail: martin.augustynek@vsb.cz

T. Klinkovsky
e-mail: tomas.klinkovsky@seznam.cz

Keywords

Measurement of blood glucose concentration
Glucose meters • Basic principles of analyzers
Blood glucose • Diabetes mellitus
Glucose measurement sensors

1 Introduction

In clinical practice, blood glucose is used as an important indicator of the state of the human body, primarily used in patients affected by diabetes mellitus. The glycemic value affects the physical proportions of the individual, the time of day and also the diet. The determined value is also dependent on the source of the analyzed blood. Normal blood glucose levels range from 3.3 to 6.6 mmol/l in capillary blood, 6.9–5.5 mmol/l in venous blood, and 4.2–6.4 mmol/l in blood plasma. A fall in blood pressure below 3.3 mmol/l is referred to as hypoglycemia and values above 5.5 mmol/l are called hyperglycemia. The basic division of methods used to measure blood glucose is invasiveness of measurement. Invasive methods are more accurate and include laboratory tests, glucometers, and part of a continuous subcutaneous electrode method [3]. Many invasive methods are based on the chemical reaction of the enzyme or the chemical with glucose, electrochemical or photometric techniques are used to determine the blood sugar concentration with the chemical reactions mentioned. Invasive methods are nowadays insurmountable in terms of accuracy and market accessibility. In addition, non-invasive methods would be very convenient in terms of comfort, but they are only experimental processes that are in the development phase. The most notable non-invasive method is the NIR method utilizing infrared interaction along with glucose. Further, there is reverse iontophoresis or Raman spectroscopy [3, 6].

The work focuses on the development of a suitable electrode or electrode system that would be able to measure

L. Lhotska et al. (eds.), *World Congress on Medical Physics and Biomedical Engineering 2018*,
IFMBE Proceedings 68/3, https://doi.org/10.1007/978-981-10-9023-3_145

the blood glucose concentration. This is the development of a new blood glucose measurement technique for continuous measurement where the sensor should be incorporated into the catheter and then inserted into the body of the patient. In end use, the sensor should be used to detect glucose levels, and then the heart rate should be calculated from this value.

2   Methods

Before the actual design of the electrode system, the type of electrode sensor had to be determined. It was taken into account that the conductivity of the solution, which will correspond to the glucose concentration, will subsequently be measured. Therefore, the proposed system should behave like an electrolytic sensor. General information on this type of sensor is given in [1, 2, 5, 6]. It was also very important to include in the proposal the fact that the enzyme that is specific to its chemical reactions is used for the measurement. The exact procedure for enzymatic chemical reactions is disclosed in the publications [4]. To illustrate the nature of the electrochemical reaction:

$$GLUCOSE + GOx(FAD) \rightarrow GLUKAGON + GOx(FADH_2)$$
$$GOx(FADH_2) + O_2 \rightarrow GOx(FAD) + H_2O_2$$

This part of the chemical reaction produces hydrogen peroxide, which oxidizes at a potential of

$$H_2O_2 \underset{exotherm, s=4MJ/kg}{\overset{0.6\,V.}{\rightarrow}} 2H^+ + 2e^-$$

Based on the specificity of the enzyme measurement, the electrode system design was followed. The first step was to select a suitable material for all elements of the electrode system so that the material did not enter and did not affect chemical reactions. The best material for the electrodes would be platinum, but due to availability, gold-plated electrodes that are of sufficient quality were used. The electrodes have a diameter of 0.6 mm. Next, the material into which the entire system is built will be chosen and will serve as a container for the injection of test samples. Silicone has been chosen in the form of a tube, allowing for easy installation of electrodes that are electrically non-conductive and do not affect the measurement. The tube diameter is 0.4 cm, the height above the earth electrode is 0.5 cm. The size of the space for application of the solution is $25.13 * 10^{-8}$ l. The electrodes embedded in the tube are connected by means of copper wires with pinheads, which can be connected to the measuring circuit. The material for the earth electrode is copper wires in the insulator. The size of the diameter is chosen with respect to the size of the

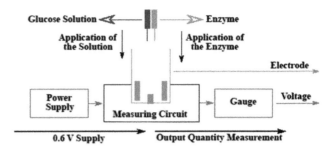

Fig. 1 Diagram of design principle of the proposed system

silicone tubing to provide a seal for the tube. Its main function is to eliminate the interference of the entire measuring system. It consists of a system of copper wires in an insulator.

To measure the blood glucose concentration, it was essential to construct a suitable circuit that would supply a required 0.6 V source voltage to the electrode and allow a suitable signal to be measured to match the glucose concentration. Since the assumed signal is assumed to be small, the circuit must be capable of capturing it. The whole is shown in Fig. 1.

The concept of conductivity, which is one of the methods for determining the concentrations of electrical conductivity of solutions, is applied in this issue. Here we take into account the given biosensor parameters, such as the cross-section and the distance that we calculate the resistance of the given system (Tables 1 and 2).

Based on the above parameters, the system resistance is calculated.

$$R = \rho * {}^l/_S \tag{1}$$

$$G = {}^1/_R[S] \tag{2}$$

$$\rho = Specific\ Resistance, l = Distance\ of\ electrodes$$
$$S = Cross\ Section\ of\ electrode;$$
$$G = k * C, C = S/l, k = Specific\ Conductivity$$

3   Results

The blood glucose level was measured under constant, unchanging conditions. For illustration and orientation in glucose analysis, blood was not used but aqueous solutions with a known glucose concentration of 3 mM, 7 mM, 12 mM, 18 mM with a volume of 0.3 ml. The measurement was carried out in two phases, the first phase being measured only with a pure glucose solution. Based on the fact that aqueous glucose solutions are made up of distilled water, we

Table 1 Overview of electrode system parameters

Electrode system	Cross section [mm]	Distance [mm]	Specific resistance [Ω cm]	Length [mm]
1	0.6	0.47	2.35 μ	4

Table 2 Overview of electrode system parameters

Electrode system	Resistance [Ω]	Conductivity [S]
1	$0.02 * 10^{-6}$	$50 * 10^{6}$

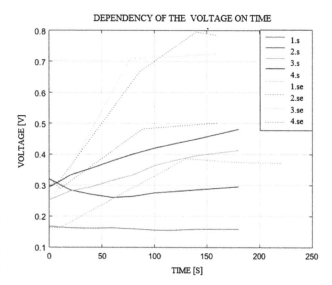

DEPENDENCY OF THE VOLTAGE ON TIME

Fig. 2 Graphical representation of the measured values using the first electrode system (Color figure online)

can assume that when the electrical voltage is applied to the solution, it will be possible to measure the current proportional to the glucose concentration in the solution. During the measurement, the value was recorded at the specified time point. In the second step, 0.01 g of the enzyme was added to the 10 s solution and the reading was read at the time point before the enzyme was added and at the time of enzyme application. Subsequently, there was a violent change in the voltage caused by the chemical reaction, and the response time was read off, the voltage value was again read off after the steady state. Each sample was measured 10 times. The voltage was recorded at certain time points in order to fully understand the nature of the measurements and the principle of chemical reactions.

A detailed description of the measured values presented in tables and graphs with statistic data analysis can be found in the literature. Tables 3 and 4 show the selected representatives of the measured groups, Table 3 characterizing the measurement of the pure glucose solution and Table 4 with the enzyme measurement. From both tables it is possible to read that the measured voltage value increases with increasing glucose concentration. Enzyme measurement is very characteristic of its course as the chemical reaction between glucose oxidase and glucose is fully applied here.

At this point, the time of the chemical reaction was measured, and the value of the stress after the steadying was then read. From these values it is again seen that the value of the measured voltage with increasing concentration increased.

It can be seen from the graph of Fig. 2 that the measured voltage measured at certain time intervals increases as the concentration of the solution analyzed at that time increases. The solid lines correspond to pure glucose solutions, is no enzyme added. The dotted lines plot the samples with the added enzyme. For each solution concentration, whether pure solution or with added enzyme, one color is selected. Measurement with and without enzyme is distinguished only by line type. The blue curves record a course of 3 mM solution analysis. In both cases these are the lowest values. Therefore, the assumption is that samples with a 3 mM glucose solution should have the lowest strain. The red lines

Table 3 Table of readings of pure glucose solution obtained by the first electrode system

Solution (mM)	[V]	20 s/[V]	40 s/[V]	60 s/[V]	80 s/[V]	100 s/[V]	120 s/[V]	3 min/[V]
3	0.168	0.17	0.173	0.182	0.19	0.195	0.199	0.22
7	0.322	0.296	0.284	0.271	0.276	0.272	0.276	0.295
12	0.253	0.281	0.295	0.316	0.334	0.364	0.381	0.413
18	0.348	0.358	0.379	0.411	0.429	0.438	0.451	0.499

Table 4 Table of readings of the pure enzyme solution with the added enzyme obtained by the first electrode system

Solution (mM)	[V]	Enzyme 10 s/[V]	Reaction [s]	Change [V] [min]	2:20 min/[V]	2:40 min/[V]
3	0.173	0.183	20	0.387(1:55)	0.392	0.385
7	0.241	0.246	16	0.492(1:17)	0.478	0.443
12	0.262	0.274	14	0.616(1:21)	0.662	0.664
18	0.286	0.294	14	0.671(1:21)	0.766	0.782

ranged from 7 mM samples. These curves give higher voltages than blue, so the assumption is confirmed again. Green includes curves plotting the measured voltage from a solution of 12 mM. Again, these are higher voltage values than in previous situations. The black curves recorded samples measured at a concentration of 18 mM. These values are visibly the highest.

4 Discussion

In this work it was a creation of a functional unit consisting of an electrode system and a measuring circuit that could determine the value of the concentration of glucose in the aqueous solution. After the construction of the measuring unit, it was possible to test it. This took place in two cycles, when the stress value of the pure glucose solution was determined with different concentrations of this substance and then the enzyme tests were performed. During the initial measurement experiments, we have immediately confirmed the assumption that electrical energy can transfer even a pure glucose solution, is without a proper chemical reaction. After this verification, the final set of experimental measurements could be switched. From the measured values that were subjected to the graphical comparison, it is clear that the values obtained by analyzing the pure glucose solution are dependent on the concentration of the glucose solution used, as the increasing concentration increases the value of the measured voltage. This fact was confirmed by the second assumption. Another important assumption was that data measured by analysis of the enzyme solution using glucose should become higher in voltage. This was confirmed during the comparison of the individual voltages that were

measured using the same glucose solution concentrations. This comparison clearly shows that the samples with the enzyme used show higher voltages. The course of these measurements is characterized by a rapid increase in voltage with a consequent rapid retreat. This phenomenon is associated with the chemical reaction and its saturation. However, when measuring the level of blood glucose, where many of the parts that are capable of transmitting electrical energy, we must accurately distinguish the electrical signal generated by the chemical reaction.

Acknowledgements The work and the contributions were supported by the project SV4507741/2101, 'Biomedicínské inženýrské systémy XIII'. This study was supported by the research project The Czech Science Foundation (GACR) No. 17-03037S, Investment evaluation of medical device development.

References

1. Baumstark, A., et al.: Lot-to-lot variability of test strips and accuracy assessment of systems for self-monitoring of blood glucose. ISO 15197. Journal of diabetes science and technology 6.5 (2012): 1076–1086.
2. Scheller, F., Schubert, F.: Biosensors. New York: Elsevier, 1992. ISBN 0-444-98783-5.
3. Clarke, S. F., Foster J. R.: A history of blood glucose meters and their role in self-monitoring of diabetes mellitus. British journal of biomedical science 69.2 (2012): 83.
4. Chang, T., Ming, S.: Biomedical applications of immobilized enzymes and proteins. Vol. 1. Springer Science & Business Media, 2013.
5. Holzinger, U., et al.: Real-time continuous glucose monitoring in critically ill patients. *Diabetes care* 33.3 (2010): 467–472.
6. Hruška, F.: Senzory: fyzikální principy, úpravy signálů, praktické použití. (2011).

Evaluation Application for Tracking and Statistical Analysis of Patient Data from Hospital Real Time Location System

David Oczka, Marek Penhaker, Lukáš Knybel, and Jan Kubíček

Abstract

In the recent time, an importance of the Real-time Locating systems (RTLS) in the hospital environment is increasingly important. Such systems are essentially utilized for the localization of either unmovable objects, or patients in the hospital environment. It is supposed that patients who underwent some trauma and surgery may have influenced their cognitive functions in a sense of lost concentration, and orientation is space. Such disorders may endanger the patient's health. In cooperation with the Trauma Center of University Hospital we have designed a localization system constituting the patients IR tags which are detected by the IR anchors placed in every hospital room. Data from such system represents the patient's movement and localization completed by sophisticated system of alarms. In this context we have developed the SW application which is connected with the RTLS database where the patient's localization data are stored. This application is fully integrated into the RTLS system. The SW application evaluates time data represents spent time in every room. The patient's data are consecutively refreshed and updated in the SW application to receive current information.

Keywords

RTLS • RF system • IR system

1 Introduction

Real-time location systems provide easy and efficient localization of objects or humans. These systems can be based on several types of technology including infrared radiation, radiofrequency radiation, GPS signal, ultrasonic waves and other or any combination of technologies listed before.

Basic model of real-time location system is consisted of an electronic tag for every tracked object, which is watched by some type of checking devices e.g. by gates. The gates are connected to some service which process and present position data to an authorized user in real time. The connection can be wireless or by wire depending on data workload. In most cases a processing service is web based and the users are connected to the system by the internet.

Real-time location systems are usually used to track goods, equipment, vehicles, pets or humans e.g. attendants of marathon. The tracking tags are very small and energy-independent to fulfill their function as long as possible [1–7].

2 Real-time Location System Overview

Real-time location system described in this paper uses a combination of infrared and radiofrequency technologies. An infrared emitter is used to determine position and a radiofrequency transmitter is used to transfer data. System is consisted of tags, infra-red anchors and receiving gates.

All positions, users, events and other data are saved in database. Every tag, gate and patient has unique identification number by which is distinguished in the system and the database. The database also containing information about which tag is held by which patient and that way assigning position data to the corresponding patient [8–11].

The system is event based and every patient movement between rooms is detected and saved to database. All events

D. Oczka (✉) · M. Penhaker · L. Knybel · J. Kubíček
Faculty of Electrical Engineering and Computer Science,
Department of Cybernetics and Biomedical Engineering, VŠB—Technical University of Ostrava, 17. Listopadu 15, 708 33
Ostrava, Poruba, Czech Republic
e-mail: david.oczka@vsb.cz

© Springer Nature Singapore Pte Ltd. 2019
L. Lhotska et al. (eds.), *World Congress on Medical Physics and Biomedical Engineering 2018*,
IFMBE Proceedings 68/3, https://doi.org/10.1007/978-981-10-9023-3_146

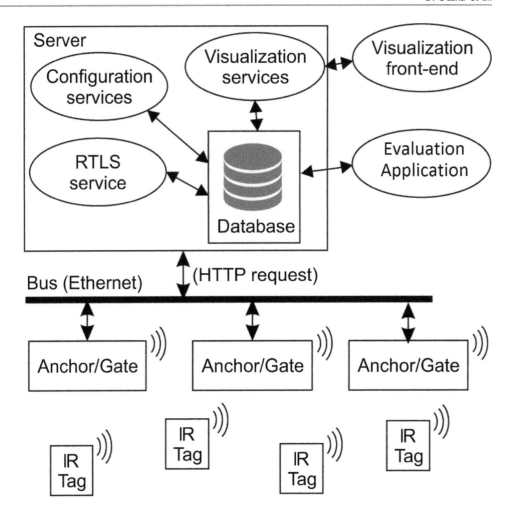

Fig. 1 Architecture of RTLS system

have assigned time stamp and room identifier, therefore every tag can be located in room and spent time in room can be calculated [12–15] (Fig. 1).

3 Design and Implementation

Software application is implemented in .NET Framework using Windows Forms platform and C# programming language. Database connection is realized by Firebird database library. The application use shared libraries of software part of Real-time location system, mentioned above in Chap. 2, to easier communication and processing of data offered by the system [16–20].

Application layout is as simple as possible and is consisted of two views: Patient listing view and Data processing view.

Patient listing view shows all available patients, which have finished their tracking session; their tracking tag in database is assigned as inactive. The view also offer sorting feature and data obtained from the system can sorted by many aspects e.g. name, surname, sex, social security

number, birth date. Detailed information about selected patient and his corresponding tag are available in the bottom of the Patient listing view. When patient or more patients are selected, their data from database are processed and shown in Data processing view (Fig. 2).

Data processing view process data of selected patients from Patient listing view and show them in time table and pie chart. In time table are shown all calculated times spent in single rooms by patient, the same data are shown in pie chart. Time data are calculated from events of patient change room. Final time is calculated as sum of differences between each event. If more patients are selected, calculated data from all patients are averaged (Fig. 3).

The application has also exporting feature. All selected and processed patient data can be exported in machine-readable and human-readable format. Machine-readable format is Comma Separated Values (CSV) format. Human-readable format is Portable Document Format (PDF).

Data from database are secured by password protection on database connection level and also authorized user login on application level. User data and credentials are protected by MD5 and AES encryption.

Fig. 2 Patient listing view

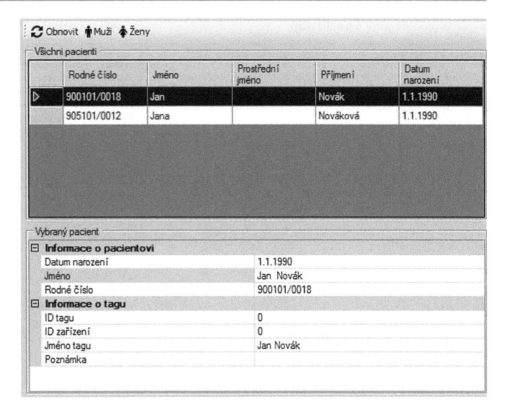

Fig. 3 Data processing view

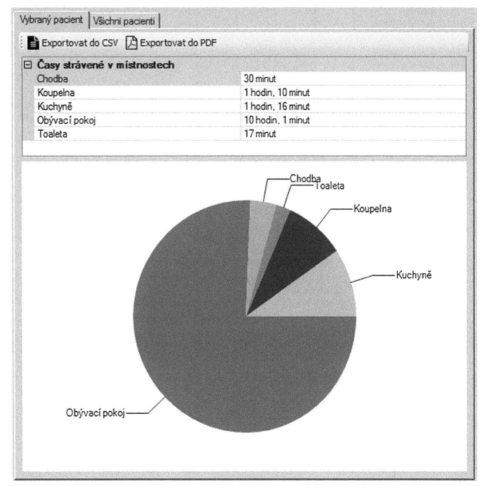

4 Discussion and Conclusion

The system is still under development, therefore database does not offer all data needed to implement new functions. There will be more features added in future to calculate other statistically useful results. The system and its collected data can reveal security risks which are not obvious and thanks to alarm features the system can alert appropriate staff to prevent an accident right in time.

Acknowledgements The work and the contributions were supported by the project SV4507741/2101, 'Biomedicínské inženýrské systémy XIII'. This study was supported by the research project The Czech Science Foundation (GACR) No. 17-03037S, Investment evaluation of medical device development.

References

1. J. Kubicek, I. Bryjova, H. Klosova, J. Stetinsky, and M. Penhaker, "The modelling of the post-operative perfusion in burns from LDI data," *Advances in Electrical and Electronic Engineering*, vol. 15, pp. 63–70, Mar 2017.

2. M. Penhaker, R. Geyerova, J. Kubicek, M. Augustynek, K. Panackova, and V. Novak, "Weariness and vigilance data mining using mobile platform assessment," in *Proceedings—2016 5th IIAI International Congress on Advanced Applied Informatics, IIAI-AAI 2016*, 2016, pp. 13–18.

3. M. Cerny and M. Penhaker, "Circadian Rhythm Monitoring in HomeCare Systems," in *13th International Conference on Biomedical Engineering, vol. 1–3*. vol. 23, C. T. Lim and J. C. H. Goh, Eds., ed, 2009, pp. 950 .

4. M. Penhaker, M. Stankus, J. Kijonka, and P. Grygarek, "Design and application of mobile embedded systems for home care applications," in *2010 2nd International Conference on Computer Engineering and Applications, ICCEA 2010*, 2010, pp. 412–416.

5. R. Hudak, J. Zivcak, T. Toth, J. Majernik, and M. Lisy, "Usage of Industrial Computed Tomography for Evaluation of Custom-Made Implants," in *Applications of Computational Intelligence in Biomedical Technology*. vol. 606, R. Bris, J. Majernik, K. Pancerz, and E. Zaitseva, Eds., ed, 2016, pp. 29–45.

6. J. Kubicek, M. Penhaker, K. Pavelova, A. Selamat, R. Hudak, and J. Majernik, "Segmentation of MRI Data to Extract the Blood Vessels Based on Fuzzy Thresholding," in *New Trends in Intelligent Information and Database Systems*. vol. 598, D. Barbucha, N. T. Nguyen, and J. Batubara, Eds., ed, 2015, pp. 43–52.

7. J. Grepl, M. Penhaker, J. Kubícek, A. Liberda, A. Selamat, J. Majerník, et al., "Real time breathing signal measurement: Current methods," *IFAC-PapersOnLine*, vol. 28, pp. 153–158, 2015.

8. R. Sabol, P. Klein, T. Ryba, L. Hvizdos, R. Varga, M. Rovnak, et al., "Novel Applications of Bistable Magnetic Microwires," *Acta Physica Polonica A*, vol. 131, pp. 1150–1152, Apr 2017.

9. R. Hudak, I. Polacek, L. Fedorova, P. Halfarova, M. Lisy, and J. Zivcak, "Mechanical Properties of Lumbar Bilateral Systems Using Two Different Spinal Rods are Comparable," in *6th European Conference of the International Federation for Medical and Biological Engineering*. vol. 45, I. Lackovic and D. Vasic, Eds., ed, 2015, pp. 285–289.

10. M. Tkacova, P. Foffova, R. Hudak, J. Svehlik, J. Zivcak, and Ieee, *Medical Thermography Application in Neuro-Vascular Deseases Diagnostics*, 2010.

11. J. Zivcak, P. Kneppo, R. Hudak, and Ieee, "Methodics of IR imaging in SCI individuals rehabilitation," in *2005 27th Annual International Conference of the IEEE Engineering in Medicine and Biology Society*, vol. 1–7, ed, 2005, pp. 6863–6866.

12. M. Penhaker, P. Klimes, J. Pindor, and D. Korpas, "Advanced Intracardial Biosignal Processing," in *Computer Information Systems and Industrial Management*, vol. 7564, A. Cortesi, N. Chaki, K. Saeed, and S. Wierzchon, Eds., ed, 2012, pp. 215–223.

13. J. Grepl, M. Penhaker, J. Kubicek, A. Liberda, and R. Mashinchi, "Real time signal detection and computer visualization of the patient respiration," in *Lecture Notes in Electrical Engineering* vol. 362, ed, 2016, pp. 783–793.

14. O. Krejcar, D. Janckulik, L. Motalova, K. Musil, and M. Penhaker, "Real Time Measurement and Visualization of ECG on Mobile Monitoring Stations of Biotelemetric System," in *Advances in Intelligent Information and Database Systems*. vol. 283, N. T. Nguyen, R. Katarzyniak, and S. M. Chen, Eds., ed, 2010, pp. 67–78.

15. O. Krejcar, M. Penhaker, D. Janckulik, and L. Motalova, "Performance test of multiplatform real time processing of biomedical signals," in *IEEE International Conference on Industrial Informatics (INDIN)*, 2010, pp. 825–830.

16. J. Majernik, L. Szerdiova, D. Schwarz, J. Zivcak, and Ieee, *Integration of Virtual Patients into Modernizing Activities of Medical Education across MEFANET*, 2016.

17. J. Majernik, P. Jarcuska, and Ieee, *Web-based delivery of medical education contents used to facilitate learning of infectology subjects*, 2014.

18. S. Conforto and T. D'Alessio, "Spectral analysis for non-stationary signals from mechanical measurements: A parametric approach," *Mechanical Systems and Signal Processing*, vol. 13, pp. 395–411, May 1999.

19. M. Goffredo, M. Schmid, S. Conforto, M. Carli, A. Neri, and T. D'Alessio, "Markerless Human Motion Analysis in Gauss-Laguerre Transform Domain: An Application to Sit-To-Stand in Young and Elderly People," *IEEE Transactions on Information Technology in Biomedicine*, vol. 13, pp. 207–216, Mar 2009.

20. M. Mancini, D. Brignani, S. Conforto, P. Mauri, C. Miniussi, and M. C. Pellicciari, "Assessing cortical synchronization during transcranial direct current stimulation: A graph-theoretical analysis," *Neuroimage*, vol. 140, pp. 57–65, Oct 2016.

Baby Cry Recognition Using Deep Neural Networks

Boon Fei Yong, Hua Nong Ting, and Kwan Hoong Ng

Abstract

Infant cry recognition is a challenging task as it is hard to determine the speech features that can allow researchers to clearly separate between different types of cries. However, baby cry is treated as a different way of communication of speech. The types of baby cry can be differentiated using Mel-Frequency Cepstral Coefficient (MFCC) with appropriate artificial intelligence model. Stacked restricted Boltzmann machine (RBN) is popular in providing few layers of neural networks to convert the high dimensional data to lower dimensional data to fine tune the input data to a better initialized weight for the neural networks. Usually RBN is used with another deep neural network to form the deep belief networks (DBN), and the studies in this direction is heading towards the convolutional-RBN variant. The study on RBN to pre-train Convolutional neural networks (CNN) without convolution function in the RBN meanwhile is scarce due to the Back propagation and principal component analysis can be applied directly to the CNN. In this paper, we describe the hybrid system between RBN and CNN for learning class specific features for baby cry recognition using the feature of Mel-Frequency Cepstral Coefficient. We archived an 78.6% of accuracy on 5 types of baby cries by validating the proposed model on baby cry recognition.

Keywords

Infant cry recognition • Restricted boltzmann machine Convolution neural networks

1 Introduction

Infant's cry is the primary form of communication which reflecting the infant's physiological function to the adults. Detecting the types of infant's cry can help to understands the current situation of the infants and help to detect early pathological diseases [1]. Scientists believe that infant cries are like the adults and therefore the studies of infant's cry recognition should be referenced to the adult speech recognition. The acoustics information of infant's cry had shown significant differences among different types of cries and this will help in the infant's cry recognition using machine learning [2]. Many infant's cry recognition studies are being done in past and they can be categorized into two main categories, statistical analysis and classification studies using the acoustics parameters and machine learning related infant's cry recognition studies. In this paper we are going to propose the use of machine learning method in formulating a model that can be used to classify different types of infant's cry.

The used type of machine learning in adult speech recognition can be repeated on the infant's cry recognition studies provided there are clear boundary of the acoustic features of different type of cries. Thus, the infant's cry recognition experiment processes are same as the adult speech recognition study. RBN and CNN are two famous deep learning neural networks models that are been used in image and speech recognition [3]. The acoustic features of infant's cry are first converted to the mathematical model which is closed to the human ear perception call the Mel-Frequency Cepstral Coefficient (MFCC) [4]. RBN is having the distributed hidden state to model and classify different infant's cry. This can make sure the acoustic

B. F. Yong · H. N. Ting (✉)
Faculty of Engineering, Biomedical Engineering Department, University of Malaya, Kuala Lumpur, Malaysia
e-mail: tinghn@um.edu.my

K. H. Ng
Department of Biomedical Imaging, University of Malaya, Kuala Lumpur, Malaysia

L. Lhotska et al. (eds.), *World Congress on Medical Physics and Biomedical Engineering 2018*,
IFMBE Proceedings 68/3, https://doi.org/10.1007/978-981-10-9023-3_147

features feed into the RBN are correlated with the different types of infant's cries so that the RBN can position the high level raw data into correlated lower level data which can be feed into the CNN. By having no connection between the hidden layers, the RBN largely breakdown the raw MFCC into only visible-hidden connections which means the infant's cry recognition problem is converted to the bipartite graph problem which can be solved using the Gibbs Sampling in RBN or other mathematical models [5]. By stacking the RBN into few layers, the deep belief net formed can be linked to the CNN. CNN is use in image recognition because its ability to handle well the dimensionality of the raw data of images which is usually formed by curves and boundary. However, CNN is not suitable to mode the MFCC local frequency directly because of there is a convolution layer in this CNN which is having the filter to convolve the input signals with a limited bandwidth. To solve this problem, we proposed to use the RBN as a pre-training to the CNN so that this MFCC acts like a static spectrum that can be correlated transformed and represents the infant cry in correlated locally process spectrum [6].

In the next section, the infant's cry acoustic features used in this study are explained with the processes of feature extraction from raw data into MFCC. Then this is follow by the explanation of the detail of RBN-CNN hybrid system architecture in handling the MFCC. In the Sect. 4, the experiment process and the results are presented. Then, the paper is followed by discussion and conclusion.

2 Features Extraction

The infant's cry samples were collected from the University Malaya Medical Centre with the help of the medical staffs. The samples were recorded with sampling rate of 16 kHz with 8-bit resolutions by putting the Olympus sound recorder 5 cm away from the crying babies' mouth. A total of 500 infant cry sounds were collected with 100 cry samples for every type of the cry. The infants were less than 2 weeks old. All the files are recorded as WAV files format. The five types of cry were pain, cold, hungry, diapers changed and discomfort (Another cry reason which is unknown) were collected separately. The pain cry samples were collected when the infants received their routine injections. Hunger cry samples were collected after before feeding. Cold cry samples were collected when the infant were having their routine bathing session. Diaper change cry samples were collected when the diaper was changed and for other unknown cries occur spontaneously they were classified as cry due to discomfort. Each cry sample corresponded to only one type or reason of reason.

MFCC is one of the most popular features used in the machine learning classification related problems. The MFCC is originated from the Mel scale filter bank which is simulating the human auditory function. For this study, only the voiced part of the infant's cry is taking into the conversion of the raw data to MFCC. The samples are parsing through hamming window and then only the voiced part are consider into the analysis window. The total window size used is 145 ms, with 25 ms static moving window overlapped with 10 ms of window size. This created total number of 13 frames per sample $((145-25)/10 +1)$. Each of the frame is being converted using the short time Fourier transform of the logarithm of the power spectrum to form the Mel Scale filter bank. Then they are converted into the coefficient by the discrete Fourier transform. Each of the frame will takes its original 13 orders of MFCC with its first and second derivatives forming the total of 507 (39×13) input nodes that will be fed into the RBN. By taking the overlapping window into consideration the warping function can represent the changes in the static window so that the cry samples are well represented into trainable form. There are total of 100 samples for each of the cries which form the total of 500 training and testing samples. They are divided into 5 folds of cross validation set which will helps in the generalization of the experiment which is described next.

3 RBM-CNN Hybrid System Architecture

3.1 RBM-CNN Hybrid System

The spectral variability of the input signal is modeled using the RBM model. In this study, we use a RBM that is only with connections between visible nodes and the hidden nodes. There is no visible-visible connection and hidden-hidden connection in the RBM. Each of the visible-hidden connections are represented by an energy function with its weights and bias value. Since infant's cry is considered as a time series problem, the type of RBM used is the conditional RBM because the ability to model sequential data by grouping the visible nodes into one group, and then perform the conditioning time step to move from one group to another. By doing so, the dynamic of the infant's cry can be well modeled before feed into the next layer of the hidden layer. Here the total number of group of frames are 13, where each group contains 39 nodes (represented by each frame of the window that is converted to the MFCC). Since there is no connection within the hidden units, all the hidden nodes are conditionally independent to each other. For the experiment, we are evaluating different numbers of hidden nodes which are 200, 500, 1000 or 2000. The numbers of hidden nodes will change the structure of the CNN but the method of grouping of the input nodes of the CNN are similar for all different numbers of hidden nodes with different scaling applied on the CNN input.

The CNN layer consists of the convolution layer and the max pooling layer. The RBM output is connected to the input layer of the CNN which is the convolution layer. For simplicity we use one layer of input and one hidden layer. The convolution layer applies the filter to the output layer of the RBN where each of the filter generalized along the input space. In this study we evaluated the convolution layer with filter size of 4 bands with total filters of 100 per band. The 200, 500, 1000 or 2000 output nodes used in the RBM are convoluted with the 100 filters. Every band has its own shared weight which is connected to the hidden layer of the convolution layer, each of the band is convoluted to the hidden layer with the shared weight which reduce the total numbers of nodes in the hybrid system and thus decrease the overfitting that may generates in the CNN layer. For max pooling layer, we are evaluating the use of different band sizes (1–8) of max pooling to see which is the most suitable for the hybrid system. The output layer of the CNN is a layer with five nodes which also equivalent to the total number of the infant's cry. When each bands are pooled towards the output layer, the lower number in the top layer provides a convergence in the training of the CNN.

The entire architecture of the RBM-CNN hybrid is shown in Fig. 1. **V** and **h** are visible and hidden weight connections respectively, **B** is the bias added to the connections.

3.2 Training

Every connection of the visible-hidden are using the same energy function with the softmax to model the joint probability function of the infant's cry [7]. The energy connection function is as the following.

$$E(\mathrm{v}, \mathrm{h}) = \sum_{i=1}^{V} \sum_{j=1}^{H} \omega_{ij} h_j v_{ij} - \sum_{j=1}^{H} h_{ij} \\ - \sum_{i=1}^{V} \frac{(v_i - b_i)^2}{2} \tag{1}$$

Probabilities of hidden layers are formulated as

$$p(\mathrm{v}, \mathrm{h}) = -\frac{e^{E(\mathrm{v}, \mathrm{h})}}{z} \tag{2}$$

where v and h represent the visible nodes and hidden nodes respectively; i and j represent the weight connections for visible node and hidden node respectively; b represents the bias; z is the probability constant. To train the RBM formed efficiently since the hidden layer is conditionally independent and the structure of conditional RBM where we group the visible nodes into different groups, we are going to use Gibbs Sampling with alternating sampling value of one visible layer group to the hidden layer and then repeating it

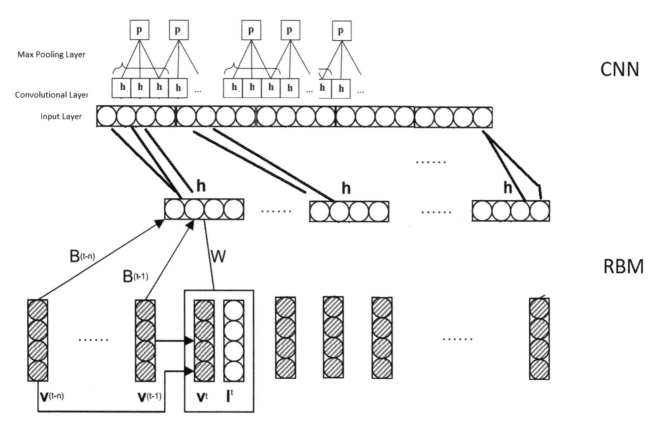

Fig. 1 RBN-CNN architecture

Table 1 Comparison of different architecture set up of RBN-CNN in infant's cry recognition

Method	Accuracy (%)
RBM200-CNN	68.3
RBM500-CNN	74.2
RBM1000-CNN	74.5
RBM2000-CNN	78.6

Table 2 Confusion matrix of the best recognition accuracy (82%)

	Hungry	Pain	Cold	Diaper-change	Discomfort
Hungry	78	4	5	6	7
Pain	5	87	2	3	3
Cold	4	4	80	6	6
Diaper-change	3	1	1	85	10
Discomfort	6	1	5	8	80

to the other visible layer group of the energy function using the following distribution.

$$p(\text{v}, \text{h}, m) = \alpha \left(\sum_{j=1}^{H} \omega_{ij} h_j v_{ij} \right) + m \qquad (3)$$

The weights and bias of the energy function are updated by solving the equation two using the stochastics gradient descent on the negative log likelihood as following where α is the learning rate (only use 0.01 in this study) and m is the momentum to smoothen the weight and bias update. Compared to other studies that used RBM for classification that use convergence divergence (CD) approximation of the gradient on the visible-hidden connections, the CD formula that used in this study is a simplified version where the samples are generalized by the Gibbs sampler to formulate the infant's cry dynamic data into the real value feature vectors. Last to speed up the training and avoid the RBM overfitting, the following generative and discriminative optimizer is added to the training process.

For the CNN layer, all the connections are trained using the stochastic gradient descent with Adam optimizer and dropout rate of 30%. All visible and hidden layers are connected in shared weights so the update of the weights are occurred in parallel to all the shared weights [8].

4 Results and Discussions

The infant's cry recognition average results are shown as Table 1.

The results show a trend of increasing numbers of hidden layer nodes in the RBM improved the accuracy. The best result is RBM with 2000 hidden nodes with CNN band size of 4 which achieved the average accuracy of 78.6% on a five-fold cross validation. From this setup, the highest accuracy out of the five folds cross validation is 82%. The confusion matrix of the best recognition accuracy is shown in Table 2. We can see diaper change and discomfort cry are two types of cry that slightly confused each other. However, we also noticed that changing the band size from 1 to 8 does

not bring any significant changes to the result (variation less than 1%).

5 Conclusions and Future Works

In this paper, we proposed the hybrid system of using Restricted Boltzmann Machine and Convolutional Neural Networks to perform infant's cry recognition on the newborn infants. The results are encouraging and it shows that CNN can be used for infant's cry recognition although we are using the MFCC as the features. To be more efficient, future works should be focused on using other speech features to replace MFCC as CNN require features which are locally correlated in time and frequency. Although we have proven the use of RBM and the conversion of MFCC into the form of spectrum are workable for the infant's cry recognition, we think that other speech features such as chromatogram, Mel spectrum, linear predictive cepstral coefficients, or other non-discrete cosine converted features will be more suitable for the CNN. The experiment should be repeated on larger size of infant's cry database and extends to the use in the pathological analysis and recognition studies.

References

1. O. Reyes-Galaviz and C. A. Reyes-Garcia, "A System for the Processing of Infant Cry to Recognize Pathologies in Recently Born Babies with Neural Networks," in *Conference Speech and Computer*, 2004.
2. J. Garcia and C. A. Reyes Garcia, "Mel-frequency cepstrum coefficients extraction from infant cry for classification of normal and pathological cry with feed-forward neural networks," in *Neural Networks, 2003. Proceedings of the International Joint Conference on*, 2013.
3. O. Abdel-Hamid, M. Abdel-Hamid, H. Jiang and G. Penn, "Applying Convolutional Neural Networks concepts to hybrid NN-HMM model for speech recognition," 2012.
4. A. Zabidi, W. Mansor, Y. K. Lee and F. Y. Abdul Rahman, "Mel-frequency cepstrum coefficient analysis of infant cry with hypothyroidism," in *Signal Processing & Its Applications, 2009. CSPA 2009. 5th International Colloquium on*, 2009.

5. G. Hinton, "A Practical Guide to Training Restricted Boltzmann Machines," in *Neural Networks: Tricks of the Trade*, Berlin, Springer, 2012, pp. 599–619.

6. N. jaitly and G. Hinton, "Learning a better representation of speech soundwaves using restricted boltzmann machines," in *Acoustics, Speech and Signal Processing (ICASSP), 2011 IEEE International Conference on*, 2011.

7. H. Chen and A. Murray, "Continuous restricted Boltzmann machine with an implementable training algorithm," in *IEE Proceedings-Vision, Image and Signal Processing, 2003*, 2003.

8. O. Abdel-Hamid, M. Abdel-Hamid, J. Hui, L. Deng, G. Penn and D. Yu, "Convolutional Neural Networks for Speech Recognition," *IEEE Transactions on Audio, Speech, and Language Processing*, vol. 20, no. 10, pp. 1–11, 2014.

Part XX

Clinical Engineering Innovation Leading to Improved Clinical Outcomes

Healthcare Technology Management (HTM) by Japanese Clinical Engineers: The Importance of CEs in Hospitals in Japan

Jun Yoshioka, Keiko Fukuta, Hiroki Igeta, Takeshi Ifuku, and Takashi Honma

Abstract

Japanese clinical engineer (CE) is a significant and unique profession compared with other nations with its dual clinical and technology focus and national licensing. The CE system of licensing was established in May 1987 under the Clinical Engineers Act. CEs are required to complete 3 to 4 years in designated schools and pass a national examination. It is a professional medical position responsible for the operation and maintenance of life-support and non-life-support medical device systems under the direction of physicians. In Japan, CEs support and operate various life-support medical devices. Technology developments have led to significant improvements in performance, making devices easier to break and requiring specialized maintenance. Some of our healthcare technology management (HTM) initiatives include: 1. Rental equipment: oversee use and conduct in-house testing and repair, avoiding faulty units. 2. Ventilator maintenance: a multi-year track record of assessing and replacing defective parts in-house, contributing to prompt repairs and reduced costs. 3. Battery-equipped devices: created a more efficient system for charge management. 4. Intermittent pneumatic compression device dedicated tester: reducing the incidence of thromboses and embolism in patients.

Keywords

Japanese clinical engineer • National licensing Healthcare technology management

1 Introduction

It is important to ensure appropriate quality control of higher risk medical equipment, such as mechanical ventilators and anesthesia machines [1–3]. In the past in Japan, doctors and nurses performed maintenance for ventilators [4], with faulty devices repaired by the distributor. However, since the late 1990s, clinical engineers—who have a certificate of completion and have met and completed established course requirements—now perform this maintenance. We would like to introduce the benefits of having CEs in the hospital, how they can reduce the exposure of potential harm to patients.

2 What Are Japanese Clinical Engineers

In Japan, Clinical Engineer is a paramedical profession. CEs are medical technologists who work in the clinical field in hospitals and maintain and operate various life-support medical devices such as mechanical ventilators, hemodialysis units, heart-lung machines, and other medical devices. The demand for CEs is high, and the number has increased steadily since the specialty was introduced in 1987. As of January 2018, there are approximately 40,000 CEs working in Japan. CEs are required to complete 3 to 4 years of study in designated schools and pass a national examination. CEs practice under the direction of doctors. The CE's familiarity with medical equipment allows them to train doctors and nurses in its best use, a critical element for avoiding user error.

3 Department of Clinical Engineering, Yamagata University Hospital

The CE department was created in our hospital in 2011. There are currently sixteen CEs working in a broad variety of clinical areas such as in operating rooms, dialysis room,

J. Yoshioka (✉)
Clinical Engineering Services, Yamagata University Hospital, Yamagata, Japan
e-mail: jyoshioka@med.id.yamagata-u.ac.jp

J. Yoshioka · K. Fukuta · H. Igeta · T. Ifuku · T. Honma
International Exchange Committee, Japan Association for Clinical Engineers, Tokyo, Japan

© Springer Nature Singapore Pte Ltd. 2019
L. Lhotska et al. (eds.), *World Congress on Medical Physics and Biomedical Engineering 2018*,
IFMBE Proceedings 68/3, https://doi.org/10.1007/978-981-10-9023-3_148

cardiac catheterization room, ICU, HCU, NICU, emergency room and hospital wards. The main CE activities include extracorporeal circulation, blood purification, hyperbaric oxygenation, HTM, and education. We currently have a large number of devices—a total of 2100, including mechanical ventilators, anesthetic machine, hemodialysis, heart-lung machines, PCPS (percutaneous cardiopulmonary support), IABP (intra aortic balloon pumping), defibrillators, pacemakers, incubators, foot pumps, continuous low pressure suction units, infusion pumps, syringe pumps, patient monitoring, electric scalpels, and endoscope devices except for the radiological equipment.

4 Safety Management

How do you know if medical equipment is reliable, accurate, and safe? Maintaining this equipment is a key safety issue. The stage of technology development is a critical aspect of safety management.

- **Earlier**: During the 1980s, general-purpose medical equipment was in use, with a simpler operating principles and few circuit boards and/or sensors. The equipment was rugged and did not break down easily. Operator error accounted for most of the sudden malfunctions.
- **Present day**: However, over the last two decades, technology development has led to significant improvements in the high performance medical equipment. Modern critical care medical equipment has numerous CPU boards and sensors, and more failure points. Thus, increased specialized maintenance of medical equipment is necessary. In our hospital, medical equipment is maintained using specialized analyzers/test equipment for incubators, defibrillators, ventilators, infusion devices, external pacemakers, electrical safety, electrosurgery, and others.
- **Key CE services**: (1) Overseeing equipment utilization and performing in-house repairs has avoided the use of faulty rental units. (2) Replacing non-functioning parts in-house has contributed to prompt service response and reduced repair costs. (3) Use of specialized test equipment has contributed to the discovery of previously undetected malfunctions in the new generation of high performance medical equipment. The CE department has decreased medical equipment failure, lowering potentially harmful risks to patients. (4) However, proactive maintenance does not preclude or prevent the sudden problems that can occur in daily practice. Clinicians using medical equipment must continue to exercise vigilance and good clinical judgment in order to ensure patient safety during care delivery; CEs partner with them to lower risk.

5 The Benefits of Having Clinical Engineers

- Ventilator example: The involvement of CEs in maintaining ventilators has ensured safe management [5]. Mechanical ventilator failures expose patients to unacceptable risks and maintaining consistent ventilator safety is very important. We examined the usefulness of maintaining ventilators by CEs using a specialized analyzer.

Table 1 shows the year-to-year comparisons of inspection times, cases of suspected ventilator failure, number of ventilators, minor problems, and failures. Tests to verify device accuracy were performed 2,430 times during the period from January 2004 to December 2010. There were a total of 151 (0.07%) cases of suspected ventilator failure. The number of faulty ventilators was 90 (0.04%) for ventilator volume, 39 (0.02%) for oxygen concentration, and 22 (0.01%) for malfunctions.

Faulty ventilators were repaired by calibration and by replacing sensors, circuit boards, and other components. The number of ventilators in use has increased each year. However, minor problems during daily practice that need to be handled on the spot have been reduced because of after-use maintenance of ventilators by CEs. These minor problems included oxygen or flow sensor calibration as a result of an inability to maintain parameters within acceptable limits, start-up problems, dead batteries, breathing circuit settings, faulty alarms, etc. Major failures, which necessitated a change out of the ventilator, have been reduced because of after-use maintenance of ventilators by CEs.

Table 2 shows the details regarding failures and whether a ventilator was repaired in-house or was sent to a distributor. In our hospital, patients experienced no long-term squealer, deaths, or serious injuries associated with failure during this study period.

The use of the PTS-2000 calibration analyzer has contributed to the discovery of previously undetected malfunctions in the new generation of high-performance mechanical ventilators. In this way, evaluating ventilation and carrying out in-house repairs has proved to be effective for obviating the chance of renting faulty units. Clinical engineering has decreased medical device failure which exposes patients to potentially harmful risks.

The most important item should be the checking of mechanical ventilators by CEs. CEs are certainly specialists in medical devices, and their involvement in maintaining mechanical ventilators is logical for the hospital, prompt, and most importantly safe. CEs fill an essential role for the safe operation of mechanical ventilators, and, more importantly, the CEs technology provides safe maintenance for many hospitals.

Table 1 Year-to-year comparisons of inspection times, cases of suspected ventilator failure, number of ventilators, minor problems, and failures

Year	2004	2005	2006	2007	2008	2009	2010	Total
Inspection times	275	290	250	339	388	430	466	2430
Cases of suspected mechanical ventilators failure	11	9	13	28	28	28	34	151
·Ventilatory volume	0	0	5	21	18	20	26	90
·Oxygen concentration	5	5	3	5	9	6	7	39
·Malfunctions	6	4	5	2	1	2	2	22
Number of mechanical ventilators	15	18	21	2 4	28	3 0	30	–
Minor problems	62	60	49	47	25	10	12	265
Failures	17	11	5	3	0	4	0	40

Case of suspected mechanical ventilator failure: There failures were discovers by using the PTS-2000 calibrator at the time of after-use inspections
Minor problems and failures: The incidents of mechanical ventilator that occurred while in use in general wards or the ICU

Table 2 Details of mechanical ventilator failures

Faulty parts		Result of accuracy test		Repair		
				Successful calibrated	Parts exchanges	
Ventilatory volume	Flow sensor	Decrease in tidal volume	24	12	12	☆
		Increase in tidal volume	4	2	2	☆
		Calibration error/breakdown	62	–	62	☆
			90	14	76	
Oxygen concentration	Oxygen sensor	Decrease in oxygen concentration	6	3	3	☆
		Increase in oxygen concentration	32	6	27	☆
			39	9	30	
Malfunctions	Circuit board	Ventilation defects	10	–	10	○
	Flow trigger sensor	Auto triggering	1	–	1	○
	Pressure trigger board	Auto triggering	1	–	1	○
	Battery and filter	Weak battery/deteriorated filter	10	–	10	☆
			22	–	22	

☆In-house: Parts could be replaced in-house by CEs
○Distributor: When in-house repair by CEs was impossible due to a serious problem, the ventilator was sent to the distributor for repair

- Development of the VOLT BANK

We are developing a better system for more efficient charge management of battery-equipped medical equipment [6]. Battery-power in medical devices is often an important function for mobile devices, yet not always checked by users. The VOLT BANK (Fig. 1-TAKASHIN, Japan) was developed by our institution to address this concern.

A charge-control box was created with various functions, and a special rack was incorporated. In the 100 V cutoff function, when the batteries are fully charged, the equipment automatically turns off the power supply of 100 V, in the order that they finish charging.

A maximum of 36 medical devices can be simultaneously charged and stored. This rack was thought to enable construction of a safe, smoothly operating system for battery-equipped medical equipment and efficient battery-charge management that prevents overcharge and electrical discharge.

- Development of the IPCD tester

Inspection to ensure the safe use of devices is an essential part of the daily checks vital to safety in the clinical field. Only basic checks of external appearance and operation checks were possible during maintenance control of intermittent pneumatic compression devices until now, and no specific tester had been available. The vinyl tube that connects the device to pressurized sleeves worn on the lower limbs can break easily and ways to conduct a thorough leak check have been in demand. An intermittent pneumatic compression device tester (IPCD Tester: Fig. 2-TAKASHIN, Japan) was

Fig. 1 Volt Bank, construction of sage operating system and efficient battery-charge management for battery-equipped medical equipment

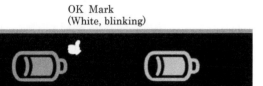

OK Mark
(White, blinking)

COMPLETION Lamp
(Green, Lighting)

CHARGING Lamp
(Orange, Lighting)

Fig. 2 IPDC tester

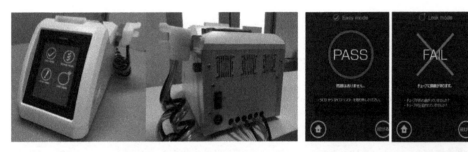

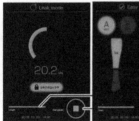

Fig. 3 Obtaining inspection results with high reliability by IPCD Tester

	Air pressure		System errors		Leak			
Pass	94		95		150			
Fail	5		4		48			
Location	In-house	Manufacture	Manufacture		In-house			
	2	3	2	2				
Dam-age parts					Connector		Tube	
					18		30	
Cause of damage	Air leak from air flow passage	Compressor malfunction	Pressure transducer malfunction, Compressor malfunction	Un-known	Break	Crack	Crack	Pin-hole
					3	15	8	22
Repair	Reconnected the air tube inside of SCD	Compressor Modulation2, Replacement 1	Pressure transducer Calibration1, Compressor Modulation1	Non				

co-developed and marketed in collaboration with a local company, and then applied to existing tests at our hospital [7]. This tester not only checks the air pressure value of the compressor (or "heart") and alarm function of the intermittent pneumatic compression device, but even checks the tube connector. Consequently, efficiency measurement, assessment of system errors, and testing of the easily broken sequential compression device (SCD) connector tube in intermittent pneumatic compression devices is now possible.

Introducing IPCD testers to existing tests allows detection and repair of issues in intermittent pneumatic compression devices. Lending out of broken devices can be avoided before it happens, thus reducing the incidence of thromboses and embolism in patients (Fig. 3).

Objects 99 intermittent pneumatic compression devices that is maintained by CEs at our hospital.

6 Conclusions

The combined user and maintenance role of CEs is unique in Japan. On one hand, as medical technologists operating high risk devices alongside clinicians, CEs decrease equipment user errors. On the other hand, the involvement of CEs in device maintenance demonstrably provides safer care delivery and cost-effective equipment management. Having these unique CE capabilities in hospitals also create value in other current and emerging aspects of safety management. In conclusion, clinical engineering in Japan has had a significant impact on improving patient safety.

Acknowledgements We recognize members of ACCE, USA for their helpful and constructive comments on this article. The Japan Association for Clinical Engineers also appreciates ACCE for their faithful service to their country during the tragic 2011 Tsunami.

Conflict of Interest The authors declare that they have no conflict of interest.

References

1. Chapkowski S, Pucilo NP, Cane RD et al. (1984) Impact on ventilator-check time of an in-circuit computerized respiratory monitoring system. Respir Care 29:144–146.
2. Association of Anaesthetists of Great Britain and Ireland(AAGBI), Hartle A, Anderson E. (2012) Checking anaesthetic equipment 2012: association of anaesthetists of Great Britain and Ireland. Anaesthesia 67:660–668.
3. Bourgain JL, Baguenard P, Puizillout JM et al. (1999) The breakdown of anesthesia equipment survey. Ann Fr Anesth Reanim 18:303–308.
4. Chapkowski S, Pucilo NP, Cane RD et al. (1984) Impact on ventilator-check time of an in-circuit computerized respiratory monitoring system. Respir Care 29:144–146.
5. Yoshioka J, Kawamae K, Nakane M. (2014) Healthcare Technology Management (HTM) of mechanical ventilators by clinical engineers. J of Intensive Care 2:27.
6. Yoshioka J, Murakami M, Sadahiro M et al. (2015) Development of VOLT BANK: Construction of a safe operating system and efficient battery-charge management for battery-equipped medical equipment. J of Clinical Engineering 40:90–96.
7. Yoshioka J, Ishiyama S, Miharu M et al. (2017) Development and operational usefulness of an intermittent pneumatic compression device dedicated tester. The Japanese J of Medical Instrumentation 87:308–317 (In Japanese).

Partial Findings of the Clinical Engineering Body of Knowledge and Body of Practice Survey

S. J. Calil and L. N. Nascimento

Abstract

Clinical Engineering has been fundamental to health care for decades, providing expertise in the interaction between medical devices and the health care system. Because the skills and activities required from clinical engineers around the world are not homogeneous, the Clinical Engineering Division at IFMBE decided to promote a global survey to identify the body of knowledge and body of practices they adopt. The survey was aimed at collecting data about employers and professional status, background knowledge, activity responsibilities, and the time spent in the multiple classes of activities. Survey results suggest the profession is still associated to certain traditional characteristics, such as the predominance of professionals with background in electrical, electronic, or mechanical engineering and the prevalence of hospitals and clinics as employers. The questionnaire seems adequate to reveal which skills and activities are considered the most relevant by clinical engineers, but more responses are required before a solid Body of Knowledge and Body of Practice can be defined.

Keywords

Clinical engineering • Body of knowledge
Body of practice

1 Introduction

Clinical Engineering (CE) has been a frontline ally of health care for decades, providing technical expertise for multiple levels of the health care system, such as medical device manufacturers, health care institutions, regulatory agencies, and service delivery organizations.

S. J. Calil (✉) · L. N. Nascimento
School of Electrical and Computer Engineering, University of Campinas, Campinas, Brazil
e-mail: Calil.saide@gmail.com

Considering the perceived differences in knowledge and practices of clinical engineers around the world, the Division declared that one of its mission statements is [to] "define and promote an international body of knowledge, skills and competences on which the profession of clinical engineering can be practiced in various clinical setting" [1].

In 2005, The Clinical Engineering Division of the International Federation for Medical and Biological Engineering (IFMBE/CED) developed a survey to identify the activities developed by Clinical Engineers worldwide. The aim was to understand CE profession in different regions/countries and if there was a common set, if any, of CE activities. The results showed there is a common set of activities but the number of CEs developing each activity is highly dependent on the region/countries [2]. At that time, the conclusion was the need to obtain additional data and to deeper explore such findings.

Recently, a joint effort between the Global Task Force and IFMBE/CED, launched a second survey. This time however was not only to identify the activities practiced by CEs worldwide (Called Book of Practice—BoP), but also to understand the kind of knowledge CEs need to develop their work (called Book of Knowledge—BoK). The final goal is to define a set of subjects that will help any teaching unit to revise and develop its academic curricula aiming to train clinical engineers.

The objective of this article is to present part of the results of this last survey aimed at identifying the BoP and BoP clinical engineers adopt around the world.

2 Materials and Methods

A total of 574 invitations in English were sent to respond the survey.

The kind of questions and format of this survey was based on a similar survey carried out by the American College of Clinical Engineering to understand the profile and practices of CEs working in USA and Canada [3].

© Springer Nature Singapore Pte Ltd. 2019
L. Lhotska et al. (eds.), *World Congress on Medical Physics and Biomedical Engineering 2018*,
IFMBE Proceedings 68/3, https://doi.org/10.1007/978-981-10-9023-3_149

The survey developed and described here was divided into five parts.

The first part, "Contact Information", asked for identification and general information about the respondent, such as: name, company, country, and e-mail address. Here, only the respondent country was mandatory.

The second part of the survey, "Job Information", was developed to acquire data about employer and background. It included questions about the type of employer, about how respondents describe their profession, the primary nature of their current position, about their educational background (Engineering or other), and about CE certification.

The third section of the survey, "Knowledge", presented a list of 28 background knowledge topics and asked the respondents to rate how important (Minor, Moderate, or High Importance) they are for the development of their activities.

The fourth part, "Responsibilities", presented eight classes of activities (Technology Management, Service Delivery Management, Product Development Management, IT/Telecommunications, Education, Facilities Management, Risk Management/ Safety, and General Management). A list of multiple skills related to each one of these classes was then presented and respondents were asked to rate how important (No, Minor, Moderate, or High importance) each skill is to develop each responsibility.

The final and fifth session, "Work Activities", asked the percentage of time the respondents dedicate to each of those eight classes of activities.

For a better interpretation of the processed data and generation of the radar charts, weights were assigned to the multiple responsibility responses: where the answers required three rating levels, 0 (zero) was assigned for "Minor", 1 for "Moderate" and 2 for "High" importance; where 4 rating levels were required, 0 (zero) Zero was assigned for "No", 1 for "Minor", 2 for "Moderate", and 3 for "High".

3 Results

Among several strategies to process and present the data, here it was gather the answers in 7 world regions, according to the country of the respondent. Hence, from the 574 invitations to respond the survey, 199 responses were received from 35 countries. From those; 35% came from Latin America, 20% from Oceania, 14% from Asia, 11% from Middle East, 10% from Europe, 6% from Africa and 4% from USA and Canada.

The second part of the questionnaire showed that almost half of the respondents work in hospitals or clinics (48.2%); followed by 12.5% in government agencies; 9% in the health system, and 6.5% at the Academia. Except for the option

"Other", none of the remaining employer categories reaches 5% of the responses. These categories include private practice consultants, medical equipment manufacturers, medical equipment vendors, and standards development organizations. For 48.2% of the respondents, their profession is best described as "Clinical Engineer", followed by 18.1% as "Healthcare Technology Manager", 13.1% as "Biomedical Equipment Technician", 5.5% as "Clinical Systems Engineer", and 4.0% as "Medical Equipment Planner"—the questionnaire provided definitions for each class and included an "Other" option for open responses.

The main answers for the nature of the respondents' present positions were "Management" (45.7%), "Service Delivery" (15.6%), and "Professional Support" (14.1%).

Around 65% of respondents indicated they had degrees in electrical/electronic/mechanical engineering and 11.5% (extracted from the open responses) indicated they had degrees in biomedical engineering. No other background presented a relevant number of responses.

Regarding Clinical Engineering Certification, 47.7% of the respondents answered there was no certification in their region and were not certified elsewhere; 26.1% have a certification system in their region and were certified; 16.1% that there was certification, but were not certified; and 10.1% have no certification system in their region, but were certified elsewhere.

The results of the third section, "Knowledge" is summarized and presented in Fig. 1, a single example to clarify the "radar chart"; while the knowledge about Respiratory Therapy is quite relevant to CEs in African region (6% of the respondents), one cannot say the same for CEs in Europe (10% of the respondents).

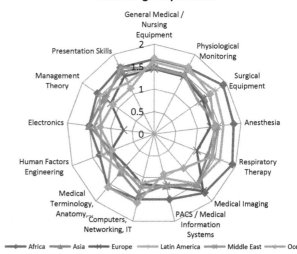

Fig. 1 Background knowledge importance per region. Only topics with 75% or more "Moderate" or "High" importance ratings (Minor = 0, Moderate =1 and High =2)

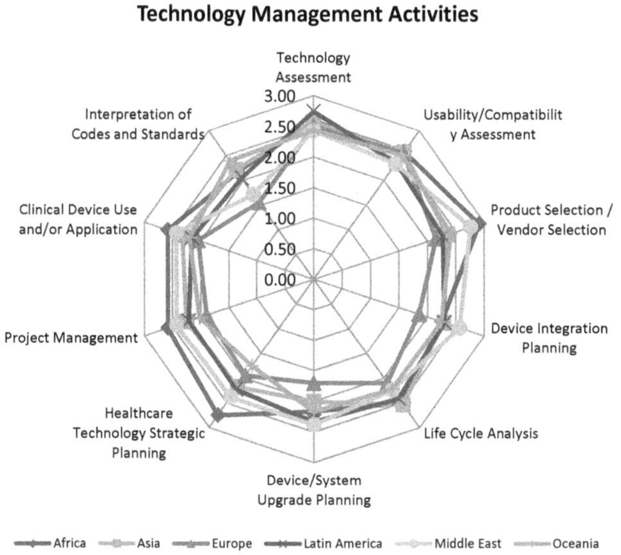

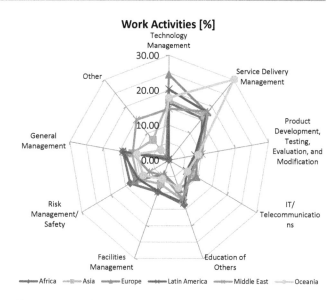

Fig. 2 Technology management responsibilities skill ratings per region

Fig. 3 Overall time spent in each group of activities

As the amount of data obtained in the fourth section (Responsibilities) is quite extensive to be presented in this paper, only one of the eight groups of activities is presented here—Technology Management. Figure 2 shows the rates given for 10 skills related to Technology Management, per region. It should be noted that a set of 20 skills was presented in the survey, but, to make the answers readable, only the skills that received "moderate" or "high" by at least 70% of the respondents were considered.

An example to better understand the radar chart: while the Responsibility for Interpretation of Codes and Standards skill rating is between Minor and Moderate for European CEs (10% of the respondents), it is above Moderate for Latin American CEs (35% of the respondents).

The results regarding the amount of time dedicated to each of the 8 general activities by each respondent per region, explored by fifth section, is presented in Fig. 3. While CEs in Oceania (20% of the respondents) employ more than 30% of their time to Service Delivery activities, CEs in the Middle Eastern region (11% of the respondents) employ less than 20% of theirs.

4 Discussion

The low number of respondents from the 574 invitations is probably due to two main reasons: the survey extension and the absence of translations to other languages.

Because the number of responses varied a lot among regions and even countries within each region, it is still early to discuss specifics, but certain tendencies can already be identified.

First, electrical/electronic/mechanical engineers (65%) who work in hospitals and clinics (48.2%) as managers are still a relevant group.

Second, the answers related to the "Clinical Engineering Certification", revealed that a number of respondents (10.1%) went beyond the limits of their region to get certified elsewhere, indicating a certain interest among the CE community in strengthening and add quality to the profession.

Third, certain topics not traditionally related to Clinical Engineering (such as IT and Human Factors Engineering) are considered quite important background knowledge by the respondents. This reveals that digital technologies and safety requirements using this kind of knowledges are being adopted by health services all over the world and CEs feel the need of such knowledge.

And fourth, overall activity importance seems to vary among regions, but the relative importance ratings between certain groups of activities are somewhat similar (education tends to be considered more important than IT or Facilities Management).

5 Conclusion

The amount of information revealed by this survey can be extremely important to understand not only what is happening in the regions but also in each country. The crossing of data can reveal, for instance, the need of knowledge of CEs working in different environment/country/region. Despite the relatively low number of responses, the results already forecast the set of knowledge needed worldwide by CEs.

The translation of the questionnaire to other languages, followed by another call for responses, might help collecting enough data to draw a solid profile of the clinical engineering profession.

A larger volume of responses more evenly distributed among regions might help identifying regional particularities and confirm which skills and activities should be part of the body of knowledge and practices for clinical engineers.

Compliance with Ethical Requirements The authors have no conflict of interest to declare.

References

1. CED Homepage, http://cedglobal.org/, last accessed 2018/01/25.
2. Calil, S. J., Nascimento, L. N., & Painter, F. R.: Findings of the worldwide clinical engineering survey conducted by the clinical engineering division of the international federation for medicine and biological engineering. In: *11th Mediterranean Conference on Medical and Biomedical Engineering and Computing 2007*, pp. 1085–1088. Springer, Berlin, Heidelberg (2007).
3. Subhan, A. 2015 American College of Clinical Engineering Body of Knowledge Survey Results. Journal of Clinical Engineering, 42(3), 105–106 (2017).

Global Clinical Engineering Innovation, Overview and New Perspectives

Mario Castañeda and Thomas Judd

Abstract

Health Technology (HT) is vital to health care and wellness programs. The dependence on HT services and the expectation for novel approaches has never been greater. Patients, payers, and administrators are demanding innovative HT and optimal services. Clinical Engineers (CEs) are critical members of the healthcare team and are responsible for current and emerging strategies for HT management. But is their role recognized? An IFMBE Clinical Engineering Division (CED) survey determined it was generally unrecognized and collected further data with a landmark survey of success stories. 400 stories from 125 countries qualified as evidence-based CE contributions. A subset of innovation stories were subsequently extracted to see how innovation approaches and solutions could be effectively shared with stakeholders. These stories demonstrated significant benefits from HT innovation and evidenced a compelling case for CEs to embrace innovation as brand for their work and a path to enhance recognition within the global health community.

Keywords

Clinical engineering • Innovation • Recognition
Influence

1 Introduction

The clinical engineering (CE) profession has grown alongside important and visible professions like physicians and other patient caregivers. In the decade of the 1970s, the nascent patient care technology that CEs began to manage included discreet medical devices. Visibility and influence for CEs was enhanced by the unique and valuable skill set of the clinical engineer and their close relationship with physicians. The few clinical engineers available were involved in the planning, acquisition, maintenance, and use of medical devices.

A snapshot from the past illustrates the special relationship clinical engineers had with physicians. From time to time, a clinical engineer could receive requests from physicians such as coming up with a way to measure the pressure of a heart chamber to confirm a diagnosis. The CE would assemble a set of devices that included a blood pressure monitor to connect to the catheter with a transducer tip to be inserted into the patient's heart, an EKG monitor; and a defibrillator. The procedure was conducted in a sterile room with a fluoroscope in a dynamic mode to indicate to the radiologist the catheter's progress along the patient's vessels. The team for this intervention included the CE who scrubbed, sat next to the large arrangement of devices, and ensured all systems were appropriately connected, calibrated, and functioning.

Moving forward just a few decades, the incredible growth and complexity of health technology along with the increased demand for patient services began changing the health delivery scene. Digital and information technology (IT) changed the health care team relationships. New players in the patient space like the health IT professionals brought another perspective to the health care team. The proliferation of user devices and networks moved health IT to center front in visibility to the policy and decision makers. Meanwhile clinical engineers were concentrating on medical devices mostly at operational levels.

Gaining visibility and enhancing the value to our organizations and stakeholders is imperative. Innovation is on the current priority list of most health care organizations. Focusing on this aspect is of high value to Clinical Engineers (CEs). This paper describes some examples of how to get visibility with our executives and grow in our careers.

M. Castañeda
Healthitek Inc, San Rafael, CA 94903, USA

T. Judd (✉)
IFMBE CED Secretary, Marietta, GA 30068, USA
e-mail: judd.tom@gmail.com

© Springer Nature Singapore Pte Ltd. 2019
L. Lhotska et al. (eds.), *World Congress on Medical Physics and Biomedical Engineering 2018*,
IFMBE Proceedings 68/3, https://doi.org/10.1007/978-981-10-9023-3_150

2 Methodology

2.1 Background

The 1st Global CE Summit, under IFMBE/CED, was organized in 2015 to determine the common international challenges to the CE profession. At the 2nd Global CE Summit in September 2017, these concerns were reviewed and updated. At both events, attendees agreed to address the most pertinent barriers:

- lack of professional recognition and influence, and
- lack of sufficient education and training for entering the field and for professional development.

The Summits' action plans included first, data collection identifying if CE contributions qualify as improvement to world health and wellness, and whether these stories can it be substantiated through evidence-based (reproducible, data-driven) records.

The results of this timely research were the base to further the quest for options to address the barriers identified in the Summits. Over four hundred stories from around the globe were collected and validated. Among the stories, there were many that displayed a high degree of innovation. Innovation is a highly valued process and outcome, and the various projects clearly validated the involvement and contribution of clinical engineers in the health care community. Following this line of thought, 78 "Success Stories" with innovation were extracted from the main body of research to start on the path of aligning with innovation to tell our stories.

2.2 Definitions

Innovation often occurs at the beginning of the HT life cycle where new ideas can offer solutions to current problems faced by healthcare providers or their patients. Clinical engineers are well positioned to understand the current problems and guide different or new approaches to resolve them. Innovation, in the CED data collection effort, means to demonstrate the team approach to solving problems all the way from a concept and building a prototype and continuing with clinical trials and demonstration of compliance with standards, regulations, and intended outcomes.

Innovation then is a pathway for clinical engineers to achieve improvement for both professional challenges outlined above. Be an innovator! We CEs already are … we just tell our stories. We need to be able to learn from all kinds of sources—both within healthcare and outside; and use 'systems thinking' to collaborate with others. The C-Suite in health care thrives on innovation.

Innovation for our Success Stories was defined as: "through provision of new HT solutions, adaptation of existing, or a combination to address several issues." [1]. Summaries from CED's 2016 and 2017 data collection of evidence-based Innovation stories are outlined below: (1) CED's HT Resources [1] document provided to the World Health Organization (WHO's) World Health Assembly in May 2016, WHO's May 2017 Third Global Forum on Medical Devices [2]; (2), CED's September 2017 Brazil II ICEHTMC [3] (S), and Others [4] from 2016–2017 IFMBE published sources (O).

2.3 Success Stories

2.3.1 2016 = 17 Total; 13 + Countries

USA	India	Ethiopia	Australia	Uruguay
MG Hospital re test for superbugs [1]	Sharma, AMTZ—Local production of HT and MOH innov [1]	Optimal design of O2 concentrators [1]	New blood/fluid warner design [1]	Simini, CE driving facility design [1]
WHO: LRC innov [1], HT ebola [1]	**Bangladesh:** HT policy and HTM [1]	**South Africa:** Local production of HT in Africa [1]	**Italy:** Robotic surgery [1]	**Colombia:** Bus. opps in HT; Castaneda [1]
WHO, Tech specs O2 concentrators [1]	**Malaysia:** Biomechanics for Amputee [1]	**Tanzania:** MCH (maternal child health) rural HT [1]	**Peru:** Heavy metals detection [1]	**Canada:** Province respiratory outreach [1]

2.3.2 2017 = 61 More; 16 + Countries

WHO/Global	India	USA	UK	Brazil	Colombia
G-PATH (20) MCH [2]; G-WHO assistive HT	G-Prasant, GANDI—needs driven innov. [2]	G-MGH: LMIC Inn [2]; Africa post-part hem [2]	G-Un. Oxford child Pneu. diag [2]	G-Orthostatic chair [2], S-Loc. HT via WiFi [3]	O-Cipro pharm model. and circuits [4]
G-Priority devices, WHO HQ and EMRO [2]	G-Hypothermia alert for newborns-Bempu [2]	G-Early detection brst canc—Un WA [2]	G-Endo GI canc scrg leeds Un [2].	G-Photometric test gestational age [2]	O-ECG signal modeling [4]
G-UNICEF: how we drive innovations [2]	G-Prevent apnea prematurity-bempu [2]	G-Field test neo photothera. Lit Spar [2]	**Norway, Denmark**	G-Prem light detect [2], S-Remote Eq. Mon [3].	O-Mechanical knee modeling [4]
G-EPFL Dig. X-ray [2], G-IARC Therm. Coag [2]	G-Remote monitor. critical infants-Bempu [2]	G-Test pre-eclampsia Un. OSU-Geneva [2]	O-Tiss engr impr. monitoring [4]	S-travel ECG telem [3]; S-dig surgical video [3]	O-Mech ventilation from pesticides [4]
G-e-stethoscope child Pneu diag, Un.Gen [2]; G-Emerg. Care	G- AMTZ 2 -HT Policy impr. Svc del. [2]	**Senegal:** G-O2 concentrators [2], G-Inn. HT inf diseases [2]	**Italy:** G-Phototh. and transfusions [2]	S-Flow anal bld pmp [3]; O-Respiratory control with exercise [2]	O-Parkinsn EP study [4] O-perm.mag drug del [4]
G-UNICEF devices for MCH, LaBarre [2], G-WHO TB diagnoses [2]	**Bangladesh:** G-jaundice screening-harvard [2]	**IFMBE:** G-BME and HTA, Pecchia [2]	O-HTA trends [4], O-HTA tablet for dig pathologies [4]	S-BME aid diagnose pathologies [3], S-Hi flow nasal therapy [3]	O-Respiratory system simulation [4], O-Parkin. EEG study [4]
G-WHO: Dig Hlth Inv [2]; G-Innov LRC 2	**China:** S-renal GRF estimation [3];	G-CED Inn, david [2], G-CE-IT	**Croatia:** O-Diab Pts	**Mexico:** G-Hand orthosis [2]	**Chile:** S-Clin Sim prior to

(continued)

WHO/Global	India	USA	UK	Brazil	Colombia
[2]; G-NCD kit refugees [2]	**Australia**: G-O2 storage [2]	innovation, castaneda, judd [2]	remote mon. [4]		new Facility opening [3].

2.4 Discussion

From the validated success stories there were 78 stories that met de criteria for innovations. Additionally, they were grouped in four categories and selected ones were presented at a Special Session on CE Innovation at the 2018 Prague World BME/CE (IUPESM) World Congress: (Table 1)

A useful grouping of innovations is the simplified classification of groups into technical, administrative, product and process [5]. Further refining yields the following groups.

- Structure of health care delivery
- Process of care delivery
- Outcomes of health care delivery approach
- Individual medical device/HT system design improvement

To position the importance and need for the innovation programs presented within the four groups, it is useful to review the global healthcare needs and priorities. One of the credible sources in health care is the annual Deloitte Global Healthcare Outlook [6]. Information contained in the report is reviewed by a large number of stakeholder governments, C-suites, and industry among others.

According to the report Global health care spending is projected to increase at an annual rate of 4.1% in 2017–2021, up from just 1.3% in 2012–2016. Although the battle against communicable diseases is far from over, countries are making headway through improved sanitation, better living conditions, and wider access to health care and vaccinations. The estimated number of malaria deaths worldwide fell to 429,000 in 2015, down from nearly 1 million in 2000. The number of AIDS-related deaths dropped from 2.3 million in 2005 to an estimated 1.1 million in 2015, due largely to the successful rollout of treatment.

Rapid urbanization, sedentary lifestyles, changing diets, and rising obesity levels are fueling an increase in chronic diseases—most prominently, cancer, heart disease, and diabetes—even in developing markets. China and India have the largest number of diabetes sufferers in the world, at around 114 million and 69 million, respectively. Globally; the number is expected to rise from the current 415 million to 642 million by 2040.

3 Gaining Visibility

The highest contributors to solving the problems and issues of an organization enjoy visibility of their immediate superiors and clients. Increasing the level and value of the solutions increases the visibility and reputation of clinical engineers as problem solvers. One indication that a clinical engineer has been perceived by their employers as a valuable and innovative contributor in an organization is when the engineer gets invited to meetings of other peer departments who are seeking for solution [7]. The reputation of a CE innovator brings positive visibility.

How to be aligned with the priorities of a field and organization begins with the awareness of the current problems and priorities. With the advent of Internet information about healthcare technology is readily available. Global reports and analysis from public [8] and private organizations provided analysis and priorities that are widely read and acknowledged. For example, Deloitte points out their consensus of global health care priorities for 2018:

- Creating a positive margin in an uncertain and changing health economy
- Strategically moving from volume to value
- Responding to health policy and complex regulations
- Investing in exponential technologies to reduce costs, increase access, and improve care

Table 1 Example of health technology innovation stories

Structure	Process	Outcome	Device
India AMTZ (In-country HT production)	**Canada** (Provincial respiratory outreach)	**Malaysia** (Biomechanics for Amputee)	**USA** (Mass Gen Hosp. Superbug Test)
UNICEF (NGO focused on Maternal child health-MCH HT)	**Brazil** (Travel ECG telemedicine)	**Croatia** (Diabetes control remote patient monitoring)	**Ethiopia** (O2 concentrator)
PATH (20-country NGO focused on MCH HT projects)	**Bangladesh** (HT Policy and HTM)	**Colombia** (Mechanical ventilation after Pesticides)	**Australia** (Blood/fluid warmer)
WHO (Global agency focused on various priority HT areas)	**Italy** (Phototherapy and transfusions)	**Peru** (Heavy metals detection)	**Norway, Denmark** (tissue engr monitoring)
Ministries of Health-MOH (Sponsored HT projects)	**Chile** (Clinical simulations prior to facility opening)	**China** (Renal GRF estimation)	**Mexico** (Hand orthotics)

- Engaging with consumers and improving the patient experience
- Shaping the workforce of the future

Being aware of global priorities and specifically on priorities that are of special significance to your organization and clients present an opportunity to contribute with innovative solutions. Innovation focused on immediate needs is welcomed at all level of organization. A gap analysis-oriented methodology is very useful [9]. The gap is the difference between the potential and the actual status.

Relentless thinking about solutions that add value fosters a culture of innovation. Culture encompasses many factors such as the collective consensus of what is useful and valuable, what is rewarded or punished, and what is beautiful, etc. A culture of innovation can be fostered by employers and clients. However it starts with the individual who internalizes inquiry about status quo and has the discipline of adding several perspectives to solving the issues at hand.

Embracing and internalizing innovation is a basic step. Producing outcomes that elegantly a creatively support the goals and priorities of the organization is a path to visibility. One organizing frame to evaluate the level of success of an innovative outcome is to rate the outcome in terms of uniqueness and value [10]. The desired outcome is to have a unique and valuable process or product. Any other combination of uniqueness and value provides a grade for the outcome. Not unique and not valuable process or product is therefore at the bottom of the scale.

4   Conclusions

4.1   Perspective

As a result, here are some practical observations and steps forward:

1. Innovation is cool, is wanted, is admired, and everybody in health care (government, private sector, industry, and academia) has acknowledged this positive item.
2. We have a survey that validates the two most pressing issues for clinical engineers: lack of recognition and influence, and lack of educational pathways to the top.
3. We have this one-of-a-kind global clinical engineering (CE) project that collected and validated success stories around the world. Wow, there is some serious innovation work in these success stories.
4. We will acknowledge and show this work at an Innovation Track at the IUPESM World Congress.

5. As we are looking for wide (outside our peers) recognition, and innovation is a language more widely recognized, can we use these innovation success stories to communicate our work with wide audiences?
6. Let us submit that CE innovation is one of our recognition tickets because it swiftly reaches beyond our peer group.
7. Further, let us submit that embracing innovation as a part of our professional DNA can associate a CE with a positive brand of creativity, problem solving, and usefulness to the organization and clients.
8. Let us continue to write paper and craft presentations to introduce the excellent CE innovation work around the globe and propose a CE culture adjustment that transplants innovation to all of our issue/resolution processes.

5   Summary

To transform the perception of clinical engineers by policy and decision makers, first there needs to be an alignment of values and priorities on both sides, and secondly, there needs to be meaningful communication. While the alignment of values is more natural and easier in a health care team, the priorities may be quite different.

In the current environment, the concept of innovation is understood and desired by leadership. When clinical engineers are perceived as innovators, they would be noticed and acknowledged for their culture and results. To complete the transformation the second part, meaningful communication, requires that clinical engineers also embrace a process for reporting and updating their superiors and clients of their innovation accomplishments and challenges. Professional recognition for clinical engineers will come from our peers, employers, industry, and other stakeholders when we are identified as members of a culture of innovation, and we all can speak and understand a common language—we have 78 opportunities (stories) to do it. Influence is the corollary of a valued profession.

References

1. IFMBE Clinical Engineering Division (CED), *Health Technologies Resource*, May 2016, http://cedglobal.org/global-ce-success-stories/.
2. World Health Organization Third Global Forum on Medical Devices, May 2017, http://www.who.int/medical_devices/global_forum/3rd_gfmd/en/.
3. IFMBE CED 2nd International Clinical Engineering and Health Technology Management Congress (II ICEHTMC) Proceedings, September 2017, http://cedglobal.org/icehtmc2017-proceedings/.

4. Other IFMBE related CE Papers, http://cedglobal.org/global-ce-stories-other/.

5. "Health Technology Innovation Adoption" Trugul U Daim et al., page 43 (Springer).

6. Deloitte 2017 and 2018 Global Health Care Sector Outlook. © 2017 and © 2018 Publications. Deloitte Touche Tohmatsu Limited.

7. Kaiser Permanente *Clinical Technology Department* Innovation teams 2010.

8. World Health Organization (WHO) and Pan-American Health Organization (PAHO) various HT publications.

9. Reverse Innovation, Vijay Govindarajan, Chris Trimble, Harvard Business Press.

10. Lessons from Guy Kawasaki, Forbes 1/6/2016 by Denise Lee Yohn.

Part XXI

Recent Advances in EEG Signal Processing

Feature Extraction and Visualization of MI-EEG with L-MVU Algorithm

Ming-ai Li, Hong-wei Xi, and Yan-jun Sun

Abstract

The feature extraction of Motor Imagery Electroencephalography (MI-EEG), as a key technique of brain computer interface system, has attracted increasing attention in recent years. Because of the high temporal resolution of MI-EEG, researchers are usually bedeviled by the curse of dimensionality. Some manifold learning approaches, such as Isometric Mapping (ISOMAP) and Local Linear Embedding (LLE) etc., have been applied to dimension reduction of MI-EEG by modeling the nonlinear intrinsic structure embedded in the original high-dimensional data. However, these methods are difficulty to exactly represent the nonlinear manifold, affecting the classification accuracy. The Maximum Variance Unfolding (MVU) can solve this problem, but it is unsuitable for online application due to the computation complexity. In this paper, a novel feature extraction approach is proposed based on the Landmark version of Maximum Variance Unfolding (L-MVU). First, the MI-EEG signals are preprocessed according to the event-related desynchronization (ERD) and event-related synchronization (ERS). Then, L-MVU is used to extract the nonlinear features, and a joint optimization of parameters is performed by using the traversing method. Finally, a back-propagation neural network is selected to classify the features. Based on a public dataset, some experiments are conducted, and the experiment results show that L-MVU can preserve more information and perfectly extract the nonlinear nature of original MI-EEG, and reduce the redundant and irrelevant information by introducing the landmark points as well, yielding a higher classification accuracy and a lower computation cost. Furthermore, the proposed method has a better effect on feature visualization with an obvious clustering distribution.

Keywords

Motor imagery electroencephalography
Feature extraction • Dimension reduction
Landmark maximum variance unfolding

1 Introduction

Brain-Computer-Interface (BCI) is defined as a technique to provide direct communication and control pathway between human's brain and external devices [1, 2]. The BCI has been applicated to numerous fields such as rehabilitation therapy, Virtual Reality (VR), disease diagnosis and so on [3–5]. The MI-EEG signals contains a large amount of physiological information and more important is that they have a close correlation with the state of consciousness. Consequently, the recognition of MI-EEG signal is a key point in BCI-based rehabilitation therapy. And extracting the feature of MI-EEG is top priority of a BCI system to obtain a better classification effect.

MI-EEG is a nonlinear, non-stationary, high temporal resolution signal and has the individual differences. The approach of feature extraction has to cope with its characteristics. In the last few years, a large number of feature extraction methods have been proposed. The time domain methods are the earliest proposed methods. Limited by its low frequent resolution, it is difficult to get better classification effect when it is applicated in MI-EEG signal. Wavelet Transform (WT) is the most widely applicated in the feature extraction of MI-EEG. However, because of the

M. Li (✉) · H. Xi · Y. Sun
Faculty of Information Technology,
Beijing University of Technology, Beijing, 100124, China
e-mail: limingai@bjut.edu.cn

H. Xi
e-mail: S201602123@emails.bjut.edu.cn

Y. Sun
e-mail: sunyj@bjut.edu.cn

M. Li
Beijing Key Laboratory of Computational Intelligence and Intelligent System, Beijing, China

© Springer Nature Singapore Pte Ltd. 2019
L. Lhotska et al. (eds.), *World Congress on Medical Physics and Biomedical Engineering 2018*,
IFMBE Proceedings 68/3, https://doi.org/10.1007/978-981-10-9023-3_151

irrelevant information contained in wavelet coefficients, WT can't extract the accurate feature of MI-EEG signal either. As a nonlinear and high temporal resolution signal, it is complicated to process MI-EEG signal due to its high dimension and nonlinear structure. Recently, the arising of Manifold Learning (ML) provides a better new way to process MI-EEG signal. ML can recover the structure of lower dimensional manifolds from high-dimensional data and can help us obtain the corresponding nonlinear embedded coordinates that are regarded as a meaningful representation of reduced the dimension of data [6]. Many of ML algorithms have been proposed based on different instinct. And most of them has been applied in the dimension reduction of EEG signal. Krivov et al. [7] involve Riemannian geometry in the space of symmetric and positive-definite matrices to measure distances between covariance matrices in more accurate fashion. Then the ISOMAP algorithm is applied to the Riemannian pairwise distances to locate manifold, corresponding to human EEG signals, and arrange points, corresponding to covariance matrices, in lowdimensional space, preserving geodesic distances. Finally, linear discriminant analysis is applied for classification. On the public dataset of four classification problem, the ISOMAP obtain the accuracy equal to Common Spatial Pattern (CSP). Yin et al. [8] first use the LLE technique elicited the mental workload (MWL) indicators from different cortical regions. Then, the support vector clustering (SVC) approach is used to find the clusters of these MWL indicators and thereby to detect MWL variations. It has been demonstrated that the proposed framework can lead to acceptable computational accuracy. Gramfort et al. [9] apply the Laplacian Eigenmaps (LE) on the EEG data successfully. They prove that it provides a powerful approach to visualize and understand the underlying structure of evoked potentials or multi-trial time series. From these studies, we can discover that the approaches based on ML have been applicated in feature extraction and visualization of EEG signal. Nevertheless, most of these methods only applied in the data dimension reduction of the original signal rather than using the low dimension feature to classify the motor imagery task. As the matter of fact, during the process of dimension reduction, these methods only preserve the distance of the local area. It leads to many of useful information is missed. The MVU algorithm can preserve more information of the local area [10]. However, limited by the high computational complexity and large storage requirement, it is difficult to use MVU on massive data processing. In this paper, we adopt L-MVU to reduce the computational complexity and storage requirement. Firstly, we choose the optimal time block of the original MI-EEG signal what the task is imagine left/right hands movement. Then we apply L-MVU in the chosen time block to complete the feature extraction and visualization of MI-EEG signal.

Finally, we use the back-propagation neural network classifier to classify the features obtain by L-MVU. By means of the joint-optimal against the parameter of L-MVU, we obtain the high classify accuracy and acceptable computation complexity.

The rest of the paper is organized as follows: In Sect. 2, the feature extraction method based on L-MVU is briefly introduced. Section 3 describes the specific experimental process and the results of the experiment. Section 4 is the conclusion and future work.

2 Methods

MVU is a typical ML algorithm based on full spectral technique. As for the high dimensional input data, it can produce the faithful output data of low dimension. The "faithful" means the output data contains the key information of the input data [10, 11]. But limited by the high computation complexity, MVU can't process the dimension reduction of large data set. The L-MVU is proposed to solve this problem by introducing landmark points to decompose the kernel matrix in MVU [12].

In this paper, based on L-MVU, we proposed a novel feature extraction method of MI-EEG. More details steps of this method is as follows.

Step 1: Based on the ERS/ERD physiological phenomenon of the MI-EEG signal, we chosen the optimal time block $O[min, max]$ by analyzing instantaneous power spectrum of C3 and C4 conductor. For a given $x(i,j)$, which represents the jth data in trial i, the average, the average power $P(j)$ can be calculated as:

$$P(j) = \frac{1}{N}\sum\nolimits_{i=1}^{N} x^2(i,j) \tag{1}$$

where the N is the number of trials. Because the ERS/ERD is more obvious between C3 and C4, the results of imagine left/right hand movement instantaneous power spectra are computed only on C3 and C4.

Step 2: According to the optimal time block $O[min, max]$, we computed the difference value of C3 and C4 conductor. For a given $x(i,j)$ defined in Step 1, difference value of C3 and C4 conductor $d(i,j)$ can be calculated as:

$$d(i,j) = x(i,j)_{C3} - x(i,j)_{C4}, j \in O \tag{2}$$

Then we obtain the (max-min) dimension vector $\vec{D}_i = (d(i,min)d(i,min+1)...d(i,max))^T$

Step 3: we use the $\vec{D}_i$ computed in Step 2 as the high dimension input data of L-MVU. Then we formulate a semidefinite program (SDP) as follows:

Maximize trace (QLQ^T) subject to:

I. $(QLQ^T)_{ii} - 2(QLQ^T)_{ij} + (QLQ^T)_{jj} \leq ||\vec{D}_i - \vec{D}_j||^2$

for all (i,j) with $\eta_{ij} = 1$

II. $\sum_{ij} (QLQ^T)_{ij} = 0$

III. $L \geq 0$

$$(3)$$

where the L is m × m submatrix of inner products between landmarks (with m ≪ n, and n = max − min), the Q is an n × m transformation derived from solving a sparse set of linear equation, and $\eta_{ij} \in (0, 1)$ denote weather input $\vec{D}_i$ and $\vec{D}_j$ are k-nearest neighbors. The landmark points are chosen randomly of the input data points. The transfer matrix Q is computed by the weight matrix W in LLE algorithm. The connection between objective function and low dimension output data $\vec{y}_i$ is as follows:

$$(QLQ^T)_{ij} \approx \vec{y}_i \cdot \vec{y}_i \qquad (4)$$

where the $\vec{y} \in R^{1 \times d}$ are the features of MI-EEG.

To verify the effectiveness of our method, back-propagation neural network is selected to classify the features.

3 Experimental Research

3.1 Data Set Description

The experimental dataset was from BCI Competition 2003 provided by BCI Lab, Graz University of Technology. The dataset was composed of 280 trials, of which 140 were for training and 140 were used for testing images of left/right hand movements. The data were sampled at 128 Hz. Three MI-EEG channels were measured over a C3, CZ, and C4 conductor, using AgCl as an electrode. The placement of the electrode obeys the 10–20 electrode system.

3.2 Feature Extraction and Visualization

First, we choose the optimal time block of the original MI-EEG. The results of imagine left/right hand movement instantaneous power spectra are presented in Fig. 1.

We can discover that in the time range from 3.5 to 7 s is the most variable between C3 and C4, so we choose the data in this time block to applicate in the next experiment processing.

As mentioned in the Sect. 2, L-MVU transforms the optimization problem of feature extraction to be a SDP. Therefore, we should solve this SDP. In this paper, CSDP 6.2.1 solver is used to solve SDP [13]. Then a joint optimization of parameters is performed. By using the traversing methods, we have chosen the optimal value of parameters of L-MVU. The four types of parameters are the expected dimension d, the nearest neighbor r (used to derive locally linear reconstructions), the nearest neighbor k (used to generate distance constraints in the SDP) and the numbers of landmarks m. Figure 2 illustrates the segmental results of the parameter optimization (r = 52, m = 11). We can discover that the highest classification rate appeared when we chose k = 20 and m = 11. The feature visualization results of ML methods are showed in Fig. 3. The ML methods include ISOMAP, LLE, LE, Landmark ISOMAP (L-ISOMAP), MVU, L-MVU, Local Tangent Space Alignment (LTSA), Linear LTSA (LLTSA), Diffusion Mapping (DM) and Hessian LLE (HLLE). The blue points represent the task of imagine left hand movement, and the yellow points represent the task of imagine right hand movement. From Fig. 3, we can see clearly that MVU and L-MVU has a better effect on feature visualization with an obvious clustering distribution compare with other ML methods.

3.3 Classification Accuracy and Computation Complexity

On the dataset we introduced in 3.1 and the same experimental conditions, we have completed the experiments, in which ML methods mentioned in Fig. 3 are applied to extract the features of original MI-EEG signal. Figure 4 shows the classification accuracy of 10-fold cross validation by using these methods including MVU and L-MVU. Furthermore, Table 1 shows the embedded time consumption of these ML methods on the same data set used in this paper. The experimental environment is the windows 10 64-bit

Fig. 1 Instantaneous power spectrum of left/right hand movement

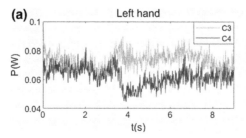
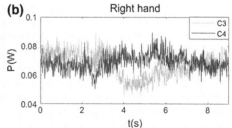

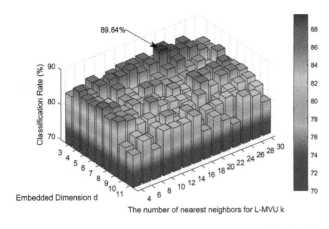

Fig. 2 Results of the parameter optimization about d and k (r = 52, m = 11)

operating system; the CPU is Intel(R) Xeon(R) E5-2683 v3; the memory is 16 GB; the software is MATLAB R2017a.

We can conclude from Fig. 4 that MVU and L-MVU show its superiority on classification accuracy. This is because these two algorithms preserve more information from high dimension data. This information includes the angle between two input points. The experimental results indicate that this information is useful for our classification task.

From Table 1 we discover that MVU waste too much time on producing low dimension data, this will make significance influence on the online experiments. Therefore, we use L-MVU solve this problem basically. More surprising is that L-MVU gains 1% advance in average classification rate. This is actually reasonable. We choose landmark points to

Fig. 3 Feature visualization results of ML methods

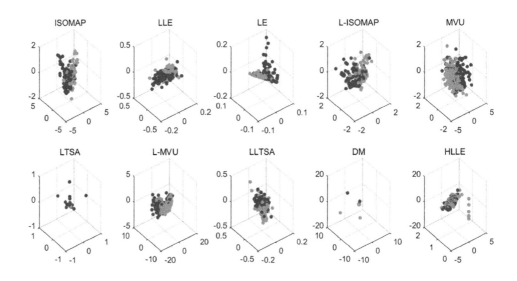

Fig. 4 Average classifier rate of multiple ML methods

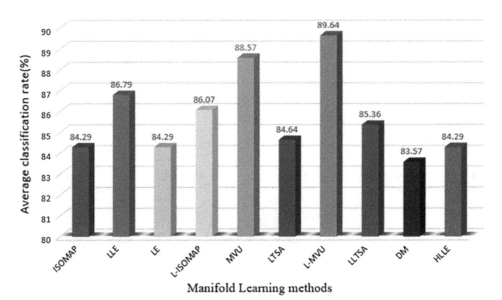

Table 1 Embedded time consumption of multiple ML methods

Method	Time(s)	Method	Time(s)
ISOMAP	2.64	LTSA	2.29
LLE	2.448	L-MVU	*6.184*
LE	0.176	LLTSA	1.201
L-ISOMAP	0.35	DM	0.05
MVU	*465*	HLLE	11.148

lower the computation complexity. Simultaneously, this process reduces the redundant information contains in the input data. Table 1 proved L-MVU can basically meet the requirement of online experiments.

4 Conclusions

In this paper, a novel feature extraction method is proposed based on L-MVU algorithm for MI-EEG. The original MI-EEG signal whose time block is chosen by instantaneous power spectra analysis is used as the input high dimension data of the L-MVU algorithm. By means of adjusts parameters of L-MVU algorithm, we find the optimal parameter of data set used in this paper. Compare to other ML algorithm, our feature extraction method performs better on the classifier accuracy. The computation complexity is also better than its traditional version. It makes our methods could basically meet the request of the online recognition of larger data set.

In the future work, the proposed method could improve in three aspects. First, many ML algorithm has incremental version, such as ISOMAP [14] and LLE [15], but either MVU or L-MVU has not yet, that lead the problems that if we want to obtain the low dimension data of new test data, we must use the train data to reproduce the whole data points. Therefore, the incremental version of L-MVU is necessary. Second, the landmark points of L-MVU are chosen randomly. If we find a criterion for choosing landmark points, the experiment result could be better. Finally, in this paper we use L-MVU only to obtain the nonlinear structure of MI-EEG. However, we ignore the time-frequency characteristic of MI-EEG. Combined the time-frequency methods with L-MVU algorithm is also the work we have to do next [16, 17].

Acknowledgements This work was financially supported by the National Natural Science Foundation of China (No. 81471770, No. 61672070) and the Natural Science Foundation of Beijing (No. 7132021).

References

1. Pfurtscheller, Gert, et al. "Brain-computer interfaces for communication and control. " Supplements to Clinical Neurophysiology 57.5(2002):607.
2. Vaughan, T. M., and J. R. Wolpaw. "The Third International Meeting on Brain-Computer Interface Technology: making a difference. " IEEE Transactions on Neural Systems & Rehabilitation Engineering A Publication of the IEEE Engineering in Medicine & Biology Society 14.2(2006):126.
3. Daly, J. J., and J. R. Wolpaw. "Brain-computer interfaces in neurological rehabilitation." Lancet Neurology 7.11(2008):1032.
4. Lotte, Fabien, et al. Combining BCI with Virtual Reality: Towards New Applications and Improved BCI. Towards Practical Brain-Computer Interfaces. 2013:197–220.
5. Alsaggaf E, Kamel M. Using EEGs to Diagnose Autism Disorder by Classification Algorithm. Life Sci J 2014;11(6):305–308.
6. Seung, H. Sebastian, and D. D. Lee. "The Manifold Ways of Perception." Science 290.5500(2000):2268–9.
7. Krivov, Egor, and M. Belyaev. "Dimensionality reduction with isomap algorithm for EEG covariance matrices." International Winter Conference on Brain-Computer Interface 2016:1–4.
8. Yin, Z., and J. Zhang. "Identification of temporal variations in mental workload using locally-linear-embedding-based EEG feature reduction and support-vector-machine-based clustering and classification techniques. " Comput Methods Programs Biomed 115.3(2014):119–34.
9. Gramfort, Alexandre, and M. Clerc. "Low Dimensional Representations of MEG/EEG Data Using Laplacian Eigenmaps." Joint Meeting of the, International Symposium on Noninvasive Functional Source Imaging of the Brain and Heart and the International Conference on Functional Biomedical Imaging, 2007. Nfsi-Icfbi IEEE, 2007:169–172.
10. Weinberger, Killan Q., and L. K. Saul. "An introduction to nonlinear dimensionality reduction by maximum variance unfolding." National Conference on Artificial Intelligence AAAI Press, 2006:1683–1686.
11. Weinberger, Kilian Q., and L. K. Saul. "Unsupervised learning of image manifolds by semidefinite programming." Computer Vision and Pattern Recognition, 2004. CVPR 2004. Proceedings of the 2004 IEEE Computer Society Conference on IEEE, 2004:988–995.
12. Weinberger, Kilian Q., B. D. Packer, and L. K. Saul. "Nonlinear dimensionality reduction by semidefinite programming and kernel matrix." Tenth International Workshop on Artificial Intelligence and Statistics 2005.
13. Brian Borchers. "CSDP, A C library for semidefinite programming." Optimization Methods & Software 11.1–4(1999):613–623.
14. Law, M. H. C., and A. K. Jain. "Incremental nonlinear dimensionality reduction by manifold learning." IEEE Transactions on Pattern Analysis & Machine Intelligence 28.3(2006):377–91.
15. Kouropteva, Olga, O. Okun, and M. Pietikäinen. "Incremental Locally Linear Embedding Algorithm." Pattern Recognition 38.10 (2005):1764–1767.
16. Day, To This. "Applying a Locally Linear Embedding Algorithm for Feature Extraction and Visualization of MI-EEG." 2016.2 (2016):1–9.
17. Li, Ming Ai, et al. "Adaptive Feature Extraction of Motor Imagery EEG with Optimal Wavelet Packets and SE-Isomap." Applied Sciences 7.4(2017):390.

Influence of Parameter Choice on the Detection of High-Dimensional Functional Networks

Britta Pester, Karl-Jürgen Bär, and Lutz Leistritz

Abstract

The detection of directed interactions within networks derived from spatially highly resolved data, such as functional magnetic resonance imaging (fMRI) has been a challenging task for the last years. Commonly this is solved by restricting the analysis to a small number of representative network nodes (e.g. fMRI voxels), to regions of interest (e.g. brain areas) or by using dimension reduction methods like principal or independent component analysis. Recently, these problems have successfully been encountered by combining multivariate autoregressive models and parallel factor analysis. This approach involves a cascade of analysis steps, entailing a number of parameters that have to be chosen carefully. Yet, the question of an appropriate choice of analysis parameters has not been clarified so far—in particular for temporally varying models. In this work we fill this gap. Synthetic data with known underlying ground truth structure are generated to evaluate the correctness of results in dependence on the parameter choice. Resting state fMRI data are used to assess the influence of the involved parameters in the clinical application. We found that model residuals offer a good means for determining appropriate filter algorithm parameters; the model order should be chosen according to two aspects: the well-established information criteria and the fit between Fourier and estimated spectra.

Keywords

Large scale granger causality • Resting state fMRI Time-variant multivariate autoregressive models

1 Introduction

The human brain is a complex network, exchanging an immense mass of information between remote neuronal areas. Therefore, a thorough investigation of brain processes does not only require the analysis of brain activity but also the consideration brain connectivity [1]. In other words, two regions being active at the same time do not necessarily transfer information between each other. For this reason many approaches have successfully developed in order to quantify the extent of connectivity between different brain regions. Prominent examples are dynamic causal modelling, transfer entropy, Granger causality, directed transfer function and partial directed coherence [1–3].

However, spatially high-resolved data lead to two problems: first, the number of spatial nodes (in this work: fMRI voxels) by far exceed the number of temporal samples (in this work: fMRI volumes). Second, from a practical point of view, the computational capacities are exhausted due to the high network dimensionality: in the fMRI case, the number of network nodes reaches ten thousands up to a hundreds of thousands; in addition, the number of possible connections quadratically rises with the number of network nodes. This makes any conventional connectivity analysis unfeasible. In most cases, the analysis is limited to time series derived from a smaller set of selected or aggregated voxels. Another alternative is the application of dimension reduction methodologies as for example independent or principal component analysis (PCA) [4, 5].

A new approach has been proposed in [6]. Here, a PCA dimensionality reduction from high-dimensional (HD) into low-dimensional (LD) space is combined with a following multivariate autoregressive (MVAR) estimation which is

B. Pester (✉) · L. Leistritz
Institute of Medical Statistics, Computer Sciences and Documentation, Jena, University Hospital, Friedrich Schiller University Jena, Jena, Germany
e-mail: Britta.Pester@med.uni-jena.de
URL: http://www.imsid.uniklinikum-jena.de/IMSID.html

B. Pester
Chair of Clinical Psychology and Behavioural Neuroscience, Technische Universität Dresden, Dresden, Germany

K.-J. Bär
Department of Psychiatry, Jena University Hospital, Friedrich Schiller University Jena, Jena, Germany

© Springer Nature Singapore Pte Ltd. 2019
L. Lhotska et al. (eds.), *World Congress on Medical Physics and Biomedical Engineering 2018*,
IFMBE Proceedings 68/3, https://doi.org/10.1007/978-981-10-9023-3_152

transferred back into HD space. This finally enables the calculation of a highly resolved network, i.e. the quantification of directed connectivity from voxel to voxel.

What has been missing so far is an in-depth consideration of the necessary analysis parameters. The proposed methodology requires many analysis configurations, such as settings of the estimation algorithm or the proportion of retained variance after the PCA dimension reduction. Here, we successively vary the involved parameters and show the influence and reciprocal effects of parameter choice on the quality of network identification.

2 Material

In this work, we followed two complementary approaches: first, simulated data with known ground truth structure were generated in order to evaluate the correctness of results in dependence on the parameter choice. Second, resting state fMRI data of 154 healthy subjects were used to assess the influence of the involved parameters in a clinical application.

2.1 Synthetic Data

Simulated time series were realized as time-variant MVAR processes, where the model coefficients were chosen according to pre-defined ground truth networks. These networks were designed in such a way that the network nodes form four non-overlapping clusters, so-called modules [7]. This means that the expected value for an intra-module connection is considerably higher than that for an extra-module connection. The number of network nodes was set to $D = 50$ and the number of temporal samples to $N = 1000$, providing a good balance between network size and temporal resolution [8]. At sample $n = 500$, ground truth changed from one network into another, which enables the generation of time series based on a temporally varying model.

2.2 Resting State fMRI Data

To evaluate the influence of analysis parameters in practice, data from a resting state fMRI experiment conducted by the Department of Psychiatry and Psychotherapy, Jena University Hospital were used [9]. Pseudonymized data of 154 subjects were acquired using the 12 channel head coil at the 3T MRI scanner (MAGNETOM TIM Trio, Siemens). The experiment included a resting state fMRI scan with a subsequent high-resolution, anatomical T1-weighted MR scan. A to-

tal of $N = 240$ volumes were acquired; each consisting of 45 transversal slices covering the whole brain, deliberately including the lower brainstem [9].

3 Methods

3.1 Applied Analysis Steps

The herein applied methodologies are based on time-variant multivariate autoregressive models (tvMVAR) [10]. This tvMVAR approach has been further developed to the *large scale* MVAR model (lsMVAR) that can be used to estimate time-variant approximations of high-dimensional data [8]. Despite the benefit of time variance, this approach offers the possibility to apply any tvMVAR-based connectivity measure in high dimensions, including frequency-selective approaches.

The initial step of the lsMVAR approach is a reduction from HD space comprising D (D large) network nodes to LD space with C (C small) network nodes by means of PCA. Let $\mathbf{x} \in \mathbb{R}^{C \times N}$ be the LD matrix containing the C retained principle components of N temporal samples derived from HD data. Then, consider the LD tvMVAR model of order p for $\mathbf{x}$:

$$x(n) = \sum_{r=1}^{p} \mathbf{B}^r(n)x(n-r) + e(n), \quad n = p+1,\ldots,N, \quad (1)$$

with LD model parameters $\mathbf{B}^r \in \mathbb{R}^{C \times C}$ and LD model residuals $e(n) \in \mathbb{R}^C$. Then, the whole model can be projected back onto D-dimensional space by a left multiplication of the pseudoinverse of the (truncated) mixing matrix $\mathbf{W} \in \mathbb{R}^{C \times D}$:

$$\mathbf{W}^+ x(n) = \mathbf{W}^+ \left(\sum_{r=1}^{p} \mathbf{B}^r x(n-r) + e(n) \right) \quad (2)$$

which can be rearranged to

$$\underbrace{\mathbf{W}^+ x(n)}_{:=\tilde{y}(n)} = \sum_{r=1}^{p} \underbrace{\mathbf{W}^+ \mathbf{B}^r \mathbf{W}}_{:=\mathbf{A}^r} \underbrace{\mathbf{W}^+ x(n-r)}_{:=\tilde{y}(n-1)} + \underbrace{\mathbf{W}^+ e(n)}_{:=\tilde{e}(n)} \in \mathbb{R}^D,$$

$$(3)$$

with approximated HD data $\tilde{y}(n)$, HD model parameters $\mathbf{A}^r$ and HD residuals $\tilde{e}(n)$. This offers the opportunity for the estimation of *time-variant* MVAR models. In this work, connectivity was assessed by means of time-variant partial directed coherence (PDC) which has the benefit that directed connectivity can be quantified under consideration of various frequencies [1].

3.2 Involved Parameters

The lsMVAR approach requires four parameters that are involved in three different analysis steps:

- TvMVAR parameters were estimated by means of the Kalman Filter [11]. This time-variant approach requires two constants: c_1 regulates the adaption of the covariance matrix; c_2 defines the step-width of the random walk that is used to update the tvMVAR parameters.
- The tvMVAR model p has to be determined. This parameter defines the number of temporal samples in the past that are considered for the estimation of the current value.
- PCA dimension reduction demands an a priori definition of the number of retained PCA components C. This value determines the proportion of variance explanation after PCA, i.e. the higher C, the higher the explanation of variance.

4 Results

4.1 Synthetic Data

Simulations offer the possibility to clearly decide whether the results are correct or not. To assess the goodness of fit between ground truth and computed networks, we used the Cohen's kappa [12]. It quantifies the agreement between two raters; in this case between the derived networks and the known ground truth networks. The results of our simulations can be summarized as follows:

- A quite reasonable possibility for choosing the *Kalman filter parameters* c_1, c_2 is to consider the tvMVAR model residuals. In our simulations, this approach has proven to be useful: synthetic data showed that a high Kappa coefficient—and thus a high accordance between GTNs and PDC networks—corresponds to low mean squared model residuals. Therefore, surveying the model residuals offers a suitable possibility for an adequate choice of c_1, c_2.
- The determination of the *tvMVAR model order* p has proven to be not that clear. Conventional information criteria like Akaike's and Bayesian information criterion [10] provide a first recommendation by establishing a balance between goodness of fit and number of parameters that have to be estimated. However, for frequency-selective approaches like PDC it is important whether the model order is suitable to properly reproduce the frequency spectrum of original data. We found that

the best way is to initially use the information criteria to obtain a first impression of a reasonable region for the choice of p; then, Fourier spectra of real time series should be checked against those of estimated MVAR-based data.

- The successive variation of the *number of retained components* C did not show surprising results for simulated data. Figure 1a shows the performance in dependence on C by means of the area under the receiver operating characteristic curve [13]; clearly, the accordance of PDC networks with GTNs rises with increasing C. Cohen's kappa in dependence on the explained variance is represented in Fig. 1b; again, a higher explanation of variance leads to a deteriorated agreement between GTNs and lsMVAR-driven networks.

4.2 FMRI Data

Individual fMRI connectivity patterns heavily differed between subjects. However, it turned out that in despite of this variation, the influence of parameter choice was similar for the whole group. Therefore, we show the results for one exemplary subject.

First, all parameters were chosen according to the suggestions described in Sect. 4.1, then they were kept fix and successively one parameter has been systematically varied.

- The variation of *Kalman filter parameters* c_1, c_2 showed that the model residuals slightly rise with increasing c_1 while they intensively decrease with increasing c_2. That means: a faster adaption of the covariance matrix and a lower the step width of the random walk lead to a better model fit. As a consequence, it does not appear to be useful to solely consider the model residuals but also whether the adaption of the estimator is satisfying, which of course requires a certain experience of the user in the application of the method.
- According to Akaike's and Bayesian information criterion the *model order* was suggested to be set to $p = 8$. As described in Sect. 4.1, it is important to also consider the spectral properties; we found that for our data, $p \geq 11$ should be preferred with the aim to adequately separate connectivity patterns regarding the frequency domain. The reason is that for $p < 11$, time-frequency-maps of PDC are quite smeared, getting clearer with increasing p, while for $p > 11$, there is hardly any further improvement regarding this point. As an example, Fig. 2a shows the connection from the locus coeruleus complex (LC) to the nucleus raphes dorsalis (DRN), which have proven to be connected during resting state situations [9]. In our

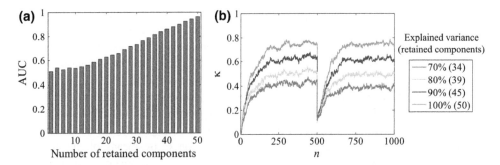

Fig. 1 Performance of lsMVAR-based PDC analysis. Panel **a** shows the temporal mean of the AUC values in dependence on the number of retained components $C = 1, \ldots, 50$. In panel **b**, the temporal dynamics of Cohen's kappa for 70, 80, 90 and 100% explained variance are depicted

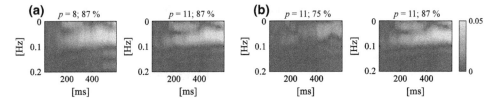

Fig. 2 PDC results of the connection from LC to DRN. Subplot **a** provides a comparison between two different model orders $p = 8$ (suggested choice based on information criteria) and $p = 11$

(data-driven optimum). Analogously, subplot **b** shows the PDC results for two different proportions of explained variance, 75 and 87%

time-variant, frequency-selective analysis approach, the order $p = 8$ suggested by the information criteria is not enough to separate the connection in low frequencies (around 0.06 Hz) emerging during the second half (Fig. 2a, left panel) in a clear manner as compared to $p = 11$ (Fig. 2a, right panel).

- A similar situation is when the number of *retained components* C has to be chosen. Whenever C is increased, the model gets more accurate; on the other hand it has to be considered that a high number of components leads to high computational efforts. Therefore, a good strategy is to inspect the results in dependence on the explained variance and identify a proper balance between explained variance and adequate computational efforts. For our data, we found that an explained variance of around 87% provides a good compromise. Again, the rationale is that this choice represents the point, where for higher values of C the detected networks hardly vary, while for smaller C the derived networks immensely differ when C is changed. This property is by far more pronounced for the choice of C as compared to the choice of p. Figure 2b demonstrates this property: analogous to Fig. 2a, it illustrates the PDC time-frequency maps of the connection from LC to DRN. In the left panel, the map for 75% variance explanation is shown and on the left for 87%. The most striking difference occurs in the lower frequency domain: for 75%, high connections are indicated around 0.03 Hz,

while for 87% it is around 0.06 Hz. Notably, this 0.06 Hz connection remains for higher variance explanation than 87%, this is why that point provides a suitable indicator for a proper choice of C.

5 Discussion and Future Work

Any newly introduced method requires a substantial evaluation to justify the application to real-world data. Nonetheless, in addition it is important to test and understand the influence and mutual effects of analysis configurations in order to avoid misinterpretations due to inappropriate parameter settings. For conventional PDC there has been in-depth work on that aspect based on simulations and EEG data, providing recommendations for the application of this method [14].

However, for the recently proposed lsMVAR approach this point has not been investigated yet. The lsMVAR methodology combines a PCA dimension reduction with tvMVAR modelling and involves four important parameters: two parameters that control the adaption of the estimation algorithm; the tvMVAR model order, defining the number steps in the past that are included for the estimation of the current value; and the number of retained PCA components which corresponds to the proportion of explained variance.

Based on the analysis of synthetic data, we found that model residuals yield a useful indication for a suitable setting of the Kalman filter control parameters. A combination between common information criteria and the consideration of frequency spectra give support in choosing an appropriate tvMVAR model order p. Not surprisingly, a higher number of retained components C leads to a better agreement between GTNs and PDC networks.

For the resting state fMRI data, we found that the choice of Kalman filter parameters based on model residuals is not advisable. Besides surveying the residuals, a sufficient expertise is necessary to find an appropriate compromise between fast adaption and smoothness of the estimated model. The model order p should be chosen in two steps: first, information criteria should be applied to get an appropriate initial value. Second, the results in this range should be inspected with regard to the fit between Fourier and estimated spectra, in order to find out whether this value is adequate. Finally we found that the most impact on the results was made by the explained variance after PCA (i.e. number of retained components C). Similar to p, C should successively be varied to identify the setting when the results of higher C only differ to a small extent.

So far, we inspected the results from a methodological point of view. After finding an appropriate parameter choice, the next step will be to investigate the results in addition to methodological questions—can the lsMVAR approach provide new insights into the default mode network [15]? Furthermore, what has not been done yet is to take advantage of the possibility to explore temporally varying experimental setups [16]. Finally, a comparison between groups is of great interest [17], which however will be a big challenge due to the high number of output data.

Acknowledgements This study was supported by the German Research Foundation (DFG Pe 2546/1-1).

References

1. K. Sameshima and L. A. Baccalá, *Methods in brain connectivity inference through multivariate time series analysis.* 1em plus 0.5em minus 0.4em CRC press, 2014.
2. L. Lee, K. Friston, and B. Horwitz, "Large-scale neural models and dynamic causal modelling," *Neuroimage*, vol. 30, no. 4, pp. 1243–1254, 2006.
3. A. K. Seth, "Causal connectivity of evolved neural networks during behavior," *Network-Computation in Neural Systems*, vol. 16, no. 1, pp. 35–54, 2005. [Online]. Available: <Go to ISI > ://WOS:000233575000003 = 0pt.
4. S. M. Smith, A. Hyvärinen, G. Varoquaux, K. L. Miller, and C. F. Beckmann, "Group-pca for very large fMRI datasets," *NeuroImage*, vol. 101, pp. 738–749, 2014.
5. I. Jolliffe, *Principal component analysis.* 1em plus 0.5em minus 0.4em Wiley Online Library, 2002.
6. C. Schmidt, B. Pester, N. Schmid-Hertel, H. Witte, A. Wismüller, and L. Leistritz, "A multivariate granger causality concept towards full brain functional connectivity," *PloS one*, vol. 11, no. 4, p. e0153105, 2016.
7. V. D. Blondel, J. L. Guillaume, R. Lambiotte, and E. Lefebvre, "Fast unfolding of communities in large networks," *Journal of Statistical Mechanics-Theory and Experiment*, 2008.
8. B. Pester, "Novel approaches for exploring highly resolved brain connectivity: development, evaluation and practical application," Thesis, Institute of Medical Statistics, Computer Sciences and Documentation, 2016.
9. K.-J. Baer, F. de la Cruz, A. Schumann, S. Koehler, H. Sauer, H. Critchley, and G. Wagner, "Functional connectivity and network analysis of midbrain and brainstem nuclei," *Neuroimage*, vol. 134, pp. 53–63, 2016.
10. H. Lütkepohl, *New introduction to multiple time series analysis.* 1em plus 0.5em minus 0.4em Springer Science & Business Media, 2005.
11. T. Milde, L. Leistritz, A. Astolfi, W. H. Miltner, T. Weiss, F. Babiloni, and H. Witte, "A new kalman filter approach for the estimation of high-dimensional time-variant multivariate ar models and its application in analysis of laser-evoked brain potentials," *Neuroimage*, vol. 50, no. 3, pp. 960–969, 2010.
12. J. Cohen, "A coefficient of agreement for nominal scales," *Educational and psychological measurement*, vol. 20, no. 1, pp. 37–46, 1960.
13. L. Sachs, *Angewandte Statistik: Anwendung statistischer Methoden.* 1em plus 0.5em minus 0.4em Springer-Verlag, 2013.
14. L. Leistritz, B. Pester, A. Doering, K. Schiecke, F. Babiloni, L. Astolfi, and H. Witte, "Time-variant partial directed coherence for analysing connectivity: a methodological study," *Philosophical Transactions of the Royal Society of London A: Mathematical, Physical and Engineering Sciences*, vol. 371, no. 1997, p. 20110616, 2013.
15. R. Hindriks, M. H. Adhikari, Y. Murayama, M. Ganzetti, D. Mantini, N. K. Logothetis, and G. Deco, "Can sliding-window correlations reveal dynamic functional connectivity in resting-state fMRI?" *Neuroimage*, vol. 127, pp. 242–256, 2016.
16. J. Liu, B. A. Duffy, D. Bernal-Casas, Z. Fang, and J. H. Lee, "Comparison of fMRI analysis methods for heterogeneous bold responses in block design studies," *NeuroImage*, vol. 147, pp. 390–408, 2017.
17. L. Lim, H. Hart, M. A. Mehta, A. Simmons, K. Mirza, and K. Rubia, "Neural correlates of error processing in young people with a history of severe childhood abuse: an fMRI study," *American Journal of Psychiatry*, vol. 172, no. 9, pp. 892–900, 2015.

Extraction of Diagnostic Information from Phonocardiographic Signal Using Time-Growing Neural Network

Arash Gharehbaghi, Ankica Babic, and Amir A. Sepehri

Abstract

This paper presents an original method for extracting medical information from a heart sound recording, so called Phonocardiographic (PCG) signal. The extracted information is employed by a binary classifier to distinguish between stenosis and regurgitation murmurs. The method is based on using our original neural network, the Time-Growing Neural Network (TGNN), in an innovative way. Children with an obstruction on their semilunar valve are considered as the patient group (PG) against a reference group (RG) of children with a regurgitation in their atrioventricular valve. PCG signals were collected from 55 children, 25/30 from the PG/RG, who referred to the Children Medical Center of Tehran University. The study was conducted according to the guidelines of Good Clinical Practices and the Declaration of Helsinki. Informed consents were obtained for all the patients prior to the data acquisition. The accuracy and sensitivity of the method was estimated to be 85% and 80% respectively, exhibiting a very good performance to be used as a part of decision support system. Such a decision support system can improve the screening accuracy in primary healthcare centers, thanks to the innovative use of TGNN.

Keywords

Intelligent phonocardiography
Time-growing neural network
Deep time-growing neural network

1 Introduction

The fact that screening accuracy of pediatric heart disease is still insufficient in primary healthcare centers, has been reported in a number of the studies [1, 2]. It is evident that a huge amount of the medical costs on the global healthcare system can be prevented by improving the screening accuracy, especially when it comes with the pediatric heart disease, in which as many as 70% of the healthy children can have a sort of innocent murmurs. Many healthy children are sent to the hospital for the cardiac investigation because of the innocent murmur. This brings an extensive unnecessary expenses to the healthcare system, and also stress to the families. On the other hand, a number of the diseased children are overlooked by the practitioners due to the complexities in heart sound auscultation. Our previous studies introduced a non-invasive and inexpensive approach for such a screening, which we called "intelligent phonocardiography" [3, 4]. The Intelligent PhonoCardioGraphy (IPCG) is indeed a computerized phonocardiography, supported by the intelligent machine learning algorithms for the decision making. Although the current practical usage of IPCG is for the screening purpose, our recent studies revealed that further diagnostic information can be provided by this approach when sophisticated processing algorithms are employed, i.e. advanced machine learning methods [5–7]. One of the important challenges of this approach is discrimination between two groups of heart disease, obstructive abnormalities on a semilunar valve, aortic or pulmonary valve, and regurgitation from an atrioventricular valve, mitral or tricuspid. Both of the groups, create systolic murmur, but with different characteristic in terms of the

A. Gharehbaghi (✉)
Department of Innovation, Design and Technology, Mälardalen University, Västerås, Sweden
e-mail: arash.ghareh.baghi@mdh.se

A. Babic
Department of Biomedical Engineering, Linköping University, Linköping, Sweden

A. Babic
Department of Information Science and Media Studies, University of Bergen, Bergen, Norway

A. A. Sepehri
CAPIS Biomedical Research and Development Centre, Mon, Belgium

time-frequency distribution. However, an inexperienced practitioner can easily mix them up, especially in mild cases. A consequence of this misconception could improper disease management, as the obstructive defects can require different follow-up routines due to the higher risk of the left ventricular hypertrophy comparing to the regurgitation defects. It is therefore, a priority for the approach to develop sophisticated methods to discriminate between these two groups of the disease, even though other possibilities of IPCG have been recently investigated [8–11]. This paper presents an original method for classifying heart murmurs, based on our deep learning method for characterizing cyclic time series [12]. The proposed provides a robust classification by modifying the deep learning method in a way to be sophisticated for this application in which a precise learning of systolic murmurs is objective. Meanwhile, complexities of our previous method is drastically reduces, to avoid the over-fitting problem exist in learning with small data size. The experimental results show that the method can efficiently profile the systolic murmur caused by the obstructive defects. However, there is no theoretical obligation to employ the method for other case studies. Simplicity of the testing phase provide the possibility to install the resulting method on the mobile or web technology.

2 Materials

2.1 The Tools

A set of the WelchAllyn Meditron Anlyzer system was used for synchronous recording of heart sounds and electrocardiogram (ECG) signals in conjunction with a DELL laptop. Each signal contains 10 s of the recording with sampling rate of 44,100 Hz and a resolution of 16bit resolution. The heart sounds were recorded from the thoracic apical site. The ECG was used for timing and localization of the systolic segment. We used the Meditron software for data collection, but all the processing algorithms were implemented under the MATLAB platform.

2.2 The Patient Population

The pediatric referrals to the echocardiographic lab at the Children Medical Center hospital, Tehran University of medical science, Iran, participated in this study. All the participants underwent echocardiography as well as other complementary tests including electrocardiogram and chest X-ray, according to the guideline of the hospital. Patients with aortic or pulmonary stenosis were selected as the

Patient Group (PG) against a Reference Group (RG), composed of those ones with mitral or tricuspid regurgitation. Table 1 lists the patient population.

All the referrals gave their informed consent according to the Good Clinical Practice, and the study complied with rules of the World Medical Association and the Declaration of Helsinki. The study was approved by the institutional committee of ethics.

3 Methods

3.1 The Heart Sound Processing

Heart sound signals were filtered and down-sampled to 2 kHz before being segmented in which the systolic intervals between first and second heart sounds are extracted for processing. The method is illustrated in Fig. 1.

The systolic segments were divided into three successive slices with equal length. Then, three different fashions of time growing neural network, the forward, the backward and the mid-growing scheme of time growing neural network is employed for the feature extraction. Details of the growing schemes are found in [12]. For each scheme, three temporal frames are employed, for spectral calculations. A unique discriminative frequency band is found for each temporal frame, based on using the Fisher criteria and discriminant analysis. The spectral energy of the signal over each discriminative band is independently calculated for each frame using periodogram. A multi layer perceptron neural network performs the nonlinear mapping to the discriminative vectors of 4 dimension. The neural network has 6 input nodes of the spectral energies. The hidden and output layers contain 10 and 4 neurons, respectively. Each neuron of the output layer is indicative for one class of the patients of the dataset (see Table 1). The neural network was trained by using back propagation error method. A support vector machine performs the binary classification, discrimination between the PG and RG.

3.2 Statistical Evaluation

The repeated random sub sampling was employed to evaluate the method using accuracy P_{ac} as the performance measure:

$$P_{ac} = 100 \frac{N_{TP} + N_{TN}}{N_{TP} + N_{TN} + N_{FP} + N_{FN}}$$

where N_{TN} and N_{TP} are the number of the correctly classified individuals from the RG and the PG, respectively. The N_{FP}

Table 1 The characteristics of the participating patients

	Aortic stenosis	Pulmonary stenosis	Mitral regurgitation	Tricuspid regurgitation
Number of patients	13	12	15	15
The age range	1–8	1–10	4–18	5–9 years

and N_{FN} are the number of the incorrectly classified individuals from the PG and the RG, respectively. In this method, 15 individuals from the PG, and 20 individuals from the RG are randomly selected for training, and the rest for testing the method. The performance measure is then, calculated. This procedure is repeated several time and the average of the P_{ac} is calculated.

4 Results

Figure 2 represents one cycles of the four classes of the signals in our dataset.

The systolic murmur is seen in all the cases. However, differentiation between the classes is not easy, even for a skilled person. The repeated random sub sampling with 100 iterations was applied to the method and the average accuracy was estimated to be 85% with a sensitivity of 80%. In order to find an understanding about the structural risk of the method, the A-Test method is employed. In this method, k-fold validation is applied to the method, using different validation index K, and average value of the classification error is calculated as a measure of the structural risk. To this end, 2-fold validation is firstly applied to method and the classification error is calculated. This is repeated with

different values of K, (3-fold...), and the corresponding classification are plotted. The plot shows stability of the classification method against the training data. Details of the A-Test are found in [12]. Figure 3 demonstrates the variation of the classification error according to the A-Test.

The A-Test shows a relatively stable valve for the classification error, with an average classification error of 17.4% ± 5.5%. The minimum value of the classification error, is as small as 11%, showing a high capacity for the method to learn the patterns with a higher size of the training data.

5 Discussion

This paper proposed an important application of the intelligent phonocardiography, in assessing origin of the underlying systolic murmur. Such an assessment helps the practitioners to chose a proper management for the patients. For instance, aortic stenosis can remain asymptomatically over a decade after which the lifetime is drastically decreased. It is therefore, important for the patients to be managed properly. Art of the approach is its easiness and inexpensiveness that avails healthcare to those who are of real need by reducing the unnecessary medical investigations. The intelligent phonocardiography takes advantage of the advanced machine learning methods, which allows efficient learning by modifying our deep learning method. This modification includes the use of three scheme of the growing sectors to profile the systolic segment in an efficient way. This novel aspect is considered as an elaborative feature of the method, that provides the flexibility to effectively learn temporal and spectral contents of the signal, as reflected by the A-Test method. This is especially important for assessing

Fig. 1 An illustration of the method. More details of the calculations can be found in [12]

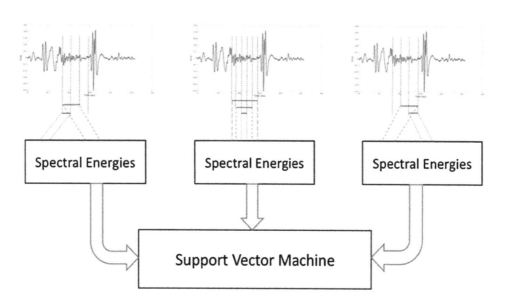

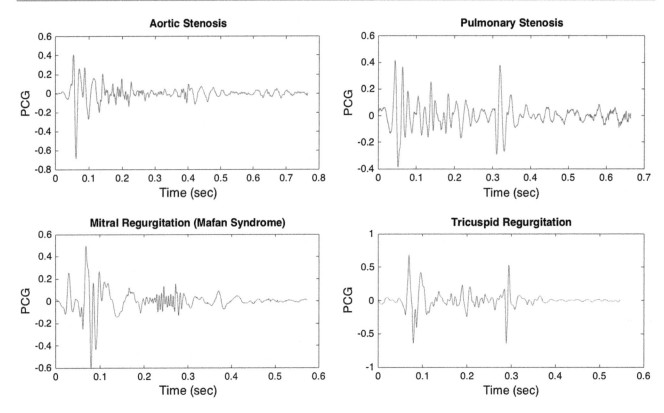

Fig. 2 One sample of each class of the signals

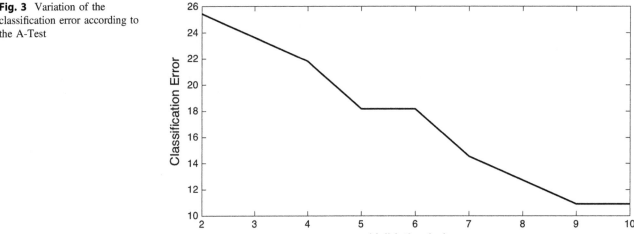

Fig. 3 Variation of the classification error according to the A-Test

mild valvular obstruction in which the between-class dissimilarity is marginal. It can be considered as an important contribution of the study, comparing to the existing methods for processing heart sound signal [13–16]. Our proposed method has the potential to be rather improved by including a more comprehensive training data, in favor of using the time-growing neural network which associate a high learning capacity to the approach.

6 Conclusions

This paper revealed an important application of intelligent phonocardiography in discriminating between two groups of heart abnormalities, obstruction on the semilunar valve and regurgitation from the atrioventricular valves. Both the groups introduce systolic murmur, both the patients need different managements, especially for the obstructive defects where management is rather important. Time-growing neural network was modified in an innovative way, to exhibit less complexities and lower structural risk. The statistical validation shows an acceptable performance for the method in such a difficult classification, that is not easy to be performed by only relying on the conventional auscultation even for the skilled practitioners.

Acknowledgements The authors would like to thank Prof. A. Kocharian for his valuable cooperation in data collection. This study was supported by the CAPIS Inc., Mons, Belgium, and also by the KKS financed research profile in embedded sensor systems for health at Mälardalen University, Västerås, Sweden.

Conflict of Interest The authors declare that they have no conflict of interest.

References

1. Watrous, R. L., Thompson, W. R., Ackerma, S. J.: The impact of computer-assisted auscultation on physician referrals of asymptomatic patients with heart murmurs. Clinical Cardiology, 31(2), 79–83 (2008).
2. Watrous, R. L.: Computer-aided auscultation of the heart: From anatomy and physiology to diagnostic decision support. In: Engineering in Medicine and Biology Society, pp. 140–143, IEEE (2006).
3. Gharehbaghi, A., Ekman, I., Ask, P., Nylander, E., Janerot-Sjöberg, B.: Assessment of aortic valve stenosis using intelligent phonocardiography. International Journal of Cardiology, (198), 58–60 (2015).
4. Gharehbaghi, A., Dutoit, T., Sepehri, A. A., Kocharian, A., Lindén, M.: A novel method for screening children with isolated bicuspid aortic valve. Cardiovascular Engineering and Technology, 6(4), 546–556 (2015).
5. Gharehbaghi, A., Ask, P., Nylander, E., Janerot-Sjöberg, B., Ekman, I., Lindén, M., Babic, A.: A hybrid model for diagnosing sever aortic stenosis in asymptomatic patients using phonocardiogram. In: World Congress on Medical Physics and Biomedical Engineering, Toronto, Canada, pp. 1006–1009, Springer International Publishing (2015).
6. Gharehbaghi, A., Sepehri, A. A., Kocharian, A., Lindén, M.: An intelligent method for discrimination between aortic and pulmonary stenosis using phonocardiogram. In: World Congress on Medical Physics and Biomedical Engineering, Toronto, Canada, pp. 1010–1013, Springer International Publishing (2015).
7. Gharehbaghi, A., Sepehri, A. A., Lindén, M., Babic, A.: Intelligent phonocardiography for screening ventricular septal defect using time growing neural network. In: Mantas, J., Hasman, G., Gallos, G. (eds.) Informatics Empowers Healthcare Transformation, pp. 108–111. IOS Press (2017).
8. Akay, Y. M., Akay, M., Welkowitz, W., Semmlow, J. L., Kostis, J. B.: Noninvasive acoustical detection of coronary artery disease: a comparative study of signal processing methods. IEEE Transactions on Biomedical Engineering, 40 (6), 571–578 (1993).
9. Gharehbaghi, A., Sepehri, A. A., Lindén, M., Babic, A.: A Hybrid Machine Learning Method for Detecting Cardiac Ejection Murmurs. In: EMBEC & NBC 2017, pp. 787–790 Springer Singapore (2017).
10. Gharehbaghi, A., Lindén, M., Babic, A.: A Decision Support System for Cardiac Disease Diagnosis Based on Machine Learning Methods. In: Informatics for Health: Connected Citizen-Led Wellness and Population Health, pp. 235–238 IOS Press (2017).
11. Welkowitz, W., Akay, M., Wang, J. Z., Semmlow, J., Kotis, J.: A Model for Distributed Coronary Artery Flow with Phonocardiographis Verification. Cardiac Electrophysiology, Circulation and Transport, (121) 261–272 (1991).
12. Gharehbaghi, A., Lindén, M.: A Deep Machine Learning Method for Classifying Cyclic Time Series of Biological Signals Using Time-Growing Neural Network, IEEE Transcations on Neural Networks and Learning Systems, Volume: PP, Issue: 99, pp. 1–14.
13. DeGroff, C. G., et al.: Artificial neural network-based method of screening heart murmurs in children. Circultion, (103) 2711–2716 (2001).
14. Sinha, R. K., et al. > Backpropagation artificial neural network classifier to detect changes in heart sound due to mitral valve regurgitation. Journal of Medical Systems, (31) 205–209 (2007).
15. Ari, S., et al.: In search of an optimization technique for artificial neural network to classify abnormal heart sounds. Applied Soft Computing, (9) 330–340 (2009).
16. Dokure, Z., et al.: Heart sound classification using wavelet transform and incremental self-organizing map. Digital Signal Processing, (18) 951–959 (2008).

Physiological Data Monitoring of Members of Air Forces During Training on Simulators

Jiri Kacer, Vaclav Krivanek, Ludek Cicmanec, Patrik Kutilek, Jan Farlik, Jan Hejda, Slavka Viteckova, Petr Volf, Karel Hana, and Pavel Smrcka

Abstract

Many complex situations can be induced to the members of air forces during training on simulators, which may result in mentally vigorous situations or even overload. The aim of the paper is to describe the current state and our contribution to development of systems for measurement of the physiological data of basic member of air force including mission commander, pilots, air traffic controllers and ground support staff. The reason for physiological data monitoring is to test the possibility of usage them to estimate the physical and psychological state of the team members. The base for the design of physiological data monitoring was the FlexiGuar system, originally developed at the FBMI CTU. The core of simulators for training of military personnel in aviation was Lockheed Martin's Prepar3D simulation software. Two airplane cockpits were used as simulators for training of two pilots, air traffic control simulator, i.e. a control tower simulator, and an airport ground station for the preparation of aviation ground staff. The proposed systems are used for simultaneous measurement of the working performance and physiological data of members of the four-member team during their training. The physiological data, heart rate, body temperature, movement activity and perspiration intensity, are transferred to the commander visualization unit for further evaluation. Designed systems and methods could help to monitor, on the base of physiological data and data from simulators, the stress load of team members.

Keywords

Physiological data • Stress load • Air force
Training • Simulators

J. Kacer · V. Krivanek (✉) · L. Cicmanec · J. Farlik
Faculty of Military Technology, University of Defence,
Kounicova 65, Brno, Czech Republic
e-mail: vaclav.krivanek@unob.cz

J. Kacer
e-mail: jiri.kacer@unob.cz

L. Cicmanec
e-mail: ludek.cicmanec@unob.cz

J. Farlik
e-mail: jan.farlik@unob.cz

P. Kutilek · J. Hejda · S. Viteckova · P. Volf · K. Hana · P. Smrcka
Faculty of Biomedical Engineering, Czech Technical University in
Prague, Sitna sq., 3105 Kladno, Czech Republic
e-mail: kutilek@fbmi.cvut.cz

J. Hejda
e-mail: jan.hejda@fbmi.cvut.cz

S. Viteckova
e-mail: slavka.viteckova@fbmi.cvut.cz

P. Volf
e-mail: petr.volf@fbmi.cvut.cz

K. Hana
e-mail: hana@fbmi.cvut.cz

P. Smrcka
e-mail: smrcka@fbmi.cvut.cz

1 Introduction

Intensive research is currently underway on the development of for monitoring health state of employees in aviation [1], including in military aviation [2]. The objective of health state measurement is to use this information on employee health state to increase safety, i.e. to supplement the information used in the monitoring or sensor subsystem of the control system, which will be able to intervene to control [3, 4]. However, in the development of prospective systems, pilot health state monitoring is only assumed. The health state monitoring of other aviation personnel, such as air traffic controllers (ATC) and ground support staff, is given

© Springer Nature Singapore Pte Ltd. 2019
L. Lhotska et al. (eds.), *World Congress on Medical Physics and Biomedical Engineering 2018*,
IFMBE Proceedings 68/3, https://doi.org/10.1007/978-981-10-9023-3_154

little attention. Also, the monitoring of the health status of all key members in aviation is nowhere suggested or presented.

The aim of this article is to describe the current state and our contribution to development of systems for measurement of the physiological data of basic member of air force including mission commander, pilots, air traffic controllers and ground support staff. The reason for physiological data monitoring is to test the possibility of usage them to estimate the physical and psychological state and eventually for identification of dangerous situations in the staffing of the military mission.

2   Methods

Based on the above mentioned requirements, methods of direct measurement of the physical and psychological state of the staff members can be designed. Methods may provide both behavioural and cognitive/emotional physiological measures to assess the performance of members and unit as a whole [5]. All these methods assume a direct measurement of the physiological indicators of the health state of staff.

2.1   Participants

Twenty-eight soldiers (aged 21.1 (SD 2.2)) were recruited for measurement. Soldiers were cadets of University of Defence which is the only military institution of higher education of the Czech Armed Forces. Students were future members of air forces preparing for the profession of pilot and air traffic controllers. However, they had only minimal or zero experience with aircraft piloting. Cadets were subjected to diagnostic evaluation focused on detailed disease history, a neurologic examination, and routine laboratory testing. The study was performed in accordance with the Helsinki Declaration. The study protocol was approved by the local Ethics Committee of the Faculty of Biomedical Engineering of the Czech Technical University (CTU) in Prague. The subjects were measured on same days.

2.2   Measurement Equipment

Measurement systems can be divided into two groups: simulators for training and system for monitoring physiological data.

The base for the design of physiological data monitoring was the FlexiGuar system, originally developed at the Faculty of Biomedical Engineering, Czech Technical University in Prague [6]. The FlexiGuard system is modular biotelemetric system for real-time monitoring of special military

units. However, this system and its previous use assumed the measurement of individuals of a homogeneous group, i.e. individuals performing an identical task with the same difficulty during a mission. In our case, the application assumes measurement of a non-homogeneous group, i.e. parallel monitoring of each member of the special team individually. The systems consists of a set of sensors for monitoring body temperature, heart rate, acceleration and humidity [6, 7]. The modular sensing units records the measured data and send them wireless to the visualization unit. All data are transmitted wirelessly to visualisation unit. The limit values of the measured physiological data for each subject can be individually set in accordance with the expected values of a particular subject undergoing a specific task within the mission. These values are determined and set by the expert before the measurement. The physiological data (heart rate, body temperature, movement activity and perspiration intensity) and their comparison with the limit values are transferred to the commander visualization unit for further evaluation.

Simulators for training can be divided into three groups: airplane cockpit for training of pilots, air traffic control simulator (i.e. a control tower) for training of air traffic controller ATC, airport ground station simulator for the preparation of member of ground support. In our case we used two airplane cockpits, one ATC simulator and one airport ground station simulator. The core of all simulators for training of military personnel in aviation was Lockheed Martin's Prepar3D simulation software. It is necessary to emphasize one essential fact, simulators are based on commonly commercially available software and hardware. The workplace is compatible with Microsoft Flight Simulator X. The workplaces consist of two basic elements: the visualization system and the virtual console. Visualization systems are based on commercially available monitors of computers. Virtual consoles contain elements (rudder pedals, centre stick, etc.) to represent the cores of workplaces of individual members of air force. HW interface is built on a modular system of IO Cards Simulation System. IO Cards Simulation System is linked via the Flight Simulator Universal Inter-Process Communication to the Microsoft Flight Simulator X software. The VR Group Ltd. (Prague, Czech Republic) is the manufacturer of system of simulators. All simulators are located in separate rooms. Communication between simulators is ensured through an intranet. The simulators allow us to store the following data: latitude, longitude, heading, altitude, pitch, bank, airspeed, flaps position, vertical speed, etc.

The data from simulators for training and system for monitoring physiological data were synchronized by the trainer's command i.e. physical activity recorded by both group of systems. Systems collected data at a sampling rate of 1 Hz.

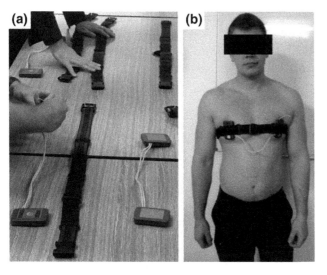

Fig. 1 Preparation (**a**) and application (**b**) of modular sensing units

2.3 Test Procedure

The proposed systems are used for simultaneous measurement of the working performance and physiological data of members of the four-member team (two pilots, one ATC staff and one ground staff member) during their training.

Twenty-eight subjects were divided into seven groups of four members. Thus, four subjects were measured at the same time. Before each measurement, four portable systems for the monitoring physiological data were placed on the trunk of each soldier in accordance with [6, 7], see Fig. 1.

Then, two airplane cockpit simulators, one air traffic control simulator and one airport ground station simulator were used for training of military mission. The instructor proposed a list of tasks that all team members had to perform. It was all about ten types of different tasks. The commander-instructor monitored the performance of tasks by individual members and their physiological data, see Fig. 2. All data was stored for the following evaluation.

2.4 Method of Data Processing

The recordings of the data obtained during performing the tasks of each group on simulators are approximately a half of an hour long to 45 min. The records of physiological data and data from simulators can be processed to study the health conditions of staff members. This process was done using a custom-designed MatLab program based on the functions of the MatLab software.

Fig. 2 Concept of the designed physiological data monitoring of members of air forces during training on simulators

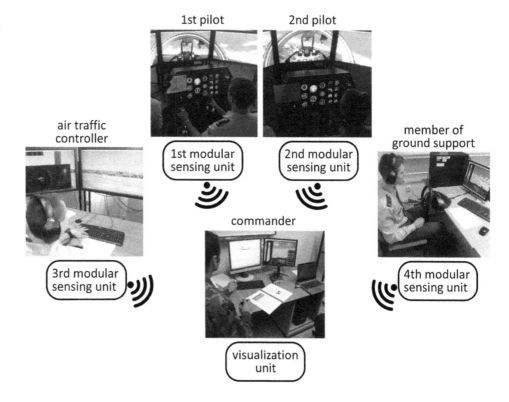

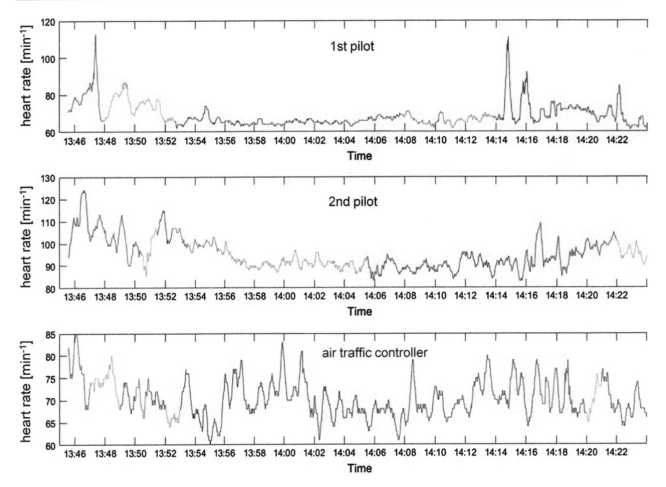

Fig. 3 Example of the output from user sw: real-time heart rate graphs of selected team members during training with color differentiation of performed individual tasks required by instructor (Color figure online)

3 Results

Preliminary results showed that there are dependencies in the development of physiological parameters of individual subjects in the team, although the correlation does not show causality. These dependencies are strong among pilots and controlling traffic controller, see Figs. 3 and 4, with less dependency identified in member of grout support. This is especially noticeable at the beginning of the start measurement, where the highest heart rate value is identified for all three members. Heart rate values are slowly decreasing after the start of aircrafts. Values of heart rate are gradually rising again when landing. The reason for this is that the start and landing of the most challenging phases of the flight for cadets-pilots, and the finding is that air traffic controllers are also responding to this, though not so much. Similar outputs are found for all measured teams. Preliminary results also show an increasing temperature of the body surface of all three members of team during the task on the simulator.

Thus, the preliminary results showed higher correlations among the subjects above mentioned.

4 Discussion

Authors present how the modules of monitoring system were designed, training simulators adapted, and measurement and tests were performed. Testing of the functionality of the methodology took place on the training means of the army of the Czech Republic. In the case of carrying out the measurements on seven groups (i.e. 28 soldiers) of members of air forces during training on simulators, the sensor and simulator networks worked without any bigger problems. Preliminary results show higher correlations in biomedical data among the team members. From the results, we can assume that the monitoring system and training simulators allow us to study physical and psychological state and eventually identification of dangerous situations in the staffing of the military mission. However, this needs to be

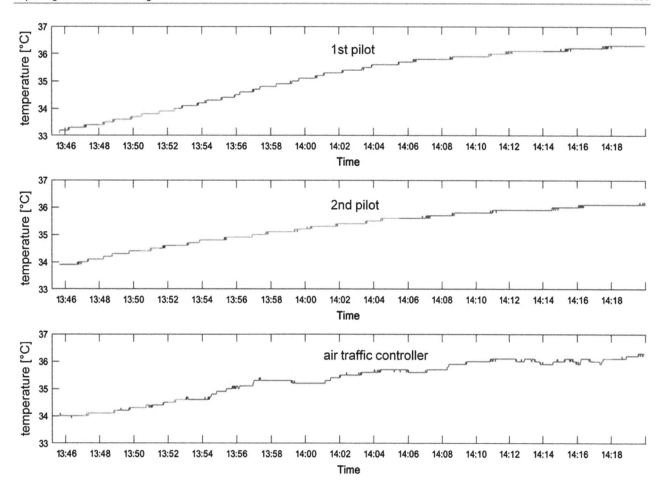

Fig. 4 Example of the output from user sw: real-time temperature graphs of selected team members during training with color differentiation of individual tasks (Color figure online)

further explored in future studies. We can say that low body temperature and heart rate could refer to better physical and psychological state and vice versa [7]. Thus, the systems and technique can offer information which may help to monitor the stress load level and operational preparedness of members of air forces. The assumption of research is future use of the technique in practice.

5 Conclusion

The proposed methodology and measurements were tested on 28 members of Air Force of the Czech Republic. Described preliminary findings demonstrate the ability of the proposed systems and measurement techniques to identify differences in the health state of team members of Air Force during training. The method can be used for quantifying in mental and physical state of the team members. Next goal is to verify the method on more subjects measured over more complex training missions, and integrate the system into practice.

Acknowledgements This work was done in the framework of research project SGS17/108/OHK4/1T/17 sponsored by Czech Technical University in Prague as well as by the Czech Republic Ministry of Defence (University of Defence development program "Research of sensor and control systems to achieve battlefield information superiority").

Conflict of Interest The authors declare that they have no conflict of interest.

References

1. Kacer J.: Modelling of the Pilot Behavior. In: ICMT 2017—6th International Conference on Military Technologies, pp. 477–480. University of Defense: Brno (2017).
2. Boril, J., Jirgl, M., Jalovecky, R.: Use of flight simulators in analyzing pilot behavior. In: Proceedings of the IFIP Advances in Information and Communication Technology, pp. 255–263. Springer: Berlin (2016).
3. Rerucha, V., Krupka, Z.; The pilot-aircraft intelligent interface concept. In: Proceedings of 5th International conference on application of electrical engineering, pp. 1204–1207. WSEAS: Athens (2006).

4. Boril, J., Zaplatilek, K., Jalovecky, R.: Analog filter realization for human—Machine interaction in aerospace. In: Proceedings of IEEE International conference on the science of electrical engineering, pp. 1–5. IEEE: Piscataway (2017).

5. Granholm E., Steinhauer, S.: Pupillometric measures of cognitive and emotional processes. International Journal of Psychophysiology, 52(1), pp. 1–6 (2004).

6. Schlenker, J., Socha, V., Smrčka, P., Hána, K., Begera, V., Kutilek, P., Hon, Z., Kašpar, J., Kučera, L., Mužík, J., Veselý, T., Vítězník, M.: FlexiGuard: Modular Biotelemetry System for Military Applications. In: Proceedings of the 5th International Conference on Military Technologies—ICMT'15. University of Defence: Brno (2015).

7. Kutilek, P., Volf, P., Viteckova, S., Smrcka, P., Krivanek, V., Lhotska, L., Hana, K., Doskocil, R., Navratil, L., Hon, Z., Stefek, A.: Wearable Systems for Monitoring the Health Condition of Soldiers: Review and Application. In: ICMT 2017—6th International Conference on Military Technologies. University of Defence: Brno (2017).

Application of Smart Sock System for Testing of Shoe Cushioning Properties

Alexander Oks, Alexei Katashev, Peteris Eizentals, Zane Pavare, and Darta Balcuna

Abstract

Appropriate choice of shoes with required cushioning characteristics is rather an urgent problem for people from very different groups, such as sportsmen, elderly people, people with foot disorders and locomotion problems. Present research is devoted to further development of wireless DAids™ Pressure Sock System and its application for shoe cushioning estimation. In particular, a new version of pressure sensors with improved sensitivity and working range is designed and tested. Based on above-mentioned developments, the possibility of shoe cushioning testing using DAid™ Pressure Sock System was studied. For this purpose, gait records of several test subjects who used sets of shoes with different cushioning properties, as well as bare walking, were made. Data analysis showed that the developed system gives the possibility to recognize different shoe cushioning. Several approaches to data processing to increase the sensitivity of such recognition are discussed. The comparison showed the potential ability of the developed system to test wirelessly shoe cushioning in real outdoor conditions. Such ability also provides the possibility to monitor and estimate degradation of cushioning quality of shoes under deterioration and environment.

Keywords

Textile sensors • Smart socks • Cushioning control

A. Oks (✉)
Riga Technical University, Institute of Design and Technology, Riga, Latvia
e-mail: aleksandrs.okss@rtu.lv

A. Katashev · P. Eizentals
Riga Technical University, Institute of Biomedical Engineering and Nanotechnologies, Riga, Latvia
e-mail: aleksejs.katasevs@rtu.lv

P. Eizentals
e-mail: peteris.eizentals@gmail.com

Z. Pavare · D. Balcuna
Medical Education Technology Centre, Riga Stradins University, Riga, Latvia

1 Introduction

It is known that cushioning ability of shoes is one of the factors which have essential influence on lower feet loading level. For example, according to [1] midsole degradation of running shoes can lead to increase of heel impact loading forces up to 20–30%. These data were obtained by using force platforms. Data obtained with smart insoles [2] showed an increment of loading impact forces due to midsole degradation up to 100%. So, appropriate choice of shoes midsoles and periodic monitoring of their cushioning characteristics can be rather an urgent problem for people from very different groups, such as sportsmen, elderly people, people with foot and locomotion disorders.

Existing devices which can provide shoe cushioning estimation are quite complicated and expensive and most of them are only for indoor application and cannot be used for outdoor shoe tests. Moreover, as clear from results [1, 2], force platforms can give strongly underestimated data comparatively with in-shoe pressure monitoring.

Reported in [3, 4] DAid™ Pressure Sock System is in-shoe low cost device developed for outdoor monitoring of human gate and running. Comprehensive tests of this system showed its applicable accuracy for monitoring of temporal parameters of locomotion. Present paper is devoted to analysis of the possibility to use DAid™ Pressure Sock System for shoe cushioning ability estimation. Different types of sensor designs are developed and compared to improve system sensitivity and accuracy for plantar pressure measurement. Two data analysis methods of shoe cushioning estimation are proposed and compared as well.

2 Pressure Sensors Tests

2.1 Materials and Methods

Three types of pressure sensors ("filled", "ribbed" and "curved line", referred further as type A, B and C, correspondently) (Fig. 1) were designed and produced using

© Springer Nature Singapore Pte Ltd. 2019
L. Lhotska et al. (eds.), *World Congress on Medical Physics and Biomedical Engineering 2018*,
IFMBE Proceedings 68/3, https://doi.org/10.1007/978-981-10-9023-3_155

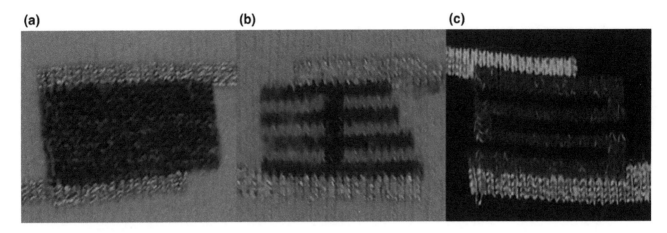

Fig. 1 Knitted pressure sensors

Fig. 2 Dependences $R_{min}/R(p)$

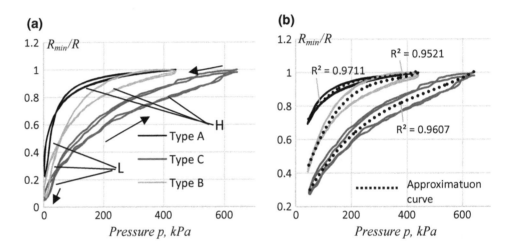

commercially available knitting machines. Five sensors of each type were tested by using complex of Zwick/Roell Z2.5 and Agelent 34970A devices. This complex has provided circling loading of sensors and simultaneous monitoring of their electrical resistance. Loading range was from 2 to 50 N; loading rate—25 N/s. Each test consisted of five loading-unloading cycles.

Effective squares of pressure platforms were 1.142 m² for A and B types and 0.782 m² for C sensor type. Thus, provided loading pressure was up to 440 kPa for A and B types and up to 640 kPa for C type. As the result, dependences of electrical resistance R from applied pressure p were obtained.

2.2 Test Results and Analysis

Obtained characteristics $R(p)$ had the same main features as were described in [3]: high decrement of resistance in low pressure zone ($p < 50$ kPa) and low decrement for $p > 50$ kPa.

It is known [5], that region of maximal plantar pressures during human gate and running corresponds to $p > 200$ kPa, i.e. to high pressure zone. So, to increase the sensitivity of measurement in this zone, it was assumed to use inverted dependences $1/R(p)$ instead of direct data $R(p)$ in further analysis. Examples of normalized inverted characteristics $R_{min}/R(p)$ for all the sensor types are shown in Fig. 2a for both fourth and fifth loading cycles. It can be seen, that dependences $1/R(p)$ of all sensor types represent hysteresis loops, which consist of two subloops ("low" and "high" pressure subloops, marked as L and H, correspondently. Arrows show loading and unloading branches of the loops.) It can be also seen, that sensors demonstrate quite good repeatability under cyclic loading. Exclusion of several first "warming" loading cycles from data analysis provides essential increase of data repeatability. To achieve the best approximation possible for $p > 100$ kPa data only H subloops were used (see Fig. 2b). Analysis showed that these subloops can be approximated by third order polynomial functions with $R^2_{A,B,C}$-squared values not lower than $R^2_{A,B,C} < 0.95$. Obtained approximation

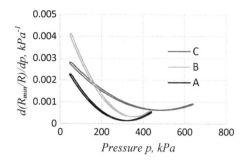

Fig. 3 Relative sensitivity of knitted pressure sensors

functions were used to compare sensors sensitivity to pressure variations and to estimate possible measurement errors. Figure 3 shows relative sensitivity graphs $d(R_{min}/R)/dp$ of studied sensors. It can be seen, that in the high pressures range the best sensitivity to pressure variation has type C. On the contrary, in lower pressures range type B is more sensitive. The lowest sensitivity in all range of pressures has type A.

Estimation of measurement errors due to assumed approximation of hysteresis loops showed, that the highest relative error corresponds to the middle parts of the loops for all type of sensors. Value of maximal error varies from $\pm 25\%$ for type A to $\pm 15\%$ for type C. To improve accuracy of measurements special hysteresis loop models and algorithms of hysteresis compensation can be used [6].

3 Monitoring of Shoe Cushioning Properties

3.1 Test Protocol

Three male volunteers had participated in three walk test trial series: first trial—walking without shoes ("bare" walking—only smart socks had been put on the feet) and second shoed trials—walking in shoes with different midsole stiffness (hard and soft). These trials will be referred further as "B"-bare, "H"-hard and "S"-soft. Midsole stiffness differs for each volunteer, because they used their own shoes, comfortable for them. So, midsole stiffness of shoes used for S trial by volunteer 1 can be harder or equal to the same of H trial of second volunteer, and so on. Each trial had included 6 repeated series of walking with the length of 40–46 strides each. All trials were made with the same walking speed, which was monitored by using metronome device.

3.2 Materials and Methods

Plantar loading forces were measured during walking trials by using DAid™ Pressure Sock System with sensor placement on the sole as follows [3]: toe part—under first and fifth metatarsal; midfoot—inside under foot arc and symmetrically

outside; heel part—in the middle of the heel. Shoes cushioning was tested by measurement of midsole's heel parts stiffness. Measurements were made using Zwick Z 100 device.

To study the possibility of midsole cushioning estimation by using smart socks, two approaches of data processing and analysis were used. The first one—Peak Value analysis (PVA) was based on defining and comparison of the relative average $\overline{a_k}$ of peak values of pressure forces for B, H and S trials. The second one—Loading Rate Analysis (LRA)—on comparison of average loading rate of each sensor or sensor group for B, H and S trials. Calculations for AP and LRA were made using the following formulas:

For PVA:

$$\left(\overline{A_k}\right)_{B,H,S} = \left(\frac{1}{n}\sum_{i=1}^{n} A_{ik}\right)_{B,H,S} \tag{1}$$

$$\left(\overline{A_{sum}}\right)_{B,H,S} = \left(\frac{1}{5}\sum_{i=1}^{5} A_k\right)_{B,H,S} \tag{2}$$

$$\left(\overline{a_k}\right)_{B,H,S} = \frac{\left(\overline{A_k}\right)_{B,H,S}}{\left(\overline{A_{sum}}\right)_B}, \tag{3}$$

where $k = 1\ldots5$—sensor number, n—number of strides in a trial, A_{ik}—peak value of pressure force.

For LRA:

$$\left(u'_{ik}\right)_{B,H,S} = \left(\frac{u_{ik} - \min_{a \le i \le b} u_{ik}}{\max_{a \le i \le b} u_{ik} - \min_{a \le i \le b} u_{ik}}\right)_{B,H,S} \tag{4}$$

$$a = i - 0.5w \tag{5}$$

$$b = i + 0.5w \tag{6}$$

$$\left(v_{ik}^{max}\right)_{B,H,S} = \left(\max\left(\frac{du'_{ik}}{dt}\right)\right)_{B,H,S} \tag{7}$$

and then using Eqs. (1)–(3), replacing there A_{ik} with v_{ik}^{max}.

In Eqs. (4)–(6) u_{ik} is inverted signal of i-th stride from k-th sensor, w is the normalization window size.

3.3 Results and Analysis

Data of measurements of shoe midsoles stiffness are presented in Table 1. Further, for convenience, results from H trial of volunteer with number n will be marked as nH, from

Table 1 Shoes midsole average stiffness, kN/m

Trial/Volunteer	1	2	3
H	110	300	170
S	71	100	110

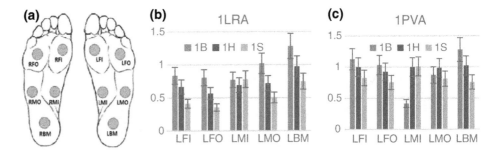

Fig. 4 Sensor placement schema (**a**) and loading histograms for volunteer 1, left foot (**b**, **c**)

his S trial—as nS, etc. Also following abbreviations for sensors are used: L, R—left and right foot; F, M, B—front, middle, back; I, M, O—inside, middle, outside. So, for example, LFI is the mark of frontal inside sensor of the left foot (Fig. 4a) Using PVA and LRA, histograms of $\overline{(\overline{a_k})}_{B,H,S}$ and $\overline{(v_k^{max})}_{B,H,S}$ values were built for each sensor separately and, also, by zones (toe, midfoot and heel). Examples of such histograms built for separate sensors are shown in Fig. 4b, c.

It can be seen, that tested system "fills" even local changes of lower foot cushioning: essential difference in force amplitude peak values and sole loading rates performs between B, H and S trials. One can also see, that PVA and LRA give similar results for toe and heel parts of foot (see LFI, LFO and LBM sensors). For midfoot part results are different. It is explained by specificity of sensors placement and features of midfoot loading. So, to recognize local cushioning ability of shoes the results from both PVA and LRA must be compared. Grouping of sensors gives the possibility to test quality of midsoles cushioning by specific zones.

4 Conclusions

New types of knitted pressure sensors are designed and tested. It is shown that highest sensitivity has "curved line" type sensor. Maximal amplitude measurement error with third order polynomial approximation of sensor characteristic is not exceeding 15%. Possibility to use DAid™ Pres-

sure Sock System for testing shoe midsole cushioning quality is shown. Methods of data processing for cushioning quality estimation and comparison are proposed.

Acknowledgements This research is co-financed by the ESF within the project "Synthesis of textile surface coating modified in nano-level and energetically independent measurement system integration in smart clothing with functions of medical monitoring", Project implementation agreement No. 1.1.1.1./16/A/020. The authors would like to thank A. Linarts for the help during sensor tests.

Conflict of Interest The authors declare that there is no any conflict of interest.

References

1. Schwanitz S, Odenwald S.: Long term cushioning properties of running shoes, The Engineering of Sport, Vol. 2. 7th ed., Springer-Verlag, 95–100 (2008).
2. Verdejo R, Mills NJ: Heel-shoe interactions and the durability of EVA foam running-shoe midsoles. J. Biomech. 37, 1379–1386 (2004).
3. Oks A., Katashev A et al.: Development of Smart Sock System for Gate Analysis and Foot Pressure Control.: IFMBE Proceedings, Vol. 57, pp. 466–469. Springer (2016).
4. Okss, A., Kataševs, A., Bernans, E., Abolins, V. A.: Comparison of the Accuracy of Smart Sock System for Monitoring of Temporal Parameters of Locomotion. OP Conf. Ser.: Mater. Sci. Eng., Vol. 254, pp. 1–6 (2017).
5. Castro M, Abreo S. et al.: In-Shoe Plantar Pressures and Ground Reaction Forces During Overweight Adults' Overground Walking. Research Quarterly for Exercise and Sport, 85(2), 188–197 (2014).
6. Visone C.: Hysteresis modelling and compensation for smart sensors and actuators. J. Phys.: Conf. Ser. (138), 1–24 (2008).

Wireless Assistance System During Episodes of Freezing of Gait by Means Superficial Electrical Stimulation

B. Barzallo, C. Punin, C. Llumiguano, and M. Huerta

Abstract

Parkinson's disease in an advanced stage presents the symptom of freezing of gait, approximately 80% of patients may have a freeze after 17 years of illness, provoking falls and injuries in the 60% of them, the medication is obsolete before this symptom. In search of new methodologies and instruments to help improve the lifestyle of these patients, a non-invasive wireless system is proposed to detect freezing and restart walking by means of superficial electrical stimulation during an episode. A sensor based on a triaxial accelerometer placed on the posterior secondary nerve was used to acquire and store the data of the right lower extremity during the presence of a freezing episode. The data was processed on a smartphone with Android operative system using the tool of discrete wavelet transform developed in Java. The transcutaneous electrical stimulation is applied near the posterior tibial nerve of the lower extremities to continue the walk, which presents better results compared to the vibratory stimulation presented in an earlier version from the authors. The results show feasible diagnostic tests for the validation of the system, such as precision, sensitivity and specificity.

Keywords

FOG • Parkinson's disease • DWT • Variance
IMU • TENS

B. Barzallo · C. Punin (✉) · M. Huerta
Universidad Politécnica Salesiana, Cuenca, Ecuador
e-mail: bpunin@ups.edu.ec

B. Barzallo
e-mail: bbarzallo@ups.edu.ec

M. Huerta
e-mail: mhuerta@ups.edu.ec

C. Llumiguano
Hospital Vozandes Quito, Quito, Ecuador
e-mail: carlos.llumiguano@yahoo.com

1 Introduction

The Parkinson's disease (PD) is a chronic and progressive neurodegenerative disorder, produced by the lack of dopamine in the brain that causes difficulty to move, dopamine is the reason for the controlled movement, coordinated and balanced development of the muscles [1–3]. The clinical symptoms of PD include non-motor deficits: depression, anxiety and problems related to memory, and motor ones: resting tremor, bradykinesia, muscle stiffness and freezing of gait (FOG) [4, 5]. The freezing infers in the daily activities and the social relations of the patients that present these episodes of FOG, increasing the risk to fall and affecting their quality of life.

In search of help so that patients can leave these episodes through the use of external devices, in Ecuador Punin et al. 2017 [6] develop a system to detect and avoid episodes of non-invasive FOG through the use of inertial measurement unit (IMU) systems and a vibratory stimulation of a micro motor placed in the lower extremities near the posterior tibial nerve, the results of the system before the occurrence of an episode of FOG gave a specificity of 86.66% and a sensitivity of 60.61% in the detection, while the effectiveness of resumption of walking after detection was 80%. In [7] a portable assistant was developed for the online detection of FOG with an auditory feedback or vibration to resume walking, with 88.6% in sensitivity and 92.4% specificity. Tay with other researchers [8], developed a real-time PD monitoring and biofeedback system using IMU sensors, it is able to detect a FOG episode and activate an audio and vibration feedback to prevent or reduce freezing. Regarding of the Transcutaneous Electrical Nerve Stimulation (TENS), due to the non-invasive approach, it has become a well-known method to allow motor functions, TENS is perceived to the underlying area level where the electrodes are placed, such as sensation of fibrillary contractions; the range of frequencies that are acceptable to TENS is of 2 to 200 Hz [9], the impulses are of short length and with a high excitabilidad nervous, of high voltage and low intensity.

Departing from the find on this stimulation, such investigations are presented like: the development of an exoskeleton that uses the electrical stimulation as a therapy to incite the march in paraplegic persons [10]. Also the multichannel TENS system developed in [11] stimulates the entire area of the group of electrodes assembly by first applying regulated pulses of current of 8 mA.

This article proposes a wireless system consisting on inertial measurement units that record the walk to be processed by Discrete Wavelet Transform (DWT) and detect episodes of FOG with the calculation of changes in variance, when anomalies in the signal manifest, TENS is activated with the objective of restart the walk. The system consists of two devices located in the lower extremities, connected wirelessly together with an application in a smartphone that processes, stores and displays the data of the motor activity, the system was subjected to diagnostic tests such as specificity, sensitivity and effectiveness, for its validation.

2 Freezing of Gait (FOG)

Is defined as an episodic phenomenon commonly seen as an advanced symptom in PD, it is a sudden motor block with short and transient movements, especially before starting the march, during turns, walking in line straight, going through narrow places like a door or finding obstacles [12]. FOG events are transient, usually last a few seconds and tend to increase in frequency and disability as the disease progresses. A freezing band has a frequency component that varies from 3 to 8 Hz [6, 10, 13]. One of the important characteristics of FOG is that the foot or toe does not leave the ground or just peel off the ground. The episodic and unpredictable nature of FOG can be quantified and evaluated for several minutes. There are three different subtypes of FOG, we have the well-known clinical declaration that it is a patient who suddenly becomes unable to begin to walk or cannot keep on advancing (akinesia) [14]. The second FOG subtype is related to a final movement absence, while the third FOG subtype consists on advancing dragging the feet with very short steps [15, 16].

3 Development

3.1 Acquisition

The study involved 5 men and 15 women with Parkinson's disease, 16 had FOG and 4 did not with an age range of 57 ± 24 years. Participants go through a closed test of 10 m length, which includes curves and straight trajectories with narrow sections, designed with the intention of encouraging

the appearance of FOG. During the route, the patients carried a triaxial accelerometer, placed in the posterior sural nerve that acquires the data at a frequency of 20 Hz by means of an Arduino Pro Mini.

3.2 Processing

The discretization allows to represent a signal in terms of elemental functions accompanied by coefficients [17, 18], then it is possible to represent a signal as a summation of wavelet functions or mother $\psi(t)$ and scale functions $\emptyset(t)$, this is manifested in (1). To obtain a discretized signal, the DWT is applied, which provides specific frequency information and time information at low and high frequencies, its equation is represented in (2).

$$f(t) = \sum_k \sum_j c_{j,k} \emptyset(t) + \sum_k \sum_j d_{j,k} \psi(t) \qquad (1)$$

$$DWT(j,k) = \frac{1}{\sqrt{[2]^j}} \int_{-\infty}^{\infty} x(t) \psi\left(\frac{t - 2^j k}{2^j}\right) dt \qquad (2)$$

In (1) and (2), the x(t) is the input signal, ψ the mother wavelet, c is the vector of approximation coefficients, d is the vector of detail coefficients and j and k are called scaling and time location (displacement), respectively.

Daubechies 4 (DB4) is used as a wavelet mother, due to their characteristics: asymmetric, orthogonal, Biorthogonal, also has the highest number of times of dispersion, which implies more smoothly, since the wavelet function is the best adapted for the analysis of biosignals [19, 20]. In the first step of DWT, the input signal passes simultaneously through the high and low pass filters and its outputs are called approximation coefficients (A1) and detail (D1) of the first level that have half the bandwidth frequency of the original signal, for the next decomposition the same process is applied again to the A1, being its resulting the approximation coefficients (A2) and detail (D2) of the second level, this is replicated until completing 4 levels of decomposition (A3, D3 and A4, D4).

From wavelet coefficient vectors, D2 and D3 are selected for the reconstruction of the signal, since they contain details of the input signal between 5–10 Hz and 2.5–5 Hz, respectively, ranges where FOG episodes may occur [21]. The Wavelet Transformation (WT) consists of a signal in scaled and displaced versions of a wavelet, called a mother wavelet, hence is feasible to use the calculation of the variance to estimate changes in the signals [22–24].

The variance algorithm has the following process: calculation of the variance by sections [24], points estimation of the change of variance and location and segmentation in the

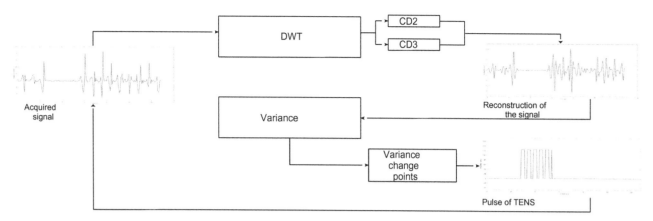

Fig. 1 Block diagram of the system

original signal [23, 25]. The variance is calculated from the reconstructed data vector of the DWT, then the variance change estimation detects the amount of variations in its results to be used as segmentation values in the original signal, signaling the presence of FOG during the walk. In Fig. 1 the process diagram, that involves the DWT and the variance is illustrated [25].

3.3 Transmission Protocols

The transmission protocols are designed as a loop. The tri-axial accelerometer sends the data through the I2C protocol to the microprocessor, it is preprocessed and sent via Bluetooth V2.0 to a previously linked smartphone. The data is stored and processed; part of its results is a logical state (1 or 0), that is sent to the Microprocessor by the same Bluetooth link, where it is used for the control of the stimulation communicating by radiofrequency to 433 MHz.

3.4 Storage, Visualization and Stimulation

The acceleration data, the coefficients of the DWT, the reconstructed signal, the number of detected FOG episodes and the start and end time of the episodes are stored every 10 s in a .csv file inside the internal memory of the Smartphone through an Android application. The results are organized to be used with new processing and extraction of signal characteristics, being accessible to the user and specialist. The application is linked via Bluetooth with the devices, it graphs the real-time data of the acceleration, reflects the state of the stimulus (On/Off) and the manual control of it.

In the presence of FOG episodes, Transcutaneous Electrical Nerve Stimulation (TENS) is applied, a method widely used in physiotherapy. Its purpose is to transmit electric

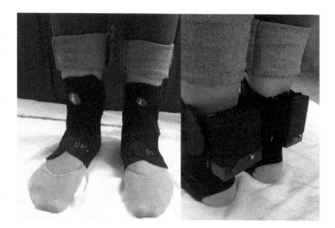

Fig. 2 Devices placed on the patient

current superficially so that the nerves of the lower extremities react to an external stimulus that leads to the resumption of the march. Electrodes placed on the posterior tibial nerve and the sural nerve send square wave electric impulses of 3 V/150 mA at a frequency of 2 Hz, as shown in Fig. 2, allowing exogenous neurostimulation to occur.

4 Results Analysis

In the Fig. 3 are the signals of two patients who have Parkinson's disease in a range of age between 70 and 84 years. As we can see these signals are non-stationary, the patient that presents PD but does not manifest FOG, as shown in the Fig. 3a, has a continuous walk with an acceleration of 28 m/s^2 as the highest peak. While the patient that does present episode of FOG, as shown in the Fig. 3b, has its highest peak in 18 m/s^2; an abrupt deceleration is also noted with tendency to the resting state (9.81 m/s^2) that starts at 2.3 s until 5.8 s, which is the range of time that this episode of FOG lasts.

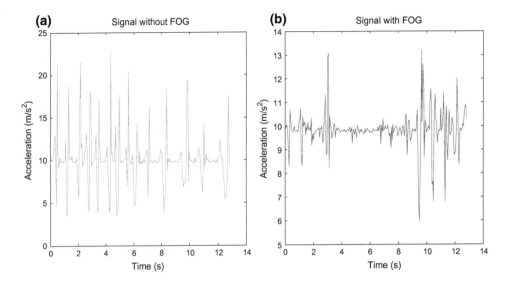

Fig. 3 Comparation of signals of patients with PD: (**a**) with FOG and (**b**) without FOG

The frequency band that presides a FOG is from 3 to 8 Hz [6, 14–16], when performing the wavelet decomposition of the signal acquired we have compensation of the coefficients D2 and D3, located within this band; it is how the reconstruction of the signal with these coefficients is done to determine the characteristics of an episode of FOG. Figure 4a shows a signal with an episode of FOG, its wavelet reconstruction is represented in Fig. 4b, that has maximum oscillating amplitude of 0.3 m/s^2 with respect to the rest, persisting in a period of 3.5 s. For this reason, a variance calculation is made to establish the instants of changes in the signal, thus was obtain the start time of a FOG and the system detects the beginning of this episode and feeds back by superficial electrical stimulation until the

start of the gait. In Fig. 4c shows the electrical pulse of 125 mA, this pulse is activated at the start of FOG (red line) until leaving this episode (purple line), the pulses are carried out every 2 Hz, resuming the march.

A rapid response to the stimulus may be due to a hypersensitivity, common in the PD, which indicates a decrease in the duration of the episodes before the application of the stimulus (see Table 1), with the group of younger patients being the ones that take longer to resume the march (5.069 s) and the older ones are those who respond faster to stimulation (2.307 s). In addition, the results of the specialist and the system were calculated and compared: number of episodes of FOG, mean duration of FOG, mean error in start (MEST) and end time (MEET). In contrast to the opinion of the

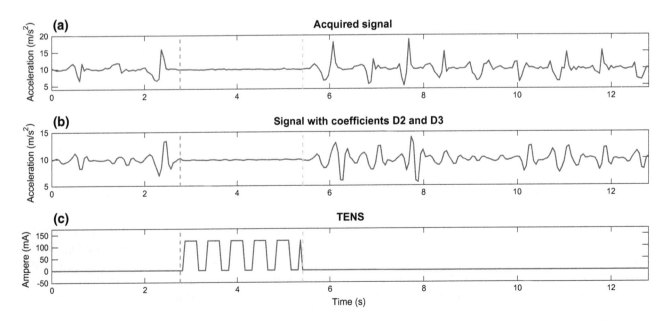

Fig. 4 (**a**) Signal acquired with episode of FOG, (**b**) Signal processed with coefficients D2 and D3 and (**c**) Stimulus waveform applied

Table 1 Results of system tests by age groups of patients

Groups	Patients	FOG: Specialist	FOG: System	MEST (s): Specialist/system	MEET (s): Specialist/system	Mean duration of FOG (s): System
G1: 40–54	6	4	3	0.368	1.919	5.069
G2: 55–69	9	9	7	0.210	0.669	2.542
G3: 70–84	5	4	4	0.327	0.250	2.307
Total	20	17	14	0.283	0.918	3.115

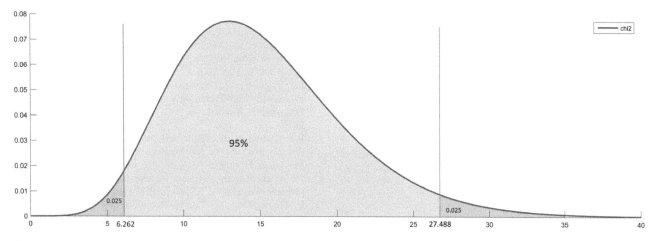

Fig. 5 Chi square- reliability intervals

neurologist specialist, the system presents a MEST and MEET of FOG episodes of 0.283 s and 0.918 s, respectively.

From the results of the tests to the 20 patients is obtained: The true positives (TP = 14), true negatives (TN = 4), false positives (FP = 1) and false negatives (FN = 2). The above data calculates a sensitivity of 87.5%, specificity of 88.89% and effectiveness of 82.35%, showing an improvement compared to its previous version [6], of 21.74%, 2.23% and 2.35%, respectively, in which DWT, calculation of energies and vibratory stimulation, differentiating also in that the current system was tested in a larger sample.

Assuming that the duration of the FOG of the entire population represents an exponential decay distribution and choosing the patients who presented FOG (16 patients), the chi-square test for variance is used, defined in (3), with the aim of verify if the variance of the population is equal to the variance of patient groups (G1, G2, G3).

$$\chi^2 = \frac{(n-1)S^2}{\sigma^2} \qquad (3)$$

The Eq. (3), where n is the sample size, S^2 is the variance of the samples and σ^2 is the variance of the population, was used under the following hypotheses:

- H0: $\sigma^2 = 3.074\,s^2$ (null hypotheses)
- H1: $\sigma^2 \neq 3.074\,s^2$ (alternative hypotheses)

Knowing the 15 degrees of freedom (16 patients–1) of the population distribution, selecting a level of significance of 0.05 (divided into 0.025 and 0.025, for a bilateral analysis) and taking into account the exponential decay distribution, was obtained from the table of chi-square distribution [26, 27] the critical lower and upper values, corresponding to 6.262 and 27.488, respectively, as seen in Fig. 5.

From (3) was calculated the chi-square of Group 1 (χ^2_{G1}), Group 2 (χ^2_{G2}) and Group 3 (χ^2_{G3}). Then, because $\chi^2_{0.025} = 6.262 < \chi^2_{G1} = 8.532, \chi^2_{G2} = 13.337,$ $\chi^2_{G3} = 6.985 < \chi^2_{0.975} = 27.488$, that there is insufficient evidence that the population variance is different from $3.074\,s^2$.

5 Conclusions

The WT has the property of invariance with respect to the translations or changes of the signal, which does not affect the reconstruction of the signal as a function of wavelet coefficients and the processing of the signal by the DWT allows extracting characteristics of the signal in low frequency as it is a non-stationary signal. Based on the above, it is feasible to use the calculation of the variance in the reconstructed signal with the coefficients D2 and D3 to estimate changes in the segments of the signals that represent an episode of FOG.

The duration of an episode of FOG is distinct, even for the same patient, since it depends on the scenario, medication, emotional and physical conditions of the patient. Patients of G3 perceive the electrical stimulus more effectively; this is due to age specific percentages of permeability.

The manifestation of FOG is related to the degree of Parkinson's disease, the greater the degree of disease, the greater the possibility that the individual will develop episodes of FOG. During the study, all patients who demonstrated or claimed that they usually exhibit FOG have a degree of disease 3 (determined by the specialist) along with a diagnosis of the disease greater than 10 years and an average age of 62 years.

Acknowledgements The authors gratefully acknowledge the support of the Project named *"SISMO-NEURO: Análisis del Movimiento Corporal en Enfermedades Neurológicas" (Analysis of the body movement in neurological diseases)* of Universidad Politécnica Salesiana from Ecuador.

Conflict of Interest Statement The authors have no conflict of interest.

References

1. K. Devi Das, A. J. Saji and C. S. Kumar: Frequency analysis of gait signals for detection of neurodegenerative diseases. 2017 International Conference on Circuit, Power and Computing Technologies (ICCPCT), pp. 1–6. Kollam (2017).
2. Enfermedades neurodegenerativas: MedlinePlus en español, Medlineplus.gov, 2017.
3. M. S. Baby, A. J. Saji and C. S. Kumar: Parkinsons disease classification using wavelet transform based feature extraction of gait data. 2017 International Conference on Circuit, Power and Computing Technologies (ICCPCT), pp. 1–6. Kollam (2017).
4. O. Hornykiewicz, "Biochemical aspects of Parkinson's disease," Neurology, vol. 51, no. 2 Suppl 2, pp. S2–S9, 1998.
5. P. Brodal, The Central Nervous System: Structure and Function, 3rd ed. Oxford University Press, 2003.
6. C. Punin, B. Barzallo, M. Huerta, A. Bermeo, M. Bravo and C. Llumiguano: Wireless devices to restart walking during an episode of FOG on patients with Parkinson's disease. 2017 IEEE Second Ecuador Technical Chapters Meeting (ETCM), Salinas, 2017, pp. 1–6.
7. S. Mazilu et al., "Online detection of freezing of gait with smartphones and machine learning techniques," 2012 6th International Conference on Pervasive Computing Technologies for Healthcare (PervasiveHealth) and Workshops, San Diego, CA, 2012, pp. 123–130.
8. A. Tay et al., "Freezing of Gait (FoG) detection for Parkinson Disease," 2015 10th Asian Control Conference (ASCC), Kota Kinabalu, 2015, pp. 1–6.
9. D. Graupe, "EMG pattern analysis for patient-responsive control of FES in paraplegics for walker-supported walking," in IEEE Transactions on Biomedical Engineering, vol. 36, no. 7, pp. 711–719, July 1989.
10. M. R. Popovic, "Transcutaneous Electrical Stimulation Technology for Functional Electrical Therapy Applications," 2006 International Conference of the IEEE Engineering in Medicine and Biology Society, New York, NY, 2006, pp. 2142–2145.
11. Keller, M. Lawrence, A. Kuhn and M. Morari, "New Multi-Channel Transcutaneous Electrical Stimulation Technology for Rehabilitation," 2006 International Conference of the IEEE Engineering in Medicine and Biology Society, New York, NY, 2006, pp. 194–197.
12. C. Punin, B. Barzallo, M. Huerta, A. Bermeo, M. Bravo and C. Llumiguano: Wireless system for detection of FOG in patients with Parkinson's Disease. 2017 Global Medical Engineering Physics Exchanges/Pan American Health Care Exchanges (GMEPE/PAHCE), Tuxtla Gutierrez, 2017, pp. 1–4.
13. K. Niazmand et al., "Freezing of Gait detection in Parkinson's disease using accelerometer based smart clothes," 2011 IEEE Biomedical Circuits and Systems Conference (BioCAS), San Diego, CA, 2011, pp. 201–204.
14. M. Macht et al., "Predictors of freezing in Parkinson's disease: A survey of 6,620 patients," Movement Disorders, vol. 22, no. 7, pp. 953–956, May 2007.
15. E.E. Tripoliti et al., "Automatic detection of freezing og gait events in patients with Parkinson's disease", Computer Methods and Programs in Biomedicine, vol 110, insuue 1, pp 12–26, April 2013.
16. B.R. Bloem et al., "Falls and freezing of gait in Parkinson's disease: a review of two interconnected, episodic phenomena", Movement Disorders, 2004, pp. 871–884.
17. M. Wickerhauser, Adapted wavelet analysis from theory to software. Piscataway, NJ [u.a.]: IEEE Press [u.a.], 1996.
18. E. Gómez-Luna, J. Cuartas-Bermúdez and E. Marles-Sáenz, "Obtención de la fase de la impedancia eléctrica usando transformada Wavelet y transformada de Fourier de señales transitorias. Parte 1: Análisis teórico", DYNA, vol. 84, no. 201, p. 138, 2017.
19. N. Ghassemi, F. Marxreiter, C. Pasluosta, P. Kugler, J. Schlachetzki, A. Schramm, B. Eskofier and J. Klucken, "Combined accelerometer and EMG analysis to differentiate essential tremor from Parkinson's disease", 2016 38th Annual International Conference of the IEEE Engineering in Medicine and Biology Society (EMBC), 2016.
20. D. Jeong, Y. Kim, I. Song, Y. Chung and J. Jeong, "Wavelet Energy and Wavelet Coherence as EEG Biomarkers for the Diagnosis of Parkinson's Disease-Related Dementia and Alzheimer's Disease", Entropy, vol. 18, no. 1, p. 8, 2015.
21. S. Saraswat, G. Srivastava and S. Shukla, "Decomposition of ECG Signals Using Discrete Wavelet Transform for Wolff Parkinson White Syndrome Patients", 2016 International Conference on Micro-Electronics and Telecommunication Engineering (ICMETE), 2016.
22. S. Nanda, W. Lin, M. Lee and R. Chen, "A quantitative classification of essential and Parkinson's tremor using wavelet transform and artificial neural network on sEMG and accelerometer signals", 2015 IEEE 12th International Conference on Networking, Sensing and Control, 2015.
23. C. Chatfield, The Analysis of Time Series. Hoboken: Taylor and Francis, 2013.
24. D. Joshi, A. Khajuria and P. Joshi, "An automatic non-invasive method for Parkinson's disease classification", Computer Methods and Programs in Biomedicine, vol. 145, pp. 135–145, 2017.
25. F. Li, Z. Tian, Y. Xiao and Z. Chen, "Variance change-point detection in panel data models", Economics Letters, vol. 126, pp. 140–143, 2015.
26. F. Carlborg, Introduction to statistics. [Glenview, Ill.]: Scott, Foresman, 1968.
27. H. Lancaster: The Chi-Squared Distribution. 1st edn. Wiley: New York (1969).

Practical Performance Assessment of Dry Electrodes Under Skin Moisture for Wearable Long-Term Cardiac Rhythm Monitoring Systems

Antonio Bosnjak and Omar J. Escalona

Abstract

The use of wearable dry sensors for long term recordings of electrocardiographic bipolar leads located in comfortable areas of the body, is a requirement for detecting certain heart rhythms. Knowledge of the skin-electrode electrical performance of dry electrodes is necessary when seeking to improve various processing stages for signal quality enhancement. In this paper, methods for the assessment of skin-electrode impedance (Zse) of dry electrodes and its modelling are presented. We need to know the behavior of dry electrodes when they are moistened with skin sweat, either at the time of exercise or when it comes to warm climates, under the following posed hypothesis: the impedance magnitude of dry electrodes under study would be significantly lower after they have been moistened with sweat, and comparatively could reach levels of impedance characteristics presented by standard pre-gelled ECG electrodes. Measurements were carried out on selected dry-electrode materials such as silver, stainless steel, AgCl (dry), polyurethane and iron (Fe). These presented, |Zse| values between 500 kΩ and 1 MΩ within the main ECG frequency range (1–100 Hz), under no sweat conditions, and values of few kiloohms under artificial sweat conditions. However, in spite of the sweat conditions, open bandwidth ECG traces were of similar quality and stability, within tolerance; with dry AgCl electrode material presenting the best ECG trace performance.

Keywords

Dry electrodes • Moistened skin-electrode impedance Spectroscopy • Effect of sweat • Arm ECG Wearable monitoring devices • Heart rhythm

1 Introduction

Some patients need to be continuously monitored due to the nature of their specific condition. For example, a surgical procedure is sometimes used for long term rhythm monitoring and involves an implantable loop recorder. In this procedure, a device is positioned on the chest wall, under the skin, and the patient can be monitored for two years or longer. However, there are some inconvenient issues associated with this procedure, the device is expensive, and its use involves risks such as infection and anesthetic reactions among others [1]. Recently, a growing research interest has arisen in the area of wearable physiological measurement systems [2]. The development of devices that use high-performance electrodes for such demanding ambulatory applications remains a research challenge [3, 4].

1.1 Electrical Bioimpedance

Mathematically, the complex impedance Z (in Eq. 1) is represented by its resistive component R (the real part), and the reactance X (imaginary part), or alternatively, as its inverse parameter, the admittance Y (in Eq. 2) integrated by its conductance G (real part), and its susceptance B (imaginary part). These two parameters are also used either for serial type models (impedance) or parallel type models (admittance):

$$Z = R + j \cdot X \qquad (1)$$

$$Y = \frac{1}{Z} = G + j \cdot B \qquad (2)$$

A. Bosnjak
Centro de Procesamiento de Imágenes, Universidad de Carabobo, Valencia, Venezuela

O. J. Escalona (✉)
Engineering Research Institute. Ulster University, Newtownabbey, UK
e-mail: oj.escalona@ulster.ac.uk

© Springer Nature Singapore Pte Ltd. 2019
L. Lhotska et al. (eds.), *World Congress on Medical Physics and Biomedical Engineering 2018*,
IFMBE Proceedings 68/3, https://doi.org/10.1007/978-981-10-9023-3_157

The impedance of the skin (or admittance) depends on many factors. And it is frequently described numerically in the frequency domain by the Cole Eq. (3) [5].

$$Z = R_\infty + \frac{R_0 - R_\infty}{1 + (j\omega\tau_z)^\alpha} \quad (3)$$

where R_0 is the impedance at frequency 0 and is represented on the right, in a Nyquist plot as shown in Fig. 1, as the highest impedance of the electrode-skin. R_∞ is represented on the left in the Nyquist plot, corresponding to the lowest impedance of the electrode-skin interface. Between these two points, the impedance travels from right to left, on a crushed semicircle whose center is below the X (real axis). The parameter 'α' is a dimensionless numerical constant, and it is precisely its value which determines how flattened the semi-circle curve.

2 Methods

The method for determining circuit parameters is represented by the block diagram shown in Fig. 2. Measurements of impedance phase and magnitude, obtained with the impedance/gain-phase analyzer, were used as input to a curve fitting process. The goal was to fit, to experimental data, a polynomial function with coefficients derived from the circuit model [6]. In order to validate that the method was working correctly, a circuit with known values of resistors and capacitors was assembled on a protoboard and

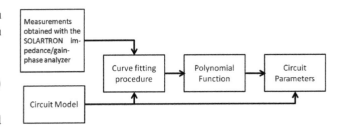

Fig. 2 Curve fitting and estimation of circuit parameters block diagram

using the measurements of impedance phase, parameters of the test circuit were obtained with an error lesser than 3% on average.

Assambo et al. [7] proposed a circuit based on a double time constant model similar as shown in Fig. 3. The impedance function of the circuit is:

$$\frac{V_1 - V_2}{I_1} = \frac{R_5}{1 + j\omega R_5 C_5} + R_{tissue} + \frac{R_4}{1 + j\omega R_4 C_4} + R_3 \quad (4)$$

which can be simplified by defining the following resistor:

$$R_{t3} = R_{tissue} + R_3 \quad (5)$$

and the following time constants:

$$\tau_5 = R_5 C_5 \quad \text{and} \quad \tau_4 = R_4 C_4 \quad (6)$$

Once simplified using R_{t3} the impedance function magnitude can be expressed as:

Fig. 1 A schematic diagram of a typical impedance locus for a biological tissue

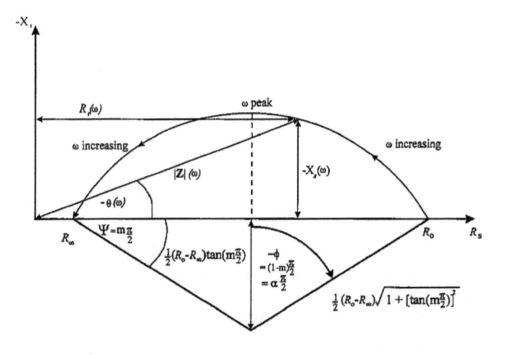

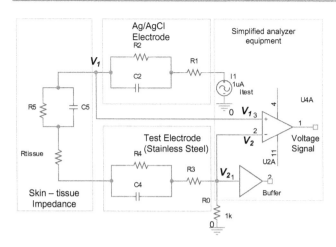

Fig. 3 Schematic representation of the circuit setup for measurements assuming a double time constant model

$$Z_{SE} = \frac{R_5(1+j\omega\tau_4) + R_4(1+j\omega\tau_5) + R_{t3}(1+j\omega\tau_4)(1+j\omega\tau_5)}{1 - \omega^2\tau_5\tau_4 + j(\omega\tau_4 + \omega\tau_5)}$$

(7)

In the above expression for the transfer function, the phase angle is important because it can be used as a measure of how close are the experimental plot, output by the impedance analyzer, and the analytical curve. Computing the arctangent from Eq. (7) the following relation can be obtained:

$$\phi(w) = Arg(Z_{SE})$$
$$= \tan^{-1}\left(\frac{\omega[R_5\tau_5 + R_4\tau_4] + \omega^3\tau_5\tau_4[R_4\tau_5 + R_5\tau_4]}{R_T + \omega^2[\tau_4^2 R_{5t} + \tau_5^2 R_{4t}] + \omega^4 R_{t3}(\tau_5\tau_4)^2}\right)$$

(8)

where:

$$R_{4t} = R_4 + R_{t3}; \quad R_{5t} = R_5 + R_{t3}; \quad R_T = R_4 + R_5 + R_{t3}$$

(9)

3 Results

The modelling approach described in the Methods section was used with the purpose of performing impedance spectroscopy measurements using a Solartron Impedance

Analyser [6] on five different types of dry electrodes that are used experimentally in electrocardiography and electroencephalography. Photographic illustration of the dry electrode materials used for the study are shown in Fig. 4. These were: (1) AgCl stud (Ringtrode, BQEL1). (2) Iron (Fe), (3) Polyurethane—multipin—A, (4) Polyurethane—multipin—B and (5) Silver disc (Ag).

In this work the experiment consisted of the following: (1) the impedance of each of the electrode materials shown in Fig. 4, as initially placed in its condition of dry electrodes (without sweat) was measured. (2) Each of the electrodes was moistened with physiological solution (medical intravenous saline solution, at 0.9% concentration) and their impedance spectroscopy measure. It is well known that the impedance of the wetted electrodes (few kΩ) is lower than that of the dry electrodes (near to MΩ). Our objective is to determine quantitatively how fast its impedance decreases to a stable level, and if it decreases to a comparable levels of the impedance reported for standard pre-gelled ECG electrodes.

As it can be observed in Fig. 5, after moistening the iron electrode (Fe) with artificial sweat used (saline solution, 0.9%), it significantly decreases its impedance profile to a few kΩ magnitude, and thus being able to reach values similar to the impedance values of a standard pre-gelled disposable electrode. The variation in time of the impedance of a standard pre-gelled electrode type were evidenced experimentally in this study, and compared with each of the dry electrode materials in Fig. 4, after moistening them. There were no overall significant difference observed in all cases.

In Tables 1 and 2, R_1 corresponds to the Resistance at the infinite frequency R_∞ (see Eq. 3) in the Nyquist impedance locus, whereas $R_2 = R_0 - R_1$, where R_0 (Eq. 3) is the resistance value at zero frequency and corresponds to the X-axis cutoff.

The impedance of each one of the electrodes after being wetted with the artificial sweat decreases significantly, and in most cases the parameters decreases less than ten times. The most significant decrease in the parameters can be observed in the Iron electrode (Fe). This dry electrode is a poor conductor since its impedance is in the order of 1.9 MΩ, but

Fig. 4 Dry electrodes used for the measurements

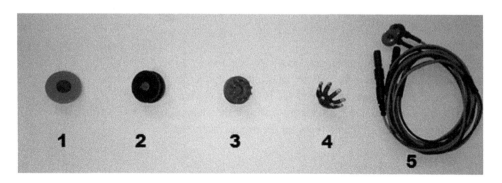

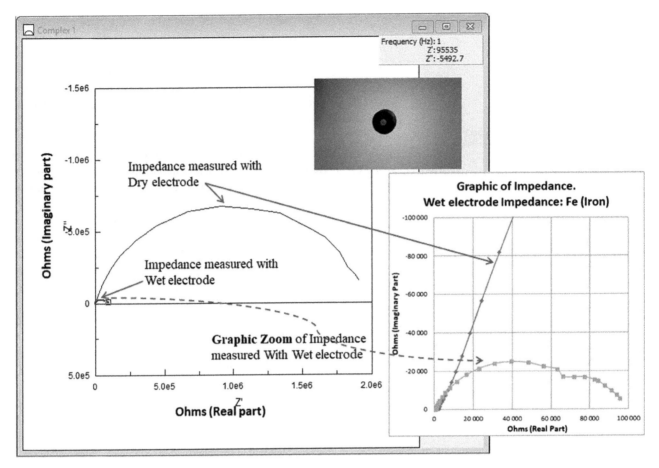

Fig. 5 Nyquist plot showing comparatively the variation of the complex impedance for the particular case of iron material dry electrode (Fe). The upper graph is the dry-electrode (without artificial sweat), while the lower graph corresponds to the complex impedance of electrode after being moistened with artificial sweat

Table 1 Resulting circuit parameters obtained from Nyquist plot, for five dry electrodes

Type of Electrode	R_1 (Ω)	R_2	CPE—T Pseudo Capacitance	CPE —P α
AgCl Ringtrode	746.2	260.4 kΩ	1.263 e-8	0.8556
Iron (Fe)	2398	1.939 MΩ	8.114 e-9	0.7808
Polyurethane multipin A	277.3	201.08 kΩ	2.715 e-8	0.8437
Polyurethane multipin B	1826	830.41 kΩ	1.927 e-8	0.8241
Silver Disc (Ag)	464.7	533.47 kΩ	1.509 e-8	0.8868

when it gets wet the impedance drastically decreases 25 times less to be located in 75 kΩ. So, an electrode that looks bad when it is dry, once moistened his behaves is equivalently to a pre-gelled electrode. Figure 6 shows the change that occurs in three different types of electrodes such as:

(a) Iron (Fe), (b) Polyurethane Multipin B, and (c) Silver Disc (Ag). In all of these electrodes a significant decrease of the impedance is observed. We concluded according to the graph of Fig. 6, that the impedance of the wet electrode decrease between 10 and 25 times, with respect to the dry electrode.

Table 2 Resulting circuit parameters obtained using moisture electrodes with artificial sweat

Type of Electrode	R_1 (Ω)	R_2 (k Ω)	CPE—T Pseudo Capacitance	CPE— P α
AgCl Ringtrode	670	100.07	6.3133 e-8	0.80171
Iron (Fe)	931	75.22	3.8243 e-8	0.79239
Polyurethane multipin A	440.3	45.31	1.57 e-7	0.78575
Polyurethane multipin B	1803	37.11	3.2892 e-7	0.72440
Silver Disc (Ag)	438	80.01	6.3283 e-8	0.84147

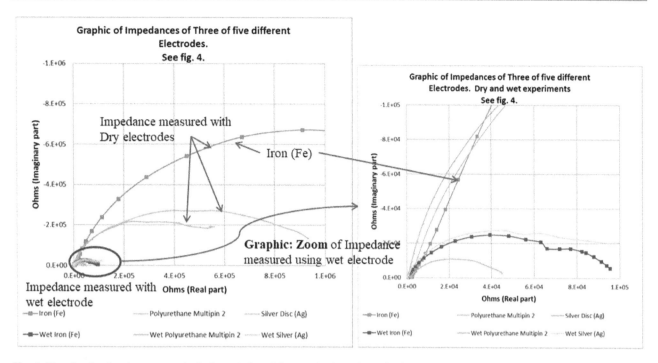

Fig. 6 Nyquist plot showing comparatively the variation of the complex impedance for three different electrodes: Iron (Fe), Polyurethane multipin B, Silver Disc (Ag). The graph on the right is a Zoom of impedances when each one of these electrodes are moisture with artificial sweat

4 Discussion and Conclusions

Although considerable amount of research effort on high-performance electrode materials, destined to be used without the application of a gel substance, continues increasingly, their electrical skin interface performance needs to be evaluated both under absolute dry conditions and under the reality of naturally sweating skin; particularly if the wearable long-term ECG monitoring device is to be located along the arm. This study has demonstrated the drastic changes of a low performance dry-electrode material: iron (Fe), presenting impedance magnitude values of around 1.9 MΩ (Fig. 5) before the influence of "sweat", to just 40 kΩ after they are moistened.

Acknowledgements This research is supported by funding from the European Union (EU): H2020-MSCA-RISE Programme (WASTCArD Project, Grant #645759). Prof Omar Escalona is supported by funds equally from the Ulster Garden Villages Ltd. and the McGrath Trust, UK.

References

1. W.D. Lynn, O.J. Escalona and D.J. McEneaney, "ECG monitoring techniques using advanced signal recovery and arm worn sensors".

IEEE International Conference on Bioinformatics and Biomedicine, IEEE BIBM 2014, pp. 51–55.

2. O. Escalona and M. Mendoza, "Electrocardiographic Waveforms Fitness Check Device Technique for Sudden Cardiac Death Risk Screening". IEEE-EMBC 2016, pp. 3453–3456.

3. A. Bosnjak, P. Linares, J. McLaughlin, O.J. Escalona. "Characterizing dry electrodes impedance by parametric modeling for arm wearable long-term cardiac rhythm monitoring". In Computing in Cardiology 2017, 44:1–4. https://doi.org/10.22489/cinc.2017.130-461.

4. Y.H Chen, M.O. Beeck, L. Vanderheyden, et al. "Soft, Comfortable Polymer Dry Electrodes for High Quality ECG and EEG Recording". *Sensors*, Vol 14, 2014, pp. 23758–780.

5. S. Grimnes, G., Martinsen. "Bioimpedance and Bioelectricity (Basics)", Third Edition, Elsevier and Academic Press (2015).

6. A. Bosnjak, A. Kennedy, P. Linares, M. Borges, J.A.D. McLaughlin, O.J. Escalona, "Performance assessment of dry electrodes for wearable long term cardiac rhythm monitoring: skin-electrode impedance spectroscopy". In Proceedings of the Annual International Conference of the IEEE-EMBS 8037209, pp. 1861–1864 (2017).

7. C. Assambo, R. Dozio, A. Baba and M. J. Burke "Determination of the Parameters of the Skin Electrode Impedance Model for ECG Measurement". Proc. 6th Int. Conf. on Electronics, Hardware, Wireless and Optical Communications, Corfu, pp. 540–318 (2007).

A Multimodal Machine Learning Approach to Omics-Based Risk Stratification in Coronary Artery Disease

Eleni I. Georga, Nikolaos S. Tachos, Antonis I. Sakellarios, Gualtiero Pelosi, Silvia Rocchiccioli, Oberdan Parodi, Lampros K. Michalis, and Dimitrios I. Fotiadis

Abstract

This study aims at developing a personalized model for coronary artery disease (CAD) risk stratification based on machine learning modelling of non-imaging data, i.e. clinical, molecular, cellular, inflammatory, and omics data. A multimodal architectural approach is proposed whose generalization capability, with respect to CAD stratification, is currently evaluated. Different data fusion techniques are investigated, ranging from early to late integration methods, aiming at designing a predictive model capable of representing genotype-phenotype interactions pertaining to CAD development. An initial evaluation of the discriminative capacity of the feature space with respect to a binary classification problem (No CAD, CAD), although not complete, shows that: (i) kernel-based classification provides more accurate results as compared with neural network-based and decision tree-based modelling, and (ii) appropriate input refinement by feature ranking has the potential to increase the sensitivity of the model.

Keywords

Coronary artery disease • Patient risk stratification Multi-modal machine learning

E. I. Georga (✉) · D. I. Fotiadis
Unit of Medical Technology and Intelligent Information Systems, Department of Materials Science and Engineering, University of Ioannina, Ioannina, Greece
e-mail: egewrga@gmail.com

N. S. Tachos · A. I. Sakellarios · D. I. Fotiadis
Department of Biomedical Research, Institute of Molecular Biology and Biotechnology, Foundation for Research and Technology–Hellas (FORTH), Ioannina, Greece

G. Pelosi · S. Rocchiccioli · O. Parodi
Institute of Clinical Physiology, National Research Council, Pisa, Italy

L. K. Michalis
Department of Cardiology, Medical School, University of Ioannina, Ioannina, Greece

1 Introduction

Coronary artery disease (CAD) is a multi-factorial disease characterized by the accumulation of lipids into the arterial wall and the subsequent inflammatory response [1, 2]. The phenotype of disease progression is affected by several factors, including clinical risk factors (e.g. gender, smoking, hyperlipidaemia, hypertension, diabetes) as well as molecular, biohumoral and biomechanical factors (e.g. low endothelial shear stress). CAD diagnosis is validated through invasive coronary angiography (CA); however, different invasive [e.g. intravascular ultrasound (IVUS), optical coherence tomography (OCT)] and non-invasive imaging modalities [e.g. computed tomography angiography [CTA], magnetic resonance imaging (MRI)] are nowadays available to visualize the vessel wall, quantify the plaque burden and characterize the type of the atherosclerotic plaque.

Predicting the risk of CAD constitutes a widely-studied problem from the perspective of statistical modelling. The majority of existing risk models, such as the Framingham risk score (FRS) [3], the Systematic COronary Risk Evaluation (SCORE) [4] and the QRISK [5], postulate a Cox proportional hazard regression or logistic regression model of relatively few traditional predictors of the disease, focusing on CAD or cardiovascular disease (CVD). In spite of the reported good discrimination ability of parametric linear regression models, a recent systematic review demonstrated the paucity of external validation and head-to-head comparisons, the poor reporting of their technical characteristics as well as the variability in outcome variables, predictors and prediction horizons, which limits their applicability in evidence-based decision making in healthcare [6]. Precision medicine suggests individualized dynamic predictive modelling approaches not being hypotheses-driven [7–9]. Moreover, the increasing availability of electronic health records (EHRs), personal health records (PHRs) and omics big data give rise to multiscale multi-parametric predictive big data analytics in

L. Lhotska et al. (eds.), *World Congress on Medical Physics and Biomedical Engineering 2018*,
IFMBE Proceedings 68/3, https://doi.org/10.1007/978-981-10-9023-3_158

personalized medicine in cardiovascular research and clinical practice [10–12].

The purpose of this study is to design and develop a machine learning-based model effectively integrating multiple categories of biological data towards precise risk stratification in coronary artery disease. Herein, we outline the formulation of the problem, present the main components of the model architecture, and investigate the predictive power of the currently available feature set.

2 CAD Risk Stratification Methodology

2.1 Problem Formulation

CAD risk stratification is formulated as a multiclass classification problem, representing the severity of the disease as a nonlinear parametric function of a confined set of features $f(x) = C_i, x = [x_1,\ldots,x_d], i = 1,\ldots,k$. The utilized feature set is provided in Table 1. Three dominant classes $C_i, i = 1,\ldots,k$ have been defined, namely "No CAD", "Non Obstructive CAD", and "Obstructive CAD", with a $\geq 50\%$

Table 1 Description of the feature set

Category	Features
Demographics	Age, gender
Risk factors	Family history of CAD, hypertension, diabetes, dyslipidaemia, smoking, obesity, metabolic syndrome
Molecular systemic variables	Alanine aminotransferase, alkaline phosphatase, aspartate aminotransferase, creatinine, gamma-glutamyl transferase, glucose, HDL, high-sensitivity C-reactive protein, interleukin-6, LDL, leptin, total cholesterol, triglycerides, uric acid
Symptoms	Typical angina, atypical angina, non angina chest pain, other symptoms, no symptoms
Exposome	Alcohol consumption, vegetable consumption, physical activity, home environment, exposition to pollutants
Inflammatory markers	ICAM1, VCAM1
Monocyte markers	$CCR2_{val1}$, $CCR2_{val2}$, $CCR5_{val1}$, $CCR5_{val2}$, $CD11b_{val1}$, $CD11b_{val2}$, $CD14(++/+)_{val1}$, $CD14(++/+)_{val2}$, CD14 ++/CD16 +/ CCR2 + $_{val1}$, CD14 ++/CD16−/ CCR2 + $_{val1}$, CD14 +/CD16 ++/CCR2− $_{val1}$, $CD163_{val1}$, $CD163_{val2}$, $CD16_{val1}$, $CD16_{val2}$, $CD18_{val1}$, $CD18_{val2}$, $CX3CR1_{val1}$, $CX3CR1_{val2}$, $CXCR4_{val1}$, $CXCR4_{val2}$, $HLA-DR_{val1}$, $HLA-DR_{val2}$, MONOCYTE COUNT$_{val2}$
Omics data	Lipid profile, MRNA sequencing, exome sequencing

val1: % of +, val2: RFI

diameter stenosis in at least one main coronary artery vessel, as assessed by computed tomography coronary angiography (CTCA), characterizing patients with obstructive CAD.

2.2 Multimodal Machine-Learning CAD Stratification Model

A multimodal architecture was specified relying on two processing layers which are defined according to late or intermediate data integration strategies [13]. First, the following feature classes (or views) were defined: (View 1) demographics, (View 2) clinical data, risk factors, symptoms, (View 3) molecular variables (i.e. biohumoral, inflammatory markers and lipids profile), (View 4) gene expression data, (View 5) exposome, and (View 6) monocytes. As it is shown in Fig. 1, late data integration consists in the construction of: (i) an ensemble of decision tree-based prediction models (i.e. random forests, boosted decision trees) for each data view, whose individual decisions are effectively merged using simple mechanisms (e.g. weighted voting), or (ii) a multimodal deep neural network comprising of appropriate deep learning subnetworks for each separate data view and, unifying their output into higher network layers.

Intermediate data integration is based on multiple kernel learning (Fig. 2. Kernel matrices are computed for each data view, and then they are combined, through a parametric linear function, in order to generate the final kernel matrix. Kernel-based classification (i.e. support vector machine, relevance vector machine) is subsequently applied to predict CAD risk stratification.

The skeleton and individual modules of the integrative model (i.e. merging mechanisms, machine learning algorithms, metric learning, regularization, and feature extraction) are implemented in R.

3 Results

Currently, the dataset is confined to demographics, risk factors, biohumoral markers and symptoms, which led us to concatenate all features into a single vector. In particular, three machine learning algorithms have been examined; a parametric model [i.e. feed-forward neural network (FFNN)], a non-parametric kernel-based model [i.e. support vector machine (SVM)] and an ensemble model [i.e. random forest (RF)]. In addition, the discriminative capacity of the available data categories has been evaluated via: (i) a knowledge-based approach consisting in the a priori definition of 3 input cases (Case 1: Demographics, Risk Factors; Case 2: Demographics, Risk Factors, Symptoms; Case 3 Demographics, Risk Factors, Symptoms, Molecular Systemic Variables), and (ii) feature ranking according to the

Fig. 1 Integrative CAD risk stratification model

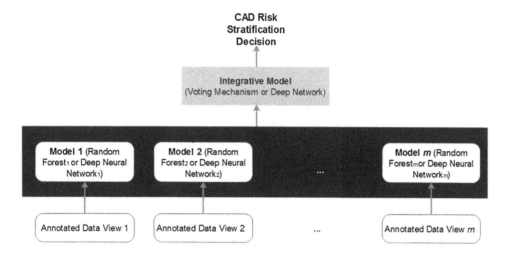

Fig. 2 Integrative multiple kernel-based CAD risk stratification model

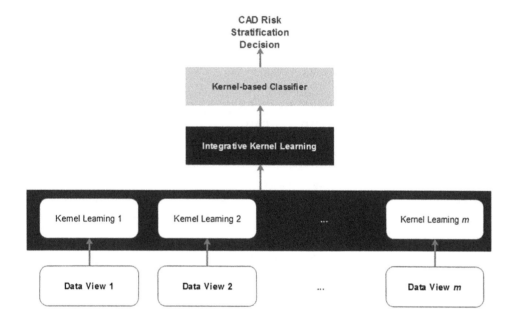

InfoGain criterion accompanied by a forward selection procedure (Case 4).

Table 2 reports classification results on 101 patients (No CAD: $n = 25$, Age: 58.36 ± 7.45; Mild to Severe CAD: $n = 76$, Age: 63.61 ± 7.43) by 10-fold cross-validation. The gradual improvement of accuracy with the enhancement of the input space is apparent, with proper customization of the input via feature ranking ($d = 20$) better balancing the sensitivity to specificity ratio. SVM outperforms FFNN and RF resulting in an overall accuracy 85.1% and a nearly perfect sensitivity (98.7%), whereas specificity remains low (44.0%), presumably due to the class imbalance in the dataset. The confusion matrices corresponding to SVM output in Case 3 and Case 4 are reported in Tables 3 and 4, respectively.

4 Discussion and Conclusions

CAD diagnosis is currently performed according to well-known screening strategies (i.e. CA, IVUS, OCT, CTA, MRI), whereas CVD risk can be assessed by linear regres-sion models of clinical, laboratory and anthropometric features, assuming linearity as well as time-invariance of the underly-ing input-output relationships. Non-linearity is addressed by black-box parameterizations (neural networks and kernel-based models) or more transparent architectures (de-cision trees, dynamic |Bayesian networks) or ensembles of classification models (random forests), which feature space, however, resembles that of linear approaches (i.e. established risk factors). The generalization capability of the existing

Table 2 Classification performance

	FFNN			SVM			RF		
	Acc.	Se.	Sp.	Acc.	Se.	Sp.	Acc.	Se.	Sp.
Case 1	66.3	78.9	28.0	77.2	97.4	16.0	73.3	85.5	36.0
Case 2	70.3	81.6	36.0	81.2	94.7	40.0	75.2	88.2	36.0
Case 3	74.3	84.2	44.0	84.2	97.4	44.0	77.2	97.4	16.0
Case 4	78.2	90.8	40.0	85.1	98.7	44.0	81.2	92.1	48.0

Acc.: Accuracy, *Se.*: Sensitivity, *Sp.*: Specificity

Table 3 Confusion matrix of SVM results in Case 3

		Prediction	
		No CAD	Mild to severe CAD
Annotation	No CAD	11	14
	Mild to severe CAD	2	74

Table 4 Confusion matrix of SVM Results in Case 4

		Prediction	
		No CAD	Mild to severe CAD
Annotation	No CAD	11	14
	Mild to severe CAD	1	75

machine learning models for the diagnosis of CAD or the estimation of eventful or asymptomatic CAD progression is promising; however, new knowledge coming from big data sources (e.g. molecular, cellular, inflammatory and omics data) requires more integrative machine learning solutions.

In this study, a new machine-learning approach to CAD risk stratification has been proposed relying on multimodal data integration. Its deployment and evaluation are ongoing by: (i) integrating new features concerning the lipid profile, the exome and mRNA sequencing, the exposome, and inflammatory and monocyte markers, and (ii) selecting the most effective multimodal predictive modelling scheme. Moreover, the multi-class classification problem is going to be refined by considering established risk scores of coronary atherosclerosis combining markers of stenosis severity, plaque location and composition, as assessed by computed tomography angiography.

Acknowledgements This work is funded by the European Commission: Project SMARTOOL, "Simulation Modeling of coronary ARTery disease: a tool for clinical decision support—SMARTool" GA number: 689068.

Compliance with Ethical Standards

Conflict of Interest: The authors declare that they have no conflict of interest.
Ethical approval: "All procedures performed in studies involving human participants were in accordance with the ethical standards of the institutional and/or national research committee and with the 1964 Helsinki declaration and its later amendments or comparable ethical standards."
Informed consent: "Informed consent was obtained from all individual participants included in the study."

References

1. Stone, P.H., et al., *Prediction of progression of coronary artery disease and clinical outcomes using vascular profiling of endothelial shear stress and arterial plaque characteristics: the PREDICTION Study*. Circulation, 2012. 126(2): p. 172–81.
2. Sakellarios, A., et al., *Prediction of atherosclerotic disease progression using LDL transport modelling: a serial computed tomographic coronary angiographic study*. European Heart Journal: Cardiovascular Imaging, 2017. 18(1): p. 11–18.
3. D'Agostino, R.B., Sr., et al., *General cardiovascular risk profile for use in primary care: the Framingham Heart Study*. Circulation, 2008. 117(6): p. 743–53.
4. Conroy, R.M., et al., *Estimation of ten-year risk of fatal cardiovascular disease in Europe: the SCORE project*. Eur Heart J, 2003. 24(11): p. 987–1003.
5. Hippisley-Cox, J., et al., *Derivation, validation, and evaluation of a new QRISK model to estimate lifetime risk of cardiovascular disease: cohort study using QResearch database*. BMJ, 2010. 341: p. c6624.
6. Damen, J.A., et al., *Prediction models for cardiovascular disease risk in the general population: systematic review*. BMJ, 2016. 353: p. i2416.
7. Weng, S.F., et al., *Can machine-learning improve cardiovascular risk prediction using routine clinical data?* PLOS ONE, 2017. 12 (4): p. e0174944.
8. Choi, E., et al., *Using recurrent neural network models for early detection of heart failure onset*. Journal of the American Medical Informatics Association: JAMIA, 2017. 24(2): p. 361–370.
9. Motwani, M., et al., *Machine learning for prediction of all-cause mortality in patients with suspected coronary artery disease: a 5-year multicentre prospective registry analysis*. European Heart Journal, 2017. 38(7): p. 500–507.
10. Goldstein, B.A., A.M. Navar, and R.E. Carter, *Moving beyond regression techniques in cardiovascular risk prediction: applying machine learning to address analytic challenges*. European Heart Journal, 2017. 38(23): p. 1805–1814.
11. Rumsfeld, J.S., K.E. Joynt, and T.M. Maddox, *Big data analytics to improve cardiovascular care: promise and challenges*. Nat Rev Cardiol, 2016. 13(6): p. 350–9.
12. Groeneveld, P.W. and J.S. Rumsfeld, *Can Big Data Fulfill Its Promise?* Circ Cardiovasc Qual Outcomes, 2016. 9(6): p. 679–682.
13. Li, Y., F.X. Wu, and A. Ngom, *A review on machine learning principles for multi-view biological data integration*. Brief Bioinform, 2016.

Author Index

Pichugin, Vladimir, 139
Pilt, Kristjan, 13
Pisanelli, Domenico M., 795
Pit'hová, Pavlína, 753
Plata-Guao, Viena Sofia, 295
Pojtinger, Stefan, 589
Pokorny, Tomas, 775
Pollo, B., 693
Polyaka, Nataliya, 139
Ponce, Sergio Damián, 347
Ponce, Sergio, 195
Potocnak, Tomas, 149
Priscilla, Longo, 355
Provaznik, Ivo, 149, 155
Punin, C., 865

Q

Qamhiyeh, Sima, 599
Quiroga-Torres, Daniel Alejandro, 295

R

Růžička, Filip, 105
Råglund, Jari, 183
Rama Raju, Venkateshwarla, 47, 65
Rantaniva, Teppo, 183
Remoto, Randal Zandro, 509
Repaka, Ramjee, 663
Ribeiro, Jonatas Magno Tavares, 341
Richter, Aleš, 239
Rivas, Rossana, 325, 723
Roberti, Martín, 195
Rocchiccioli, Silvia, 879
Rocha, Humberto, 413
Rocha, Sandra Luz , 383
Rocha, Sandra N.L., 627
Rodrigues, Carlos, 761
Rodríguez-Dueñas, William Ricardo, 273, 295
Rogalewicz, Vladimír, 107
Rojas, Jefferson Steven Sarmiento, 273
Román, Esteban Hernández San, 407
Romanov, A.V., 651
Rubio, Débora, 347
Rudžianskas, Viktoras, 643
Russomando, A., 693
Rykhalskiy, O.Y., 651

S

Sagbay, Giovanni, 201, 221
Saiti, Kyriaki, 753
Saito, N., 565
Sakama, M., 565
Sakellarios, Antonis I., 879
Salsac, Anne-Virginie, 135
Salvi de Souza, Giordana, 699
Samnick, Samuel, 685
Samra, Rafael S., 627
Sanchez-Nieto, Beatriz, 677
Santamaría-Vázquez, Eduardo, 41
Santos, Fernanda Stephanie, 533
Santos, João, 499
Santos, Natalia Aurora, 227
Sarian, L. O. Z., 371
Sarasanandarajah, Siva, 477

Sarmiento-Rojas, Jefferson S., 295
Satou, Y., 565
Schadt, Fabian, 685
Schaefers, Gregor, 599
Scherer, Daniel, 69
Schiariti, M., 693
Schlect, David, 443
Schlett, Paul, 77
Schuemer, Gustavo Ariel, 347
SchÜller, Andreas, 589
Selčan, Miroslav, 377
Sepehri, Amir A., 849
Serrano, Benjamin, 469
Sharkeev, Yurii, 139
Shen, Chun-Chiang, 655
Shevchenko, A.D., 651
Shimizu, M., 565
Shimono, Tetsunori, 417
Shiraishi, Yasuyuki, 717
Shi, Shawn, 253
Sieger, Tomáš, 105
Silva, L.A., 367
Simões, Marco, 113
Simonelli, Francesca, 387
Simon, Luc, 633
Singh, Sundeep, 663
Siroky, David, 249
Skopalik, Josef, 149, 155
Smrcka, Pavel, 189, 855
Sneiders, M., 3
Soejoko, D.S., 571
Solfaroli Camillocci, E., 693
Sorelli, Michele, 173
Souza, F., 493
Sparapani, Alexis, 195
Štechová, Kateřina, 753, 757
Steklá, Michaela, 377
Stephens, Holly, 553
Strazza, Annachiara, 719
Štveráková, Markéta, 377
Sucena, Ana, 595
Sukupova, Lucie, 541
Sun, Yan-jun, 835
Susilo, A., 571
Svarca, A., 619
Svoboda, Ondrej, 149, 155
Sylvander, Steven, 451, 545
Syskov, Alexey, 71

T

Tabuchi, Akihiko, 525, 537
Tachos, Nikolaos S., 879
Takahashi, Yusuke, 121
Takemura, Akihiro, 487, 505, 581
Tanaka, T., 565
Tangboonduangjit, Puangpen, 481
Tanki, Nobuyoshi, 525, 537
Teke, Tony, 421
Tello, Marcos, 235
Terini, Ricardo, 519
Terzaghi, F., 317
Tesarik, Jan, 771, 775, 779, 785
Testa, Francesco, 387
Thoelen, Ronald, 27
Ting, Hua Nong, 809

Lightning Source UK Ltd.
Milton Keynes UK
UKHW05f0840110718
325537UK00003B/54/P